JG juxtaglomerular
JGA juxtaglomerular apparatus

K potassium (K$^+$ potassium ion)
kcal kilocalorie
kg kilogram
km/h kilometer per hour
k_p permeability constant

L liter, lumbar
L tube length
lb pound
LDH lactate dehydrogenase
LDL low-density lipoprotein
LH luteinizing hormone
l_o optimal length
LSD lysergic acid diethylamide
LTD long-term depression
LTP long-term potentiation

m meter, milli-
M molar, myosin
M* activated myosin
M phase mitosis phase of cell cycle
MAC membrane attack complex
MAP mean arterial pressure
mEq milliequivalent
MES microsomal enzyme system
mg milligram
Mg magnesium (Mg^{2+} magnesium ion)
MHC major histocompatibility complex
mi mile
mi/h miles per hour
MIS Müllerian inhibiting substance
min minute
miu milli international unit
ml milliliter
mM millimolar
mmol millimol
mm millimeter
mmHg millimeters of mercury
mol mole
mOsm milliosmolar
mOsmol milliosmol
mRNA messenger RNA
ms millisecond
mV millivolt
μg microgram
μl microliter
μm micrometer
μM micromolar
μmol micromol
μV microvolt

n any whole number
N nitrogen
Na sodium (Na$^+$ sodium ion)
NAD$^+$ nicotinamide adenine dinucleotide

NE norepinephrine
NFP net filtration pressure
ng nanogram
—NH$_2$ amino group (—NH$_3^+$ ionized amino group)
NH$_3$ ammonia
NH$_4^+$ ammonium ion
NIDDM noninsulin-dependent diabetes mellitus
NK cell natural killer cell
nm nanometer
nM nanomolar
nmol nanomol
NO nitric oxide
NPY neuropeptide Y
NREM nonrapid eye movement
NSAIDs nonsteroidal anti-inflammatory drugs

O$_2$ oxygen
O$_2$·$^-$ superoxide anion
—OH$^-$ hydroxyl group
OH· hydroxyl radical
1,25-(OH)$_2$D$_3$ 1,25-dihydroxyvitamin D$_3$
Osm osmolar

p pico
P product
P partial pressure, pressure, permeability, plasma concentration of a substance
PAH para-aminohippurate
P_{alv} alveolar pressure
P_{atm} atmospheric pressure
P_{BS} Bowman's space pressure
P_{GC} glomerular capillary pressure
PF platelet factor
pg picogram
PGA prostaglandin of the A type
PGE prostaglandin of the E type
PGE$_2$ prostaglandin E$_2$
PGI$_2$ prostacyclin, prostaglandin I$_2$
PHI peptide histidine isoleucine
PHM peptide histidine methionine
P_i inorganic phosphate
PIH prolactin inhibiting hormone
P_{ip} intrapleural pressure
PIP$_2$ phosphatidylinositol bisphosphate
pM picomolar
PMDD premenstrual dysphoric disorder
PMS premenstrual syndrome
PRF prolactin releasing factor
PRG primary response gene
P_s plasma concentration of substance s

R remainder of molecule
R resistance
r inside radius of tube

REM rapid eye movement
RNA ribonucleic acid
RQ respiratory quotient
rRNA ribosomal RNA

s second, sacral
S substrate, substance
S phase synthesis phase of cell cycle
SA sinoatrial
SAD seasonal affective disorder
SE substrate-enzyme complex
SERM selective estrogen receptor modulator
—SH sulfhydryl group
SO$_4^{2-}$ sulfate ion
SP systolic pressure
SR sarcoplasmic reticulum
SRY sex-determining region on the Y chromosome
SS somatostatin
SSRIs serotonin-specific reuptake inhibitors
STD sexually transmitted disease
SV stroke volume

T thymine, thoracic
T$_3$ triiodothyronine
T$_4$ thyroxine
TENS transcutaneous electric nerve stimulation
t-PA tissue plasminogen activator
T tubule transverse tubule
TBW total body water
TFPI tissue factor pathway inhibitor
TH thyroid hormones
TIA transient ischemic attack
T_m transport maximum
TNF tumor necrosis factor
TPR total peripheral resistance
TRH thyrotropin releasing hormone
tRNA transfer RNA
TSH thyroid-stimulating hormone

U uracil
U urine concentration of a substance
UTP uracil triphosphate

V volume, volume of urine per unit time
VIP vasoactive intestinal peptide
V_L lung volume
VLDL very low density lipoprotein
$\dot{V}_{O_2}$**max** maximal oxygen consumption
vWF von Willebrand factor

W work

x general term for any substance

eighth edition

HUMAN Physiology

The Mechanisms of Body Function

Arthur Vander
Professor Emeritus of Physiology
University of Michigan

James Sherman
Associate Professor of Physiology
University of Michigan

Dorothy Luciano
Formerly of the Department of Physiology
University of Michigan

Boston   Burr Ridge, IL   Dubuque, IA   Madison, WI   New York   San Francisco St. Louis
Bangkok Bogotá Caracas Lisbon London Madrid
Mexico City   Milan   New Delhi   Seoul   Singapore   Sydney   Taipei   Toronto

McGraw-Hill Higher Education

A Division of The **McGraw-Hill** Companies

HUMAN PHYSIOLOGY:
THE MECHANISMS OF BODY FUNCTION, EIGHTH EDITION

Published by McGraw-Hill, an imprint of The McGraw-Hill Companies, Inc., 1221 Avenue of the Americas, New York, NY 10020. Copyright © 2001, 1998, 1994, 1990, 1985, 1980, 1975, 1970 by The McGraw-Hill Companies, Inc. All rights reserved. No part of this publication may be reproduced or distributed in any form or by any means, or stored in a database or retrieval system, without the prior written consent of The McGraw-Hill Companies, Inc., including, but not limited to, in any network or other electronic storage or transmission, or broadcast for distance learning.

Some ancillaries, including electronic and print components, may not be available to customers outside the United States.

This book is printed on acid-free paper.

3 4 5 6 7 8 9 0 VNH/VNH 0 9 8 7 6 5 4 3 2 1

ISBN 0-07-290801-7

Vice president and editor-in-chief: *Kevin T. Kane*
Publisher: *Michael D. Lange*
Sponsoring editor: *Kristine Tibbetts*
Developmental editor: *Patrick F. Anglin*
Marketing managers: *Michelle Watnick/Heather K. Wagner*
Senior project manager: *Peggy J. Selle*
Senior production supervisor: *Mary E. Haas*
Design manager: *Stuart D. Paterson*
Cover/interior designer: *Christopher Reese*
Cover image: *Photo Researchers*
Photo research coordinator: *John C. Leland*
Photo research: *Mary Reeg Photo Research*
Senior supplement coordinator: *Audrey A. Reiter*
Compositor: *York Graphic Services, Inc.*
Typeface: *10/12 Palatino*
Printer: *Von Hoffmann Press, Inc.*

The credits section for this book begins on page 779 and is considered an extension of the copyright page.

Library of Congress Cataloging-in-Publication Data

Vander, Arthur J., 1933–
    Human physiology: the mechanisms of body function / Arthur Vander, James
Sherman, Dorothy Luciano.—8th ed.
        p. cm.
    Includes bibliographical references and index.
    ISBN 0-07-290801-7
    1. Human physiology.   I. Sherman, James H., 1936–   II. Luciano, Dorothy S.   III. Title.
    [DNLM: 1. Physiological Processes. QT 104 V228h 2001]
QP34.5 .V36 2001
612—dc21   00–020039
                                                          CIP

www.mhhe.com

Contents in Brief

iii

Contents

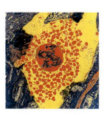

Chapter 4
Protein Activity and Cellular Metabolism   53

Chapter 5
Genetic Information and Protein Synthesis   91

Chapter 6
Movement of Molecules Across Cell Membranes 115

Part Two
Biological Control Systems

Chapter 7
Homeostatic Mechanisms and Cellular Communication 143

Chapter 8
Neural Control Mechanisms 175

Chapter 9
The Sensory Systems 227

Chapter 11
Muscle 291

Chapter 10
Principles of Hormonal Control Systems 263

Chapter 12
Control of Body Movement   333

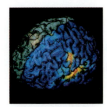

Chapter 13
Consciousness and Behavior   351

Part Three
Coordinated Body Functions

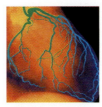

Chapter 14
Circulation   373

Chapter 15
Respiration   463

Chapter 16
The Kidneys and Regulation of Water and Inorganic Ions   505

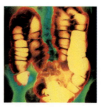

Chapter 17
**The Digestion and Absorption
of Food** 553

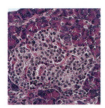

Chapter 18
Regulation of Organic Metabolism, Growth, and Energy Balance 593

Chapter 19
Reproduction 635

Chapter 20
Defense Mechanisms of the Body   687

Preface

Goals and Orientation

The purpose of this book remains what it was in the first seven editions: to present the fundamental principles and facts of human physiology in a format that is suitable for undergraduate students, regardless of academic backgrounds or fields of study: liberal arts, biology, nursing, pharmacy, or other allied health professions. The book is also suitable for dental students, and many medical students have also used previous editions to lay the foundation for the more detailed coverage they receive in their courses.

The most significant feature of this book is its clear, up-to-date, accurate explanations of **mechanisms,** rather than the mere description of facts and events. Because there are no limits to what can be covered in an introductory text, it is essential to reinforce over and over, through clear explanations, that physiology can be understood in terms of basic themes and principles. As evidenced by the very large number of flow diagrams employed, the book emphasizes understanding based on the ability to think in **clearly defined chains of causal links.** This approach is particularly evident in our emphasis of the dominant theme of human physiology and of this book—**homeostasis** as achieved through the coordinated function of **homeostatic control systems.**

To repeat, we have attempted to explain, integrate, and synthesize information rather than simply to describe, so that students will achieve a working knowledge of physiology, not just a memory bank of physiological facts. Since our aim has been to tell a coherent story, rather than to write an encyclopedia, we have been willing to devote considerable space to the logical development of difficult but essential concepts; examples are second messengers (Chapter 7), membrane potentials (Chapter 8), and the role of intrapleural pressure in breathing (Chapter 15).

In keeping with our goals, the book progresses from the cell to the body, utilizing information and principles developed previously at each level of complexity. One example of this approach is as follows: the characteristics that account for protein specificity are presented in Part One (Chapter 4), and this concept is used there to explain the "recognition" process exhibited by enzymes. It is then used again in Part Two (Chapter 7) for membrane receptors, and again in Part Three (Chapter 20) for antibodies. In this manner, the student is helped to see the basic foundations upon which more complex functions such as homeostatic neuroendocrine and immune responses are built.

Another example: Rather than presenting, in a single chapter, a gland-by-gland description of all the hormones, we give a description of the basic principles of endocrinology in Chapter 10, but then save the details of individual hormones for later chapters. This permits the student to focus on the functions of the hormones in the context of the homeostatic control systems in which they participate.

Alternative Sequences

Given the inevitable restrictions of time, our organization permits a variety of sequences and approaches to be adopted. Chapter 1 should definitely be read first as it introduces the basic themes that dominate the book. Depending on the time available, the instructor's goals, and the students' backgrounds in physical science and cellular and molecular biology, the chapters of Part One can be either worked through systematically at the outset or be used more selectively as background reading in the contexts of Parts Two and Three.

In Part Two, the absolutely essential chapters are, in order, Chapters 7, 8, 10, and 11, for they present the basic concepts and facts relevant to homeostasis, intercellular communication, signal transduction, nervous and endocrine systems, and muscle. This material, therefore, is critical for an understanding of Part Three.

We believe it is best to begin the coordinated body functions of Part Three with circulation (Chapter 14), but otherwise the chapters of Part Three, as well as Chapters 9, 12, and 13 of Part Two, can be rearranged and used or not used to suit individual instructor's preferences and time availability.

Revision Highlights

There were two major goals for this revision: (1) to redo the entire illustration program (and give the general layout of the book a "face-lift") for greater teaching effectiveness, clarity, consistency, and esthetic appeal; and (2) to update all material and assure the greatest accuracy possible.

Illustration Program

Almost all the figures have been redone to some extent, ranging from a complete redrawing of the figure to simply changing the labeling of graph axes for greater clarity. Figures 20–1 and 20–10 (Figure 20–9 in the previous edition) provide examples of how a more realistic three-dimensional perspective has been added to many of the figures, and Figure 20–13 (Figure 20–12 in the previous edition) shows how the picturing of complex events has been improved. Also, even when a specific part of the text has not required revision, we have added some new figures (for example, Figure 20–7) to illustrate the text, particularly in the case of material we know to be difficult.

Of course, the extensive use of flow diagrams, which we introduced in our first edition, has been continued. Conventions, which have been expanded in this edition, are used in these diagrams throughout the book to enhance learning. Look, for example, at Figure 16–28. The beginning and ending boxes of the flow diagram are in green, and the beginning is further clarified by the use of a "Begin" logo. Blue three-dimensional boxes are used to denote events that occur inside organs and tissues (identified by bold-faced underlined labels in the upper right of the boxes), so that the reader can easily pick out the anatomic entities that participate in the sequences of events. The participation of hormones in the sequences stand out by the placing of changes in their plasma concentrations in reddish/orange boxes. Similarly, changes in urinary excretion are shown in yellow boxes. All other boxes are purple. Thus, color is used in these diagrams for particular purposes, not just for the sake of decoration.

Other types of color coding are also now used consistently throughout the book. Thus, to take just a few examples, there are specific colors for the extracellular fluid, the intracellular fluid, muscle, particular molecules (the two strands of DNA, for example), and the lumen of the renal tubules and GI tract. Even a quick perusal of Chapter 20 will reveal how consistent use of different colors for the different types of lymphocytes, as well as macrophages, should help learning.

Updating of Material

Once again, we have considerably rewritten material to improve clarity of presentation. In addition, as noted above, most figures have been extensively redone, and new figures have been added (only a few of these are listed below). Finally, as a result of new research or in response to suggestions by our colleagues, many topics have either been significantly altered or added for the first time in this edition; the following is a partial list of these topics.

Chapter 1 Introductory section: "The Scope of Human Physiology"

Chapter 2 New figures: Hemoglobin molecule, DNA double helix base pairings, purine-pyrimidine hydrogen bond pairings

Chapter 3 Cholesterol in membrane function
Procedures for studying cell organelles
Endosomes
Peroxisomes

Chapter 5 Mitochondrial DNA
Preinitiation complex
Factors altering the activity of specific cell proteins
Protein delivery and entry into mitochondria
Regulation of cell division at checkpoints in mitotic cycle

Chapter 6 Patch clamping
Primary active-transport mechanisms
Digitalis and inhibition of Na,K-ATPase
Cystic fibrosis chloride channel
Endocytosis
New figures illustrating transporter conformational changes

Chapter 7 Paracrine/autocrine agents
Melatonin and brain pacemakers
Receptors as tyrosine kinases and guanylyl cyclase
JAK kinases and receptors
Phospholipase, diacylglycerol, and inositol trisphosphate
Calcium-induced calcium release
Receptor inactivation

Chapter 8 Regeneration of neurons
Comparison of voltage-gated sodium and potassium channels
Information on neurotransmitters
Functional anatomy of the central nervous system

Chapter 9 Pain
Olfaction

Chapter 10 Diagnosis of the site of a hormone abnormality

Chapter 11 Passive elastic properties and role of titan
Factors causing fatigue
Role of nitric oxide in relaxing smooth muscle

Chapter 12 Cortical control of motor behavior
Parkinson's disease
Effect of the corticospinal pathways on local-level neurons
Walking

Chapter 13 Electroencephalogram
Sleep
Binding problem
Emotions

Schizophrenia
Serotonin-specific reuptake inhibitors (SSRIs)
Learning and memory, and their neural bases

Chapter 14 Erythropoietin mechanism of action
Anti-angiogenic factors in treatment of cancer
Capillary filtration coefficient
Shock
Static exercise and blood pressure
Aging and heart rate
Drug therapy for hypertension, heart failure, and coronary artery disease
Dysfunctional endothelium in atherosclerosis
Homocysteine, folate, and vitamin E in atherosclerosis
Coronary stents
Nitric oxide and peripheral veins
Platelet receptors for fibrinogen
Therapy of stroke with t-PA

Chapter 15 Pulmonary vessels and gravitational/physical forces
Hemoglobin cooperativity
Carbon monoxide and oxygen carriage
Emphysema

Chapter 16 Mesangial cells and glomerular filtration coefficient
Channels, transporters, and genetic renal diseases
Micturition, including role of sympathetic neurons
Aquaporins
Medullary circulation and urinary concentration
Pressure natriuresis
Calcitonin
Bisphosphonates and osteoporosis

Chapter 17 Colipase and fat digestion
HCl secretion and inhibitory role of somatostatin
Intestinal fluid secretion and absorption

Chapter 18 Inhibition of glucagon secretion by insulin
Roles of HDL and LDL
IGF-I and fetal growth
IGF-II
Mechanism of calorigenic effect of thyroid hormones
Leptin effects on hypothalamus and anterior pituitary
Overweight and obesity
Fever and neural pathways from liver
Endogenous cryogens

Chapter 19 Dehydroepiandrosterone (DHEA)
Viagra (mechanism of action)
Therapy of prostate cancer with blockers of dihydrotestosterone formation
Mechanism of dominant follicle selection and function
Mechanism of corpus luteum regression
Estrogen effect in males
Cause of premenstrual tension, syndrome, and dysphoric disorder
Estrogen, learning, and Alzheimer's disease
Oxytocin and sperm transport

Parturition and placental corticotropin releasing hormone
Postcoital contraception
Lack of crossing-over in X and Y chromosomes
ACTH and onset of puberty
Leptin and onset of puberty
Tamoxifen and selective estrogen receptor modulators (SERMs)

Chapter 20 Carbohydrates and lipids as nonspecific markers on foreign cells
C-reactive protein and other nonspecific opsonins
Apoptosis of immune cells
Mechanism by which diversity arises in lymphocytes
Tumor necrosis factor and lymphocyte activation
Roles of acute phase proteins
Mechanisms of immune tolerance
Psychological stress and disease

Also, our coverage of pathophysiology, everyday applications of physiology, exercise physiology, and molecular biology have again been expanded.

Despite many additions, a ruthless removal of material no longer deemed essential has permitted us to maintain the text size unchanged from the previous edition.

Finally, *The Dynamic Human* CD-ROM is correlated to several figures. A Dynamic Human (dancing man) icon appears in appropriate figure legends. The *WCB Life Science Animations* Videotape Series is also correlated to several figure legends, and videotape icons appear in relevant figure legends.

Study Aids

A variety of pedagogical aids are utilized:

1. **Bold-faced key terms** throughout each chapter. Clinical terms are designated by *bold-faced italics.*
2. The illustration program is described earlier in the preface.
3. Summary tables. We have increased the number of reference and summary tables in this edition. Some summarize small or moderate amounts of information (for example, the summary of the major hormones influencing growth in Table 18–6), whereas others bring together large amounts of information that may be scattered throughout the book (for example, the reference figure of liver functions in Chapter 17). In several places, mini-glossaries are included as reference tables in the text (for example, the list of immune-system cells and chemical mediators in Chapter 20). Because the tables complement the figures, these two learning aids taken

together provide a rapid means of reviewing the most important material in a chapter.

4. End-of-section or chapter study aids
   a. Extensive summaries in outline form
   b. Key-term lists of all bold-faced words in the section/chapter (excluding the clinical terms)
   c. Comprehensive review questions in essay format. These review questions, in essence, constitute a complete list of learning objectives.
   d. Clinical term lists of all bold-face italicized words in the chapter. This serves to remind the student of how the physiology has been applied to clinical examples in the chapter.
   e. Thought questions that challenge the student to go beyond the memorization of facts to solve problems, often presented as case histories or experiments. Complete Answers to Thought Questions are given in Appendix A.

The chapter summaries, key-term definition lists, and review questions appear at the ends of the sections in those chapters that are broken into sections. These aids appear at the ends of nonsectioned chapters. Clinical term lists and thought questions are always at the ends of chapters.

5. A very extensive glossary, with pronunciation guides, is provided in Appendix B.
6. Appendixes C and D present, respectively, English-metric interconversions and Electrophysiology equations. Appendix E is an outline index of exercise physiology.
7. A complete alphabetized list of all abbreviations used in the text is given on the endpapers (the insides of the book's covers).

Supplements

1. *Essential Study Partner* (007-235897-1). This CD-ROM is an interactive study tool packed with hundreds of animations and learning activities, including quizzes, and interactive diagrams. A self-quizzing feature allows students to check their knowledge of a topic before moving on to a new module. Additional unit exams give students the opportunity to review coverage after completing entire units. A large number of anatomical supplements are also included. The ESP is packaged free with textbooks.
2. *Online Learning Center* (http://www.mhhe.com/biosci/ap/vander8e/). Students and instructors gain access to a world of opportunities through this Web site. Students will find quizzes, activities, links, suggested readings, and much more. Instructors will find all the enhancement

tools needed for teaching on-line, or for incorporating technology in the traditional course.

3. *The Student Study Guide* is now available as part of the Online Learning Center. Written by Donna Van Wynsberghe of the University of Wisconsin—Milwaukee, it contains a large variety of study aids, including learning hints and many test questions with answers.
4. *Instructor's Manual and Test Item File* (007-290803-3) by Sharon Russell of the University of California—Berkeley contains suggestions for teaching, as well as a complete test item file.
5. *MicroTest III testing software.* Available in Windows (007-290805-X) and Macintosh (007-290804-1). A computerized test generator for use with the text allows for quick creation of tests based on questions from the test item file and requires no programming experience.
6. *Overhead transparencies* (007-290806-8). A set of 200 full-color transparencies representing the most important figures from the book is available to instructors.
7. *McGraw-Hill Visual Resource Library* (007-290807-6). A CD-ROM containing all of the line art from the text with an easy-to-use interface program enabling the user to quickly move among the images, show or hide labels, and create a multimedia presentation.

Other Materials Available from McGraw-Hill

8. *The Dynamic Human* CD-ROM (0697-38935-9) illustrates the important relationships between anatomical structures and their functions in the human body. Realistic computer visualization and three-dimensional visualizations are the premier features of this CD-ROM. Various figures throughout this text are correlated to modules of *The Dynamic Human*. See pages xxvi–xxvii for a detailed listing of figures.
9. *The Dynamic Human Videodisc* (0-667-38937-5) contains all the animations (200+) from the CD-ROM. A bar code directory is also available.
10. *Life Science Animations Videotape Series* is a series of five videotapes containing 53 animations that cover many of the key physiological processes. Another videotape containing similar animations is also available, entitled *Physiological Concepts of Life Science*. Various figures throughout this text are correlated to animations from the *Life Science Animations*. See pages xxvii–xxviii for a detailed listing of figures.

Tape 1: Chemistry, The Cell, Energetics (0-697-25068-7)

Tape 2: Cell Division, Heredity, Genetics, Reproduction and Development (0-697-25069-5)

Tape 3: Animal Biology I (0-697-25070-9)

Tape 4: Animal Biology II (0-697-25071-7)

Tape 5: Plant Biology, Evolution, and Ecology (0-697-26600-1)

Tape 6: Physiological Concepts of Life Science (0-697-21512-1)

11. *Life Science Animations 3D* CD-ROM (007-234296-X). More than 120 animations that illustrate key biological processes are available at your fingertips on this exciting CD-ROM. This CD contains all of the animations found on the *Essential Study Partner* and much more. The animations can be imported into presentation programs, such as PowerPoint. Imagine the benefit of showing the animations during lecture.

12. *Life Science Animations 3D Videotape* (007-290652-9). Featuring 42 animations of key biologic processes, this tape contains 3D animations and is fully narrated. Various figures throughout this text are correlated to video animations. See page xxviii for a detailed listing of figures.

13. *Life Science Living Lexicon* CD-ROM (0-697-37993-0 hybrid) contains a comprehensive collection of life science terms, including definitions of their roots, prefixes, and suffixes as well as audio pronunciations and illustrations. The Lexicon is student-interactive, featuring quizzing and notetaking capabilities.

14. *The Virtual Physiology Lab* CD-ROM (0-697-37994-9 hybrid) containing 10 dry labs of the most common and important physiology experiments.

15. *Anatomy and Physiology Videodisc* (0-697-27716-X) is a four-sided videodisc containing more than 30 animations of physiological processes, as well as line art and micrographs. A bar code directory is also available.

16. *Anatomy and Physiology Video Series* consists of the following:
    a. Internal Organs and the Circulatory System of the Cat (0-697-13922-0)
    b. Blood Cell Counting, Identification & Grouping (0-697-11629-8)
    c. Introduction to the Human Cadaver and Prosection (0-697-11177-6)
    d. Introduction to Cat Dissection: Musculature (0-697-11630-1)

17. *Study Cards for Anatomy and Physiology* (007-290818-1) by Van De Graaff, et al., is a boxed set of 300 3-by-5 inch cards. It serves as a well-organized and illustrated synopsis of the structure and function of the human body. The Study Cards offer a quick and effective way for students to review human anatomy and physiology.

18. *Coloring Guide to Anatomy and Physiology* (0-697-17109-4) by Robert and Judith Stone emphasizes learning through the process of color association. The Coloring Guide provides a thorough review of anatomical and physiological concepts.

19. *Atlas of the Skeletal Muscles* (0-697-13790-2) by Robert and Judith Stone is a guide to the structure and function of human skeletal muscles. The illustrations help students locate muscles and understand their actions.

20. *Laboratory Atlas of Anatomy and Physiology* (0-697-39480-8) by Eder, et al., is a full-color atlas containing histology, human skeletal anatomy, human muscular anatomy, dissections, and reference tables.

21. *Case Histories in Human Physiology*, third edition, by Donna Van Wynesberghe and Gregory Cooley is a web-based workbook that stimulates analytical thinking through case studies and problem solving; includes an instructor's answer key. (www.mhhe.com/biosci/ap/vanwyn/).

22. *Survey of Infectious and Parasitic Diseases* (0-697-27535-3) by Kent M. Van De Graaff is a black-and-white booklet that presents the essential information on 100 of the most common and clinically significant diseases.

Acknowledgments

We are grateful to those colleagues who read one or more chapters during various stages of this revision:

Jennifer Carr Burtwistle
Northeast Community College

Nicholas G. Despo
Thiel College

Jean-Pierre Dujardin
The Ohio State University

David A. Gapp
Hamilton College

H. Maurice Goodman
University of Massachusetts Medical School

David L. Hammerman
Long Island University

Dona Housh
University of Nebraska Medical Center

Sarah N. Jerome
University of Central Arkansas

Fred Karsch
University of Michigan

Stephanie Burdine King
Wood College

Steven L. Kunkel
University of Michigan Medical School

Michael G. Levitzky
Louisiana State University Medical Center

Joseph V. Martin
Rutgers University

John L. McCarthy
Southern Methodist University

Kerry McDonald
University of Missouri

Philip Nelson
Barstow College

C. S. Nicoll
University of California, Berkeley

Colleen J. Nolan
St. Mary's University

David Quadagno
Florida State University

Sharon M. Russell
University of California, Berkeley

Allen F. Sanborn
Barry University

David J. Saxon
Morehead State University

Amanda Starnes
Emory University

Edward K. Stauffer
University of Minnesota

Leeann Sticker
Northwestern State University of Louisiana

James D. Stockand
Emory University

Richard Stripp
Arnold and Marie Schwartz College of Pharmacy,
Long Island University

Donna Van Wynsberghe
University of Wisconsin-Milwaukee

Samuel J. Velez
Dartmouth College

Benjamin Walcott
SUNY at Stony Brook

Curt Walker
Dixie College

R. Douglas Watson
University of Alabama at Birmingham

Scott Wells
Missouri Southern State College

Eric P. Widmaier
Boston University

Judy Williams
Southeastern Oklahoma State University

John Q. Zhang
Sherman College of Straight Chiropractic

Their advice was very useful in helping us to be accurate and balanced in our coverage. We hope that they will be understanding of the occasions when we did not heed their advice, and we are, of course, solely responsible for any errors that have crept in. We would like to express our appreciation to Kris Tibbetts, Sponsoring Editor; Pat Anglin, Developmental Editor; and Peggy Selle, Project Manager.

To our parents, and to Judy, Peggy, and Joe without whose understanding it would have been impossible

human Physiology

**The Mechanisms of
Body Function**

Visual Tour

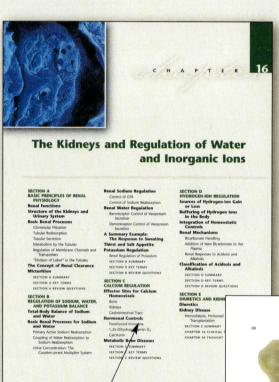

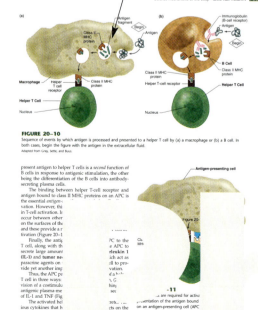

Beautifully Rendered Full-color Art

Almost all of the figures have been redone in this edition, ranging from a complete redrawing of the figure to simple labeling changes. A realistic three-dimensional perspective has been added to many of the figures for greater clarity and understanding of the concept.

Chapter Outline

Before you begin a chapter, it is important to have a broad overview of what it covers. Each chapter has an outline that permits you to see at a glance how the chapter is organized and what major topics are included.

Color-coded Illustrations

Color-coding is effectively used to promote learning. For example, there are specific colors for the extracellular fluid, the intracellular fluid, muscle, and the lumen of the renal tubules and GI tract.

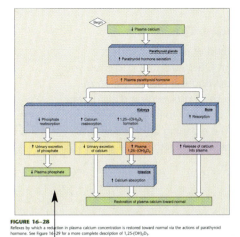

FIGURE 16–28
Reflexes by which a reduction in plasma calcium concentration is restored toward normal via the actions of parathyroid hormone. See Figure 16-29 for a more complete description of 1,25-(OH)₂D₃.

Flow Diagrams

Long a hallmark of this book, extensive use of flow diagrams have been continued and expanded in this edition. A bookmark has been included with your book to give a further explanation.

Summary Tables

Some summary tables summarize small or moderate amounts of information whereas others bring together large amounts of information that may be scattered throughout the book. The tables complement the accompanying figures to provide a rapid means of reviewing the most important material in a chapter.

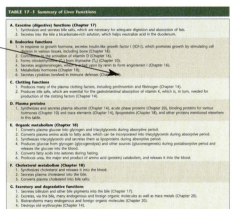

Thought Questions

At the end of each chapter are Thought Questions that challenge you to go beyond the memorization of facts to solve problems and encourage you to stop and think more deeply about the meaning or broader significance of what you have just read.

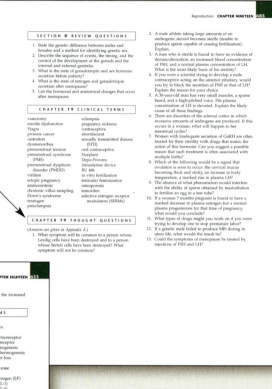

SECTION D REVIEW QUESTIONS

1. State the genetic difference between males and females and a method for identifying genetic sex.
2. Describe the sequence of events, the timing, and the control of the development of the gonads and the internal and external genitalia.
3. What is the state of gonadotropin and sex hormone secretion before puberty?
4. What is the state of estrogen and gonadotropin secretion after menopause?
5. List the hormonal and anatomical changes that occur after menopause.

CHAPTER 19 CLINICAL TERMS

vasectomy	eclampsia
erectile dysfunction	pregnancy sickness
Viagra	contraceptive
prostate cancer	abortifacient
castration	sexually transmitted disease (STD)
dysmenorrhea	oral contraceptive
premenstrual tension	Norplant
premenstrual syndrome (PMS)	Depo-Provera
premenstrual dysphoric disorder (PMDD)	intrauterine device
virilism	RU 486
ectopic pregnancy	in vitro fertilization
amniocentesis	testicular feminization
chorionic villus sampling	osteoporosis
Down's syndrome	tamoxifen
teratogen	selective estrogen receptor modulators (SERMs)
preeclampsia	

CHAPTER 19 THOUGHT QUESTIONS

(Answers are given in Appendix A.)

1. What symptom will be common to a person whose Leydig cells have been destroyed and to a person whose Sertoli cells have been destroyed? What symptom will not be common?
2. A male athlete taking large amounts of an androgenic steroid becomes sterile (unable to produce sperm capable of causing fertilization). Explain.
3. A man who is sterile is found to have no evidence of demasculinization, an increased blood concentration of FSH, and a normal plasma concentration of LH. What is the most likely basis of his sterility?
4. If you were a scientist trying to develop a male contraceptive acting on the anterior pituitary, would you try to block the secretion of FSH or that of LH? Explain the reason for your choice.
5. A 30-year-old man has very small muscles, a sparse beard, and a high-pitched voice. His plasma concentration of LH is elevated. Explain the likely cause of all these findings.
6. There are disorders of the adrenal cortex in which excessive amounts of androgens are produced. If this occurs in a woman, what will happen to her menstrual cycles?
7. Women with inadequate secretion of GnRH are often treated for their sterility with drugs that mimic the action of this hormone. Can you suggest a possible reason that such treatment is often associated with multiple births?
8. Which of the following would be a signal that ovulation is soon to occur: the cervical mucus becoming thick and sticky, an increase in body temperature, a marked rise in plasma LH?
9. The absence of what phenomenon would interfere with the ability of sperm obtained by masturbation to fertilize an egg in a test tube?
10. If a woman 7 months pregnant is found to have a marked decrease in plasma estrogen but a normal plasma progesterone for that time of pregnancy, what would you conclude?
11. What types of drugs might you work on if you were trying to develop or to stop premature labor?
12. If a genetic male failed to produce MIS during in utero life, what would the result be?
13. Could the symptoms of menopause be treated by injections of FSH and LH?

Regulation of Total-Body Energy Stores

I. Energy storage as fat can be positive or negative when the metabolic rate is less than or greater than, respectively, the energy content of ingested food.
   a. Energy storage is regulated mainly by reflex adjustment of food intake.
   b. In addition, the metabolic rate increases or decreases to some extent when food intake is chronically increased or decreased, respectively.
II. Food intake is controlled by leptin, secreted by adipose-tissue cells, and a variety of satiety factors, as summarized in Figure 18–17.
III. Being overweight or obese, the result of an imbalance between food intake and metabolic rate, increases the risk of many diseases.

Regulation of Body Temperature

I. Core body temperature shows a circadian rhythm, being highest during the day and lowest at night.
II. The body exchanges heat with the external environment by radiation, conduction, convection, and evaporation of water from the body surface.
III. The hypothalamus and other brain areas contain the integrating centers for temperature-regulating reflexes, and both peripheral and central thermoreceptors participate in these reflexes.
IV. Body temperature is regulated by altering heat production and/or heat loss so as to change total body heat content.
   a. Heat production is altered by increasing muscle tone, shivering, and voluntary activity.
   b. Heat loss by radiation, conduction, and convection depends on the difference between the skin surface and the environment.
   c. In response to cold, skin temperature is decreased by decreasing skin blood flow through reflex stimulation of the sympathetic nerves to the skin. In response to heat, skin temperature is increased by inhibiting these nerves.
   d. Behavioral responses such as putting on more clothes also influence heat loss.
   e. Evaporation of water occurs all the time as insensible loss from the skin and respiratory lining. Additional water for evaporation is supplied by sweat, stimulated by the sympathetic nerves to the sweat glands.
   f. Increased heat production is essential for temperature regulation at environmental temperatures below the thermoneutral zone, and sweating is essential at temperatures above this zone.
V. Temperature acclimatization to heat is achieved by an earlier onset of sweating, an increased volume of sweat, and a decreased sodium concentration of the sweat.
VI. Fever is due to a resetting of the temperature set point so that heat production is increased and heat loss is decreased in order to raise body temperature to the new set point and keep it there. The stimulus is endogenous pyrogen, which is interleukin 1 and other peptides as well.

VII. The hyperthermia of exercise is due to the increased heat produced by the muscles.

SECTION C KEY TERMS

external work	convection
internal work	wind-chill index
total energy expenditure	evaporation
kilocalorie (kcal)	peripheral thermoreceptor
metabolic rate	central thermoreceptor
basal metabolic rate (BMR)	shivering thermogenesis
caloric effect	nonshivering thermogenesis
food-induced thermogenesis	insensible water loss
leptin	sweat gland
satiety signal	fever
body mass index (BMI)	endogenous pyrogen (EP)
homeothermic	interleukin 1 (IL-1)
radiation	interleukin 6 (IL-6)
conduction	endogenous cryogens
	hyperthermia

SECTION C REVIEW QUESTIONS

1. State the formula relating total energy expenditure, heat produced, external work, and energy storage.
2. What two hormones alter the basal metabolic rate?
3. State the equation for total-body energy balance. Describe the three possible states of balance with regard to energy storage.
4. What happens to the basal metabolic rate after a person has either lost or gained weight?
5. List five satiety signals.
6. List three beneficial effects of exercise in a weight-loss program.
7. Compare and contrast the four mechanisms for heat loss.
8. Describe the control of skin blood vessels during exposure to cold or heat.
9. With a diagram, summarize the reflex responses to heat or cold. What are the dominant mechanisms for temperature regulation in the thermoneutral zone and in temperatures below and above this range?
10. What changes are exhibited by a heat-acclimatized person?
11. Summarize the sequence of events leading to a fever and contrast this to the sequence leading to hyperthermia during exercise.

CHAPTER 18 CLINICAL TERMS

diabetes mellitus	sulfonylureas
insulin-dependent diabetes mellitus (IDDM)	fasting hypoglycemia
	atherosclerosis
noninsulin-dependent diabetes mellitus (NIDDM)	cancer
	oncogene
diabetic ketoacidosis	giantism
insulin resistance	dwarfism
	acromegaly

Chapter Summary

A summary, in outline form, at the end of each chapter reinforces your mastery of the chapter content.

Answers to Thought Questions

Complete answers to Thought Questions are given in Appendix A.

Glossary

A very extensive Glossary, with pronunciation guides, is provided in Appendix B.

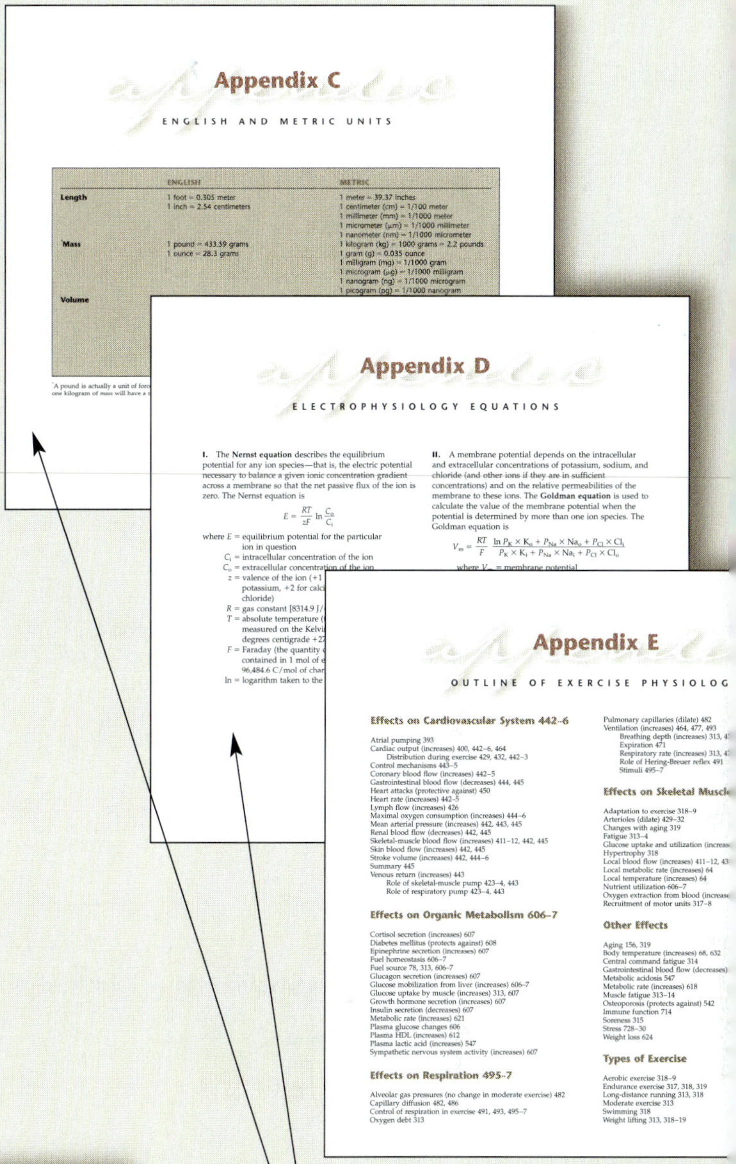

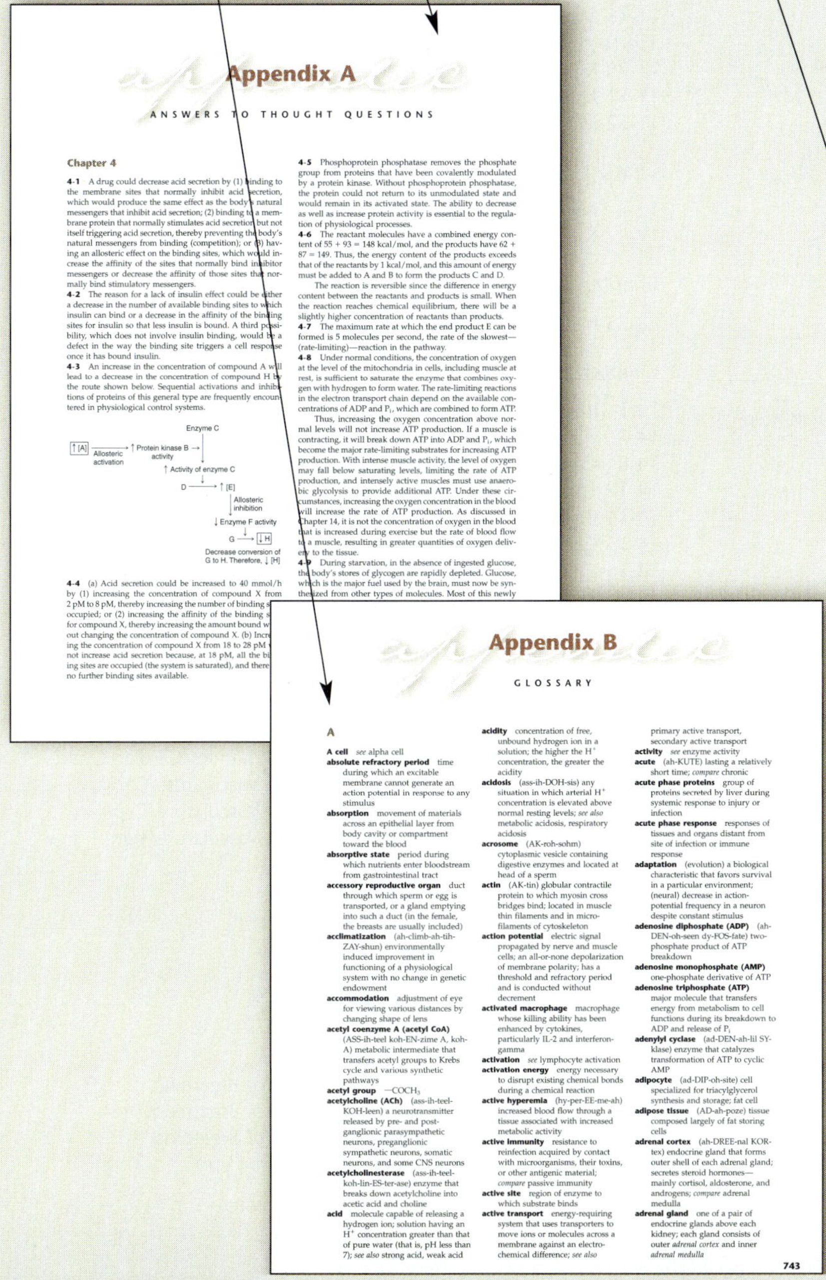

Appendix C

ENGLISH AND METRIC UNITS

	ENGLISH	METRIC
Length	1 foot = 0.305 meter 1 inch = 2.54 centimeters	1 meter = 39.37 inches 1 centimeter (cm) = 1/100 meter 1 millimeter (mm) = 1/1000 meter 1 micrometer (μm) = 1/1000 millimeter 1 nanometer (nm) = 1/1000 micrometer
Mass	1 pound = 433.59 grams 1 ounce = 28.3 grams	1 kilogram (kg) = 1000 grams = 2.2 pounds 1 gram = 0.035 ounce 1 milligram (mg) = 1/1000 gram 1 microgram (μg) = 1/1000 milligram 1 nanogram (ng) = 1/1000 microgram 1 picogram (pg) = 1/1000 nanogram
Volume		

Appendix D

ELECTROPHYSIOLOGY EQUATIONS

I. The **Nernst equation** describes the equilibrium potential for any ion species—that is, the electric potential necessary to balance a given ionic concentration gradient across a membrane so that the net passive flux of the ion is zero. The Nernst equation is

$$E = \frac{RT}{zF} \ln \frac{C_o}{C_i}$$

where E = equilibrium potential for the particular ion in question
C_i = intracellular concentration of the ion
C_o = extracellular concentration of the ion
z = valence of the ion (+1 potassium, +2 for calcium, chloride)
R = gas constant [8314.9 J/...]
T = absolute temperature measured on the Kelvin degrees centigrade +273
F = Faraday (the quantity contained in 1 mol of... 96,484.6 C/mol of char...
$\ln$ = logarithm taken to the...

II. A membrane potential depends on the intracellular and extracellular concentrations of potassium, sodium, and chloride (and other ions if they are in sufficient concentrations) and on the relative permeabilities of the membrane to these ions. The **Goldman equation** is used to calculate the value of the membrane potential when the potential is determined by more than one ion species. The Goldman equation is

$$V_m = \frac{RT}{F} \ln \frac{P_K \times K_o + P_{Na} \times Na_o + P_{Cl} \times Cl_i}{P_K \times K_i + P_{Na} \times Na_i + P_{Cl} \times Cl_o}$$

where V_m = membrane potential

Appendix E

OUTLINE OF EXERCISE PHYSIOLOGY

Effects on Cardiovascular System 442–6

Atrial pumping 393
Cardiac output (increases) 400, 442–6, 464
  Distribution during exercise 429, 432, 442–3
Control mechanisms 443–5
Coronary blood flow (increases) 442–5
Gastrointestinal blood flow (decreases) 444, 445
Heart attacks (protective against) 450
Heart rate (increases) 442–5
Lymph flow (increases) 426
Maximal oxygen consumption (increases) 444–6
Mean arterial pressure (increases) 442, 443, 445
Renal blood flow (decreases) 442, 445
Skeletal-muscle blood flow (increases) 411–12, 442, 445
Skin blood flow (increases) 442, 445
Stroke volume (increases) 442, 444–6
Summary 445
Venous return (increases) 443
  Role of skeletal-muscle pump 423–4, 443
  Role of respiratory pump 423–4, 443

Pulmonary capillaries (dilate) 482
Ventilation (increases) 464, 477, 493
  Breathing depth (increases) 313, 4...
  Expiration 471
  Respiratory rate (increases) 313...
  Role of Hering-Breuer reflex in...
  Stimuli 495–7

Effects on Skeletal Muscle

Adaptation to exercise 318–9
Arterioles (dilate) 429–32
Changes with aging 319
Fatigue 313–4
Glucose uptake and utilization (increase...
Hypertrophy 318
Local blood flow (increases) 411–12, 43...
Local metabolic rate (increased) 64
Local temperature (increases) 442
Nutrient utilization 606–7
Oxygen extraction from blood (increase...
Recruitment of motor units 317–8

Effects on Organic Metabolism 606–7

Cortisol secretion (increases) 607
Diabetes mellitus (protects against) 608
Epinephrine secretion (increases) 607
Fuel homeostasis 606–7
Fuel source 78, 313, 606–7
Glucagon secretion (increases) 607
Glucose mobilization from liver (increases) 606–7
Glucose uptake by muscle (increases) 313, 607
Growth hormone secretion (increases) 607
Insulin secretion (decreases) 607
Metabolic rate (increases) 621
Plasma glucose change 606
Plasma HDL (increases) 612
Plasma lactic acid (increases) 547
Sympathetic nervous system activity (increases) 607

Other Effects

Aging 156, 319
Body temperature (increases) 68, 632
Central command fatigue 314
Gastrointestinal blood flow (decreases...
Metabolic acidosis 547
Metabolic rate (increases) 618
Muscle fatigue 313–14
Osteoporosis (protects against) 542
Immune function 714
Soreness 315
Stress 728–30
Weight loss 624

Effects on Respiration 495–7

Alveolar gas pressures (no change in moderate exercise) 482
Capillary diffusion 482, 486
Control of respiration in exercise 491, 493, 495–7
Oxygen debt 313

Types of Exercise

Aerobic exercise 318–9
Endurance exercise 317, 318, 319
Long-distance running 313, 318
Moderate exercise 315
Swimming 318
Weight lifting 313, 318–19

Appendix A

ANSWERS TO THOUGHT QUESTIONS

Chapter 4

4-1 A drug could decrease acid secretion by (1) binding to the membrane sites that normally inhibit acid secretion, which would produce the same effect as the body's natural messengers that inhibit acid secretion; (2) binding to a membrane protein that normally stimulates acid secretion but not itself triggering acid secretion, thereby preventing the body's natural messengers from binding (competition); or (3) having an allosteric effect on the binding sites, which would increase the affinity of the sites that normally bind inhibitor messengers or decrease the affinity of those sites that normally bind stimulatory messengers.

4-2 The reason for a lack of insulin effect could be either a decrease in the number of available binding sites to which insulin can bind or a decrease in the affinity of the binding sites for insulin so that less insulin is bound. A third possibility, which does not involve insulin binding, would be a defect in the way the binding site triggers a cell response once it has bound insulin.

4-3 An increase in the concentration of compound A will lead to a decrease in the concentration of compound H by the route shown below. Sequential activations and inhibitions of proteins of this general type are frequently encountered in physiological control systems.

Enzyme C
↑ [A] → Allosteric activation → ↑ Protein kinase B activity
↑ Activity of enzyme C
↓ D → ↑ [E]
Allosteric inhibition
↓ Enzyme F activity
↓ G → ↑ [H]
Decrease conversion of G to H. Therefore, ↓ [H]

4-4 (a) Acid secretion could be increased to 40 mmol/h by (1) increasing the concentration of compound X from 2 pM to 8 pM, thereby increasing the number of binding sites occupied; or (2) increasing the affinity of the binding sites for compound X, thereby increasing the amount bound without changing the concentration of compound X. (b) Increasing the concentration of compound X from 18 to 28 pM will not increase acid secretion because, at 18 pM, all the binding sites are occupied (the system is saturated), and there are no further binding sites available.

4-5 Phosphoprotein phosphatase removes the phosphate group from proteins that have been covalently modulated by a protein kinase. Without phosphoprotein phosphatase, the protein could not return to its unmodulated state and would remain in its activated state. The ability to decrease as well as increase protein activity is essential to the regulation of physiological processes.

4-6 The reactant molecules have a combined energy content of $55 + 93 = 148$ kcal/mol, and the products have $62 + 87 = 149$. Thus, the energy content of the products exceeds that of the reactants by 1 kcal/mol, and this amount of energy must be added to A and B to form the products C and D.

The reaction is reversible since the difference in energy content between the reactants and products is small. When the reaction reaches chemical equilibrium, there will be a slightly higher concentration of reactants than products.

4-7 The maximum rate at which the end product E can be formed is 5 molecules per second, the rate of the slowest—(rate-limiting)—reaction in the pathway.

4-8 Under normal conditions, the concentration of oxygen at the level of the mitochondria in cells, including muscle at rest, is sufficient to saturate the enzyme that combines oxygen with hydrogen to form water. The rate-limiting reactions in the electron transport chain depend on the available concentrations of ADP and P_i, which are combined to form ATP. Thus, increasing the oxygen concentration above normal levels will not increase ATP production. If a muscle is contracting, it will break down ATP into ADP and P_i, which become the major rate-limiting substrates for increasing ATP production. With intense muscle activity, the level of oxygen may fall below saturating levels, limiting the rate of ATP production, and intensely active muscles must use anaerobic glycolysis to provide additional ATP. Under these circumstances, increasing the oxygen concentration in the blood will increase the rate of ATP production. As discussed in Chapter 14, it is not the concentration of oxygen in the blood that is increased during exercise but the rate of blood flow to a muscle, resulting in greater quantities of oxygen delivery to the tissue.

4-9 During starvation, in the absence of ingested glucose, the body's stores of glycogen are rapidly depleted. Glucose, which is the major fuel used by the brain, must now be synthesized from other types of molecules. Most of this newly

Appendix B

GLOSSARY

A

A cell *see* alpha cell

absolute refractory period time during which an excitable membrane cannot generate an action potential in response to any stimulus

absorption movement of materials across an epithelial layer from body cavity or compartment toward the blood

absorptive state period during which nutrients enter bloodstream from gastrointestinal tract

accessory reproductive organ duct through which sperm or egg is transported, or a gland emptying into such a duct (in the female, the breasts are usually included)

acclimatization (ah-climb-ah-tih-ZAY-shun) environmentally induced improvement in functioning of a physiological system with no change in genetic endowment

accommodation adjustment of eye for viewing various distances by changing shape of lens

acetyl coenzyme A (acetyl CoA) (ASS-ih-teel koh-EN-zime A, koh-A) metabolic intermediate that transfers acetyl groups to Krebs cycle and various synthetic pathways

acetyl group —COCH₃

acetylcholine (ACh) (ass-ih-teel-KOH-leen) a neurotransmitter released by pre- and post-ganglionic parasympathetic neurons, preganglionic sympathetic neurons, and some CNS neurons

acetylcholinesterase (ass-ih-teel-koh-lin-ES-ter-ase) enzyme that breaks down acetylcholine into acetic acid and choline

acid molecule capable of releasing a hydrogen ion; solution having an H⁺ concentration greater than that of pure water (that is, pH less than 7); *see also* strong acid, weak acid

acidity concentration of free, unbound hydrogen ion in a solution; the higher the H⁺ concentration, the greater the acidity

acidosis (ass-ih-DOH-sis) any situation in which arterial H⁺ concentration is elevated above normal resting levels; *see also* metabolic acidosis, respiratory acidosis

acrosome (AK-roh-sohm) cytoplasmic vesicle containing digestive enzymes and located at head of a sperm

actin (AK-tin) globular contractile protein to which myosin cross bridges bind; located in muscle thin filaments and microfilaments of cytoskeleton

action potential electric signal propagated by nerve and muscle cells; an all-or-none depolarization of membrane polarity; has a threshold and refractory period and is conducted without decrement

activated macrophage macrophage whose killing ability has been enhanced by cytokines, particularly IL-2 and interferon-gamma

activation *see* lymphocyte activation

activation energy energy necessary to disrupt existing chemical bonds during a chemical reaction

active hyperemia (hy-per-EE-me-ah) increased blood flow through a tissue associated with increased metabolic activity

active immunity resistance to reinfection acquired by contact with microorganisms, their toxins, or other antigenic material; *compare* passive immunity

active site region of enzyme to which substrate binds

active transport energy-requiring system that uses transporters to move ions or molecules across a membrane against an electrochemical difference; *see also*

primary active transport, secondary active transport

activity *see* enzyme activity

acute (ah-KUTE) lasting a relatively short time; *compare* chronic

acute phase proteins group of proteins secreted by liver during systemic response to injury or infection

acute phase response responses of tissues and organs distant from site of infection or immune response

adaptation (evolution) a biological characteristic that favors survival in a particular environment; (neural) decrease in action-potential frequency in a neuron despite constant stimulus

adenosine diphosphate (ADP) (ah-DEN-oh-seen dy-FOS-fate) two-phosphate product of ATP breakdown

adenosine monophosphate (AMP) one-phosphate derivative of ATP

adenosine triphosphate (ATP) major molecule that transfers energy from metabolism to cell functions during its breakdown to ADP and release of P_i

adenylyl cyclase (ad-DEN-ah-lil SY-klase) enzyme that catalyzes transformation of ATP to cyclic AMP

adipocyte (ad-DIP-oh-site) cell specialized for triacylglycerol synthesis and storage; fat cell

adipose tissue (AD-ah-poze) tissue composed largely of fat storing cells

adrenal cortex (ah-DREE-nal KOR-tex) endocrine gland that forms outer shell of each adrenal gland; secretes steroid hormones—mainly cortisol, aldosterone, and androgens; *compare* adrenal medulla

adrenal gland one of a pair of endocrine glands above each kidney; each gland consists of outer *adrenal cortex* and inner *adrenal medulla*

Appendixes

Appendix C presents English-metric interconversions, Appendix D features Electrophysiology equations, and Appendix E is an outline index of Exercise Physiology.

Correlations

15-25 Respiratory/Explorations/Oxygen Transport
Respiratory/Explorations/Gas Exchange
15-27 Respiratory/Explorations/Oxygen Transport
Respiratory/Explorations/Gas Exchange

Chapter 16

16-1 Urinary/Anatomy/Gross Anatomy
16-2 Urinary/Anatomy/Nephron Anatomy
16-3 Urinary/Anatomy/3D Viewer: Nephron
Urinary/Anatomy/Nephron Anatomy
16-4 Urinary/Anatomy/Kidney Anatomy
16-6 Urinary/Explorations/Urine Formation
16-11 Urinary/Explorations/Urine Formation

Chapter 17

17-1 Digestive/Anatomy/3D Viewer: Digestive Anatomy
Digestive/Anatomy/Gross Anatomy
17-3 Digestive/Anatomy/3D Viewer: Digestive Anatomy
Digestive/Anatomy/Gross Anatomy
Digestive/Explorations/Digestion

17-4 Digestive/Anatomy/3D Viewer: Digestive Anatomy
Digestive/Anatomy/Gross Anatomy
17-7 Digestive/Histology/Duodenal Villi
17-11 Digestive/Explorations/Digestion
17-12 Digestive/Explorations/Digestion
17-14 Digestive/Explorations/Oral Cavity
17-15 Digestive/Anatomy/Gross Anatomy
17-16 Digestive/Anatomy/Gross Anatomy
17-17 Digestive/Histology/Fundic Stomach
17-21 Digestive/Explorations/Digestion
17-22 Digestive/Explorations/Digestion
17-25 Digestion/Anatomy/Gross Anatomy
17-33 Digestion/Anatomy/Gross Anatomy

Chapter 18

18-7 Endocrine/Clinical Applications/Diabetes
18-9 Endocrine/Clinical Applications/Diabetes
18-14 Skeletal/Explorations/Cross section of a Long Bone
18-21 Immune/Explorations/Non-specific Immunity

Life Science Animations Correlation Guide

Chapter 3

3-4	Tape 1	Concept 2	Journey into a Cell
3-12	Tape 1	Concept 2	Journey into a Cell
3-13	Tape 1	Concept 2	Journey into a Cell
3-14	Tape 1	Concept 2	Journey into a Cell

Chapter 4

4-16	Tape 1	Concept 11	ATP as an Energy Carrier
4-17	Tape 1	Concept 11	ATP as an Energy Carrier
4-18	Tape 1	Concept 11	ATP as an Energy Carrier
4-19	Tape 1	Concept 5	Glycolysis
4-22	Tape 1	Concept 6	Oxidative Respiration
	Tape 6	Concept 5	Electron Transport and Oxidative Phosphorylation
4-23	Tape 1	Concept 6	Oxidative Respiration
	Tape 6	Concept 5	Electron Transport and Oxidative Phosphorylation
	Tape 1	Concept 7	Electron Transport Chain and the Production of ATP
4-24	Tape 1	Concept 6	Oxidative Respiration
	Tape 6	Concept 5	Electron Transport and Oxidative Phosphorylation
	Tape 1	Concept 7	Electron Transport Chain and the Production of ATP

Chapter 5

5-3	Tape 2	Concept 16	Transcription of a Gene
5-4	Tape 2	Concept 17	Protein Synthesis
5-6	Tape 2	Concept 17	Protein Synthesis
5-9	Tape 2	Concept 16	Transcription of a Gene
5-10	Tape 1	Concept 4	Cellular Secretion
	Tape 6	Concept 4	Cellular Secretion
5-11	Tape 2	Concept 15	DNA Replication
5-12	Tape 2	Concept 12	Mitosis
5-13	Tape 2	Concept 12	Mitosis

Chapter 6

6-11	Tape 6	Concept 3	Active Transport
6-13	Tape 6	Concept 3	Active Transport
6-19	Tape 6	Concept 2	Osmosis
6-21	Tape 1	Concept 3	Endocytosis
6-22	Tape 1	Concept 3	Endocytosis
6-23	Tape 1	Concept 3	Active Transport
6-24	Tape 1	Concept 3	Active Transport

Chapter 7

7-14	Tape 6	Concept 12	Cyclic AMP Action
	Tape 3	Concept 28	Peptide Hormone Action (cAMP)
7-17	Tape 6	Concept 12	Cyclic AMP Action
	Tape 3	Concept 28	Peptide Hormone Action (cAMP)

Chapter 8

8-3	Tape 3	Concept 22	Formation of Myelin Sheath
8-16	Tape 6	Concept 7	Temporal and Spatial Summation
8-24	Tape 3	Concept 24	Signal Integration
8-33	Tape 6	Concept 8	Synaptic Transmission

Chapter 9

9-5	Tape 3	Concept 24	Signal Integration
9-25	Tape 6	Concept 9	Visual Accommodation
9-34	Tape 3	Concept 27	Organ of Corti
9-35	Tape 3	Concept 27	Organ of Corti
9-36	Tape 3	Concept 27	Organ of Corti
9-37	Tape 3	Concept 27	Organ of Corti
9-38	Tape 3	Concept 27	Organ of Corti
9-39	Tape 3	Concept 26	Organ of Static Equilibrium
9-42	Tape 3	Concept 26	Organ of Static Equilibrium

Chapter 10

10-2	Tape 1	Concept 4	Cellular Secretion
	Tape 6	Concept 4	Cellular Secretion

Chapter 11

11-4	Tape 3	Concept 29	Levels of Muscle Structure
11-8	Tape 3	Concept 30	Sliding Filament Model
11-12	Tape 3	Concept 30	Sliding Filament Model
11-13	Tape 3	Concept 31	Regulation of Muscle Contraction
11-16	Tape 3	Concept 31	Regulation of Muscle Contraction
11-26	Tape 1	Concept 6	Oxidative Respiration
	Tape 6	Concept 5	Electron Transport Chain and Oxidative Phosphorylation
11-37	Tape 3	Concept 31	Regulation of Muscle Contraction

Chapter 12

12-3	Tape 3	Concept 24	Signal Integration
12-8	Tape 3	Concept 25	Reflex Arc

Chapter 14

14-7	Tape 4	Concept 37	Blood Circulation
14-12	Tape 4	Concept 37	Blood Circulation
14-14	Tape 4	Concept 37	Blood Circulation
14-20	Tape 4	Concept 38	Production of Electrocardiogram
14-21	Tape 4	Concept 38	Production of Electrocardiogram
14-24	Tape 4	Concept 32	Cardiac Cycle and Production of Sounds
14-25	Tape 4	Concept 32	Cardiac Cycle and Production of Sounds
	Tape 4	Concept 38	Production of Electrocardiogram

Chapter 17

17-9	Tape 4	Concept 36	Digestion of Lipids
17-10	Tape 4	Concept 36	Digestion of Lipids
17-11	Tape 4	Concept 36	Digestion of Lipids
17-21	Tape 4	Concept 35	Digestion of Proteins
17-22	Tape 4	Concept 35	Digestion of Proteins
17-32	Tape 4	Concept 33	Peristalsis

Life Science 3D Animations Correlation Guide

Chapter 2

2-22	Tape 1	Module 13	Structure of DNA
2-23	Tape 1	Module 13	Structure of DNA
2-24	Tape 1	Module 13	Structure of DNA

Chapter 4

4-8	Tape 1	Module 7	Enzyme Action
4-22	Tape 1	Module 9	Electron Transport Chain
4-23	Tape 1	Module 9	Electron Transport Chain
4-24	Tape 1	Module 9	Electron Transport Chain

Chapter 5

5-3	Tape 2	Module 18	Transcription
5-4	Tape 2	Module 19	Translation
5-6	Tape 2	Module 19	Translation
5-9	Tape 2	Module 18	Transcription
5-10	Tape 1	Module 3	Cellular Secretion
5-11	Tape 2	Module 14	DNA Replication
5-12	Tape 2	Module 10	Mitosis
5-13	Tape 2	Module 10	Mitosis

Chapter 6

6-1	Tape 1	Module 4	Diffusion
6-12	Tape 1	Module 6	Sodium/Potassium Pump
6-18	Tape 1	Module 4	Diffusion
6-19	Tape 1	Module 5	Osmosis

Chapter 8

8-13	Tape 1	Module 6	Sodium/Potassium Pump
8-18	Tape 5	Module 39	Action Potential

Chapter 10

10-2	Tape 1	Module 3	Cellular Secretion
10-7	Tape 5	Module 41	Hormone Action

Chapter 11

11-8	Tape 5	Module 40	Muscle Contraction
11-12	Tape 5	Module 40	Muscle Contraction
11-16	Tape 5	Module 40	Muscle Contraction

Chapter 15

15-6	Tape 5	Module 37	Gas Exchange
15-8	Tape 1	Module 2	Boyle's Law
15-25	Tape 5	Module 37	Gas Exchange

Chapter 16

16-6	Tape 5	Module 38	Kidney Function
16-14	Tape 5	Module 38	Kidney Function

The Online
Learning Center
Your Password
to Success

http://www.mhhe.com/biosci/ap/vander8e/.

Human Physiology: The Mechanisms of Body Function presents a world of opportunities in physiology. A multitude of learning and teaching tools are at your fingertips on our web site. All the student resources available on the Online Learning Center are included with your purchase of a new copy of the text.

Student Resources

Animations
Extensive Quizzing
Histology Exercise
Chapter Outlines
Message Board
Web Links
Additional Readings
Clinical Applications
Internet Activities
Case Studies
Art Labeling Exercises
Student Study Guide

Instructor Resources

Instructor's Manual
Online Periodicals
Classroom Activities
Chapter Outlines
Page Out: Course Web Site
  Development Center
Message Board
Web Links

**Imagine the advantages of having so many
learning and teaching tools
all in one place—all at your fingertips—FREE.**

Essential Study Partner CD-ROM

A free study partner that engages, investigates, and reinforces what you are learning from your textbook. You'll find the **Essential Study Partner** for *Human Physiology* to be a complete, interactive student study tool packed with hundreds of animations and learning activities. From quizzes to interactive diagrams, you'll find that there has never been a better study partner to ensure the mastery of core concepts. Best of all, it's **FREE** with your new textbook purchase.

The unit pop-up menu is accessible at anytime within the program. Clicking on the current unit will bring up a menu of other units available in the program.

The topic menu contains an interactive list of the available topics. Clicking on any of the listings within this menu will open your selection and will show the specific concepts presented within this topic. Clicking any of the concepts will move you to your selection. You can use the UP and DOWN arrow keys to move through the topics.

To the right of the arrows is a row of icons that represent the number of screens in a concept. There are three different icons, each representing different functions that a screen in that section will serve. The screen that is currently displayed will highlight yellow and visited ones will be checked.

Along the bottom of the screen you will find various navigational aids. At the left are arrows that allow you to page forward and backward through text screens or interactive exercise screens. You can also use the LEFT and RIGHT arrows on your keyboard to perform the same function.

The film icon represents an animation screen.

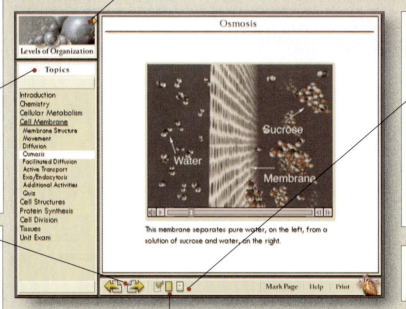

The activity icon represents an interactive learning activity.

The page icon represents a page of informational text.

The heart icon allows you to access additional controls, navigation, and exit.

The "Audio" button opens up the controls for the audio. You may turn the sound on or off and adjust the volume.

The Bookmarks button lets you access the pages that you have marked with the "Mark Page" button.

"Search" will help you locate specific content in the CD.

"Text Guide" will find the correlation between areas in the textbook and screens in the CD.

Use the "Glossary" to find definitions of underlined words.

"Internet" will launch your web browser and take you to the McGraw-Hill web site.

"Exit" will leave the program.

The "Mark Page" button allows you to tag a screen. The tagged screen then appears in the Bookmarks.

"Help" will bring you to the guided help walkthrough.

"Print" will print the current screen's content area.

Each topic ends with a quiz for review of the material. The quiz has a "Review Topic" feature for additional study.

New Version 2.0!

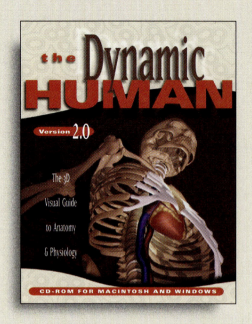

by McGraw-Hill and Engineering Animation, Inc.

1998 • CD-ROMs (two CD set) for Macintosh and Windows • ISBN 0-697-38935-9

This powerful new version of *The Dynamic Human* CD-ROM builds upon the best in anatomy and physiology technology. **Version 2.0** features added detail and depth— expanding upon a variety of topics and incorporating the World Wide Web.

Features

- The World Wide Web provides the ideal opportunity to implement and greatly extend the powers of this updated CD-ROM. The all-new *Dynamic Human Version 2.0* web site features an immense amount of supplemental information, including:
  - approximately 20 to 30 readings per body system
  - extensive links to additional informational sites on the World Wide Web
  - supplemental visuals, including animations, video, and vivid illustrations
  - correlation guides for many McGraw-Hill texts
  - on-line self-testing
- An updated design and interface make navigation much simpler with easy-to-use pop-up menus.
- Each body system can be studied through four standard content areas: anatomy, explorations, clinical concepts, and histology.
- A self-quizzing section is available for each body system.
- Powerful 3D images and even more animations enhance the thorough content and draw your students into the wonder of anatomy and physiology.

Contents

Human Body • Skeletal System • Muscular System • Nervous System • Endocrine System • Cardiovascular System • Lymphatic System • Digestive System • Respiratory System • Urinary System • Reproductive System

System Requirements

IBM/PC or compatible
486/66 or better (Pentium recommended)
Windows 95 or newer
16 MB RAM or better
640 x 480 x 256 color monitor
CD-ROM drive
(transfer rate of 300 kbs or better)
SoundBlaster compatible audio card
Mouse

Macintosh
Power PC
System 7.1 or newer
16 MB RAM or better
640 x 480 x 256 color monitor
CD-ROM drive
(transfer rate of 300 kbs or better)
Mouse

This distinctive icon appears in appropriate figure legends of the eighth edition of *Human Physiology: The Mechanisms of Body Function* by Vander, Sherman, Luciano, and correlates information to specific modules found on the new Dynamic Human CD-ROM, Version 2.0.

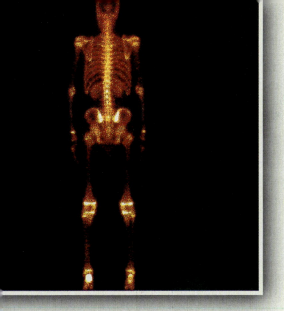

A Framework for Human Physiology

One cannot meaningfully analyze the complex activities of the human body without a framework upon which to build, a set of viewpoints to guide one's thinking. It is the purpose of this chapter to provide such an orientation to the subject of human physiology.

The Scope of Human Physiology

Stated most simply and broadly, **physiology** is the study of how living organisms work. As applied to human beings, its scope is extremely broad. At one end of the spectrum, it includes the study of individual molecules—for example, how a particular protein's shape and electrical properties allow it to function as a channel for sodium ions to move into or out of a cell. At the other end, it is concerned with complex processes that depend on the interplay of many widely separated organs in the body—for example, how the brain, heart, and several glands all work together to cause the excretion of more sodium in the urine when a person has eaten salty food.

What makes physiologists unique among biologists is that they are always interested in function and integration—how things work together at various levels of organization and, most importantly, in the entire organism. Thus, even when physiologists study parts of organisms, all the way down to individual molecules, the intention is always ultimately to have whatever information is gained applied to the function of the whole body. As the nineteenth-century physiologist Claude Bernard put it: "After carrying out an analysis of phenomena, we must . . . always reconstruct our physiological synthesis, so as to see the joint action of all the parts we have isolated . . . ."

In this regard, a very important point must be made about the present status and future of physiology. It is easy for a student to gain the impression from a textbook that almost everything is known about the subject, but nothing could be farther from the truth for physiology. Many areas of function are still only poorly understood (for example, how the workings of the brain produce the phenomena we associate with the word "mind").

Indeed, we can predict with certainty a coming explosion of new physiological information and understanding. One of the major reasons is as follows. As you will learn in Chapters 4 and 5, proteins are molecules that are associated with practically every function performed in the body, and the directions for the synthesis of each type of protein are coded into a unique gene. Presently, only a fraction of all the body's proteins has been identified, and the roles of these known proteins in normal body function and disease often remain incompletely understood. But recently, with the revolution in molecular biology, it has become possible to add or eliminate a particular gene from a living organism (Chapter 5) in order to better study the physiological significance of the protein for which that gene codes. Moreover, the gaining of new physiological information of this type will expand enormously as the Human Genome Project (Chapter 5) continues its task of identifying all of the estimated 50,000 to 100,000 genes in the body, most of these genes coding for proteins whose functions are unknown.

Finally, a word should be said about the interaction of physiology and medicine. Disease states can be viewed as physiology "gone wrong," or **pathophysiology,** and for this reason an understanding of physiology is absolutely essential for the study and practice of medicine. Indeed, many physiologists are themselves actively engaged in research on the physiological bases of a wide range of diseases. In this text, we will give many examples of pathophysiology, always to illustrate the basic physiology that underlies the disease.

Mechanism and Causality

The **mechanist view** of life, the view taken by physiologists, holds that all phenomena, no matter how complex, can ultimately be described in terms of physical and chemical laws. In contrast, **vitalism** is the view that some "vital force" beyond physics and chemistry is required to explain life. The mechanist view has predominated in the twentieth century because virtually all information gathered from observation and experiment has agreed with it.

Physiologists should not be misunderstood when they sometimes say that "the whole is greater than the sum of its parts." This statement in no way implies a vital force but rather recognizes that *integration* of an enormous number of individual physical and chemical events occurring at all levels of organization is required for biological systems to function.

A common denominator of physiological processes is their contribution to survival. Unfortunately, it is easy to misunderstand the nature of this relationship. Consider, for example, the statement, "During exercise a person sweats because the body *needs* to get rid of the excess heat generated." This type of statement is an example of **teleology,** the explanation of events in terms of purpose, but it is not an explanation at all in the scientific sense of the word. It is somewhat like saying, "The furnace is on because the house needs to be heated." Clearly, the furnace is on not because it senses in some mystical manner the house's "needs," but because the temperature has fallen below the thermostat's set point and the electric current in the connecting wires has turned on the heater.

Of course, sweating really does serve a useful purpose during exercise because the excess heat, if not eliminated, might cause sickness or even death. But this is totally different from stating that a need to avoid injury *causes* the sweating. The cause of the sweating is a sequence of events initiated by the increased heat generation: increased heat generation → increased blood temperature → increased activity of specific nerve cells in the brain → increased activity of a series of nerve cells → increased production of sweat by the sweat-gland cells. Each step occurs by means of physicochemical changes in the cells involved. In science, to explain a phenomenon is to reduce it to a causally linked sequence of physicochemical events. This is the scientific meaning of causality, of the word "because."

This is a good place to emphasize that causal chains can be not only long, as in the example just cited, but also multiple. In other words, one should not assume the simple relationship of one cause, one effect. We shall see that multiple factors often must interact to elicit a response. To take an example from medicine, cigarette smoking can cause lung cancer, but the likelihood of cancer developing in a smoker depends on a variety of other factors, including the way that person's body processes the chemicals in cigarette smoke, the rate at which damaged molecules are repaired, and so on.

That a phenomenon is beneficial to a person, while not explaining the *mechanism* of the phenomenon, is of obvious interest and importance. Evolution is the key to understanding why most body activities do indeed appear to be purposeful, since responses that have survival value undergo natural selection. Throughout this book we emphasize how a particular process contributes to survival, but the reader must never confuse the survival value of a process with the explanation of the mechanisms by which the process occurs.

A Society of Cells

Cells: The Basic Units of Living Organisms

The simplest structural units into which a complex multicellular organism can be divided and still retain the functions characteristic of life are called **cells.** One of the unifying generalizations of biology is that certain fundamental activities are common to almost all cells and represent the minimal requirements for maintaining cell integrity and life. Thus, for example, a hu-

man liver cell and an amoeba are remarkably similar in their means of exchanging materials with their immediate environments, of obtaining energy from organic nutrients, of synthesizing complex molecules, of duplicating themselves, and of detecting and responding to signals in their immediate environment.

Each human organism begins as a single cell, a fertilized egg, which divides to create two cells, each of which divides in turn, resulting in four cells, and so on. If cell multiplication were the only event occurring, the end result would be a spherical mass of identical cells. During development, however, each cell becomes specialized for the performance of a particular function, such as producing force and movement (muscle cells) or generating electric signals (nerve cells). The process of transforming an unspecialized cell into a specialized cell is known as **cell differentiation,** the study of which is one of the most exciting areas in biology today. As described in Chapter 5, all cells in a person have the same genes; how then is one unspecialized cell instructed to differentiate into a nerve cell, another into a muscle cell, and so on? What are the external chemical signals that constitute these "instructions," and how do they affect various cells differently? For the most part, the answers to these questions are unknown.

In addition to differentiating, cells migrate to new locations during development and form selective adhesions with other cells to produce multicellular structures. In this manner, the cells of the body are arranged in various combinations to form a hierarchy of organized structures. Differentiated cells with similar properties aggregate to form **tissues** (nerve tissue, muscle tissue, and so on), which combine with other types of tissues to form **organs** (the heart, lungs, kidneys, and so on), which are linked together to form **organ systems** (Figure 1–1).

About 200 distinct kinds of cells can be identified in the body in terms of differences in structure and function. When cells are classified according to the broad types of function they perform, however, four categories emerge: (1) muscle cells, (2) nerve cells, (3) epithelial cells, and (4) connective-tissue cells. In each of these functional categories, there are several cell types that perform variations of the specialized function. For example, there are three types of muscle cells—skeletal, cardiac, and smooth—which differ from each other in shape, in the mechanisms controlling their contractile activity, and in their location in the various organs of the body.

Muscle cells are specialized to generate the mechanical forces that produce force and movement. They may be attached to bones and produce movements of the limbs or trunk. They may be attached to skin, as for example, the muscles producing facial

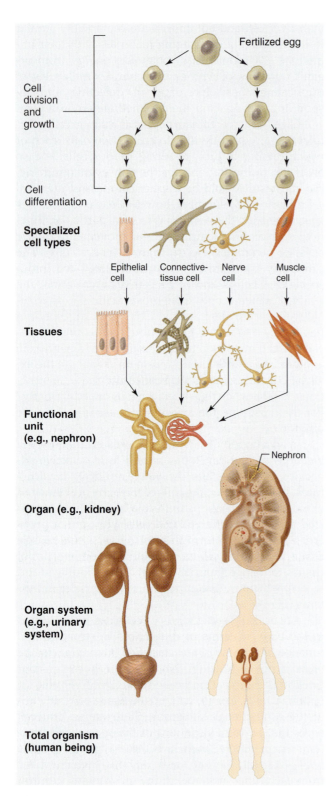

FIGURE 1–1
Levels of cellular organization.

expressions. They may enclose hollow cavities so that their contraction expels the contents of the cavity, as in the pumping of the heart. Muscle cells also surround many of the tubes in the body—blood vessels, for example—and their contraction changes the diameter of these tubes.

Nerve cells are specialized to initiate and conduct electric signals, often over long distances. A signal may initiate new electric signals in other nerve cells, or it may stimulate secretion by a gland cell or contraction of a muscle cell. Thus, nerve cells provide a major means of controlling the activities of other cells. The incredible complexity of nerve-cell connections and activity underlie such phenomena as consciousness and perception.

Epithelial cells are specialized for the selective secretion and absorption of ions and organic molecules. They are located mainly at the surfaces that either cover the body or individual organs or else line the walls of various tubular and hollow structures within the body. Epithelial cells, which rest on a homogeneous extracellular protein layer called the **basement membrane,** form the boundaries between compartments and function as selective barriers regulating the exchange of molecules across them. For example, the epithelial cells at the surface of the skin form a barrier that prevents most substances in the **external environment**—the environment surrounding the body—from entering the body through the skin. Epithelial cells are also found in glands that form from the invagination of epithelial surfaces.

Connective-tissue cells, as their name implies, have as their major function connecting, anchoring, and supporting the structures of the body. These cells typically have a large amount of material between them. Some connective-tissue cells are found in the loose meshwork of cells and fibers underlying most epithelial layers; other types include fat-storing cells, bone cells, and red blood cells and white blood cells.

Tissues

Most specialized cells are associated with other cells of a similar kind to form tissues. Corresponding to the four general categories of differentiated cells, there are four general classes of tissues: (1) **muscle tissue,** (2) **nerve tissue,** (3) **epithelial tissue,** and (4) **connective tissue.** It should be noted that the term "tissue" is used in different ways. It is formally defined as an aggregate of a single type of specialized cell. However, it is also commonly used to denote the general cellular fabric of any organ or structure, for example, kidney tissue or lung tissue, each of which in fact usually contains all four classes of tissue.

We will emphasize later in this chapter that the immediate environment of each individual cell in the body is the extracellular fluid. Actually this fluid is interspersed within a complex **extracellular matrix** consisting of a mixture of protein molecules (and, in some cases, minerals) specific for any given tissue. The matrix serves two general functions: (1) It provides a scaffold for cellular attachments, and (2) it transmits to the cells information, in the form of chemical messengers, that helps regulate their migration, growth, and differentiation.

The proteins of the extracellular matrix consist of **fibers**—ropelike **collagen fibers** and rubberband-like **elastin fibers**—and a mixture of other proteins that contain chains of complex sugars (carbohydrates). In some ways, the extracellular matrix is analogous to reinforced concrete. The fibers of the matrix, particularly collagen, which constitutes one-third of all bodily proteins, are like the reinforcing iron mesh or rods in the concrete, and the carbohydrate-containing protein molecules are the surrounding cement. However, these latter molecules are not merely inert "packing material," as in concrete, but function as adhesion/recognition molecules between cells and as important links in the communication between extracellular messenger molecules and cells.

Organs and Organ Systems

Organs are composed of the four kinds of tissues arranged in various proportions and patterns: sheets, tubes, layers, bundles, strips, and so on. For example, the kidneys consist of (1) a series of small tubes, each composed of a single layer of epithelial cells; (2) blood vessels, whose walls contain varying quantities of smooth muscle and connective tissue; (3) nerve-cell extensions that end near the muscle and epithelial cells; (4) a loose network of connective-tissue elements that are interspersed throughout the kidneys and also form enclosing capsules; and (5) extracellular fluid and matrix.

Many organs are organized into small, similar subunits often referred to as **functional units,** each performing the function of the organ. For example, the kidneys' 2 million functional units are termed nephrons (which contain the small tubes mentioned in the previous paragraph), and the total production of urine by the kidneys is the sum of the amounts formed by the individual nephrons.

Finally we have the organ system, a collection of organs that together perform an overall function. For example, the kidneys, the urinary bladder, the tubes leading from the kidneys to the bladder, and the tube leading from the bladder to the exterior constitute the urinary system. There are 10 organ systems in the body. Their components and functions are given in Table 1–1.

To sum up, the human body can be viewed as a complex society of differentiated cells structurally and functionally combined and interrelated to carry out the functions essential to the survival of the entire organism. The individual cells constitute the basic units of this society, and almost all of these cells individually exhibit the fundamental activities common to all forms of life. Indeed, many of the cells can be removed and maintained in test tubes as free-living organisms (this is termed *in vitro,* literally "in glass," as opposed to *in vivo,* meaning "within the body").

There is a paradox in this analysis: How is it that the functions of the organ systems are essential to the survival of the body when each individual cell seems capable of performing its own fundamental activities? As described in the next section, the resolution of this paradox is found in the isolation of most of the cells of the body from the external environment and in the existence of an internal environment.

The Internal Environment and Homeostasis

An amoeba and a human liver cell both obtain their energy by breaking down certain organic nutrients. The chemical reactions involved in this intracellular process are remarkably similar in the two types of cells and involve the utilization of oxygen and the production of carbon dioxide. The amoeba picks up oxygen directly from the fluid surrounding it (its external environment) and eliminates carbon dioxide into the same fluid. But how can the liver cell and all other internal parts of the body obtain oxygen and eliminate carbon dioxide when, unlike the amoeba, they are not in direct contact with the external environment—the air surrounding the body?

Figure 1–2 summarizes the exchanges of matter that occur in a person. Supplying oxygen is the function both of the respiratory system, which takes up oxygen from the external environment, and of the circulatory system, which distributes the oxygen to all parts of the body. In addition, the circulatory system carries the carbon dioxide generated by all the cells of the body to the lungs, which eliminate it to the exterior. Similarly, the digestive and circulatory systems working together make nutrients from the external environment available to all the body's cells. Wastes other than carbon dioxide are carried by the circulatory system from the cells that produced them to the kidneys and liver, which excrete them from the body. The kidneys also regulate the amounts of water and many essential minerals in the body. The nervous and hormonal systems coordinate and control the activities of all the other organ systems.

Thus the overall effect of the activities of organ systems is to create within the body an environment in

TABLE 1–1 Organ Systems of the Body

System	Major Organs or Tissues	Primary Functions
Circulatory	Heart, blood vessels, blood (Some classifications also include lymphatic vessels and lymph in this system.)	Transport of blood throughout the body's tissues
Respiratory	Nose, pharynx, larynx, trachea, bronchi, lungs	Exchange of carbon dioxide and oxygen; regulation of hydrogen-ion concentration
Digestive	Mouth, pharynx, esophagus, stomach, intestines, salivary glands, pancreas, liver, gallbladder	Digestion and absorption of organic nutrients, salts, and water
Urinary	Kidneys, ureters, bladder, urethra	Regulation of plasma composition through controlled excretion of salts, water, and organic wastes
Musculoskeletal	Cartilage, bone, ligaments, tendons, joints, skeletal muscle	Support, protection, and movement of the body; production of blood cells
Immune	White blood cells, lymph vessels and nodes, spleen, thymus, and other lymphoid tissues	Defense against foreign invaders; return of extracellular fluid to blood; formation of white blood cells
Nervous	Brain, spinal cord, peripheral nerves and ganglia, special sense organs	Regulation and coordination of many activities in the body; detection of changes in the internal and external environments; states of consciousness; learning; cognition
Endocrine	All glands secreting hormones: Pancreas, testes, ovaries, hypothalamus, kidneys, pituitary, thyroid, parathyroid, adrenal, intestinal, thymus, heart, and pineal, and endocrine cells in other locations	Regulation and coordination of many activities in the body
Reproductive	Male: Testes, penis, and associated ducts and glands Female: Ovaries, uterine tubes, uterus, vagina, mammary glands	Production of sperm; transfer of sperm to female Production of eggs; provision of a nutritive environment for the developing embryo and fetus; nutrition of the infant
Integumentary	Skin	Protection against injury and dehydration; defense against foreign invaders; regulation of temperature

which all cells can survive and function. This fluid environment surrounding each cell is called the **internal environment.** The internal environment is not merely a theoretical physiological concept. It can be identified quite specifically in anatomical terms. The body's internal environment is the **extracellular fluid** (literally, fluid outside the cells), which bathes each cell.

In other words, the environment in which each cell lives is not the external environment surrounding the entire body but the local extracellular fluid surrounding that cell. It is from this fluid that the cells receive oxygen and nutrients and into which they excrete wastes. A multicellular organism can survive only as long as it is able to maintain the composition of its internal environment in a state compatible with the sur-

vival of its individual cells. In 1857, Claude Bernard clearly described the central importance of the extracellular fluid: *"It is the fixity of the internal environment that is the condition of free and independent life. . . . All the vital mechanisms, however varied they may be, have only one object, that of preserving constant the conditions of life in the internal environment."*

The relative constancy of the internal environment is known as **homeostasis.** Changes do occur, but the magnitudes of these changes are small and are kept within narrow limits. As emphasized by the twentieth-century American physiologist, Walter B. Cannon, such stability can be achieved only through the operation of carefully coordinated physiological processes. The activities of the cells, tissues, and organs must be

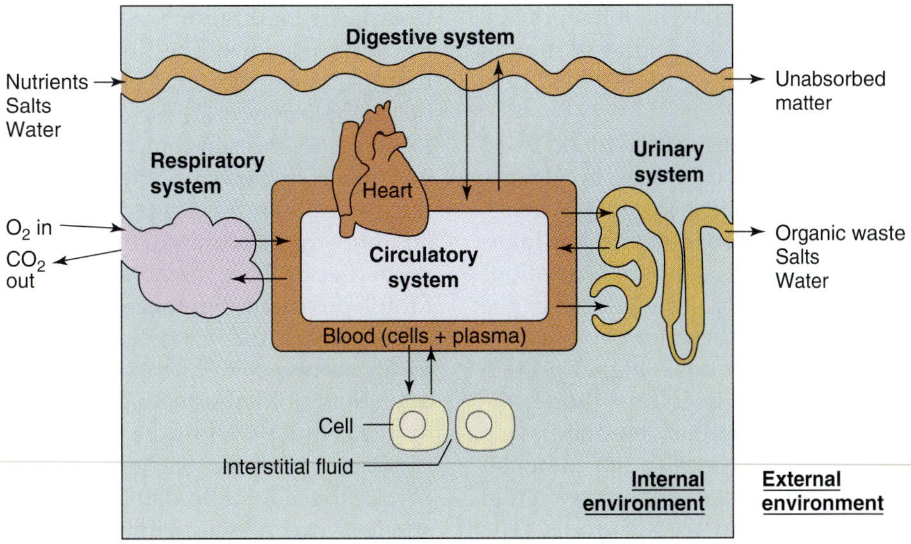

FIGURE 1–2

Exchanges of matter occur between the external environment and the circulatory system via the digestive, respiratory, and urinary systems. Extracellular fluid (plasma and interstitial fluid) is the internal environment of the body. The external environment is the air surrounding the body.

regulated and integrated with each other in such a way that any change in the extracellular fluid initiates a reaction to minimize the change.

A collection of body components that functions to keep a physical or chemical property of the internal environment relatively constant is termed a **homeostatic control system.** As will be described in detail in Chapter 7, such a system must detect changes in the magnitude of the property, relay this information to an appropriate site for integration with other incoming information, and elicit a "command" to particular cells to alter their rates of function in such a way as to restore the property toward its original value.

The description at the beginning of this chapter of how sweating is brought about in response to increased heat generation during exercise is an example of a homeostatic control system in operation; the sweating (more precisely, the evaporation of the sweat) removes heat from the body and keeps the body temperature relatively constant even though more heat is being produced by the exercising muscles.

Here is another example: A mountaineer who ascends to high altitude suffers a decrease in the concentration of oxygen in his or her blood because of the decrease in the amount of oxygen in inspired air; the nervous system detects this change in the blood and increases its signals to the skeletal muscles responsible for breathing. The result is that the mountaineer breathes more rapidly and deeply, and the increase in the amount of air inspired helps keep the blood oxy-

gen concentration from falling as much as it otherwise would.

We emphasized at the beginning of this chapter the intimate relationship between physiology and medicine. Another way of putting it is that physicians, for the most part, diagnose and treat disease-induced disruptions of homeostasis.

To summarize, the activities of every individual cell in the body fall into two categories: (1) Each cell performs for itself all those fundamental basic cellular processes—movement of materials across its membrane, extraction of energy, protein synthesis, and so on—that represent the minimal requirements for maintaining its own individual integrity and life; and (2) each cell simultaneously performs one or more specialized activities that, in concert with the activities performed by the other cells of its tissue or organ system, contribute to the survival of the body by maintaining the stable internal environment required by all cells.

Body-Fluid Compartments

To repeat, the internal environment can be equated with the extracellular fluid. It was not stated earlier that extracellular fluid exists in two locations—surrounding cells and inside blood vessels. Approximately 80 percent of the extracellular fluid surrounds all the body's cells except the blood cells. Because it lies "between cells," this 80 percent of the extracellular fluid is known as **interstitial fluid.** The remaining 20 percent of the extracellular fluid is the fluid portion of the blood, the **plasma,** in which the various blood cells are suspended.

As the blood (plasma plus suspended blood cells) flows through the smallest of blood vessels in all parts of the body, the plasma exchanges oxygen, nutrients, wastes, and other metabolic products with the interstitial fluid. Because of these exchanges, concentrations of dissolved substances are virtually identical in the plasma and interstitial fluid, except for protein concentration. With this major exception—higher protein concentration in plasma than in interstitial fluid—the entire extracellular fluid may be considered to have a homogeneous composition. In contrast, the composition of the extracellular fluid is very different from that of the **intracellular fluid,** the fluid inside the cells. (The actual differences will be presented in Chapter 6, Table 6–1.)

In essence, the fluids in the body are enclosed in "compartments." The volumes of the body-fluid compartments are summarized in Figure 1–3 in terms of water, since water is by far the major component of the fluids. Water accounts for about 60 percent of normal body weight. Two-thirds of this water (28 L in a typical normal 70-kg person) is intracellular fluid. The remaining one-third (14 L) is extracellular and as described above, 80 percent of this extracellular fluid is interstitial fluid (11 L) and 20 percent (3 L) is plasma.

Compartmentalization is an important general principle in physiology. (We shall see in Chapter 3 that the inside of cells is also divided into compartments.)

Compartmentalization is achieved by barriers between the compartments. The properties of the barriers determine which substances can move between contiguous compartments. These movements in turn account for the differences in composition of the different compartments. In the case of the body-fluid compartments, the intracellular fluid is separated from the extracellular fluid by membranes that surround the cells; the properties of these membranes and how they account for the profound differences between intracellular and extracellular fluid are described in Chapter 6. In contrast, the two components of extracellular fluid—the interstitial fluid and the blood plasma—are separated by the cellular wall of the smallest blood vessels, the capillaries. How this barrier normally keeps 80 percent of the extracellular fluid in the interstitial compartment and restricts proteins mainly to the plasma is described in Chapter 14.

This completes our introductory framework. With it in mind, the overall organization and approach of this book should easily be understood. Because the fundamental features of cell function are shared by virtually all cells and because these features constitute the foundation upon which specialization develops, we devote Part 1 of the book to an analysis of basic cell physiology.

Part 2 provides the principles and information required to bridge the gap between the functions of individual cells and the integrated systems of the body. Chapter 7 describes the basic characteristics of homeostatic control systems and the required cellular communications. The other chapters of Part 2 deal with the

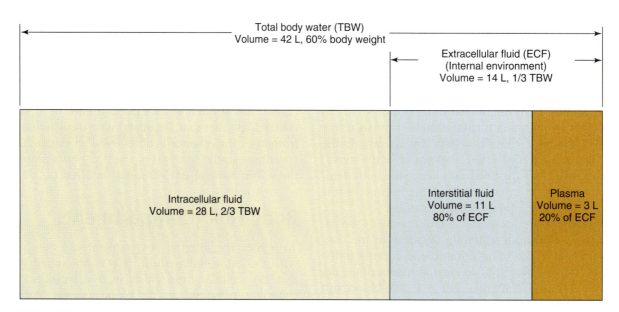

FIGURE 1–3

Fluid compartments of the body. Volumes are for an average 70-kg (154-lb) person. TBW = total body water; ECF = extracellular fluid.

specific components of the body's control systems: nerve cells, muscle cells, and gland cells.

Part 3 describes the coordinated functions (circulation, respiration, and so on) of the body, emphasizing how they result from the precisely controlled and integrated activities of specialized cells grouped together in tissues and organs. The theme of these descriptions is that each function, with the obvious exception of reproduction, serves to keep some important aspect of the body's internal environment relatively constant. Thus, homeostasis, achieved by homeostatic control systems, is the single most important unifying idea to be kept in mind in Part 3.

SUMMARY

The Scope of Human Physiology

I. Physiology is the study of how living organisms work. Physiologists are unique among biologists in that they are always interested in function.
II. Disease states are physiology "gone wrong" (pathophysiology).

Mechanism and Causality

I. The mechanist view of life, the view taken by physiologists, holds that all phenomena can be described in terms of physical and chemical laws.
II. Vitalism holds that some additional force is required to explain the function of living organisms.

A Society of Cells

I. Cells are the simplest structural units into which a complex multicellular organism can be divided and still retain the functions characteristic of life.
II. Cell differentiation results in the formation of four categories of specialized cells.
   a. Muscle cells generate the mechanical activities that produce force and movement.
   b. Nerve cells initiate and conduct electric signals.
   c. Epithelial cells selectively secrete and absorb ions and organic molecules.
   d. Connective-tissue cells connect, anchor, and support the structures of the body.
III. Specialized cells associate with similar cells to form tissues: muscle tissue, nerve tissue, epithelial tissue, and connective tissue.
IV. Organs are composed of the four kinds of tissues arranged in various proportions and patterns; many organs contain multiple small, similar functional units.
V. An organ system is a collection of organs that together perform an overall function.

The Internal Environment and Homeostasis

I. The body's internal environment is the extracellular fluid surrounding cells.
II. The function of organ systems is to maintain the internal environment relatively constant—

homeostasis. This is achieved by homeostatic control systems.
III. Each cell performs the basic cellular processes required to maintain its own integrity plus specialized activities that help achieve homeostasis.

Body-Fluid Compartments

I. The body fluids are enclosed in compartments.
   a. The extracellular fluid is composed of the interstitial fluid (the fluid between cells) and the blood plasma. Of the extracellular fluid, 80 percent is interstitial fluid, and 20 percent is plasma.
   b. Interstitial fluid and plasma have essentially the same composition except that plasma contains a much higher concentration of protein.
   c. Extracellular fluid differs markedly in composition from the fluid inside cells—the intracellular fluid.
   d. Approximately one-third of body water is in the extracellular compartment, and two-thirds is intracellular.
II. The differing compositions of the compartments reflect the activities of the barriers separating them.

KEY TERMS

physiology	muscle tissue
pathophysiology	nerve tissue
mechanist view	epithelial tissue
vitalism	connective tissue
teleology	extracellular matrix
cell	fiber
cell differentiation	collagen fiber
tissue	elastin fiber
organ	functional unit
organ system	internal environment
muscle cell	extracellular fluid
nerve cell	homeostasis
epithelial cell	homeostatic control system
basement membrane	interstitial fluid
external environment	plasma
connective-tissue cell	intracellular fluid

REVIEW QUESTIONS

1. Describe the levels of cellular organization and state the four types of specialized cells and tissues.
2. List the 10 organ systems of the body and give one-sentence descriptions of their functions.
3. Contrast the two categories of functions performed by every cell.
4. Name two fluids that constitute the extracellular fluid. What are their relative proportions in the body, and how do they differ from each other in composition?
5. State the relative volumes of water in the body-fluid compartments.

Chemical Composition of the Body

Atoms and molecules are the chemical units of cell structure and function. In this chapter we describe the distinguishing characteristics of the major chemicals in the human body. The specific roles of these substances will be discussed in subsequent chapters. This chapter is, in essence, an expanded glossary of chemical terms and structures, and like a glossary, it should be consulted as needed.

Atoms

The units of matter that form all chemical substances are called **atoms.** The smallest atom, hydrogen, is approximately 2.7 billionths of an inch in diameter. Each type of atom—carbon, hydrogen, oxygen, and so on—is called a **chemical element.** A one- or two-letter symbol is used as a shorthand identification for each element. Although slightly more than 100 elements exist in the universe, only 24 (Table 2–1) are known to be essential for the structure and function of the human body.

The chemical properties of atoms can be described in terms of three subatomic particles—**protons, neutrons,** and **electrons.** The protons and neutrons are confined to a very small volume at the center of an atom, the **atomic nucleus,** whereas the electrons revolve in orbits at various distances from the nucleus. This miniature solar-system model of an atom is an oversimplification, but it is sufficient to provide a conceptual framework for understanding the chemical and physical interactions of atoms.

Each of the subatomic particles has a different electric charge: Protons have one unit of positive charge, electrons have one unit of negative charge, and neutrons are electrically neutral (Table 2–2). Since the protons are located in the atomic nucleus, the nucleus has a net positive charge equal to the number of protons it contains. The entire atom has no net electric charge, however, because the number of negatively charged electrons orbiting the nucleus is equal to the number of positively charged protons in the nucleus.

Atomic Number

Every atom of each chemical element contains a specific number of protons, and it is this number that distinguishes one type of atom from another. This number is known as the **atomic number.** For example, hydrogen, the simplest atom, has an atomic number of 1, corresponding to its single proton; calcium has an atomic number of 20, corresponding to its 20 protons. Since an atom is electrically neutral, the atomic number is also equal to the number of electrons in the atom.

Atomic Weight

Atoms have very little mass. A single hydrogen atom, for example, has a mass of only 1.67×10^{-24} g. The **atomic weight** scale indicates an atom's mass relative to the mass of other atoms. This scale is based upon assigning the carbon atom a mass of 12. On this scale,

TABLE 2–1 Essential Chemical Elements in the Body

Element	Symbol
MAJOR ELEMENTS: 99.3% OF TOTAL ATOMS	
Hydrogen	H (63%)
Oxygen	O (26%)
Carbon	C (9%)
Nitrogen	N (1%)
MINERAL ELEMENTS: 0.7% OF TOTAL ATOMS	
Calcium	Ca
Phosphorus	P
Potassium	K (Latin *kalium*)
Sulfur	S
Sodium	Na (Latin *natrium*)
Chlorine	Cl
Magnesium	Mg
TRACE ELEMENTS: LESS THAN 0.01% OF TOTAL ATOMS	
Iron	Fe (Latin *ferrum*)
Iodine	I
Copper	Cu (Latin *cuprum*)
Zinc	Zn
Manganese	Mn
Cobalt	Co
Chromium	Cr
Selenium	Se
Molybdenum	Mo
Fluorine	F
Tin	Sn (Latin *stannum*)
Silicon	Si
Vanadium	V

TABLE 2–2 Characteristics of Major Subatomic Particles

Particle	Mass Relative to Electron Mass	Electric Charge	Location in Atom
Proton	1836	+1	Nucleus
Neutron	1839	0	Nucleus
Electron	1	−1	Orbiting the nucleus

a hydrogen atom has an atomic weight of approximately 1, indicating that it has one-twelfth the mass of a carbon atom; a magnesium atom, with an atomic weight of 24, has twice the mass of a carbon atom.

Since the atomic weight scale is a *ratio* of atomic masses, it has no units. The unit of atomic mass is known as a dalton. One dalton (d) equals one-twelfth the mass of a carbon atom. Thus, carbon has an atomic weight of 12, and a carbon atom has an atomic mass of 12 daltons.

Although the number of neutrons in the nucleus of an atom is often equal to the number of protons, many chemical elements can exist in multiple forms, called **isotopes,** which differ in the number of neutrons they contain. For example, the most abundant form of the carbon atom, ^{12}C, contains 6 protons and 6 neutrons, and thus has an atomic number of 6. Protons and neutrons are approximately equal in mass; therefore, ^{12}C has an atomic weight of 12. The radioactive carbon isotope ^{14}C contains 6 protons and 8 neutrons, giving it an atomic number of 6 but an atomic weight of 14.

One **gram atomic mass** of a chemical element is the amount of the element in grams that is equal to the numerical value of its atomic weight. Thus, 12 g of carbon (assuming it is all ^{12}C) is 1 gram atomic mass of carbon. *One gram atomic mass of any element contains the same number of atoms.* For example, 1 g of hydrogen contains 6×10^{23} atoms, and 12 g of carbon, whose atoms have 12 times the mass of a hydrogen atom, also has 6×10^{23} atoms.

Atomic Composition of the Body

Just four of the body's essential elements (Table 2–1)—hydrogen, oxygen, carbon, and nitrogen—account for over 99 percent of the atoms in the body.

The seven essential mineral elements are the most abundant substances dissolved in the extracellular and intracellular fluids. Most of the body's calcium and phosphorus atoms, however, make up the solid matrix of bone tissue.

The 13 essential **trace elements** are present in extremely small quantities, but they are nonetheless essential for normal growth and function. For example, iron plays a critical role in the transport of oxygen by the blood. Additional trace elements will likely be added to this list as the chemistry of the body becomes better understood.

Many other elements, in addition to the 24 listed in Table 2–1, can be detected in the body. These elements enter in the foods we eat and the air we breathe but are not essential for normal body function and may even interfere with normal body chemistry. For example, ingested arsenic has poisonous effects.

Molecules

Two or more atoms bonded together make up a **molecule.** For example, a molecule of water contains two hydrogen atoms and one oxygen atom, which can be represented by H_2O. The atomic composition of glucose, a sugar, is $C_6H_{12}O_6$, indicating that the molecule contains 6 carbon atoms, 12 hydrogen atoms, and 6 oxygen atoms. Such formulas, however, do not indicate how the atoms are linked together in the molecule.

Covalent Chemical Bonds

The atoms in molecules are held together by chemical bonds, which are formed when electrons are transferred from one atom to another or are shared between two atoms. The strongest chemical bond between two atoms, a **covalent bond,** is formed when one electron in the outer electron orbit of each atom is shared between the two atoms (Figure 2–1). The atoms in most molecules found in the body are linked by covalent bonds.

The atoms of some elements can form more than one covalent bond and thus become linked simultaneously to two or more other atoms. Each type of atom forms a characteristic number of covalent bonds, which depends on the number of electrons in its outermost orbit. The number of chemical bonds formed by the four most abundant atoms in the body are hydrogen, one; oxygen, two; nitrogen, three; and carbon, four. When the structure of a molecule is diagramed, each covalent bond is represented by a line indicating a pair of shared electrons. The covalent bonds of the four elements mentioned above can be represented as

$$H- \qquad -O- \qquad -\overset{|}{N}- \qquad -\overset{|}{\underset{|}{C}}-$$

A molecule of water H_2O can be diagramed as

$$H-O-H$$

In some cases, two covalent bonds—a double bond—are formed between two atoms by the sharing of two electrons from each atom. Carbon dioxide (CO_2) contains two double bonds:

$$O=C=O$$

Note that in this molecule the carbon atom still forms four covalent bonds and each oxygen atom only two.

Molecular Shape

When atoms are linked together, molecules with various shapes can be formed. Although we draw diagrammatic structures of molecules on flat sheets of paper, molecules are actually three-dimensional. When

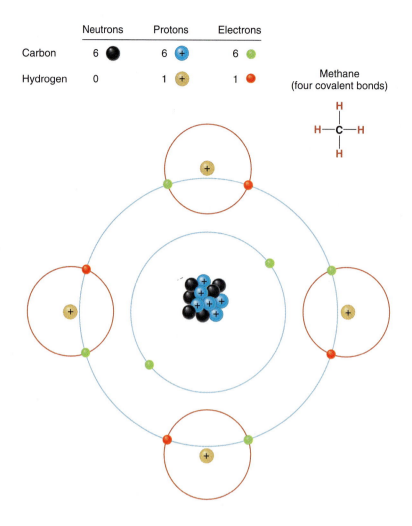

	Neutrons	Protons	Electrons
Carbon	6	6	6
Hydrogen	0	1	1

Methane
(four covalent bonds)

FIGURE 2–1

Each of the four hydrogen atoms in a molecule of methane (CH_4) forms a covalent bond with the carbon atom by sharing its one electron with one of the electrons in carbon. Each shared pair of electrons—one electron from the carbon and one from a hydrogen atom—forms a covalent bond.

more than one covalent bond is formed with a given atom, the bonds are distributed around the atom in a pattern that may or may not be symmetrical (Figure 2–2).

Molecules are not rigid, inflexible structures. Within certain limits, the shape of a molecule can be changed without breaking the covalent bonds linking its atoms together. A covalent bond is like an axle around which the joined atoms can rotate. As illustrated in Figure 2–3, a sequence of six carbon atoms can assume a number of shapes as a result of rotations around various covalent bonds. As we shall see, the three-dimensional shape of molecules is one of the major factors governing molecular interactions.

Ions

A single atom is electrically neutral since it contains equal numbers of negative electrons and positive protons. If, however, an atom gains or loses one or more electrons, it acquires a net electric charge and becomes an **ion.** For example, when a sodium atom (Na), which has 11 electrons, loses 1 electron, it becomes a sodium ion (Na^+) with a net positive charge; it still has 11 protons, but it now has only 10 electrons. On the other hand, a chlorine atom (Cl), which has 17 electrons, can gain an electron and become a chloride ion (Cl^-) with a net negative charge—it now has 18 electrons but only 17 protons. Some atoms can gain or lose more than 1 electron to become ions with two or even three units of net electric charge (for example, calcium Ca^{2+}).

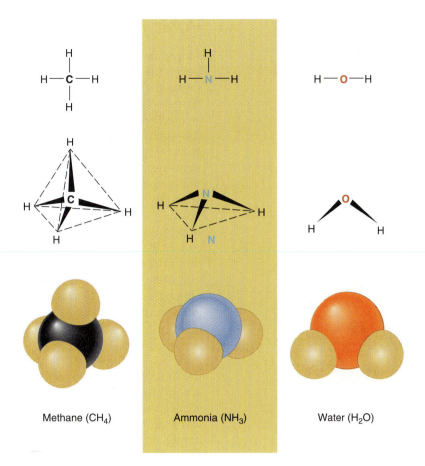

FIGURE 2–2
Geometric configuration of covalent bonds around the carbon, nitrogen, and oxygen atoms bonded to hydrogen atoms.

Hydrogen atoms and most mineral and trace element atoms readily form ions. Table 2–3 lists the ionic forms of some of these elements. Ions that have a net positive charge are called **cations,** while those that have a net negative charge are called **anions.** Because of their ability to conduct electricity when dissolved in water, the ionic forms of the seven mineral elements are collectively referred to as **electrolytes.**

The process of ion formation, known as ionization, can occur in single atoms or in atoms that are covalently linked in molecules. Within molecules two commonly encountered groups of atoms that undergo ionization are the **carboxyl group** ($-COOH$) and the **amino group** ($-NH_2$). The shorthand formula when indicating only a portion of a molecule can be written as $R-COOH$ or $R-NH_2$, where R signifies the remaining portion of the molecule. The carboxyl group ionizes when the oxygen linked to the hydrogen captures the hydrogen's only electron to form a carboxyl ion ($R-COO^-$) and releases a hydrogen ion (H^+):

$$R-COOH \rightleftharpoons R-COO^- + H^+$$

The amino group can bind a hydrogen ion to form an ionized amino group ($R-NH_3^+$):

$$R-NH_2 + H^+ \rightleftharpoons R-NH_3^+$$

The ionization of each of the above groups can be reversed, as indicated by the double arrows; the ionized carboxyl group can combine with a hydrogen ion to form an un-ionized carboxyl group, and the ionized amino group can lose a hydrogen ion and become an un-ionized amino group.

Free Radicals

The electrons that revolve around the nucleus of an atom occupy regions known as orbitals, each of which can be occupied by two electrons. An atom is most stable when each orbital is occupied by two electrons. An atom containing a single electron in its outermost orbital is known as a **free radical,** as are molecules containing such atoms. Most free radicals react rapidly with other atoms, thereby filling the unpaired orbital; thus free radicals normally exist for only brief periods of time before combining with other atoms.

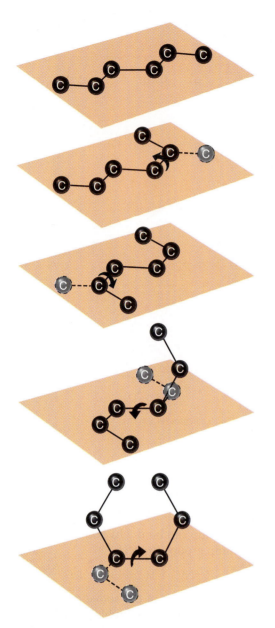

FIGURE 2–3
Changes in molecular shape occur as portions of a molecule rotate around different carbon-to-carbon bonds, transforming this molecule's shape, for example, from a relatively straight chain into a ring.

Free radicals are diagramed with a dot next to the atomic symbol. Examples of biologically important free radicals are superoxide anion, $O_2 \cdot {}^-$; hydroxyl radical, $OH \cdot$; and nitric oxide, $NO \cdot$. Note that a free radical configuration can occur in either an ionized or an un-ionized atom. A number of free radicals play important roles in the normal and abnormal functioning of the body.

Polar Molecules

As we have seen, when the electrons of two atoms interact, the two atoms may share the electrons equally, forming a covalent bond that is electrically neutral. Alternatively, one of the atoms may completely capture an electron from the other, forming two ions. Between these two extremes are bonds in which the electrons are not shared equally between the two atoms, but instead reside closer to one atom of the pair. This atom thus acquires a slight negative charge, while the other atom, having partly lost an electron, becomes slightly positive. Such bonds are known as **polar covalent bonds** (or, simply, polar bonds) since the atoms at each end of the bond have an opposite electric charge. For example, the bond between hydrogen and oxygen in a **hydroxyl group** (—OH) is a polar covalent bond in which the oxygen is slightly negative and the hydrogen slightly positive:

$$(-) \ (+)$$
$$R—O—H$$

(Polar bonds will be diagramed with parentheses around the charges, as above.) The electric charge associated with the ends of a polar bond is considerably less than the charge on a fully ionized atom. (For example, the oxygen in the polarized hydroxyl group has only about 13 percent of the negative charge associated with the oxygen in an ionized carboxyl group, $R—COO^-$.) Polar bonds do not have a *net* electric charge, as do ions, since they contain equal amounts of negative and positive charge.

Atoms of oxygen and nitrogen, which have a relatively strong attraction for electrons, form polar bonds with hydrogen atoms; in contrast, bonds between carbon and hydrogen atoms and between two carbon atoms are electrically neutral (Table 2–4).

Different regions of a single molecule may contain nonpolar bonds, polar bonds, and ionized groups. Molecules containing significant numbers of polar bonds or ionized groups are known as **polar molecules,** whereas molecules composed predominantly of electrically neutral bonds are known as **nonpolar molecules.** As we shall see, the physical characteristics of these two classes of molecules, especially their solubility in water, are quite different.

Hydrogen Bonds

The electrical attraction between the hydrogen atom in a polar bond in one molecule and an oxygen or nitrogen atom in a polar bond of another molecule—or within the same molecule if the bonds are sufficiently separated from each other—forms a **hydrogen bond.** This type of bond is very weak, having only about 4 percent of the strength of the polar bonds linking the

TABLE 2–3 Most Frequently Encountered Ionic Forms of Elements

Atom	Chemical Symbol	Ion	Chemical Symbol	Electrons Gained or Lost
Hydrogen	H	Hydrogen ion	H^+	1 lost
Sodium	Na	Sodium ion	Na^+	1 lost
Potassium	K	Potassium ion	K^+	1 lost
Chlorine	Cl	Chloride ion	Cl^-	1 gained
Magnesium	Mg	Magnesium ion	Mg^{2+}	2 lost
Calcium	Ca	Calcium ion	Ca^{2+}	2 lost

TABLE 2–4 Examples of Nonpolar and Polar Bonds, and Ionized Chemical Groups

Nonpolar Bonds	$-\overset{\textstyle\vert}{\underset{\textstyle\vert}{C}}-H$	Carbon-hydrogen bond
	$-\overset{\textstyle\vert}{\underset{\textstyle\vert}{C}}-\overset{\textstyle\vert}{\underset{\textstyle\vert}{C}}-$	Carbon-carbon bond
Polar Bonds	$\overset{(-)\ (+)}{R-O-H}$	Hydroxyl group (R—OH)
	$\overset{(-)\ (+)}{R-S-H}$	Sulfhydryl group (R—SH)
	$R-\overset{\textstyle H\,(+)}{\underset{(-)}{N}}-R$	Nitrogen-hydrogen bond
Ionized Groups	$R-\overset{\textstyle O}{\overset{\|\|}{C}}-O^-$	Carboxyl group (R—COO$^-$)
	$R-\overset{\textstyle H}{\underset{\textstyle H}{\overset{\|}{\underset{\|}{N^+}}}}-H$	Amino group (R—NH$_3{}^+$)
	$R-O-\overset{\textstyle O}{\underset{\textstyle O^-}{\overset{\|\|}{\underset{\|}{P}}}}-O^-$	Phosphate group (R—PO$_4{}^{2-}$)

hydrogen and oxygen within a water molecule (H_2O). Hydrogen bonds are represented in diagrams by dashed or dotted lines to distinguish them from covalent bonds (Figure 2–4). Hydrogen bonds between and within molecules play an important role in molecular interactions and in determining the shape of large molecules.

Water

Hydrogen is the most numerous atom in the body, and water is the most numerous molecule. Out of every 100 molecules, 99 are water. The covalent bonds linking the two hydrogen atoms to the oxygen atom in a water molecule are polar. Therefore, the oxygen in water has a slight negative charge, and each hydrogen has a slight positive charge. The positively polarized regions near the hydrogen atoms of one water molecule are electrically attracted to the negatively polarized regions of the oxygen atoms in adjacent water molecules by hydrogen bonds (Figure 2–4).

At body temperature, water exists as a liquid because the weak hydrogen bonds between water molecules are continuously being formed and broken. If the temperature is increased, the hydrogen bonds are

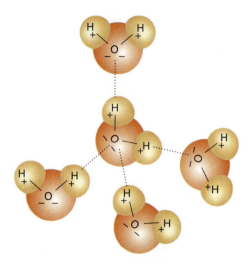

FIGURE 2–4

Five water molecules. Note that polarized covalent bonds link the hydrogen and oxygen atoms within each molecule and that hydrogen bonds occur between adjacent molecules. Hydrogen bonds are represented in diagrams by dashed or dotted lines, and covalent bonds by solid lines.

broken more readily, and molecules of water escape into the gaseous state; however, if the temperature is lowered, hydrogen bonds are broken less frequently so that larger and larger clusters of water molecules are formed until at 0° C water freezes into a continuous crystalline matrix—ice.

Water molecules take part in many chemical reactions of the general type:

$$R_1—R_2 + H—O—H \rightarrow R_1—OH + H—R_2$$

In this reaction the covalent bond between R_1 and R_2 and the one between a hydrogen atom and oxygen in water are broken, and the hydroxyl group and hydrogen atom are transferred to R_1 and R_2, respectively. Reactions of this type are known as hydrolytic reactions, or **hydrolysis.** Many large molecules in the body are broken down into smaller molecular units by hydrolysis.

Solutions

Substances dissolved in a liquid are known as **solutes,** and the liquid in which they are dissolved is the **solvent.** Solutes dissolve in a solvent to form a **solution.** Water is the most abundant solvent in the body, accounting for 60 percent of the total body weight. A majority of the chemical reactions that occur in the body involve molecules that are dissolved in water, either in the intracellular or extracellular fluid. However, not all molecules dissolve in water.

Molecular Solubility

In order to dissolve in water, a substance must be electrically attracted to water molecules. For example, table salt (NaCl) is a solid crystalline substance because of the strong electrical attraction between positive sodium ions and negative chloride ions. This strong attraction between two oppositely charged ions is known as an **ionic bond.** When a crystal of sodium chloride is placed in water, the polar water molecules are attracted to the charged sodium and chloride ions (Figure 2–5). The ions become surrounded by clusters of water molecules, allowing the sodium and chloride ions to separate from the salt crystal and enter the water—that is, to dissolve.

Molecules having a number of polar bonds and/or ionized groups will dissolve in water. Such molecules are said to be **hydrophilic,** or "water-loving." Thus, the presence in a molecule of ionized groups, such as carboxyl and amino groups, or of polar groups, such as hydroxyl groups, promotes solubility in water. In contrast, molecules composed predominantly of carbon and hydrogen are insoluble in water since their electrically neutral covalent bonds are not attracted to water molecules. These molecules are **hydrophobic,** or "water-fearing."

When nonpolar molecules are mixed with water, two phases are formed, as occurs when oil is mixed with water. The strong attraction between polar molecules "squeezes" the nonpolar molecules out of the water phase. Such a separation is never 100 percent complete, however, and very small amounts of nonpolar solutes remain dissolved in the water phase.

Molecules that have a polar or ionized region at one end and a nonpolar region at the opposite end are called **amphipathic**—consisting of two parts. When mixed with water, amiphipathic molecules form clusters, with their polar (hydrophilic) regions at the surface of the cluster where they are attracted to the surrounding water molecules. The nonpolar (hydrophobic) ends are oriented toward the interior of the cluster (Figure 2–6). Such an arrangement provides the maximal interaction between water molecules and the polar ends of the amphipathic molecules. Nonpolar molecules can dissolve in the central nonpolar regions of these clusters and thus exist in aqueous solutions in far higher amounts than would otherwise be possible based on their low solubility in water. As we shall see, the orientation of amphipathic molecules plays an important role in the structure of cell membranes and in both the absorption of nonpolar molecules from the gastrointestinal tract and their transport in the blood.

Concentration

Solute **concentration** is defined as the amount of the solute present in a unit volume of solution. One measure of the amount of a substance is its mass given in

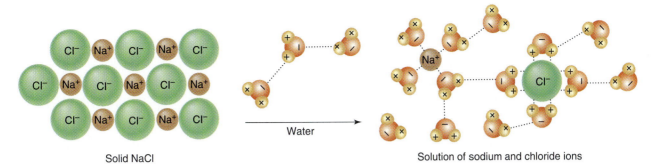

FIGURE 2–5

The ability of water to dissolve sodium chloride crystals depends upon the electrical attraction between the polar water molecules and the charged sodium and chloride ions.

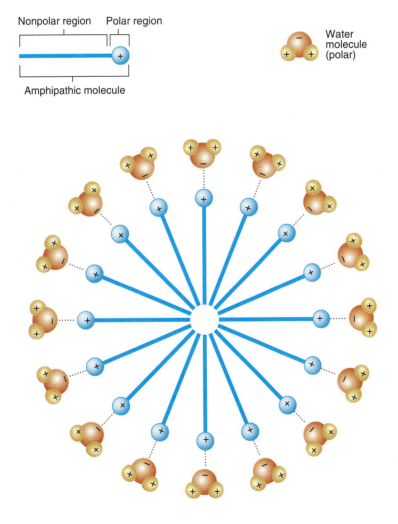

FIGURE 2–6

In water, amphipathic molecules aggregate into spherical clusters. Their polar regions form hydrogen bonds with water molecules at the surface of the cluster.

grams. The unit of volume in the metric system is a liter (L). (One liter equals 1.06 quarts. See Appendix C for metric and English units.) Smaller units are the milliliter (ml, or 0.001 liter) and the microliter (μl, or 0.001 ml). The concentration of a solute in a solution can then be expressed as the number of grams of the substance present in one liter of solution (g/L).

A comparison of the concentrations of two different substances on the basis of the number of *grams* per liter of solution does not directly indicate how many *molecules* of each substance are present. For example, 10 g of compound X, whose molecules are heavier than those of compound Y, will contain fewer molecules than 10 g of compound Y. Concentrations in units of grams per liter are most often used when the chemical structure of the solute is unknown. When the structure of a molecule is known, concentrations are usually expressed as moles per liter, which provides a unit of concentration based upon the number of solute molecules in solution, as described next.

The **molecular weight** of a molecule is equal to the sum of the atomic weights of all the atoms in the molecule. For example, glucose ($C_6H_{12}O_6$) has a molecular weight of 180 ($6 \times 12 + 12 \times 1 + 6 \times 16 = 180$). One **mole** (abbreviated mol) of a compound is the amount of the compound in grams equal to its molecular weight. A solution containing 180 g of glucose (1 mol) in 1 L of solution is a 1 molar solution of glucose (1 mol/L). If 90 g of glucose are dissolved in enough water to produce 1 L of solution, the solution will have a concentration of 0.5 mol/L. Just as 1 gram atomic mass of any element contains the same number of atoms, 1 mol (1 gram molecular mass) of any molecule will contain the same number of molecules— 6×10^{23}. Thus, a 1 mol/L solution of glucose contains the same number of solute molecules per liter as a 1 mol/L solution of urea or any other substance.

The concentrations of solutes dissolved in the body fluids are much less than 1 mol/L. Many have concentrations in the range of millimoles per liter (1 mmol/L = 0.001 mol/L), while others are present in even smaller concentrations—micromoles per liter (1 μmol/L = 0.000001 mol/L) or nanomoles per liter (1 nmol/L = 0.000000001 mol/L).

Hydrogen Ions and Acidity

As mentioned earlier, a hydrogen atom has a single proton in its nucleus orbited by a single electron. A hydrogen ion (H^+), formed by the loss of the electron, is thus a single free proton. Hydrogen ions are formed when the proton of a hydrogen atom in a molecule is released, leaving behind its electron. Molecules that release protons (hydrogen ions) in solution are called **acids**, for example:

$$HCl \longrightarrow H^+ + Cl^-$$
$$\text{hydrochloric acid} \qquad \text{chloride}$$

$$H_2CO_3 \rightleftharpoons H^+ + HCO_3^-$$
$$\text{carbonic acid} \qquad \text{bicarbonate}$$

$$CH_3-\underset{\underset{H}{|}}{\overset{\overset{OH}{|}}{C}}-COOH \rightleftharpoons H^+ + CH_3-\underset{\underset{H}{|}}{\overset{\overset{OH}{|}}{C}}-COO^-$$
$$\text{lactic acid} \qquad \text{lactate}$$

Conversely, any substance that can accept a hydrogen ion (proton) is termed a **base**. In the reactions above, bicarbonate and lactate are bases since they can combine with hydrogen ions (note the double arrows in the two reactions). It is important to distinguish between the un-ionized acid and ionized base forms of these molecules and to note that separate terms are used for the acid forms, lactic acid and carbonic acid, and the bases derived from the acids, lactate and bicarbonate. By combining with hydrogen ions, bases lower the hydrogen-ion concentration of a solution.

When hydrochloric acid is dissolved in water, 100 percent of its atoms separate to form hydrogen and chloride ions, and these ions do not recombine in solution (note the one-way arrow above). In the case of lactic acid, however, only a fraction of the lactic acid molecules in solution release hydrogen ions at any instant. Therefore, if a 1 mol/L solution of hydrochloric acid is compared with a 1 mol/L solution of lactic acid, the hydrogen-ion concentration will be lower in the lactic acid solution than in the hydrochloric acid solution. Hydrochloric acid and other acids that are 100 percent ionized in solution are known as **strong acids**, whereas carbonic and lactic acids and other acids that do not completely ionize in solution are **weak acids**. The same principles apply to bases.

It must be understood that the hydrogen-ion concentration of a solution refers only to the hydrogen ions that are free in solution and not to those that may be bound, for example, to amino groups ($R-NH_3^+$). The **acidity** of a solution refers to the *free* (unbound) hydrogen-ion concentration in the solution; the higher the hydrogen-ion concentration, the greater the acidity. The hydrogen-ion concentration is frequently expressed in terms of the **pH** of a solution, which is defined as the negative logarithm to the base 10 of the hydrogen-ion concentration (the brackets around the symbol for the hydrogen ion in the formula below indicate concentration):

$$pH = -\log [H^+]$$

Thus, a solution with a hydrogen-ion concentration of 10^{-7} mol/L has a pH of 7, whereas a more acidic solution with a concentration of 10^{-6} mol/L has a pH of

6. Note that as the acidity *increases,* the pH *decreases;* a change in pH from 7 to 6 represents a tenfold increase in the hydrogen-ion concentration.

Pure water, due to the ionization of some of the molecules into H^+ and OH^-, has a hydrogen-ion concentration of 10^{-7} mol/L (pH = 7.0) and is termed a **neutral solution. Alkaline solutions** have a lower hydrogen-ion concentration (a pH higher than 7.0), while those with a higher hydrogen-ion concentration (a pH lower than 7.0) are **acidic solutions.** The extracellular fluid of the body has a hydrogen-ion concentration of about 4×10^{-8} mol/L (pH = 7.4), with a normal range of about pH 7.35 to 7.45, and is thus slightly alkaline. Most intracellular fluids have a slightly higher hydrogen-ion concentration (pH 7.0 to 7.2) than extracellular fluids.

As we saw earlier, the ionization of carboxyl and amino groups involves the release and uptake, respectively, of hydrogen ions. These groups behave as weak acids and bases. Changes in the acidity of solutions containing molecules with carboxyl and amino groups alter the net electric charge on these molecules by shifting the ionization reaction to the right or left.

$$R-COO^- + H^+ \rightleftharpoons R-COOH$$

For example, if the acidity of a solution containing lactate is increased by adding hydrochloric acid, the concentration of lactic acid will increase and that of lactate will decrease.

If the electric charge on a molecule is altered, its interaction with other molecules or with other regions within the same molecule is altered, and thus its functional characteristics are altered. In the extracellular fluid, hydrogen-ion concentrations beyond the tenfold pH range of 7.8 to 6.8 are incompatible with life if maintained for more than a brief period of time. Even small changes in the hydrogen-ion concentration can produce large changes in molecular interactions, as we shall see.

Classes of Organic Molecules

Because most naturally occurring carbon-containing molecules are found in living organisms, the study of these compounds became known as organic chemistry. (Inorganic chemistry is the study of noncarbon-containing molecules.) However, the chemistry of living organisms, **biochemistry,** now forms only a portion of the broad field of organic chemistry.

One of the properties of the carbon atom that makes life possible is its ability to form four covalent bonds with other atoms, in particular with other carbon atoms. Since carbon atoms can also combine with hydrogen, oxygen, nitrogen, and sulfur atoms, a vast number of compounds can be formed with relatively few chemical elements. Some of these molecules are extremely large **(macromolecules),** being composed of thousands of atoms. Such large molecules are formed by linking together hundreds of smaller molecules (subunits) and are thus known as **polymers** (many small parts). The structure of macromolecules depends upon the structure of the subunits, the number of subunits linked together, and the position along the chain of each type of subunit.

Most of the organic molecules in the body can be classified into one of four groups: carbohydrates, lipids, proteins, and nucleic acids (Table 2–5).

Carbohydrates

Although carbohydrates account for only about 1 percent of the body weight, they play a central role in the chemical reactions that provide cells with energy. Carbohydrates are composed of carbon, hydrogen, and oxygen atoms in the proportions represented by the general formula $Cn(H_2O)n$, where *n* is any whole number. It is from this formula that the class of molecules gets its name, **carbohydrate**—water-containing (hydrated) carbon atoms. Linked to most of the carbon atoms in a carbohydrate are a hydrogen atom and a hydroxyl group:

$$H-\overset{|}{\underset{|}{C}}-OH$$

The presence of numerous hydroxyl groups makes carbohydrates readily soluble in water.

Most carbohydrates taste sweet, and it is among the carbohydrates that we find the substances known as sugars. The simplest sugars are the **monosaccharides** (single-sweet), the most abundant of which is **glucose,** a six-carbon molecule ($C_6H_{12}O_6$) often called "blood sugar" because it is the major monosaccharide found in the blood.

There are two ways of representing the linkage between the atoms of a monosaccharide, as illustrated in Figure 2–7. The first is the conventional way of drawing the structure of organic molecules, but the second gives a better representation of their three-dimensional shape. Five carbon atoms and an oxygen atom form a ring that lies in an essentially flat plane. The hydrogen and hydroxyl groups on each carbon lie above and below the plane of this ring. If one of the hydroxyl groups below the ring is shifted to a position above the ring, as shown in Figure 2–8, a different monosaccharide is produced.

Most monosaccharides in the body contain five or six carbon atoms and are called **pentoses** and **hexoses,** respectively. Larger carbohydrates can be formed by

TABLE 2–5 Major Categories of Organic Molecules in the Body

Category	Percent of Body Weight	Majority of Atoms	Subclass	Subunits
Carbohydrates	1	C, H, O	Monosaccharides (sugars) Polysaccharides	Monosaccharides
Lipids	15	C, H	Triacylglycerols Phospholipids Steroids	3 fatty acids + glycerol 2 fatty acids + glycerol + phosphate + small charged nitrogen molecule
Proteins	17	C, H, O, N	Peptides Proteins	Amino acids Amino acids
Nucleic acids	2	C, H, O, N	DNA RNA	Nucleotides containing the bases adenine, cytosine, guanine, thymine, the sugar deoxyribose, and phosphate Nucleotides containing the bases adenine, cytosine, guanine, uracil, the sugar ribose, and phosphate

linking a number of monosaccharides together. Carbohydrates composed of two monosaccharides are known as **disaccharides. Sucrose,** or table sugar (Figure 2–9), is composed of two monosaccharides, glucose and fructose. The linking together of most monosaccharides involves the removal of a hydroxyl group from one monosaccharide and a hydrogen atom from the other, giving rise to a molecule of water and linking the two sugars together through an oxygen atom. Conversely, hydrolysis of the disaccharide breaks this linkage by adding back the water and thus uncoupling the two monosaccharides. Additional disaccharides

frequently encountered are maltose (glucose-glucose), formed during the digestion of large carbohydrates in the intestinal tract, and lactose (glucose-galactose), present in milk.

When many monosaccharides are linked together to form polymers, the molecules are known as **polysaccharides.** Starch, found in plant cells, and **glycogen** (Figure 2–10), present in animal cells and often called "animal starch," are examples of polysaccharides. Both of these polysaccharides are composed of thousands of glucose molecules linked together in long chains, differing only in the degree of branching along the

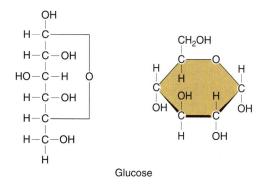

FIGURE 2–7

Two ways of diagraming the structure of the monosaccharide glucose.

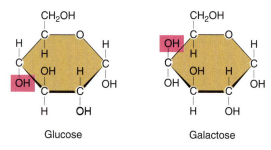

FIGURE 2–8

The structural difference between the monosaccharides glucose and galactose has to do with whether the hydroxyl group at the position indicated lies below or above the plane of the ring.

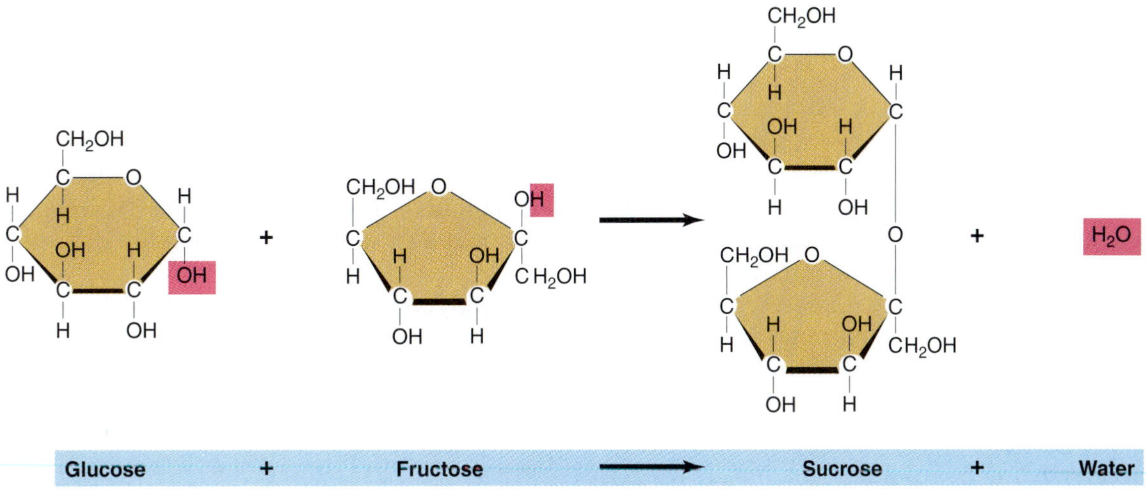

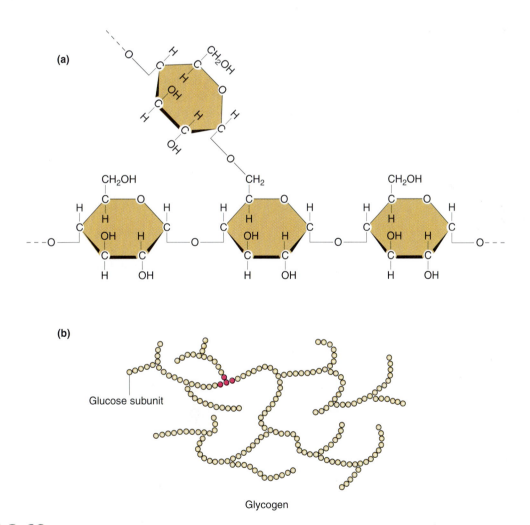

FIGURE 2–9

Sucrose (table sugar) is a disaccharide formed by the linking together of two monosaccharides, glucose and fructose.

FIGURE 2–10

Many molecules of glucose linked end-to-end and at branch points form the branched-chain polysaccharide glycogen, shown in diagrammatic form in (a). The four red subunits in (b) correspond to the four glucose subunits in (a).

chain. Hydrolysis of these polysaccharides leads to release of the glucose subunits.

Lipids

Lipids are molecules composed predominantly of hydrogen and carbon atoms. Since these atoms are linked by neutral covalent bonds, lipids are nonpolar and thus have a very low solubility in water. It is the physical property of insolubility in water that characterizes this class of organic molecules. Lipids, which account for about 40 percent of the organic matter in the average body (15 percent of the body weight), can be divided into four subclasses: fatty acids, triacylglycerols, phospholipids, and steroids.

Fatty Acids A **fatty acid** consists of a chain of carbon and hydrogen atoms with a carboxyl group at one end (Figure 2–11). Because fatty acids are synthesized in the body by the linking together of two-carbon fragments, most fatty acids have an even number of carbon atoms, with 16- and 18-carbon fatty acids being the most common. When all the carbons in a fatty acid are linked by single covalent bonds, the fatty acid is said to be a **saturated fatty acid**. Some fatty acids contain one or more double bonds, and these are known as **unsaturated fatty acids.** If one double bond is present, the acid is said to be **monounsaturated**, and if

there is more than one double bond, **polyunsaturated** (Figure 2–11).

Some fatty acids can be altered to produce a special class of molecules that regulate a number of cell functions. As described in more detail in Chapter 7, these modified fatty acids—collectively termed **eicosanoids**—are derived from the 20-carbon, polyunsaturated fatty acid **arachidonic acid.**

Triacylglycerols Triacylglycerols (also known as triglycerides) constitute the majority of the lipids in the body, and it is these molecules that are generally referred to simply as "fat." Triacylglycerols are formed by the linking together of **glycerol**, a three-carbon carbohydrate, with three fatty acids (Figure 2–11). Each of the three hydroxyl groups in glycerol is linked to the carboxyl group of a fatty acid by the removal of a molecule of water.

The three fatty acids in a molecule of triacylglycerol need not be identical; therefore, a variety of fats can be formed with fatty acids of different chain lengths and degrees of saturation. Animal fats generally contain a high proportion of saturated fatty acids, whereas vegetable fats contain more unsaturated fatty acids. Hydrolysis of triacylglycerols releases the fatty acids from glycerol, and these products can then be metabolized to provide energy for cell functions.

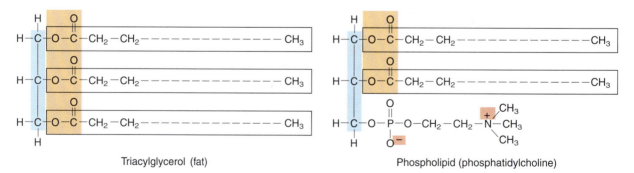

FIGURE 2–11

Glycerol and fatty acids are the major subunits that combine to form triacylglycerols and phospholipids.

Phospholipids Phospholipids are similar in overall structure to triacylglycerols, with one important difference. The third hydroxyl group of glycerol, rather than being attached to a fatty acid, is linked to phosphate. In addition, a small polar or ionized nitrogen-containing molecule is usually attached to this phosphate (Figure 2–11). These groups constitute a polar (hydrophilic) region at one end of the phospholipid, whereas the fatty acid chains provide a nonpolar (hydrophobic) region at the opposite end. Therefore, phospholipids are amphipathic. In water, they become organized into clusters, with their polar ends attracted to the water molecules.

Steroids Steroids have a distinctly different structure from that of the other subclasses of lipid molecules. Four interconnected rings of carbon atoms form the skeleton of all steroids (Figure 2–12). A few hydroxyl groups, which are polar, may be attached to this ring structure, but they are not numerous enough to make a steroid water-soluble. Examples of steroids are cholesterol, cortisol from the adrenal glands, and female (estrogen) and male (testosterone) sex hormones secreted by the gonads.

Proteins

The term **"protein"** comes from the Greek *proteios* ("of the first rank"), which aptly describes their importance. These molecules, which account for about 50 percent of the organic material in the body (17 percent of the body weight), play critical roles in almost every physiological process. Proteins are composed of carbon, hydrogen, oxygen, nitrogen, and small amounts of other elements, notably sulfur. They are macromolecules, often containing thousands of atoms, and like most large molecules, they are formed by the linking together of a large number of small subunits to form long chains.

Amino Acid Subunits The subunits of proteins are **amino acids;** thus, proteins are polymers of amino acids. Every amino acid except proline has an amino ($-NH_2$) and a carboxyl ($-COOH$) group linked to the terminal carbon in the molecule:

$$\begin{array}{c} H \\ | \\ R-C-COOH \\ | \\ NH_2 \end{array}$$

The third bond of this terminal carbon is linked to a hydrogen and the fourth to the remainder of the molecule, which is known as the **amino acid side chain** (R in the formula). These side chains are relatively small, ranging from a single hydrogen to 9 carbons.

The proteins of all living organisms are composed of the same set of 20 different amino acids, corresponding to 20 different side chains. The side chains may be nonpolar (8 amino acids), polar (7 amino acids), or ionized (5 amino acids) (Figure 2–13).

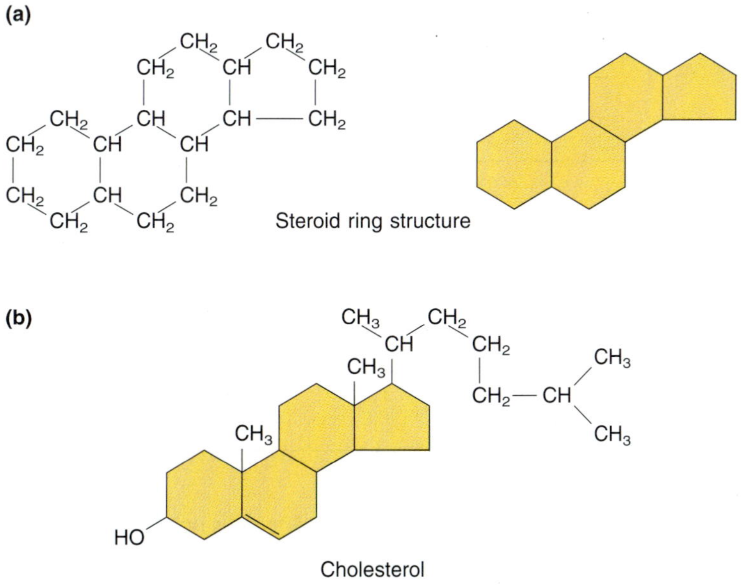

FIGURE 2–12

(a) Steroid ring structure, shown with all the carbon and hydrogen atoms in the rings and again without these atoms to emphasize the overall ring structure of this class of lipids. (b) Different steroids have different types and numbers of chemical groups attached at various locations on the steroid ring, as shown by the structure of cholesterol.

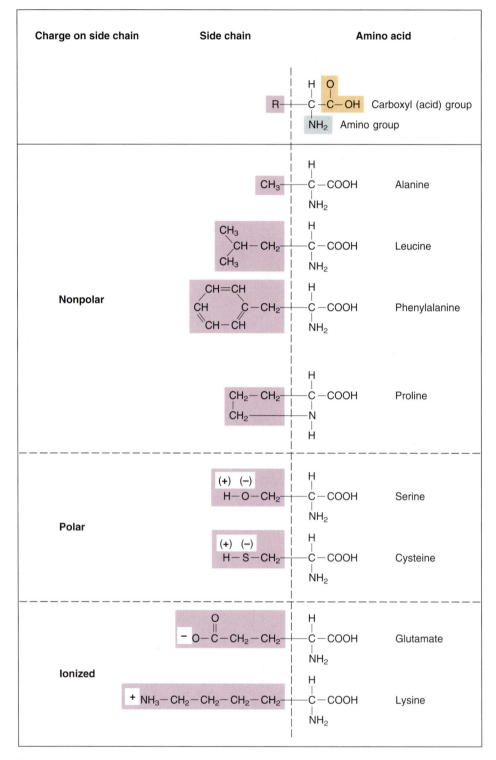

FIGURE 2–13

Structures of 8 of the 20 amino acids found in proteins. Note that proline does not have a free amino group, but it can still form a peptide bond.

Polypeptides Amino acids are joined together by linking the carboxyl group of one amino acid to the amino group of another. In the process, a molecule of water is formed (Figure 2–14). The bond formed between the amino and carboxyl group is called a **peptide bond,** and it is a polar covalent bond.

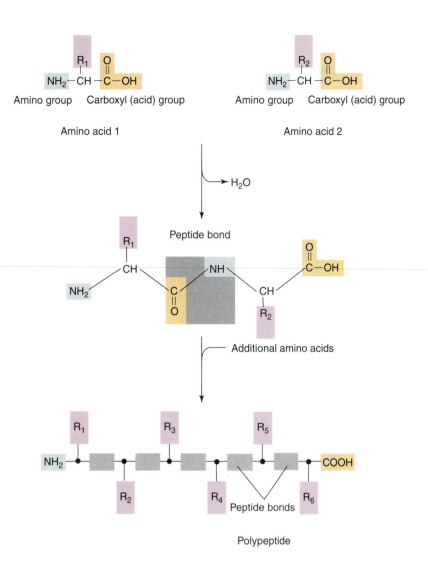

FIGURE 2–14
Linkage of amino acids by peptide bonds to form a polypeptide.

Note that when two amino acids are linked together, one end of the resulting molecule has a free amino group, and the other has a free carboxyl group. Additional amino acids can be linked by peptide bonds to these free ends. A sequence of amino acids linked by peptide bonds is known as a **polypeptide.** The peptide bonds form the backbone of the polypeptide, and the side chain of each amino acid sticks out from the side of the chain. If the number of amino acids in a polypeptide is 50 or less, the molecule is known as a **peptide;** if the sequence is more than 50 amino acid units, it is known as a protein. The number 50 is arbitrary but has become the convention for distinguishing between large and small polypeptides.

One or more monosaccharides can be covalently attached to the side chains of specific amino acids (serine and threonine) to form a class of proteins known as **glycoproteins.**

Primary Protein Structure   Two variables determine the primary structure of a polypeptide: (1) the number of amino acids in the chain, and (2) the specific type of amino acid at each position along the chain (Figure 2–15). Each position along the chain can be occupied by any one of the 20 different amino acids. Let us consider the number of different peptides that can be formed that have a sequence of three amino acids. Any one of the 20 different amino acids may occupy the first position in the sequence, any one of the 20 the second position, and any one of the 20 the third position, for a total of $20 \times 20 \times 20 = 20^3 = 8000$ possible sequences of three amino acids. If the peptide is 6 amino acids in length, $20^6 = 64,000,000$ possible combinations can be formed. Peptides that are only 6 amino acids long are still very small compared to proteins, which may have sequences of 1000 or more amino acids. Thus, with 20 different amino acids, an almost

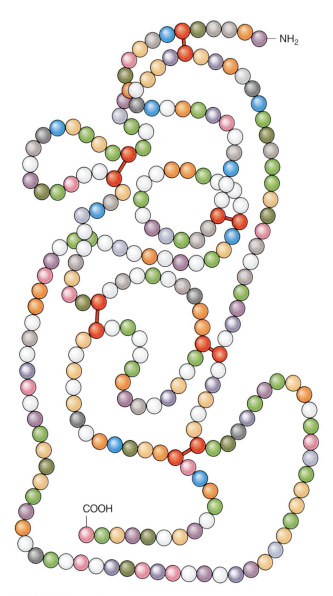

tate around their peptide bonds, a polypeptide chain is flexible and can be bent into a number of shapes, just as a string of beads can be twisted into many configurations. The three-dimensional shape of a molecule is known as its **conformation** (Figure 2–16). The conformations of peptides and proteins play a major role in their functioning, as we shall see in Chapter 4.

Four factors determine the conformation of a polypeptide chain once the amino acid sequence has been formed: (1) hydrogen bonds between portions of the chain or with surrounding water molecules; (2) ionic bonds between polar and ionized regions along the chain; (3) **van der Waals forces,** which are very weak forces of attraction between nonpolar (hydrophobic) regions in close proximity to each other; and (4) covalent bonds linking the side chains of two amino acids (Figure 2–17).

An example of the attractions between various regions along a polypeptide chain is the hydrogen bond that can occur between the hydrogen linked to the nitrogen atom in one peptide bond and the double-bonded oxygen in another peptide bond (Figure 2–18). Since peptide bonds occur at regular intervals along a polypeptide chain, the hydrogen bonds between them tend to force the chain into a coiled conformation known as an **alpha helix.** Hydrogen bonds can also form between peptide bonds when extended regions

FIGURE 2–15

The position of each type of amino acid in a polypeptide chain and the total number of amino acids in the chain distinguish one polypeptide from another. The polypeptide illustrated contains 223 amino acids with different amino acids represented by different-colored circles. The bonds between various regions of the chain (red to red) represent covalent disulfide bonds between cysteine side chains.

unlimited variety of polypeptides can be formed by altering both the amino acid sequence and the total number of amino acids in the chain.

Protein Conformation   A polypeptide is analogous to a string of beads, each bead representing one amino acid (Figure 2–15). Moreover, since amino acids can ro-

FIGURE 2–16

Conformation (shape) of the protein molecule myoglobin. Each dot corresponds to the location of a single amino acid.
Adopted from Albert L. Lehninger.

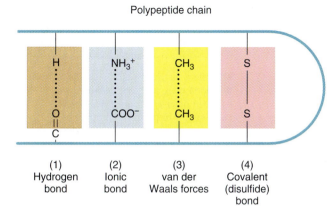

Polypeptide chain

(1)	(2)	(3)	(4)
Hydrogen bond	Ionic bond	van der Waals forces	Covalent (disulfide) bond

FIGURE 2–17

Factors that contribute to the folding of polypeptide chains and thus to their conformation are (1) hydrogen bonds between side chains or with surrounding water molecules, (2) ionic bonds between polar or ionized side chains, (3) van der Waals forces between nonpolar side chains, and (4) covalent bonds between side chains.

of a polypeptide chain run approximately parallel to each other, forming a relatively straight, extended region known as a **beta sheet** (Figure 2–19). However, for several reasons, a given region of a polypeptide chain may not assume either a helical or beta sheet conformation. For example, the sizes of the side chains and the ionic bonds between oppositely charged side chains can interfere with the repetitive hydrogen bonding required to produce these shapes. These irregular regions are known as loop conformations and occur in

regions linking the more regular helical and beta sheet patterns (Figure 2–19).

Covalent bonds between certain side chains can also distort the regular folding patterns. For example, the side chain of the amino acid cysteine contains a sulfhydryl group (R—SH), which can react with a sulfhydryl group in another cysteine side chain to produce a **disulfide bond** (R—S—S—R), which links the two amino acid side chains together (Figure 2–20). Disulfide bonds form covalent bonds between portions of a polypeptide chain, in contrast to the weaker hydrogen and ionic bonds, which are more easily broken. Table 2–6 provides a summary of the types of bonding forces that contribute to the conformation of polypeptide chains. These same bonds are also involved in other intermolecular interactions, which will be described in later chapters.

A number of proteins are composed of more than one polypeptide chain and are known as **multimeric proteins** (many parts). The same factors that influence the conformation of a single polypeptide also determine the interactions between the polypeptides in a multimeric protein. Thus, the chains can be held together by interactions between various ionized, polar, and nonpolar side chains, as well as by disulfide covalent bonds between the chains.

The polypeptide chains in a multimeric protein may be identical or different. For example, hemoglobin, the protein that transports oxygen in the blood, is a multimeric protein with four polypeptide chains, two of one kind and two of another (Figure 2-21).

The primary structures (amino acid sequences) of a large number of proteins are known, but

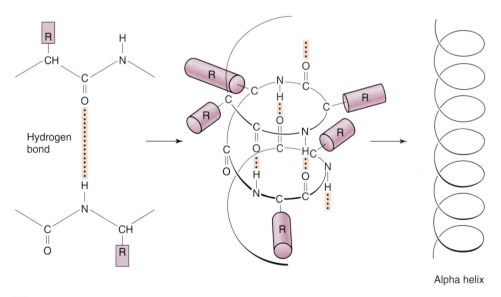

Alpha helix

FIGURE 2–18

Hydrogen bonds between regularly spaced peptide bonds can produce a helical conformation in a polypeptide chain.

TABLE 2–6 Bonding Forces Between Atoms and Molecules

Bond	Strength	Characteristics	Examples
Hydrogen	Weak	Electrical attraction between polarized bonds, usually hydrogen and oxygen	Attractions between peptide bonds forming the alpha helix structure of proteins and between polar amino acid side chains contributing to protein conformation; attractions between water molecules
Ionic	Strong	Electrical attraction between oppositely charged ionized groups	Attractions between ionized groups in amino acid side chains contributing to protein conformation; attractions between ions in a salt
van der Waals	Very weak	Attraction between nonpolar molecules and groups when very close to each other	Attractions between nonpolar amino acids in proteins contributing to protein conformation; attractions between lipid molecules
Covalent	Very strong	Shared electrons between atoms Nonpolar covalent bonds share electrons equally while in polar bonds the electrons reside closer to one atom in the pair	Most bonds linking atoms together to form molecules

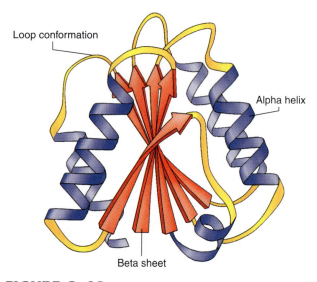

Loop conformation

Alpha helix

Beta sheet

FIGURE 2–19

A ribbon diagram illustrating the pathway followed by the backbone of a single polypeptide chain. Helical regions (blue) are coiled, beta sheets (red) of parallel chains are shown as relatively straight arrows, and loop conformations (yellow) connect the various helical and beta sheet regions. Beginning at the end of the chain labeled "Beta sheet," there is a continuous chain of amino acids that passes through various conformations.

three-dimensional conformations have been determined for only a few. Because of the multiple factors that can influence the folding of a polypeptide chain, it is not yet possible to predict accurately the conformation of a protein from its primary amino acid sequence.

Nucleic Acids

Nucleic acids account for only 2 percent of the body's weight, yet these molecules are extremely important because they are responsible for the storage, expression, and transmission of genetic information. It is the expression of genetic information (in the form of specific proteins) that determines whether one is a human being or a mouse, or whether a cell is a muscle cell or a nerve cell.

There are two classes of nucleic acids, **deoxyribonucleic acid (DNA)** and **ribonucleic acid (RNA).** DNA molecules store genetic information coded in the sequence of their subunits, whereas RNA molecules are involved in the decoding of this information into instructions for linking together a specific sequence of amino acids to form a specific polypeptide chain. The mechanisms of gene expression and protein synthesis will be described in Chapter 5.

Both types of nucleic acids are polymers and are therefore composed of linear sequences of repeating subunits. Each subunit, known as a **nucleotide,** has

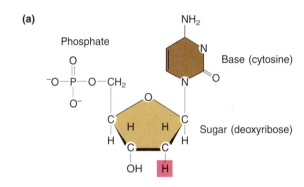

FIGURE 2–20
Formation of a disulfide bond between the side chains of two cysteine amino acids links two regions of the polypeptide together. The hydrogen atoms on the sulfhydryl groups of the cysteines are transferred to another molecule, X, during the formation of the disulfide bond.

three components: a phosphate group, a sugar, and a ring of carbon and nitrogen atoms known as a base because it can accept hydrogen ions (Figure 2–22). The phosphate group of one nucleotide is linked to the sugar of the adjacent nucleotide to form a chain, with the bases sticking out from the side of the phosphate-sugar backbone (Figure 2–23).

DNA The nucleotides in DNA contain the five-carbon sugar **deoxyribose** (hence the name "deoxyribonucleic acid"). Four different nucleotides are present in DNA, corresponding to the four different bases that can be linked to deoxyribose. These bases are divided into

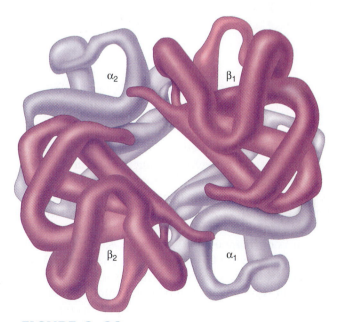

FIGURE 2–21
Hemoglobin, a multimeric protein composed of two identical α chains and two identical β chains. (The heme groups attached to each globin chain are not shown.)

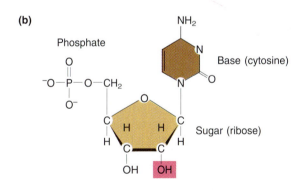

FIGURE 2–22
Nucleotide subunits of DNA and RNA. Nucleotides are composed of a sugar, a base, and phosphate. (a) Deoxyribonucleotides present in DNA contain the sugar deoxyribose. (b) The sugar in ribonucleotides, present in RNA, is ribose, which has an OH at the position that lacks this group in deoxyribose.

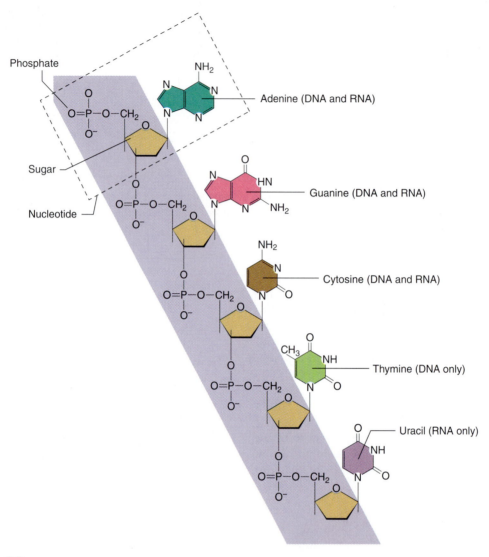

Phosphate

Adenine (DNA and RNA)

Sugar

Nucleotide

Guanine (DNA and RNA)

Cytosine (DNA and RNA)

Thymine (DNA only)

Uracil (RNA only)

FIGURE 2–23

Phosphate-sugar bonds link nucleotides in sequence to form nucleic acids. Note that the pyrimidine base thymine is only found in DNA, and uracil is only present in RNA.

two classes: (1) the **purine** bases, **adenine** (A) and **guanine** (G), which have double (fused) rings of nitrogen and carbon atoms, and (2) the **pyrimidine** bases, **cytosine** (C) and **thymine** (T), which have only a single ring (Figure 2–23).

A DNA molecule consists of not one but two chains of nucleotides coiled around each other in the form of a double helix (Figure 2–24). The two chains are held together by hydrogen bonds between a purine base on one chain and a pyrimidine base on the opposite chain. The ring structure of each base lies in a flat plane perpendicular to the phosphate-sugar backbone, appearing as steps on a spiral staircase. This base pairing maintains a constant distance between the sugar-phosphate backbones of the two chains as they coil around each other.

Specificity is imposed on the base pairings by the location of the hydrogen-bonding groups in the four bases (Figure 2–25). Three hydrogen bonds are formed between the purine guanine and the pyrimidine cytosine (G—C pairing), while only two hydrogen bonds can be formed between the purine adenine and the pyrimidine thymine (A—T pairing). As a result, G is always paired with C, and A with T. In Chapter 5 we shall see how this specificity provides the mechanism for duplicating and transferring genetic information.

RNA RNA molecules differ in only a few respects from DNA (Table 2–7): (1) RNA consists of a single (rather than a double) chain of nucleotides; (2) in RNA, the sugar in each nucleotide is **ribose** rather than deoxyribose; and (3) the pyrimidine base thymine in

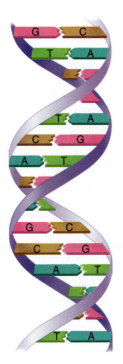

FIGURE 2–24

Base pairings between a purine and pyrimidine base link the two polynucleotide strands of the DNA double helix.

TABLE 2–7 Comparison of DNA and RNA Composition

	DNA	RNA
Nucleotide sugar	Deoxyribose	Ribose
Nucleotide bases		
Purines	Adenine	Adenine
	Guanine	Guanine
Pyrimidines	Cytosine	Cytosine
	Thymine	Uracil
Number of chains	Two	One

DNA is replaced in RNA by the pyrimidine base **uracil** (U) (Figure 2–23), which can base-pair with the purine adenine (A—U pairing). The other three bases, adenine, guanine, and cytosine, are the same in both DNA and RNA. Although RNA contains only a single chain of nucleotides, portions of this chain can bend back upon itself and undergo base pairing with nucleotides in the same chain or in other molecules of DNA or RNA.

SUMMARY

Atoms

I. Atoms are composed of three subatomic particles: positive protons and neutral neutrons, both located in the nucleus, and negative electrons revolving around the nucleus.

II. The atomic number is the number of protons in an atom, and because atoms are electrically neutral, it is also the number of electrons.

III. The atomic weight of an atom is the ratio of the atom's mass relative to that of a carbon-12 atom.

IV. One gram atomic mass is the number of grams of an element equal to its atomic weight. One gram atomic mass of any element contains the same number of atoms—6×10^{23}.

V. The 24 elements essential for normal body function are listed in Table 2–1.

Molecules

I. Molecules are formed by linking atoms together.

II. A covalent bond is formed when two atoms share a pair of electrons. Each type of atom can form a characteristic number of covalent bonds: hydrogen forms one; oxygen, two; nitrogen, three; and carbon, four.

III. Molecules have characteristic shapes, which can be altered within limits by the rotation of their atoms around covalent bonds.

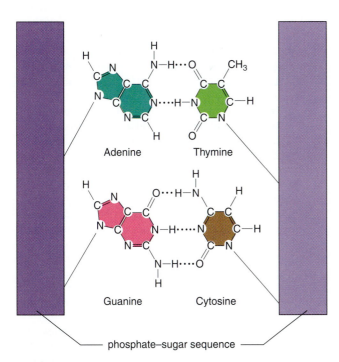

FIGURE 2–25

Hydrogen bonds between the nucleotide bases in DNA determine the specificity of base pairings: adenine with thymine and guanine with cytosine.

Ions

When an atom gains or loses one or more electrons, it acquires a net electric charge and becomes an ion.

Free Radicals

Free radicals are atoms or molecules that contain atoms having an unpaired electron in their outer electron orbital.

Polar Molecules

I. In polar covalent bonds, one atom attracts the bonding electrons more than the other atom of the pair.
II. The electrical attraction between hydrogen and an oxygen or nitrogen atom in a separate molecule or different region of the same molecule forms a hydrogen bond.
III. Water, a polar molecule, is attracted to other water molecules by hydrogen bonds.

Solutions

I. Substances dissolved in a liquid are solutes, and the liquid in which they are dissolved is the solvent. Water is the most abundant solvent in the body.
II. Substances that have polar or ionized groups dissolve in water by being electrically attracted to the polar water molecules.
III. In water, amphipathic molecules form clusters with the polar regions at the surface and the nonpolar regions in the interior of the cluster.
IV. The molecular weight of a molecule is the sum of the atomic weights of all its atoms. One mole of any substance is its molecular weight in grams and contains 6×10^{23} molecules.
V. Substances that release a hydrogen ion in solution are called acids. Those that accept a hydrogen ion are bases.
   a. The acidity of a solution is determined by its free hydrogen-ion concentration; the greater the hydrogen-ion concentration, the greater the acidity.
   b. The pH of a solution is the negative logarithm of the hydrogen-ion concentration. As the acidity of a solution increases, the pH decreases. Acid solutions have a pH less than 7.0, whereas alkaline solutions have a pH greater than 7.0.

Classes of Organic Molecules

I. Carbohydrates are composed of carbon, hydrogen, and oxygen in the proportions $Cn(H_2O)n$.
   a. The presence of the polar hydroxyl groups makes carbohydrates soluble in water.
   b. The most abundant monosaccharide in the body is glucose ($C_6H_{12}O_6$), which is stored in cells in the form of the polysaccharide glycogen.
II. Most lipids lack polar and ionized groups, a characteristic that makes them insoluble in water.
   a. Triacylglycerols (fats) are formed when fatty acids are linked to each of the three hydroxyl groups in glycerol.
   b. Phospholipids contain two fatty acids linked to two of the hydroxyl groups in glycerol, with the third hydroxyl linked to phosphate, which in turn is linked to a small charged or polar compound. The polar and ionized groups at one end of phospholipids make these molecules amphipathic.
   c. Steroids are composed of four interconnected rings, often containing a few hydroxyl and other groups.
III. Proteins, macromolecules composed primarily of carbon, hydrogen, oxygen, and nitrogen, are polymers of 20 different amino acids.
   a. Amino acids have an amino ($-NH_2$) and a carboxyl ($-COOH$) group linked to their terminal carbon atom.
   b. Amino acids are linked together by peptide bonds between the carboxyl group of one amino acid and the amino group of the next.
   c. The primary structure of a polypeptide chain is determined by (1) the number of amino acids in sequence, and (2) the type of amino acid at each position.
   d. The factors that determine the conformation of a polypeptide chain are summarized in Figure 2–17.
   e. Hydrogen bonds between peptide bonds along a polypeptide force much of the chain into an alpha helix.
   f. Covalent disulfide bonds can form between the sulfhydryl groups of cysteine side chains to hold regions of a polypeptide chain close to each other.
   g. Multimeric proteins have multiple polypeptide chains.
IV. Nucleic acids are responsible for the storage, expression, and transmission of genetic information.
   a. Deoxyribonucleic acid (DNA) stores genetic information.
   b. Ribonucleic acid (RNA) is involved in decoding the information in DNA into instructions for linking amino acids together to form proteins.
   c. Both types of nucleic acids are polymers of nucleotides, each containing a phosphate group, a sugar, and a base of carbon, hydrogen, oxygen, and nitrogen atoms.
   d. DNA contains the sugar deoxyribose and consists of two chains of nucleotides coiled around each other in a double helix. The chains are held together by hydrogen bonds between purine and pyrimidine bases in the two chains.
   e. Base pairings in DNA always occur between guanine and cytosine and between adenine and thymine.
   f. RNA consists of a single chain of nucleotides, containing the sugar ribose and three of the four bases found in DNA. The fourth base in RNA is the pyrimidine uracil rather than thymine. Uracil base-pairs with adenine.

KEY TERMS

atom
chemical element
proton
neutron
electron
atomic nucleus
atomic number
atomic weight
isotope
gram atomic mass
trace element
molecule
covalent bond
ion
cation
anion
electrolyte
carboxyl group
amino group
free radical
polar covalent bond
hydroxyl group
polar molecule
nonpolar molecule
hydrogen bond
hydrolysis
solute
solvent
solution
ionic bond
hydrophilic
hydrophobic
amphipathic
concentration
molecular weight
mole
acid
base
strong acid
weak acid
acidity
pH
neutral solution
alkaline solution
acidic solution
biochemistry
macromolecule
polymer

carbohydrate
monosaccharide
glucose
pentose
hexose
disaccharide
sucrose
polysaccharide
glycogen
lipid
fatty acid
saturated fatty acid
unsaturated fatty acid
monounsaturated fatty acid
polyunsaturated fatty acid
eicosanoid
arachidonic acid
triacylglycerol
glycerol
phospholipid
steroid
protein
amino acid
amino acid side chain
peptide bond
polypeptide
peptide
glycoprotein
conformation
van der Waals forces
alpha helix
beta sheet
disulfide bond
multimeric protein
nucleic acid
deoxyribonucleic acid (DNA)
ribonucleic acid (RNA)
nucleotide
deoxyribose
purine
adenine
guanine
pyrimidine
cytosine
thymine
ribose
uracil

REVIEW QUESTIONS

1. Describe the electric charge, mass, and location of the three major subatomic particles in an atom.
2. Which four kinds of atoms are most abundant in the body?
3. Describe the distinguishing characteristics of the three classes of essential chemical elements found in the body.
4. How many covalent bonds can be formed by atoms of carbon, nitrogen, oxygen, and hydrogen?
5. What property of molecules allows them to change their three-dimensional shape?
6. Describe how an ion is formed.
7. Draw the structures of an ionized carboxyl group and an ionized amino group.
8. Define a free radical.
9. Describe the polar characteristics of a water molecule.
10. What determines a molecule's solubility or lack of solubility in water?
11. Describe the organization of amphipathic molecules in water.
12. What is the molar concentration of 80 g of glucose dissolved in sufficient water to make 2 L of solution?
13. What distinguishes a weak acid from a strong acid?
14. What effect does increasing the pH of a solution have upon the ionization of a carboxyl group? An amino group?
15. Name the four classes of organic molecules in the body.
16. Describe the three subclasses of carbohydrate molecules.
17. To which subclass of carbohydrates do each of the following molecules belong: glucose, sucrose, and glycogen?
18. What properties are characteristic of lipids?
19. Describe the subclasses of lipids.
20. Describe the linkage between amino acids to form a polypeptide chain.
21. What is the difference between a peptide and a protein?
22. What two factors determine the primary structure of a polypeptide chain?
23. Describe the types of interactions that determine the conformation of a polypeptide chain.
24. Describe the structure of DNA and RNA.
25. Describe the characteristics of base pairings between nucleotide bases.

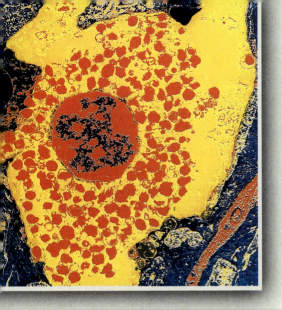

Cell Structure

As we learned in Chapter 1, cells are the structural and functional units of all living organisms. The word "cell" means "a small chamber" (like a jail cell). The human body is composed of trillions of cells, each a microscopic compartment (Figure 3–1). In this chapter, we describe the structures found in most of the body's cells and state their functions. Subsequent chapters describe how these structures perform their functions.

The cells of a mouse, a human being, and an elephant are all approximately the same size. An elephant is large because it has more cells, not because it has larger cells. A majority of the cells in a human being have diameters in the range of 10 to 20 μm, although cells as small as 2 μm and as large as 120 μm are present. A cell 10 μm in diameter is about one-tenth the size of the smallest object that can be seen with the naked eye; a microscope must therefore be used to observe cells and their internal structure.

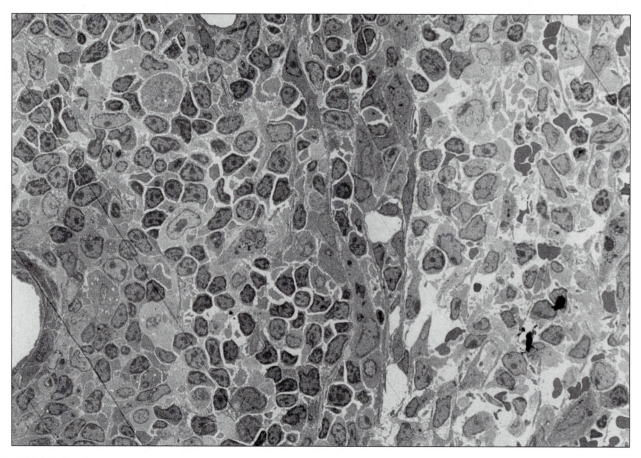

FIGURE 3–1
Cellular organization of tissues, as illustrated by a portion of spleen. Oval, clear spaces in the micrograph are blood vessels.
From Johannes A. G. Rhodin, "Histology, A Text & Atlas," Oxford University Press, New York, 1974.

Microscopic Observations of Cells

The smallest object that can be resolved with a microscope depends upon the wavelength of the radiation used to illuminate the specimen—the shorter the wavelength, the smaller the object that can be seen. With a **light microscope,** objects as small as 0.2 μm in diameter can be resolved, whereas an **electron micro-** scope, which uses electron beams instead of light rays, can resolve structures as small as 0.002 μm. A greater resolution is achieved with an electron microscope because electrons behave as waves with much shorter wavelengths than those of visible light. Typical sizes of cells and cellular components are illustrated in Figure 3–2.

Although *living* cells can be observed with a light microscope, this is not possible with an electron microscope. To form an image with an electron beam,

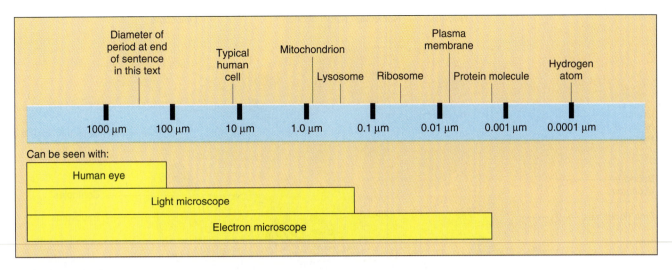

FIGURE 3–2
Sizes of cell structures, plotted on a logarithmic scale. Typical sizes are indicated.

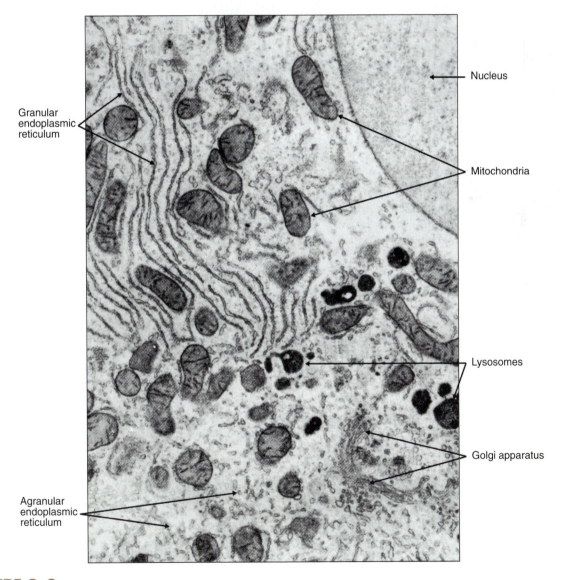

FIGURE 3–3
Electron micrograph of a thin section through a portion of a rat liver cell.

From K. R. Porter in T. W. Goodwin and O. Lindberg (eds.), "Biological Structure and Function," vol. I, Academic Press, Inc., New York, 1961.

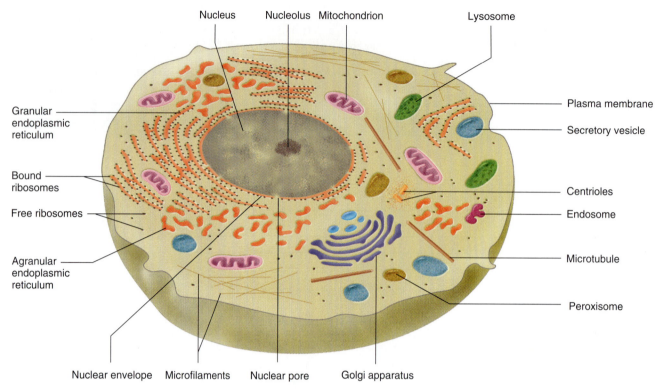

Nucleus   Nucleolus   Mitochondrion   Lysosome

Granular endoplasmic reticulum

Bound ribosomes

Free ribosomes

Agranular endoplasmic reticulum

Plasma membrane

Secretory vesicle

Centrioles

Endosome

Microtubule

Peroxisome

Nuclear envelope   Microfilaments   Nuclear pore   Golgi apparatus

FIGURE 3–4

Structures found in most human cells.

most of the electrons must pass through the specimen, just as light passes through a specimen in a light microscope. However, electrons can penetrate only a short distance through matter; therefore, the observed specimen must be very thin. Cells to be observed with an electron microscope must be cut into sections on the order of 0.1 μm thick, which is about one-hundredth of the thickness of a typical cell.

Because electron micrographs (such as Figure 3–3) are images of very thin sections of a cell, they can often be misleading. Structures that appear as separate objects in the electron micrograph may actually be continuous structures that are connected through a region lying outside the plane of the section. As an analogy, a thin section through a ball of string would appear as a collection of separate lines and disconnected dots even though the piece of string was originally continuous.

Two classes of cells, **eukaryotic cells** and **prokaryotic cells,** can be distinguished by their structure. The cells of the human body, as well as those of other multicellular animals and plants, are eukaryotic (true-nucleus) cells. These cells contain a nuclear membrane surrounding the cell nucleus and numerous other membrane-bound structures. Prokaryotic cells, for example, bacteria, lack these membranous structures. This chapter describes the structure of eukaryotic cells only.

Compare an electron micrograph of a section through a cell (Figure 3–3) with a diagrammatic illustration of a typical human cell (Figure 3–4). What is immediately obvious from both figures is the extensive structure inside the cell. Cells are surrounded by a limiting barrier, the **plasma membrane,** which covers the cell surface. The cell interior is divided into a number of compartments surrounded by membranes. These membrane-bound compartments, along with some particles and filaments, are known as **cell organelles** (little organs). Each cell organelle performs specific functions that contribute to the cell's survival.

The interior of a cell is divided into two regions: (1) the **nucleus,** a spherical or oval structure usually near the center of the cell, and (2) the **cytoplasm,** the region outside the nucleus (Figure 3–5). The cytoplasm contains two components: (1) cell organelles and (2) the fluid surrounding the organelles known as the **cytosol** (cytoplasmic solution). The term **intracellular fluid** refers to *all* the fluid inside a cell—in other words, cytosol plus the fluid inside all the organelles, including the nucleus. The chemical compositions of the fluids in these cell organelles differ from that of the cytosol. The cytosol is by far the largest intracellular fluid compartment.

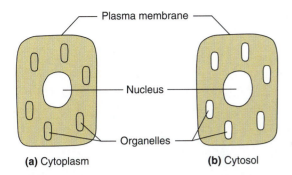

Plasma membrane

Nucleus

Organelles

(a) Cytoplasm   **(b)** Cytosol

FIGURE 3–5
Comparison of cytoplasm and cytosol. (a) Cytoplasm
(colored area) is the region of the cell outside the nucleus.
(b) Cytosol (colored area) is the fluid portion of the
cytoplasm outside the cell organelles.

TABLE 3–1 Functions of Cell Membranes
1. Regulate the passage of substances into and out of cells and between cell organelles and cytosol
2. Detect chemical messengers arriving at the cell surface
3. Link adjacent cells together by membrane junctions
4. Anchor cells to the extracellular matrix

Membranes

Membranes form a major structural element in cells. Although membranes perform a variety of functions, their most universal role is to act as a selective barrier to the passage of molecules, allowing some molecules to cross while excluding others. The plasma membrane regulates the passage of substances into and out of the cell, whereas the membranes surrounding cell organelles allow selective movement of substances between the organelles and the cytosol. One of the advantages of restricting the movements of molecules across membranes is confining the products of chemical reactions to specific cell organelles. As we shall see in Chapter 6, the hindrance offered by a membrane to the passage of substances can be altered to allow increased or decreased flow of molecules or ions across the membrane in response to various signals.

The plasma membrane, in addition to acting as a selective barrier, plays an important role in detecting chemical signals from other cells and in anchoring cells to adjacent cells and to the extracellular matrix of connective-tissue proteins (Table 3–1).

Membrane Structure

All membranes consist of a double layer of lipid molecules in which proteins are embedded (Figure 3–6). The major membrane lipids are **phospholipids.** As described in Chapter 2, these are amphipathic molecules: one end has a charged region, and the remainder of the molecule, which consists of two long fatty acid chains, is nonpolar. The phospholipids in cell membranes are organized into a bimolecular layer with the nonpolar fatty acid chains in the middle. The polar regions of the phospholipids are oriented toward the surfaces of the membrane as a result of their attraction to the polar water molecules in the extracellular fluid and cytosol.

No chemical bonds link the phospholipids to each other or to the membrane proteins, and therefore, each molecule is free to move independently of the others. This results in considerable random lateral movement of both membrane lipids and proteins parallel to the surfaces of the bilayer. In addition, the long fatty acid chains can bend and wiggle back and forth. Thus, the lipid bilayer has the characteristics of a fluid, much like a thin layer of oil on a water surface, and this makes the membrane quite flexible. This flexibility, along with the fact that cells are filled with fluid, allows cells to undergo considerable changes in shape without disruption of their structural integrity. Like a piece of cloth, membranes can be bent and folded but cannot be stretched without being torn.

The plasma membrane also contains **cholesterol** (about one molecule of cholesterol for each molecule of phospholipid), whereas intracellular membranes contain very little cholesterol. Cholesterol, a steroid, is slightly amphipathic because of a single polar hydroxyl group (see Figure 2–12) on its nonpolar ring structure. Therefore, cholesterol, like the phospholipids, is inserted into the lipid bilayer with its polar region at a bilayer surface and its nonpolar rings in the interior in association with the fatty acid chains. Cholesterol associates with certain classes of plasma membrane phospholipids and proteins, forming organized clusters that function in the pinching off of portions of the plasma membrane to form vesicles that deliver their contents to various intracellular organelles, as described in Chapter 6.

There are two classes of membrane proteins: integral and peripheral. **Integral membrane proteins** are closely associated with the membrane lipids and cannot be extracted from the membrane without disrupting the lipid bilayer. Like the phospholipids, the integral proteins are amphipathic, having polar amino acid side chains in one region of the molecule and nonpolar side chains clustered together in a separate region. Because they are amphipathic, integral proteins are arranged in the membrane with the same orientation as amphipathic lipids—the polar regions are at the surfaces in association with polar water molecules, and

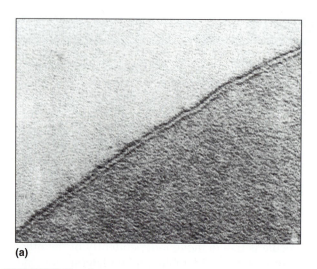

(a)

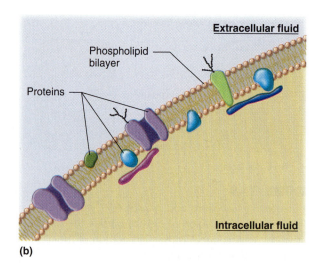

(b)

FIGURE 3–6

(a) Electron micrograph of a human red-cell plasma membrane. Cell membranes are 6 to 10 nm thick, too thin to be seen without the aid of an electron microscope. In an electron micrograph, a membrane appears as two dark lines separated by a light interspace. The dark lines correspond to the polar regions of the proteins and lipids, whereas the light interspace corresponds to the nonpolar regions of these molecules. (b) Arrangement of the proteins and lipids in a membrane.

From J. D. Robertson in Michael Locke (ed.), "Cell Membranes in Development," Academic Press, Inc., New York, 1964.

the nonpolar regions are in the interior in association with nonpolar fatty acid chains (Figure 3–7). Like the membrane lipids, many of the integral proteins can move laterally in the plane of the membrane, but others are immobilized because they are linked to a network of peripheral proteins located primarily at the cytosolic surface of the membrane.

Most integral proteins span the entire membrane and are referred to as **transmembrane proteins.** Most of these transmembrane proteins cross the lipid bilayer several times (Figure 3–8). These proteins have polar regions connected by nonpolar segments that associate with the nonpolar regions of the lipids in the membrane interior. The polar regions of transmembrane proteins may extend far beyond the surfaces of the lipid bilayer. Some transmembrane proteins form channels through which ions or water can cross the membrane, whereas others are associated with the transmission of chemical signals across the membrane or the anchoring of extracellular and intracellular protein filaments to the plasma membrane.

Peripheral membrane proteins are not amphipathic and do not associate with the nonpolar regions of the lipids in the interior of the membrane. They are located at the membrane surface where they are bound to the polar regions of the integral membrane proteins (see Figure 3–7). Most of the peripheral proteins are on the cytosolic surface of the plasma membrane where they are associated with cytoskeletal elements that influence cell shape and motility.

The extracellular surface of the plasma membrane contains small amounts of carbohydrate covalently linked to some of the membrane lipids and proteins. These carbohydrates consist of short, branched chains of monosaccharides that extend from the cell surface into the extracellular fluid where they form a fuzzy, "sugar-coated" layer known as the **glycocalyx.** These surface carbohydrates play important roles in enabling cells to identify and interact with each other.

FIGURE 3–7

Arrangement of integral and peripheral membrane proteins in association with a bimolecular layer of phospholipids.

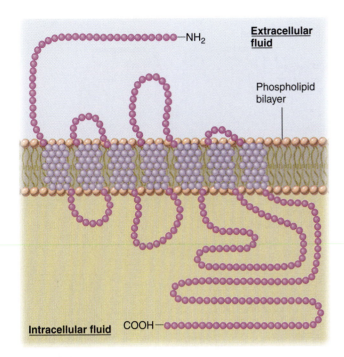

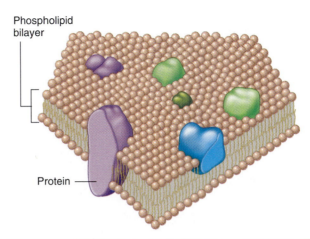

FIGURE 3–9

Fluid-mosaic model of cell membrane structure.

Redrawn from S. J. Singer and G. L. Nicholson, Science, 175:723. Copyright 1972 by the American Association for the Advancement of Science.

FIGURE 3–8

A typical transmembrane protein with multiple hydrophobic segments traversing the lipid bilayer. Each transmembrane segment is composed of nonpolar amino acids spiraled in an alpha helical conformation.

The lipids in the outer half of the bilayer differ somewhat in kind and amount from those in the inner half, and, as we have seen, the proteins or portions of proteins on the outer surface differ from those on the inner surface. Many membrane functions are related to these asymmetries in chemical composition between the two surfaces of a membrane.

All membranes have the general structure described above, which has come to be known as the **fluid-mosaic model** in which membrane proteins float in a sea of lipid (Figure 3–9). However, the proteins and, to a lesser extent, the lipids (the distribution of cholesterol, for example) in the plasma membrane are different from those in organelle membranes. Thus, the special functions of membranes, which depend primarily on the membrane proteins, may differ in the various membrane-bound organelles and in the plasma membranes of different types of cells.

Membrane Junctions

In addition to providing a barrier to the movements of molecules between the intracellular and extracellular fluids, plasma membranes are involved in interactions between cells to form tissues. Some cells, particularly those in the blood, are not anchored to other cells, but are suspended in a fluid–the blood plasma in the case of blood cells. Most cells, however, are packaged into tissues and are not free to move around the body. But even in tissues there is usually a space between the plasma membranes of adjacent cells. This space is filled with extracellular fluid and provides the pathway for substances to pass between cells on their way to and from the blood.

The forces that organize cells into tissues and organs are poorly understood, but they depend, at least in part, on the ability of certain transmembrane proteins in the plasma membrane, known as **integrins,** to bind to specific proteins in the extracellular matrix and to membrane proteins on adjacent cells. Integrins also transmit signals from the extracellular matrix to the cell interior that can influence cell shape and growth.

Many cells are physically joined at discrete locations along their membranes by specialized types of junctions known as desmosomes, tight junctions, and gap junctions. **Desmosomes** (Figure 3–10a) consist of a region between two adjacent cells where the apposed plasma membranes are separated by about 20 nm and have a dense accumulation of protein at the cytoplasmic surface of each membrane and in the space between the two membranes. Protein fibers extend from the cytoplasmic surface of desmosomes into the cell and are linked to other desmosomes on the opposite side of the cell. Desmosomes function to hold adjacent cells firmly together in areas that are subject to considerable stretching, such as in the skin. The specialized area of the membrane in the region of a desmosome is usually disk-shaped, and these membrane junctions could be likened to rivets or spot-welds.

A second type of membrane junction, the **tight junction,** (Figure 3–10b) is formed when the extracellular

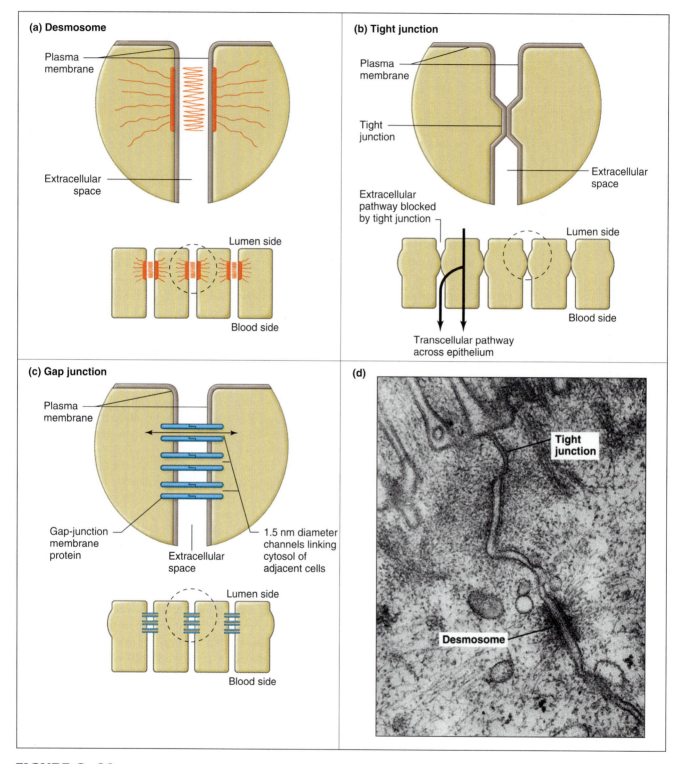

FIGURE 3–10

Three types of specialized membrane junctions: (a) desmosome, (b) tight junction, and (c) gap junction. (d) Electron micrograph of two intestinal epithelial cells joined by a tight junction near the luminal surface and a desmosome below the tight junction.

Electron micrograph from M. Farquhar and G.E. Palade, J. Cell. Biol., 17:375–412 (1963).

surfaces of two adjacent plasma membranes are joined together so that there is no extracellular space between them. Unlike the desmosome, which is limited to a disk-shaped area of the membrane, the tight junction occurs in a band around the entire circumference of the cell.

Most epithelial cells are joined by tight junctions. For example, epithelial cells cover the inner surface of the intestinal tract, where they come in contact with the digestion products in the cavity of the tract. During absorption, the products of digestion move across the epithelium and enter the blood. This transfer could take place theoretically by movement either through the extracellular space between the epithelial cells or through the epithelial cells themselves. For many substances, however, movement through the extracellular space is blocked by the tight junctions, and organic nutrients are required to pass through the cells, rather than between them. In this way, the selective barrier properties of the plasma membrane can control the types and amounts of absorbed substances. The ability of tight junctions to impede molecular movement between cells is not absolute. Ions and water can move through these junctions with varying degrees of ease in different epithelium. Figure 3–10d shows both a tight junction and a desmosome near the luminal border between two epithelial cells.

A third type of junction, the **gap junction,** consists of protein channels linking the cytosols of adjacent cells (Figure 3–10c). In the region of the gap junction, the two opposing plasma membranes come within 2 to 4 nm of each other, which allows specific proteins from the two membranes to join, forming small, protein-lined channels linking the two cells. The small diameter of these channels (about 1.5 nm) limits what can pass between the cytosols of the connected cells to small molecules and ions, such as sodium and potassium, and excludes the exchange of large proteins. A variety of cell types possess gap junctions, including the muscle cells of the heart and smooth-muscle cells where, as we shall see in Chapter 11, they play a very important role in the transmission of electrical activity between the cells. In other cases, gap junctions coordinate the activities of adjacent cells by allowing chemical messengers to move from one cell to another.

Cell Organelles

The contents of cells can be released by grinding a tissue against rotating glass surfaces (homogenization) or using various chemical methods to break the plasma membrane. The cell organelles thus released can then be isolated by subjecting the homogenate to ultracentrifugation in which the mixture is spun at very high speeds producing centrifugal forces thousands of times that of gravity. Cell organelles of different sizes and density settle out at various rates and, by controlling the speed and time of centrifugation, various fractions can be separated. Examination of these fractions in the electron microscope allows identification of the type of cell organelle they contain by comparison with similar structures found in intact cells. These isolated cell organelles can then be studied to learn their chemical composition and metabolic functions.

Nucleus

Almost all cells contain a single nucleus, the largest of the membrane-bound cell organelles. A few specialized cells, for example, skeletal-muscle cells, contain multiple nuclei, while the mature red blood cell has none. The primary function of the nucleus is the storage and the transmission of genetic information to the next generation of cells. This information coded in molecules of DNA is also used to synthesize the proteins that determine the structure and function of the cell (Chapter 5).

Surrounding the nucleus is a barrier, the **nuclear envelope,** composed of two membranes. At regular intervals along the surface of the nuclear envelope, the two membranes are joined to each other, forming the rims of circular openings known as **nuclear pores** (Figure 3–11). Molecules of RNA that determine the structure of proteins synthesized in the cytoplasm move between the nucleus and cytoplasm through these nuclear pores. Proteins that modulate the expression of various genes in DNA move into the nucleus through these pores. The movement of very large molecules, such as RNA and proteins, is selective—that is, restricted to specific macromolecules. An energy-dependent process that alters the diameter of the pore in response to specific signals is involved in the transfer process.

Within the nucleus, DNA, in association with proteins, forms a fine network of threads known as **chromatin;** the threads are coiled to a greater or lesser degree, producing the variations in the density of the nuclear contents seen in electron micrographs (Figure 3–11). At the time of cell division, the chromatin threads become tightly condensed, forming rodlike bodies known as **chromosomes.**

The most prominent structure in the nucleus is the **nucleolus,** a densely staining filamentous region without a membrane. It is associated with specific regions of DNA that contain the genes for forming the particular RNA found in cytoplasmic organelles called ribosomes (see next page). It is in the nucleolus that this RNA and the protein components of ribosomal subunits are assembled; these subunits are then transferred through the nuclear pores to the cytoplasm, where they combine to form functional ribosomes.

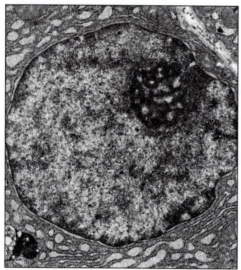

Nucleus

Structure: Largest organelle. Round or oval body located near the cell center. Surrounded by a nuclear envelope composed of two membranes. Envelope contains nuclear pores through which messenger molecules pass between the nucleus and the cytoplasm. No membrane-bound organelles are present in the nucleus, which contains coiled strands of DNA known as chromatin. These condense to form chromosomes at the time of cell division.

Function: Stores and transmits genetic information in the form of DNA. Genetic information passes from the nucleus to the cytoplasm, where amino acids are assembled into proteins.

Nucleolus

Structure: Densely stained filamentous structure within the nucleus. Consists of proteins associated with DNA in regions where information concerning ribosomal proteins is being expressed.

Function: Site of ribosomal RNA synthesis. Assembles RNA and protein components of ribosomal subunits, which then move to the cytoplasm through nuclear pores.

FIGURE 3–11
Nucleus.
Electron micrograph courtesy of K. R. Porter.

Ribosomes

Ribosomes are the protein factories of a cell. On ribosomes, protein molecules are synthesized from amino acids, using genetic information carried by RNA messenger molecules from DNA in the nucleus. Ribosomes are large particles, about 20 nm in diameter, composed of about 70 proteins and several RNA molecules (Chapter 5). Ribosomes are either bound to the organelle called granular endoplasmic reticulum (described next) or are found free in the cytoplasm.

The proteins synthesized on the free ribosomes are released into the cytosol, where they perform their functions. The proteins synthesized by ribosomes attached to the granular endoplasmic reticulum pass into the lumen of the reticulum and are then transferred to

yet another organelle, the Golgi apparatus. They are ultimately secreted from the cell or distributed to other organelles.

Endoplasmic Reticulum

The most extensive cytoplasmic organelle is the network of membranes that forms the **endoplasmic reticulum** (Figure 3–12). These membranes enclose a space that is continuous throughout the network. (The continuity of the endoplasmic reticulum is not obvious when examining a single electron micrograph because only a portion of the network is present in any one section.)

Two forms of endoplasmic reticulum can be distinguished: **granular** (rough-surfaced) and **agranular** (smooth-surfaced). As noted on the next page, the

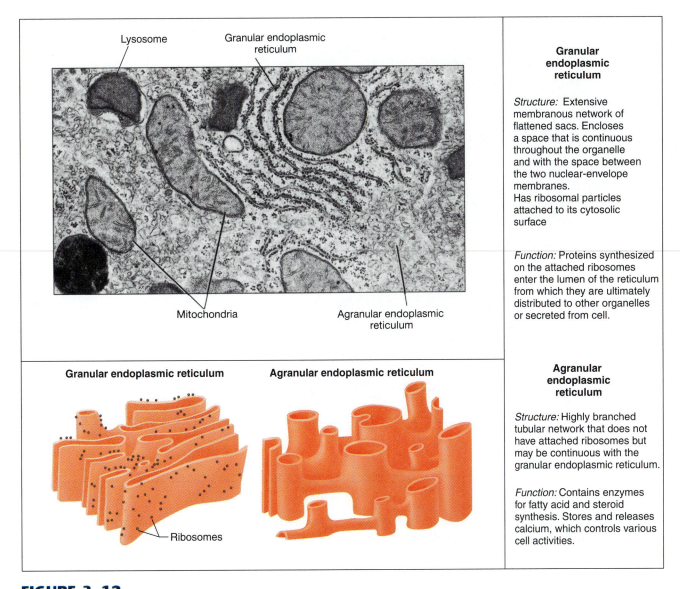

Lysosome

Granular endoplasmic reticulum

Mitochondria

Agranular endoplasmic reticulum

Granular endoplasmic reticulum

Agranular endoplasmic reticulum

Ribosomes

Granular endoplasmic reticulum

Structure: Extensive membranous network of flattened sacs. Encloses a space that is continuous throughout the organelle and with the space between the two nuclear-envelope membranes. Has ribosomal particles attached to its cytosolic surface

Function: Proteins synthesized on the attached ribosomes enter the lumen of the reticulum from which they are ultimately distributed to other organelles or secreted from cell.

Agranular endoplasmic reticulum

Structure: Highly branched tubular network that does not have attached ribosomes but may be continuous with the granular endoplasmic reticulum.

Function: Contains enzymes for fatty acid and steroid synthesis. Stores and releases calcium, which controls various cell activities.

FIGURE 3–12

Endoplasmic reticulum.

Electron micrograph from D. W. Fawcett, "The Cell, An Atlas of Fine Structure," W. B. Saunders Company, Philadelphia, 1966.

granular endoplasmic reticulum has ribosomes bound to its cytosolic surface, and it has a flattened-sac appearance. The outer membrane of the *nuclear envelope* also has ribosomes on its surface, and the space between the two nuclear-envelope membranes is continuous with the lumen of the granular endoplasmic reticulum (see Figure 3–4). Granular endoplasmic reticulum is involved in the packaging of proteins that, after processing in the Golgi apparatus, are to be secreted by cells or distributed to other cell organelles (Chapter 6).

The agranular endoplasmic reticulum has no ribosomal particles on its surface and has a branched, tubular structure. It is the site at which lipid molecules are synthesized (Chapter 4), and it also stores and releases calcium ions involved in controlling various cell activities (Chapter 7).

Both granular and agranular endoplasmic reticulum exist in the same cell, but the relative amounts of the two types vary in different cells and even within the same cell during different periods of cell activity.

Golgi Apparatus

The **Golgi apparatus** is a series of closely opposed, flattened membranous sacs that are slightly curved, forming a cup-shaped structure (Figure 3–13). Most cells have a single Golgi apparatus located near the nucleus, although some cells may have several. Associated with

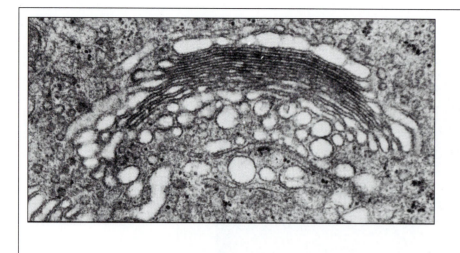

Golgi apparatus

Structure: Series of cup-shaped, closely opposed, flattened, membranous sacs; associated with numerous vesicles. Generally, a single Golgi apparatus is located in the central portion of a cell near its nucleus.

Function: Concentrates, modifies, and sorts proteins arriving from the granular endoplasmic reticulum prior to their distribution, by way of the Golgi vesicles, to other organelles or their secretion from cell.

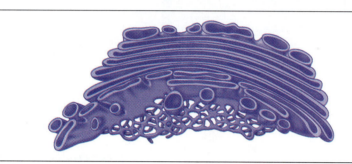

FIGURE 3–13

Golgi apparatus.

Electron micrograph from W. Bloom and D. W. Fawcett, "Textbook of Histology," 9th ed. W. B. Sanders Company, Philadelphia, 1968.

this organelle, particularly near its concave surface, are a number of approximately spherical, membrane-enclosed vesicles.

Proteins arriving at the Golgi apparatus from the granular endoplasmic reticulum undergo a series of modifications as they pass from one Golgi compartment to the next. For example, carbohydrates are linked to proteins to form glycoproteins, and the length of the protein is often shortened by removing a terminal portion of the polypeptide chain. The Golgi apparatus sorts the modified proteins into discrete classes of transport vesicles that will be delivered to various cell organelles and to the plasma membrane, where the protein contents of the vesicle are released to the outside of the cell. Vesicles containing proteins to be secreted from the cell are known as **secretory vesicles.**

Endosomes

A number of membrane-bound vesicular and tubular structures called **endosomes** lie between the plasma membrane and the Golgi apparatus. Certain types of vesicles that pinch off the plasma membrane travel to and fuse with endosomes. In turn, the endosome can pinch off vesicles that are then sent to other cell or-

ganelles or returned to the plasma membrane. Like the Golgi apparatus, endosomes are involved in sorting, modifying, and directing vesicular traffic in cells, as will be described in Chapter 6.

Mitochondria

Mitochondria (singular, *mitochondrion*) are primarily concerned with the chemical processes by which energy in the form of adenosine triphosphate (ATP) molecules is made available to cells (Chapter 4). Most of the ATP used by cells is formed in the mitochondria by a process that consumes oxygen and produces carbon dioxide.

Mitochondria are spherical or elongated, rodlike structures surrounded by an inner and an outer membrane (Figure 3–14). The outer membrane is smooth, whereas the inner membrane is folded into sheets or tubules known as **cristae,** which extend into the inner mitochondrial compartment, the **matrix.** Mitochondria are found throughout the cytoplasm. Large numbers of them, as many as 1000, are present in cells that utilize large amounts of energy, whereas less active cells contain fewer.

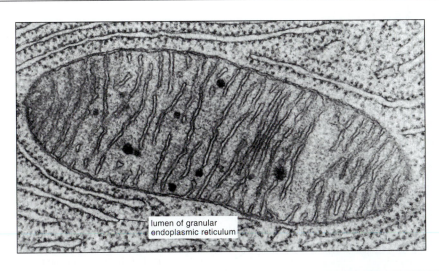

lumen of granular
endoplasmic reticulum

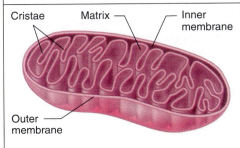

Cristae   Matrix   Inner
membrane

Outer
membrane

Mitochondrion

Structure: Rod- or oval-shaped body
surrounded by two membranes. Inner
membrane folds into matrix of the
mitochondrion, forming cristae.

Function: Major site of ATP production,
O_2 utilization, and CO_2 formation.
Contains enzymes of Krebs cycle and
oxidative phosphorylation.

FIGURE 3–14

Mitochondrion. ✗ ⬛
Electron micrograph courtesy of
K. R. Porter.

Mitochondria have small amounts of DNA that contain the genes for the synthesis of some of the mitochondrial proteins. Evidence suggests that cells gained mitochondria millions of years ago when a bacteria-like organism was engulfed by another cell and, rather than being destroyed, its metabolic functions became integrated with those of the host cell.

Lysosomes

Lysosomes are spherical or oval organelles surrounded by a single membrane (see Figure 3–4). A typical cell may contain several hundred lysosomes. The fluid within a lysosome is highly acidic and contains a variety of digestive enzymes. Lysosomes act as "cellular stomachs," breaking down bacteria and the debris from dead cells that have been engulfed by a cell. They may also break down cell organelles that have been damaged and no longer function normally. They play an especially important role in the various cells that make up the defense systems of the body (Chapter 20).

Peroxisomes

The structure of **peroxisomes** is similar to that of lysosomes—that is, both are moderately dense oval bodies enclosed by a single membrane. Like mitochondria,

peroxisomes consume molecular oxygen, although in much smaller amounts, but this oxygen is not used to store energy in ATP. Instead it undergoes reactions that remove hydrogen from various organic molecules including lipids, alcohol, and various potentially toxic ingested substances. One of the reaction products is hydrogen peroxide, H_2O_2, thus the organelle's name. Hydrogen peroxide can be toxic to cells in high concentrations, but peroxisomes can also destroy hydrogen peroxide and thus prevent its toxic effects. It has been suggested that peroxisomes represent organelles that arose when the oxygen levels in the atmosphere began to rise, protecting cells from the potentially toxic effects of oxygen.

Cytoskeleton

In addition to the membrane-enclosed organelles, the cytoplasm of most cells contains a variety of protein filaments. This filamentous network is referred to as the cell's **cytoskeleton** (Figure 3–15), and, like the bony skeleton of the body, it is associated with processes that maintain and change cell shape and produce cell movements.

There are three classes of cytoskeletal filaments, based on their diameter and the types of protein they contain. In order of size, starting with the thinnest, they

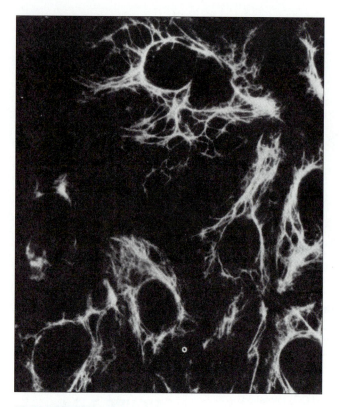

FIGURE 3–15

Cells stained to show the intermediate filament components of the cytoskeleton.

From Roy A. Quinlan, et al., Annals of the New York Academy of Sciences, Vol. 455, New York, 1985.

are (1) microfilaments, (2) intermediate filaments, and (3) microtubules (Figure 3–16). Microfilaments and microtubules can be assembled and disassembled rapidly, allowing a cell to alter these components of its cytoskeletal framework according to changing requirements. In contrast, intermediate filaments, once assembled, are less readily disassembled.

Microfilaments, which are composed of the contractile protein **actin,** make up a major portion of the cytoskeleton in all cells. **Intermediate filaments** are most extensively developed in regions of cells that are subject to mechanical stress (for example, in association with desmosomes).

Microtubules are hollow tubes about 25 nm in diameter, whose subunits are composed of the protein **tubulin.** They are the most rigid of the cytoskeletal filaments and are present in the long processes of nerve cells, where they provide the framework that maintains the processes' cylindrical shape. Microtubules radiate from a region of the cell known as the **centrosome,** which surrounds two small cylindrical bodies, **centrioles,** composed of nine sets of fused microtubules. The centrosome is a cloud of amorphous material that regulates the formation and elongation of microtubules. During cell division the centrosome generates the microtubular spindle fibers used in chromosome separation. Microtubules and microfilaments have also been implicated in the movements of organelles within the cytoplasm. These fibrous elements form the tracks along which organelles are propelled by contractile proteins attached to the surface of the organelles.

Cilia, the hairlike motile extensions on the surfaces of some epithelial cells, have a central core of microtubules organized in a pattern similar to that found in

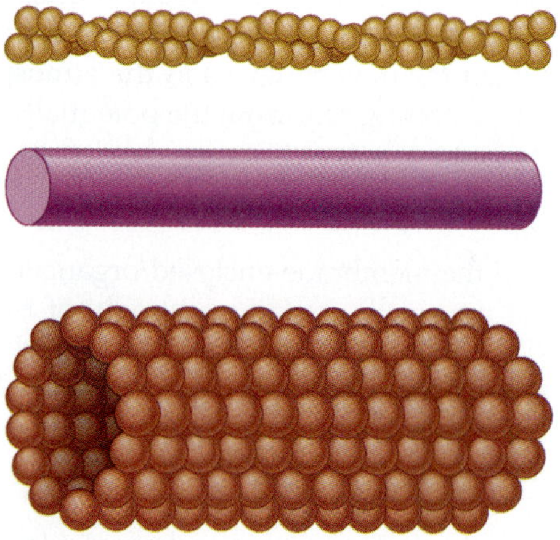

Cytoskeletal filaments	Diameter (nm)	Protein subunit
Microfilament	7	Actin
Intermediate filament	10	Several proteins
Microtubule	25	Tubulin

FIGURE 3–16

Cytoskeletal filaments associated with cell shape and motility.

the centrioles. These microtubules, in combination with a contractile protein, produce movements of the cilia. In hollow organs that are lined with ciliated epithelium, the cilia wave back and forth, propelling the luminal contents along the surface of the epithelium.

SUMMARY

Microscopic Observations of Cells

I. All living matter is composed of cells.
II. There are two types of cells: prokaryotic cells (bacteria) and eukaryotic cells (plant and animal cells).

Membranes

I. Every cell is surrounded by a plasma membrane.
II. Within each eukaryotic cell are numerous membrane-bound compartments, nonmembranous particles, and filaments, known collectively as cell organelles.
III. A cell is divided into two regions, the nucleus and the cytoplasm, the latter composed of the cytosol and cell organelles other than the nucleus.
IV. The membranes that surround the cell and cell organelles regulate the movements of molecules and ions into and out of the cell and its compartments.
    a. Membranes consist of a bimolecular lipid layer, composed of phospholipids in which proteins are embedded.
    b. Integral membrane proteins are amphipathic proteins that often span the membrane, whereas peripheral membrane proteins are confined to the surfaces of the membrane.
V. Three types of membrane junctions link adjacent cells.
    a. Desmosomes link cells that are subject to considerable stretching.
    b. Tight junctions, found primarily in epithelial cells, limit the passage of molecules through the extracellular space between the cells.
    c. Gap junctions form channels between the cytosols of adjacent cells.

Cell Organelles

I. The nucleus transmits and expresses genetic information.
    a. Threads of chromatin, composed of DNA and protein, condense to form chromosomes when a cell divides.
    b. Ribosomal subunits are assembled in the nucleolus.
II. Ribosomes, composed of RNA and protein, are the sites of protein synthesis.
III. The endoplasmic reticulum is a network of flattened sacs and tubules in the cytoplasm.
    a. Granular endoplasmic reticulum has attached ribosomes and is primarily involved in the packaging of proteins that are to be secreted by the cell or distributed to other organelles.
    b. Agranular endoplasmic reticulum is tubular, lacks ribosomes, and is the site of lipid synthesis and calcium accumulation and release.
IV. The Golgi apparatus modifies and sorts the proteins that are synthesized on the granular endoplasmic reticulum and packages them into secretory vesicles.
V. Endosomes are membrane-bound vesicles that fuse with vesicles derived from the plasma membrane and bud off vesicles that are sent to other cell organelles.
VI. Mitochondria are the major cell sites that consume oxygen and produce carbon dioxide in chemical processes that transfer energy to ATP, which can then provide energy for cell functions.
VII. Lysosomes digest particulate matter that enters the cell.
VIII. Peroxisomes use oxygen to remove hydrogen from organic molecules and in the process form hydrogen peroxide.
IX. The cytoplasm contains a network of three types of filaments that form the cytoskeleton: (1) microfilaments, (2) intermediate filaments, and (3) microtubules.

KEY TERMS

light microscope	chromosome
electron microscope	nucleolus
eukaryotic cell	ribosome
prokaryotic cell	endoplasmic reticulum
plasma membrane	granular endoplasmic
cell organelle	reticulum
nucleus	agranular endoplasmic
cytoplasm	reticulum
cytosol	Golgi apparatus
intracellular fluid	secretory vesicle
phospholipid	endosomes
cholesterol	mitochondria
integral membrane protein	mitochondrial cristae
transmembrane protein	mitochondrial matrix
peripheral membrane	lysosome
protein	peroxisome
glycocalyx	cytoskeleton
fluid-mosaic model	microfilament
integrin	actin
desmosome	intermediate filament
tight junction	microtubule
gap junction	tubulin
nuclear envelope	centrosome
nuclear pore	centriole
chromatin	cilia

REVIEW QUESTIONS

1. In terms of the size and number of cells, what makes an elephant larger than a mouse?
2. Identify the location of cytoplasm, cytosol, and intracellular fluid within a cell.

3. Identify the classes of organic molecules found in cell membranes.
4. Describe the orientation of the phospholipid molecules in a membrane.
5. Which plasma membrane components are responsible for membrane fluidity?
6. Describe the location and characteristics of integral and peripheral membrane proteins.
7. Describe the structure and function of the three types of junctions found between cells.
8. What function is performed by the nucleolus?
9. Describe the location and function of ribosomes.
10. Contrast the structure and functions of the granular and agranular endoplasmic reticulum.
11. What function is performed by the Golgi apparatus?
12. What functions are performed by endosomes?
13. Describe the structure and primary function of mitochondria.
14. What functions are performed by lysosomes and peroxisomes?
15. List the three types of filaments associated with the cytoskeleton. Identify the structures in cells that are composed of microtubules.

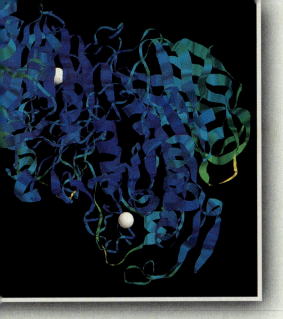

C H A P T E R **4**

Protein Activity and Cellular Metabolism

Proteins are associated with practically every function performed by living cells. One fact is crucial for an understanding of protein function, and thus the functioning of a living organism: Each protein has a unique shape or conformation that enables it to bind specific molecules to a portion of its surface known as a protein binding site. We begin this chapter with a discussion of the properties of protein binding sites that apply to all proteins, and we see how these properties are involved in one class of protein functions—the ability of enzymes to accelerate specific chemical reactions. We then apply this information to a description of the multitude of chemical reactions known as metabolism.

PROTEIN BINDING SITES

Binding Site Characteristics

The ability of various molecules and ions to bind to specific sites on the surface of a protein forms the basis for the wide variety of protein functions. A **ligand** is any molecule or ion that is bound to the surface of a protein by one of the following forces: (1) electrical attractions between oppositely charged ionic or polarized groups on the ligand and the protein, or (2) weaker attractions due to van der Waals forces between nonpolar regions on the two molecules (Chapter 2). Note that this binding does not involve covalent bonds. The region of a protein to which a ligand binds is known as a **binding site.** A protein may contain several binding sites, each specific for a particular ligand.

Chemical Specificity

The force of electrical attraction between oppositely charged regions on a protein and a ligand decreases markedly as the distance between them increases. The even weaker van der Waals forces act only between nonpolar groups that are very close to each other. Therefore, for a ligand to bind to a protein, the ligand must be close to the protein surface. This proximity occurs when the shape of the ligand is complementary to the shape of the protein binding site, such that the two fit together like pieces of a jigsaw puzzle (Figure 4–1).

The binding between a ligand and a protein may be so specific that a binding site can bind only one type of ligand and no other. Such selectivity allows a protein to "identify" (by binding) one particular molecule in a solution containing hundreds of different molecules. This ability of a protein binding site to bind specific ligands is known as **chemical specificity,** since the binding site determines the type of chemical that is bound.

In Chapter 2 we described how the conformation of a protein is determined by the location of the vari-ous amino acids along the polypeptide chain. Accordingly, proteins with different amino acid sequences have different shapes and therefore differently shaped binding sites, each with its own chemical specificity. As illustrated in Figure 4–2, the amino acids that interact with a ligand at a binding site need not be adjacent to each other along the polypeptide chain since the folding of the protein may bring various segments of the molecule into juxtaposition.

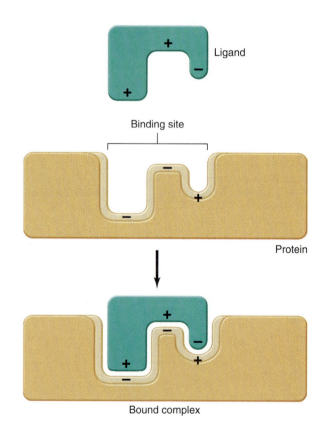

FIGURE 4–1

Complementary shapes of ligand and protein binding site determine the chemical specificity of binding.

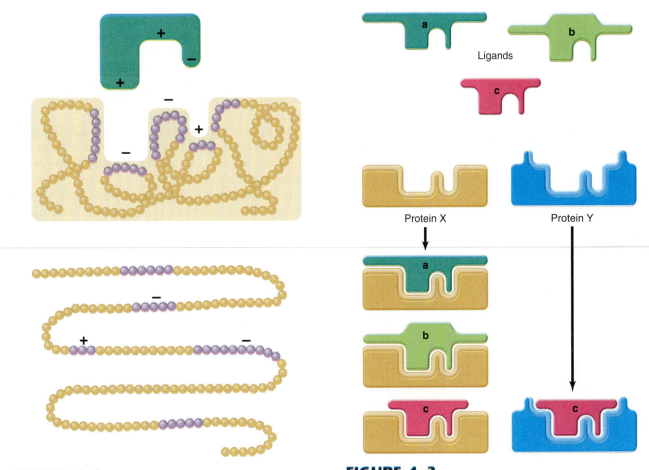

FIGURE 4–2

Amino acids that interact with the ligand at a binding site need not be at adjacent sites along the polypeptide chain, as indicated in this model showing the three-dimensional folding of a protein. The unfolded polypeptide chain is shown below.

FIGURE 4–3

Protein X is able to bind all three ligands, which have similar chemical structures. Protein Y, because of the shape of its binding site, can bind only ligand c. Protein Y, therefore, has a greater chemical specificity than protein X.

Although some binding sites have a chemical specificity that allows them to bind only one type of ligand, others are less specific and thus are able to bind a number of related ligands. For example, three different ligands can combine with the binding site of protein X in Figure 4–3 since a portion of each ligand is complementary to the shape of the binding site. In contrast, protein Y has a greater—that is, more limited—chemical specificity and can bind only one of the three ligands.

Affinity

The strength of ligand-protein binding is a property of the binding site known as **affinity.** The affinity of a binding site for a ligand determines how likely it is that a bound ligand will leave the protein surface and return to its unbound state. Binding sites that tightly bind a ligand are called high-affinity binding sites;

those to which the ligand is weakly bound are low-affinity binding sites.

Affinity and chemical specificity are two distinct, although closely related, properties of binding sites. Chemical specificity, as we have seen, depends only on the shape of the binding site, whereas affinity depends on the strength of the attraction between the protein and the ligand. Thus, different proteins may be able to bind the same ligand—that is, may have the same chemical specificity—but may have different affinities for that ligand. For example, a ligand may have a negatively charged ionized group that would bind strongly to a site containing a positively charged amino acid side chain but would bind less strongly to a binding site having the same shape but no positive charge (Figure 4–4). In addition, the closer the surfaces of the ligand and binding site are to each other, the stronger the attractions. Hence, the more closely the ligand

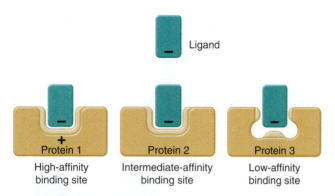

FIGURE 4–4
Three binding sites with the same chemical specificity for a ligand but different affinities.

shape matches the binding site shape, the greater the affinity. In other words, shape can influence affinity as well as chemical specificity.

Saturation

An equilibrium is rapidly reached between unbound ligands in solution and their corresponding protein binding sites such that at any instant some of the free ligands become bound to unoccupied binding sites and some of the bound ligands are released into solution. A single binding site is either occupied or unoccupied. The term **saturation** refers to the fraction of to-tal binding sites that are occupied at any given time. When all the binding sites are occupied, the population of binding sites is 100 percent saturated. When half the available sites are occupied, the system is 50 percent saturated, and so on. A *single* binding site would also be 50 percent saturated if it were occupied by a ligand 50 percent of the time.

The percent saturation of a binding site depends upon two factors: (1) the concentration of unbound ligand in the solution, and (2) the affinity of the binding site for the ligand.

The greater the ligand concentration, the greater the probability of a ligand molecule encountering an unoccupied binding site and becoming bound. Thus, the percent saturation of binding sites increases with increasing ligand concentration until all the sites become occupied (Figure 4–5). Assuming that the ligand is a molecule that exerts a biological effect when it is bound to a protein, the magnitude of the effect would also increase with increasing numbers of bound ligands until all the binding sites were occupied. Further increases in ligand concentration would produce no further effect since there would be no additional sites to be occupied. To generalize, a continuous increase in the magnitude of a chemical stimulus (ligand concentration) that exerts its effects by binding to proteins will produce an increased biological response up to the point at which the protein binding sites are 100 percent saturated.

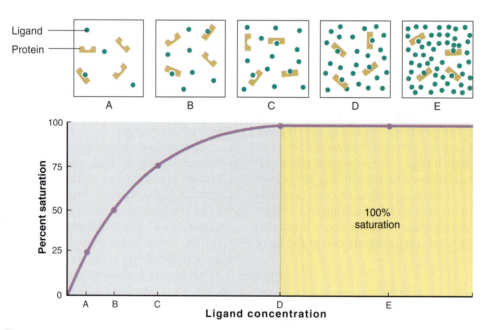

FIGURE 4–5
Increasing ligand concentration increases the number of binding sites occupied—that is, increases the percent saturation. At 100 percent saturation, all the binding sites are occupied, and further increases in ligand concentration do not increase the amount bound.

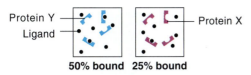

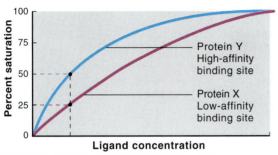

FIGURE 4–6

When two different proteins, X and Y, are able to bind the same ligand, the protein with the higher-affinity binding site (protein Y) has more bound sites at any given ligand concentration until 100 percent saturation.

The second factor determining the percent of binding site saturation is the affinity of the binding site. Collisions between molecules in a solution and a protein containing a bound ligand can dislodge a loosely bound ligand, just as tackling a football player may cause a fumble. If a binding site has a high affinity for a ligand, even a low ligand concentration will result in a high degree of saturation since, once bound to the site, the ligand is not easily dislodged. A low-affinity site, on the other hand, requires a much higher concentration of ligand to achieve the same degree of saturation (Figure 4–6). One measure of binding-site affinity is the ligand concentration necessary to produce 50 percent saturation; the lower the ligand concentration required to bind to half the binding sites, the greater the affinity of the binding site (Figure 4–6).

Competition

As we have seen, more than one type of ligand can bind to certain binding sites (see Figure 4–3). In such cases **competition** occurs between the ligands for the same binding site. In other words, the presence of multiple ligands able to bind to the same binding site affects the percentage of binding sites occupied by any one ligand. If two competing ligands, A and B, are present, increasing the concentration of A will increase the amount of A that is bound, thereby decreasing the number of sites available to B, and decreasing the amount of B that is bound.

As a result of competition, the biological effects of one ligand may be diminished by the presence of another. For example, many drugs produce their effects by competing with the body's natural ligands for bind-

ing sites. By occupying the binding sites, the drug decreases the amount of natural ligand that can be bound.

Regulation of Binding Site Characteristics

Because proteins are associated with practically everything that occurs in a cell, the mechanisms for controlling these functions center on the control of protein activity. There are two ways of controlling protein activity: (1) Changing protein shape, which alters its binding of ligands, and (2) regulating protein synthesis and degradation, which determines the types and amounts of proteins in a cell. The first type of regulation—control of protein shape—is discussed in this section, and the second—protein synthesis and degradation—in Chapter 5.

Since a protein's shape depends on electrical attractions between charged or polarized groups in various regions of the protein (Chapter 2), a change in the charge distribution along a protein or in the polarity of the molecules immediately surrounding it will alter its shape. The two mechanisms used by cells to selectively alter protein shape are known as allosteric modulation and covalent modulation. Before describing these mechanisms, however, it should be emphasized that only certain key proteins are regulated by modulation. Most proteins are not subject to either of these types of modulation.

Allosteric Modulation

Whenever a ligand binds to a protein, the attracting forces between the ligand and the protein alter the protein's shape. For example, as a ligand approaches a binding site, these attracting forces can cause the surface of the binding site to bend into a shape that more closely approximates the shape of the ligand's surface.

Moreover, as the shape of a binding site changes, it produces changes in the shape of *other* regions of the protein, just as pulling on one end of a rope (the polypeptide chain) causes the other end of the rope to move. Therefore, when a protein contains *two* binding sites, the noncovalent binding of a ligand to one site can alter the shape of the second binding site and, hence, the binding characteristics of that site. This is termed **allosteric** (other shape) **modulation** (Figure 4–7a), and such proteins are known as **allosteric proteins.**

One binding site on an allosteric protein, known as the **functional site** (also termed the active site), carries out the protein's physiological function. The other binding site is the **regulatory site,** and the ligand that binds to this site is known as a **modulator molecule**

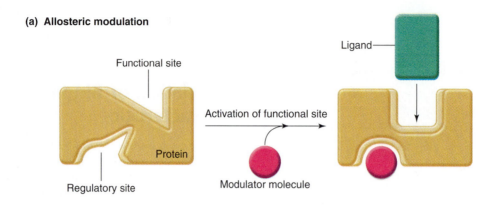

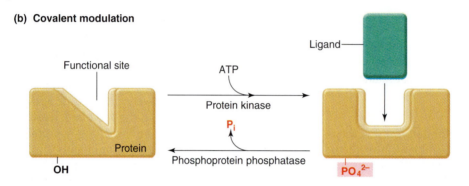

FIGURE 4–7

(a) Allosteric modulation and (b) covalent modulation of a protein's functional binding site.

since its binding to the regulatory site allosterically modulates the shape, and thus the activity, of the functional site.

The regulatory site to which modulator molecules bind is the equivalent of a molecular switch that controls the functional site. In some allosteric proteins, the binding of the modulator molecule to the regulatory site turns *on* the functional site by changing its shape so that it can bind the functional ligand. In other cases, the binding of a modulator molecule turns *off* the functional site by preventing the functional site from binding its ligand. In still other cases, binding of the modulator molecule may decrease or increase the affinity of the functional site. For example, if the functional site is 50 percent saturated at a particular ligand concentration, the binding of a modulator molecule that increases the affinity of the functional site may increase its saturation to 75 percent.

To summarize, the activity of a protein can be increased without changing the concentration of either the protein or the functional ligand. By controlling the concentration of the modulator molecule, and thus the percent saturation of the regulatory site, the functional activity of an allosterically regulated protein can be increased or decreased.

We have spoken thus far only of interactions between regulatory and functional binding sites. There is, however, a way that functional sites can influence each other in certain proteins. These proteins are composed of more than one polypeptide chain held together by electrical attractions between the chains. There may be only one binding site, a functional binding site, on each chain. The binding of a functional ligand to one of the chains, however, can result in an alteration of the functional binding sites in the other chains. This happens because the change in shape of the chain with the bound ligand induces a change in the shape of the other chains. The interaction between the functional binding sites of a multimeric (more than one polypeptide chain) protein is known as **cooperativity.** It can result in a progressive increase in the affinity for ligand binding as more and more of the sites become occupied. Such an increase occurs, for example, in the binding of oxygen to hemoglobin, a protein composed of four polypeptide chains, each containing one binding site for oxygen (Chapter 15).

Covalent Modulation

The second way to alter the shape and therefore the activity of a protein is by the covalent bonding of charged chemical groups to some of the protein's side chains. This is known as **covalent modulation.** In most cases, a phosphate group, which has a net negative charge, is covalently attached by a chemical reaction called **phosphorylation,** in which a phosphate group is transferred from one molecule to another. Phosphorylation of one of the side chains of certain amino acids in a protein introduces a negative charge into that region of the protein. This charge alters the distribution of electric forces in the protein and produces a change in protein conformation (Figure 4–7b). If the conformational change affects a binding site, it changes the binding site's properties. Although the mechanism is completely different, the *effects* produced by covalent modulation are the same as those of allosteric modulation—that is, a functional binding site may be turned on or off or the affinity of the site for its ligand may be altered. To reiterate, unlike allosteric modulation, which involves *noncovalent binding* of modulator molecules, covalent modulation requires chemical reactions in which *covalent bonds* are formed.

Most chemical reactions in the body are mediated by a special class of proteins known as enzymes, whose properties will be discussed in Section B of this chapter. For now, suffice it to say that enzymes accelerate the rate at which reactant molecules (called substrates) are converted to different molecules called products. Two enzymes control a protein's activity by covalent modulation: One adds phosphate, and one removes it. Any enzyme that mediates protein phosphorylation is called a **protein kinase.** These enzymes catalyze the transfer of phosphate from a molecule of adenosine triphosphate (ATP) (discussed in Section B of this chapter) to a hydroxyl group present on the side chain of certain amino acids:

$$\text{Protein} + \text{ATP} \xrightarrow{\text{protein kinase}} \text{Protein}-\text{PO}_4^{2-} + \text{ADP}$$

The protein and ATP are the substrates for protein kinase, and the phosphorylated protein and adenosine diphosphate (ADP) are the products of the reaction.

There is also a mechanism for removing the phosphate group and returning the protein to its original shape. This dephosphorylation is accomplished by a second enzyme known as **phosphoprotein phosphatase.**

$$\text{Protein}-\text{PO}_4^{2-} + \text{H}_2\text{O} \xrightarrow[\text{phosphatase}]{\text{phosphoprotein}} \text{Protein} + \text{HPO}_4^{2-}$$

The activity of the protein will depend on the relative activity of the kinase and phosphatase that con-

trol the extent of the protein's phosphorylation. There are many protein kinases, each with specificities for different proteins, and several kinases may be present in the same cell. The chemical specificities of the phosphoprotein phosphatases are broader, and a single enzyme can dephosphorylate many different phosphorylated proteins.

An important interaction between allosteric and covalent modulation results from the fact that protein kinases are themselves allosteric proteins whose activity can be controlled by modulator molecules. Thus, the process of covalent modulation is itself indirectly regulated by allosteric mechanisms. In addition, some allosteric proteins can also be modified by covalent modulation.

In Chapter 7 we will describe how cell activities can be regulated in response to signals that alter the concentrations of various modulator molecules that in turn alter specific protein activities via allosteric and covalent modulations. Table 4–1 summarizes the factors influencing protein function.

TABLE 4–1 Factors that Influence Protein Function

I. Changing protein shape
  a. Allosteric modulation
  b. Covalent modulation
    i. Protein kinase activity
    ii. Phosphoprotein phosphatase activity

II. Changing protein concentration
  a. Protein synthesis
  b. Protein degradation

SECTION A SUMMARY

Binding Site Characteristics

I. Ligands bind to proteins at sites with shapes complementary to the ligand shape.
II. Protein binding sites have the properties of chemical specificity, affinity, saturation, and competition.

Regulation of Binding Site Characteristics

I. Protein function in a cell can be controlled by regulating either the shape of the protein or the amounts of protein synthesized and degraded.
II. The binding of a modulator molecule to the regulatory site on an allosteric protein alters the shape of the functional binding site, thereby altering

its binding characteristics and the activity of the protein. The activity of allosteric proteins is regulated by varying the concentrations of their modulator molecules.

III. Protein kinase enzymes catalyze the addition of a phosphate group to the side chains of certain amino acids in a protein, changing the shape of the protein's functional binding site and thus altering the protein's activity by covalent modulation. A second enzyme is required to remove the phosphate group, returning the protein to its original state.

SECTION A KEY TERMS

ligand	functional site
binding site	regulatory site
chemical specificity	modulator molecule
affinity	cooperativity
saturation	covalent modulation
competition	phosphorylation
allosteric modulation	protein kinase
allosteric protein	phosphoprotein phosphatase

SECTION A REVIEW QUESTIONS

1. List the four characteristics of a protein binding site.
2. List the types of forces that hold a ligand on a protein surface.
3. What characteristics of a binding site determine its chemical specificity?
4. Under what conditions can a single binding site have a chemical specificity for more than one type of ligand?
5. What characteristics of a binding site determine its affinity for a ligand?
6. What two factors determine the percent saturation of a binding site?
7. Describe the mechanism responsible for competition in terms of the properties of binding sites.
8. Describe two ways of controlling protein activity in a cell.
9. How is the activity of an allosteric protein modulated?
10. How does regulation of protein activity by covalent modulation differ from that by allosteric modulation?

SECTION B

ENZYMES AND CHEMICAL ENERGY

Thousands of chemical reactions occur each instant throughout the body; this coordinated process of chemical change is termed **metabolism** (Greek, change). Metabolism includes the synthesis and breakdown of organic molecules required for cell structure and function and the release of chemical energy used for cell functions. The synthesis of organic molecules by cells is called **anabolism,** and their breakdown, **catabolism.**

The body's organic molecules undergo continuous transformation as some molecules are broken down while others of the same type are being synthesized. Chemically, no person is the same at noon as at 8 o'-clock in the morning since during even this short period much of the body's structure has been torn apart and replaced with newly synthesized molecules. In adults the body's composition is in a steady state in which the anabolic and catabolic rates for the synthesis and breakdown of most molecules are equal.

Chemical Reactions

Chemical reactions involve (1) the breaking of chemical bonds in reactant molecules, followed by (2) the making of new chemical bonds to form the product molecules. In the chemical reaction in which carbonic acid is transformed into carbon dioxide and water, for example, two of the chemical bonds in carbonic acid are broken, and the product molecules are formed by establishing two new bonds between different pairs of atoms:

$$H-O-\overset{\overset{\displaystyle O}{\|}}{C}-O-H \longrightarrow O=C + H-O-H$$
Broken Broken   Formed Formed

$$H_2CO_3 \longrightarrow CO_2 + H_2O + 4\,kcal/mol$$
Carbonic acid   Carbon dioxide   Water

Since the energy contents of the reactants and products are usually different, and because energy can neither be created nor destroyed, energy must either be added or released during most chemical reactions. For example, the breakdown of carbonic acid into carbon dioxide and water occurs with the release of 4 kcal of energy per mole of products formed since carbonic acid has a higher energy content (155 kcal/mol) than the sum of the energy contents of carbon dioxide and water ($94 + 57 = 151$ kcal/mol).

The energy that is released appears as heat, the energy of increased molecular motion, which is measured in units of calories. One **calorie** (1 cal) is the amount of heat required to raise the temperature of 1 g of water $1°$ on the Celsius scale. Energies associated

TABLE 4–2 Determinants of Chemical Reaction Rates
1. Reactant concentrations (higher concentrations: faster reaction rate)
2. Activation energy (higher activation energy: slower reaction rate)
3. Temperature (higher temperature: faster reaction rate)
4. Catalyst (increases reaction rate)

with most chemical reactions are several thousand calories per mole and are reported as **kilocalories** (1 kcal = 1000 cal).

Determinants of Reaction Rates

The rate of a chemical reaction (in other words, how many molecules of product are formed per unit time) can be determined by measuring the change in the concentration of reactants or products per unit of time. The faster the product concentration increases or the reactant concentration decreases, the greater rate of the reaction. Four factors (Table 4–2) influence the reaction rate: reactant concentration, activation energy, temperature, and the presence of a catalyst.

The lower the concentration of reactants, the slower the reaction simply because there are fewer molecules available to react. Conversely, the higher the concentration of reactants, the faster the reaction rate.

Given the same initial concentrations of reactants, however, all reactions do not occur at the same rate. Each type of chemical reaction has its own characteristic rate, which depends upon what is called the activation energy for the reaction. In order for a chemical reaction to occur, reactant molecules must acquire enough energy—the **activation energy**—to enter an activated state in which chemical bonds can be broken and formed. The activation energy does not affect the difference in energy content between the reactants and final products since the activation energy is released when the products are formed.

How do reactants acquire activation energy? In most of the metabolic reactions we will be considering, activation energy is obtained when reactants collide with other molecules. If the activation energy required for a reaction is large, then the probability of a given reactant molecule acquiring this amount of energy will be small, and the reaction rate will be slow. Thus, the higher the activation energy, the slower the rate of a chemical reaction.

Temperature is the third factor influencing reaction rates. The higher the temperature, the faster molecules move and thus the greater their impact when they collide. Therefore, one reason that increasing the temperature increases a reaction rate is that reactants have a better chance of acquiring sufficient activation energy from a collision. In addition, faster-moving molecules will collide more frequently.

A **catalyst** is a substance that interacts with a reactant in such a manner that it alters the distribution of energy between the chemical bonds of the reactant, the result being a decrease in the activation energy required to transform the reactant into product. Since less activation energy is required, a reaction will proceed at a faster rate in the presence of a catalyst. The chemical composition of a catalyst is not altered by the reaction, and thus a single catalyst molecule can be used over and over again to catalyze the conversion of many reactant molecules to products. Furthermore, a catalyst does not alter the difference in the energy contents of the reactants and products.

Reversible and Irreversible Reactions

Every chemical reaction is in theory reversible. Reactants are converted to products (we will call this a "forward reaction"), and products are converted to reactants (a "reverse reaction"). The overall reaction is a **reversible reaction:**

$$\text{Reactants} \underset{\text{reverse}}{\overset{\text{forward}}{\rightleftharpoons}} \text{Products}$$

As a reaction progresses, the rate of the forward reaction will decrease as the concentration of reactants decreases. Simultaneously the rate of the reverse reaction will increase as the concentration of the product molecules increases. Eventually the reaction will reach a state of **chemical equilibrium** in which the forward and reverse reaction rates are equal. At this point there will be no further change in the concentrations of reactants or products even though reactants will continue to be converted into products and products converted to reactants.

Consider our previous example in which carbonic acid breaks down into carbon dioxide and water. The products of this reaction, carbon dioxide and water, can also recombine to form carbonic acid:

$$\underset{94\,\text{kcal/mol}}{CO_2} + \underset{57\,\text{kcal/mol}}{H_2O} + 4\,\text{kcal/mol} \rightleftharpoons \underset{155\,\text{kcal/mol}}{H_2CO_3}$$

Since carbonic acid has a greater energy content than the sum of the energies contained in carbon dioxide and water, energy must be added to the latter molecules in order to form carbonic acid. (This 4 kcal of energy is *not* activation energy but is an integral part of the energy balance.) This energy can be obtained, along with the activation energy, through collisions with other molecules.

TABLE 4–3 Characteristics of Reversible and Irreversible Chemical Reactions

Reversible Reactions	A + B $\rightleftharpoons$ C + D + small amount of energy At chemical equilibrium, product concentrations are only slightly higher than reactant concentrations.
Irreversible Reactions	E + F $\longrightarrow$ G + H + large amount of energy At chemical equilibrium, almost all reactant molecules have been converted to product.

When chemical equilibrium has been reached, the concentration of products need not be equal to the concentration of reactants even though the forward and reverse reaction rates are equal. The ratio of product concentration to reactant concentration at equilibrium depends upon the amount of energy released (or added) during the reaction. The greater the energy released, the smaller the probability that the product molecules will be able to obtain this energy and undergo the reverse reaction to reform reactants. Therefore, in such a case, the ratio of product to reactant concentration at chemical equilibrium will be large. For example, when carbonic acid breaks down to form carbon dioxide and water, the amount of energy released is 4 kcal per mol, and the ratio of product to reactant molecules at equilibrium is about 1000 to 1. If there is no difference in the energy contents of reactants and products, their concentrations will be equal at equilibrium.

Thus, although all chemical reactions are reversible to some extent, reactions that release large quantities of energy are said to be **irreversible reactions** in the sense that almost all of the reactant molecules have been converted to product molecules when chemical equilibrium is reached. It must be emphasized that the energy released in a reaction determines the degree to which the reaction is reversible or irreversible. This energy is *not* the activation energy and it does *not* determine the reaction rate, which is governed by the four factors discussed earlier. The characteristics of reversible and irreversible reactions are summarized in Table 4–3.

Law of Mass Action

The concentrations of reactants and products play a very important role in determining not only the rates of the forward and reverse reactions but also the direction in which the *net* reaction proceeds—whether products or reactants are accumulating at a given time.

Consider the following reversible reaction that has reached chemical equilibrium:

$$A + B \underset{\text{reverse}}{\overset{\text{forward}}{\rightleftharpoons}} C + D$$

$$\underset{\text{Reactants}}{} \qquad \underset{\text{Products}}{}$$

If at this point we increase the concentration of one of the reactants, the rate of the forward reaction will increase and lead to increased product formation. In contrast, increasing the concentration of one of the product molecules will drive the reaction in the reverse direction, increasing the formation of reactants. The direction in which the net reaction is proceeding can also be altered by *decreasing* the concentration of one of the participants. Thus, decreasing the concentration of one of the products drives the net reaction in the forward direction since it decreases the rate of the reverse reaction without changing the rate of the forward reaction.

These effects of reaction and product concentrations on the direction in which the net reaction proceeds are known as the **law of mass action.** Mass action is often a major determining factor controlling the direction in which metabolic pathways proceed since reactions in the body seldom come to chemical equilibrium as new reactant molecules are being added and product molecules are simultaneously being removed by other reactions.

Enzymes

Most of the chemical reactions in the body, if carried out in a test tube with only reactants and products present, would proceed at very low rates because they have high activation energies. In order to achieve the high reaction rates observed in living organisms, catalysts are required to lower the activation energies. These particular catalysts are called enzymes (meaning "in yeast" since the first enzymes were discovered in yeast cells). Enzymes are protein molecules, so an **enzyme** can be defined as a protein catalyst. (Although some RNA molecules possess catalytic activity, the number of reactions they catalyze is very small, and we shall restrict the term "enzyme" to protein catalysts.)

To function, an enzyme must come into contact with reactants, which are called **substrates** in the case of enzyme-mediated reactions. The substrate becomes bound to the enzyme, forming an enzyme-substrate complex, which breaks down to release products and enzyme. The reaction between enzyme and substrate can be written:

$$S + E \rightleftharpoons ES \rightleftharpoons P + E$$

Substrate Enzyme   Enzyme-   Product   Enzyme
                    substrate
                    complex

At the end of the reaction, the enzyme is free to undergo the same reaction with additional substrate molecules. The overall effect is to accelerate the conversion of substrate into product, with the enzyme acting as a catalyst. Note that an enzyme increases both the forward and reverse rates of a reaction and thus does not change the chemical equilibrium that is finally reached.

The interaction between substrate and enzyme has all the characteristics described previously for the binding of a ligand to a binding site on a protein—specificity, affinity, competition, and saturation. The region of the enzyme to which the substrate binds is known as the enzyme's **active site** (a term equivalent to "binding site"). The shape of the enzyme in the region of the active site provides the basis for the enzyme's chemical specificity since the shape of the active site is complementary to the substrate's shape (Figure 4–8).

There are approximately 4000 different enzymes in a typical cell, each capable of catalyzing a different chemical reaction. Enzymes are generally named by adding the suffix *-ase* to the name of either the substrate or the type of reaction catalyzed by the enzyme.

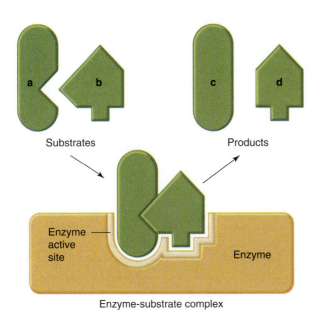

FIGURE 4–8

Binding of substrate to the active site of an enzyme catalyzes the formation of products.

TABLE 4–4 Characteristics of Enzymes
1. An enzyme undergoes no net chemical change as a consequence of the reaction it catalyzes.
2. The binding of substrate to an enzyme's active site has all the characteristics—chemical specificity, affinity, competition, and saturation—of a ligand binding to a protein.
3. An enzyme increases the rate of a chemical reaction but does not cause a reaction to occur that would not occur in its absence.
4. An enzyme increases both the forward and reverse rates of a chemical reaction and thus does not change the chemical equilibrium that is finally reached. It only increases the rate at which equilibrium is achieved.
5. An enzyme lowers the activation energy of a reaction but does not alter the net amount of energy that is added to or released by the reactants in the course of the reaction.

For example, the reaction in which carbonic acid is broken down into carbon dioxide and water is catalyzed by the enzyme carbonic anhydrase.

The catalytic activity of an enzyme can be extremely large. For example, a single molecule of carbonic anhydrase can catalyze the conversion of about 100,000 substrate molecules to products in 1 s. The major characteristics of enzymes are listed in Table 4–4.

Cofactors

Many enzymes are inactive in the absence of small amounts of other substances known as **cofactors.** In some cases, the cofactor is a trace metal (Chapter 2), such as magnesium, iron, zinc, or copper, and its binding to an enzyme alters the enzyme's conformation so that it can interact with the substrate (this is a form of allosteric modulation). Since only a few enzyme molecules need be present to catalyze the conversion of large amounts of substrate to product, very small quantities of these trace metals are sufficient to maintain enzymatic activity.

In other cases, the cofactor is an organic molecule that directly participates as one of the substrates in the reaction, in which case the cofactor is termed a **coenzyme.** Enzymes that require coenzymes catalyze reactions in which a few atoms (for example, hydrogen, acetyl, or methyl groups) are either removed from or added to a substrate. For example:

$$R-2H + Coenzyme \xrightarrow{\text{enzyme}} R + Coenzyme-2H$$

What makes a coenzyme different from an ordinary substrate is the fate of the coenzyme. In our example, the two hydrogen atoms that are transferred to

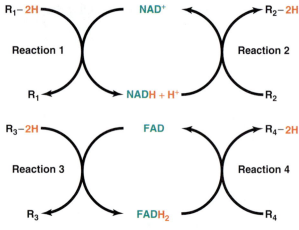

FIGURE 4–9

The coenzymes NAD^+ and FAD are used to transfer two hydrogen atoms from one reaction to a second reaction. In the process, the hydrogen-free forms of the coenzymes are regenerated.

the coenzyme can then be transferred from the coenzyme to another substrate with the aid of a second enzyme. This second reaction converts the coenzyme back to its original form so that it becomes available to accept two more hydrogen atoms (Figure 4–9). A single coenzyme molecule can be used over and over again to transfer molecular fragments from one reaction to another. Thus, as with metallic cofactors, only small quantities of coenzymes are necessary to maintain the enzymatic reactions in which they participate.

Coenzymes are derived from several members of a special class of nutrients known as **vitamins.** For example, the coenzymes **NAD^+** (nicotinamide adenine dinucleotide) and **FAD** (flavine adenine dinucleotide) (Figure 4–9) are derived from the B-vitamins niacin and riboflavin, respectively. As we shall see, they play major roles in energy metabolism by transferring hydrogen from one substrate to another.

Regulation of Enzyme-Mediated Reactions

The rate of an enzyme-mediated reaction depends on substrate concentration and on the concentration and activity (a term defined below) of the enzyme that catalyzes the reaction. Since body temperature is normally maintained nearly constant, changes in temperature are not used directly to alter the rates of metabolic reactions. Increases in body temperature can occur during a fever, however, and around muscle tissue during exercise, and such increases in temperature increase the rates of all metabolic reactions in the affected tissues.

Substrate Concentration

Substrate concentration may be altered as a result of factors that alter the supply of a substrate from outside a cell. For example, there may be changes in its blood concentration due to changes in diet or rate of substrate absorption from the intestinal tract. In addition, a substrate's entry into the cell through the plasma membrane can be controlled by mechanisms that will be discussed in Chapter 6. Intracellular substrate concentration can also be altered by cellular reactions that either utilize the substrate, and thus lower its concentration, or synthesize the substrate, and thereby increase its concentration.

The rate of an enzyme-mediated reaction increases as the substrate concentration increases, as illustrated in Figure 4–10, until it reaches a maximal rate, which remains constant despite further increases in substrate concentration. The maximal rate is reached when the enzyme becomes saturated with substrate—that is, when the active binding site of every enzyme molecule is occupied by a substrate molecule.

Since coenzymes function as substrates in certain enzyme reactions, changes in coenzyme concentration also affect reaction rates, as occurs with ordinary substrates.

Enzyme Concentration

At any substrate concentration, including saturating concentrations, the rate of an enzyme-mediated reaction can be increased by increasing the enzyme concentration. In most metabolic reactions, the substrate concentration is much greater than the concentration of enzyme available to catalyze the reaction. Therefore, if the number of enzyme molecules is doubled, twice

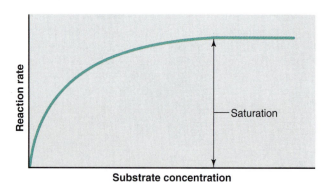

FIGURE 4–10

Rate of an enzyme-catalyzed reaction as a function of substrate concentration.

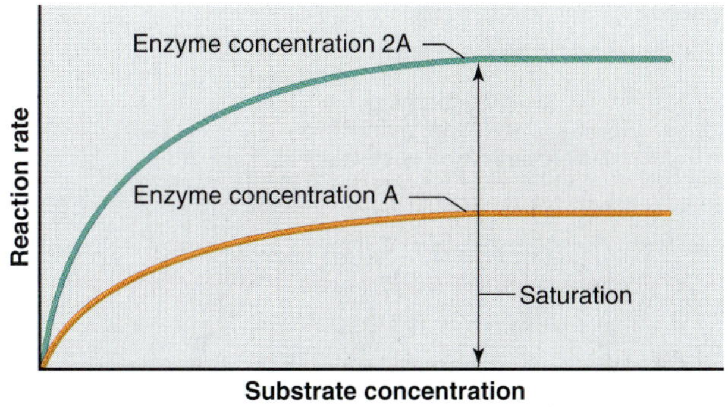

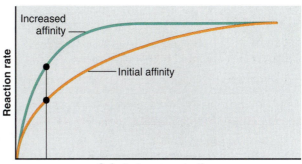

FIGURE 4–11

Rate of an enzyme-catalyzed reaction as a function of substrate concentration at two enzyme concentrations, A and 2A. Enzyme concentration 2A is twice the enzyme concentration of A, resulting in a reaction that proceeds twice as fast at any substrate concentration.

FIGURE 4–12

At a constant substrate concentration, increasing the affinity of an enzyme for its substrate by allosteric or covalent modulation increases the rate of the enzyme-mediated reaction. Note that increasing the enzyme's affinity does not increase the *maximal* rate of the enzyme-mediated reaction.

as many active sites will be available to bind substrate, and twice as many substrate molecules will be converted to product (Figure 4–11). Certain reactions proceed faster in some cells than in others because more enzyme molecules are present.

In order to change the concentration of an enzyme, either the rate of enzyme synthesis or the rate of enzyme breakdown must be altered. Since enzymes are proteins, this involves changing the rates of protein synthesis or breakdown (to be discussed in Chapter 5). Regardless of whether altered synthesis or altered breakdown is involved, changing the concentration of enzymes is a relatively slow process, generally requiring several hours to produce noticeable changes in reaction rates.

Enzyme Activity

In addition to changing the rate of enzyme-mediated reactions by changing the *concentration* of either substrate or enzyme, the rate can be altered by changing **enzyme activity.** A change in enzyme activity occurs when the properties of the enzyme's active site are altered by either allosteric or covalent modulation. Such modulation alters the rate at which the binding site converts substrate to product, the affinity of the binding site for substrate, or both.

Figure 4–12 illustrates the effect of increasing the affinity of an enzyme's active site without changing the substrate or enzyme concentration. Provided the substrate concentration is less than the saturating concentration, the increased affinity of the enzyme's binding site results in an increased number of active sites bound to substrate, and thus an increase in the reaction rate.

The regulation of metabolism through the control of enzyme activity is an extremely complex process since, in many cases, the activity of an enzyme can be altered by more than one agent (Figure 4–13). The modulator molecules that *allosterically* alter enzyme activities are product molecules of other cellular reactions. The result is that the overall rates of metabolism can be adjusted to meet various metabolic demands, as will be illustrated in the next section. In contrast, *covalent* modulation of enzyme activity is mediated by protein kinase enzymes that are themselves activated by various chemical signals received by the cell, for example, from a hormone. (The regulatory mechanisms controlling the activity of protein kinase enzymes will be described in Chapter 7.)

Figure 4–14 summarizes the factors that regulate the rate of an enzyme-mediated reaction.

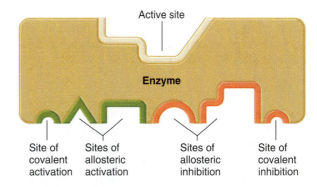

FIGURE 4–13

On a single enzyme, multiple sites can modulate enzyme activity and hence the reaction rate by allosteric and covalent activation or inhibition.

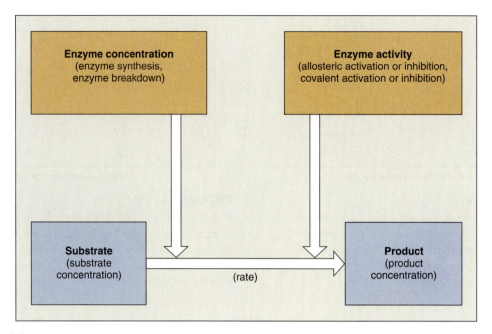

FIGURE 4–14
Factors that affect the rate of enzyme-mediated reactions.

Multienzyme Metabolic Pathways

The sequence of enzyme-mediated reactions leading to the formation of a particular product is known as a **metabolic pathway.** For example, the 19 reactions that convert glucose to carbon dioxide and water constitute the metabolic pathway for glucose catabolism. Each reaction produces only a small change in the structure of the substrate (see, for example, Figures 4–19 and 4–22). By such a sequence of small steps, a complex chemical structure, such as glucose, can be transformed to the relatively simple molecular structures, carbon dioxide and water.

Consider a metabolic pathway containing four enzymes (e_1, e_2, e_3, and e_4) and leading from an initial substrate A to the end product E, through a series of intermediates, B, C, and D:

$$A \underset{}{\overset{e_1}{\rightleftharpoons}} B \underset{}{\overset{e_2}{\rightleftharpoons}} C \underset{}{\overset{e_3}{\rightleftharpoons}} D \overset{e_4}{\longrightarrow} E$$

(The irreversibility of the last reaction is of no consequence for the moment.) By mass action, increasing the concentration of A will lead to an increase in the concentration of B (provided e_1 is not already saturated with substrate), and so on until eventually there is an increase in the concentration of the end product E.

Since different enzymes have different concentrations and activities, it would be extremely unlikely that the reaction rates of all these steps would be exactly the same. Thus, one step is likely to be slower than all the others. This step is known as the **rate-limiting reaction** in a metabolic pathway. None of the reactions that occur later in the sequence, including the formation of end product, can proceed more rapidly than the rate-limiting reaction since their substrates are being supplied by the previous steps. By regulating the concentration or activity of the rate-limiting enzyme, the rate of flow through the whole pathway can be increased or decreased. Thus, it is not necessary to alter all the enzymes in a metabolic pathway to control the rate at which the end product is produced.

Rate-limiting enzymes are often the sites of allosteric or covalent regulation. For example, if enzyme e_2 is rate limiting in the pathway described above, and if the end product E inhibits the activity of e_2, **end-product inhibition** occurs (Figure 4–15). As the concentration of the product increases, the inhibition of product formation increases. Such inhibition is frequently found in synthetic pathways where the formation of end product is effectively shut down when it is not being utilized, preventing excessive accumulation of the end product.

Control of enzyme activity also can be critical for *reversing* a metabolic pathway. Consider the pathway we have been discussing, ignoring the presence of end-product inhibition of enzyme e_2. The pathway consists of three reversible reactions mediated by e_1, e_2, and e_3,

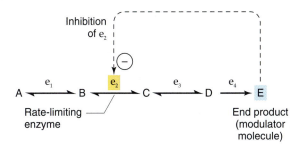

FIGURE 4–15

End-product inhibition of the rate-limiting enzyme in a metabolic pathway. The end product E becomes the modulator molecule that produces inhibition of enzyme e_2.

followed by an irreversible reaction mediated by enzyme e_4. E can be converted into D, however, if the reaction is coupled to the simultaneous breakdown of a molecule that releases large quantities of energy. In other words, an irreversible step can be "reversed" by an alternative route, using a second enzyme and its substrate to provide the large amount of required energy. Two such high-energy irreversible reactions are indicated by bowed arrows to emphasize that two separate enzymes are involved in the two directions:

$$A \underset{e_1}{\rightleftarrows} B \underset{e_2}{\rightleftarrows} C \underset{e_3}{\rightleftarrows} D \underset{\substack{Y \quad e_5 \quad X}}{\overset{e_4}{\rightleftarrows}} E$$

By controlling the concentration and/or activities of e_4 and e_5, the direction of flow through the pathway can be regulated. If e_4 is activated and e_5 inhibited, the flow will proceed from A to E, whereas inhibition of e_4 and activation of e_5 will produce flow from E to A.

Another situation involving the differential control of several enzymes arises when there is a branch in a metabolic pathway. A single metabolite, C, may be the substrate for more than one enzyme, as illustrated by the pathway

$$A \underset{e_1}{\rightleftarrows} B \underset{e_2}{\rightleftarrows} C \overset{\substack{e_3 \\}}{\underset{e_6}{\diagdown}} \begin{matrix} D \underset{e_4}{\rightleftarrows} E \\ \\ F \underset{e_7}{\rightleftarrows} G \end{matrix}$$

Altering the concentration and/or activities of e_3 and e_6 regulates the flow of metabolite C through the two branches of the pathway.

When one considers the thousands of reactions that occur in the body and the permutations and combinations of possible control points, the overall result is staggering. The details of regulating the many metabolic pathways at the enzymatic level are beyond the scope of this book. In the remainder of this chapter, we consider only (1) the overall characteristics of the pathways by which cells obtain energy, and (2) the major pathways by which carbohydrates, fats, and proteins are broken down and synthesized.

ATP

The functioning of a cell depends upon its ability to extract and use the chemical energy in organic molecules. For example, when, in the presence of oxygen, a cell breaks down 1 mol of glucose to carbon dioxide and water, 686 kcal of energy is released. Some of this energy appears as heat, but a cell cannot use heat energy to perform its functions. The remainder of the energy is transferred to another molecule that can in turn transfer it to yet another molecule or to energy-requiring processes. In all cells, from bacterial to human, the primary molecule to which energy from the breakdown of fuel molecules—carbohydrates, fats, and proteins—is transferred and which then transfers this energy to cell functions is the nucleotide **adenosine triphosphate (ATP)** (Figure 4–16). (As we shall see in subsequent chapters, other nucleotide triphosphates, such as GTP, are also used to transfer energy in special cases.) For the moment we will disregard how ATP is formed from fuel molecules and focus on its energy release.

The chemical reaction (referred to as ATP hydrolysis) that removes the terminal phosphate group from ATP is accompanied by the release of a large amount of energy, 7 kcal/mol:

$$ATP + H_2O \longrightarrow ADP + P_i + H^+ + 7 \text{ kcal/mol}$$

The products of the reaction are adenosine diphosphate (ADP), inorganic phosphate (P_i) and H^+. Note that 7 kcal of energy is released when *one mol* 6×10^{23} molecules) of ATP is hydrolyzed, not just one molecule.

The energy derived from the hydrolysis of ATP is used by energy-requiring processes in cells for (1) the production of force and movement, as in muscle contraction (Chapter 11); (2) active transport across membranes (Chapter 6); and (3) synthesis of the organic molecules used in cell structures and functions.

We must emphasize that cells use ATP not to *store* energy but rather to *transfer* it. ATP is an energy-carrying molecule that transfers relatively small amounts of energy from fuel molecules to the cell

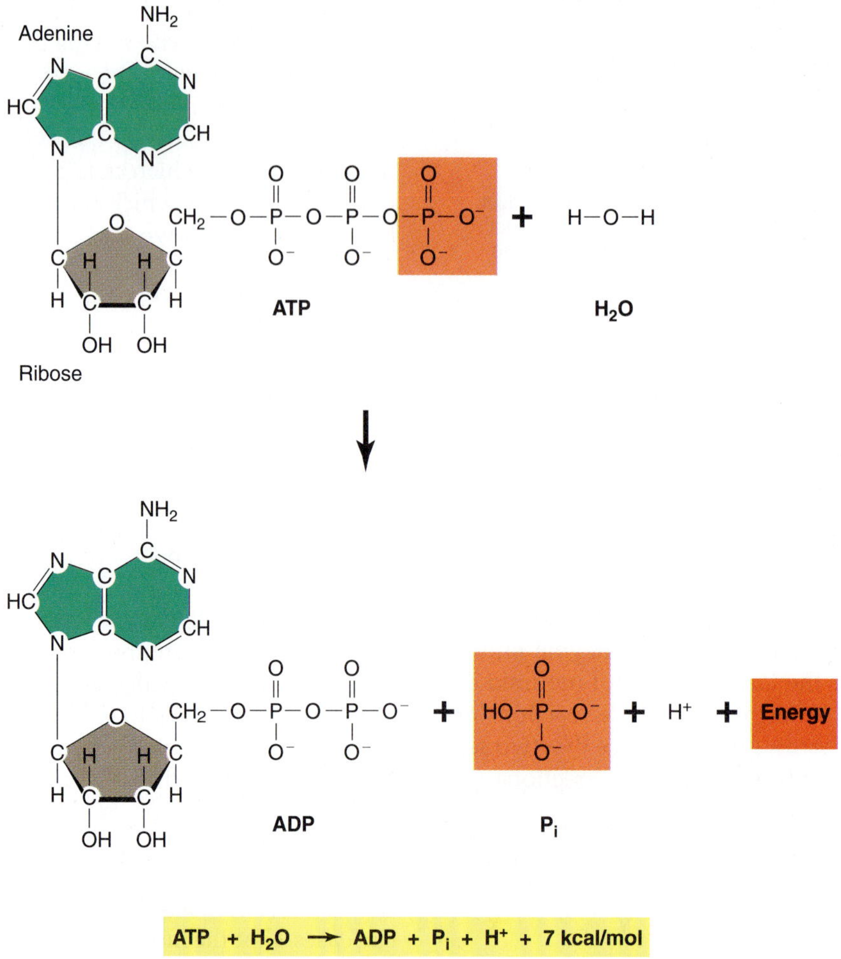

FIGURE 4–16
Chemical structure of ATP. Its breakdown to ADP and P_i is accompanied by the release of 7 kcal of energy per mol.

processes that require energy. ATP is often referred to as the energy currency of the cell. By analogy, if the amount of usable energy released by the catabolism of one molecule of glucose were equivalent to a $10 bill, then the energy released by the hydrolysis of one molecule of ATP would be worth about a quarter. The energy-requiring machinery of a cell uses only quarters—it will not accept $10 bills. Transferring energy to ATP is the cell's way of making change. However, the amount of energy released in a reaction is the same whether it is released all at once (as in combustion) or in small steps, as occurs physiologically.

Energy is continuously cycled through ATP in a cell. A typical ATP molecule may exist for only a few seconds before it is broken down to ADP and P_i, with the released energy used to perform a cell function. Equally rapidly, the products of ATP hydrolysis, ADP and P_i, are converted back into ATP through coupling to reactions that release energy during the catabolism of carbohydrates, fats, or proteins (Figure 4–17).

$$ADP + P_i + 7 \text{ kcal/mol} \longrightarrow ATP + H_2O$$
$$\text{(From catabolism}$$
$$\text{of fuel molecules)}$$

The total amount of ATP in the body is sufficient to maintain the resting functions of the tissues for only about 90 s. Thus, energy must be continuously transferred from fuel molecules to ATP.

Only about 40 percent of the energy released by the catabolism of fuel molecules is transferred to ATP, the remaining 60 percent appearing as heat, which is used to maintain the high body temperature found in birds and mammals. Increased metabolic activity, as occurs during exercise, releases increased amounts of heat, producing an elevation in body temperature.

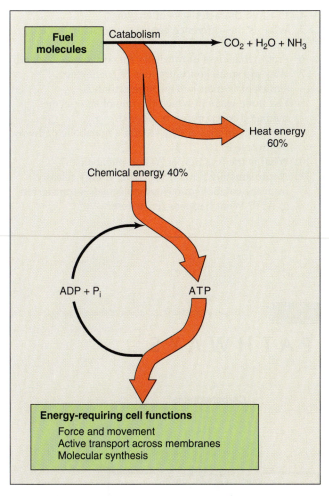

FIGURE 4–17

Flow of chemical energy from fuel molecules to ATP and heat, and from ATP to energy-requiring cell functions.

V. Catalysts increase the rate of a reaction by lowering the activation energy.

VI. The characteristics of reversible and irreversible reactions are listed in Table 4–3.

VII. The net direction in which a reaction proceeds can be altered, according to the law of mass action, by increases or decreases in the concentrations of reactants or products.

Enzymes

I. Nearly all chemical reactions in the body are catalyzed by enzymes, the characteristics of which are summarized in Table 4–4.

II. Some enzymes require small concentrations of cofactors for activity.

   a. The binding of trace metal cofactors maintains the conformation of the enzyme's binding site so that it is able to bind substrate.

   b. Coenzymes, derived from vitamins, transfer small groups of atoms from one substrate to another. The coenzyme is regenerated in the course of these reactions and can be used over and over again.

Regulation of Enzyme-Mediated Reactions

The rates of enzyme-mediated reactions can be altered by changes in temperature, substrate concentration, enzyme concentration, and enzyme activity. Enzyme activity is altered by allosteric or covalent modulation.

Multienzyme Metabolic Pathways

I. The rate of product formation in a metabolic pathway can be controlled by allosteric or covalent modulation of the enzyme mediating the rate-limiting reaction in the pathway. The end product often acts as a modulator molecule, inhibiting the rate-limiting enzyme's activity.

II. An "irreversible" step in a metabolic pathway can be reversed by the use of two enzymes, one for the forward reaction and one for the reverse direction via another, energy-yielding reaction.

ATP

In all cells, energy from the catabolism of fuel molecules is transferred to ATP. The hydrolysis of ATP to ADP and P_i then transfers this energy to cell functions.

SECTION **B** KEY TERMS	
metabolism	chemical equilibrium
anabolism	irreversible reaction
catabolism	law of mass action
calorie	enzyme
kilocalorie	substrate
activation energy	active site
catalyst	cofactor
reversible reaction	coenzyme

SECTION **B** SUMMARY

In adults, the rates at which organic molecules are continuously synthesized (anabolism) and broken down (catabolism) are approximately equal.

Chemical Reactions

I. The difference in the energy content of reactants and products is the amount of energy (measured in calories) that is released or added during a reaction.

II. The energy released during a chemical reaction either is released as heat or is transferred to other molecules.

III. The four factors that can alter the rate of a chemical reaction are listed in Table 4–2.

IV. The activation energy required to initiate the breaking of chemical bonds in a reaction is usually acquired through collisions with other molecules.

vitamin
NAD$^+$
FAD
enzyme activity
metabolic pathway

rate-limiting reaction
end-product inhibition
adenosine triphosphate
(ATP)

SECTION B REVIEW QUESTIONS

1. How do molecules acquire the activation energy required for a chemical reaction?
2. List the four factors that influence the rate of a chemical reaction and state whether increasing the factor will increase or decrease the rate of the reaction.
3. What characteristics of a chemical reaction make it reversible or irreversible?

4. List five characteristics of enzymes.
5. What is the difference between a cofactor and a coenzyme?
6. From what class of nutrients are coenzymes derived?
7. Why are small concentrations of coenzymes sufficient to maintain enzyme activity?
8. List three ways in which the rate of an enzyme-mediated reaction can be altered.
9. How can an irreversible step in a metabolic pathway be reversed?
10. What is the function of ATP in metabolism?
11. Approximately how much of the energy released from the catabolism of fuel molecules is transferred to ATP? What happens to the rest?

SECTION C

METABOLIC PATHWAYS

Three distinct but linked metabolic pathways are used by cells to transfer the energy released from the breakdown of fuel molecules of ATP. They are known as glycolysis, the Krebs cycle, and oxidative phosphorylation (Figure 4–18). In the following section, we will describe the major characteristics of these three pathways in terms of the location of the pathway enzymes in a cell, the relative contribution of each pathway to ATP production, the sites of carbon dioxide formation and oxygen utilization, and the key molecules that enter and leave each pathway.

In this last regard, several facts should be noted in Figure 4–18. First, glycolysis operates only on carbohydrates. Second, all the categories of nutrients—carbohydrates, fats, and proteins—contribute to ATP production via the Krebs cycle and oxidative phosphorylation. Third, mitochondria are essential for the Krebs cycle and oxidative phosphorylation. Finally one important generalization to keep in mind is that glycolysis can occur in either the presence or absence of oxygen, whereas both the Krebs cycle and oxidative phosphorylation require oxygen as we shall see.

Cellular Energy Transfer

Glycolysis

Glycolysis (from the Greek *glycos*, sugar, and *lysis*, breakdown) is a pathway that partially catabolizes carbohydrates, primarily glucose. It consists of 10 enzymatic reactions that convert a six-carbon molecule of glucose into two three-carbon molecules of **pyruvate**, the ionized form of pyruvic acid (Figure 4–19). The

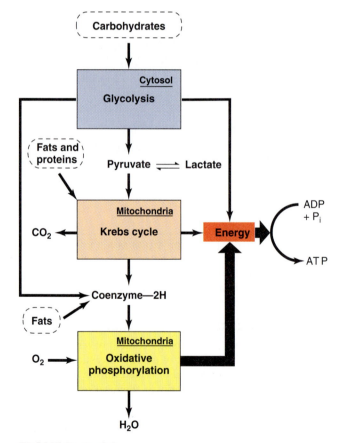

FIGURE 4–18

Pathways linking the energy released from the catabolism of fuel molecules to the formation of ATP.

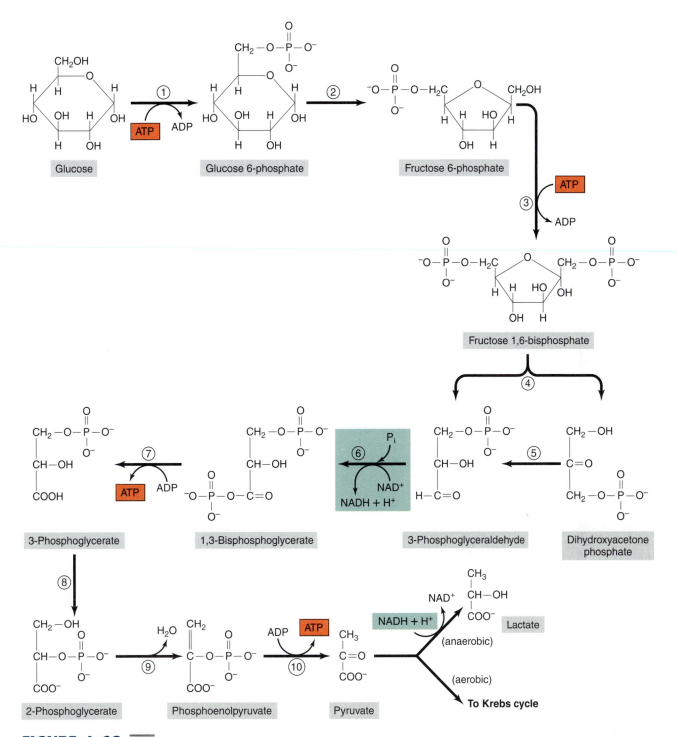

FIGURE 4–19

Glycolytic pathway. Under anaerobic conditions, there is a net synthesis of two molecules of ATP for every molecule of glucose that enters the pathway. Note that at the pH existing in the body, the products produced by the various glycolytic steps exist in the ionized, anionic form (pyruvate, for example). They are actually produced as acids (pyruvic acid, for example) that then ionize.

reactions produce a net gain of two molecules of ATP and four atoms of hydrogen, two of which are transferred to NAD^+ and two are released as hydrogen ions:

$$\text{Glucose} + 2\text{ ADP} + 2\text{ P}_i + 2\text{ NAD}^+ \longrightarrow \qquad (4\text{–}1)$$
$$2\text{ Pyruvate} + 2\text{ ATP} + 2\text{ NADH} + 2\text{ H}^+ + 2\text{ H}_2\text{O}$$

These 10 reactions, *none of which utilizes molecular oxygen*, take place in the cytosol. Note (Figure 4–19) that all the intermediates between glucose and the end product pyruvate contain one or more ionized phosphate groups. As we shall learn in Chapter 6, plasma membranes are impermeable to such highly ionized molecules, and thus these molecules remain trapped within the cell.

Note that the early steps in glycolysis (reactions 1 and 3) each *use*, rather than produce, one molecule of ATP, to form phosphorylated intermediates. In addition, note that reaction 4 splits a six-carbon intermediate into two three-carbon molecules, and reaction 5 converts one of these three-carbon molecules into the other so that at the end of reaction 5 we have two molecules of 3-phosphoglyceraldehyde derived from one molecule of glucose. Keep in mind, then, that from this point on, *two* molecules of each intermediate are involved.

The first *formation* of ATP in glycolysis occurs during reaction 7 when a phosphate group is transferred to ADP to form ATP. Since, as stressed above, two intermediates exist at this point, reaction 7 produces *two* molecules of ATP, one from each of them. In this reaction, the mechanism of forming ATP is known as **substrate-level phosphorylation** since the phosphate group is transferred from a substrate molecule to ADP. As we shall see, this mechanism is quite different from that used during oxidative phosphorylation, in which *free* inorganic phosphate is coupled to ADP to form ATP.

A similar substrate-level phosphorylation of ADP occurs during reaction 10, where again two molecules of ATP are formed. Thus, reactions 7 and 10 generate a total of four molecules of ATP for every molecule of glucose entering the pathway. There is a net gain, however, of only two molecules of ATP during glycolysis because two molecules of ATP were used in reactions 1 and 3.

The end product of glycolysis, pyruvate, can proceed in one of two directions, depending on the availability of molecular oxygen, which, as we stressed earlier, is *not* utilized in any of the glycolytic reactions themselves. If oxygen is present—that is, if **aerobic** conditions exist—pyruvate can enter the Krebs cycle and be broken down into carbon dioxide, as described in the next section. In contrast, in the absence of oxygen (**anaerobic** conditions), pyruvate is converted to **lactate** (the ionized form of lactic acid) by a single enzyme-mediated reaction. In this reaction (Figure 4–20) two hydrogen atoms derived from NADH + H$^+$ are transferred to each molecule of pyruvate to form lactate, and NAD$^+$ is regenerated. These hydrogens had originally been transferred to NAD$^+$ during reaction 6 of glycolysis, so the coenzyme NAD$^+$ shuttles hydrogen between the two reactions during anaerobic glycolysis. The *overall* reaction for anaerobic glycolysis is

$$\text{Glucose} + 2\ \text{ADP} + 2\ P_i \longrightarrow$$
$$2\ \text{Lactate} + 2\ \text{ATP} + 2\ H_2O \quad (4\text{–}2)$$

As stated in the previous paragraph, under aerobic conditions pyruvate is not converted to lactate but rather enters the Krebs cycle. Therefore, the mechanism

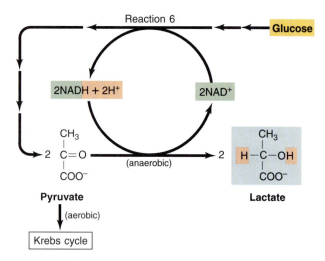

FIGURE 4–20

Under anaerobic conditions, the coenzyme NAD$^+$ utilized in the glycolytic reaction 6 (see Figure 4–19) is regenerated when it transfers its hydrogen atoms to pyruvate during the formation of lactate.

just described for regenerating NAD$^+$ from NADH + H$^+$ by forming lactate does not occur. (Compare Equations 4–1 and 4–2.) Instead, as we shall see, H$^+$ and the hydrogens of NADH are transferred to oxygen during oxidative phosphorylation, regenerating NAD$^+$ and producing H$_2$O.

In most cells, the amount of ATP produced by glycolysis from one molecule of glucose is much smaller than the amount formed under aerobic conditions by the other two ATP-generating pathways—the Krebs cycle and oxidative phosphorylation. There are special cases, however, in which glycolysis supplies most, or even all, of a cell's ATP. For example, erythrocytes contain the enzymes for glycolysis but have no mitochondria, which, as we have said, are required for the other pathways. All of their ATP production occurs, therefore, by glycolysis. Also, certain types of skeletal muscles contain considerable amounts of glycolytic enzymes but have few mitochondria. During intense muscle activity, glycolysis provides most of the ATP in these cells and is associated with the production of large amounts of lactate. Despite these exceptions, most cells do not have sufficient concentrations of glycolytic enzymes or enough glucose to provide, by glycolysis alone, the high rates of ATP production necessary to meet their energy requirements and thus are unable to function for long under anaerobic conditions.

Our discussion of glycolysis has focused upon glucose as the major carbohydrate entering the glycolytic pathway. However, other carbohydrates such as fructose, derived from the disaccharide sucrose (table sugar), and galactose, from the disaccharide lactose

TABLE 4–5 Characteristics of Glycolysis

Entering substrates	Glucose and other monosaccharides
Enzyme location	Cytosol
Net ATP production	2 ATP formed directly per molecule of glucose entering pathway Can be produced in the absence of oxygen (anaerobically)
Coenzyme production	2 NADH + 2 H⁺ formed under aerobic conditions
Final products	Pyruvate—under aerobic conditions Lactate—under anaerobic conditions
Net reaction	
Aerobic:	Glucose + 2 ADP + 2 P$_i$ + 2 NAD⁺ $\longrightarrow$ 2 pyruvate + 2 ATP + 2 NADH + 2 H⁺ + 2 H$_2$O
Anaerobic:	Glucose + 2 ADP + 2 P$_i$ $\longrightarrow$ 2 lactate + 2 ATP + 2 H$_2$O

(milk sugar), can also be catabolized by glycolysis since these carbohydrates are converted into several of the intermediates that participate in the early portion of the glycolytic pathway.

In some microorganisms (yeast cells, for example), pyruvate is converted under anaerobic conditions to carbon dioxide and alcohol (CH_3CH_2OH) rather than to lactate. This process is known as fermentation and forms the basis for the production of alcohol from cereal grains rich in carbohydrates.

Table 4–5 summarizes the major characteristics of glycolysis.

Krebs Cycle

The **Krebs cycle,** named in honor of Hans Krebs, who worked out the intermediate steps in this pathway (also known as the **citric acid cycle** or **tricarboxylic acid cycle**), is the second of the three pathways involved in fuel catabolism and ATP production. It utilizes molecular fragments formed during carbohydrate, protein, and fat breakdown, and it produces carbon dioxide, hydrogen atoms (half of which are bound to coenzymes), and small amounts of ATP. The enzymes for this pathway are located in the inner mitochondrial compartment, the matrix.

The primary molecule entering at the beginning of the Krebs cycle is **acetyl coenzyme A (acetyl CoA):**

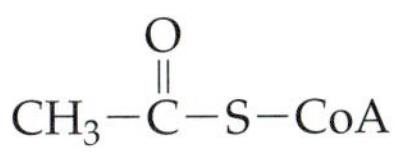

$$CH_3-\overset{\overset{\textstyle O}{\textstyle \|}}{C}-S-CoA$$

Coenzyme A (CoA) is derived from the B vitamin pantothenic acid and functions primarily to transfer acetyl groups, which contain two carbons, from one molecule to another. These acetyl groups come either from pyruvate, which, as we have just seen, is the end prod-

uct of aerobic glycolysis, or from the breakdown of fatty acids and some amino acids, as we shall see in a later section.

Pyruvate, upon entering mitochondria from the cytosol, is converted to acetyl CoA and CO_2 (Figure 4–21). Note that this reaction produces the first molecule of CO_2 formed thus far in the pathways of fuel catabolism, and that hydrogen atoms have been transferred to NAD⁺.

The Krebs cycle begins with the transfer of the acetyl group of acetyl CoA to the four-carbon molecule, oxaloacetate, to form the six-carbon molecule, citrate (Figure 4–22). At the third step in the cycle a molecule of CO_2 is produced, and again at the fourth step. Thus, two carbon atoms entered the cycle as part of the acetyl group attached to CoA, and two carbons (although not the same ones) have left in the form of CO_2. Note also that the oxygen that appears in the CO_2 is not derived from molecular oxygen but from the carboxyl groups of Krebs-cycle intermediates.

In the remainder of the cycle, the four-carbon molecule formed in reaction 4 is modified through a series of reactions to produce the four-carbon molecule oxaloacetate, which becomes available to accept another acetyl group and repeat the cycle.

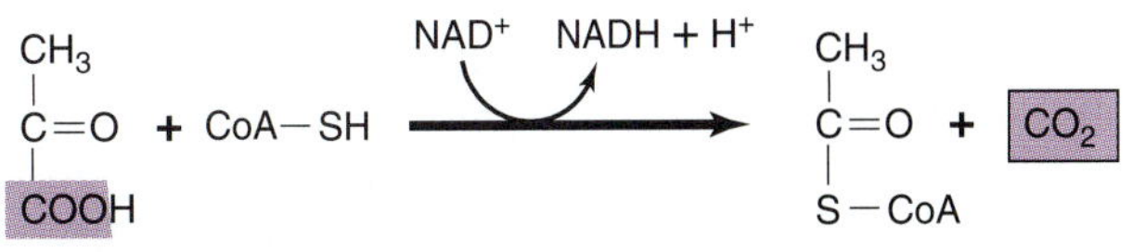

FIGURE 4–21

Formation of acetyl coenzyme A from pyruvic acid with the formation of a molecule of carbon dioxide.

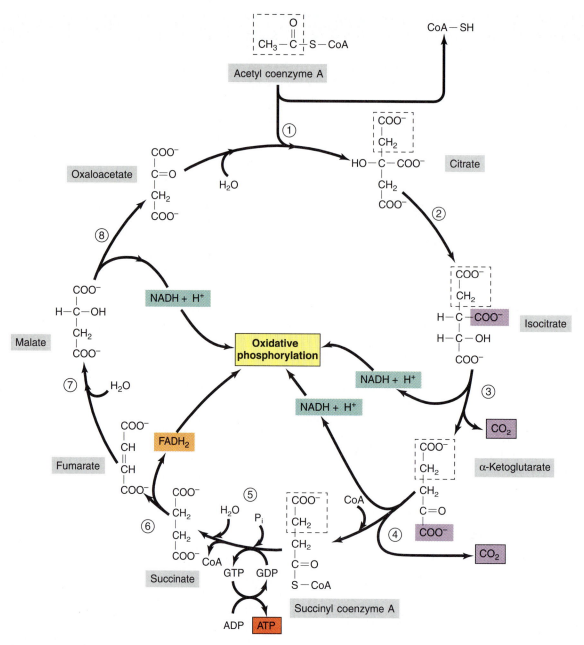

FIGURE 4–22
The Krebs-cycle pathway. Note that the carbon atoms in the two molecules of CO_2 produced by a turn of the cycle are not the same two carbon atoms that entered the cycle as an acetyl group (identified by the dashed boxes in this figure).

Now we come to a crucial fact: In addition to producing carbon dioxide, intermediates in the Krebs cycle generate hydrogen atoms, most of which are transferred to the coenzymes NAD^+ and FAD to form NADH and $FADH_2$. This hydrogen transfer to NAD^+ occurs in each of steps 3, 4, and 8, and to FAD in reaction 6. These hydrogens will be transferred from the coenzymes, along with the free H^+, to oxygen in the next stage of fuel metabolism—oxidative phosphorylation. Since oxidative phosphorylation is necessary for

regeneration of the hydrogen-free form of these coenzymes, *the Krebs cycle can operate only under aerobic conditions.* There is no pathway in the mitochondria that can remove the hydrogen from these coenzymes under anaerobic conditions.

So far we have said nothing of how the Krebs cycle contributes to the formation of ATP. In fact, the Krebs cycle *directly* produces only one high-energy nucleotide triphosphate. This occurs during reaction 5 in which inorganic phosphate is transferred to guanosine

TABLE 4–6 Characteristics of the Krebs Cycle

Entering substrate	Acetyl coenzyme A—acetyl groups derived from pyruvate, fatty acids, and amino acids Some intermediates derived from amino acids
Enzyme location	Inner compartment of mitochondria (the mitochondrial matrix)
ATP production	1 GTP formed directly, which can be converted into ATP Operates only under aerobic conditions even though molecular oxygen is not used directly in this pathway
Coenzyme production	3 NADH + 3 H$^+$ and 2 FADH$_2$
Final products	2 CO$_2$ for each molecule of acetyl coenzyme A entering pathway Some intermediates used to synthesize amino acids and other organic molecules required for special cell functions
Net reaction	Acetyl CoA + 3 NAD$^+$ + FAD + GDP + P$_i$ + 2 H$_2$O $\longrightarrow$ 2 CO$_2$ + CoA + 3 NADH + 3 H$^+$ + FADH$_2$ + GTP

diphosphate (GDP) to form guanosine triphosphate (GTP). The hydrolysis of GTP, like that of ATP, can provide energy for some energy-requiring reactions. In addition, the energy in GTP can be transferred to ATP by the reaction

$$GTP + ADP \rightleftharpoons GDP + ATP$$

This reaction is reversible, and the energy in ATP can be used to form GTP from GDP when additional GTP is required for protein synthesis (Chapter 5) and signal transduction (Chapter 7).

To reiterate, the formation of ATP from GTP is the only mechanism by which ATP is formed within the Krebs cycle. Why, then, is the Krebs cycle so important? Because the hydrogen atoms transferred to coenzymes during the cycle (plus the free hydrogen ions generated) are used in the next pathway, oxidative phosphorylation, to form large amounts of ATP.

The net result of the catabolism of one acetyl group from acetyl CoA by way of the Krebs cycle can be written:

$$Acetyl\ CoA + 3\ NAD^+ + FAD + GDP + P_i + 2\ H_2O \longrightarrow$$
$$2\ CO_2 + CoA + 3\ NADH + 3\ H^+ + FADH_2 + GTP \quad (4\text{--}3)$$

One more point should be noted: Although the major function of the Krebs cycle is to provide hydrogen atoms to the oxidative-phosphorylation pathway, some of the intermediates in the cycle can be used to synthesize organic molecules, especially several types of amino acids, required by cells. Oxaloacetate is one of the intermediates used in this manner. When a molecule of oxaloacetate is removed from the Krebs cycle in the process of forming amino acids, however, it is not available to combine with the acetate fragment of acetyl CoA at the beginning of the cycle. Thus, there must be a way of replacing the oxaloacetate and other Krebs-cycle intermediates that are consumed in syn-

thetic pathways. Carbohydrates provide one source of oxaloacetate replacement by the following reaction, which converts pyruvate into oxaloacetate.

$$Pyruvate + CO_2 + ATP \longrightarrow$$
$$Oxaloacetate + ADP + P_i \quad (4\text{--}4)$$

Certain amino acid derivatives, as we shall see, can also be used to form oxaloacetate and other Krebs-cycle intermediates.

Table 4–6 summarizes the characteristics of the Krebs cycle reactions.

Oxidative Phosphorylation

Oxidative phosphorylation provides the third, and quantitatively most important, mechanism by which energy derived from fuel molecules can be transferred to ATP. The basic principle behind this pathway is simple: The energy transferred to ATP is derived from the energy released when hydrogen ions combine with molecular oxygen to form water. The hydrogen comes from the NADH + H$^+$ and FADH$_2$ coenzymes generated by the Krebs cycle, by the metabolism of fatty acids (see below), and, to a much lesser extent, during aerobic glycolysis. The net reaction is

$$\tfrac{1}{2} O_2 + NADH + H^+ \longrightarrow H_2O + NAD^+ + 53\ kcal/mol$$

The proteins that mediate oxidative phosphorylation are embedded in the inner mitochondrial membrane unlike the enzymes of the Krebs cycle, which are soluble enzymes in the mitochondrial matrix. The proteins for oxidative phosphorylation can be divided into two groups: (1) those that mediate the series of reactions by which hydrogen ions are transferred to molecular oxygen, and (2) those that couple the energy released by these reactions to the synthesis of ATP.

Most of the first group of proteins contain iron and copper cofactors, and are known as **cytochromes** (because in pure form they are brightly colored). Their structure resembles the red iron-containing hemoglobin molecule, which binds oxygen in red blood cells. The cytochromes form the components of the **electron transport chain,** in which two electrons from the hydrogen atoms are initially transferred either from $NADH + H^+$ or $FADH_2$ to one of the elements in this chain. These electrons are then successively transferred to other compounds in the chain, often to or from an iron or copper ion, until the electrons are finally transferred to molecular oxygen, which then combines with hydrogen ions (protons) to form water. These hydrogen ions, like the electrons, come from the free hydrogen ions and the hydrogen-bearing coenzymes, having been released from them early in the transport chain when the electrons from the hydrogen atoms were transferred to the cytochromes.

Importantly, in addition to transferring the coenzyme hydrogens to water, this process regenerates the hydrogen-free form of the coenzymes, which then become available to accept two more hydrogens from intermediates in the Krebs cycle, glycolysis, or fatty acid pathway (as described below). Thus, the electron transport chain provides the *aerobic* mechanism for regenerating the hydrogen-free form of the coenzymes,

whereas, as described earlier, the *anaerobic* mechanism, which applies only to glycolysis, is coupled to the formation of lactate.

At each step along the electron transport chain, small amounts of energy are released, which in total account for the full 53 kcal/mol released from a direct reaction between hydrogen and oxygen. Because this energy is released in small steps, it can be linked to the synthesis of several molecules of ATP, each of which requires only 7 kcal/mol.

ATP is formed at three points along the electron transport chain. The mechanism by which this occurs is known as the **chemiosmotic hypothesis.** As electrons are transferred from one cytochrome to another along the electron transport chain, the energy released is used to move hydrogen ions (protons) from the matrix into the compartment between the inner and outer mitochondrial membranes (Figure 4–23), thus producing a source of potential energy in the form of a hydrogen-ion gradient across the membrane. At three points along the chain, a protein complex forms a channel in the inner mitochondrial membrane through which the hydrogen ions can flow back to the matrix side and in the process transfer energy to the formation of ATP from ADP and P_i. $FADH_2$ has a slightly lower chemical energy content than does $NADH + H^+$ and enters the electron transport chain at a point

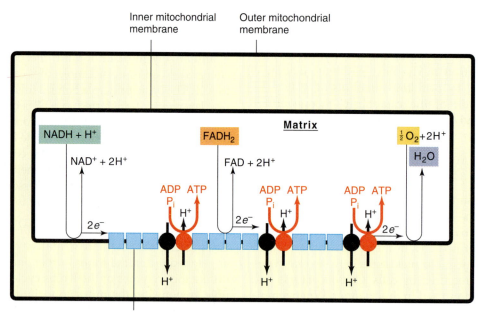

FIGURE 4–23

ATP is formed during oxidative phosphorylation by the flow of hydrogen ions across the inner mitochondrial membrane. Two or three molecules of ATP are produced per pair of electrons donated, depending on the point at which a particular coenzyme enters the electron transport chain.

TABLE 4–7 Characteristics of Oxidative Phosphorylation

Entering substrates	Hydrogen atoms obtained from NADH + H$^+$ and FADH$_2$ formed (1) during glycolysis, (2) by the Krebs cycle during the breakdown of pyruvate and amino acids, and (3) during the breakdown of fatty acids Molecular oxygen
Enzyme location	Inner mitochondrial membrane
ATP production	3 ATP formed from each NADH + H$^+$ 2 ATP formed from each FADH$_2$
Final products	H$_2$O—one molecule for each pair of hydrogens entering pathway.
Net reaction	$\frac{1}{2}$ O$_2$ + NADH + H$^+$ + 3 ADP + 3 P$_i$ $\longrightarrow$ H$_2$O + NAD$^+$ + 3 ATP

beyond the first site of ATP generation (Figure 4–23). Thus, the transfer of its electrons to oxygen produces only two ATP rather than the three formed from NADH + H$^+$.

To repeat, the majority of the ATP formed in the body is produced during oxidative phosphorylation as a result of processing hydrogen atoms that originated largely from the Krebs cycle, during the breakdown of carbohydrates, fats, and proteins. The mitochondria, where the oxidative phosphorylation and the Krebs-cycle reactions occur, are thus considered the power-houses of the cell. In addition, as we have just seen, it is within these organelles that the majority of the oxygen we breathe is consumed, and the majority of the carbon dioxide we expire is produced.

Table 4–7 summaries the key features of oxidative phosphorylation.

Reactive Oxygen Species

As we have just seen, the formation of ATP by oxidative phosphorylation involves the transfer of electrons and hydrogen to molecular oxygen. Several highly reactive transient oxygen derivatives can also be formed during this process—**hydrogen peroxide** and the free radicals **superoxide anion** and **hydroxyl radical.**

$$O_2 \xrightarrow{\;e^-\;} O_2^- \bullet \xrightarrow[2\,H^+]{\;e^-\;} H_2O_2 \xrightarrow{\;e^-\;}$$

Superoxide anion   Hydrogen peroxide

$$OH^- + OH \bullet \xrightarrow{\;e^-\;} 2\,OH^- \xrightarrow[2\,H^+]{} 2\,H_2O$$

Hydroxyl radical

Although most of the electrons transferred along the electron transport chain go into the formation of water, small amounts can combine with oxygen to

form reactive oxygen species. These species can react with and damage proteins, membrane phospholipids, and nucleic acids. Such damage has been implicated in the aging process and in inflammatory reactions to tissue injury. Some cells use these reactive molecules to kill invading bacteria, as described in Chapter 20.

Reactive oxygen molecules are also formed by the action of ionizing radiation on oxygen and by reactions of oxygen with heavy metals such as iron. Cells contain several enzymatic mechanisms for removing these reactive oxygen species and thus providing protection from their damaging effects.

Carbohydrate, Fat, and Protein Metabolism

Having described the three pathways by which energy is transferred to ATP, we now consider how each of the three classes of fuel molecules—carbohydrates, fats, and proteins—enters the ATP-generating pathways. We also consider the synthesis of these fuel molecules and the pathways and restrictions governing their conversion from one class to another. These anabolic pathways are also used to synthesize molecules that have functions other than the storage and release of energy. For example, with the addition of a few enzymes, the pathway for fat synthesis is also used for synthesis of the phospholipids found in membranes.

Carbohydrate Metabolism

Carbohydrate Catabolism In the previous sections, we described the major pathways of carbohydrate catabolism: the breakdown of glucose to pyruvate or lactate by way of the glycolytic pathway, and the metabolism of pyruvate to carbon dioxide and water by way of the Krebs cycle and oxidative phosphorylation.

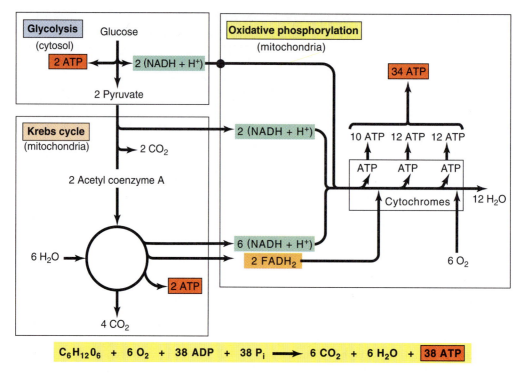

FIGURE 4–24

Pathways of aerobic glucose catabolism and their linkage to ATP formation.

The amount of energy released during the catabolism of glucose to carbon dioxide and water is 686 kcal/mol of glucose:

$$C_6H_{12}O_6 + 6 O_2 \longrightarrow 6 H_2O + 6 CO_2 + 686 \text{ kcal/mol}$$

As noted earlier, about 40 percent of this energy is transferred to ATP. Figure 4–24 illustrates the points at which ATP is formed during glucose catabolism. As we have seen, a net gain of two ATP molecules occurs by substrate-level phosphorylation during glycolysis, and two more are formed during the Krebs cycle from GTP, one from each of the two molecules of pyruvate entering the cycle. The major portion of ATP molecules produced in glucose catabolism—34 ATP per molecule—is formed during oxidative phosphorylation from the hydrogens generated at various steps during glucose breakdown.

To reiterate, in the absence of oxygen, only 2 molecules of ATP can be formed by the breakdown of glucose to lactate. This yield represents only 2 percent of the energy stored in glucose. Thus, the evolution of aerobic metabolic pathways greatly increased the amount of energy available to a cell from glucose catabolism. For example, if a muscle consumed 38 molecules of ATP during a contraction, this amount of ATP could be supplied by the breakdown of 1 molecule of glucose in the presence of oxygen or 19 molecules of glucose under anaerobic conditions.

It is important to note, however, that although only 2 molecules of ATP are formed per molecule of glucose under anaerobic conditions, large amounts of ATP can still be supplied by the glycolytic pathway if large amounts of glucose are broken down to lactate. This is not an efficient utilization of fuel energy, but it does permit continued ATP production under anaerobic conditions, such as occur during intense exercise (Chapter 11).

Glycogen Storage A small amount of glucose can be stored in the body to provide a reserve supply for use when glucose is not being absorbed into the blood from the intestinal tract. It is stored as the polysaccharide **glycogen,** mostly in skeletal muscles and the liver.

Glycogen is synthesized from glucose by the pathway illustrated in Figure 4–25. The enzymes for both glycogen synthesis and glycogen breakdown are located in the cytosol. The first step in glycogen synthesis, the transfer of phosphate from a molecule of ATP to glucose, forming glucose 6-phosphate, is the same as the first step in glycolysis. Thus, glucose 6-phosphate can either be broken down to pyruvate or used to form glycogen.

Note that, as indicated by the bowed arrows in Figure 4–25, different enzymes are used to synthesize and break down glycogen. The existence of two pathways

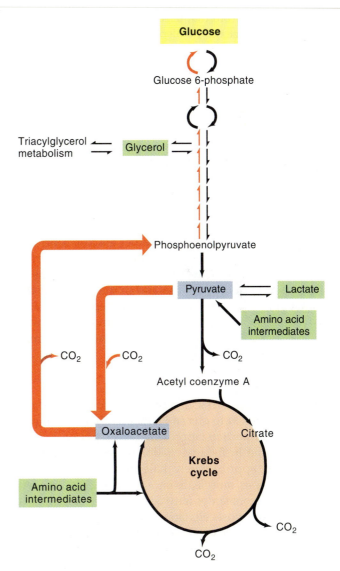

amino acids. This process of generating new molecules of glucose is known as **gluconeogenesis.** The major substrate in gluconeogenesis is pyruvate, formed from lactate and from several amino acids during protein breakdown. In addition, as we shall see, glycerol derived from the hydrolysis of triacylglycerols can be converted into glucose via a pathway that does not involve pyruvate.

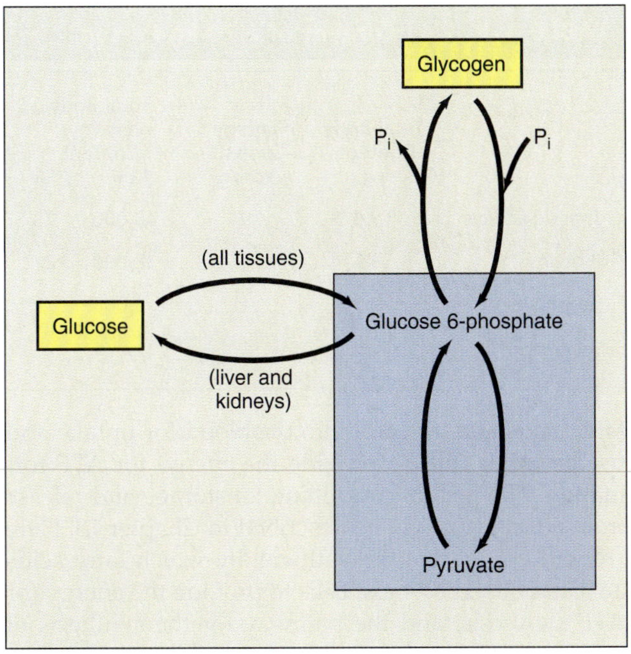

The pathway for gluconeogenesis in the liver and kidneys (Figure 4–26) makes use of many but not all of the enzymes used in glycolysis because most of these reactions are reversible. However, reactions 1, 3

FIGURE 4–25

Pathways for glycogen synthesis and breakdown. Each bowed arrow indicates one or more irreversible reactions that requires different enzymes to catalyze the reaction in the forward and reverse direction.

containing enzymes that are subject to both covalent and allosteric modulation provides a mechanism for regulating the flow of glucose to and from glycogen. When an excess of glucose is available to a liver or muscle cell, the enzymes in the glycogen synthesis pathway are activated by the chemical signals described in Chapter 18, and the enzyme that breaks down glycogen is simultaneously inhibited. This combination leads to the net storage of glucose in the form of glycogen.

When less glucose is available, the reverse combination of enzyme stimulation and inhibition occurs, and net breakdown of glycogen to glucose 6-phosphate occurs. Two paths are available to this glucose 6-phosphate: (1) In most cells, including skeletal muscle, it enters the glycolytic pathway where it is catabolized to provide the energy for ATP formation; (2) in liver (and kidney cells), glucose 6-phosphate can be converted to free glucose by removal of the phosphate group, and glucose is then able to pass out of the cell into the blood, for use as fuel by other cells (Chapter 18).

Glucose Synthesis   In addition to being formed in the liver from the breakdown of glycogen, glucose can be synthesized in the liver and kidneys from intermediates derived from the catabolism of glycerol and some

FIGURE 4–26

Gluconeogenic pathway by which pyruvate, lactate, glycerol, and various amino acid intermediates can be converted into glucose in the liver. Note the points at which each of these precursors, supplied by the blood, enters the pathway.

and 10 (see Figure 4–19) are irreversible, and additional enzymes are required, therefore, to form glucose from pyruvate. Pyruvate is converted to phosphoenolpyruvate by a series of mitochondrial reactions in which CO_2 is added to pyruvate to form the four-carbon Krebs-cycle intermediate oxaloacetate. [In addition to being an important intermediary step in gluconeogenesis, this reaction (Equation 4–4) provides a pathway for replacing Krebs-cycle intermediates, as described earlier.] An additional series of reactions leads to the transfer of a four-carbon intermediate derived from oxaloacetate out of the mitochondria and its conversion to phosphoenolpyruvate in the cytosol. Phosphoenolpyruvate then reverses the steps of glycolysis back to the level of reaction 3, in which a different enzyme from that used in glycolysis is required to convert fructose 1,6-bisphosphate to fructose 6-phosphate. From this point on, the reactions are again reversible, leading to glucose 6-phosphate, which can be converted to glucose in the liver and kidneys or stored as glycogen. Since energy is released during the glycolytic breakdown of glucose to pyruvate in the form of heat and ATP generation, energy must be added to reverse this pathway. A total of six ATP are consumed in the reactions of gluconeogenesis per molecule of glucose formed.

Many of the same enzymes are used in glycolysis and gluconeogenesis, so the question arises: What controls the direction of the reactions in these pathways? What conditions determine whether glucose is broken down to pyruvate or whether pyruvate is converted into glucose? The answer lies in the concentrations of glucose or pyruvate in a cell and in the control of the enzymes involved in the irreversible steps in the pathway, a control exerted via various hormones that alter the concentrations and activities of these key enzymes (Chapter 18).

Fat Metabolism

Fat Catabolism   Triacylglycerol (fat) consists of three fatty acids linked to glycerol (Chapter 2). Fat accounts for the major portion (approximately 80 percent) of the energy stored in the body (Table 4–8). Under resting conditions, approximately half the energy used by such tissues as muscle, liver, and kidneys is derived from the catabolism of fatty acids.

Although most cells store small amounts of fat, the majority of the body's fat is stored in specialized cells known as **adipocytes.** Almost the entire cytoplasm of these cells is filled with a single large fat droplet. Clusters of adipocytes form **adipose tissue,** most of which is in deposits underlying the skin. The function of adipocytes is to synthesize and store triacylglycerols during periods of food uptake and then, when food is not being absorbed from the intestinal tract, to release

TABLE 4–8 Fuel Content of a 70-kg Person

	Total-Body Content, kg	Energy Content, kcal/g	Total-Body Energy Content kcal	%
Triacylglycerols	15.6	9	140,000	78
Proteins	9.5	4	38,000	21
Carbohydrates	0.5	4	2,000	1

fatty acids and glycerol into the blood for uptake and use by other cells to provide the energy for ATP formation. The factors controlling fat storage and release from adipocytes will be described in Chapter 18. Here we will emphasize the pathway by which fatty acids are catabolized by most cells to provide the energy for ATP synthesis, and the pathway for the synthesis of fatty acids from other fuel molecules.

Figure 4–27 shows the pathway for fatty acid catabolism, which is achieved by enzymes present in the mitochondrial matrix. The breakdown of a fatty acid is initiated by linking a molecule of coenzyme A to the carboxyl end of the fatty acid. This initial step is accompanied by the *breakdown* of ATP to AMP and two P_i.

The coenzyme-A derivative of the fatty acid then proceeds through a series of reactions, known as **beta oxidation,** which split off a molecule of acetyl coenzyme A from the end of the fatty acid and transfer two pairs of hydrogen atoms to coenzymes (one pair to FAD and the other to NAD^+). The hydrogen atoms from the coenzymes then enter the oxidative-phosphorylation pathway to form ATP.

When an acetyl coenzyme A is split from the end of a fatty acid, another coenzyme A is added (ATP is not required for this step), and the sequence is repeated. Each passage through this sequence shortens the fatty acid chain by two carbon atoms until all the carbon atoms have been transferred to coenzyme A molecules. As we saw, these molecules then enter the Krebs cycle to produce CO_2 and ATP via the Krebs cycle and oxidative phosphorylation.

How much ATP is formed as a result of the total catabolism of a fatty acid? Most fatty acids in the body contain 14 to 22 carbons, 16 and 18 being most common. The catabolism of one 18-carbon saturated fatty acid yields 146 ATP molecules. In contrast, as we have seen, the catabolism of one glucose molecule yields a maximum of 38 ATP molecules. Thus, taking into account the difference in molecular weight of the fatty acid and glucose, the amount of ATP formed from the catabolism of a gram of fat is about $2\frac{1}{2}$ times greater

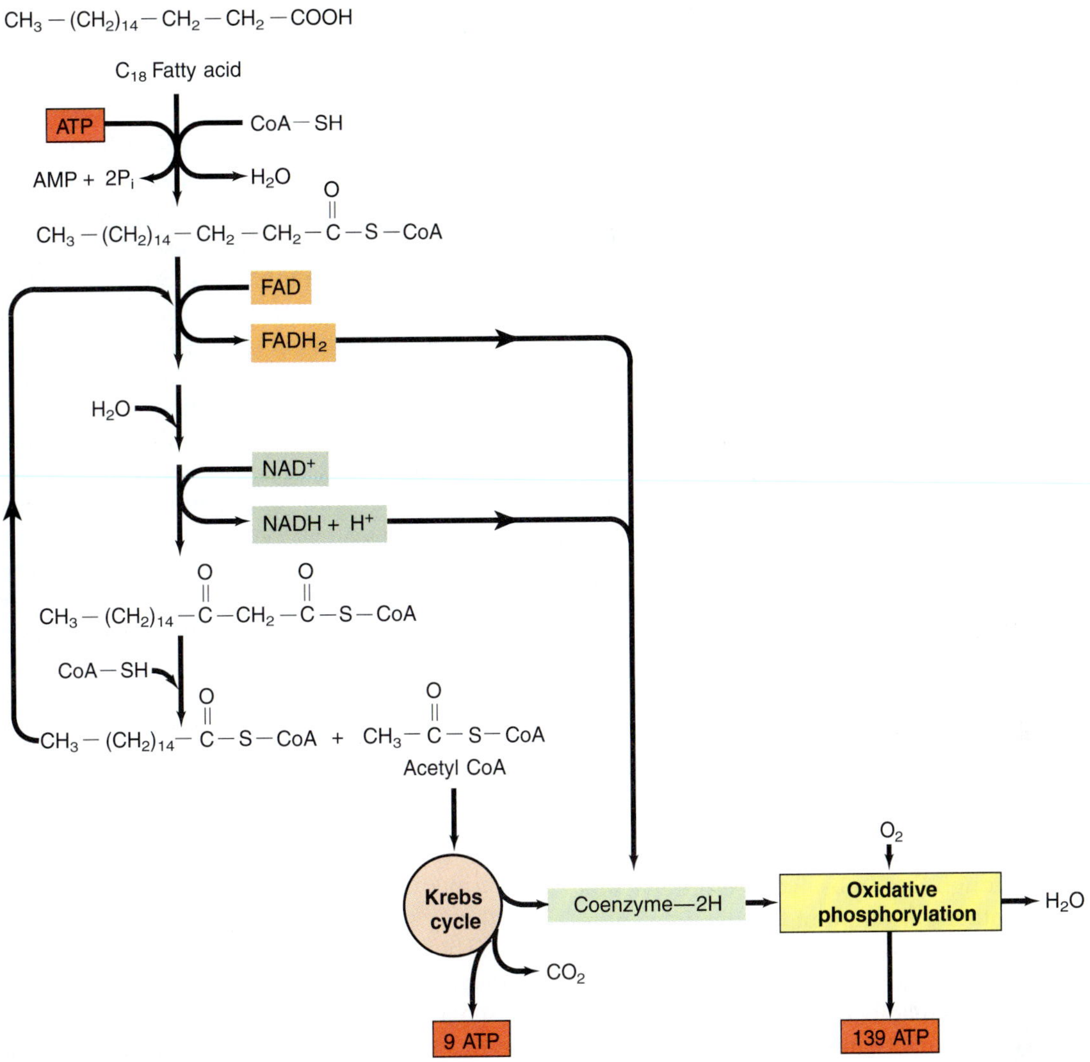

$$CH_3 - (CH_2)_{14} - CH_2 - CH_2 - COOH$$

C_{18} Fatty acid

ATP

$AMP + 2P_i$ CoA — SH

        H_2O

$$CH_3 - (CH_2)_{14} - CH_2 - CH_2 - \overset{\overset{\displaystyle O}{\|}}{C} - S - CoA$$

FAD

FADH$_2$

H_2O

NAD$^+$

NADH + H$^+$

$$CH_3 - (CH_2)_{14} - \overset{\overset{\displaystyle O}{\|}}{C} - CH_2 - \overset{\overset{\displaystyle O}{\|}}{C} - S - CoA$$

CoA — SH

$$CH_3 - (CH_2)_{14} - \overset{\overset{\displaystyle O}{\|}}{C} - S - CoA \;+\; CH_3 - \overset{\overset{\displaystyle O}{\|}}{C} - S - CoA$$

Acetyl CoA

Krebs cycle → Coenzyme — 2H → **Oxidative phosphorylation** → H_2O

O_2

→ CO_2

9 ATP   139 ATP

FIGURE 4–27

Pathway of fatty acid catabolism, which takes place in the mitochondria. The energy equivalent of two ATP is consumed at the start of the pathway.

than the amount of ATP produced by catabolizing 1 gram of carbohydrate. If an average person stored most of his or her fuel as carbohydrate rather than fat, body weight would have to be approximately 30 percent greater in order to store the same amount of usable energy, and the person would consume more energy moving this extra weight around. Thus, a major step in fuel economy occurred when animals evolved the ability to store fuel as fat. In contrast, plants store almost all their fuel as carbohydrate (starch).

Fat Synthesis   The synthesis of fatty acids occurs by reactions that are almost the reverse of those that degrade them. However, the enzymes in the synthetic pathway are in the cytosol, whereas (as we have just seen) the enzymes catalyzing fatty acid breakdown are

in the mitochondria. Fatty acid synthesis begins with cytoplasmic acetyl coenzyme A, which transfers its acetyl group to another molecule of acetyl coenzyme A to form a four-carbon chain. By repetition of this process, long-chain fatty acids are built up two carbons at a time, which accounts for the fact that all the fatty acids synthesized in the body contain an even number of carbon atoms.

Once the fatty acids are formed, triacylglycerol can be synthesized by linking fatty acids to each of the three hydroxyl groups in glycerol, more specifically, to a phosphorylated form of glycerol called **α-glycerol phosphate.** The synthesis of triacylglycerol is carried out by enzymes associated with the membranes of the smooth endoplasmic reticulum.

Compare the molecules produced by glucose catabolism with those required for synthesis of both fatty acids and α-glycerol phosphate. First, acetyl coenzyme A, the starting material for fatty acid synthesis, can be formed from pyruvate, the end product of glycolysis. Second, the other ingredients required for fatty acid synthesis—hydrogen-bound coenzymes and ATP—are produced during carbohydrate catabolism. Third, α-glycerol phosphate can be formed from a glucose intermediate. It should not be surprising, therefore, that much of the carbohydrate in food is converted into fat and stored in adipose tissue shortly after its absorption from the gastrointestinal tract. Mass action resulting from the increased concentration of glucose intermediates, as well as the specific hormonal regulation of key enzymes, promotes this conversion, as will be described in Chapter 18.

It is very important to note that fatty acids, or more specifically the acetyl coenzyme A derived from fatty acid breakdown, cannot be used to synthesize *new* molecules of glucose. The reasons for this can be seen by examining the pathways for glucose synthesis (see Figure 4–26). First, because the reaction in which pyruvate is broken down to acetyl coenzyme A and carbon dioxide is irreversible, acetyl coenzyme A cannot be converted into pyruvate, a molecule that could lead to the production of glucose. Second, the equivalent of the two carbon atoms in acetyl coenzyme A are converted into two molecules of carbon dioxide during their passage through the Krebs cycle before reaching oxaloacetate, another takeoff point for glucose synthesis, and therefore cannot be used to synthesize *net* amounts of oxaloacetate.

Thus, glucose can readily be converted into fat, but the fatty acid portion of fat cannot be converted to glucose. However, the three-carbon glycerol backbone of fat can be converted into an intermediate in the gluconeogenic pathway and thus give rise to glucose, as mentioned earlier.

Protein and Amino Acid Metabolism

In contrast to the complexities of protein synthesis, described in Chapter 5, protein catabolism requires only a few enzymes, termed **proteases,** to break the peptide bonds between amino acids. Some of these enzymes split off one amino acid at a time from the ends of the protein chain, whereas others break peptide bonds between specific amino acids within the chain, forming peptides rather than free amino acids.

Amino acids can be catabolized to provide energy for ATP synthesis, and they can also provide intermediates for the synthesis of a number of molecules other than proteins. Since there are 20 different amino acids, a large number of intermediates can be formed, and there are many pathways for processing them. A few basic types of reactions common to most of these pathways can provide an overview of amino acid catabolism.

Unlike most carbohydrates and fats, amino acids contain nitrogen atoms (in their amino groups) in addition to carbon, hydrogen, and oxygen atoms. Once the nitrogen-containing amino group is removed, the remainder of most amino acids can be metabolized to intermediates capable of entering either the glycolytic pathway or the Krebs cycle.

The two types of reactions by which the amino group is removed are illustrated in Figure 4–28. In the first reaction, **oxidative deamination,** the amino group gives rise to a molecule of ammonia (NH_3) and is replaced by an oxygen atom derived from water to form

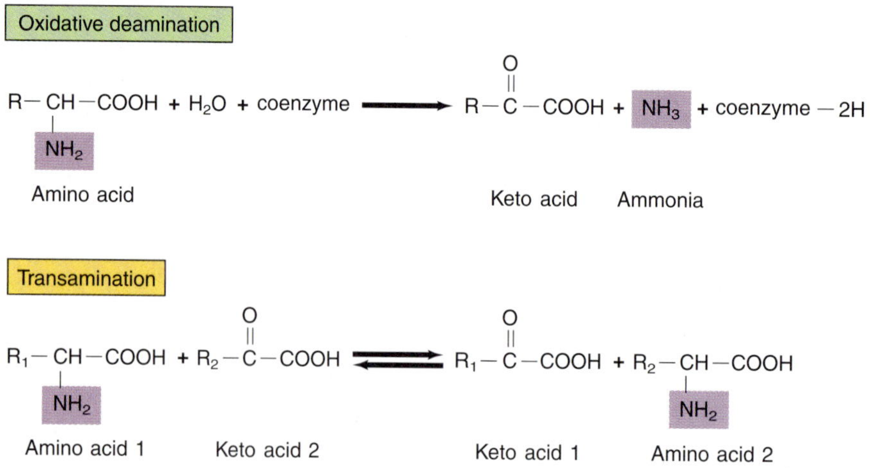

FIGURE 4–28
Oxidative deamination and transamination of amino acids.

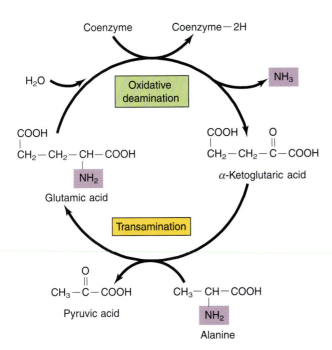

FIGURE 4–29

Oxidative deamination and transamination of the amino acids glutamic acid and alanine lead to keto acids that can enter the carbohydrate pathways.

a **keto acid,** a categorical name rather than the name of a specific molecule. The second means of removing an amino group is known as **transamination** and involves transfer of the amino group from an amino acid to a keto acid. Note that the keto acid to which the amino group is transferred becomes an amino acid. The nitrogen derived from amino groups can also be used by cells to synthesize other important nitrogen-containing molecules, such as the purine and pyrimidine bases found in nucleic acids.

Figure 4–29 illustrates the oxidative deamination of the amino acid glutamic acid and the transamination of the amino acid alanine. Note that the keto acids formed are intermediates either in the Krebs cycle (α ketoglutaric acid) or glycolytic pathway (pyruvic acid). Once formed, these keto acids can be metabolized to produce carbon dioxide and form ATP, or they can be used as intermediates in the synthetic pathway leading to the formation of glucose. As a third alternative, they can be used to synthesize fatty acids after their conversion to acetyl coenzyme A by way of pyruvic acid. Thus, amino acids can be used as a source of energy, and some can be converted into carbohydrate and fat.

As we have seen, the oxidative deamination of amino acids yields ammonia. This substance, which is highly toxic to cells if allowed to accumulate, readily passes through cell membranes and enters the blood,

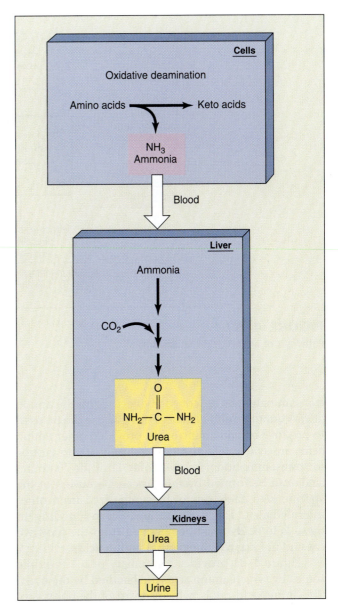

FIGURE 4–30

Formation and excretion of urea, the major nitrogenous waste product of protein catabolism.

which carries it to the liver (Figure 4–30). The liver contains enzymes that can link two molecules of ammonia with carbon dioxide to form **urea.** Thus, urea, which is relatively nontoxic, is the major nitrogenous waste product of protein catabolism. It enters the blood from the liver and is excreted by the kidneys into the urine. Two of the 20 amino acids also contain atoms of sulfur, which can be converted to sulfate, SO_4^{2-}, and excreted in the urine.

Thus far, we have discussed mainly amino acid *catabolism;* now we turn to amino acid synthesis. The keto acids pyruvic acid and α-ketoglutaric acid can be derived from the breakdown of glucose; they can then be

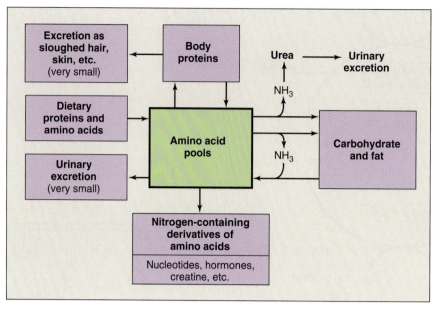

FIGURE 4–31
Pathways of amino acid metabolism.

transaminated, as described above, to form the amino acids glutamate and alanine. Thus glucose can be used to produce certain amino acids, provided other amino acids are available in the diet to supply amino groups for transamination. However, only 11 of the 20 amino acids can be formed by this process because 9 of the specific keto acids cannot be synthesized from other intermediates. The 9 amino acids corresponding to these keto acids must be obtained from the food we eat and are known as **essential amino acids.**

Figure 4–31 provides a summary of the multiple routes by which amino acids are handled by the body. The amino acid pools, which consist of the body's total free amino acids, are derived from (1) ingested protein, which is degraded to amino acids during digestion in the intestinal tract, (2) the synthesis of nonessential amino acids from the keto acids derived from carbohydrates and fat, and (3) the continuous breakdown of body proteins. These pools are the source of amino acids for the resynthesis of body protein and a host of specialized amino acid derivatives, as well as for conversion to carbohydrate and fat. A very small quantity of amino acids and protein is lost from the body via the urine, skin, hair, fingernails, and in women, the menstrual fluid. The major route for the loss of amino acids is not their excretion but rather their deamination, with ultimate excretion of the nitrogen atoms as urea in the urine. The terms **negative nitrogen balance** and **positive nitrogen balance** refer to whether there is a net loss or gain, respectively, of amino acids in the body over any period of time.

If any of the essential amino acids are missing from the diet, a negative nitrogen balance—that is, loss greater than gain—always results. The proteins that require a missing essential amino acid cannot be synthesized, and the other amino acids that would have been incorporated into these proteins are metabolized. This explains why a dietary requirement for protein cannot be specified without regard to the amino acid composition of that protein. Protein is graded in terms of how closely its relative proportions of essential amino acids approximate those in the average body protein. The highest quality proteins are found in animal products, whereas the quality of most plant proteins is lower. Nevertheless, it is quite possible to obtain adequate quantities of all essential amino acids from a mixture of plant proteins alone.

Fuel Metabolism Summary

Having discussed the metabolism of the three major classes of organic molecules, we can now briefly review how each class is related to the others and to the process of synthesizing ATP. Figure 4–32, which is an expanded version of Figure 4–18, illustrates the major pathways we have discussed and the relations of the common intermediates. All three classes of molecules can enter the Krebs cycle through some intermediate, and thus all three can be used as a source of energy for the synthesis of ATP. Glucose can be converted into fat or into some amino acids by way of common intermediates such as pyruvate, oxaloacetate, and acetyl coenzyme A. Similarly, some amino acids can be

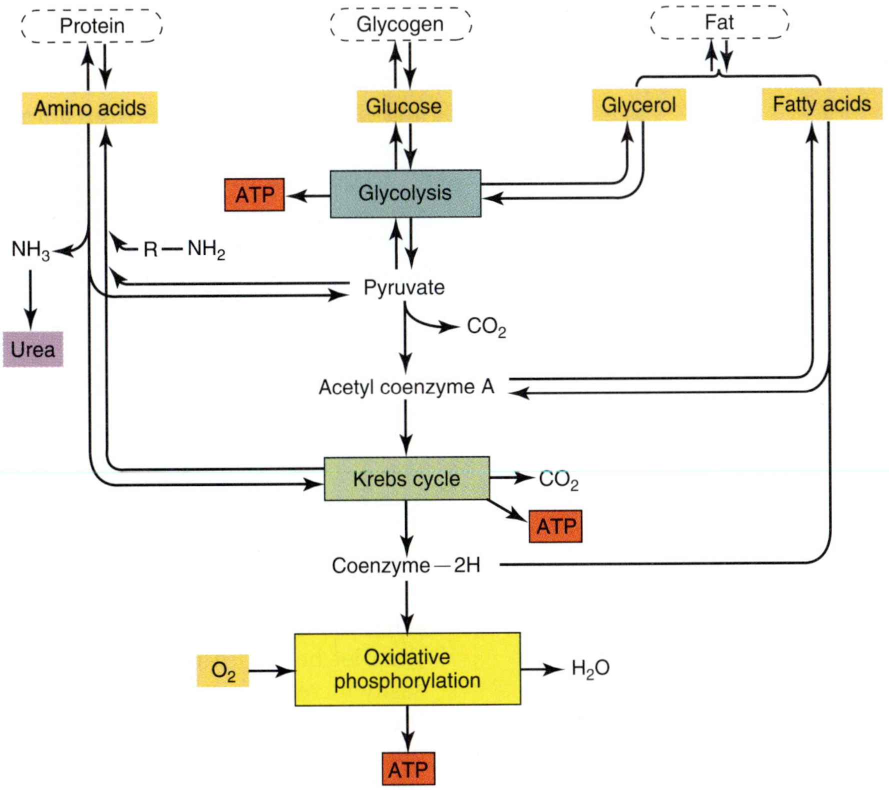

FIGURE 4–32

Interrelations between the pathways for the metabolism of carbohydrate, fat, and protein.

converted into glucose and fat. Fatty acids cannot be converted into glucose because of the irreversibility of the reaction converting pyruvate to acetyl coenzyme A, but the glycerol portion of triacylglycerols can be converted into glucose. Fatty acids can be used to synthesize portions of the keto acids used to form some amino acids. Metabolism is thus a highly integrated process in which all classes of molecules can be used, if necessary, to provide energy, and in which each class of molecule can provide the raw materials required to synthesize most but not all members of other classes.

Essential Nutrients

About 50 substances required for normal or optimal body function cannot be synthesized by the body or are synthesized in amounts inadequate to keep pace with the rates at which they are broken down or excreted. Such substances are known as **essential nutrients** (Table 4–9). Because they are all removed from the body at some finite rate, they must be continually supplied in the foods we eat.

It must be emphasized that the term "essential nutrient" is reserved for substances that fulfill *two* criteria: (1) they must be essential for health, and (2) they

must not be synthesized by the body in adequate amounts. Thus, glucose, although "essential" for normal metabolism, is not classified as an essential nutrient because the body normally can synthesize all it needs, from amino acids, for example. Furthermore, the quantity of an essential nutrient that must be present in the diet in order to maintain health is not a criterion for determining if the substance is essential. Approximately 1500 g of water, 2 g of the amino acid methionine, but only about 1 mg of the vitamin thiamine are required per day.

Water is an essential nutrient because far more of it is lost in the urine and from the skin and respiratory tract than can be synthesized by the body. (Recall that water is formed as an end product of oxidative phosphorylation as well as from several other metabolic reactions.) Therefore, to maintain water balance, water intake is essential.

The mineral elements provide an example of substances that cannot be synthesized or broken down but are continually lost from the body in the urine, feces, and various secretions. The major minerals must be supplied in fairly large amounts, whereas only small quantities of the trace elements are required.

We have already noted that 9 of the 20 amino acids are essential. Two fatty acids, linoleic and linolenic

TABLE 4–9 Essential Nutrients

Water

Mineral Elements
 7 major mineral elements (see Table 2–1)
 13 trace elements (see Table 2–1)

Essential Amino Acids
 Isoleucine
 Leucine
 Lysine
 Methionine
 Phenylalanine
 Threonine
 Tryptophan
 Tyrosine
 Valine

Essential Fatty Acids
 Linoleic
 Linolenic

Vitamins
 Water-soluble vitamins
  B₁: thiamine
  B₂: riboflavin
  B₆: pyridoxine
  B₁₂: cobalamine
  Niacin   } Vitamin B complex
  Pantothenic acid
  Folic acid
  Biotin
  Lipoic acid
  Vitamin C
 Fat-soluble vitamins
  Vitamin A
  Vitamin D
  Vitamin E
  Vitamin K

Other Essential Nutrients
 Inositol
 Choline
 Carnitine

be composed of eight substances now known as the vitamin B complex. Plants and bacteria have the enzymes necessary for vitamin synthesis, and it is by eating either plants or meat from animals that have eaten plants that we get our vitamins.

The vitamins as a class have no particular chemical structure in common, but they can be divided into the **water-soluble vitamins** and the **fat-soluble vitamins.** The water-soluble vitamins form portions of coenzymes such as NAD^+, FAD, and coenzyme A. The fat-soluble vitamins (A, D, E, and K) in general do not function as coenzymes. For example, vitamin A (retinol) is used to form the light-sensitive pigment in the eye, and lack of this vitamin leads to night blindness. The specific functions of each of the fat-soluble vitamins will be described in later chapters.

The catabolism of vitamins does not provide chemical energy, although some of them participate as coenzymes in chemical reactions that release energy from other molecules. Increasing the amount of vitamins in the diet beyond a certain minimum does not necessarily increase the activity of those enzymes for which the vitamin functions as a coenzyme. Only very small quantities of coenzymes participate in the chemical reactions that require them and increasing the concentration above this level does not increase the reaction rate.

The fate of large quantities of ingested vitamins varies depending upon whether the vitamin is water-soluble or fat-soluble. As the amount of water-soluble vitamins in the diet is increased, so is the amount excreted in the urine; thus the accumulation of these vitamins in the body is limited. On the other hand, fat-soluble vitamins can accumulate in the body because they are poorly excreted by the kidneys and because they dissolve in the fat stores in adipose tissue. The intake of very large quantities of fat-soluble vitamins can produce toxic effects.

A great deal of research is presently being done concerning the health consequences of taking large amounts of different vitamins, amounts much larger than one would ever normally ingest in food. Many claims have been made for the beneficial effects of this practice—the use of vitamins as drugs—but most of these claims remain unsubstantiated. On the other hand, it is now clear that ingesting large amounts of certain vitamins does indeed have proven health-promoting effects; most notably, the ingestion of large amounts of vitamin E (400 International Units per day) is protective against both heart disease and multiple forms of cancer, the most likely explanation of these effects being that vitamin E is an antioxidant and thus scavenges toxic free radicals. (See also the section on aging in Chapter 7.)

acid, which contain a number of double bonds and serve important roles in chemical messenger systems, are also essential nutrients. Three additional essential nutrients—inositol, choline, and carnitine—have functions that will be described in later chapters but do not fall into any common category other than being essential nutrients. Finally, the class of essential nutrients known as vitamins deserves special attention.

Vitamins

Vitamins are a group of 14 organic essential nutrients that are required in very small amounts in the diet. The exact chemical structures of the first vitamins to be discovered were unknown, and they were simply identified by letters of the alphabet. Vitamin B turned out to

Cellular Energy Transfer

I. The end products of glycolysis under aerobic conditions are ATP and pyruvate, whereas ATP and lactate are the end products under anaerobic conditions.

    a. Carbohydrates are the only major fuel molecules that can enter the glycolytic pathway, enzymes for which are located in the cytosol.

    b. During anaerobic glycolysis, hydrogen atoms are transferred to NAD^+, which then transfers them to pyruvate to form lactate, thus regenerating the original coenzyme molecule.

    c. During aerobic glycolysis, $NADH + H^+$ transfers hydrogen atoms to the oxidative-phosphorylation pathway.

    d. The formation of ATP in glycolysis is by substrate-level phosphorylation, a process in which a phosphate group is transferred from a phosphorylated metabolic intermediate directly to ADP.

II. The Krebs cycle, the enzymes of which are in the matrix of the mitochondria, catabolizes molecular fragments derived from fuel molecules and produces carbon dioxide, hydrogen atoms, and ATP.

    a. Acetyl coenzyme A, the acetyl portion of which is derived from all three types of fuel molecules, is the major substrate entering the Krebs cycle. Amino acids can also enter at several sites in the cycle by being converted to cycle intermediates.

    b. During one rotation of the Krebs cycle, two molecules of carbon dioxide are produced, and four pairs of hydrogen atoms are transferred to coenzymes. Substrate-level phosphorylation produces one molecule of GTP, which can be converted to ATP.

III. Oxidative phosphorylation forms ATP from ADP and P_i, using the energy released when molecular oxygen ultimately combines with hydrogen atoms to form water.

    a. The enzymes for oxidative phosphorylation are located on the inner membrane of mitochondria.

    b. Hydrogen atoms derived from glycolysis, the Krebs cycle, and the breakdown of fatty acids are delivered, most bound to coenzymes, to the electron transport chain, which regenerates the hydrogen-free forms of the coenzymes NAD^+ and FAD by transferring the hydrogens to molecular oxygen to form water.

    c. The reactions of the electron transport chain produce a hydrogen-ion gradient across the inner mitochondrial membrane. The flow of hydrogen ions back across the membrane provides the energy for ATP synthesis.

    d. Small amounts of reactive oxygen species, which can damage proteins, lipids, and nucleic acids, are formed during electron transport.

Carbohydrate, Fat, and Protein Metabolism

I. The aerobic catabolism of carbohydrates proceeds through the glycolytic pathway to pyruvate, which enters the Krebs cycle and is broken down to carbon dioxide and to hydrogens, which are transferred to coenzymes.

    a. About 40 percent of the chemical energy in glucose can be transferred to ATP under aerobic conditions; the rest is released as heat.

    b. Under aerobic conditions, 38 molecules of ATP can be formed from 1 molecule of glucose: 34 from oxidative phosphorylation, 2 from glycolysis, and 2 from the Krebs cycle.

    c. Under anaerobic conditions, 2 molecules of ATP are formed from 1 molecule of glucose during glycolysis.

II. Carbohydrates are stored as glycogen, primarily in the liver and skeletal muscles.

    a. Two different enzymes are used to synthesize and break down glycogen. The control of these enzymes regulates the flow of glucose to and from glycogen.

    b. In most cells, glucose 6-phosphate is formed by glycogen breakdown and is catabolized to produce ATP. In liver and kidney cells, glucose can be derived from glycogen and released from the cells into the blood.

III. New glucose can be synthesized (gluconeogenesis) from some amino acids, lactate, and glycerol via the enzymes that catalyze reversible reactions in the glycolytic pathway. Fatty acids cannot be used to synthesize new glucose.

IV. Fat, stored primarily in adipose tissue, provides about 80 percent of the stored energy in the body.

    a. Fatty acids are broken down, two carbon atoms at a time, in the mitochondrial matrix by beta oxidation, to form acetyl coenzyme A and hydrogen atoms, which combine with coenzymes.

    b. The acetyl portion of acetyl coenzyme A is catabolized to carbon dioxide in the Krebs cycle, and the hydrogen atoms generated there, plus those generated during beta oxidation, enter the oxidative-phosphorylation pathway to form ATP.

    c. The amount of ATP formed by the catabolism of 1 g of fat is about $2\frac{1}{2}$ times greater than the amount formed from 1 g of carbohydrate.

    d. Fatty acids are synthesized from acetyl coenzyme A by enzymes in the cytosol and are linked to α-glycerol phosphate, produced from carbohydrates, to form triacylglycerols by enzymes in the smooth endoplasmic reticulum.

V. Proteins are broken down to free amino acids by proteases.

    a. The removal of amino groups from amino acids leaves keto acids, which can either be catabolized via the Krebs cycle to provide energy for the synthesis of ATP or be converted into glucose and fatty acids.

b. Amino groups are removed by (1) oxidative deamination, which gives rise to ammonia, or by (2) transamination, in which the amino group is transferred to a keto acid to form a new amino acid.

c. The ammonia formed from the oxidative deamination of amino acids is converted to urea by enzymes in the liver and then excreted in the urine by the kidneys.

VI. Some amino acids can be synthesized from keto acids derived from glucose, whereas others cannot be synthesized by the body and must be provided in the diet.

Essential Nutrients

I. Approximately 50 essential nutrients, listed in Table 4–9, are necessary for health but cannot be synthesized in adequate amounts by the body and must therefore be provided in the diet.

II. A large intake of water-soluble vitamins leads to their rapid excretion in the urine, whereas large intakes of fat-soluble vitamins lead to their accumulation in adipose tissue and may produce toxic effects.

SECTION C KEY TERMS

glycolysis	glycogen
pyruvate	gluconeogenesis
substrate-level phosphorylation	adipocyte
aerobic	adipose tissue
anaerobic	beta oxidation
lactate	α-glycerol phosphate
Krebs cycle	protease
citric acid cycle	oxidative deamination
tricarboxylic acid cycle	keto acid
acetyl coenzyme A (acetyl CoA)	transamination
oxidative phosphorylation	urea
cytochrome	essential amino acid
electron transport chain	negative nitrogen balance
chemiosmotic hypothesis	positive nitrogen balance
hydrogen peroxide	essential nutrient
superoxide anion	water-soluble vitamin
hydroxyl radical	fat-soluble vitamin

SECTION C REVIEW QUESTIONS

1. What are the end products of glycolysis under aerobic and anaerobic conditions?
2. To which molecule are the hydrogen atoms in $NADH + H^+$ transferred during anaerobic glycolysis? During aerobic glycolysis?
3. What are the major substrates entering the Krebs cycle, and what are the products formed?
4. Why does the Krebs cycle operate only under aerobic conditions even though molecular oxygen is not used in any of its reactions?

5. Identify the molecules that enter the oxidative-phosphorylation pathway and the products that are formed.
6. Where are the enzymes for the Krebs cycle located? The enzymes for oxidative phosphorylation? The enzymes for glycolysis?
7. How many molecules of ATP can be formed from the breakdown of one molecule of glucose under aerobic conditions? Under anaerobic conditions?
8. Describe the origin and effects of reactive oxygen molecules.
9. Describe the pathways by which glycogen is synthesized and broken down by cells.
10. What molecules can be used to synthesize glucose?
11. Why can't fatty acids be used to synthesize glucose?
12. Describe the pathways used to catabolize fatty acids to carbon dioxide.
13. Why is it more efficient to store fuel as fat than as glycogen?
14. Describe the pathway by which glucose is converted into fat.
15. Describe the two processes by which amino groups are removed from amino acids.
16. What can keto acids be converted into?
17. What is the source of the nitrogen atoms in urea, and in what organ is urea synthesized?
18. Why is water considered an essential nutrient whereas glucose is not?
19. What is the consequence of ingesting large quantities of water-soluble vitamins? Fat-soluble vitamins?

CHAPTER 4 THOUGHT QUESTIONS

(Answers are given in Appendix A.)

1. A variety of chemical messengers that normally regulate acid secretion in the stomach bind to proteins in the plasma membranes of the acid-secreting cells. Some of these binding reactions lead to increased acid secretion, and others to decreased secretion. In what ways might a drug that causes decreased acid secretion be acting on these cells?
2. In one type of diabetes, the plasma concentration of the hormone insulin is normal, but the response of the cells to which insulin usually binds is markedly decreased. Suggest a reason for this in terms of the properties of protein binding sites.
3. Given the following substances in a cell and their effects on each other, predict the change in compound H that will result from an increase in compound A, and diagram this sequence of changes.

   Compound A is a modulator molecule that allosterically activates protein B.
   Protein B is a protein kinase enzyme that activates protein C.
   Protein C is an enzyme that converts substrate D to product E.
   Compound E is a modulator molecule that allosterically inhibits protein F.

Protein F is an enzyme that converts substrate G to product H.

4. Shown below is the relation between the amount of acid secreted and the concentration of compound X, which stimulates acid secretion in the stomach by binding to a membrane protein.

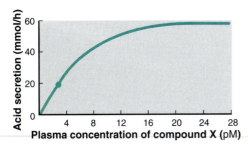

At a plasma concentration of 2 pM, compound X produces an acid secretion of 20 mmol/h.

   a. Specify two ways in which acid secretion by compound X could be increased to 40 mmol/h.

   b. Why will increasing the concentration of compound X to 28 pM not produce more acid secretion than increasing the concentration of X to 18 pM.

5. How would protein regulation be affected by a mutation that causes the loss of phosphoprotein phosphatase from cells?

6. How much energy is added to or released from a reaction in which reactants A and B are converted to products C and D if the energy content, in kilocalories per mole, of the participating molecules is: A = 55, B = 93, C = 62, and D = 87? Is this reaction reversible or irreversible? Explain.

7. In the following metabolic pathway, what is the rate of formation of the end product E if substrate A is present at a saturating concentration? The maximal rates (products formed per second) of the individual steps are indicated.

$$A \xrightarrow{30} B \xrightarrow{5} C \xrightarrow{20} D \xrightarrow{40} E$$

8. If the concentration of oxygen in the blood delivered to a muscle is increased, what effect will this have on the rate of ATP production by the muscle?

9. During prolonged starvation, when glucose is not being absorbed from the gastrointestinal tract, what molecules can be used to synthesize new glucose?

10. Why does the catabolism of fatty acids occur only under aerobic conditions?

11. Why do certain forms of liver disease produce an increase in the blood levels of ammonia?

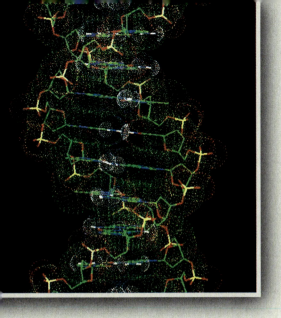

Genetic Information and Protein Synthesis

Whether an organism is a human being or a mouse, has blue eyes or black, has light skin or dark, is determined by the types of proteins the organism synthesizes. Moreover, the properties of muscle cells differ from those of nerve cells and epithelial cells because of the types of proteins present in each cell type and the functions performed by these proteins.

The information for synthesizing the cell's proteins is contained in the hereditary material in each cell coded into DNA molecules. Given that different cell types synthesize different proteins and that the specifications for these proteins are coded in DNA, one might be led to conclude that different cell types contain different DNA molecules. However, this is not the case. All cells in the body, with the exception of sperm or egg cells, receive the same genetic information

when DNA molecules are duplicated and passed on to daughter cells at the time of cell division. Therefore, cells differ in structure and function because only a portion of the total genetic information common to all cells is used by any given cell to synthesize proteins.

This chapter describes: (1) how genetic information is used to synthesize proteins, (2) some of the factors that govern the selective expression of genetic information, (3) the process by which DNA molecules are replicated and their genetic information passed on to daughter cells during cell division, and (4) how altering the genetic message—mutation—can lead to the class of diseases known as inherited disorders as well as to cancers.

Genetic Code

Molecules of DNA contain information, coded in the sequence of nucleotides, for the synthesis of proteins. A sequence of DNA nucleotides containing the information that specifies the amino acid sequence of a single polypeptide chain is known as a **gene.** A gene is thus a unit of hereditary information. A single molecule of DNA contains many genes.

The total genetic information coded in the DNA of a typical cell in an organism is known as its **genome.** The human genome contains between 50,000 and 100,000 genes, the information required for producing 50,000 to 100,000 different proteins. Currently, scientists from around the world are collaborating in the Human Genome Project to determine the nucleotide sequence of the human genome that will involve locating the position of the approximately 3 billion nucleotides that make up the human genome.

It is easy to misunderstand the relationship between genes, DNA molecules, and chromosomes. In all human cells (other than the eggs or sperm), there are 46 separate DNA molecules in the cell nucleus, each molecule containing many genes. Each DNA molecule is packaged into a single chromosome composed of DNA and proteins, so there are 46 chromosomes in each cell. A **chromosome** contains not only its DNA molecule, but a special class of proteins called histone proteins, or simply **histones.** The cell's nucleus is a marvel of packaging; the very long DNA molecules, having lengths a thousand times greater than the diameter of the nucleus, fit into the nucleus by coiling around clusters of histones at frequent intervals to form complexes known as **nucleosomes.** There are

about 25 million of these complexes on the chromosomes, resembling beads on a string.

Although DNA contains the information specifying the amino acid sequences in proteins, it does not itself participate *directly* in the assembly of protein molecules. Most of a cell's DNA is in the nucleus (a small amount is in the mitochondria), whereas most protein synthesis occurs in the cytoplasm. The transfer of information from DNA to the site of protein synthesis is the function of RNA molecules, whose synthesis is governed by the information coded in DNA. Genetic information flows from DNA to RNA and then to protein (Figure 5–1). The process of transferring genetic information from DNA to RNA in the nucleus is known as **transcription;** the process that uses the coded information in RNA to assemble a protein in the cytoplasm is known as **translation.**

$$DNA \xrightarrow{\text{transcription}} RNA \xrightarrow{\text{translation}} Protein$$

As described in Chapter 2, a molecule of DNA consists of two chains of nucleotides coiled around each other to form a double helix (see Figure 2–24). Each DNA nucleotide contains one of four bases—adenine (A), guanine (G), cytosine (C), or thymine (T)—and each of these bases is specifically paired by hydrogen bonds with a base on the opposite chain of the double helix. In this base pairing, A and T bond together and G and C bond together. Thus, both nucleotide chains contain a specifically ordered sequence of bases, one chain being complementary to the other. This specificity of base pairing, as we shall see, forms the basis of the transfer of information from DNA to RNA and of the duplication of DNA during cell division.

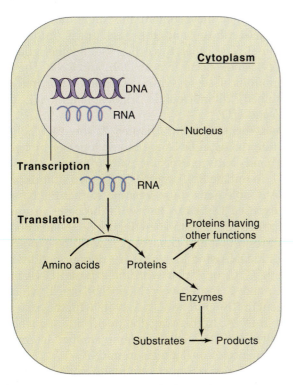

FIGURE 5–1

The expression of genetic information in a cell occurs through the transcription of coded information from DNA to RNA in the nucleus, followed by the translation of the RNA information into protein synthesis in the cytoplasm. The proteins then perform the functions that determine the characteristics of the cell.

The genetic language is similar in principle to a written language, which consists of a set of symbols, such as A, B, C, D, that form an alphabet. The letters are arranged in specific sequences to form words, and the words are arranged in linear sequences to form sentences. The genetic language contains only four letters, corresponding to the bases A, G, C, and T. The genetic words are three-base sequences that specify particular amino acids—that is, each word in the genetic language is only three letters long. This is termed a triplet code. The sequence of three-letter code words (triplets) along a gene in a single strand of DNA specifies the sequence of amino acids in a polypeptide chain (Figure 5–2). Thus, a gene is equivalent to a sentence, and the genetic information in the human genome is equivalent to a book containing 50,000 to 100,000 sentences. Using a single letter (A, T, C, G) to specify each of the four bases in the DNA nucleotides, it will require about 550,000 pages, each equivalent to this text page to print the nucleotide sequence of the human genome.

The four bases in the DNA alphabet can be arranged in 64 different three-letter combinations to form 64 code words ($4 \times 4 \times 4 = 64$). Thus, this code actually provides more than enough words to code for the 20 different amino acids that are found in proteins. This means that a given amino acid is usually specified by more than one code word. For example, the four DNA triplets C–C–A, C–C–G, C–C–T, and C–C–C all specify the amino acid glycine. Only 61 of the 64 possible code words are used to specify amino acids. The code words that do not specify amino acids are known as **"stop" signals.** They perform the same function as does a period at the end of a sentence—they indicate that the end of a genetic message has been reached.

The genetic code is a universal language used by all living cells. For example, the code words for the amino acid tryptophan are the same in the DNA of a bacterium, an amoeba, a plant, and a human being. Although the same code words are used by all living cells, the messages they spell out—the sequences of code words that code for a specific protein—vary from gene to gene in each organism. The universal nature of the genetic code supports the concept that all forms of life on earth evolved from a common ancestor.

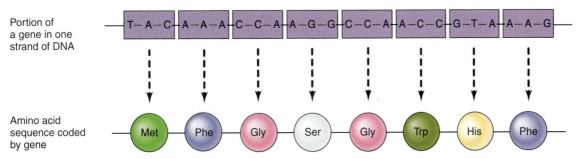

FIGURE 5–2

The sequence of three-letter code words in a gene determines the sequence of amino acids in a polypeptide chain. The names of the amino acids are abbreviated. Note that more than one three-letter code sequence can indicate the same amino acid; for example, the amino acid phenylalanine (Phe) is coded by two triplet codes, A–A–A and A–A–G.

Before we turn to the specific mechanisms by which the DNA code is used in protein synthesis, an important clarification and qualification is required. As noted earlier, the information coded in genes is always first transcribed into RNA. As we shall see in the next section there are several classes of RNA—messenger RNA, ribosomal RNA, transfer RNA, and small nuclear RNAs. Only messenger RNA *directly* codes for the amino acid sequences of proteins even though the other RNA classes participate in the overall process of protein synthesis. For this reason, the customary definition of a gene as the sequence of DNA nucleotides that specifies the amino acid sequence of a protein is true only for those genes that are transcribed into messenger RNA. The vast majority of genes are of this type, but it should at least be noted that the genes that code for the other classes of RNA do not technically fit this definition.

Protein Synthesis

To repeat, the first step in using the genetic information in DNA to synthesize a protein is called transcription, and it involves the synthesis of an RNA molecule containing coded information that corresponds to the information in a single gene. As noted above, several classes of RNA molecules take part in protein synthesis; the class of RNA molecules that specifies the amino acid sequence of a protein and carries this message from DNA to the site of protein synthesis in the cytoplasm is known as **messenger RNA (mRNA).**

Transcription: mRNA Synthesis

As described in Chapter 2, ribonucleic acids are single-chain polynucleotides whose nucleotides differ from DNA in that they contain the sugar ribose (rather than deoxyribose) and the base uracil (rather than thymine). The other three bases—adenine, guanine, and cytosine—occur in both DNA and RNA. The pool of subunits used to synthesize mRNA are free (uncombined) ribonucleotide triphosphates: ATP, GTP, CTP, and UTP.

As mentioned in Chapter 2, the two polynucleotide chains in DNA are linked together by hydrogen bonds between specific pairs of bases—A–T and C–G. To initiate RNA synthesis, the two strands of the DNA double helix must separate so that the bases in the exposed DNA can pair with the bases in free ribonucleotide triphosphates (Figure 5–3). Free ribonucleotides containing U bases pair with the exposed A bases in DNA, and likewise, free ribonucleotides containing G, C, or A pair with the exposed DNA bases C, G, and T, respectively. Note that uracil, which is present in RNA but not DNA, pairs with the base adenine in DNA. In this way, the nucleotide sequence in one strand of DNA acts as a template that determines the sequence of nucleotides in mRNA.

The aligned ribonucleotides are joined together by the enzyme **RNA polymerase,** which hydrolyses the nucleotide triphosphates, releasing two of the terminal phosphate groups, and joining the remaining phosphate in covalent linkage to the ribose of the adjacent nucleotide.

Since DNA consists of *two* strands of polynucleotides, both of which are exposed during transcription, it should theoretically be possible to form two different RNA molecules, one from each strand. However, only one of the two potential RNAs is ever formed. Which of the two DNA strands is used as the **template strand** for RNA synthesis from a particular gene is

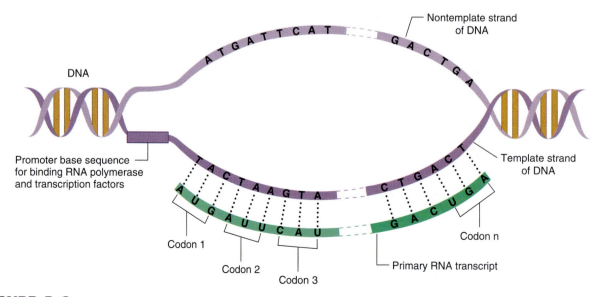

FIGURE 5–3

Transcription of a gene from the template strand of DNA to a primary RNA transcript.

determined by a specific sequence of DNA nucleotides called the **promoter,** which is located near the beginning of the gene on the strand that is to be transcribed (Figure 5–3). It is to this promoter region that RNA polymerase binds. Thus, for any given gene, only one strand is used, and that is the strand with the promoter region at the beginning of the gene. However, different transcribed genes may be located on either of the two strands of the DNA double helix.

To repeat, transcription of a gene begins with the binding of RNA polymerase to the promoter region of that gene. This initiates the separation of the two strands of DNA. RNA polymerase moves along the template strand, joining one ribonucleotide at a time (at a rate of about 30 nucleotides per second) to the growing RNA chain. Upon reaching a "stop" signal specifying the end of the gene, the RNA polymerase releases the newly formed RNA transcript. After the RNA transcript is released, a series of 100 to 200 adenine nucleotides is added to its end, forming a poly A tail that acts as a signal to allow RNA to move out of the nucleus and bind to ribosomes in the cytoplasm.

In a given cell, the information in only 10 to 20 percent of the genes present in DNA is transcribed into RNA. Genes are transcribed only when RNA polymerase can bind to their promoter sites. Various mechanisms, described later in this chapter, are used by cells either to block or to make accessible the promoter region of a particular gene to RNA polymerase. Such regulation of gene transcription provides a means of controlling the synthesis of specific proteins and thereby the activities characteristic of a particular type of differentiated cell.

It must be emphasized that the base sequence in the RNA transcript is *not identical* to that in the template strand of DNA, since the RNA's formation depends on the pairing between *complementary,* not identical, bases (Figure 5–3). A three-base sequence in RNA that specifies one amino acid is called a **codon.** Each codon is *complementary* to a three-base sequence in DNA. For example, the base sequence T–A–C in the template strand of DNA corresponds to the codon A–U–G in transcribed RNA.

Although the entire sequence of nucleotides in the template strand of a gene is transcribed into a complementary sequence of nucleotides known as the **primary RNA transcript,** only certain segments of the gene actually code for sequences of amino acids. These regions of the gene, known as **exons** (expression regions), are separated by noncoding sequences of nucleotides known as **introns** (intervening sequences). It is estimated that as much as 75–95 percent of human DNA is composed of intron sequences that do not contain protein-coding information. What role, if any, such large amounts of "nonsense" DNA may perform is unclear.

Before passing to the cytoplasm, a newly formed primary RNA transcript must undergo splicing (Figure 5–4) to remove the sequences that correspond to the DNA introns and thereby form the continuous sequence of exons that will be translated into protein (only after this splicing is the RNA termed messenger RNA).

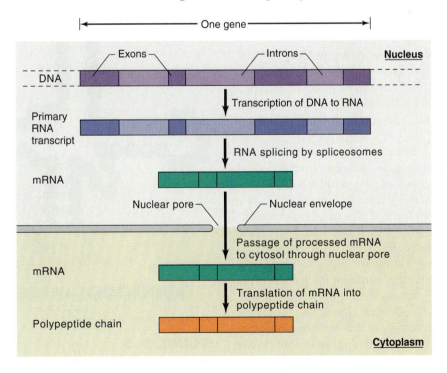

FIGURE 5–4

Spliceosomes remove the noncoding intron-derived segments from a primary RNA transcript and link the exon-derived segments together to form the mRNA molecule that passes through the nuclear pores to the cytosol. The lengths of the intron- and exon-derived segments represent the relative lengths of the base sequences in these regions.

Splicing occurs in the nucleus and is performed by a complex of proteins and small nuclear RNAs known as a **spliceosome.** The spliceosome identifies specific nucleotide sequences at the beginning and end of each intron-derived segment in the primary RNA transcript, removes the segment, and splices the end of one exon-derived segment to the beginning of another to form mRNA with a continuous coding sequence. Moreover, in some cases, during the splicing process the exon-derived segments from a single gene can be spliced together in different sequences, or some exon-derived segments can be deleted entirely. These processes result in the formation of different mRNA sequences from the same gene and give rise, in turn, to proteins with slightly different amino acid sequences.

Translation: Polypeptide Synthesis

After splicing, the mRNA moves through the pores in the nuclear envelope into the cytoplasm. Although the nuclear pores allow the diffusion of small molecules and ions between the nucleus and cytoplasm, they have specific energy-dependent mechanisms for the selective transport of large molecules such as proteins and RNA.

In the cytoplasm, mRNA binds to a ribosome, the cell organelle that contains the enzymes and other components required for the translation of mRNA's coded message into protein. Before describing this assembly process, we will examine the structure of a ribosome and the characteristics of two additional classes of RNA involved in protein synthesis.

Ribosomes and rRNA   As described in Chapter 3, ribosomes are small granules in the cytoplasm, either suspended in the cytosol (free ribosomes) or attached to the surface of the endoplasmic reticulum (bound ribosomes). A typical cell may contain 10 million ribosomes.

A ribosome is a complex particle composed of about 80 different proteins in association with a class of RNA molecules known as **ribosomal RNA (rRNA).** The genes for rRNA are transcribed from DNA in a process similar to that for mRNA except that a different RNA polymerase is used. Ribosomal RNA transcription occurs in the region of the nucleus known as the nucleolus. Ribosomal proteins, like other proteins, are synthesized in the cytoplasm from the mRNAs specific for them. These proteins then move back through nuclear pores to the nucleolus where they combine with newly synthesized rRNA to form two ribosomal subunits, one large and one small. These subunits are then individually transported to the cytoplasm where they combine to form a functional ribosome during protein translation.

Transfer RNA   How do individual amino acids identify the appropriate codons in mRNA during the process of translation? By themselves, free amino acids do not have the ability to bind to the bases in mRNA codons. This process of identification involves the third major class of RNA, known as **transfer RNA (tRNA).** Transfer RNA molecules are the smallest (about 80 nucleotides long) of the major classes of RNA. The single chain of tRNA loops back upon itself, forming a structure resembling a cloverleaf with three loops (Figure 5–5).

Like mRNA and rRNA, tRNA molecules are synthesized in the nucleus by base-pairing with DNA nucleotides at specific tRNA genes and then move to the cytoplasm. The key to tRNA's role in protein synthesis is its ability to combine with both a specific amino acid and a codon in ribosome-bound mRNA specific for that amino acid. This permits tRNA to act as the link between an amino acid and the mRNA codon for that amino acid.

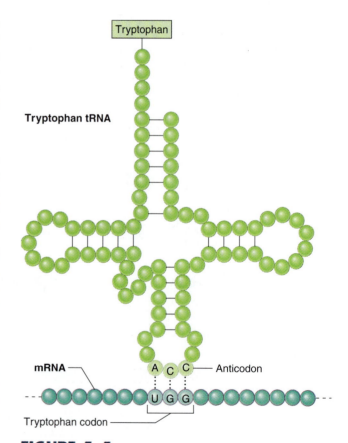

FIGURE 5–5

Base-pairing between the anticodon region of a tRNA molecule and the corresponding codon region of an mRNA molecule.

A tRNA molecule is covalently linked to a specific amino acid by an enzyme known as aminoacyl-tRNA synthetase. There are 20 different aminoacyl-tRNA synthetases, each of which catalyzes the linkage of a specific amino acid to a specific type of tRNA. The next step is to link the tRNA, bearing its attached amino acid, to the mRNA codon for that amino acid. As one might predict, this is achieved by base-pairing between tRNA and mRNA. A three-nucleotide sequence at the end of one of the loops of tRNA can base-pair with a complementary codon in mRNA. This tRNA three-letter code sequence is appropriately termed an **anti-codon.** Figure 5–5 illustrates the binding between mRNA and a tRNA specific for the amino acid tryptophan. Note that tryptophan is covalently linked to one end of tRNA and does not bind to either the anti-codon region of tRNA or the codon region of mRNA.

Protein Assembly The process of assembling a polypeptide chain based on an mRNA message involves three stages—initiation, elongation, and termination. Synthesis is initiated by the binding of a tRNA containing the amino acid methionine to the small ribosomal subunit. A number of proteins known as **initiation factors** are required to establish an initiation complex, which positions the methionine-containing tRNA opposite the mRNA codon that signals the start site at which assembly is to begin. The large ribosomal subunit then binds, enclosing the mRNA between the two subunits. This initiation phase is the slowest step in protein assembly, and the rate of protein synthesis can be regulated by factors that influence the activity of initiation factors.

Following the initiation process, the protein chain is elongated by the successive addition of amino acids (Figure 5–6). A ribosome has two binding sites for tRNA. Site 1 holds the tRNA linked to the portion of the protein chain that has been assembled up to this point, and site 2 holds the tRNA containing the next amino acid to be added to the chain. Ribosomal enzymes catalyze the linkage of the protein chain to the newly arrived amino acid. Following the formation of the peptide bond, the tRNA at site 1 is released from the ribosome, and the tRNA at site 2—now linked to the peptide chain—is transferred to site 1. The ribosome moves down one codon along the mRNA, making room for the binding of the next amino acid–tRNA molecule. This process is repeated over and over as amino acids are added to the growing peptide chain (at an average rate of two to three per second). When

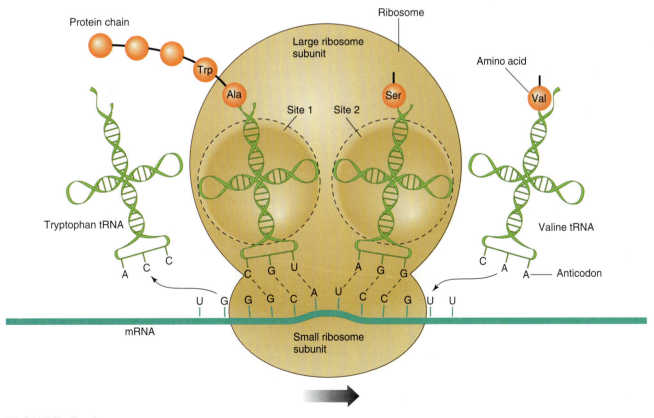

FIGURE 5–6
Sequence of events during the synthesis of a protein by a ribosome.

the ribosome reaches a termination sequence in mRNA specifying the end of the protein, the link between the polypeptide chain and the last tRNA is broken, and the completed protein is released from the ribosome.

Messenger-RNA molecules are not destroyed during protein synthesis, so they may be used to synthesize many protein molecules. Moreover, while one ribosome is moving along a particular strand of mRNA, a second ribosome may become attached to the start site on that same mRNA and begin the synthesis of a second identical protein molecule. Thus, a number of ribosomes, as many as 70, may be moving along a single strand of mRNA, each at a different stage of the translation process (Figure 5–7).

Molecules of mRNA do not, however, remain in the cytoplasm indefinitely. Eventually they are broken down into nucleotides by cytoplasmic enzymes. Therefore, if a gene corresponding to a particular protein ceases to be transcribed into mRNA, the protein will no longer be formed after its cytoplasmic mRNA molecules are broken down.

For small proteins, the folding that gives the protein its characteristic three-dimensional shape occurs spontaneously as the polypeptide chain emerges from the ribosome. Large proteins have a folding problem because their final conformation may depend upon interactions with portions of the molecule that have not yet emerged from the ribosome. In addition, a large segment of unfolded protein tends to aggregate with other proteins, which inhibits its proper folding. These problems are overcome by a complex of proteins known as **chaperones,** which form a small, hollow chamber into which the emerging protein chain is inserted. Within the confines of the chaperone, the

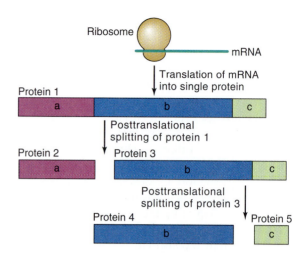

FIGURE 5–8

Posttranslational splitting of a protein can result in several smaller proteins, each of which may perform a different function. All these proteins are derived from the same gene.

polypeptide chain is able to complete its folding. The chaperones thus provide an isolated environment where protein folding can occur without interference.

Once a polypeptide chain has been assembled, it may undergo posttranslational modifications to its amino acid sequence. For example, the amino acid methionine that is used to identify the start site of the assembly process is cleaved from the end of most proteins. In some cases, other specific peptide bonds within the polypeptide chain are broken, producing a number of smaller peptides, each of which may perform a different function. For example, as illustrated in Figure 5–8, five different proteins can be derived from the same mRNA as a result of posttranslational cleavage. The same initial polypeptide may be split at different points in different cells depending on the specificity of the hydrolyzing enzymes present.

Carbohydrates and lipid derivatives are often covalently linked to particular amino acid side chains. These additions may protect the protein from rapid degradation by proteolytic enzymes or act as signals to direct the protein to those locations in the cell where it is to function. The addition of a fatty acid to a protein, for example, can lead to the anchoring of the protein to a membrane as the nonpolar portion of the fatty acid becomes inserted into the lipid bilayer.

The steps leading from DNA to a functional protein are summarized in Table 5–1.

Although 99 percent of eukaryotic DNA is located in the nucleus, a small amount is present in mitochondria. Mitochondrial DNA, like bacterial DNA, does not contain introns and is circular. These characteristics support the hypothesis that mitochondria arose during an early stage of evolution when an

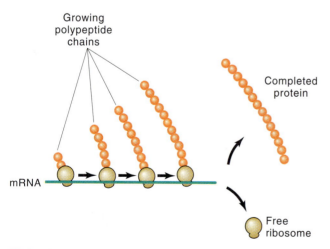

FIGURE 5–7

Several ribosomes can simultaneously move along a strand of mRNA, producing the same protein in different states of assembly.

TABLE 5–1 Events Leading from DNA to Protein Synthesis

Transcription

1. RNA polymerase binds to the promoter region of a gene and separates the two strands of the DNA double helix in the region of the gene to be transcribed.

2. Free ribonucleotide triphosphates base-pair with the deoxynucleotides in the template strand of DNA.

3. The ribonucleotides paired with this strand of DNA are linked by RNA polymerase to form a primary RNA transcript containing a sequence of bases complementary to the template strand of the DNA base sequence.

4. RNA splicing removes the intron-derived regions in the primary RNA transcript, which contain noncoding sequences, and splices together the exon-derived regions, which code for specific amino acids, producing a molecule of mRNA.

Translation

5. The mRNA passes from the nucleus to the cytoplasm, where one end of the mRNA binds to the small subunit of a ribosome.

6. Free amino acids are linked to their corresponding tRNAs by aminoacyl-tRNA synthetase.

7. The three-base anticodon in an amino acid–tRNA complex pairs with its corresponding codon in the region of the mRNA bound to the ribosome.

8. The amino acid on the tRNA is linked by a peptide bond to the end of the growing polypeptide chain (see Figure 5–6).

9. The tRNA that has been freed of its amino acid is released from the ribosome.

10. The ribosome moves one codon step along mRNA.

11. Steps 7 to 10 are repeated over and over until a termination sequence is reached, and the completed protein is released from the ribosome.

12. Chaperone proteins guide the folding of some proteins into their proper conformation.

13. In some cases, the protein undergoes posttranslational processing in which various chemical groups are attached to specific side chains and/or the protein is split into several smaller peptide chains.

anaerobic cell ingested an aerobic bacterium that ultimately became what we know today as mitochondria. Mitochondria also have the machinery, including ribosomes, for protein synthesis. However, the mitochondrial DNA contains the genes for only 13 mitochondrial proteins and a few of the rRNA and tRNA genes. Therefore, additional components are required for protein synthesis by the mitochondria, and most of the mitochondrial proteins are coded by nuclear DNA genes. These components are synthesized in the cytoplasm and then transported into the mitochondria.

Regulation of Protein Synthesis

As noted earlier, in any given cell only a small fraction of the genes in the human genome are ever transcribed into mRNA and translated into proteins. Of this fraction, a small number of genes are *continuously* being transcribed into mRNA, but the transcription of other genes is regulated and can be turned on or off in response either to signals generated within the cell or to external signals received by the cell. In order for a gene to be transcribed, RNA polymerase must be able to bind to the promoter region of the gene and be in an activated configuration.

Transcription of most genes is regulated by a class of proteins known as **transcription factors,** which act as gene switches, interacting in a variety of ways to activate or repress the initiation process that takes place at the promoter region of a particular gene. The influence of a transcription factor on transcription is not necessarily all or none, on or off; it may have the effect of slowing or speeding up the initiation of the transcription process. The transcription factors, along with accessory proteins, form a **preinitiation complex** at the promoter which is required to carry out the process of separating the DNA strands, removing any blocking nucleosomes in the region of the promoter, activating the bound RNA polymerase, and moving the complex along the template strand of DNA. Some transcription factors bind to regions of DNA that are far removed from the promoter region of the gene whose transcription they regulate. In this case, the DNA containing the bound transcription factor forms a loop that brings the transcription factor into contact with the promoter region where it may activate or repress transcription (Figure 5–9).

Many genes contain regulatory sites that can be influenced by a common transcription factor; thus there does not need to be a different transcription factor for every gene. In addition, more than one transcription factor may interact in controlling the transcription of a given gene.

Since transcription factors are proteins, the activity of a particular transcription factor—that is, its ability to bind to DNA or to other regulatory proteins—can be turned on or off by allosteric or covalent modulation in response to signals either received by a cell or generated within it. Thus, specific genes can be regulated in response to specific signals. These signaling mechanisms will be discussed in Chapter 7.

To summarize, the rate of a protein's synthesis can be regulated at various points: (1) gene transcription into mRNA; (2) the initiation of protein assembly on a ribosome; and (3) mRNA degradation in the cytoplasm.

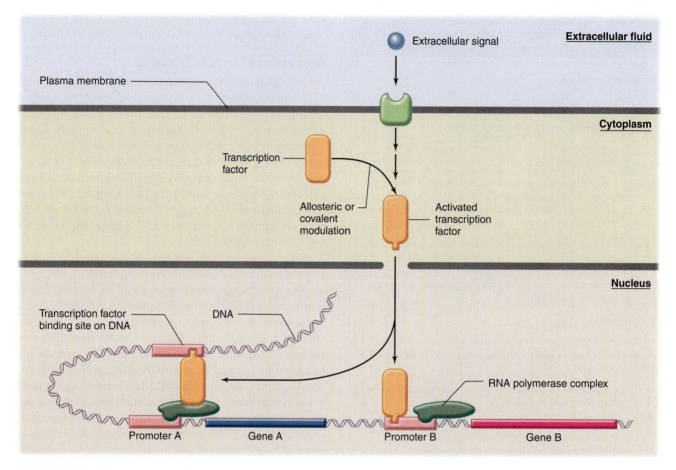

FIGURE 5–9

Transcription of gene B is modulated by the binding of an activated transcription factor directly to the promoter region. In contrast, transcription of gene A is modulated by the same transcription factor which, in this case, binds to a region of DNA considerably distant from the promoter region.

Protein Degradation

We have thus far emphasized protein synthesis, but an important fact is that the concentration of a particular protein in a cell at a particular time depends not only upon its rate of synthesis but upon its rates of degradation and/or secretion.

Different proteins are degraded at different rates. In part this depends on the structure of the protein, with some proteins having a higher affinity for certain proteolytic enzymes than others. A denatured (unfolded) protein is more readily digested than a protein with an intact conformation. Proteins can be targeted for degradation by the attachment of a small peptide, **ubiquitin,** to the protein. This peptide directs the protein to a protein complex known as a **proteosome,** which unfolds the protein and breaks it down into small peptides.

In summary, there are many steps between a gene in DNA and a fully active protein at which the rate of protein synthesis or the final active form of the protein can be altered (Table 5–2). By controlling these various steps, the total amount of a specific protein in a particular cell can be regulated by signals as described in Chapter 7.

Protein Secretion

Most proteins synthesized by a cell remain in the cell, providing structure and function for the cell's survival. Some proteins, however, are secreted into the extracellular fluid, where they act as signals to other cells or provide material for forming the extracellular matrix to which tissue cells are anchored. Since proteins are large, charged molecules that cannot diffuse through cell membranes (as will be described in more

TABLE 5–2 Factors that Alter the Amount and Activity of Specific Cell Proteins

Process Altered	Mechanism of Alteration
1. Transcription of DNA	Activation or inhibition by transcription factors
2. Splicing of RNA	Activity of enzymes in spliceosome
3. mRNA degradation	Activity of RNAase
4. Translation of mRNA	Activity of initiating factors on ribosomes
5. Protein degradation	Activity of proteosomes
6. Allosteric and covalent modulation	Signal ligands, protein kinases, and phosphatases

detail in Chapter 6), special mechanisms are required to insert them into or move them through membranes.

Proteins destined to be secreted from a cell or become integral membrane proteins are recognized during the early stages of protein synthesis. For such proteins, the first 15 to 30 amino acids that emerge from the surface of the ribosome act as a recognition signal, known as the **signal sequence,** or signal peptide.

The signal sequence binds to a complex of proteins known as a signal recognition particle, which temporarily inhibits further growth of the polypeptide chain on the ribosome. The signal recognition particle then binds to a specific membrane protein on the surface of the granular endoplasmic reticulum. This binding restarts the process of protein assembly, and the growing polypeptide chain is fed through a protein complex in the endoplasmic reticulum membrane into the lumen of the reticulum (Figure 5–10). Upon completion of protein assembly, proteins that are to be secreted end up in the lumen of the granular endoplasmic reticulum. Proteins that are destined to function as integral membrane proteins remain embedded in the reticulum membrane.

Within the lumen of the endoplasmic reticulum, enzymes remove the signal sequence from most proteins, and so this portion is not present in the final protein. In addition, carbohydrate groups are linked to various side chains in the proteins; almost all proteins secreted from the cell are glycoproteins.

Following these modifications, portions of the reticulum membrane bud off, forming vesicles that contain the newly synthesized proteins. These vesicles migrate to the Golgi apparatus (Figure 5–10) and fuse with the Golgi membranes. Vesicle budding, movement through the cytosol, and fusion with the Golgi membranes require the interaction of a number of proteins that initiate the budding process, serve as molecular motors that transport vesicles along microtubules, and provide the docking signals to direct the vesicles to the appropriate membrane. These processes require chemical energy derived from the hydrolysis of ATP and GTP.

Within the Golgi apparatus, the protein undergoes still further modification. Some of the carbohydrates that were added in the granular endoplasmic reticulum are now removed and new groups added. These new carbohydrate groups function as labels that can be recognized when the protein encounters various binding sites during the remainder of its trip through the cell.

While in the Golgi apparatus, the many different proteins that have been funneled into this organelle become sorted out according to their final destination. This sorting involves the binding of regions of a particular protein to specific proteins in the Golgi membrane that are destined to form vesicles targeted to a particular destination.

Following modification and sorting, the proteins are packaged into vesicles that bud off the surface of the Golgi membrane. Some of the vesicles travel to the plasma membrane where they fuse with the membrane and release their contents to the extracellular fluid, a process known as exocytosis (Chapter 6). Other vesicles dock and fuse with lysosome membranes, delivering digestive enzymes to the interior of this organelle. The specific interactions governing the formation and distribution of these vesicles from the Golgi apparatus are similar in mechanism to those involved in vesicular shuttling between the endoplasmic reticulum and the Golgi apparatus. Specific proteins on the surface of a vesicle are recognized by specific docking proteins on the surface of the membranes with which the vesicle finally fuses.

In contrast to this entire story, if a protein does not have a signal sequence, synthesis continues on a free ribosome until the completed protein is released into the cytosol. These proteins are not secreted but are destined to function within the cell. Many remain in the cytosol where they function, for example, as enzymes in various metabolic pathways. Others are targeted to particular cell organelles; for example, ribosomal proteins are directed to the nucleus where they combine with rRNA before returning to the cytosol as part of the ribosomal subunits. The specific location of a protein is determined by binding sites on the protein that bind to specific sites at the protein's destination. For example, in the case of the ribosomal proteins, they bind to sites on the nuclear pores that control access to the nucleus.

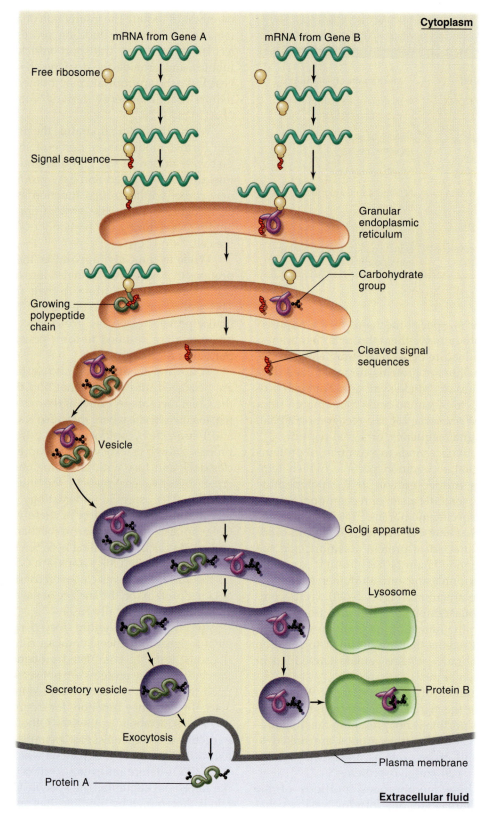

FIGURE 5–10

Pathway of proteins destined to be secreted by cells or transferred to lysosomes.

As we described earlier, although some mitochondrial proteins can be synthesized within the mitochondria from mitochondrial DNA genes, most mitochondrial proteins are coded by nuclear genes and are synthesized in the cytosol on free ribosomes. To gain access to the mitochondrial matrix, these proteins bind to recognition sites on the mitochondrial membrane; their folded conformation is unfolded, and they are fed through a pore complex into the mitochondrial matrix, a process similar to inserting a bound ribosomal protein into the lumen of the endoplasmic reticulum. In the mitochondrial matrix, the protein refolds into its functional conformation. A similar process delivers proteins to the lumen of peroxisomes.

Replication and Expression of Genetic Information

The set of genes present in each cell of an individual is inherited from the father and mother at the time of fertilization of an egg by a sperm. The egg and sperm cell each contain 23 molecules of DNA associated with histone proteins in chromosomes. Each of the 23 chromosomes contains a different set of genes, some containing more genes than others along its single continuous DNA molecule. Twenty-two of the 23 chromosomes contain genes that produce the proteins that govern most cell structures and functions and are known as **autosomes.** The remaining chromosome, known as the **sex chromosome,** contains genes whose expression determines the development of male or female gender. The 22 autosomes in the egg and those in the sperm contain corresponding genes. For example, a chromosome in the egg contains a gene for hemoglobin that is homologous to a similar gene in one of the sperm's chromosomes.

When the egg and sperm unite, the resulting fertilized egg contains 46 chromosomes—44 autosomes and 2 sex chromosomes. With the exception of the genes on the sex chromosomes, each cell of an individual contains 22 pairs of homologous genes. Of each pair, one chromosome was inherited from the mother and one from the father, with each potentially able to code for the same type of protein.

The development of an individual is determined by the controlled expression of the set of genes inherited at the time of conception. Growth occurs through the successive division of cells to form the trillions of cells that make up the adult human body. Each time a cell divides, the 46 DNA molecules in the 46 chromosomes must be replicated, and identical DNA copies passed on to each of the two new cells, termed **daughter cells.** Thus every cell in the body, with the exception of the reproductive cells, contains an identical set

of 46 DNA molecules, and therefore an identical set of genes. (See Chapter 19 for a discussion of the special processes associated with the formation of the reproductive cells in which the number of chromosomes is reduced from 46 to 23.) What makes one cell different from another depends on the differential expression of various sets of genes in this gene pool common to every cell.

Replication of DNA

DNA is the only molecule in a cell able to duplicate itself without information from some other cell component. In contrast, as we have seen, RNA can only be formed using the information present in DNA, protein formation uses the information in mRNA, and all other molecules use protein enzymes to determine the structure of the products formed.

DNA replication is, in principle, similar to the process whereby RNA is synthesized. During DNA replication (Figure 5–11), the two strands of the double helix separate, and *each* exposed strand acts as a template to which free *deoxyribonucleotide* triphosphates can base-pair, A with T and G with C. An enzyme, **DNA polymerase,** then links the free nucleotides together at a rate of about 50 nucleotides per second as it moves along the strand, forming a new strand complementary to each template strand of DNA. The end result is two identical molecules of DNA. In each molecule, one strand of nucleotides, the template strand, was present in the original DNA molecule, and one strand has been newly synthesized.

This description of DNA synthesis provides an overview of the basic elements of the process, but the individual steps are considerably more complex. A number of proteins in addition to DNA polymerase are required. Some of these proteins determine where along the DNA strand replication will begin, others open the DNA helix so that it can be copied, while still others prevent the tangling that can occur as the helix unwinds and rewinds.

A special problem arises as the replication process approaches the end of the DNA molecule. The complex of proteins that carry out the replication sequence is in part anchored to a portion of the DNA molecule that lies ahead of the site at which the two strands separate during replication. If a DNA molecule ended at the very end of the last gene, this gene could not be copied during DNA replication because there are no more downstream sites to anchor the replication complex.

This problem is overcome by an enzyme that adds to the ends of DNA a chain of nucleotides composed of several hundred to several thousand repeats of the six-nucleotide sequence TTAGGG. This terminal repetitive segment is known as a **telomere,** and the enzyme that catalyzes the formation of a telomere is

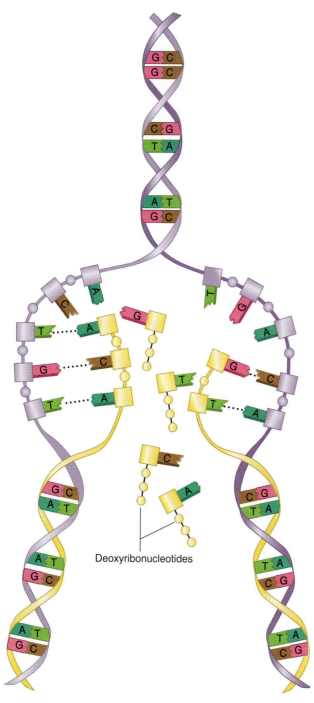

Deoxyribonucleotides

FIGURE 5–11

Replication of DNA involves the pairing of free deoxyribonucleotides with the bases of the separated DNA strands, giving rise to two new identical DNA molecules, each containing one old and one new polynucleotide strand.

telomerase. In the absence of telomerase, each replication of DNA results in a shorter molecule because of failure to replicate the ends of DNA.

Cells that continue to divide throughout the life of an organism contain telomerase, as do cancer cells and the cells that give rise to sperm and egg cells. The presence of telomerase allows cells to restore their telomeres after each cell division, thus preventing shortening of their DNA. However, many cells do not express telomerase, and each replication of DNA leads to a loss of coded genetic information. It is hypothesized that the telomeres serve as a biological clock that sets the number of divisions a cell can undergo and still remain viable.

In order to form the approximately 40 trillion cells of the adult human body, a minimum of 40 trillion individual cell divisions must occur. Thus, the DNA in the original fertilized egg must be replicated at least 40 trillion times. Actually, many more than 40 trillion divisions occur during the growth of a fertilized egg into an adult human being since many cells die during development and are replaced by the division of existing cells.

If a secretary were to type the same manuscript 40 trillion times, one would expect to find some typing errors. Therefore, it is not surprising to find that during the duplication of DNA, errors occur that result in an altered sequence of bases and a change in the genetic message. What is amazing is that DNA can be duplicated so many times with relatively few errors.

A mechanism called **proofreading** corrects most errors in the base sequence as it is being duplicated and is largely responsible for the low error rate observed during DNA replication. If an incorrect free nucleotide has become temporarily paired with a base in the template strand of DNA (for example, C pairing with A rather than its appropriate partner G), the DNA polymerase somehow "recognizes" this abnormal pairing and will not proceed in the linking of nucleotides until the abnormal pairing has been replaced. Note that in performing this proofreading, the DNA polymerase needs to identify only two configurations, the normal A–T and G–C pairing; any other combination halts polymerase activity. In this manner each nucleotide, as it is inserted into the new DNA chain, is checked for its appropriate complementarity to the base in the template strand.

Cell Division

Starting with a single fertilized egg, the first cell division produces 2 cells. When these daughter cells divide, they each produce 2 cells, giving a total of 4. These 4 cells produce a total of 8, and so on. Thus, starting from a single cell, 3 division cycles will produce 8 cells (2^3), 10 division cycles will produce $2^{10} = 1024$ cells, and 20 division cycles will produce $2^{20} = 1,048,576$ cells. If the development of the human body involved only cell division and growth without any cell death, only about 46 division cycles would be needed to produce all the cells in the adult body. However, large numbers of cells die during the course of development, and even in the adult many cells survive only a few days and are continually replaced by the division of existing cells.

The time between cell divisions varies considerably in different types of cells, with the most rapidly growing cells dividing about once every 24 h. During most of this period, there is no visible evidence that the cell will divide. For example, in a 24-h division cycle, changes in cell structure begin to appear 23 h after the last division. The period between the end of one division and the appearance of the structural changes that indicate the beginning of the next division is known as **interphase.** Since the physical process of dividing one cell into two cells takes only about 1 h, the cell spends most of its time in interphase, and most of the cell properties described in this book are properties of interphase cells.

One very important event related to subsequent cell division does occur during interphase, namely, the replication of DNA, which begins about 10 h before the first visible signs of division and lasts about 7 h. This period of the cell cycle is known as the S phase (synthesis) (Figure 5–12). Following the end of DNA synthesis, there is a brief interval, G_2 (second gap), before the signs of cell division begin. The period from the end of cell division to the beginning of the S phase is the G_1 (first gap) phase of the cell cycle.

In terms of the capacity to undergo cell division, there are two classes of cells in the adult body. Some cells proceed continuously through one cell cycle after another, while others seldom or never divide once they have differentiated. The first group consists of the stem cells, which provide a continuous supply of cells that form the specialized cells to replace those (such as blood cells, skin cells, and the cells lining the intestinal tract) that are continuously lost. The second class includes a number of differentiated, specialized cell types, such as nerve and striated-muscle cells, that rarely or never divide once they have differentiated. Also included in this second class are cells that leave the cell cycle and enter a phase known as G_0 (Figure 5–12) in which the process that initiates DNA replication is blocked. A cell in the G_0 phase, upon receiving an appropriate signal, can reenter the cell cycle, begin replicating DNA, and proceed to divide.

Cell division involves two processes: nuclear division, or **mitosis,** and cytoplasmic division, or **cytokinesis.** Although mitosis and cytokinesis are separate events, the term mitosis is often used in a broad sense to include the subsequent cytokinesis, and so the two events constitute the M phase (mitosis) of the cell cycle. Nuclear division that is not followed by cytokinesis produces multinucleated cells found in the liver, placenta, and some embryonic cells and cancer cells.

When a DNA molecule replicates, the result is two identical chains termed **sister chromatids,** which initially are joined together at a single point called the **centromere** (Figure 5–13). As a cell begins to divide, each chromatid pair becomes highly coiled and condensed, forming a visible, rod-shaped body, a chromosome. In the condensed state prior to division, each of the 46 chromosomes, each consisting of 2 chromatids, can be identified microscopically by its length and position of its centromere.

As the duplicated chains condense, the nuclear membrane breaks down, and the chromosomes become linked in the region of their centromeres to spindle fibers (Figure 5–13c). The **spindle fibers,** composed of microtubules, are formed in the region of the cell known as the **centrosome.** The centrosome, which contains two centrioles (described in Chapter 2) and associated proteins, is required for microtubule assembly.

When a cell enters the mitotic phase of the cell cycle, the two centrioles divide, and a pair of centrioles migrates to opposite sides of the cell, thus establishing the axis of cell division. One centrosome will pass to each of the daughter cells during cytokinesis. Some of the spindle fibers extend between the two

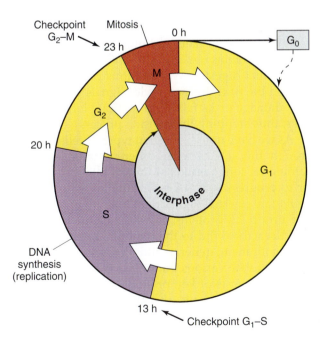

FIGURE 5–12

Phases of the cell cycle with approximate elapsed time in a cell that divides every 24 h. A cell may leave the cell cycle and enter the G_0 phase where division ceases unless the cell receives a specific signal to reenter the cycle.

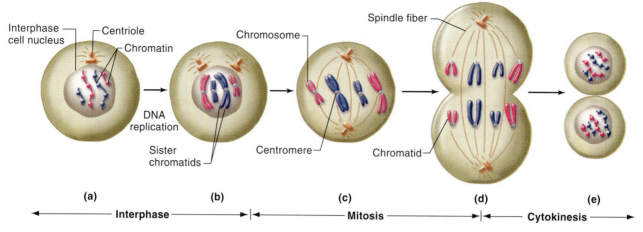

FIGURE 5-13

Mitosis and cytokinesis. (Only 4 of the 46 chromosomes in a human cell are illustrated.) (a) During interphase, chromatin exists in the nucleus as long, extended chains. The chains are partially coiled around clusters of histone proteins, producing a beaded appearance. (b) Prior to the onset of mitosis, DNA replicates, forming two identical sister chromatids that are joined at the centromere. A second pair of centrioles is formed at this time. As mitosis begins, the chromatids become highly condensed and (c) become attached to spindle fibers. (d) The two chromatids of each chromosome separate and move toward opposite poles of the cell (e) as the cell divides (cytokinesis) into two daughter cells.

centrosomes, while others connect the centrioles to the chromosomes. The spindle fibers and centrosomes constitute the **mitotic apparatus.**

As mitosis proceeds, the sister chromatids of each chromosome separate at the centromere and move toward opposite centrioles (Figure 5–13d). Cytokinesis begins as the sister chromatids separate. The cell begins to constrict along a plane perpendicular to the axis of the mitotic apparatus, and constriction continues until the cell has been pinched in half, forming the two daughter cells (Figure 5–13e), each having half the volume of the parent cell. Following cytokinesis, in each daughter cell, the spindle fibers dissolve, a nuclear envelope forms, and the chromatids uncoil.

The forces producing the movements associated with mitosis and cytokinesis are generated by (1) contractile proteins similar to those producing the forces generated by muscle cells (described in Chapter 11) and (2) the chemical kinetics associated with the elongation and shrinkage of microtubular filaments.

There are two critical checkpoints in the cell cycle, at which special events must occur in order for a cell to progress to the next phase (see Figure 5–12). One is at the boundary between G_1 and S. For example, if some of the chromosomes have not completed their DNA replication during S phase, the cell will not begin mitosis until the replication is complete. The second checkpoint is between G_2 and M. Thus if DNA has been damaged, by x-rays for example, the cell will not enter M phase until the DNA has been repaired.

Two classes of proteins are the major players in timing cell division and the progression through these checkpoints—**cell division cycle kinases** (cdc kinases) and **cyclins.** Cyclins act as modulator molecules to activate the cdc kinases. The concentration of cyclins progressively increases during interphase and then rapidly falls during mitosis. Once activated, the kinase enzymes phosphorylate, and thus activate or inhibit a variety of proteins necessary for division, including an enzyme that digests cyclin and thus prepares the cell to begin the next division cycle. Signals generated by DNA damage or its failure to replicate inhibit cdc kinases, thus stopping the division process.

As we have noted, different types of cells progress through the cell cycle at different rates, some remaining for long periods of time in interphase. In order to progress to DNA replication, most cells must receive an external signal delivered by one or more of a group of proteins known as **growth factors.** Growth factors bind to their specific receptors in the cell membrane to generate intracellular signals; these signals activate various transcription factors that control the synthesis of key proteins involved in the division process and the checkpoint mechanisms. At least 50 growth factors have been identified. Many are secreted by one cell and stimulate other specific cell types to divide; others stimulate division in the cell that secretes them. Growth factors also influence various aspects of metabolism and cell differentiation. In the absence of the appropriate growth factor, most cells will not divide.

Mutation

Any alteration in the nucleotide sequence that spells out a genetic message in DNA is known as a *mutation.* Certain chemicals and various forms of ionizing radiation, such as x-rays, cosmic rays, and atomic radiation, can break the chemical bonds in DNA. This can result in the loss of segments of DNA or the incorporation of the wrong base when the broken bonds are reformed. Environmental factors that increase the rate of mutation are known as *mutagens.* Even in the absence of environmental mutagens, the mutation rate is never zero. In spite of proofreading, some errors are made during the replication of DNA, and some of the normal compounds present in cells, particularly reactive oxygen species, can damage DNA, leading to mutations.

Types of Mutations   The simplest type of mutation, known as a point mutation, occurs when a single base is replaced by a different one. For example, the base sequence C–G–T is the DNA code word for the amino acid alanine. If guanine (G) is replaced by adenine (A), the sequence becomes C–A–T, which is the code for valine. If, however, cytosine (C) replaces thymine (T), the sequence becomes C–G–C, which is another code for alanine, and the amino acid sequence transcribed from the mutated gene would not be altered. On the other hand, if an amino acid code is mutated to one of the three termination code words, the translation of the mRNA message will cease when this code word is reached, resulting in the synthesis of a shortened, typically nonfunctional protein.

Assume that a mutation has altered a single code word in a gene, for example, alanine C–G–T changed to valine C–A–T, so that it now codes for a protein with one different amino acid. What effect does this mutation have upon the cell? The answer depends upon where in the gene the mutation has occurred. Although proteins are composed of many amino acids, the properties of a protein often depend upon a very

small region of the total molecule, such as the binding site of an enzyme. If the mutation does not alter the conformation of the binding site, there may be little or no change in the protein's properties. On the other hand, if the mutation alters the binding site, a marked change in the protein's properties may occur. Thus, if the protein is an enzyme, a mutation may change its affinity for a substrate or render the enzyme totally inactive. To take another situation, if the mutation occurs within an intron segment of a gene, it will have no effect upon the amino acid sequence coded by the exon segments (unless it alters the ability of the intron segment to undergo normal splicing from the primary RNA transcript).

In a second general category of mutation, single bases or whole sections of DNA are deleted or added. Such mutations may result in the loss of an entire gene or group of genes or may cause the misreading of a sequence of bases. Figure 5–14 shows the effect of removing a single base on the reading of the genetic code. Since the code is read in three-base sequences, the removal of one base not only alters the code word containing that base, but also causes a misreading of all subsequent bases by shifting the reading sequence. Addition of an extra base causes a similar misreading of all subsequent code words, which often results in a protein having an amino acid sequence that does not correspond to any functional protein.

What effects do these various types of mutation have upon the functioning of a cell? If a mutated, nonfunctional enzyme is in a pathway supplying most of a cell's chemical energy, the loss of the enzyme's function *could* lead to the death of the cell. The story is more complex, however, since the cell contains a second gene for this enzyme on its homologous chromosome, one which has not been mutated and is able to form an active enzyme. Thus, little or no change in cell function would result from this mutation. If both genes had mutations that rendered their products inactive, then no functional enzyme would be formed, and the cell

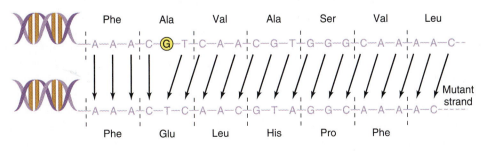

FIGURE 5–14

A deletion mutation caused by the loss of a single base G in one of the two DNA strands causes a misreading of all code words beyond the point of the mutation.

would die. In contrast, if the active enzyme were involved in the synthesis of a particular amino acid, and if the cell could also obtain that amino acid from the extracellular fluid, the cell function would not be impaired by the absence of the enzyme.

To generalize, a mutation may have any one of three effects upon a cell: (1) It may cause no noticeable change in cell function; (2) it may modify cell function, but still be compatible with cell growth and replication; or (3) it may lead to cell death.

With one exception—cancer, to be described later—the malfunction of a single cell, other than a sperm or egg, as a result of mutation usually has no significant effect because there are many cells performing the same function in the individual. Unfortunately, the story is different when the mutation occurs in a sperm or egg. In this case, the mutation will be passed on to *all* the cells in the body of the new individual. Thus, mutations in a sperm or egg cell do not affect the individual in which they occur but do affect, often catastrophically, the child produced by these cells. Moreover, these mutations may be passed on to some individuals in future generations descended from the individual carrying the mutant gene.

DNA Repair Mechanisms Cells possess a number of enzymatic mechanisms for repairing DNA that has been altered. These repair mechanisms all depend on the damage occurring in only one of the two DNA strands, so that the undamaged strand can provide the correct code for rebuilding the damaged strand. A repair enzyme identifies an abnormal region in one of the DNA strands and cuts out the damaged segment. DNA polymerase then rebuilds the segment after basepairing with the undamaged strand just as it did during DNA replication. If adjacent regions in both strands of DNA are damaged, a permanent mutation is created that cannot be repaired by these mechanisms.

This repair mechanism is particularly important for long-lived cells, such as skeletal muscle cells, that do not divide and therefore do not replicate their DNA. This means that the same molecule of DNA must continue to function and maintain the stability of its genetic information for as long as the cell lives, which could be as long as 100 years. One aspect of aging may be related to the accumulation of unrepaired mutations in these long-lived cells.

Mutations and Evolution Mutations contribute to the evolution of organisms. Although most mutations result in either no change or an impairment of cell function, a very small number may alter the activity of a protein in such a way that it is more, rather than less, active, or they may introduce an entirely new type of protein activity into a cell. If an organism carrying such a mutant gene is able to perform some function more effectively than an organism lacking the mutant gene, it has a better chance of reproducing and passing on the mutant gene to its descendants. On the other hand, if the mutation produces an organism that functions less effectively than organisms lacking the mutation, the organism is less likely to reproduce and pass on the mutant gene. This is the principle of **natural selection.** Although any one mutation, if it is able to survive in the population, may cause only a very slight alteration in the properties of a cell, given enough time, a large number of small changes can accumulate to produce very large changes in the structure and function of an organism.

The Gene Pool Given the fact that there are billions of people living on the surface of the earth, all carrying genes encoded in DNA and subject to mutation, any given gene is likely to have a slightly different sequence in some individuals as a result of these ongoing mutations. These variants of the same gene are known as **alleles,** and the number of different alleles for a particular gene in the population is known as the **gene pool.** At conception, one allele of each gene from the father and one allele from the mother are present in the fertilized egg. If both alleles of the gene are identical, the individual is said to be **homozygous** for that particular gene, but if the two alleles differ, the individual is **heterozygous.**

The set of alleles present in an individual is referred to as the individual's **genotype.** With the exception of the genes in the sex chromosomes, both of the homologous genes inherited by an individual can be transcribed and translated into proteins, given the appropriate signals. The expression of the genotype into proteins produces a specific structural or functional activity that is recognized as a particular trait in the individual and is known as the person's **phenotype.** For example, blue eyes and black eyes represent the phenotypes of the genes involved in the formation of eye pigments.

A particular phenotype is said to be **dominant** when only one of the two inherited alleles is required to express the trait, and **recessive** when both inherited alleles must be the same—that is, the individual must be homozygous for the trait to be present. For example, black eye color is inherited as a dominant trait, while blue eyes are a recessive trait. If an individual receives an allele of the gene controlling black eye pigment from either parent, the individual will have black eyes. A single copy of the allele for black eye color is sufficient to express the proteins forming black eye pigment. In contrast, the expression of the blue-eyed phenotype occurs only when both alleles in the individual code for a protein able to form the blue-eyed pigment. Although *genes* are often described as dominant or recessive, it is the activity or lack of activity of the *proteins* expressed by the genes that determines the phenotypic characteristics observed.

Genetic Disease   Many diseases are referred to as "genetic"—that is, due to abnormal structure or function resulting from the inheritance of mutant genes, rather than the result of microbial infections, toxic agents, or improper nutrition. Over 4000 diseases are linked to genetic abnormalities, and these diseases are currently a major cause of infant mortality. Genetic diseases can be inherited as either a dominant or recessive trait. Let us look at a few examples.

Familial hypercholesterolemia is an autosomal dominant disease affecting 1 in 500 individuals. These individuals have elevated blood levels of cholesterol because of a defect in a plasma-membrane protein involved in cholesterol removal from the blood and are, therefore, at increased risk of developing heart disease. Inheritance of only a single mutant allele from either the mother or father is sufficient to produce this condition.

Cystic fibrosis, an autosomal recessive disease, is the most common lethal genetic disease among Caucasians, with a prevalence of about 1 in 2000 births. Because of a defective mechanism for the transfer of fluid across epithelial membranes (to be discussed in Chapter 6), various ducts in the lungs, intestines, and reproductive tract become obstructed, with the most serious complications generally developing in the lungs and leading to death from respiratory failure. An individual must inherit a mutant allele from *both* parents in order for this recessive disease to be expressed. Individuals who are heterozygous, having only one copy of the mutant allele, do not show the symptoms of the disease because a single copy of the normal allele is sufficient to produce the protein required to maintain epithelial fluid transport. However, such individuals are carriers who are able to transmit the mutant allele to their offspring.

Familial hypercholesterolemia and cystic fibrosis are examples of *single gene diseases,* as are sickle-cell anemia, hemophilia, and muscular dystrophy. Two other recognized classes of genetic disease are *chromosomal* and *polygenic diseases,* both of which require the expression or lack of expression of *multiple* genes to produce the phenotypic trait. Chromosomal diseases are the result of the addition or deletion of chromosomes or portions of chromosomes during the process of reducing the 46 chromosomes to 23 during the formation of egg and sperm cells (to be discussed in Chapter 19). The classic example of a chromosomal disease is *Down's syndrome (trisomy 21),* in which the fertilized egg has an extra copy, or translocation, of chromosome 21. This abnormality occurs in approximately 1 of every 800 births and is characterized by retardation of growth and mental function. Other forms of chromosomal abnormalities are the major cause of spontaneous abortions or miscarriages.

Polygenic diseases result from the interaction of multiple mutant genes, any one of which by itself produces little or no effect, but when present with other mutant genes produces disease. This category of genetic disease is involved in most forms of the major diseases in our modern society, such as diabetes, hypertension, and cancer.

Cancer

Like the inherited genetic diseases described previously, cancer results from gene mutations. However, with a few exceptions, cancer is not an *inherited* genetic disease that depends on mutations in the reproductive cells. Rather, most cancers arise from mutations that can occur in any cell at anytime. As noted earlier, most mutations in a single nonreproductive cell have no effect upon the overall functioning of an organism, even if they lead to the death of that particular cell. If, however, mutations result in the failure of the control systems that regulate cell division, a cell with a capacity for uncontrolled growth, a *cancer cell,* may form and lead to the full-blown disease.

Cancer is the second leading cause of death in America after heart disease, with approximately 25 percent of all deaths due to cancer. Fifty percent of cancers occur in three organs—lung (28 percent), colon (13 percent), and breast (9 percent). About 90 percent of cancers develop in epithelial cells and are known as *carcinomas.* Those derived from connective tissue and muscle cells are called *sarcomas,* and those from white blood cells are *leukemias* and *lymphomas.*

The abnormal replication of cells forms a growing mass of tissue known as a *tumor.* If the cells remain localized and do not invade surrounding tissues, the tumor is said to be a *benign tumor.* If, however, the tumor cells grow into the surrounding tissues, disrupting their functions, or spread to other regions of the body via the circulation, a process known as *metastasis,* the tumor is said to be a *malignant tumor* (used synonymously with *cancer*) and may lead to the death of the individual.

The transformation of a normal cell into a cancer cell is a multistep process that involves altering not only the mechanisms that regulate cell replication but also those that control the invasiveness of the cell and its ability to subvert the body's defense mechanisms. (As will be discussed in Chapter 20, the body's defense system is normally able to detect and destroy most cancerous cells when they first appear.) A cancer cell does not arise in its fully malignant form from a single mutation but progresses through various stages as a result of successive mutations. The incidence of cancer increases with age as a result of the accumulation of these mutations. Some of the early stages of transformation result in changes in the cell's morphology, known as *dysplasia,* a precancerous state that can be detected by microscopic examination. At this stage, the cell has not yet acquired a capacity for unlimited replication or an ability to invade surrounding tissues.

As mentioned earlier, a number of agents—termed mutagens—in the environment can damage DNA, increasing the mutation rate. Mutagens that increase the probability of a cancerous transformation in a cell are known as *carcinogens;* examples of carcinogens are the chemicals in tobacco smoke, radiation, certain microbes, and some synthetic chemicals in our food, water, and air. Some of these carcinogens act directly on DNA, while others are converted in the body into compounds that damage DNA. It is estimated that approximately 90 percent of all cancers require the participation of environmental factors, some of which have been added to the environment by our modern lifestyle.

A growing number of genes have been identified that contribute to the cancerous state when they mutate. These cancer-related genes fall into two classes: dominant and recessive. The dominant cancer-producing genes are called *oncogenes* (Greek, *onkos,* mass, tumor; the branch of medicine that deals with cancer is known as oncology). Oncogenes arise as mutations of normal genes known as *proto-oncogenes.* For example, some oncogenes code for abnormal forms of cell surface receptors that bind growth factors, producing a state in which the altered receptor produces a continuous growth signal in the absence of bound growth factor. The oncogenes are considered dominant since only one of the two homologous proto-oncogenes needs to be mutated for the mutation to contribute to the cancerous state.

The second class of genes involved in cancer are genes known as *tumor suppressor genes.* In their unmutated state, these genes code for proteins that inhibit various steps in cell replication. In the absence or malfunction of these proteins, cell replication cannot be inhibited by the normal signals that regulate growth. Mutation of one of the pair of alleles of tumor suppressor genes inactivates its function, but leaves a normal gene on the homologous chromosome that can still suppress tumor development. It is only when both alleles have been mutated that a cell may become cancerous. Thus, this type of cancer phenotype is recessive.

One of the most frequently encountered mutations in cancer cells is a tumor suppressor gene that codes for a phosphoprotein known as **p53** (because it has a molecular mass of 53,000 daltons). Normally, p53 functions as a transcription factor that stimulates transcription of a gene that codes for a protein that inhibits the cdc kinase required for progression of a cell from the G_1 to the S phase of the mitotic cycle. The concentration of p53 increases in cells that have suffered damage to their DNA and acts to prevent the replication of these damaged cells, including cells that have undergone cancerous mutations at other gene sites. Mutation of both homologous copies of p53 results in the loss of a cell's ability to inhibit the proliferation of damaged cells and thus provides one step in the progression to a fully malignant cancer cell. Cells carrying one copy of a mutated p53 are at increased risk of progressing to a cancerous state if the remaining normal gene becomes mutated.

Although most cancers are not directly inherited, the risk of developing cancer can be increased if, for example, one mutant p53 gene is inherited and is therefore present in all cells of the body. Because cells contain multiple control systems to regulate various stages of cell proliferation, disruption of one system, although it may produce a precancerous state, is not usually sufficient to form a fully malignant cell.

If a cancer is detected in the early stages of its growth, before it has metastasized, the tumor may be removed by surgery. Once it has metastasized to other organs, curative surgery is no longer possible. Drugs and radiation can be used to inhibit cell multiplication and destroy malignant cells, both before and after metastasis, although these treatments unfortunately also damage the growth of normal cells. Some cancer cells retain the ability to respond to normal growth signals, such as the growth of breast tissue in response to the hormone estrogen. Blocking the action of the hormones on hormone-dependent tumor cells can inhibit their growth. Chapter 20 describes therapies that utilize the weapons of the immune system. The development of more selective drugs and the mechanisms for targeting them to cancer cells is one of the benefits that may arise from the field of genetic engineering.

Genetic Engineering

Since the discovery of the structure of DNA in the early 1950s, techniques have been developed that enable scientists not only to determine the base sequence of a particular DNA molecule but to modify that sequence by the addition or deletion of specific bases, altering in a controlled manner the message encoded by the DNA. Through the use of these techniques, it may become possible to successfully replace mutated genes in specific cells with normal genes. We end this chapter with a discussion of some of the ways in which DNA can be studied and manipulated.

In order to manipulate a gene, it must first be identified among the many thousands of genes in the genome, isolated in sufficient quantities to allow a determination of its base sequence, and finally inserted back into a living cell. Several methods are available for performing each of these steps.

One of the key factors in solving each of these problems is a class of bacterial enzymes called *restriction nucleases,* which bind to specific sequences in DNA that are four to six nucleotides long and are called restriction sites. The enzyme cuts each of the two strands of DNA at these sites. Since there are numerous restriction sites located along a large molecule of DNA and a number of restriction nucleases with different binding site specificities, the use of multiple enzymes produces a number of small DNA fragments of varying lengths, some of which may contain the complete sequence of a gene while most contain only a fragment of a gene. This reduction in the size of the DNA fragments allows various procedures to be performed that cannot be carried out on the very large molecules of intact DNA.

One application of restriction nucleases is in a procedure known as ***DNA fingerprinting,*** which can be used in an attempt to identify a particular individual (for example, a person alleged to be involved in a crime) by analysis of blood or tissue fragments found at the scene of the crime. The DNA from these tissue samples is subjected to digestion by restriction nucleases, producing fragments of varying lengths. These fragments are then separated by a technique called gel electrophoresis, in which the fragments are placed at one end of a gel and subjected to an electrical current that causes the fragments to move along the gel at rates dependent on their electrical charge and size, separating the fragments into bands at different positions along the gel. Since no two individuals, with the exception of identical twins, have inherited the same combination of alleles and thus DNA sequences, different individuals produce different-sized restriction fragments. Comparing the pattern of the sample with the pattern from the tissue of a suspect can then be used to establish the probability that the two samples came from the same individual.

Restriction enzymes also provide a way to cut and paste genes between different DNA molecules. This results from the way in which restriction nucleases break the two strands of DNA. The two strands are broken at slightly different points (Figure 5–15) such that the end of one strand has a short, exposed sequence of

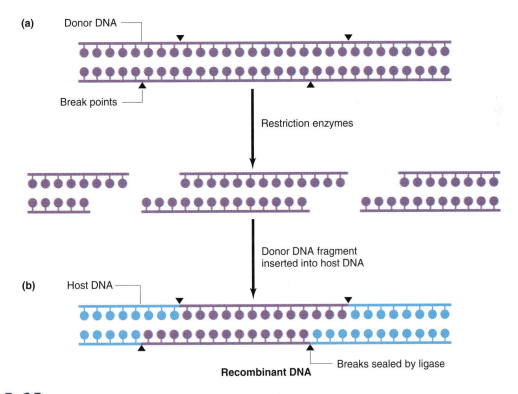

FIGURE 5–15

The basis of gene transfection. (a) Bacterial restriction enzymes break the two strands of DNA at different points, producing ends with exposed bases that are complementary to each other. (b) A segment of DNA containing one or more genes from one organism (donor) can be inserted into the DNA of another organism (host) by using the same restriction enzymes to produce complementary breaks in the host DNA, to which the donor DNA can bind.

bases that is complementary to the exposed strand on the other side of the break. This produces DNA fragments with "sticky" ends that can undergo base pairing. If a particular fragment that contains a gene or gene segment of interest can be identified and isolated, it can be inserted into another molecule of DNA, allowing the exposed ends of the fragment to hybridize with the exposed ends of the DNA that have been treated with the same nuclease. An enzyme known as a ligase can then be used to covalently link together the cut ends, resulting in the insertion of the DNA fragment into a second molecule of DNA. This technique can be used to insert DNA from one organism into the DNA of another, a procedure known as *transfection.* The organisms into which the DNA has been transfected are known as *transgenic organisms.*

A major problem occurs at the point where fragments of DNA must be introduced into a living cell because large molecules, such as DNA fragments, do not readily cross cell membranes. To overcome this problem, DNA fragments are inserted into the DNA of viruses that are able to enter host cells, carrying the modified DNA with them.

Replication of the transfected DNA inserted into bacteria produces additional copies of the DNA, or *cloned DNA,* each time the bacterium divides, that can be isolated in sufficient quantities to determine its sequence. Bacterial DNA, however, does not have introns, and so bacteria lack the spliceosomes required to delete intron-derived segments from DNA. Thus, bacteria are unable to use the transfected intron-containing DNA of eukaryotic organisms to form protein. DNA segments lacking introns, which are known as *cDNA,* or complementary DNA, can be derived from the isolated spliced mRNA that lacks introns. Using a viral enzyme called reverse transcriptase, the isolated mRNA can serve as a template for the synthesis of a complementary DNA strand. This cDNA can be transfected into bacteria that can then use it to form protein.

The transfection of a human gene, in its cDNA form, into bacteria can be used to produce large quantities of human proteins. For example, the gene for human insulin can be transfected into bacteria where it is transcribed into the protein insulin, which can then be isolated from the transfected bacteria and made available for the treatment of some forms of diabetes in which the patients are unable to synthesize insulin (to be discussed in Chapter 18).

Another genetic engineering procedure used in the study of DNA includes the formation of transgenic animals, primarily mice, in which a particular gene in the reproductive cells has been inactivated or deleted, forming a *knockout organism.* The effects of the absence of the gene's expression can be observed in the offspring of these mice, which provides insights into the normal function of the absent protein. Transgenic

techniques can also be used to form cells that overproduce a particular protein.

It is hoped that the techniques of genetic engineering will one day be able to selectively replace mutant genes in humans with normal genes and thus provide a cure for genetic diseases.

SUMMARY

Genetic Code

I. Genetic information is coded in the nucleotide sequences of DNA molecules. A single gene contains either (a) the information that, via mRNA, determines the amino acid sequence in a specific protein, or (b) the information for forming rRNA, tRNA, or small nuclear RNAs, which assists in protein assembly.

II. Genetic information is transferred from DNA to mRNA in the nucleus (transcription), and then mRNA passes to the cytoplasm, where its information is used to synthesize protein (translation).

III. The words in the DNA genetic code consist of a sequence of three nucleotide bases that specify a single amino acid. The sequence of three-letter code words along a gene determines the sequence of amino acids in a protein. More than one code word can specify a given amino acid.

Protein Synthesis, Degradation, and Secretion

I. Table 5–1 summarizes the steps leading from DNA to protein synthesis.

II. Transcription involves the formation of a primary RNA transcript by base-pairing with the template strand of DNA containing a single gene and the removal of intron-derived segments by spliceosomes to form mRNA, which moves to the cytoplasm.

III. Translation of mRNA occurs on the ribosomes in the cytoplasm by base-pairing between the anticodons in tRNAs linked to single amino acids, with the corresponding codons in mRNA.

IV. Chaperones help fold large proteins into their proper conformation as they are released from ribosomes.

V. Protein transcription factors activate or repress the transcription of specific genes by binding to regions of DNA that interact with the promoter region of a gene.

VI. The concentration of a particular protein in a cell depends on: (1) the rate of its gene's transcription, (2) the rate of initiating protein assembly on a ribosome, (3) the rate at which mRNA is degraded, (4) the rate of protein digestion by enzymes associated with proteosomes, and (5) the rate of secretion, if any, of the protein from the cell.

VII. Proteins secreted by cells pass through the sequence of steps illustrated in Figure 5–10. Targeting of a protein to the secretory pathway depends on the signal sequence of amino acids that first emerge from a ribosome during synthesis.

Replication and Expression of Genetic Information

I. Human cells contain 46 chromosomes, consisting of 44 autosomes and 2 sex chromosomes. At conception, 22 homologous chromosomes and one sex chromosome are supplied to the fertilized egg by each parent.

II. When a cell divides, the DNA molecule in each of the 46 chromosomes is replicated, one copy passing to each daughter cell, so that both receive the same complete set of genetic instructions.

III. DNA replication involves base-pairing of the exposed bases in each of the two unwound strands of DNA with free deoxyribonucleotide triphosphate bases. DNA polymerase joins the nucleotides together, forming two molecules of DNA, one from each of the original DNA strands.

IV. Telomeres are added to the ends of replicating DNA in some cells. In the absence of telomeres, the length of DNA decreases with each replication.

V. Proofreading mechanisms help prevent the introduction of errors during DNA replication.

VI. Cell division, consisting of nuclear division (mitosis) and cytoplasmic division (cytokinesis) lasts about 1 h. The period between divisions, known as interphase, is divided into 3 phases—G_1, S, and G_2.

VII. During the S phase of interphase, DNA replicates, forming two identical sister chromatids joined by a centromere.

VIII. In mitosis:

   a. The chromatin condenses into highly coiled chromosomes.

   b. The centromeres of each chromosome become attached to spindle fibers extending from the centrioles, which have migrated to opposite poles of the nucleus.

   c. The two chromatids of each chromosome separate and move toward opposite poles of the cell as the cell divides into two daughter cells. Following cell division, the condensed chromatids uncoil into their extended interphase form.

IX. Entry into the S and M phases of the cell cycle is controlled by cell division cycle kinases. These enzymes are activated by a rising concentration of cyclin proteins, which are then rapidly destroyed as the cell passes through each of these checkpoints.

X. Extracellular growth factors act on cells to produce intracellular signals that regulate the rate of cell proliferation.

XI. Mutagens alter DNA molecules, resulting in the addition or deletion of nucleotides or segments of DNA. The result is an altered DNA sequence known as a mutation.

   a. A mutation may (1) cause no noticeable change in cell function, (2) modify cell function but still be compatible with cell growth and replication, or (3) lead to the death of the cell.

   b. Mutations occurring in egg or sperm cells are passed on to all the cells of a new individual and possibly to some individuals in future generations.

XII. DNA repair mechanisms are important in preventing the accumulation of mutations, particularly in long-lived cells that do not divide.

XIII. A number of different forms of each gene, called alleles, exist in the population. A homozygous individual has two identical alleles for a particular gene, while a heterozygous individual has two different alleles of the gene.

XIV. The phenotypic traits produced by genes are described as dominant if only one copy of the two inherited alleles is sufficient to produce the trait and are recessive if the same allele must be inherited from both parents for the trait to be expressed.

XV. Genetic diseases are the result of inherited mutated genes. They can result from single gene mutations, for example, losses or additions of chromosomal segments, or they can be polygenic, when more than one mutated gene is required for the disease to be expressed.

Cancer

I. Cancer cells are characterized by their capacity for unlimited multiplication and their ability to metastasize to other parts of the body, forming multiple tumor sites.

   a. Mutations in proto-oncogenes and tumor suppressor genes can lead to cancer. In their unmutated state, these genes code for proteins that function at various stages in the control systems that regulate cell replication.

   b. More than one mutation is necessary to cause the transformation of a normal cell into a cancer cell.

Genetic Engineering

I. With the use of bacterial restriction nucleases, segments of DNA can be cut from the DNA of one cell and inserted into the DNA of another cell—transfection—forming a transgenic organism.

II. Transfection of human genes into bacteria provides a mechanism for producing large quantities of the expressed protein, which can be isolated and used to treat disease (for example, the production of insulin).

III. Analysis of the pattern of tissue DNA fragments formed by nuclease digestion is the basis of DNA fingerprinting used to identify a specific individual.

IV. Experimental techniques that lead to the selective removal or inactivation of a specific gene produce a knockout organism that can be used to study the functional consequences of the loss of the gene's activity.

KEY TERMS	
gene	codon
genome	primary RNA transcript
chromosome	exon
histone	intron
nucleosome	spliceosome
transcription	ribosomal RNA (rRNA)
translation	transfer RNA (tRNA)
"stop" signal	anticodon
messenger RNA (mRNA)	initiation factor
RNA polymerase	chaperone
template strand	transcription factor
promoter	preinitiation complex

ubiquitin
proteosome
signal sequence
autosome
sex chromosome
daughter cells
DNA polymerase
telomere
telomerase
proofreading
interphase
mitosis
cytokinesis
sister chromatid
centromere
spindle fiber

centrosome
mitotic apparatus
cell division cycle kinase
cyclin
growth factor
natural selection
allele
gene pool
homozygous
heterozygous
genotype
phenotype
dominant
recessive
p53

25. Describe the mechanism of DNA repair.
26. State the difference between a homozygote and a heterozygote in terms of alleles.
27. Describe the difference between a phenotype that is inherited as a dominant or recessive trait.
28. Describe the characteristics of a cancer cell.
29. Describe the difference between a benign and a malignant tumor.
30. Describe the difference between an oncogene and a tumor suppressor gene.
31. Describe the properties of bacterial restriction nucleases and their role in gene transfection.
32. Describe the process of DNA fingerprinting.
33. How does the base sequence in a cDNA molecule differ from the base sequence in the gene from which it is derived?

REVIEW QUESTIONS

1. Summarize the direction of information flow during protein synthesis.
2. Describe how the genetic code in DNA specifies the amino acid sequence in a protein.
3. List the four nucleotides found in mRNA.
4. Describe the main events in the transcription of genetic information in DNA into mRNA.
5. State the difference between an exon and an intron.
6. What is the function of a spliceosome?
7. Identify the site of ribosomal subunit assembly.
8. Describe the role of tRNA in protein assembly.
9. Describe the events of protein translation that occur on the surface of a ribosome.
10. What is the function of a chaperone?
11. Describe the effects of transcription factors on gene transcription.
12. List the factors that regulate the concentration of a protein in a cell.
13. What is the function of the signal sequence of a protein? How is it formed, and where is it located?
14. Describe the pathway that leads to the secretion of proteins from cells.
15. Describe the functions of the Golgi apparatus.
16. Describe the structure of chromatin, and state the number and types of chromosomes found in a human cell.
17. Describe the mechanism by which DNA is replicated.
18. What is a telomere, and what is its function?
19. Summarize the main events of mitosis and cytokinesis.
20. Describe the role of cell division cycle kinases and cyclins in controlling cell division.
21. Describe the function of growth factors.
22. Describe several ways in which the genetic message can be altered by mutation.
23. How will the deletion of a single base in a gene affect the protein synthesized?
24. List the three general types of effects that a mutation can have on a cell's function.

CLINICAL TERMS

mutation
mutagen
familial hypercho-
  lesterolemia
cystic fibrosis
single gene disease
chromosomal disease
polygenic disease
Down's syndrome
  (trisomy 21)
cancer cell
carcinoma
sarcoma
leukemia
lymphoma
tumor

benign tumor
metastasis
malignant tumor
dysplasia
carcinogen
oncogene
proto-oncogene
tumor suppressor gene
restriction nuclease
DNA fingerprinting
transfection
transgenic organism
cloned DNA
cDNA
knockout organism

THOUGHT QUESTIONS

(Answers are given in Appendix A)

1. A base sequence in a portion of one strand of DNA is A–G–T–G–C–A–A–G–T–C–T. Predict:
   a. the base sequence in the complementary strand of DNA.
   b. The base sequence in RNA transcribed from the sequence shown.
2. The triplet code in DNA for the amino acid histidine is G–T–A. Predict the mRNA codon for this amino acid and the tRNA anticodon.
3. If a protein contains 100 amino acids, how many nucleotides will be present in the gene that codes for this protein?
4. Why do chemical agents that inhibit the polymerization of tubulin (Chapter 3) inhibit cell division?
5. Why are drugs that inhibit the replication of DNA potentially useful in the treatment of cancer? What are some of the limitations of such drugs?

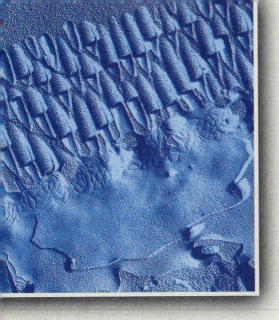

Movement of Molecules Across Cell Membranes

As we saw in Chapter 3, the contents of a cell are separated from the surrounding extracellular fluid by a thin layer of lipids and protein—the **plasma membrane.** In addition, membranes associated with mitochondria, endoplasmic reticulum, lysosomes, the Golgi apparatus, and the nucleus divide the intracellular fluid into several membrane-bound compartments. The movements of molecules and ions both between the various cell organelles and the cytosol, and between the cytosol and the extracellular fluid, depend on

the properties of these membranes. The rates at which different substances move through membranes vary considerably and in some cases can be controlled—increased or decreased—in response to various signals. This chapter focuses upon the transport functions of membranes, with emphasis on the plasma membrane. There are several mechanisms by which substances pass through membranes, and we begin our discussion of these mechanisms with the physical process known as diffusion.

Diffusion

The molecules of any substance, be it solid, liquid, or gas, are in a continuous state of movement or vibration, and the warmer a substance is, the faster its molecules move. The average speed of this "thermal motion" also depends upon the mass of the molecule. At body temperature, a molecule of water moves at about 2500 km/h (1500 mi/h), whereas a molecule of glucose, which is 10 times heavier, moves at about 850 km/h. In solutions, such rapidly moving molecules cannot travel very far before colliding with other molecules. They bounce off each other like rubber balls, undergoing millions of collisions every second. Each collision alters the direction of the molecule's movement, and the path of any one molecule becomes unpredictable. Since a molecule may at any instant be moving in any direction, such movement is said to be "random," meaning that it has no preferred direction of movement.

The random thermal motion of molecules in a liquid or gas will eventually distribute them uniformly throughout the container. Thus, if we start with a solution in which a solute is more concentrated in one region than another (Figure 6–1a), random thermal motion will redistribute the solute from regions of higher concentration to regions of lower concentration until the solute reaches a uniform concentration throughout the solution (Figure 6–1b). This movement of molecules from one location to another solely as a result of their random thermal motion is known as **diffusion.**

Many processes in living organisms are closely associated with diffusion. For example, oxygen, nutrients, and other molecules enter and leave the smallest blood vessels (capillaries) by diffusion, and the movement of many substances across plasma membranes and organelle membranes occurs by diffusion.

(a)

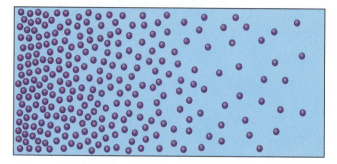

(b)

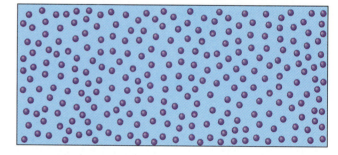

FIGURE 6–1

Molecules initially concentrated in one region of a solution (a) will, due to their random thermal motion, undergo a net diffusion from the region of higher to the region of lower concentration until they become uniformly distributed throughout the solution (b).

Magnitude and Direction of Diffusion

The diffusion of glucose between two compartments of equal volume separated by a permeable barrier is illustrated in Figure 6–2. Initially glucose is present in compartment 1 at a concentration of 20 mmol/L, and there is no glucose in compartment 2. The random movements of the glucose molecules in compartment

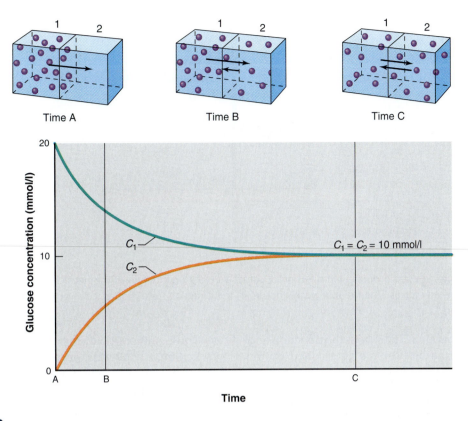

FIGURE 6–2

Diffusion of glucose between two compartments of equal volume separated by a barrier permeable to glucose. Initially, time A, compartment 1 contains glucose at a concentration of 20 mmol/L, and no glucose is present in compartment 2. At time B, some glucose molecules have moved into compartment 2, and some of these are moving back into compartment 1. The length of the arrows represents the magnitudes of the one-way movements. At time C, diffusion equilibrium has been reached, the concentrations of glucose are equal in the two compartments (10 mmol/l), and the net movement is zero.

In the graph at the bottom of the figure, the blue line represents the concentration in compartment 1 (C_1), and the orange line represents the concentration in compartment 2 (C_2).

1 carry some of them into compartment 2. The amount of material crossing a surface in a unit of time is known as a **flux.** This one-way flux of glucose from compartment 1 to compartment 2 depends on the concentration of glucose in compartment 1. If the number of molecules in a unit of volume is doubled, the flux of molecules across each surface of the unit will also be doubled, since twice as many molecules will be moving in any direction at a given time.

After a short time, some of the glucose molecules that have entered compartment 2 will randomly move back into compartment 1 (Figure 6–2, time B). The magnitude of the glucose flux from compartment 2 to compartment 1 depends upon the concentration of glucose in compartment 2 at any time.

The **net flux** of glucose between the two compartments at any instant is the difference between the two one-way fluxes. It is the net flux that determines the net gain of molecules by compartment 2 and the net loss from compartment 1.

Eventually the concentrations of glucose in the two compartments become equal at 10 mmol/L. The two

one-way fluxes are then equal in magnitude but opposite in direction, and the net flux of glucose is zero (Figure 6–2, time C). The system has now reached **diffusion equilibrium.** No further change in the glucose concentration of the two compartments will occur, since equal numbers of glucose molecules will continue to diffuse in both directions between the two compartments.

Several important properties of diffusion can be reemphasized using this example. Three fluxes can be identified at any surface—the two one-way fluxes occurring in opposite directions from one compartment to the other, and the net flux, which is the difference between them (Figure 6–3). The net flux is the most important component in diffusion since it is the net amount of material transferred from one location to another. Although the movement of individual molecules is random, *the net flux always proceeds from regions of higher concentration to regions of lower concentration.* For this reason, we often say that substances move "downhill" by diffusion. The greater the difference in concentration between any two regions, the greater the

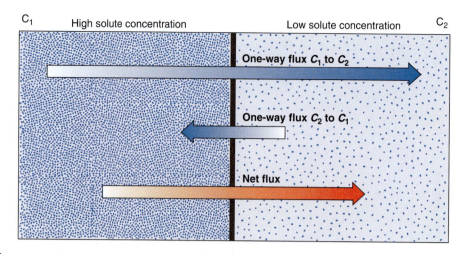

C_1 High solute concentration   Low solute concentration C_2

One-way flux C_1 to C_2

One-way flux C_2 to C_1

Net flux

FIGURE 6–3

The two one-way fluxes occurring during the diffusion of solute across a boundary and the net flux, which is the difference between the two one-way fluxes. The net flux always occurs in the direction from higher to lower concentration.

magnitude of the net flux. Thus, both the direction and the magnitude of the net flux are determined by the concentration difference.

At any concentration difference, however, the magnitude of the net flux depends on several additional factors: (1) temperature—the higher the temperature, the greater the speed of molecular movement and the greater the net flux; (2) mass of the molecule—large molecules (for example, proteins) have a greater mass and lower speed than smaller molecules (for example, glucose) and thus have a smaller net flux; (3) surface area—the greater the surface area between two regions, the greater the space available for diffusion and thus the greater the net flux; and (4) medium through which the molecules are moving—molecules diffuse more rapidly in air than in water because collisions are less frequent in a gas phase, and as we shall see, when a membrane is involved, its chemical composition influences diffusion rates.

Diffusion Rate versus Distance

The distance over which molecules diffuse is an important factor in determining the rate at which they can reach a cell from the blood or move throughout the interior of a cell after crossing the plasma membrane. Although individual molecules travel at high speeds, the number of collisions they undergo prevents them from traveling very far in a straight line. Diffusion times increase in proportion to the *square* of the distance over which the molecules diffuse. It is for this reason, for example, that it takes glucose approximately 3.5 s to reach 90 percent of diffusion equilibrium at a point 10 μm away from a source of glucose, such as the blood, but it would take over 11 years to reach the same concentration at a point 10 cm away from the source.

Thus, although diffusion equilibrium can be reached rapidly over distances of cellular dimensions, it takes a very long time when distances of a few centimeters or more are involved. For an organism as large as a human being, the diffusion of oxygen and nutrients from the body surface to tissues located only a few centimeters below the surface would be far too slow to provide adequate nourishment. Accordingly, the circulatory system provides the mechanism for rapidly moving materials over large distances (by blood flow using a mechanical pump, the heart), with diffusion providing movement over the short distance between the blood and tissue cells.

The rate at which diffusion is able to move molecules *within* a cell is one of the reasons that cells must be small. A cell would not have to be very large before diffusion failed to provide sufficient nutrients to its central regions. For example, the center of a 20-μm diameter cell reaches diffusion equilibrium with extracellular oxygen in about 15 ms, but it would take 265 days to reach equilibrium at the center of a cell the size of a basketball.

Diffusion through Membranes

The rate at which a substance diffuses across a plasma membrane can be measured by monitoring the rate at which its intracellular concentration approaches diffusion equilibrium with its concentration in the extracellular fluid. Let us assume that since the volume of extracellular fluid is large, its solute concentration will remain essentially constant as the substance diffuses into the small intracellular volume (Figure 6–4). As with all diffusion processes, the net flux F of material across the membrane is from the region of higher concentration (the extracellular solution in this case) to the region of lower concentration (the intracellular fluid).

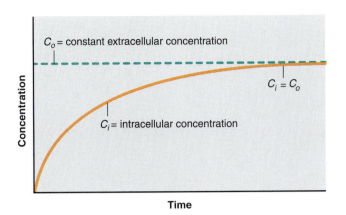

FIGURE 6–4

The increase in intracellular concentration as a substance diffuses from a constant extracellular concentration until diffusion equilibrium ($C_i = C_o$) is reached across the plasma membrane of a cell.

The magnitude of the net flux is directly proportional to the difference in concentration across the membrane ($C_o - C_i$), the surface area of the membrane A, and the membrane **permeability constant k_p:**

$$F = k_p A(C_o - C_i)$$

The numerical value of the permeability constant k_p is an experimentally determined number for a particular type of molecule at a given temperature, and it reflects the ease with which the molecule is able to move through a given membrane. In other words, the greater the permeability constant, the larger the net flux across the membrane for any given concentration difference and membrane surface area.

The rates at which molecules diffuse across membranes, as measured by their permeability constants, are a thousand to a million times smaller than the diffusion rates of the same molecules through a water layer of equal thickness. In other words, a membrane acts as a barrier that considerably slows the diffusion of molecules across its surface. The major factor limiting diffusion across a membrane is its lipid bilayer.

Diffusion through the Lipid Bilayer When the permeability constants of different organic molecules are examined in relation to their molecular structures, a correlation emerges. Whereas most polar molecules diffuse into cells very slowly or not at all, nonpolar molecules diffuse much more rapidly across plasma membranes—that is, they have large permeability constants. The reason is that nonpolar molecules can dissolve in the nonpolar regions of the membrane—regions occupied by the fatty acid chains of the membrane phospholipids. In contrast, polar molecules

have a much lower solubility in the membrane lipids. Increasing the lipid solubility of a substance (decreasing the number of polar or ionized groups it contains) will increase the number of molecules dissolved in the membrane lipids and thus increase its flux across the membrane. Oxygen, carbon dioxide, fatty acids, and steroid hormones are examples of nonpolar molecules that diffuse rapidly through the lipid portions of membranes. Most of the organic molecules that make up the intermediate stages of the various metabolic pathways (Chapter 4) are ionized or polar molecules, often containing an ionized phosphate group, and thus have a low solubility in the lipid bilayer. Most of these substances are retained within cells and organelles because they cannot diffuse across the lipid barrier of membranes.

Diffusion of Ions through Protein Channels Ions such as Na^+, K^+, Cl^-, and Ca^{2+} diffuse across plasma membranes at rates that are much faster than would be predicted from their very low solubility in membrane lipids. Moreover, different cells have quite different permeabilities to these ions, whereas nonpolar substances have similar permeabilities when different cells are compared. The fact that artificial lipid bilayers containing no protein are practically impermeable to these ions indicates that it is the protein component of the membrane that is responsible for these permeability differences.

As we have seen (Chapter 3), integral membrane proteins can span the lipid bilayer. Some of these proteins form **channels** through which ions can diffuse across the membrane. A single protein may have a conformation similar to that of a doughnut, with the hole in the middle providing the channel for ion movement. More often, several proteins aggregate, each forming a subunit of the walls of a channel (Figure 6–5). The diameters of protein channels are very small, only slightly larger than those of the ions that pass through them. The small size of the channels prevents larger, polar, organic molecules from entering the channel.

Ion channels show a selectivity for the type of ion that can pass through them. This selectivity is based partially on the channel diameter and partially on the charged and polar surfaces of the protein subunits that form the channel walls and electrically attract or repel the ions. For example, some channels (K channels) allow only potassium ions to pass, others are specific for sodium (Na channels), and still others allow both sodium and potassium ions to pass (Na,K channels). For this reason, two membranes that have the same permeability to potassium because they have the same number of K channels may have quite different permeabilities to sodium because they contain different numbers of Na channels.

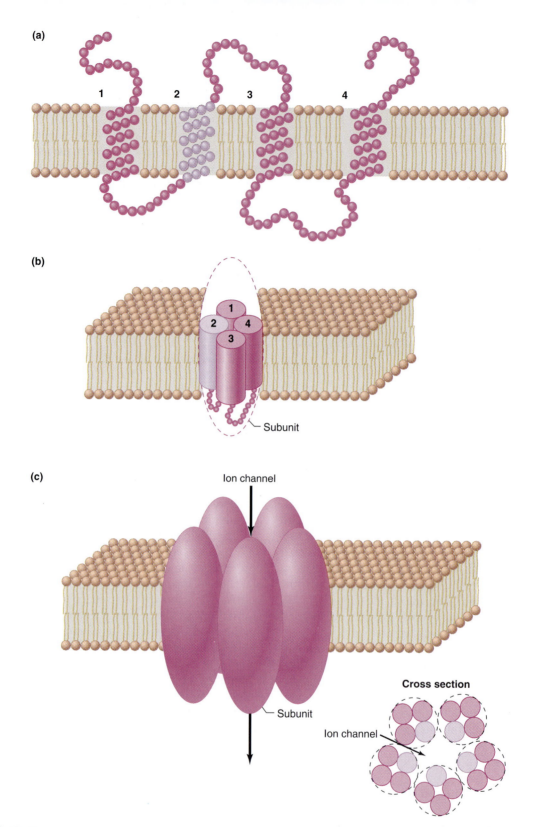

FIGURE 6–5

Model of an ion channel composed of five polypeptide subunits. (a) A channel subunit consisting of an integral membrane protein containing four transmembrane segments (1, 2, 3, and 4), each of which has an alpha helical configuration within the membrane. Although this model has only four transmembrane segments, some channel proteins have as many as 12. (b) The same subunit as in (a) shown in three dimensions within the membrane with the four transmembrane helices aggregated together. (c) The ion channel consists of five of the subunits illustrated in b, which form the sides of the channel. As shown in cross section, the helical transmembrane segment (a,2) (light purple) of each subunit forms the sides of the channel opening. The presence of ionized amino acid side chains along this region determines the selectivity of the channel to ions. Although this model shows the five subunits as being identical, many ion channels are formed from the aggregation of several different types of subunit polypeptides.

Role of Electric Forces on Ion Movement Thus far we have described the direction and magnitude of solute diffusion across a membrane in terms of the solute's concentration difference across the membrane, its solubility in the membrane lipids, the presence of membrane ion channels, and the area of the membrane. When describing the diffusion of ions, since they are charged, one additional factor must be considered: the presence of electric forces acting upon the ions.

There exists a separation of electric charge across plasma membranes, known as a **membrane potential** (Figure 6–6), the origin of which will be described in Chapter 8. The membrane potential provides an electric force that influences the movement of ions across the membrane. Electric charges of the same sign, both positive or both negative, repel each other, while opposite charges attract. For example, if the inside of a cell has a net negative charge with respect to the outside, as it does in most cells, there will be an electric force attracting positive ions into the cell and repelling negative ions. Even if there were no difference in ion concentration across the membrane, there would still be a net movement of positive ions into and negative ions out of the cell because of the membrane potential. Thus, the direction and magnitude of *ion* fluxes across membranes depend on both the concentration difference and the electrical difference (the membrane potential). These two driving forces are collectively known as the **electrochemical gradient,** also termed the electrochemical difference across a membrane.

It is important to recognize that the two forces that make up the electrochemical gradient may oppose each other. Thus, the *membrane potential* may be driving potassium ions, for example, in one direction across the membrane, while the *concentration difference* for potassium is driving these ions in the opposite direction. The net movement of potassium in this case would be determined by the magnitudes of the two opposing forces—that is, by the electrochemical gradient across the membrane.

Regulation of Diffusion through Ion Channels Ion channels can exist in an open or closed state (Figure 6–7), and changes in a membrane's permeability to ions can occur rapidly as a result of the opening or closing of these channels. The process of opening and closing ion channels is known as **channel gating,** like the opening and closing of a gate in a fence. A single ion channel may open and close many times each second, suggesting that the channel protein fluctuates between two (or more) conformations. Over an extended period of time, at any given electrochemical gradient, the total number of ions that pass through a channel depends on how frequently the channel opens and how long it stays open.

In the 1980s, a technique was developed to allow investigators to monitor the properties of single ion channels. The technique, known as **patch clamping,** involves placing the tip of a glass pipette on a small region of a cell's surface and applying a slight suction so that the membrane patch becomes sealed to the edges of the pipette and remains attached when the pipette is withdrawn. Since ions carry an electric charge, the flow of ions through an ion channel in the membrane patch produces an electric current that can be monitored. Investigators found that the current flow was intermittent, corresponding to the opening and closing of the ion channel, and that the current magnitude was a measure of the channel permeability. By adding possible inhibitors or stimulants to the solution in the pipette (or to the bath fluid, which is now in contact with the intracellular surface of the membrane patch), one can analyze the effects of these agents in modifying the frequency and duration of channel opening. Patch clamping thus allows investigators to follow the behavior of a single channel over time.

Three factors can alter the channel protein conformations, producing changes in the opening frequency or duration: (1) As described in Chapter 7, the binding of specific molecules to channel proteins may directly or indirectly produce either an allosteric or covalent change in the shape of the channel protein; such channels are termed **ligand-sensitive channels,** and the ligands that influence them are often chemical messengers. (2) Changes in the membrane potential can cause movement of the charged regions on a channel protein, altering its shape—**voltage-gated channels**

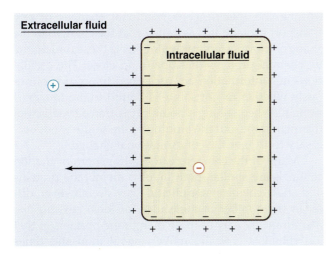

FIGURE 6–6

The separation of electric charge across a plasma membrane (the membrane potential) provides the electric force that drives positive ions into a cell and negative ions out.

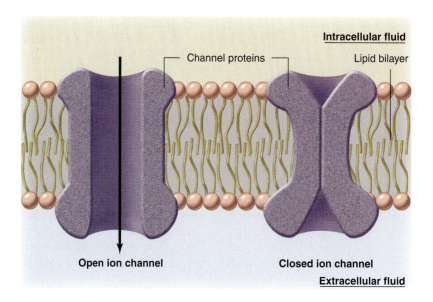

Intracellular fluid

Channel proteins

Lipid bilayer

Open ion channel

Closed ion channel

Extracellular fluid

FIGURE 6–7

As a result of conformational changes in the proteins forming an ion channel, the channel may be open, allowing ions to diffuse across the membrane, or may be closed.

(voltage-sensitive channels). (3) Stretching the membrane may affect the conformation of some channel proteins—**mechanosensitive channels.** A single channel may be affected by more than one of these factors.

A particular type of ion may pass through several different types of channels. For example, a membrane may contain ligand-sensitive potassium channels (K channels), voltage-sensitive K channels, and mechanosensitive K channels. Moreover, the same membrane may have several types of voltage-sensitive K channels, each responding to a different range of membrane voltage, or several types of ligand-sensitive K channels, each responding to a different chemical messenger. The roles of these gated channels in cell communication and electrical activity will be discussed in Chapters 7 through 9.

Mediated-Transport Systems

Although diffusion through channels accounts for some of the transmembrane movement of ions, it does not account for all. Moreover, there are a number of other molecules, including amino acids and glucose, that are able to cross membranes yet are too polar to diffuse through the lipid bilayer and too large to diffuse through ion channels. The passage of these molecules and the nondiffusional movements of ions are mediated by integral membrane proteins known as **transporters** (or carriers). Movement of substances through a membrane by these **mediated-transport** systems depends on conformational changes in these transporters.

The transported solute must first bind to a specific site on a transporter (Figure 6–8), a site that is exposed to the solute on one surface of the membrane. A portion of the transporter then undergoes a change in shape, exposing this same binding site to the solution on the opposite side of the membrane. The dissociation of the substance from the transporter binding site completes the process of moving the material through the membrane. Using this mechanism, molecules can move in either direction, getting on the transporter on one side and off at the other.

The diagram of the transporter in Figure 6–8 is only a model, since we have little information concerning the specific conformational changes of any transport protein. It is assumed that the changes in the shape of transporters are analogous to those undergone by channel proteins that open and close. The oscillations in conformation are presumed to occur continuously whether or not solute is bound to the transport protein. When solute is bound, it is transferred across the membrane, but the binding of the solute is not necessary to trigger the conformational change.

Many of the characteristics of transporters and ion channels are similar. Both involve membrane proteins and show chemical specificity. They do, however, differ in the number of molecules (or ions) crossing the membrane by way of these membrane proteins in that ion channels typically move several thousand times more ions per unit time than do transporters. In part, this reflects the fact that for each molecule transported across the membrane, a transporter must change its shape, while an open ion channel can support a continuous flow of ions without a change in conformation.

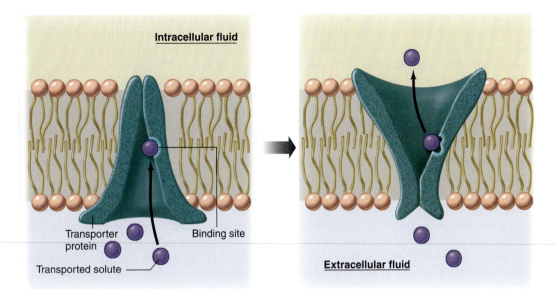

Intracellular fluid

Transporter protein — Binding site

Transported solute

Extracellular fluid

FIGURE 6–8

Model of mediated transport. A change in the conformation of the transporter exposes the transporter binding site first to one surface of the membrane then to the other, thereby transferring the bound solute from one side of the membrane to the other. This model shows net mediated transport from the extracellular fluid to the inside of the cell. In many cases, the net transport is in the opposite direction.

There are many types of transporters in membranes, each type having binding sites that are specific for a particular substance or a specific class of related substances. For example, although both amino acids and sugars undergo mediated transport, a protein that transports amino acids does not transport sugars, and vice versa. Just as with ion channels, the plasma membranes of different cells contain different types and numbers of transporters and thus exhibit differences in the types of substances transported and their rates of transport.

Three factors determine the magnitude of the solute flux through a mediated-transport system: (1) the extent to which the transporter binding sites are saturated, which depends on both the solute concentration and the affinity of the transporters for the solute, (2) the number of transporters in the membrane—the greater the number of transporters, the greater the flux at any level of saturation, and (3) the rate at which the conformational change in the transport protein occurs. The flux through a mediated-transport system can be altered by changing any of these three factors.

For any transported solute there is a finite number of specific transporters in a given membrane at any particular moment. As with any binding site, as the concentration of the ligand (the solute to be transported, in this case) is increased, the number of occupied binding sites increases until the transporters become saturated—that is, until all the binding sites become occupied. When the transporter binding sites are saturated, the maximal flux across the membrane

has been reached, and no further increase in solute flux will occur with increases in solute concentration. Contrast the solute flux resulting from mediated transport with the flux produced by diffusion through the lipid portion of a membrane (Figure 6–9). The flux due to diffusion increases in direct proportion to the increase in extracellular concentration, and there is no limit

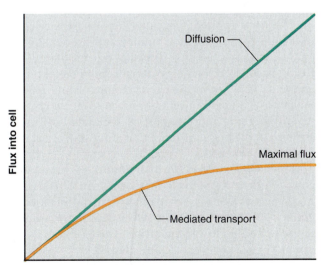

Extracellular solute concentration

FIGURE 6–9

The flux of molecules diffusing into a cell across the lipid bilayer of a plasma membrane (blue line) increases continuously in proportion to the extracellular concentration, whereas the flux of molecules through a mediated-transport system (orange line) reaches a maximal value.

since diffusion does not involve binding to a fixed number of sites. (At very high ion concentrations, however, diffusion through ion channels may approach a limiting value because of the fixed number of channels available just as there is an upper limit to the rate at which a crowd of people can pass through a single open doorway.)

When the transporters are saturated, the maximal transport flux depends upon the rate at which the conformational changes in the transporters can transfer their binding sites from one surface to the other. This rate is much slower than the rate of ion diffusion through ion channels.

Thus far we have described mediated transport as though all transporters had similar properties. In fact, two types of mediated transport can be distinguished—**facilitated diffusion** and **active transport.** Facilitated diffusion uses a transporter to move solute *downhill* from a higher to a lower concentration across a membrane (as in Figure 6–8), whereas active transport uses a transporter that is coupled to an energy source to move solute *uphill* across a membrane—that is, *against* its electrochemical gradient.

Facilitated Diffusion

"Facilitated diffusion" is an unfortunate term since the process it denotes does not involve diffusion. The term arose because the end results of both diffusion and facilitated diffusion are the same. In both processes, the net flux of an uncharged molecule across a membrane always proceeds from higher to lower concentration and continues until the concentrations on the two sides of the membrane become equal. At this point in facilitated diffusion, equal numbers of molecules are binding to the transporter at the outer surface of the cell and moving into the cell as are binding at the inner surface and moving out. Neither diffusion nor facilitated diffusion is coupled to energy derived from metabolism, and thus they are incapable of moving solute from a lower to a higher concentration across a membrane.

Among the most important facilitated-diffusion systems in the body are those that move glucose across plasma membranes. Without such glucose transporters, cells would be virtually impermeable to glucose, which is a relatively large, polar molecule. One might expect that as a result of facilitated diffusion the glucose concentration inside cells would become equal to the extracellular concentration. This does not occur in most cells, however, because glucose is metabolized to glucose 6-phosphate almost as quickly as it enters. Thus, the intracellular glucose concentration remains lower than the extracellular concentration, and there is a continuous net flux of glucose into cells.

Several distinct transporters are known to mediate the facilitated diffusion of glucose across cell membranes. Each transporter is coded by a different gene, and these genes are expressed in different types of

cells. The transporters differ in the affinity of their binding sites for glucose, their maximal rates of transport when saturated, and the modulation of their transport activity by various chemical signals, such as the hormone insulin. As discussed in Chapter 18, although glucose enters all cells by means of glucose transporters, insulin affects only the type of glucose transporter expressed primarily in muscle and adipose tissue. Insulin increases the number of these glucose transporters in the membrane and, hence, the rate of glucose movement into cells.

Active Transport

Active transport differs from facilitated diffusion in that it uses energy to move a substance *uphill* across a membrane—that is, against the substance's electrochemical gradient (Figure 6–10). As with facilitated diffusion, active transport requires binding of a substance to the transporter in the membrane. Because these transporters move the substance *uphill,* they are often referred to as "pumps." As with facilitated-diffusion transporters, active-transport transporters exhibit specificity and saturation—that is, the flux via the transporter is maximal when all transporter binding sites are saturated.

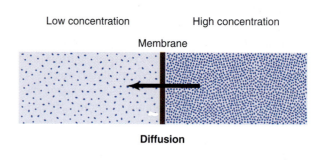

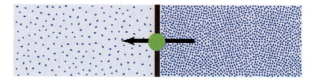

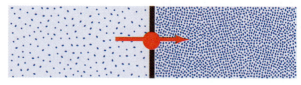

FIGURE 6–10

Direction of net solute flux crossing a membrane by: (1) diffusion (high to low concentration), (2) facilitated diffusion (high to low concentration), and active transport (low to high concentration).

The net movement from lower to higher concentration and the maintenance of a higher steady-state concentration on one side of a membrane can be achieved only by the continuous input of energy into the active-transport process. This energy can (1) alter the affinity of the binding site on the transporter such that it has a higher affinity when facing one side of the membrane than when facing the other side; or (2) alter the rates at which the binding site on the transporter is shifted from one surface to the other.

To repeat, in order to move molecules from a lower concentration (lower energy state) to a higher concentration (higher energy state), energy must be added. Therefore, active transport must be coupled to the simultaneous flow of some energy source from a higher energy level to a lower energy level. Two means of coupling an energy flow to transporters are known: (1) the direct use of ATP in **primary active transport,** and (2) the use of an ion concentration difference across a membrane to drive the process in **secondary active transport.**

Primary Active Transport　The hydrolysis of ATP by a transporter provides the energy for primary active transport. The transporter is an enzyme (an ATPase) that catalyzes the breakdown of ATP and, in the process, phosphorylates itself. Phosphorylation of the transporter protein (covalent modulation) changes the affinity of the transporter's solute binding site. Figure 6–11 illustrates the sequence of events leading to the active transport (that is, transport from low to higher concentration) of a solute into a cell. (1) Initially, the binding site for the transported solute is exposed to

the extracellular fluid and has a high affinity because the protein has been phosphorylated on its intracellular surface by ATP. This phosphorylation occurs only when the transporter is in the conformation shown on the left side of the figure. (2) The transported solute in the extracellular fluid binds to the high-affinity binding site. Random thermal oscillations repeatedly expose the binding site to one side of the membrane, then to the other, independent of the protein's phosphorylation. (3) Removal of the phosphate group from the transporter decreases the affinity of the binding site, leading to (4) the release of the transported solute into the intracellular fluid. When the low-affinity site is returned to the extracellular face of the membrane by the random oscillation of the transporter (5), it is in a conformation which again permits phosphorylation, and the cycle can be repeated.

To see why this will lead to movement from low to higher concentration (that is, uphill movement), consider the flow of solute through the transporter at a point in time when the concentration is equal on the two sides of the membrane. More solute will be bound to the high-affinity site at the extracellular surface of the membrane than to the low-affinity site on the intracellular surface. Thus more solute will move in than out when the transporter oscillates between sides.

The major primary active-transport proteins found in most cells are (1) Na,K-ATPase; (2) Ca-ATPase; (3) H-ATPase; and (4) H,K-ATPase.

Na,K-ATPase is present in all plasma membranes. The pumping activity of this primary active-transport protein leads to the characteristic distribution of high intracellular potassium and low intracellular sodium

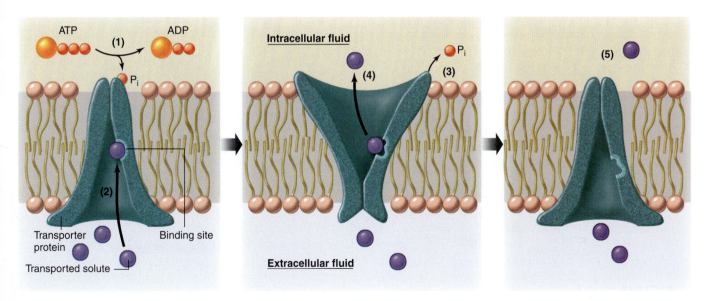

FIGURE 6–11
Primary active-transport model. Changes in the binding site affinity for a transported solute are produced by phosphorylation and dephosphorylation of the transporter (covalent modulation) as it oscillates between two conformations. See text for the numbered sequence of events occurring during transport.

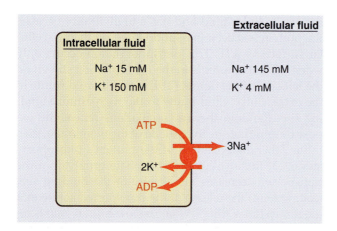

FIGURE 6–12

The primary active transport of sodium and potassium ions in opposite directions by the Na,K-ATPase in plasma membranes is responsible for the low sodium and high potassium intracellular concentrations. For each ATP hydrolyzed, three sodium ions are moved out of a cell, and two potassium ions are moved in.

relative to their respective extracellular concentrations (Figure 6–12). For each molecule of ATP that is hydrolyzed, this transporter moves three sodium ions *out* of a cell and two potassium ions *in*. This results in the net transfer of positive charge to the outside of the cell, and thus this transport process is not electrically neutral, a point to which we will return in Chapter 8.

Ca-ATPase is found in the plasma membrane and several organelle membranes, including the membranes of the endoplasmic reticulum. In the plasma membrane, the direction of active calcium transport is from cytosol to extracellular fluid. In organelle membranes, it is from cytosol into the organelle lumen. Thus active transport of calcium out of the cytosol, via Ca-ATPase, is one reason that the cytosol of most cells has a very low calcium concentration, about 10^{-7} mol/L compared with an extracellular calcium concentration of 10^{-3} mol/L, 10,000 times greater (a second reason will be given below).

H-ATPase is in the plasma membrane and several organelle membranes, including the inner mitochondrial and lysosomal membranes. In the plasma membrane, the H-ATPase moves hydrogen ions out of cells.

H,K-ATPase is in the plasma membranes of the acid-secreting cells in the stomach and kidneys, where it pumps one hydrogen ion out of the cell and moves one potassium in for each molecule of ATP hydrolyzed. (This pump is thus electrically neutral in contrast to the other three ATPases.)

Secondary Active Transport Secondary active transport is distinguished from primary active transport by its use of an ion concentration gradient across a membrane as the energy source. The flow of ions from a higher concentration (higher energy state) to a lower concentration (lower energy state) provides energy for the uphill movement of the actively transported solute.

In addition to having a binding site for the actively transported solute, the transport protein in a secondary active-transport system also has a binding site for an ion (Figure 6–13). This ion is usually sodium, but

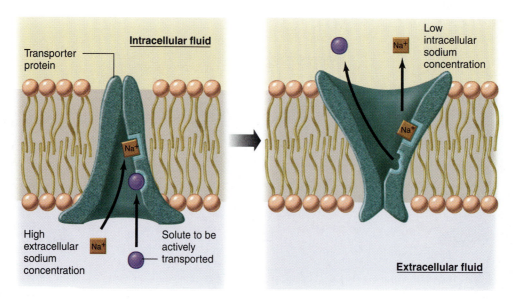

FIGURE 6–13

Secondary active transport model. The binding of a sodium ion to the transporter produces an allosteric alteration in the affinity of the solute binding site at the extracellular surface of the membrane. The absence of sodium binding at the intracellular surface, due to the low intracellular sodium concentration, reverses these changes, producing a low-affinity binding site for the solute, which is then released.

in some cases it can be another ion such as bicarbonate, chloride, or potassium. The binding of an ion to the secondary active transporter produces similar changes in the transporter as occur in primary active transport, namely, (1) altering the affinity of the binding site for the transported solute, or (2) altering the rate at which the binding site on the transport protein is shifted from one surface to the other. Note, however, that during primary active transport, the transport protein is altered by covalent modulation resulting from the covalent linkage of phosphate from ATP to the transport protein; in secondary active transport, the changes are brought about through allosteric modulation as a result of ion binding (Chapter 4).

There is a very important indirect link between the secondary active transporters that utilize sodium and the primary active sodium transporter, the Na,K-ATPase. Recall that the intracellular concentration of sodium is much lower than the extracellular sodium concentration because of the primary active transport of sodium out of the cell by Na,K-ATPase. Because of the low intracellular sodium concentration, few of the sodium binding sites on the secondary active-transport protein are occupied at the intracellular surface of the transporter. This difference provides the basis for the asymmetry in the transport fluxes, leading to the uphill movement of the transported solute. At the same time, the sodium ion that binds to the transporter at the extracellular surface moves *downhill* into the cell when the transporter undergoes its conformational change.

To summarize, the creation of a sodium concentration gradient across the plasma membrane by the *primary* active transport of sodium is a means of indirectly "storing" energy that can then be used to drive *secondary* active-transport pumps linked to sodium. Ultimately, however, the energy for secondary active transport is derived from metabolism in the form of the ATP that is used by the Na,K-ATPase to create the sodium concentration gradient. If the production of ATP were inhibited, the primary active transport of sodium would cease, and the cell would no longer be able to maintain a sodium concentration gradient across the membrane. This in turn would lead to a failure of the secondary active-transport systems that depend on the sodium gradient for their source of energy. Between 10 and 40 percent of the ATP produced by a cell, under resting conditions, is used by the Na,K-ATPase to maintain the sodium gradient, which in turn drives a multitude of secondary active-transport systems.

As discussed in Chapter 4, the energy stored in an ion concentration gradient across a membrane can also be used to *synthesize* ATP from ADP and P_i. Electron transport through the cytochrome chain produces a hydrogen-ion concentration gradient across the inner mitochondrial membrane. The movement of hydrogen ions down this gradient provides the energy that is coupled to the synthesis of ATP during oxidative phosphorylation—the chemiosmotic hypothesis.

As noted earlier, the net movement of sodium by a secondary active-transport protein is always from high extracellular concentration into the cell, where the concentration of sodium is lower. Thus, in secondary active transport, the movement of sodium is always *downhill*, while the net movement of the actively transported solute on the same transport protein is *uphill*, moving from lower to higher concentration. The movement of the actively transported solute can be either into the cell (in the same direction as sodium), in which case it is known as **cotransport**, or out of the cell (opposite the direction of sodium movement), which is called **countertransport** (Figure 6–14). The terms "symport" and "antiport" are also used to refer to the processes of cotransport and countertransport, respectively.

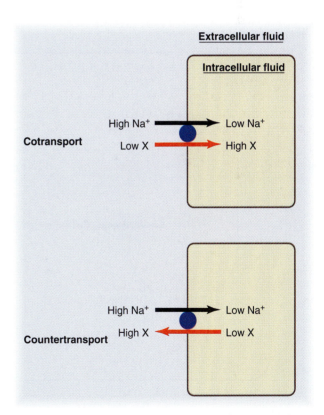

FIGURE 6–14

Cotransport and countertransport during secondary active transport driven by sodium. Sodium ions always move *down* their concentration gradient into a cell, and the transported solute always moves *up* its gradient. Both sodium and the transported solute X move in the same direction during cotransport but in opposite directions during countertransport.

A variety of organic molecules and a few ions are moved across membranes by sodium-coupled secondary active transport. For example, in most cells, amino acids are actively transported into the cell by cotransport with sodium ions, attaining intracellular concentrations 2 to 20 times higher than in the extracellular fluid.

An example of the secondary active transport of ions is provided by calcium. In addition to the previously described primary active transport of calcium from cytosol to extracellular fluid and organelle interior via Ca-ATPase, in many membranes there are also Na-Ca countertransporters (or Na-Ca "exchangers") that use the downhill movement of sodium ions into a cell to pump calcium ions out. Figure 6–15 illustrates the dependence of cytosolic calcium concentration on the several pathways that can move calcium ions into or out of the cytosol: calcium channels, Ca-ATPase pumps, and Na-Ca countertransport. Alterations in the movement of calcium through any of these pathways will lead to a change in cytosolic calcium concentra-

tion and, as a result, alter the cellular activities that are dependent on cytosolic calcium, as will be described in subsequent chapters.

For example, the mechanism of action of a group of drugs, including digitalis, that are used to strengthen the contraction of the heart (Chapter 14) involves several of these transport processes. These drugs inhibit the Na,K-ATPase pumps in the plasma membranes of the heart muscle, leading to an increase in cytosolic sodium concentration. This decreases the gradient for sodium diffusion into the cell, thereby decreasing calcium exit from the cell via sodium-calcium exchange and increasing cytosolic calcium concentration, which acts in muscle cells on the mechanisms that increase the force of contraction.

A large number of genetic diseases result from defects in the various proteins that form ion channels and transport proteins. These mutations can produce malfunctioning of the electrical properties of nerve and muscle cells, and the absorptive and secretory properties

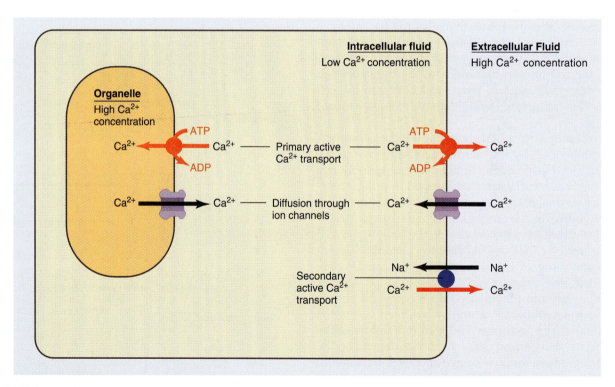

FIGURE 6–15

Pathways affecting cytosolic calcium concentration. The active transport of calcium, both by primary Ca-ATPase pumps and by secondary active calcium countertransport with sodium, moves calcium ions out of the cytosol. Calcium channels allow net diffusion of calcium into the cytosol from both the extracellular fluid and cell organelles. Cytosolic calcium concentration is the resultant of all these processes. The symbols in this diagram will be used throughout this book to represent primary active transport and secondary active transport. The red arrow indicates the direction of the actively transported solute. Black arrows denote downhill movement. Diffusion of ions through channels will use the channel symbol.

TABLE 6-1 Composition of Extracellular and Intracellular Fluids		
	Extracellular Concentration, mM	Intracellular Concentration,* mM
Na^+	145	15
K^+	4	150
Ca^{2+}	1	1.5
Mg^{2+}	1.5	12
Cl^-	110	10
HCO_3^-	24	10
P_i	2	40
Amino acids	2	8
Glucose	5.6	1
ATP	0	4
Protein	0.2	4

*The intracellular concentrations differ slightly from one tissue to another, but the concentrations shown above are typical of most cells. The intracellular concentrations listed above may not reflect the *free* concentration of the substance in the cytosol since some may be bound to proteins or confined within cell organelles. For example, the free cytosolic concentration of calcium is only about 0.0001 mM.

of epithelial cells lining the intestinal tract, kidney, and lung airways. Cystic fibrosis provides one example; others will be discussed in later chapters. Cystic fibrosis, as mentioned earlier, is the result of a defective membrane channel through which chloride ions move from cells into the extracellular fluid. Failure to secrete adequate amounts of chloride ions decreases the fluid secreted by the epithelial cells that is necessary to prevent the build up of mucus, which if allowed to thicken, leads to the eventual obstruction of the airways, pancreatic ducts, and male genital ducts.

In summary the distribution of substances between the intracellular and extracellular fluid is often unequal (Table 6–1) due to the presence in the plasma membrane of primary and secondary active transporters, ion channels, and the membrane potential.

Table 6–2 provides a summary of the major characteristics of the different pathways by which substances move through cell membranes, while Figure 6–16 illustrates the variety of commonly encountered channels and transporters associated with the movement of substances across a typical plasma membrane.

TABLE 6-2 Major Characteristics of Pathways by which Substances Cross Membranes					
	DIFFUSION		MEDIATED TRANSPORT		
	Through Lipid Bilayer	Through Protein Channel	Facilitated Diffusion	Primary Active Transport	Secondary Active Transport
Direction of net flux	High to low concentration	High to low concentration	High to low concentration	Low to high concentration	Low to high concentration
Equilibrium or steady state	$C_o = C_i$	$C_o = C_i*$	$C_o = C_i$	$C_o \neq C_i$	$C_o \neq C_i$
Use of integral membrane protein	No	Yes	Yes	Yes	Yes
Maximal flux at high concentration (saturation)	No	No	Yes	Yes	Yes
Chemical specificity	No	Yes	Yes	Yes	Yes
Use of energy and source	No	No	No	Yes: ATP	Yes: ion gradient (often Na)
Typical molecules using pathway	Nonpolar: O_2, CO_2, fatty acids	Ions: Na^+, K^+, Ca^{2+}	Polar: glucose	Ions: Na^+, K^+, Ca^{2+}, H^+	Polar: amino acids, glucose, some ions

*In the presence of a membrane potential, the intracellular and extracellular ion concentrations will not be equal at equilibrium.

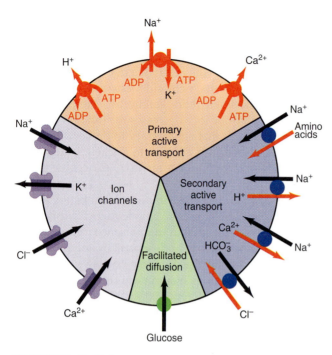

FIGURE 6–16

Movements of solutes across a typical plasma membrane involving membrane proteins. A specialized cell may contain additional transporters and channels not shown in this figure. Many of these membrane proteins can be modulated by various signals leading to a controlled increase or decrease in specific solute fluxes across the membrane.

The addition of a solute to water lowers the concentration of water in the solution compared to the concentration of pure water. For example, if a solute such as glucose is dissolved in water, the concentration of water in the resulting solution is less than that of pure water. A given volume of a glucose solution contains fewer water molecules than an equal volume of pure water since each glucose molecule occupies space formerly occupied by a water molecule (Figure 6–17). In quantitative terms, a liter of pure water weighs about 1000 g, and the molecular weight of water is 18. Thus, the concentration of water in pure water is 1000/18 = 55.5 M. The decrease in water concentration in a solution is approximately equal to the concentration of added solute. In other words, one solute molecule will displace one water molecule. The water concentration in a 1 M glucose solution is therefore approximately 54.5 M rather than 55.5 M. Just as adding water to a solution will dilute the solute, adding solute to a solution will "dilute" the water. *The greater the solute concentration, the lower the water concentration.*

It is essential to recognize that the degree to which the water concentration is decreased by the addition of solute depends upon the *number* of particles (molecules or ions) of solute in solution (the solute concentration) and not upon the chemical nature of the solute.

Osmosis

Water is a small, polar molecule, about 0.3 nm in diameter, that diffuses across most cell membranes very rapidly. One might expect that, because of its polar structure, water would not penetrate the nonpolar lipid regions of membranes. Artificial phospholipid bilayers are somewhat permeable to water, indicating that this small polar molecule can diffuse, at least to some extent, through the membrane lipid layer. Most plasma membranes, however, have a permeability to water that is 10 times greater than that of an artificial lipid membrane. The reason is that a group of membrane proteins known as **aquaporins** form channels through which water can diffuse. The concentration of these water channels differs in different membranes, and in some cells the number of aquaporin channels, and thus the permeability of the membrane to water, can be altered in response to various signals.

The net diffusion of water across a membrane is called **osmosis**. As with any diffusion process, there must be a concentration difference in order to produce a net flux. How can a difference in water concentration be established across a membrane?

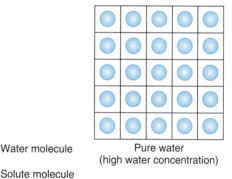

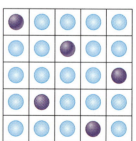

FIGURE 6–17

The addition of solute molecules to pure water lowers the water concentration in the solution.

For example, 1 mol of glucose in 1 L of solution decreases the water concentration to approximately the same extent as does 1 mol of an amino acid, or 1 mol of urea, or 1 mol of any other molecule that exists as a single particle in solution. On the other hand, a molecule that ionizes in solution decreases the water concentration in proportion to the number of ions formed. Hence, 1 mol of sodium chloride in solution gives rise to 1 mol of sodium ions and 1 mol of chloride ions, producing 2 mol of solute particles, which lowers the water concentration twice as much as 1 mol of glucose. By the same reasoning, a 1 M $MgCl_2$ solution lowers the water concentration three times as much as a 1 M glucose solution.

Since the water concentration in a solution depends upon the number of solute particles, it is useful to have a concentration term that refers to the total concentration of solute particles in a solution, regardless of their chemical composition. The total solute concentration of a solution is known as its **osmolarity.** One **osmol** is equal to 1 mol of solute particles. Thus, a 1 M solution of glucose has a concentration of 1 Osm (1 osmol per liter), whereas a 1 M solution of sodium chloride contains 2 osmol of solute per liter of solution. A liter of solution containing 1 mol of glucose and 1 mol of sodium chloride has an osmolarity of 3 Osm. A solution with an osmolarity of 3 Osm may contain 1 mol of glucose and 1 mol of sodium chloride, or 3 mol of glucose, or 1.5 mol of sodium chloride, or any other combination of solutes as long as the total solute concentration is equal to 3 Osm.

Although osmolarity refers to the concentration of solute particles, it is essential to realize that it also determines the *water concentration* in the solution since the *higher* the osmolarity, the *lower* the water concentration. The concentration of water in any two solutions having the same osmolarity is the same since the total number of solute particles per unit volume is the same.

Let us now apply these principles governing water concentration to the diffusion of water across membranes. Figure 6–18 shows two 1-L compartments separated by a membrane permeable to *both* solute and water. Initially the concentration of solute is 2 Osm in compartment 1 and 4 Osm in compartment 2. This difference in solute concentration means there is also a difference in water concentration across the membrane: 53.5 M in compartment 1 and 51.5 M in compartment 2. Therefore, there will be a net diffusion of water from the higher concentration in 1 to the lower concentration in 2, and of solute in the opposite direction, from 2 to 1. When diffusion equilibrium is reached, the two compartments will have identical solute and water concentrations, 3 Osm and 52.5 M, respectively. One mol of water will have diffused from

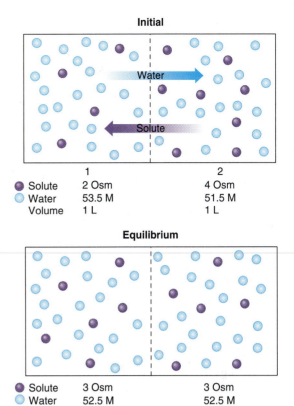

Initial

	1	2
● Solute	2 Osm	4 Osm
○ Water	53.5 M	51.5 M
Volume	1 L	1 L

Equilibrium

● Solute	3 Osm	3 Osm
○ Water	52.5 M	52.5 M
Volume	1 L	1 L

FIGURE 6–18

Between two compartments of equal volume, the net diffusion of water and solute across a membrane permeable to both leads to diffusion equilibrium of both, with no change in the volume of either compartment.

compartment 1 to compartment 2, and 1 mol of solute will have diffused from 2 to 1. Since 1 mol of solute has replaced 1 mol of water in compartment 1, and vice versa in compartment 2, there is no change in the *volume* of either compartment.

If the membrane is now replaced by one that is permeable to water but impermeable to solute (Figure 6–19), the same *concentrations* of water and solute will be reached at equilibrium as before, but there will now be a change in the *volumes* of the compartments. Water will diffuse from 1 to 2, but there will be no solute diffusion in the opposite direction because the membrane is impermeable to solute. Water will continue to diffuse into compartment 2, therefore, until the water concentrations on the two sides become equal. The solute concentration in compartment 2 decreases as it is diluted by the incoming water, and the solute in compartment 1 becomes more concentrated as water moves out. When the water reaches diffusion equilibrium, the osmolarities of the compartments will be equal, and thus the solute concentrations must also be equal. To reach this state of equilibrium, enough

Initial

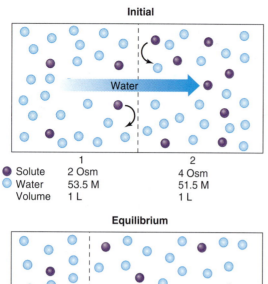

	1	2
● Solute	2 Osm	4 Osm
○ Water	53.5 M	51.5 M
Volume	1 L	1 L

Equilibrium

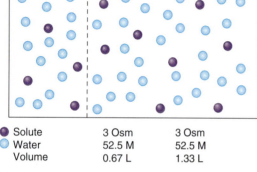

● Solute	3 Osm	3 Osm
○ Water	52.5 M	52.5 M
Volume	0.67 L	1.33 L

FIGURE 6–19

The movement of water across a membrane that is permeable to water but not permeable to solute leads to an equilibrium state in which there is a change in the volumes of the two compartments due to the net diffusion of water (0.33 L in this case) from compartment 1 to 2. (We will assume that the membrane in this example stretches as the volume of compartment 2 increases so that no significant change in compartment pressure occurs.)

water must pass from compartment 1 to 2 to increase the volume of compartment 2 by one-third and decrease the volume of compartment 1 by an equal amount. Note that it is the presence of a membrane impermeable to solute that leads to the volume changes associated with osmosis.

We have treated the two compartments in our example as if they were infinitely expandable, so that the net transfer of water does not create a pressure difference across the membrane. This is essentially the situation that occurs across plasma membranes. In contrast, if the walls of compartment 2 could not expand, the movement of water into compartment 2 would raise the pressure in compartment 2, which would oppose further net water entry. Thus the movement of water into compartment 2 can be prevented by the application of a pressure to compartment 2. This leads to a crucial definition: When a *solution* containing nonpenetrating solutes is separated from pure water by a membrane, the pressure that must be applied to the *solution* to prevent the net flow of water into the *solution* is termed the **osmotic pressure** of the *solution*. The greater the osmolarity of a solution, the greater its osmotic pressure. It is important to recognize that the osmotic pressure of a solution does not push water molecules into the solution. Rather it is the amount of pressure that would have to be applied to the solution to *prevent* the net flow of water into the solution. Like osmolarity, the osmotic pressure of a solution is a measure of the solution's water concentration—the lower the water concentration, the higher the osmotic pressure.

Extracellular Osmolarity and Cell Volume

We can now apply the principles learned about osmosis to cells, which meet all the criteria necessary to produce an osmotic flow of water across a membrane. Both the intracellular and extracellular fluids contain water, and cells are surrounded by a membrane that is very permeable to water but impermeable to many substances (**nonpenetrating solutes**).

About 85 percent of the extracellular solute particles are sodium and chloride ions, which can diffuse into the cell through protein channels in the plasma membrane or enter the cell during secondary active transport. As we have seen, however, the plasma membrane contains Na,K-ATPase pumps that actively move sodium ions out of the cell. Thus, sodium moves into cells and is pumped back out, behaving as if it never entered in the first place; that is, extracellular sodium behaves like a nonpenetrating solute. Also, secondary active-transport pumps and the membrane potential move chloride ions out of cells as rapidly as they enter, with the result that extracellular chloride ions also behave as if they were nonpenetrating solutes.

Inside the cell, the major solute particles are potassium ions and a number of organic solutes. Most of the latter are large polar molecules unable to diffuse through the plasma membrane. Although potassium ions can diffuse out of a cell through potassium channels, they are actively transported back by the Na,K-ATPase pump. The net effect, as with extracellular sodium and chloride, is that potassium behaves as if it were a nonpenetrating solute, but in this case one confined to the intracellular fluid. Thus, sodium and chloride outside the cell and potassium and organic solutes inside the cell behave as nonpenetrating solutes on the two sides of the plasma membrane.

The osmolarity of the extracellular fluid is normally about 300 mOsm. Since water can diffuse across plasma membranes, the water in the intracellular and extracellular fluids will come to diffusion equilibrium. At equilibrium, therefore, the osmolarities of the intracellular and extracellular fluids are the same—300 mOsm. Changes in extracellular osmolarity can cause cells to shrink or swell as a result of the movements of water across the plasma membrane.

If cells are placed in a solution of nonpenetrating solutes having an osmolarity of 300 mOsm, they will neither swell nor shrink since the water concentrations in the intra- and extracellular fluid are the same, and the solutes cannot leave or enter. Such solutions are said to be **isotonic** (Figure 6–20), defined as having the same concentration of *nonpenetrating* solutes as normal extracellular fluid. Solutions containing less than 300 mOsm of nonpenetrating solutes (**hypotonic** solutions) cause cells to swell because water diffuses into the cell from its higher concentration in the extracellular fluid. Solutions containing greater than 300 mOsm of nonpenetrating solutes (**hypertonic** solutions) cause cells to shrink as water diffuses out of the cell into the fluid with the lower water concentration. Note that the concentration of *nonpenetrating* solutes in a solution, not the total osmolarity, determines its tonicity—hypotonic, isotonic, or hypertonic. Penetrating solutes do not contribute to the tonicity of a solution.

In contrast, another set of terms—**isoosmotic, hyperosmotic,** and **hypoosmotic**—denotes simply the osmolarity of a solution relative to that of normal extracellular fluid without regard to whether the solute is penetrating or nonpenetrating. The two sets of terms are therefore not synonymous. For example, a 1-L solution containing 300 mOsmol of nonpenetrating NaCl and 100 mOsmol of urea, which can cross plasma membranes, would have a total osmolarity of 400 mOsm and would be hyperosmotic. It would, however, also be an isotonic solution, producing no change in the equilibrium volume of cells immersed in it. The reason is that urea will diffuse into the cells and reach the same concentration as the urea in the extracellular solution, and thus both the intracellular and extracellular solutions will have the same osmolarity (400 mOsm). Therefore, there will be no difference in the water concentration across the membrane and thus no change in cell volume.

Table 6–3 provides a comparison of the various terms used to describe the osmolarity and tonicity of solutions. All hypoosmotic solutions are also hypotonic, whereas a hyperosmotic solution can be hypertonic, isotonic, or hypotonic.

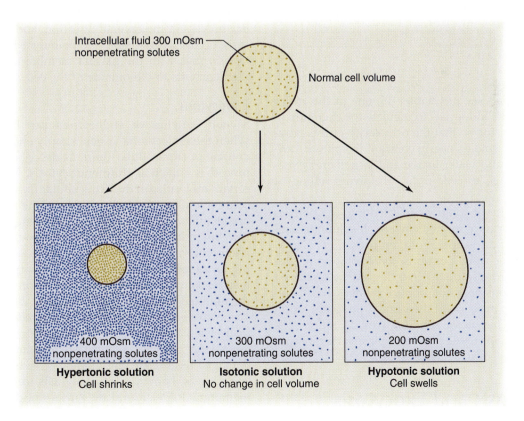

Intracellular fluid 300 mOsm nonpenetrating solutes

Normal cell volume

400 mOsm nonpenetrating solutes

Hypertonic solution
Cell shrinks

300 mOsm nonpenetrating solutes

Isotonic solution
No change in cell volume

200 mOsm nonpenetrating solutes

Hypotonic solution
Cell swells

FIGURE 6–20
Changes in cell volume produced by hypertonic, isotonic, and hypotonic solutions.

TABLE 6–3	Terms Referring to Both the Osmolarity and Tonicity of Solutions
Isotonic	A solution containing 300 mOsmol/L of nonpenetrating solutes, regardless of the concentration of membrane-penetrating solutes that may be present
Hypertonic	A solution containing greater than 300 mOsmol/L of nonpenetrating solutes, regardless of the concentration of membrane-penetrating solutes that may be present
Hypotonic	A solution containing less than 300 mOsmol/L of nonpenetrating solutes, regardless of the concentration of membrane-penetrating solutes that may be present
Isoosmotic	A solution containing 300 mOsmol/L of solute, regardless of its composition of membrane-penetrating and nonpenetrating solutes
Hyperosmotic	A solution containing greater than 300 mOsmol/L of solutes, regardless of the composition of membrane-penetrating and nonpenetrating solutes
Hypoosmotic	A solution containing less than 300 mOsmol/L of solutes, regardless of the composition of membrane-penetrating and nonpenetrating solutes

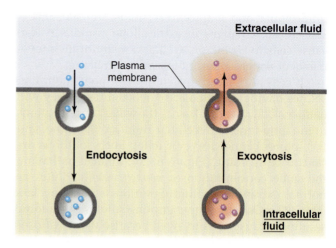

FIGURE 6–21
Endocytosis and exocytosis.

As we shall see in Chapter 16, one of the major functions of the kidneys is to regulate the excretion of water in the urine so that the osmolarity of the extracellular fluid remains nearly constant in spite of variations in salt and water intake and loss, thereby preventing damage to cells from excessive swelling or shrinkage.

The tonicity of solutions injected into the body is of great importance in medicine. Such solutions usually consist of an isotonic solution of NaCl (150 mM NaCl—isotonic saline) or an isotonic solution of glucose (5% dextrose solution). Injecting a drug dissolved in such solutions does not produce changes in cell volume, whereas injection of the same drug dissolved in pure water, a hypotonic solution, would produce cell swelling, perhaps to the point that plasma membranes would rupture, destroying cells.

Endocytosis and Exocytosis

In addition to diffusion and mediated transport, there is another pathway by which substances can enter or leave cells, one that does not require the molecules to pass through the structural matrix of the plasma membrane. When living cells are observed under a light microscope, regions of the plasma membrane can be seen to fold into the cell, forming small pockets that pinch off to produce intracellular, membrane-bound vesicles that enclose a small volume of extracellular fluid. This process is known as **endocytosis** (Figure 6–21). A similar process in the reverse direction, known as **exocytosis,** occurs when membrane-bound vesicles in the *cytoplasm* fuse with the plasma membrane and release their contents to the outside of the cell.

Endocytosis

Several varieties of endocytosis can be identified. When the endocytotic vesicle simply encloses a small volume of extracellular fluid, as described above, the process is known as **fluid endocytosis.** In other cases, certain molecules in the extracellular fluid bind to specific proteins on the outer surface of the plasma membrane and are carried into the cell along with the extracellular fluid when the membrane invaginates. This is known as **adsorptive endocytosis.** In addition to taking in trapped extracellular fluid, adsorptive endocytosis leads to a selective concentration in the vesicle of the material bound to the membrane. Both fluid and adsorptive endocytosis are often referred to as **pinocytosis** (cell drinking). A third type of endocytosis occurs when large particles, such as bacteria and debris from damaged tissues, are engulfed by cells. In this form of endocytosis, known as **phagocytosis** (cell eating), the membrane folds around the surface of the particle so that little extracellular fluid is enclosed within the vesicle. While most cells undergo pinocytosis, only a few special cells carry out phagocytosis (Chapter 20).

Endocytosis of any kind requires metabolic energy and is associated with the binding of specific "coating" proteins that form a shell around the newly forming vesicle on its cytoplasmic surface. After the vesicle separates from the plasma membrane, the coating proteins that helped to form it are removed, and the vesicle membrane now fuses with the membranes of intracellular organelles, adding the contents of the vesicle to the lumen of that organelle. The passage of material from one membrane-bound organelle to another involves the formation of vesicles from one organelle and the fusion with the second. These processes of intracellular budding and fusion are similar to endo- and exocytotic events occurring at the plasma membrane and involve some of the same proteins to mediate vesicle formation and fusion with other membranes.

What is the fate of most endocytotic vesicles once they enter the cell? After separating from the plasma membrane, they fuse with a series of vesicles and tubular elements known as **endosomes,** which lie between the plasma membrane and the Golgi apparatus (Figure 6–22). Like the Golgi apparatus, the endosomes perform a sorting function, distributing the contents of the vesicle and its membrane to various locations. Most of the contents of endocytotic vesicles are passed from the endosomes to lysosomes, organelles that contain digestive enzymes that break down large molecules such as proteins, polysaccharides, and nucleic acids. The fusion of endosomal vesicles with the lysosomal membrane exposes the contents of the vesicle to these digestive enzymes. The phagocytosis of bacteria and their destruction by the lysosomal digestive enzymes is one of the body's major defense mechanisms against germs (Chapter 20).

Some endocytotic vesicles pass through the cytoplasm and fuse with the plasma membrane on the opposite side of the cell, releasing their contents to the extracellular space. This provides a pathway for the transfer of large molecules, such as proteins, across the epithelial cells that separate two compartments (for example, blood and interstitial fluid). A similar process allows small amounts of protein to be moved across the intestinal epithelium.

Each episode of endocytosis removes a small portion of the membrane from the cell surface. In cells that have a great deal of endocytotic activity, more than 100 percent of the plasma membrane may be internalized in an hour, yet the membrane surface area remains constant. This is because the membrane is replaced at about the same rate by vesicle membrane that fuses with the plasma membrane during *exocytosis.* Some of the plasma-membrane proteins taken into the cell during endocytosis are stored in the membranes of endosomes, and upon receiving the appropriate signal can be returned to the plasma membrane

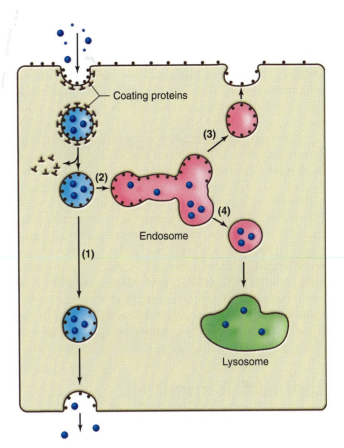

FIGURE 6–22

Fate of endocytotic vesicles. Pathway 1 transfers extracellular materials from one side of the cell to the other. Pathway 2 leads to fusion with endosomes, from which point the membrane components may be recycled to the plasma membrane (3) or the contents of the vesicle may be transferred to lysosomes for digestion (4).

during exocytosis when the endosomal vesicles fuse with the plasma membrane.

Exocytosis

Exocytosis performs two functions for cells: (1) It provides a way to replace portions of the plasma membrane that have been removed by endocytosis and, in the process, to add new membrane components as well, and (2) it provides a route by which membrane-impermeable molecules, such as protein hormones, synthesized by cells can be released (secreted) into the extracellular fluid.

How are substances that are to be secreted by exocytosis packaged into vesicles? The entry of newly formed proteins into the lumen of the endoplasmic reticulum and the protein's processing through the Golgi apparatus were described in Chapter 5. From the Golgi apparatus, the proteins to be secreted travel to the plasma membrane in vesicles from which they can be released into the extracellular fluid by exocytosis. Very high concentrations of various organic molecules, such as neurotransmitters (Chapter 8), can be achieved within vesicles by a combination of mediated transport across the vesicle membrane followed by the binding of the transported substances to proteins within the vesicle.

The secretion of substances by exocytosis is triggered in most cells by stimuli that lead to an increase in cytosolic calcium concentration in the cell. As described in Chapter 7, these stimuli open calcium channels in either the plasma membrane and/or the membranes of intracellular organelles. The resulting increase in cytosolic calcium concentration activates proteins required for the vesicle membrane to fuse with the plasma membrane and release the vesicle contents into the extracellular fluid. Material stored in secretory vesicles is available for rapid secretion in response to a stimulus, without delays that might occur if the material had to be synthesized after the arrival of the stimulus.

Epithelial Transport

Epithelial cells line hollow organs or tubes and regulate the absorption or secretion of substances across these surfaces. One surface of an epithelial cell generally faces a hollow or fluid-filled chamber, and the plasma membrane on this side is referred to as the **luminal membrane** (also known as the apical, or mucosal, membrane) of the epithelium. The plasma membrane on the opposite surface, which is usually adjacent to a network of blood vessels, is referred to as the **basolateral membrane** (also known as the serosal membrane).

There are two pathways by which a substance can cross a layer of epithelial cells: (1) by diffusion *between* the adjacent cells of the epithelium—the **paracellular pathway,** and (2) by movement into an epithelial cell across either the luminal or basolateral membrane, diffusion through the cytosol, and exit across the opposite membrane. This is termed the **transcellular pathway.**

Diffusion through the paracellular pathway is limited by the presence of tight junctions between adjacent cells, since these junctions form a seal around the luminal end of the epithelial cells (Chapter 3). Although small ions and water are able to diffuse to some degree through tight junctions, the amount of paracellular diffusion is limited by the tightness of the junctional seal and the relatively small area available for diffusion. The leakiness of the paracellular pathway varies in different types of epithelium, with some being very leaky and others very tight.

During transcellular transport, the movement of molecules through the plasma membranes of epithe-

lial cells occurs via the pathways (diffusion and mediated transport) already described for movement across membranes in general. However, the transport and permeability characteristics of the luminal and basolateral membranes are not the same. These two membranes contain different ion channels and different transporters for mediated transport. As a result of these differences, substances can undergo a net movement from a low concentration on one side of an epithelium to a higher concentration on the other side, or in other words, to undergo active transport across the overall epithelial layer. Examples of such transport are the absorption of material from the gastrointestinal tract into the blood, the movement of substances between the kidney tubules and the blood during urine formation, and the secretion of salts and fluid by glands.

Figures 6–23 and 6–24 illustrate two examples of active transport across an epithelium. Sodium is actively transported across most epithelia from lumen to blood side in absorptive processes, and from blood

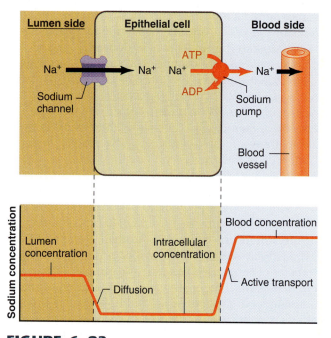

FIGURE 6–23

Active transport of sodium across an epithelial cell. The transepithelial transport of sodium always involves primary active transport out of the cell across one of the plasma membranes. (For clarity in this and the next two figures, the entrance of potassium via Na,K-ATPase transporters is not shown.) The movement of sodium into the cell across the plasma membrane on the opposite side is always downhill. Sometimes, as in this example, it is by diffusion through sodium channels, whereas in other epithelia this downhill movement is by means of a secondary active transporter. Shown below the cell is the concentration profile of the transported solute across the epithelium.

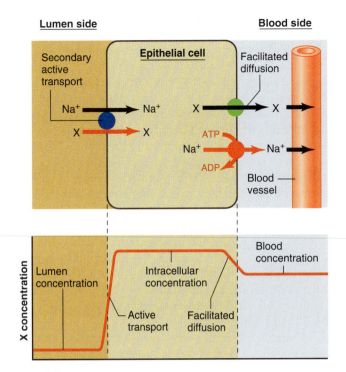

FIGURE 6–24

The transepithelial transport of most organic solutes (X) involves their movement into a cell through a secondary active transport driven by the downhill flow of sodium. The organic substance then moves out of the cell at the blood side down a concentration gradient, by means of facilitated diffusion. Shown below the cell is the concentration profile of the transported solute across the epithelium.

cell to a lower concentration in the extracellular fluid on the blood side. The concentration of the substance may be considerably higher on the blood side than in the lumen since the blood-side concentration can approach equilibrium with the high intracellular concentration created by the luminal membrane entry step.

Although water is not actively transported across cell membranes, net movement of water across an epithelium can be achieved by osmosis as a result of the active transport of solutes, especially sodium, across the epithelium. The active transport of sodium, as described above, results in a decrease in the sodium concentration on one side of an epithelial layer (the luminal side in our example) and an increase on the other. These changes in solute concentration are accompanied by changes in the water concentration on the two sides since a change in solute concentration, as we have seen, produces a change in water concentration. The water concentration difference produced will cause water to move by osmosis from the low-sodium to the high-sodium side of the epithelium (Figure 6–25). Thus, net movement of solute across an epithelium is accompanied by a flow of water in the same direction. If the epithelial cells are highly permeable to water, large net movements of water can occur with very small differences in osmolarity.

side to lumen during secretion. In our example, the movement of sodium from the lumen into the epithelial cell occurs by diffusion through sodium channels in the luminal membrane (Figure 6–23). Sodium diffuses into the cell because the intracellular concentration of sodium is kept low by the active transport of sodium out of the cell across the basolateral membrane on the opposite side, where all of the Na,K-ATPase pumps are located. In other words, sodium moves downhill into the cell and then uphill out of it. The net result is that sodium can be moved from lower to higher concentration across the epithelium.

Figure 6–24 illustrates the active absorption of organic molecules across an epithelium. In this case, entry of an organic molecule X across the luminal plasma membrane occurs via a secondary active transporter linked to the downhill movement of sodium into the cell. In the process, X moves from a lower concentration in the luminal fluid to a higher concentration in the cell. Exit of the substance across the basolateral membrane occurs by facilitated diffusion, which moves the material from its higher concentration in the

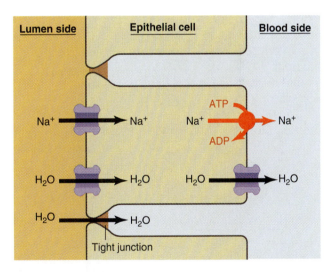

FIGURE 6–25

Net movements of water across an epithelium are dependent on net solute movements. The active transport of sodium across the cells, into the surrounding interstitial spaces, produces an elevated osmolarity in this region and a decreased osmolarity in the lumen. This leads to the osmotic flow of water across the epithelium in the same direction as the net solute movement. The water diffuses through protein water channels in the membrane and across the tight junctions between the epithelial cells.

Glands

Glands secrete specific substances into the extracellular fluid or the lumen of ducts in response to appropriate stimuli. The mechanisms of glandular secretion depend upon the principles of membrane diffusion, mediated transport, and exocytosis described in this chapter.

Glands are formed during embryonic development by the infolding of the epithelial layer of an organ's surface. Many glands remain connected by ducts to the epithelial surfaces from which they were formed, while others lose this connection and become isolated clusters of cells. The first type of gland is known as an **exocrine gland,** and its secretions flow through the ducts and are discharged into the lumen of an organ or, in the case of the skin glands, onto the surface of the skin (Figure 6–26). Sweat glands and salivary glands are examples of exocrine glands.

In the second type of gland, known as an **endocrine gland,** or ductless gland, secretions are released directly into the interstitial fluid surrounding the gland cells (Figure 6–26). From this point, the ma-

terial diffuses into the blood, which carries it throughout the body. The endocrine glands secrete a major class of chemical messengers, the **hormones,** and in practical usage the term "endocrine gland" has come to be synonymous with "hormone-secreting gland." However, it should be noted that there are "ductless glands" that secrete nonhormonal, organic substances into the blood. For example, the liver secretes glucose, amino acids, fats, and proteins into the blood. The substances secreted by such nonendocrine glands serve as nutrients for other cells or perform special functions in the blood, but they do not act as messengers and therefore are not hormones.

The substances secreted by glands fall into two chemical categories: (1) organic materials that are, for the most part, synthesized by the gland cells, and (2) salts and water, which are moved from one extracellular compartment to another across the glandular epithelium. Ultimately this salt and water come from the blood supplying the tissue.

Organic molecules are secreted by gland cells via all the pathways already described: diffusion in the case of lipid-soluble materials, mediated transport for some polar materials, and exocytosis for very large molecules such as proteins. Salts are actively transported across glandular epithelia, producing changes in the extracellular osmolarity, which in turn causes the osmotic flow of water. In exocrine glands, as the secreted fluid passes along the ducts connecting the gland to the luminal surface, the composition of the fluid originally secreted may be altered as a result of absorption or secretion by the duct cells (Figure 6–26). Often the composition of the secreted material at the end of the duct varies with the rate of secretion, reflecting the amount of time the fluid remains in the duct, where it can be modified.

Most glands undergo a low, basal rate of secretion, which can be greatly augmented in response to an appropriate signal, usually a nerve impulse, hormone, or a locally generated chemical messenger. The mechanism of the increased secretion is again one of altering some portion of the secretory pathway. This may involve (1) increasing the rate at which a secreted organic substance is synthesized by activating the appropriate enzymes, (2) providing the calcium signal for exocytosis of already synthesized material, or (3) altering the pumping rates of transporters or opening ion channels.

The volume of fluid secreted by an exocrine gland can be increased by increasing the sodium pump activity or by controlling the opening of sodium channels in the plasma membrane. The increased movement of sodium across the epithelium increases the sodium concentration in the lumen, which in turn increases the flow of water by osmosis. The greater the solute transfer, the greater the water flow.

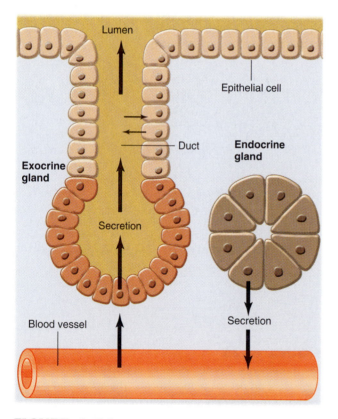

FIGURE 6–26

Glands are composed of epithelial cells. Exocrine-gland secretions enter ducts, whereas hormones or other substances secreted by endocrine (ductless) glands diffuse into the blood.

Diffusion

I. Diffusion is the movement of molecules from one location to another by random thermal motion.
   a. The net flux between two compartments always proceeds from higher to lower concentration.
   b. Diffusion equilibrium is reached when the concentrations of the diffusing substance in the two compartments become equal.

II. The magnitude of the net flux F across a membrane is directly proportional to the concentration difference across the membrane $C_o - C_i$, the surface area of the membrane A, and the membrane permeability constant k_p.

III. Nonpolar molecules diffuse through the lipid portions of membranes much more rapidly than do polar or ionized molecules because nonpolar molecules can dissolve in the lipids in the membrane.

IV. Mineral ions diffuse across membranes by passing through ion channels formed by integral membrane proteins.
   a. The diffusion of ions across a membrane depends on both the concentration gradient and the membrane potential.
   b. The flux of ions across a membrane can be altered by opening or closing ion channels.

Mediated-Transport Systems

I. The mediated transport of molecules or ions across a membrane involves binding of the transported solute to a transporter protein in the membrane. Changes in the conformation of the transporter move the binding site to the opposite side of the membrane, where the solute dissociates from the protein.
   a. The binding sites on transporters exhibit chemical specificity, affinity, and saturation.
   b. The magnitude of the flux through a mediated-transport system depends on the degree of transporter saturation, the number of transporters in the membrane, and the rate at which the conformational change in the transporter occurs.

II. Facilitated diffusion is a mediated-transport process that moves molecules from higher to lower concentration across a membrane by means of a transporter until the two concentrations become equal. Metabolic energy is not required for this process.

III. Active transport is a mediated-transport process that moves molecules against an electrochemical gradient across a membrane by means of a transporter and requires an input of energy.
   a. Active transport is achieved either by altering the affinity of the binding site so that it differs on the two sides of the membrane or by altering the rate at which the protein changes its conformation from one side of the membrane to the other.
   b. Primary active transport uses the phosphorylation of the transporter by ATP to drive the transport process.
   c. Secondary active transport uses the binding of ions (often sodium) to the transporter to drive the transport process.
   d. In secondary active transport, the downhill flow of an ion is linked to the uphill movement of a second solute either in the same direction as the ion (cotransport) or in the opposite direction of the ion (countertransport).

Osmosis

I. Water crosses membranes by (1) diffusing through the lipid bilayer, and (2) diffusing through protein channels in the membrane.

II. Osmosis is the diffusion of water across a membrane from a region of higher water concentration to a region of lower water concentration. The osmolarity—total solute concentration in a solution—determines the water concentration: The higher the osmolarity of a solution, the lower the water concentration.

III. Osmosis across a membrane permeable to water but impermeable to solute leads to an increase in the volume of the compartment on the side that initially had the higher osmolarity, and a decrease in the volume on the side that initially had the lower osmolarity.

IV. Application to a solution of sufficient pressure will prevent the osmotic flow of water into the solution from a compartment of pure water. This pressure is called the osmotic pressure. The greater the osmolarity of a solution, the greater its osmotic pressure. Net water movement occurs from a region of lower osmotic pressure to one of higher osmotic pressure.

V. The osmolarity of the extracellular fluid is about 300 mOsm. Since water comes to diffusion equilibrium across cell membranes, the intracellular fluid has an osmolarity equal to that of the extracellular fluid.
   a. Na^+ and Cl^- ions are the major effectively nonpenetrating solutes in the extracellular fluid, whereas K^+ ions and various organic solutes are the major effectively nonpenetrating solutes in the intracellular fluid.
   b. The terms used to describe the osmolarity and tonicity of solutions containing different compositions of penetrating and nonpenetrating solutes are given in Table 6–3.

Endocytosis and Exocytosis

I. During endocytosis, regions of the plasma membrane invaginate and pinch off to form vesicles that enclose a small volume of extracellular material.
   a. The three classes of endocytosis are (1) fluid endocytosis, (2) adsorptive endocytosis, and (3) phagocytosis.
   b. Most endocytotic vesicles fuse with endosomes, which in turn transfer the vesicle contents to lysosomes where they are digested by lysosomal enzymes.

II. Exocytosis, which occurs when intracellular vesicles fuse with the plasma membrane, provides a means of adding components to the plasma membrane and a route by which membrane-impermeable molecules, such as proteins synthesized by cells, can be released into the extracellular fluid.

Epithelial Transport

I. Molecules can cross an epithelial layer of cells by two pathways: (1) through the extracellular spaces between the cells—the paracellular pathway, and (2) through the cell, across both the luminal and basolateral membranes as well as the cell's cytoplasm—the transcellular pathway.

II. In epithelial cells, the permeability and transport characteristics of the luminal and basolateral plasma membranes differ, resulting in the ability of the cells to actively transport a substance between the fluid on one side of the cell and the fluid on the opposite side of the cell.

III. The active transport of sodium through an epithelium increases the osmolarity on one side of the cell and decreases it on the other, causing water to move by osmosis in the same direction as the transported sodium.

IV. Glands are composed of epithelial cells that secrete water and solutes in response to stimulation.
   a. There are two categories of glands: (1) exocrine glands, which secrete into ducts, and (2) endocrine glands (ductless glands), which secrete hormones and other substances into the extracellular fluid, from which they diffuse into the blood.
   b. The secretions of glands consist of (1) organic substances that have been synthesized by the gland, and (2) salts and water, which have been transported across the gland cells from the blood.

KEY TERMS

plasma membrane	osmolarity
diffusion	osmol
flux	osmotic pressure
net flux	nonpenetrating solute
diffusion equilibrium	isotonic
permeability constant, k_p	hypotonic
channel	hypertonic
membrane potential	isoosmotic
electrochemical gradient	hyperosmotic
channel gating	hypoosmotic
patch clamping	endocytosis
ligand-sensitive channel	exocytosis
voltage-gated channel	fluid endocytosis
mechanosensitive channel	adsorptive endocytosis
transporter	pinocytosis
mediated transport	phagocytosis
facilitated diffusion	endosome
active transport	luminal membrane
primary active transport	basolateral membrane
secondary active transport	paracellular pathway
cotransport	transcellular pathway
countertransport	exocrine gland
aquaporin	endocrine gland
osmosis	hormone

REVIEW QUESTIONS

1. What determines the direction in which net diffusion of a nonpolar molecule will occur?
2. In what ways can the net solute flux between two compartments separated by a permeable membrane be increased?
3. Why are membranes more permeable to nonpolar molecules than to most polar and ionized molecules?
4. Ions diffuse across cell membranes by what pathway?
5. When considering the diffusion of ions across a membrane, what driving force, in addition to the ion concentration gradient, must be considered?
6. What factors can alter the opening and closing of protein channels in a membrane?
7. Describe the mechanism by which a transporter of a mediated-transport system moves a solute from one side of a membrane to the other.
8. What determines the magnitude of flux across a membrane in a mediated-transport system?
9. What characteristics distinguish diffusion from facilitated diffusion?
10. What characteristics distinguish facilitated diffusion from active transport?
11. Contrast the mechanism by which energy is coupled to a transporter during (a) primary active transport and (b) secondary active transport.
12. Describe the direction in which sodium ions and a solute transported by secondary active transport move during cotransport and countertransport.
13. How can the concentration of water in a solution be decreased?
14. If two solutions having different osmolarities are separated by a water-permeable membrane, why will there be a change in the volumes of the two compartments if the membrane is impermeable to the solutes, but no change in volume if the membrane is permeable to solute?
15. To which solution must pressure be applied to prevent the osmotic flow of water across a membrane separating a solution of higher osmolarity and a solution of lower osmolarity?
16. Why do sodium and chloride ions in the extracellular fluid and potassium ions in the intracellular fluid behave as if they are nonpenetrating solutes?
17. What is the osmolarity of the extracellular fluid? Of the intracellular fluid?
18. What change in cell volume will occur when a cell is placed in a hypotonic solution? In a hypertonic solution?
19. Under what conditions will a hyperosmotic solution be isotonic?
20. Endocytotic vesicles deliver their contents to which parts of a cell?
21. How do the mechanisms for actively transporting glucose and sodium across an epithelium differ?
22. By what mechanism does the active transport of sodium lead to the osmotic flow of water across an epithelium?
23. What is the difference between an endocrine gland and an exocrine gland?

THOUGHT QUESTIONS

(Answers are given in Appendix A.)

1. In two cases (A and B), the concentrations of solute X in two 1-L compartments separated by a membrane through which X can diffuse are

   CONCENTRATION OF X, mM

Case	Compartment 1	Compartment 2
A	3	5
B	32	30

   a. In what direction will the net flux of X take place in case A and in case B?
   b. When diffusion equilibrium is reached, what will be the concentration of solute in each compartment in case A and in case B?
   c. Will A reach diffusion equilibrium faster, slower, or at the same rate as B?

2. When the extracellular concentration of the amino acid alanine is increased, the net flux of the amino acid leucine into a cell is decreased. How might this observation be explained?

3. If a transporter that mediates active transport of a substance has a lower affinity for the transported substance on the extracellular surface of the plasma membrane than on the intracellular surface, in what direction will there be a net transport of the substance across the membrane? (Assume that the rate of transporter conformational change is the same in both directions.)

4. Why will inhibition of ATP synthesis by a cell lead eventually to a decrease and, ultimately, cessation in secondary active transport?

5. Given the following solutions, which has the lowest water concentration? Which two have the same osmolarity?

   CONCENTRATION, mM

Solution	Glucose	Urea	NaCl	CaCl$_2$
A	20	30	150	10
B	10	100	20	50
C	100	200	10	20
D	30	10	60	100

6. Assume that a membrane separating two compartments is permeable to urea but not permeable to NaCl. If compartment 1 contains 200 mmol/L of NaCl and 100 mmol/L of urea, and compartment 2 contains 100 mmol/L of NaCl and 300 mmol/L of urea, which compartment will have increased in volume when osmotic equilibrium is reached?

7. What will happen to cell volume if a cell is placed in each of the following solutions?

   CONCENTRATION, mM

Solution	NaCl (nonpenetrating)	Urea (penetrating)
A	150	100
B	100	150
C	200	100
D	100	50

8. Characterize each of the solutions in question 7 as to whether it is isotonic, hypotonic, hypertonic, isoosmotic, hypoosmotic, or hyperosmotic.

9. By what mechanism might an increase in intracellular sodium concentration lead to an increase in exocytosis?

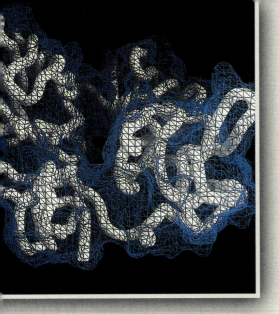

Homeostatic Mechanisms and Cellular Communication

Part 2 of this book provides the information needed to bridge the gap between the cell physiology of Part 1 and the coordinated body functions of Part 3. Section A of this chapter begins that process by amplifying the concept of homeostasis first presented in Chapter 1 and describing the general characteristics and components of homeostatic control systems. It also presents several processes, such as biological rhythms, that are related to and influence homeostasis. In these discussions, many specific examples will be used purely for the purpose of illustration—for example, certain features of temperature regulation. The reader should recognize that such information will be presented again in its more specific context in later chapters.

The operation of control systems requires that cells be able to communicate with each other, often over long distances. Much of this *inter*cellular communication is mediated by chemical messengers. Accordingly, Section B describes how these messengers interact with their target cells and how these interactions trigger *intra*cellular chains of chemical events that lead to the cell's response. Throughout Section B, the reader should carefully distinguish *inter*cellular and *intra*cellular chemical messengers and communication.

HOMEOSTATIC CONTROL SYSTEMS

General Characteristics

As described in Chapter 1, the activities of cells, tissues, and organs must be regulated and integrated with each other in such a way that any change in the extracellular fluid initiates a reaction to minimize the change. **Homeostasis** denotes the *relatively* stable conditions of the internal environment that result from these compensating regulatory responses performed by **homeostatic control systems.**

Consider the regulation of body temperature. Our subject is a resting, lightly clad man in a room having a temperature of 20°C and moderate humidity. His internal body temperature is 37°C, and he is losing heat to the external environment because it is at a lower temperature. However, the chemical reactions occurring within the cells of his body are producing heat at a rate equal to the rate of heat loss. Under these conditions, the body undergoes no net gain or loss of heat, and the body temperature remains constant. The system is said to be in a **steady state,** defined as a system in which a particular variable (temperature, in this case) is not changing but energy (in this case, heat) must be added continuously to maintain this variable constant. (Steady states differ from equilibrium situations, in which a particular variable is not changing but no input of energy is required to maintain the constancy.) The steady-state temperature in our example is known as the **set point** (also termed the operating point) of the thermoregulatory system.

This example illustrates a crucial generalization about homeostasis: Stability of an internal environmental variable is achieved by the balancing of inputs and outputs. In this case, the variable (body temperature) remains constant because metabolic heat production (input) equals heat loss from the body (output).

Now we lower the temperature of the room rapidly, say to 5°C, and keep it there. This immediately increases the loss of heat from our subject's warm skin, upsetting the dynamic balance between heat gain and loss. The body temperature therefore starts to fall. Very rapidly, however, a variety of homeostatic responses occur to limit the fall. These are summarized in Figure 7–1. *The reader is urged to study Figure 7–1 and its legend carefully because the figure is typical of those used throughout the remainder of the book to illustrate homeostatic systems, and the legend emphasizes several conventions common to such figures.*

The first homeostatic response is that blood vessels to the skin narrow, reducing the amount of warm blood flowing through the skin and thus reducing heat loss. At a room temperature of 5°C, however, blood vessel constriction cannot completely eliminate the extra heat loss from the skin. Our subject curls up in order to reduce the surface area of the skin available for heat loss. This helps a bit, but excessive heat loss still continues, and body temperature keeps falling, although at a slower rate. He has a strong desire to put on more clothing—"voluntary" behavioral responses are often crucial events in homeostasis—but no clothing is available. Clearly, then, if excessive heat loss (output) cannot be prevented, the only way of restoring the balance between heat input and output is to increase input, and this is precisely what occurs. He begins to shiver, and the chemical reactions responsible

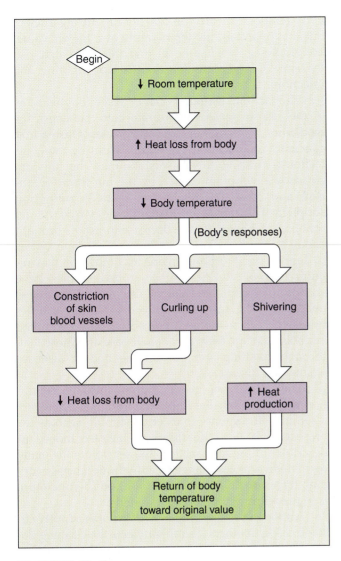

FIGURE 7–1

The homeostatic control system maintains a relatively constant body temperature when room temperature decreases. This flow diagram is typical of those used throughout the remainder of this book to illustrate homeostatic systems, and several conventions should be noted. (See also the legend for Figure 7–4.) The "begin" sign indicates where to start. The arrows next to each term within the boxes denote increases or decreases. The arrows connecting any two boxes in the figure denote cause and effect; that is, an arrow can be read as "causes" or "leads to." (For example, decreased room temperature "leads to" increased heat loss from the body.) In general, one should add the words "tends to" in thinking about these cause-and-effect relationships. For example, decreased room temperature tends to cause an increase in heat loss from the body, and curling up tends to cause a decrease in heat loss from the body. Qualifying the relationship in this way is necessary because variables like heat production and heat loss are under the influence of many factors, some of which oppose each other.

for the skeletal muscular contractions that constitute shivering produce large quantities of heat.

Indeed, heat production may transiently exceed heat loss so that body temperature begins to go back toward the value existing before the room temperature was lowered (Figure 7–2). It eventually stabilizes at a temperature *a bit below this original value;* at this new steady state, heat input and heat output are both higher than their original values but are once again equal to each other.

The thermoregulatory system just described is an example of a **negative-feedback** system, in which an increase or decrease in the variable being regulated brings about responses that tend to move the variable in the direction opposite ("negative" to) the direction of the original change. Thus, in our example, the *decrease* in body temperature led to responses that tended to *increase* the body temperature—that is, move it toward its original value.

Negative-feedback control systems are the most common homeostatic mechanisms in the body, but there is another type of feedback known as **positive feedback** in which an initial disturbance in a system sets off a train of events that increase the disturbance even further. Thus positive feedback does not favor stability and often abruptly displaces a system away from its normal set point. As we shall see, several

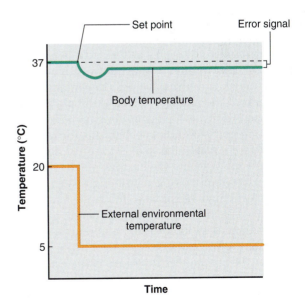

FIGURE 7–2

Changes in internal body temperature during exposure to a low external environmental temperature. As long as the environmental perturbation persists, the homeostatic responses do not return the regulated variable completely to its original value. The deviation from the original value is called the error signal.

important positive-feedback relationships occur in the body, contractions of the uterus during labor being one example.

Note that in our thermoregulatory example the negative-feedback system did not bring the person's temperature back completely to its original value. This illustrates another important generalization about homeostasis: Homeostatic control systems do not maintain *complete* constancy of the internal environment in the face of continued change in the external environment, but can only minimize changes. This is the reason we have said that homeostatic systems maintain the internal environment *relatively* stable. The explanation is that as long as the initiating event (exposure to cold, in our example) continues, some change in the regulated variable (the decrease in body temperature, in our example) must persist to serve as a signal to maintain the homeostatic responses. (This last statement will be qualified below.) Such a persisting signal is termed an **error signal** (Figure 7–2). This situation applies only when the initiating event continues; thus, in our example if the external temperature eventually goes back up to its original value, the homeostatic systems will be able to restore body temperature completely back to its original value.

It is essential to recognize that normally a control system does not overcompensate (that is, drive the system beyond the normal set point to create another physiological imbalance).

Inherent in the concept of error signals is another generalization about homeostasis: Even in reference to one individual, thus ignoring variation among persons, any regulated variable in the body cannot be assigned a single "normal" value but has a more-or-less narrow range of normal values, depending on the magnitude of the changes in the external conditions and the sensitivity of the responding homeostatic system. The more precise the mechanisms for regulating a variable are—that is, the smaller the error signal need be to drive the system—the narrower is the range. For example, the temperature-regulating systems of the body are extremely sensitive so that body temperature normally varies by only about 1°C even in the face of marked changes in the external environment or heat production during exercise.

As we have seen, perturbations in the external environment can displace a variable from its preexisting set point. In addition, the set points for many regulated variables can be physiologically altered or *reset*; that is, the values that the homeostatic control systems are "trying" to keep relatively constant can be altered. A common example is fever, the increase in body temperature that occurs in response to infection and that is analogous to raising the setting of your house's thermostat. The homeostatic control systems regulating body temperature are still functioning during a fever, but they maintain the temperature at a higher value. We shall see in Chapter 20 that this regulated rise in body temperature is adaptive for fighting the infection.

The fact that set points can be reset adaptively, as in the case of fever, raises important challenges for medicine, as another example illustrates. Plasma iron concentration decreases significantly during many infections. Until recently it was assumed that this decrease is a symptom caused by the infectious organism and that it should be treated with iron supplements. In fact, just the opposite is true: As described in Chapter 20, the decrease in iron is brought about by the body's defense mechanisms and serves to deprive the infectious organisms of the iron they require to replicate. Several controlled studies have shown that iron replacement can make the illness much worse. Clearly it is crucial to distinguish between those deviations of homeostatically controlled variables that are truly part of a disease and those that, through resetting, are part of the body's defenses against the disease.

The examples of fever and plasma iron concentration may have left the impression that set points are reset only in response to external stimuli, such as the presence of bacteria, but this is not the case. Indeed, as described in the next section, the set points for many regulated variables change on a rhythmical basis every day; for example, the set point for body temperature is higher during the day than at night.

Although the resetting of a set point is adaptive in some cases, in others it simply reflects the clashing demands of different regulatory systems. This brings us to one more generalization: It is not possible for everything to be maintained relatively constant by homeostatic control systems. In our example, body temperature was kept relatively constant, but only because large changes in skin blood flow and skeletal-muscle contraction were brought about by the homeostatic control system. Moreover, because so many properties of the internal environment are closely interrelated, it is often possible to keep one property relatively constant only by moving others farther from their usual set point. This is what we meant by "clashing demands."

The generalizations we have given concerning homeostatic control systems are summarized in Table 7–1. One additional point is that, as is illustrated by the regulation of body temperature, *multiple* systems frequently control a *single* parameter. The adaptive value of such redundancy is that it provides much greater fine-tuning and also permits regulation to occur even when one of the systems is not functioning properly because of disease.

TABLE 7–1 Some Important Generalizations About Homeostatic Control Systems

1. Stability of an internal environmental variable is achieved by balancing inputs and outputs. It is not the absolute magnitudes of the inputs and outputs that matter but the balance between them.

2. In negative-feedback systems, a change in the variable being regulated brings about responses that tend to move the variable in the direction opposite the original change—that is, back toward the initial value (set point).

3. Homeostatic control systems cannot maintain complete constancy of any given feature of the internal environment. Therefore, any regulated variable will have a more-or-less narrow range of normal values depending on the external environmental conditions.

4. The set point of some variables regulated by homeostatic control systems can be reset—that is, physiologically raised or lowered.

5. It is not possible for everything to be maintained relatively constant by homeostatic control systems. There is a hierarchy of importance, such that the constancy of certain variables may be altered markedly to maintain others at relatively constant levels.

Feedforward Regulation

Another type of regulatory process frequently used in conjunction with negative-feedback systems is feedforward. Let us give an example of feedforward and then define it. The temperature-sensitive nerve cells that trigger negative-feedback regulation of body temperature when body temperature begins to fall are located *inside* the body. In addition, there are temperature-sensitive nerve cells in the skin, and these cells, in effect, monitor *outside* temperature. When outside temperature falls, as in our example, these nerve cells immediately detect the change and relay this information to the brain, which then sends out signals to the blood vessels and muscles, resulting in heat conservation and increased heat production. In this manner, compensatory thermoregulatory responses are activated *before* the colder outside temperature can cause the internal body temperature to fall. Thus, **feedforward** regulation anticipates changes in a regulated variable such as internal body temperature, improves the speed of the body's homeostatic responses, and minimizes fluctuations in the level of the variable being regulated—that is, it reduces the amount of deviation from the set point.

In our example, feedforward control utilizes a set of "external environmental" detectors. It is likely, however, that most feedforward control is the result of a different phenomenon—learning. The first times they occur, early in life, perturbations in the external environment probably cause relatively large changes in regulated internal environmental factors, and in responding to these changes the central nervous system learns to anticipate them and resist them more effectively. A familiar form of this is learning to ride a bicycle with minimal swaying. Learning of this type probably explains many situations in which the error signals observed are extremely small or even undetectable despite profound perturbations in the environment.

Components of Homeostatic Control Systems

Reflexes

The thermoregulatory system we used as an example in the previous section, and many of the body's other homeostatic control systems, belong to the general category of stimulus-response sequences known as reflexes. Although in some reflexes we are aware of the stimulus and/or the response, many reflexes regulating the internal environment occur without any conscious awareness.

In the most narrow sense of the word, a **reflex** is a specific involuntary, unpremeditated, unlearned "built-in" response to a particular stimulus. Examples of such reflexes include pulling one's hand away from a hot object or shutting one's eyes as an object rapidly approaches the face. There are also many responses, however, that appear to be automatic and stereotyped but are actually the result of learning and practice. For example, an experienced driver performs many complicated acts in operating a car. To the driver these motions are, in large part, automatic, stereotyped, and unpremeditated, but they occur only because a great deal of conscious effort was spent learning them. We term such reflexes **learned,** or **acquired.** In general, most reflexes, no matter how basic they may appear to be, are subject to alteration by learning; that is, there is often no clear distinction between a basic reflex and one with a learned component.

The pathway mediating a reflex is known as the **reflex arc,** and its components are shown in Figure 7–3.

A **stimulus** is defined as a detectable change in the internal or external environment, such as a change in temperature, plasma potassium concentration, or blood pressure. A **receptor** detects the environmental change; we referred to the receptor as a "detector" earlier. A stimulus acts upon a receptor to produce a signal that is relayed to an integrating center. The pathway traveled by the signal between the receptor and the integrating center is known as the **afferent pathway** (the general term "afferent" means "to carry to," in this case, to the integrating center).

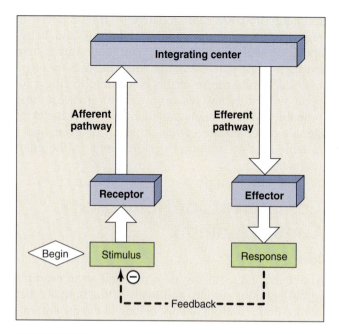

FIGURE 7–3

General components of a reflex arc that functions as a negative-feedback control system. The response of the system has the effect of counteracting or eliminating the stimulus. This phenomenon of negative feedback is emphasized by the minus sign in the dashed feedback loop.

An **integrating center** often receives signals from many receptors, some of which may be responding to quite different types of stimuli. Thus, the output of an integrating center reflects the net effect of the total afferent input; that is, it represents an integration of numerous bits of information.

The output of an integrating center is sent to the last component of the system, a device whose change in activity constitutes the overall response of the system. This component is known as an **effector.** The information going from an integrating center to an effector is like a command directing the effector to alter its activity. The pathway along which this information travels is known as the **efferent pathway** (the general term "efferent" means "to carry away from," in this case, away from the integrating center).

Thus far we have described the reflex arc as the sequence of events linking a stimulus to a response. If the response produced by the effector causes a decrease in the magnitude of the stimulus that triggered the sequence of events, then the reflex leads to negative feedback and we have a typical homeostatic control system. Not all reflexes are associated with such feedback. For example, the smell of food stimulates the secretion of a hormone by the stomach, but this hormone does not eliminate the smell of food (the stimulus).

To illustrate the components of a negative-feedback homeostatic reflex arc, let us use Figure 7–4 to apply these terms to thermoregulation. The temperature receptors are the endings of certain nerve cells in various parts of the body. They generate electric signals in the nerve cells at a rate determined by the temperature. These electric signals are conducted by the nerve fibers—the afferent pathway—to a specific part of the brain—the integrating center for temperature regulation. The integrating center, in turn, determines the signals sent out along those nerve cells that cause skeletal muscles and the muscles in skin blood vessels to contract. The nerve fibers to the muscles are the efferent pathway, and the muscles are the effectors. The dashed arrow and the ⊖ indicate the negative-feedback nature of the reflex.

Almost all body cells can act as effectors in homeostatic reflexes. There are, however, two specialized classes of tissues—muscle and gland—that are the major effectors of biological control systems. The physiology of glands is described in Chapter 6, that of muscle in Chapter 11.

Traditionally, the term "reflex" was restricted to situations in which the receptors, afferent pathway, integrating center, and efferent pathway were all parts of the nervous system, as in the thermoregulatory reflex. Present usage is not so restrictive, however, and recognizes that the principles are essentially the same when a blood-borne chemical messenger known as a **hormone,** rather than a nerve fiber, serves as the efferent pathway, or when a hormone-secreting gland (termed an **endocrine gland**) serves as the integrating center. Thus, in the thermoregulation example, the integrating center in the brain not only sends signals by way of nerve fibers, as shown in Figure 7–4, but also causes the release of a hormone that travels via the blood to many cells, where it produces an increase in the amount of heat produced by these cells. This hormone therefore also serves as an efferent pathway in thermoregulatory reflexes.

Accordingly, in our use of the term "reflex," we include hormones as reflex components. Moreover, depending on the specific nature of the reflex, the integrating center may reside either in the nervous system or in an endocrine gland. In addition, an endocrine gland may act as both receptor and integrating center in a reflex; for example, the endocrine-gland cells that secrete the hormone insulin themselves detect changes in the plasma glucose concentration.

In conclusion, many reflexes function in a homeostatic manner to keep a physical or chemical variable of the body relatively constant. One can analyze any such system by answering the questions listed in Table 7–2.

<table>
<tr><td>

TABLE 7–2 Questions to Be Asked About Any Homeostatic Reflex

1. What is the variable (for example, plasma potassium concentration, body temperature, blood pressure) that is maintained relatively constant in the face of changing conditions?

2. Where are the receptors that detect changes in the state of this variable?

3. Where is the integrating center to which these receptors send information and from which information is sent out to the effectors, and what is the nature of these afferent and efferent pathways?

4. What are the effectors, and how do they alter their activities so as to maintain the regulated variable near the set point of the system?

</td></tr>
</table>

Local Homeostatic Responses

In addition to reflexes, another group of biological responses is of great importance for homeostasis. We shall call them **local homeostatic responses.** They are initiated by a change in the external or internal environment (that is, a stimulus), and they induce an alteration of cell activity with the net effect of counteracting the stimulus. Like a reflex, therefore, a local response is the result of a sequence of events proceeding from a stimulus. Unlike a reflex, however, the entire sequence occurs only in the area of the stimulus. For example, damage to an area of skin causes cells in the damaged area to release certain chemicals that help the local defense against further injury. The significance of local responses is that they provide individual areas of the body with mechanisms for local self-regulation.

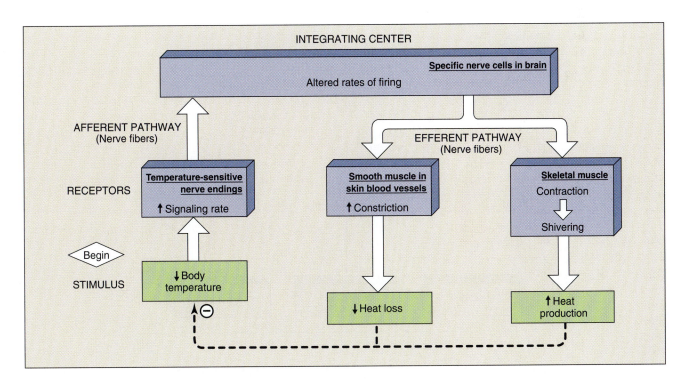

FIGURE 7–4

Reflex for minimizing the decrease in body temperature that occurs on exposure to a reduced external environmental temperature. This figure provides the internal components for the reflex shown in Figure 7–1. The dashed arrow and the ⊖ indicate the negative-feedback nature of the reflex, denoting that the reflex responses cause the decreased body temperature to return toward normal. Two flow-diagram conventions in addition to those described in Figure 7–1 are shown in this figure: (1) Blue 3-dimensional boxes always denote events that are occurring in anatomical structures (labelled in bold, underlined type at the upper right of the box); and (2) the phenomenon of negative feedback is denoted by a circled minus sign at the end of a dashed arrow.

Intercellular Chemical Messengers

Essential to reflexes and local homeostatic responses, and therefore to homeostasis, is the ability of cells to communicate with one another. In the vast majority of cases, this communication *between* cells—*inter*cellular communication—is performed by chemical messengers. There are three categories of such messengers: hormones, neurotransmitters, and paracrine agents (Figure 7–5).

A hormone functions as a chemical messenger that enables the hormone-secreting cell to communicate with the cell acted upon by the hormone—its **target cell**—with the blood acting as the delivery service. Most nerve cells communicate with each other or with effector cells by means of chemical messengers called **neurotransmitters.** Thus, one nerve cell alters the activity of another by releasing from its ending a neurotransmitter that diffuses through the extracellular fluid separating the two nerve cells and acts upon the second cell. Similarly, neurotransmitters released from

nerve cells into the extracellular fluid in the immediate vicinity of effector cells constitute the controlling input to the effector cells.

As described more fully in Chapter 10, the chemical messengers released by certain nerve cells act neither on adjacent nerve cells nor on adjacent effector cells but rather enter the bloodstream to act on target cells elsewhere in the body. For this reason, these messengers are properly termed hormones (or neurohormones), not neurotransmitters.

Chemical messengers participate not only in reflexes but also in local responses. Chemical messengers involved in local communication between cells are known as **paracrine agents.**

Paracrine/Autocrine Agents

Paracrine agents are synthesized by cells and released, once given the appropriate stimulus, into the extracellular fluid. They then diffuse to neighboring cells, some of which are their target cells. (Note that, given this broad definition, neurotransmitters theoretically

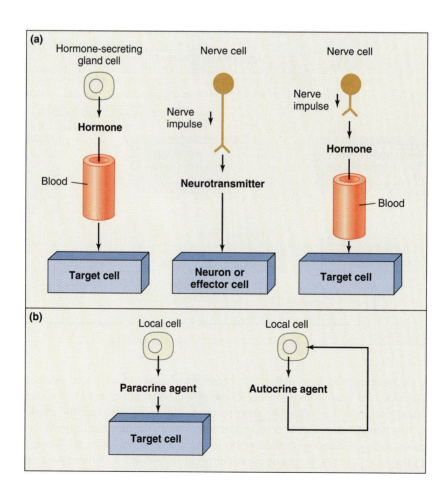

FIGURE 7–5

Categories of chemical messengers. (a) Reflexes. Note that chemical messengers that are secreted by nerve cells and act on adjacent nerve cells or effector cells are termed neurotransmitters, whereas those that enter the blood and act on distant effector cells (synonymous with target cells) are classified as hormones (also termed neurohormones). (b) Local homeostatic responses. With the exception of autocrine agents, all messengers act between cells—that is, *inter*cellularly.

could be classified as a subgroup of paracrine agents, but by convention they are not.) Paracrine agents are generally inactivated rapidly by locally existing enzymes so that they do not enter the bloodstream in large quantities.

There is one category of local chemical messengers that are not *inter*cellular messengers—that is, they do not communicate *between* cells. Rather, the chemical is secreted by a cell into the extracellular fluid and then acts upon the very cell that secreted it. Such messengers are termed **autocrine agents** (Figure 7–5). Frequently a messenger may serve both paracrine and autocrine functions simultaneously—that is, molecules of the messenger released by a cell may act locally on adjacent cells as well as on the same cell that released the messenger.

One of the most exciting developments in physiology today is the identification of a seemingly endless number of paracrine/autocrine agents and the extremely diverse effects they exert. Their structures span the gamut from a simple gas (nitric oxide) to fatty acid derivatives (the eicosanoids, see below) to peptides and amino acid derivatives. They tend to be secreted by multiple cell types in many tissues and organs. According to their structures and functions, they can be gathered into families; for example, one such family constitutes the "growth factors," encompassing more than 50 distinct molecules, each of which is highly effective in stimulating certain cells to divide and/or differentiate.

Stimuli for the release of paracrine/autocrine agents are also extremely varied, including not only local chemical changes (for example, in the concentration of oxygen), but neurotransmitters and hormones as well. In these two latter cases, the paracrine/autocrine agent often serves to oppose the effects induced locally by the neurotransmitter or hormone. For example, the neurotransmitter norepinephrine strongly constricts blood vessels in the kidneys, but it simultaneously causes certain kidney cells to secrete paracrine agents that cause the same vessels to dilate. This provides a local negative feedback, in which the paracrine agents keep the action of norepinephrine from becoming too intense.

In other cases, a neurotransmitter or hormone may stimulate the local release of a paracrine/autocrine agent that actually is responsible for causing the cellular response to that neurotransmitter or hormone. For example, most of the growth-promoting effects of growth hormone on bone are not exerted *directly* on bone cells by this hormone; rather, growth hormone stimulates the release from the bone cells of a paracrine/autocrine agent that then stimulates the bone growth.

Eicosanoids The general approach of this text is to describe the specific chemical messengers in the context of the functions they influence. However, one set of paracrine/autocrine agents exerts such a wide variety of effects in virtually every tissue and organ system that it is best described separately at this point. These are the **eicosanoids**, a family of substances produced from the polyunsaturated fatty acid **arachidonic acid,** which is present in plasma-membrane phospholipids. The eicosanoids include the **cyclic endoperoxides**, the **prostaglandins**, the **thromboxanes**, and the **leukotrienes** (Figure 7–6).

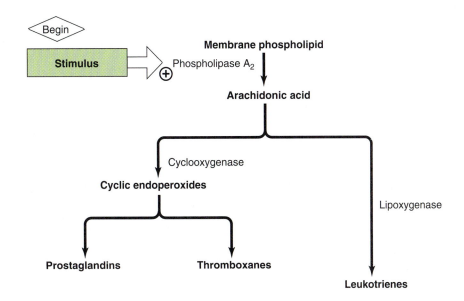

FIGURE 7–6

Pathways for the synthesis of eicosanoids. Phospholipase A_2 is the one enzyme common to the formation of all the eicosanoids; it is the site at which stimuli act. Anti-inflammatory steroids inhibit phospholipase A_2. The step mediated by cyclooxygenase is inhibited by aspirin and other nonsteroidal anti-inflammatory drugs (termed NSAIDs).

The synthesis of eicosanoids begins when an appropriate stimulus—hormone, neurotransmitter, paracrine agent, drug, or toxic agent—activates an enzyme, **phospholipase A_2,** in the plasma membrane of the stimulated cell. As shown in Figure 7–6, this enzyme splits off arachidonic acid from the membrane phospholipids, and the arachidonic acid can then be metabolized by two pathways. One pathway is initiated by an enzyme called **cyclooxygenase (COX)** and leads ultimately to formation of the cyclic endoperoxides, prostaglandins, and thromboxanes. The other pathway is initiated by the enzyme **lipoxygenase** and leads to formation of the leukotrienes. Within both of these pathways, synthesis of the various specific eicosanoids is enzyme-mediated. Accordingly, beyond phospholipase A_2, the eicosanoid-pathway enzymes found in a particular cell determine which eicosanoids the cell synthesizes in response to a stimulus.

Each of the major eicosanoid subdivisions contains more than one member, as indicated by the use of the plural in referring to them (prostaglandins, for example). On the basis of structural differences, the different molecules within each subdivision are designated by a letter—for example, PGA and PGE for prostaglandins of the A and E types—which then may be further subdivided—for example, PGE_2.

Once synthesized in response to a stimulus, the eicosanoids are not stored to any extent but are released immediately and act locally; accordingly, the eicosanoids are categorized as paracrine and autocrine agents. After they act, they are quickly metabolized by local enzymes to inactive forms. The eicosanoids exert a bewildering array of effects, many of which we will describe in future chapters.

Finally, a word about drugs that influence the eicosanoid pathway since these are perhaps the most commonly used drugs in the world today. At the top of the list must come *aspirin,* which inhibits cyclooxygenase and, therefore, blocks the synthesis of the endoperoxides, prostaglandins, and thromboxanes. It and the new drugs that also block cyclooxygenase are collectively termed *nonsteroidal anti-inflammatory drugs (NSAIDs).* Their major uses are to reduce pain, fever, and inflammation. The term "nonsteroidal" distinguishes them from the *adrenal steroids* (Chapters 10 and 20) that are used in large doses as anti-inflammatory drugs; these steroids inhibit phospholipase A_2 and thus block the production of all eicosanoids.

Conclusion

A point of great importance must be emphasized here to avoid later confusion: A nerve cell, endocrine gland cell, or other cell type may all secrete the same chemical messenger. Thus, a particular messenger may sometimes function as a neurotransmitter, as a hormone, or as a paracrine/autocrine agent.

All types of intercellular communication described so far in this section involve secretion of a chemical messenger into the extracellular fluid. However, there are two important types of chemical communication between cells that do not require such secretion. In the first type, which occurs via gap junctions (Chapter 3), chemicals move from one cell to an adjacent cell without ever entering the extracellular fluid. In the second type, the chemical messenger is not actually released from the cell producing it but rather is located in the plasma membrane of that cell; when the cell encounters another cell type capable of responding to the message, the two cells link up via the membrane-bound messenger. This type of signaling (sometimes termed "juxtacrine") is of particular importance in the growth and differentiation of tissues as well as in the functioning of cells that protect the body against microbes and other foreign cells (Chapter 20).

Processes Related to Homeostasis

A variety of seemingly unrelated processes, such as biological rhythms and aging, have important implications for homeostasis and are discussed here to emphasize this point.

Acclimatization

The term **adaptation** denotes a characteristic that favors survival in specific environments. Homeostatic control systems are inherited biological adaptations. An individual's ability to respond to a particular environmental stress is not fixed, however, but can be enhanced, with no change in genetic endowment, by prolonged exposure to that stress. This type of adaptation—the improved functioning of an already existing homeostatic system—is known as **acclimatization.**

Let us take sweating in response to heat exposure as an example and perform a simple experiment. On day 1 we expose a person for 30 min to a high temperature and ask her to do a standardized exercise test. Body temperature rises, and sweating begins after a certain period of time. The sweating provides a mechanism for increasing heat loss from the body and thus tends to minimize the rise in body temperature in a hot environment. The volume of sweat produced under these conditions is measured. Then, for a week, our subject enters the heat chamber for 1 or 2 h per day and exercises. On day 8, her body temperature and sweating rate are again measured during the same exercise test performed on day 1; the striking finding is that the subject begins to sweat earlier and much more profusely than she did on day 1. Accordingly, her body temperature does not rise to nearly the same degree.

The subject has become acclimatized to the heat; that is, she has undergone an adaptive change induced by repeated exposure to the heat and is now better able to respond to heat exposure.

The precise anatomical and physiological changes that bring about increased capacity to withstand change during acclimatization are highly varied. Typically, they involve an increase in the number, size, or sensitivity of one or more of the cell types in the homeostatic control system that mediates the basic response.

Acclimatizations are usually completely reversible. Thus, if the daily exposures to heat are discontinued, the sweating rate of our subject will revert to the preacclimatized value within a relatively short time. If an acclimatization is induced very early in life, however, at the **critical period** for development of a structure or response, it is termed a **developmental acclimatization** and may be irreversible. For example, the barrel-shaped chests of natives of the Andes Mountains represent not a genetic difference between them and their lowland compatriots but rather an irreversible acclimatization induced during the first few years of their lives by their exposure to the low-oxygen environment of high altitude. The altered chest size remains even though the individual moves to a lowland environment later in life and stays there. Lowland persons who have suffered oxygen deprivation from heart or lung disease during their early years show precisely the same chest shape.

Biological Rhythms

A striking characteristic of many body functions is the rhythmical changes they manifest. The most common type is the **circadian rhythm,** which cycles approximately once every 24 h. Waking and sleeping, body temperature, hormone concentrations in the blood, the excretion of ions into the urine, and many other functions undergo circadian variation (Figure 7–7). Other cycles have much longer periods, the menstrual cycle (approximately 28 days) being the most well known.

What have biological rhythms to do with homeostasis? They add yet another "anticipatory" component to homeostatic control systems, in effect a feedforward system operating without detectors. The negative-feedback homeostatic responses we described earlier in this chapter are *corrective* responses, in that they are initiated *after* the steady state of the individual has been perturbed. In contrast, biological rhythms enable homeostatic mechanisms to be utilized immediately and automatically by activating them at times when a challenge is *likely* to occur but before it actually does occur. For example, there is a rhythm in the urinary excretion of potassium such that excretion is high during the day and low at night. This makes

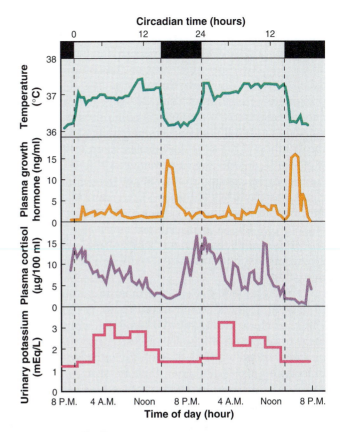

FIGURE 7–7

Circadian rhythms of several physiological variables in a human subject with room lights on (open bars at top) for 16 h and off (black bars at top) for 8 h. As is usual in dealing with rhythms, we have used a 24-h clock in which both 0 and 24 designate midnight and 12 designates noon. Cortisol is a hormone secreted by the adrenal glands.

Adapted from Moore-Ede and Sulzman.

sense since we ingest potassium in our food during the day, not at night when we are asleep. Therefore, the total amount of potassium in the body fluctuates less than if the rhythm did not exist.

A crucial point concerning most body rhythms is that they are *internally* driven. Environmental factors do not drive the rhythm but rather provide the timing cues important for **entrainment** (that is, setting of the actual hours) of the rhythm. A classic experiment will clarify this distinction.

Subjects were put in experimental chambers that completely isolated them from their usual external environment. For the first few days, they were exposed to a 24 h rest-activity cycle in which the room lights were turned on and off at the same time each day. Under these conditions, their sleep-wake cycles were 24 h long. Then, all environmental time cues were eliminated, and the individuals were allowed to control the lights themselves. Immediately, their sleep-wake patterns began to change. On the average, bedtime began about 30 min later *each day* and so did wake-up time. Thus a sleep-wake cycle persisted in the complete absence of environmental cues, and such a rhythm is called a **free-running rhythm.** In this case it was approximately 25 h rather than 24. This indicates that cues are required to entrain a circadian rhythm to 24 h.

One more point should be mentioned: By altering the duration of the light-dark cycles, sleep-wake cycles can be entrained to any value between 23 and 27 h, but shorter or longer durations cannot be entrained; instead, the rhythm continues to free-run. Because of this, people whose work causes them to adopt sleep-wake cycles longer than 27 h are never able to make the proper adjustments and achieve stable rhythms. The result is symptoms similar to those of jet lag, to be described later.

The light-dark cycle is the most important environmental time cue in our lives but not the only one. Others include external environmental temperature, meal timing, and many social cues. Thus, if several people were undergoing the experiment just described in isolation from each other, their free-runs would be somewhat different, but if they were all in the same room, social cues would entrain all of them to the same rhythm.

Environmental time cues also function to **phase-shift** rhythms—in other words, to reset the internal clock. Thus if one jets west or east to a different time zone, the sleep-wake cycle and other circadian rhythms slowly shift to the new light-dark cycle. These shifts take time, however, and the disparity between external time and internal time is one of the causes of the symptoms of *jet lag*—disruption of sleep, gastrointestinal disturbances, decreased vigilance and attention span, and a general feeling of malaise.

Similar symptoms occur in workers on permanent or rotating night shifts. Such individuals generally do not adapt to these schedules even after several years because they are exposed to the usual outdoor light-dark cycle (normal indoor lighting is too dim to function as a good entrainer). In recent experiments, night-shift workers were exposed to extremely bright indoor lighting while they worked and 8 h of total darkness during the day when they slept. This schedule produced total adaptation to the night-shift work within 5 days.

What is the neural basis of body rhythms? In the part of the brain called the hypothalamus is a specific collection of nerve cells (the suprachiasmatic nucleus) that function as the principal **pacemaker** (time clock) for circadian rhythms. How it "keeps time" independent of any external environmental cues is not really understood, but it probably involves the rhythmical turning on and off of critical genes in the pacemaker cells. Indeed, just such genes—one has been named *clock*—and the proteins that they code have recently been discovered in the mouse pacemaker—that is, in the cells of the mouse's suprachiasmatic nucleus.

The pacemaker receives input from the eyes and many other parts of the nervous system, and these inputs mediate the entrainment effects exerted by the external environment. In turn, the pacemaker sends out neural signals to other parts of the brain, which then influence the various body systems, activating some and inhibiting others. One output of the pacemaker is to the **pineal gland,** an offshoot of the brain that secretes the hormone **melatonin** (Chapter 10); these neural signals from the pacemaker cause the pineal to secrete melatonin during darkness but not to secrete it during daylight. It has been hypothesized, therefore, that melatonin may act as an important "middleman" to influence other organs either directly or by altering the activity of the parts of the brain that control these organs. Studies to determine whether administration of melatonin at specific times can reduce the symptoms of jet lag remain inconclusive (the same can be said for virtually every other proposed treatment for jet lag).

It should not be surprising that rhythms have effects on the body's resistance to various stresses and responses to different drugs. Also, certain diseases have characteristic rhythms. For example, *heart attacks* are almost twice as common in the first hours after waking, and *asthma* frequently flares at night. Insights about these rhythms have already been incorporated into therapy; for example, once-a-day timed-release pills for asthma are taken at night and deliver a high dose of medication between midnight and 6 A.M.

Regulated Cell Death: Apoptosis

It is obvious that the proliferation and differentiation of cells are important for the development and maintenance of homeostasis in multicellular organisms. Only recently, however, have physiologists come to appreciate the contribution of another characteristic shared by virtually all cells—the ability to self-destruct by activation of an intrinsic cell suicide program. This type of cell death, termed **apoptosis,** plays important roles in the sculpting of a developing organism and in the elimination of undesirable cells (for example, cells that have become cancerous), but it is particularly crucial for regulating the number of cells in a tissue or organ. Thus, the control of cell number within each cell lineage is normally determined by a balance between cell proliferation and cell death, both of which are regulated processes. For example, white blood cells called neutrophils are programmed to die by apoptosis 24 hours after they are produced in the bone marrow.

Apoptosis occurs by controlled autodigestion of the cell contents. Within a cell, endogenous enzymes are activated that break down the cell nucleus and its DNA, as well as other cell organelles. Importantly, the plasma membrane is maintained as the cell dies so that the cell contents are not dispersed. Instead the apoptotic cell sends out chemical messengers that attract neighboring phagocytic cells (cells that "eat" matter or other cells), which engulf and digest the dying or dead cell. In this way the leakage of breakdown products, many of which are toxic, from apoptotic cells is prevented. Apoptosis is, therefore, very different from the death of a cell due to externally imposed injury; in that case (termed necrosis) the plasma membrane is disrupted, and the cell swells and releases its cytoplasmic material, inducing an inflammatory response, as described in Chapter 20.

The fact that virtually all normal cells contain the enzymes capable of carrying out apoptosis means that these enzymes must normally remain inactive if the cell is to survive. In most tissues this inactivity is maintained by the constant supply to the cell of a large number of chemical "survival signals" provided by neighboring cells, hormones, and the extracellular matrix. In other words, most cells are programmed to commit suicide if survival signals are not received from the internal environment. For example, prostate-gland cells undergo apoptosis when the influence on them of testosterone, the male sex hormone, is removed. In addition, there are other chemical signals, some exogenous to the organism (for example, certain viruses and bacterial toxins) and some endogenous (for example, certain messengers released by nerve cells and white blood cells) that can inhibit or override survival signals and induce the cell to undergo apoptosis.

It is very likely that abnormal inhibition of appropriate apoptosis may contribute to diseases, like cancer, characterized by excessive numbers of cells. At the other end of the spectrum, too high a rate of apoptosis probably contributes to degenerative diseases, such as that of bone in the disease called *osteoporosis.* The hope is that therapies designed to enhance or decrease apoptosis, depending on the situation, would ameliorate these diseases.

Aging

The physiological manifestations of aging are a gradual deterioration in the function of virtually all tissues and organ systems and in the capacity of the body's homeostatic control systems to respond to environmental stresses.

Aging represents the operation of a distinct process that is distinguishable from those diseases, such as heart disease, frequently associated with aging. Aging is typified by a decrease in the number of cells in the body, due to some combination of decreased cell division and increased cell death, and by malfunction of many of the cells that remain. The immediate cause of these changes is probably an interference in the function of the cells' macromolecules, particularly DNA. The crucial question is, what causes the interference?

With regard to the decreased cell division typical of aging, it is likely that cells have a built-in limit to the number of times they can divide. As described in Chapter 5, this limit is set by the fact that the DNA of the cell loses a portion of its terminal segment—the telomere—each time it replicates prior to cell division; therefore, after a certain number of divisions, a cell's telomere is completely gone, and the DNA of that cell will no longer be able to replicate.

In addition to this basic limitation, there are almost certainly other factors—both genetic and environmental—that act on cells' macromolecules to influence the ability of cells to divide and function. For example, there is progressive accumulation of damage to macromolecules as a result of the toxic effects of free radicals produced during oxidative metabolism, and other reactive oxygen molecules.

Whatever the precise factors, studies in humans (for example, of twins) indicate that about one-third of the variability in life span among individuals can be ascribed to their genes, and the remaining two-thirds to their differing environments. What kinds of genes (other than the one coding for telomerase) might be most likely to influence aging? The strongest candidates are genes that code for proteins that regulate the processes of cellular and macromolecular maintenance and repair. In keeping with this view is the fact that a rare inherited disease, *Werner's syndrome*, which is characterized by premature aging, is caused by mutation of a single gene, one that is critical for normal DNA replication or repair. Other kinds of genes that should, in theory, be important for aging are those that code for proteins that are important in cellular responses to various forms of stress or contribute to preventing or ameliorating the effects of toxic molecules such as free radicals.

It is difficult to sort out the extent to which any particular age-related change in physiological function is due to aging itself and the extent to which it is secondary to disease and lifestyle changes. For example, until recently it was believed that the functioning of the nervous system markedly deteriorates as a result of aging per se, but this view is incorrect. It was based on studies of the performance of individuals with age-related diseases. Studies of people without such diseases do document changes, including loss of memory, increased difficulty in learning new tasks, decrease in the speed of processing by the brain, and loss of brain mass, but these changes are relatively modest. Most brain functions considered to underlie intelligence seem to remain relatively intact.

Can the aging process be inhibited or at least slowed down? One factor—physical exercise—has been shown to prolong life in human beings, although it is not clear that it does so by altering the aging process itself. In various chapters we shall mention the beneficial effects of exercise in various diseases (coronary heart disease and diabetes mellitus, for example), and the explanations for these benefits are usually logical in terms of the pathophysiology of the diseases. But the remarkable finding has been that persons who participate in aerobic sports activities of even moderate intensity have a significantly lower risk of dying *from any cause.* Moreover, such persons are much less likely to develop life-disturbing disabilities. It is this nonspecific aspect of exercise's benefits that raises the question as to whether being physically fit somehow alters the aging process itself rather than simply the diseases associated with aging.

A second approach—marked restriction of calories, but with enough protein, fat, vitamins, and minerals provided to prevent malnutrition—is the only intervention that has consistently been shown to prolong life in experimental animals. This type of caloric restriction increases not only the *average* life span of the animals but also the maximum span—that is, the lifetime of the longest-surviving members of the group. It is this increase in maximum life span that indicates that caloric restriction is influencing the basic aging process, not simply postponing the major diseases that are common late in life (caloric restriction does that, too). How it delays aging—one theory is that it reduces the formation of free radicals—and the relevance of these findings, if any, to human beings is still unclear.

Balance in the Homeostasis of Chemicals

Many homeostatic systems are concerned with the balance between the addition to and removal from the body of a chemical substance. Figure 7–8 is a generalized schema of the possible pathways involved in such balance. The **pool** occupies a position of central importance in the balance sheet. It is the body's readily available quantity of the particular substance and is frequently identical to the amount present in the extracellular fluid. The pool receives substances from and contributes them to all the pathways.

The pathways on the left of the figure are sources of net gain to the body. A substance may enter the body through the gastrointestinal (GI) tract or the lungs. Alternatively, a substance may be synthesized within the body from other materials.

The pathways on the right of the figure are causes of net loss from the body. A substance may be lost in the urine, feces, expired air, or menstrual fluid, as well as from the surface of the body as skin, hair, nails, sweat, and tears. The substance may also be chemically altered and thus removed by metabolism.

The central portion of the figure illustrates the distribution of the substance within the body. The substance may be taken from the pool and accumulated in storage depots (for example, the accumulation of fat in adipose tissue). Conversely, it may leave the storage depots to reenter the pool. Finally, the substance may be incorporated reversibly into some other molecular structure, such as fatty acids into membranes or iodine into thyroxine. Incorporation is reversible in that the substance is liberated again whenever the more complex structure is broken down. This pathway is distinguished from storage in that the incorporation of the substance into the other molecules produces new molecules with specific functions.

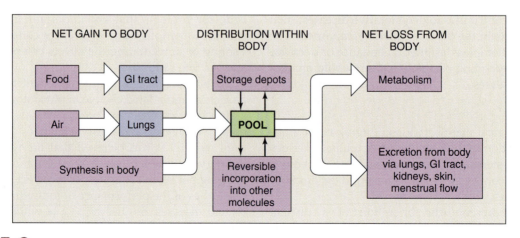

FIGURE 7–8
Balance diagram for a chemical substance.

It should be recognized that not every pathway of this generalized schema is applicable to every substance. For example, mineral electrolytes such as sodium cannot be synthesized, do not normally enter through the lungs, and cannot be removed by metabolism.

The orientation of Figure 7–8 illustrates two important generalizations concerning the balance concept: (1) During any period of time, total-body balance depends upon the relative rates of net gain and net loss to the body; and (2) the pool concentration depends not only upon the total amount of the substance in the body, but also upon exchanges of the substance *within* the body.

For any chemical, three states of total-body balance are possible: (1) Loss exceeds gain, so that the total amount of the substance in the body is decreasing, and the person is said to be in **negative balance;** (2) gain exceeds loss, so that the total amount of the substance in the body is increasing, and the person is said to be in **positive balance;** and (3) gain equals loss, and the person is in **stable balance.**

Clearly a stable balance can be upset by alteration of the amount being gained or lost in any single pathway in the schema; for example, severe negative water balance can be caused by increased sweating. Conversely, stable balance can be restored by homeostatic control of water intake and output.

Let us take sodium balance as another example. The control systems for sodium balance have as their targets the kidneys, and the systems operate by inducing the kidneys to excrete into the urine an amount of sodium approximately equal to the amount ingested daily. In this example, we assume for simplicity that all sodium loss from the body occurs via the urine. Now imagine a person with a daily intake and excretion of 7 g of sodium—a moderate intake for most Americans—and a stable amount of sodium in her body (Figure 7–9). On day 2 of our experiment, the subject changes her diet so that her daily sodium consumption rises to 15 g—a fairly large but commonly observed intake—*and remains there indefinitely.* On this same day, the kidneys excrete into the urine somewhat more than 7 g of sodium, but not all the ingested 15 g. The result is that some excess sodium is retained in the body on that day—that is, the person is in positive sodium balance. The kidneys do somewhat better on day 3, but it is probably not until day 4 or 5 that they are excreting 15 g. From this time on, output from the body once again equals input, and sodium balance is once again stable. (The delay of several days before stability is reached is quite typical for the kidneys' handling of sodium, but should not be assumed to apply to other homeostatic responses, most of which are much more rapid.)

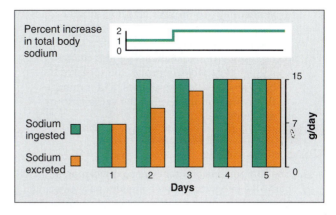

FIGURE 7–9

Effects of a continued change in the amount of sodium ingested on sodium excretion and total-body sodium balance. Stable sodium balance is reattained by day 4 but with some gain of total-body sodium.

But , and this is an important point, although again in stable balance, the woman has perhaps 2 percent more sodium in her body than was the case when she was in stable balance ingesting 7 g. It is this 2 percent extra body sodium that constitutes the continuous error signal to the control systems driving the kidneys to excrete 15 g/day rather than 7 g/day. [Recall the generalization (Table 7–1, no. 3) that homeostatic control systems cannot maintain complete constancy of the internal environment *in the face of continued change in the perturbing event* since some change in the regulated variable (body sodium content in our example) must persist to serve as a signal to maintain the compensating responses.] An increase of 2 percent does not seem large, but it has been hypothesized that this small gain might facilitate the development of high blood pressure (**hypertension**) in some persons.

SECTION A SUMMARY

General Characteristics of Homeostatic Control Systems

I. Homeostasis denotes the stable conditions of the internal environment that result from the operation of compensatory homeostatic control systems.
   a. In a negative-feedback control system, a change in the variable being regulated brings about responses that tend to push the variable in the direction opposite to the original change. Negative feedback minimizes changes from the set point of the system, leading to stability.
   b. In a positive-feedback system, an initial disturbance in the system sets off a train of events that increases the disturbance even further.

c. Homeostatic control systems minimize changes in the internal environment but cannot maintain complete constancy.

d. Feedforward regulation anticipates changes in a regulated variable, improves the speed of the body's homeostatic responses, and minimizes fluctuations in the level of the variable being regulated.

Components of Homeostatic Control Systems

I. The components of a reflex arc are receptor, afferent pathway, integrating center, efferent pathway, and effector. The pathways may be neural or hormonal.

II. Local homeostatic responses are also stimulus-response sequences, but they occur only in the area of the stimulus, neither nerves nor hormones being involved.

III. Intercellular communication is essential to reflexes and local responses and is achieved by neurotransmitters, hormones, and paracrine agents. Less common is intercellular communication through either gap junctions or cell-bound messengers.

IV. The eicosanoids are a widespread family of messenger molecules derived from arachidonic acid. They function mainly as paracrine and autocrine agents in local responses.

a. The first step in production of the eicosanoids is the splitting-off of arachidonic acid from plasma membrane phospholipids by the action of phospholipase A_2.

b. There are two pathways from arachidonic acid, one mediated by cyclooxygenase and leading to the formation of prostaglandins and thromboxanes, and the other mediated by lipoxygenase and leading to the formation of leukotrienes.

Processes Related to Homeostasis

I. Acclimatization is an improved ability to respond to an environmental stress.

a. The improvement is induced by prolonged exposure to the stress with no change in genetic endowment.

b. If acclimatization occurs early in life, it may be irreversible and is known as a developmental acclimatization.

II. Biological rhythms provide a feedforward component to homeostatic control systems.

a. The rhythms are internally driven by brain pacemakers, but are entrained by environmental cues, such as light, which also serve to phase-shift (reset) the rhythms when necessary.

b. In the absence of cues, rhythms free-run.

III. Apoptosis, regulated cell death, plays an important role in homeostasis by helping to regulate cell numbers and eliminating undesirable cells.

IV. Aging is associated with a decrease in the number of cells in the body and with a disordered functioning of many of the cells that remain.

a. It is a process distinct from the diseases associated with aging.

b. Its physiological manifestations are a deterioration in organ-system function and in the capacity to respond homeostatically to environmental stresses.

V. The balance of substances in the body is achieved by a matching of inputs and outputs. Total body balance of a substance may be negative, positive, or stable.

SECTION A KEY TERMS

homeostasis	eicosanoid
homeostatic control system	arachidonic acid
	cyclic endoperoxide
steady state	prostaglandin
set point	thromboxane
negative feedback	leukotriene
positive feedback	phospholipase A_2
error signal	cyclooxygenase (COX)
feedforward	lipoxygenase
reflex	adaptation
learned reflex	acclimatization
acquired reflex	critical period
reflex arc	developmental
stimulus	acclimatization
receptor (in reflex)	circadian rhythm
afferent pathway	entrainment
integrating center	free-running rhythm
effector	phase-shift
efferent pathway	pacemaker
hormone	pineal gland
endocrine gland	melatonin
local homeostatic response	apoptosis
	pool
target cell	negative balance
neurotransmitter	positive balance
paracrine agent	stable balance
autocrine agent	

SECTION A REVIEW QUESTIONS

1. Describe five important generalizations about homeostatic control systems.

2. Contrast negative-feedback systems and positive-feedback systems.

3. Contrast feedforward and negative feedback.

4. How do error signals develop, and why are they essential for maintaining homeostasis?

5. List the components of a reflex arc.

6. What is the basic difference between a local homeostatic response and a reflex?

7. List the general categories of intercellular messengers.

8. Describe two types of intercellular communication that do not depend on extracellular chemical messengers.

9. Draw a figure illustrating the various pathways for eicosanoid synthesis.

10. Describe the conditions under which acclimatization occurs. In what period of life might an acclimatization be irreversible? Are acclimatizations passed on to a person's offspring?
11. Under what conditions do circadian rhythms become free-running?
12. How do phase shifts occur?
13. What are the important environmental cues for entrainment of body rhythms?
14. What are the physiological manifestations of aging?
15. Draw a figure illustrating the balance concept in homeostasis.
16. What are the three possible states of total-body balance of any chemical?

MECHANISMS BY WHICH CHEMICAL MESSENGERS CONTROL CELLS

Receptors

The vast majority of homeostatic systems require cell-to-cell communication via chemical messengers. The first step in the action of any intercellular chemical messenger is the binding of the messenger to specific target-cell proteins known as **receptors.** In the general language of Chapter 4, a chemical messenger is a "ligand," and the receptor is a "binding site." The binding of a messenger to a receptor then initiates a sequence of events in the cell leading to the cell's response to that messenger.

The term "receptor" can be the source of confusion because the same word is used to denote the "detectors" in a reflex arc, as described earlier in this chapter. The reader must keep in mind the fact that "receptor" has two totally distinct meanings, but the context in which the term is used usually makes it quite clear which is meant.

What is the nature of the receptors with which intercellular chemical messengers combine? They are proteins (or glycoproteins) located either in the cell's plasma membrane or inside the cell, mainly in the nucleus. The plasma membrane is the much more common location, applying to the very large number of messengers that are lipid-insoluble and so do not traverse the lipid-rich plasma membrane. In contrast, the much smaller number of lipid-soluble messengers pass through membranes (mainly by diffusion but, in some cases, by mediated transport as well) to bind to their receptors inside the cell.

Plasma-membrane receptors are transmembrane proteins; that is, they span the entire membrane thickness. A typical plasma-membrane receptor is illustrated in Figure 7–10. Like other transmembrane proteins, a plasma-membrane receptor has segments within the membrane, one or more segments extending out from the membrane into the extracellular fluid, and other segments extending into the intracellular fluid. It is to the extracellular portions that the mes-

senger binds. Like other transmembrane proteins, a receptor is often composed of two or more nonidentical subunits bound together.

It is the combination of chemical messenger and receptor that initiates the events leading to the cell's response. The existence of receptors explains a very important characteristic of intercellular communication—**specificity** (see Table 7–3 for a glossary of terms concerning receptors). Although a chemical messenger (hormone, neurotransmitter, paracrine/autocrine agent, or plasma-membrane–bound messenger) may come into contact with many different cells, it influences only certain cells and not others. The explanation is that cells differ in the types of receptors they contain. Accordingly, only certain cell types, frequently just one, possess the receptor required for combination with a given chemical messenger (Figure 7–11). In many cases, the receptors for a group of messengers are closely related structurally; thus, for example, endocrinologists refer to "superfamilies" of hormone receptors.

Where different types of cells possess the same receptors for a particular messenger, the responses of the various cell types to that messenger may differ from each other. For example, the neurotransmitter norepinephrine causes the smooth muscle of blood vessels to contract, but via the same type of receptor, norepinephrine causes endocrine cells in the pancreas to secrete less insulin. In essence, then, the receptor functions as a molecular "switch" that elicits the cell's response when "switched on" by the messenger binding to it. Just as identical types of switches can be used to turn on a light or a radio, a single type of receptor can be used to produce quite different responses in different cell types.

Similar reasoning explains a more surprising phenomenon: A single cell may contain several different receptor types *for a single messenger*. Combination of the messenger with one of these receptor types may produce a cellular response quite different from,

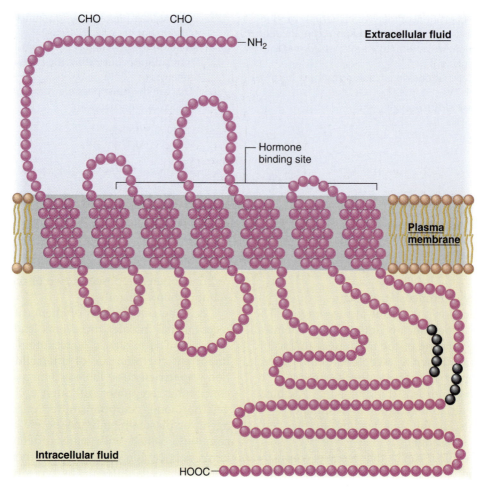

FIGURE 7–10

Structure of a human receptor that binds the hormone epinephrine. The seven clusters of amino acids in the plasma membrane represent hydrophobic portions of the protein's alpha helix. Note that the binding site for the hormone includes several of the segments that extend into the extracellular fluid. The amino acids denoted by black circles represent sites at which the receptor can be phosphorylated, and thereby regulated, by intracellular substances.

Adapted from Dohlman et al.

indeed sometimes opposite to, that produced when the messenger combines with the other receptors. For example, as we shall see in Chapter 14, there are two distinct types of receptors for the hormone epinephrine in the smooth muscle of certain blood vessels, and this hormone can cause either contraction or relaxation of the muscle depending on the relative degrees of binding to the two different types. The degree to which the molecules of a particular messenger bind to different receptor types in a single cell is determined by the affinity of the different receptor types for the messenger.

It should not be inferred from these descriptions that a cell has receptors for only one messenger. In fact, a single cell usually contains many different receptors for different chemical messengers.

Other characteristics of messenger-receptor interactions are **saturation** and **competition.** These phenomena were described in Chapter 4 for ligands binding to binding sites on proteins and are fully applicable here. In most systems, a cell's response to a messenger increases as the extracellular concentration of messenger increases, because the number of receptors occupied by messenger molecules increases. There is an upper limit to this responsiveness, however, because only a finite number of receptors are available, and they become saturated at some point.

Competition is the ability of different messenger molecules that are very similar in structure to compete with each other for a receptor. Competition occurs physiologically with closely related messengers, and it also underlies the action of many drugs. If researchers

TABLE 7–3 A Glossary of Terms Concerning Receptors

Receptor	A specific protein in either the plasma membrane or interior of a target cell with which a chemical messenger combines.
Specificity	The ability of a receptor to bind only one type or a limited number of structurally related types of chemical messengers.
Saturation	The degree to which receptors are occupied by a messenger. If all are occupied, the receptors are fully saturated; if half are occupied, the saturation is 50 percent, and so on.
Affinity	The strength with which a chemical messenger binds to its receptor.
Competition	The ability of different molecules very similar in structure to combine with the same receptor.
Antagonist	A molecule that competes for a receptor with a chemical messenger normally present in the body. The antagonist binds to the receptor but does not trigger the cell's response.
Agonist	A chemical messenger that binds to a receptor and triggers the cell's response; often refers to a drug that mimics a normal messenger's action.
Down-regulation	A decrease in the total number of target-cell receptors for a given messenger in response to chronic high extracellular concentration of the messenger.
Up-regulation	An increase in the total number of target-cell receptors for a given messenger in response to a chronic low extracellular concentration of the messenger.
Supersensitivity	The increased responsiveness of a target cell to a given messenger, resulting from up-regulation.

or physicians wish to interfere with the action of a particular messenger, they can administer competing molecules, if available, that bind to the receptors for that messenger without activating them. This prevents the messenger from binding and does not trigger the cell's response. Such drugs are known as **antagonists** with regard to the usual chemical messenger. For example,

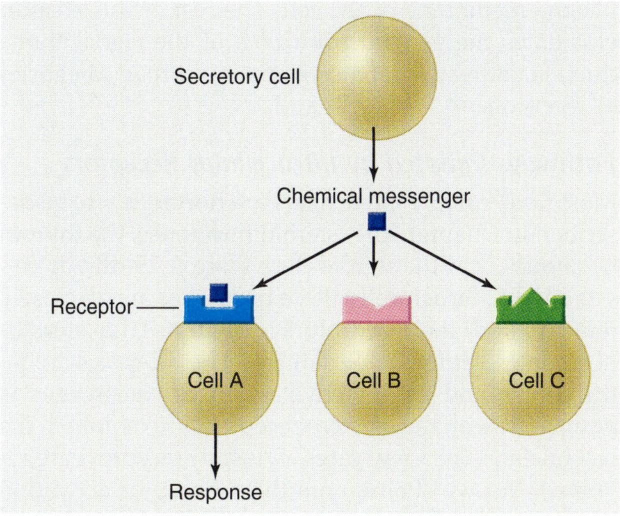

FIGURE 7–11

Specificity of receptors for chemical messengers. Only cell A has the appropriate receptor for this chemical messenger and, therefore, is a target cell for the messenger.

so-called beta-blockers, used in the treatment of high blood pressure and other diseases, are drugs that antagonize the ability of epinephrine and norepinephrine to bind to one of their receptors—the beta-adrenergic receptor (Chapter 8). On the other hand, some drugs that bind to a particular receptor type do trigger the cell's response exactly as if the true chemical messenger had combined with the receptor; such drugs are known as **agonists** and are used to mimic the messenger's action. For example, the decongestant drug ephedrine mimics the action of epinephrine.

Regulation of Receptors

Receptors are themselves subject to physiological regulation. The number of receptors a cell has (or the affinity of the receptors for their specific messenger) can be increased or decreased, at least in certain systems. An important example of such regulation is the phenomenon of **down-regulation.** When a high extracellular concentration of a messenger is maintained for some time, the total number of the target-cell's receptors for that messenger may decrease—that is, downregulate. Down-regulation has the effect of reducing the target cells' responsiveness to frequent or intense stimulation by a messenger and thus represents a local negative-feedback mechanism. For example, a prolonged high plasma concentration of the hormone insulin, which stimulates glucose uptake by its target cells, causes down-regulation of its receptors, and this acts to dampen the ability of insulin to stimulate glucose uptake.

Change in the opposite direction (up-regulation) also occurs. Cells exposed for a prolonged period to very low concentrations of a messenger may come to have many more receptors for that messenger, thereby developing increased sensitivity to it. For example, days after the nerves to a muscle are cut, thereby eliminating the neurotransmitter released by those nerves, the muscle will contract in response to amounts of experimentally injected neurotransmitter much smaller than those to which an innervated muscle can respond.

Up-regulation and down-regulation are made possible because there is a continuous degradation and synthesis of receptors. The main cause of down-regulation of plasma-membrane receptors is as follows: The binding of a messenger to its receptor can stimulate the internalization of the complex; that is, the messenger-receptor complex is taken into the cell by endocytosis (an example of so-called receptor-mediated endocytosis); this increases the rate of receptor degradation inside the cell. Thus, at high hormone concentrations, the number of plasma-membrane receptors of that type gradually decreases.

The opposite events also occur and contribute to up-regulation: The cell may contain stores of receptors in the membrane of intracellular vesicles, and these are available for insertion into the membrane via exocytosis (Chapter 6).

Another important mechanism of up-regulation and down-regulation is alteration of the expression of the genes that code for the receptors.

Down-regulation and up-regulation are physiological responses, but there are also many disease processes in which the number of receptors or their affinity for messenger becomes abnormal. The result is unusually large or small responses to any given level of messenger. For example, the disease called *myasthenia gravis* is due to destruction of the skeletal muscle receptors for acetylcholine, the neurotransmitter that normally causes contraction of the muscle in response to nerve stimulation; the result is muscle weakness or paralysis.

Signal Transduction Pathways

What are the sequences of events by which the binding of a chemical messenger (hormone, neurotransmitter, or paracrine/autocrine agent) to a receptor causes the cell to respond to the messenger?

The combination of messenger with receptor causes a change in the conformation of the receptor. This event, known as **receptor activation,** is always the initial step leading to the cell's ultimate responses to the messenger. These responses can take the form of changes in: (1) the permeability, transport properties, or electrical state of the cell's plasma membrane;

(2) the cell's metabolism; (3) the cell's secretory activity; (4) the cell's rate of proliferation and differentiation; and (5) the cell's contractile activity.

Despite the seeming variety of these five types of ultimate responses, there is a common denominator: They are all due directly to alterations of particular cell proteins. Let us take a few examples of messenger-induced responses, all of which are described fully in subsequent chapters. Generation of electric signals in nerve cells reflects the altered conformation of membrane proteins constituting ion channels through which ions can diffuse between extracellular fluid and intracellular fluid. Changes in the rate of glucose secretion by the liver reflect the altered activity and concentration of enzymes in the metabolic pathways for glucose synthesis. Muscle contraction results from the altered conformation of contractile proteins.

To repeat, receptor activation by a messenger is only the first step leading to the cell's ultimate response (contraction, secretion, and so on). The sequences of events, however, between receptor activation and the responses may be very complicated and are termed **signal transduction pathways.** The "signal" is the receptor activation, and "transduction" denotes the process by which a stimulus is transformed into a response. The important question is: *How does receptor activation influence the cell's internal proteins, which are usually critical for the response but may be located far from the receptor?*

Signal transduction pathways differ at the very outset for lipid-soluble and lipid-insoluble messengers since, as described earlier, the receptors for these two broad chemical classes of messenger are in different locations—the former inside the cell and the latter in the plasma membrane of the cell. The rest of this chapter elucidates the general principles of the signal transduction pathways initiated by the two broad categories of receptors.

Pathways Initiated by Intracellular Receptors

Most lipid-soluble messengers are hormones (to be described in Chapter 10)—steroid hormones, the thyroid hormones, and the steroid derivative, 1,25-dihydroxy-vitamin D_3. Structurally these hormones are all closely related, and their receptors constitute the steroid-hormone receptor "superfamily." The receptors are intracellular and are inactive when no messenger is bound to them; the inactive receptors are mainly in the cell nucleus. (In a few cases, cytosolic receptors are involved, but we shall ignore this.) Receptor activation leads to altered rates of gene transcription, the sequence of events being as follows.

The messenger diffuses across the cell's plasma membrane and nuclear membrane to enter the nucleus and bind to the receptor there (Figure 7–12). The

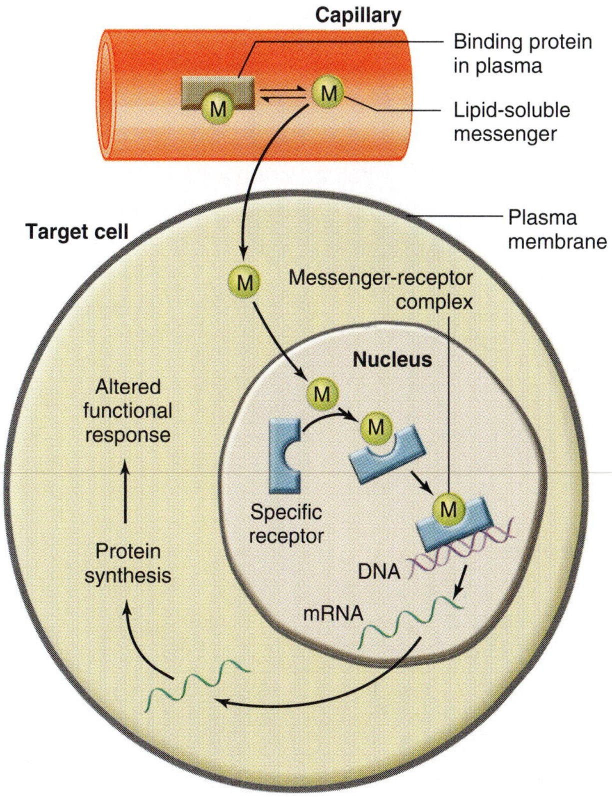

FIGURE 7–12

Mechanism of action of lipid-soluble messengers. This figure shows the receptor for these messengers as being in the nucleus. In some cases, the unbound receptor is in the cytosol rather than the nucleus, in which case the binding occurs there, and the messenger-receptor complex moves into the nucleus.

receptor, activated by the binding of hormone to it, then functions in the nucleus as a **transcription factor,** defined in Chapter 5 as any regulatory protein that directly influences gene transcription. The receptor binds to a specific sequence near a gene in DNA, termed a response element, and increases the rate of that gene's transcription into mRNA. The mRNA molecules formed enter the cytosol and direct the synthesis, on ribosomes, of the protein encoded by the gene. The result is an increase in cellular concentration of the protein or its rate of secretion, and this accounts for the cell's ultimate response to the messenger. For example, if the protein encoded by the gene is an enzyme, the cell's response is an increase in the rate of the reaction catalyzed by that enzyme.

Two other points should be mentioned. First, more than one gene may be subject to control by a single receptor type, and second, in some cases the transcription of the gene(s) is *decreased* by the activated receptor rather than increased.

TABLE 7–4 Classification of Receptors Based on Their Locations and the Signal Transduction Pathways They Use

1. INTRACELLULAR RECEPTORS (Figure 7–12) (for lipid-soluble messengers) Function in the nucleus as transcription factors to alter the rate of transcription of particular genes.
2. PLASMA-MEMBRANE RECEPTORS (Figure 7–13) (for lipid-insoluble messengers)
   a. Receptors that themselves function as ion channels.
   b. Receptors that themselves function as enzymes.
   c. Receptors that are bound to and activate cytoplasmic JAK kinases.
   d. Receptors that activate G proteins, which in turn act upon effector proteins—either ion channels or enzymes—in the plasma membrane.

Pathways Initiated by Plasma-Membrane Receptors

On the basis of the signal transduction pathways they initiate, plasma-membrane receptors can be classified into the types listed in Table 7–4 and illustrated in Figure 7–13.

Three notes on general terminology are essential for this discussion. First, the intercellular chemical messengers (hormones, neurotransmitters, and paracrine/autocrine agents), which reach the cell from the extracellular fluid and bind to their specific receptors, are often referred to as **first messengers. Second messengers** are nonprotein substances that enter the cytoplasm or are enzymatically generated there as a result of plasma-membrane receptor activation and diffuse throughout the cell to transmit signals. They serve as chemical relays from the plasma membrane to the biochemical machinery inside the cell.

The third essential general term is **protein kinase.** As described in Chapter 4, protein kinase is the name for *any* enzyme that phosphorylates other proteins by transferring to them a phosphate group from ATP. Introduction of the phosphate group changes the conformation and/or activity of the recipient protein, often itself an enzyme. There are many distinct protein kinases, each type being able to phosphorylate only certain proteins. The important point is that a variety of protein kinases are involved in signal transduction pathways. These pathways may involve long and complex series of reactions in which a particular inactive protein kinase is activated by phosphorylation and then catalyses the phosphorylation of another inactive protein kinase, and so on. At the ends of these sequences, the ultimate phosphorylation of key proteins (transporters, metabolic enzymes, ion channels, contractile proteins, and so on) underlies the cell's response to the original first messenger.

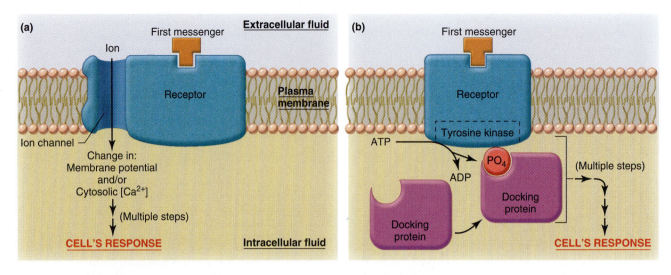

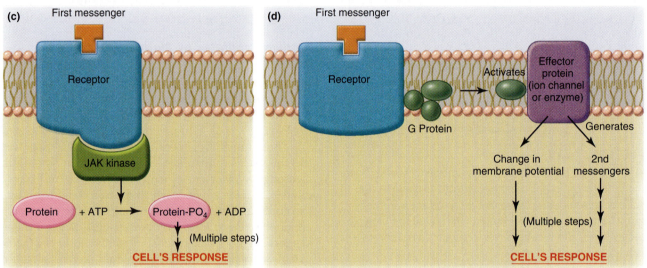

FIGURE 7–13

Mechanisms of action of lipid-insoluble messengers (noted as "first messengers" in this and subsequent figures). (a) Signal transduction mechanism in which the receptor complex itself contains an ion channel. (b) Signal transduction mechanism in which the receptor itself functions as an enzyme, usually a tyrosine kinase. (c) Signal transduction mechanism in which the receptor activates a JAK kinase in the cytoplasm. (d) Signal transduction mechanism involving G proteins.

As described in Chapter 4, other enzymes do the reverse of protein kinases; that is, they dephosphorylate proteins. These enzymes, termed protein phosphatases, also participate in signal transduction pathways, but their roles are much less well understood than those of the protein kinases and will not be described here.

Receptors That Function as Ion Channels In the first type of plasma-membrane receptor listed in Table 7–4 (Figure 7–13a), the protein that acts as the receptor itself constitutes an ion channel, and activation of the receptor by a first messenger causes the channel to open. The opening results in an increase in the net diffusion across the plasma membrane of the ion or ions specific to the channel. As we shall see in Chapter 8, such a change in ion diffusion is usually associated with a change in the membrane potential, and this electric signal is often the essential event in the cell's response to the messenger. In addition, as described later in this chapter, when the channel is a calcium channel, its opening results in an increase, by diffusion, in the cytosolic calcium concentration, another essential event in the signal transduction pathway for many receptors.

Receptors That Function as Enzymes The receptors in the second category of plasma-membrane receptors listed in Table 7–4 (Figure 7–13b) have intrinsic enzyme activity. With one major exception (discussed below), the many receptors that possess intrinsic enzyme activity are all protein kinases. Of these, the great majority phosphorylate specifically the portions of proteins that contain the amino acid tyrosine; accordingly, they are termed **tyrosine kinases.** Keep in mind that (1) tyrosine kinases simply constitute one category of protein kinases, and that (2) the label does not denote any particular enzyme but is a generic term that defines the type of function—phosphorylating tyrosine groups—these protein kinases perform.

The sequence of events for receptors with intrinsic tyrosine kinase activity is as follows. The binding of a specific messenger to the receptor changes the conformation of the receptor so that its enzymatic portion, located on the cytoplasmic side of the plasma membrane, is activated. This results in autophosphorylation of the receptor; that is, the receptor phosphorylates its own tyrosine groups! The newly created phosphotyrosines on the cytoplasmic portion of the receptor then serve as "docking" sites for cytoplasmic proteins that have a high affinity for those phosphotyrosines displayed by that particular receptor. The bound docking proteins then bind other proteins, which leads to a cascade of signaling pathways within the cell. The common denominator of these pathways is that, at one or more points in their sequences, they all involve activation of cytoplasmic proteins by phosphorylation.

The number of kinases that mediate these phosphorylations can be very large, and their names constitute a veritable alphabet soup—RAF, MEK, MAPKK, and so on. In all this complexity, it is easy to lose track of the point that the end result of all these pathways is the activation or synthesis of molecules, usually proteins, that ultimately mediate the response of the cell to the messenger. The receptors with intrinsic tyrosine kinase activity all bind first messengers that influence cell proliferation and differentiation.

As stated above, there is one major exception to the generalization that plasma-membrane receptors with inherent enzyme activity function as protein kinases. In this exception, the receptor functions as a **guanylyl cyclase** to catalyse the formation, in the cytoplasm, of a molecule known as **cyclic GMP (cGMP).** In turn, cGMP functions as a second messenger to activate a particular protein kinase, **cGMP-dependent protein kinase,** which phosphorylates particular proteins that then mediate the cell's response to the original messenger. This signal transduction pathway is used by only a small number of messengers and should not be confused with the much more important cAMP system to be described in a later section.

(Also, we will see in Chapter 8 that in certain cells, guanylyl cyclase enzymes are present *in the cytoplasm;* in these cases, a first messenger—nitric oxide—diffuses into the cell and combines with the guanylyl cyclase there to trigger the formation of cGMP.)

Receptors that Interact with Cytoplasmic JAK Kinases To repeat, in the previous category, the receptor itself has intrinsic enzyme activity. In contrast, in the present category of receptors (Table 7–4 and Figure 7–13c), the enzymatic activity—again tyrosine kinase activity—resides not in the receptor but in a family of *separate cytoplasmic* kinases, termed **JAK kinases,** which are bound to the receptor. (The term "JAK" has several derivations, including "just another kinase.") In these cases, the receptor and its associated JAK kinase function as a unit; the binding of a first messenger to the receptor causes a conformational change in the receptor that leads to activation of the JAK kinase. Different receptors associate with different members of the JAK kinase family, and the different JAK kinases phosphorylate different target proteins, many of which act as transcription factors. The result of these pathways is the synthesis of new proteins, which mediate the cell's response to the first messenger.

Receptors that Interact with G Proteins The fourth category of plasma-membrane receptors in Table 7–4 (Figure 7–13d) is by far the largest, including hundreds of distinct receptors. Bound to the receptor is a protein located on the inner (cytosolic) surface of the plasma membrane and belonging to the family of proteins known as **G proteins.** The binding of a first messenger to the receptor changes the conformation of the receptor. This change causes one of the three subunits of the G protein to link up with still another plasma-membrane protein, either an ion channel or an enzyme. These ion channels and enzymes are termed **plasma-membrane effector proteins** since they mediate the next steps in the sequences of events leading to the cell's response.

In essence, then, a G protein serves as a switch to "couple" a receptor to an ion channel or an enzyme in the plasma membrane. The G protein may cause the ion channel to open, with resulting generation of electric signals or, in the case of calcium channels, changes in the cytosolic calcium concentration. Alternatively, the G protein may activate or inhibit the membrane enzyme with which it interacts; these are enzymes that, when activated, cause the generation, inside the cell, of second messengers.

There are three subfamilies of plasma-membrane G proteins, each with multiple distinct members, and a single receptor may be associated with more than one type of G protein. Moreover, some G proteins may couple to more than one type of plasma-membrane effector protein. Thus, a first-messenger-activated receptor, via its G-protein couplings, can call into action a variety of plasma-membrane effector proteins—ion channels and enzymes—which in turn induce a variety of cellular events.

To illustrate some of the major points concerning G proteins, plasma-membrane effector proteins, second messengers, and protein kinases, the next two sections describe the two most important effector-protein enzymes—adenylyl cyclase and phospholipase C—regulated by G proteins and the subsequent portions of the signal transduction pathways in which they participate.

Before doing so, however, we would like to emphasize that the plasma-membrane G proteins activated by receptors encompass only a subset of G proteins, a term that includes all those proteins that, regardless of location and function, share a particular chemical characteristic (the ability to bind certain guanine nucleotides). In contrast to G proteins coupled to receptors is a class of small (one-subunit), mainly cytoplasmic G proteins (with names like Ras, Rho, and Rac). These G proteins play an important role in the signal transduction pathways from tyrosine kinase receptors, but they do not interact directly with either the receptor or membrane-bound effector molecules.

Adenylyl Cyclase and Cyclic AMP In this pathway, activation of the receptor (Figure 7–14) by the binding of the first messenger (for example, the hormone epinephrine) allows the receptor to activate its associated G protein, in this example known as G_s (the subscript s denotes "stimulatory"). This causes G_s to activate its effector protein, the membrane enzyme called **adenylyl cyclase** (also termed adenylate cyclase). The activated adenylyl cyclase, whose enzymatic site is located on the cytosolic surface of the plasma membrane, then catalyzes the conversion of some cytosolic ATP molecules to cyclic 3',5'-adenosine monophosphate, called simply **cyclic AMP (cAMP)** (Figure 7–15). Cyclic AMP then acts as a second messenger (Figure 7–14). It diffuses throughout the cell to trigger the sequences of events leading to the cell's ultimate response to the first messenger. The action of cAMP is eventually terminated by its breakdown to noncyclic AMP, a reaction catalyzed by the enzyme **phosphodiesterase** (Figure 7–15). This enzyme is also subject to physiological control so that the cellular concentration of cAMP can be changed either by altering the rate of its messenger-mediated generation or the rate of its phosphodiesterase-mediated breakdown.

What does cAMP actually do inside the cell? It binds to and activates an enzyme known as **cAMP-dependent protein kinase** (also termed protein

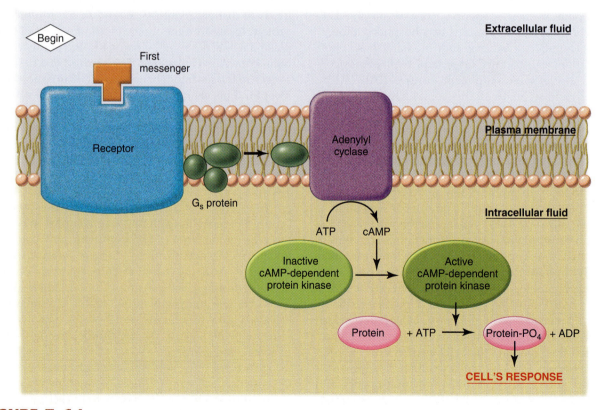

FIGURE 7–14

Cyclic AMP second-messenger system. Not shown in the figure is the existence of another regulatory protein, G_i, with which certain receptors can react to cause inhibition of adenylyl cyclase.

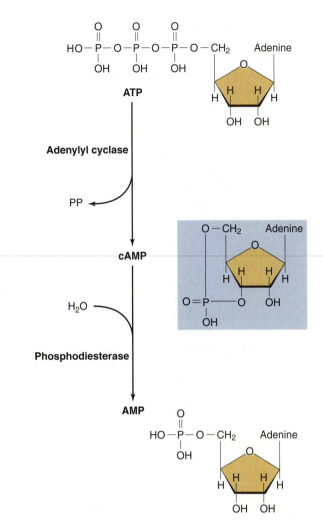

FIGURE 7–15
Structure of ATP, cAMP, and AMP, the last resulting from enzymatic alteration of cAMP.

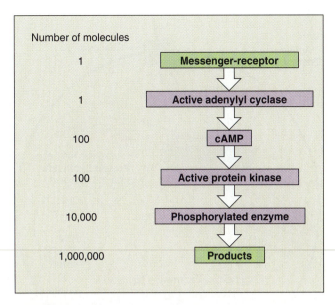

FIGURE 7–16
Example of amplification in the cAMP system.

kinase A) (Figure 7–14). As emphasized, protein kinases phosphorylate other proteins—often enzymes—by transferring a phosphate group to them. The change in the activity of those proteins phosphorylated by cAMP-dependent protein kinase brings about the response of the cell (secretion, contraction, and so on). Again we emphasize that each of the various protein kinases that participate in the multiple signal transduction pathways described in this chapter has its own specific substrates.

In essence, then, the activation of adenylyl cyclase by a G protein initiates a chain, or "cascade," of events in which proteins are converted in sequence from inactive to active forms. Figure 7–16 illustrates the benefit of such a cascade. While it is active, a single enzyme molecule is capable of transforming into product not one but many substrate molecules, let us say 100. Therefore, one active molecule of adenylyl cyclase may catalyze the generation of 100 cAMP molecules. At

each of the two subsequent enzyme-activation steps in our example, another hundredfold amplification occurs. Therefore, the end result is that a single molecule of the first messenger could, in this example, cause the generation of 1 million product molecules. This fact helps to explain how hormones and other messengers can be effective at extremely low extracellular concentrations. To take an actual example, one molecule of the hormone epinephrine can cause the generation and release by the liver of 10^8 molecules of glucose.

How can activation of a single molecule, cAMP-dependent protein kinase, by cAMP be an event common to the great variety of biochemical sequences and cell responses initiated by cAMP-generating first messengers? The major answer is that cAMP-dependent protein kinase has a large number of distinct substrates—it can phosphorylate a large number of different proteins (Figure 7–17). Thus, activated cAMP-dependent protein kinase can exert multiple actions within a single cell and different actions in different cells. For example, epinephrine acts via the cAMP pathway on fat cells to cause both glycogen breakdown (mediated by one phosphorylated enzyme) and triacylglycerol breakdown (mediated by another phosphorylated enzyme).

It must be emphasized that whereas phosphorylation mediated by cAMP-dependent protein kinase *activates* certain enzymes, it *inhibits* others. For example, the enzyme catalyzing the rate-limiting step in glycogen synthesis is inhibited by phosphorylation, and this fact explains how epinephrine inhibits glycogen synthesis at the same time that it stimulates glycogen breakdown by activating the enzyme that catalyzes the latter response.

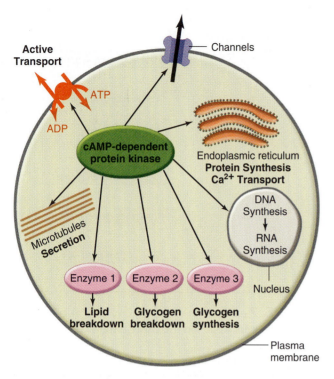

FIGURE 7–17

The variety of cellular responses induced by cAMP is due mainly to the fact that activated cAMP-dependent protein kinase can phosphorylate many different proteins, activating them or inhibiting them. In this figure, the protein kinase is shown phosphorylating eight different proteins—a microtubular protein, an ATPase, an ion channel, a protein in the endoplasmic reticulum, a protein involved in DNA synthesis, and three enzymes.

Not mentioned so far is the fact that receptors for some first messengers, upon activation by their messengers, cause adenylyl cyclase to be *inhibited,* resulting in less, rather than more, generation of cAMP. This occurs because these receptors are associated with a different G protein, known as G_i (the subscript i denotes "inhibitory"), and activation of G_i causes inhibition of adenylyl cyclase. The result is to decrease the concentration of cAMP in the cell and, thereby, to decrease the phosphorylation of key proteins inside the cell.

Phospholipase C, Diacylglycerol, and Inositol Trisphosphate In this system, the relevant G protein (termed G_q), activated by a first-messenger–bound receptor, activates a plasma-membrane effector enzyme called **phospholipase C.** This enzyme catalyzes the breakdown of a plasma-membrane phospholipid known as phosphatidylinositol bisphosphate, abbreviated PIP_2, to **diacylglycerol (DAG)** and **inositol**

trisphosphate (IP₃)** (Figure 7–18). Both DAG and IP_3 then function as second messengers but in very different ways.

DAG activates a particular protein kinase known as **protein kinase C,** which then phosphorylates a large number of other proteins, leading to the cell's response.

IP_3, in contrast to DAG, does not exert its second messenger role by directly activating a protein kinase. Rather, IP_3, after entering the cytosol, binds to calcium channels on the outer membranes of the endoplasmic reticulum and opens them. Because the concentration of calcium is much higher in the endoplasmic reticulum than in the cytosol, calcium diffuses out of this organelle into the cytosol, significantly increasing cytosolic calcium concentration. This increased calcium concentration then continues the sequence of events leading to the cell's response to the first messenger. We will pick up this thread in a later section.

Control of Ion Channels by G Proteins A comparison of Figures 7–13d and 7–17 emphasizes one more important feature of G-protein function—its ability to both directly and indirectly gate ion channels. As shown in Figure 7–13d and described earlier, an ion channel can be the effector protein for a G protein. This situation is known as *direct G-protein gating* of plasma-membrane ion channels because the G protein interacts directly with the channel (the term "gating" denotes control of the opening or closing of a channel). All the events occur in the plasma membrane and are independent of second messengers. Now look at Figure 7–17, and you will see that cAMP-dependent protein kinase can phosphorylate a plasma-membrane ion channel, thereby causing it to open. Since, as we have seen, the sequence of events leading to activation of cAMP-dependent protein kinase proceeds through a G protein, it should be clear that the opening of this channel is indirectly dependent on that G protein. To generalize, the *indirect gating* of ion channels by G proteins utilizes a second-messenger pathway for the opening (or closing) of the channel. Not just cAMP-dependent protein kinase but protein kinases involved in other signal transduction pathways can participate in reactions leading to such indirect gating. Table 7–5 summarizes the three ways we have described by which receptor activation by a first messenger leads to opening or closing of ion channels.

Calcium as a Second Messenger The calcium ion (Ca^{2+}) functions as a second messenger in a great variety of cellular responses to stimuli, both chemical (first messenger) and electrical. The physiology of calcium as a second messenger requires an analysis of two broad questions: (1) How do stimuli cause the cytosolic calcium concentration to increase? (2) How does the increased calcium concentration elicit the cells' responses? Note that, for simplicity, our two questions are phrased in terms of an *increase* in cytosolic concentration. There are, in fact, first messengers that elicit a *decrease* in cytosolic calcium concentration and therefore a decrease in calcium's second-messenger effects. Now for the answer to the first question.

The regulation of cytosolic calcium concentration is described in Chapter 6. In brief, by means of active-transport systems in the plasma membrane and cell organelles, Ca^{2+} is maintained at an extremely low concentration in the cytosol. Accordingly, there is always a large electrochemical gradient favoring diffusion of calcium into the cytosol via calcium channels in both the plasma membrane and endoplasmic reticulum. A stimulus to the cell can alter this steady state by influencing the active-transport systems and/or the ion channels, resulting in a change in cytosolic calcium concentration.

The most common ways that receptor activation by a first messenger increases the cytosolic Ca^{2+} concentration have already been presented in this chapter and are summarized in the top part of Table 7–6.

TABLE 7–5 Summary of Mechanisms by Which Receptor Activation Influences Channels
1. The ion channel is part of the receptor.
2. A G protein directly gates the channel.
3. A G protein gates the channel indirectly via a second messenger.

The previous paragraph dealt with *receptor-initiated* sequences of events. This is a good place, however, to emphasize that there are calcium channels in the plasma membrane that are opened directly by an *electric* stimulus to the membrane (Chapter 6). Calcium can act as a second messenger, therefore, in response not only to chemical stimuli acting via receptors, but to electric stimuli acting via voltage-gated calcium channels as well. Moreover, extracellular calcium entering the cell via these channels can, in certain cells, bind to calcium-sensitive channels in the endoplasmic reticulum and open them. In this manner,

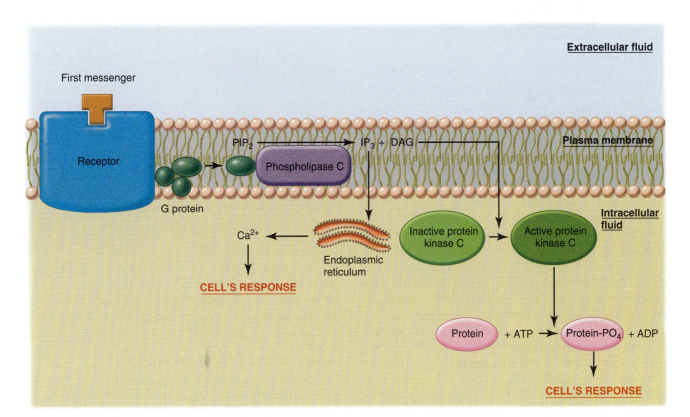

FIGURE 7–18

Mechanism by which an activated receptor stimulates the enzymatically mediated breakdown of PIP$_2$ to yield of IP$_3$ and DAG. IP$_3$ then causes the release of calcium ions from the endoplasmic reticulum, and DAG activates a particular protein kinase known as protein kinase C.

TABLE 7–6 Calcium as a Second Messenger

Common mechanisms by which stimulation of a cell leads to an increase in cytosolic Ca^{2+} concentration:

1. Receptor activation
   a. Plasma-membrane calcium channels open in response to a first messenger; the receptor itself may contain the channel, or the receptor may activate a G protein that opens the channel via a second messenger.
   b. Calcium is released from the endoplasmic reticulum; this is mediated by second messengers, particularly IP_3 and calcium entering from the extracellular fluid.
   c. Active calcium transport out of the cell is inhibited by a second messenger.

2. Opening of voltage-sensitive calcium channels

Major mechanisms by which an increase in cytosolic Ca^{2+} concentration induces the cell's responses:

1. Calcium binds to calmodulin. On binding calcium, the calmodulin changes shape, which allows it to activate or inhibit a large variety of enzymes and other proteins. Many of these enzymes are protein kinases.

2. Calcium combines with calcium-binding intermediary proteins other than calmodulin. These proteins then act in a manner analogous to calmodulin.

3. Calcium combines with and alters response proteins directly, without the intermediation of any specific calcium-binding protein.

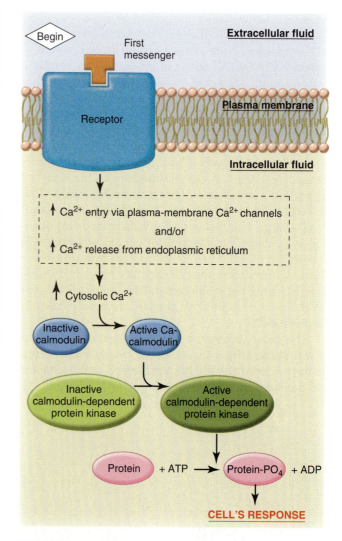

FIGURE 7–19
Calcium, calmodulin, and the calmodulin-dependent protein kinase system. (There are multiple calmodulin-dependent protein kinases.) The mechanisms for increasing cytosolic calcium concentration are summarized in Table 7–6.

a small amount of extracellular calcium entering the cell can function as a second messenger to release a much larger amount of calcium from the endoplasmic reticulum. This is termed "calcium-induced calcium release." Thus, depending on the cell and the signal— first messenger or an electrical impulse—the major second messenger that releases calcium from the endoplasmic reticulum can be either IP_3 or calcium itself (item 1b in the top of Table 7–6).

Now we turn to the question of how the increased cytosolic calcium concentration elicits the cells' responses (bottom of Table 7–6). The common denominator of calcium's actions is its ability to bind to various cytosolic proteins, altering their conformation and thereby activating their function. One of the most important of these is a protein found in virtually all cells and known as **calmodulin** (Figure 7–19). On binding with calcium, calmodulin changes shape, and this allows calcium-calmodulin to activate or inhibit a large variety of enzymes and other proteins, many of which are protein kinases. Activation or inhibition of **calmodulin-dependent protein kinases** leads, via phosphorylation, to activation or inhibition of proteins involved in the cell's ultimate responses to the first messenger.

Calmodulin is not, however, the only intracellular protein influenced by calcium binding. For example, as we shall see in Chapter 11, calcium binds to the protein troponin in certain types of muscle to initiate contraction.

Receptors and Gene Transcription
As described earlier in this chapter, the receptors for lipid-soluble messengers, once activated by hormone binding, act in the nucleus as transcription factors to increase or decrease the rate of gene transcription. We now emphasize that there are many other transcription factors inside cells and that the signal transduction pathways initiated by *plasma-membrane* receptors

often result in the activation, by phosphorylation, of these transcription factors. Thus, many first messengers that bind to plasma-membrane receptors can also alter gene transcription via second messengers. For example, at least three of the proteins phosphorylated by cAMP-dependent protein kinase function as transcription factors.

Some of the genes influenced by transcription factors activated in response to first messengers are known collectively as **primary response genes,** or **PRGs** (also termed immediate-early genes). In many cases, especially those involving first messengers that influence the proliferation or differentiation of their target cells, the story does not stop with a PRG and the protein it encodes. In these cases, the protein encoded by the PRG is itself a transcription factor for other genes (Figure 7–20). Thus, an initial transcription factor activated in the signal transduction pathway causes the synthesis of a different transcription factor, which in turn causes the synthesis of additional proteins, ones particularly important for the long-term biochemical events required for cellular proliferation and differentiation. A great deal of research is being done on the transcription factors encoded by PRGs because of their relevance for the abnormal growth and differentiation typical of cancer.

Cessation of Activity in Signal Transduction Pathways

A word is needed about how signal transduction pathways are shut off. As expected, the key event is usually the cessation of receptor activation. Because organic second messengers are rapidly inactivated (for example, cAMP by phosphodiesterase) or broken down intracellularly, and because calcium is continuously being pumped out of the cell or back into the endoplasmic reticulum, increases in the cytosolic concentrations of all these components are transient events and continue only as long as the receptor is being activated by a first messenger.

A major way that receptor activation ceases is by a decrease in the concentration of first messenger molecules in the region of the receptor. This occurs as the first messenger is metabolized by enzymes in the vicinity, taken up by adjacent cells, or simply diffuses away.

In addition, receptors can be inactivated in two other ways: (1) The receptor becomes chemically altered (usually by phosphorylation), which lowers its affinity for a first messenger, and so the messenger is released; and (2) removal of plasma-membrane receptors occurs when the combination of first messenger and receptor is taken into the cell by endocytosis. The processes described here are physiologically controlled. For example, in many cases the inhibitory phosphorylation of a receptor is mediated by a protein

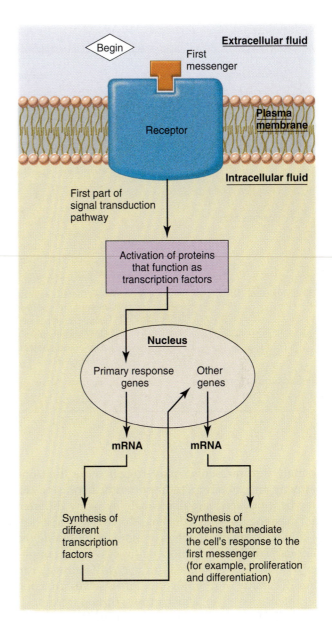

FIGURE 7–20
Role of multiple transcription factors and primary response genes in mediating protein synthesis in response to a first messenger binding to a plasma-membrane receptor.

kinase in the signal transduction pathway triggered by first-messenger binding to that very receptor; thus, this receptor inactivation constitutes a negative feedback.

This concludes our description of the basic principles of signal transduction pathways. It is essential to recognize that the pathways do not exist in isolation but may be active simultaneously in a single cell, exhibiting complex interactions. This is possible because

TABLE 7–7 Reference Table of Important Second Messengers

Substance	Source	Effects
Calcium	Enters cell through plasma-membrane ion channels or is released from endoplasmic reticulum	Activates calmodulin and other calcium-binding proteins; calcium-calmodulin activates calmodulin-dependent protein kinases
Cyclic AMP (cAMP)	A G protein activates plasma-membrane adenylyl cyclase, which catalyzes formation of cAMP from ATP	Activates cAMP-dependent protein kinase (protein kinase A)
Cyclic GMP (cGMP)	Generated from guanosine triphosphate in a reaction catalyzed by a plasma-membrane receptor with guanylyl cyclase activity	Activates cGMP-dependent protein kinase (protein kinase G)
Diacylglycerol (DAG)	A G protein activates plasma-membrane phospholipase C, which catalyzes generation of DAG and IP_3 from plasma membrane phosphatidylinositol bisphosphate (PIP_2)	Activates protein kinase C
Inositol trisphosphate (IP_3)	See DAG above	Releases calcium from endoplasmic reticulum

a single first messenger may trigger more than one pathway and, much more importantly, because a cell may be influenced simultaneously by many different first messengers—often dozens. Moreover, a great deal of cross-talk can occur at one or more levels among the various signal transduction pathways. For example, active molecules generated in the cAMP pathway can alter the ability of receptors that, themselves, function as protein kinases to activate transcription factors.

Why should signal transduction pathways be so diverse and complex? The only way to achieve controlled distinct effects by a cell in the face of the barrage of multiple first messengers, each often having more than one ultimate effect, is to have diverse pathways with branch points at which one pathway can be enhanced and another reduced.

The biochemistry and physiology of plasma-membrane signal transduction pathways are among the most rapidly expanding fields in biology, and most of this information, beyond the basic principles we have presented, exceeds the scope of this book. For example, the protein kinases we have identified are those that are closest in the various sequences to the original receptor activation; in fact, as noted earlier there are often cascades of protein kinases in the remaining portions of the pathways. Moreover, there are a host of molecules other than protein kinases that play "helper" roles.

Finally, for reference purposes, Table 7–7 summarizes the biochemistry of the second messengers described in this chapter.

SECTION B SUMMARY

Receptors

I. Receptors for chemical messengers are proteins located either inside the cell or, much more commonly, in the plasma membrane. The binding of a messenger by a receptor manifests specificity, saturation, and competition.

II. Receptors are subject to physiological regulation by their own messengers. This includes down-regulation and up-regulation.

Signal Transduction Pathways

I. Binding a chemical messenger activates a receptor, and this initiates one or more signal transduction pathways leading to the cell's response.

II. Lipid-soluble messengers bind to receptors inside the target cell, and the activated receptor acts in the nucleus as a transcription factor to alter the rate of transcription of specific genes, resulting in a change in the concentration or secretion of the proteins coded by the genes.

III. Lipid-insoluble messengers bind to receptors on the plasma membrane. The pathways induced by activation of the receptor often involve second messengers and protein kinases.

a. The receptor may contain an ion channel, which opens, resulting in an electric signal in the membrane and, when calcium channels are involved, an increase in the cytosolic calcium concentration.

b. The receptor may itself act as an enzyme. With one exception, the enzyme activity is that of a protein kinase, usually a tyrosine kinase. The exception is the receptor that functions as a guanylyl cyclase to generate cyclic GMP.

c. The receptor may activate a cytosolic JAK kinase associated with it.

d. The receptor may interact with an associated plasma-membrane G protein, which in turn interacts with plasma-membrane effector proteins—ion channels or enzymes.

e. Very commonly, the receptor may activate, via a G_s protein, or inhibit, via a G_i protein, the membrane effector enzyme adenylyl cyclase, which catalyzes the conversion of cytosolic ATP to cyclic AMP. Cyclic AMP acts as a second messenger to activate intracellular cAMP-dependent protein kinase, which phosphorylates proteins that mediate the cell's ultimate responses to the first messenger.

f. The receptor may activate, via a G protein, the plasma-membrane enzyme phospholipase C, which catalyzes the formation of diacylglycerol (DAG) and inositol trisphosphate (IP_3). DAG activates protein kinase C, and IP_3 acts as a second messenger to release calcium from the endoplasmic reticulum.

IV. The receptor, via a G protein, may directly open or close an adjacent ion channel. This differs from indirect G-protein gating of channels, in which a second messenger acts upon the channel.

V. The calcium ion is one of the most widespread second messengers.

a. An activated receptor can increase cytosolic calcium concentration by causing certain calcium channels in the plasma membrane and/or endoplasmic reticulum to open. Voltage-gated calcium channels can also influence cytosolic calcium concentration.

b. Calcium binds to one of several intracellular proteins, most often calmodulin. Calcium-activated calmodulin activates or inhibits many proteins, including calmodulin-dependent protein kinases.

VI. The signal transduction pathways triggered by activated plasma-membrane receptors may influence genetic expression by activating transcription factors. In some cases, the primary response genes influenced by these transcription factors code for still other transcription factors. This is particularly true in pathways initiated by first messengers that stimulate their target cell's proliferation or differentiation.

VII. Cessation of receptor activity occurs by decreased first messenger molecule concentration and when the receptor is chemically altered or internalized, in the case of plasma-membrane receptors.

SECTION B REVIEW QUESTIONS

1. What is the chemical nature of receptors? Where are they located?
2. Explain why different types of cells may respond differently to the same chemical messenger.
3. Describe how the metabolism of receptors can lead to down-regulation or up-regulation.
4. What is the first step in the action of a messenger on a cell?
5. Describe the signal transduction pathway used by lipid-soluble messengers.
6. Classify plasma-membrane receptors according to the signal transduction pathways they initiate.
7. What is the result of opening a membrane ion channel?
8. Contrast receptors that have intrinsic enzyme activity with those associated with cytoplasmic JAK kinases.
9. Describe the role of plasma-membrane G proteins.
10. Draw a diagram describing the adenylyl cyclase-cAMP system.
11. Draw a diagram illustrating the phospholipase C/DAG/IP_3 system.
12. Contrast direct and indirect gating of ion channels by G proteins.
13. What are the two general mechanisms by which first messengers elicit an increase in cytosolic calcium concentration? What are the sources of the calcium in each mechanism?
14. How does the calcium-calmodulin system function?
15. Describe the manner in which activated plasma-membrane receptors influence gene expression.

CHAPTER 7 CLINICAL TERMS

aspirin

nonsteroidal anti-
  inflammatory drugs
  (NSAIDs)

adrenal steroids

jet lag

heart attack

asthma

Werner's syndrome

hypertension

myasthenia gravis

CHAPTER 7 THOUGHT QUESTIONS

(Answers are given in Appendix A.)

1. A person's plasma potassium concentration (a homeostatically regulated variable) is 4 mmol/L when she is eating 150 mmol of potassium per day. One day she doubles her potassium intake and continues to eat that amount indefinitely. At the new steady state, do you think her plasma potassium concentration is more likely to be 8, 4.4, or 4 mmol/L? (The answer to this question requires no knowledge about potassium, only the ability to reason about homeostatic control systems.)

2. Eskimos have a remarkable ability to work in the cold without gloves and not suffer decreased skin blood flow. Does this prove that there is a genetic difference between Eskimos and other people with regard to this characteristic?

3. Patient A is given a drug that blocks the synthesis of all eicosanoids, whereas patient B is given a drug that blocks the synthesis of leukotrienes but none of the other eicosanoids. What are the enzymes most likely blocked by these drugs?

4. Certain nerves to the heart release the neurotransmitter norepinephrine. If these nerves are removed in experimental animals, the heart becomes extremely sensitive to the administration of a drug that is an agonist of norepinephrine. Explain why, in terms of receptor physiology.

5. A particular hormone is known to elicit, completely by way of the cyclic AMP system, six different responses in its target cell. A drug is found that eliminates one of these responses but not the other five. Which of the following, if any, could the drug be blocking: the hormone's receptors, G_s protein, adenylyl cyclase, or cyclic AMP?

6. If a drug were found that blocked all calcium channels directly linked to G proteins, would this eliminate the role of calcium as a second messenger?

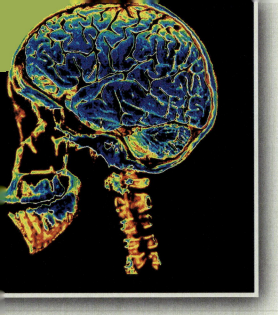

Neural Control Mechanisms

In order to coordinate the functions of the trillions of cells of the human body, two control systems exist. One, the endocrine system, is a collection of blood-borne messengers that works slowly, while the other, the nervous system, is a rapid control system. Together they regulate most internal functions and organize and control the activities we know collectively as human behavior. These activities include not only such easily observed acts as smiling and walking but also experiences such as feeling angry, being motivated, having an idea, or remembering a long-past event. Such experiences, which we attribute to the "mind," are related to the integrated activities of nerve cells in as yet unidentified ways.

The various structures of the nervous system are intimately interconnected, but for convenience they are divided into two parts: (1) the **central nervous system (CNS)**, composed of the brain and spinal cord; and (2) the **peripheral nervous system**, consisting of the nerves, which extend between the brain or spinal cord and the body's muscles, glands, and sense organs (Figure 8–1). For example, branches of the peripheral nervous system go between the base of the spine and the tips of the toes and, although they are not shown in Figure 8–1, between the base of the brain and the internal organs.

In this chapter, we are concerned with the components common to all neural mechanisms: the structure of individual nerve cells, the mechanisms underlying neural function, and the basic organization and major divisions of the nervous system.

NEURAL TISSUE

The basic unit of the nervous system is the individual nerve cell, or **neuron.** Nerve cells operate by generating electric signals that pass from one part of the cell to another part of the same cell and by releasing chemical messengers—**neurotransmitters**—to communicate with other cells. Neurons serve as **integrators** because their output reflects the balance of inputs they receive from the thousands or even hundreds of thousands of other neurons that impinge upon them.

Structure and Maintenance of Neurons

Neurons occur in a variety of sizes and shapes; nevertheless, as shown in Figure 8–2, most of them contain four parts: (1) a cell body, (2) dendrites, (3) an axon, and (4) axon terminals.

As in other types of cells, a neuron's **cell body** contains the nucleus and ribosomes and thus has the genetic information and machinery necessary for protein synthesis. The **dendrites** form a series of highly branched outgrowths from the cell body. They and the cell body receive most of the inputs from other neurons, the dendrites being vastly more important in this role than the cell body. The branching dendrites (some neurons may have as many as 400,000!) increase the cell's receptive surface area and thereby increase its capacity to receive signals from a myriad of other neurons.

The **axon,** sometimes also called a **nerve fiber,** is a single long process that extends from the cell body to its target cells. In length, axons can be a few micrometers or a meter or more. The portion of the axon closest to the cell body plus the part of the cell body where the axon is joined are known as the **initial segment,** or axon hillock. The initial segment is the "trigger zone" where, in most neurons, the electric signals are generated that then propagate away from the cell body along the axon or, sometimes, back along the dendrites. The main axon may have branches, called **collaterals,** along its course; near the ends both the main axon and its collaterals undergo further branching (Figure 8–2). The greater the degree of branching of the axon and axon collaterals, the greater the cell's sphere of influence.

Each branch ends in an **axon terminal,** which is responsible for releasing neurotransmitters from the axon. These chemical messengers diffuse across an extracellular gap to the cell opposite the terminal. Alternatively, some neurons release their chemical messengers from a series of bulging areas along the axon known as **varicosities.** Different parts of nerve cells serve different functions because of the segregated distribution of various membrane-bound channels and pumps as well as other molecules and organelles.

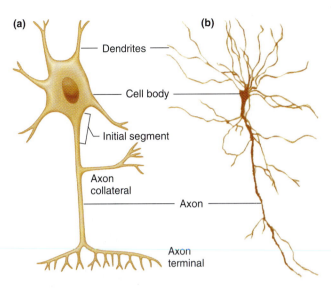

FIGURE 8–2

(a) Diagrammatic representation of a neuron. The proportions shown here are misleading because the axon may be 5000 to 10,000 times longer than the cell body is wide. This neuron is a common type, but there are several other types, one of which has no axon. (b) A neuron as observed through a microscope. The axon terminals cannot be seen at this magnification.

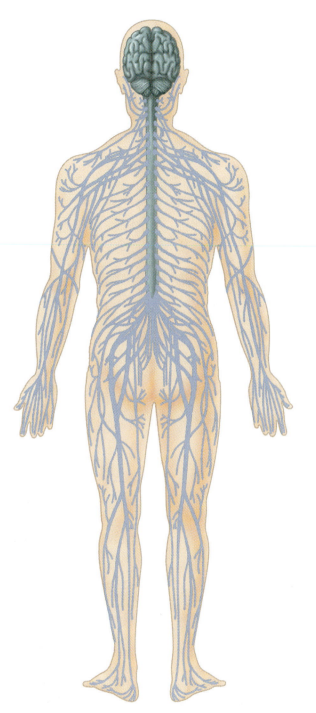

FIGURE 8–1

The central nervous system (green) and the peripheral nervous system (blue). Some of the peripheral nerves connect with the brain (these nerves are not shown) and others with the spinal cord.

The axons of some neurons are covered by **myelin** (Figure 8–3), which consists of 20 to 200 layers of highly modified plasma membrane wrapped around the axon by a nearby supporting cell. In the central nervous system these myelin-forming cells are the **oligodendroglia** (a type of neuroglia, or simply glial cell to be described later in the chapter), and in the peripheral nervous system they are the **Schwann cells.** The spaces between adjacent sections of myelin where the axon's plasma membrane is exposed to extracellular fluid are the **nodes of Ranvier.** The myelin sheath speeds up conduction of the electric signals along the axon and conserves energy, as will be discussed later.

Various organelles and materials must be moved as much as one meter from the cell body, where they are made, to the axon and its terminals in order to maintain the structure and function of the cell axon. This movement is termed **axon transport.** The substances and organelles being moved are linked by proteins to microtubules in the cell body and axon. The microtubules serve as the "rails" along which the transport occurs. The linking proteins act both as the "motors" of axon transport and as ATPase enzymes, providing energy from split ATP to the "motors."

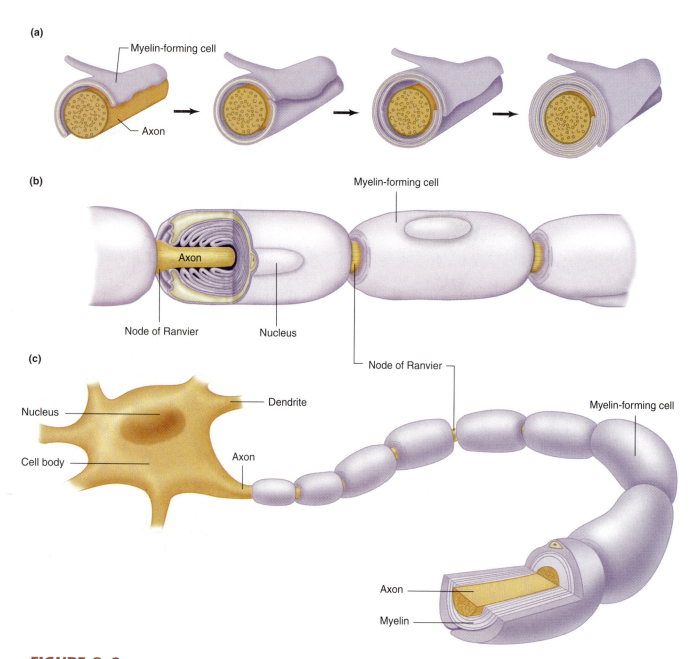

(a)

Myelin-forming cell

Axon

(b)

Myelin-forming cell

Axon

Node of Ranvier

Nucleus

Node of Ranvier

(c)

Nucleus

Dendrite

Myelin-forming cell

Cell body

Axon

Axon

Myelin

FIGURE 8–3

(a) Cross section of an axon in successive stages of myelinization. The myelin-forming cell may migrate around the axon, trailing successive layers of its plasma membrane or, as shown here, it may add to its tip, which lies against the axon, so that the tip is pushed around the axon, burrowing under the layers of myelin that are already formed. The latter process must be used in the central nervous system where each myelin-forming cell may send branches to as many as 40 axons. (b) Adjacent myelin-forming cells are each separated by a small space, the node of Ranvier. (c) A myelinated neuron.

Part a redrawn from Meyer-Franke and Barres.

Axon transport of certain materials also occurs in the opposite direction, from the axon terminals to the cell body. By this route, growth factors and other chemical signals picked up at the terminals can affect the neuron's morphology, biochemistry, and connectivity.

This is also the route by which certain harmful substances, such as tetanus toxin and herpes and polio viruses, taken up by the peripheral axon terminals can enter the central nervous system.

Functional Classes of Neurons

Neurons can be divided into three functional classes: afferent neurons, efferent neurons, and interneurons. **Afferent neurons** convey information from the tissues and organs of the body *into* the central nervous system, **efferent neurons** transmit electric signals from the central nervous system *out* to effector cells (particularly muscle or gland cells or other neurons), and **interneurons** connect neurons *within* the central nervous system (Figure 8–4). As a rough estimate, for each afferent neuron entering the central nervous system, there are 10 efferent neurons and 200,000 interneurons. Thus, by far most of the neurons in the central nervous system are interneurons.

At their peripheral ends (the ends farthest from the central nervous system), afferent neurons have **sensory receptors,** which respond to various physical or chemical changes in their environment by causing electric signals to be generated in the neuron. The receptor region may be a specialized portion of the plasma membrane or a separate cell closely associated with the neuron ending. (Recall from Chapter 7 that the term "receptor" has two distinct meanings, the one defined here and the other referring to the specific proteins with which a chemical messenger combines to exert its effects on a target cell; both types of receptors will be referred to frequently in this chapter.) Afferent neurons propagate electric signals from their receptors into the brain or spinal cord.

Afferent neurons are atypical in that they have only a single process, usually considered to be an axon. Shortly after leaving the cell body, the axon divides. One branch, the peripheral process, ends at the receptors; the other branch, the central process, enters the central nervous system to form junctions with other neurons. Note in Figure 8–4 that for afferent neurons both the cell body and the long peripheral process of the axon are *outside* the central nervous system, and only a part of the central process enters the brain or spinal cord.

The cell bodies and dendrites of efferent neurons are within the central nervous system, but the axons extend out into the periphery. The axons of both the afferent and efferent neurons, except for the small part in the brain or spinal cord, form the **nerves** of the peripheral nervous system. Note that a nerve fiber is a *single* axon, and a nerve is a *bundle* of axons bound together by connective tissue.

Interneurons lie entirely within the central nervous system. They account for over 99 percent of all neurons and have a wide range of physiological properties, shapes, and functions. The number of interneurons interposed between certain afferent and

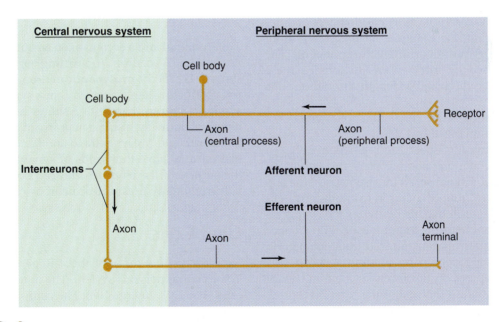

FIGURE 8–4

Three classes of neurons. The dendrites are not shown. The arrows indicate the direction of transmission of neural activity. The stylized neurons in this figure show the conventions that we will use throughout this book for the different parts of neurons. As discussed later, there are efferent components of the peripheral nervous system that consist of two neurons, not one as shown here.

TABLE 8–1 Three Classes of Neurons

I. Afferent neurons
A. Transmit information into the central nervous system from receptors at their peripheral endings
B. Cell body and the long peripheral process of the axon are in the peripheral nervous system; only the short central process of the axon enters the central nervous system
C. Have no dendrites

II. Efferent neurons
A. Transmit information out of the central nervous system to effector cells, particularly muscles, glands, or other neurons
B. Cell body, dendrites, and a small segment of the axon are in the central nervous system; most of the axon is in the peripheral nervous system

III. Interneurons
A. Function as integrators and signal changers
B. Integrate groups of afferent and efferent neurons into reflex circuits
C. Lie entirely within the central nervous system
D. Account for 99 percent of all neurons

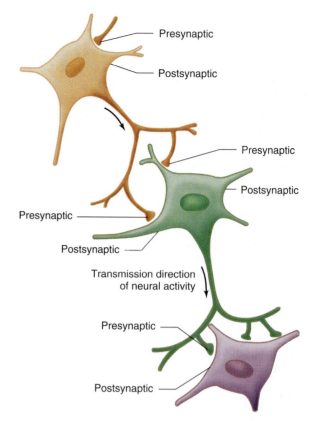

FIGURE 8–5

A neuron postsynaptic to one cell can be presynaptic to another.

efferent neurons varies according to the complexity of the action. The knee-jerk reflex elicited by tapping below the kneecap has no interneurons—the afferent neurons end directly on the efferent neurons. In contrast, stimuli invoking memory or language may involve millions of interneurons.

Interneurons can serve as signal changers or gate-keepers, changing, for example, an excitatory input into an inhibitory output or into no output at all. The mechanisms used by interneurons to achieve these functions will be discussed at length throughout this chapter.

Characteristics of the three functional classes of neurons are summarized in Table 8–1.

The anatomically specialized junction between two neurons where one neuron alters the activity of another is called a **synapse.** At most synapses, the signal is transmitted from one neuron to another by neurotransmitters, a term that also includes the chemicals by which efferent neurons communicate with effector cells. The neurotransmitters released from one neuron alter the receiving neuron by binding with specific membrane receptors on the receiving neuron. (Once again, do not confuse this use of the term "receptor" with the sensory receptors mentioned above that are at the peripheral ends of afferent neurons.)

Most synapses occur between the axon terminal of one neuron and the dendrite or cell body of a second neuron. In certain areas, however, synapses also occur between two dendrites or between a dendrite and a cell body to modulate the input to a cell, or between an

axon terminal and a second axon terminal to modulate its output. A neuron conducting signals toward a synapse is called a **presynaptic neuron,** whereas a neuron conducting signals away from a synapse is a **postsynaptic neuron.** Figure 8–5 shows how, in a multineuronal pathway, a single neuron can be postsynaptic to one cell and presynaptic to another. A postsynaptic neuron may have thousands of synaptic junctions on the surface of its dendrites and cell body, so that signals from many presynaptic neurons can affect it.

Glial Cells

Neurons account for only about 10 percent of the cells in the central nervous system. The remainder are **glial cells** (also called neuroglia). The neurons branch more extensively than glia do, however, and therefore neurons occupy about 50 percent of the volume of the brain and spinal cord.

Glial cells physically and metabolically support neurons and, as noted earlier, some glia, the oligodendroglia, form the myelin covering of CNS axons. A second type of glial cell, the **astroglia,** helps regulate the composition of the extracellular fluid in the central nervous system by removing potassium ions and neurotransmitters around synapses. Astroglia sustain the neurons metabolically—for example, by providing glucose and removing ammonia. In development of the embryo, astroglia guide neurons as they migrate, and they stimulate the neurons' growth by secreting growth factors. In addition, astroglia have many neuron-like characteristics, for example, they have ion channels, receptors for certain neurotransmitters and the enzymes for processing them, and the capability of generating weak electrical responses. Thus, in addition to all their other roles, it is speculated that astroglia may take part in information signaling in the brain. A third type of glia, the **microglia,** perform immune functions in the central nervous system.

Schwann cells, the glial cells of the peripheral nervous system, have most of the properties of the central nervous system glia. As mentioned earlier, Schwann cells produce the myelin sheath of peripheral nerve fibers.

Neural Growth and Regeneration

The elaborate networks of nerve-cell processes that characterize the nervous system are remarkably similar in all human beings and depend upon the outgrowth of specific axons to specific targets.

Development of the nervous system in the embryo begins with a series of divisions of precursor cells that can develop into neurons or glia. After the last cell division, each neuronal daughter cell differentiates, migrates to its final location, and sends out processes that will become its axon and dendrites. A specialized enlargement, the growth cone, forms the tip of each extending axon and is involved in finding the correct route and final target for the process.

As the axon grows, it is guided along the surfaces of other cells, most commonly glial cells. Which particular route is followed depends largely on attracting, supporting, deflecting, or inhibiting influences exerted by several types of molecules. Some of these molecules, such as cell adhesion molecules, reside on the membranes of the glia and embryonic neurons. Others are soluble **neurotropic factors** (growth factors for neural tissue) in the extracellular fluid surrounding the growth cone or its distant target.

Once the target of the advancing growth cone is reached, synapses are formed. The synapses are active, however, before their final maturation occurs, and this early activity, in part, determines their final use. During these intricate early stages of neural development, which occur during all trimesters of pregnancy and into infancy, alcohol and other drugs, radiation,

malnutrition, and viruses can exert effects that cause permanent damage to the developing fetal nervous system.

A normal, although unexpected, aspect of development of the nervous system occurs after growth and projection of the axons. Many of the newly formed neurons and synapses *degenerate.* In fact, as many as 50 to 70 percent of neurons die by apoptosis in some regions of the developing nervous system! Exactly why this seemingly wasteful process occurs is unknown although neuroscientists speculate that in this way connectivity in the nervous system is refined, or "fine tuned."

Although the basic shape and location of existing neurons in the mature central nervous system do not change, the creation and removal of synaptic contacts begun during fetal development continue, albeit at a slower pace, throughout life as part of normal growth, learning, and aging. Division of neuron precursors is largely complete before birth, and after early infancy new neurons are formed at a slower pace to replace those that die.

Severed axons can repair themselves, however, and significant function regained, provided that the damage occurs *outside* the central nervous system and does not affect the neuron's cell body. After repairable injury, the axon segment now separated from the cell body degenerates. The proximal part of the axon (the stump still attached to the cell body) then gives rise to a growth cone, which grows out to the effector organ so that in some cases function is restored.

In contrast, severed axons *within* the central nervous system attempt sprouting, but no significant regeneration of the axon occurs across the damaged site, and there are no well-documented reports of significant function return. Either some basic difference of central nervous system neurons or some property of their environment, such as inhibitory factors associated with nearby glia, prevents their functional regeneration.

In humans, however, spinal injuries typically crush rather than cut the tissue, leaving the axons intact. In this case, a primary problem is self-destruction (apoptosis) of the nearby oligodendroglia, because when these cells die and their associated axons lose their myelin coat, the axons cannot transmit information effectively.

Researchers are attempting a variety of measures to provide an environment that will support axonal regeneration in the central nervous system. They are creating tubes to support regrowth of the severed axons, redirecting the axons to regions of the spinal cord that lack the growth-inhibiting factors, preventing apoptosis of the oligodendrocytes so myelin can be maintained, and supplying neurotropic factors that support recovery of the damaged tissue.

Attempts are also being made to restore function to damaged or diseased brains by the implantation of precursor cells that will develop into new neurons that will replace missing neurotransmitters or neurotropic factors. Alternatively, pieces of fetal brain or tissues from the patient that produce the needed neurotransmitters or growth factors are implanted. For example, the adrenal medulla, which is part of the adrenal glands, synthesizes and secretes chemicals similar to some of the neurotransmitters found in the brain. When pieces of a patient's own adrenal medulla are inserted into damaged parts of the brain, the pieces continue to secrete these chemicals and provide the missing neurotransmitters.

We now turn to the mechanisms by which neurons and synapses function, beginning with the electrical properties that underlie all these events.

SECTION A SUMMARY

I. The nervous system is divided into two parts: The central nervous system (CNS) comprises the brain and spinal cord, and the peripheral nervous system consists of nerves extending from the CNS.

Structure and Maintenance of Neurons

I. The basic unit of the nervous system is the nerve cell, or neuron.
II. The cell body and dendrites receive information from other neurons.
III. The axon (nerve fiber), which may be covered with sections of myelin separated by nodes of Ranvier, transmits information to other neurons or effector cells.

Functional Classes of Neurons

I. Neurons are classified in three ways:
  a. Afferent neurons transmit information into the CNS from receptors at their peripheral endings.
  b. Efferent neurons transmit information out of the CNS to effector cells.
  c. Interneurons lie entirely within the CNS and form circuits with other interneurons or connect afferent and efferent neurons.
II. Information is transmitted across a synapse by neurotransmitters, which are released by a

presynaptic neuron and combine with receptors on a postsynaptic neuron.

Glial Cells

I. The CNS also contains glial cells, which help regulate the extracellular fluid composition, sustain the neurons metabolically, form myelin, serve as guides for developing neurons, and provide immune functions.

Neural Growth and Regeneration

I. Neurons develop from precursor cells, migrate to their final location, and send out processes to their target cells.
II. Cell division to form new neurons is markedly slowed after birth.
III. After degeneration of a severed axon, damaged peripheral neurons may regrow the axon to their target organ. Damaged neurons of the CNS do not regenerate or restore significant function.

SECTION A KEY TERMS

central nervous system (CNS)	Schwann cell
peripheral nervous system	node of Ranvier
neuron	axon transport
neurotransmitter	afferent neuron
integrator	efferent neuron
cell body	interneuron
dendrite	sensory receptor
axon	nerve
nerve fiber	synapse
initial segment	presynaptic neuron
collateral	postsynaptic neuron
axon terminal	glial cell
varicosity	astroglia
myelin	microglia
oligodendroglia	neurotropic factor

SECTION A REVIEW QUESTIONS

1. Describe the direction of information flow through a neuron and also through a network consisting of afferent neurons, efferent neurons, and interneurons.
2. Contrast the two uses of the word "receptor."

SECTION B

MEMBRANE POTENTIALS

Basic Principles of Electricity

As discussed in Chapter 6, with the exception of water the major chemical substances in the extracellular fluid are sodium and chloride ions, whereas the

intracellular fluid contains high concentrations of potassium ions and ionized nondiffusible molecules, particularly proteins, with negatively charged side chains and phosphate compounds. Electrical phenomena resulting from the distribution of these charged

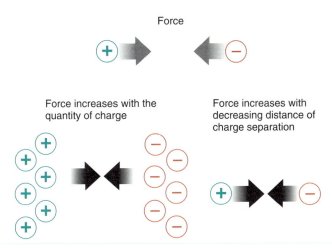

FIGURE 8–6

The electric force of attraction between positive and negative charges increases with the quantity of charge and with decreasing distance between charges.

particles occur at the cell's plasma membrane and play a significant role in cell integration and communication, the two major functions of the nervous system.

Like charges repel each other; that is, positive charge repels positive charge, and negative charge repels negative charge. In contrast, an electric force draws oppositely charged substances together (Figure 8–6).

Separated electric charges of opposite sign have the potential of doing work if they are allowed to come together. This potential is called an **electric potential** or, because it is determined by the difference in the amount of charge between two points, a **potential difference,** which we shall often shorten to **potential.** The units of electric potential are volts, but since the total charge that can be separated in most biological systems is very small, the potential differences are small and are measured in millivolts (1 mV = 0.001 V).

The movement of electric charge is called a **current.** The electric force between charges tends to make them flow, producing a current. If the charges are of opposite sign, the current brings them toward each other; if the charges are alike, the current increases the separation between them. The amount of charge that moves—in other words, the current—depends on the potential difference between the charges and on the nature of the material through which they are moving. The hindrance to electric charge movement is known as **resistance.** The relationship between current I, voltage E (for electric potential), and resistance R is given by **Ohm's law:**

$$I = E/R$$

Materials that have a high electrical resistance are known as insulators, whereas materials that have a low resistance are conductors.

Water that contains dissolved ions is a relatively good conductor of electricity because the ions can carry the current. As we have seen, the intracellular and extracellular fluids contain numerous ions and can therefore carry current. Lipids, however, contain very few charged groups and cannot carry current. Therefore, the lipid layers of the plasma membrane are regions of high electrical resistance separating two water compartments—the intracellular fluid and the extracellular fluid—of low resistance.

The Resting Membrane Potential

All cells under resting conditions have a potential difference across their plasma membranes oriented with the inside of the cell negatively charged with respect to the outside (Figure 8–7a). This potential is the **resting membrane potential.**

(a)

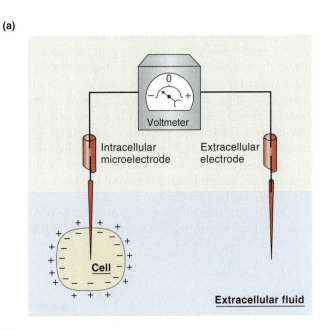

(b)

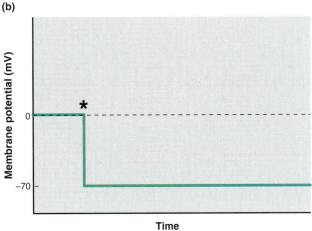

FIGURE 8–7

(a) Apparatus for measuring membrane potentials. (b) The potential difference across a plasma membrane as measured by an intracellular microelectrode. The asterisk indicates the time the electrode entered the cell.

By convention, extracellular fluid is assigned a voltage of zero, and the polarity (positive or negative) of the membrane potential is stated in terms of the sign of the excess charge on the inside of the cell. For example, if the intracellular fluid has an excess of negative charge and the potential difference across the membrane has a magnitude of 70 mV, we say that the membrane potential is −70 mV.

The magnitude of the resting membrane potential varies from about −5 to −100 mV, depending upon the type of cell; in neurons, it is generally in the range of −40 to −75 mV (Figure 8–7b). The membrane potential of some cells can change rapidly in response to stimulation, an ability of key importance in their functioning.

The resting membrane potential exists because there is a tiny excess of negative ions inside the cell and an excess of positive ions outside. The excess negative charges inside are electrically attracted to the excess positive charges outside the cell, and vice versa. Thus, the excess charges (ions) collect in a thin shell tight against the inner and outer surfaces of the plasma membrane (Figure 8–8), whereas the bulk of the intracellular and extracellular fluids are neutral. Unlike the diagrammatic representation in Figure 8–8, the number of positive and negative charges that have to be separated across a membrane to account for the potential is an infinitesimal fraction of the total number of charges in the two compartments.

The magnitude of the resting membrane potential is determined mainly by two factors (a third factor will

TABLE 8–2 Distribution of Major Ions Across the Plasma Membrane of a Typical Nerve Cell		
	Concentration, mmol/L	
Ion	Extracellular	Intracellular
Na$^+$	150	15
Cl$^-$	110	10
K$^+$	5	150

be given later): (1) differences in specific ion concentrations in the intracellular and extracellular fluids, and (2) differences in membrane permeabilities to the different ions, which reflect the number of open channels for the different ions in the plasma membrane. The rest of this section analyzes how these two factors operate.

The concentrations of sodium, potassium, and chloride ions in the extracellular fluid and in the intracellular fluid of a typical nerve cell are listed in Table 8–2. Although this table appears to contradict our earlier assertion that the bulk of the intra- and extracellular fluids are electrically neutral, there are many other ions, including Mg^{2+}, Ca^{2+}, H^+, HCO_3^-, HPO_4^{2-}, SO_4^{2-}, amino acids, and proteins, in both fluid compartments. Of the mobile ions, sodium, potassium, and chloride ions are present in the highest concentrations, and the membrane permeabilities to these ions are restricted, although, as we shall see, to different degrees. Sodium and potassium generally play the most important roles in generating the resting membrane potential. Note that the sodium and chloride concentrations are lower inside the cell than outside, and that the potassium concentration is greater inside the cell. As we described in Chapter 6, the concentration differences for sodium and potassium are due to the action of a plasma-membrane active-transport system that pumps sodium out of the cell and potassium into it. We will see later the reason for the chloride distribution.

To understand how such concentration differences for sodium and potassium create membrane potentials, let us consider the situation in Figure 8–9. The assumption in this model is that the membrane contains potassium channels but no sodium channels. *Initially*, compartment 1 contains 0.15 M NaCl, and compartment 2 contains 0.15 M KCl. There is no potential difference across the membrane because the two compartments contain equal numbers of positive and negative ions; that is, they are electrically neutral. The positive ions are different—sodium versus potassium, but the *total* numbers of positive ions in the two compartments are the same, and each positive ion is balanced by a chloride ion.

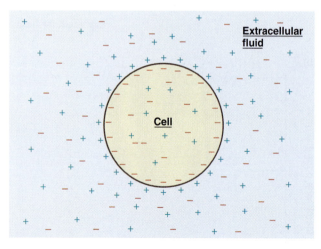

FIGURE 8–8

The excess of positive charges outside the cell and the excess of negative charges inside collect tight against the plasma membrane. In reality, these excess charges are only an extremely small fraction of the total number of ions inside and outside the cell.

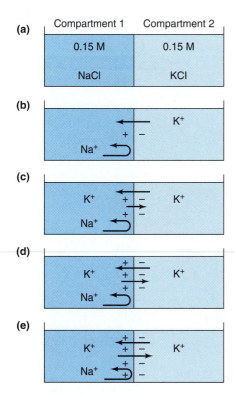

FIGURE 8–9

Generation of a diffusion potential across a membrane that contains only potassium channels. Arrows represent ion movements.

This initial state will not, however, last. Because of the potassium channels, potassium will diffuse down its concentration gradient from compartment 2 into compartment 1. After a few potassium ions have moved into compartment 1, that compartment will have an excess of positive charge, leaving behind an excess of negative charge in compartment 2. Thus, a potential difference has been created across the membrane.

Now we introduce a second factor that can cause net movement of ions across a membrane: an electrical potential. As compartment 1 becomes increasingly positive and compartment 2 increasingly negative, the membrane potential difference begins to influence the movement of the potassium ions. They are attracted by the negative charge of compartment 2 and repulsed by the positive charge of compartment 1.

As long as the flux due to the potassium concentration gradient is greater than the flux due to the membrane potential, there will be net movement of potassium from compartment 2 to compartment 1, and the membrane potential will progressively increase. However, eventually the membrane potential will become negative enough to produce a flux equal but opposite the flux due to the concentration gradient. The

membrane potential at which these two fluxes become equal in magnitude but opposite in direction is called the **equilibrium potential** for that type of ion—in this case, potassium. At the equilibrium potential for an ion, there is no net movement of the ion because the opposing fluxes are equal, and the potential will undergo no further change.

The value of the equilibrium potential for any type of ion depends on the concentration gradient for that ion across the membrane. If the concentrations on the two sides were equal, the flux due to the concentration gradient would be zero, and the equilibrium potential would also be zero. The larger the concentration gradient, the larger the equilibrium potential because a larger electrically driven movement of ions will be required to balance the larger movement due to the concentration difference.

If the membrane separating the two compartments is replaced with one that contains only sodium channels, a parallel situation will occur (Figure 8–10). A sodium equilibrium potential will eventually be established, but compartment 2 will be *positive* with respect to compartment 1, at which point net movement of sodium will cease. Again, at the equilibrium potential the movement of ions due to the concentration gradient is equal but opposite to the movement due to the electrical gradient.

Thus, the equilibrium potential for one ion species can be different in magnitude and direction from those for other ion species, depending on the concentration gradients for each ion. (Given the concentration gradient for any ion, the equilibrium potential for that ion can be calculated by means of the Nernst equation, Appendix D.)

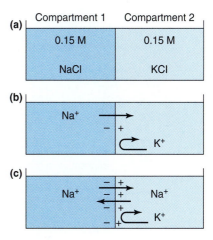

FIGURE 8–10

Generation of a diffusion potential across a membrane that contains only sodium channels. Arrows represent ion movements.

Our examples were based on the membrane being permeable to only one ion at a time. When more than one ion species can diffuse across the membrane, the permeabilities and concentration gradients for all the ions must be considered when accounting for the membrane potential. For a given concentration gradient, the greater the membrane permeability to an ion species, the greater the contribution that ion species will make to the membrane potential. (Given the concentration gradients and membrane permeabilities for several ion species, the potential of a membrane permeable to these species can be calculated by the Goldman equation, Appendix D.)

It is not difficult to move from these hypothetical examples to a nerve cell at rest where (1) the potassium concentration is much greater inside than outside (Figure 8–11a) and the sodium concentration profile is just the opposite (Figure 8–12a); and (2) the plasma membrane contains 50 to 75 times as many open potassium channels as open sodium channels.

Given the actual potassium and sodium concentration differences, one can calculate that the potassium equilibrium potential will be approximately −90 mV (Figure 8–11b) and the sodium equilibrium potential about +60 mV (Figure 8–12b). However, since the membrane is permeable, to some extent, to both

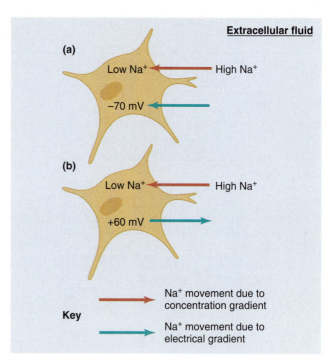

FIGURE 8–12

Forces acting on sodium when the membrane of a neuron is at (a) the resting potential (−70 mV, inside negative), and (b) the sodium equilibrium potential (+60 mV, inside positive).

potassium and sodium, the resting membrane potential cannot be equal to either of these two equilibrium potentials. The resting potential will be much closer to the potassium equilibrium potential because the membrane is so much more permeable to potassium than to sodium.

In other words, a potential is generated across the plasma membrane largely because of the movement of potassium out of the cell down its concentration gradient through open potassium channels, so that the inside of the cell becomes negative with respect to the outside. To repeat, the experimentally measured resting membrane potential is *not* equal to the potassium equilibrium potential, because a small number of sodium channels are open in the resting state, and some sodium ions continually move into the cell, canceling the effect of an equivalent number of potassium ions simultaneously moving out.

An actual resting membrane potential when recorded is about −70 mV, a typical value for neurons, and neither sodium nor potassium is at its equilibrium potential. Thus, there is net movement through ion channels of sodium into the cell and potassium out. The concentration of intracellular sodium and potassium ions does not change, however, because active-transport mechanisms in the plasma membrane utilize

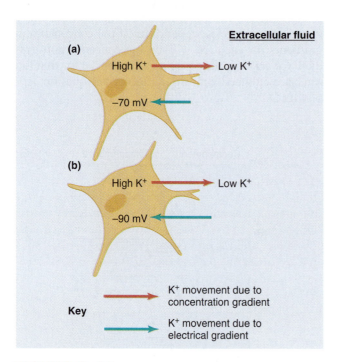

FIGURE 8–11

Forces acting on potassium when the membrane of a neuron is at (a) the resting potential (−70 mV, inside negative), and (b) the potassium equilibrium potential (−90 mV, inside negative).

energy derived from cellular metabolism to pump the sodium back out of the cell and the potassium back in. Actually, the pumping of these ions is linked because they are both transported by the Na,K-ATPase pumps in the membrane (Chapter 6).

In a resting cell, the number of ions moved by the pump equals the number of ions that move in the opposite direction through membrane channels down their concentration and/or electrical gradients. Therefore the concentrations of sodium and potassium in the cell do not change. As long as the concentration gradients remain stable and the ion permeabilities of the plasma membrane do not change, the electric potential across the resting membrane will also remain constant.

Thus far, we have described the membrane potential as due purely and directly to the passive movement of ions down their electrical and concentration gradients, the concentration gradients having been established by membrane pumps. There is, however, as mentioned in Chapter 6, another component to the membrane potential that reflects the *direct* separation of charge across the membrane by the transport of ions by the membrane Na,K-ATPase pumps. These pumps actually move three sodium ions out of the cell for every two potassium ions that they bring in. This unequal transport of positive ions makes the inside of the cell more negative than it would be from ion diffusion alone. A pump that moves net charge across the membrane contributes directly to the membrane potential and is known as an **electrogenic pump.**

In most cells (but by no means all), the electrogenic contribution to the membrane potential is quite small. It must be reemphasized, however, that even when the electrogenic contribution of the Na,K-ATPase pump is small, the pump always makes an essential *indirect* contribution to the membrane potential because it maintains the concentration gradients down which the ions diffuse to produce most of the charge separation that makes up the potential.

Figure 8–13 summarizes the information we have been presenting. This figure may mistakenly be seen to present a conflict: The *development* of the resting membrane potential depends predominantly on the diffusion of potassium out of the cell, yet in the steady state, sodium diffusion into the cell, indicated by the black Na$^+$ arrow in Figure 8–13, is greater than potassium diffusion out of the cell. The reason is that although there are relatively few open sodium channels, sodium has a much larger electrochemical force acting upon it—that is, it is far from its equilibrium potential. The greater diffusion of sodium into the cell than potassium out compensates for the fact that the membrane pump moves three sodium ions out of the cell

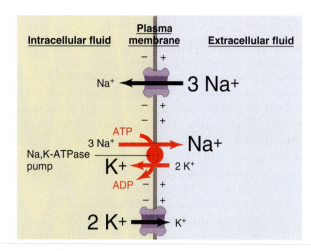

FIGURE 8–13

Movements of sodium and potassium ions across the plasma membrane of a resting neuron in the steady state. The passive movements (black arrows) are exactly balanced by the active transport (red arrows) of the ions in the opposite direction.

for every two potassium ions that are moved in. Figure 8–13 shows ion movements *once steady state has been achieved,* not during its achievement.

We have not yet dealt with chloride ions. The plasma membranes of many cells have chloride channels but do not contain chloride-ion pumps. Therefore, in these cells chloride concentrations simply shift until the equilibrium potential for chloride is equal to the resting membrane potential. In other words, the negative membrane potential moves chloride out of the cell, and the chloride concentration outside the cell becomes higher than that inside. This concentration gradient produces a diffusion of chloride back into the cell that exactly opposes the movement out because of the electric potential.

In contrast, some cells have a non-electrogenic active transport system that moves chloride *out* of the cell. In these cells, the membrane potential is not at the chloride equilibrium potential, and net chloride diffusion *into* the cell contributes to the excess negative charge inside the cell; that is, net chloride diffusion makes the membrane potential more negative than it would otherwise be.

We noted earlier that most of the negative charge in neurons is accounted for not by chloride ions but by negatively charged organic molecules, such as proteins and phosphate compounds. Unlike chloride, however, these molecules do not readily cross the plasma membrane but remain inside the cell, where their charge contributes to the total negative charge within the cell.

Graded Potentials and Action Potentials

Transient changes in the membrane potential from its resting level produce electric signals. Such changes are the most important way that nerve cells process and transmit information. These signals occur in two forms: graded potentials and action potentials. Graded potentials are important in signaling over short distances, whereas action potentials are the long-distance signals of nerve and muscle membranes.

The terms "depolarize," "repolarize," and "hyperpolarize" are used to describe the direction of changes in the membrane potential relative to the resting potential (Figure 8–14). The membrane is said to be **depolarized** when its potential is less negative (closer to zero) than the resting level. **Overshoot** refers to a reversal of the membrane potential polarity—that is, when the inside of a cell becomes positive relative to the outside. When a membrane potential that has been depolarized returns toward the resting value, it is said to be **repolarizing.** The membrane is **hyperpolarized** when the potential is more negative than the resting level.

Graded Potentials

Graded potentials are changes in membrane potential that are confined to a relatively small region of the plasma membrane and die out within 1 to 2 mm of their site of origin. They are usually produced by some specific change in the cell's environment acting on a specialized region of the membrane, and they are called "graded potentials" simply because the magnitude of the potential change can vary (is graded). We shall encounter a number of graded potentials, which are given various names related to the location of the potential or to the function it performs: receptor potential, synaptic potential, and pacemaker potential (Table 8–3).

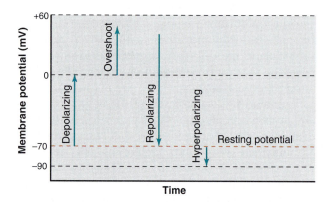

FIGURE 8–14

Depolarizing, repolarizing, hyperpolarizing, and overshoot changes in membrane potential.

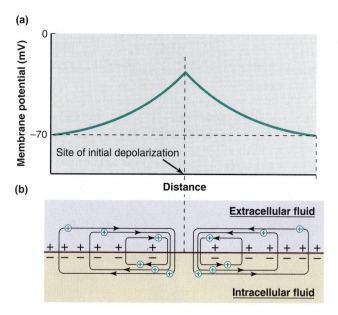

FIGURE 8–15

The membrane potential of a cell can be depolarized by using a stimulating current generator, and the potential can be recorded by a pair of electrodes, one inside the cell and the other in the extracellular fluid, as in Figure 8–7. (a) Membrane potential is closer to the resting potential with increasing distance from the depolarization site. (b) Local current surrounding the depolarized region produces depolarization of adjacent regions.

Whenever a graded potential occurs, charge flows between the place of origin of the potential and adjacent regions of the plasma membrane, which are still at the resting potential. In Figure 8–15a, a small region of a membrane has been depolarized by a stimulus and therefore has a potential less negative than adjacent areas. Inside the cell (Figure 8–15b), positive charge (positive ions) will flow through the intracellular fluid away from the depolarized region and toward the more negative, resting regions of the membrane. Simultaneously, outside the cell, positive charge will flow from the more positive region of the resting membrane toward the less positive region just created by the depolarization. The greater the potential change, the greater the currents. By convention, the direction in which positive ions move is designated the direction of the current, although negatively charged ions simultaneously move in the opposite direction. In fact, the local current is carried by ions such as K^+, Na^+, Cl^-, and HCO_3^-.

Note that this local current moves positive charges toward the depolarization site along the *outside* of the membrane and away from the depolarization site along the *inside* of the membrane. Thus it produces a decrease in the amount of charge separation (depolarization) in the membrane sites *adjacent to* the originally depolarized region.

TABLE 8–3 A Miniglossary of Terms Describing the Membrane Potential

Potential = potential difference	The voltage difference between two points.
Membrane potential = transmembrane potential	The voltage difference between the inside and outside of a cell.
Equilibrium potential	The voltage difference across a membrane that produces a flux of a given ion species that is equal but opposite the flux due to the concentration gradient affecting that same ion species.
Resting membrane potential = resting potential	The steady transmembrane potential of a cell that is not producing an electric signal.
Graded potential	A potential change of variable amplitude and duration that is conducted decrementally; it has no threshold or refractory period.
Action potential	A brief all-or-none depolarization of the membrane, reversing polarity in neurons; it has a threshold and refractory period and is conducted without decrement.
Synaptic potential	A graded potential change produced in the postsynaptic neuron in response to release of a neurotransmitter by a presynaptic terminal; it may be depolarizing (an excitatory postsynaptic potential or EPSP) or hyperpolarizing (an inhibitory postsynaptic potential or IPSP).
Receptor potential	A graded potential produced at the peripheral endings of afferent neurons (or in separate receptor cells) in response to a stimulus.
Pacemaker potential	A spontaneously occurring graded potential change that occurs in certain specialized cells.

Depending upon the initiating event, graded potentials can occur in either a depolarizing or hyperpolarizing direction (Figure 8–16a), and their magnitude is related to the magnitude of the initiating event (Figure 8–16b). Moreover, local current flows much like water flows through a leaky hose. Charge is lost across the membrane because the membrane is permeable to ions, just as water is lost from the leaky hose. The result is that the magnitude of the current decreases with the distance from the initial site of the potential change, just as water flow decreases the farther along the leaky hose you are from the faucet (Figure 8–17). In fact, plasma membranes are so leaky to ions that local currents die out almost completely within a few millimeters of their point of origin. There is another way of saying the same thing: Local current is **decremental;** that is, its amplitude decreases with increasing distance from the site of origin of the potential. The resulting change in membrane potential from resting level therefore also decreases with the distance from the potential's site of origin (Figures 8–15a and 8–16c).

Because the electric signal decreases with distance, graded potentials (and the local current they generate) can function as signals only over very short distances (a few millimeters). Nevertheless, graded potentials are the only means of communication used by some neurons and, as we shall see, play very important roles in the initiation and integration of the long-distance signals by neurons and some other cells.

Action Potentials

Action potentials are very different from graded potentials. They are rapid, large alterations in the membrane potential during which time the membrane potential may change 100 mV, from −70 to +30 mV, and then repolarize to its resting membrane potential (Figure 8–18a). Nerve and muscle cells as well as some endocrine, immune, and reproductive cells have plasma membranes capable of producing action potentials. These membranes are called **excitable membranes,** and their ability to generate action potentials is known as **excitability.** Whereas all cells are capable of conducting graded potentials, only excitable membranes can conduct action potentials. The propagation of action potentials is the mechanism used by the nervous system to communicate over long distances.

How does an excitable membrane make rapid changes in its membrane potential? How is an action potential propagated along an excitable membrane? These questions are discussed in the following sections.

Ionic Basis of the Action Potential Action potentials can be explained by the concepts already developed for describing the origins of resting membrane potentials. We have seen that the magnitude of the resting membrane potential depends upon the concentration gradients of and membrane permeabilities to different ions, particularly sodium and potassium. This is true for the action potential as well: The action

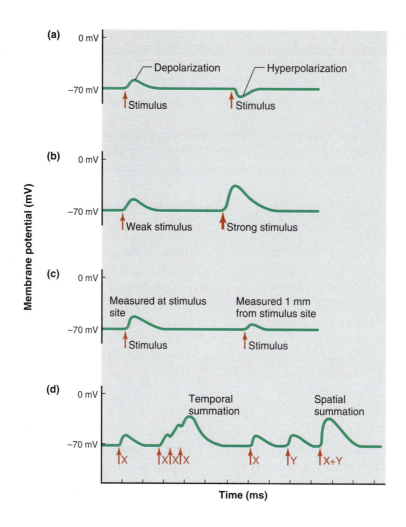

FIGURE 8–16

Graded potentials can be recorded under experimental conditions in which the stimulus or recording conditions can be varied. Such experiments show that graded potentials (a) can be depolarizing or hyperpolarizing, (b) can vary in size, (c) are conducted decrementally, and (d) can be summed. Temporal and spatial summation will be discussed later in the chapter. ▭

potential results from a transient change in membrane ion permeability, which allows selected ions to move down their concentration gradients.

In the resting state, the open channels in the plasma membrane are predominantly those that are permeable to potassium (and chloride) ions. Very few sodium-ion channels are open, and the resting potential is there-

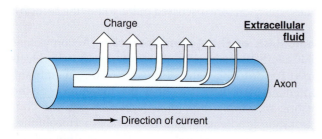

FIGURE 8–17

Leakage of charge across the plasma membrane reduces the local current at sites farther along the membrane.

fore close to the potassium equilibrium potential. During an action potential, however, the membrane permeabilities to sodium and potassium ions are markedly altered. (A review of voltage-gated ion channels, Chapter 6, may be helpful at this time.)

The depolarizing phase of the action potential is due to the opening of voltage-gated sodium channels, which increases the membrane permeability to sodium ions several hundredfold (purple line in Figure 8–18b). This allows more sodium ions to move into the cell. During this period, therefore, more positive charge enters the cell in the form of sodium ions than leaves in the form of potassium ions, and the membrane depolarizes. It may even overshoot, becoming positive on the inside and negative on the outside of the membrane. In this phase, the membrane potential approaches but does not quite reach the sodium equilibrium potential (+60 mV).

Action potentials in nerve cells last only about 1 ms and typically show an overshoot. (They may last much longer in certain types of muscle cells.) The

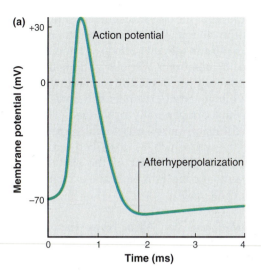

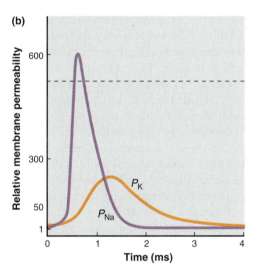

FIGURE 8–18

The changes during an action potential in (a) membrane potential and (b) membrane permeability (*P*) to sodium (purple) and potassium (orange) ions.

channels have closed, some of the voltage-gated potassium channels are still open, and in nerve cells there is generally a small hyperpolarization of the membrane potential beyond the resting level (**afterhyperpolarization,** Figure 8–18a). The differences between voltage-gated sodium and potassium channels are summarized in Table 8–4. Chloride permeability does not change during the action potential.

One might think that large movements of ions across the membrane are required to produce such large changes in membrane potential. Actually, the number of ions that cross the membrane during an action potential is extremely small compared to the total number of ions in the cell, producing only infinitesimal changes in the intracellular ion concentrations. Yet if this tiny number of additional ions crossing the membrane with each action potential were not eventually moved back across the membrane, the concentration gradients of sodium and potassium would gradually disappear, and action potentials could no longer be generated. As might be expected, cellular accumulation of sodium and loss of potassium are prevented by the continuous action of the membrane Na,K-ATPase pumps.

What is achieved by letting sodium move into the neuron and then pumping it back out? Sodium movement down its electrochemical gradient into the cell creates the electric signal necessary for communication between parts of the cell, and pumping sodium out maintains the concentration gradient so that, in response to a new stimulus, sodium will again enter the cell and create another signal.

Mechanism of Ion-channel Changes   In the above section, we described the various phases of the action potential as due to the opening and/or closing of voltage-gated ion channels. What causes these changes? The very first part of the depolarization, as we shall see later, is due to local current. Once depolarization starts, the depolarization itself causes voltage-gated sodium channels to open. In light of our discussion of the ionic basis of membrane potentials, it is very easy to confuse the cause-and-effect relationships of this last statement. Earlier we pointed out that an increase in sodium permeability *causes* mem-

membrane potential returns so rapidly to its resting level because: (1) the sodium channels that opened during the depolarization phase undergo inactivation near the peak of the action potential, which causes them to close; and (2) voltage-gated potassium channels, which open more slowly than sodium channels, open in response to the depolarization. The timing of these two events can be seen in Figure 8–18b.

Closure of the sodium channels alone would restore the membrane potential to its resting level since potassium flux out would then exceed sodium flux in. However, the process is speeded up by the simultaneous increase in potassium permeability. Potassium diffusion out of the cell is then much greater than the sodium diffusion in, rapidly returning the membrane potential to its resting level. In fact, after the sodium

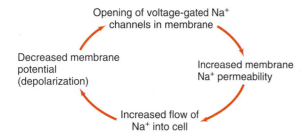

FIGURE 8–19

Positive-feedback relation between membrane depolarization and increased sodium permeability, which leads to the rapid depolarizing phase of the action potential.

brane depolarization; now we are saying that depolarization *causes* an increase in sodium permeability. Combining these two distinct causal relationships yields the positive-feedback cycle (Figure 8–19) responsible for the depolarizing phase of the action potential: The initial depolarization opens voltage-gated sodium channels so that the membrane permeability to sodium increases. Because of increased sodium permeability, sodium diffuses into the cell; this addition of positive charge to the cell further depolarizes the membrane, which in turn opens still more voltage-gated sodium channels, which produces a still greater increase in sodium permeability, etc. Many cells that have graded potentials cannot form action potentials because they have no voltage-gated sodium channels.

The potassium channels that open during an action potential are also voltage-gated. In fact, their opening is triggered by the same depolarization that opens the sodium channels, but the potassium channel opening is slightly delayed.

What about the inactivation of the voltage-gated sodium channels that opened during the rising phase of the action potential? This is the result of a voltage-induced change in the conformation of the proteins that constitute the channel, which closes the channel after its brief opening.

The generation of action potentials is prevented by *local anesthetics* such as procaine (Novocaine) and lidocaine (Xylocaine) because these drugs bind to the voltage-gated sodium channels and block them, preventing their opening in response to depolarization. Without action potentials, graded signals generated in the periphery—in response to injury, for example—cannot reach the brain and give rise to the sensation of pain.

Some animals produce toxins that work by interfering with nerve conduction in the same way that local anesthetics do. For example, the puffer fish produces an extremely potent toxin, tetrodotoxin, that binds to voltage-gated sodium channels and prevents the sodium component of the action potential.

Although we have discussed only sodium and potassium channels, in certain areas of neurons and in various nonneural cells *calcium* channels open in response to membrane depolarization. In some of the nonneural cells, calcium diffusion into the cell through these voltage-gated channels generates action potentials, which are generally prolonged. The inward calcium diffusion also raises calcium concentration within the cell, which, as described in Chapter 7, constitutes an essential part of the signal transduction pathway that couples membrane excitability to events within these cells.

Threshold and the All-or-None Response   Not all membrane depolarizations in excitable cells trigger the positive-feedback relationship that leads to an action potential.

The event that initiates the membrane depolarization provides an ionic current that adds positive charge to the inside of the cell, causing the initial depolarization from the resting membrane potential. As the depolarization begins, however, potassium efflux *increases* above its resting rate because the inside negativity, which tends to keep potassium in the cell, is weaker. Moreover, *initial* movement of sodium into the cell decreases because of this same lessened negativity; however, also in response to the depolarization, voltage-gated sodium channels open, increasing sodium permeability, which enhances sodium influx. All in all, at this stage potassium exit still exceeds sodium entry. But as the stimulus continues to add current (positive charge) to the inside of the cell, the depolarization increases, and more and more voltage-gated sodium channels open, allowing the influx of sodium ions to increase. Once the point is reached that the sodium influx exceeds potassium efflux, the positive-feedback cycle takes off and an action potential occurs. From this moment on, the membrane events are independent of the initial disturbing event and are driven entirely by the membrane properties.

In other words, action potentials occur only when the *net* movement of positive charge through ion channels is inward. The membrane potential at which this occurs is called the **threshold potential**, and stimuli that are just strong enough to depolarize the membrane to this level are **threshold stimuli** (Figure 8–20).

The threshold of most excitable membranes is about 15 mV less negative than the resting membrane potential. Thus, if the resting potential of a neuron is -70 mV, the threshold potential may be -55 mV. At depolarizations less than threshold, outward potassium movement still exceeds sodium entry, and the positive-feedback cycle cannot get started despite the increase in sodium entry. In such cases, the membrane will return to its resting level as soon as the stimulus

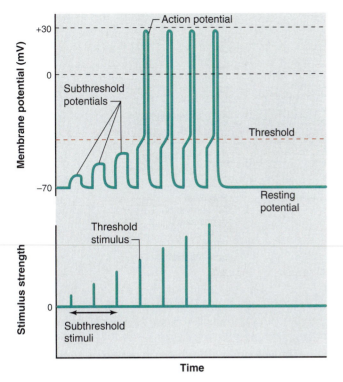

FIGURE 8–20

Changes in the membrane potential with increasing strength of depolarizing stimulus. When the membrane potential reaches threshold, action potentials are generated. Increasing the stimulus strength above threshold level does not cause larger action potentials. (The afterhyperpolarization has been omitted from this figure for clarity, and the absolute value of threshold is not indicated because it varies from cell to cell.)

is removed, and no action potential is generated. These weak depolarizations are **subthreshold potentials,** and the stimuli that cause them are **subthreshold stimuli.**

Stimuli of *more than* threshold magnitude also elicit action potentials, but as can be seen in Figure 8–20, the action potentials resulting from such stimuli have exactly the same amplitude as those caused by threshold stimuli. This is because once threshold is reached, membrane events are no longer dependent upon stimulus strength. Rather, the depolarization generates an action potential because the positive-feedback cycle is operating. Action potentials either occur maximally or they do not occur at all. Another way of saying this is that action potentials are **all-or-none.** The actual shape and amplitude of the action potential depends on the membrane conditions existing at a given time. For example, if the extracellular sodium concentration changes, the shape of the action potential will change.

The firing of a gun is a mechanical analogy that shows the principle of all-or-none behavior. The magnitude of the explosion and the velocity at which the bullet leaves the gun do not depend on how hard the trigger is squeezed. Either the trigger is pulled hard enough to fire the gun, or it is not; the gun cannot be fired halfway.

Because of its all-or-none nature, a single action potential cannot convey information about the magnitude of the stimulus that initiated it. How then does one distinguish between a loud noise and a whisper, a light touch and a pinch? This information, as we shall see, depends upon the number and pattern of action potentials transmitted per unit of time and not upon their magnitude.

Refractory Periods During the action potential, a second stimulus, no matter how strong, will not produce a second action potential, and the membrane is said to be in its **absolute refractory period.** This occurs because the voltage-gated sodium channels enter a closed, inactive state at the peak of the action potential. The membrane must repolarize before the sodium channel proteins return to the state in which they can be opened again by depolarization.

Following the absolute refractory period, there is an interval during which a second action potential can be produced, but only if the stimulus strength is considerably greater than usual. This is the **relative refractory period,** which can last 10 to 15 ms or longer in neurons and coincides roughly with the period of afterhyperpolarization. During the relative refractory period, there is lingering inactivation of the voltage-gated sodium channels, and an increased number of potassium channels are open. If a depolarization exceeds the increased threshold or outlasts the relative refractory period, additional action potentials will be fired.

The refractory periods limit the number of action potentials that can be produced by an excitable membrane in a given period of time. They also increase the reliability of neural signaling because they help limit extra impulses. Most nerve cells respond at frequencies of up to 100 action potentials per second, and some may produce much higher frequencies for brief periods. Finally, the refractory periods are key in determining the direction of action potential propagation, as will be discussed in the following section.

Action-Potential Propagation As we have seen, the inside of the cell becomes positive with respect to the outside at the site of an action potential. This area of the membrane is also positive with respect to other regions where the membrane is still at its resting potential. The difference in potentials between the active and resting regions causes ions to flow, and this local current depolarizes the membrane adjacent to the action-potential site to *its* threshold potential. The sodium positive-feedback cycle takes over, and a new action potential occurs there.

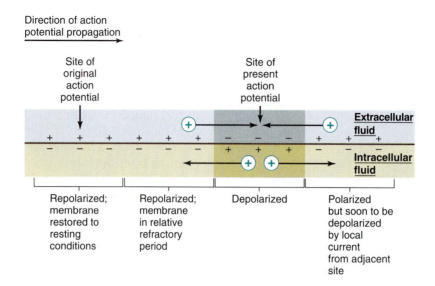

Direction of action
potential propagation →

Site of original action potential

Site of present action potential

Extracellular fluid

Intracellular fluid

Repolarized; membrane restored to resting conditions

Repolarized; membrane in relative refractory period

Depolarized

Polarized but soon to be depolarized by local current from adjacent site

FIGURE 8–21
Propagation of an action potential along a plasma membrane.

The new action potential then produces local currents of its own, which depolarize the region adjacent to it, producing yet another action potential at the next site, and so on to cause **action-potential propagation** along the length of the membrane. Thus, there is a sequential opening and closing of sodium and potassium channels along the membrane. It is like lighting a trail of gunpowder; the action potential doesn't move but "sets off" a new action potential in the region of the axon just ahead of it. Because each action potential depends on the sodium-feedback cycle of the membrane where the action potential is occurring, the action potential arriving at the end of the membrane is virtually identical in form to the initial one. Thus, action potentials are not conducted decrementally as are graded potentials.

Because the membrane areas that have just undergone an action potential are refractory and cannot immediately undergo another, the only direction of action potential propagation is away from a region of membrane that has recently been active (Figure 8–21).

If the membrane through which the action potential must travel is not refractory, excitable membranes are able to conduct action potentials in either direction, the direction of propagation being determined by the stimulus location. For example, the action potentials in skeletal-muscle cells are initiated near the middle of these cylindrical cells and propagate toward the two ends. In most nerve cells, however, action potentials are initiated *physiologically* at one end of the cell (for reasons to be described in the next section) and propagate toward the other end. The propagation ceases when the action potential reaches the end of an axon.

The velocity with which an action potential propagates along a membrane depends upon fiber diameter and whether or not the fiber is myelinated. The larger the fiber diameter, the faster the action potential propagates. This is because a large fiber offers less resistance to local current; more ions will flow in a given time, bringing adjacent regions of the membrane to threshold faster.

Myelin is an insulator that makes it more difficult for charge to flow between intracellular and extracellular fluid compartments. Because there is less "leakage" of charge across the myelin, the graded potential can spread farther along the axon. Moreover, the concentration of voltage-gated sodium channels in the myelinated region of axons is low. Therefore, action potentials occur only at the nodes of Ranvier where the myelin coating is interrupted and the concentration of voltage-gated sodium channels is high. Thus, action potentials literally jump from one node to the next as they propagate along a myelinated fiber, and for this reason such propagation is called **saltatory conduction** (Latin, *saltare*, to leap).

Propagation via saltatory conduction is faster than propagation in nonmyelinated fibers of the same axon diameter because less charge leaks out through the myelin-covered sections of the membrane (Figure 8–22). More charge arrives at the node adjacent to the active node, and an action potential is generated there sooner than if the myelin were not present. Moreover, because ions cross the membrane only at the nodes of Ranvier, the membrane pumps need restore fewer ions. Myelinated axons are therefore metabolically more

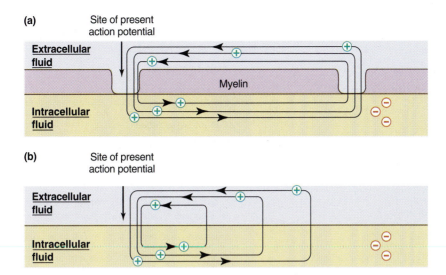

(a)

Site of present action potential

Extracellular fluid

Myelin

Intracellular fluid

(b)

Site of present action potential

Extracellular fluid

Intracellular fluid

FIGURE 8–22

Current during an action potential in (a) a myelinated and (b) an unmyelinated axon.

cost-effective than unmyelinated ones. Thus, myelin adds efficiency in speed and metabolic cost, and it saves room in the nervous system because the axons can be thinner.

Conduction velocities range from about 0.5 m/s (1 mi/h) for small-diameter, unmyelinated fibers to about 100 m/s (225 mi/h) for large-diameter, myelinated fibers. At 0.5 m/s, an action potential would travel the distance from the head to the toe of an average-sized person in about 4 s; at a velocity of 100 m/s, it takes about 0.02 s.

Initiation of Action Potentials In our description of action potentials thus far, we have spoken of "stimuli" as the initiators of action potentials. How are action potentials actually initiated in various types of neurons?

In afferent neurons, the initial depolarization to threshold is achieved by a graded potential—here called a **receptor potential**, which is generated in the sensory receptors at the peripheral ends of the neurons. These are the ends farthest from the central nervous system, and where the nervous system functionally encounters the outside world. In all other neurons, the depolarization to threshold is due either to a graded potential generated by synaptic input to the neuron or to a spontaneous change in the neuron's membrane potential, known as a **pacemaker potential.** How synaptic potentials are produced is the subject of the next section. The production of receptor potentials is discussed in Chapter 9.

Spontaneous generation of pacemaker potentials occurs in the absence of any identifiable external stimulus and is an inherent property of certain neurons (and other excitable cells, including certain smooth-muscle and cardiac-muscle cells). In these cells, the activity of different types of ion channels in the plasma membrane causes a graded depolarization of the membrane—the pacemaker potential. If threshold is reached, an action potential occurs; the membrane then repolarizes and again begins to depolarize (Figure 8–23). There is no stable, resting membrane potential in such cells because of the continuous change in membrane permeability. The rate at which the membrane depolarizes to threshold determines the

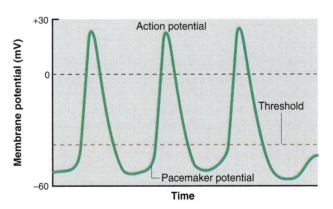

FIGURE 8–23

Action potentials resulting from pacemaker potentials.

TABLE 8–5 Differences between Graded Potentials and Action Potentials

Graded Potential	Action Potential
Amplitude varies with conditions of the initiating event	All-or-none once membrane is depolarized to threshold, amplitude is independent of initiating event
Can be summed	Cannot be summed
Has no threshold	Has a threshold that is usually about 15 mV depolarized relative to the resting potential
Has no refractory period	Has a refractory period
Is conducted decrementally; that is, amplitude decreases with distance	Is conducted without decrement; the depolarization is amplified to a constant value at each point along the membrane
Duration varies with initiating conditions	Duration constant for a given cell type under constant conditions
Can be a depolarization or a hyperpolarization	Is a depolarization (with overshoot in neurons)
Initiated by environmental stimulus (receptor), by neurotransmitter (synapse), or spontaneously	Initiated by a graded potential
Mechanism depends on ligand-sensitive channels or other chemical or physical changes	Mechanism depends on voltage-gated channels

action-potential frequency. Pacemaker potentials are implicated in many rhythmical behaviors, such as breathing, the heartbeat, and movements within the walls of the stomach and intestines.

The differences between graded potentials and action potentials are listed in Table 8–5.

SECTION B SUMMARY

The Resting Membrane Potential

I. Membrane potentials are generated mainly by diffusion of ions and are determined by (a) the ionic concentration differences across the membrane, and (b) the membrane's relative permeabilities to different ions.
   a. Plasma-membrane Na,K-ATPase pumps maintain intracellular sodium concentration low and potassium high.
   b. In almost all resting cells, the plasma membrane is much more permeable to potassium than to sodium, so the membrane potential is close to the potassium equilibrium potential—that is, the inside is negative relative to the outside.
   c. The Na,K-ATPase pumps also contribute directly a small component of the potential because they are electrogenic.

Graded Potentials and Action Potentials

I. Neurons signal information by graded potentials and action potentials (APs).

II. Graded potentials are local potentials whose magnitude can vary and that die out within 1 or 2 mm of their site of origin.

III. An AP is a rapid change in the membrane potential during which the membrane rapidly depolarizes and repolarizes. In neurons, the potential reverses and the membrane becomes positive inside. APs provide long-distance transmission of information through the nervous system.
   a. APs occur in excitable membranes because these membranes contain voltage-gated sodium channels, which open as the membrane depolarizes, causing a positive feedback toward the sodium equilibrium potential.
   b. The AP is ended as the sodium channels close and additional potassium channels open, which restores the resting conditions.
   c. Depolarization of excitable membranes triggers APs only when the membrane potential exceeds a threshold potential.
   d. Regardless of the size of the stimulus, if the membrane reaches threshold, the APs generated are all the same size.
   e. A membrane is refractory for a brief time even though stimuli that were previously effective are applied.
   f. APs are propagated without any change in size from one site to another along a membrane.
   g. In myelinated nerve fibers, APs manifest saltatory conduction.
   h. APs can be initiated by receptors at the ends of afferent neurons, at synapses, or in some cells, by pacemaker potentials.

```
SECTION  B  REVIEW  QUESTIONS
```

1. Describe how negative and positive charges interact.
2. Contrast the abilities of intracellular and extracellular fluids and membrane lipids to conduct electric current.
3. Draw a simple cell; indicate where the concentrations of Na^+, K^+, and Cl^- are high and low and the electric-potential difference across the membrane when the cell is at rest.
4. Explain the conditions that give rise to the resting membrane potential. What effect does membrane permeability have on this potential? What is the role of Na,K-ATPase membrane pumps in the membrane potential? Is this role direct or indirect?
5. Which two factors involving ion diffusion determine the magnitude of the resting membrane potential?
6. Explain why the resting membrane potential is not equal to the potassium equilibrium potential.
7. Draw a graded potential and an action potential on a graph of membrane potential versus time. Indicate zero membrane potential, resting membrane potential, and threshold potential; indicate when the membrane is depolarized, repolarizing, and hyperpolarized.
8. List the differences between graded potentials and action potentials.
9. Describe the ionic basis of an action potential; include the role of voltage-gated channels and the positive-feedback cycle.
10. Explain threshold and the relative and absolute refractory periods in terms of the ionic basis of the action potential.
11. Describe the propagation of an action potential. Contrast this event in myelinated and unmyelinated axons.
12. List three ways in which action potentials can be initiated in neurons.

```
SECTION C
```

SYNAPSES

As defined earlier, a synapse is an anatomically specialized junction between two neurons, at which the electrical activity in one neuron, the presynaptic neuron, influences the electrical (or metabolic) activity in the second, postsynaptic neuron. Anatomically, synapses include parts of the presynaptic and postsynaptic neurons and the extracellular space between these two cells. According to the latest estimate, there are approximately 10^{14} (100 quadrillion!) synapses in the CNS.

When active, synapses can increase or decrease the likelihood that the postsynaptic neuron will fire action potentials by producing a brief, graded potential there. The membrane potential of a postsynaptic neuron is brought closer to threshold at an **excitatory synapse,** and it is either driven farther from threshold or stabilized at its present level at an **inhibitory synapse.**

Thousands of synapses from many different presynaptic cells can affect a single postsynaptic cell (**convergence**), and a single presynaptic cell can send branches to affect many other postsynaptic cells (**divergence,** Figure 8–24). Convergence allows information from many sources to influence a cell's activity; divergence allows one information source to affect multiple pathways.

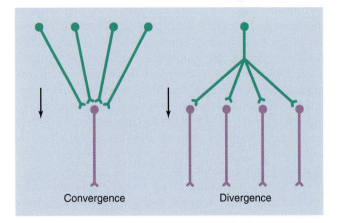

FIGURE 8–24

Convergence of neural input from many neurons onto a single neuron, and divergence of output from a single neuron onto many others. Presynaptic neurons are shown in green and postsynaptic neurons in purple. Arrows indicate the direction of transmission of neural activity.

The level of excitability of a postsynaptic cell at any moment (that is, how close its membrane potential is to threshold) depends on the number of synapses active at any one time and the number that are excitatory or inhibitory. If the membrane of the postsynaptic neuron reaches threshold, it will generate action potentials that are propagated along its axon to the terminal branches, which influence the excitability of *other* cells by divergence.

Functional Anatomy of Synapses

There are two types of synapses: **electric** and **chemical.** At electric synapses, the plasma membranes of the pre- and postsynaptic cells are joined by gap junctions (Chapter 3). These allow the local currents resulting from arriving action potentials to flow directly across the junction through the connecting channels in either direction from one neuron to the neuron on the other side of the junction, depolarizing the membrane to threshold and thus initiating an action potential in the second cell. Although numerous in cardiac and smooth muscles, electric synapses are relatively rare in the mammalian nervous system, and we shall henceforth discuss only the much more common, chemical synapse.

Figure 8–25 shows the structure of a single typical chemical synapse. The axon of the presynaptic neuron ends in a slight swelling, the axon terminal, and the postsynaptic membrane under the axon terminal appears denser. Note that in actuality the size and shape of the pre- and postsynaptic elements can vary greatly (Figure 8–26). A 10- to 20-nm extracellular space, the **synaptic cleft,** separates the pre- and postsynaptic neurons and prevents *direct* propagation of the current from the presynaptic neuron to the postsynaptic cell. Instead, signals are transmitted across the synaptic cleft by means of a chemical messenger—a neurotransmitter—released from the presynaptic axon terminal. Sometimes more than one neurotransmitter may be simultaneously released from an axon, in which case the additional neurotransmitter is called a **cotransmitter.** These neurotransmitters have different receptors in the postsynaptic cell.

The neurotransmitter in terminals is stored in membrane-bound **synaptic vesicles,** some of which are docked at specialized regions of the synaptic membrane. When an action potential in the presynaptic neuron reaches the end of the axon and depolarizes the axon terminal, voltage-gated calcium channels in the membrane open, and calcium diffuses from the extracellular fluid into the axon terminal near the docked vesicles. The calcium ions induce a series of reactions that allow some of the docked vesicles to fuse with the presynaptic plasma membrane and liberate their contents into the synaptic cleft by the process of exocytosis.

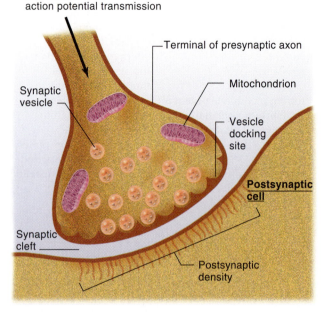

(a)

FIGURE 8–25

(a) Diagram of a synapse. Some vesicles are docked at the presynaptic membrane ready for release. The postsynaptic membrane is distinguished microscopically by "postsynaptic density," which contains proteins associated with the receptors. (b) An enlargement showing synaptic specialization.

Part b redrawn from Walmsley et al.

Once released from the presynaptic axon terminal, neurotransmitter and cotransmitter, if there is one, diffuse across the cleft. A fraction of these molecules bind to receptors on the plasma membrane of the postsynaptic cell (the fate of the others will be described later). The activated receptors themselves may contain an ion channel, or they may act indirectly, via a G protein, on separate ion channels. In either case, the result of the binding of neurotransmitter to receptor is the opening or closing of specific ion channels in the postsynaptic plasma membrane. These channels belong, therefore, to the class of ligand-sensitive channels whose function is controlled by receptors, as discussed in Chapter 7, and are distinct from voltage-gated channels. (Exceptions to this generalization—the activation of metabolic pathways rather than ion channels—will be discussed later.)

Although Figure 8–26 shows a few exceptions, in general the neurotransmitter is stored on the presynaptic side of the synaptic cleft, whereas receptors for the neurotransmitters are on the postsynaptic side. Therefore, most chemical synapses operate in only one direction. One-way conduction across synapses causes action potentials to be transmitted along a given multineuronal pathway in one direction.

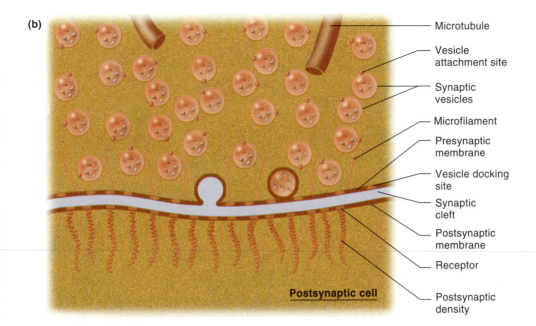

(b)

- Microtubule
- Vesicle attachment site
- Synaptic vesicles
- Microfilament
- Presynaptic membrane
- Vesicle docking site
- Synaptic cleft
- Postsynaptic membrane
- Receptor
- Postsynaptic density

Postsynaptic cell

Because of the sequence of events involved, there is a very brief synaptic delay—as short as 0.2 sec—between the arrival of an action potential at a presynaptic terminal and the membrane-potential changes in the postsynaptic cell.

Neurotransmitter binding to the receptor is a transient event, and as with any binding site, the bound ligand—in this case, the neurotransmitter—is in equilibrium with the unbound form. Thus, if the concentration of unbound neurotransmitter in the synaptic cleft is decreased, the number of occupied receptors will decrease. The ion channels in the postsynaptic membrane return to their resting state when the neurotransmitter is no longer bound. Unbound neurotransmitters are removed from the synaptic cleft when they (1) are actively transported back into the axon terminal or, in some cases into nearby glial cells; (2) diffuse away from the receptor site; or (3) are enzymatically transformed into ineffective substances, some of which are transported back into the axon terminal for reuse.

The two kinds of chemical synapses—excitatory and inhibitory—are differentiated by the effects of the neurotransmitter on the postsynaptic cell. Whether the effect is inhibitory or excitatory depends on the type of signal transduction mechanism brought into operation when the neurotransmitter binds to a receptor and the type of channel the receptor influences.

Excitatory Chemical Synapses

At an excitatory synapse, the postsynaptic response to the neurotransmitter is a depolarization, bringing the membrane potential closer to threshold. The usual effect of the activated receptor on the postsynaptic

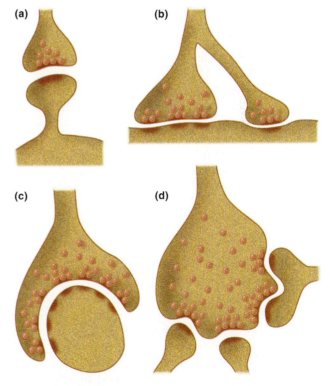

(a)   **(b)**

(c)   **(d)**

FIGURE 8–26

Synapses appear in many forms as demonstrated here in views (a) to (d). The presynaptic fiber contains synaptic vesicles.

Redrawn from Walmsley et al.

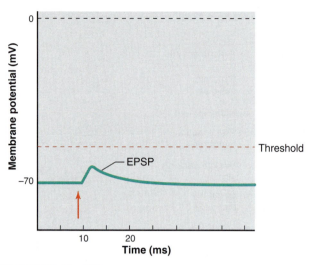

FIGURE 8–27
Excitatory postsynaptic potential (EPSP). Stimulation of the presynaptic neuron is marked by the arrow.

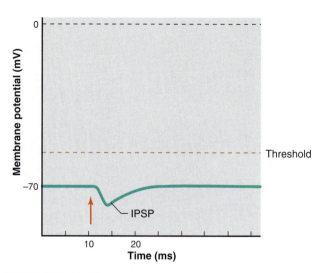

FIGURE 8–28
Inhibitory postsynaptic potential (IPSP). Stimulation of the presynaptic neuron is marked by the arrow.

membrane at such synapses is to open postsynaptic-membrane ion channels that are permeable to sodium, potassium, and other small, positively charged ions. These ions then are free to move according to the electrical and chemical gradients across the membrane.

There is both an electrical and a concentration gradient driving sodium into the cell, while for potassium, the electrical gradient is opposed by the concentration gradient. Opening channels nonspecifically to all small positively charged ions, therefore, results in the simultaneous movement of a relatively small number of potassium ions out of the cell and a larger number of sodium ions into the cell. Thus, the *net* movement of positive ions is into the postsynaptic cell, and this slightly depolarizes it. This potential change is called an **excitatory postsynaptic potential** (**EPSP,** Figure 8–27).

The EPSP is a graded potential that spreads decrementally away from the synapse by local current. Its only function is to bring the membrane potential of the postsynaptic neuron closer to threshold.

Inhibitory Chemical Synapses

At inhibitory synapses, the potential change in the postsynaptic neuron is a hyperpolarizing graded potential called an **inhibitory postsynaptic potential** (**IPSP,** Figure 8–28). Alternatively, there may be no IPSP but rather *stabilization* of the membrane potential at its existing value. In either case, activation of an inhibitory synapse lessens the likelihood that the postsynaptic cell will depolarize to threshold and generate an action potential.

At an inhibitory synapse, the activated receptors on the postsynaptic membrane open chloride or, sometimes, potassium channels; sodium channels are not

affected. In cells that actively transport chloride ions out of the cell, the chloride equilibrium potential (-80 mV) is more negative than the resting potential. Therefore, as chloride channels open, more chloride enters the cell, producing a hyperpolarization—that is, an IPSP. In cells that do not actively transport chloride, the equilibrium potential for chloride is equal to the resting membrane potential. A rise in chloride-ion permeability therefore does not change the membrane potential but does increase chloride's influence on the membrane potential. This in turn makes it more difficult for other ion types to change the potential and results in a stabilization of the membrane at the resting level without producing a hyperpolarization.

Increased potassium permeability, when it occurs in the postsynaptic cell, also produces an IPSP. Earlier it was noted that if a cell membrane were permeable only to potassium ions, the resting membrane potential would equal the potassium equilibrium potential; that is, the resting membrane potential would be -90 mV instead of -70 mV. Thus, with an increased potassium permeability, more potassium ions leave the cell and the membrane moves closer to the potassium equilibrium potential, causing a hyperpolarization.

Activation of the Postsynaptic Cell

A feature that makes postsynaptic integration possible is that in most neurons one excitatory synaptic event by itself is not enough to cause threshold to be reached in the postsynaptic neuron. For example, a single EPSP may be only 0.5 mV, whereas changes of about 15 mV

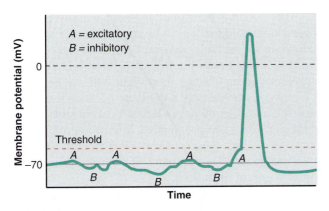

FIGURE 8–29

Intracellular recording from a postsynaptic cell during episodes when (A) excitatory synaptic activity predominates and the cell is facilitated, and (B) inhibitory synaptic activity dominates.

are necessary to depolarize the neuron's membrane to threshold. This being the case, an action potential can be initiated only by the combined effects of many excitatory synapses.

Of the thousands of synapses on any one neuron, probably hundreds are active simultaneously or close enough in time so that the effects can add together. The membrane potential of the postsynaptic neuron at any moment is, therefore, the resultant of all the synaptic activity affecting it at that time. There is a depolarization of the membrane toward threshold when excitatory synaptic input predominates, and either a hyperpolarization or stabilization when inhibitory input predominates (Figure 8–29).

Let us perform a simple experiment to see how EPSPs and IPSPs interact (Figure 8–30). Assume there are three synaptic inputs to the postsynaptic cell: The synapses from axons A and B are excitatory, and the

synapse from axon C is inhibitory. There are laboratory stimulators on axons A, B, and C so that each can be activated individually. An electrode is placed in the cell body of the postsynaptic neuron and connected to record the membrane potential. In Part 1 of the experiment, we shall test the interaction of two EPSPs by stimulating axon A and then, after a short time, stimulating it again. Part 1 of Figure 8–30 shows that no interaction occurs between the two EPSPs. The reason is that the change in membrane potential associated with an EPSP is fairly short-lived. Within a few milliseconds (by the time we stimulate axon A for the second time), the postsynaptic cell has returned to its resting condition.

In Part 2, we stimulate axon A for the second time *before* the first EPSP has died away; the second synaptic potential adds to the previous one and creates a greater depolarization than from one input alone. This is called **temporal summation** since the input signals arrive at the same cell at different *times.* The potentials summate because there are a greater number of open ion channels and, therefore, a greater flow of positive ions into the cell. In Part 3, axon B is stimulated alone to determine its response, and then axons A and B are stimulated simultaneously. The two EPSPs that result also summate in the postsynaptic neuron; this is called **spatial summation** since the two inputs occurred at different *locations* on the same cell. The interaction of multiple EPSPs through ongoing spatial and temporal summation can increase the inward flow of positive ions and bring the postsynaptic membrane to threshold so that action potentials are initiated (Part 4).

So far we have tested only the patterns of interaction of excitatory synapses. Since EPSPs and IPSPs are due to oppositely directed local currents, they tend to cancel each other, and there is little or no change in membrane potential (Figure 8–30, Part 5). Inhibitory potentials can also show spatial and temporal summation.

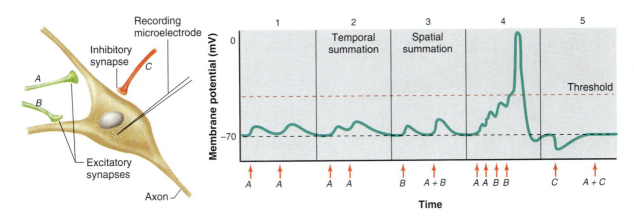

FIGURE 8–30

Interaction of EPSPs and IPSPs at the postsynaptic neuron. Arrows indicate time of stimulation.

Via the local current mechanisms described earlier, the plasma membrane of the entire postsynaptic cell body and the initial segment reflect the changes at the postsynaptic membrane. The membrane of a large area of the cell becomes slightly depolarized during activation of an excitatory synapse and slightly hyperpolarized or stabilized during activation of an inhibitory synapse, although these graded potentials will decrease with distance from the synaptic junction (Figure 8–31).

In the previous examples, we referred to the threshold of the postsynaptic neuron as though it were the same for all parts of the cell. However, different parts of the neuron have different thresholds. In many cells the initial segment has a lower threshold (that is, much closer to the resting potential) than the threshold of the cell body and dendrites. In these cells the initial segment reaches threshold first whenever enough EPSPs summate, and the resulting action potential is then propagated from this point down the axon (and, sometimes, back over the cell body and dendrites).

The fact that the initial segment usually has the lowest threshold explains why the location of individual synapses on the postsynaptic cell is important. A synapse located near the initial segment will produce a greater voltage change there than will a synapse on the outermost branch of a dendrite because it will expose the initial segment to a larger local current. In fact, some dendrites use propagated action potentials over portions of their length to convey information about the synaptic events occurring at their endings to the initial segment of the cell.

Postsynaptic potentials last much longer than action potentials. In the event that cumulative EPSPs cause the initial segment to still be depolarized to threshold after an action potential has been fired and the refractory period is over, a second action potential will occur. In fact, as long as the membrane is depolarized to threshold, action potentials will continue to arise. Neuronal responses at synapses almost always occur in bursts of action potentials rather than as single isolated events.

Synaptic Effectiveness

Individual synaptic events—whether excitatory or inhibitory—have been presented as though their effects are constant and reproducible. Actually, the variability in postsynaptic potentials following any particular presynaptic input is enormous. The effectiveness of a given synapse can be influenced by both presynaptic and postsynaptic mechanisms.

First, a presynaptic terminal does not release a constant amount of neurotransmitter every time it is activated. One reason for this variation involves calcium concentration. Calcium that has entered the terminal during previous action potentials is pumped out of the cell or (temporarily) into intracellular organelles. If calcium removal does not keep up with entry, as can occur during high-frequency stimulation, calcium concentration in the terminal, and hence the amount of neurotransmitter released upon subsequent stimulation, will be greater than usual. The greater the amount of neurotransmitter released, the greater the number of ion channels opened (or closed) in the postsynaptic membrane, and the larger the amplitude of the EPSP or IPSP in the postsynaptic cell.

The neurotransmitter output of some presynaptic terminals is also altered by activation of membrane receptors in the terminals themselves. These *presynaptic* receptors are often associated with a second synaptic ending known as an axon-axon synapse, or **presynaptic synapse,** in which an axon terminal of one neuron ends on an axon terminal of another. For example, in Figure 8–32 the neurotransmitter released by A combines with receptors on B, resulting in a change in the

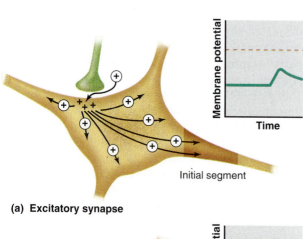

(a) Excitatory synapse

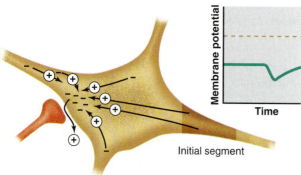

(b) Inhibitory synapse

FIGURE 8–31

Comparison of excitatory and inhibitory synapses, showing current direction through the postsynaptic cell following synaptic activation. (a) Current through the postsynaptic cell is away from the excitatory synapse, depolarizing the initial segment. (b) Current through the postsynaptic cell hyperpolarizes the initial segment.

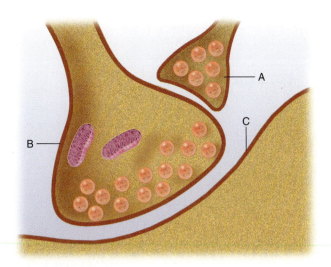

FIGURE 8–32
A presynaptic (axon-axon) synapse between axon terminal A and axon terminal B. C is the final postsynaptic cell body.

amount of neurotransmitter released from B in response to action potentials. Thus, neuron A has no *direct* effect on neuron C, but it has an important influence on the ability of B to influence C. Neuron A is said to be exerting a *presynaptic* effect on the synapse between B and C. Depending upon the nature of the neurotransmitter released from A and the type of receptors activated by that neurotransmitter on B, the presynaptic effect may decrease the amount of neurotransmitter released from B **(presynaptic inhibition)** or increase it **(presynaptic facilitation).**

Presynaptic synapses such as A in Figure 8–32 can alter the calcium concentration in axon terminal B or even affect neurotransmitter synthesis there. If the calcium concentration increases, the number of vesicles releasing neurotransmitter from B increases; decreased calcium reduces the number of vesicles that are releasing transmitter. Presynaptic synapses are important because they selectively control one specific input to the postsynaptic neuron C.

Some receptors on the presynaptic terminal are not associated with axon-axon synapses. Rather they are activated by neurotransmitters or other chemical messengers released by nearby neurons or glia or even the axon terminal itself. In the last case, the receptors are called **autoreceptors** and provide an important feedback mechanism by which the neuron can regulate its own neurotransmitter output. In most cases, the released neurotransmitter acts on autoreceptors to decrease its own release, thereby providing negative-feedback control.

Postsynaptic mechanisms for varying synaptic effectiveness also exist. For example, as described in Chapter 7, there are many types and subtypes of receptors for each kind of neurotransmitter. The different receptor types operate by different signal transduction mechanisms and have different—sometimes even opposite—effects on the postsynaptic mechanisms they influence. Moreover, a given signal transduction mechanism may be regulated by multiple neurotransmitters, and the various second-messenger systems affecting a channel may interact with each other.

Recall, too, from Chapter 7 that the number of receptors is not constant, varying with up- and down-regulation, for example. Also, the ability of a given receptor to respond to its neurotransmitter can change. Thus, in some systems a receptor responds once and then temporarily fails to respond despite the continued presence of the receptor's neurotransmitter, a phenomenon known as receptor desensitization.

Imagine the complexity when a cotransmitter (or several cotransmitters) is released with the neurotransmitter to act upon postsynaptic receptors and maybe upon presynaptic receptors as well! Clearly, the possible variations in transmission at even a single synapse are great, and the functions of a given neurotransmitter can be extremely difficult to identify.

Modification of Synaptic Transmission by Drugs and Disease
The great majority of drugs that act on the nervous system do so by altering synaptic mechanisms and thus synaptic effectiveness. All the synaptic mechanisms labeled in Figure 8–33 are vulnerable.

The long-term effects of drugs are sometimes difficult to predict because the imbalances produced by the initial drug action are soon counteracted by feedback mechanisms that normally regulate the processes. For example, if a drug interferes with the action of a neurotransmitter by inhibiting the rate-limiting enzyme in its synthetic pathway, the neurons may respond by increasing the rate of precursor transport into the axon terminals to maximize the use of any enzyme that is available.

Recall from Chapter 7 that drugs that bind to a receptor and produce a response similar to the normal activation of that receptor are called **agonists,** and drugs that bind to the receptor but are unable to activate it are **antagonists.** By occupying the receptors, antagonists prevent binding of the normal neurotransmitter when it is released at the synapse. Specific agonists and antagonists can affect receptors on both pre- and postsynaptic membranes.

Diseases can also affect synaptic mechanisms. For example, the toxin produced by the bacillus *Clostridium tetani* **(tetanus toxin)** is a protease that destroys certain proteins in the synaptic-vesicle docking mechanism of neurons that provide inhibitory synaptic input to the neurons supplying skeletal muscles. The toxin of the *Clostridium botulinum* bacilli and the venom of the black widow spider also affect neurotransmitter

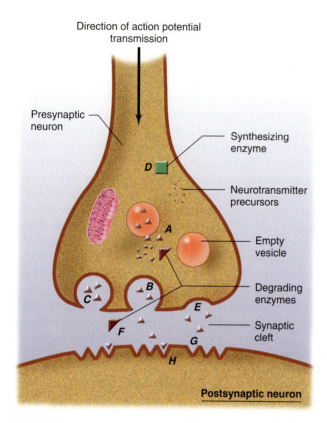

Direction of action potential transmission

Presynaptic neuron

Synthesizing enzyme

D

Neurotransmitter precursors

A

Empty vesicle

B

C

Degrading enzymes

E

F

G

Synaptic cleft

H

Postsynaptic neuron

FIGURE 8–33

Drug actions at synapses: (A) Increase leakage of neurotransmitter from vesicle to cytoplasm, exposing it to enzyme breakdown, (B) increase transmitter release, (C) block transmitter release, (D) inhibit transmitter synthesis, (E) block transmitter reuptake, (F) block enzymes that metabolize transmitter, (G) bind to receptor to block (antagonist) or mimic (agonist) transmitter action, and (H) inhibit or facilitate second-messenger activity.

Redrawn from DRUGS AND THE BRAIN by Solomon H. Snyder. Copyright © 1986 by Scientific American Books, Inc. Reprinted by permission of W. H. Freeman and Company.

TABLE 8–6 Factors that Determine Synaptic Effectiveness

I. Presynaptic factors
  A. Availability of neurotransmitter
    1. Availability of precursor molecules
    2. Amount (or activity) of the rate-limiting enzyme in the pathway for neurotransmitter synthesis
  B. Axon terminal membrane potential
  C. Axon terminal calcium
  D. Activation of membrane receptors on presynaptic terminal
    1. Presynaptic (axon-axon) synapses
    2. Autoreceptors
    3. Other receptors
  E. Certain drugs and diseases, which act via the above mechanisms A–D

II. Postsynaptic factors
  A. Immediate past history of electrical state of postsynaptic membrane (that is, facilitation or inhibition from temporal or spatial summation)
  B. Effects of other neurotransmitters or neuromodulators acting on postsynaptic neuron
  C. Certain drugs and diseases

III. General factors
  A. Area of synaptic contact
  B. Enzymatic destruction of neurotransmitter
  C. Geometry of diffusion path
  D. Neurotransmitter reuptake

---

release from synaptic vesicles by interfering with docking proteins, but they act on axon terminals of neurons different from those affected by tetanus toxin.

Table 8–6 summarizes the factors that determine synaptic effectiveness.

Neurotransmitters and Neuromodulators

We have emphasized the role of neurotransmitters in eliciting EPSPs and IPSPs. However, certain chemical messengers elicit complex responses that cannot be simply described as EPSPs or IPSPs. The word "modulation" is used for these complex responses, and the messengers that cause them are called **neuromodulators.** The distinctions between neuromodulators and neurotransmitters are, however, far from clear. In fact, certain neuromodulators are often synthesized by the presynaptic cell and co-released with the neurotransmitter. To add to the complexity, certain hormones, paracrine agents, and messengers used by the immune system serve as neuromodulators.

Neuromodulators often modify the *postsynaptic* cell's response to specific neurotransmitters, amplifying or dampening the effectiveness of ongoing synaptic activity. Alternatively, they may change the *presynaptic* cell's synthesis, release, reuptake, or metabolism of a transmitter. In other words, they alter the effectiveness of the synapse.

In general, the receptors for *neurotransmitters* influence ion channels that directly affect excitation or inhibition of the postsynaptic cell. These mechanisms operate within milliseconds. Receptors for *neuromodulators,* on the other hand, more often bring about changes in metabolic processes in neurons, often via G proteins coupled to second-messenger systems. Such changes, which can occur over minutes, hours, or even days, include alterations in enzyme activity or, by way of influences on DNA transcription, in protein synthesis. Thus, neurotransmitters are involved in rapid communication, whereas neuromodulators

TABLE 8–7 Classes of Some of the Chemicals Known or Presumed to be Neurotransmitters or Neuromodulators

1. **Acetylcholine (ACh)**

2. **Biogenic amines**

   Catecholamines

      Dopamine (DA)

      Norepinephrine (NE)

      Epinephrine (Epi)

   Serotonin (5-hydroxytryptamine, 5-HT)

   Histamine

3. **Amino acids**

   Excitatory amino acids; for example, glutamate

   Inhibitory amino acids; for example, gamma-aminobutyric acid (GABA)

4. **Neuropeptides;** for example, the endogenous opioids

5. **Miscellaneous**

   Gases; for example, nitric oxide

   Purines; for example, adenosine and ATP

tend to be associated with slower events such as learning, development, motivational states, or even some sensory or motor activities.

Table 8–7 lists the major categories of substances generally accepted as neurotransmitters or neuromodulators. A huge amount of information has accumulated concerning the synthesis, metabolism, and mechanisms of action of these messengers—material well beyond the scope of this book. The following sections will therefore present only some basic generalizations about certain of the neurotransmitters presently deemed most important. For simplicity's sake, we use the term "neurotransmitter" in a general sense, realizing that sometimes the messenger may more appropriately be described as a neuromodulator. A note on terminology should also be included here: Neurons are often referred to as "-ergic," where the missing prefix is the type of neurotransmitter released by the neuron. For example, "dopaminergic" applies to neurons that release the neurotransmitter dopamine.

Acetylcholine

Acetylcholine (ACh) is synthesized from choline and acetyl coenzyme A in the cytoplasm of synaptic terminals and stored in synaptic vesicles. After it is released and activates receptors on the postsynaptic membrane, the concentration of ACh at the postsynaptic membrane is reduced (thereby stopping receptor activation) by the enzyme **acetylcholinesterase.** This enzyme is located on the pre- and postsynaptic membranes and rapidly destroys ACh, releasing choline. The choline is then transported back into the axon terminals where it is reused in the synthesis of new ACh. The ACh concentration at the receptors is also reduced by simple diffusion away from the site and eventual breakdown of the molecule by an enzyme in the blood.

Acetylcholine is a major neurotransmitter in the peripheral nervous system, and it is also present in the brain. Fibers that release ACh are called **cholinergic** fibers. The cell bodies of the brain's cholinergic neurons are concentrated in relatively few areas, but their axons are widely distributed.

Some ACh receptors respond not only to acetylcholine but to the drug nicotine and, therefore, have come to be known as **nicotinic receptors.** The nicotinic receptor is an excellent example of a receptor that itself contains an ion channel; in this case the channel is selective for positively charged ions.

Nicotinic receptors in the brain are important in cognitive functions. For example, one cholinergic system that employs nicotinic receptors plays a major role in attention, learning, and memory by reinforcing the ability to detect and respond to meaningful stimuli. Neurons associated with this system degenerate in people with *Alzheimer's disease,* a brain disease that is usually age-related and is the most common cause of declining intellectual function in late life, affecting 10 to 15 percent of people over age 65, and 50 percent of people over age 85. Because of the degeneration of cholinergic neurons, this disease is associated with a decreased amount of ACh in certain areas of the brain and even the loss of the postsynaptic neurons that would have responded to it. These defects and those in other neurotransmitter systems that are affected in this disease are related to the declining language and perceptual abilities, confusion, and memory loss that characterize Alzheimer's victims. The exact causes of this degeneration are unknown.

Other cholinergic receptors are stimulated not only by acetylcholine but by the mushroom poison muscarine; therefore, they are called **muscarinic receptors.** These receptors couple with G proteins, which then alter the activity of a number of different enzymes and ion channels.

Biogenic Amines

The **biogenic amines** are neurotransmitters that are synthesized from amino acids and contain an amino group ($R–NH_2$). The most common biogenic amines are dopamine, norepinephrine, serotonin, and histamine. Epinephrine, another biogenic amine, is not a common neurotransmitter in the central nervous system but is the major *hormone* secreted by the adrenal medulla. Norepinephrine is an important neurotransmitter in both the central and peripheral components of the nervous system.

Catecholamines Dopamine, norepinephrine (NE), and **epinephrine** all contain a catechol ring (a six-carbon ring with two adjacent hydroxyl groups) and an amine group; thus they are called **catecholamines.** The catecholamines are formed from the amino acid tyrosine and share the same basic synthetic pathway (Figure 8–34), which begins with the uptake of tyrosine by the axon terminals. Depending on the enzymes present in the terminals, any one of the three catecholamines may be ultimately released. Synthesis and release of the catecholamines from the presynaptic terminals are strongly modulated by autoreceptors on the presynaptic terminals.

After activation of the receptors on the postsynaptic cell, the catecholamine concentration in the synaptic cleft declines, mainly because the catecholamine is actively transported back into the axon terminal. The catecholamine neurotransmitters are also broken down in both the extracellular fluid and the axon terminal by enzymes such as monoamine oxidase. Monoamine oxidase inhibitors, which increase the brain extracellular concentration of the catecholamine neurotransmitters, are used in the treatment of diseases such as depression, as will be discussed in Chapter 13.

Within the central nervous system, the cell bodies of the catecholamine-releasing neurons lie in parts of the brain called the brainstem and hypothalamus, and although relatively few in number, their axons branch greatly and may go to virtually all parts of the brain and spinal cord. The catecholamines exert a much greater influence in the central nervous system than the number of neurons alone would suggest, possibly because of their neuromodulator-like effects on post-synaptic neurons. These neurotransmitters play essential roles in states of consciousness, mood, motivation, directed attention, movement, blood-pressure regulation, and hormone release, all functions that will be covered in later chapters.

During the early experiments on norepinephrine and epinephrine, norepinephrine was mistakenly taken to be epinephrine, and epinephrine was called by its British name "adrenaline." Consequently, nerve fibers that release epinephrine or norepinephrine came to be called **adrenergic** fibers. Norepinephrine-releasing fibers are also called **noradrenergic.**

There are two major classes of receptors for norepinephrine and epinephrine: **alpha-adrenergic receptors** and **beta-adrenergic receptors** (these are also called alpha-adrenoceptors and beta-adrenoceptors). The major way of distinguishing between the two classes of receptors is that they are influenced by different drugs. Both alpha- and beta-adrenergic receptors can be subdivided still further (alpha$_1$ and alpha$_2$, for example), again according to the drugs that influence them and their second-messenger systems.

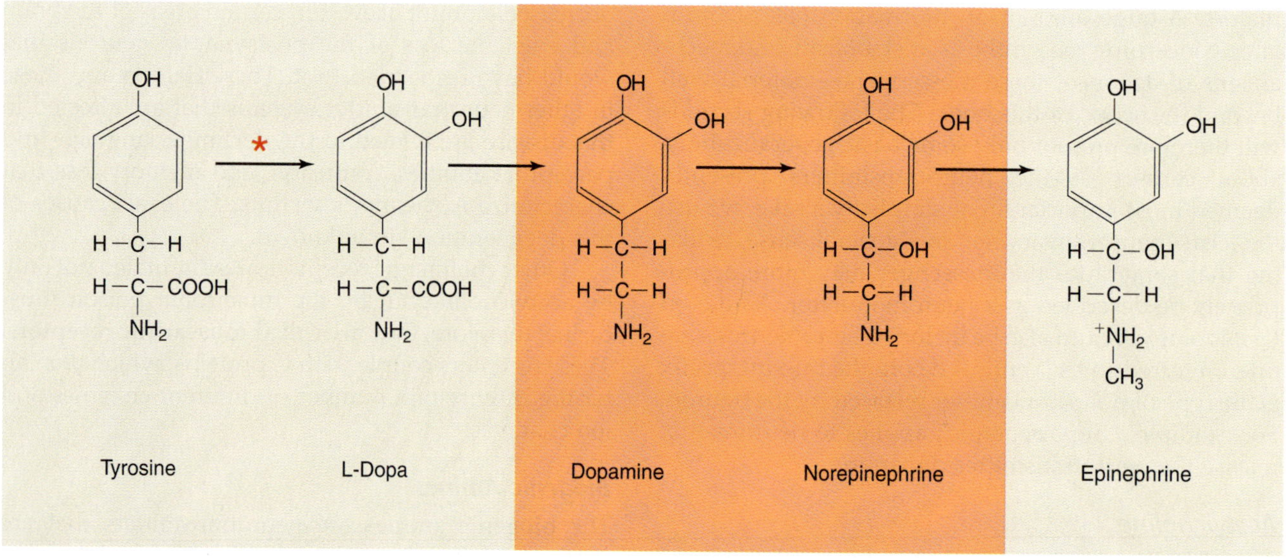

FIGURE 8–34
Catecholamine biosynthetic pathway. The red asterisk indicates the site of action of tyrosine hydroxylase, the rate-limiting enzyme; the colored screen indicates the more common CNS catecholamine neurotransmitters.

Serotonin While not a catecholamine, **serotonin** (5-hydroxytryptamine, or 5-HT) is an important biogenic amine. It is produced from tryptophan, an essential amino acid. Its effects generally have a slow onset, indicating that it works as a neuromodulator. Serotonin-releasing neurons innervate virtually every structure in the brain and spinal cord and operate via at least 16 different receptor types.

In general, serotonin has an excitatory effect on pathways that are involved in the control of muscles, and an inhibitory effect on pathways that mediate sensations. The activity of serotonergic neurons is lowest or absent during sleep and highest during states of alert wakefulness. In addition to their contributions to motor activity and sleep, serotonergic pathways also function in the regulation of food intake, reproductive behavior, and emotional states such as mood and anxiety.

Serotonin is also present in many nonneural cells (for example, blood platelets and certain cells of the immune system and digestive tract). In fact, the brain contains only 1 to 2 percent of the body's serotonin.

Amino Acid Neurotransmitters

In addition to the neurotransmitters that are synthesized from amino acids, several amino acids themselves function as neurotransmitters. Although the amino acid neurotransmitters chemically fit the category of biogenic amines, neurophysiologists traditionally put them into a category of their own. The amino acid neurotransmitters are by far the most prevalent neurotransmitters in the central nervous system, and they affect virtually all neurons there.

Two so-called **excitatory amino acids, glutamate** and **aspartate,** serve as neurotransmitters at the vast majority of excitatory synapses in the central nervous system. In fact, most excitatory synapses in the brain release glutamate. The excitatory amino acids function in learning, memory, and neural development. They are also implicated in epilepsy, Alzheimer's and Parkinson's diseases, and the neural damage that follows strokes, brain trauma, and other conditions of low oxygen availability. One of the family of glutamate receptors is the site of action of a number of mind-altering drugs, such as phencyclidine ("angel dust").

GABA (gamma-aminobutyric acid) and the amino acid **glycine** are the major inhibitory neurotransmitters in the central nervous system. (GABA is not one of the 20 amino acids used to build proteins, but because it is a modified form of glutamate, it is classified with the amino acid neurotransmitters.) Drugs such as Valium that reduce anxiety, guard against seizures, and induce sleep enhance the action of GABA.

Neuropeptides

The **neuropeptides** are composed of two or more amino acids linked together by peptide bonds. Some 85 neuropeptides have been found, but their physiological roles are often unknown. It seems that evolution has selected the same chemical messengers for use in widely differing circumstances, and many of the neuropeptides had been previously identified in nonneural tissue where they function as hormones or paracrine agents. They generally retain the name they were given when first discovered in the nonneural tissue.

The neuropeptides are formed differently from other neurotransmitters, which are synthesized in the axon terminals by very few enzyme-mediated steps. The neuropeptides, in contrast, are derived from large precursor proteins, which in themselves have little, if any, inherent biological activity. The synthesis of these precursors is directed by mRNA and occurs on ribosomes, which exist only in the cell body and large dendrites of the neuron, often a considerable distance from axon terminals or varicosities where the peptides are released.

In the cell body, the precursor protein is packaged into vesicles, which are then moved by axon transport into the terminals or varicosities where the vesicle contents are cleaved by specific peptidases. Many of the precursor proteins contain multiple peptides, which may be different or copies of one peptide. Neurons that release one or more of the peptide neurotransmitters are collectively called **peptidergic.** In many cases, neuropeptides are cosecreted with another type of neurotransmitter and act as neuromodulators.

Certain neuropeptides, termed **endogenous opioids—beta-endorphin,** the **dynorphins,** and the **enkephalins**—have attracted much interest because their receptors are the sites of action of opiate drugs such as morphine and codeine. The opiate drugs are powerful *analgesics* (that is, they relieve pain without loss of consciousness), and the endogenous opioids undoubtedly play a role in regulating pain. The opioids have been implicated in the runner's "second wind," when the athlete feels a boost of energy and a decrease in pain and effort, and in the general feeling of well-being experienced after a bout of strenuous exercise, the so-called jogger's high. There is also evidence that the opioids play a role in eating and drinking behavior, in regulation of the cardiovascular system, and in mood and emotion.

Substance P, another of the neuropeptides, is a transmitter released by afferent neurons that relay sensory information into the central nervous system.

Miscellaneous

Surprisingly, at least one gas—**nitric oxide**—serves as a neurotransmitter. Gases are not released from presynaptic vesicles, nor do they bind to postsynaptic plasma-membrane receptors. They simply diffuse from their sites of origin in one cell into the intracellular fluid of nearby cells. Nitric oxide serves as a messenger between some neurons and between neurons and effector cells. It is produced in one cell from the amino acid arginine (in a reaction catalyzed by nitric oxide synthase), and it binds to and activates guanylyl cyclase in the recipient cell, thereby increasing the concentration of the second-messenger cyclic GMP in that cell (Chapter 7).

Nitric oxide plays a role in a bewildering array of neurally mediated events—learning, development, drug tolerance, penile erection, and sensory and motor modulation, to name a few. Paradoxically, it is also implicated in neural damage that results, for example, from the stoppage of blood flow to the brain or from a head injury. In later chapters we shall see that nitric oxide is produced not only in the central and peripheral nervous systems but by a variety of nonneural cells as well and plays an important paracrine role in the cardiovascular and immune systems, among others.

Another surprise is that **ATP,** the molecule that serves as an important energy source (Chapter 4) is also a neurotransmitter, as is **adenine,** the purine base from which ATP is formed. Like glutamate, ATP is a very fast acting excitatory transmitter.

Neuroeffector Communication

Thus far we have described the effects of neurotransmitters released at synapses. Many neurons of the peripheral nervous system end, however, not at synapses on other neurons but at neuroeffector junctions on muscle and gland cells. The neurotransmitters released by these efferent neurons' terminals or varicosities provide the link by which electrical activity of the nervous system is able to regulate effector cell activity.

The events that occur at neuroeffector junctions are similar to those at a synapse. The neurotransmitter is released from the efferent neuron upon the arrival of an action potential at the neuron's axon terminals or varicosities. The neurotransmitter then diffuses to the surface of the effector cell, where it binds to receptors on that cell's plasma membrane. The receptors may be directly under the axon terminal or varicosity, or they may be some distance away so that the diffusion path followed by the neurotransmitter is tortuous and long. The receptors on the effector cell may be associated with ion channels that alter the membrane potential of the cell, or they may be coupled via a G protein, to enzymes that result in the formation of second messengers in the effector cell. The response (altered muscle contraction or glandular secretion) of the effector cell to these changes will be described in later chapters. As we shall see in the next section, the major neurotransmitters released at neuroeffector junctions are acetylcholine and norepinephrine.

SECTION C SUMMARY

I. An excitatory synapse brings the membrane of the postsynaptic cell closer to threshold. An inhibitory synapse hyperpolarizes the postsynaptic cell or stabilizes it at its resting level.
II. Whether a postsynaptic cell fires action potentials depends on the number of synapses that are active and whether they are excitatory or inhibitory.

Functional Anatomy of Synapses

I. A neurotransmitter, which is stored in synaptic vesicles in the presynaptic axon terminal, carries the signal from a pre- to a postsynaptic neuron. Depolarization of the axon terminal raises the calcium concentration within the terminal, which causes the release of neurotransmitter into the synaptic cleft.
II. The neurotransmitter diffuses across the synaptic cleft and binds to receptors on the postsynaptic cell; the activated receptors usually open ion channels.
   a. At an excitatory synapse, the electrical response in the postsynaptic cell is called an excitatory postsynaptic potential (EPSP). At an inhibitory synapse, it is an inhibitory postsynaptic potential (IPSP).
   b. Usually at an excitatory synapse, channels in the postsynaptic cell that are permeable to sodium, potassium, and other small positive ions are opened; at inhibitory synapses, channels to chloride and/or potassium are opened.
   c. The postsynaptic cell's membrane potential is the result of temporal and spatial summation of the EPSPs and IPSPs at the many active excitatory and inhibitory synapses on the cell.

Activation of the Postsynaptic Cell

I. Action potentials are generally initiated by the temporal and spatial summation of many EPSPs.

Synaptic Effectiveness

I. Synaptic effects are influenced by pre- and postsynaptic events, drugs, and diseases (Table 8–6).

Neurotransmitters and Neuromodulators

I. In general, neurotransmitters cause EPSPs and IPSPs, and neuromodulators cause, via second messengers, more complex metabolic effects in the postsynaptic cell.
II. The actions of neurotransmitters are usually faster than those of neuromodulators.

III. A substance can act as a neurotransmitter at one type of receptor and as a neuromodulator at another.

IV. The major classes of known or suspected neurotransmitters and neuromodulators are listed in Table 8–7.

Neuroeffector Communication

I. The junction between a neuron and an effector cell is called a neuroeffector junction.

II. The events at a neuroeffector junction (release of neurotransmitter into an extracellular space, diffusion of neurotransmitter to the effector cell, and binding with a receptor on the effector cell) are similar to those at a synapse.

SECTION C KEY TERMS

excitatory synapse	spatial summation
inhibitory synapse	presynaptic synapse
convergence	presynaptic inhibition
divergence	presynaptic facilitation
electric synapse	autoreceptor
chemical synapse	neuromodulator
synaptic cleft	acetylcholine (ACh)
cotransmitter	acetylcholinesterase
synaptic vesicle	cholinergic
excitatory postsynaptic	nicotinic receptor
potential (EPSP)	muscarinic receptor
inhibitory postsynaptic	biogenic amine
potential (IPSP)	dopamine

temporal summation	norepinephrine (NE)
epinephrine	glycine
catecholamine	neuropeptide
adrenergic	peptidergic
noradrenergic	endogenous opioid
alpha-adrenergic receptor	beta-endorphin
beta-adrenergic receptor	dynorphin
serotonin	enkephalin
excitatory amino acid	substance P
glutamate	nitric oxide
aspartate	ATP
GABA (gamma- aminobutyric acid)	adenine

SECTION C REVIEW QUESTIONS

1. Contrast the postsynaptic mechanisms of excitatory and inhibitory synapses.
2. Explain how synapses allow neurons to act as integrators; include the concepts of facilitation, temporal and spatial summation, and convergence in your explanation.
3. List at least eight ways in which the effectiveness of synapses may be altered.
4. Discuss differences between neurotransmitters and neuromodulators.
5. Discuss the relationship between dopamine, norepinephrine, and epinephrine.

SECTION D

STRUCTURE OF THE NERVOUS SYSTEM

We shall now survey the anatomy and broad functions of the major structures of the nervous system; future chapters will describe these functions in more detail. First, we must deal with some potentially confusing terminology. Recall that a long extension from a *single* neuron is called an axon or a nerve fiber and that the term "nerve" refers to a group of *many* nerve fibers that are traveling together to the same general location in the peripheral nervous system. There are no nerves in the *central* nervous system. Rather, a group of nerve fibers traveling together in the central nervous system is called a **pathway,** a **tract,** or, when it links the right and left halves of the central nervous system, a **commissure.**

Information can pass through the central nervous system along two types of pathways: (1) **long neural pathways,** in which neurons with long axons carry information directly between the brain and spinal cord or between large regions of the brain, and (2) **multineuronal** or **multisynaptic pathways** (Figure 8–35). As their name suggests, the multineuronal pathways are made up of many neurons and many synaptic connections. Since synapses are the sites where new

information can be integrated into neural messages, there are many opportunities for neural processing along the multineuronal pathways. The long pathways, on the other hand, consist of chains of only a few sequentially connected neurons. Because the long pathways contain few synapses, there is little opportunity for alteration in the information they transmit.

The cell bodies of neurons having similar functions are often clustered together. Groups of neuron cell bodies in the peripheral nervous system are called **ganglia** (singular, *ganglion*), and in the central nervous system they are called **nuclei** (singular, *nucleus*), not to be confused with cell nuclei.

Central Nervous System: Spinal Cord

The spinal cord lies within the bony vertebral column (Figure 8–36). It is a slender cylinder of soft tissue about as big around as the little finger. The central butterfly-shaped area (in cross section) of **gray matter**

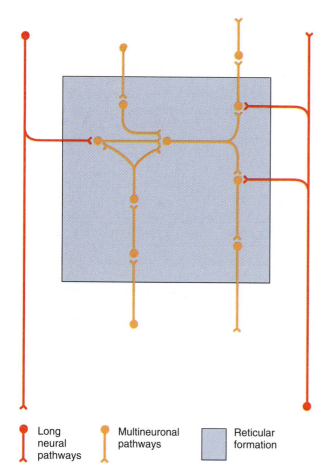

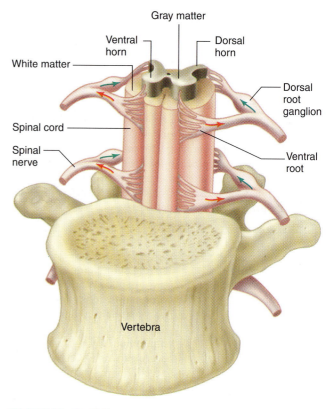

FIGURE 8–36

Section of the spinal cord, ventral view. The arrows indicate the direction of transmission of neural activity. ✗

FIGURE 8–35

Long neural pathways and multineuronal (multisynaptic) pathways and their relationship to the reticular formation.

is composed of interneurons, the cell bodies and dendrites of efferent neurons, the entering fibers of afferent neurons, and glial cells. It is called gray matter because there are more cells than myelinated fibers, and the cells appear gray.

The gray matter is surrounded by **white matter,** which consists of groups of myelinated axons of interneurons. These groups of axons, called fiber tracts or pathways, run longitudinally through the cord, some descending to relay information from the brain to the spinal cord, others ascending to transmit information to the brain. Pathways also transmit information between different levels of the spinal cord.

Groups of afferent fibers that enter the spinal cord from the peripheral nerves enter on the dorsal side of the cord (the side nearest the back of the body) via the **dorsal roots** (Figure 8–36). Small bumps on the dorsal roots, the **dorsal root ganglia,** contain the cell bodies of the afferent neurons. The axons of efferent neurons leave the spinal cord on the ventral side (nearest the front surface of the body) via the **ventral roots.** A short distance from the cord, the dorsal and ventral roots

from the same level combine to form a **spinal nerve,** one on each side of the spinal cord. The 31 pairs of spinal nerves are designated by the four vertebral levels from which they exit: cervical, thoracic, lumbar, and sacral (Figure 8–37).

Central Nervous System: Brain

During development, the central nervous system forms from a long tube. As the anterior part of the tube, which becomes the brain, folds during its continuing formation, four different regions become apparent. These regions become the four subdivisions of the brain: **cerebrum, diencephalon, brainstem,** and **cerebellum** (Figure 8–38). The cerebrum and diencephalon together constitute the **forebrain.** The brainstem consists of the **midbrain, pons,** and **medulla oblongata.** The brain also contains four interconnected cavities, the **cerebral ventricles,** which are filled with circulating cerebrospinal fluid (Figure 8–39), to be discussed more fully later in this chapter.

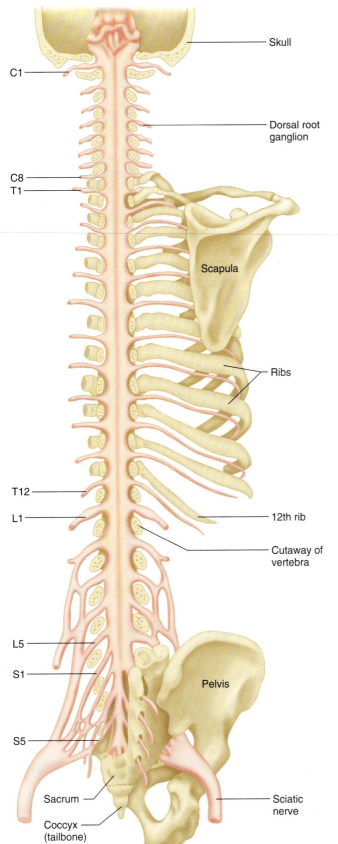

C1

Skull

Dorsal root
ganglion

C8
T1

Scapula

Ribs

T12
L1

12th rib

Cutaway of
vertebra

L5

S1

Pelvis

S5

Sacrum

Coccyx
(tailbone)

Sciatic
nerve

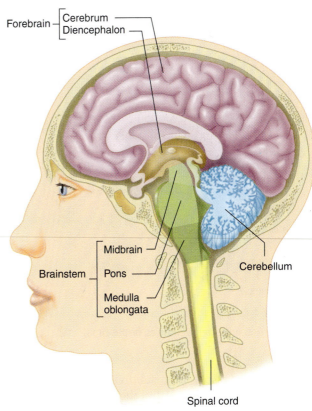

Forebrain — Cerebrum
Diencephalon

Brainstem — Midbrain
Pons
Medulla
oblongata

Cerebellum

Spinal cord

FIGURE 8–38
The spinal cord and divisions of the brain.

FIGURE 8–37
Dorsal view of the spinal cord. Parts of the skull and
vertebrae have been cut away. In general, the eight cervical
nerves (C) control the muscles and glands and receive
sensory input from the neck, shoulder, arm, and hand. The
12 thoracic nerves (T) are associated with the chest and
abdominal walls. The five lumbar nerves (L) are associated
with the hip and leg, and the five sacral nerves (S) are
associated with the genitals and lower digestive tract.

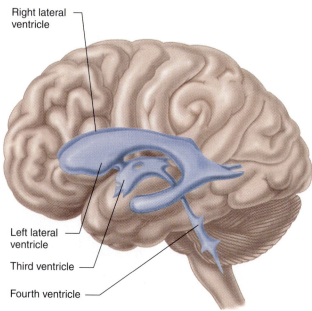

Right lateral ventricle

Left lateral ventricle

Third ventricle

Fourth ventricle

FIGURE 8–39

The four interconnected ventricles of the brain.

Overviews of the brain subdivisions are included here and later in Table 8–9, but their functions are discussed more fully in remaining chapters of the book, particularly Chapters 9, 12, and 13.

Brainstem

All the nerve fibers that relay signals between the spinal cord, forebrain, and cerebellum pass through the brainstem. Running through the core of the brainstem and consisting of loosely arranged neuron cell bodies intermingled with bundles of axons is the **reticular formation,** which is the one part of the brain absolutely essential for life. It receives and integrates input from all regions of the central nervous system and processes a great deal of neural information. The reticular formation is involved in motor functions, cardiovascular and respiratory control, and the mechanisms that regulate sleep and wakefulness and focus attention. Most of the biogenic amine neurotransmitters are released from the axons of cells in the reticular formation and, because of the far-reaching projections of these cells, affect all levels of the nervous system.

Some reticular formation neurons send axons for considerable distances up or down the brainstem and beyond, to most regions of the brain and spinal cord. This pattern explains the very large scope of influence that the reticular formation has over other parts of the central nervous system and explains the widespread effects of the biogenic amines.

The pathways that convey information from the reticular formation to the upper portions of the brain affect wakefulness and the direction of attention to specific events by selectively facilitating neurons in some areas of the brain while inhibiting others. The fibers that descend from the reticular formation to the spinal cord influence activity in both efferent and afferent neurons. There is considerable interaction between the reticular pathways that go up to the forebrain, down to the spinal cord, and to the cerebellum. For example, all three components function in controlling muscle activity.

The reticular formation encompasses a large portion of the brainstem, and many areas within the reticular formation serve distinct functions. For example, some reticular-formation neurons are clustered together, forming brainstem nuclei and integrating centers. These include the cardiovascular, respiratory, swallowing, and vomiting centers, all of which are discussed in subsequent chapters. The reticular formation also has nuclei important in eye-movement control and the reflex orientation of the body in space.

In addition, the brainstem contains nuclei involved in processing information for 10 of the 12 pairs of **cranial nerves.** These are the peripheral nerves that connect with the brain and innervate the muscles, glands, and sensory receptors of the head, as well as many organs in the thoracic and abdominal cavities (Table 8–8).

Cerebellum

The cerebellum consists of an outer layer of cells, the cerebellar cortex (don't confuse this with the cerebral cortex, described below), and several deeper cell clusters. Although the cerebellum does not initiate voluntary movements, it is an important center for coordinating movements and for controlling posture and balance. In order to carry out these functions, the cerebellum receives information from the muscles and joints, skin, eyes and ears, viscera, and the parts of the brain involved in control of movement. Although the cerebellum's function is almost exclusively motor, it is implicated in some forms of learning.

Forebrain

The larger component of the forebrain (see Figure 8–38), the cerebrum, consists of the right and left **cerebral hemispheres** as well as certain other structures on the underside of the brain. The central core of the forebrain is formed by the diencephalon.

The cerebral hemispheres (Figure 8–40) consist of the **cerebral cortex,** an outer shell of gray matter covering myelinated fiber tracts, which form the white matter. This in turn overlies cell clusters, which are also gray matter and are collectively termed the **subcortical nuclei.** The fiber tracts consist of the many nerve fibers that bring information into the cerebrum, carry information out, and connect different areas within a

TABLE 8–8 The Cranial Nerves

Name	Fibers	Comments
I. Olfactory	Afferent	Carries input from receptors in olfactory (smell) neuroepithelium.
II. Optic	Afferent	Carries input from receptors in eye.p
III. Oculomotor	Efferent	Innervates skeletal muscles that move eyeball up, down, and medially and raise upper eyelid; innervates smooth muscles that constrict pupil and alter lens shape for near and far vision.
	Afferent	Transmits information from receptors in muscles.
IV. Trochlear	Efferent	Innervates skeletal muscles that move eyeball downward and laterally.
	Afferent	Transmits information from receptors in muscle.
V. Trigeminal	Efferent	Innervates skeletal chewing muscles.
	Afferent	Transmits information from receptors in skin; skeletal muscles of face, nose, and mouth; and teeth sockets.
VI. Abducens	Efferent	Innervates skeletal muscles that move eyeball laterally.
	Afferent	Transmits information from receptors in muscle.
VII. Facial	Efferent	Innervates skeletal muscles of facial expression and swallowing; innervates nose, palate, and lacrimal and salivary glands.
	Afferent	Transmits information from taste buds in front of tongue and mouth.
VIII. Vestibulocochlear	Afferent	Transmits information from receptors in ear.
IX. Glossopharyngeal	Efferent	Innervates skeletal muscles involved in swallowing and parotid salivary gland.
	Afferent	Transmits information from taste buds at back of tongue and receptors in auditory-tube skin.
X. Vagus	Efferent	Innervates skeletal muscles of pharynx and larynx and smooth muscle and glands of thorax and abdomen.
	Afferent	Transmits information from receptors in thorax and abdomen.
XI. Accessory	Efferent	Innervates neck skeletal muscles.
XII. Hypoglossal	Efferent	Innervates skeletal muscles of tongue.

hemisphere. The cortex layers of the two cerebral hemispheres, although largely separated by a longitudinal division, are connected by a massive bundle of nerve fibers known as the **corpus callosum** (Figure 8–40).

The cortex of each cerebral hemisphere is divided into four **lobes: the frontal, parietal, occipital,** and **temporal** (Figure 8–41). Although it averages only 3 mm in thickness, the cortex is highly folded, which results in an area for cortical neurons that is four times larger than it would be if unfolded, yet does not appreciably increase the volume of the brain. The cells of the cerebral cortex are organized in six layers. The cortical neurons are of two basic types: pyramidal cells (named for the shape of their cell bodies) and nonpyramidal cells. The pyramidal cells form the major output cells of the cortex, sending their axons to other parts of the cortex and to other parts of the central nervous system.

The cerebral cortex is the most complex integrating area of the nervous system. It is where basic afferent information is collected and processed into meaningful perceptual images, and where the ultimate refinement of control over the systems that govern the movement of the skeletal muscles occurs. Nerve fibers enter the cortex predominantly from the diencephalon, specifically from a region known as the thalamus (see below), other regions of the cortex, and the reticular formation of the brainstem. Some of the input fibers convey information about specific events in the environment, whereas others have as their function controlling levels of cortical excitability, determining states of arousal, and directing attention to specific stimuli.

The subcortical nuclei are heterogeneous groups of gray matter that lie deep within the cerebral hemispheres. Predominant among them are the **basal ganglia,** which play an important role in the control of movement and posture and in more complex aspects

(a)

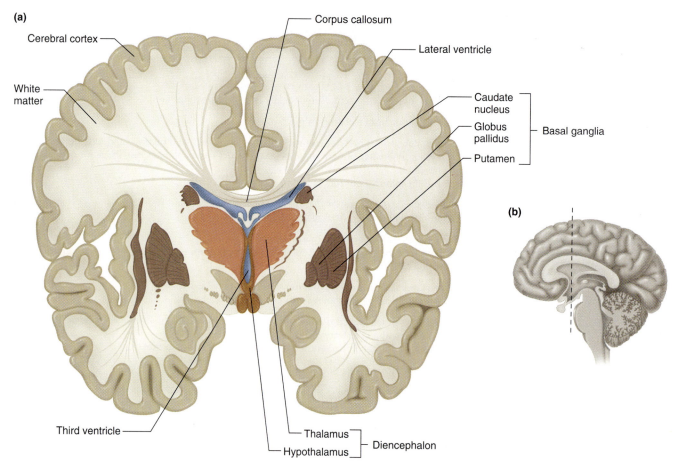

FIGURE 8–40

(a) Coronal (side-to-side) section of the brain. (b) The dashed line indicates the location of the cross section in a.

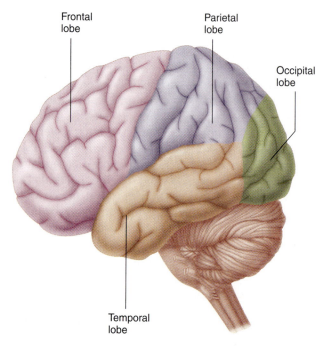

FIGURE 8–41

A lateral view of the brain. The outer layer of the forebrain (the cortex) is divided into four lobes, as shown.

of behavior. (Note that the name "basal ganglia" is an exception to the generalization that ganglia are neuronal cell clusters that lie *outside* the central nervous system.)

The diencephalon, which is divided in two by the slitlike third ventricle, is the second component of the forebrain. It contains two major parts: the thalamus and the hypothalamus (see Figure 8–40). The **thalamus** is a collection of several large nuclei that serve as synaptic relay stations and important integrating centers for most inputs to the cortex. It also plays a key role in nonspecific arousal and focused attention.

The **hypothalamus** (see Figure 8–40) lies below the thalamus and is on the undersurface of the brain. Although it is a tiny region that accounts for less than 1 percent of the brain's weight, it contains different cell groups and pathways that form the master command center for neural and endocrine coordination. Indeed, the hypothalamus is the single most important control area for homeostatic regulation of the internal environment and behaviors having to do with preservation of the individual—for example, eating and drinking—

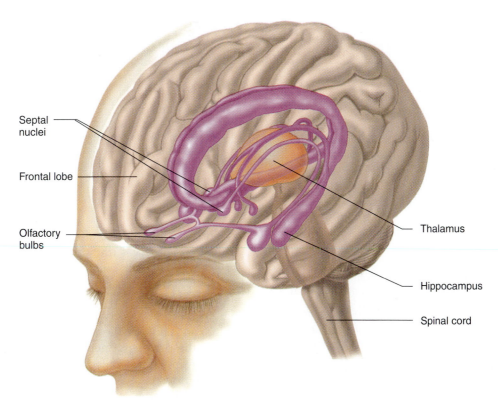

Septal nuclei

Frontal lobe

Olfactory bulbs

Thalamus

Hippocampus

Spinal cord

FIGURE 8–42

Structures of the limbic system are shown shaded in violet in this partially transparent view of the brain.

Redrawn from BRAIN, MIND, AND BEHAVIOR by Floyd E. Bloom and Arlyne Lazerson. Copyright 1985, 1988 by Educational Broadcasting Corporation.

Reprinted by permission of W. H. Freeman and Company.

and preservation of the species—reproduction. The hypothalamus lies directly above the pituitary gland, an important endocrine structure, to which it is attached by a stalk (Chapter 10).

Thus far we have described discrete anatomical areas of the forebrain. Some of these forebrain areas, consisting of both gray and white matter, are also classified together in a functional system, termed the **limbic system.** This interconnected group of brain structures includes portions of frontal-lobe cortex, temporal lobe, thalamus, and hypothalamus, as well as the circuitous fiber pathways that connect them (Figure 8–42). Besides being connected with each other, the parts of the limbic system are connected with many other parts of the central nervous system. Structures within the limbic system are associated with learning, emotional experience and behavior, and a wide variety of visceral and endocrine functions. In fact, much of the output of the limbic system is coordinated by the hypothalamus into behavioral and endocrine responses.

The functions of the major parts of the brain are listed in Table 8–9.

Peripheral Nervous System

Nerve fibers in the peripheral nervous system transmit signals between the central nervous system and receptors and effectors in all other parts of the body. As noted earlier, the nerve fibers are grouped into bundles called nerves. The peripheral nervous system consists of 43 pairs of nerves: 12 pairs of cranial nerves and 31 pairs that connect with the spinal cord as the spinal nerves. The cranial nerves and a summary of the information they transmit were listed in Table 8–8. In general, of the spinal nerves, eight cervical nerves control the muscles and glands and receive sensory input from the neck, shoulder, arm, and hand. The 12 thoracic nerves are associated with the chest and abdominal walls. The five lumbar nerves are associated with the hip and leg, and the five sacral nerves are associated with the genitals and lower digestive tract. (A single pair of coccygeal nerves brings the total to 31 pair.)

Each nerve fiber is surrounded by a Schwann cell. Some of the fibers are wrapped in layers of Schwann-cell

TABLE 8–9 Summary of Functions of the Major Parts of the Brain

I. Brainstem
  A. Contains all the fibers passing between the spinal cord, forebrain, and cerebellum
  B. Contains the reticular formation and its various integrating centers, including those for cardiovascular and respiratory activity (Chapters 14 and 15)
  C. Contains nuclei for cranial nerves III through XII

II. Cerebellum
  A. Coordinates movements, including those for posture and balance (Chapter 12)
  B. Participates in some forms of learning (Chapter 13)

III. Forebrain
  A. Cerebral hemispheres
    1. Contain the cerebral cortex, which participates in perception (Chapter 9), the generation of skilled movements (Chapter 12), reasoning, learning, and memory (Chapter 13)
    2. Contain subcortical nuclei, including those that participate in coordination of skeletal-muscle activity (Chapter 12)
    3. Contain interconnecting fiber pathways
  B. Thalamus
    1. Is a synaptic relay station for sensory pathways on their way to the cerebral cortex (Chapter 9)
    2. Participates in control of skeletal-muscle coordination (Chapter 12)
    3. Plays a key role in awareness (Chapter 13)
  C. Hypothalamus
    1. Regulates anterior pituitary gland (Chapter 10)
    2. Regulates water balance (Chapter 16)
    3. Participates in regulation of autonomic nervous system (Chapters 8 and 18)
    4. Regulates eating and drinking behavior (Chapter 18)
    5. Regulates reproductive system (Chapters 10 and 19)
    6. Reinforces certain behaviors (Chapter 13)
    7. Generates and regulates circadian rhythms (Chapters 7, 9, 10, and 18)
    8. Regulates body temperature (Chapter 18)
    9. Participates in generation of emotional behavior (Chapter 13)
  D. Limbic system
    1. Participates in generation of emotions and emotional behavior (Chapter 13)
    2. Plays essential role in most kinds of learning (Chapter 13)

TABLE 8–10 Divisions of the Peripheral Nervous System

I. Afferent division

II. Efferent division
  A. Somatic nervous system
  B. Autonomic nervous system
    1. Sympathetic division
    2. Parasympathetic division
    3. Enteric division

contain both afferent and efferent fibers, whereas some of the cranial nerves (the optic nerves from the eyes, for example) contain only afferent fibers.

As noted earlier, afferent neurons convey information from sensory receptors at their peripheral endings to the central nervous system. The long part of their axon is outside the central nervous system and is part of the peripheral nervous system. Afferent neurons are sometimes called primary afferents or first-order neurons because they are the first cells entering the central nervous system in the synaptically linked chains of neurons that handle incoming information.

Recall that efferent neurons carry signals out from the central nervous system to muscles or glands. The efferent division of the peripheral nervous system is more complicated than the afferent, being subdivided into a **somatic nervous system** and an **autonomic nervous system.** These terms are somewhat misleading because they suggest additional nervous systems distinct from the central and peripheral systems. Keep in mind that the terms together denote the efferent division of the peripheral nervous system.

The simplest distinction between the somatic and autonomic systems is that the neurons of the somatic division innervate skeletal muscle, whereas the autonomic neurons innervate smooth and cardiac muscle, glands, and neurons in the gastrointestinal tract. Other differences are listed in Table 8–11.

The somatic portion of the efferent division of the peripheral nervous system is made up of all the nerve fibers going from the central nervous system to skeletal-muscle cells. The cell bodies of these neurons are located in groups in the brainstem or spinal cord. Their large diameter, myelinated axons leave the central nervous system and pass without any synapses to skeletal-muscle cells. The neurotransmitter released by these neurons is acetylcholine. Because activity in the somatic neurons leads to contraction of the innervated skeletal-muscle cells, these neurons are called **motor neurons.** Excitation of motor neurons leads only to the *contraction* of skeletal-muscle cells; there are no somatic neurons that inhibit skeletal muscles.

membrane, and these tightly wrapped membranes form a myelin sheath (see Figure 8–3). Other fibers are unmyelinated.

A nerve contains nerve fibers that are the axons of efferent neurons or afferent neurons or both. Accordingly, fibers in a nerve may be classified as belonging to the **efferent** or the **afferent division** of the peripheral nervous system (Table 8–10). All the spinal nerves

TABLE 8–11 Peripheral Nervous System: Somatic and Autonomic Divisions
Somatic   1. Consists of a single neuron between central nervous system and skeletal-muscle cells   2. Innervates skeletal muscle   3. Can lead only to muscle excitation   **Autonomic**   1. Has two-neuron chain (connected by a synapse) between central nervous system and effector organ   2. Innervates smooth and cardiac muscle, glands, and GI neurons   3. Can be either excitatory or inhibitory

Autonomic Nervous System

The efferent innervation of all tissues other than skeletal muscle is by way of the autonomic nervous system. A special case occurs in the gastrointestinal tract, where autonomic neurons innervate a nerve network in the wall of the intestinal tract. This network, termed the **enteric nervous system,** will be described in Chapter 17.

In the autonomic nervous system, parallel chains, each with two neurons, connect the central nervous system and the effector cells (Figure 8–43). (This is in contrast to the single neuron of the somatic system.) The first neuron has its cell body in the central nervous system. The synapse between the two neurons is outside the central nervous system, in a cell cluster called an **autonomic ganglion.** The nerve fibers passing between the central nervous system and the ganglia are called **preganglionic fibers;** those passing between the ganglia and the effector cells are **postganglionic fibers.** There is the potential for integration in the autonomic ganglia because of the convergence and divergence of the pathways there.

Anatomical and physiological differences within the autonomic nervous system are the basis for its further subdivision into **sympathetic** and **parasympa-** thetic components (see Table 8–10). The nerve fibers of the sympathetic and parasympathetic components leave the central nervous system at different levels— the sympathetic fibers from the thoracic (chest) and lumbar regions of the spinal cord, and the parasympathetic fibers from the brain and the sacral portion of the spinal cord (lower back, Figure 8–44). Therefore, the sympathetic division is also called the thoracolumbar division, and the parasympathetic is called the craniosacral division.

The two divisions also differ in the location of ganglia. Most of the sympathetic ganglia lie close to the spinal cord and form the two chains of ganglia—one on each side of the cord—known as the **sympathetic trunks** (Figure 8–44). Other sympathetic ganglia, called collateral ganglia—the celiac, superior mesenteric, and inferior mesenteric ganglia—are in the abdominal cavity, closer to the innervated organ (Figure 8–44). In contrast, the parasympathetic ganglia lie within the organs innervated by the postganglionic neurons or very close to the organs.

The anatomy of the sympathetic nervous system can be confusing. Preganglionic sympathetic *fibers* leave the spinal cord only between the first thoracic and third lumbar segments, whereas sympathetic *trunks* extend the entire length of the cord, from the cervical levels high in the neck down to the sacral levels. The ganglia in the extra lengths of sympathetic trunks receive preganglionic fibers from the thoracolumbar regions because some of the preganglionic fibers, once in the sympathetic trunks, turn to travel upward or downward for several segments before forming synapses with postganglionic neurons (Figure 8–45, numbers 1 and 4). Other possible paths taken by the sympathetic fibers are shown in Figure 8–45, numbers 2, 3, and 5.

The anatomical arrangements in the sympathetic nervous system to some extent tie the entire system together so it can act as a single unit, although small segments of the system can still be regulated independently. The parasympathetic system, in contrast, is

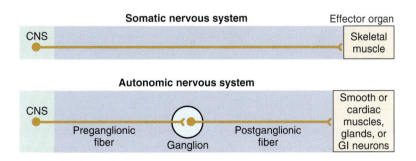

FIGURE 8–43
Efferent division of the peripheral nervous system. Overall plan of the somatic and autonomic nervous systems.

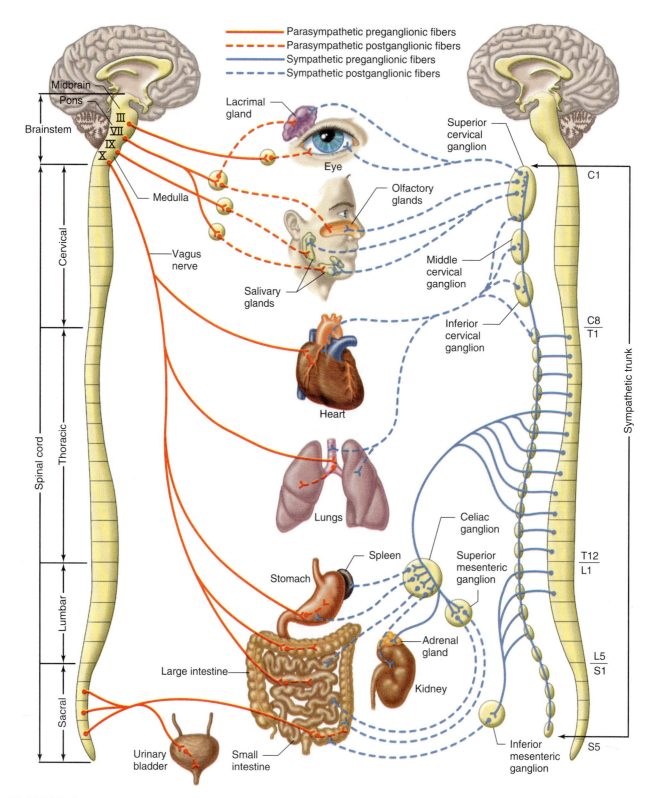

FIGURE 8-44

The parasympathetic (left) and sympathetic (right) divisions of the autonomic nervous system. The celiac, superior mesenteric, and inferior mesenteric ganglia are collateral ganglia. Only one sympathetic trunk is indicated, although there are two, one on each side of the spinal cord. Not shown are the fibers passing to the liver, blood vessels, genitalia and skin glands.

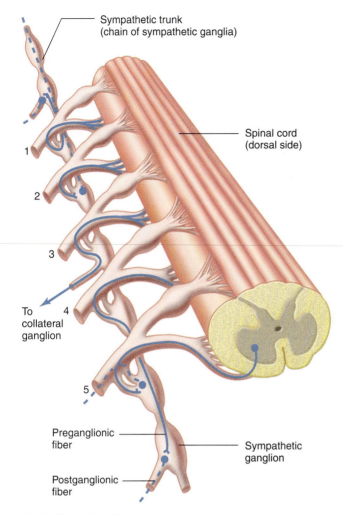

Sympathetic trunk
(chain of sympathetic ganglia)

Spinal cord
(dorsal side)

1

2

3

To
collateral
ganglion

4

5

Preganglionic
fiber

Sympathetic
ganglion

Postganglionic
fiber

FIGURE 8–45

Relationship between a sympathetic trunk and spinal cord
(1 through 5) with the various courses that preganglionic
sympathetic fibers (solid lines) take through the sympathetic
trunk. Dashed lines represent postganglionic fibers. A mirror
image of this exists on the opposite side of the spinal cord.

TABLE 8–12 Classes of Receptors for Acetylcholine, Norepinephrine, and Epinephrine
I. Receptors for acetylcholine
a. Nicotinic receptors
On postganglionic neurons in the autonomic ganglia
At neuromuscular junctions of skeletal muscle
On some central nervous system neurons
b. Muscarinic receptors
On smooth muscle
On cardiac muscle
On gland cells
On some central nervous system neurons
On some neurons of autonomic ganglia (although the great majority of receptors at this site are nicotinic)
II. Receptors for norepinephrine and epinephrine
On smooth muscle
On cardiac muscle
On gland cells
On some central nervous system neurons

---

some sympathetic postganglionic endings. Moreover, one or more cotransmitters are usually stored and released with the autonomic transmitters; these include ATP, dopamine, and several of the neuropeptides. These all, however, play a relatively small role.

In addition to the classical autonomic neurotransmitters just described, there is a widespread network of postganglionic fibers recognized as nonadrenergic and noncholinergic. These fibers use nitric oxide and other neurotransmitters to mediate some forms of blood vessel dilation and to regulate various gastrointestinal, respiratory, urinary, and reproductive functions.

Many of the drugs that stimulate or inhibit various components of the autonomic nervous system affect receptors for acetylcholine and norepinephrine. Recall that there are several types of receptors for each neurotransmitter (Table 8–12). The great majority of acetylcholine receptors in the autonomic ganglia are nicotinic receptors. In contrast, the acetylcholine receptors on smooth-muscle, cardiac-muscle, and gland cells are muscarinic receptors. To complete the story of the peripheral cholinergic receptors, it should be emphasized that the cholinergic receptors on skeletal-muscle fibers, innervated by the *somatic* motor neurons, not autonomic neurons, are nicotinic receptors.

One set of postganglionic neurons in the sympathetic division never develops axons; instead, upon activation by preganglionic axons, the cells of this "ganglion" release their transmitters into the bloodstream (Figure 8–46). This "ganglion," called

---

made up of relatively independent components. Thus, overall autonomic responses, made up of many small parts, are quite variable and finely tailored to the specific demands of any given situation.

In both sympathetic and parasympathetic divisions, acetylcholine is the major neurotransmitter released between pre- and postganglionic fibers in autonomic ganglia (Figure 8–46). In the parasympathetic division, acetylcholine is also the major neurotransmitter between the postganglionic fiber and the effector cell. In the sympathetic division, norepinephrine is usually the major transmitter between the postganglionic fiber and the effector cell. We say "major" and "usually" because acetylcholine is also released by

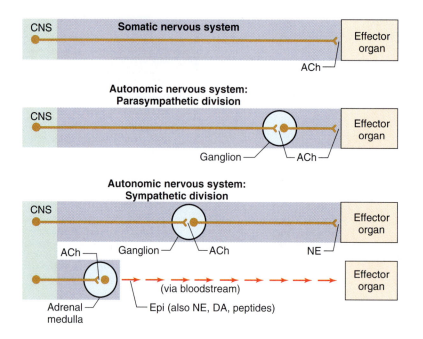

FIGURE 8–46

Transmitters used in the various components of the peripheral efferent nervous system. In a few cases (to be described later), sympathetic neurons release a transmitter other than norepinephrine. Notice that the first neuron exiting the central nervous system—whether in the somatic or the autonomic nervous system—releases acetylcholine. ACh, acetylcholine; NE, norepinephrine; Epi, epinephrine; DA, dopamine.

the **adrenal medulla,** therefore functions as an endocrine gland whose secretion is controlled by sympathetic preganglionic nerve fibers. It releases a mixture of about 80 percent epinephrine and 20 percent norepinephrine into the blood (plus small amounts of other substances, including dopamine, ATP, and neuropeptides). These catecholamines, properly called hormones rather than neurotransmitters in this circumstance, are transported via the blood to effector cells having receptors sensitive to them. The receptors may be the same adrenergic receptors that are located near the release sites of sympathetic postganglionic neurons and normally activated by the norepinephrine released from these neurons, or the receptors may be located at places that are not near the neurons and therefore activated only by the circulating epinephrine or norepinephrine.

Table 8–13 is a reference list of the effects of autonomic nervous system activity, which will be described in subsequent chapters. Note that the heart and many glands and smooth muscles are innervated by both sympathetic and parasympathetic fibers; that is, they receive **dual innervation.** Whatever effect one division has on the effector cells, the other division usually has the opposite effect. (Several exceptions to this rule are indicated in Table 8–13.) Moreover, the two divisions are usually activated reciprocally; that is, as the activity

of one division is increased, the activity of the other is decreased. Dual innervation by nerve fibers that cause opposite responses provides a very fine degree of control over the effector organ.

A useful generalization is that the sympathetic system increases its response under conditions of physical or psychological stress. Indeed, a full-blown sympathetic response is called the **fight-or-flight response,** describing the situation of an animal forced to challenge an attacker or run from it. All resources are mobilized: heart rate and blood pressure increase; blood flow to the skeletal muscles, heart, and brain increase; the liver releases glucose; and the pupils dilate. Simultaneously, activity of the gastrointestinal tract and blood flow to the skin are decreased by inhibitory sympathetic effects.

The two divisions of the autonomic nervous system rarely operate independently, and autonomic responses generally represent the regulated interplay of both divisions. Autonomic responses usually occur without conscious control or awareness, as though they were indeed autonomous (in fact, the autonomic nervous system has been called the "involuntary" nervous system). However, it is wrong to assume that this is always the case, for it has been shown that discrete visceral or glandular responses can be learned and thus, to this extent, voluntarily controlled.

TABLE 8–13 Some Effects of Autonomic Nervous System Activity

Effector Organ	Receptor Type*	Sympathetic Effect	Parasympathetic Effect†
Eyes			
Iris muscles	Alpha	Contracts radial muscle (widens pupil)	Contracts sphincter muscle (makes pupil smaller)
Ciliary muscle	Beta	Relaxes (flattens lens for far vision)	Contracts (allows lens to become more convex for near vision)
Heart			
SA node	Beta	Increases heart rate	Decreases heart rate
Atria	Beta	Increases contractility	Decreases contractility
AV node	Beta	Increases conduction velocity	Decreases conduction velocity
Ventricles	Beta	Increases contractility	Decreases contractility slightly
Arterioles			
Coronary	Alpha	Constricts	—
	Beta	Dilates	
Skin	Alpha	Constricts	—‡
Skeletal muscle	Alpha	Constricts	—
	Beta	Dilates	
Abdominal viscera	Alpha	Constricts	—
	Beta	Dilates	
Salivary glands	Alpha	Constricts	Dilates
Veins	Alpha	Constricts	—
	Beta	Dilates	
Lungs			
Bronchial muscle	Beta	Relaxes	Contracts
Bronchial glands	Alpha	Inhibits secretion	Stimulates secretion
	Beta	Stimulates secretion	
Salivary glands	Alpha	Stimulates watery secretion	Stimulates watery secretion
	Beta	Stimulates enzyme secretion	
Stomach			
Motility, tone	Alpha and Beta	Decreases	Increases
Sphincters	Alpha	Contracts	Relaxes
Secretion		Inhibits (?)	Stimulates
Intestine			
Motility	Alpha and Beta	Decreases	Increases
Sphincters	Alpha	Contracts (usually)	Relaxes (usually)
Secretion	Alpha	Inhibits	Stimulates
Gallbladder	Beta	Relaxes	Contracts
Liver	Alpha and Beta	Glycogenolysis and gluconeogenesis	—
Pancreas			
Exocrine glands	Alpha	Inhibits secretion	Stimulates secretion
Endocrine glands	Alpha	Inhibits secretion	—
	Beta	Stimulates secretion	

(continued)

TABLE 8–13 Some Effects of Autonomic Nervous System Activity (cont.)

Effector Organ	Sympathetic		Parasympathetic Effect†
	Receptor Type*	Effect	
Fat cells	Alpha and Beta	Increases fat breakdown	—
Kidneys	Beta	Increases renin secretion	—
Urinary bladder			
Bladder wall	Beta	Relaxes	Contracts
Sphincter	Alpha	Contracts	Relaxes
Uterus	Alpha	Contracts in pregnancy	Variable
	Beta	Relaxes	
Reproductive tract (male)	Alpha	Ejaculation	Erection
Skin			
Muscles causing hair erection	Alpha	Contracts	—
Sweat glands	Alpha	Localized secretion	Generalized secretion
Lacrimal glands	Alpha	Secretion	Secretion

Table adapted from "Goodman and Gilman's The Pharmacological Basis of Therapeutics," Joel G. Hardman, Lee E. Limbird, Perry B. Molinoff, Raymond W. Ruddon, and Alfred Goodman Gilman, eds., 9th edn., McGraw-Hill, New York, 1996.

*Note that in many effector organs, there are both alpha-adrenergic and beta-adrenergic receptors. Activation of these receptors may produce either the same or opposing effects. For simplicity, except for the arterioles and a few other cases, only the dominant sympathetic effect is given when the two receptors oppose each other.

†These effects are all mediated by muscarinic receptors.

‡A dash means these cells are not innervated by this branch of the autonomic nervous system or that these nerves do not play a significant physiological role.

Blood Supply, Blood-Brain Barrier Phenomena, and Cerebrospinal Fluid

As mentioned earlier, the brain lies within the skull, and the spinal cord within the vertebral column. Between the soft neural tissues and the bones that house them are three types of membranous coverings called **meninges:** the dura mater next to the bone, the arachnoid in the middle, and the pia mater next to the nervous tissue. A space, the subarachnoid space, between the arachnoid and pia is filled with **cerebrospinal fluid (CSF).** The meninges and their specialized parts protect and support the central nervous system, and they produce, circulate, and absorb the cerebrospinal fluid. (As described later, a portion of the cerebrospinal fluid is also formed in the cerebral ventricles.)

The cerebrospinal fluid circulates through the interconnected ventricular system to the brainstem, where it passes through small openings out to a space between the meninges on the surface of the brain and spinal cord (Figure 8–47). Aided by circulatory, respiratory, and postural pressure changes, the fluid ultimately flows to the top of the outer surface of the brain, where most of it enters the bloodstream through one-

way valves in large veins. Thus, the central nervous system literally floats in a cushion of cerebrospinal fluid. Since the brain and spinal cord are soft, delicate tissues with a consistency similar to Jello, they are somewhat protected by the shock-absorbing fluid from sudden and jarring movements. If the flow is obstructed, cerebrospinal fluid accumulates, causing **hydrocephalus** ("water on the brain"). In severe untreated cases, the resulting elevation of pressure in the ventricles leads to compression of the brain's blood vessels, which may lead to inadequate blood flow to the neurons, neuronal damage, and mental retardation.

Under normal conditions, glucose is the only substrate metabolized by the brain to supply its energy requirements, and most of the energy from the oxidative breakdown of glucose is transferred to ATP. The brain's glycogen stores being negligible, it is completely dependent upon a continuous blood supply of glucose and oxygen. In fact, the most common form of brain damage is caused by a stoppage of the blood supply to a region of the brain. When neurons in the region are without a blood supply and deprived of nutrients and oxygen for even a few minutes, they cease to function and die. This neuronal death results in a *stroke.*

Although the adult brain makes up only 2 percent of the body weight, it receives 12 to 15 percent of the

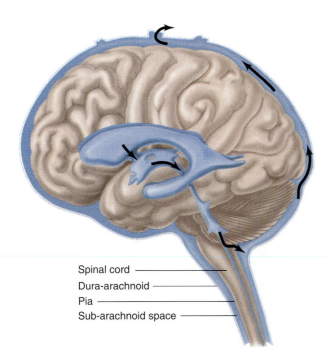

Spinal cord
Dura-arachnoid
Pia
Sub-arachnoid space

FIGURE 8–47

The ventricular system of the brain and the distribution of the cerebrospinal fluid, shown in blue.

total blood supply, which supports its high oxygen utilization. If the blood flow to a region of the brain is reduced to 10 to 25 percent of its normal level, energy stores are depleted, energy-dependent membrane ion pumps fail, membrane ion gradients decrease, the membranes depolarize, and extracellular potassium concentrations increase.

The exchange of substances between blood and extracellular fluid in the central nervous system is different from the more-or-less unrestricted diffusion of nonprotein substances from blood to extracellular fluid in the other organs of the body. A complex group of **blood-brain barrier** mechanisms closely control both the kinds of substances that enter the extracellular fluid of the brain and the rates at which they enter. These mechanisms minimize the ability of many harmful substances to reach the neurons, but they also reduce the access of the immune system to the brain.

The blood-brain barrier, which comprises the cells that line the smallest blood vessels in the brain, has both anatomical structures, such as tight junctions, and physiological transport systems that handle different classes of substances in different ways. For example, substances that dissolve readily in the lipid components of the plasma membranes enter the brain quickly. Therefore, the extracellular fluid of the brain and spinal cord is a product of, but chemically different from, the blood.

The blood-brain barrier accounts for some drug actions, too, as can be seen from the following scenario: Morphine differs chemically from heroin only in that morphine has two hydroxyl groups whereas heroin has two acetyl groups ($-COCH_3$). This small difference renders heroin more lipid soluble and able to cross the blood-brain barrier more readily than morphine. As soon as heroin enters the brain, however, enzymes remove the acetyl groups from heroin and change it to morphine. The morphine, insoluble in lipid, is then effectively trapped in the brain where it continues to exert its effect. Other drugs that have rapid effects in the central nervous system because of their high lipid solubility are the barbiturates, nicotine, caffeine, and alcohol.

Many substances that do not dissolve readily in lipids, such as glucose and other important substrates of brain metabolism, nonetheless enter the brain quite rapidly by combining with membrane transport proteins in the cells that line the smallest brain blood vessels. Similar transport systems also move substances *out* of the brain and into the blood, preventing the buildup of molecules that could interfere with brain function.

In addition to its blood supply, the central nervous system is perfused by the cerebrospinal fluid. The cerebrospinal fluid is secreted into the ventricles by epithelial cells that cover the **choroid plexuses,** which form part of the lining of the four ventricles. A barrier is present here, too, between the blood in the capillaries of the choroid plexuses and the cerebrospinal fluid, and cerebrospinal fluid is a selective secretion. For example, potassium and calcium concentrations are slightly lower in cerebrospinal fluid than in plasma, whereas the sodium and chloride concentrations are slightly higher. The choroid plexuses also trap toxic heavy metals such as lead, thus affording a degree of protection to the brain from these substances.

The cerebrospinal fluid and the extracellular fluid of the brain are, over time, in diffusion equilibrium. Thus, the extracellular environment of the brain and spinal cord neurons is regulated by restrictive, selective barrier mechanisms in the capillaries of the brain and choroid plexuses.

SECTION D SUMMARY

I. Inside the skull and vertebral column, the brain and spinal cord are enclosed in and protected by the meninges.

Central Nervous System: Spinal Cord

I. The spinal cord is divided into two areas: central gray matter, which contains nerve cell bodies and dendrites; and white matter, which surrounds the gray matter and contains myelinated axons organized into ascending or descending tracts.

II. The axons of the afferent and efferent neurons form the spinal nerves.

Central Nervous System: Brain

I. The brain is divided into six regions: cerebrum, diencephalon, midbrain, pons, medulla oblongata, and cerebellum.

II. The midbrain, pons, and medulla oblongata form the brainstem, which contains the reticular formation.

III. The cerebellum plays a role in posture, movement, and some kinds of memory.

IV. The cerebrum, made up of right and left cerebral hemispheres, and the diencephalon together form the forebrain. The cerebral cortex forms the outer shell of the cerebrum and is divided into parietal, frontal, occipital, and temporal lobes.

V. The diencephalon contains the thalamus and hypothalamus.

VI. The limbic system is a set of deep forebrain structures associated with learning and emotions.

Peripheral Nervous System

I. The peripheral nervous system consists of 43 paired nerves—12 pairs of cranial nerves and 31 pairs of spinal nerves. Most nerves contain axons of both afferent and efferent neurons.

II. The efferent division of the peripheral nervous system is divided into somatic and autonomic parts. The somatic fibers innervate skeletal-muscle cells and release the neurotransmitter acetylcholine.

III. The autonomic nervous system innervates cardiac and smooth muscle, glands, and gastrointestinal-tract neurons. Each autonomic pathway consists of a preganglionic neuron with its cell body in the CNS and a postganglionic neuron with its cell body in an autonomic ganglion outside the CNS.

  a. The autonomic nervous system is divided into sympathetic and parasympathetic components. The preganglionic neurons in both sympathetic and parasympathetic divisions release acetylcholine; the postganglionic parasympathetic neurons release mainly acetylcholine; and the postganglionic sympathetics release mainly norepinephrine.

  b. The receptors that respond to acetylcholine are classified as nicotinic and muscarinic, and those that respond to norepinephrine or epinephrine as alpha- and beta-adrenergic types.

  c. The adrenal medulla is a hormone-secreting part of the sympathetic nervous system and secretes mainly epinephrine.

  d. Many effector organs innervated by the autonomic nervous system receive dual innervation.

Blood Supply, Blood-Brain Barrier Phenomena, and Cerebrospinal Fluid

I. Brain tissue depends on a continuous supply of glucose and oxygen for metabolism.

II. The brain ventricles and the space within the meninges are filled with cerebrospinal fluid, which is formed in the ventricles.

III. The chemical composition of the extracellular fluid of the CNS is closely regulated by the blood-brain barrier.

SECTION D KEY TERMS	
pathway	temporal lobe
tract	basal ganglia
commissure	thalamus
long neural pathway	hypothalamus
multineuronal pathway	limbic system
multisynaptic pathway	efferent division of the
ganglia	peripheral nervous
nuclei	system
gray matter	afferent division of the
white matter	peripheral nervous
dorsal root	system
dorsal root ganglia	somatic nervous system
ventral root	autonomic nervous system
spinal nerve	motor neuron
cerebrum	enteric nervous system
diencephalon	autonomic ganglion
brainstem	preganglionic fiber
cerebellum	postganglionic fiber
forebrain	sympathetic division of the
midbrain	autonomic nervous
pons	system
medulla oblongata	parasympathetic division of
cerebral ventricle	the autonomic nervous
reticular formation	system
cranial nerve	sympathetic trunk
cerebral hemisphere	adrenal medulla
cerebral cortex	dual innervation
subcortical nuclei	fight-or-flight response
corpus callosum	meninges
frontal lobe	cerebrospinal fluid (CSF)
parietal lobe	blood-brain barrier
occipital lobe	choroid plexuses

SECTION D REVIEW QUESTIONS

1. Make an organizational chart showing the central nervous system, peripheral nervous system, brain, spinal cord, spinal nerves, cranial nerves, forebrain, brainstem, cerebrum, diencephalon, midbrain, pons, medulla oblongata, and cerebellum.

2. Draw a cross section of the spinal cord showing the gray and white matter, dorsal and ventral roots, dorsal root ganglion, and spinal nerve. Indicate the general location of pathways.

3. List two functions of the thalamus.

4. List the functions of the hypothalamus.

5. Make a peripheral nervous system chart indicating the relationships among afferent and efferent divisions, somatic and autonomic nervous systems, and sympathetic and parasympathetic divisions.

6. Contrast the somatic and autonomic divisions of the efferent nervous system; mention at least three characteristics of each.

7. Name the neurotransmitter released at each synapse or neuroeffector junction in the somatic and autonomic systems.
8. Contrast the sympathetic and parasympathetic components of the autonomic nervous system; mention at least four characteristics of each.
9. Explain how the adrenal medulla can affect receptors on various effector organs despite the fact that its cells have no axons.
10. The chemical composition of the CNS extracellular fluid is different from that of blood. Explain how this difference is achieved.

| CHAPTER 8 CLINICAL TERMS |

local anesthetic   Alzheimer's disease
agonist   analgesic
antagonist   hydrocephalus
tetanus toxin   stroke

| CHAPTER 8 THOUGHT QUESTIONS |

(Answers are given in Appendix A)

1. Neurons are treated with a drug that instantly and permanently stops the Na,K-ATPase pumps. Assume for this question that the pumps are not electrogenic. What happens to the resting membrane potential immediately and over time?
2. Extracellular potassium concentration in a person is increased with no change in intracellular potassium concentration. What happens to the resting potential and the action potential?
3. A person has received a severe blow to the head but appears to be all right. Over the next weeks, however, he develops loss of appetite, thirst, and sexual capacity, but no loss in sensory or motor function. What part of the brain do you think may have been damaged?
4. A person is taking a drug that causes, among other things, dryness of the mouth and speeding of the heart rate but no impairment of the ability to use the skeletal muscles. What type of receptor does this drug probably block? (Table 8–12 will help you answer this.)
5. Some cells are treated with a drug that blocks chloride channels, and the membrane potential of these cells becomes slightly depolarized (less negative). From these facts, predict whether the plasma membrane of these cells actively transports chloride and, if so, in what direction.
6. If the enzyme acetylcholinesterase were blocked with a drug, what malfunctions would occur?
7. The compound tetraethylammonium (TEA) blocks the voltage-gated changes in potassium permeability that occur during an action potential. After administration of TEA, what changes would you expect in the action potential? In the afterhyperpolarization?

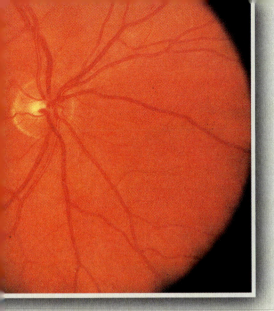

C H A P T E R 9

The Sensory Systems

GENERAL PRINCIPLES

Awareness of our internal and external world is brought about by the neural mechanisms that process afferent information. The initial step of this processing is the transformation of stimulus energy first into graded potentials—the receptor potentials—and then into action potentials in nerve fibers. The pattern of action potentials in particular nerve fibers is a code that provides information about the world even though, as is frequently the case with symbols, the action potentials differ vastly from what they represent.

A **sensory system** is a part of the nervous system that consists of sensory receptors that receive stimuli from the external or internal environment, the neural pathways that conduct information from the receptors to the brain, and those parts of the brain that deal primarily with processing the information.

Information processed by a sensory system may or may not lead to conscious awareness of the stimulus. Regardless of whether the information reaches consciousness, it is called **sensory information.** If the information does reach consciousness, it can also be called a **sensation.** A person's understanding of the sensation's meaning is called **perception.** For example, feeling pain is a sensation, but my awareness that my tooth hurts is a perception. Perceptions are the result of the neural processing of sensory information. At present we have little understanding of the final stages in the processing by which patterns of action potentials become sensations or perceptions.

Intuitively, it might seem that sensory systems operate like familiar electrical equipment, but this is true only up to a point. As an example, let us compare telephone transmission with our auditory (hearing) sensory system. The telephone changes sound waves into electric impulses, which are then transmitted along wires to the receiver. Thus far the analogy holds. (Of course, the mechanisms by which electric currents and action potentials are transmitted are quite different, but this does not affect our argument.) The telephone receiver then changes the coded electric impulses *back into sound waves.* Here is the crucial difference, for our brain does not physically translate the code into sound. Rather, the coded information itself or some correlate of it is what we perceive as sound.

Receptors

Neural activity is initiated at the border between the nervous system and the outside world by **sensory receptors.** Since some receptors respond to changes in the internal environment, the "outside world" with

regard to the sensory receptors can also mean, for example, distension of a blood vessel in our body.

Information about the external world and about the body's internal environment exists in different energy forms—pressure, temperature, light, sound waves, and so on. Receptors at the peripheral ends of afferent neurons change these energy forms into graded potentials that can initiate action potentials, which travel into the central nervous system. The receptors are either specialized endings of afferent neurons themselves (Figure 9–1a) or separate cells that affect the ends of afferent neurons (Figure 9–1b).

To avoid confusion in the remainder of this chapter, the reader must recall from Chapter 7 that the term "receptor" has two completely different meanings. One meaning is that of "sensory receptor," as just defined. The second usage is for the individual proteins in the plasma membrane or inside the cell to which specific chemical messengers bind, triggering an intracellular signal transduction pathway that culminates in the cell's response. The potential confusion between these two meanings is magnified by the fact that the stimuli for some sensory receptors (for example, those involved in taste and smell) are chemicals that bind to protein receptors in the plasma membrane of the sensory receptor. If you are in doubt as to which meaning is intended, add the adjective "sensory" or "protein" to see which makes sense in the context.

To repeat, regardless of the original form of the energy, information from sensory receptors linking the nervous system with the outside world must be translated into the language of graded potentials or action potentials. The energy that impinges upon and activates a sensory receptor is known as a **stimulus.** The process

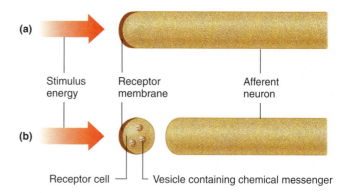

(a)

Stimulus energy   Receptor membrane   Afferent neuron

(b)

Receptor cell   Vesicle containing chemical messenger

FIGURE 9–1

Sensory receptors. The sensitive membrane that responds to a stimulus is either (a) an ending of an afferent neuron or (b) on a separate cell adjacent to an afferent neuron (highly schematized).

by which a stimulus—a photon of light, say, or the mechanical stretch of a tissue—is transformed into an electrical response is known as **stimulus transduction.**

There are many types of sensory receptors, each of which is specific; that is, each type responds much more readily to one form of energy than to others. The type of energy to which a receptor responds in normal functioning is known as its **adequate stimulus.**

Specificity exists at still another level. Within the general energy type that serves as a receptor's adequate stimulus, a particular receptor responds best (that is, at lowest threshold) to only a very narrow range of stimulus energies. For example, individual receptors in the eye respond best to photic energy of one range of light wavelengths.

Most sensory receptors are exquisitely sensitive to their specific energy form. For example, some olfactory receptors respond to as few as three or four odor molecules in the inspired air, and visual receptors can respond to a single photon, the smallest quantity of light.

Virtually all sensory receptors, however, can be activated by several forms of energy if the intensity is sufficiently high. For example, the receptors of the eye normally respond to light, but they can be activated by an intense mechanical stimulus, like a poke in the eye. Note, however, that one experiences the sensation of *light* in response to a poke in the eye. Regardless of how the receptor is stimulated, any given receptor gives rise to only one sensation.

The Receptor Potential

The transduction process in all sensory receptors involves the opening or closing of ion channels that receive—either directly or through a second-messenger system—information about the outside world. The ion channels occur in a specialized receptor membrane and not on ordinary plasma membranes. The gating of these ion channels allows a change in the ion fluxes across the receptor membrane, which in turn produces a change in the membrane potential there. This change in potential is a graded potential called a **receptor potential.** The different mechanisms by which ion channels are affected in the various types of sensory receptors are described throughout this chapter.

The specialized receptor membrane where the initial ion-channel changes occur, unlike the axonal plasma membrane, does not generate action potentials. Instead, local current from the receptor membrane flows a short distance along the axon to a region where the membrane can generate action potentials. In myelinated afferent neurons, this region is usually at the first node of Ranvier of the myelin sheath (Figure 9–2).

In the case where the receptor membrane is on a separate cell, the receptor potential there causes the release of neurotransmitter, which diffuses across the extracellular cleft between the receptor cell and the afferent neuron and binds to specific sites on the afferent neuron. Thus, this junction is like a synapse. The combination of neurotransmitter with its binding sites on the afferent neuron generates a graded potential in the neuron's end analogous to either an excitatory postsynaptic potential or, in some cases, an inhibitory postsynaptic potential.

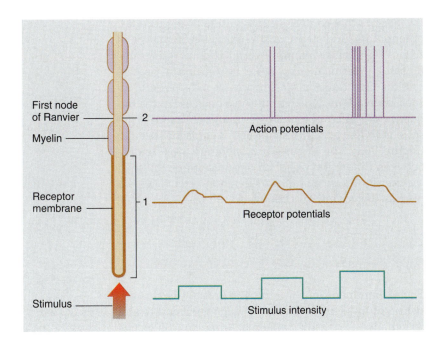

FIGURE 9–2

An afferent neuron with a receptor ending. The receptor potential arises at the nerve ending 1, and the action potential arises at the first node of the myelin sheath 2.

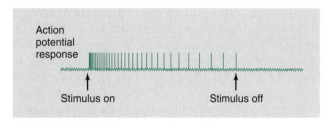

FIGURE 9–3

Action potentials in a single afferent nerve fiber showing adaptation to a stimulus of constant strength.

As is true of all graded potentials, the magnitude of a receptor potential (or a graded potential in the axon adjacent to the receptor cell) decreases with distance from its origin. However, if the amount of depolarization at the first node in the afferent neuron is large enough to bring the membrane there to threshold, action potentials are initiated, which then propagate along the nerve fiber. The only function of the graded potential is to trigger action potentials. (See Figure 8–16 to review the properties of graded potentials.)

As long as the afferent neuron remains depolarized to or above threshold, action potentials continue to fire and propagate along the afferent neuron. Moreover, for complex reasons, an increase in the graded-potential magnitude causes an increase in the action-potential frequency in the afferent neuron (up to the limit imposed by the neuron's refractory period). Although the graded-potential magnitude determines action-potential *frequency,* it does not determine action-potential *magnitude.* Since the action potential is all-or-none, its magnitude is independent of the strength of the initiating stimulus.

Factors that control the magnitude of the receptor potential include stimulus strength, rate of change of stimulus strength, temporal summation of successive receptor potentials (see Figure 8–16), and a process called **adaptation.** This last process is a decrease in receptor sensitivity, which results in a decrease in the frequency of action potentials in an afferent neuron despite maintenance of the stimulus at constant strength (Figure 9–3). The degrees of adaptation vary widely between different types of sensory receptors. We shall see the significance of these differences later when we discuss the coding of stimulus duration.

Neural Pathways in Sensory Systems

The afferent neurons form the first link in a chain consisting of three or more neurons connected end to end by synapses. A bundle of parallel, three-neuron chains together form a **sensory pathway.** The chains in a given pathway run parallel to each other in the central nervous system and, with one exception, carry information to the part of the cerebral cortex responsible for conscious recognition of the information. Sensory pathways are also called **ascending pathways** because they go "up" to the brain.

Sensory Units

A single afferent neuron with all its receptor endings makes up a **sensory unit.** In a few cases, the afferent neuron has a single receptor, but generally the peripheral end of an afferent neuron divides into many fine branches, each terminating at a receptor.

The portion of the body that, when stimulated, leads to activity in a particular afferent neuron is called the **receptive field** for that neuron (Figure 9–4). Receptive fields of neighboring afferent neurons overlap so that stimulation of a single point activates several sensory units; thus, activation at a single sensory unit almost never occurs. As we shall see, the degree of overlap varies in different parts of the body.

Ascending Pathways

The central processes of the afferent neurons enter the brain or spinal cord and synapse upon interneurons there. The central processes diverge to terminate on several, or many, interneurons (Figure 9–5a) and converge so that the processes of many afferent neurons terminate upon a single interneuron (Figure 9–5b). The interneurons upon which the afferent neurons synapse are termed second-order neurons, and these in turn synapse with third-order neurons, and so on, until the information (coded action potentials) reaches the cerebral cortex.

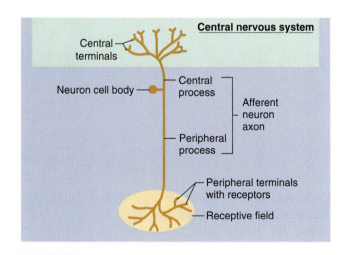

FIGURE 9–4

Sensory unit and receptive field.

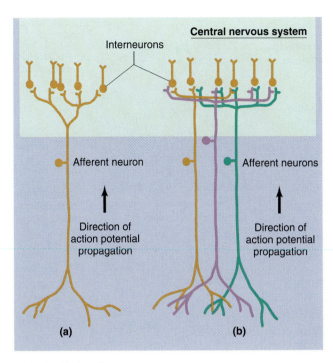

FIGURE 9–5

(a) Divergence of afferent neuron terminals. (b) Convergence of input from several afferent neurons onto single interneurons.

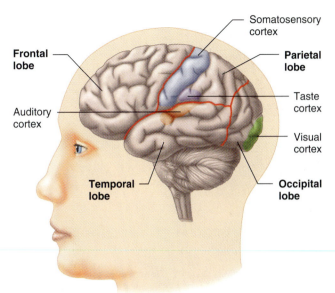

FIGURE 9–6

Primary sensory areas of the cerebral cortex.

Some of the sensory pathways convey information about only a single type of sensory information. Thus, one pathway is influenced only by information from mechanoreceptors, whereas another is influenced only by information from thermoreceptors. The ascending pathways in the spinal cord and brain that carry information about single types of stimuli are known as the **specific ascending pathways.** The specific pathways pass to the brainstem and thalamus, and the final neurons in the pathways go from there to different areas of the cerebral cortex (Figure 9–6). (The olfactory pathways are an exception because they go to parts of the limbic system rather than to the thalamus and because they terminate in the limbic system.) By and large, the specific pathways cross to the side of the central nervous system that is opposite to the location of their sensory receptors so that information from receptors on the right side of the body is transmitted to the left cerebral hemisphere and vice versa.

The specific ascending pathways that transmit information from **somatic receptors**—that is, the receptors in the framework or outer walls of the body, including skin, skeletal muscle, tendons, and joints—go to the **somatosensory cortex,** a strip of cortex that lies in the parietal lobe of the brain just behind the junction of the parietal and frontal lobes (Figure 9–6).

The specific pathways from the eyes go to a different primary cortical receiving area, the **visual cortex,** which is in the occipital lobe, and the specific pathways from the ears go to the **auditory cortex,** which is in the temporal lobe (Figure 9–6). Specific pathways from the taste buds pass to a cortical area adjacent to the face region of the somatosensory cortex. As we have indicated, the pathways serving olfaction have no representation in the cerebral cortex.

Finally, the processing of afferent information does not end in the primary cortical receiving areas but continues from these areas to association areas of the cerebral cortex.

In contrast to the specific ascending pathways, neurons in the **nonspecific ascending pathways** are activated by sensory units of several different types (Figure 9–7) and therefore signal general information. In other words, they indicate that *something* is happening, without specifying just what or where. A given second-order neuron in a nonspecific pathway may respond, for example, to input from several afferent neurons, each activated by a different stimulus, such as maintained skin pressure, heating, and cooling. Such pathway neurons are called **polymodal neurons.** The nonspecific pathways, as well as collaterals from the specific pathways, end in the brainstem reticular formation and regions of the thalamus and cerebral cortex that are not highly discriminative, but are important in the control of alertness and arousal.

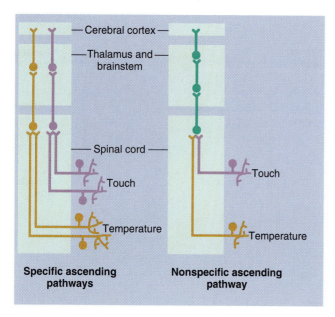

FIGURE 9–7
Diagrammatic representation of two specific sensory pathways and a nonspecific sensory pathway.

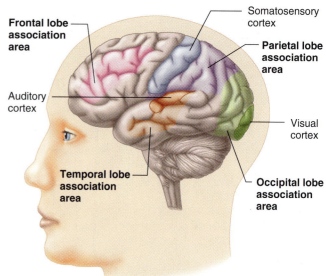

FIGURE 9–8
Areas of association cortex.

Association Cortex and Perceptual Processing

The **cortical association areas** (Figure 9–8) are brain areas that lie outside the primary cortical sensory or motor areas but are adjacent to them. The association areas are not considered part of the sensory pathways but rather play a role in the progressively more complex analysis of incoming information.

Although neurons in the earlier stages of the sensory pathways are associated with perception, information from the primary sensory cortical areas is elaborated after it is relayed to a cortical association area. The region of association cortex closest to the primary sensory cortical area processes the information in fairly simple ways and serves basic sensory-related functions. Regions farther from the primary sensory areas process the information in more complicated ways, including, for example, greater input from areas of the brain serving arousal, attention, memory, and language. Some of the neurons in these latter regions also receive input concerning two or more other types of sensory stimuli. Thus, an association-area neuron receiving input from both the visual cortex and the "neck" region of the somatosensory cortex might be concerned with integrating visual information with sensory information about head position so that, for example, a tree is understood to be vertical even though the viewer's head is tipped sideways.

Fibers from neurons of the parietal and temporal lobes go to association areas in the frontal lobes that are part of the limbic system. Through these connections, sensory information can be invested with emotional and motivational significance.

Further perceptual processing involves not only arousal, attention, learning, memory, language, and emotions, but also comparing the information presented via one type of sensation with that of another. For example, we may hear a growling dog, but our perception of the event and our emotional response vary markedly, depending upon whether our visual system detects the sound source to be an angry animal or a loudspeaker.

Factors That Affect Perception

We put great trust in our sensory-perceptual processes despite the inevitable modifications we know to exist. Some of the following factors are known to affect our perceptions of the real world:

1. Afferent information is influenced by sensory receptor mechanisms (for example by adaptation), and by processing of the information along afferent pathways.
2. Factors such as emotions, personality, experience, and social background can influence perceptions so that two people can witness the same events and yet perceive them differently.
3. Not all information entering the central nervous system gives rise to conscious sensation. Actually, this is a very good thing because many unwanted signals are generated by the extreme

sensitivity of our sensory receptors. For example, under ideal conditions the rods of the eye can detect the flame of a candle 17 mi away. The hair cells of the ear can detect vibrations of an amplitude much lower than those caused by blood flow through the ears' blood vessels and can even detect molecules in random motion bumping against the ear drum. It is possible to detect one action potential generated by a certain type of mechanoreceptor. Although these receptors are capable of giving rise to sensations, much of their information is canceled out by receptor or central mechanisms, which will be discussed later. In other receptors' afferent pathways, information is not canceled out—it simply does not feed into parts of the brain that give rise to a conscious sensation. For example, stretch receptors in the walls of some of the largest blood vessels monitor blood pressure as part of reflex regulation of this pressure, but people have no conscious awareness of their blood pressure.

4. We lack suitable receptors for many energy forms. For example, we cannot directly detect ionizing radiation and radio or television waves.

5. Damaged neural networks may give faulty perceptions as in the bizarre phenomenon known as **phantom limb**, in which a limb that has been lost by accident or amputation is experienced as though it were still in place. The missing limb is perceived to be the "site" of tingling, touch, pressure, warmth, itch, wetness, pain, and even fatigue, and it is felt as though it were still a part of "self." It seems that the sensory neural networks in the central nervous system that exist genetically in everyone and are normally triggered by receptor activation are, instead, in the case of phantom limb, activated independently of peripheral input. The activated neural networks continue to generate the usual sensations, which are perceived as arising from the missing receptors. Moreover, somatosensory cortex undergoes marked reorganization after the loss of input from a part of the body so that a person whose arm has been amputated may perceive a touch on the cheek as though it were a touch on the phantom arm; because of the reorganization, the arm area of somatosensory cortex receives input normally directed to the face somatosensory area.

6. Some drugs alter perceptions. In fact, the most dramatic examples of a clear difference between the real world and our perceptual world can be found in illusions and drug- and disease-induced hallucinations, where whole worlds can be created.

In summary, for perception to occur, the three processes involved—transducing stimulus energy into action potentials by the receptor, transmitting data through the nervous system, and interpreting data—cannot be separated. Sensory information is processed at each synapse along the afferent pathways and at many levels of the central nervous system, with the more complex stages receiving input only after it has been processed by the more elementary systems. This hierarchical processing of afferent information along individual pathways is an important organizational principle of sensory systems. As we shall see, a second important principle is that information is processed by *parallel* pathways, each of which handles a limited aspect of the neural signals generated by the sensory transducers. A third principle is that information at each stage along the pathway is modified by "top-down" influences serving emotions, attention, memory, and language. Every synapse along the afferent pathway adds an element of organization and contributes to the sensory experience so that what we perceive is not a simple—or even an absolutely accurate—image of the stimulus that originally activated our receptors.

We turn now to how the particular characteristics of a stimulus are coded by the various receptors and sensory pathways.

Primary Sensory Coding

The sensory systems code four aspects of a stimulus: stimulus type, intensity, location, and duration.

Stimulus Type

Another term for stimulus type (heat, cold, sound, or pressure, for example) is stimulus **modality**. Modalities can be divided into submodalities: Cold and warm are submodalities of temperature, whereas salt, sweet, bitter, and sour are submodalities of taste. The type of sensory receptor activated by a stimulus plays the primary role in coding the stimulus modality.

As mentioned earlier, a given receptor type is particularly sensitive to one stimulus modality—the adequate stimulus—because of the signal transduction mechanisms and ion channels incorporated in the receptor's plasma membrane. For example, receptors for vision contain pigment molecules whose shape is transformed by light; these receptors also have intracellular mechanisms by which changes in the pigment molecules alter the activity of membrane ion channels and generate a neural signal. Receptors in the skin have neither light-sensitive molecules nor plasma-membrane ion channels that can be affected by them; thus, receptors in the eyes respond to light and those in the skin do not.

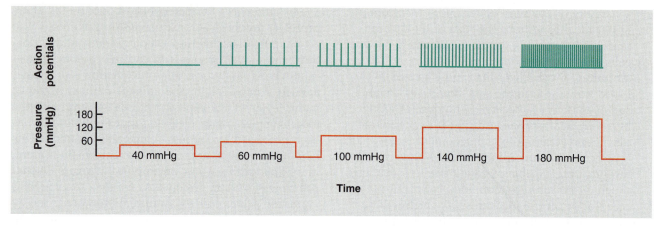

FIGURE 9–9

Action potentials from an afferent fiber leading from the pressure receptors of a single sensory unit as the receptors are subjected to pressures of different magnitudes.

All the receptors of a single afferent neuron are preferentially sensitive to the same type of stimulus. For example, they are all sensitive to cold or all to pressure. Adjacent sensory units, however, may be sensitive to different types of stimuli. Since the receptive fields for different modalities overlap, a single stimulus, such as an ice cube on the skin, can give rise simultaneously to the sensations of touch and temperature.

Stimulus Intensity

How is a strong stimulus distinguished from a weak one when the information about both stimuli is relayed by action potentials that are all the same size? The *frequency* of action potentials in a single receptor is one way, since as described earlier, increased stimulus strength means a larger receptor potential and a higher frequency of action-potential firing.

In addition to an increased firing rate from individual receptors, receptors on other branches of the same afferent neuron also begin to respond. The action potentials generated by these receptors propagate along the branches to the main afferent nerve fiber and add to the train of action potentials there. Figure 9–9 is a record of an experiment in which increased stimulus intensity to the receptors of a sensory unit is reflected in increased action-potential frequency in its afferent nerve fiber.

In addition to increasing the firing frequency in a single afferent neuron, stronger stimuli usually affect a larger area and activate similar receptors on the endings of *other* afferent neurons. For example, when one touches a surface lightly with a finger, the area of skin in contact with the surface is small, and only receptors in that skin area are stimulated. Pressing down firmly increases the area of skin stimulated. This "calling in" of receptors on additional afferent neurons is known as **recruitment.**

Stimulus Location

A third type of information to be signaled is the location of the stimulus—in other words, where the stimulus is being applied. (It should be noted that in vision, hearing, and smell, stimulus location is interpreted as arising from the site from which the stimulus *originated* rather than the place on our body where the stimulus was actually *applied.* For example, we interpret the sight and sound of a barking dog as occurring in that furry thing on the other side of the fence rather than in a specific region of our eyes and ears. More will be said of this later; we deal here with the senses in which the stimulus is located to a site on the body.)

The main factor coding stimulus location is the site of the stimulated receptor. The precision, or **acuity,** with which one stimulus can be located and differentiated from an adjacent one depends upon the amount of convergence of neuronal input in the specific ascending pathways. The greater the convergence, the less the acuity. Other factors affecting acuity are the size of the receptive field covered by a single sensory unit and the amount of overlap of nearby receptive fields. For example, it is easy to discriminate between two adjacent stimuli (two-point discrimination) applied to the skin on a finger, where the sensory units are small and the overlap considerable. It is harder to do so on the back, where the sensory units are large and widely spaced. Locating sensations from internal organs is less precise than from the skin because there are fewer afferent neurons in the internal organs and each has a larger receptive field.

It is fairly simple to see why a stimulus to a neuron that has a small receptive field can be located more precisely than a stimulus to a neuron with a large receptive field (Figure 9–10). The fact is, however, that even in the former case one cannot distinguish exactly where within the receptive field of a single neuron a

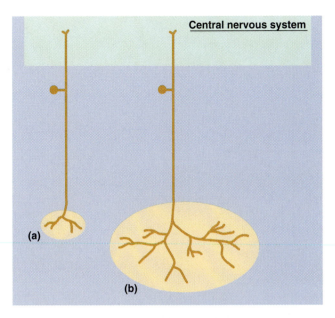

FIGURE 9–10

The information from neuron a indicates the stimulus location more precisely than does that from neuron b because a's receptive field is smaller.

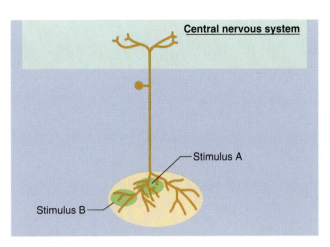

FIGURE 9–11

Two stimulus points, A and B, in the receptive field of a single afferent neuron. The density of nerve endings around area A is greater than around B, and the frequency of action potentials in response to a stimulus in area A will be greater than the response to a similar stimulus in B.

stimulus has been applied; one can only tell that the afferent neuron has been activated. In this case, receptive-field overlap aids stimulus localization even though, intuitively, overlap would seem to "muddy" the image. Let us examine in the next two paragraphs how this works.

An afferent neuron responds most vigorously to stimuli applied at the center of its receptive field because the receptor density—that is, the number of receptors in a given area—is greatest there. The response decreases as the stimulus is moved toward the receptive-field periphery. Thus, a stimulus activates more receptors and generates more action potentials if it occurs at the center of the receptive field (point A in Figure 9–11). The firing frequency of the afferent neuron is also related to stimulus strength, however, and a high frequency of impulses in the single afferent nerve fiber of Figure 9–11 could mean either that a moderately intense stimulus was applied to the center at A or that a strong stimulus was applied to the periphery at B. Thus, neither the intensity nor the location of the stimulus can be detected precisely with a single afferent neuron.

Since the receptor endings of different afferent neurons overlap, however, a stimulus will trigger activity in more than one sensory unit. In Figure 9–12, neurons A and C, stimulated near the edge of their receptive fields where the receptor density is low, fire at a lower frequency than neuron B, stimulated at the center of its receptive field. In the group of sensory

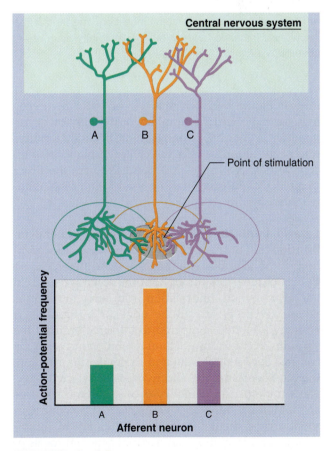

FIGURE 9–12

A stimulus point falls within the overlapping receptive fields of three afferent neurons. Note the difference in receptor response (that is, the action-potential frequency in the three neurons) due to the difference in receptor distribution under the stimulus (low receptor density in A and C, high in B).

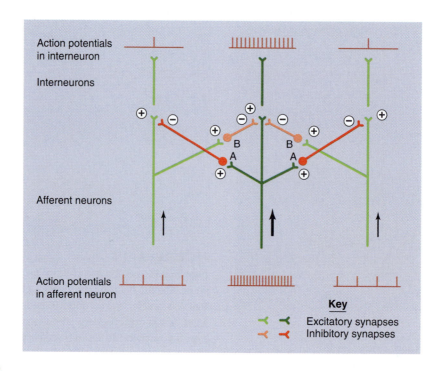

FIGURE 9–13

Afferent pathways showing lateral inhibition. The central fiber at the beginning of the pathway (bottom of figure) is firing at the highest frequency and inhibits, via inhibitory neurons A, the lateral neurons more strongly than the lateral pathways inhibit it, via inhibitory neurons B.

units in Figure 9–12, a high action-potential frequency in neuron B occurring simultaneously with lower frequencies in A and C permits a more accurate localization of the stimulus near the center of neuron B's receptive field. Once this location is known, the firing frequency of neuron B can be used to indicate stimulus intensity.

Lateral Inhibition The phenomenon of **lateral inhibition** is, however, far more important in localization of the stimulus site than are the different sensitivities of receptors throughout the receptive field. In lateral inhibition, information from afferent neurons whose receptors are at the edge of a stimulus is strongly inhibited compared to information from the stimulus's center. Thus, lateral inhibition increases the contrast between relevant and irrelevant information, thereby increasing the effectiveness of selected pathways and focusing sensory-processing mechanisms on "important" messages. Figure 9–13 shows one neuronal arrangement that accomplishes lateral inhibition. Lateral inhibition can occur at different levels of the sensory pathways but typically happens at an early stage.

Lateral inhibition can be demonstrated by pressing the tip of a pencil against your finger. With your eyes closed, you can localize the pencil point precisely, even though the region around the pencil tip is also

indented and mechanoreceptors within this region are activated (Figure 9–14). Exact localization occurs because the information from the peripheral regions is removed by lateral inhibition.

Lateral inhibition is utilized to the greatest degree in the pathways providing the most accurate localization. For example, movement of skin hairs, which we can locate quite well, activates pathways that have significant lateral inhibition, but temperature and pain, which we can locate only poorly, activate pathways that use lateral inhibition to a lesser degree.

Stimulus Duration

Receptors differ in the way they respond to a constantly maintained stimulus—that is, in the way they undergo adaptation.

The response—the action-potential frequency—at the beginning of the stimulus indicates the stimulus strength, but after this initial response, the frequency differs widely in different types of receptors. Some receptors respond very rapidly at the stimulus onset, but, after their initial burst of activity, fire only very slowly or stop firing all together during the remainder of the stimulus. These are the **rapidly adapting receptors;** they are important in signaling rapid change (for example, vibrating or moving stimuli). Some receptors adapt so rapidly that they fire only a single action potential at the onset of a stimulus—an on response—

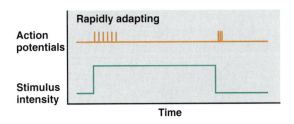

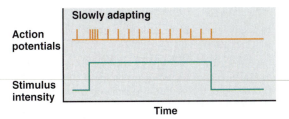

FIGURE 9–15

Rapidly and slowly adapting receptors. The top line in each graph indicates the action-potential firing of the afferent nerve fiber from the receptor, and the bottom line, application of the stimulus.

FIGURE 9–14

(a) A pencil tip pressed against the skin depresses surrounding tissue. Receptors are activated under the pencil tip and in the adjacent tissue. (b) Because of lateral inhibition, the central area of excitation is surrounded by an area where the afferent information is inhibited. (c) The sensation is localized to a more restricted region than that in which mechanoreceptors are actually stimulated.

while others respond at the beginning of the stimulus and again at its removal—so-called on-off responses. The rapid fading of the sensation of clothes pressing on one's skin is due to rapidly adapting receptors.

Slowly adapting receptors maintain their response at or near the initial level of firing regardless of the stimulus duration (Figure 9–15). These receptors signal slow changes or prolonged events, such as occur in the joint and muscle receptors that participate in the maintenance of upright posture when standing or sitting for long periods of time.

Central Control of Afferent Information

All sensory signals are subject to extensive control at the various synapses along the ascending pathways before they reach higher levels of the central nervous system. Much of the incoming information is reduced or even abolished by inhibition from collaterals from other neurons in ascending pathways (lateral inhibition, discussed earlier) or by pathways descending

from higher centers in the brain. The reticular formation and cerebral cortex, in particular, control the input of afferent information via descending pathways. The inhibitory controls may be exerted directly by synapses on the axon terminals of the primary afferent neurons (an example of presynaptic inhibition) or indirectly via interneurons that affect other neurons in the sensory pathways (Figure 9–16).

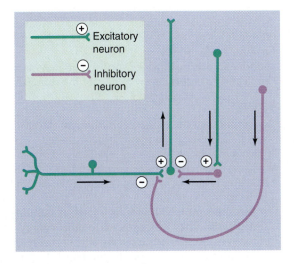

FIGURE 9–16

Descending pathways may control sensory information by directly inhibiting the central terminals of the afferent neuron (an example of presynaptic inhibition) or via an interneuron that affects the ascending pathway by inhibitory synapses. Arrows indicate the direction of action-potential transmission.

TABLE 9–1 Principles of Sensory System Organization

1. Specific sensory receptor types are sensitive to certain modalities and submodalities.

2. A specific sensory pathway codes for a particular modality or submodality.

3. The ascending pathways are crossed so that sensory information is generally processed by the side of the brain opposite the side of the body that was stimulated.

4. In addition to other synaptic relay points, all ascending pathways, except for those involved in smell, synapse in the thalamus on their way to the cortex.

5. Information is organized such that initial cortical processing of the various modalities occurs in different parts of the brain.

6. Ascending pathways are subject to descending controls.

In some cases (for example, in the pain pathways), the afferent input is continuously inhibited to some degree. This provides the flexibility of either removing the inhibition (disinhibition) so as to allow a greater degree of signal transmission or of increasing the inhibition so as to block the signal more completely.

We conclude our general introduction to sensory system pathways and coding with a summary of the general principles of the organization of the sensory systems (Table 9–1). We now present the individual systems.

SECTION A SUMMARY

I. Sensory processing begins with the transformation of stimulus energy into graded potentials and then into action potentials in nerve fibers.

II. Information carried in a sensory system may or may not lead to a conscious awareness of the stimulus.

Receptors

I. Receptors translate information from the external world and internal environment into graded potentials, which then generate action potentials.
   a. Receptors may be either specialized endings of afferent neurons or separate cells at the end of the neurons.
   b. Receptors respond best to one form of stimulus energy, but they may respond to other energy forms if the stimulus intensity is abnormally high.
   c. Regardless of how a specific receptor is stimulated, activation of that receptor always leads to perception of one sensation. Not all receptor activations lead, however, to conscious sensations.

II. The transduction process in all sensory receptors involves—either directly or indirectly—the opening or closing of ion channels in the receptor. Ions then flow across the membrane, causing a receptor potential.
   a. Receptor-potential magnitude and action-potential frequency increase as stimulus strength increases.
   b. Receptor-potential magnitude varies with stimulus strength, rate of change of stimulus application, temporal summation of successive receptor potentials, and adaptation.

Neural Pathways in Sensory Systems

I. A single afferent neuron with all its receptor endings is a sensory unit.
   a. Afferent neurons, which usually have more than one receptor of the same type, are the first neurons in sensory pathways.
   b. The area of the body that, when stimulated, causes activity in a sensory unit or other neuron in the ascending pathway of that unit is called the receptive field for that neuron.

II. Neurons in the specific ascending pathways convey information to specific primary receiving areas of the cerebral cortex about only a single type of stimulus.

III. Nonspecific ascending pathways convey information from more than one type of sensory unit to the brainstem reticular formation and regions of the thalamus that are not part of the specific ascending pathways.

Association Cortex and Perceptual Processing

I. Information from the primary sensory cortical areas is elaborated after it is relayed to a cortical association area.
   a. The primary sensory cortical area and the region of association cortex closest to it process the information in fairly simple ways and serve basic sensory-related functions.
   b. Regions of association cortex farther from the primary sensory areas process the sensory information in more complicated ways.
   c. Processing in the association cortex includes input from areas of the brain serving other sensory modalities, arousal, attention, memory, language, and emotions.

Primary Sensory Coding

I. The type of stimulus perceived is determined in part by the type of receptor activated. All receptors of a given sensory unit respond to the same stimulus modality.

II. Stimulus intensity is coded by the rate of firing of individual sensory units and by the number of sensory units activated.

III. Perception of the stimulus location depends on the size of the receptive field covered by a single sensory unit and on the overlap of nearby receptive fields. Lateral inhibition is a means by which ascending pathways emphasize wanted information and increase sensory acuity.

IV. Stimulus duration is coded by slowly adapting receptors.

V. Information coming into the nervous system is subject to control by both ascending and descending pathways.

SECTION **A** REVIEW QUESTIONS

1. Distinguish between a sensation and a perception.
2. Describe the general process of transduction in a receptor that is a cell separate from the afferent neuron. Include in your description the following terms: specificity, stimulus, receptor potential, neurotransmitter, graded potential, and action potential.
3. List several ways in which the magnitude of a receptor potential can be varied.
4. Describe the relationship between sensory information processing in the primary cortical sensory areas and in the cortical association areas.
5. List several ways in which sensory information can be distorted.
6. How does the nervous system distinguish between stimuli of different types?
7. How is information about stimulus intensity coded by the nervous system?
8. Make a diagram showing how a specific ascending pathway relays information from peripheral receptors to the cerebral cortex.

SECTION B

SPECIFIC SENSORY SYSTEMS

Somatic Sensation

Sensation from the skin, muscles, bones, tendons, and joints is termed **somatic sensation** and is initiated by a variety of somatic receptors (Figure 9–17). Some respond to mechanical stimulation of the skin, hairs, and underlying tissues, whereas others respond to temperature or chemical changes. Activation of somatic receptors gives rise to the sensations of touch, pressure, warmth, cold, pain, and awareness of the position of the body parts and their movement. The receptors for visceral sensations, which arise in certain organs of the thoracic and abdominal cavities, are the same types as the receptors that give rise to somatic sensations. Some organs, such as the liver, have no sensory receptors at all.

Each sensation is associated with a specific receptor type. In other words, there are distinct receptors for heat, cold, touch, pressure, limb position or movement, and pain. After entering the central nervous system, the afferent nerve fibers from the somatic receptors synapse on neurons that form the specific ascending pathways going primarily to the somatosensory cortex via the brainstem and thalamus. They also synapse on interneurons that give rise to the nonspecific pathways. For reference, the location of some important ascending pathways is shown in a cross section of the spinal cord (Figure 9–18a), and two are diagrammed as examples in Figure 9–18b and c.

Note that the pathways cross from the side where the afferent neurons enter the central nervous system to the opposite side either in the spinal cord (Figure 9–18b) or brainstem (Figure 9–18c). Thus, the sensory pathways from somatic receptors on the left side of the body go to the somatosensory cortex of the right cerebral hemisphere, and vice versa.

In the somatosensory cortex, the endings of the axons of the specific somatic pathways are grouped according to the location of the receptors giving rise to the pathways (Figure 9–19). The parts of the body that are most densely innervated—fingers, thumb, and lips—are represented by the largest areas of the somatosensory cortex. There are qualifications, however, to this seemingly precise picture: The sizes of the areas can be modified with changing sensory experience, and there is considerable overlap of the body-part representations.

Touch-Pressure

Stimulation of the variety of mechanoreceptors in the skin (see Figure 9–17) leads to a wide range of touch-pressure experiences—hair bending, deep pressure, vibrations, and superficial touch, for example. These mechanoreceptors are highly specialized nerve endings encapsulated in elaborate cellular structures. The details of the mechanoreceptors vary, but generally the nerve endings are linked to collagen-fiber networks within the capsule. These networks transmit the

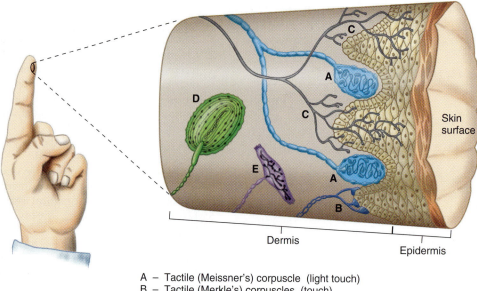

A – Tactile (Meissner's) corpuscle (light touch)
B – Tactile (Merkle's) corpuscles (touch)
C – Free terminal (pain)
D – Lamellated (Pacinian) corpuscle (deep pressure)
E – Ruffini corpuscle (warmth)

FIGURE 9–17

Skin receptors. Some nerve fibers have free endings not related to any apparent receptor structure. Thicker, myelinated axons, on the other hand, end in receptors that have a complex structure. (Not drawn to scale; for example, Pacinian corpuscles are actually four to five times larger than Meissner's corpuscles.)

mechanical tension in the capsule to ion channels in the nerve endings and activate them.

The skin mechanoreceptors adapt at different rates, about half adapting rapidly (that is, they fire only when the stimulus is changing), and the others adapting slowly. Activation of rapidly adapting receptors gives rise to the sensations of touch, movement, and vibration, whereas slowly adapting receptors give rise to the sensation of pressure.

In both categories, some receptors have small, well-defined receptive fields and are able to provide precise information about the contours of objects indenting the skin. As might be expected, these receptors are concentrated at the fingertips. In contrast, other receptors have large receptive fields with obscure boundaries, sometimes covering a whole finger or a large part of the palm. These receptors are not involved in detailed spatial discrimination but signal information about vibration, skin stretch, and joint movement.

Sense of Posture and Movement

The senses of posture and movement are complex. The major receptors responsible for these senses are the muscle-spindle stretch receptors, which occur in skeletal muscles and respond both to the absolute magnitude of muscle stretch and to the rate at which the stretch occurs (to be described in Chapter 12). The senses of posture and movement are also supported

by vision and the vestibular organs (the "sense organs of balance," described later). Mechanoreceptors in the joints, tendons, ligaments, and skin also play a role. The term **kinesthesia** refers to the sense of movement at a joint.

Temperature

There are two types of thermoreceptors in the skin, each of which responds to a limited range of temperature. Warmth receptors respond to temperatures between 30 and 43°C with an increased discharge rate upon warming, whereas receptors for cold are stimulated by small decreases in temperature. It is not known how heat or cold alter the endings of the thermosensitive afferent neurons to generate receptor potentials.

Pain

A stimulus that causes (or is on the verge of causing) tissue damage usually elicits a sensation of pain. Receptors for such stimuli are known as **nociceptors.** They respond to intense mechanical deformation, excessive heat, and many chemicals, including neuropeptide transmitters, bradykinin, histamine, cytokines, and prostaglandins, several of which are released by damaged cells. These substances act by combining with specific ligand-sensitive ion channels on the nociceptor plasma membrane.

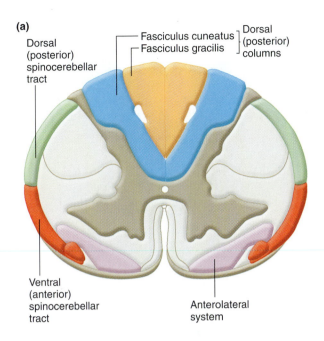

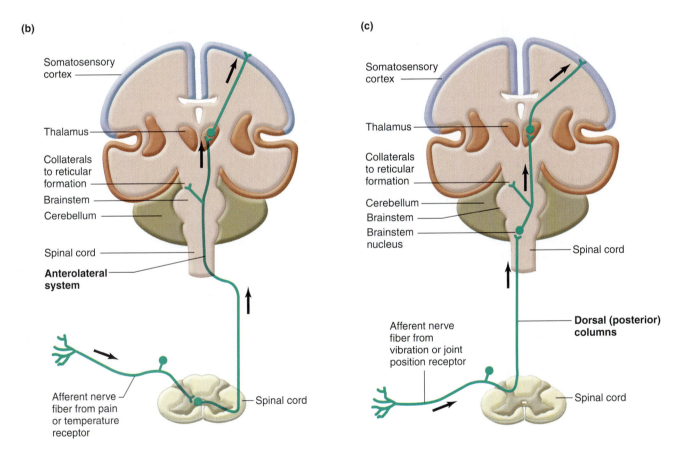

FIGURE 9–18

(a) A reference cross section of the spinal cord showing the relative locations of the major ascending fiber tracts. (b) The anterolateral system. (c) The dorsal columns. Information carried over collaterals to the reticular formation in (b) and (c) contribute to alertness and arousal mechanisms.

Parts b and c adapted from Gardner.

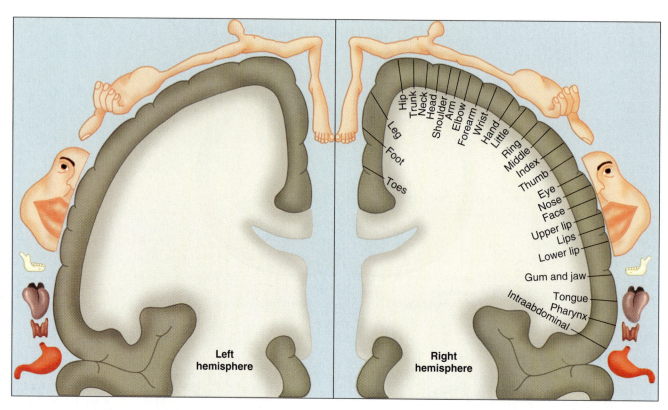

FIGURE 9–19

Location of pathway terminations for different parts of the body in somatosensory cortex, although there is actually much overlap between the cortical regions. The left half of the body is represented on the right hemisphere of the brain, and the right half of the body is represented on the left hemisphere.

Several of these chemicals are secreted by cells of the immune system (described in Chapter 20) that have moved into the injured area. In fact, there is a great deal of interaction between substances released from the damaged tissue, cells of the immune system, and nearby afferent pain neurons. All three of these—the tissue, immune cells, and afferent neurons themselves—release substances that affect the nociceptors and are, in turn, affected by these substances.

Pain differs significantly from the other somatosensory modalities. After transduction of the first noxious stimuli into action potentials in the afferent neuron, a series of changes occur in components of the pain pathway—including the ion channels in the nociceptors themselves—that alter the way these components respond to subsequent stimuli, a process referred to as sensitization. When these changes result in an increased sensitivity to painful stimuli, it is known as *hyperalgesia* and can last for hours after the original stimulus is over. Thus, the pain experienced in response to stimuli occurring even a short time after the original stimulus (and the reactions to that pain) can be very different from the pain experienced initially. Moreover, probably more than any other type of

sensation, pain can be altered by past experiences, suggestion, emotions (particularly anxiety), and the simultaneous activation of other sensory modalities. Thus, the level of pain experienced is not solely a physical property of the stimulus.

The primary afferents having nociceptor endings synapse on interneurons after entering the central nervous system (glutamate and the neuropeptide, substance P, are among the neurotransmitters released at these synapses). Some of these interneurons form the ascending anterolateral system, the pathway on one side of the spinal cord receiving information from receptors on the opposite side of the body (see Figure 9–18b). These pathways transmit information that leads to both the localization of pain and its sensory and emotional components.

The activation of interneurons by incoming nociceptive afferents may lead to the phenomenon of *referred pain,* in which the sensation of pain is experienced at a site other than the injured or diseased part. For example, during a heart attack, pain is often experienced in the left arm. Referred pain occurs because both visceral and somatic afferents often converge on the same interneurons in the pain pathway. Excitation of the somatic afferent fibers is the more usual source of afferent discharge, so we "refer" the location of receptor activation to the somatic source even though, in the case of visceral pain, the perception is incorrect.

Analgesia is the selective suppression of pain without effects on consciousness or other sensations. Electrical stimulation of specific areas of the central nervous system can produce a profound reduction in pain, a phenomenon called ***stimulation-produced analgesia,*** by inhibiting pain pathways. This occurs because descending pathways that originate in these brain areas selectively inhibit the transmission of information originating in nociceptors. The descending axons end at lower brainstem and spinal levels on interneurons in the pain pathways as well as on the synaptic terminals of the afferent nociceptor neurons themselves. Some of the neurons in these inhibitory pathways release or are sensitive to certain endogenous opioids (Chapter 8). Thus, infusion of morphine, which binds to and stimulates opioid receptors, into the spinal cord at the level of entry of the active nociceptor fibers can provide relief in many cases of intractable pain. This is separate from morphine's effect on the brain.

Transcutaneous electric nerve stimulation (TENS), in which the painful site itself or the nerves leading from it are stimulated by electrodes placed on the surface of the skin, can be useful in lessening pain. TENS works because the stimulation of nonpain, low-threshold afferent fibers (for example, the fibers from touch receptors) leads to inhibition of neurons in the pain pathways. We often apply our own type of TENS therapy when we rub or press hard on a painful area.

Under certain circumstances, the ancient Chinese therapy, ***acupuncture,*** prevents or alleviates pain. During acupuncture analgesia, needles are introduced into specific parts of the body to stimulate afferent fibers, which causes analgesia. Endogenous opioid neurotransmitters are involved in acupuncture analgesia.

Stimulation-produced analgesia, TENS, and acupuncture work by exploiting the body's built-in mechanisms that control pain.

Vision

The eyes are composed of an optical portion, which focuses the visual image on the receptor cells, and a neural component, which transforms the visual image into a pattern of neural discharges.

Light

The receptors of the eye are sensitive only to that tiny portion of the vast spectrum of electromagnetic radiation that we call visible light (Figure 9–20). Radiant energy is described in terms of wavelengths and

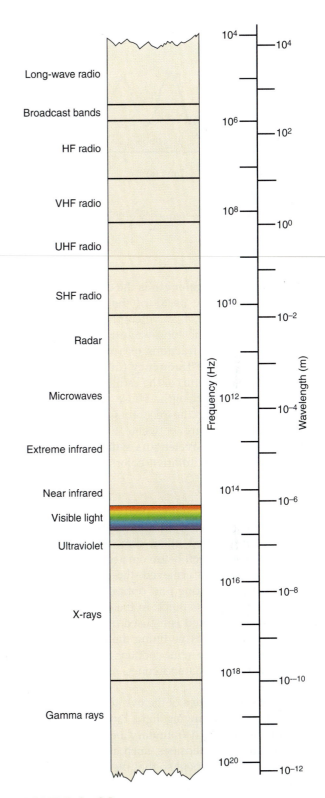

FIGURE 9–20
Electromagnetic spectrum.

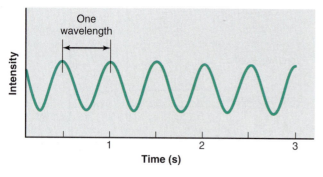

FIGURE 9–21

Properties of a wave. The frequency of this wave is 2 Hz (cycles/s).

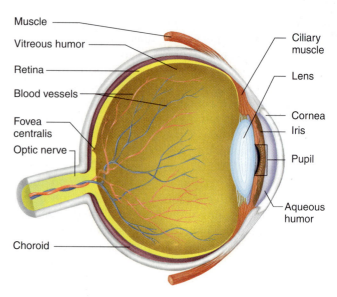

FIGURE 9–22

The human eye. The blood vessels depicted run along the back of the eye between the retina and vitreous humor, not through the vitreous humor.

frequencies. The **wavelength** is the distance between two successive wave peaks of the electromagnetic radiation (Figure 9–21). Wavelengths vary from several kilometers at the long-wave radio end of the spectrum to minute fractions of a millimeter at the gamma-ray end. The **frequency** (in hertz, the number of cycles per second) of the radiation wave varies inversely with wavelength. Those wavelengths capable of stimulating the receptors of the eye—the **visible spectrum**—are between 400 and 700 nm. Light of different wavelengths within this band is perceived as having different colors.

The Optics of Vision

The light wave can be represented by a line drawn in the direction in which the wave is traveling. Light waves are propagated in all directions from every point of a visible object. Before an accurate image of a point on the object is achieved, these divergent light waves must pass through an optical system that focuses them back into a point. In the eye, the image of the object being viewed is focused upon the **retina**, a thin layer of neural tissue lining the back of the eyeball (Figure 9–22). The retina contains the light-sensitive receptor cells, the rods and cones, as well as several types of neurons.

The **lens** and **cornea** of the eye are the optical systems that focus impinging light rays into an image upon the retina. At a boundary between two substances of different densities, such as the cornea and the air, light rays are bent so that they travel in a new direction. The cornea plays a larger quantitative role than the lens in focusing light rays because the rays are bent more in passing from air into the cornea than they are when passing into and out of the lens or any other transparent structure of the eye.

The surface of the cornea is curved so that light rays coming from a single point source hit the cornea at different angles and are bent different amounts,

directing the light rays back to a point after emerging from the lens. The image is focused on a specialized area known as the **fovea centralis** (Figure 9–22), the area of the retina that gives rise to the greatest visual clarity. The image on the retina is upside down relative to the original light source (Figure 9–23), and it is also reversed right to left.

Light rays from objects close to the eye strike the cornea at greater angles and must be bent more in order to reconverge on the retina. Although, as noted above, the cornea performs the greater part quantitatively of focusing the visual image on the retina, all *adjustments* for distance are made by changes in lens shape. Such changes are part of the process known as **accommodation.**

The shape of the lens is controlled by the **ciliary muscle** and the tension it applies to the **zonular fibers,** which attach this smooth muscle to the lens (Figure 9–24). To focus on distant objects, the zonular fibers pull the lens into a flattened, oval shape. When their pull is removed for near vision, the natural elasticity of the lens causes it to become more spherical. This more spherical shape provides additional bending of the light rays, which is important to focus near objects on the retina. The ciliary muscle, which is stimulated by parasympathetic nerves, is circular, like a sphincter, so that it draws nearer to the lens as it contracts and therefore removes tension on the zonular fibers, resulting in accommodation for viewing near objects (Figure 9–25). Accommodation also includes other mechanisms that move the lens slightly toward the back of the eye, turn the eyes inward toward the nose

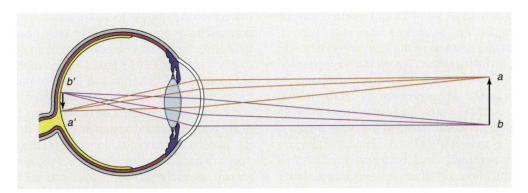

FIGURE 9–23

Refraction (bending) of light by the lens system of the eye. For simplicity, we show light refraction only at the surface of the cornea where the greatest refraction occurs. Refraction also occurs in the lens and at other sites in the eye. ✗

(convergence), and constrict the pupil. The sequence of events for accommodation is reversed when distant objects are viewed.

The cells that make up most of the lens lose their internal membranous organelles early in life and are thus transparent, but they lack the ability to replicate. The only lens cells that retain the capacity to divide are on the surface of the lens, and as new cells are formed, older cells come to lie deeper within the lens. With increasing age, the central part of the lens becomes denser and stiffer and acquires a coloration that progresses from yellow to black.

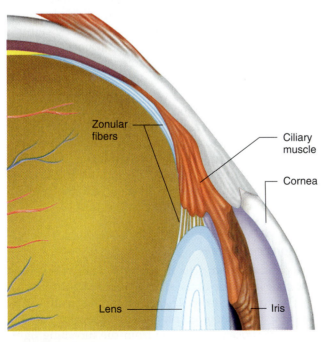

FIGURE 9–24

Ciliary muscle, zonular fibers, and lens of the eye. ✗

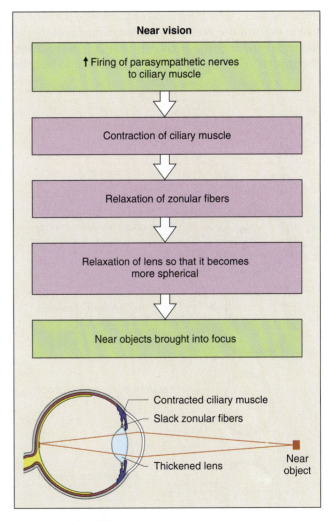

FIGURE 9–25

Accommodation of the lens for near vision. Although refraction is shown only at the surface of the cornea, the change in refraction during accommodation is a function of the lens, not the cornea. ✗

Since the lens must be elastic to assume a more spherical shape during accommodation for near vision, the increasing stiffness of the lens that occurs with aging makes accommodation for near vision increasingly difficult. This condition, known as *presbyopia,* is a normal part of the aging process and is the reason that people around 45 years of age may have to begin wearing reading glasses or bifocals for close work.

The changes in lens color that occur with aging are responsible for *cataract,* which is an opacity of the lens and one of the most common eye disorders. Early changes in lens color do not interfere with vision, but vision is impaired as the process slowly continues. The opaque lens can be removed surgically. With the aid of an implanted artificial lens or compensating eyeglasses, effective vision can be restored, although the ability to accommodate is lost.

Cornea and lens shape and eyeball length determine the point where light rays reconverge. Defects in vision occur if the eyeball is too long in relation to the focusing power of the lens (Figure 9–26). In this case, the images of near objects fall on the retina, but the images of far objects focus at a point in front of the retina. This is a *nearsighted,* or *myopic,* eye, which is unable to see distant objects clearly. In contrast, if the eye is too short for the lens, images of distant objects are focused on the retina but those of near objects are focused behind it. This eye is *farsighted,* or *hyperopic,* and near vision is poor. The use of corrective lenses for near- and farsighted vision is shown in Figure 9–26.

Defects in vision also occur where the lens or cornea does not have a smoothly spherical surface, a condition known as *astigmatism.* These surface imperfections can usually be compensated for by eyeglasses.

The lens separates two fluid-filled chambers in the eye, the anterior chamber, which contains aqueous humor, and the posterior chamber, which contains the more viscous vitreous humor (see Figure 9–22). These two fluids are colorless and permit the transmission of light from the front of the eye to the retina. The aqueous humor is formed by special vascular tissue that overlies the ciliary muscle. In some instances, the aqueous humor is formed faster than it is removed, which results in increased pressure within the eye. **Glaucoma,** the leading cause of irreversible blindness, is a disease in which the axons of the optic nerve die, but it is often associated with increased pressure within the eye.

The amount of light entering the eye is controlled by muscles in the ringlike, pigmented tissue known as the **iris** (see Figure 9–22), the color being of no importance as long as the tissue is sufficiently opaque to prevent the passage of light. The hole in the center of the iris through which light enters the eye is the **pupil.** The iris is composed of smooth muscle, which is innervated by autonomic nerves. Stimulation of sympathetic nerves to the iris enlarges the pupil by causing the radially arranged muscle fibers to contract. Stimulation of parasympathetic fibers to the iris makes the pupil smaller by causing the sphincter muscle fibers, which circle around the pupil, to contract.

These neurally induced changes occur in response to light-sensitive reflexes. Bright light causes a decrease in the diameter of the pupil, which reduces the amount of light entering the eye and restricts the light to the central part of the lens for more accurate vision. Conversely, the iris enlarges in dim light, when maximal illumination is needed. Changes also occur as a result of emotion or pain.

Photoreceptor Cells

The photoreceptor cells in the retina are called **rods** and **cones** because of the shapes of their light-sensitive tips. Note in Figure 9–27 that the light-sensitive portion of the photoreceptor cells—the tips of the rods and cones—faces *away* from the incoming light, and the light must pass through all the cell layers of the retina before reaching the photoreceptors and

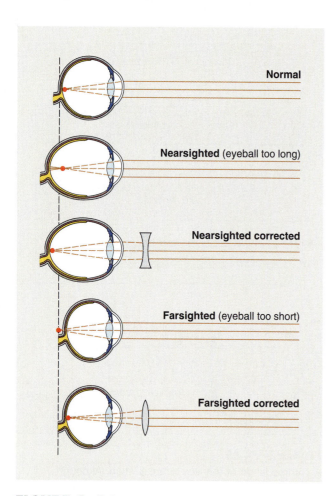

FIGURE 9–26

Nearsightedness, farsightedness, and their correction.

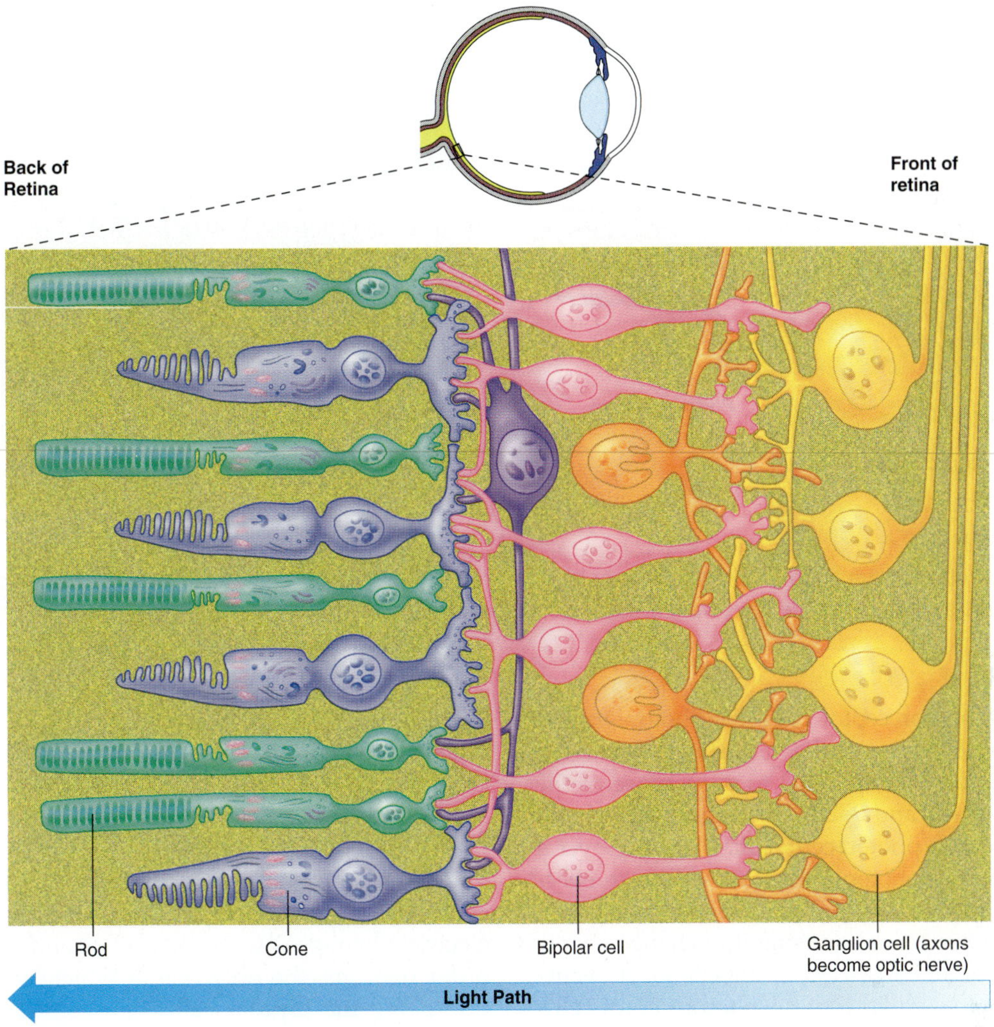

FIGURE 9–27

Organization of the retina. Light enters through the cornea, passes through the aqueous humor, pupil, vitreous humor, and the front surface of the retina before reaching the photoreceptors. The membranes that contain the photoreceptors form discrete discs in the rods but are continuous with the plasma membrane in the cones, which accounts for the comblike appearance of these latter cells. Two other neuron types, depicted here in purple and orange, provide lateral inhibition between neurons of the retina.

Redrawn from Dowling and Boycott.

stimulating them. A pigmented layer (the choroid), which lies behind the retina (see Figure 9–22), absorbs light and prevents its reflection back to the rods and cones, which would cause the visual image to be blurred. The rods are extremely sensitive and respond to very low levels of illumination, whereas the cones are considerably less sensitive and respond only when the light is brighter than, for example, twilight.

The photoreceptors contain molecules called **photopigments,** which absorb light. There are four different photopigments in the retina, one (**rhodopsin**) in the rods and one in each of the three cone types. Each photopigment contains an opsin and a chromophore.

Opsin is a collective term for a group of integral membrane proteins, one of which surrounds and binds a **chromophore** molecule (Figure 9–28). The chromophore, which is the actual light-sensitive part of the photopigment, is the same in each of the four photopigments and is **retinal,** a derivative of vitamin A. The opsin differs in each of the four photopigments. Since each type of opsin binds to the chromophore in a different way and filters light differently, each of the four photopigments absorbs light most effectively at a different part of the visible spectrum. For example, one photopigment absorbs wavelengths in the range of red light best, whereas another absorbs green light best.

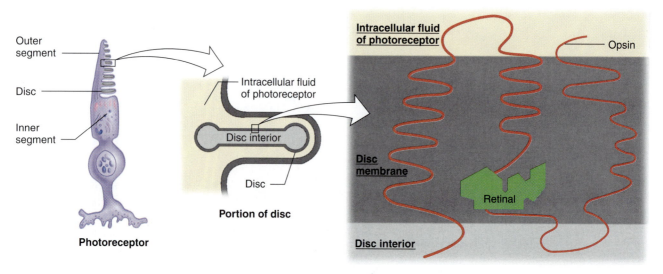

FIGURE 9–28

The arrangement of the opsin and retinal (the chromophore) in the membrane of the photoreceptor discs of a cone. The opsin actually crosses the membrane seven times, not three as shown here.

Within the photoreceptor cells, the photopigments lie in specialized membranes that are arranged in highly ordered stacks, or discs, parallel to the surface of the retina (Figures 9–27 and 9–28). The repeated layers of membranes in each photoreceptor may contain over a billion molecules of photopigment, providing an effective trap for light.

Light activates retinal, causing it to change shape. This change triggers a cascade of biochemical events that lead to *hyperpolarization* of the photoreceptor cell's plasma membrane and, thereby, *decreased* release of neurotransmitter (glutamate) from the cell. Note that in the case of photoreceptors the response of the cell to a stimulus (light) is a hyperpolarizing receptor potential and a decrease in neurotransmitter release. The decrease in neurotransmitter then causes the bipolar cells, which synapse with the photoreceptor cell, to undergo a hyperpolarization in membrane potential.

After its activation by light, retinal changes back to its resting shape by several mechanisms that do not depend on light but are enzyme mediated. Thus, in the dark, retinal has its resting shape, the photoreceptor cell is partially depolarized, and *more* neurotransmitter is being released.

When one steps back from a place of bright sunlight into a darkened room, dark adaptation, a temporary "blindness," takes place. In the low levels of illumination of the darkened room, vision can only be supplied by the rods, which have greater sensitivity than the cones. During the exposure to bright light, however, the rods' rhodopsin has been completely activated. It cannot respond fully again until it is restored to its resting state, a process requiring some tens of minutes. Dark adaptation occurs, in part, as enzymes regenerate the initial form of rhodopsin, which can respond to light.

Neural Pathways of Vision

The neural pathways of vision begin with the rods and cones. These photoreceptors communicate by way of electrical synapses with each other and with second-order neurons, the only one of which we shall mention being the **bipolar cell** (Figures 9–27 and 9–29). The bipolar cells synapse (still within the retina) both upon neurons that pass information horizontally from one part of the retina to another and upon the **ganglion cells.** Via these latter synapses, the ganglion cells are caused to respond differentially to the various characteristics of visual images, such as color, intensity, form, and movement. Thus, a great deal of information processing takes place at this early stage of the sensory pathway.

The distinct characteristics of the visual image are transmitted through the visual system along multiple, parallel pathways by two types of ganglion cells, each type concerned with different aspects of the visual stimulus. Parallel processing of information continues all the way to and within the cerebral cortex, to the highest stages of visual neural networks. (Note that in this discussion of the visual pathway, the terms "respond" and "response" denote not a *direct* response to a light stimulus—only the rods and cones show such responses—but rather to the synaptic input reaching the relevant pathway neuron as a consequence of the original stimulus to the rods and cones.)

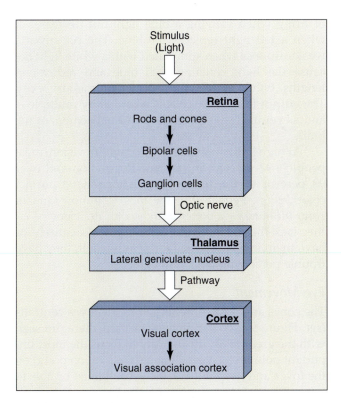

FIGURE 9–29
Diagrammatic representation of the visual pathways.

Ganglion cells are the first cells in the visual system to respond to activation by producing action potentials, whereas the rods and cones and almost all other retinal neurons produce only graded potentials. The axons of the ganglion cells form the output from the retina—the optic nerve, cranial nerve II. The two optic nerves meet at the base of the brain to form the optic chiasm, where some of the fibers cross to the opposite side of the brain, providing both cerebral hemispheres with input from each eye.

Optic nerve fibers project to several structures in the brain, the largest number passing to the thalamus (specifically to the lateral geniculate nucleus, Figure 9–29), where the information from the different ganglion cell types is kept distinct. In addition to the input from the retina, many neurons of the lateral geniculate nucleus also receive input from the brainstem reticular formation and input relayed *back* from the visual cortex. These nonretinal inputs can control the transmission of information from the retina to the visual cortex and may be involved in the ability to shift attention between vision and the other sensory modalities.

The lateral geniculate nucleus sends action potentials to the visual cortex, the primary visual area of the cerebral cortex (see Figure 9–6). Different aspects of visual information are carried in parallel pathways and

are processed simultaneously in a number of independent ways in different parts of the cerebral cortex before they are reintegrated to produce the conscious sensation of sight and the perceptions associated with it. The cells of the visual pathways are organized to handle information about line, contrast, movement, and color. They do not, however, form a picture in the brain. Rather, they form a spatial and temporal pattern of electrical activity.

We mentioned that a substantial number of fibers of the visual pathway project to regions of the brain other than the visual cortex. For example, visual information is transmitted to the **suprachiasmatic nucleus,** which lies just above the optic chiasm and functions as a "biological clock," as described in Chapter 7. Information about diurnal cycles of light intensity is used to entrain this neuronal clock. Other visual information is passed to the brainstem and cerebellum, where it is used in the coordination of eye and head movements, fixation of gaze, and change in pupil size.

Color Vision

The colors we perceive are related to the wavelengths of light that are reflected, absorbed, or transmitted by the pigments in the objects of our visual world. For example, an object appears red because shorter wavelengths, which would be perceived as blue, are absorbed by the object, while the longer wavelengths, perceived as red, are reflected from the object to excite the photopigment of the retina most sensitive to red. Light perceived as white is a mixture of all wavelengths, and black is the absence of all light.

Color vision begins with activation of the photopigments in the *cone* receptor cells. Human retinas have three kinds of cones, which contain red-, green-, or blue-sensitive photopigments. As their names imply, these pigments absorb and hence respond optimally to light of different wavelengths. Because the red pigment is actually more sensitive to the wavelengths that correspond to yellow, this pigment is sometimes called the yellow photopigment.

Although each type of cone is excited most effectively by light of one particular wavelength, it responds to other wavelengths as well. Thus, for any given wavelength, the three cone types are excited to different degrees (Figure 9–30). For example, in response to light of 531-nm wavelengths, the green cones respond maximally, the red cones less, and the blue cones not at all. Our sensation of color depends upon the relative outputs of these three types of cone cells and their comparison by higher-order cells in the visual system.

The pathways for color vision follow those described in Figure 9–29. Ganglion cells of one type respond to a broad band of wavelengths. In other words,

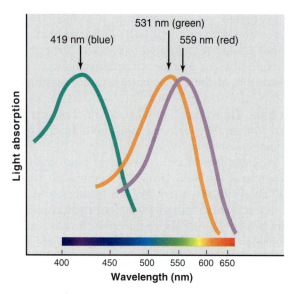

FIGURE 9–30

The sensitivities of the photopigments in the three types of cones in the normal human retina. Action-potential frequency in the optic nerve is directly related to absorption of light by a photopigment.

they receive input from all three types of cones, and they signal not specific color but general brightness. Ganglion cells of a second type code specific colors. These latter cells are also called **opponent color cells** because they have an excitatory input from one type of cone receptor and an inhibitory input from another. For example, the cell in Figure 9–31 increases its rate

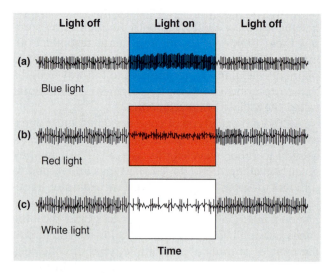

FIGURE 9–31

Response of a single opponent color ganglion cell to blue, red, and white lights.

Redrawn from Hubel and Wiesel.

of firing when viewing a blue light but decreases it when a red light replaces the blue. The cell gives a weak response when stimulated with a white light because the light contains both blue and red wavelengths. Other more complicated patterns also exist. The output from these cells is recorded by multiple—and as yet unclear—strategies in visual centers of the brain.

At high light intensities, as in daylight vision, most people—92 percent of the male population and over 99 percent of the female population—have normal color vision. People with the most common kind of *color blindness*—a better term is color deficiency—either lack the red or green cone pigments entirely or have them in an abnormal form; as a result, they have trouble perceiving red versus green.

Eye Movement

The cones are most concentrated in the fovea centralis (see Figure 9–22), and images focused there are seen with the greatest acuity. In order to focus the most important point in the visual image (the fixation point) on the fovea and keep it there, the eyeball must be able to move. Six skeletal muscles attached to the outside of each eyeball (Figure 9–32) control its movement. These muscles perform two basic movements, fast and slow.

The fast movements, called **saccades,** are small, jerking movements that rapidly bring the eye from one fixation point to another to allow search of the visual field. In addition, saccades move the visual image over the receptors, thereby preventing adaptation. Saccades also occur during certain periods of sleep when the eyes are closed, and may be associated with "watching" the visual imagery of dreams.

Slow eye movements are involved both in tracking visual objects as they move through the visual field and during compensation for movements of the head. The control centers for these compensating movements obtain their information about head movement from the vestibular system, which will be described shortly. Control systems for the other slow movements of the eyes require the continuous feedback of visual information about the moving object.

Hearing

The sense of hearing is based on the physics of sound and the physiology of the external, middle, and inner ear, the nerves to the brain, and the brain parts involved in processing acoustic information.

Sound

Sound energy is transmitted through a gaseous, liquid, or solid medium by setting up a vibration of the medium's molecules, air being the most common

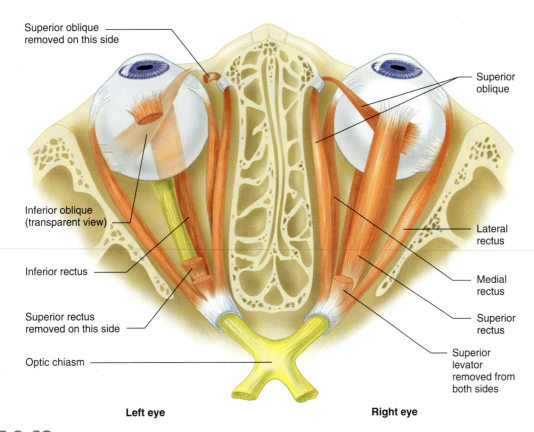

FIGURE 9–32

A superior view of the muscles that move the eyes to direct the gaze and give convergence.

medium. When there are no molecules, as in a vacuum, there can be no sound. Anything capable of creating a disturbance of molecules—for example, vibrating objects—can serve as a sound source. Figure 9–33 demonstrates the basic principles using a tuning fork. The disturbance of air molecules that makes up the sound wave consists of zones of compression, in which the molecules are close together and the pressure is increased, alternating with zones of rarefaction, where the molecules are farther apart and the pressure is lower (Figure 9–33a through d).

A sound wave measured over time (Figure 9–33e) consists of rapidly alternating pressures that vary continuously from a high during compression of molecules, to a low during rarefaction, and back again. The difference between the pressure of molecules in zones of compression and rarefaction determines the wave's amplitude, which is related to the loudness of the sound; the greater the amplitude, the louder the sound. The frequency of vibration of the sound source (that is, the number of zones of compression or rarefaction in a given time) determines the pitch we hear; the faster the vibration, the higher the pitch. The sounds heard most keenly by human ears are those from sources vibrating at frequencies between 1000 and 4000 Hz (hertz, or cycles per second), but the entire range of frequencies audible to human beings extends from 20 to 20,000 Hz. Sound waves with sequences of pitches are generally perceived as musical, the complexity of the individual waves giving the sound its characteristic quality, or timbre.

We can distinguish about 400,000 different sounds. For example, we can distinguish the note A played on a piano from the same note on a violin. We can also selectively *not* hear sounds, tuning out the babble of a party to concentrate on a single voice.

Sound Transmission in the Ear

The first step in hearing is the entrance of sound waves into the **external auditory canal** (Figure 9–34). The shapes of the outer ear (the pinna, or auricle) and the external auditory canal help to amplify and direct the sound. The sound waves reverberate from the sides and end of the external auditory canal, filling it with the continuous vibrations of pressure waves.

The **tympanic membrane** (eardrum) is stretched across the end of the external auditory canal, and air molecules push against the membrane, causing it to vibrate at the same frequency as the sound wave. Under higher pressure during a zone of compression, the tympanic membrane bows inward. The distance the membrane moves, although always very small, is a

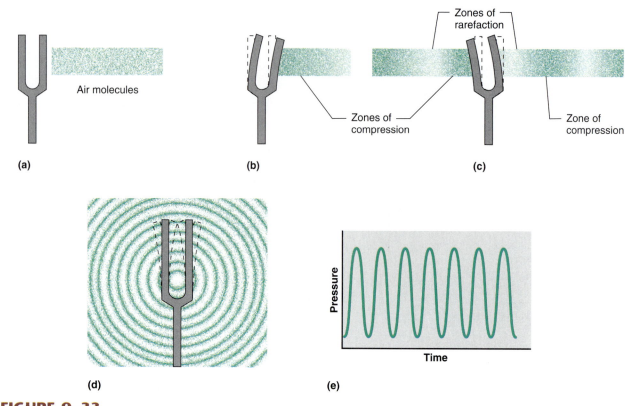

FIGURE 9–33

Formation of sound waves from a vibrating tuning fork.

function of the force with which the air molecules hit it and is related to the sound pressure and therefore its loudness. During the subsequent zone of rarefaction, the membrane returns to its original position. The exquisitely sensitive tympanic membrane responds to all the varying pressures of the sound waves, vibrating slowly in response to low-frequency sounds and rapidly in response to high-frequency ones.

The tympanic membrane separates the external auditory canal from the **middle ear cavity,** an air-filled cavity in the temporal bone of the skull. The pressures in the external auditory canal and middle ear cavity are normally equal to atmospheric pressure. The middle ear cavity is exposed to atmospheric pressure through the **auditory** (eustachian) **tube,** which connects the middle ear to the pharynx. The slitlike ending of this tube in the pharynx is normally closed, but muscle movements open the tube during yawning, swallowing, or sneezing, and the pressure in the middle ear equilibrates with atmospheric pressure. A difference in pressure can be produced with sudden changes in altitude (as in an ascending or descending elevator or airplane), when the pressure outside the ear and in the ear canal changes while the pressure in the middle ear remains constant because the auditory tube is closed. This pressure difference can stretch the tympanic membrane and cause pain.

The second step in hearing is the transmission of sound energy from the tympanic membrane through the middle-ear cavity to the **inner ear.** The inner ear, called the **cochlea,** is a *fluid-filled,* spiral-shaped passage in the temporal bone. The temporal bone also houses other passages, including the semicircular canals, which contain the sensory organs for balance and movement. These passages are connected to the cochlea but will be discussed later.

Because liquid is more difficult to move than air, the sound pressure transmitted to the inner ear must be amplified. This is achieved by a movable chain of three small bones, the **malleus, incus,** and **stapes** (Figure 9–35); these bones act as a piston and couple the motions of the tympanic membrane to the **oval window,** a membrane covered opening separating the middle and inner ear (Figure 9–36).

The *total* force of a sound wave applied to the tympanic membrane is transferred to the oval window, but because the oval window is much smaller than the tympanic membrane, the *force per unit area* (that is, the pressure) is increased 15 to 20 times. Additional advantage is gained through the lever action of the middle-ear bones. The amount of energy transmitted to the inner ear can be lessened by the contraction of two small skeletal muscles in the middle ear that alter the tension of the tympanic membrane and the

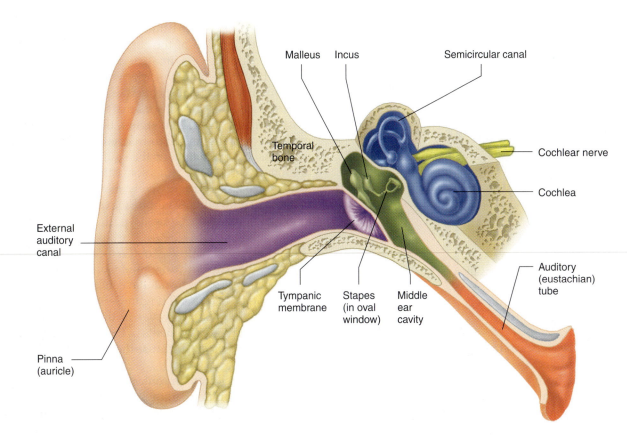

FIGURE 9–34

The human ear. In this and the following drawing, violet indicates the outer ear, green the middle ear, and blue the inner ear. The malleus, incus, and stapes are bones even though they are colored green in this figure to indicate that they are components of the middle ear compartment. Actually, the auditory tube is closed except during movements of the pharynx, such as swallowing or yawning.

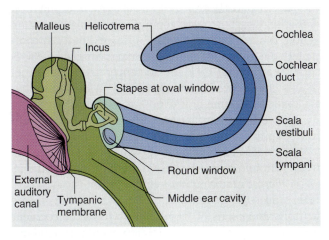

FIGURE 9–35

Relationship between the middle ear bones and the cochlea. Movement of the stapes against the membrane covering the oval window sets up pressure waves in the fluid-filled scala vestibuli. These waves cause vibration of the cochlear duct and the basilar membrane. Some of the pressure is transmitted around the helicotrema directly into the scala tympani.

Redrawn from Kandel and Schwartz.

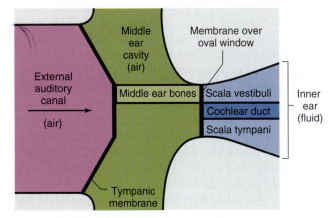

FIGURE 9–36

Diagrammatic representation showing that the middle ear bones act as a piston against the fluid of the inner ear.

Redrawn from von Bekesy.

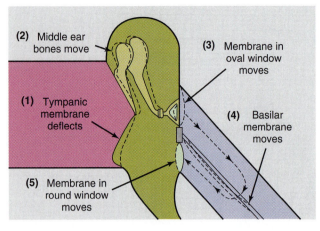

(2) Middle ear bones move

(3) Membrane in oval window moves

(1) Tympanic membrane deflects

(4) Basilar membrane moves

(5) Membrane in round window moves

FIGURE 9–37

Transmission of sound vibrations through the middle and inner ear.

Redrawn from Davis and Silverman.

position of the stapes in the oval window. These muscles help to protect the delicate receptor apparatus of the inner ear from continuous intense sound stimuli and improve hearing over certain frequency ranges.

The entire system described thus far has been concerned with the transmission of sound energy into the cochlea, where the receptor cells are located. The cochlea is almost completely divided lengthwise by a fluid-filled membranous tube, the **cochlear duct,** which follows the cochlear spiral (see Figure 9–35). On either side of the cochlear duct are fluid-filled compartments: the **scala vestibuli,** which is on the side of the cochlear duct that ends at the oval window; and the **scala tympani,** which is below the cochlear duct and ends in a second membrane-covered opening to the middle ear, the round window. The scala vestibuli and scala tympani meet at the end of the cochlear duct at the helicotrema (see Figure 9–35).

Sound waves in the ear canal cause in-and-out movement of the tympanic membrane, which moves the chain of middle-ear bones against the membrane covering the oval window, causing it to bow into the scala vestibuli and back out (Figure 9–37), creating waves of pressure there. The wall of the scala vestibuli is largely bone, and there are only two paths by which the pressure waves can be dissipated. One path is to the helicotrema, where the waves pass around the end of the cochlear duct into the scala tympani and back to the round-window membrane, which is then bowed out into the middle ear cavity. However, most of the pressure is transmitted from the scala vestibuli across the cochlear duct.

One side of the cochlear duct is formed by the **basilar membrane** (Figure 9–38), upon which sits the **organ of Corti,** which contains the ear's sensitive

receptor cells. Pressure differences across the cochlear duct cause vibration of the basilar membrane.

The region of maximal displacement of the vibrating basilar membrane varies with the frequency of the sound source. The properties of the membrane nearest the middle ear are such that this region vibrates most easily—that is, undergoes the greatest movement, in response to high-frequency (high-pitched) tones. As the frequency of the sound is lowered, vibration waves travel out along the membrane for greater distances. Progressively more distant regions of the basilar membrane vibrate maximally in response to progressively lower tones.

Hair Cells of the Organ of Corti

The receptor cells of the organ of Corti, the **hair cells,** are mechanoreceptors that have hairlike **stereocilia** protruding from one end (Figure 9–38c). The hair cells transform the pressure waves in the cochlea into receptor potentials. Movements of the basilar membrane stimulate the hair cells because they are attached to the membrane.

The stereocilia are in contact with the overhanging **tectorial membrane** (Figure 9–38c), which projects inward from the side of the cochlea. As the basilar membrane is displaced by pressure waves, the hair cells move in relation to the tectorial membrane, and, consequently, the stereocilia are bent. Whenever the stereocilia bend, ion channels in the plasma membrane of the hair cell open, and the resulting ion movements depolarize the membrane and create a receptor potential.

Efferent nerve fibers from the brainstem regulate the activity of certain of the hair cells and dampen their response, which protects them. Despite this protective action, the hair cells are easily damaged or even completely destroyed by exposure to high-intensity noises such as amplified rock music concerts, engines of jet planes, and revved-up motorcycles. Lesser noise levels also cause damage if exposure is chronic.

Hair cell depolarization leads to release of the neurotransmitter glutamate (the same neurotransmitter released by photoreceptor cells), which binds to and activates protein binding sites on the terminals of the 10 or so *afferent* neurons that synapse upon the hair cell. This causes the generation of action potentials in the neurons, the axons of which join to form the cochlear nerve (a component of cranial nerve VIII). The greater the energy (loudness) of the sound wave, the greater the frequency of action potentials generated in the afferent nerve fibers. Because of its position on the basilar membrane, each hair cell and, therefore, the nerve fibers that synapse upon it respond to a limited range of sound frequency and intensity, and they respond best to a single frequency.

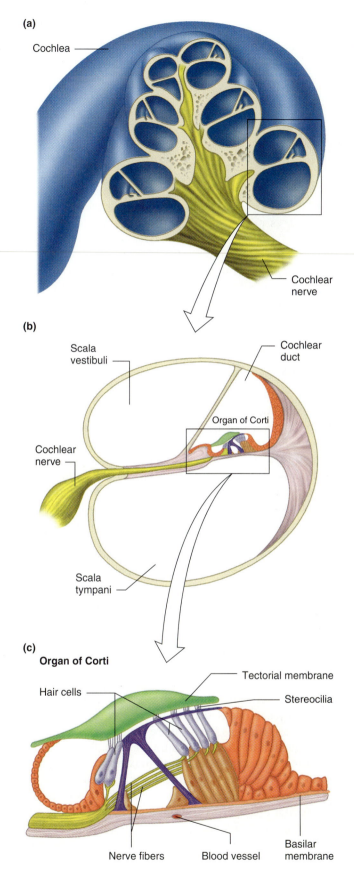

(a)

Cochlea

Cochlear nerve

(b)

Scala vestibuli

Cochlear duct

Organ of Corti

Cochlear nerve

Scala tympani

(c)

Organ of Corti

Tectorial membrane

Hair cells

Stereocilia

Nerve fibers   Blood vessel   Basilar membrane

Neural Pathways in Hearing

Cochlear nerve fibers enter the brainstem and synapse with interneurons there, fibers from both ears often converging on the same neuron. Many of these interneurons are influenced by the different arrival times and intensities of the input from the two ears. The different arrival times of low-frequency sounds and the difference in intensities of high-frequency sounds are used to determine the direction of the sound source. If, for example, a sound is louder in the right ear or arrives sooner at the right ear than at the left, we assume that the sound source is on the right. The shape of the outer ear (the pinna, see Figure 9–34) and movements of the head are also important in localizing the source of a sound.

From the brainstem, the information is transmitted via a multineuron pathway to the thalamus and on to the auditory cortex (see Figure 9–6). The neurons responding to different pitches (frequencies) are arranged along the auditory cortex in an orderly manner in much the same way that signals from different regions of the body are represented at different sites in the somatosensory cortex. Different areas of the auditory system are further specialized, some neurons responding best to complex sounds such as those used in verbal communication, whereas others signal the location, movement, duration, or loudness of a sound.

Electronic devices can help compensate for damage to the intricate middle ear, cochlea, or neural structures. *Hearing aids* amplify incoming sounds, which then pass via the ear canal to the same cochlear mechanisms used by normal sound. When substantial damage has occurred, however, and hearing aids cannot correct the deafness, electronic devices known as *cochlear implants* may restore functional hearing. In response to sound, cochlear implants directly stimulate the cochlear nerve with tiny electric currents so that sound signals are transmitted directly to the auditory pathways, bypassing the cochlea.

FIGURE 9–38

Cross section of the membranes and compartments of the inner ear with detailed view of the hair cells and other structures on the basilar membrane as shown with increasing magnification in views (a), (b), and (c).

Redrawn from Rasmussen.

Vestibular System

Changes in the motion and position of the head are detected by hair cells in the **vestibular apparatus** of the inner ear (Figure 9–39), a series of fluid-filled membranous tubes that connect with each other and with the cochlear duct. The vestibular apparatus consists of three membranous **semicircular ducts** and two saclike swellings, the **utricle** and **saccule,** all of which lie in tunnels (canals) in the temporal bone on each side of the head. The bony canals of the inner ear in which the vestibular apparatus and cochlea are housed have such a complicated shape that they are sometimes called the **labyrinth.** The canals that house the semicircular ducts are the **semicircular canals** (a term often used to denote the semicircular ducts).

The Semicircular Canals

The semicircular ducts detect angular acceleration during *rotation* of the head along three perpendicular axes. The three axes of the semicircular canals and the ducts within them are those activated while nodding the head up and down as in signifying "yes," shaking the head from side to side as in signifying "no," and tipping the head so the ear touches the shoulder (Figure 9–40).

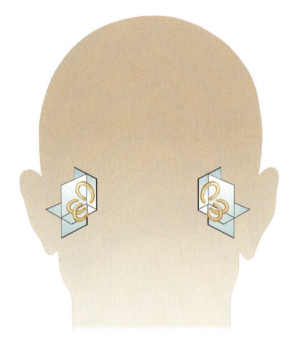

FIGURE 9–40

Relation of the two sets of semicircular canals.

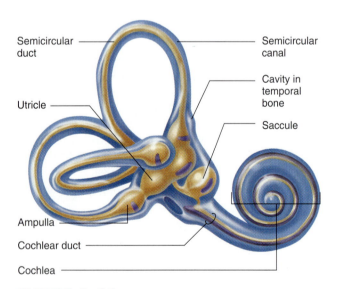

Semicircular duct

Semicircular canal

Cavity in temporal bone

Utricle

Saccule

Ampulla

Cochlear duct

Cochlea

FIGURE 9–39

A tunnel in the temporal bone contains a fluid-filled membranous duct system, part of which can be seen in this cutaway view. The semicircular duct, utricle, and saccule make up the vestibular apparatus. This duct system is connected to the cochlear duct. The purple region on the ducts indicates the locations of the hair (receptor) cells.

Redrawn from Hudspeth.

Receptor cells of the semicircular ducts like those of the organ of Corti, contain hairlike stereocilia. These stereocilia are closely ensheathed by a gelatinous mass, the **cupula,** which extends across the lumen of each semicircular duct at the **ampulla,** a slight bulge in the wall of each duct (Figure 9–41). Whenever the head is moved, the bony semicircular canal, its enclosed duct, and the attached bodies of the hair cells all move with it. The fluid filling the duct, however, is not attached to the skull, and because of inertia, tends to retain its original position (that is, to be "left behind"). Thus, the moving ampulla is pushed against the stationary fluid, which causes bending of the stereocilia and alteration in the rate of release of a chemical transmitter from the hair cells. This transmitter activates the nerve terminals synapsing with the hair cells.

The speed and magnitude of rotational head movements determine the direction in which the stereocilia are bent and the hair cells stimulated. Neurotransmitter is released from the hair cells at rest, and the release changes from this resting rate according to the direction in which the hairs are bent. Each hair cell receptor has one direction of maximum neurotransmitter release, and when its stereocilia are bent in this direction, the receptor cell depolarizes. When the stereocilia are bent in the opposite direction, the cell hyperpolarizes (Figure 9–42). The frequency of action potentials in the afferent nerve fibers that synapse with the hair cells is related to both the amount of force bending the stereocilia on the receptor cells and to the direction in which this force is applied.

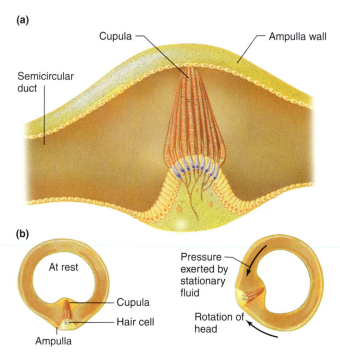

(a) Cupula — Ampulla wall

Semicircular duct

(b)

At rest

Pressure exerted by stationary fluid

Cupula

Hair cell

Rotation of head

Ampulla

FIGURE 9–41

(a) Organization of a cupula and ampulla. (b) Relation of the cupula to the ampulla when the head is at rest and when it is accelerating.

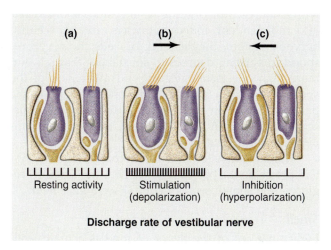

(a) **(b)** **(c)**

Resting activity | Stimulation (depolarization) | Inhibition (hyperpolarization)

Discharge rate of vestibular nerve

FIGURE 9–42

Relationship between position of hairs and action-potential firing in afferent neurons. (a) Resting activity. (b) Movement of hairs in one direction increases the action-potential frequency in the afferent nerve activated by the hair cell. (c) Movement in the opposite direction decreases the rate relative to the resting state.

Redrawn from Wersall, Gleisner, and Lundquist.

At a constant velocity, the duct fluid begins to move at the same rate as the rest of the head, and the stereocilia slowly return to their resting position. For this reason, the hair cells are stimulated only during *changes* in the rate of rotation (that is, during acceleration or deceleration) of the head.

The Utricle and Saccule

The utricle and saccule provide information about *linear*—up and down, back and forth—acceleration and changes in head position relative to the forces of gravity. Here, too, the receptor cells are mechanoreceptors sensitive to the displacement of projecting hairs. The patch of hair cells in the utricle is nearly horizontal in a standing person, and that in the saccule is vertical.

The stereocilia projecting from the hair cells are covered by a gelatinous substance in which tiny stones, or otoliths, are embedded; the otoliths, which are calcium carbonate crystals, make the gelatinous substance heavier than the surrounding fluid. In response to a change in position, the gelatinous otolithic material moves according to the forces of gravity and pulls against the hair cells so that the stereocilia on the hair cells are bent and the receptor cells stimulated.

Vestibular Information and Dysfunction

Information about hair cell stimulation is relayed from the vestibular apparatus to the brainstem via the vestibular branch of cranial nerve VIII (the same cranial nerve that carries acoustic information). It is transmitted via a multineuronal pathway to a system of vestibular centers in the parietal lobe. Vestibular information is integrated with information from the joints, tendons, and skin, leading to the sense of posture and movement.

Vestibular information is used in three ways. One is to control the eye muscles so that, in spite of changes in head position, the eyes can remain fixed on the same point. *Nystagmus* is a large, jerky, back-and-forth movement of the eyes that can occur in response to unusual vestibular input in normal people but can also be a pathological sign.

The second use of vestibular information is in reflex mechanisms for maintaining upright posture. The vestibular apparatus plays a role in the support of the head during movement, orientation of the head in space, and reflexes accompanying locomotion. Very few postural reflexes, however, depend exclusively on input from the vestibular system despite the fact that the vestibular organs are sometimes called the sense organs of balance.

The third use of vestibular information is in providing conscious awareness of the position and acceleration of the body, perception of the space surrounding the body, and memory of spatial information.

Unexpected inputs from the vestibular system and other sensory systems such as those caused by a stroke, irritation of the labyrinths by infection, or even loose particles of calcium carbonate in the semicircular ducts can induce *vertigo,* defined as an illusion of movement—usually spinning—that is often accompanied by feelings of nausea and lightheadedness. This can also occur when there is a mismatch in information from the various sensory systems as, for example, when one is in a high place looking down to the ground: The visual input indicates that you are floating in space while the vestibular system signals that you are not moving at all. *Motion sickness* also involves the vestibular system, occurring when unfamiliar patterns of linear and rotational acceleration are experienced and adaptation to them has not occurred.

Ménière's disease involves the vestibular system and is associated with episodes of abrupt and often severe dizziness, ringing in the ears, and bouts of hearing loss. It is due to an increased fluid pressure in the membranous duct system of the inner ear. The dizziness occurs because the inputs from the two ears are not balanced, either because only one ear is affected or because the two are affected to different degrees.

Chemical Senses

Receptors sensitive to specific chemicals are **chemoreceptors.** Some of these respond to chemical changes in the internal environment, two examples being the oxygen and hydrogen-ion receptors in certain large blood vessels (Chapter 15). Others respond to external chemical changes, and in this category are the receptors for taste and smell, which affect a person's appetite, saliva flow, gastric secretions, and avoidance of harmful substances.

Taste

The specialized sense organs for taste are the 10,000 or so **taste buds** that are found primarily on the tongue. The receptor cells are arranged in the taste buds like the segments of an orange (Figure 9–43). A long narrow process on the upper surface of each receptor cell extends into a small pore at the surface of the taste bud, where the process is bathed by the fluids of the mouth.

Many chemicals can generate the sensation of taste, but taste sensations (modalities) are traditionally divided into four basic groups: sweet, sour, salty, and bitter, each group having a distinct transductional system. For example, salt taste begins with sodium entry into the cell through plasma-membrane ion channels, which depolarizes the plasma membrane; depolarization causes neurotransmitter release from the receptor cell at synapses with afferent nerve fibers. In contrast, molecules such as carbohydrates interact with plasma-

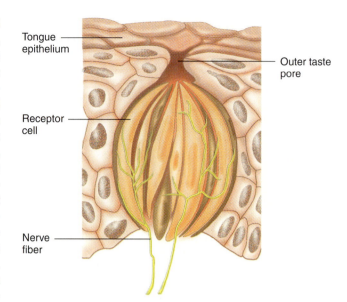

FIGURE 9–43

Structure and innervation of a taste bud. Unlike the taste bud here, each taste bud contains 100 or so receptor cells. The different types of cells are actually receptor cells in different stages of development.

membrane receptors that regulate second-messenger cascades. It is likely that different mechanisms occur in separate cell types.

Although more than one afferent fiber synapses with each receptor cell, the taste system is organized into independent coded pathways into the central nervous system. Single receptor cells, however, respond in varying degrees to substances that fall into more than one taste category and awareness of the specific taste of a substance depends also upon the pattern of firing in a group of neurons. For example, sensations of pain ("hot" spices), texture, and temperature contribute to taste.

The pathways for taste in the central nervous system project to the parietal cortex, near the "mouth" region of the somatosensory cortex (see Figure 9–6).

Smell

Eighty percent of the flavor of food is actually contributed by the sense of smell, or **olfaction,** as is attested by the common experience that food lacks taste when one has a head cold. The odor of a substance is directly related to its chemical structure. Since we are able to recognize and identify hundreds of different odors with a great deal of accuracy, neural circuits that deal with olfaction must encode information about different chemical structures, store (learn) the different code patterns that represent the different structures, and at a later time recognize a particular neural code to identify the odor.

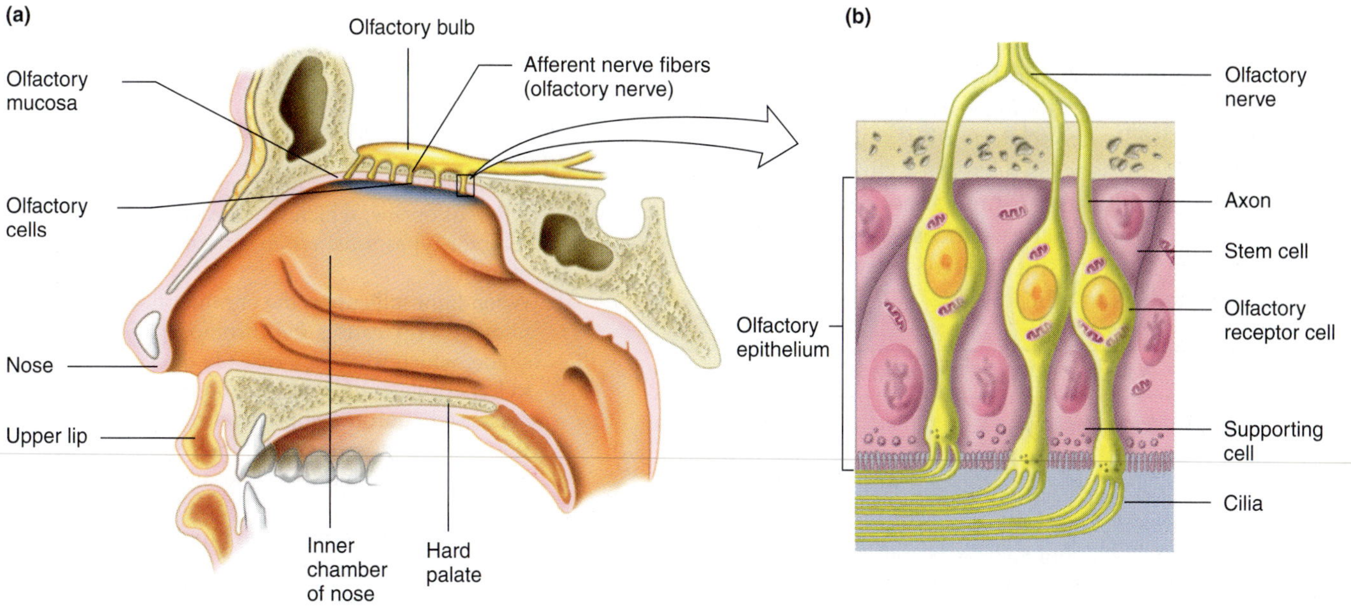

(a)

Olfactory bulb

Olfactory mucosa

Afferent nerve fibers (olfactory nerve)

Olfactory cells

Nose

Upper lip

Inner chamber of nose

Hard palate

(b)

Olfactory nerve

Axon

Stem cell

Olfactory receptor cell

Olfactory epithelium

Supporting cell

Cilia

FIGURE 9–44

(a) Location and (b) enlargement of a portion of the mucosa showing the structure of the olfactory receptor cells. In addition to these cells, the olfactory epithelium contains stem cells, which give rise to new receptors, and supporting cells.

The olfactory receptor cells, the first cells in the pathways that give rise to the sense of smell, lie in a small patch of membrane, the **olfactory epithelium,** in the upper part of the nasal cavity (Figure 9–44a). These cells are specialized afferent neurons that have a single enlarged dendrite that extends to the surface of the epithelium. Several long nonmotile cilia extend out from the tip of the dendrite and lie along the surface of the olfactory epithelium (Figure 9–44b) where they are bathed in mucus. The cilia contain the receptor proteins (binding sites) for olfactory stimuli. The axons of the neurons form the olfactory nerve, which is cranial nerve I.

For an odorous substance (that is, an **odorant**) to be detected, molecules of the substance must first diffuse into the air and pass into the nose to the region of the olfactory epithelium. Once there, they dissolve in the mucus that covers the epithelium and then bind to specific odorant receptors on the cilia. Proteins in the mucus may interact with the odorant molecules, transport them to the receptors, and facilitate their binding to the receptors.

Although there are many thousands of olfactory receptor cells, each contains one, or at most a few, of the 1000 or so different plasma-membrane odorant receptor types, each of which responds only to a specific chemically related group of odorant molecules. Each odorant has characteristic chemical groups that distinguish it from other odorants, and each of these groups activates a different plasma-membrane odorant receptor type. Thus, the identity of a particular odorant is determined by the activation of a precise combination of plasma-membrane receptors, each of which is contained in a distinct group of olfactory receptor cells.

The axons of the olfactory receptor cells synapse in the brain structures known as olfactory bulbs, which lie on the undersurface of the frontal lobes. Axons from olfactory receptor cells sharing a common receptor specificity synapse together on certain olfactory-bulb neurons, thereby maintaining the specificity of the original stimulus. In other words, specific odorant receptor cells activate only certain olfactory-bulb neurons, thereby allowing the brain to determine which receptors have been stimulated. The codes used to transmit olfactory information probably use both spatial (which neurons are firing?) and temporal (what is the timing of the action-potential responses?) components.

Information is passed from the olfactory bulbs to olfactory cortex, which is in the limbic system (see Figure 8–42), a part of the brain intimately associated with emotional, food-getting, and sexual behavior. There, different odors elicit different patterns of electrical activity in a variety of cortical areas, allowing humans to discriminate between some 10,000 different odorants even though they have only 1000 different olfactory receptor types.

Olfactory discrimination varies with attentiveness: hunger—sensitivity is greater in hungry subjects; gender—women in general have keener olfactory sensitivities than men; smoking—decreased sensitivity has been repeatedly associated with smoking; age—the ability to identify odors decreases with age, and a large percentage of elderly persons cannot detect odors at all; and state of the olfactory mucosa—as we have mentioned, the sense of smell decreases when the mucosa is congested, as in a head cold.

SECTION B SUMMARY

Somatic Sensation

I. Sensory function of the skin and underlying tissues is served by a variety of receptors sensitive to one (or a few) stimulus types.

II. Information about somatic sensation enters both specific and nonspecific ascending pathways. The specific pathways cross to the opposite side of the brain.

III. The somatic sensations include touch-pressure, the senses of posture and movement, temperature, and pain.

 a. Rapidly adapting mechanoreceptors of the skin give rise to sensations such as vibration, touch, and movement, whereas slowly adapting ones give rise to the sensation of pressure.

 b. Skin receptors having small receptive fields are involved in fine spatial discrimination, whereas receptors having larger receptive fields signal less spatially precise touch-pressure sensations.

 c. A major receptor type responsible for the senses of posture and kinesthesia is the muscle-spindle stretch receptor.

 d. Cold receptors are sensitive to decreasing temperature; warmth receptors signal information about increasing temperature.

 e. Tissue damage and immune cells release chemical agents that stimulate specific receptors that give rise to the sensation of pain.

 f. Stimulation-produced analgesia, transcutaneous nerve stimulation (TENS), and acupuncture control pain by blocking transmission in the pain pathways.

Vision

I. Light is defined by its wavelength or frequency.

II. The light that falls on the retina is focused by the cornea and lens.

 a. Lens shape is changed to permit viewing near or distant objects (accommodation) so that they are focused on the retina.

 b. Stiffening of the lens with aging interferes with accommodation. Cataracts decrease the amount of light transmitted through the lens.

 c. An eyeball too long or too short relative to the focusing power of the lens causes nearsighted or farsighted vision, respectively.

III. The photopigments of the rods and cones are made up of a protein component (opsin) and a chromophore (retinal).

 a. The rods and each of the three cone types have different opsins, which make each of the four receptor types sensitive to different wavelengths of light.

 b. When light falls upon the chromophore, the photic energy causes the chromophore to change shape, which triggers a cascade of events leading to hyperpolarization of the photoreceptors and decreased neurotransmitter release from them. When exposed to darkness, the rods and cones are depolarized and therefore release more neurotransmitter.

IV. The rods and cones synapse on bipolar cells, which synapse on ganglion cells.

 a. Ganglion-cell axons form the optic nerves, which lead into the brain.

 b. The optic-nerve fibers from half of each retina cross to the opposite side of the brain in the optic chiasm. The fibers from the optic nerves terminate in the lateral geniculate nuclei of the thalamus, which send fibers to the visual cortex.

 c. Visual information is also relayed to areas of the brain dealing with biological rhythms.

V. Coding in the visual system occurs along parallel pathways, in which different aspects of visual information, such as color, form, movement, and depth, are kept separate from each other.

VI. The colors we perceive are related to the wavelength of light. Different wavelengths excite one of the three cone photopigments most strongly.

 a. Certain ganglion cells are excited by input from one type of cone cell and inhibited by input from a different cone type.

 b. Our sensation of color depends on the output of the various opponent-color cells and the processing of this output by brain areas involved in color vision.

VII. Six skeletal muscles control eye movement to scan the visual field for objects of interest, keep the fixation point focused on the fovea centralis despite movements of the object or the head, prevent adaptation of the photoreceptors, and move the eyes during accommodation.

Hearing

I. Sound energy is transmitted by movements of pressure waves.

 a. Sound wave frequency determines pitch.

 b. Sound wave amplitude determines loudness.

II. The sequence of sound transmission is as follows:

 a. Sound waves enter the external auditory canal and press against the tympanic membrane, causing it to vibrate.

b. The vibrating membrane causes movement of the three small middle-ear bones; the stapes vibrates against the oval-window membrane.

c. Movements of the oval-window membrane set up pressure waves in the fluid-filled scala vestibuli, which cause vibrations in the cochlear duct wall, setting up pressure waves in the fluid there.

d. These pressure waves cause vibrations in the basilar membrane, which is located on one side of the cochlear duct.

e. As this membrane vibrates, the hair cells of the organ of Corti move in relation to the tectorial membrane.

f. Movement of the hair cells' stereocilia stimulates the hair cells to release neurotransmitter, which activates receptors on the peripheral ends of the afferent nerve fibers.

III. Each part of the basilar membrane vibrates maximally in response to one particular sound frequency.

Vestibular System

I. A vestibular apparatus lies in the temporal bone on each side of the head and consists of three semicircular ducts, a utricle, and a saccule.

II. The semicircular ducts detect angular acceleration during rotation of the head, which causes bending of the stereocilia on their hair cells.

III. Otoliths in the gelatinous substance of the utricle and saccule move in response to changes in linear acceleration and the position of the head relative to gravity, and stimulate the stereocilia on the hair cells.

Chemical Senses

I. The receptors for taste lie in taste buds throughout the mouth, principally on the tongue. Different types of taste receptors operate by different mechanisms.

II. Olfactory receptors, which are part of the afferent olfactory neurons, lie in the upper nasal cavity.

a. Odorant molecules, once dissolved in the mucus that bathes the olfactory receptors, bind to specific receptors (protein binding sites). Each olfactory receptor cell has one of the 1000 different receptor types.

b. Olfactory pathways go to the limbic system.

---

suprachiasmatic nucleus	organ of Corti
opponent color cell	hair cell
saccade	stereocilia
external auditory canal	tectorial membrane
tympanic membrane	vestibular apparatus
middle ear cavity	semicircular duct
auditory tube	utricle
inner ear	saccule
cochlea	labyrinth
malleus	semicircular canal
incus	cupula
stapes	ampulla
oval window	chemoreceptor
cochlear duct	taste bud
scala vestibuli	olfaction
scala tympani	olfactory epithelium
basilar membrane	odorant

SECTION B REVIEW QUESTIONS

1. Describe the similarities between pain and the other somatic sensations. Describe the differences.
2. List the structures through which light must pass before it reaches the photopigment in the rods and cones.
3. Describe the events that take place during accommodation for far vision.
4. What changes take place in neurotransmitter release from the rods or cones when they are exposed to light?
5. Beginning with the ganglion cells of the retina, describe the visual pathway.
6. List the sequence of events that occur between entry of a sound wave into the external auditory canal and the firing of action potentials in the cochlear nerve.
7. Describe the anatomical relationship between the cochlea and the cochlear duct.
8. What is the relationship between head movement and cupula movement in a semicircular canal?
9. What causes the release of neurotransmitter from the utricle and saccule receptor cells?
10. In what ways are the sensory systems for taste and olfaction similar? In what ways are they different?

SECTION B KEY TERMS

somatic sensation	zonular fiber
kinesthesia	iris
nociceptor	pupil
wavelength	rod
frequency	cone
visible spectrum	photopigment
retina	rhodopsin
lens	opsin
cornea	chromophore
fovea centralis	retinal
accommodation	bipolar cell
ciliary muscle	ganglion cell

CHAPTER 9 CLINICAL TERMS

phantom limb	myopic
hyperalgesia	farsighted
referred pain	hyperopic
analgesia	astigmatism
stimulation-produced analgesia	glaucoma
	color blindness
transcutaneous electric nerve stimulation (TENS)	hearing aid
	cochlear implant
acupuncture	nystagmus
presbyopia	vertigo
cataract	motion sickness
nearsighted	Ménière's disease

CHAPTER 9 THOUGHT QUESTIONS

(Answers are given in Appendix A.)

1. Describe several mechanisms by which pain could theoretically be controlled medically or surgically.

2. At what two sites would central nervous system injuries interfere with the perception of heat applied to the right side of the body? At what single site would a central nervous system injury interfere with the perception of heat applied to either side of the body?

3. What would vision be like after a drug has destroyed all the cones in the retina?

4. Damage to what parts of the cerebral cortex could explain the following behaviors? (a) A person walks into a chair placed in her path. (b) The person does not walk into the chair but does not know what the chair can be used for.

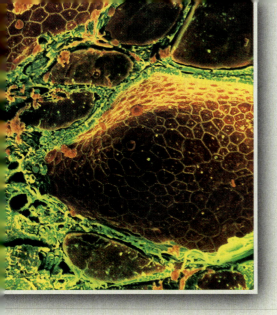

Principles of Hormonal Control Systems

The endocrine system is one of the body's two major communication systems, the nervous system being the other. The **endocrine system** consists of all those glands, termed **endocrine glands,** that secrete hormones. As described in Chapter 6, endocrine glands are also called ductless glands to distinguish them from all other (exocrine) glands. **Hormones,** as noted in Chapter 7, are chemical messengers that enter the *blood* which carries them from endocrine glands to the cells upon which they act. The cells influenced by a particular hormone are the **target cells** for that hormone.

Table 10–1 summarizes, for reference and orientation, the endocrine glands, the hormones they secrete, and the major functions the hormones control. The endocrine system differs from most of the other organ systems of the body in that the various glands are not anatomically continuous; however, they do form a system in the functional sense. The reader may be puzzled to see listed as endocrine glands some organs—the heart, for instance—that clearly have other functions. The explanation is that, in addition to the cells that carry out the organ's other functions, the organ also contains cells that secrete hormones. This illustrates the fact that organs are made up of different types of cells.

Note also in Table 10–1 that the hypothalamus, a part of the brain, is considered part of the endocrine system too. This is because the chemical messengers released by certain neuron terminals in both the hypothalamus and its extension, the posterior pituitary, do not function as neurotransmitters affecting adjacent cells but rather enter the blood, which carries them to their sites of action.

Table 10–1 demonstrates that there are a large number of endocrine glands and hormones. One way of describing the physiology of the individual hormones is to present all relevant material, gland by gland, in a single chapter. In keeping with our emphasis on hormones as messengers in homeostatic control mechanisms, however, we have chosen to describe the physiology of specific hormones and the glands that secrete them in subsequent chapters, in the context of the control systems in which they participate. For example, the pancreatic hormones are described in Chapter 18, which is on organic metabolism, parathyroid hormone in Chapter 16 in the context of calcium metabolism, and so on.

The aims of the present chapter are therefore limited to presenting: (1) the general principles of endocrinology—that is, a structural and functional analysis of hormones in general that transcends individual glands; and (2) an analysis of the hypothalamus-pituitary hormonal system. The control systems for the hormones of this particular system are so interconnected that they are best described as a unit to lay the foundation for subsequent descriptions in other chapters.

Before turning to these presentations, however, several additional general points should be made concerning Table 10–1. One phenomenon evident from this table is that a single gland may secrete multiple hormones. The usual pattern in such cases is that a single cell type secretes only one hormone, so that multiple hormone secretion reflects the presence of different types of endocrine cells in the same gland. In a few cases, however, a single cell may secrete more than one hormone (for example, the secretion of follicle-stimulating hormone and luteinizing hormone by the anterior pituitary).

Another point of interest illustrated by Table 10–1 is that a particular hormone may be produced by more than one type of endocrine gland. For example, somatostatin is secreted by endocrine cells in both the gastrointestinal tract and the pancreas and is also one of the hormones secreted by the hypothalamus.

Finally, as emphasized in Chapter 7, a chemical messenger secreted by an endocrine gland cell is often also secreted by other cell types and serves in these other locations as a neurotransmitter or paracrine/autocrine agent. For example, prolactin is secreted not only by the anterior pituitary but by at least 12 other cell types where it is thought to exert a huge number of paracrine/autocrine functions.

TABLE 10–1 Summary of the Hormones

Site Produced (Endocrine Gland)	Hormone	Major Function* Is Control of:
Adipose tissue cells	Leptin	Food intake; metabolic rate
Adrenal:		
Adrenal cortex	Cortisol	Organic metabolism; response to stress; immune system
	Androgens	Sex drive in women
	Aldosterone	Sodium, potassium, and acid excretion by kidneys
Adrenal medulla	Epinephrine	Organic metabolism; cardiovascular function; response to stress
	Norepinephrine	
Gastrointestinal tract	Gastrin	Gastrointestinal tract; liver; pancreas; gallbladder
	Secretin	
	Cholecystokinin	
	Glucose-dependent insulinotropic peptide (GIP)†	
	Motilin	
Gonads:		
Ovaries: female	Estrogen	Reproductive system; breasts; growth and development; influences gametes
	Progesterone	
	Inhibin	FSH secretion
	Relaxin	? Relaxation of cervix and pubic ligaments
Testes: male	Testosterone	Reproductive system; growth and development; sex drive; influences gametes
	Inhibin	FSH secretion
	Müllerian-inhibiting hormone	Regression of Müllerian ducts
Heart	Atrial natriuretic factor (ANF, atriopeptin)	Sodium excretion by kidneys; blood pressure
Hypothalamus	Hypophysiotropic hormones:	Secretion of hormones by the anterior pituitary
	Corticotropin releasing hormone (CRH)	Secretion of adrenocorticotropic hormone (stimulation)
	Thyrotropin releasing hormone (TRH)	Secretion of thyroid-stimulating hormone (stimulation)
	Growth hormone releasing hormone (GHRH)	Secretion of growth hormone (stimulation)
	Somatostatin (SS)	Secretion of growth hormone (inhibition)
	Gonadotropin releasing hormone (GnRH)	Secretion of luteinizing hormone and follicle-stimulating hormone (stimulation)
	Dopamine (DA, also called prolactin-inhibiting hormone, PIH)	Secretion of prolactin (inhibition)
	Posterior pituitary hormones	See posterior pituitary
Kidneys	Renin (an enzyme that generates angiotensin)	Aldosterone secretion; blood pressure
	Erythropoietin	Erythrocyte production
	1,25-dihydroxyvitamin D_3	Plasma calcium
Leukocytes, macrophages, endothelial cells, and fibroblasts	Cytokines‡ (these include the interleukins, colony-stimulating factors, interferons, tumor necrosis factors)	Immune defenses
Liver and other cells	Insulin-like growth factors (IGF-I and II)	Cell division and growth
Pancreas	Insulin	Organic metabolism; plasma glucose
	Glucagon	
	Somatostatin	
Parathyroids	Parathyroid hormone (PTH, PH, parathormone)	Plasma calcium and phosphate
Pineal	Melatonin	? Sexual maturity; body rhythms

TABLE 10–1 Summary of the Hormones (*continued*)

Site Produced (Endocrine Gland)	Hormone	Major Function* Is Control of:
Pituitary glands:		
Anterior pituitary	Growth hormone (GH, somatotropin)	Growth, mainly via secretion of IGF-I; protein, carbohydrate, and lipid metabolism
	Thyroid-stimulating hormone (TSH, thyrotropin)	Thyroid gland
	Adrenocorticotropic hormone (ACTH, corticotropin)	Adrenal cortex
	Prolactin	Breast growth and milk synthesis; may be permissive for certain reproductive functions in the male
	Gonadotropic hormones: Follicle-stimulating hormone (FSH) Luteinizing hormone (LH)	Gonads (gamete production and sex hormone secretion)
	β-lipotropin and β-endorphin	Unknown
Posterior pituitary[§]	Oxytocin	Milk let-down; uterine motility
	Vasopressin (antidiuretic hormone, ADH)	Water excretion by the kidneys; blood pressure
Placenta	Chorionic gonadotropin (CG)	Secretion by corpus luteum
	Estrogens	See Gonads: ovaries
	Progesterone	See Gonads: ovaries
	Placental lactogen	Breast development; organic metabolism
Thymus	Thymopoietin	T-lymphocyte function
Thyroid	Thyroxine (T$_4$) Triiodothyronine (T$_3$)	Metabolic rate; growth; brain development and function
	Calcitonin	Plasma calcium
Multiple cell types	Growth factors[‡] (e.g., epidermal growth factor)	Growth and proliferation of specific cell types

°This table does not list all functions of the hormones.
†The names and abbreviations in parentheses are synonyms.
‡Some classifications include the cytokines under the category of growth factors.

§The posterior pituitary stores and secretes these hormones; they are made in the hypothalamus.

Hormone Structures and Synthesis

Hormones fall into three chemical classes: (1) amines, (2) peptides and proteins, and (3) steroids.

Amine Hormones

The **amine hormones** are all derivatives of the amino acid tyrosine. They include the thyroid hormones, epinephrine and norepinephrine (produced by the adrenal medulla), and dopamine (produced by the hypothalamus).

Thyroid Hormones The **thyroid gland** is located in the lower part of the neck wrapped around the front of the trachea (windpipe). It is composed of many spherical structures called follicles, each consisting of a single layer of epithelial cells surrounding an extracellular central space. This space is filled with a glycoprotein called **thyroglobulin.** The follicles secrete the two iodine-containing amine hormones—

thyroxine (T$_4$) and triiodothyronine (T$_3$) (Figure 10–1), collectively known as the **thyroid hormones (TH),** Parafollicular cells, which are located between follicles, secrete a third hormone—a peptide called calcitonin; this hormone does not contain iodine and is not included in the term "thyroid hormones."

Iodine is an essential element that functions as a component of T$_4$ and T$_3$. Most of the iodine ingested in food is absorbed into the blood from the gastrointestinal tract by active transport; in the process it is converted to the ionized form, iodide. Iodide is actively transported from the blood into the thyroid follicular cells. Once in the cells, the iodide is converted back to iodine, which is then coupled to the side chains of tyrosine molecules that had previously been incorporated into thyroglobulin precursor. The result is the formation of thyroglobulin, which is stored in the central space of the follicles.

During hormone secretion, thyroglobulin is moved into the follicular cells by endocytosis and, after fusion with lysosomes, is digested to release the thyroid

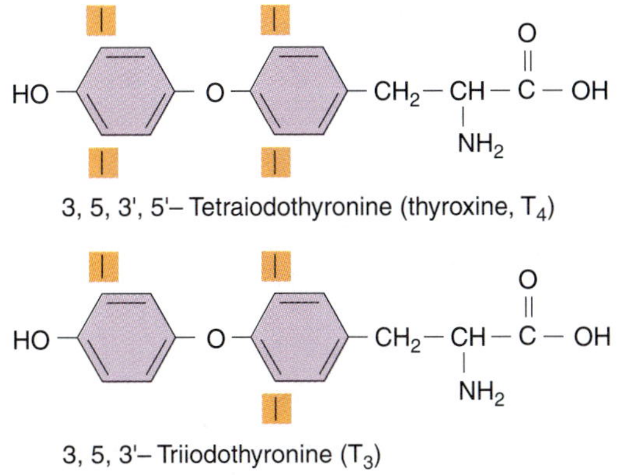

FIGURE 10–1

Chemical structures of the thyroid hormones, thyroxine and triiodothyronine. The two molecules differ by only one iodine atom, a difference noted in the abbreviations T_3 and T_4.

hormones, which cross the cells' plasma membranes to enter the blood. T_4 is secreted in much larger amounts than is T_3. However, the plasma membranes of many cell types contain enzymes that convert most of this T_4 into T_3 by removal of one iodine atom, and this constitutes the major source of plasma T_3. This is an important point because T_3 is a much more active hormone than is T_4. Indeed it is likely that T_4 has little or no action unless it is converted into T_3. Thus, persons with low peripheral deiodination because of defective enzymes can show evidence of thyroid-hormone deficiency even though they have normal or elevated plasma concentrations of T_4.

Virtually every tissue in the body is affected by the thyroid hormones. These effects, which are described in Chapter 18, include regulation of oxygen consumption, growth, and brain development and function.

Adrenal Medullary Hormones and Dopamine
There are two adrenal glands, one on the top of each kidney. Each **adrenal gland** comprises two distinct endocrine glands, an inner **adrenal medulla,** which secretes amine hormones, and a surrounding **adrenal cortex,** which secretes steroid hormones. As described in Chapter 8, the adrenal medulla is really a modified sympathetic ganglion whose cell bodies do not have axons but instead release their secretions into the blood, thereby fulfilling a criterion for an endocrine gland.

The adrenal medulla secretes mainly two amine hormones, **epinephrine (E)** and **norepinephrine (NE).** Recall from Chapter 8 that these molecules constitute, with dopamine, the chemical family of **catecholamines.** The structures and pathways for synthesis of the catecholamines were described in Chapter 8, when they were dealt with as neurotransmitters. In humans, the adrenal medulla secretes approximately four times more epinephrine than norepinephrine. Epinephrine and norepinephrine exert actions similar to those of the sympathetic nerves. These effects are described in various chapters and summarized in Chapter 20 in the section on stress.

The adrenal medulla also secretes small amounts of dopamine and several substances other than catecholamines, but whether any of these adrenal secretions other than epinephrine and norepinephrine actually serve hormonal functions is unknown. In contrast, as described below, the dopamine secreted by certain cells in the hypothalamus definitely functions as a hormone.

Peptide Hormones
The great majority of hormones are either peptides or proteins. They range in size from small peptides having only three amino acids to small proteins (some of which are glycoproteins). For convenience, we shall follow a common practice of endocrinologists and refer to all these hormones as **peptide hormones.**

In many cases, they are initially synthesized on the ribosomes of the endocrine cells as larger proteins known as preprohormones, which are then cleaved to **prohormones** by proteolytic enzymes in the granular endoplasmic reticulum (Figure 10–2). The prohormone is then packaged into secretory vesicles by the Golgi apparatus. In this process, the prohormone is cleaved to yield the active hormone and other peptide chains found in the prohormone. Therefore, when the cell is stimulated to release the contents of the secretory vesicles by exocytosis, the other peptides are cosecreted with the hormone. In certain cases they, too, may exert hormonal effects. In other words, instead of just one peptide hormone, the cell may be secreting multiple peptide hormones that differ in their effects on target cells.

One more point about peptide hormones: As mentioned in Chapters 7 and 8, many peptides serve as both neurotransmitters (or neuromodulators) and as hormones. For example, most of the hormones secreted by the endocrine glands in the gastrointestinal tract (for example, cholecystokinin) are also produced by neurons in the brain where they function as neurotransmitters.

Steroid Hormones
The third family of hormones is the steroids, the lipids whose ringlike structure was described in Chapter 2. **Steroid hormones** are produced by the adrenal cortex and the **gonads** (testes and ovaries) as well as by the placenta during pregnancy. Examples are shown in Figure 10–3. In addition, the hormone 1,25-dihydroxyvitamin D_3, the active form of vitamin D, is a steroid derivative.

Cholesterol is the precursor of all steroid hormones. The cells producing these hormones synthesize some of their own cholesterol, but most is provided to them from the plasma, as described in Chapter 18. The many biochemical steps in steroid synthesis beyond cholesterol involve small changes in the molecules and are mediated by specific enzymes. The steroids produced by a particular cell depend, therefore, on the types and concentrations of enzymes present. Because steroids are highly lipid-soluble, once they are synthesized they simply diffuse across the plasma membrane of the

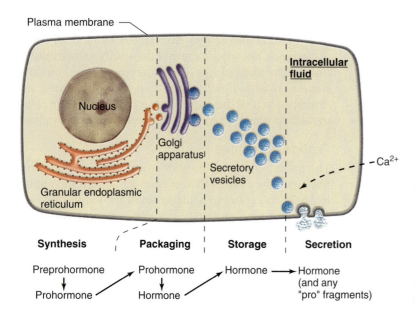

FIGURE 10–2

Typical synthesis and secretion of peptide hormones. In some cells, the calcium that causes exocytosis is released from the endoplasmic reticulum by action of a second messenger rather than entering from the extracellular fluid.

Adapted from Hedge et al.

steroid-producing cell and enter the interstitial fluid and then the blood where they mainly bind to certain plasma proteins.

The next sections describe the pathways for steroid synthesis by the adrenal cortex and gonads. Those for the placenta are somewhat unusual and are discussed in Chapter 19.

Hormones of the Adrenal Cortex Steroid synthesis by the adrenal cortex is illustrated in Figure 10–4. The five hormones normally secreted in physiologically significant amounts by the adrenal cortex are aldosterone, cortisol, corticosterone, dehydroepiandrosterone (DHEA), and androstenedione. **Aldosterone** is known as a **mineralocorticoid** because its effects are on salt (mineral) balance, mainly on the kidneys' handling of sodium, potassium, and hydrogen ions. **Cortisol** and corticosterone are called **glucocorticoids** because they have important effects on the metabolism of glucose and other organic nutrients. Cortisol is by far the more important of the two glucocorticoids in humans, and so we shall deal only with it in future discussions. In addition to its effects on organic metabolism (described in Chapter 18), cortisol exerts many other effects, including facilitation of the body's responses to stress and regulation of the immune system (Chapter 20).

Dehydroepiandrosterone (DHEA) and androstenedione belong to the class of hormones known as **androgens,** which also includes the major male sex hormone, testosterone, produced by the testes. All androgens have actions similar to those of testosterone. Because the adrenal androgens are much less potent than testosterone, they are of little physiological significance in the adult male; they do, however, play roles in the adult female, and in both sexes at puberty, as described in Chapter 19. The use of DHEA as a "supplement" is also discussed in Chapter 19.

The adrenal cortex is not a homogeneous gland but is composed of three distinct layers (Figure 10–5). The outer layer—the zona glomerulosa—possesses very high concentrations of the enzymes required to convert corticosterone to aldosterone but lacks the enzymes required for the formation of cortisol and androgens. Accordingly, this layer synthesizes and secretes aldosterone but not the other major adrenal cortical hormones. In contrast, the zona fasciculata and zona reticularis have just the opposite enzyme profile. They, therefore, secrete no aldosterone but much cortisol and androgen.

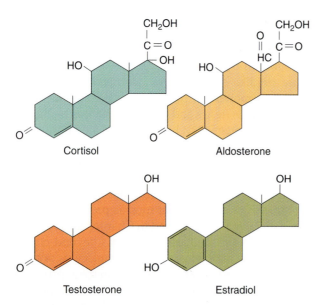

FIGURE 10–3

Structures of representative steroid hormones.

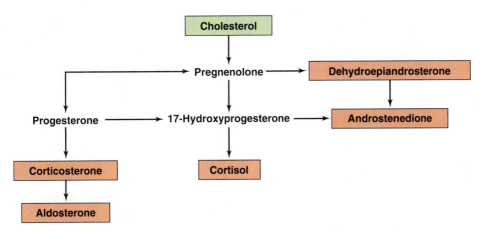

FIGURE 10-4

Simplified flow sheet for synthesis of steroid hormones by the adrenal cortex; several intermediate steps have been left out. The various steps are mediated by specific enzymes. The five hormones shown in boxes are the major hormones secreted. Dehydroepiandrosterone (DHEA) and androstenedione are androgens—that is, testosterone-like hormones. Cortisol and corticosterone are glucocorticoids, and aldosterone is a mineralocorticoid.

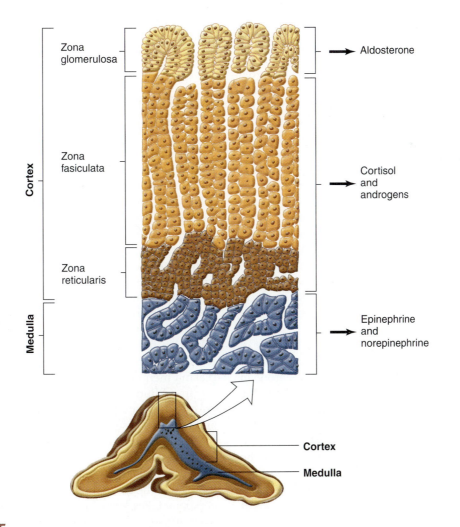

FIGURE 10-5

Section through an adrenal gland showing both the medulla and cortex, as well as the hormones they secrete.

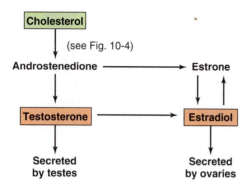

FIGURE 10–6

Gonadal production of steroids. Only the ovaries have high concentrations of the enzymes (aromatase) required to produce the estrogens estrone and estradiol.

In certain disease states, the adrenal cortex may secrete decreased or increased amounts of various steroids. For example, an absence of the enzymes for the formation of cortisol by the adrenal cortex can result in the shunting of the cortisol precursors into the androgen pathway. In a woman, the result of the large increase in androgen secretion would be *masculinization,* in addition to the effects of cortisol deficiency (Chapter 20).

Hormones of the Gonads   Compared to the adrenal cortex, the gonads have very different concentrations of key enzymes in their steroid pathways. Endocrine cells in both testes and ovaries lack the enzymes needed to produce aldosterone and cortisol. They possess high concentrations of enzymes in the androgen pathways leading to androstenedione, as in the adrenal cortex. In addition, the endocrine cells in the testes contain a high concentration of the enzyme that converts androstenedione to **testosterone,** which is therefore the major androgen secreted by the testes (Figure 10–6). The ovarian endocrine cells that synthesize the major female sex hormone, **estradiol,** have a high concentration of the enzyme (aromatase) required to go one step further—that is, to transform androgens to estradiol (Figure 10–6). Accordingly, estradiol, rather than androgens, is secreted by the ovaries.

Very small amounts of testosterone do leak out of ovarian endocrine cells, however, and very small amounts of estradiol are produced from testosterone in the testes. Moreover, following their secretion into the blood by the gonads and the adrenal cortex, steroid hormones may undergo further interconversion in either the blood or other organs. For example, testosterone is converted to estradiol in some of its target cells. Thus, the major male and female sex hormones—testosterone and estradiol, respectively—are not unique to males and females, although, of course, the relative concentrations of the hormones are quite different in the two sexes.

Finally, certain ovarian endocrine cells secrete another major steroid hormone, **progesterone.**

Hormone Transport in the Blood

Peptide and catecholamine hormones are water-soluble. Therefore, with the exception of a few peptides, these hormones are transported simply dissolved in plasma (Table 10–2). In contrast, the steroid hormones and the thyroid hormones circulate in the blood largely bound to plasma proteins.

Even though the steroid and thyroid hormones exist in plasma mainly bound to large proteins, small concentrations of these hormones do exist dissolved in the plasma. The dissolved, or free, hormone is in equilibrium with the bound hormone:

Free hormone + Binding protein $\rightleftharpoons$

Hormone-protein complex

The total hormone concentration in plasma is the sum of the free and bound hormone. It is important to realize, however, that only the *free* hormone can diffuse across capillary walls and encounter its target cells. Accordingly, the concentration of the free hormone is what is physiologically important rather than the concentration of the total hormone, most of which is bound. As we shall see, the degree of protein binding also influences the rate of metabolism and the excretion of the hormone.

Hormone Metabolism and Excretion

A hormone's concentration in the plasma depends not only upon its rate of secretion by the endocrine gland but also upon its rate of removal from the blood, either by excretion or by metabolic transformation. The liver and the kidneys are the major organs that excrete or metabolize hormones.

The liver and kidneys, however, are not the only routes for eliminating hormones. Sometimes the hormone is metabolized by the cells upon which it acts. Very importantly, in the case of peptide hormones, endocytosis of hormone-receptor complexes on plasma membranes enables cells to remove the hormones rapidly from their surfaces and catabolize them intracellularly. The receptors are then often recycled to the plasma membrane.

In addition, catecholamine and peptide hormones are excreted rapidly or attacked by enzymes in the blood and tissues. These hormones therefore tend to remain in the bloodstream for only brief periods—minutes to an hour. In contrast, because protein-bound hormones are less vulnerable to excretion or metabolism by enzymes, removal of the circulating steroid and thyroid hormones generally takes longer, often several hours (with thyroid hormone remaining in the plasma for days).

In some cases, metabolism of the hormone after its secretion *activates* the hormone rather than inactivates

TABLE 10–2 Categories of Hormones

Types	Major Form in Plasma	Location of Receptors	Signal Transduction Mechanisms	Rate of Excretion/Metabolism
Peptides and catecholamines	Free	Plasma membrane	Receptors alter: Channels intrinsic to the receptors; Enzymatic activity intrinsic to the receptor; Enzymatic activity of cytoplasmic JAK kinases associated with the receptor; G proteins in the plasma membrane. These control plasma-membrane channels or enzymes that generate second messengers (cAMP, DAG, IP$_3$).	Fast (minutes to an hour)
Steroids and thyroid hormones	Protein-bound	Cell interior	Receptors directly alter gene transcription	Slow (hours to days)

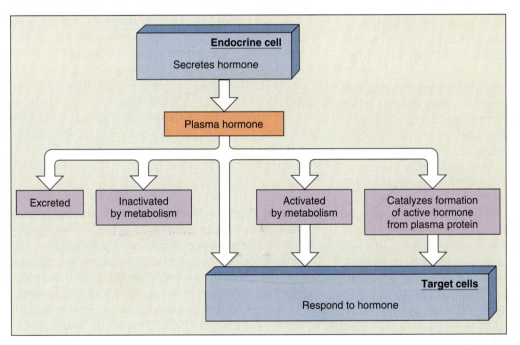

FIGURE 10–7

Possible fates and actions of a hormone following its secretion by an endocrine cell. Not all paths apply to all hormones.

it. In other words, the secreted hormone may be relatively or completely unable to act upon a target cell until metabolism transforms it into a substance that can act. We have already seen one example of hormone activation—the conversion of circulating T$_4$ to the far more active T$_3$. Another example is provided by testosterone, which is converted either to estradiol or dihydrotestosterone in certain of its target cells. These molecules, rather than testosterone itself, then bind to receptors inside the target cell and elicit the cell's response.

There is another kind of "activation" that applies to a few hormones. Instead of the hormone itself being activated after secretion, it acts enzymatically on a completely different plasma protein to split off a peptide that functions as the active hormone. The best known example of this is the renin-angiotensin system, described in Chapters 14 and 16; renin is not technically a hormone but an enzyme that participates in the generation of the hormone angiotensin.

Figure 10–7 summarizes the fates of hormones after their secretion.

Mechanisms of Hormone Action

Hormone Receptors

Because they travel in the blood, hormones are able to reach virtually all tissues. Yet the body's response to a hormone is not all-inclusive, but is highly specific, involving only the target cells for that hormone. The ability to respond depends upon the presence on (or in) the target cells of specific receptors for those hormones.

As emphasized in Chapter 7, the response of a target cell to a chemical messenger is the final event in a sequence that begins when the messenger binds to specific cell receptors. As described in that chapter, the receptors for peptide hormones and catecholamines are proteins located in the plasma membranes of the target cells. In contrast, the receptors for steroid hormones and the thyroid hormones are proteins located mainly *inside* the target cells.

Hormones can influence the ability of target cells to respond by regulating hormone receptors. Basic concepts of receptor modulation (up-regulation and down-regulation) were described in Chapter 7. In the context of hormones, **up-regulation** is an increase in the number of a hormone's receptors, often resulting from a prolonged exposure to a low concentration of the hormone. **Down-regulation** is a decrease in receptor number, often from exposure to high concentrations of the hormone.

Hormones can down-regulate or up-regulate not only their own receptors but the receptors for other hormones as well. If one hormone induces a loss of a second hormone's receptors, the result will be a reduction of the second hormone's effectiveness; in such cases, the one hormone is said to antagonize the action of the other. On the other hand, a hormone may induce an increase in the number of receptors for a second hormone. In this case the effectiveness of the second hormone is increased.

This latter phenomenon, in some cases, underlies the important hormone-hormone interaction known as permissiveness. In general terms, **permissiveness** means that hormone A must be present for the full strength of hormone B's effect. A low concentration of hormone A is usually all that is needed for this permissive effect, which is due to A's positive effect on B's receptors. For example (Figure 10–8), epinephrine causes a large release of fatty acids from adipose tissue, but only in the presence of permissive amounts of thyroid hormone. The major reason is that thyroid hormone facilitates the synthesis of receptors for epinephrine in adipose tissue and so the tissue becomes much more sensitive to epinephrine. It should be noted, however, that receptor alteration does not explain all cases of permissiveness; often the explanation is not known.

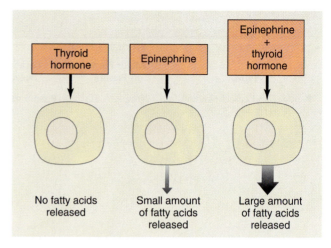

FIGURE 10–8

Ability of thyroid hormone to "permit" epinephrine-induced release of fatty acids from adipose-tissue cells. The thyroid hormones exert this effect by causing an increased number of epinephrine receptors on the cell.

Events Elicited by Hormone-Receptor Binding

The events initiated by the binding of a hormone to its receptor—that is, the mechanisms by which the hormone elicits a cellular response—are one or more of the signal transduction pathways that apply to all chemical messengers, as described in Chapter 7 (see Figures 7–12 and 7–13). In other words, there is nothing unique about the mechanisms initiated by hormones as compared to those utilized by neurotransmitters and paracrine/autocrine agents, and so they are only briefly reviewed at this point (see Table 10–2).

Effects of Peptide Hormones and Catecholamines
As stated above, the receptors for peptide hormones and the catecholamine hormones are located on the outer surface of the target cell's plasma membrane. When activated by hormone binding, the receptors trigger one or more of the signal transduction pathways described for plasma-membrane receptors in Chapter 7 (see Figure 7–13). That is, the activated receptors directly influence: (1) ion channels that are part of the receptors; (2) enzyme activity that is part of the receptor; (3) activity of cytoplasmic JAK kinases associated with the receptor; or (4) G proteins coupled in the plasma membrane to effector proteins—ion channels and enzymes—that generate second messengers. The opening or closing of ion channels changes the electrical potential across the membrane and, when a calcium channel is involved, changes the cytosolic concentration of this important ionic second messenger. The changes in enzyme activity produce—most

commonly by phosphorylation catalysed by protein kinase enzymes—changes in the conformation and hence the activity of various cellular proteins. In some cases the signal transduction pathways also lead to activation (or inhibition) of particular genes, causing a change in the rate of synthesis of the proteins coded for by these genes.

Effects of Steroid and Thyroid Hormones Structurally, the steroid hormones, the thyroid hormones, and the steroid derivative 1,25-dihydroxyvitamin D_3 are all closely related, and their receptors, which are intracellular, constitute the steroid-hormone receptor superfamily. As described in Chapter 7 (see Figure 7–12), the binding of hormone to one of these receptors leads to activation (or in some cases, inhibition) of particular genes, causing a change in the rate of synthesis of the proteins coded for by those genes. The ultimate result of changes in the concentrations of these proteins is an enhancement or inhibition of particular processes carried out by the cell, or a change in the rate of protein secretion by the cell.

Surprisingly, in addition to having intracellular receptors, some target cells also have *plasma-membrane receptors* for certain of the steroid hormones. In such cases the signal-transduction pathways initiated by the plasma-membrane receptors elicit rapid nongenomic cell responses while the intracellular receptors mediate a delayed response, requiring new protein synthesis.

Pharmacological Effects of Hormones

Administration of very large quantities of a hormone for medical purposes may have effects that are never seen in a normal healthy person. They are called ***pharmacological effects,*** and they can also occur in diseases when excessive amounts of hormones are secreted. Pharmacological effects are of great importance in medicine, since hormones are often used in large doses as therapeutic agents. Perhaps the most common example is that of the adrenal cortical hormone cortisol, which is administered in large amounts to suppress allergic and inflammatory reactions.

Inputs That Control Hormone Secretion

Most hormones are released in short bursts, with little or no release occurring between bursts. Accordingly, the plasma concentrations of hormones may fluctuate rapidly over brief time periods. Hormones also manifest 24-h cyclical variations in their secretory rates, the circadian patterns being different for different hormones. Some are clearly linked to sleep; for example, growth hormone's secretion is markedly increased

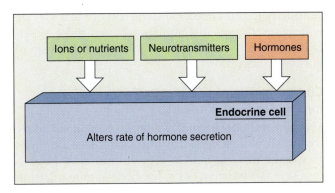

FIGURE 10–9
Inputs that act directly on endocrine-gland cells to stimulate or inhibit hormone secretion.

during the early period of sleep and is quite low or absent during the rest of the day and night. The mechanisms underlying these cycles are ultimately traceable to cyclical variations in the activity of neural pathways involved in the hormone's release.

Hormone secretion is controlled mainly by three types of inputs to endocrine cells (Figure 10–9): (1) changes in the plasma concentrations of mineral ions or organic nutrients; (2) neurotransmitters released from neurons impinging on the endocrine cell; and (3) another hormone (or, in some cases, a paracrine/autocrine agent) acting on the endocrine cell. There is actually a fourth type of input—chemical and physical factors in the lumen of the gastrointestinal tract—but it applies only to the hormones secreted by the gastrointestinal tract and will be described in Chapter 17.

Before we look more closely at each category, it must be stressed that, in many cases, hormone secretion is influenced by more than one input. For example, insulin secretion is controlled by the extracellular concentrations of glucose and other nutrients, by both sympathetic and parasympathetic neurons to the insulin-secreting endocrine cells, and by several hormones acting on these cells. Thus, endocrine cells, like neurons, may be subject to multiple, simultaneous, often opposing inputs, and the resulting output—the rate of hormone secretion—reflects the integration of all these inputs.

One more point should be made to avoid misunderstanding. The term "secretion" applied to a hormone denotes its synthesis and release from the cell. Some inputs to endocrine cells specifically stimulate or inhibit only synthesis, for example, by altering the expression of the gene for that hormone, with changes in release occurring as a secondary result. In contrast, other inputs directly influence only the actual release of the hormone from the cell, and some inputs influence both synthesis and release. For simplicity in this chapter and the rest of the book, we will generally not distinguish between these possibilities when we refer to stimulation or inhibition of hormone "secretion."

Control by Plasma Concentrations of Mineral Ions or Organic Nutrients

There are multiple hormones whose secretion is directly controlled, at least in part, by the plasma concentrations of specific mineral ions or organic nutrients. In each case, a major function of the hormone is to regulate, in a negative-feedback manner, the plasma concentration of the ion or nutrient controlling its secretion. For example, insulin secretion is stimulated by an elevated plasma glucose concentration, and the additional insulin then causes, by several actions, the plasma glucose concentration to decrease (Figure 10–10).

Control by Neurons

The adrenal medulla behaves like a sympathetic ganglion and thus is stimulated by sympathetic preganglionic fibers. In addition to its control of the adrenal medulla, the autonomic nervous system has influences on other endocrine glands (Figure 10–11b). Both parasympathetic and sympathetic inputs to these other glands may occur, some inhibitory and some stimulatory. Examples are the secretion of insulin and the gastrointestinal hormones.

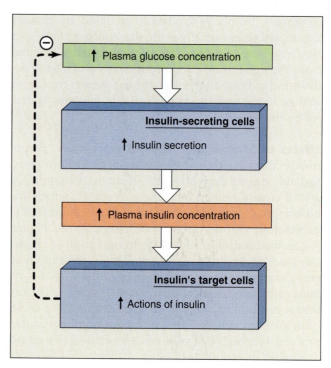

FIGURE 10–10

Example of how the direct control of hormone secretion by the plasma concentration of a substance, either organic nutrient or mineral ion, results in the negative-feedback control of the substance's plasma concentration.

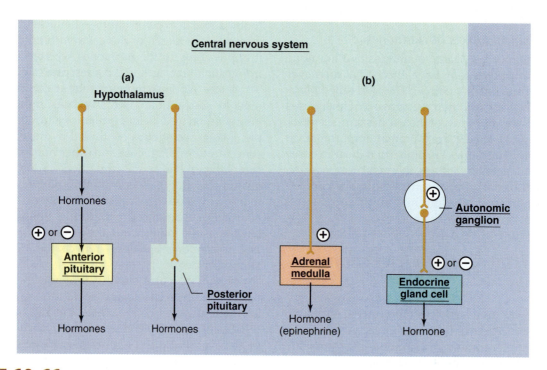

FIGURE 10–11

Pathways by which the nervous system influences hormone secretion. (a) Certain neurons in the hypothalamus, some of which terminate in the posterior pituitary, secrete hormones. The secretion of hypothalamic hormones from the posterior pituitary and the effects of other hypothalamic hormones on the anterior pituitary are described later in this chapter. (b) The autonomic nervous system controls hormone secretion by the adrenal medulla and many other endocrine glands.

Thus far, our discussion of neural control of hormone release has been limited to the role of the *autonomic* nervous system (Figure 10–11b). However, one large group of hormones—those secreted by the hypothalamus and its extension, the posterior pituitary—are under the direct control not of autonomic neurons but of neurons in the brain itself (Figure 10–11a). This category will be described in detail later in this chapter.

Control by Other Hormones

In many cases, the secretion of a particular hormone is directly controlled by the blood concentration of another hormone. As we shall see, there are often complex sequences in which the only function of the first few hormones in the sequence is to stimulate the secretion of the next one. A hormone that stimulates the secretion of another hormone is often referred to as a **tropic hormone.** The tropic hormones usually stimulate not only secretion but the growth of the stimulated gland as well.

Control Systems Involving the Hypothalamus and Pituitary

The **pituitary gland,** or hypophysis, lies in a pocket (the sella turcica) of the sphenoid bone at the base of the brain (Figure 10–12), just below the brain area called the **hypothalamus.** The pituitary is connected to the hypothalamus by the infundibulum, a stalk containing nerve fibers and small blood vessels. In adult human beings, the pituitary gland is composed of two adjacent lobes—the **anterior pituitary** (toward the front of the head) and the **posterior pituitary** (toward the back of the head)—each of which is a more or less distinct gland. (In many mammalian species an intermediate lobe is found between the anterior and posterior portions of the pituitary, but this is not the case in humans.) The anterior pituitary and its part of the stalk are termed the adenohypophysis, and the posterior pituitary with its part of the stalk is the neurohypophysis.

The posterior pituitary is an outgrowth of the hypothalamus and is neural tissue. The axons of two well-defined clusters of hypothalamic neurons (the supraoptic and paraventricular nuclei) pass down the infundibulum and end within the posterior pituitary in close proximity to capillaries (the smallest of blood vessels) (Figure 10–12b).

In contrast to the neural connections between the hypothalamus and *posterior* pituitary, there are no important neural connections between the hypothalamus and *anterior* pituitary. There is, however, an unusual *blood-vessel* connection (Figure 10–12b). The capillaries at the base of the hypothalamus (the **median eminence**) recombine to form the **hypothalamo-pituitary**

portal vessels—the term "portal" denotes vessels that connect distinct capillary beds. The hypothalamo-pituitary portal vessels pass down the stalk connecting the hypothalamus and pituitary and enter the anterior pituitary where they drain into a second capillary bed, the anterior pituitary capillaries. Thus, the hypothalamo-pituitary portal vessels offer a local route for blood flow directly from the hypothalamus to the anterior pituitary.

Posterior Pituitary Hormones

We stressed above that the posterior pituitary is really a neural extension of the hypothalamus (Figure 10–12). The term "posterior pituitary hormones" is somewhat of a misnomer, therefore, since the hormones are not synthesized in the posterior pituitary itself but in the hypothalamus, specifically in the cell bodies of the hypothalamic neurons whose axons pass down the infundibulum and end in the posterior pituitary. Only one hormone is produced by any particular neuron in the two relevant hypothalamic nuclei. Enclosed in small vesicles, the hormone moves down the neural axons to accumulate at the axon terminals in the posterior pituitary. Stimuli (either hormones or neurotransmitters) act to generate action potentials in the neurons; these action potentials propagate to the axon terminals and trigger the release of the hormone by exocytosis. The hormone then enters the capillaries to be carried away by the blood returning to the heart.

The two posterior pituitary hormones are **oxytocin** and **vasopressin** (the latter is also known as **antidiuretic hormone,** or **ADH**). Vasopressin participates in the control of water excretion by the kidneys and of blood pressure (Chapters 16 and 14). Oxytocin acts on smooth muscle cells in the breasts and uterus (Chapter 19); its functions, if any, in the male are uncertain.

Vasopressin and oxytocin are also produced in other areas of the brain and serve in those sites as neurotransmitters or neuromodulators.

The Hypothalamus and Anterior Pituitary

Hypothalamic neurons different from those that produce the hormones released from the *posterior* pituitary, secrete hormones that control, in large part, the secretion of all the *anterior* pituitary hormones. These hypothalamic hormones are collectively termed **hypophysiotropic hormones** (recall that another name for the pituitary is hypophysis); they are also commonly called hypothalamic releasing hormones. One more word about terminology is appropriate: "Hypophysiotropic hormones" denotes only those hormones from the *hypothalamus* that influence the anterior pituitary. We shall see later that nonhypothalamic hormones can also influence the anterior pituitary, but they are not categorized as hypophysiotropic hormones.

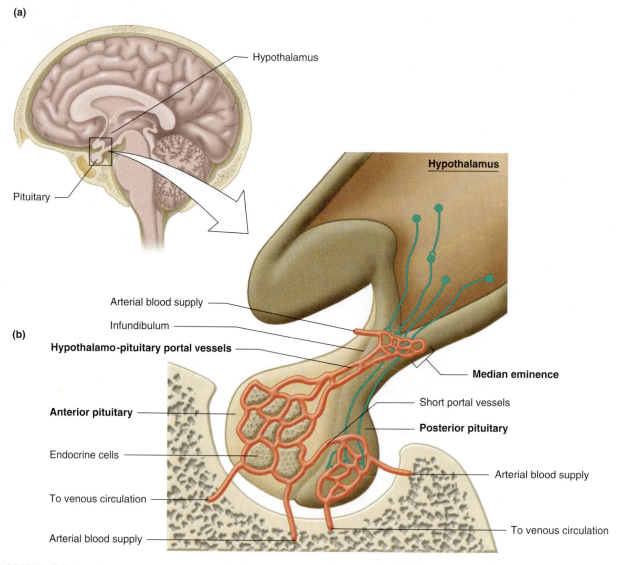

(a)

Hypothalamus

Pituitary

(b)

Hypothalamus

Arterial blood supply

Infundibulum

Hypothalamo-pituitary portal vessels

Median eminence

Anterior pituitary

Short portal vessels

Posterior pituitary

Endocrine cells

To venous circulation

Arterial blood supply

Arterial blood supply

To venous circulation

FIGURE 10–12

(a) Relation of the pituitary gland to the brain and hypothalamus. (b) Neural and vascular connections between the hypothalamus and pituitary. Some hypothalamic neurons run down the infundibulum to end in the posterior pituitary, whereas others end in the median eminence. Almost the entire blood supply to the anterior pituitary comes via the hypothalamo-pituitary portal vessels, which originate in the median eminence. (The short portal vessels, which originate in the posterior pituitary, carry only a small fraction of the blood leaving the posterior pituitary and supply only a small fraction of the blood received by the anterior pituitary.)

With one exception, each of the hypophysiotropic hormones is the first in a *three*-hormone sequence: (1) A hypophysiotropic hormone controls the secretion of (2) an anterior pituitary hormone, which controls the secretion of (3) a hormone from some other endocrine gland (Figure 10–13). This last hormone then acts on its target cells. As we shall see later, the adaptive value of such chains is that they permit a variety of important hormonal feedbacks. We begin our description of these sequences in the middle—that is, with the anterior pituitary hormones—because the names and actions of the hypophysiotropic hormones are all based on the names of the anterior pituitary hormones.

Anterior Pituitary Hormones As shown in Table 10–1, the anterior pituitary secretes at least eight hormones, but only six have well-established functions. All peptides, these six "classical" hormones are **follicle-stimulating hormone (FSH), luteinizing hormone (LH), growth hormone (GH), thyroid-stimulating hormone** (TSH, thyrotropin), **prolactin,** and **adrenocorticotropic hormone** (ACTH, corticotropin). Each of the last four is probably secreted by a distinct cell type in the anterior pituitary, whereas FSH and LH, collectively termed **gonadotropic hormones** (or gonadotropins) because they stimulate the gonads, are both secreted by the same cells.

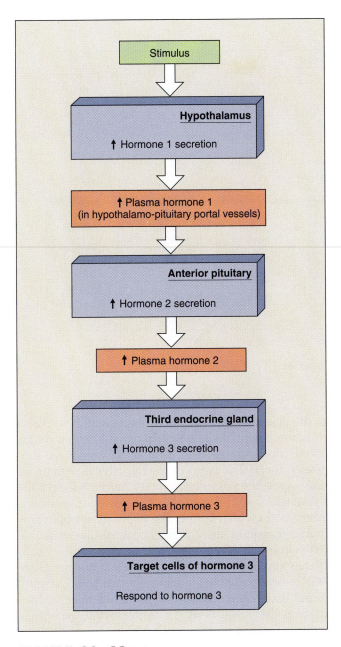

FIGURE 10–13
Typical sequential pattern by which a hypophysiotropic hormone (hormone 1 from the hypothalamus) controls the secretion of an anterior pituitary hormone, which in turn controls the secretion of a hormone by a third endocrine gland. The hypothalamo-pituitary portal vessels are illustrated in Figure 10–12.

What about the other two peptides—beta-lipotropin and beta-endorphin—secreted by the anterior pituitary? Their functions are not presently known, but some possibilities are described in Chapters 9 and 13.

The target organs and major functions of the six classical anterior pituitary hormones are summarized in Figure 10–14. Note that the only major function of two of the six is to stimulate secretion of other hormones by their target cells (and to maintain the growth and function of these cells): thyroid-stimulating hormone induces secretion of thyroxine and triiodothyronine from the thyroid; adrenocorticotropic hormone, meaning "hormone that stimulates the adrenal cortex," stimulates the secretion of cortisol by that gland.

Three other anterior pituitary hormones also stimulate secretion of another hormone but have an additional function as well. Follicle-stimulating hormone and luteinizing hormone stimulate secretion of the sex hormones by the gonads—estradiol and progesterone from the ovaries, or testosterone from the testes—but in addition, they regulate the growth and development of ova and sperm. Growth hormone stimulates the liver and many other body cells to secrete a growth-promoting peptide hormone known as **insulin-like growth factor I (IGF-I),** and in addition, growth hormone exerts direct effects on the metabolism of protein, carbohydrate, and lipid by various organs and tissues (Chapter 18).

Prolactin is unique among the anterior pituitary hormones in that it does not exert major control over the secretion of a hormone by another endocrine gland. Rather, its most important action is to stimulate development of the mammary glands and milk production by direct effects upon the breasts. In the male, prolactin may facilitate several components of reproductive function.

It should be emphasized that many, perhaps all, of the anterior pituitary hormones are also secreted by cells elsewhere in the body and may, in those other sites, exert functions (as neurotransmitters, neuromodulators, or paracrine/autocrine agents) quite different from those summarized in Figure 10–14. For example, ACTH is an important neurotransmitter in several brain areas.

Hypophysiotropic Hormones As stated above, secretion of the anterior pituitary hormones is largely regulated by hormones produced by the hypothalamus and collectively called hypophysiotropic hormones. These hormones are secreted by neurons that originate in diverse areas of the hypothalamus and terminate in the median eminence around the capillaries that are the origins of the hypothalamo-pituitary portal vessels. The generation of action potentials in these neurons causes them to release their hormones, which enter the capillaries and are carried by the hypothalamo-pituitary portal vessels to the anterior pituitary (Figure 10–15). There they act upon the various anterior pituitary cells to control their hormone secretions.

Thus, these hypothalamic neurons secrete hormones in a manner identical to that described previously for the hypothalamic neurons whose axons end in the posterior pituitary. In both cases the hormones are made in hypothalamic neurons, pass down axons

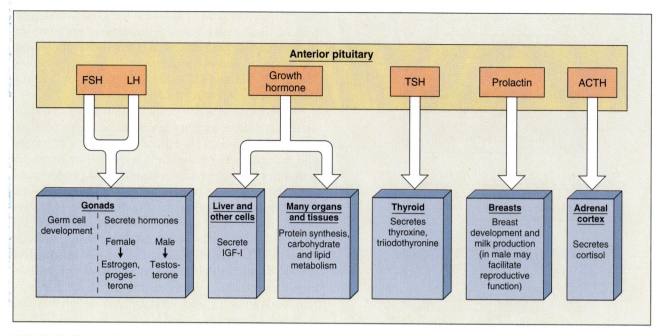

FIGURE 10–14

Targets and major functions of the six classical anterior pituitary hormones.

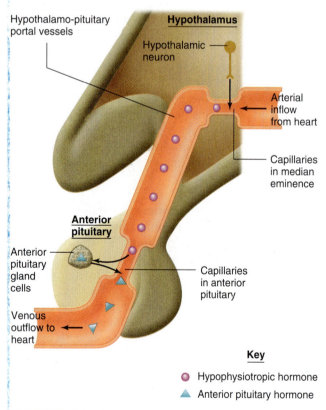

FIGURE 10–15

Hormone secretion by the anterior pituitary is controlled by hypophysiotropic hormones released by hypothalamic neurons and reaching the anterior pituitary by way of the hypothalamo-pituitary portal vessels.

to the neuron terminals, and are released in response to action potentials in the neurons. The crucial differences, however, between the two systems need emphasizing: (1) The axons of the hypothalamic neurons that secrete the posterior pituitary hormones leave the hypothalamus and end in the posterior pituitary, whereas those that secrete the hypophysiotropic hormones remain in the hypothalamus, ending in its median eminence. (2) Most of the posterior pituitary capillaries into which the posterior pituitary hormones are secreted immediately drain into the main bloodstream, which carries the hormones to the heart to be distributed to the entire body. In contrast, the hypophysiotropic hormones enter capillaries in the median eminence of the hypothalamus that do not directly join the main bloodstream, but empty into the hypothalamo-pituitary portal vessels, which carry them to the anterior pituitary (from which the blood then returns to the heart). Thus, the anterior pituitary is uniquely exposed to plasma concentrations of the hypophysiotropic hormones much higher than those existing in the general bloodstream. The hypothalamo-pituitary portal system allows these hormones to act preferentially on the anterior pituitary without influencing other organs and tissues that may have receptors for them.

There are multiple discrete hypophysiotropic hormones, each secreted by a particular group of hypothalamic neurons and influencing the release of one or, in at least one case, two of the anterior pituitary

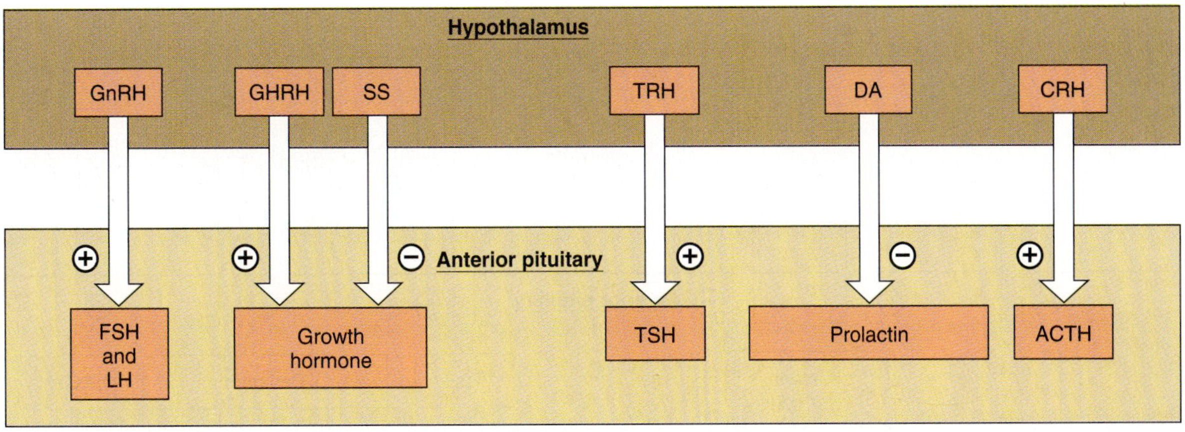

Major known hypophysiotropic hormones	Major effect on anterior pituitary
Corticotropin releasing hormone (CRH) $\longrightarrow$	Stimulates secretion of ACTH
Thyrotropin releasing hormone (TRH)* $\longrightarrow$	Stimulates secretion of TSH
Growth hormone releasing hormone (GHRH) $\longrightarrow$	Stimulates secretion of GH
Somatostatin (SS, also known as growth hormone $\longrightarrow$ release inhibiting hormone, GIH)	Inhibits secretion of GH
Gonadotropin releasing hormone (GnRH) $\longrightarrow$	Stimulates secretion of LH and FSH
Dopamine (DA, also known as prolactin-inhibiting $\longrightarrow$ hormone, PIH)‡	Inhibits secretion of prolactin

*TRH can also stimulate the release of prolactin, but whether this occurs physiologically is unclear.

‡Dopamine is a catecholamine; all the other hypophysiotropic hormones are peptides.

FIGURE 10–16

The effects of definitely established hypophysiotropic hormones on the anterior pituitary. The hypophysiotropic hormones reach the anterior pituitary via the hypothalamo-pituitary portal vessels.

hormones. For simplicity, Figure 10–16 and the text of this chapter summarize only those hypophysiotropic hormones that are known to play important physiological roles and whose structures have been identified. These hypophysiotropic hormones may also exert effects in addition to those described here, but these effects have not been established with certainty under physiological conditions.

Each hypophysiotropic hormone is named for the anterior pituitary hormone whose secretion it controls. Thus, secretion of ACTH (corticotropin) is stimulated by **corticotropin releasing hormone (CRH),** secretion of growth hormone is stimulated by **growth hormone releasing hormone (GHRH),** secretion of thyroid-stimulating hormone (thyrotropin) is stimulated by **thyrotropin releasing hormone (TRH),** and secretion of both luteinizing hormone and follicle-stimulating hormone (the gonadotropins) is stimulated by **gonadotropin releasing hormone (GnRH).**

Note, however, in Figure 10–16, that two of the hypophysiotropic hormones do not *stimulate* release of an anterior pituitary hormone but rather *inhibit* its release.

One of them inhibits secretion of growth hormone and is most commonly called **somatostatin (SS).** The other, **dopamine** (also termed, in this location, **prolactin-inhibiting hormone, PIH**), inhibits secretion of prolactin.

As shown in Figure 10–16, growth hormone is controlled by *two* hypophysiotropic hormones—somatostatin, which inhibits its release, and growth hormone releasing hormone, which stimulates it. The rate of growth hormone secretion depends, therefore, upon the relative amounts of the opposing hormones released by the hypothalamic neurons, as well as upon the relative sensitivities of the anterior pituitary to them. Such dual controls may also exist for prolactin, but the identity of the hypothesized "prolactin releasing hormone" (not shown in Figure 10–16) and its importance in humans remain uncertain; this will be discussed further in Chapter 19.

With one exception, all the known hypophysiotropic hormones are peptides that also occur in locations other than the hypothalamus, particularly in other areas of the central nervous system, where they function as neurotransmitters or neuromodulators,

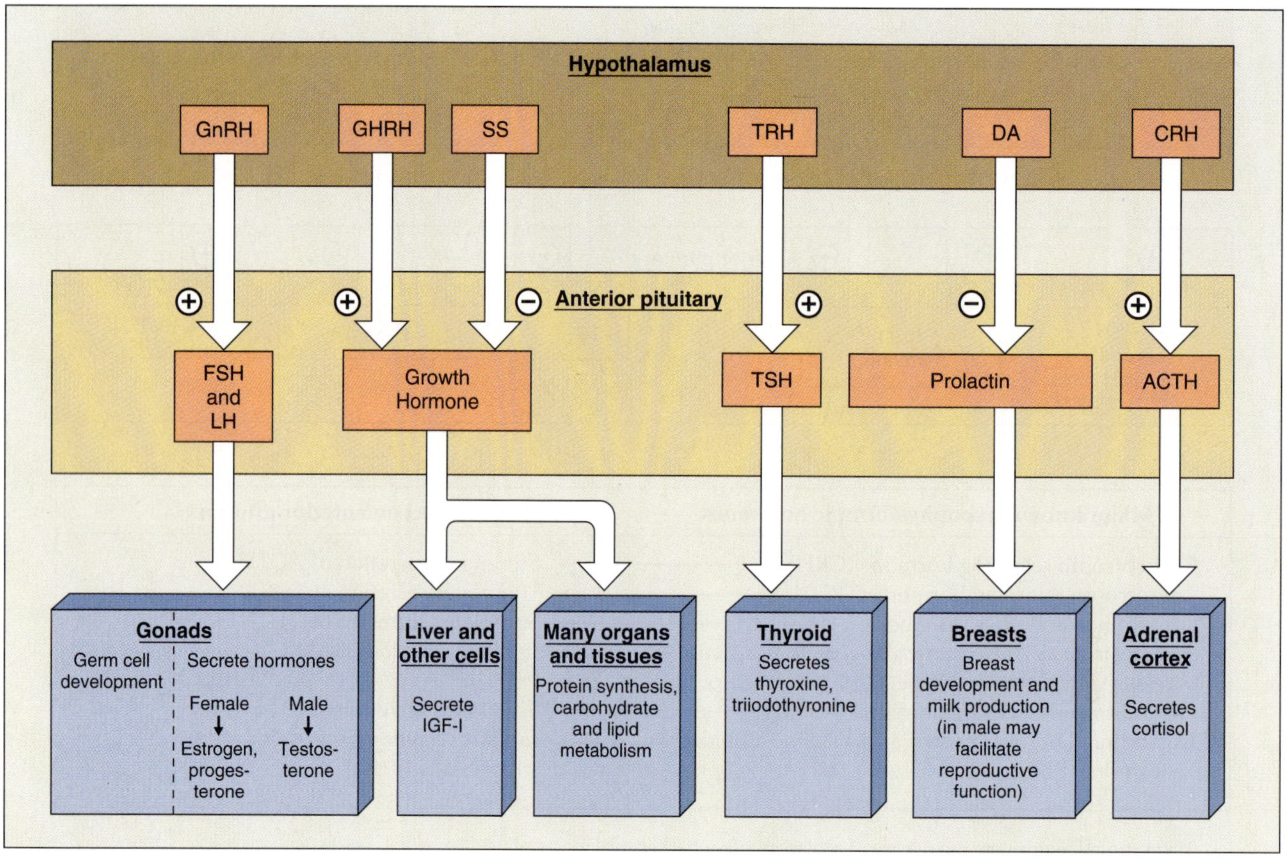

FIGURE 10–17

A combination of Figures 10–14 and 10–16 summarizes the hypothalamic-anterior-pituitary system.

and in the gastrointestinal tract. The one hypophysiotropic hormone that is not a peptide is dopamine. As described earlier, this substance is an amine and a member of the catecholamine family.

Figure 10–17 summarizes the information presented in Figures 10–14 and 10–16 to demonstrate the full sequence of hypothalamic control of endocrine function.

Given that the hypophysiotropic hormones control anterior pituitary function, we must now ask: What controls secretion of the hypophysiotropic hormones? Some of the neurons that secrete hypophysiotropic hormones may possess spontaneous activity, but the firing of most of them requires neural and hormonal input. First, let us deal with the neural input.

Neural Control of Hypophysiotropic Hormones

Neurons of the hypothalamus receive synaptic input, both stimulatory and inhibitory, from virtually all areas of the central nervous system, and specific neural pathways influence secretion of the individual hypophysiotropic hormones. A large number of neurotransmit-

ters (for example, the catecholamines and serotonin) are released at the synapses on the hormone-secreting hypothalamic neurons, and this explains why the secretion of the hypophysiotropic hormones can be altered by drugs that influence these neurotransmitters.

Figure 10–18 illustrates one example of the role of neural input to the hypothalamus. Corticotropin releasing hormone (CRH) from the hypothalamus stimulates the anterior pituitary to secrete ACTH, which in turn stimulates the adrenal cortex to secrete cortisol. A wide variety of stresses, both physical and emotional, act via neural pathways to the hypothalamus to increase CRH secretion, and, hence, ACTH and cortisol secretion, markedly above basal values. Thus, stress is the common denominator of reflexes leading to increased cortisol secretion. Cortisol then functions to facilitate an individual's response to stress. Even in an unstressed person, however, the secretion of cortisol varies in a highly stereotyped manner during a 24-h period because neural rhythms within the central nervous system also impinge upon the hypothalamic neurons that secrete CRH.

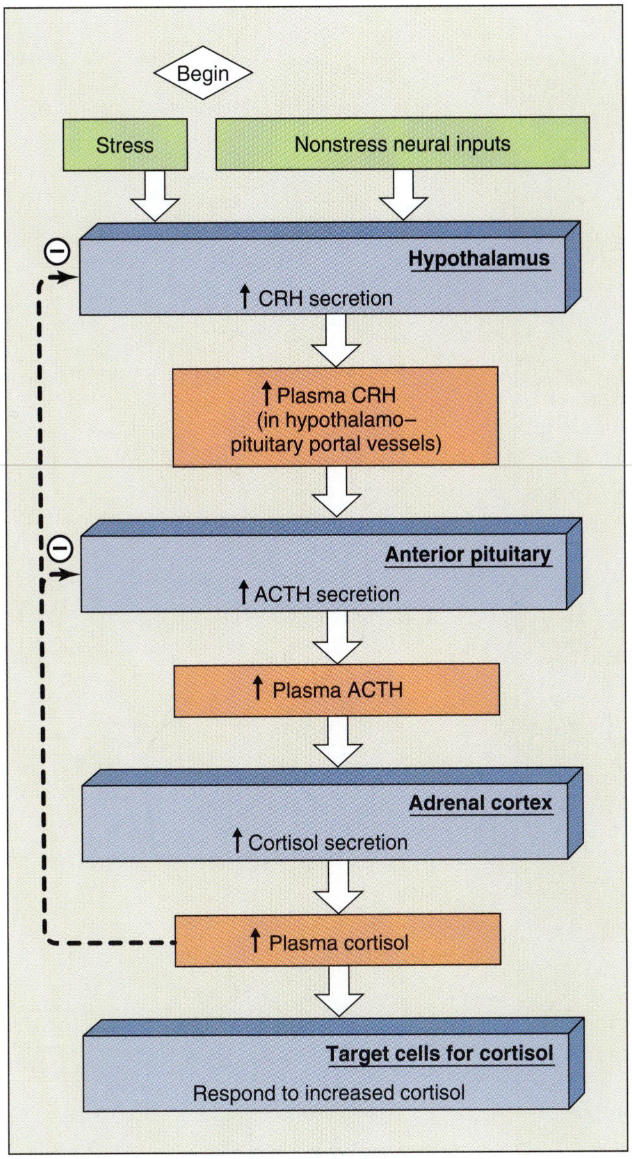

FIGURE 10–18

CRH-ACTH-cortisol sequence. The "stress" input to the hypothalamus is via neural pathways. Cortisol exerts a negative-feedback control over the system by acting on (1) the hypothalamus to inhibit CRH secretion and (2) the anterior pituitary to reduce responsiveness to CRH.

Hormonal Feedback Control of the Hypothalamus and Anterior Pituitary A prominent feature of each of the hormonal sequences initiated by a hypophysiotropic hormone is negative feedback exerted upon the hypothalamo-pituitary system by one or more of the hormones in its sequence. For example, in the CRH-ACTH-cortisol sequence (Figure 10–18), the final hormone, cortisol, acts upon the hypothalamus to

reduce secretion of CRH by causing a decrease in the frequency of action potentials in the neurons secreting CRH. In addition, cortisol acts directly on the anterior pituitary to reduce the response of the ACTH-secreting cells to CRH. Thus, by a double-barreled action, cortisol exerts a negative-feedback control over its own secretion.

Such a system is effective in dampening hormonal responses—that is, in limiting the extremes of hormone secretory rates. For example, when a painful stimulus elicits increased secretion, in turn, of CRH, ACTH, and cortisol, the resulting elevation in plasma cortisol concentration feeds back to inhibit the hypothalamus and anterior pituitary. Therefore, cortisol secretion does not rise as much as it would without these negative feedbacks.

Another adaptive function of these negative-feedback mechanisms is that they maintain the plasma concentration of the final hormone in a sequence relatively constant whenever a disease-induced primary change occurs in the secretion or metabolism of that hormone. An example of this is shown in Figure 10–19 for cortisol.

Another example is provided by the TRH-TSH-TH system. The thyroid hormones (TH) exert a feedback inhibition on the hypothalamo-pituitary system (mainly by decreasing the response of anterior pituitary TSH-secreting cells to the stimulatory effects of TRH). Since iodine is essential for the synthesis of TH, individuals with iodine deficiencies tend to have a deficient production of TH. The resulting decrease in plasma TH concentration relieves some of the feedback inhibition TH exerts on the pituitary. Therefore, more TSH is secreted in response to the TRH coming from the hypothalamus, and the increased plasma TSH stimulates the thyroid gland to enlarge and to utilize more efficiently whatever iodine is available (the enlarged gland is known as *iodine-deficient goiter*). In this manner, plasma TH concentration can be kept quite close to normal.

The situations described above for cortisol and the thyroid hormones, in which the hormone secreted by the third endocrine gland in a sequence exerts a negative-feedback effect over the anterior pituitary and/or hypothalamus, is known as a **long-loop negative feedback** (Figure 10–20). This type of feedback exists for each of the five three-hormone sequences initiated by a hypophysiotropic hormone.

Long-loop feedback does not exist for prolactin since this is one anterior pituitary hormone that does not have major control over another endocrine gland—that is, it does not participate in a three-hormone sequence. Nonetheless, there is negative feedback in

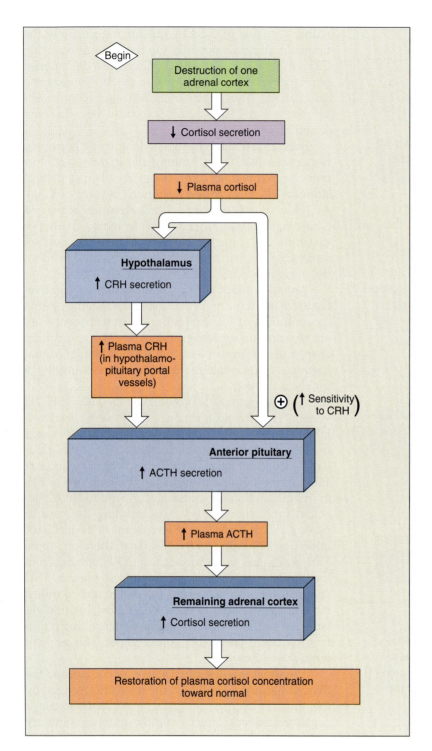

FIGURE 10–19
How negative feedback in a hormonal sequence helps maintain the plasma concentration of the final hormone when disease-induced changes in hormone secretion occur. The same analysis would apply if the original reduction in plasma cortisol were due to excessive metabolism of cortisol rather than deficient secretion.

the prolactin system, for this hormone itself acts upon the hypothalamus to *stimulate* the secretion of dopamine, which then, you will recall, *inhibits* the secretion of prolactin. The influence of an anterior pituitary hormone on the hypothalamus is known as a

short-loop negative feedback (Figure 10–20). Like prolactin, several other anterior pituitary hormones, including growth hormone, also exert such feedback on the hypothalamus.

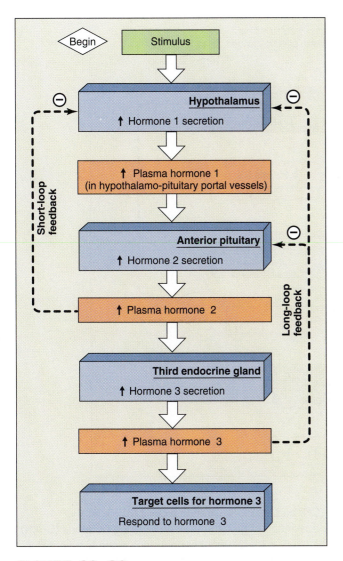

FIGURE 10–20

Short-loop and long-loop feedbacks. Long-loop feedback is exerted on the hypothalamus and/or anterior pituitary by the third hormone in the sequence. Short-loop feedback is exerted by the anterior pituitary hormone on the hypothalamus.

The Role of "Nonsequence" Hormones on the Hypothalamus and Anterior Pituitary It must be emphasized that there are many stimulatory and inhibitory hormonal influences on the hypothalamus and/or anterior pituitary other than those that fit the feedback patterns just described. In other words, a hormone that is not itself in a particular hormonal sequence may nevertheless exert important influences on the secretion of the hypophysiotropic or anterior pituitary hormones in that sequence. For example,

estrogen markedly enhances the secretion of prolactin by the anterior pituitary, even though estrogen secretion is not controlled by prolactin. Thus, one should not view the sequences we have been describing as isolated units.

A Summary Example: Control of Growth Hormone Secretion The three-hormone sequences beginning in the hypothalamus can be extremely complex, incorporating multiple sites of feedback, both long-loop and short-loop, as well as other hormones not in the sequence. Purely for the sake of illustrating this complexity, we describe here the control of growth hormone secretion (Figure 10–21), building upon the information already presented in this chapter.

Recall from Figure 10–16 that the secretion of GH by the anterior pituitary is controlled mainly by two hormones from the hypothalamus: (1) somatostatin, which inhibits GH secretion; and (2) GHRH, which stimulates it. With such a dual control system, the rate of GH secretion at any moment reflects the relative amounts of simultaneous stimulation by GHRH and inhibition by somatostatin. For example, very little growth hormone is secreted during the day in non-stressed persons who are eating normally, but whether this is due to a very low secretion of GHRH or a very high secretion of somatostatin is not yet clear. Similarly, a large number of physiological states (exercise, stress, fasting, a low plasma glucose concentration, and sleep) stimulate growth hormone secretion by decreasing the secretion of somatostatin and/or increasing that of GHRH.

As shown in Figure 10–14 and 10–17, growth hormone stimulates the secretion of another hormone, IGF-I, from the liver and many other target cells of GH. This makes possible a variety of feedbacks by which an increase in plasma GH results in inhibition of GH secretion (Figure 10–21): (1) a short-loop negative feedback exerted by GH on the hypothalamus and (2) a long-loop negative feedback exerted by IGF-I on the hypothalamus. Both of these feedbacks operate by inhibiting secretion of GHRH and/or stimulating secretion of somatostatin, the result in either case being less stimulation of GH secretion. (3) IGF-I also acts directly on the pituitary to inhibit the stimulatory effect of GHRH on GH secretion; again, the result is decreased GH secretion. Finally, the secretion of growth hormone is influenced by several other hormones not in the sequence, for example, by the sex hormones. Some of these hormones may affect GH secretion indirectly by altering the release of somatostatin and/or GHRH from the hypothalamus, whereas others may act directly upon the anterior pituitary.

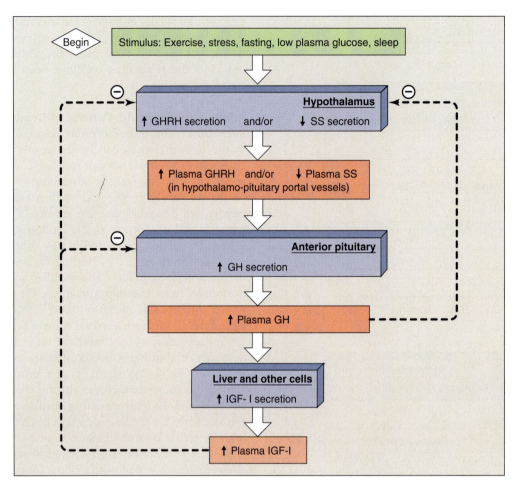

FIGURE 10–21

Hormonal pathways controlling the secretion of growth hormone (GH) and insulin-like growth factor I (IGF-I). At the hypothalamus, the minus sign ⊖ denotes that the input inhibits the secretion of growth hormone releasing hormone (GHRH) and/or stimulates the release of somatostatin (SS). Not shown in the figure is that several hormones not in the sequence (for example, the thyroid hormones) influence growth hormone secretion via effects on the hypothalamus and/or anterior pituitary.

Candidate Hormones

Many substances, termed **candidate hormones,** are suspected of being hormones in humans but are not considered classical hormones for one of two reasons: Either (1) their functions have not been conclusively documented; or (2) they have well-documented functions as paracrine/autocrine agents, but it is not certain that they ever reach additional target cells via the blood, an essential criterion for classification as a hormone. This second category includes certain of the eicosanoids (Chapter 7) and a large number of what are called growth factors (Chapter 18) that are secreted by multiple cell types and stimulate specific cells to undergo cell division and differentiation.

A substance that fits the first category described above is the amino acid derivative **melatonin,** which

is synthesized from serotonin. This candidate hormone is produced by the **pineal gland,** an outgrowth from the roof of the diencephalon of the brain (shown but not labeled in Figure 8–38). The exact functions of melatonin in humans are uncertain, but this hormone probably plays an important role in the setting of the body's circadian rhythms and in sleep (Chapter 7). Its secretion is stimulated by sympathetic neurons that constitute the last link in a neuronal chain primarily triggered by receptors in the eyes; darkness stimulates melatonin secretion, and light inhibits it. Melatonin secretion, therefore, undergoes a marked 24-h cycle, being high at night and low during the day. [Environmental lighting does not *cause* the circadian rhythm, which is of internal origin, but *entrains* it (see Chapter 7)]. Melatonin's ability to reduce the symptoms of *jet lag* when administered in small amounts at the proper time, its relationship to *seasonal affective disorder* ("winter depression"), its potential use as a "natural" sleeping pill, its ability to scavenge damaging free radicals, and its possible role in the control of the reproductive system are all being studied.

Types of Endocrine Disorders

Most endocrine disorders fall into one of four categories: (1) too little hormone (*hyposecretion*); (2) too much hormone (*hypersecretion*); (3) reduced response of the target cells (*hyporesponsiveness*); and (4) increased response of the target cells (*hyperresponsiveness*). In the first two categories, the phrases "too little hormone" and "too much hormone" here mean too little or too much for any given physiological situation. For example, as we shall see, insulin secretion decreases during fasting, and this decrease is an adaptive physiological response, not too little insulin. In contrast, insulin secretion should increase after eating, and if its increase is less than normal, this is too little hormone secretion.

Hyposecretion

An endocrine gland may be secreting too little hormone because the gland is not able to function normally. This is termed *primary hyposecretion.* Examples of primary hyposecretion include (1) genetic absence of a steroid-forming enzyme in the adrenal cortex, leading to decreased cortisol secretion, and (2) dietary deficiency of iodine leading to decreased secretion of thyroid hormones. There are many other causes—infections, toxic chemicals, and so on—all having the common denominator of damaging the endocrine gland.

In contrast to primary hyposecretion, a gland may be secreting too little hormone not because the gland is abnormal but because there is not enough of its tropic hormone. This is termed *secondary hyposecretion.* For example, there may be nothing wrong with the thyroid gland, but it may be secreting too little thyroid hormone because the secretion of TSH by the anterior pituitary is abnormally low. Thus, the hyposecretion by the thyroid gland in this case is secondary to inadequate secretion by the anterior pituitary.

This example raises the next question applicable to any of the other anterior pituitary hormones as well: Is the hyposecretion of TSH primary (that is, due to a defect in the anterior pituitary), or is it secondary to a hypothalamic defect causing too little secretion of TRH? If the latter were true, then we would have the following sequence: hypothalamic defect → primary hyposecretion of TRH → secondary hyposecretion of TSH → *tertiary hyposecretion* of TH.

In diagnosing the presence of hyposecretion, a basic measurement to be made is the concentration of the hormone in either plasma or, for some hormones, urine. The finding of a low concentration will not distinguish between primary and secondary hyposecretion, however. To do this, the concentration of the relevant tropic hormone must also be measured. Thus, in our example, if the hyposecretion of TH is secondary to hyposecretion of TSH, then the plasma concentrations of both will be decreased. If the hyposecretion of TH is primary, then TH concentration will be decreased and TSH concentration will be *increased* because of less negative-feedback inhibition by TH over TSH secretion.

Another diagnostic approach is to attempt to stimulate the gland in question by administering either its tropic hormone or some other substance known to elicit increased secretion. The increase in hormone secretion elicited by the stimulus will be normal if the original hyposecretion is secondary, but less than normal if primary hyposecretion is the problem.

The most common means of treating hormone hyposecretion is to administer the hormone that is missing or present in too small amounts. In cases of secondary hyposecretion, there is a choice since at least two hormones are involved. In our example, the TH deficiency resulting from primary hyposecretion of TSH could theoretically be eliminated by administering either TH or TSH.

Hypersecretion

A hormone can also undergo either *primary hypersecretion* (the gland is secreting too much of the hormone on its own) or *secondary hypersecretion* (there is excessive stimulation of the gland by its tropic hormone). One of the most common causes of primary hypersecretion is the presence of a hormone-secreting endocrine-cell tumor.

The diagnosis of primary versus secondary hypersecretion of a particular hormone is analogous to that of hyposecretion. The concentrations of the hormone and, if relevant, its tropic hormone are measured in plasma or urine. If both concentrations are elevated, then the hormone in question is being secondarily hypersecreted. For example, if both TSH and TH are increased in plasma, then the increased TH must be secondary to increased TSH. If the hypersecretion is primary, there will be a decreased concentration of the tropic hormone because of negative feedback by the high concentration of the hormone being hypersecreted. Again as with hyposecretion, one can get hypersecretion of a hypophysiotropic hormone, leading to secondary hypersecretion of an anterior pituitary hormone, leading to *tertiary hypersecretion* of the peripheral endocrine gland.

When an endocrine tumor is the cause of hypersecretion, it can often be removed surgically or destroyed with radiation. In many cases, hypersecretion can also be blocked by drugs that inhibit the hormone's synthesis. Alternatively, the situation can be treated with drugs that do not alter the hormone's secretion but instead block the hormone's actions on its target cells.

Hyporesponsiveness and Hyperresponsiveness

In some cases, the endocrine system may be dysfunctioning even though there is nothing wrong with hormone secretion. The problem is that the target cells do not respond normally to the hormone. This condition is termed hyporesponsiveness (Table 10–3). An important example of a disease resulting from hyporesponsiveness is the major form of *diabetes mellitus* ("sugar diabetes"), in which the target cells of the hormone insulin are hyporesponsive to this hormone.

One cause of hyporesponsiveness is deficiency of receptors for the hormone. For example, certain men have a genetic defect manifested by the *absence of receptors for dihydrotestosterone,* the form of testosterone active in many target cells. In such men, these cells are unable to bind dihydrotestosterone, and the result is lack of development of certain male characteristics, just as though the hormone were not being produced.

In a second type of hyporesponsiveness, the receptors for a hormone may be normal, but some event occurring after the hormone binds to receptors may be defective. For example, the activated receptor might be unable to stimulate formation of cyclic AMP or open a plasma-membrane channel.

A third cause of hyporesponsiveness applies to hormones that require metabolic activation by some other tissue after secretion. There may be a lack or deficiency of the enzymes that catalyze the activation. For example, some men secrete testosterone normally and have normal receptors for dihydrotestosterone but are missing the enzyme that converts testosterone to dihydrotestosterone.

In situations characterized by hyporesponsiveness to a hormone, the plasma concentration of the hormone in question is normal or elevated, but the response of target cells to administered hormone is diminished.

Finally, hyperresponsiveness to a hormone can also occur and cause problems. For example, the thyroid hormone causes an up-regulation of certain receptors for epinephrine; therefore, hypersecretion of the thyroid hormones (hyperthyroidism) causes, in turn, a hyperresponsiveness to epinephrine. One result of this is the increased heart rate typical of persons with hyperthyroidism.

TABLE 10–3 Use of Plasma Hormone Measurements to Diagnose the Problem in a Person with Symptoms of Hyperthyroidism

Plasma Concentrations	Diagnosis
↑TH, ↓TSH	Primary hypersecretion of TH (primary problem is in thyroid gland)
↑TH, ↑TSH	Secondary hypersecretion of TH (primary problem is in hypothalamus or anterior pituitary)
Normal TH	Hyperresponsiveness to TH (problem is in target cells for TH)

SUMMARY

I. The endocrine system is one of the body's two major communications systems. It consists of all the glands that secrete hormones, which are chemical messengers carried by the blood from the endocrine glands to target cells elsewhere in the body.

Hormone Structures and Synthesis

I. The amine hormones are the iodine-containing thyroid hormones—thyroxine and triiodothyronine—and the catecholamines secreted by the adrenal medulla and the hypothalamus.
II. The majority of hormones are peptides, many of which are synthesized as larger molecules, which are then cleaved.
III. Steroid hormones are produced from cholesterol by the adrenal cortex and the gonads, and by the placenta during pregnancy.
  a. The most important steroid hormones produced by the adrenal cortex are the mineralocorticoid aldosterone, the glucocorticoid cortisol, and two androgens.
  b. The ovaries produce mainly estradiol and progesterone, and the testes mainly testosterone.

Hormone Transport in the Blood

I. Peptide hormones and catecholamines circulate dissolved in the plasma water, but steroid and thyroid hormones circulate mainly bound to plasma proteins.

Hormone Metabolism and Excretion

I. The liver and kidneys are the major organs that remove hormones from the plasma by metabolizing or excreting them.
II. The peptide hormones and catecholamines are rapidly removed from the blood, whereas the steroid and thyroid hormones are removed more slowly.
III. After their secretion, some hormones are metabolized to more active molecules in their target cells or other organs.

Mechanisms of Hormone Action

I. The great majority of receptors for steroid and thyroid hormones are inside the target cells; those for the peptide hormones and catecholamines are on the plasma membrane.

II. Hormones can cause up-regulation and down-regulation of their own receptors and those of other hormones. The induction of one hormone's receptors by another hormone increases the first hormone's effectiveness and may be essential to permit the first hormone to exert its effects.

III. Receptors activated by peptide hormones and catecholamines utilize one or more of the signal transduction pathways available to plasma-membrane receptors; the result is altered membrane potential or activity of proteins in the cell.

IV. Intracellular receptors activated by steroid and thyroid hormones function as transcription factors, combining with DNA in the nucleus and inducing the transcription of DNA into mRNA; the result is increased synthesis of particular proteins.

V. In pharmacological doses, hormones can have effects not seen under ordinary circumstances.

Inputs That Control Hormone Secretion

I. The secretion of a hormone may be controlled by the plasma concentration of an ion or nutrient that the hormone regulates, by neural input to the endocrine cells, and by one or more hormones.

II. The autonomic nervous system is the neural input controlling many hormones, but the hypothalamic and posterior pituitary hormones are controlled by neurons in the brain.

Control Systems Involving the Hypothalamus and Pituitary

I. The pituitary gland, comprising the anterior pituitary and the posterior pituitary, is connected to the hypothalamus by a stalk containing nerve axons and blood vessels.

II. Specific axons, whose cell bodies are in the hypothalamus, terminate in the posterior pituitary and release oxytocin and vasopressin.

III. The anterior pituitary secretes growth hormone (GH), thyroid-stimulating hormone (TSH), adrenocorticotropic hormone (ACTH), prolactin, and two gonadotropic hormones—follicle-stimulating hormone (FSH) and luteinizing hormone (LH). The functions of these hormones are summarized in Figure 10–14.

IV. Secretion of the anterior pituitary hormones is controlled mainly by hypophysiotropic hormones secreted into capillaries in the median eminence of the hypothalamus and reaching the anterior pituitary via the portal vessels connecting the hypothalamus and anterior pituitary. The actions of the hypophysiotropic hormones on the anterior pituitary are summarized in Figure 10–16.

V. The secretion of each hypophysiotropic hormone is controlled by neuronal and hormonal input to the hypothalamic neurons producing it.

a. In each of the three-hormone sequences beginning with a hypophysiotropic hormone, the third hormone exerts a long-loop negative-feedback effect on the secretion of the hypothalamic and/or anterior pituitary hormone.

b. The anterior pituitary hormone may exert a short-loop negative-feedback inhibition of the hypothalamic releasing hormone(s) controlling it.

c. Hormones not in a particular sequence can also influence secretion of the hypothalamic and/or anterior pituitary hormones in that sequence.

Candidate Hormones

I. Substances that are suspected of functioning as hormones but have not yet been proven to do so are called candidate hormones.

II. Melatonin is a candidate hormone secreted with a 24-h rhythm by the pineal gland; it probably exerts effects on the body's circadian rhythms.

Types of Endocrine Disorders

I. Endocrine disorders may be classified as hyposecretion, hypersecretion, and target-cell hyporesponsiveness or hyperresponsiveness.

a. Primary disorders are those in which the defect is in the cells that secrete the hormone.

b. Secondary disorders are those in which there is too much or too little tropic hormone.

c. Hyporesponsiveness is due to an alteration in the receptors for the hormone, to disordered postreceptor events, or to failure of normal metabolic activation of the hormone in those cases requiring such activation.

II. These disorders can be distinguished by measurements of the hormone and any tropic hormones under both basal conditions and during experimental stimulation of the hormone's secretion.

KEY TERMS

endocrine system	steroid hormone
endocrine gland	gonad
hormone	aldosterone
target cell	mineralocorticoid
amine hormone	cortisol
thyroid gland	glucocorticoid
thyroglobulin	androgen
thyroxine (T$_4$)	testosterone
triiodothyronine (T$_3$)	estradiol
thyroid hormones (TH)	progesterone
iodine	up-regulation
adrenal gland	down-regulation
adrenal medulla	permissiveness
adrenal cortex	tropic hormone
epinephrine (E)	pituitary gland
norepinephrine (NE)	hypothalamus
catecholamine	anterior pituitary
peptide hormone	posterior pituitary
prohormone	median eminence

hypothalamo-pituitary
  portal vessels
oxytocin
vasopressin
antidiuretic hormone
  (ADH)
hypophysiotropic hormones
follicle-stimulating hormone
  (FSH)
luteinizing hormone (LH)
growth hormone (GH)
thyroid-stimulating hormone
  (TSH)
prolactin
adrenocorticotropic hormone
  (ACTH)
gonadotropic hormones
insulin-like growth factor
  I (IGF-I)

corticotropin releasing
  hormone (CRH)
growth hormone releasing
  hormone (GHRH)
thyrotropin releasing
  hormone (TRH)
gonadotropin releasing
  hormone (GnRH)
somatostatin (SS)
dopamine (prolactin-
  inhibiting hormone, PIH)
long-loop negative
  feedback
short-loop negative
  feedback
candidate hormones
melatonin
pineal gland

CHAPTER 10 REVIEW QUESTIONS

1. What are the three general chemical classes of hormones?
2. What essential nutrient is needed for synthesis of the thyroid hormones?
3. Which catecholamine is secreted in the largest amount by the adrenal medulla?
4. What are the major hormones produced by the adrenal cortex? By the testes? By the ovaries?
5. Which classes of hormones are carried in the blood mainly as unbound, dissolved hormone? Mainly bound to plasma proteins?
6. Do protein-bound hormones cross capillary walls?
7. Which organs are the major sites of hormone excretion and metabolic transformation?
8. How do the rates of metabolism and excretion differ for the various classes of hormones?
9. What must some hormones undergo after their secretion to become activated?
10. Contrast the locations of receptors for the various classes of hormones.
11. How do hormones influence the concentrations of their own receptors and those of other hormones? How does this explain permissiveness in hormone action?
12. Describe the sequence of events when peptide or catecholamine hormones bind to their receptors.
13. Describe the sequence of events when steroid or thyroid hormones bind to their receptors.
14. What are the direct inputs to endocrine glands controlling hormone secretion?
15. How does control of hormone secretion by plasma mineral ions and nutrients achieve negative-feedback control of these substances?

16. What roles does the autonomic nervous system play in controlling hormone secretion?
17. What groups of hormones receive input from neurons located in the brain rather than in the autonomic nervous system?
18. Describe the anatomical relationships between the hypothalamus and the pituitary.
19. Name the two posterior pituitary hormones and describe their site of synthesis and mechanism of release.
20. List all six well-established anterior pituitary hormones and their functions.
21. List the major hypophysiotropic hormones and the hormone whose release each controls.
22. What kinds of inputs control secretion of the hypophysiotropic hormones?
23. Diagram the CRH-ACTH-cortisol system.
24. What is the difference between long-loop and short-loop negative feedback in the hypothalamo-anterior pituitary system?
25. How would you distinguish between primary and secondary hyposecretion of a hormone? Between hyposecretion and hyporesponsiveness?

CLINICAL TERMS

masculinization of a female
pharmacological effect
iodine-deficient goiter
jet lag
seasonal affective disorder
hyposecretion
hypersecretion
hyporesponsiveness
hyperresponsiveness

primary hyposecretion
secondary hyposecretion
tertiary hyposecretion
primary hypersecretion
secondary hypersecretion
tertiary hypersecretion
diabetes mellitus
absence of receptors for
  dihydrotestosterone

THOUGHT QUESTIONS

(Answers are given in Appendix A.)

1. In an experimental animal, the sympathetic preganglionic fibers to the adrenal medulla are cut. What happens to the plasma concentration of epinephrine at rest and during stress?
2. During pregnancy there is an increase in the production (by the liver) and, hence, the plasma concentration of the major plasma binding protein for the thyroid hormones (TH). This causes a sequence of events involving feedback that results in an increase in the plasma concentration of TH, but no evidence of hyperthyroidism. Describe the sequence of events.
3. A child shows the following symptoms: deficient growth; failure to show sexual development;

decreased ability to respond to stress. What is the most likely cause of all these symptoms?

4. If all the neural connections between the hypothalamus and pituitary were severed, the secretion of which pituitary hormones would be affected? Which pituitary hormones would not be affected?

5. An antibody to a peptide combines with the peptide and renders it nonfunctional. If an animal were given an antibody to somatostatin, the secretion of which anterior pituitary hormone would change and in what direction?

6. A drug that blocks the action of norepinephrine is injected directly into the hypothalamus of an experimental animal, and the secretion rates of several anterior pituitary hormones are observed to change. How is this possible, since norepinephrine is not a hypophysiotropic hormone?

7. A person is receiving very large doses of a cortisol-like drug to treat her arthritis. What happens to her secretion of cortisol?

8. A person with symptoms of hypothyroidism (for example, sluggishness and intolerance to cold) is found to have abnormally low plasma concentrations of T_4, T_3, and TSH. After an injection of TRH, the plasma concentrations of all three hormones increase. Where is the site of the defect leading to the hypothyroidism?

Muscle

The ability to use chemical energy to produce force and movement is present to a limited extent in most cells. It is, however, in muscle cells that this process has become dominant. The primary function of these specialized cells is to generate the forces and movements used in the regulation of the internal environment and to produce movements in the external environment. In humans, the ability to communicate, whether by speech, writing, or artistic expression, also depends on muscle contractions. Indeed, it is only by controlling the activity of muscles that the human mind ultimately expresses itself.

Three types of muscle tissue can be identified on the basis of structure, contractile properties, and control mechanisms: (1) **skeletal muscle**, (2) **smooth muscle**, and (3) **cardiac muscle.** Most skeletal muscle, as the name implies, is attached to bone, and its contraction is responsible for supporting and moving the skeleton. The contraction of skeletal muscle is initiated by impulses in the neurons to the muscle and is usually under voluntary control.

Sheets of smooth muscle surround various hollow organs and tubes, including the stomach, intestines, urinary bladder, uterus, blood vessels, and airways in the lungs. Contraction of the smooth muscle surrounding hollow organs may propel the luminal contents through the organ, or it may regulate internal flow by changing the tube diameter. In addition, small bundles of smooth-muscle cells are attached to the hairs of the skin and iris of the eye. Smooth-muscle contraction is controlled by the autonomic nervous system, hormones, autocrine/paracrine agents, and other local chemical signals. Some smooth muscles contract spontaneously, however, even in the absence of such signals. In contrast to skeletal muscle, smooth muscle is not normally under voluntary control.

Cardiac muscle is the muscle of the heart. Its contraction propels blood through the circulatory system. Like smooth muscle, it is regulated by the autonomic nervous system, hormones, and autocrine/paracrine agents, and certain portions of it can undergo spontaneous contractions.

Although there are significant differences in these three types of muscle, the force-generating mechanism is similar in all of them. Skeletal muscle will be described first, followed by a discussion of smooth muscle. Cardiac muscle, which combines some of the properties of both skeletal and smooth muscle, will be described in Chapter 14 in association with its role in the circulatory system.

SKELETAL MUSCLE

Structure

A single skeletal-muscle cell is known as a **muscle fiber.** Each muscle fiber is formed during development by the fusion of a number of undifferentiated, mononucleated cells, known as **myoblasts,** into a single cylindrical, multinucleated cell. Skeletal muscle differentiation is completed around the time of birth, and these differentiated fibers continue to increase in size during growth from infancy to adult stature, but no new fibers are formed from myoblasts. Adult skeletal-muscle fibers have diameters between 10 and 100 μm and lengths that may extend up to 20 cm.

If skeletal-muscle fibers are destroyed after birth as a result of injury, they cannot be replaced by the division of other existing muscle fibers. New fibers can be formed, however, from undifferentiated cells known as **satellite cells,** which are located adjacent to the muscle fibers and undergo differentiation similar to that followed by embryonic myoblasts. This capacity for forming new skeletal-muscle fibers is considerable but will not restore a severely damaged muscle to full strength. Much of the compensation for a loss of muscle tissue occurs through an increase in the size of the remaining muscle fibers.

The term **muscle** refers to a number of muscle fibers bound together by connective tissue (Figure 11–1). The relationship between a single muscle fiber and a muscle is analogous to that between a single neuron and a nerve, which is composed of the axons of many neurons. Muscles are usually linked to bones by bundles of collagen fibers known as **tendons,** which are located at each end of the muscle.

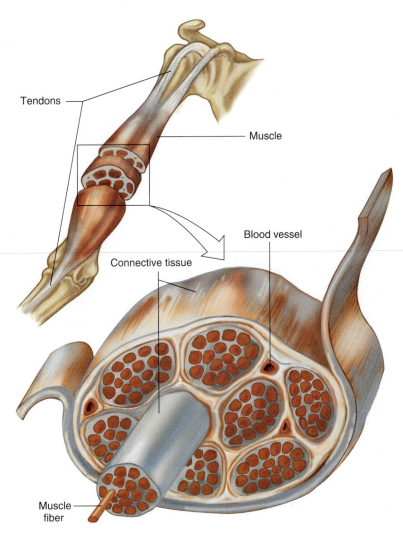

FIGURE 11–1

Organization of cylindrical skeletal-muscle fibers in a muscle that is attached to bones by tendons.

In some muscles, the individual fibers extend the entire length of the muscle, but in most, the fibers are shorter, often oriented at an angle to the longitudinal axis of the muscle. The transmission of force from muscle to bone is like a number of people pulling on a rope, each person corresponding to a single muscle fiber and the rope corresponding to the connective tissue and tendons.

Some tendons are very long, with the site of tendon attachment to bone far removed from the end of the muscle. For example, some of the muscles that move the fingers are in the forearm, as one can observe by wiggling one's fingers and feeling the movement of the muscles in the lower arm. These muscles are connected to the fingers by long tendons.

The most striking feature seen when observing skeletal- or cardiac-muscle fibers through a light microscope (Figure 11–2) is a series of light and dark bands perpendicular to the long axis of the fiber. Because of this characteristic banding, both types are known as **striated muscle** (Figure 11–3). Smooth-muscle cells do not show a banding pattern. The striated pattern in skeletal and cardiac fibers results from the arrangement of numerous thick and thin filaments in the cytoplasm into approximately cylindrical bundles (1 to 2 μm in diameter) known as **myofibrils** (Figure 11–4). Most of the cytoplasm of a fiber is filled with myofibrils, each of which extends from one end of the fiber to the other and is linked to the tendons at the ends of the fiber.

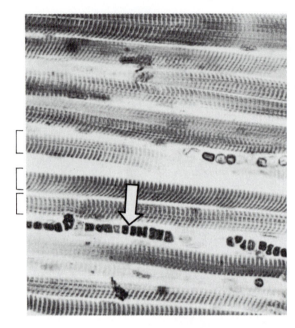

FIGURE 11–2

Skeletal-muscle fibers viewed through a light microscope. Each bracket at the left indicates one muscle fiber. Arrow indicates a blood vessel containing red blood cells.

From Edward K. Keith and Michael H. Ross, "Atlas of Descriptive Histology," Harper & Row, New York, 1968.

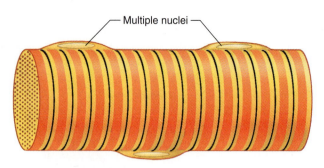

Skeletal-muscle fiber

Cardiac-muscle fiber

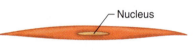

Smooth-muscle fiber

FIGURE 11–3

The three types of muscle fibers. Note the differences in fiber diameter. (The single nucleus in the center of the cardiac-muscle fiber is not shown.)

The thick and thin filaments in each myofibril (Figures 11–4 and 11–5) are arranged in a repeating pattern along the length of the myofibril. One unit of this repeating pattern is known as a **sarcomere** (Greek, *sarco*, muscle; *mere*, small). The **thick filaments** are composed almost entirely of the contractile protein **myosin.** The **thin filaments** (which are about half the diameter of the thick filaments) contain the contractile protein **actin,** as well as to two other proteins—troponin and tropomyosin—that play important roles in regulating contraction, as we shall see.

The thick filaments are located in the middle of each sarcomere, where their orderly parallel arrangement produces a wide, dark band known as the **A band** (Figure 11–4). Each sarcomere contains two sets of thin filaments, one at each end. One end of each thin filament is anchored to a network of interconnecting proteins known as the **Z line,** whereas the other end overlaps a portion of the thick filaments. Two successive Z lines define the limits of one sarcomere. Thus, thin filaments from two adjacent sarcomeres are anchored to the two sides of each Z line.

A light band, known as the **I band** (Figure 11–4), lies between the ends of the A bands of two adjacent sarcomeres and contains those portions of the thin filaments that do not overlap the thick filaments. It is bisected by the Z line.

Two additional bands are present in the A-band region of each sarcomere (Figure 11–5). The **H zone** is a narrow light band in the center of the A band. It corresponds to the space between the opposing ends of the two sets of thin filaments in each sarcomere; hence, only thick filaments, specifically their central parts, are found in the H zone. A narrow, dark band in the center of the H zone is known as the **M line** and corresponds to proteins that link together the central region of the thick filaments. In addition, filaments composed of the protein **titin** extend from the Z line to the M line and are linked to both the M-line proteins and the thick filaments. Both the M-line linkage between thick filaments and the titin filaments act to maintain the regular array of thick filaments in the middle of each sarcomere. Thus, neither the thick nor the thin filaments are free-floating.

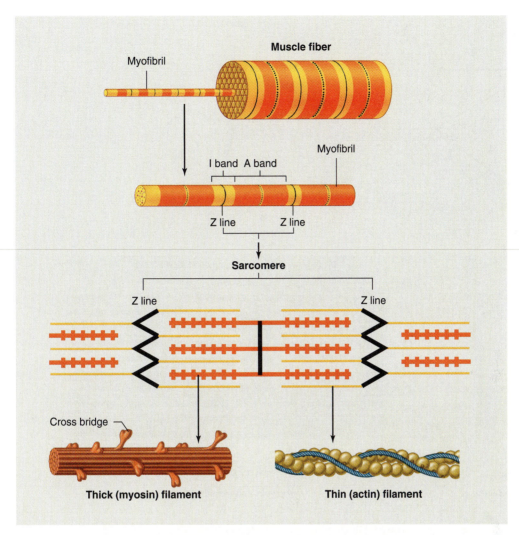

FIGURE 11–4

Arrangement of filaments in a skeletal-muscle fiber that produces the striated banding pattern. 🏃 ▭

A cross section through the A bands (Figure 11–6), shows the regular, almost crystalline, arrangement of overlapping thick and thin filaments. Each thick filament is surrounded by a hexagonal array of six thin filaments, and each thin filament is surrounded by a triangular arrangement of three thick filaments. Altogether there are twice as many thin as thick filaments in the region of filament overlap.

The space between overlapping thick and thin filaments is bridged by projections known as **cross bridges.** These are portions of myosin molecules that extend from the surface of the thick filaments toward the thin filaments (Figures 11–4 and 11–7). During muscle contraction, the cross bridges make contact with the thin filaments and exert force on them.

Molecular Mechanisms of Contraction

The term **contraction,** as used in muscle physiology, does not necessarily mean "shortening"; rather it refers only to the turning on of the force-generating sites—the cross bridges—in a muscle fiber. Following contraction, the mechanisms that initiate force generation are turned off, and tension declines, allowing **relaxation** of the muscle fiber.

Sliding-Filament Mechanism

When force generation produces shortening of a skeletal-muscle fiber, the overlapping thick and thin

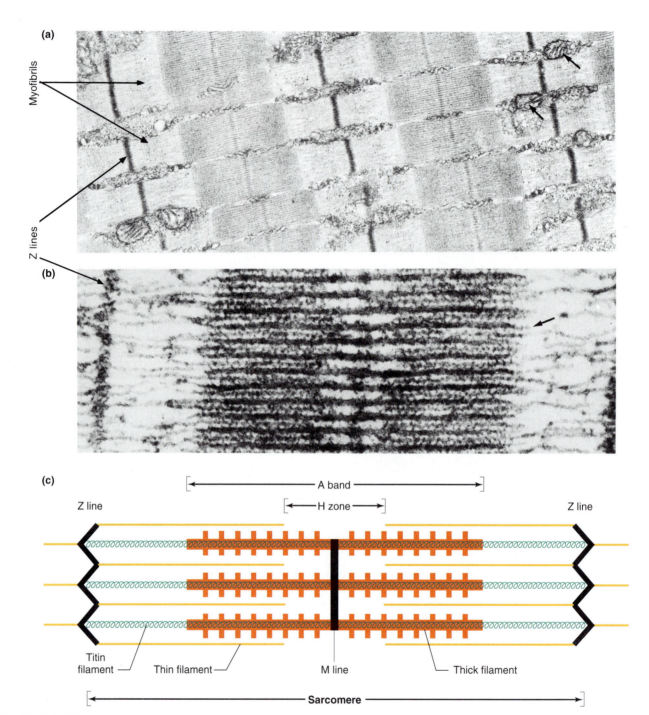

FIGURE 11–5

(a) Numerous myofibrils in a single skeletal-muscle fiber (arrows in upper right corner indicate mitochondria between the myofibrils). (b) High magnification of a sarcomere within a myofibril (arrow at the right of A band indicates end of a thick filament). (c) Arrangement of the thick and thin filaments in the sarcomere shown in b. 𝔛

(a)

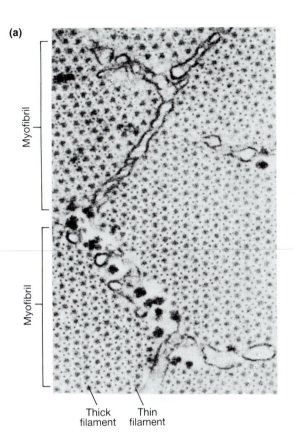

Myofibril

Myofibril

Thick filament Thin filament

(b)

Thick filament — Thin filament —

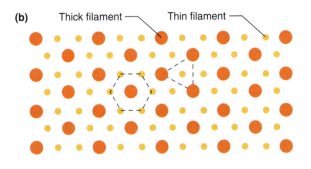

FIGURE 11–6

(a) Electron micrograph of a cross section through several myofibrils in a single skeletal-muscle fiber.

From H. E. Huxley, J. Mol. Biol., 37:507–520 (1968).

(b) Hexagonal arrangements of the thick and thin filaments in the overlap region in a single myofibril. Six thin filaments surround each thick filament, and three thick filaments surround each thin filament.

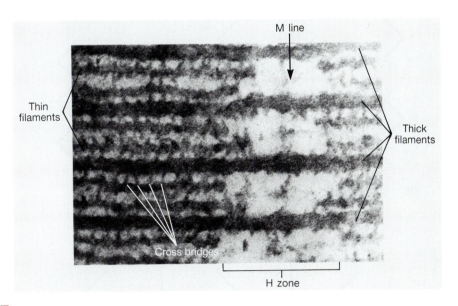

M line

Thin filaments

Thick filaments

Cross bridges

H zone

FIGURE 11–7

High-magnification electron micrograph in the filament-overlap region near the middle of a sarcomere. Cross bridges between the thick and thin filaments can be seen at regular intervals along the filaments.

From H. E. Huxley and J. Hanson, in G. H. Bourne (ed.), "The Structure and Function of Muscle," Vol. 1, Academic Press, New York, 1960.

filaments in each sarcomere move past each other, propelled by movements of the cross bridges. During this shortening of the sarcomeres, there is no change in the lengths of either the thick or thin filaments (Figure 11–8). This is known as the **sliding-filament mechanism** of muscle contraction.

During shortening, each cross bridge attached to a thin filament moves in an arc much like an oar on a boat. This swiveling motion of many cross bridges forces the thin filaments at either end of the A band toward the center of the sarcomere, thereby shortening the sarcomere (Figure 11–9). One stroke of a cross bridge produces only a very small movement of a thin filament relative to a thick filament. As long as a muscle fiber remains "turned on," however, each cross bridge repeats its swiveling motion many times, resulting in large displacements of the filaments.

Let us look more closely at these events. A muscle fiber's ability to generate force and movement depends on the interactions of the two so-called contractile proteins—myosin in the thick filaments and actin in the thin filaments—and energy provided by ATP.

An actin molecule is a globular protein composed of a single polypeptide that polymerizes with other actins to form two intertwined helical chains (Figure 11–10) that make up the core of a thin filament. Each actin molecule contains a binding site for myosin. The myosin molecule, on the other hand, is composed of two large polypeptide heavy chains and four smaller light chains. These polypeptides combine to form a molecule that consists of two globular heads (containing heavy and light chains) and a long tail formed by the two intertwined heavy chains (Figure 11–11b). The tail of each myosin molecule lies along the axis of the thick filament, and the two globular heads extend out to the sides, forming the cross bridges. Each globular head contains two binding sites, one for actin and one for ATP. The ATP binding site also serves as an enzyme—an ATPase that hydrolyzes the bound ATP.

The myosin molecules in the two ends of each thick filament are oriented in opposite directions, such that their tail ends are directed toward the center of the filament (Figure 11–11a). Because of this arrangement, the power strokes of the cross bridges move the attached thin filaments at the two ends of the sarcomere toward the center during shortening (see Figure 11–9).

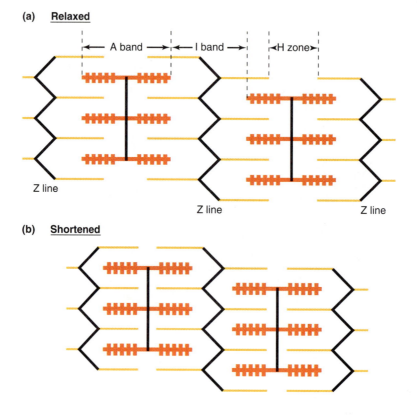

(a) **Relaxed**

A band — I band — H zone

Z line   Z line   Z line

(b) **Shortened**

FIGURE 11–8

The sliding of thick filaments past overlapping thin filaments produces shortening with no change in thick or thin filament length. The I band and H zone have, however, decreased.

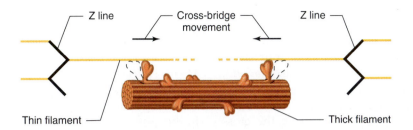

FIGURE 11–9

Cross bridges in the thick filaments bind to actin in the thin filaments and undergo a conformational change that propels the thin filaments toward the center of a sarcomere. (Only 2 of the approximately 200 cross bridges in each thick filament are shown.)

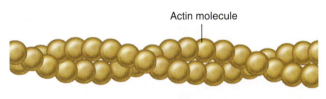

Thin filament

FIGURE 11–10

Two intertwined helical chains of actin molecules form the primary structure of the thin filaments.

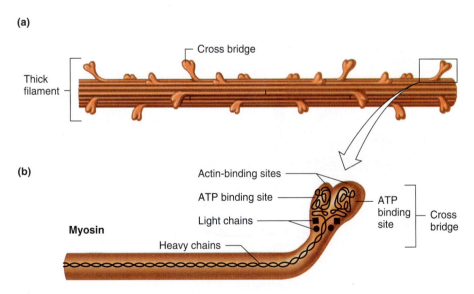

FIGURE 11–11

(a) The heavy chains of myosin molecules form the core of a thick filament. The myosin molecules are oriented in opposite directions in either half of a thick filament. (b) Structure of a myosin molecule. The two globular heads of each myosin molecule extend from the sides of a thick filament forming a cross bridge.

The sequence of events that occurs between the time a cross bridge binds to a thin filament, moves, and then is set to repeat the process is known as a **cross-bridge cycle.** Each cycle consists of four steps: (1) attachment of the cross bridge to a thin filament, (2) movement of the cross bridge, producing tension in the thin filament, (3) detachment of the cross bridge from the thin filament, and (4) energizing the cross bridge so that it can again attach to a thin filament and repeat the cycle. Each cross bridge undergoes its own cycle of movement independently of the other cross bridges, and at any one instant during contraction only a portion of the cross bridges overlapping a thin filament are attached to the thin filaments and producing

tension, while others are in a detached portion of their cycle.

The chemical and physical events during the four steps of a cross-bridge cycle are illustrated in Figure 11–12. At the conclusion (step 4) of the preceding cycle, the ATP bound to myosin is split, releasing chemical energy which results in a conformational change in the cross bridge. This produces an energized form of myosin (M*) to which the products of ATP hydrolysis, ADP and inorganic phosphate (P_i), are still bound. This storage of energy in myosin is analogous to the storage of potential energy in a stretched spring.

$$\text{Step 4} \qquad A + M \cdot ATP \xrightarrow[\text{ATP hydrolysis}]{} A + M^* \cdot ADP \cdot P_i$$

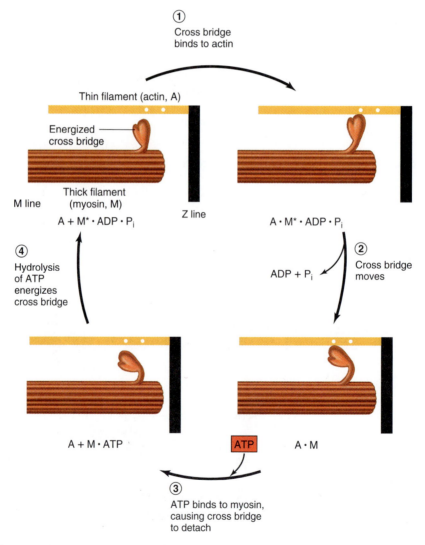

① Cross bridge binds to actin

Thin filament (actin, A)

Energized cross bridge

M line

Thick filament (myosin, M)

$A + M^* \cdot ADP \cdot P_i$

Z line

$A \cdot M^* \cdot ADP \cdot P_i$

④ Hydrolysis of ATP energizes cross bridge

② Cross bridge moves

$ADP + P_i$

$A + M \cdot ATP$

ATP

$A \cdot M$

③ ATP binds to myosin, causing cross bridge to detach

FIGURE 11–12

Chemical and mechanical changes during the four stages of a cross-bridge cycle. In a resting muscle fiber, contraction begins with the binding of a cross bridge to actin in a thin filament—step 1. (M* represents an energized myosin cross bridge.)

A new cross-bridge cycle begins with the binding of an energized myosin cross bridge to actin (A) in a thin filament (step 1):

Step 1 $A + M^* \cdot ADP \cdot P_i \longrightarrow A \cdot M^* \cdot ADP \cdot P_i$
                                      Actin
                                      binding

The binding of energized myosin to actin triggers the release of the strained conformation of the energized bridge, which produces the movement of the bound cross bridge (step 2) and the release of ADP and P_i:

Step 2 $A \cdot M^* \cdot ADP \cdot P_i \longrightarrow A \cdot M + ADP + P_i$
                                 Cross-bridge
                                 movement

This sequence of energy storage and release by myosin is analogous to the operation of a mousetrap: Energy is stored in the trap by cocking the spring (ATP hydrolysis) and released after springing the trap (binding to actin).

During the cross-bridge movement, myosin is bound very firmly to actin, and this linkage must be broken in order to allow the cross bridge to be re-energized and repeat the cycle. The binding of a molecule of ATP to myosin breaks the link between actin and myosin (step 3):

Step 3 $A \cdot M + ATP \longrightarrow A + M \cdot ATP$
                              Cross-bridge
                              dissociation from actin

The dissociation of actin and myosin by ATP is an example of allosteric regulation of protein activity. The binding of ATP at one site on myosin decreases myosin's affinity for actin bound at another site. Thus, ATP is acting as a modulator molecule controlling the binding of actin to myosin. Note that ATP is not split in this step; that is, it is not acting as an energy source but only as a modulator molecule that produces an allosteric modulation of the myosin head that weakens the binding of myosin to actin.

Then, following the dissociation of actin and myosin, the ATP bound to myosin is split (step 4), thereby re-forming the energized state of myosin, which can now reattach to a new site on the actin filament and repeat the cycle. Note that the release of energy by the hydrolysis of ATP (step 4) and the movement of the cross bridge (step 2) are not simultaneous events.

To summarize, ATP performs two distinct roles in the cross-bridge cycle: (1) The energy released from ATP *hydrolysis* ultimately provides the energy for cross-bridge movement, and (2) ATP *binding* (not hydrolysis) to myosin breaks the link formed between actin and myosin during the cycle, allowing the cycle to be repeated.

The importance of ATP in dissociating actin and myosin during step 3 of a cross-bridge cycle is illustrated by **rigor mortis,** the stiffening of skeletal muscles that begins several hours after death and is complete after about 12 h. The ATP concentration in cells, including muscle cells, declines after death because the nutrients and oxygen required by the metabolic pathways to form ATP are no longer supplied by the circulation. In the absence of ATP, nonenergized cross bridges can bind to actin, but the subsequent movement of the cross bridge and the breakage of the link between actin and myosin do not occur because these events require ATP. The thick and thin filaments become bound to each other by immobilized cross bridges, producing a rigid condition in which the thick and thin filaments cannot be passively pulled past each other. The stiffness of rigor mortis disappears about 48 to 60 h after death as a result of the disintegration of muscle tissue.

Roles of Troponin, Tropomyosin, and Calcium in Contraction

Since every muscle fiber contains all the ingredients necessary for cross-bridge activity (actin, myosin, and ATP) the question arises: Why are muscles not in a continuous state of contractile activity? The answer is that in a resting muscle fiber, the cross bridges are prevented from interacting with actin by two proteins, **troponin** and **tropomyosin,** which, as noted earlier, are located on thin filaments (Figure 11–13).

Tropomyosin is a rod-shaped molecule composed of two intertwined polypeptides with a length approximately equal to that of seven actin molecules. Chains of tropomyosin molecules are arranged end to end along the actin thin filament. These tropomyosin molecules partially cover the myosin-binding site on each actin molecule, thereby preventing the cross bridges from making contact with actin. Each tropomyosin molecule is held in this blocking position by troponin, a smaller, globular protein that is bound to both tropomyosin and actin. One molecule of troponin binds to each molecule of tropomyosin and regulates the access to myosin-binding sites on the seven actin molecules in contact with tropomyosin.

Having described the system that prevents cross-bridge activity and thus keeps a muscle fiber in a resting state, we can now ask: What enables cross bridges to bind to actin and begin cycling? For this to occur, tropomyosin molecules must be moved away from their blocking positions on actin. This happens when calcium binds to specific binding sites on troponin (not tropomyosin). The binding of calcium produces a change in the shape of troponin, which through troponin's linkage to tropomyosin,

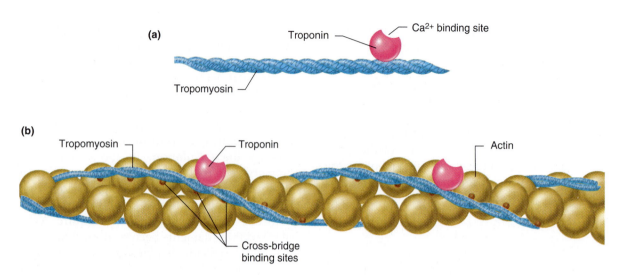

FIGURE 11–13

(a) Molecule of troponin bound to a molecule of tropomyosin. (b) Two chains of tropomyosin on a thin filament regulate access of cross bridges to binding sites on actin.

drags tropomyosin away from the myosin-binding site on each actin molecule. Conversely, removal of calcium from troponin reverses the process, turning off contractile activity.

Thus, cytosolic calcium-ion concentration determines the number of troponin sites occupied by calcium, which in turn determines the number of actin sites available for cross-bridge binding. Changes in cytosolic calcium concentration are controlled by electrical events in the muscle plasma membrane, which we now discuss.

Excitation-Contraction Coupling

Excitation-contraction coupling refers to the sequence of events by which an action potential in the plasma membrane of a muscle fiber leads to cross-bridge activity by the mechanisms just described. The skeletal-muscle plasma membrane is an excitable membrane capable of generating and propagating action potentials by mechanisms similar to those described for nerve cells (Chapter 8). An action potential in a skeletal-muscle fiber lasts 1 to 2 ms and is completed before any signs of mechanical activity begin (Figure 11–14). Once begun, the mechanical activity following an action potential may last 100 ms or more. The electrical activity in the plasma membrane does not *directly* act upon the contractile proteins but instead produces a state of increased cytosolic calcium concentration, which continues to activate the contractile apparatus long after the electrical activity in the membrane has ceased.

In a resting muscle fiber, the concentration of free, ionized calcium in the cytosol surrounding the thick and thin filaments is very low, about 10^{-7} mol/L. At this low calcium concentration, very few of the calcium-binding sites on troponin are occupied, and thus cross-bridge activity is blocked by tropomyosin. Following an action potential, there is a rapid increase in cytosolic calcium concentration, and calcium binds to troponin, removing the blocking effect of tropomyosin and allowing cross-bridge cycling. The source of the increased cytosolic calcium is the **sarcoplasmic reticulum** within the muscle fiber.

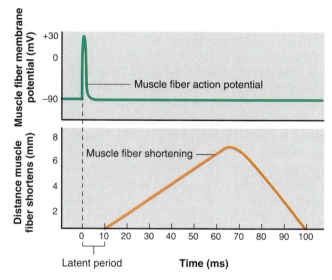

FIGURE 11–14

Time relations between a skeletal-muscle fiber action potential and the resulting shortening and relaxation of the muscle fiber.

Sarcoplasmic Reticulum The sarcoplasmic reticulum in muscle is homologous to the endoplasmic reticulum found in most cells and forms a series of sleevelike structures around each myofibril (Figure 11–15), one segment surrounding the A band and another the I band. At the end of each segment there are two enlarged regions, known as **lateral sacs** that are connected to each other by a series of smaller tubular elements. The lateral sacs store the calcium that is released following membrane excitation.

(a)

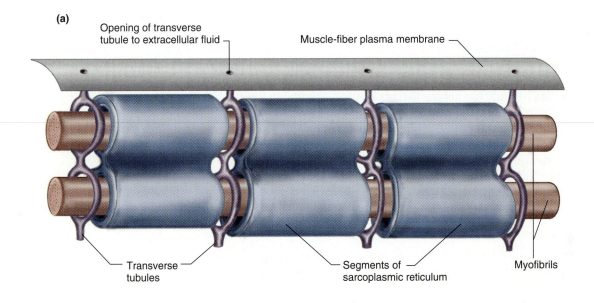

Opening of transverse tubule to extracellular fluid

Muscle-fiber plasma membrane

Transverse tubules

Segments of sarcoplasmic reticulum

Myofibrils

(b)

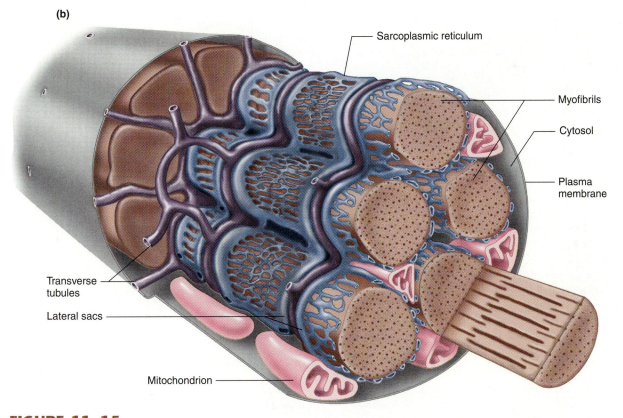

Sarcoplasmic reticulum

Myofibrils

Cytosol

Plasma membrane

Transverse tubules

Lateral sacs

Mitochondrion

FIGURE 11–15

(a) Diagrammatic representation of the sarcoplasmic reticulum, the transverse tubules, and the myofibrils. (b) Anatomical structure of transverse tubules and sarcoplasmic reticulum in a single skeletal-muscle fiber.

A separate tubular structure, the **transverse tubule (T tubule),** crosses the muscle fiber at the level of each A-I junction, passing between adjacent lateral sacs and eventually joining the plasma membrane. The lumen of the T tubule is continuous with the extracellular fluid surrounding the muscle fiber. The membrane of the T tubule, like the plasma membrane, is able to propagate action potentials. Once initiated in the plasma membrane, an action potential is rapidly conducted over the surface of the fiber and into its interior by way of the T tubules. The action potential in a T tubule adjacent to the lateral sacs activates voltage-gated proteins in the T-tubule membrane that are physically or chemically linked to calcium-release channels in the membrane of the lateral sacs. Depolarization of the T tubule by an action potential thus leads to the opening of the calcium channels in the lateral sacs,

allowing calcium to diffuse from the calcium-rich lumen of the lateral sacs into the cytosol. The rise in cytosolic calcium concentration is normally enough to turn on all the cross bridges in the fiber.

A contraction continues until calcium is removed from troponin, and this is achieved by lowering the calcium concentration in the cytosol back to its pre-release level. The membranes of the sarcoplasmic reticulum contain primary active-transport proteins, Ca-ATPases, that pump calcium ions from the cytosol back into the lumen of the reticulum. As we just saw, calcium is released from the reticulum upon arrival of an action potential in the T tubule, but the pumping of the released calcium back into the reticulum requires a much longer time. Therefore, the cytosolic calcium concentration remains elevated, and the contraction continues for some time after a single action potential.

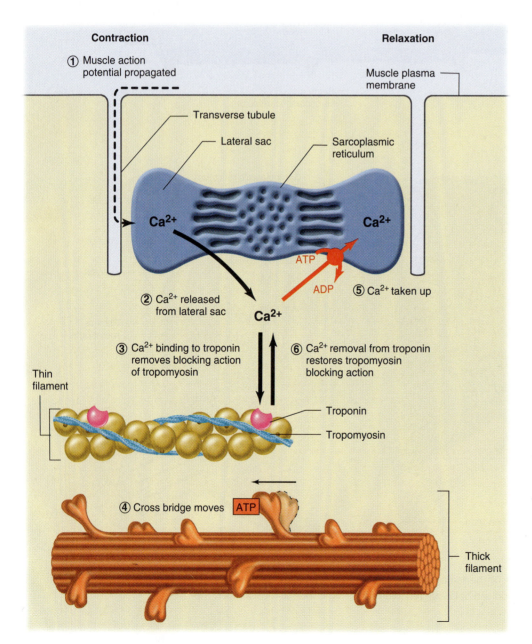

FIGURE 11–16

Release and uptake of calcium by the sarcoplasmic reticulum during contraction and relaxation of a skeletal-muscle fiber.

TABLE 11–1	Functions of ATP in Skeletal-Muscle Contraction

1. Hydrolysis of ATP by myosin energizes the cross bridges, providing the energy for force generation.

2. Binding of ATP to myosin dissociates cross bridges bound to actin, allowing the bridges to repeat their cycle of activity.

3. Hydrolysis of ATP by the Ca-ATPase in the sarcoplasmic reticulum provides the energy for the active transport of calcium ions into the lateral sacs of the reticulum, lowering cytosolic calcium to pre-release levels, ending the contraction, and allowing the muscle fiber to relax.

(a) Single motor unit

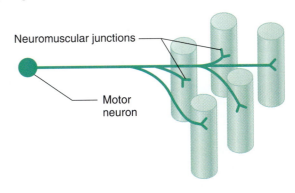

(b) Two motor units

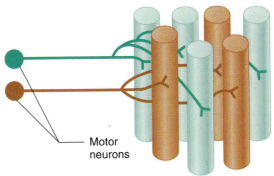

FIGURE 11–17

(a) Single motor unit consisting of one motor neuron and the muscle fibers it innervates. (b) Two motor units and their intermingled fibers in a muscle.

To reiterate, just as contraction results from the release of calcium ions stored in the sarcoplasmic reticulum, so contraction ends and relaxation begins as calcium is pumped back into the reticulum (Figure 11–16). ATP is required to provide the energy for the calcium pump, and this is the third major role of ATP in muscle contraction (Table 11–1).

Membrane Excitation: The Neuromuscular Junction

We have just seen that an action potential in the plasma membrane of a skeletal-muscle fiber is the signal that triggers contraction. The next question we must ask then is: How are these action potentials initiated? Stimulation of the nerve fibers to a skeletal muscle is the only mechanism by which action potentials are initiated in this type of muscle. As we shall learn, there are additional mechanisms for activating cardiac- and smooth-muscle contraction.

The nerve cells whose axons innervate skeletal-muscle fibers are known as **motor neurons** (or somatic efferent neurons), and their cell bodies are located in either the brainstem or the spinal cord. The axons of motor neurons are myelinated and are the largest-diameter axons in the body. They are therefore able to propagate action potentials at high velocities, allowing signals from the central nervous system to be transmitted to skeletal-muscle fibers with minimal delay.

Upon reaching a muscle, the axon of a motor neuron divides into many branches, each branch forming a single junction with a muscle fiber. A single motor neuron innervates many muscle fibers, but each muscle fiber is controlled by a branch from only one motor neuron. A motor neuron plus the muscle fibers it innervates is called a **motor unit** (Figure 11–17a). The muscle fibers in a single motor unit are located in one muscle, but they are scattered throughout the muscle and are not adjacent to each other (Figure 11–17b). When an action potential occurs in a motor neuron, all the muscle fibers in its motor unit are stimulated to contract.

The myelin sheath surrounding the axon of each motor neuron ends near the surface of a muscle fiber, and the axon divides into a number of short processes that lie embedded in grooves on the muscle-fiber surface. The region of the muscle-fiber plasma membrane that lies directly under the terminal portion of the axon has special properties and is known as the **motor end plate.** The junction of an axon terminal with the motor end plate is known as a **neuromuscular junction** (Figure 11–18).

The axon terminals of a motor neuron contain vesicles similar to the vesicles found at synaptic junctions between two neurons. The vesicles contain the neurotransmitter **acetylcholine (ACh).** When an action potential in a motor neuron arrives at the axon terminal, it depolarizes the nerve plasma membrane, opening voltage-sensitive calcium channels and allowing calcium ions to diffuse into the axon terminal from the extracellular fluid. This calcium binds to proteins that enable the membranes of acetylcholine-containing vesicles to fuse with the nerve plasma membrane thereby releasing acetylcholine into the extracellular cleft separating the axon terminal and the motor end plate.

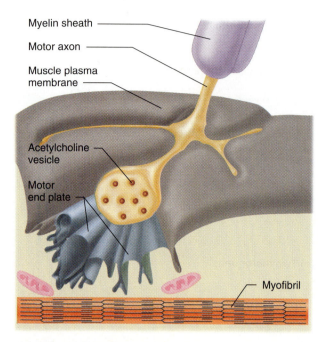

FIGURE 11–18

Neuromuscular junction. The motor axon terminals are embedded in grooves in the muscle fiber's surface. ✗

ACh diffuses from the axon terminal to the motor end plate where it binds to receptors [of the nicotinic type (Chapter 8)]. The binding of ACh opens an ion channel in each receptor protein. Both sodium and potassium ions can pass through these channels. Because of the differences in electrochemical gradients across the plasma membrane (Chapter 8), more sodium moves in than potassium out, producing a local depolarization of the motor end plate known as an **end-plate potential (EPP).** Thus, an EPP is analogous to an EPSP (excitatory postsynaptic potential) at a synapse (Chapter 8).

The magnitude of a single EPP is, however, much larger than that of an EPSP because neurotransmitter is released over a larger surface area, binding to many more receptors and hence opening many more ion channels. For this reason, one EPP is normally more than sufficient to depolarize the muscle plasma membrane adjacent to the end-plate membrane, by local current flow, to its threshold potential, initiating an action potential. This action potential is then propagated over the surface of the muscle fiber by the same mechanism described in Chapter 8 for the propagation of action potentials along axon membranes (Figure 11–19). Most neuromuscular junctions are located near the middle of a muscle fiber, and newly generated muscle action potentials propagate from this region in both directions toward the ends of the fiber.

To repeat, every action potential in a motor neuron normally produces an action potential in each muscle fiber in its motor unit. This is quite different from synaptic junctions, where multiple EPSPs must occur, undergoing temporal and spatial summation, in order for threshold to be reached and an action potential elicited in the postsynaptic membrane.

A second difference between synaptic and neuromuscular junctions should be noted. As we saw in Chapter 8, at some synaptic junctions, IPSPs (inhibitory

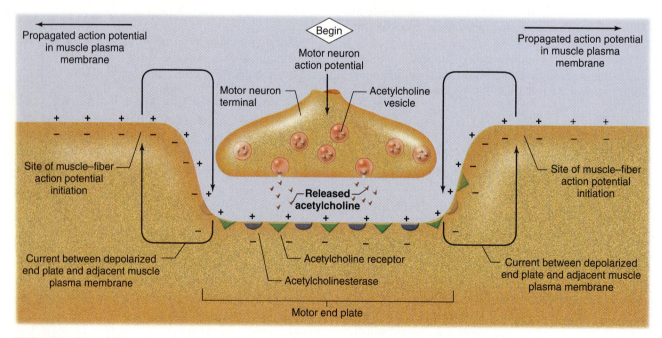

FIGURE 11–19

Events at a neuromuscular junction that lead to an action potential in the muscle-fiber plasma membrane. ✗

TABLE 11–2 **Sequence of Events Between a Motor Neuron Action Potential and Skeletal-Muscle Fiber Contraction**

1. Action potential is initiated and propagates in motor neuron axon.

2. Action potential triggers release of ACh from axon terminals at neuromuscular junction.

3. ACh diffuses from axon terminals to motor end plate in muscle fiber.

4. ACh binds to receptors on motor end plate, opening Na^+, K^+ ion channels.

5. More Na^+ moves into the fiber at the motor end plate than K^+ moves out, depolarizing the membrane, producing the end-plate potential (EPP).

6. Local currents depolarize the adjacent plasma membrane to its threshold potential, generating an action potential that propagates over the muscle fiber surface and into the fiber along the transverse tubules.

7. Action potential in transverse tubules triggers release of Ca^{2+} from lateral sacs of sarcoplasmic reticulum.

8. Ca^{2+} binds to troponin on the thin filaments, causing tropomyosin to move away from its blocking position, thereby uncovering cross-bridge binding sites on actin.

9. Energized myosin cross bridges on the thick filaments bind to actin:

$$A + M^* \cdot ADP \cdot P_i \longrightarrow A \cdot M^* \cdot ADP \cdot P_i$$

10. Cross-bridge binding triggers release of the strained conformational state of myosin, producing an angular movement of each cross bridge:

$$A \cdot M^* \cdot ADP \cdot P_i \longrightarrow A \cdot M + ADP + P_i$$

11. ATP binds to myosin, breaking linkage between actin and myosin and thereby allowing cross bridges to dissociate from actin:

$$A \cdot M + ATP \longrightarrow A + M \cdot ATP$$

12. ATP bound to myosin is split, energizing the myosin cross bridge:

$$M \cdot ATP \longrightarrow M^* \cdot ADP \cdot P_i$$

13. Cross bridges repeat steps 9 to 12, producing movement of thin filaments past thick filaments. Cycles of cross-bridge movement continue as long as Ca^{2+} remains bound to troponin.

14. Cytosolic Ca^{2+} concentration decreases as Ca^{2+} is actively transported into sarcoplasmic reticulum by Ca-ATPase.

15. Removal of Ca^{2+} from troponin restores blocking action of tropomyosin, the cross-bridge cycle ceases, and the muscle fiber relaxes.

postsynaptic potentials) are produced. They hyperpolarize or stabilize the postsynaptic membrane and decrease the probability of its firing an action potential. In contrast, inhibitory potentials do not occur in human skeletal muscle; all neuromuscular junctions are excitatory.

In addition to receptors for ACh, the surface of the motor end plate contains the enzyme **acetylcholinesterase,** which breaks down ACh, just as occurs at ACh-mediated synapses in the nervous system. ACh bound to receptors is in equilibrium with free ACh in the cleft between the nerve and muscle membranes. As the concentration of free ACh falls because of its breakdown by acetylcholinesterase, less ACh is available to bind to the receptors. When the receptors no longer contain bound ACh, the ion channels in the end plate close. The depolarized end plate returns to its resting potential and can respond to the subsequent arrival of ACh released by another nerve action potential.

Table 11–2 summarizes the sequence of events that lead from an action potential in a motor neuron to the contraction and relaxation of a skeletal-muscle fiber.

There are many ways by which events at the neuromuscular junction can be modified by disease or drugs. For example, the deadly South American arrowhead poison *curare* binds strongly to the ACh receptors, but it does not open their ion channels and is not destroyed by acetylcholinesterase. When a receptor is occupied by curare, ACh cannot bind to the receptor. Therefore, although the motor nerves still conduct normal action potentials and release ACh, there is no resulting EPP in the motor end plate and hence no contraction. Since the skeletal muscles responsible for breathing, like all skeletal muscles, depend upon neuromuscular transmission to initiate their contraction, curare poisoning can lead to death by asphyxiation. Drugs similar to curare are used in small amounts to prevent muscular contractions during certain types of surgical procedures when it is necessary to immobilize the surgical field. Patients are artificially ventilated in order to maintain respiration until the drug has been removed from the system.

Neuromuscular transmission can also be blocked by inhibiting acetylcholinesterase. Some organophosphates, which are the main ingredients in certain pesticides and "nerve gases" (the latter developed for biological warfare), inhibit this enzyme. In the presence of such agents, ACh is released normally upon the arrival of an action potential at the axon terminal and binds to the end-plate receptors. The ACh is not destroyed, however, because the acetylcholinesterase is inhibited. The ion channels in the end plate therefore remain open, producing a maintained depolarization of the end plate and the muscle plasma membrane adjacent to the end plate. A skeletal-muscle membrane maintained in a depolarized state cannot generate action potentials because the voltage-gated sodium channels in the membrane have entered an inactive state, which requires repolarization to remove. Thus, the muscle does not contract in response to subsequent nerve stimulation, and the result is skeletal-muscle paralysis and death from asphyxiation.

A third group of substances, including the toxin produced by the bacterium *Clostridium botulinum*, blocks the release of acetylcholine from nerve terminals. Botulinum toxin is an enzyme that breaks down a protein required for the binding and fusion of ACh vesicles with the plasma membrane of the axon terminal. This toxin, which produces the food poisoning called **botulism**, is one of the most potent poisons known because of the very small amount necessary to produce an effect.

Mechanics of Single-Fiber Contraction

The force exerted on an object by a contracting muscle is known as muscle **tension,** and the force exerted on the muscle by an object (usually its weight) is the **load.** Muscle tension and load are opposing forces. Whether or not force generation leads to fiber shortening depends on the relative magnitudes of the tension and the load. In order for muscle fibers to shorten, and thereby move a load, muscle tension must be greater than the opposing load.

When a muscle develops tension but does not shorten (or lengthen), the contraction is said to be **isometric** (constant length). Such contractions occur when the muscle supports a load in a constant position or attempts to move an otherwise supported load that is greater than the tension developed by the muscle. A contraction in which the muscle shortens, while the load on the muscle remains constant, is said to be **isotonic** (constant tension).

A third type of contraction is a **lengthening contraction** (eccentric contraction). This occurs when an unsupported load on a muscle is greater than the tension being generated by the cross bridges. In this situation, the load pulls the muscle to a longer length in spite of the opposing force being produced by the cross bridges. Such lengthening contractions occur when an object being supported by muscle contraction is lowered, such as occurs when you sit down from a standing position or walk down a flight of stairs. It must be emphasized that in these situations the lengthening of muscle fibers is not an active process produced by the contractile proteins, but a consequence of the external forces being applied to the muscle. In the absence of external lengthening forces, a fiber will only *shorten* when stimulated; it will never lengthen. All three types of contractions—isometric, isotonic, and lengthening—occur in the natural course of everyday activities.

During each type of contraction the cross bridges repeatedly go through the four steps of the cross-bridge cycle illustrated in Figure 11–12. During step 2 of an isotonic contraction, the cross bridges bound to actin move to their angled positions, causing shortening of the sarcomeres. In contrast, during an isometric contraction, the bound cross bridges are unable to move the thin filaments because of the load on the muscle fiber, but they do exert a force on the thin filaments—isometric tension. During a lengthening contraction, the cross bridges in step 2 are pulled backward toward the Z lines by the load while still bound to actin and exerting force. The events of steps 1, 3, and 4 are the same in all three types of contractions. Thus, the chemical changes in the contractile proteins during each type of contraction are the same. The end result (shortening, no length change, or lengthening) is determined by the magnitude of the load on the muscle.

Contraction terminology applies to both single fibers and whole muscles. We first describe the mechanics of single-fiber contractions and later discuss the factors controlling the mechanics of whole-muscle contraction.

Twitch Contractions

The mechanical response of a single muscle fiber to a single action potential is known as a **twitch.** Figure 11–20a shows the main features of an isometric twitch. Following the action potential, there is an interval of a few milliseconds, known as the **latent period,** before the tension in the muscle fiber begins to increase. During this latent period, the processes associated with excitation-contraction coupling are occurring. The time interval from the beginning of tension development at the end of the latent period to the peak tension is the **contraction time.** Not all skeletal-muscle fibers have the same twitch contraction time. Some fast fibers have contraction times as short as 10 ms, whereas slower fibers may take 100 ms or longer. The duration of the

(a) Isometric contraction

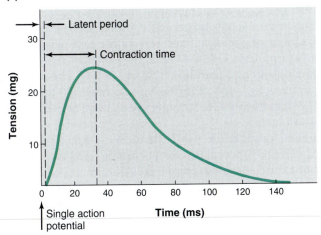

(b) Isotonic contraction

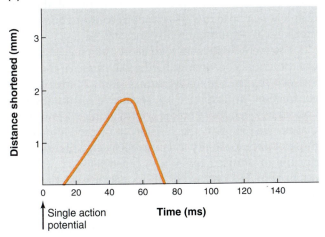

FIGURE 11–20

(a) An isometric twitch of a skeletal-muscle fiber following a single action potential. (b) An isotonic twitch of a skeletal-muscle fiber following a single action potential.

contraction time depends on the time that cytosolic calcium remains elevated so that cross bridges can continue to cycle. It is most closely related to the Ca-ATPase activity in the sarcoplasmic reticulum; activity is greater in fast-twitch fibers and less in slow-twitch fibers.

Comparing isotonic and isometric twitches in the same muscle fiber, one can see from Figure 11–20b that the latent period in an isotonic twitch is longer than that in an isometric contraction, while the duration of the mechanical event—shortening—is briefer in an isotonic twitch than the duration of force generation in an isometric twitch.

Moreover, the characteristics of an isotonic twitch depend upon the magnitude of the load being lifted (Figure 11–21): (1) at heavier loads, the latent period is longer, and (2) the velocity of shortening (distance shortened per unit of time), the duration of the twitch, and the distance shortened are all slower or shorter.

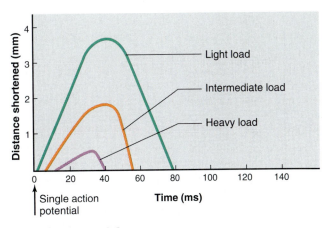

FIGURE 11–21

Isotonic twitches with different loads. The distance shortened, velocity of shortening, and duration of shortening all decrease with increased load, whereas the time from stimulation to the beginning of shortening increases with increasing load.

Let us look more closely at the sequence of events in an isotonic twitch. Following excitation, the cross bridges begin to develop force, but shortening does not begin until the muscle tension just exceeds the load on the fiber. Thus, before shortening, there is a period of *isometric* contraction during which the tension increases. The heavier the load, the longer it takes for the tension to increase to the value of the load, when shortening will begin. If the load on a fiber is increased, eventually a load is reached that the muscle is unable to lift, the velocity and distance of shortening will be zero, and the contraction will become completely isometric.

Load-Velocity Relation

It is a common experience that light objects can be moved faster than heavy objects. That is, the velocity at which a muscle fiber shortens decreases with increasing loads (Figure 11–22). The shortening velocity is maximal when there is no load and is zero when the load is equal to the maximal isometric tension. At loads greater than the maximal isometric tension, the fiber will *lengthen* at a velocity that increases with load, and at very high loads the fiber will break.

The shortening velocity is determined by the rate at which individual cross bridges undergo their cyclical activity. Because one ATP is split during each cross-bridge cycle, the rate of ATP splitting determines the shortening velocity. Increasing the load on a bridge, for complex reasons, decreases the rate of ATP hydrolysis and thus the velocity of shortening.

Frequency-Tension Relation

Since a single action potential in a skeletal-muscle fiber lasts 1 to 2 ms but the twitch may last for 100 ms, it is

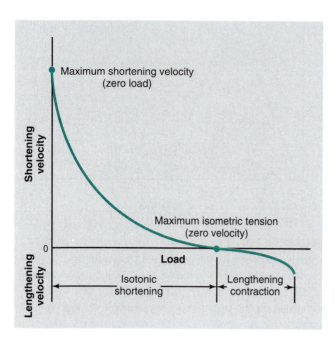

FIGURE 11–22

Velocity of skeletal-muscle fiber shortening and lengthening as a function of load. Note that the force on the cross bridges during a lengthening contraction is greater than the maximum isometric tension.

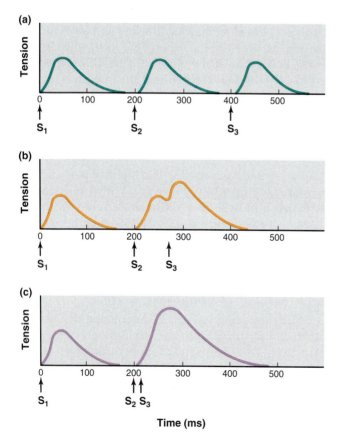

FIGURE 11–23

Summation of isometric contractions produced by shortening the time between stimuli S_2 and S_3.

possible for a second action potential to be initiated during the period of mechanical activity. Figure 11–23 illustrates the tension generated during isometric contractions of a muscle fiber in response to three successive stimuli. In Figure 11–23a, the isometric twitch following the first stimulus S_1 lasts 150 ms. The second stimulus S_2, applied to the muscle fiber 200 ms after S_1 when the fiber has completely relaxed, causes a second identical twitch, and a third stimulus S_3, equally timed, produces a third identical twitch. In Figure 11–23b, the interval between S_1 and S_2 remains 200 ms, but a third stimulus is applied 60 ms after S_2, when the mechanical response resulting from S_2 is beginning to decrease but has not yet ended. Stimulus S_3 induces a contractile response whose peak tension is greater than that produced by S_2. In Figure 11–23c, the interval between S_2 and S_3 is further reduced to 10 ms, and the resulting peak tension is even greater. Indeed, the mechanical response to S_3 is a smooth continuation of the mechanical response already induced by S_2.

The increase in muscle tension from successive action potentials occurring during the phase of mechanical activity is known as **summation.** A maintained contraction in response to repetitive stimulation is known as a **tetanus** (tetanic contraction). At low stimulation frequencies, the tension may oscillate as the muscle fiber partially relaxes between stimuli, producing an unfused tetanus. A fused tetanus, with no oscillations, is produced at higher stimulation frequencies (Figure 11–24).

As the frequency of action potentials increases, the level of tension increases by summation until a maximal fused tetanic tension is reached, beyond which tension no longer increases with further increases in stimulation frequency. This maximal tetanic tension is about three to five times greater than the isometric twitch tension. Since different muscle fibers have different contraction times, the stimulus frequency that will produce a maximal tetanic tension differs from fiber to fiber.

Summation can be explained by events occurring in the muscle fiber. The explanation requires one new piece of information: A muscle contains passive elastic elements (portions of the thick and thin filaments and tendons) that are in series with the contractile (force-generating) elements. These series elastic elements act like springs through which the active force generated by the cross bridges must pass to be applied to the load. Therefore, the time course of the rise in tension during an isometric contraction includes the time required to stretch the series elastic elements.

The tension produced by a muscle fiber at any instant depends upon (1) the number of cross bridges bound to actin and undergoing step 2 of the cross-bridge cycle in each sarcomere, (2) the force produced by each cross bridge and, (3) the amount of time the

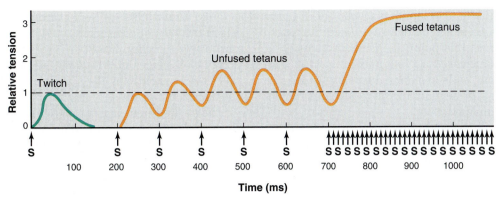

FIGURE 11–24

Isometric contractions produced by multiple stimuli at 10 stimuli per second (unfused tetanus) and 100 stimuli per second (fused tetanus), as compared with a single twitch.

cross bridges remain active. A single action potential in a skeletal-muscle fiber releases enough calcium to saturate troponin, and all the myosin-binding sites on the thin filaments are therefore *initially* available. But the binding of energized cross bridges to these sites (step 1 of the cross-bridge cycle) takes time and, as noted above, the stretching of the series elastic elements by the cross bridges also takes time. For these reasons, even though all the binding sites are available initially during a single twitch, maximal tension is not developed instantaneously. Moreover, almost immediately after the release of calcium, it begins to be pumped back into the sarcoplasmic reticulum, and the calcium concentration begins to fall from its initial high value, causing more and more of the myosin-binding sites on actin to become unavailable for cross-bridge binding. Thus, during a single twitch, the cross bridges do not remain active long enough for the series elastic element to be stretched to the maximal tension the cross bridge can exert.

In contrast, during a tetanic contraction, the successive action potentials each release calcium from the sarcoplasmic reticulum before all the calcium from the previous action potential has been pumped back into the reticulum. This results in a maintained elevated cytosolic calcium concentration and prevents a decline in the number of available binding sites on the thin filaments. Under these conditions, the maximum number of binding sites remains available and the maintained cross-bridge cycling has time to stretch the series elastic elements, thereby transmitting maximal tension to the ends of the fiber.

Length-Tension Relation

The springlike characteristics of the protein titin, which is attached to the Z line at one end and the thick filaments at the other, as described earlier, are responsible for most of the *passive* elastic properties of relaxed muscles. With increased stretch, the passive tension in a relaxed fiber increases, not from active cross-bridge

movements but from elongation of the titin filaments. If the stretched fiber is released, its length will return to an equilibrium length, much like releasing a stretched rubber band. The critical point for this section is that, on top of this increased passive tension due to stretching, the amount of active tension developed by a muscle fiber during contraction, and thus its strength, can be altered by changing the length of the fiber before contraction. One can stretch a muscle fiber to various lengths and measure the magnitude of the active tension generated in response to stimulation at each length (Figure 11–25). The length at which the fiber develops the greatest isometric active tension is termed the **optimal length, l_o.**

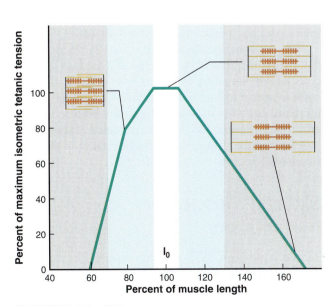

FIGURE 11–25

Variation in active isometric tetanic tension with muscle-fiber length. The blue band represents the range of length changes that can occur physiologically in the body while muscles are attached to bones.

When a muscle fiber length is 60 percent of l_o, the fiber develops no tension when stimulated. As length is increased from this point, the isometric tension at each length is increased up to a maximum at l_o. Further lengthening leads to a *drop* in tension. At lengths of 175 percent l_o or beyond, the fiber develops no tension when stimulated.

When all the skeletal muscles in the body are relaxed, the lengths of most fibers are near l_o and thus near the optimal lengths for force generation. The length of a relaxed fiber can be altered by the load on the muscle or the contraction of other muscles that stretch the relaxed fibers, but the extent to which the relaxed length can be changed is limited by the muscle's attachments to bones. It rarely exceeds a 30 percent change from l_o and is often much less. Over this range of lengths, the ability to develop tension never falls below about half of the tension that can be developed at l_o (Figure 11–25).

The relationship between fiber length and the fiber's capacity to develop active tension during contraction can be partially explained in terms of the sliding-filament mechanism. Stretching a relaxed muscle fiber pulls the thin filaments past the thick filaments, changing the amount of overlap between them. Stretching a fiber to 1.75 l_o pulls the filaments apart to the point where there is no overlap. At this point there can be no cross-bridge binding to actin and no development of tension. Between 1.75 l_o and l_o, more and more filaments overlap, and the tension developed upon stimulation increases in proportion to the increased number of cross bridges in the overlap region. Filament overlap is greatest at l_o, allowing the maximal number of cross bridges to bind to the thin filaments, thereby producing maximal tension.

The tension decline at lengths less than l_o is the result of several factors. For example, (1) the overlapping sets of thin filaments from opposite ends of the sarcomere may interfere with the cross bridges' ability to bind and exert force, and (2) for unknown reasons, the affinity of troponin for calcium decreases at short fiber lengths, resulting in fewer accessible sites on the thin filaments for cross-bridge binding.

Skeletal-Muscle Energy Metabolism

As we have seen, ATP performs three functions directly related to muscle-fiber contraction and relaxation (see Table 11–1). In no other cell type does the rate of ATP breakdown increase so much from one moment to the next as in a skeletal muscle fiber (20 to several hundredfold depending on the type of muscle fiber) when it goes from rest to a state of contractile

activity. The small supply of preformed ATP that exists at the start of contractile activity would only support a few twitches. If a fiber is to sustain contractile activity, molecules of ATP must be produced by metabolism as rapidly as they are broken down during the contractile process.

There are three ways a muscle fiber can form ATP during contractile activity (Figure 11–26): (1) phosphorylation of ADP by **creatine phosphate,** (2) oxidative phosphorylation of ADP in the mitochondria, and (3) substrate-level phosphorylation of ADP by the glycolytic pathway in the cytosol.

Phosphorylation of ADP by creatine phosphate (CP) provides a very rapid means of forming ATP at the onset of contractile activity. When the chemical bond between creatine (C) and phosphate is broken, the amount of energy released is about the same as that released when the terminal phosphate bond in ATP is broken. This energy, along with the phosphate group, can be transferred to ADP to form ATP in a reversible reaction catalyzed by creatine kinase:

$$\text{CP} + \text{ADP} \xrightleftharpoons[]{\substack{\text{Creatine} \\ \text{kinase}}} \text{C} + \text{ATP}$$

Although creatine phosphate is a high-energy molecule, its energy cannot be released by myosin to drive cross-bridge activity. During periods of rest, muscle fibers build up a concentration of creatine phosphate approximately five times that of ATP. At the beginning of contraction, when the concentration of ATP begins to fall and that of ADP to rise owing to the increased rate of ATP breakdown by myosin, mass action favors the formation of ATP from creatine phosphate. This transfer of energy is so rapid that the concentration of ATP in a muscle fiber changes very little at the start of contraction, whereas the concentration of creatine phosphate falls rapidly.

Although the formation of ATP from creatine phosphate is very rapid, requiring only a single enzymatic reaction, the amount of ATP that can be formed by this process is limited by the initial concentration of creatine phosphate in the cell. If contractile activity is to be continued for more than a few seconds, the muscle must be able to form ATP from the other two sources listed above. The use of creatine phosphate at the start of contractile activity provides the few seconds necessary for the slower, multienzyme pathways of oxidative phosphorylation and glycolysis to increase their rates of ATP formation to levels that match the rates of ATP breakdown.

At moderate levels of muscular activity, most of the ATP used for muscle contraction is formed by oxidative phosphorylation, and during the first 5 to 10 min of such exercise, muscle glycogen is the major fuel

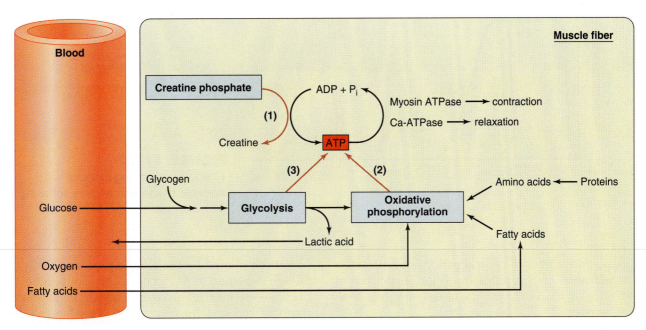

FIGURE 11–26
The three sources of ATP production during muscle contraction: (1) creatine phosphate, (2) oxidative phosphorylation, and (3) glycolysis.

contributing to oxidative phosphorylation. For the next 30 min or so, blood-borne fuels become dominant, blood glucose and fatty acids contributing approximately equally; beyond this period, fatty acids become progressively more important, and glucose utilization decreases.

If the intensity of exercise exceeds about 70 percent of the maximal rate of ATP breakdown, however, glycolysis contributes an increasingly significant fraction of the total ATP generated by the muscle. The glycolytic pathway, although producing only small quantities of ATP from each molecule of glucose metabolized, can produce large quantities of ATP when enough enzymes and substrate are available, and it can do so in the absence of oxygen. The glucose for glycolysis can be obtained from two sources: the blood or the stores of glycogen within the contracting muscle fibers. As the intensity of muscle activity increases, a greater fraction of the total ATP production is formed by anaerobic glycolysis, with a corresponding increase in the production of lactic acid (which dissociates to yield lactate ions and hydrogen ions).

At the end of muscle activity, creatine phosphate and glycogen levels in the muscle have decreased, and to return a muscle fiber to its original state, these energy-storing compounds must be replaced. Both processes require energy, and so a muscle continues to consume increased amounts of oxygen for some time after it has ceased to contract, as evidenced by the fact that one continues to breathe deeply and rapidly for a

period of time immediately following intense exercise. This elevated consumption of oxygen following exercise repays what has been called the **oxygen debt**—that is, the increased production of ATP by oxidative phosphorylation following exercise that is used to restore the energy reserves in the form of creatine phosphate and glycogen.

Muscle Fatigue

When a skeletal-muscle fiber is repeatedly stimulated, the tension developed by the fiber eventually decreases even though the stimulation continues (Figure 11–27). This decline in muscle tension as a result of previous contractile activity is known as **muscle fatigue.** Additional characteristics of fatigued muscle are a decreased shortening velocity and a slower rate of relaxation. The onset of fatigue and its rate of development depend on the type of skeletal-muscle fiber that is active and on the intensity and duration of contractile activity.

If a muscle is allowed to rest after the onset of fatigue, it can recover its ability to contract upon restimulation (Figure 11–27). The rate of recovery depends upon the duration and intensity of the previous activity. Some muscle fibers fatigue rapidly if continuously stimulated but also recover rapidly after a brief rest. This is the type of fatigue (high-frequency fatigue) that accompanies high-intensity, short-duration exercise, such as weight lifting. In contrast, low-frequency fatigue develops more slowly with low-intensity, long-duration exercise, such as long-distance running,

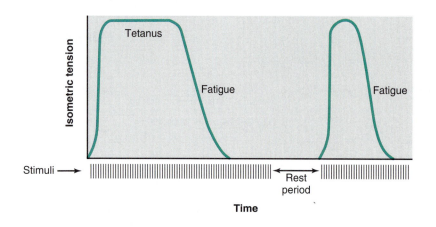

FIGURE 11–27

Muscle fatigue during a maintained isometric tetanus and recovery following a period of rest.

during which there are cyclical periods of contraction and relaxation, and requires much longer periods of rest, often up to 24 h, before the muscle achieves complete recovery.

It might seem logical that depletion of energy in the form of ATP would account for fatigue, but the ATP concentration in fatigued muscle is found to be only slightly lower than in a resting muscle, and not low enough to impair cross-bridge cycling. If contractile activity were to continue without fatigue, the ATP concentration could decrease to the point that the cross bridges would become linked in a rigor configuration, which is very damaging to muscle fibers. Thus, muscle fatigue may have evolved as a mechanism for preventing the onset of rigor.

Multiple factors can contribute to the fatigue of skeletal muscle. Fatigue from high-intensity, short-duration exercise occurs primarily because of a failure of the muscle action potential to be conducted into the fiber along the T tubules and thus a failure to release calcium from the sarcoplasmic reticulum. The conduction failure results from the build up of potassium ions in the small volume of the T tubule with each of the initial action potentials, which leads to a partial depolarization of the membrane and eventually failure to produce action potentials in the T-tubular membrane. Recovery is rapid with rest as the accumulated potassium diffuses out of the tubule, restoring excitability.

With low-intensity, long-duration exercise a number of processes have been implicated in fatigue, but no single process can completely account for the fatigue from this type of exercise. One of the major factors is the build up of lactic acid. Since the hydrogen-ion concentration can alter protein conformation and thus protein activity, the acidification of the muscle alters a number of muscle proteins, including actin and myosin, as well as proteins involved in calcium release. Recovery from this kind of fatigue probably requires protein synthesis to replace those proteins that

have been altered by the fatigue process. Finally, although depletion of ATP is not a cause of fatigue, the decrease in muscle glycogen, which is supplying much of the fuel for contraction, correlates closely with fatigue onset.

Another type of fatigue quite different from muscle fatigue is due to failure of the appropriate regions of the cerebral cortex to send excitatory signals to the motor neurons. This is called **central command fatigue,** and it may cause an individual to stop exercising even though the muscles are not fatigued. An athlete's performance depends not only on the physical state of the appropriate muscles but also upon the "will to win"—that is, the ability to initiate central commands to muscles during a period of increasingly distressful sensations.

Types of Skeletal-Muscle Fibers

All skeletal-muscle fibers do not have the same mechanical and metabolic characteristics. Different types of fibers can be identified on the basis of (1) their maximal velocities of shortening—fast and slow fibers—and (2) the major pathway used to form ATP—oxidative and glycolytic fibers.

Fast and slow fibers contain myosin isozymes that differ in the maximal rates at which they split ATP, which in turn determine the maximal rate of cross-bridge cycling and hence the fibers' maximal shortening velocity. Fibers containing myosin with high ATPase activity are classified as **fast fibers,** and those containing myosin with lower ATPase activity are **slow fibers.** Although the rate of cross-bridge cycling is about four times faster in fast fibers than in slow fibers, the force produced by both types of cross bridges is about the same.

The second means of classifying skeletal-muscle fibers is according to the type of enzymatic machinery available for synthesizing ATP. Some fibers contain numerous mitochondria and thus have a high capacity for oxidative phosphorylation. These fibers are classified as **oxidative fibers.** Most of the ATP produced by such fibers is dependent upon blood flow to deliver oxygen and fuel molecules to the muscle, and these fibers are surrounded by numerous small blood vessels. They also contain large amounts of an oxygen-binding protein known as **myoglobin,** which increases the rate of oxygen diffusion within the fiber and provides a small store of oxygen. The large amounts of myoglobin present in oxidative fibers give the fibers a dark-red color, and thus oxidative fibers are often referred to as **red muscle fibers.**

In contrast, **glycolytic fibers** have few mitochondria but possess a high concentration of glycolytic enzymes and a large store of glycogen. Corresponding to their limited use of oxygen, these fibers are surrounded by relatively few blood vessels and contain little myoglobin. The lack of myoglobin is responsible for the pale color of glycolytic fibers and their designation as **white muscle fibers.**

On the basis of these two characteristics, three types of skeletal-muscle fibers can be distinguished:

1. **Slow-oxidative fibers** (type I) combine low myosin-ATPase activity with high oxidative capacity.
2. **Fast-oxidative fibers** (type IIa) combine high myosin-ATPase activity with high oxidative capacity.
3. **Fast-glycolytic fibers** (type IIb) combine high myosin-ATPase activity with high glycolytic capacity.

Note that the fourth theoretical possibility—slow-glycolytic fibers—is not found.

In addition to these biochemical differences, there are also size differences, glycolytic fibers generally having much larger diameters than oxidative fibers (Figure 11–28). This fact has significance for tension development. The number of thick and thin filaments per unit of cross-sectional area is about the same in all types of skeletal-muscle fibers. Therefore, the larger the diameter of a muscle fiber, the greater the total number of thick and thin filaments acting in parallel to produce force, and the greater the maximum tension it can develop (greater strength). Accordingly, the average glycolytic fiber, with its larger diameter, develops more

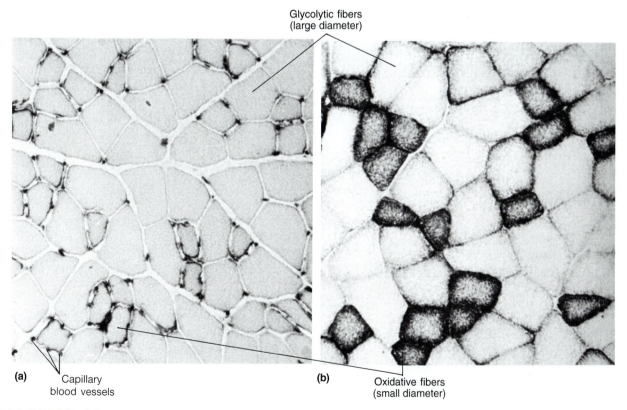

(a) Capillary blood vessels

Glycolytic fibers (large diameter)

(b) Oxidative fibers (small diameter)

FIGURE 11–28

Cross sections of skeletal muscle. (a) The capillaries surrounding the muscle fibers have been stained. Note the large number of capillaries surrounding the small-diameter oxidative fibers. (b) The mitochondria have been stained indicating the large numbers of mitochondria in the small-diameter oxidative fibers.

Courtesy of John A. Faulkner.

TABLE 11–3 Characteristics of the Three Types of Skeletal-Muscle Fibers

	Slow-Oxidative Fibers	Fast-Oxidative Fibers	Fast-Glycolytic Fibers
Primary source of ATP production	Oxidative phosphorylation	Oxidative phosphorylation	Glycolysis
Mitochondria	Many	Many	Few
Capillaries	Many	Many	Few
Myoglobin content	High (red muscle)	High (red muscle)	Low (white muscle)
Glycolytic enzyme activity	Low	Intermediate	High
Glycogen content	Low	Intermediate	High
Rate of fatigue	Slow	Intermediate	Fast
Myosin-ATPase activity	Low	High	High
Contraction velocity	Slow	Fast	Fast
Fiber diameter	Small	Intermediate	Large
Motor unit size	Small	Intermediate	Large
Size of motor neuron innervating fiber	Small	Intermediate	Large

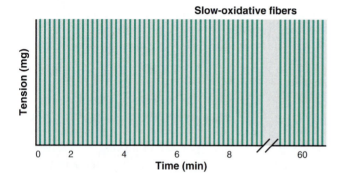

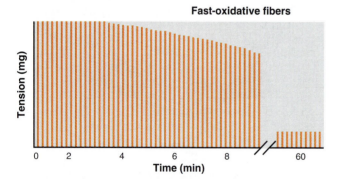

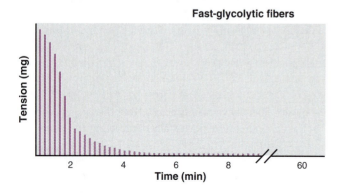

tension when it contracts than does an average oxidative fiber.

These three types of fibers also differ in their capacity to resist fatigue. Fast-glycolytic fibers fatigue rapidly, whereas slow-oxidative fibers are very resistant to fatigue, which allows them to maintain contractile activity for long periods with little loss of tension. Fast-oxidative fibers have an intermediate capacity to resist fatigue (Figure 11–29).

The characteristics of the three types of skeletal-muscle fibers are summarized in Table 11–3.

Whole-Muscle Contraction

As described earlier, whole muscles are made up of many muscle fibers organized into motor units. All the muscle fibers in a single motor unit are of the same fiber type. Thus, one can apply the fiber type designation to the motor unit and refer to slow-oxidative motor units, fast-oxidative motor units, and fast-glycolytic motor units.

Most muscles are composed of all three motor unit types interspersed with each other (Figure 11–30). No muscle has only a single fiber type. Depending on the proportions of the fiber types present, muscles can differ considerably in their maximal contraction speed,

FIGURE 11–29

The rate of fatigue development in the three fiber types. Each vertical line is the contractile response to a brief tetanic stimulus and relaxation. The contractile responses occurring between about 9 min and 60 min are not shown on the figure.

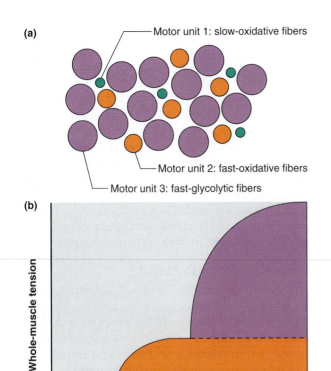

(a)

Motor unit 1: slow-oxidative fibers

Motor unit 2: fast-oxidative fibers

Motor unit 3: fast-glycolytic fibers

(b)

Whole-muscle tension

0

Time

Motor unit 1 recruited

Motor unit 2 recruited

Motor unit 3 recruited

FIGURE 11–30

(a) Diagram of a cross section through a muscle composed of three types of motor units. (b) Tetanic muscle tension resulting from the successive recruitment of the three types of motor units. Note that motor unit 3, composed of fast-glycolytic fibers, produces the greatest rise in tension because it is composed of the largest-diameter fibers and contains the largest number of fibers per motor unit.

TABLE 11–4 Factors Determining Muscle Tension
I. Tension developed by each individual fiber
a. Action-potential frequency (frequency-tension relation)
b. Fiber length (length-tension relation)
c. Fiber diameter
d. Fatigue
II. Number of active fibers
a. Number of fibers per motor unit
b. Number of active motor units

shortening velocity. The conditions that determine the amount of tension developed in a single fiber have been discussed previously and are summarized in Table 11–4.

The number of fibers contracting at any time depends on: (1) the number of fibers in each motor unit (motor unit size), and (2) the number of active motor units.

Motor unit size varies considerably from one muscle to another. The muscles in the hand and eye, which produce very delicate movements, contain small motor units. For example, one motor neuron innervates only about 13 fibers in an eye muscle. In contrast, in the more coarsely controlled muscles of the back and legs, each motor unit is large, containing hundreds and in some cases several thousand fibers. When a muscle is composed of small motor units, the total tension produced by the muscle can be increased in small steps by activating additional motor units. If the motor units are large, large increases in tension will occur as each additional motor unit is activated. Thus, finer control of muscle tension is possible in muscles with small motor units.

The force produced by a single fiber, as we have seen earlier, depends in part on the fiber diameter—the greater the diameter, the greater the force. We have also noted that fast-glycolytic fibers have the largest diameters. Thus, a motor unit composed of 100 fast-glycolytic fibers produces more force that a motor unit composed of 100 slow-oxidative fibers. In addition, fast-glycolytic motor units tend to have more muscle fibers. For both of these reasons, activating a fast-glycolytic motor unit will produce more force than activating a slow-oxidative motor unit.

The process of increasing the number of motor units that are active in a muscle at any given time is called **recruitment**. It is achieved by increasing the excitatory synaptic input to the motor neurons. The greater the number of active motor neurons, the more motor units recruited, and the greater the muscle tension.

Motor neuron size plays an important role in the recruitment of motor units (the size of a motor neuron refers to the diameter of the nerve cell body, which is

strength, and fatigability. For example, the muscles of the back and legs, which must be able to maintain their activity for long periods of time without fatigue while supporting an upright posture, contain large numbers of slow-oxidative and fast-oxidative fibers. In contrast, the muscles in the arms may be called upon to produce large amounts of tension over a short time period, as when lifting a heavy object, and these muscles have a greater proportion of fast-glycolytic fibers.

We will now use the characteristics of single fibers to describe whole-muscle contraction and its control.

Control of Muscle Tension

The total tension a muscle can develop depends upon two factors: (1) the amount of tension developed by each fiber, and (2) the number of fibers contracting at any time. By controlling these two factors, the nervous system controls whole-muscle tension, as well as

usually correlated with the diameter of its axon, and does not refer to the size of the motor unit the neuron controls). Given the same number of sodium ions entering a cell at a single excitatory synapse in a large and in a small motor neuron, the small neuron will undergo a greater depolarization because these ions will be distributed over a smaller membrane surface area. Accordingly, given the same level of synaptic input, the smallest neurons will be recruited first—that is, will begin to generate action potentials first. The larger neurons will be recruited only as the level of synaptic input increases. Since the smallest motor neurons innervate the slow-oxidative motor units (see Table 11–3), these motor units are recruited first, followed by fast-oxidative motor units, and finally, during very strong contractions, by fast-glycolytic motor units (Figure 11–30).

Thus, during moderate-strength contractions, such as are used in most endurance types of exercise, relatively few fast-glycolytic motor units are recruited, and most of the activity occurs in oxidative fibers, which are more resistant to fatigue. The large fast-glycolytic motor units, which fatigue rapidly, begin to be recruited when the intensity of contraction exceeds about 40 percent of the maximal tension that can be produced by the muscle.

In conclusion, the neural control of whole-muscle tension involves both the frequency of action potentials in individual motor units (to vary the tension generated by the fibers in that unit) and the recruitment of motor units (to vary the number of active fibers). Most motor neuron activity occurs in bursts of action potentials, which produce tetanic contractions of individual motor units rather than single twitches. Recall that the tension of a single fiber increases only three- to fivefold when going from a twitch to a maximal tetanic contraction. Therefore, varying the frequency of action potentials in the neurons supplying them provides a way to make only three- to fivefold adjustments in the tension of the recruited motor units. The force a whole muscle exerts can be varied over a much wider range than this, from very delicate movements to extremely powerful contractions, by the recruitment of motor units. Thus recruitment provides the primary means of varying tension in a whole muscle. Recruitment is controlled by the central commands from the motor centers in the brain to the various motor neurons (Chapter 12).

Control of Shortening Velocity

As we saw earlier, the velocity at which a *single* muscle fiber shortens is determined by (1) the load on the fiber and (2) whether the fiber is a fast fiber or a slow fiber. Translated to a *whole* muscle, these characteristics become (1) the load on the whole muscle and (2) the types of motor units in the muscle. For the whole muscle, however, recruitment becomes a third

very important factor, one that explains how the shortening velocity can be varied from very fast to very slow even though the load on the muscle remains constant. Consider, for the sake of illustration, a muscle composed of only two motor units of the same size and fiber type. One motor unit by itself will lift a 4-g load more slowly than a 2-g load because the shortening velocity decreases with increasing load. When both units are active and a 4-g load is lifted, each motor unit bears only half the load, and its fibers will shorten as if it were lifting only a 2-g load. In other words, the muscle will lift the 4-g load at a higher velocity when both motor units are active. Thus recruitment of motor units leads to an increase in both force and velocity.

Muscle Adaptation to Exercise

The regularity with which a muscle is used, as well as the duration and intensity of its activity, affects the properties of the muscle. If the neurons to a skeletal muscle are destroyed or the neuromuscular junctions become nonfunctional, the denervated muscle fibers will become progressively smaller in diameter, and the amount of contractile proteins they contain will decrease. This condition is known as **denervation atrophy.** A muscle can also atrophy with its nerve supply intact if the muscle is not used for a long period of time, as when a broken arm or leg is immobilized in a cast. This condition is known as **disuse atrophy.**

In contrast to the decrease in muscle mass that results from a lack of neural stimulation, increased amounts of contractile activity—in other words, exercise—can produce an increase in the size **(hypertrophy)** of muscle fibers as well as changes in their capacity for ATP production.

Since the number of fibers in a muscle remains essentially constant throughout adult life, the changes in muscle size with atrophy and hypertrophy do not result from changes in the *number* of muscle fibers but in the metabolic capacity and size of each fiber.

Exercise that is of relatively low intensity but of long duration (popularly called "aerobic exercise"), such as running and swimming, produces increases in the number of mitochondria in the fibers that are recruited in this type of activity. In addition, there is an increase in the number of capillaries around these fibers. All these changes lead to an increase in the capacity for endurance activity with a minimum of fatigue. (Surprisingly, fiber diameter decreases slightly, and thus there is a small decrease in the maximal strength of muscles as a result of endurance exercise.) As we shall see in later chapters, endurance exercise produces changes not only in the skeletal muscles but also in the respiratory and circulatory systems, changes that improve the delivery of oxygen and fuel molecules to the muscle.

In contrast, short-duration, high-intensity exercise (popularly called "strength training"), such as weight lifting, affects primarily the fast-glycolytic fibers, which are recruited during strong contractions. These fibers undergo an increase in fiber diameter (hypertrophy) due to the increased synthesis of actin and myosin filaments, which form more myofibrils. In addition, the glycolytic activity is increased by increasing the synthesis of glycolytic enzymes. The result of such high-intensity exercise is an increase in the strength of the muscle and the bulging muscles of a conditioned weight lifter. Such muscles, although very powerful, have little capacity for endurance, and they fatigue rapidly.

Exercise produces little change in the types of *myosin* enzymes formed by the fibers and thus little change in the proportions of fast and slow fibers in a muscle. As described above, however, exercise does change the rates at which *metabolic* enzymes are synthesized, leading to changes in the proportion of oxidative and glycolytic fibers within a muscle. With endurance training, there is a decrease in the number of fast-glycolytic fibers and an increase in the number of fast-oxidative fibers as the oxidative capacity of the fibers is increased. The reverse occurs with strength training as fast-oxidative fibers are converted to fast-glycolytic fibers.

The signals responsible for all these changes in muscle with different types of activity are unknown. They are related to the frequency and intensity of the contractile activity in the muscle fibers and thus to the pattern of action potentials produced in the muscle over an extended period of time.

Because different types of exercise produce quite different changes in the strength and endurance capacity of a muscle, an individual performing regular exercises to improve muscle performance must choose a type of exercise that is compatible with the type of activity he or she ultimately wishes to perform. Thus, lifting weights will not improve the endurance of a long-distance runner, and jogging will not produce the increased strength desired by a weight lifter. Most exercises, however, produce some effects on both strength and endurance.

These changes in muscle in response to repeated periods of exercise occur slowly over a period of weeks. If regular exercise is stopped, the changes in the muscle that occurred as a result of the exercise will slowly revert to their unexercised state.

The maximum force generated by a muscle decreases by 30 to 40 percent between the ages of 30 and 80. This decrease in tension-generating capacity is due primarily to a decrease in average fiber diameter. Some of the change is simply the result of diminishing physical activity with age and can be prevented by exercise programs. The ability of a muscle to adapt to exercise, however, decreases with age: The same intensity and duration of exercise in an older individual will not produce the same amount of change as in a younger person. This decreased ability to adapt to increased activity is seen in most organs as one ages (Chapter 7).

This effect of aging, however, is only partial, and there is no question that even in the elderly, exercise can produce significant adaptation. Aerobic training has received major attention because of its effect on the cardiovascular system (Chapter 14). Strength training of a modest degree, however, is also strongly recommended because it can partially prevent the loss of muscle tissue that occurs with aging. Moreover, it helps maintain stronger bones (Chapter 18).

Extensive exercise by an individual whose muscles have not been used in performing that particular type of exercise leads to muscle soreness the next day. This soreness is the result of a mild inflammation in the muscle, which occurs whenever tissues are damaged (Chapter 20). The most severe inflammation occurs following a period of lengthening contractions, indicating that the lengthening of a muscle fiber by an external force produces greater muscle damage than do either isotonic or isometric contractions. Thus, exercising by gradually lowering weights will produce greater muscle soreness than an equivalent amount of weight lifting.

The effects of anabolic steroids on skeletal-muscle growth and strength are described in Chapter 18.

Lever Action of Muscles and Bones

A contracting muscle exerts a force on bones through its connecting tendons. When the force is great enough, the bone moves as the muscle shortens. A contracting muscle exerts only a pulling force, so that as the muscle shortens, the bones to which it is attached are pulled toward each other. **Flexion** refers to the *bending* of a limb at a joint, whereas **extension** is the *straightening* of a limb (Figure 11–31). These opposing motions require at least two muscles, one to cause flexion and the other extension. Groups of muscles that produce oppositely directed movements at a joint are known as **antagonists.** For example, from Figure 11–31 it can be seen that contraction of the biceps causes flexion of the arm at the elbow, whereas contraction of the antagonistic muscle, the triceps, causes the arm to extend. Both muscles exert only a pulling force upon the forearm when they contract.

Sets of antagonistic muscles are required not only for flexion-extension, but also for side-to-side movements or rotation of a limb. The contraction of some muscles leads to two types of limb movement, depending on the contractile state of other muscles

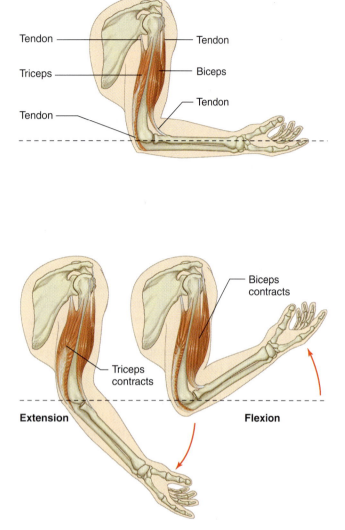

FIGURE 11–31

Antagonistic muscles for flexion and extension of the forearm. ✗

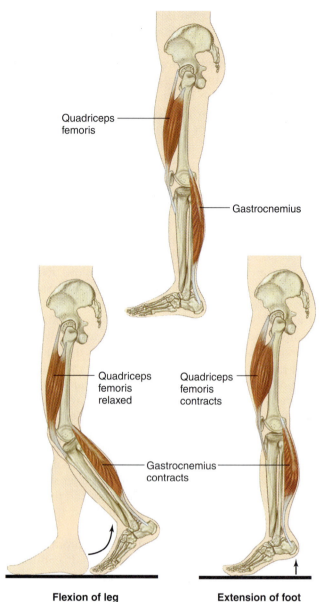

Flexion of leg   Extension of foot

FIGURE 11–32

Contraction of the gastrocnemius muscle in the calf can lead either to flexion of the leg, if the quadriceps femoris muscle is relaxed, or to extension of the foot, if the quadriceps is contracting, preventing bending of the knee joint. ✗

acting on the same limb. For example, contraction of the gastrocnemius muscle in the leg causes a flexion of the leg at the knee, as in walking (Figure 11–32). However, contraction of the gastrocnemius muscle with the simultaneous contraction of the quadriceps femoris (which causes extension of the lower leg) prevents the knee joint from bending, leaving only the ankle joint capable of moving. The foot is extended, and the body rises on tiptoe.

The muscles, bones, and joints in the body are arranged in lever systems. The basic principle of a lever is illustrated by the flexion of the arm by the biceps muscle (Figure 11–33), which exerts an upward pulling force on the forearm about 5 cm away from the elbow joint. In this example, a 10-kg weight held in the hand exerts a downward force of 10 kg about 35 cm from the elbow. A law of physics tells us that the forearm is

in mechanical equilibrium (no net forces acting on the system) when the product of the downward force (10 kg) and its distance from the elbow (35 cm) is equal to the product of the isometric tension exerted by the muscle (X), and its distance from the elbow (5 cm); that is, $10 \times 35 = 5 \times X$. Thus $X = 70$ kg. The important point is that this system is working at a mechanical disadvantage since the force exerted by the muscle (70 kg) is considerably greater than that load (10 kg) it is supporting.

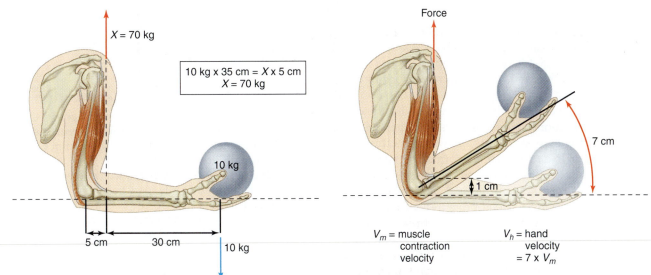

FIGURE 11–33

Mechanical equilibrium of forces acting on the forearm while supporting a 10-kg load.

FIGURE 11–34

Velocity of the biceps muscle is amplified by the lever system of the arm, producing a greater velocity of the hand. The range of movement is also amplified (1 cm of shortening by the muscle produces 7 cm of movement by the hand).

However, the mechanical disadvantage under which most muscle level systems operate is offset by increased maneuverability. In Figure 11–34, when the biceps shortens 1 cm, the hand moves through a distance of 7 cm. Since the muscle shortens 1 cm in the same amount of time that the hand moves 7 cm, the velocity at which the hand moves is seven times greater than the rate of muscle shortening. The lever system amplifies the velocity of muscle shortening so that short, relatively slow movements of the muscle produce faster movements of the hand. Thus, a pitcher can throw a baseball at 90 to 100 mi/h even though his muscles shorten at only a small fraction of this velocity.

Skeletal-Muscle Disease

A number of diseases can affect the contraction of skeletal muscle. Many of them are due to defects in the parts of the nervous system that control contraction of the muscle fibers rather than to defects in the muscle fibers themselves. For example, *poliomyelitis* is a viral disease that destroys motor neurons, leading to the paralysis of skeletal muscle, and may result in death due to respiratory failure.

Muscle Cramps Involuntary tetanic contraction of skeletal muscles produces *muscle cramps.* During cramping, nerve action potentials fire at abnormally high rates, a much greater rate than occurs during maximal voluntary contraction. The specific cause of this high activity is uncertain but is probably related to electrolyte imbalances in the extracellular fluid surrounding both the muscle and nerve fibers and changes in extracellular osmolarity, especially hypoosmolarity.

Hypocalcemic Tetany Similar in symptoms to muscular cramping is *hypocalcemic tetany,* the involuntary tetanic contraction of skeletal muscles that occurs when the extracellular calcium concentration falls to about 40 percent of its normal value. This may seem surprising since we have seen that calcium is required for excitation-contraction coupling. However, recall that this calcium is sarcoplasmic-reticulum calcium, not extracellular calcium. The effect of changes in extracellular calcium is exerted not on the sarcoplasmic-reticulum calcium, but directly on the plasma membrane. Low extracellular calcium (hypocalcemia) increases the opening of sodium channels in excitable membranes, leading to membrane depolarization and the spontaneous firing of action potentials. It is this that causes the increased muscle contractions. The mechanisms controlling the extracellular concentration of calcium ions are discussed in Chapter 16.

Muscular Dystrophy This disease is one of the most frequently encountered genetic diseases, affecting one in every 4000 boys (but much less commonly in girls) born in America. *Muscular dystrophy* is associated with the progressive degeneration of skeletal- and cardiac-muscle fibers, weakening the muscles and leading ultimately to death from respiratory or cardiac failure. While exercise strengthens normal skeletal muscle, it weakens dystrophic muscle. The symptoms become evident at about 2 to 6 years of age, and most affected individuals do not survive much beyond the age of 20.

The recessive gene responsible for a major form of muscular dystrophy has been identified on the X chromosome, and muscular dystrophy is a sex-linked recessive disease. (As described in Chapter 19, girls have two X chromosomes and boys only one. Accordingly, a girl with one abnormal X chromosome and one normal one will not develop the disease. This is why the disease is so much more common in boys.) This gene codes for a protein known as dystrophin, which is either absent or present in a nonfunctional form in patients with the disease. Dystrophin is located on the inner surface of the plasma membrane in normal muscle. It resembles other known cytoskeletal proteins and may be involved in maintaining the structural integrity of the plasma membrane or of elements within the membrane, such as ion channels, in fibers subjected to repeated structural deformation during contraction. Preliminary attempts are being made to treat the disease by inserting the normal gene into dystrophic muscle cells.

Myasthenia Gravis *Myasthenia gravis* is characterized by muscle fatigue and weakness that progressively worsens as the muscle is used. It affects about 12,000 Americans. The symptoms result from a decrease in the number of ACh receptors on the motor end plate. The release of ACh from the nerve terminals is normal, but the magnitude of the end-plate potential is markedly reduced because of the decreased number of receptors. Even in normal muscle, the amount of ACh released with each action potential decreases with repetitive activity, and thus the magnitude of the resulting EPP falls. In normal muscle, however, the EPP remains well above the threshold necessary to initiate a muscle action potential. In contrast, after a few motor nerve impulses in a myasthenia gravis patient, the magnitude of the EPP falls below the threshold for initiating a muscle action potential. As described in Chapter 20, the destruction of the ACh receptors is brought about by the body's own defense mechanisms gone awry, specifically because of the formation of antibodies to the ACh-receptor proteins.

SECTION A SUMMARY

I. There are three types of muscle—skeletal, smooth, and cardiac. Skeletal muscle is attached to bones and moves and supports the skeleton. Smooth muscle surrounds hollow cavities and tubes. Cardiac muscle is the muscle of the heart.

Structure

I. Skeletal muscles, composed of cylindrical muscle fibers (cells), are linked to bones by tendons at each end of the muscle.
II. Skeletal-muscle fibers have a repeating, striated pattern of light and dark bands due to the arrangement of the thick and thin filaments within the myofibrils.

III. Actin-containing thin filaments are anchored to the Z lines at each end of a sarcomere, while their free ends partially overlap the myosin-containing thick filaments in the A band at the center of the sarcomere.

Molecular Mechanisms of Contraction

I. When a skeletal-muscle fiber actively shortens, the thin filaments are propelled toward the center of their sarcomere by movements of the myosin cross bridges that bind to actin.
   a. The two globular heads of each cross bridge contain a binding site for actin and an enzymatic site that splits ATP.
   b. The four steps occurring during each cross-bridge cycle are summarized in Figure 11–12. The cross bridges undergo repeated cycles during a contraction, each cycle producing only a small increment of movement.
   c. The three functions of ATP in muscle contraction are summarized in Table 11–1.
II. In a resting muscle, attachment of cross bridges to actin is blocked by tropomyosin molecules that are in contact with the actin subunits of the thin filaments.
III. Contraction is initiated by an increase in cytosolic calcium concentration. The calcium ions bind to troponin, producing a change in its shape that is transmitted via tropomyosin to uncover the binding sites on actin, allowing the cross bridges to bind to the thin filaments.
   a. The rise in cytosolic calcium concentration is triggered by an action potential in the plasma membrane. The action potential is propagated into the interior of the fiber along the transverse tubules to the region of the sarcoplasmic reticulum, where it produces a release of calcium ions from the reticulum.
   b. Relaxation of a contracting muscle fiber occurs as a result of the active transport of cytosolic calcium ions back into the sarcoplasmic reticulum.
IV. Branches of a motor neuron axon form neuromuscular junctions with the muscle fibers in its motor unit. Each muscle fiber is innervated by a branch from only one motor neuron.
   a. Acetylcholine released by an action potential in a motor neuron binds to receptors on the motor end plate of the muscle membrane, opening ion channels that allow the passage of sodium and potassium ions, which depolarize the end-plate membrane.
   b. A single action potential in a motor neuron is sufficient to produce an action potential in a skeletal-muscle fiber.
V. Table 11–2 summarizes the events leading to the contraction of a skeletal-muscle fiber.

Mechanics of Single-Fiber Contraction

I. Contraction refers to the turning on of the cross-bridge cycle. Whether there is an accompanying change in muscle length depends upon the external forces acting on the muscle.

II. Three types of contractions can occur following activation of a muscle fiber: (1) an isometric contraction in which the muscle generates tension but does not change length; (2) an isotonic contraction in which the muscle shortens, moving a load; and (3) a lengthening contraction in which the external load on the muscle causes the muscle to lengthen during the period of contractile activity.

III. Increasing the frequency of action potentials in a muscle fiber increases the mechanical response (tension or shortening), up to the level of maximal tetanic tension.

IV. Maximum isometric tetanic tension is produced at the optimal sarcomere length l_o. Stretching a fiber beyond its optimal length or decreasing the fiber length below l_o decreases the tension generated.

V. The velocity of muscle-fiber shortening decreases with increases in load. Maximum velocity occurs at zero load.

Skeletal-Muscle Energy Metabolism

I. Muscle fibers form ATP by the transfer of phosphate from creatine phosphate to ADP, by oxidative phosphorylation of ADP in mitochondria, and by substrate-level phosphorylation of ADP in the glycolytic pathway.

II. At the beginning of exercise, muscle glycogen is the major fuel consumed. As the exercise proceeds, glucose and fatty acids from the blood provide most of the fuel, fatty acids becoming progressively more important during prolonged exercise. When the intensity of exercise exceeds about 70 percent of maximum, glycolysis begins to contribute an increasing fraction of the total ATP generated.

III. Muscle fatigue is caused by a variety of factors, including internal changes in acidity, glycogen depletion, and excitation-contraction coupling failure, not by a lack of ATP.

Types of Skeletal-Muscle Fibers

I. Three types of skeletal-muscle fibers can be distinguished by their maximal shortening velocities and the predominate pathway used to form ATP: slow-oxidative, fast-oxidative, and fast-glycolytic fibers.
   a. Differences in maximal shortening velocities are due to different myosin enzymes with high or low ATPase activities, giving rise to fast and slow fibers.
   b. Fast-glycolytic fibers have a larger average diameter than oxidative fibers and therefore produce greater tension, but they also fatigue more rapidly.

II. All the muscle fibers in a single motor unit belong to the same fiber type, and most muscles contain all three types.

III. Table 11–3 summarizes the characteristics of the three types of skeletal-muscle fibers.

Whole-Muscle Contraction

I. The tension produced by whole-muscle contraction depends on the amount of tension developed by each fiber and the number of active fibers in the muscle (Table 11–4).

II. Muscles that produce delicate movements have a small number of fibers per motor unit, whereas large postural muscles have much larger motor units.

III. Fast-glycolytic motor units not only have large-diameter fibers but also tend to have large numbers of fibers per motor unit.

IV. Increases in muscle tension are controlled primarily by increasing the number of active motor units in a muscle, a process known as recruitment. Slow-oxidative motor units are recruited first during weak contractions, then fast-oxidative motor units, and finally fast-glycolytic motor units during very strong contractions.

V. Increasing motor-unit recruitment increases the velocity at which a muscle will move a given load.

VI. The strength and susceptibility to fatigue of a muscle can be altered by exercise.
   a. Long-duration, low-intensity exercise increases a fiber's capacity for oxidative ATP production by increasing the number of mitochondria and blood vessels in the muscle, resulting in increased endurance.
   b. Short-duration, high-intensity exercise increases fiber diameter as a result of increased synthesis of actin and myosin, resulting in increased strength.

VII. Movement around a joint requires two antagonistic groups of muscles: one flexes the limb at the joint, and the other extends the limb.

VIII. The lever system of muscles and bones requires muscle tensions far greater than the load in order to sustain a load in an isometric contraction, but the lever system produces a shortening velocity at the end of the lever arm that is greater than the muscle-shortening velocity.

SECTION A KEY TERMS	
skeletal muscle	sliding-filament mechanism
smooth muscle	cross-bridge cycle
cardiac muscle	rigor mortis
muscle fiber	troponin
myoblast	tropomyosin
satellite cell	excitation-contraction
muscle	coupling
tendon	sarcoplasmic reticulum
striated muscle	lateral sac
myofibril	transverse tubule (T tubule)
sarcomere	motor neuron
thick filament	motor unit
myosin	motor end plate
thin filament	neuromuscular junction
actin	acetylcholine (ACh)
A band	end-plate potential (EPP)
Z line	acetylcholinesterase
I band	tension
H zone	load
M line	isometric contraction
titin	isotonic contraction
cross bridge	lengthening contraction
contraction	twitch
relaxation	latent period

contraction time
summation
tetanus
optimal length (l_o)
creatine phosphate
oxygen debt
muscle fatigue
central command fatigue
fast fiber
slow fiber
oxidative fiber
myoglobin

red muscle fiber
glycolytic fiber
white muscle fiber
slow-oxidative fiber
fast-oxidative fiber
fast-glycolytic fiber
recruitment
hypertrophy
flexion
extension
antagonist

SECTION A REVIEW QUESTIONS

1. List the three types of muscle cells and their locations.
2. Diagram the arrangement of thick and thin filaments in a striated-muscle sarcomere, and label the major bands that give rise to the striated pattern.
3. Describe the organization of myosin and actin molecules in the thick and thin filaments.
4. Describe the four steps of one cross-bridge cycle.
5. Describe the physical state of a muscle fiber in rigor mortis and the conditions that produce this state.
6. What three events in skeletal-muscle contraction and relaxation are dependent on ATP?
7. What prevents cross bridges from attaching to sites on the thin filaments in a resting skeletal muscle?
8. Describe the role and source of calcium ions in initiating contraction in skeletal muscle.
9. Describe the location, structure, and function of the sarcoplasmic reticulum in skeletal-muscle fibers.
10. Describe the structure and function of the transverse tubules.
11. Describe the events that result in the relaxation of skeletal-muscle fibers.
12. Define a motor unit and describe its structure.
13. Describe the sequence of events by which an action potential in a motor neuron produces an action potential in the plasma membrane of a skeletal-muscle fiber.
14. What is an end-plate potential, and what ions produce it?

15. Compare and contrast the transmission of electrical activity at a neuromuscular junction with that at a synapse.
16. Describe isometric, isotonic, and lengthening contractions.
17. What factors determine the duration of an isotonic twitch in skeletal muscle? An isometric twitch?
18. What effect does increasing the frequency of action potentials in a skeletal-muscle fiber have upon the force of contraction? Explain the mechanism responsible for this effect.
19. Describe the length-tension relationship in striated-muscle fibers.
20. Describe the effect of increasing the load on a skeletal-muscle fiber on the velocity of shortening.
21. What is the function of creatine phosphate in skeletal-muscle contraction?
22. What fuel molecules are metabolized to produce ATP during skeletal-muscle activity?
23. List the factors responsible for skeletal-muscle fatigue.
24. What component of skeletal-muscle fibers accounts for the differences in the fibers' maximal shortening velocities?
25. Summarize the characteristics of the three types of skeletal-muscle fibers.
26. Upon what two factors does the amount of tension developed by a whole skeletal muscle depend?
27. Describe the process of motor-unit recruitment in controlling (a) whole-muscle tension and (b) velocity of whole-muscle shortening.
28. During increases in the force of skeletal-muscle contraction, what is the order of recruitment of the different types of motor units?
29. What happens to skeletal-muscle fibers when the motor neuron to the muscle is destroyed?
30. Describe the changes that occur in skeletal muscles following a period of (a) long-duration, low-intensity exercise training; and (b) short-duration, high-intensity exercise training.
31. How are skeletal muscles arranged around joints so that a limb can push or pull?
32. What are the advantages and disadvantages of the muscle-bone-joint lever system?

SECTION B

SMOOTH MUSCLE

Having described the properties and control of skeletal muscle, we now examine the second of the three types of muscle found in the body—smooth muscle. Two characteristics are common to all smooth muscles: they lack the cross-striated banding pattern found in skeletal and cardiac fibers (hence the name "smooth" muscle), and the nerves to them are derived from the autonomic division of the nervous system rather than the somatic division. Thus, smooth muscle is not normally under direct voluntary control.

Smooth muscle, like skeletal muscle, uses cross-bridge movements between actin and myosin filaments to generate force, and calcium ions to control cross-bridge activity. However, the organization of the contractile filaments and the process of excitation-contraction coupling are quite different in these two types of muscle. Furthermore, there is considerable diversity among smooth muscles with respect to the mechanism of excitation-contraction coupling.

Structure

Each smooth-muscle fiber is a spindle-shaped cell with a diameter ranging from 2 to 10 μm, as compared to a range of 10 to 100 μm for skeletal-muscle fibers (see Figure 11–3). While skeletal-muscle fibers are multinucleate cells that are unable to divide once they have differentiated, smooth-muscle fibers have a single nucleus and have the capacity to divide throughout the life of an individual. Smooth-muscle cells can be stimulated to divide by a variety of paracrine agents, often in response to tissue injury.

Two types of filaments are present in the cytoplasm of smooth-muscle fibers (Figure 11–35): thick myosin-containing filaments and thin actin-containing filaments. The latter are anchored either to the plasma membrane or to cytoplasmic structures known as **dense bodies,** which are functionally similar to the Z lines in skeletal-muscle fibers. Note in Figure 11–35 that the filaments are oriented slightly diagonally to the long axis of the cell. When the fiber shortens, the regions of the plasma membrane between the points where actin is attached to the membrane balloon out. The thick and thin filaments are not organized into myofibrils, as in striated muscles, and there is no regular alignment of these filaments into sarcomeres, which accounts for the absence of a banding pattern (Figure 11–36). Nevertheless, smooth-muscle contraction occurs by a sliding-filament mechanism.

The concentration of myosin in smooth muscle is only about one-third of that in striated muscle, whereas the actin content can be twice as great. In spite of these differences, the maximal tension per unit of cross-sectional area developed by smooth muscles is similar to that developed by skeletal muscle.

The isometric tension produced by smooth-muscle fibers varies with fiber length in a manner qualitatively similar to that observed in skeletal muscle. There is an optimal length at which tension development is maximal, and less tension is generated at lengths shorter or longer than this optimal length. The range of muscle lengths over which smooth muscle is able to develop tension is greater, however, than it is in skeletal muscle. This property is highly adaptive since most smooth muscles surround hollow organs that undergo changes in volume with accompanying changes in the lengths of the smooth-muscle fibers in their walls. Even with relatively large increases in volume, as during the accumulation of large amounts of urine in the bladder, the smooth-muscle fibers in the wall retain some ability to develop tension, whereas such distortion might stretch skeletal-muscle fibers beyond the point of thick- and thin-filament overlap.

Contraction and Its Control

Changes in cytosolic calcium concentration control the contractile activity in smooth-muscle fibers, as in striated muscle. However, there are significant differences between the two types of muscle in the way in which calcium exerts its effects on cross-bridge activity and in the mechanisms by which stimulation leads to alterations in calcium concentration.

Cross-Bridge Activation

The thin filaments in smooth muscle do not have the calcium-binding protein troponin that mediates calcium-triggered cross-bridge activity in both skeletal and cardiac muscle. Instead, cross-bridge cycling in

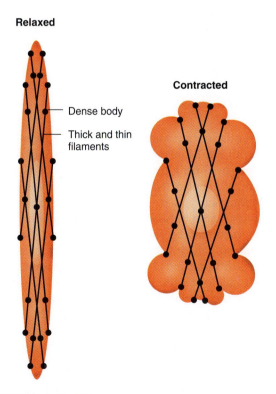

Relaxed

— Dense body

— Thick and thin filaments

Contracted

FIGURE 11–35

Thick and thin filaments in smooth muscle are arranged in slightly diagonal chains that are anchored to the plasma membrane or to dense bodies within the cytoplasm. When activated, the thick and thin filaments slide past each other causing the smooth-muscle fiber to shorten and thicken.

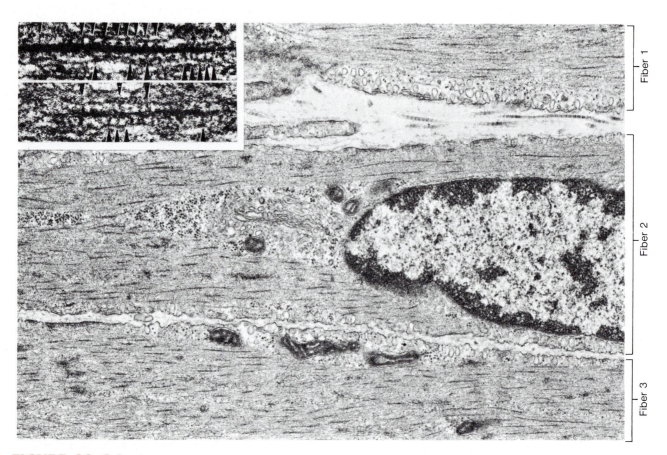

Fiber 1

Fiber 2

Fiber 3

FIGURE 11–36

Electron micrograph of portions of three smooth-muscle fibers. Higher magnification of thick filaments (insert) with arrows indicating cross bridges connecting to adjacent thin filaments.

From A. P. Somlyo, C. E. Devine, Avril V. Somlyo, and R. V. Rice, Phil. Trans. R. Soc. Lond. B, 265:223–229, (1973).

smooth muscle is controlled by a calcium-regulated enzyme that phosphorylates myosin. Only the phosphorylated form of smooth-muscle myosin is able to bind to actin and undergo cross-bridge cycling.

The following sequence of events occurs after a rise in cytosolic calcium in a smooth-muscle fiber (Figure 11–37): (1) Calcium binds to calmodulin, a calcium-binding protein that is present in most cells (Chapter 7) and whose structure is related to that of troponin. (2) The calcium-calmodulin complex binds to a protein kinase, **myosin light-chain kinase,** thereby activating the enzyme. (3) The active protein kinase then uses ATP to phosphorylate myosin light chains in the globular head of myosin. (4) The phosphorylated cross bridge binds to actin. Hence, cross-bridge activity in smooth muscle is turned on by calcium-mediated changes in the thick filaments, whereas in striated muscle, calcium mediates changes in the thin filaments.

The smooth-muscle myosin isozyme has a very low maximal rate of ATPase activity, on the order of 10 to 100 times less than that of skeletal-muscle myosin. Since the rate of ATP splitting determines the rate of cross-bridge cycling and thus shortening velocity,

smooth-muscle shortening is much slower than that of skeletal muscle. Moreover, smooth muscle does not undergo fatigue during prolonged periods of activity.

To relax a contracted smooth muscle, myosin must be dephosphorylated because dephosphorylated myosin is unable to bind to actin. This dephosphorylation is mediated by the enzyme myosin light-chain phosphatase, which is continuously active in smooth muscle during periods of rest and contraction. When cytosolic calcium rises, the rate of myosin phosphorylation by the activated kinase exceeds the rate of dephosphorylation by the phosphatase, and the amount of phosphorylated myosin in the cell increases, producing a rise in tension. When the cytosolic calcium concentration decreases, the rate of dephosphorylation exceeds the rate of phosphorylation, and the amount of phosphorylated myosin decreases, producing relaxation.

If the cytosolic calcium concentration remains elevated, the rate of ATP splitting by the cross bridges declines even though isometric tension is maintained. When a phosphorylated cross bridge is dephosphorylated while still attached to actin, it can maintain

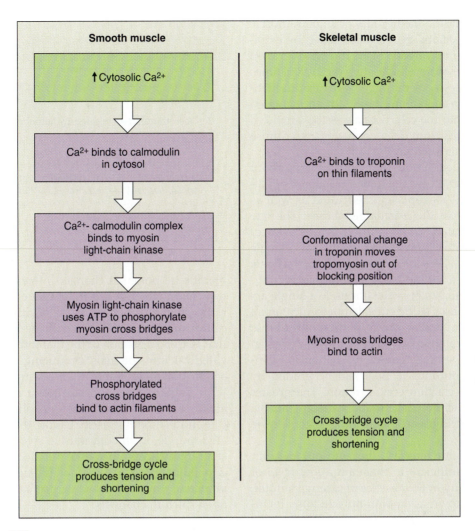

FIGURE 11–37
Pathways leading from increased cytosolic calcium to cross-bridge cycling in smooth- and skeletal-muscle fibers.

tension in a rigorlike state without movement. Dissociation of these dephosphorylated cross bridges from actin by the binding of ATP occurs at a much slower rate than dissociation of phosphorylated bridges. The net result is the ability to maintain tension for long periods of time with a very low rate of ATP consumption.

Sources of Cytosolic Calcium

Two sources of calcium contribute to the rise in cytosolic calcium that initiates smooth-muscle contraction: (1) the sarcoplasmic reticulum and (2) extracellular calcium entering the cell through plasma-membrane calcium channels. The amount of calcium contributed by these two sources differs among various smooth muscles, some being more dependent on extracellular calcium than the stores in the sarcoplasmic reticulum, and vice versa.

Let us look first at the sarcoplasmic reticulum. The total quantity of this organelle in smooth muscle is

smaller than in skeletal muscle, and it is not arranged in any specific pattern in relation to the thick and thin filaments. Moreover, there are no T tubules connected to the plasma membrane in smooth muscle. The small fiber diameter and the slow rate of contraction do not require such a rapid mechanism for getting an excitatory signal into the muscle fiber. Portions of the sarcoplasmic reticulum are located near the plasma membrane, however, forming associations similar to the relationship between T tubules and the lateral sacs in skeletal muscle. Action potentials in the plasma membrane can be coupled to the release of sarcoplasmic-reticulum calcium at these sites. In addition, second messengers released from the plasma membrane or generated in the cytosol in response to the binding of extracellular chemical messengers to plasma-membrane receptors, can trigger the release of calcium from the more centrally located sarcoplasmic reticulum.

What about extracellular calcium in excitation-contraction coupling? There are voltage-sensitive calcium channels in the plasma membranes of smooth-muscle cells, as well as calcium channels controlled by extracellular chemical messengers. Since the concentration of calcium in the extracellular fluid is 10,000 times greater than in the cytosol, the opening of calcium channels in the plasma membrane results in an increased flow of calcium into the cell. Because of the small cell size, the entering calcium does not have far to diffuse to reach binding sites within the cell.

Removal of calcium from the cytosol to bring about relaxation is achieved by the active transport of calcium back into the sarcoplasmic reticulum as well as out of the cell across the plasma membrane. The rate of calcium removal in smooth muscle is much slower than in skeletal muscle, with the result that a single twitch lasts several seconds in smooth muscle but lasts only a fraction of a second in skeletal muscle.

Moreover, whereas in skeletal muscle a single action potential releases sufficient calcium to turn on all the cross bridges in a fiber, only a portion of the cross bridges are activated in a smooth-muscle fiber in response to most stimuli. Therefore, the tension generated by a smooth-muscle fiber can be graded by varying cytosolic calcium concentration. The greater the increase in calcium concentration, the greater the number of cross bridges activated, and the greater the tension.

In some smooth muscles, the cytosolic calcium concentration is sufficient to maintain a low level of cross-bridge activity in the absence of external stimuli. This activity is known as **smooth-muscle tone.** Its intensity can be varied by factors that alter the cytosolic calcium concentration.

As in our description of skeletal muscle, we have approached the question of excitation-contraction coupling in smooth muscle backward by first describing the coupling (the changes in cytosolic calcium), and now we must ask what constitutes the excitation that elicits these changes in calcium concentration.

Membrane Activation

In contrast to skeletal muscle, in which membrane activation is dependent on a single input—the somatic neurons to the muscle—many inputs to a smooth-muscle plasma membrane can alter the contractile activity of the muscle (Table 11–5). Some of these increase contraction while others inhibit it. Moreover, at any one time, multiple inputs may be occurring, with the contractile state of the muscle dependent on the relative intensity of the various inhibitory and excitatory stimuli. All these inputs influence contractile activity by altering cytosolic calcium concentration as described in the previous section.

Some smooth muscles contract in response to membrane depolarization including action potentials, whereas others can contract in the absence of any

---

TABLE 11–5 Inputs Influencing Smooth-Muscle Contractile Activity

1. Spontaneous electrical activity in the fiber plasma membrane

2. Neurotransmitters released by autonomic neurons

3. Hormones

4. Locally induced changes in the chemical composition (paracrine agents, acidity, oxygen, osmolarity, and ion concentrations) of the extracellular fluid surrounding the fiber

5. Stretch

---

membrane potential change. Interestingly, in smooth muscles in which action potentials occur, calcium ions, rather than sodium ions, carry positive charge into the cell during the rising phase of the action potential—that is, depolarization of the membrane opens voltage-gated calcium channels, producing calcium-mediated action potentials rather than sodium-mediated ones.

Another very important point needs to be made about electrical activity and cytosolic calcium concentration in smooth muscle. Unlike the situation in striated muscle, in smooth muscle cytosolic calcium concentration can be increased (or decreased) by *graded* depolarizations (or hyperpolarizations) in membrane potential, which increase or decrease the number of open calcium channels.

Spontaneous Electrical Activity Some types of smooth-muscle fibers generate action potentials spontaneously in the absence of any neural or hormonal input. The plasma membranes of such fibers do not maintain a constant resting potential. Instead, they gradually depolarize until they reach the threshold potential and produce an action potential. Following repolarization, the membrane again begins to depolarize (Figure 11–38), so that a sequence of action potentials occurs, producing a tonic state of contractile activity. The potential change occurring during the spontaneous depolarization to threshold is known as a **pacemaker potential.** (As described in other chapters, some cardiac-muscle fibers and a few neurons in the central nervous system also have pacemaker potentials and can spontaneously generate action potentials in the absence of external stimuli.)

Nerves and Hormones The contractile activity of smooth muscles is influenced by neurotransmitters released by autonomic nerve endings. Unlike skeletal-muscle fibers, smooth-muscle fibers do not have a specialized motor end-plate region. As the axon of a postganglionic autonomic neuron enters the region

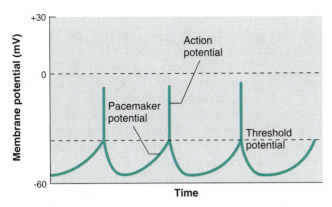

FIGURE 11–38

Generation of action potentials in a smooth-muscle fiber resulting from spontaneous depolarizations of the membrane (pacemaker potentials).

of smooth-muscle fibers, it divides into numerous branches, each branch containing a series of swollen regions known as **varicosities.** Each varicosity contains numerous vesicles filled with neurotransmitter, some of which are released when an action potential passes the varicosity. Varicosities from a single axon may be located along several muscle fibers, and a single muscle fiber may be located near varicosities belonging to postganglionic fibers of both sympathetic and parasympathetic neurons (Figure 11–39). Therefore, a number of smooth-muscle fibers are influenced by the neurotransmitters released by a single nerve fiber, and a single smooth-muscle fiber may be influenced by neurotransmitters from more than one neuron.

Whereas some neurotransmitters enhance contractile activity, others produce a *lessening* of contractile activity. Thus, in contrast to skeletal muscle, which receives only excitatory input from its motor neurons, smooth-muscle tension can be either increased or decreased by neural activity.

Moreover, a given neurotransmitter may produce opposite effects in different smooth-muscle tissues. For example, norepinephrine, the neurotransmitter released from most postganglionic sympathetic neurons, enhances contraction of vascular smooth muscle. In contrast, the same neurotransmitter produces relaxation of intestinal smooth muscle. Thus, the type of response (excitatory or inhibitory) depends not on the chemical messenger per se but on the receptor to which the chemical messenger binds in the membrane.

In addition to receptors for neurotransmitters, smooth-muscle plasma membranes contain receptors for a variety of hormones. Binding of a hormone to its receptor may lead to either increased or decreased contractile activity.

Although most changes in smooth-muscle contractile activity induced by chemical messengers are accompanied by a change in membrane potential, this is not always the case. Second messengers, for example, inositol trisphosphate, can cause the release of calcium from the sarcoplasmic reticulum, producing a contraction, without a change in membrane potential.

Local Factors Local factors, including paracrine agents, acidity, oxygen concentration, osmolarity, and the ion composition of the extracellular fluid, can also alter smooth-muscle tension. Responses to local factors provide a means for altering smooth-muscle contraction in response to changes in the muscle's immediate internal environment, which can lead to regulation that is independent of long-distance signals from nerves and hormones.

Some smooth muscles respond by contracting when they are stretched. Stretching opens mechanosensitive ion channels, leading to membrane depolarization. The resulting contraction opposes the forces acting to stretch the muscle.

On the other hand, some local factors induce smooth-muscle relaxation. Nitric oxide (NO) is one of

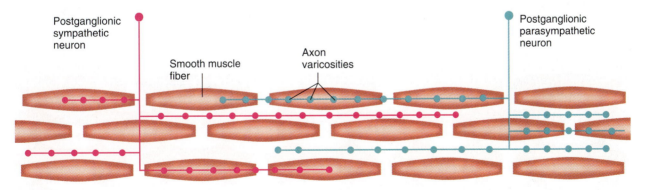

FIGURE 11–39

Innervation of smooth muscle by postganglionic autonomic neurons. Neurotransmitter is released from the varicosities along the branched axons and diffuses to receptors on muscle-fiber plasma membranes.

the most commonly encountered paracrine agents that produces smooth-muscle relaxation. NO is released from some nerve terminals as well as a variety of epithelial and endothelial cells. Because of the short life span of this reactive molecule, it acts as a paracrine agent, influencing only those cells that are very near its release site.

It is well to remember that seldom is a single agent acting on a smooth muscle, but rather the state of contractile activity at any moment depends on the simultaneous magnitude of the signals promoting contraction versus those promoting relaxation.

Types of Smooth Muscle

The great diversity of the factors that can influence the contractile activity of smooth muscles from various organs has made it difficult to classify smooth-muscle fibers. Many smooth muscles can be placed, however, into one of two groups, based on the electrical characteristics of their plasma membrane: **single-unit smooth muscles** and **multiunit smooth muscles.**

Single-Unit Smooth Muscle The muscle fibers in a single-unit smooth muscle undergo synchronous activity, both electrical and mechanical; that is, the whole muscle responds to stimulation as a single unit. This occurs because each muscle fiber is linked to adjacent fibers by gap junctions, through which action potentials occurring in one cell are propagated to other cells by local currents. Therefore, electrical activity occurring anywhere within a group of single-unit smooth-muscle fibers can be conducted to all the other connected cells (Figure 11–40).

Some of the fibers in a single-unit muscle are pacemaker cells that spontaneously generate action potentials, which are conducted by way of gap junctions into fibers that do not spontaneously generate action potentials. The majority of cells in these muscles are not pacemaker cells.

The contractile activity of single-unit smooth muscles can be altered by nerves, hormones, and local factors, using the variety of mechanisms described previously for smooth muscles in general. The extent to which these muscles are innervated varies considerably in different organs. The nerve terminals are often restricted to the regions of the muscle that contain pacemaker cells. By regulating the frequency of the pacemaker cells' action potentials, the activity of the entire muscle can be controlled.

One additional characteristic of single-unit smooth muscles is that a contractile response can often be induced by stretching the muscle. In several hollow organs—the uterus, for example—stretching the smooth muscles in the walls of the organ as a result of increases in the volume of material in the lumen initiates a contractile response.

The smooth muscles of the intestinal tract, uterus, and small-diameter blood vessels are examples of single-unit smooth muscles.

Multiunit Smooth Muscle Multiunit smooth muscles have no or few gap junctions, each fiber responds independently of its neighbors, and the muscle behaves as multiple units. Multiunit smooth muscles are richly innervated by branches of the autonomic nervous system. The contractile response of the whole muscle depends on the number of muscle fibers that are activated and on the frequency of nerve stimulation. Although stimulation of the nerve fibers to the muscle leads to some degree of depolarization and a contractile response, action potentials do not occur in most multiunit smooth muscles. Circulating hormones can increase or decrease contractile activity in multiunit smooth muscle, but stretching does not induce contraction in this type of muscle. The smooth muscle in the large airways to the lungs, in large arteries, and attached to the hairs in the skin are examples of multiunit smooth muscles.

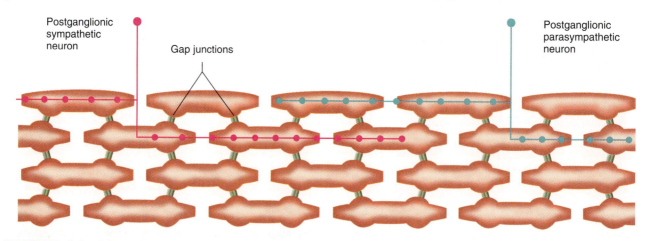

FIGURE 11–40

Innervation of a single-unit smooth muscle is often restricted to only a few fibers in the muscle. Electrical activity is conducted from fiber to fiber throughout the muscle by way of the gap junctions between the fibers.

TABLE 11–6 Characteristics of Muscle Fibers

| Characteristic | Skeletal Muscle | Smooth Muscle | | Cardiac Muscle |
		Single Unit	Multiunit	
Thick and thin filaments	Yes	Yes	Yes	Yes
Sarcomeres—banding pattern	Yes	No	No	Yes
Transverse tubules	Yes	No	No	Yes
Sarcoplasmic reticulum (SR)*	++++	+	+	++
Gap junctions between fibers	No	Yes	Few	Yes
Source of activating calcium	SR	SR and extracellular	SR and extracellular	SR and extracellular
Site of calcium regulation	Troponin	Myosin	Myosin	Troponin
Speed of contraction	Fast-slow	Very slow	Very slow	Slow
Spontaneous production of action potentials by pacemakers	No	Yes	No	Yes in certain fibers, but most not spontaneously active
Tone (low levels of maintained tension in the absence of external stimuli)	No	Yes	No	No
Effect of nerve stimulation	Excitation	Excitation or inhibition	Excitation or inhibition	Excitation or inhibition
Physiological effects of hormones on excitability and contraction	No	Yes	Yes	Yes
Stretch of fiber produces contraction	No	Yes	No	No

*Number of plus signs (+) indicates the relative amount of sarcoplasmic reticulum present in a given muscle type.

It must be emphasized that most smooth muscles do not show all the characteristics of either single-unit or multiunit smooth muscles. These two prototypes represent the two extremes in smooth-muscle characteristics, with many smooth muscles having characteristics that overlap the two groups.

Table 11–6 compares the properties of the different types of muscle. Cardiac muscle has been included for completeness although its properties are discussed in Chapter 14.

SECTION B SUMMARY

Structure

I. Smooth-muscle fibers are spindle-shaped cells that lack striations, have a single nucleus, and are capable of cell division. They contain actin and myosin filaments and contract by a sliding-filament mechanism.

Contraction and Its Control

I. An increase in cytosolic calcium leads to the binding of calcium by calmodulin. The calcium-calmodulin complex then binds to myosin light-chain kinase, activating the enzyme, which uses ATP to phosphorylate smooth-muscle myosin. Only phosphorylated myosin is able to bind to actin and undergo cross-bridge cycling.

II. Smooth-muscle myosin has a low rate of ATP splitting, resulting in a much slower shortening velocity than is found in striated muscle. However, the tension produced per unit cross-sectional area is equivalent to that of skeletal muscle.

III. Two sources of the cytosolic calcium ions initiate smooth-muscle contraction: the sarcoplasmic reticulum and extracellular calcium. The opening of calcium channels in the smooth-muscle plasma membrane and sarcoplasmic reticulum, mediated by a variety of factors, allows calcium ions to enter the cytosol.

IV. The increase in cytosolic calcium resulting from most stimuli does not activate all the cross bridges. Therefore smooth-muscle tension can be increased by agents that increase the concentration of cytosolic calcium ions.

V. Table 11–5 summarizes the types of stimuli that can initiate smooth-muscle contraction by opening or closing calcium channels in the plasma membrane or sarcoplasmic reticulum.

VI. Most, but not all, smooth-muscle cells can generate action potentials in their plasma membrane upon membrane depolarization. The rising phase of the smooth-muscle action potential is due to the influx of calcium ions into the cell through open calcium channels.

VII. Some smooth muscles generate action potentials spontaneously, in the absence of any external input, because of pacemaker potentials in the plasma membrane that repeatedly depolarize the membrane to threshold.

VIII. Smooth-muscle cells do not have a specialized end-plate region. A number of smooth-muscle fibers may be influenced by neurotransmitters released from the varicosities on a single nerve ending, and a single smooth-muscle fiber may be influenced by neurotransmitters from more than one neuron. Neurotransmitters may have either excitatory or inhibitory effects on smooth-muscle contraction.

IX. Smooth muscles can be classified broadly as single-unit or multiunit smooth muscle (Table 11–6).

SECTION B KEY TERMS

dense body	varicosity
myosin light-chain kinase	single-unit smooth muscle
smooth-muscle tone	multiunit smooth muscle
pacemaker potential	

SECTION B REVIEW QUESTIONS

1. How does the organization of thick and thin filaments in smooth-muscle fibers differ from that in striated-muscle fibers?
2. Compare the mechanisms by which a rise in cytosolic calcium concentration initiates contractile activity in skeletal- and smooth-muscle fibers.
3. What are the two sources of calcium that lead to the increase in cytosolic calcium that triggers contraction in smooth muscle?
4. What types of stimuli can trigger a rise in cytosolic calcium in smooth-muscle fibers?
5. What effect does a pacemaker potential have on a smooth-muscle cell?
6. In what ways does the neural control of smooth-muscle activity differ from that of skeletal muscle?
7. Describe how a stimulus may lead to the contraction of a smooth-muscle cell without a change in the plasma-membrane potential.
8. Describe the differences between single-unit and multiunit smooth muscles.

CHAPTER 11 CLINICAL TERMS

curare	muscle cramps
botulism	hypocalcemic tetany
denervation atrophy	muscular dystrophy
disuse atrophy	myasthenia gravis
poliomyelitis	

CHAPTER 11 THOUGHT QUESTIONS

(Answers are given in appendix A.)

1. Which of the following corresponds to the state of myosin (M) under resting conditions and in rigor mortis? (a) $M \cdot ATP$, (b) $M^* \cdot ADP \cdot P_i$, (c) $A \cdot M^* \cdot ADP \cdot P_i$, (d) $A \cdot M$.

2. If the transverse tubules of a skeletal muscle are disconnected from the plasma membrane, will action potentials trigger a contraction? Give reasons.

3. When a small load is attached to a skeletal muscle that is then tetanically stimulated, the muscle lifts the load in an isotonic contraction over a certain distance, but then stops shortening and enters a state of isometric contraction. With a heavier load, the distance shortened before entering an isometric contraction is shorter. Explain these shortening limits in terms of the length-tension relation of muscle.

4. What conditions will produce the maximum tension in a skeletal-muscle fiber?

5. A skeletal muscle can often maintain a moderate level of active tension for long periods of time, even though many of its fibers become fatigued. Explain.

6. If the blood flow to a skeletal muscle were markedly decreased, which types of motor units would most rapidly have their ability to produce ATP for muscle contraction severely reduced? Why?

7. As a result of an automobile accident, 50 percent of the muscle fibers in the biceps muscle of a patient were destroyed. Ten months later, the biceps muscle was able to generate 80 percent of its original force. Describe the changes that took place in the damaged muscle that enabled it to recover.

8. In the laboratory, if an isolated skeletal muscle is placed in a solution that contains no calcium ions, will the muscle contract when it is stimulated (1) directly by depolarizing its membrane, or (2) by stimulating the nerve to the muscle? What would happen if it were a smooth muscle?

9. The following experiments were performed on a single-unit smooth muscle in the gastrointestinal tract.
   a. Stimulating the parasympathetic nerves to the muscle produced a contraction.
   b. Applying a drug that blocks the voltage-sensitive sodium channels in most plasma membranes led to a failure to contract upon stimulating the parasympathetic nerves.
   c. Applying a drug that binds to muscarinic receptors (Chapter 8), and hence blocks the action of ACh at these receptors, did not prevent the muscle from contracting when the parasympathetic nerve was stimulated.

From these observations, what might one conclude about the mechanism by which parasympathetic nerve stimulation produces a contraction of the smooth muscle?

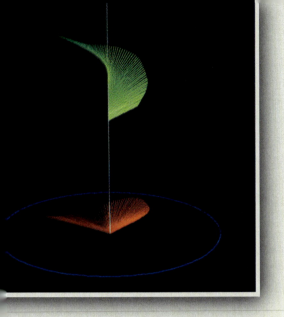

Control of Body Movement

Carrying out a coordinated movement is a complicated process involving nerves, muscles, and bones. Consider the events associated with reaching out and grasping an object. The fingers are first *extended* (straightened) to reach around the object, and then *flexed* (bent) to grasp it. The degree of extension will depend upon the size of the object (Is it a golf ball or a soccer ball?), and the force of flexion will depend upon its weight and consistency (A bowling ball or a balloon?). Simultaneously, the wrist, elbow, and shoulder are extended, and the trunk is inclined forward, the exact movements depending upon the object's position. The shoulder, elbow, and wrist are stabilized to support first the weight of the arm and hand and then the added weight of the object. Through all this, upright posture and balance are maintained despite the body's continuously shifting position.

The building blocks for these movements—as for all movements—are **motor units,** each comprising one motor neuron together with all the skeletal-muscle fibers that this neuron innervates (Chapter 11). The motor neurons are the "final common pathway" out of the central nervous system since all neural influences on skeletal muscle converge on the motor neurons and can only affect skeletal muscle through them.

All the motor neurons that supply a given muscle make up the **motor neuron pool** for the muscle. The cell bodies of the pool for a given muscle are close to each other either in the ventral horn of the spinal cord (see Figure 8–36) or in the brainstem.

Within the brainstem or spinal cord, the axons of many neurons synapse on a motor neuron to control its activity. Although no single input to a motor neuron is essential for movement of the muscle fibers it innervates, a *balanced* input from all sources is necessary to provide the precision and speed of normally coordinated actions. For example, if inhibitory synaptic input to a given motor neuron is decreased, the still-normal excitatory input to that neuron will be unopposed and the motor neuron firing will increase, which leads to excessive contraction of the muscle. This is what happens in the disease **tetanus,** where the inhibitory input to motor neurons—including those controlling the muscles of the jaw—is decreased, and all the muscles are activated. The muscles that close the jaw, however, are much stronger than those that open it, and their activity predominates. The spasms of these jaw muscles, which appear early in the disease, are responsible for the common name of the condition, **lockjaw.**

It is important to realize that movements—even simple movements such as flexing a single finger—are rarely achieved by just one muscle. Each of the myriad coordinated body movements of which a person is capable is achieved by activation, in a precise temporal order, of many motor units in various muscles.

This chapter deals with the interrelated neural inputs that converge upon motor neurons to control their activity. We present first a summary of a model of how the motor system functions and then describe each component of the model in detail.

Keep in mind throughout this section that many contractions executed by skeletal muscles, particularly the muscles involved in postural support, are isometric (Chapter 11), and even though the muscle is active during these contractions, no movement occurs. In the following discussions the general term "muscle movement" includes these isometric contractions. In addition, remember that all "information" in the nervous system is transmitted in the form of graded potentials or action potentials.

Motor Control Hierarchy

Throughout the central nervous system, the neurons involved in controlling the motor neurons to skeletal muscles can be thought of as being organized in a hierarchical fashion, each level of the hierarchy having a certain task in motor control (Figure 12–1). To begin a movement, a general "intention" such as "pick up sweater" or "write signature" or "answer telephone" is generated at the highest level of the motor control hierarchy. This highest level encompasses many regions of the brain, including those involved in memory, emotions, and motivation. Very little is known, however, as to exactly where intentions for movements are formed in the brain.

Information is relayed from these highest hierarchical neurons, referred to as the "command" neurons, to parts of the brain that make up the middle level of the motor control hierarchy. The middle-level structures specify the postures and movements needed to carry out the intended action. In our example of picking up a sweater, structures of the middle hierarchical level coordinate the commands that tilt the body and extend the arm and hand toward the sweater and shift the body's weight to maintain balance. The middle hierarchical structures are located in parts of the cerebral cortex (termed, as we shall see, the sensorimotor cortex) and in the cerebellum, subcortical nuclei, and brainstem (Figures 12–1 and 12–2a and b). These structures have extensive interconnections, as indicated by the arrows in Figure 12–1.

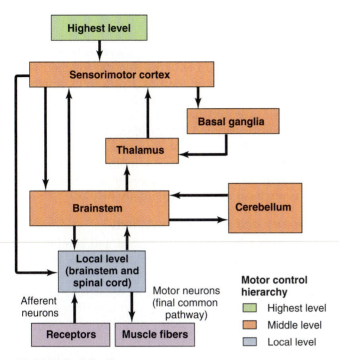

FIGURE 12–1

The conceptual hierarchical organization of the neural systems controlling body movement. All the skeletal muscles of the body are controlled by motor neurons. Sensorimotor cortex includes those parts of the cerebral cortex that act together to control skeletal-muscle activity. The middle level of the hierarchy also receives input from the vestibular apparatus and eyes (not shown in the figure).

As the neurons in the middle level of the hierarchy receive input from the command neurons, they simultaneously receive afferent information (from receptors in the muscles, tendons, joints, and skin, as well as from the vestibular apparatus and eyes) about the starting position of the body parts that are to be "commanded" to move. They also receive information about the nature of the space just outside the body into which that movement will take place. Neurons of the middle level of the hierarchy integrate all this afferent information with the signals from the command neurons to create a **motor program**—that is, the pattern of neural activity required to perform the desired movement. People can perform many slow, voluntary movements without sensory feedback, but the movements are abnormal.

The information determined by the motor program is then transmitted via **descending pathways** to the lowest level of the motor control hierarchy, the local level, at which the motor neurons to the muscles exit the brainstem or spinal cord. The local level of the hierarchy includes the motor neurons and the interneurons whose function is related to them; it is the final determinant of exactly which motor neurons will be activated to achieve the desired action and when this will happen. Note in Figure 12–1 that the descending pathways to the local level arise only in the sensorimotor cortex and brainstem; the basal ganglia, thalamus, and cerebellum exert their effects on the local level only indirectly, via the descending pathways from the cerebral cortex and brainstem.

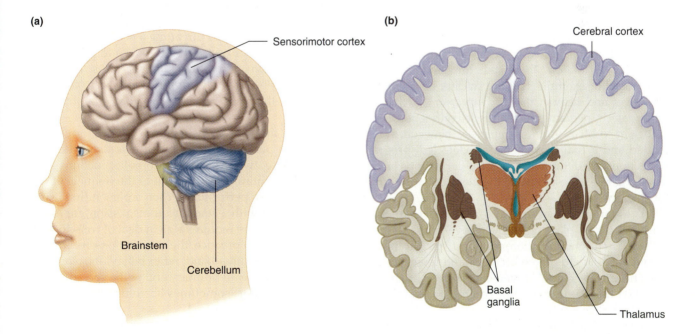

FIGURE 12–2

(a) Side view of the brain showing three of the four components of the middle level of the motor control hierarchy.
(b) Cross section of the brain showing the basal ganglia—part of the subcortical nuclei, the fourth component of the hierarchy's middle level.

TABLE 12–1 Conceptual Motor Control Hierarchy for Voluntary Movements

I. The highest level

a. Function: forms complex plans according to individual's intention and communicates with the middle level via "command neurons."

b. Structures: areas involved with memory and emotions; supplementary motor area; and association cortex. All these structures receive and correlate input from many other brain structures.

II. The middle level

a. Function: converts plans received from the highest level to a number of smaller motor programs, which determine the pattern of neural activation required to perform the movement. These programs are broken down into subprograms that determine the movements of individual joints. The programs and subprograms are transmitted, often via the cerebral cortex, through descending pathways to the lowest control level.

b. Structures: sensorimotor cortex, cerebellum, parts of basal ganglia, some brainstem nuclei.

III. The lowest level (the local level)

a. Function: specifies tension of particular muscles and angle of specific joints at specific times necessary to carry out the programs and subprograms transmitted from the middle control levels.

b. Structures: levels of brainstem or spinal cord from which motor neurons exit.

The motor programs are continuously adjusted during the course of most movements. As the initial motor program is implemented and the action gets underway, brain regions at the middle level of the hierarchy continue to receive a constant stream of updated afferent information about the movements taking place. Say, for example, that the sweater being picked up is wet and heavier than expected so that the initially determined amount of muscle contraction is not sufficient to lift it. Any discrepancies between the intended and actual movements are detected, program corrections are determined, and the corrections are relayed via the lowest level of the hierarchy to the motor neurons.

If a complex movement is repeated frequently, learning takes place and the movement becomes skilled. Then, the initial information from the middle hierarchical level is more accurate and fewer corrections need to be made. Movements performed at high speed without concern for fine control are made solely according to the initial motor program.

The structures and functions of the motor control hierarchy are summarized in Table 12–1.

We must emphasize that this hierarchical model, widely used by physiologists who work on the motor

system, is only a guide, one that requires qualification. The different areas of the brain, and neurons within each area, have so many reciprocal connections that it is often impossible to assign a specific function to a given area or group of neurons. In addition, different neurons in different areas of the brain are often active simultaneously, and neurons with similar properties are widely distributed over different regions of the brain. Nevertheless, just as researchers have found it useful to retain the notion of a motor control hierarchy despite its flaws, you the reader should also find the hierarchical model conceptually helpful.

Voluntary and Involuntary Actions

Given such a highly interconnected and complicated neuroanatomical basis for the motor system, it is difficult to use the phrase **voluntary movement** with any real precision. We shall use it, however, to refer to those actions that have the following characteristics: (1) The movement is accompanied by a conscious awareness of what we are doing and why we are doing it rather than the feeling that it "just happened," and (2) our attention is directed toward the action or its purpose.

The term "involuntary," on the other hand, describes actions that do not have these characteristics. "Unconscious," "automatic," and "reflex" are often taken to be synonyms for "involuntary," although in the motor system the term "reflex" has a more precise meaning (Chapter 7).

Despite our attempts to distinguish between voluntary and involuntary actions, almost all motor behavior involves both components, and the distinction between the two cannot be made easily. Even such a highly conscious act as threading a needle involves the unconscious postural support of the hand and forearm and inhibition of the antagonistic muscles—those muscles whose activity would oppose the intended action, in this case, the muscles that straighten the fingers.

Thus, most motor behavior is neither purely voluntary nor purely involuntary but falls somewhere between these two extremes. Moreover, actions shift along this continuum according to the frequency with which they are performed. When a person first learns to drive a car with a standard transmission, for example, shifting gears requires a great deal of conscious attention, but with practice, the same actions become automatic. On the other hand, reflex behaviors, which are all the way at the involuntary end of the spectrum, can with special effort sometimes be voluntarily modified or even prevented.

We now turn to an analysis of the individual components of the motor control system, beginning with local control mechanisms because their activity serves as a base upon which the pathways descending from the brain exert their influence. Keep in mind throughout these descriptions that motor neurons always form the "final common pathway" to the muscles.

Local Control of Motor Neurons

The local control systems are the relay points for instructions to the motor neurons from centers higher in the motor control hierarchy. In addition, the local control systems play a major role in adjusting motor unit activity to unexpected obstacles to movement and to harmful factors in the surrounding environment.

To carry out these adjustments, the local control systems use information carried by afferent fibers from sensory receptors in the muscles, tendons, joints, and skin of the body part to be moved. The afferent fibers transmit information not only to higher levels of the hierarchy, as noted earlier, but to the local level as well.

Interneurons

Most of the synaptic input to motor neurons from the descending pathways and afferent neurons does not go *directly* to motor neurons but rather to interneurons that synapse with the motor neurons. These interneurons are of several types. Some are confined to the general region of the motor neuron upon which they synapse and thus are called local interneurons. Others have processes that extend up or down short distances in the spinal cord and brainstem, or even throughout much of the length of the central nervous system. The interneurons with longer processes are important in movements that involve the coordinated interaction of, for example, a shoulder and arm or an arm and a leg.

The local interneurons are important elements of the lowest level of the motor control hierarchy, integrating inputs not only from higher centers and peripheral receptors but from other interneurons as well (Figure 12–3). They are crucial in determining which muscles are activated and when. Moreover, interneurons can act as "switches" that enable a movement to be turned on or off under the command of higher motor centers. For example, if we pick up a hot plate, a local reflex arc will be initiated by pain receptors in the skin of the hands, normally resulting in our dropping the plate. But if it contained our dinner, descending commands could inhibit the local activity, and we would hold onto the plate until we could put it down safely.

Local Afferent Input

As just noted, afferent fibers usually impinge upon the local interneurons (in one case, they synapse directly on motor neurons). The afferent fibers bring information from receptors in three areas: (1) the very muscles controlled by the motor neurons, (2) other nearby

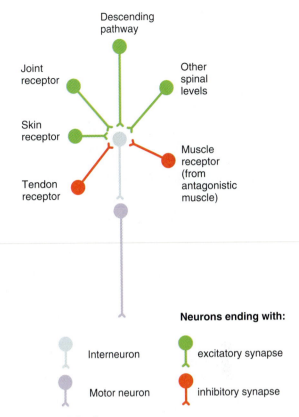

FIGURE 12–3
Convergence of axons onto a local interneuron.

muscles, and (3) the tendons, joints, and skin surrounding the muscles.

These receptors monitor the length and tension of the muscles, movement of the joints, and the effect of movements on the overlying skin. In other words, the movements themselves give rise to afferent input that, in turn, influences the movements via negative feedback. As we shall see next, their input not only provides negative-feedback control over the muscles but contributes to the conscious awareness of limb and body position as well.

Length-Monitoring Systems Absolute muscle length and changes in muscle length are monitored by stretch receptors embedded within the muscle. These receptors consist of peripheral endings of afferent nerve fibers that are wrapped around modified muscle fibers, several of which are enclosed in a connective-tissue capsule. The entire structure is called a **muscle spindle** (Figure 12–4). The modified muscle fibers within the spindle are known as **intrafusal fibers,** whereas the skeletal-muscle fibers that form the bulk of the muscle and generate its force and movement are the **extrafusal fibers.**

Within a given spindle, there are two kinds of stretch receptors: One responds best to how much the muscle has been stretched, the other to both the

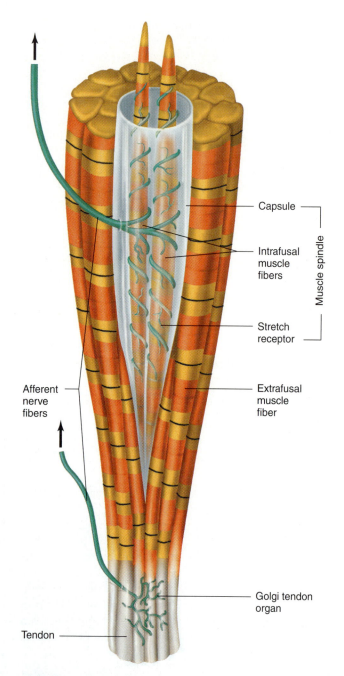

FIGURE 12–4

A muscle spindle and Golgi tendon organ. Note that the muscle spindle is parallel to the extrafusal muscle fibers. The Golgi tendon organ will be discussed later in the chapter.

Adapted from Elias, Pauly, and Burns.

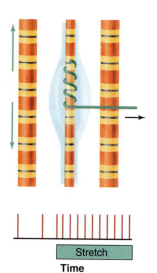

(a) Muscle stretch

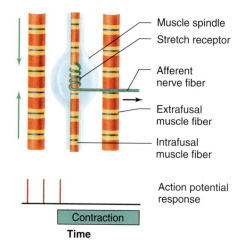

(b) Muscle contraction

FIGURE 12–5

(a) Passive stretch of the muscle activates the spindle stretch receptors and causes an increased rate of action potentials in the afferent nerve. (b) Contraction of the extrafusal fibers removes tension on the stretch receptors and lowers the rate of action potential firing. Blue arrows indicate direction of force on the muscle spindles.

magnitude of the stretch and the speed with which it occurs. Although the two kinds of stretch receptors are separate entities, they will be referred to collectively as the **muscle-spindle stretch receptors.**

The muscle spindles are parallel to the extrafusal fibers such that stretch of the muscle by an *external* force pulls on the intrafusal fibers, stretching them and

activating their receptor endings (Figure 12–5a). The more the muscle is stretched or the faster it is stretched, the greater the rate of receptor firing. In contrast, *contraction* of the extrafusal fibers and the resultant shortening of the muscle remove tension on the spindle and slow the rate of firing of the stretch receptor (Figure 12–5b).

When the afferent fibers from the muscle spindle enter the central nervous system, they divide into branches that take several different paths. In Figure 12–6, path A *directly* stimulates motor neurons that go back to the muscle that was stretched, thereby completing a reflex arc known as the **stretch reflex.**

This reflex is probably most familiar in the form of the **knee jerk,** part of a routine medical examination. The examiner taps the patellar tendon (Figure 12–6), which passes over the knee and connects extensor muscles in the thigh to the tibia in the lower leg. As the tendon is pushed in and thereby stretched by

tapping, the thigh muscles to which it is attached are stretched, and all the stretch receptors within these muscles are activated. More action potentials are generated in the afferent nerve fibers from the stretch receptors and are transmitted to the motor neurons that control these same muscles. The motor units are stimulated, the thigh muscles contract, and the patient's lower leg is extended to give the knee jerk. The proper performance of the knee jerk tells the physician that the afferent fibers, the balance of synaptic input to the motor neurons, the motor neurons themselves, the neuromuscular junctions, and the muscles are all functioning normally.

During normal movement, in contrast to the knee jerk reflex, the stretch receptors in the various muscles are rarely all activated at the same time.

Because the afferent nerve fibers mediating the stretch reflex synapse *directly* on the motor neurons without the interposition of any interneurons, the stretch reflex is called **monosynaptic.** Stretch reflexes are the only known monosynaptic reflex arcs. All other reflex arcs—including nonmuscular reflexes—are **polysynaptic,** having at least one interneuron, and usually many, between the afferent and efferent neurons.

In path B of Figure 12–6, the branches of the afferent nerve fibers from stretch receptors end on interneurons that, when activated, inhibit the motor neurons controlling antagonistic muscles; these are muscles whose contraction would interfere with the reflex response (in the knee jerk, for example, the flexor muscles of the knee are inhibited). The activation of one muscle with the simultaneous inhibition of its antagonistic muscle is called **reciprocal innervation** and is characteristic of many movements, not just the stretch reflex.

Path C in Figure 12–6 activates motor neurons of **synergistic muscles**—that is, muscles whose contraction assists the intended motion (in our example, other leg extensor muscles). The muscles activated are on the same side of the body as the receptors, and the response is therefore **ipsilateral;** a response on the opposite side of the body is **contralateral.**

In path D in Figure 12–6, the axon of the afferent neuron continues to the brainstem and synapses there with interneurons that form the next link in the pathway that conveys information about the muscle length to areas of the brain dealing with motor control. This information is especially important, as we have mentioned, during slow, controlled movements such as the performance of an unfamiliar action. Ascending paths also provide information that contributes to the conscious perception of the position of a limb.

Alpha-Gamma Coactivation As indicated in Figure 12–5b, stretch on the intrafusal fibers decreases when the muscle shortens. In this example, the spindle

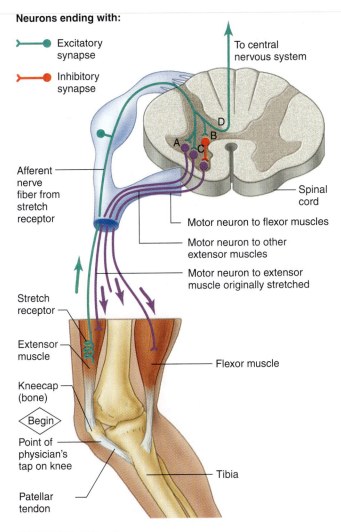

Neurons ending with:

Excitatory synapse

Inhibitory synapse

To central nervous system

Afferent nerve fiber from stretch receptor

Spinal cord

Motor neuron to flexor muscles

Motor neuron to other extensor muscles

Motor neuron to extensor muscle originally stretched

Stretch receptor

Extensor muscle

Flexor muscle

Kneecap (bone)

Begin

Point of physician's tap on knee

Tibia

Patellar tendon

FIGURE 12–6

Neural pathways involved in the knee jerk reflex. Start with the begin logo. Tapping the patellar tendon stretches the extensor muscle, causing (paths A and C) compensatory contraction of this and other extensor muscles, (path B) relaxation of flexor muscles, and (path D) information about muscle length to be sent to the brain. Arrows indicate direction of action-potential propagation.

stretch receptors go completely slack at this time, and they stop firing action potentials. In this situation, there can be no indication of any further changes in muscle length the whole time the muscle is shortening. Physiologically, to prevent this loss of information, the two ends of each intrafusal muscle fiber are stimulated to contract during the shortening of the extrafusal fibers, thus maintaining tension in the central region of the intrafusal fiber, where the stretch receptors are located (Figure 12–7). It is important to recognize that the intrafusal fibers are not large enough or strong enough to shorten a whole muscle and move joints; their sole job is to maintain tension on the spindle stretch receptors.

The intrafusal fibers contract in response to activation by motor neurons, but the motor neurons supplying them are usually not those that activate the extrafusal muscle fibers. The motor neurons controlling the extrafusal muscle fibers are larger and are classified as **alpha motor neurons,** whereas the smaller motor neurons whose axons innervate the intrafusal fibers are known as **gamma motor neurons** (Figure 12–7).

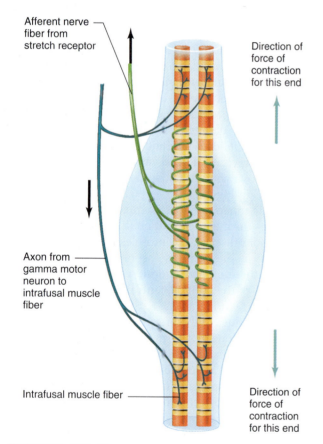

Afferent nerve fiber from stretch receptor

Direction of force of contraction for this end

Axon from gamma motor neuron to intrafusal muscle fiber

Intrafusal muscle fiber

Direction of force of contraction for this end

FIGURE 12–7

As the ends of the intrafusal fibers contract in response to gamma motor neuron activation, they pull on the center of the fiber and stretch the receptor. The black arrows indicate direction of action-potential propagation.

Both alpha and gamma motor neurons are activated by interneurons in their immediate vicinity and directly by neurons of the descending pathways. In fact, as described above, during many voluntary and involuntary movements they are **coactivated**—that is, excited at almost the same time. Coactivation ensures that information about muscle length will be continuously available to provide for adjustment during ongoing actions and to plan and program future movements.

Tension-Monitoring Systems Any given set of inputs to a given set of motor neurons can lead to various degrees of tension in the muscles they innervate, the tension depending on muscle length, the load on the muscles, and the degree of muscle fatigue. Therefore, feedback is necessary to inform the motor control systems of the tension actually achieved.

Some of this feedback is provided by vision as well as afferent input from skin, muscle, and joint receptors, but an additional receptor type specifically monitors how much tension is being exerted by the contracting motor units (or imposed on the muscle by external forces if the muscle is being stretched).

The receptors employed in this tension-monitoring system are the **Golgi tendon organs,** which are located in the tendons near their junction with the muscle (see Figure 12–4). Endings of afferent nerve fibers are wrapped around collagen bundles in the tendon, bundles that are slightly bowed in the resting state. When the attached extrafusal muscle fibers contract, they pull on the tendon, which straightens the collagen bundles and distorts the receptor endings, activating them. Thus, the Golgi tendon organs discharge in response to the tension generated by the contracting muscle and initiate action potentials that are transmitted to the central nervous system.

Branches of the afferent neuron from the Golgi tendon organ cause widespread *inhibition,* via interneurons, of the motor neurons to the contracting muscle (A in Figure 12–8) and its synergists. They also *stimulate* the motor neurons of the antagonistic muscles (B in Figure 12–8). Note that this reciprocal innervation is the opposite of that produced by the muscle-spindle afferents.

To summarize, the activity of afferent fibers from the Golgi tendon organ supplies the motor control systems (both locally and in the brain) with continuous information about the muscle's *tension.* In contrast, the spindle afferent fibers provide information about the muscle's *length.*

The Withdrawal Reflex In addition to the afferent information from the spindle stretch receptors and Golgi tendon organs of the activated muscle, other input is fed into the local motor control systems. For example, painful stimulation of the skin activates the

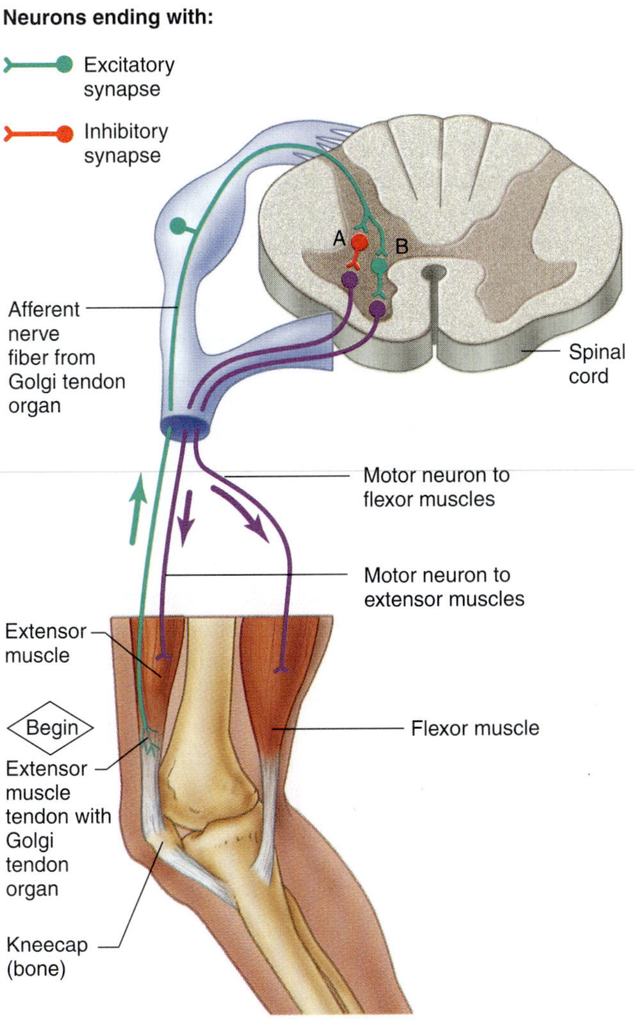

Neurons ending with:

- ⊸● Excitatory synapse
- ⊸● Inhibitory synapse

Afferent nerve fiber from Golgi tendon organ

A B

Spinal cord

Motor neuron to flexor muscles

Motor neuron to extensor muscles

Extensor muscle

Begin

Extensor muscle tendon with Golgi tendon organ

Flexor muscle

Kneecap (bone)

FIGURE 12–8

Neural pathways underlying the Golgi tendon organ component of the local control system. In this diagram, contraction of the extensor muscles causes tension in the Golgi tendon organ and increases the rate of action-potential firing in the afferent nerve fiber. By way of interneurons, this increased activity results in (path A) inhibition of the motor neuron of the extensor muscle and its synergists and (path B) excitation of flexor muscles' motor neurons. Arrows indicate direction of action-potential propagation.

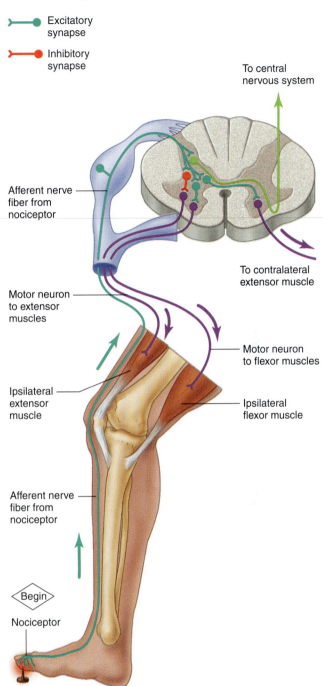

Neurons ending with:

- ⊸● Excitatory synapse
- ⊸● Inhibitory synapse

To central nervous system

Afferent nerve fiber from nociceptor

To contralateral extensor muscle

Motor neuron to extensor muscles

Motor neuron to flexor muscles

Ipsilateral extensor muscle

Ipsilateral flexor muscle

Afferent nerve fiber from nociceptor

Begin

Nociceptor

FIGURE 12–9

In response to pain, the ipsilateral flexor muscle's motor neuron is stimulated (withdrawal reflex). In the case illustrated, the opposite limb is extended (crossed-extensor reflex) to support the body's weight. Arrows indicate direction of action-potential propagation.

ipsilateral flexor motor neurons and inhibits the ipsilateral extensor motor neurons, moving the body part away from the stimulus. This is called the **withdrawal reflex** (Figure 12–9). The same stimulus causes just the opposite response on the contralateral side of the body—activation of the extensor motor neurons and inhibition of the flexor motor neurons (the **crossed-extensor reflex**). In the example in Figure 12–9, the strengthened extension of the contralateral leg means that this leg can support more of the body's weight as the hurt foot is raised from the ground by flexion.

The Brain Motor Centers and the Descending Pathways They Control

As stated earlier, the motor neurons and interneurons at the local levels of motor control are influenced by descending pathways, and these pathways are themselves controlled by various motor centers in the brain (see Figure 12–1).

Cerebral Cortex

The cerebral cortex plays a critical role in both the planning and ongoing control of voluntary movements, functioning in both the highest and middle levels of the motor control hierarchy. The term **sensorimotor cortex** is used to include all those parts of the cerebral cortex that act together in the control of muscle movement. A large number of nerve fibers that give rise to descending pathways for motor control come from two areas of sensorimotor cortex on the posterior part of the frontal lobe: the **primary motor cortex** (sometimes called simply the **motor cortex**) and the **premotor area** (Figure 12–10). Different regions of the primary motor

cortex are each concerned primarily with movements of one area of the body (Figure 12–11).

Other areas of sensorimotor cortex include the **supplementary motor cortex,** which lies mostly on the surface on the frontal lobe where the cortex folds down between the two hemispheres (see Figure 12–10b), the **somatosensory cortex,** and parts of the **parietal-lobe association cortex** (see Figures 12–10a and b).

Although these areas are distinct anatomically and functionally, they are heavily interconnected, and individual muscles or movements are represented at multiple sites. Thus, the cortical neurons that control movement form a neural network, such that many neurons participate in each single movement, and any one neuron may function in more than one movement. The neural networks can be distributed across multiple sites in parietal and frontal cortex, including the sites named in the preceding two paragraphs. The interaction of the neurons within the networks is flexible so that the neurons are capable of responding differently under different circumstances. This adaptability enhances the possibility of integration of incoming neural signals from diverse sources and the final coordination of many parts into a smooth, purposeful movement. It

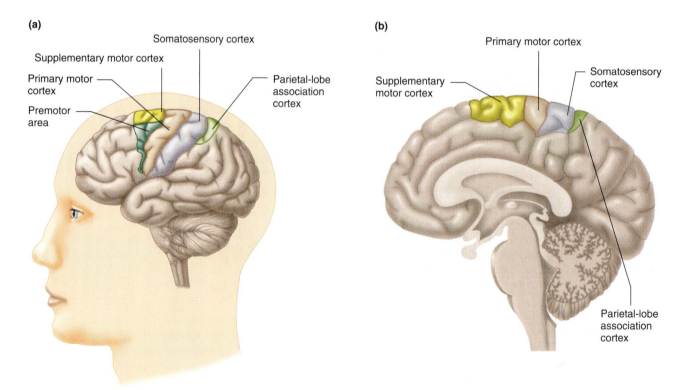

FIGURE 12–10

(a) The major motor areas of cerebral cortex. (b) Midline view of the brain showing the supplementary motor cortex, which lies in the part of the cerebral cortex that is folded down between the two cerebral hemispheres. Other cortical motor areas also extend onto this area. The premotor, supplementary motor, primary motor, somatosensory, and parietal-lobe association cortexes together make up the sensorimotor cortex.

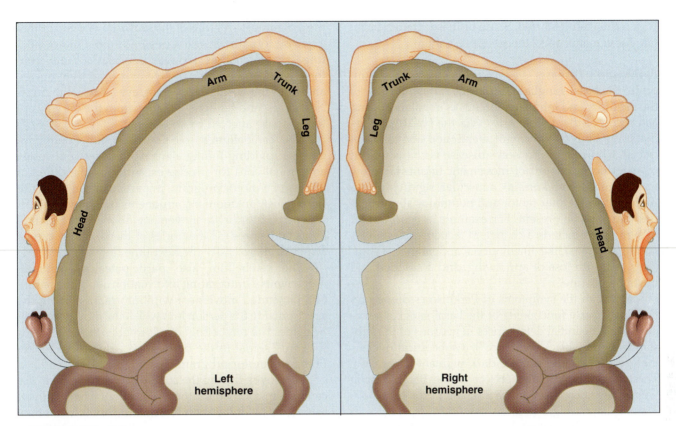

FIGURE 12–11

Representation of major body areas in primary motor cortex. Within the broad areas, however, no one area exclusively controls the movement of a single body region, and there is much overlap and duplication of cortical representation.

probably also accounts for the remarkable variety of ways in which we can approach a goal—"I can comb my hair with my right hand or my left, starting at the back of my head or the front"—and for some of the learning that occurs in all aspects of motor behavior.

We have described the various areas of sensorimotor cortex as giving rise, either directly or indirectly, to pathways descending to the motor neurons, but these areas are not the prime *initiators* of movement, and other brain areas are certainly involved. As stated earlier, we presently don't know just where or how movements are initiated except in the case of reflexes.

Association areas of the cerebral cortex play a role in motor control. For example, neurons of parietal association cortex are important in the visual control of hand action for reaching and grasping. These neurons play an important role in matching motor signals concerning the pattern of hand action with signals from the visual system concerning the three-dimensional features of the objects to be manipulated.

During activation of the cortical areas involved in motor control, subcortical mechanisms also become active, and it is to these areas of the motor control system that we now turn.

Subcortical and Brainstem Nuclei

A dozen or so highly interconnected structures lie within the cerebrum beneath the cerebral cortex and in the brainstem, and they interact with the cortex to control movements. Their influence is transmitted indirectly to the motor neurons both by pathways that go to the cerebral cortex and by descending pathways that arise from some of the brainstem nuclei.

It is not known to what extent, if any, these structures *initiate* movements, but they definitely play a prominent role in planning and monitoring them, establishing the programs that determine the specific sequence of movements needed to accomplish a desired action. Subcortical and brainstem nuclei are also important in learning skilled movements.

Prominent among the subcortical nuclei are the paired **basal ganglia** (see Figure 12–2b), which anatomically consist of a closely related group of separate nuclei. (Despite their name, these neuronal clusters are technically nuclei because they are within the central nervous system.) They form a link in some of the looping parallel circuits through which activity in the motor system is transmitted from a specific region of sensorimotor cortex to the basal ganglia, from there to the thalamus, and then back to the cortical area from which the circuit started. Some of these circuits facilitate movements and others suppress them.

Parkinson's Disease In *Parkinson's disease*, the input to the basal ganglia is diminished, the interplay of the facilitory and inhibitory circuits is unbalanced, and activation of the motor cortex (via the basal ganglia–thalamus limb of the circuit mentioned above) is reduced. Clinically, Parkinson's disease is characterized by a reduced amount of movement (*akinesia*), slow movements (*bradykinesia*), muscular rigidity, and a tremor at rest. Other motor and nonmotor abnormalities may also be present. For example, a common set of symptoms include a change in facial expression resulting in a masklike, unemotional appearance, a shuffling gait with loss of arm swing, and a stooped and unstable posture.

Although the symptoms of Parkinson's disease reflect inadequate functioning of the basal ganglia, a major part of the initial defect arises in neurons of the **substantia nigra** ("black substance"), a subcortical nucleus that gets its name from the dark pigment in its cells. These neurons normally project to the basal ganglia where they liberate dopamine from their axon terminals. The substantia nigra neurons, however, degenerate in Parkinson's disease, and the amount of dopamine they deliver to the basal ganglia is reduced, which decreases the subsequent activation of the sensorimotor cortex.

The most powerful drugs currently available for Parkinson's disease are those that mimic the action of dopamine or increase its availability. The major drug is L-dopa, a precursor of dopamine. L-dopa enters the bloodstream, crosses the blood-brain barrier, and is converted to dopamine (dopamine itself is not used as medication because it cannot cross the blood-brain barrier). The newly formed dopamine activates receptors in the basal ganglia and improves the symptoms of the disease. Another drug inhibits the brain enzyme that breaks down dopamine so that more neurotransmitter reaches the neurons in the basal ganglia. Other therapies include the electrical destruction ("lesioning") of overactive areas of the basal ganglia or stimulation of the underactive ones. Still highly controversial is the transplantation into the basal ganglia of neurons from either human fetuses or animals such as fetal pigs or cells that have been genetically engineered or taken from dopamine secreting tissues in the patient's own body. Regardless of their source, the implanted cells then synthesize the necessary dopamine as well as important growth factors.

Cerebellum

The cerebellum is behind the brainstem (see Figure 12–2a). It influences posture and movement indirectly by means of input to brainstem nuclei and (by way of the thalamus) to regions of the sensorimotor cortex that give rise to pathways that descend to the motor neurons. The cerebellum receives information both from the sensorimotor cortex (relayed via brainstem nuclei) and from the vestibular system, eyes, ears, skin, muscles, joints, and tendons—that is, from the very receptors that are affected by movement.

The cerebellum's role in motor functioning includes providing timing signals to the cerebral cortex and spinal cord for precise execution of the different phases of a motor program, in particular the timing of the agonist/antagonist components of a movement. It also helps coordinate movements that involve several joints and stores the memories of them so they can be achieved more easily the next time they are tried.

The cerebellum also participates in planning movements—integrating information about the nature of an intended movement with information about the space outside the person into which the movement will be made. The cerebellum then provides this as a "feedforward" signal to the brain areas responsible for refining the motor program.

Moreover, during the course of the movement, the cerebellum compares information about what the muscles *should* be doing with information about what they actually *are* doing. If there is a discrepancy between the intended movement and the actual one, the cerebellum sends an error signal to the motor cortex and subcortical centers to correct the ongoing program.

The role of the cerebellum in programming movements can best be appreciated when seeing the absence of this function in individuals with *cerebellar disease*, who cannot perform limb or eye movements smoothly but move with a tremor—a so-called *intention tremor* that increases as the course of the movement nears its final destination. (Note that this is different from parkinsonian patients, who have a tremor while at rest.) People with cerebellar disease also cannot start or stop movements quickly or easily, and cannot combine the movements of several joints into a single smooth, coordinated motion. The role of the cerebellum in the precision and timing of movements can be understood when one considers, for example, a tennis player who, upon seeing a ball fly over the net, anticipates its curve of flight, runs to the spot on the court where, if one swings the racquet, one can intercept the ball. People with cerebellar damage cannot achieve this level of coordinated, precise, learned movement.

Unstable posture and awkward gait are two other symptoms characteristic of cerebellar dysfunction. For example, persons with cerebellar damage walk with the feet well apart, and they have such difficulty maintaining balance that their gait appears drunken. A final symptom involves difficulty in learning new motor skills, and individuals with cerebellar dysfunction find it hard to modify movements in response to new situations. Unlike damage to areas of sensorimotor cortex, cerebellar damage does not cause paralysis or weakness.

Descending Pathways

The influence exerted by the various brain regions on posture and movement is via descending pathways to the motor neurons and the interneurons that affect these neurons. The pathways are of two types: the **corticospinal pathways,** which, as their name implies, originate in the cerebral cortex; and a second group we shall call the **brainstem pathways,** which originate in the brainstem.

Fibers from both types of descending pathways end at synapses on alpha and gamma motor neurons or on interneurons that affect the alpha motor neurons either directly or via still other interneurons. Sometimes, as mentioned earlier, these are the same interneurons that function in local reflex arcs, thereby ensuring that the descending signals are fully integrated with local information before the activity of the motor neurons is altered. In other cases, the interneurons are part of neural networks involved in posture or locomotion. The ultimate effect of the descending pathways on the alpha motor neurons may be excitatory or inhibitory.

Importantly, some of the descending fibers affect *afferent* systems. They do this via (1) presynaptic synapses on the terminals of afferent neurons as these fibers enter the central nervous system, or (2) synapses on interneurons in the ascending pathways. The overall effect of this descending input to afferent systems is to limit their influence on either the local or brain motor control areas, thereby altering the importance of a particular bit of afferent information or sharpening its focus. This descending (motor) control over ascending (sensory) information provides another example to show that there is clearly no real functional separation of the motor and sensory systems.

Corticospinal Pathway   The nerve fibers of the corticospinal pathways, as mentioned before, have their cell bodies in sensorimotor cortex and terminate in the spinal cord. The corticospinal pathways are also called the **pyramidal tracts,** or **pyramidal system** (perhaps because of their shape as they pass along the surface of the medulla oblongata or because they were formerly thought to arise solely from the pyramidal cells of the motor cortex). In the medulla oblongata near the junction of the spinal cord and brainstem, most of the corticospinal fibers cross the spinal cord to descend on the opposite side (Figure 12–12). Thus, the skeletal muscles on the left side of the body are controlled largely by neurons in the right half of the brain, and vice versa.

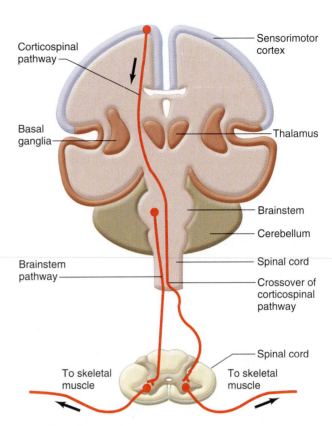

FIGURE 12–12

The corticospinal and brainstem pathways. Most of the corticospinal fibers cross in the brainstem to descend in the opposite side of the spinal cord, but the brainstem pathways are mostly uncrossed. Arrows indicate direction of action-potential propagation.

Adapted from Gardner.

As the corticospinal fibers descend through the brain from the cerebral cortex, they are accompanied by fibers of the **corticobulbar pathway** ("bulbar" means "pertaining to the brainstem"), a pathway that begins in the sensorimotor cortex and ends in the brainstem. The corticobulbar fibers control, directly or indirectly via interneurons, the motor neurons that innervate muscles of the eye, face, tongue, and throat. These fibers are the main source of control for voluntary movement of the muscles of the head and neck, whereas the corticospinal fibers serve this function for the muscles of the rest of the body. For convenience, we shall henceforth include the corticobulbar pathway in the general term "corticospinal pathways."

Convergence and divergence are hallmarks of the corticospinal pathway. For example, a great deal of convergence from different neuronal sources impinges on neurons of the sensorimotor cortex, which is not surprising when one considers the many factors that can affect motor behavior. As for the descending pathways, neurons from wide areas of the sensorimotor cortex *converge* onto single motor neurons at the local level so that multiple brain areas usually control single muscles. Also, axons of single corticospinal neurons *diverge* markedly to synapse with a number of different motor neuron populations at various levels of the spinal cord, thereby ensuring that the motor cortex can influence many different components of a movement.

This seeming "blurriness" of control appears counterintuitive when we think of the delicacy with which we can move a fingertip and when we learn that the corticospinal pathways have their greatest influence on rapid, fine movements of the distal extremities, such as those made when an object is manipulated by the fingers. After damage to the corticospinal pathways, all movements are slower and weaker, individual finger movements are absent, and it is difficult to release a grip.

Brainstem Pathways Axons from neurons in the brainstem also form pathways that descend into the spinal cord to influence motor neurons. These pathways are sometimes referred to as the **extrapyramidal system,** or indirect pathways, to distinguish them from the corticospinal (pyramidal) pathways. However, no general term is widely accepted for these pathways, and for convenience we shall refer to them collectively as the brainstem pathways.

Axons of some of the brainstem pathways cross from their side of origin in the brainstem to affect muscles on the opposite side of the body, but most remain uncrossed. In the spinal cord the fibers of the brainstem pathways descend as distinct clusters, named according to their sites of origin. For example, the vestibulospinal pathway descends to the spinal cord from the vestibular nuclei in the brainstem, whereas the reticulospinal pathway descends from neurons in the brainstem reticular formation.

The brainstem pathways are especially important in the control of upright posture, balance, and walking.

Concluding Comments on the Descending Pathways As stated above, the corticospinal neurons generally have their greatest influence over motor neurons that control muscles involved in fine, isolated movements, particularly those of the fingers and hands. The brainstem descending pathways, in contrast, are more involved with coordination of the large muscle groups used in the maintenance of upright posture, in locomotion, and in head and body movements when turning toward a specific stimulus.

There is, however, much interaction between the descending pathways. For example, some fibers of the corticospinal pathway end on interneurons that play important roles in posture, whereas fibers of the brainstem descending pathways sometimes end directly on

the alpha motor neurons to control discrete muscle movements. Because of this redundancy, loss of function resulting from damage to one system may be compensated for by the remaining system, although the compensation is generally not complete.

The distinctions between the corticospinal and brainstem descending pathways are not clear-cut. All movements, whether automatic or voluntary, require the continuous coordinated interaction of both types of pathways.

Muscle Tone

Muscle tone can be defined as the resistance of skeletal muscle to stretch as an examiner moves the limb or neck of a relaxed subject. Under such circumstances in a normal person, the resistance to passive movement is slight and uniform, regardless of the speed of the movement.

Muscle tone is due both to the viscoelastic properties of the muscles and joints and to whatever degree of alpha motor neuron activity exists. When a person is deeply relaxed, the alpha motor neuron activity probably makes no contribution to the resistance to stretch. As the person becomes increasingly alert, however, some activation of the alpha motor neurons occurs and muscle tone increases.

Abnormal Muscle Tone

Abnormally high muscle tone, called *hypertonia,* occurs in individuals with certain disease processes and is seen particularly clearly when a joint is moved passively at high speeds. The increased resistance is due to a greater-than-normal level of alpha motor neuron activity, which keeps a muscle contracted despite the individual's attempt to relax it. Hypertonia is usually found when there are disorders of the descending pathways that result in decreased inhibitory influence exerted by them on the motor neurons.

Clinically, the descending pathways—primarily the corticospinal pathways—and neurons of the motor cortex are often referred to as the "upper motor neurons" (a confusing misnomer because they are not really motor neurons at all). Abnormalities due to their dysfunction are classed, therefore, as *upper motor neuron disorders.* Thus, hypertonia indicates an upper motor neuron disorder. In this clinical classification, the alpha motor neurons—the true motor neurons—are termed lower motor neurons.

Spasticity is a form of hypertonia in which the muscles do not develop increased tone until they are stretched a bit, and after a brief increase in tone, the contraction subsides for a short time. The period of "give" occurring after a time of resistance is called the *clasp-knife phenomenon.* Spasticity may be accompanied by increased responses of motor reflexes such as the knee jerk, and by decreased coordination and strength of voluntary actions. *Rigidity* is a form of hypertonia in which the increased muscle contraction is continual and the resistance to passive stretch is constant. Two other forms of hypertonia that can occur suddenly in individual or multiple muscles are *spasms,* which are brief contractions, and *cramps,* which are prolonged and painful.

Hypotonia is a condition of abnormally low muscle tone, accompanied by weakness, atrophy (a decrease in muscle bulk), and decreased or absent reflex responses. Dexterity and coordination are generally preserved unless profound weakness is present. While hypotonia may develop after cerebellar disease, it more frequently accompanies disorders of the alpha motor neurons ("lower motor neurons"), neuromuscular junctions, or the muscles themselves. The term *flaccid,* which means "weak" or "soft," is often used to describe hypotonic muscles.

Maintenance of Upright Posture and Balance

The skeleton supporting the body is a system of long bones and a many-jointed spine that cannot stand erect against the forces of gravity without the support given by coordinated muscle activity. The muscles that maintain upright posture—that is, support the body's weight against gravity—are controlled by the brain and by reflex mechanisms that are "wired into" the neural networks of the brainstem and spinal cord.

Many of the reflex pathways previously introduced (for example, the stretch and crossed-extensor reflexes) are used in posture control.

Added to the problem of maintaining upright posture is that of maintaining balance. A human being is a very tall structure balanced on a relatively small base, and the center of gravity is quite high, being situated just above the pelvis. For stability, the center of gravity must be kept within the base of support provided by the feet (Figure 12–13). Once the center of gravity has moved beyond this base, the body will fall unless one foot is shifted to broaden the base of support. Yet people can operate under conditions of unstable equilibrium because their balance is protected by complex interacting **postural reflexes,** all the components of which we have met previously.

The afferent pathways of the postural reflexes come from three sources: the eyes, the vestibular apparatus, and the somatic receptors. The efferent pathways are the alpha motor neurons to the skeletal muscles, and the integrating centers are neuron networks in the brainstem and spinal cord.

In addition to these integrating centers, there are centers in the brain that form an internal representation of the body's geometry, its support conditions, and its orientation with respect to verticality. This internal representation serves two purposes: It serves as a reference frame for the perception of the body's position and orientation in space and for planning actions, and it contributes to stability via the motor controls involved in maintenance of upright posture.

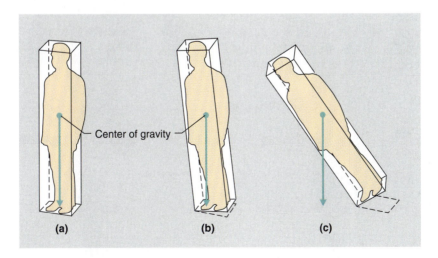

FIGURE 12–13

The center of gravity is the point in an object at which, if a string were attached at this point and pulled up, all the downward force due to gravity would be exactly balanced. (a) The center of gravity must remain within the upward vertical projections of the object's base (the tall box outlined in the drawing) if stability is to be maintained. (b) Stable conditions: The box tilts a bit, but the center of gravity remains within the base area and so the box returns to its upright position. (c) Unstable conditions: The box tilts so far that its center of gravity is not above any part of the object's base—the dashed rectangle on the floor—and the object will fall.

There are many familiar examples of using reflexes to maintain upright posture, one being the crossed-extensor reflex. As one leg is flexed and lifted off the ground, the other is extended more strongly to support the added weight of the body, and the positions of various parts of the body are shifted to move the center of gravity over the single, weight-bearing leg. This shift in the center of gravity, demonstrated in Figure 12–14, is an important component in the stepping mechanism of locomotion.

It is clear that afferent input from several sources is necessary for optimal postural adjustments, yet interfering with any one of these inputs alone does not cause a person to topple over. Blind people maintain their balance quite well with only a slight loss of precision, and people whose vestibular mechanisms have been destroyed have very little disability in everyday life as long as their visual system and somatic receptors are functioning.

The conclusion to be drawn from such examples is that the postural control mechanisms are not only effective and flexible, they are also highly adaptable.

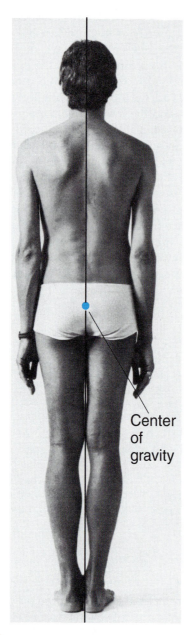 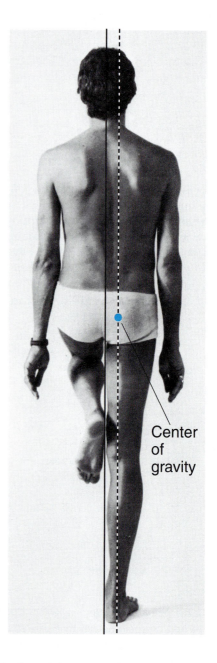

Center
of
gravity

Center
of
gravity

FIGURE 12–14

Postural changes with stepping. (Left) Normal standing posture. The line of the center of gravity falls directly between the two feet. (Right) As the left foot is raised, the whole body leans to the right so that the center of gravity shifts and is over the right foot.

Walking

Walking requires the coordination of literally hundreds of muscles, each activated to a precise degree at a precise time. Walking is initiated by allowing the body to fall forward to an unstable position and then moving one leg forward to provide support. When the extensor muscles are activated on the supported side of the body to bear the body's weight, the contralateral extensors are inhibited by reciprocal innervation to allow the nonsupporting limb to flex and swing forward. The cyclical, alternating movements of walking are brought about largely by networks of interneurons in the spinal cord at the local level. The interneuron networks coordinate the output of the various motor neuron pools that control the appropriate muscles of the arms, shoulders, trunk, hips, legs, and feet.

The network neurons rely on both plasma-membrane spontaneous pacemaker properties and patterned synaptic activity to establish their rhythms. At the same time, however, the networks are remarkably adaptable, and a single network can generate many different patterns of neural activity, depending upon its inputs. These inputs are from other local interneurons, afferent fibers, and descending pathways.

The spinal-cord neural networks can bring about the rhythmical movement of a limb in the absence of afferent information or even the descending pathways, but such afferent inputs normally contribute substantially to walking. In fact, neural activation occurs in the cerebral cortex, cerebellum, and brainstem as well as in the spinal cord during locomotion. Moreover, middle and higher levels of the motor control hierarchy are necessary for postural control, voluntary commands ("I want to jump rather than walk," so the two limbs must be operated together instead of reciprocally), and adaptations to the environment ("I am about to cross a stream on stepping stones or, perhaps, climb a ladder"). The ultimate importance of the sensorimotor cortex is attested by the fact that damage to even small areas of it can cause marked disturbances in gait.

SUMMARY

I. Skeletal muscles are controlled by their motor neurons. All the motor neurons that control a given muscle form a motor neuron pool.

Motor Control Hierarchy

I. The neural systems that control body movements can be conceptualized as being arranged in a motor control hierarchy.
   a. The highest level determines the general intention of an action.
   b. The middle level establishes a motor program and specifies the postures and movements needed to carry out the intended action, taking account of

sensory information that indicates the body's position.
   c. The lowest level ultimately determines which motor neurons will be activated.
   d. As the movement progresses, information about what the muscles are doing is fed back to the motor control centers, which make program corrections.
   e. Almost all actions have conscious and unconscious components.

Local Control of Motor Neurons

I. Most direct input to motor neurons is from local interneurons, which themselves receive input from peripheral receptors, descending pathways, and other interneurons.

II. Muscle length and the velocity of changes in length are monitored by muscle-spindle stretch receptors.
   a. Activation of these receptors initiates the stretch reflex, in which motor neurons of ipsilateral antagonists are inhibited and those of the stretched muscle and its synergists are activated.
   b. Tension on the stretch receptors is maintained during muscle contraction by gamma efferent activation to the spindle muscle fibers.
   c. Alpha and gamma motor neurons are often coactivated.

III. Muscle tension is monitored by Golgi tendon organs, which, via interneurons, activate inhibitory synapses on motor neurons of the contracting muscle and excitatory synapses on motor neurons of ipsilateral antagonists.

IV. The withdrawal reflex excites the ipsilateral flexor muscles and inhibits the ipsilateral extensors. The crossed-extensor reflex excites the contralateral extensor muscles during excitation of the ipsilateral flexors.

Brain Motor Centers and the Descending Pathways They Control

I. The location of the neurons in the motor cortex varies in general with the part of the body the neurons serve.

II. Different areas of sensorimotor cortex have different functions, but there is much overlap in activity.

III. The basal ganglia form a link in a circuit that originates in and returns to sensorimotor cortex. These subcortical nuclei facilitate some motor behaviors and inhibit others.

IV. The cerebellum coordinates posture and movement and plays a role in motor learning.

V. The corticospinal pathways pass directly from the sensorimotor cortex to motor neurons in the spinal cord (or brainstem, in the case of the corticobulbar pathways) or, more commonly, to interneurons near the motor neurons.
   a. In general, neurons on one side of the brain control muscles on the other side of the body.
   b. Corticospinal pathways serve predominately fine, precise movements.
   c. Some corticospinal fibers affect the transmission of information in afferent pathways.

VI. Other descending pathways arise in the brainstem and are involved mainly in the coordination of large groups of muscles used in posture and locomotion.
VII. There is some duplication of function between the two descending pathways.

Muscle Tone

I. Hypertonia, as seen in spasticity and rigidity, for example, usually occurs with disorders of the descending pathways.
II. Hypotonia can be seen with cerebellar disease or, more commonly, with disease of the alpha motor neurons or muscle.

Maintenance of Upright Posture and Balance

I. To maintain balance, the body's center of gravity must be maintained over the body's base.
II. The crossed-extensor reflex is a postural reflex.

Walking

I. The activity of networks of interneurons in the spinal cord brings about the cyclical, alternating movements of locomotion.
II. These pattern generators are controlled by corticospinal and brainstem descending pathways and affected by feedback and motor programs.

KEY TERMS

motor unit	Golgi tendon organ
motor neuron pool	withdrawal reflex
motor program	crossed-extensor reflex
descending pathway	sensorimotor cortex
voluntary movement	primary motor cortex
muscle spindle	motor cortex
intrafusal fiber	premotor area
extrafusal fiber	supplementary motor cortex
muscle-spindle stretch	somatosensory cortex
receptor	parietal-lobe association
stretch reflex	cortex
knee jerk	basal ganglia
monosynaptic	substantia nigra
polysynaptic	corticospinal pathway
reciprocal innervation	brainstem pathway
synergistic muscle	pyramidal tract
ipsilateral	pyramidal system
contralateral	corticobulbar pathway
alpha motor neuron	extrapyramidal system
gamma motor neuron	muscle tone
coactivated	postural reflex

REVIEW QUESTIONS

1. Describe motor control in terms of the conceptual motor control hierarchy and using the following terms: highest, middle, and lowest levels; motor program; descending pathways, and motor neuron.
2. List the characteristics of voluntary actions.
3. Picking up a book, for example, has both voluntary and involuntary components. List the components of this action and indicate whether each is voluntary or involuntary.
4. List the inputs that can converge on the interneurons active in local motor control.
5. Draw a muscle spindle within a muscle, labeling the spindle, intrafusal and extrafusal muscle fibers, stretch receptors, afferent fibers, and alpha and gamma efferent fibers.
6. Describe the components of the knee jerk reflex (stimulus, receptor, afferent pathway, integrating center, efferent pathway, effector, and response).
7. Describe the major function of alpha-gamma coactivation.
8. Distinguish among the following areas of the cerebral cortex: sensorimotor, primary motor, premotor, and supplementary motor.
9. Contrast the two major types of descending motor pathways in terms of structure and function.
10. Describe the roles that the basal ganglia and cerebellum play in motor control.
11. Explain how hypertonia might result from disease of the descending pathway.
12. Explain how hypotonia might result from lower motor neuron disease.
13. Explain the role played by the crossed-extensor reflex in postural stability.
14. Explain the role of the interneuronal networks in walking, incorporating in your discussion the following terms: interneuron, reciprocal innervation, synergist, antagonist, and feedback.

CLINICAL TERMS

tetanus	upper motor neuron disorder
lockjaw	spasticity
Parkinson's disease	clasp-knife phenomenon
akinesia	rigidity
bradykinesia	spasm
cerebellar disease	cramp
intention tremor	hypotonia
hypertonia	flaccid

THOUGHT QUESTIONS

(Answers are given in Appendix A.)

1. What changes would occur in the knee jerk reflex after destruction of the gamma motor neurons?
2. What changes would occur in the knee jerk reflex after destruction of the alpha motor neurons?
3. Draw a cross section of the spinal cord and a portion of the thigh (similar to Figure 12–6) and "wire up" and activate the neurons so the leg becomes a stiff pillar; that is, the knee does not bend.
4. We have said that hypertonia is usually considered a sign of disease of the descending motor pathways, but how might it result from abnormal function of the alpha motor neurons?

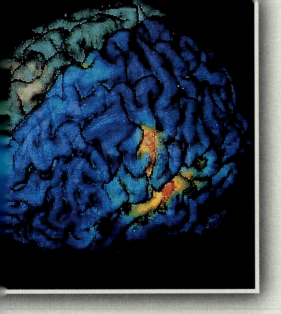

Consciousness and Behavior

The term "consciousness" includes two distinct concepts: **states of consciousness** and **conscious experiences.** The first concept refers to whether a person is awake, asleep, drowsy, and so on. "Conscious experiences" refers to those things a person is aware of—thoughts, feelings, perceptions, ideas, dreams, reasoning—during any of the states of consciousness.

States of Consciousness

A person's state of consciousness is defined in two ways: (1) by behavior, covering the spectrum from maximum attentiveness to coma, and (2) by the pattern of brain activity that can be recorded electrically. This record, known as the **electroencephalogram (EEG),** portrays the electric-potential difference between two points on the surface of the scalp. The EEG is such an important tool in identifying the different states of consciousness that we begin with it.

Electroencephalogram

Neural activity is indicated by the electric signals known as graded potentials and action potentials (Chapter 8). The electrical activity that is going on in the brain's neurons, particularly those near the surface of the brain, can be recorded from the outside of the head. Electrodes, which are wires attached to the head by a salty paste that conducts electricity, pick up electric signals generated in the brain and transmit them to a machine that records them as the EEG.

While we often think of electrical activity in neurons in terms of action potentials, action potentials usually contribute little to the EEG. Rather, EEG patterns are largely due to graded potentials, in this case summed postsynaptic potentials (Chapter 8) in the many hundreds of thousands of brain neurons that underlie the recording electrodes.

EEG patterns such as that shown in Figure 13–1, are waves, albeit complex ones, with large variations in both amplitude and frequency. The wave's *amplitude,* measured in microvolts (μV), indicates how much electrical activity of a similar type is going on beneath the recording electrodes at any given time. In other words, it indicates the degree of synchronous firing of those neurons that are generating the synaptic activity. (The properties of a wave are summarized in Figure 9–21.) The amplitude may range from 0.5 to 100 μV. Note that EEG amplitudes are about 1000 times smaller than an action potential.

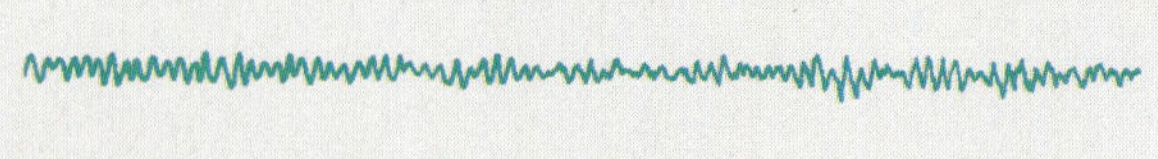

The wave's *frequency* indicates how often the wave cycles from its maximal amplitude to its minimal amplitude and back. The frequency is measured in hertz (Hz, or cycles per second), and may vary from 1 to 40 Hz or higher. Four major frequency ranges are found in EEG patterns. In general, lower EEG frequencies indicate less responsive behaviors, such as sleep, whereas higher frequencies indicate states of increased alertness.

The cause of the wavelike nature, or rhythmicity, of the EEG is not certain nor is it known exactly where in the brain it originates. It is currently thought that clusters of neurons in the thalamus play a critical role. They provide a fluctuating output through nerve fibers leading from the thalamus to the cortex. This output causes, in turn, a rhythmic pattern of synaptic activity in the neurons of the cortex. The cortical synaptic activity—not the activity of the deep thalamic structures—comprises most of a recorded EEG signal.

The synchronicity of the cortical synaptic activity (in other words, the amplitude of the EEG) does, however, indicate the degree of synchronous firing of the thalamic neuronal clusters that are generating the EEG. The purpose served by these oscillations in brain electrical activity is unknown. Theories range from the "idling hypothesis," which says that it is easier to get brain activity up and running from an "idle" as opposed to a "cold start," to the "epiphenomenon hypothesis," which says that the oscillations are simply the by-product of neuronal activity and have no functional significance at all.

The EEG is a useful clinical tool because patterns are abnormal over brain areas that are diseased or damaged (for example, by tumors, blood clots, hemorrhage, regions of dead tissue, and high or low blood sugar). Moreover, a shift from a less synchronized pattern of electrical activity (low-amplitude EEG) to a highly synchronized pattern can be a prelude to the electrical storm that signifies an epileptic seizure. *Epilepsy* is a common (1 percent of the population), neurological disease occurring in at least 40 different

FIGURE 13–1

EEG patterns are wavelike. This represents a typical EEG recorded from the parietal or occipital lobe of an awake, relaxed person. EEG wave amplitudes are generally 20–100 μV, and the durations are about 50 msec.

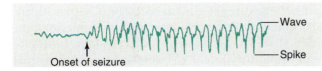

FIGURE 13-2

Spike and wave pattern in the EEG of a patient during an epileptic seizure.

forms. It is associated with abnormal synchronized discharges of cerebral neurons. These discharges are reflected in the EEG as recurrent waves having distinctive high amplitudes (up to 1000 μV) and individual spikes or combinations of spikes and waves (Figure 13–2). Epilepsy is also associated with stereotyped changes in behavior that vary according to the part of the brain affected and can include a temporary loss of consciousness.

The Waking State

Behaviorally, the waking state is far from homogeneous, comprising the infinite variety of things one can be doing. The most prominent EEG wave pattern of an awake, relaxed adult whose eyes are closed is a slow oscillation of 8 to 13 Hz, known as the **alpha rhythm** (Figure 13–3a). The alpha rhythm is recorded best over the parietal and occipital lobes and is associated with decreased levels of attention. When alpha rhythms are being generated, subjects commonly report that they feel relaxed and happy. However, people who normally experience more alpha rhythm than usual have not been shown to be psychologically different from those with less.

When people are attentive to an external stimulus or are thinking hard about something, the alpha rhythm is replaced by lower-amplitude, high-frequency oscillations, the **beta rhythm** (Figure 13–3b). This transformation is known as the **EEG arousal** and is associated with the act of paying attention to a stimulus rather than with the act of perception. For example, if people open their eyes in a completely dark room and try to see, EEG arousal occurs even though they are able to perceive nothing. With decreasing attention to repeated stimuli, the EEG pattern reverts to the alpha rhythm.

Sleep

The EEG pattern changes profoundly in sleep. As a person becomes increasingly drowsy, there is a decrease in alpha-wave amplitude and frequency. When sleep actually occurs, the EEG shifts toward slower-frequency, higher-amplitude (theta and delta) wave patterns (Figure 13–4). These EEG changes are accompanied by changes in posture, ease of arousal, threshold for sensory stimuli, and motor output.

There are two phases of sleep whose names depend on whether or not the eyes can be seen to move behind the closed eyelids: **NREM** (nonrapid eye movement) and **REM** (rapid eye movement) sleep. The EEG waves during NREM sleep are of high amplitude and low—that is, slow—frequency, and NREM sleep is also referred to as slow-wave sleep. The initial phase of sleep—NREM sleep—is itself divided into four stages, each successive stage having an EEG pattern characterized by a slower frequency and higher amplitude than the preceding one (Figure 13–4a).

Sleep begins with the progression from stage 1 to stage 4, which normally takes 30 to 45 min. The process then reverses itself, the EEG ultimately resuming the

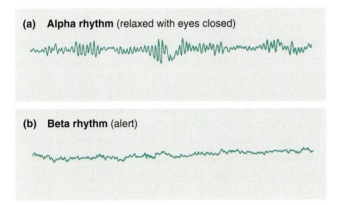

FIGURE 13-3

The (a) alpha and (b) beta rhythms of the EEG.

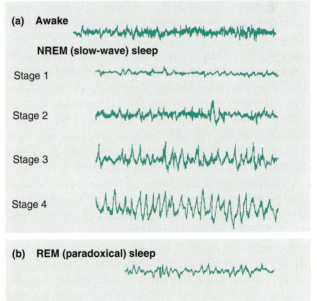

FIGURE 13-4

The EEG record of a person (a) passing from the awake state to deep sleep (stage 4) and (b) during REM sleep.

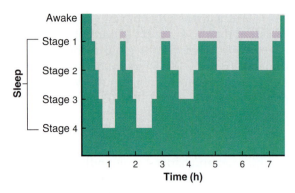

FIGURE 13–5

A typical night's sleep in an average young adult. The lavender colored lines indicate periods of REM sleep.

Adapted from Nicholi.

low-voltage, high-frequency, asynchronous pattern characteristic of the alert, awake state (Figure 13–4b). Instead of the person waking, however, the behavioral characteristics of sleep continue at this time.

REM sleep is also called **paradoxical sleep** because the sleeper is difficult to arouse despite having an EEG that is characteristic of the alert, awake state. When awakened during REM sleep, subjects generally report that they have been dreaming. This is true even in people who do not remember dreaming when they awaken later spontaneously.

If uninterrupted, sleep continues in this cyclical fashion, moving from stages 1, 2, and 3, to 4 then back up from 4 to 3, 2, and 1, where NREM sleep is punctuated by an episode of REM sleep. Continuous recordings of adults show that the average total sleep period comprises four or five such cycles, each lasting 90 to 100 min (Figure 13–5). NREM sleep constitutes 75 to 80 percent of the total sleeping time in young adults; the remainder is spent in REM sleep. The time spent in REM sleep increases toward the end of an undisturbed night. Initially, as one moves from drowsiness to stage 1 sleep, there is a considerable tension in the postural muscles, but the muscles become progressively more relaxed as NREM sleep progresses. Sleepers awakened during NREM sleep rarely report dreaming.

With several exceptions, skeletal-muscle tension, already decreased during NREM sleep, is markedly inhibited during REM sleep. Exceptions are the eye muscles, which cause the rapid bursts of saccade-like eye movements (Chapter 9), and the motor neurons to the muscles of respiration. In the disease known as *sleep apnea*, however, stimulation of the respiratory muscles temporarily ceases, sometimes hundreds of times during a night. This disease is associated with excessive—and sometimes dangerous—sleepiness during the day.

During the sleep cycle, there are many changes throughout the body, in addition to altered muscle tension. During NREM sleep, for example, there are

pulsatile releases of growth hormone (Chapter 10) and the gonadotropic hormones from the anterior pituitary, as well as decreases in blood pressure, heart rate, and respiratory rate. REM sleep is associated with an increase and irregularity in blood pressure, heart rate, and respiratory rate. Moreover, twitches of the facial muscles or limb muscles occur (despite the generalized lack of skeletal-muscle tone), as may erection of the penis and engorgement of the clitoris.

Although adults spend about one-third of their time sleeping, the functions of sleep are not known. One hypothesis is that the neural mechanisms of REM sleep facilitate the chemical and structural changes that the brain undergoes during learning and memory and, through dreams, may provide for the expression of concerns in the "subconscious."

The sleep states are summarized in Table 13–1.

Neural Substrates of States of Consciousness

Periods of sleep and wakefulness alternate about once a day; that is, they manifest a circadian rhythm consisting typically of 8 h sleep and 16 h awake. Within the sleep portion of this circadian cycle, NREM sleep and REM sleep alternate, as we have seen.

What physiological processes drive these cyclic changes in states of consciousness? Nuclei in both the brainstem and hypothalamus are involved.

Recall from Chapter 8 that neurons of certain brainstem nuclei give rise to axons that diverge to affect wide areas of the brain in a highly specific manner, forming a fiber system known as the reticular activating system. This system is, in fact, composed of several separate divisions, distinguished by their anatomical distribution and neurotransmitters. The divisions originate in different nuclei within the brainstem, and some of them send fibers to those areas of the thalamus that influence the EEG. Components of the reticular activating system that release norepinephrine, serotonin, or acetylcholine—functioning in this instance more as neuromodulators—are most involved in controlling the various states of consciousness.

One model (Figure 13–6) suggests that the waking state and REM sleep are at opposite ends of a spectrum: During waking, the aminergic neurons (those that release norepinephrine or serotonin) have the major influence, and during REM sleep the cholinergic neurons are dominant. NREM sleep, according to this model, is intermediate between the two extremes.

The aminergic neurons, which are active during the waking state, facilitate both the direction of attention to perceptions of the outer world and the enhanced motor activity that characterize awake behavior. They also inhibit certain of the cholinergic brainstem neurons. As the aminergic neurons stop firing, the cholinergic neurons, released from inhibition, increase their activity.

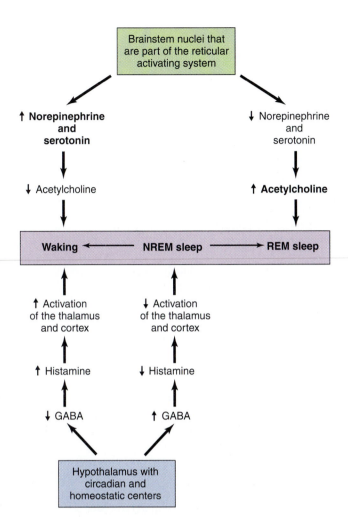

FIGURE 13–6

A model to explain the differing states of consciousness. The changes in aminergic and cholinergic influence are discussed in the text.

The switch in neurotransmitters (Figure 13–6) drives the changes that occur as we shift from the waking state through NREM sleep to REM sleep: Attention shifts to internally generated stimuli (dreams) so that we are largely insensitive to external stimuli; memory decreases (dreams are generally forgotten much faster than events we experience while awake); postural muscles lose tone so our heads nod and we slump in our chairs; and the tight rules for determining what is really real becomes relaxed, allowing the bizarre happenings of our dreams.

The basic rhythm of the sleep-awake cycle is also influenced by the biological-clock functions of the suprachiasmatic nucleus (Chapter 7) and two other hypothalamic areas (Figure 13–7). One of these areas, the preoptic area, promotes sleep, whereas increased activity of the other, in the posterior hypothalamus, leads to wakefulness. Nerve pathways connect these hypothalamic areas with those in the brainstem. The two hypothalamic areas use GABA and histamine as neurotransmitters.

In addition to these neurotransmitters, over 30 other chemical substances that affect sleep have been found in blood, urine, cerebrospinal fluid, and brain tissue. For example, two of the cytokines, a family of intercellular messengers having an important role in the body's immune defense system (Chapter 20), fluctuate in parallel with normal sleep-wake cycles. Levels of these two cytokines, known as interleukin 1 and tumor necrosis factor, increase during infections, explaining why we sleep more when we are sick.

Coma and Brain Death

The term *coma* describes a severe decrease in mental function due to structural, physiological, or metabolic impairment of the brain. A person in a coma is characterized by a sustained loss of the capacity for arousal even in response to vigorous stimulation. There is no outward behavioral expression of any mental function, the eyes are closed, and sleep-wake cycles disappear. Coma can result from extensive damage to the cerebral cortex; damage to the brainstem arousal mechanisms; interruptions of the connections between the brainstem and cortical areas; metabolic dysfunctions; brain infections; or an overdose of certain drugs, such as sedatives, sleeping pills, and, in some cases, narcotics.

TABLE 13–1 Sleep-Wakefulness Stages

Stage	Behavior	EEG (see Figures 13–3 and 13–4)
Alert wakefulness	Awake, alert with eyes open.	Beta rhythm (faster than 13 Hz).
Relaxed wakefulness	Awake, relaxed with eyes closed.	Mainly alpha rhythm (8–13 Hz) over the parietal and occipital lobes. Changes to beta rhythm in response to internal or external stimuli.
Relaxed drowsiness	Fatigued, tired, or bored; eyelids may narrow and close; head may start to droop; momentary lapses of attention and alertness. Sleepy but not asleep.	Decrease in alpha-wave amplitude and frequency.
NREM (slow-wave) sleep Stage 1	Light sleep; easily aroused by moderate stimuli or even by neck muscle jerks triggered by muscle stretch receptors as head nods; continuous lack of awareness.	Alpha waves reduced in frequency, amplitude, and percentage of time present; gaps in alpha rhythm filled with theta (4–8 Hz) and delta (slower than 4 Hz) activity.
Stage 2	Further lack of sensitivity to activation and arousal.	Alpha waves replaced by random waves of greater amplitude.
Stages 3 and 4	Deep sleep; in stage 4, activation and arousal occur only with vigorous stimulation.	Much theta and delta activity, predominant delta in stage 4.
REM (paradoxical) sleep	Deepest sleep; greatest relaxation and difficulty of arousal; begins 50–90 min after sleep onset, episodes are repeated every 60–90 min, each episode lasting about 10 min; dreaming occurs, rapid eye movements behind closed eyelids; marked increase in brain O_2 consumption.	EEG resembles that of alert awake state.

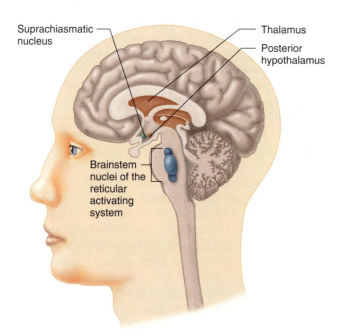

FIGURE 13–7

Brain structures involved in the sleep-wakefulness cycles. The suprachiasmatic nucleus is the site of the major "biological clock."

TABLE 13–2 Criteria for Brain Death

I. The nature and duration of the coma must be known.
  a. Known structural damage to brain or irreversible systemic metabolic disease.
  b. No chance of drug intoxication, especially from paralyzing or sedative drugs.
  c. No sign of brain function for 6 h in cases of known structural cause and when no drug or alcohol is involved; otherwise, 12–24 h without signs of brain function plus a negative drug screen.
II. Cerebral and brainstem function are absent.
  a. No response to painful stimuli administered above the spinal cord.
  b. Pupils unresponsive to light.
  c. No eye movement in response to ice-water stimulation of the vestibular reflex.
  d. Apnea (no spontaneous breathing) for 10 min.
  e. Systemic circulation may be intact.
  f. Purely spinal reflexes may be retained.
III. Supplementary (optional) criteria.
  a. Flat EEG (wave amplitudes less than 2 μV).
  b. Responses absent in vital brainstem structures.
  c. No cerebral circulation.

But a coma—even an irreversible coma—is not equivalent to death. We are left, then, with the question: When is a person actually dead? This question often has urgent medical, legal, and social consequences. For example, with the need for viable tissues for organ transplantation it becomes imperative to know just when a person is "dead" so that the organs can be removed as soon after death as possible.

Brain death is widely accepted by doctors and lawyers as the criterion for death, despite the viability of other organs. Brain death occurs when the brain no longer functions and has no possibility of functioning again.

The problem now becomes practical: How does one know when a person (for example, someone in a coma) is considered brain dead? There is general agreement that the criteria listed in Table 13–2, if met, denote brain death. Notice that the cause of a coma must be known because comas due to drug poisoning are usually reversible. Also, the criteria specify that there be no evidence of functioning neural tissues *above* the spinal cord because fragments of spinal reflexes may remain for several hours or longer after the brain is dead. The criterion for lack of spontaneous respiration (apnea) can be difficult to check because if the patient is on a respirator, it is of course inadvisable to remove him or her for the 10-min test because of the danger of further brain damage due to lack of oxygen. Therefore, apnea is diagnosed if there is no spontaneous attempt to "fight" the respirator; that is, the patient's reflexes do not drive respiration at a rate or depth different from those of the respirator.

Conscious Experiences

Having run the gamut of the states of consciousness from the awake, alert state to coma and brain death, we deal now with the conscious experiences during the awake state.

Conscious experiences are those things—either internal, such as an idea, or external—that we are aware of. The most obvious aspect of this phenomenon is sensory awareness, but there is also awareness of inner states such as fatigue and thirst and happiness. We are aware of the passing of time, of what we are presently thinking about, and of consciously recalling a fact learned in the past. We are aware of reasoning and exerting self-control, and we are aware of directing our attention to specific events. Not least, we are aware of "self."

Basic to the concept of conscious experience is the question of attention, to which we now turn.

Directed Attention

The term **directed attention** means the avoidance of distraction by irrelevant stimuli while seeking out and focusing on stimuli that are momentarily important. It is affected by both voluntary and reflex mechanisms. An example of voluntary control of directed attention familiar to students is ignoring distracting events in a busy library while studying there.

An example of reflexly directed attention occurs with the presentation of a novel stimulus to a relaxed subject showing an alpha EEG pattern. This causes the EEG to shift to the beta rhythm. If the stimulus has meaning for the individual, behavioral changes also occur. The person stops what he or she is doing and looks around, listening intently and orienting toward the stimulus source. This behavior is called the **orienting response.** If the person is concentrating hard and is not distracted by the novel stimulus, the orienting response does not occur. It is also possible to focus attention on a particular stimulus without making any behavioral response.

For attention to be directed only toward stimuli that are meaningful, the nervous system must have the means to evaluate the importance of incoming sensory information. Thus, even before we focus attention on an object in our sensory world and become aware of it, a certain amount of processing has already occurred. This so-called **preattentive processing** serves to direct our attention toward the part of the sensory world that is of particular interest and prepares the brain's perceptual processes as we direct our attention to a particular object or situation.

If a stimulus is repeated but is found to be irrelevant, the behavioral response to the stimulus progressively decreases, a process known as **habituation.** For example, when a loud bell is sounded for the first time, it may evoke an orienting response because the person might be frightened by or curious about the novel stimulus. After several ringings, however, the individual makes progressively less response and eventually may ignore the bell altogether. An extraneous stimulus of another type or the same stimulus at a different intensity can restore the orienting response.

Habituation involves a depression of synaptic transmission in the involved pathway, possibly related to prolonged inactivation of calcium channels in presynaptic axon terminals. Such inactivation results in a decreased calcium influx during depolarization and, hence, a decrease in the amount of neurotransmitter released by a terminal in response to action potentials.

Neural Mechanisms for Directed Attention Directing our attention to an object involves at least three distinct neurological processes. First, our attention must be disengaged from its present focus. Then, attention must be moved to the new focus. And, finally, attention must be engaged at the new focus. Each of these three processes occurs at a separate place in the brain.

One of the areas that plays an important role in orienting and directed attention is in the brainstem, where interaction of various sensory modalities in single cells can be detected experimentally. The receptive fields of the different modalities overlap such that, for example, a visual and auditory input from the same location in space will significantly enhance the firing rate of certain of these so-called multisensory cells, whereas the same type of stimuli originating at different places will have little effect on or even inhibit their response. Thus, weak clues add together to enhance each other's significance so we pay attention to the event, whereas a single small clue can be ignored.

There are also multisensory neurons in association areas of cerebral cortex (Chapter 8). Whereas the brainstem neurons are concerned with the *movements* associated with paying attention to a specific stimulus, the cortical multisensory neurons are more involved in the perception of the stimulus. Neuroscientists are only beginning to understand how the various areas of the attentional system interact.

The locus ceruleus, a nucleus in the brainstem pons, which projects to the parietal cortex and to many other parts of the central nervous system as well, is also implicated in directed attention. The system of fibers leading from the locus ceruleus helps determine which brain area is to gain temporary predominance in the ongoing stream of the conscious experience. Norepinephrine, the transmitter released by these neurons, acts as a neuromodulator to enhance the signals transmitted by certain sensory inputs so the difference between them and weaker signals is increased. Thus, neurons of the locus ceruleus improve information processing during directed attention.

Neural Mechanisms for Conscious Experiences

All conscious experiences are popularly attributed to the workings of the "mind," a word that conjures up the image of a nonneural "me," a phantom interposed between afferent and efferent impulses, with the implication that the mind is something more than neural activity. Most neuroscientists agree, however, that the mind represents a summation of neural activity in the brain at any given moment and does not require anything more than the brain.

Although a good definition of "mind" is not possible at this time, the term includes such actions as thinking, perceiving, making decisions, feeling, and imagining. Thus, we can conclude that "mind" refers to a "process" that gives rise to and includes conscious experiences. The truth of the matter is, however, that physiologists have only a beginning understanding of the brain mechanisms that give rise to mind or to conscious experiences.

Let us see in the rest of this section how two prominent neuroscientists, Francis Crick and Christof Koch, have speculated about this problem. They begin with the assumption that conscious experience requires neural processes—either graded potentials or action potentials—somewhere in the brain. At any moment, certain of these processes are correlated with conscious awareness and others are not. A key question here is: What is different about the processes that we are aware of?

A further assumption is that the neural activity that corresponds to a conscious experience resides not in a single anatomical cluster of "consciousness neurons" but rather in a set of neurons that are temporarily functioning together in a specific way. Since we can become aware of many different things, we further assume that this particular set of neurons can vary, shifting, for example, among parts of the brain that deal with visual or auditory stimuli, memories or new ideas, emotions or language.

Consider perception of a visual object. As was discussed in Chapter 9, different aspects of something we see are processed by different areas of the visual cortex—the object's color by one part, its motion by another, its location in the visual field by another, and its shape by still another—but we see *one* object. Not only do we perceive it; we may also know its name and function. Moreover, an object that can be seen can sometimes be heard or smelled, which requires participation of brain areas other than the visual cortex.

Neurons from the various parts of the brain that synchronously process different aspects of the information related to the object that we see are said to form a "temporary set" of neurons. It is suggested that the synchronous activity of the neurons in the temporary set leads to conscious awareness of the object we are seeing. (Instead of "leads to," perhaps we should say "corresponds to" or "is." We don't even know the appropriate term here.)

As we become aware of still other events—perhaps a memory related to this object—the set of neurons involved in the synchronous activity shifts, and a different temporary set forms. In other words, it is suggested that specific relevant neurons in many areas of the brain function together to form the unified activity that corresponds to awareness.

It is possible to record neuronal activity pulsing at a frequency of 40 to 70 Hz—faster even than the beta waves of the EEG. These oscillations may be the electrical record of a synchronous neural set, and they are the focus of a great deal of interest because they may be the clue to the "**binding problem**," namely, how the brain integrates information that occurs simultaneously in many parts of the brain into a single conscious experience. Such synchronous activity of ensembles of neurons is also implicated in the preconscious decision of whether or not stimuli are even perceived in the first place and in the planning of motor movements before their execution.

What parts of the brain might be involved in such a neuronal set? Clearly the cerebral cortex is involved, although not all parts at once, since removal of specific areas of the cortex abolish awareness of only specific types of consciousness. For example, damage to parts of the parietal lobe causes the injured person to neglect parts of the body as though they do not exist, but other parts are not neglected. Subcortical areas

such as the thalamus and basal ganglia may also be directly involved in conscious experience, but it seems that the hippocampus and cerebellum are not.

A critical question is: What binds together the functions of first one set of neurons and then another? By experience, we know that the binding must occur rapidly and that it can be very short-lived. Moreover, although its capacity at any one time may be limited, there is a huge range of combinations possible. In other words, we must postulate a mechanism that can focus our attention on a limited number of things at any one time but, over time, can bring an enormous number of things into conscious awareness.

Saying that we can bind together the activity of one set of neurons and shift the binding to a new set at a later time may be the same as saying we can focus attention on—that is, bring into conscious awareness—one object or event and then shift our focus of attention to another object or event at a later time. Thus, the mechanisms of conscious awareness and attention must be intimately related.

We include Figure 13–8 to indicate in one small way the complexity met when trying to sort out the brain's mechanisms of processing information. As confusing as it seems, the lines on this lateral view of the brain indicate only 15 percent of the possible connections in the cerebral cortex, and this is the brain of a monkey, not a human.

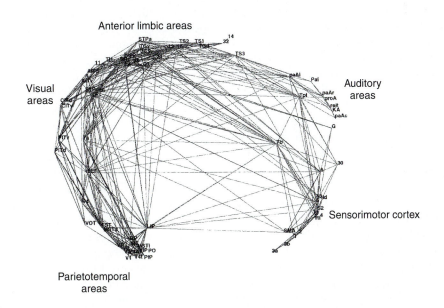

Anterior limbic areas

Visual areas

Auditory areas

Sensorimotor cortex

Parietotemporal areas

FIGURE 13–8

Possible connections for information processing in the cerebral cortex of a monkey brain. Actually, only about 15 percent of the total possible connections are represented. The small notations in the figure are not important for our purposes.

Reprinted with permission from Malcolm P. Young, "The organization of neural systems in the primate cerebral cortex," Proc. R. Soc. Biol. Sci., **252**:13–18, 1993. Fig. 3.

Motivation and Emotion

Motivation is a factor in most, if not all, behaviors, while emotions accompany many of our conscious experiences.

Motivation

Those processes responsible for the goal-directed quality of behavior are the **motivations,** or "drives" for that behavior. Motivation can lead to hormonal, autonomic, and behavioral responses. **Primary motivated behavior** is behavior related directly to homeostasis—that is, the maintenance of a relatively stable internal environment, an example being getting something to drink when one is thirsty. In such homeostatic goal-directed behavior, specific body "needs" are satisfied. Thus, in our example the correlate of need is a drop in body water concentration, and the correlate of need satisfaction is the return of the body water concentration to normal. The neurophysiological integration of much homeostatic goal-directed behavior will be discussed later (thirst and drinking, Chapter 16; food intake and temperature regulation, Chapter 18).

In many kinds of behavior, however, the relation between the behavior and the primary goal is indirect. For example, the selection of a particular drink—water or soda pop, for example—has little if any apparent relation to homeostasis. The motivation in this case is called secondary. Much of human behavior fits into this latter category and is influenced by habit, learning, intellect, and emotions—factors that can be lumped together under the term "incentives." Sometimes the primary homeostatic goals and secondary goals conflict as, for example, during a religious fast.

The concepts of reward and punishment are inseparable from motivation. Rewards are things that organisms work for or things that make the behavior that leads to them occur more often—in other words, positive reinforcers—and punishments are the opposite.

The neural system subserving reward and punishment is part of the reticular activating system, which you will recall arises in the brainstem and comprises several components. The component involved in motivation is known as the mesolimbic dopamine pathway—*meso-* because it arises in the midbrain (mesencephalon) area of the brainstem; *limbic* because it goes to areas of the limbic system, such as the prefrontal cortex, the nucleus accumbens, and the undersurface of the frontal lobe (Figure 13–9); and *dopamine* because its fibers release the neurotransmitter dopamine. The mesolimbic dopamine pathway is implicated in evaluating the availability of incentives and reinforcers (asking, "Is it worth it?", for example) and translating the evaluation into action. (We shall meet the mesolimbic dopamine pathway again later when we discuss drug dependence.)

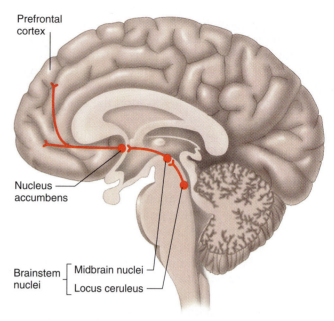

FIGURE 13–9
Schematic drawing of the mesolimbic dopamine pathway.

Much of the available information concerning the neural substrates of motivation has been obtained by studying behavioral responses of animals and, in some cases, people to rewarding or punishing stimuli. One way in which this can be done is by using the technique of **brain self-stimulation.** In this technique, an unanesthetized experimental animal (or a human undergoing neurosurgery for other reasons) regulates the rate at which electric stimuli are delivered through electrodes implanted in discrete brain areas. The experimental animal is placed in a box containing a lever it can press (Figure 13–10). If no stimulus is delivered to the brain when the bar is pressed, the animal usually presses it occasionally at random.

If, in contrast, a stimulus is delivered to the brain as a result of a bar press, a different behavior occurs, depending on the location of the electrodes. If the animal increases the bar-pressing rate above the level of random presses, the electric stimulus is by definition rewarding. If the animal decreases the press rate below the random level, the stimulus is punishing.

Thus, the rate of bar pressing with the electrode in different brain areas is taken to be a measure of the effectiveness of the reward or punishment. Different pressing rates are found for different brain regions.

Scientists expected the hypothalamus to play a role in motivation because the neural centers for the regulation of eating, drinking, temperature control, and sexual behavior are there (Chapter 8). Indeed, they found that brain self-stimulation of the lateral regions of the hypothalamus serves as a positive reward. Animals with electrodes in these areas have been known to bar-press to stimulate their brains 2000 times per

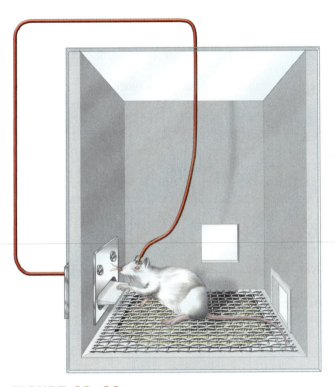

FIGURE 13–10
Apparatus for self-stimulation experiments.
Adapted from Olds.

hour continuously for 24 h until they drop from exhaustion! In fact, electric stimulation of the lateral hypothalamus is more rewarding than external rewards in that hungry rats, for example, often ignore available food for the sake of electrically stimulating their brains at that location.

Although the rewarding sites—particularly those for primary motivated behavior—are more densely packed in the lateral hypothalamus than anywhere else in the brain, self-stimulation can be obtained from a large number of brain areas. Thus, while the hypothalamus coordinates the sequenced hormonal, autonomic, and behavioral responses of motivated behaviors, motivated behaviors based on learning involve additional integrative centers, including the cortex and limbic system. Motivated behaviors also use integrating centers in the midbrain, brainstem, and spinal cord—in other words, all levels of the nervous system can be involved.

Chemical Mediators Dopamine is a major neurotransmitter in the pathway that mediates the brain reward systems and motivation. For this reason, drugs that increase synaptic activity in the dopamine pathways—amphetamine, for example, which increases the presynaptic release of dopamine and other biogenic amines—increase self-stimulation rates—that is, provide positive reinforcement. Conversely, drugs

such as chlorpromazine, an antipsychotic agent that blocks dopamine receptors and lowers activity in the catecholamine pathways, are negatively reinforcing. The catecholamines are also, as we shall see, implicated in the pathways subserving learning. This is not unexpected since rewards and punishments are believed to constitute incentives for learning.

Emotional *behavior* can be studied more easily than the anatomical systems or inner emotions because it includes responses that can be measured externally (in terms of behavior). For example, stimulation of certain regions of the lateral hypothalamus causes an experimental animal to arch its back, puff out the fur on its tail, hiss, snarl, bare its claws and teeth, flatten its ears, and strike. Simultaneously, its heart rate, blood pressure, respiration, salivation, and concentrations of plasma epinephrine and fatty acids all increase. Clearly, this behavior typifies that of an enraged or threatened animal. Moreover, the animal's behavior can be changed from savage to docile and back again simply by stimulating different areas of the limbic system, such as parts of the amygdala.

Emotion

Emotion can be considered in terms of a relation between an individual and the environment based on the individual's evaluation of the environment (is it pleasant or hostile, for example), disposition toward the environment (am I happy and attracted to the environment or fearful of it?), and the actual physical response to it. While analyzing the physiological bases of emotion, it is helpful to distinguish: (1) the anatomical sites where the emotional value of a stimulus is determined; (2) the hormonal, autonomic, and outward expressions and displays of response to the stimulus (the so-called **emotional behavior**); and (3) the conscious experience, or **inner emotions**, such as feelings of fear, love, anger, joy, anxiety, hope, and so on.

Although inner emotions seem to be handled by different areas of the brain (Figure 13–11), there is no

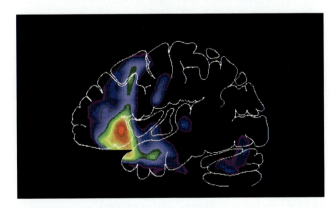

FIGURE 13–11
Computer image of neural activity during a sad thought.
Marcus E. Raichle, MD, Washington University School of Medicine.

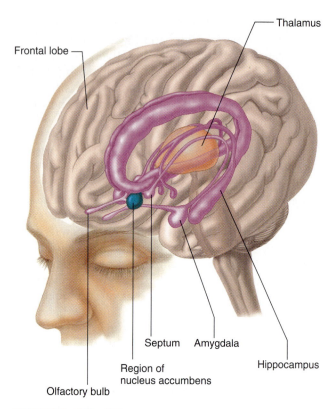

Frontal lobe

Thalamus

Septum   Amygdala

Region of
nucleus accumbens

Hippocampus

Olfactory bulb

FIGURE 13–12

Brain structures involved in directed attention, emotion,
motivation, and the affective disorders. The limbic system is
shaded purple.

From: BRAIN, MIND, AND BEHAVIOR by Floyd E. Bloom and Arlyne
Lazerson. Copyright ©1985, 1988 by Educational Broadcasting Corporation.
Reprinted by permission of W. H. Freeman and Company.

one, single "emotional system." The amygdala, a
cluster of nuclei deep in the tip of each temporal lobe
(Figure 13–12), and the region of association cortex
on the lower surface of the frontal lobe, however, are
central to most emotional states. The amygdala interacts with other parts of the brain via extensive reciprocal connections that can influence emotion in
terms of external stimuli, decision making, memory,
attention, homeostatic processes, and behavioral responses. For example, it sends output to the hypothalamus, which is central to autonomic and hormonal homeostatic processes.

Emotional behavior includes such complex behaviors as the passionate defense of a political ideol-
ogy and such simple actions as laughing, sweating,
crying, or blushing. Emotional behavior is achieved by
the autonomic and somatic nervous systems under the
influence of integrating centers such as those we just
mentioned and provides an outward sign that the
brain's "emotion systems" are activated.

The cerebral cortex plays a major role in directing
many of the motor responses during emotional behavior (for example, to approach or avoid a situation).

Moreover, it is forebrain structures, including the
cerebral cortex, that account for the modulation, direction, understanding, or even inhibition of emotional behaviors.

As for inner emotions, the limbic areas have been
stimulated in awake human beings undergoing neurosurgery. These patients reported vague feelings of
fear or anxiety during periods of stimulation to certain
areas. Stimulation of other areas induced pleasurable
sensations, which the subjects found hard to define
precisely. The cerebral cortex, however, elaborates the
conscious experience of inner emotion.

Altered States of Consciousness

The state of consciousness may be different from the
commonly experienced wakefulness, drowsiness, and
so on. Other, more bizarre situations, such as those occurring with hypnosis, mind-altering drugs, and certain diseases, are referred to as *altered states of consciousness.* These altered states are also characteristic
of psychiatric illnesses.

Schizophrenia

One of the diseases that induces altered states of consciousness is *schizophrenia,* a disease in which information is not properly regulated in the brain. The
amazingly diverse symptoms of schizophrenia include hallucinations, especially "hearing" voices, and
delusions, such as the belief that one has been chosen for a special mission or is being persecuted by
others. Schizophrenics become withdrawn, are emo-
tionally unresponsive, and experience inappropriate
moods. They also experience abnormal motor behavior, which can include total immobilization (*catato-
nia*). The symptoms occur in patterns that may or
may not overlap.

The causes of schizophrenia remain unclear. Recent studies suggest that the disease reflects a developmental disorder in which neurons migrate or mature abnormally during brain formation due to a
genetic predisposition or multiple environmental factors that may include viral infections and malnutrition
during fetal life or early childhood. The brain abnormalities involve diverse neural circuits and neurotransmitter systems that regulate basic cognitive
processes. A widely accepted explanation for schizophrenia suggests that certain dopamine pathways are
overactive. This hypothesis is supported by the facts
that the symptoms are made worse by amphetamine-
like drugs, which are dopamine agonists, and that the
most therapeutically beneficial drugs used in treating
schizophrenia block dopamine receptors.

Schizophrenia affects approximately one in every
100 people and typically appears in the late teens
or early twenties just as brain development nears

completion. Currently there is no prevention or cure for the disease, although the symptoms can often be controlled with drugs. In a small number of cases, there has been complete recovery.

The Mood Disorders: Depressions and Bipolar Disorders

The term **mood** refers to a pervasive and sustained inner emotion that affects the person's perception of the world. In addition to being part of the conscious experience of the person, it can be observed by others. In healthy people, moods can be normal, elated, or depressed, and people generally feel that they have some degree of control of their moods. That sense of control is lost, however, in the *mood disorders,* which include depressive disorders and bipolar disorders. Along with schizophrenia, the mood disorders form the major psychiatric illnesses today.

In the *depressive disorders* (*depression*), the prominent features are a pervasive sadness; a loss of energy, interest, or pleasure; anxiety; irritability; disturbed sleep; and thoughts of death or suicide. Depression can occur on its own, independent of any other illness, or it can arise secondary to other medical disorders. It is associated with decreased neuronal activity and metabolism in the anterior part of the limbic system and nearby prefrontal cortex. These same brain regions show abnormalities, albeit inconsistent ones, in bipolar disorders.

The term *bipolar disorders* describes swings between mania and depression episodes of *mania,* which are characterized by an abnormally and persistently elated mood, sometimes with euphoria (that is, an exaggerated sense of well-being), racing thoughts, excessive energy, overconfidence, and irritability.

Although the major biogenic amine neurotransmitters (norepinephrine, dopamine, and serotonin) and acetylcholine have all been implicated, the causes of the mood disorders are unknown.

Current treatment of the mood disorders emphasizes drugs and psychotherapy. The classical antidepressant drugs are of three types. The *tricyclic antidepressant drugs* such as Elavil, Desyrel, and Pamelor, interfere with serotonin and/or norepinephrine reuptake by presynaptic endings. The *monoamine oxidase inhibitors* interfere with the enzyme responsible for the breakdown of these same two neurotransmitters. A third class of antidepressant drugs, the *serotonin-specific reuptake inhibitors* (*SSRIs*), are the most widely used antidepressant drugs and include Prozac, Paxil, and Zoloft. As their name—SSRI—suggests, these drugs selectively inhibit serotonin reuptake by presynaptic terminals. In all three classes, the result is an *increased* concentration of serotonin and (except for the third class) norepinephrine in the extracellular fluid at synapses. Since the biochemical effects occur immediately but the beneficial antidepressant effects appear only after several weeks of treatment, the known biochemical effect must be only an early step in a complex, and presently unknown, sequence that leads to a therapeutic effect of these drugs.

Psychotherapy of various kinds can also be helpful in the treatment of depression. An alternative to drug therapy and psychotherapy is *electroconvulsive therapy* (*ECT*). As the name suggests, pulses of electric current are used to activate a large number of neurons in the brain simultaneously, thereby inducing a convulsion, or seizure. The patient is under anesthesia and prepared with a muscle relaxant to minimize the effects of the convulsion on the musculoskeletal system. A series of ECT treatments alters neurotransmitter function by causing a down-regulation of certain postsynaptic receptors. Another non-drug therapy used for the type of depression known as *seasonal affective disorder* (*SAD*) is *phototherapy* in which the patient is exposed to bright light for several hours per day.

A major drug used in treating patients with bipolar disorder is the chemical element, lithium, sometimes given in combination with anticonvulsant drugs. It is highly specific, normalizing both the manic and depressing moods and slowing down thinking and motor behavior without causing sedation. In addition, it decreases the severity of the swings between mania and depression that occur in the bipolar disorders. In some cases, lithium is even effective in depression not associated with manias. Lithium may interfere with the formation of members of the inositol phosphate family, thereby decreasing the postsynaptic neurons' response to neurotransmitters that utilize this signal transduction pathway (Chapter 7).

Psychoactive Substances, Dependence, and Tolerance

In the previous sections, we mentioned several drugs used to combat altered states of consciousness. Psychoactive drugs are also used as "recreational" or "street" drugs—the preferred term is now "substance"—in a deliberate attempt to elevate mood (euphorigens) and produce unusual states of consciousness ranging from meditational states to hallucinations. Virtually all the psychoactive substances exert their actions either directly or indirectly by altering neurotransmitter-receptor interactions in the biogenic amine—particularly dopamine—pathways. For example, the primary effect of cocaine comes from its ability to block the reuptake of dopamine into the presynaptic axon terminal. As mentioned in Chapter 8, psychoactive substances are often chemically similar to neurotransmitters such as

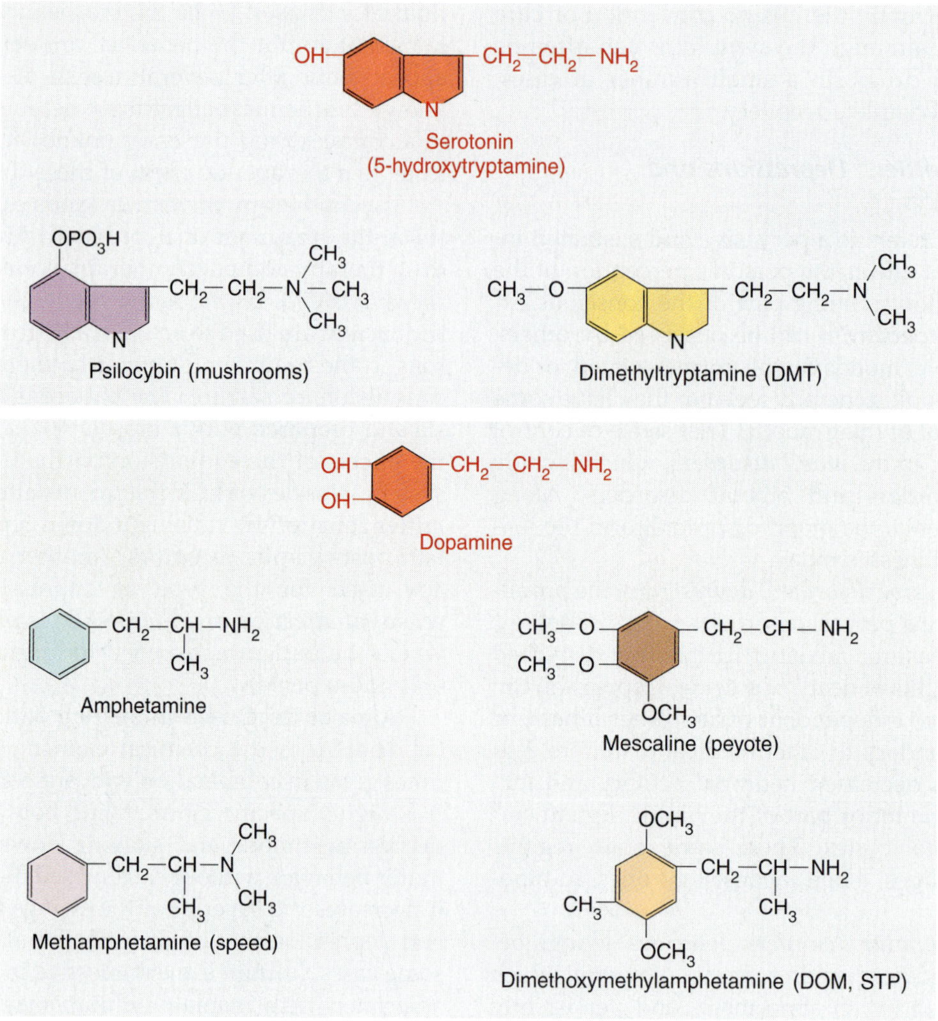

FIGURE 13–13

Molecular similarities between neurotransmitters (orange) and some euphorigens. At high doses, these euphorigens can cause hallucinations.

dopamine, serotonin, (Figure 13–13), and norepinephrine, and they interact with the receptors activated by these transmitters.

Dependence *Substance dependence,* the term now preferred for addiction, has two facets that may occur either together or independently: (1) a *psychological dependence* that is experienced as a craving for a substance and an inability to stop using the substance at will; and (2) a *physical dependence* that requires one to take the substance to avoid *withdrawal,* which is the spectrum of unpleasant physiological symptoms that occurs with cessation of substance use. Substance dependence is diagnosed if three or more of the characteristics listed in Table 13–3 occur within a 12-month period. Table 13–4 lists the dependence-producing potential of various drugs.

Several neuronal systems are involved in substance dependence, but most psychoactive substances act on the mesolimbic dopamine pathway (Figure 13–14). In addition to the actions of this system mentioned earlier in the context of motivation and emotion, the mesolimbic dopamine pathway allows a person to experience pleasure in response to pleasurable events or in response to certain substances. Although the major neurotransmitter implicated in addiction is dopamine, other neurotransmitters, including GABA, enkephalin, serotonin, and glutamate, are also involved.

According to popular opinion, people use substances primarily to feel pleasure, but some researchers are presenting a different view. They believe that in people who become substance dependent, the neurotransmitters may be handled abnormally in the mesolimbic dopamine pathway due in part to a genetic

TABLE 13–3 Diagnostic Criteria for Substance Dependence
Substance dependence is indicated when three or more of the following occur within a 12-month period.

1. Tolerance, as indicated by
  a. a need for increasing amounts of the substance to achieve the desired effect, or
  b. decreasing effects when continuing to use the same amount of the substance

2. Withdrawal, as indicated by
  a. appearance of the characteristic withdrawal symptoms upon stopping use of the substance, or
  b. use of the substance (or one closely related to it) to relieve or avoid withdrawal symptoms

3. Use of the substance in larger amounts or for longer periods of time than intended
4. Persistent desire for the substance; unsuccessful attempts to cut down or control use of the substance
5. A great deal of time is spent in activities necessary to obtain the substance, use it, or recover from its effects
6. Occupational, social, or recreational activities are given up or reduced because of substance use
7. Use of the substance is continued despite knowledge that one has a physical or psychological problem that is likely to be exacerbated by the substance

Adapted from DSM-IV, Diagnostic and Statistical Manual of Mental Disorders, 4th edn. American Psychiatric Association, Washington, D.C., 1994.

TABLE 13–4 Potential of Various Substances to Cause Dependence	
If 100 people use a substance, how many will become dependent on it?	
Nicotine	33
Heroin	25
Cocaine	16
Alcohol	15
Amphetamines	11
Marijuana	9

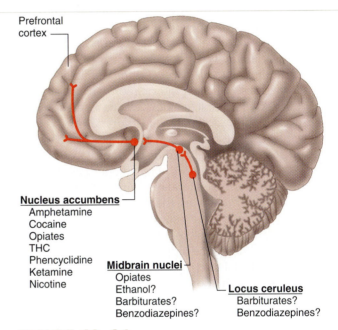

Prefrontal cortex

Nucleus accumbens
Amphetamine
Cocaine
Opiates
THC
Phencyclidine
Ketamine
Nicotine

Midbrain nuclei
Opiates
Ethanol?
Barbiturates?
Benzodiazepines?

Locus ceruleus
Barbiturates?
Benzodiazepines?

FIGURE 13–14

The sites at which various psychoactive substances are thought to work by enhancing brain reward. The mesolimbic dopamine pathway is indicated schematically by the red neurons. THC = tetrahydrocannabinol, the active ingredient in marijuana.

alteration. Thus, by using substances, addicts are self-medicating in an attempt to stabilize the activity in this pathway so they can feel normal and relaxed. In other cases, people use substances primarily to avoid the discomfort of withdrawal.

Tolerance *Tolerance* to a substance occurs when increasing doses of the substance are required to achieve effects that initially occurred in response to a smaller dose; that is, it takes more of the substance to do the same job. Moreover, tolerance can develop to one substance as a result of taking another substance, in which case, it is called **cross-tolerance.** Substance tolerance and cross-tolerance can occur with many classes of substances and are not limited to psychoactive substances.

Tolerance may develop because the presence of the substance stimulates the synthesis, especially in the microsomal enzyme system (Chapter 20), of the enzymes that degrade it. As substance concentrations increase, so do the concentrations of the enzymes that degrade it. Thus, more of the substance must be administered to produce the same plasma concentrations of the substance and hence the same initial effect.

Alternatively, tolerance can develop as a result of changes in the number and/or sensitivity of receptors that respond to the substance, the amount or activity of enzymes involved in neurotransmitter synthesis, the reuptake transport molecules, or the signal transduction pathways in the postsynaptic cell.

Learning and Memory

Learning is the acquisition and storage of information as a consequence of experience. It is measured by an increase in the likelihood of a particular behavioral response to a stimulus. Generally, rewards or punishments, as mentioned earlier, are crucial ingredients of learning, as are contact with and manipulation of the environment. **Memory** is the relatively permanent storage form of the learned information although, as we shall see, it is not a single, unitary phenomenon. Rather, the brain processes, stores, and retrieves information in different ways to suit different needs.

Memory

The term **memory encoding** defines the processes that mediate between an experience and the memory of that experience—in other words, the physiological events that lead to memory formation. This section addresses the questions: Are there different kinds of memories, where do they occur in the brain, and what happens physiologically to make them occur?

New scientific facts about memory are being generated at a tremendous pace, and the difficulty comes when one tries to fit this information into an overall, workable scheme. First, memory can be viewed in two broad categories: **Declarative memory** is involved in answering the question, "What is it, or when and where did it happen?" and is based on one's past experience. One example is the memory of having perceived an object or event and, therefore, recognizing it as familiar and maybe even knowing the specific time and place when the memory was instigated. A second example would be one's general knowledge of the world such as names and facts. The hippocampus, amygdala, and diencephalon—all parts of the limbic system—are required for the formation of declarative memories.

The second broad category of memory, **nondeclarative memory**, includes **procedural memory**, which can be thought of as the memory of how to do things. In other words, it is the memory for skilled behaviors independent of any understanding, as for example, riding a bicycle. Individuals can suffer severe deficits in declarative memory but have intact procedural memory. One case study describes a pianist who learned a new piece to accompany a singer at a concert but had no recollection the following morning of having performed the composition. He could remember *how* to play the music but could not remember having done so. The category of nondeclarative memory also includes learned emotional responses, such as fear of thunder, and the classic example of Pavlov's dog, which learned to salivate at the sound of a bell after the bell had previously been associated with food. The primary areas of the brain involved in nondeclarative memory are regions of sensorimotor cortex, the basal ganglia, and the cerebellum. As we have seen, memory can be considered as to type—declarative or nondeclarative—and the brain areas involved in its formation; it can also be looked at in terms of duration—does it last for a long or only a short time?

Working Memory Working memory, also known as primary or short-term memory, registers and retains incoming information for a short time—a matter of seconds—after its input. In other words, it is the memory that we use when we keep information consciously "in mind." Working memory makes possible a temporary impression of one's present environment in a readily accessible form and is an essential ingredient of many forms of higher mental activity. It is not surprising then that working memory is associated with a second component—the so-called executive processes—that can operate on the contents of working memory. The executive processes include such functions as selective attention, switching attention between parts of a complex task, and planning the sequence of tasks to achieve a specific goal.

Focusing attention is essential for many memory-based skills. The longer the span of attention in working memory, the better the chess player, the greater the ability to reason, and the better a student is at understanding complicated sentences and drawing inferences from texts. In fact, there is a strong correlation between working memory and standard measures of intelligence. Conversely, the specific memory deficit in victims of Alzheimer's disease, a condition marked by serious memory losses, may be in this attention-focusing component of working memory.

The Neural Basis of Learning and Memory

Not all types of memory have the same neural mechanisms or, as we have seen, involve the same parts of the brain. The **prefrontal cortex** of the frontal lobes (see Figure 13–9), is active in both components—the storage and executive processes—of working memory, but discrete regions of prefrontal cortex deal with specific kinds of information. Thus, one area of the prefrontal cortex encodes information about verbal material such as attaching names to objects or events, a different area deals with spatial memories such as how to get from one place to another, and yet another area deals with memories of events. Still other areas deal with the executive processes. Cortical areas other than the prefrontal cortex are involved in working memory as well.

Recall that the prefrontal cortex is also the destination of a major branch of the mesolimbic dopamine pathway, a pathway mentioned earlier in the context of directed attention. Directed attention is a critical component of working memory—we do not remember what we have not paid attention to—and these two mechanisms interact in prefrontal cortex. Recall, too, that dopamine is a major neurotransmitter implicated in directed attention; thus, drugs that are dopamine antagonists—drugs such as those used for schizophrenia, for example—interfere with working memory. Prefrontal cortex is also the target of a cholinergic neural system that originates in the brainstem as a component of the reticular activating system.

But what is happening during the memory formation on a cellular level? Conditions such as coma, deep anesthesia, electroconvulsive shock, and insufficient blood supply to the brain, all of which interfere with the electrical activity of the brain, interfere with working memory. Thus, it is assumed that working memory requires ongoing graded or action potentials. Working memory is interrupted when a person becomes unconscious from a blow on the head, and memories are abolished for all that happened for a variable period of time before the blow, so-called *retrograde amnesia*. (*Amnesia* is defined as the loss of memory.) Working memory is also susceptible to external interference, such as an attempt to learn conflicting information. On the other hand, long-term memory can survive deep anesthesia, trauma, or electroconvulsive shock, all of which disrupt the normal patterns of neural conduction in the brain. Thus, working memory requires electrical activity in the neurons, but the question remains: What happens next?

The problem of how exactly memories are stored in the brain is still unsolved, but some of the pieces of the puzzle are falling into place. One model for memory is **long-term potentiation (LTP),** in which certain synapses undergo a long-lasting increase in their effectiveness when they are heavily used. Long-term potentiation results from increased activation of glutamate receptors on the postsynaptic cell, which opens Ca^{2+} channels in the receptor, causing an increase of cytosolic calcium. The calcium enhances the enzymatic formation in the postsynaptic cell of **nitric oxide**, which diffuses back across the synapse to enhance the effectiveness of the synapse. An analogous process, **long-term depression (LTD),** decreases the effectiveness of synaptic contacts between neurons.

Other models for memory encoding involve feedback loops of second-messenger molecules in postsynaptic cells, which can sustain activity in these cells long after the membrane-receptor activity ceases. Moreover, it is generally accepted that memory encoding involves processes that alter gene expression and result in the synthesis of new proteins. This is achieved by a cascade of second messengers that activate genes in the cell's DNA. Yet another class of memory models is based on the idea that memory is encoded by structural changes in the synapses (for example, by an increase in the number of receptors on the postsynaptic membrane). This ability of neural tissue to change because of its activation is known as **plasticity.**

Additional Facts Concerning Learning and Memory

Certain types of learning depend not only on factors such as attention, motivation, and various neurotrans-

TABLE 13–5 General Principles about Learning and Memory
1. There are multiple memory systems in the brain.
2. Working memory requires changes in existing neural circuits, whereas long-term memory requires new protein synthesis and growth.
3. These changes may involve multiple cellular mechanisms within single neurons.
4. Second-messenger systems appear to play a role in mediating cellular changes.
5. Changes in the properties of membrane channels are often correlated with learning and memory.

Adapted from John M. Beggs et al. "Learning and Memory: Basic Mechanisms," in Michael J. Zigmond, Floyd E. Bloom, Story C. Landis, James L. Roberts, and Larry R. Squire, eds., Fundamental Neuroscience, Academic Press, San Diego, CA, 1999.

mitters but also on certain hormones. For example, the hormones epinephrine, ACTH, and vasopressin affect the retention of learned experiences. These hormones are normally released in stressful or even mildly stimulating experiences, suggesting that the hormonal consequences of our experiences affect our memories of them.

Two of the opioid peptides, enkephalin and endorphin, *interfere* with learning and memory, particularly when the lesson involves a painful stimulus. They may inhibit learning simply because they decrease the emotional (fear, anxiety) component of the painful experience associated with the learning situation, thereby decreasing the motivation necessary for learning to occur.

Memories can be encoded very rapidly, sometimes after just one trial, and they can be retained over extended periods. Information can be retrieved from memory stores after long periods of disuse, and the common notion that memory, like muscle, atrophies with lack of use is not always true. Also, unlike working memory, memory storage apparently has an unlimited capacity because people's memories never seem to be so full that they cannot learn something new. Although we have mentioned specific areas of the brain that are active in learning, we want to stress at this time the following point: Memory traces are laid down in specific neural systems throughout the brain, and different types of memory tasks utilize different systems. For in even a simple memory task, such as trying to recall a certain word from a previously seen word list, different specific parts of the brain are activated in sequence. It is as though several small "processors" are linked together in a memory system for specific memory tasks.

Some general principles about learning and memory are summarized in Table 13–5.

Cerebral Dominance and Language

The two cerebral hemispheres appear to be nearly symmetrical, but each has anatomical, chemical, and functional specializations. We have already mentioned that the left hemisphere deals with the somatosensory (Chapter 9) and motor (Chapter 12) functions of the right side of the body, and vice versa. In addition, in 90 percent of the population the left hemisphere is specialized to produce language—the conceptualization of what one wants to say or write, the neural control of the act of speaking or writing, and recent verbal memory. This is even true of the sign language used by deaf people.

Language is a complex code that includes the acts of listening, reading, and speaking. The major centers for language function are in the left hemisphere in temporal, parietal, and frontal cortex (the so-called perisylvian area, Figure 13–15a) next to the sylvian fissure, which separates the temporal lobe from the frontal and parietal lobes. Other language areas also exist in the cerebral cortex, each dealing with a separate aspect of language. For example, distinct areas are specialized for hearing, seeing, speaking, and generating words (Figure 13–15b). There are even distinct brain networks for different categories of things, such as "animals" and "tools." The cerebellum is important in speaking and writing, which involve coordinated muscle contractions. Males and females use different brain areas for language processing, probably reflecting different strategies (Figure 13–16).

Neural specialization is demonstrated by the *aphasias,* any language defect resulting from brain damage. For example, in most people, damage to the left cerebral hemisphere, but not to the right, interferes with the capacity for language manipulation, and damage to different areas of the left cerebral hemisphere affects language use differently.

Damage to the temporal region known as **Wernicke's area** (Figure 13–17) generally results in aphasias that are more closely related to *comprehension*—the individuals have difficulty understanding spoken or written language even though their hearing and vision are unimpaired, and although they may have fluent speech, their speech is incomprehensible. In contrast, damage to **Broca's area,** the language area in the frontal cortex responsible for the articulation of speech, can cause *expressive* aphasias—the individuals have difficulty carrying out the coordinated respiratory and oral movements necessary for language even though they can move their lips and tongue. They understand spoken language and know what they want to say but have trouble forming words and putting them into grammatical order.

The potential for development of language-specific mechanisms in the left hemisphere is present at birth, but the assignment of language functions to specific brain areas is fairly flexible in the early years of life.

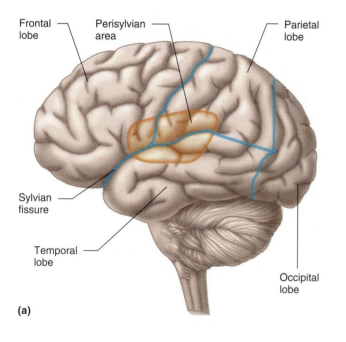

(a)

Frontal lobe
Perisylvian area
Parietal lobe
Sylvian fissure
Temporal lobe
Occipital lobe

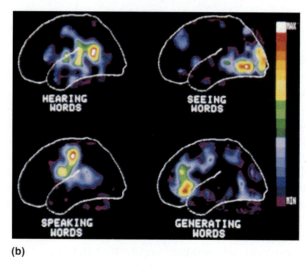

HEARING WORDS
SEEING WORDS
SPEAKING WORDS
GENERATING WORDS

(b)

FIGURE 13–15

(a) The perisylvian area, the site of many language functions, is indicated in orange. Blue lines indicate divisions of the cortex into frontal, parietal, temporal, and occipital lobes. (b) PET scans reveal changes in blood flow during various language-based activities.

Part b courtesy of Dr. Marcus E. Raichle.

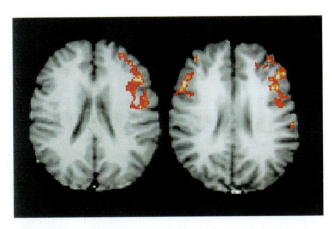

FIGURE 13–16

Images of the active areas of the brain of a male (left) and a female (right) during a language task. Note that both sides of a woman's brain are used in processing language but a man's brain is more compartmentalized.

Shaywitz, et al., 1995 NMR Research/Yale Medical School.

Thus, for example, damage to the perisylvian area of the left hemisphere during infancy or early childhood causes temporary, minor language impairment, but similar damage acquired during adulthood typically causes permanent, devastating language deficits.

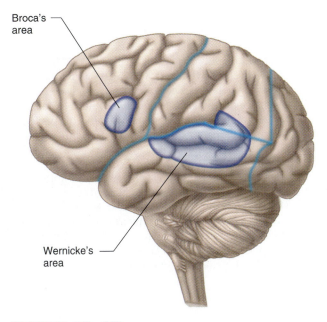

Broca's area

Wernicke's area

FIGURE 13–17

Areas found clinically to be involved in the comprehension (Wernicke's area) and motor (Broca's area) aspects of language. Blue lines indicate divisions of the cortex into frontal, parietal, temporal, and occipital lobes.

Even if the left hemisphere is traumatized in older children, functional language ability may be reestablished to some extent in the right hemisphere. Language develops in these children, however, at the expense of usual right hemisphere functions, such as spatial localization. By puberty, the transfer of language functions to the right hemisphere is less successful, and often language skills are lost permanently.

Although language skills emerge spontaneously in all normal children in all societies, there is a critical period during childhood when exposure to language is necessary for these skills to develop, just as the ability to see depends upon effective visual input early in life. The critical period is thought to end at puberty or earlier. The dramatic change at puberty in the possibility of learning language, or the ease of learning a second language, occurs as the brain attains its structural, biochemical, and functional maturity at that time.

As part of these basic language skills, the left hemisphere contains the "rules" for general grammatical principles; thus, it is much more skilled at changing verb tenses and constructing possessives than is the right hemisphere. It is also dominant for events that occur in sequences over time such as those seen in language usage as, for example, speaking the first part of a word before the last. In addition, the left hemisphere seems constantly to form theories about how the world works, to find relationships between events encountered in the world, and to assess where one stands in relation to the world. The left hemisphere has been called "the interpreter."

The right hemisphere, on the other hand, more typically handles sensory information in basic ways, such as the perception of faces and other three-dimensional objects.

Memories are handled differently in the two hemispheres, too, with verbal memories more apt to be associated with the left hemisphere, and nonverbal memories (for example, visual patterns or nonverbal memories that convey emotions) with the right. Even the emotional responses of the two hemispheres seem to be different; for example, the left hemisphere has the greater ability to understand the emotional states of oneself or others. When electroconvulsive therapy is administered in the treatment of depression, however, better effects are often obtained when the electrodes are placed over the right hemisphere. The two sides of the brain also differ in their sensitivity to psychoactive drugs. Differences between the two hemispheres are often masked by integration via the corpus callosum and other pathways that connect the two sides of the brain.

Conclusion

Mental tasks are the result of synchronized activity in vast neuronal networks made up of many functional regions of the cerebral cortex, subcortical nuclei, and brainstem. Also important are the pathways that reciprocally connect these sites and orchestrate their performance during specific tasks. In fact, the nervous system is so abundantly interconnected that it is difficult to know where any particular subsystem begins or ends.

Material in this chapter has often been highly qualified on a molecular level, largely because the answers simply are not known. For example, the scientific literature is full of statements such as, "The study of the neural basis of language is in flux," and "The understanding of learning is one of nature's most closely guarded secrets." We also caution the reader that listing brain sites that show increased activity during dependence, sleep, language, or learning does little to explain *how* these phenomena occur. For the topics discussed in this chapter, the answer to the question "How?," which makes up the stuff of physiology, is often simply, "We presently don't know." One is reminded of the lament that "We can send astronauts to the moon and space probes to Jupiter, but we can't cure the common cold"; in other words, the things most relevant to our everyday life are often the least understood.

SUMMARY

I. Consciousness includes states of consciousness and conscious experiences.

States of Consciousness

I. The electroencephalogram provides one means of defining the states of consciousness.
  a. Electric currents in the cerebral cortex due predominantly to summed postsynaptic potentials are recorded as the EEG.
  b. Slower EEG waves correlate with less responsive behaviors.
  c. Rhythm generators in the thalamus are probably responsible for the wavelike nature of the EEG.
  d. EEGs are used to diagnose brain disease and damage.
II. Alpha rhythms and, during EEG arousal, beta rhythms characterize the EEG of an awake person.
III. NREM sleep progresses from stage 1 (faster, lower-amplitude waves) through stage 4 (slower, higher-amplitude waves) and then back again, followed by an episode of REM sleep. There are generally four or five of these cycles per night.
IV. Aminergic and cholinergic brainstem centers, via their projections forward to the cerebrum as components of the reticular activating system, interact with the thalamus to regulate the sleep-wake cycles. Hypothalamic nuclei also play a role.

Conscious Experiences

I. Brain structures involved in directed attention determine which brain areas gain temporary predominance in the ongoing stream of conscious experience.
II. Conscious experiences may occur because a set of neurons temporarily functions together, the neurons comprising the set changing as the focus of attention changes.

Motivation and Emotion

I. Behaviors that satisfy homeostatic needs are primary motivated behaviors. Behavior not related to homeostasis is a result of secondary motivation.
  a. Repetition of a behavior indicates that it is rewarding, and avoidance of a behavior indicates it is punishing.
  b. The mesolimbic dopamine pathway, which goes to prefrontal cortex and parts of the limbic system, mediates emotion and motivation.
  c. Dopamine is the primary neurotransmitter in the brain pathway that mediates motivation and reward.
II. Three aspects of emotion—anatomical and physiological bases for emotion, emotional behavior, and inner emotions—can be distinguished. The limbic system integrates inner emotions and behavior.

Altered States of Consciousness

I. Hyperactivity in a brain dopaminergic system is implicated in schizophrenia.
II. The mood disorders are caused, at least in part, by disturbances in transmission at brain synapses mediated by dopamine.
III. Many psychoactive drugs, which are often chemically related to neurotransmitters, result in substance dependence, withdrawal, and tolerance. The mesolimbic dopamine pathway and the nucleus accumbens are implicated in substance abuse.

Learning and Memory

I. The brain processes, stores, and retrieves information in different ways to suit different needs.
II. Memory encoding involves cellular or molecular changes specific to different memories.
III. Declarative memories are involved in remembering facts and events. Nondeclarative memories include procedural memories, which are memories of how to do things.
IV. Prefrontal cortex and limbic regions of the temporal lobe are important brain areas for some forms of memory.
V. Memory encoding probably involves changes in second-messenger systems and protein synthesis.

Cerebral Dominance and Language

I. The two cerebral hemispheres differ anatomically, chemically, and functionally. In 90 percent of the population, the left hemisphere is superior at producing language and in performing other tasks that require rapid changes over time.

II. The development of language functions occurs in a critical period that closes at puberty.

III. After damage to the dominant hemisphere, some language function can be acquired by the opposite hemisphere—the younger the patient, the greater the transfer of function.

Conclusion

I. Many brain units are involved in the performance of even simple mental tasks.

II. Each unit is localized to a specific brain area, but, because many units are involved, widely distributed brain areas take part in mental tasks.

III. Little is known definitely about how consciousness and behavior are actually determined.

KEY TERMS

states of consciousness	emotional behavior
conscious experience	inner emotion
electroencephalogram	mood
(EEG)	learning
alpha rhythm	memory
beta rhythm	memory encoding
EEG arousal	declarative memory
NREM sleep	nondeclarative memory
REM sleep	procedural memory
paradoxical sleep	working memory
directed attention	prefrontal cortex
orienting response	long-term potentiation
preattentive processing	(LTP)
habituation	nitric oxide
binding problem	long-term depression
motivation	(LTD)
primary motivated	plasticity
behavior	Wernicke's area
brain self-stimulation	Broca's area
emotion	

REVIEW QUESTIONS

1. State the two criteria used to define one's state of consciousness.
2. What type of neural activity is recorded as the EEG?
3. Draw EEG records that show alpha and beta rhythms, the stages of NREM sleep, and REM sleep. Indicate the characteristic wave frequencies of each.
4. Distinguish NREM sleep from REM sleep.
5. Briefly describe a neural mechanism that determines the states of consciousness.

6. Name the criteria used to distinguish brain death from coma.
7. Describe the orienting response as a form of directed attention.
8. Distinguish primary from secondary motivated behavior.
9. Explain how rewards and punishments are anatomically related to emotions.
10. Explain what brain self-stimulation can tell about emotions and rewards and punishments.
11. Name the primary neurotransmitter that mediates the brain reward systems.
12. Distinguish inner emotions from emotional behavior. Name the brain areas involved in each.
13. Describe the role of the limbic system in emotions.
14. Name the major neurotransmitters involved in schizophrenia and the mood disorders.
15. Describe a mechanism that could explain tolerance and withdrawal.
16. Distinguish the types of memory.

CLINICAL TERMS

epilepsy	serotonin-specific reuptake
sleep apnea	inhibitors (SSRIs)
coma	electroconvulsive therapy
brain death	(ECT)
altered states of	seasonal affective disorder
consciousness	(SAD)
schizophrenia	phototherapy
catatonia	substance dependence
mood disorder	psychological dependence
depressive disorder	physical dependence
(depression)	withdrawal
bipolar disorder	tolerance
mania	cross-tolerance
tricyclic antidepressant drug	retrograde amnesia
monoamine oxidase	amnesia
inhibitor	aphasia

THOUGHT QUESTIONS

(Answers are given in Appendix A.)

1. Explain why patients given drugs to treat Parkinson's disease (Chapter 12) sometimes develop symptoms similar to those of schizophrenia.
2. Explain how clinical observations of individuals with various aphasias help physiologists understand the neural basis of language.

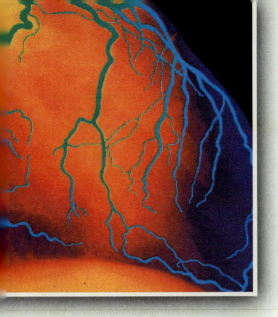

Circulation

Beyond a distance of a few cell diameters, diffusion—the random movement of substances from a region of higher concentration to one of lower concentration—is not sufficiently rapid to meet the metabolic requirements of cells. In large multicellular organisms, therefore, some mechanism other than diffusion is needed to transport molecules rapidly over the long distances between internal cells and the body's surface and between the various specialized tissues and organs. In the animal kingdom this is achieved by the **circulatory system**, which comprises the blood, the set of interconnected tubes (**blood vessels**, or **vascular system**) through which the blood flows, and a pump (the **heart**) that produces this flow. The heart and blood vessels together are termed the **cardiovascular system**.

_____

SECTION A

BLOOD

Blood is composed of cells and a liquid, called **plasma,** in which they are suspended. The cells are the **erythrocytes** (red blood cells), the **leukocytes** (white blood cells), and the **platelets,** which are not complete cells but cell fragments. More than 99 percent of blood cells are erythrocytes, which carry oxygen. The leukocytes protect against infection and cancer (Chapter 20), and the platelets function in blood clotting (Section G of this chapter). In the cardiovascular system, the constant motion of the blood keeps all the cells well dispersed throughout the plasma.

The **hematocrit** is defined as the percentage of blood volume that is occupied by erythrocytes. It is measured by centrifuging (spinning at high speed) a sample of blood, the erythrocytes being forced to the bottom of the centrifuge tube and the plasma to the top, the leukocytes and platelets forming a very thin layer between them (Figure 14–1). The normal hematocrit is approximately 45 percent in men and 42 percent in women.

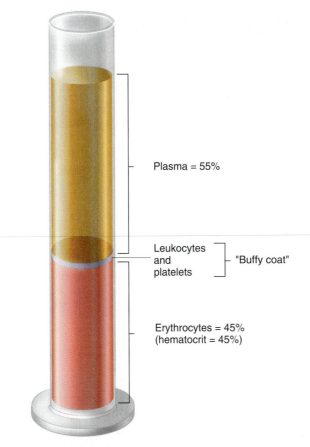

FIGURE 14–1

Measurement of the hematocrit—the percentage of blood volume that is erythrocytes—by centrifugation. The presence of a thin layer of leukocytes and platelets between the plasma and red cells explains why, in this example, the value for plasma determined by centrifugation should actually be slightly less than 55 percent.

The volume of blood in an average-sized person (70 kg; 154 lbs) is approximately 5.5 L. If we take the hematocrit to be 45 percent, then

$$\text{Erythrocyte volume} = 0.45 \times 5.5 \text{ L} = 2.5 \text{ L}$$

Since the volume occupied by the leukocytes and platelets is normally negligible, the plasma volume equals the difference between blood volume and erythrocyte volume; therefore, in our average person

$$\text{Plasma volume} = 5.5 \text{ L} - 2.5 \text{ L} = 3.0 \text{ L}$$

Plasma

Plasma, the liquid portion of the blood, consists of a large number of organic and inorganic substances dissolved in water. Table 14–1 provides, for reference, the plasma concentrations of many of these substances. The characteristic straw color of plasma is due largely to a waste product of hemoglobin breakdown called **bilirubin,** which will be described in a later section.

The **plasma proteins** constitute, by weight, most of the plasma solutes. They can be classified, according to certain physical and chemical reactions, into three broad groups: the **albumins** and **globulins,** which have many overlapping functions summarized in Table 14–1 and described in relevant chapters, and **fibrinogen,** which functions in blood clotting (Section G of this chapter). **Serum** is plasma from which fibrinogen and other proteins involved in clotting have been removed as a result of clotting. The albumins are the most abundant of the three plasma protein groups and are synthesized by the liver. It must be emphasized that the plasma proteins normally are not taken up by cells; cells use plasma amino acids, not plasma proteins, to make their own proteins. Accordingly, plasma proteins must be viewed quite differently from most of the other organic constituents of plasma, which use the plasma as a medium for transport to and from cells. In contrast, most plasma proteins perform their functions in the plasma itself or in the interstitial fluid.

In addition to the organic solutes, including proteins, nutrients, metabolic waste products, and hormones, plasma contains a variety of mineral electrolytes. Note in Table 14–1 that these ions contribute much less to the *weight* of plasma than do the proteins, but in most cases they have much higher *molar concentrations*. This is because molarity is a measure not of weight but of number of molecules or ions per unit volume (Chapter 2). Thus, there are many more ions than protein molecules, but the protein molecules are so large that a very small number of them greatly outweighs the much larger number of ions.

The Blood Cells

Erythrocytes

The major functions of erythrocytes are to carry oxygen taken in by the lungs and carbon dioxide produced by cells. Erythrocytes contain large amounts of the protein **hemoglobin** with which oxygen and, to a lesser extent, carbon dioxide reversibly combine. Oxygen binds to iron atoms (Fe) in the hemoglobin molecules. The average concentration of hemoglobin is 14 g/100 ml blood in women and 16 g/100 ml in men. Further description of hemoglobin structure and functions is given in Chapter 15, where the transport of oxygen and carbon dioxide is presented.

Erythrocytes have the shape of a biconcave disk—that is, a disk thicker at the edges than in the middle, like a doughnut with a center depression on each side

TABLE 14–1 Reference Table of Plasma Constituents

Constituent	Amount/Concentration	Major Functions
Water	93% of plasma weight	Medium for carrying all other constituents
Electrolytes (inorganic)	Total < 1% of plasma weight	Keep H_2O in extracellular compartment; act as buffers; function in membrane excitability and blood clotting
Na^+	145 mM	
K^+	4 mM	
Ca^{2+}	2.5 mM	
Mg^{2+}	1.5 mM	
H^+	0.0004 mM	
Cl^-	103 mM	
HCO_3^-	24 mM	
Phosphate (mostly HPO_4^{2-})	1 mM	
SO_4^{2-}	0.5 mM	
Proteins	Total = 7% of plasma weight, 7.3 g/100 ml (2.5 mM)	Provide nonpenetrating solutes of plasma; act as buffers; bind and transport other plasma constituents (lipids, hormones, vitamins, metals, etc.);
Albumins	4.2 g/100 ml	
Globulins	2.8 g/100 ml	clotting factors; enzymes, enzyme precursors; antibodies (immune globulins); hormones
Fibrinogen	0.3 g/100 ml	Blood clotting
Gases		
CO_2	2 ml/100 ml (1 mM)	A waste product
O_2	0.2 ml/100 ml (0.1 mM)	Oxidative metabolism
N_2	0.9 ml/100 ml (0.5 mM)	No function
Nutrients		(See Chapters 2, 4, and 18)
Glucose and other carbohydrates	100 mg/100 ml (5.6 mM)	
Total amino acids	40 mg/100 ml (2 mM)	
Total lipids	500 mg/100 ml (7.5 mM)	
Cholesterol	150–250 mg/100 ml (4–7 mM)	
Individual vitamins	0.0001–2.5 mg/100 ml (0.00005–0.1 mM)	
Individual trace elements	0.001–0.3 mg/100 ml (0.0001–0.01 mM)	
Waste products		
Urea (from protein)	34 mg/100 ml (5.7 mM)	
Creatinine (from creatine)	1 mg/100 ml (0.09 mM)	
Uric acid (from nucleic acids)	5 mg/100 ml (0.3 mM)	
Bilirubin (from heme)	0.2–1.2 mg/100 ml (0.003–0.018 mM)	
Individual hormones	0.000001–0.05 mg/100 ml (10^{-9}–10^{-6} mM)	Messengers in control systems

instead of a hole (Figure 14–2). This shape and their small size (7 μm in diameter) impart to the erythrocytes a high surface-to-volume ratio, so that oxygen and carbon dioxide can diffuse rapidly to and from the interior of the cell. The plasma membrane of erythrocytes contains specific polysaccharides and proteins that differ from person to person, and these confer upon the blood its so-called type, or group. Blood groups are described in Chapter 20, in the context of the immune responses that occur in transfusion reactions.

The site of erythrocyte production is the soft interior of bones called **bone marrow,** specifically the "red" bone marrow. With differentiation, the erythrocyte precursors produce hemoglobin but then they ultimately lose their nuclei and organelles—their machinery for protein synthesis. Young erythrocytes in the bone marrow still contain a few ribosomes, which produce a web-like (reticular) appearance when treated with special stains, an appearance that gives these young erythrocytes the name **reticulocyte.** Normally, only mature erythrocytes, which have lost these ribosomes,

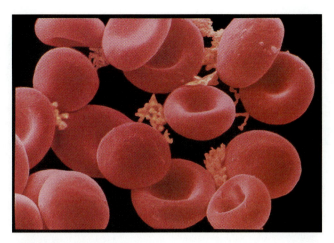

FIGURE 14–2

Electron micrograph of erythrocytes.

© Bruce Iverson

leave the bone marrow and enter the general circulation. In the presence of unusually rapid erythrocyte production, however, many reticulocytes do enter the blood, a fact of clinical diagnostic usefulness.

Because erythrocytes lack nuclei and organelles, they can neither reproduce themselves nor maintain their normal structure for very long. The average life span of an erythrocyte is approximately 120 days, which means that almost 1 percent of the body's erythrocytes are destroyed and must be replaced every day. This amounts to 250 billion cells per day! Erythrocyte destruction normally occurs in the spleen and the liver. As will be described below, most of the iron released in the process is conserved. The major breakdown product of hemoglobin is bilirubin, which, as noted above, gives plasma its color (the fate of this substance will be described in Chapter 17).

The production of erythrocytes requires the usual nutrients needed to synthesize any cell: amino acids, lipids, and carbohydrates. In addition, both iron and certain growth factors, including the vitamins folic acid and vitamin B_{12}, are essential.

Iron As noted above, **iron** is the element to which oxygen binds on a hemoglobin molecule within an erythrocyte. Small amounts of iron are lost from the body via the urine, feces, sweat, and cells sloughed from the skin. In addition, women lose an additional amount via menstrual blood. In order to remain in iron balance, the amount of iron lost from the body must be replaced by ingestion of iron-containing foods; particularly rich sources are meat, liver, shellfish, egg yolk, beans, nuts, and cereals. A significant upset of iron balance can result either in *iron deficiency,* leading to inadequate hemoglobin production, or in an excess of iron in the body, with serious toxic effects (*hemochromatosis*).

The homeostatic control of iron balance resides primarily in the intestinal epithelium, which actively absorbs iron from ingested foods. Normally, only a small fraction of ingested iron is absorbed, but more importantly, this fraction is increased or decreased, in a negative feedback manner, depending upon the state of the body's iron balance—the more iron in the body, the less ingested iron is absorbed (the mechanism is given in Chapter 17).

The body has a considerable store of iron, mainly in the liver, bound up in a protein called **ferritin.** Ferritin serves as a buffer against iron deficiency. About 50 percent of the total body iron is in hemoglobin, 25 percent is in other heme-containing proteins (mainly the cytochromes) in the cells of the body, and 25 percent is in liver ferritin. Moreover, the recycling of iron is very efficient (Figure 14–3). As old erythrocytes are destroyed in the spleen (and liver), their iron is released into the plasma and bound to an iron-transport plasma protein called **transferrin.** Almost all of this iron is delivered by transferrin to the bone marrow to be incorporated into new erythrocytes. Recirculation of erythrocyte iron is very important because it involves 20 times more iron per day than is absorbed and excreted. On a much lesser scale, nonerythrocyte cells, some of which are continuously dying and being replaced, release iron from their cytochromes into the plasma and take up iron from it, transferrin serving as a carrier.

Folic Acid and Vitamin B_{12} **Folic acid,** a vitamin found in large amounts in leafy plants, yeast, and liver, is required for synthesis of the nucleotide base thymine. It is, therefore, essential for the formation of DNA and hence for normal cell division. When this vitamin is not present in adequate amounts, impairment of cell division occurs throughout the body but is most striking in rapidly proliferating cells, including erythrocyte precursors. Thus, fewer erythrocytes are produced when folic acid is deficient.

Production of normal erythrocyte numbers also requires extremely small quantities (one-millionth of a gram per day) of a cobalt-containing molecule, **vitamin B_{12}** (also called cobalamin), since this vitamin is required for the action of folic acid. Vitamin B_{12} is found only in animal products, and strictly vegetarian diets are deficient in it. Also, as described in Chapter 17, the absorption of vitamin B_{12} from the gastrointestinal tract requires a protein called **intrinsic factor,** which is secreted by the stomach; lack of this protein also causes vitamin B_{12} deficiency.

Regulation of Erythrocyte Production In a normal person, the total volume of circulating erythrocytes remains remarkably constant because of reflexes that regulate the bone marrow's production of these cells.

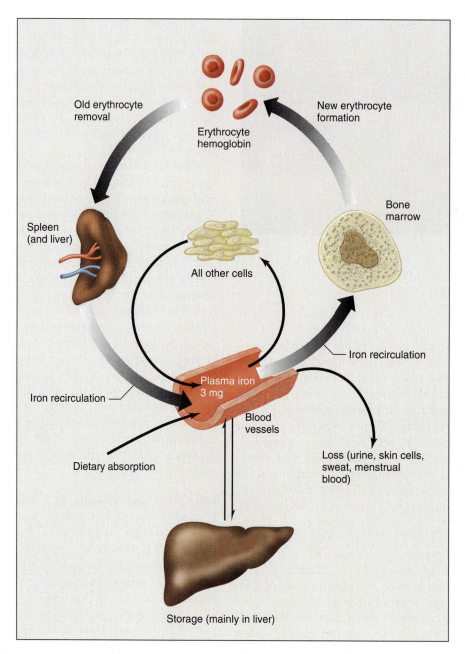

FIGURE 14–3

Summary of iron balance. The thickness of the arrows corresponds approximately to the amount of iron involved. In the steady state, the rate of gastrointestinal iron absorption equals the rate of iron loss via urine, skin, and menstrual flow.
Adapted from Crosby.

In the previous section, we stated that iron, folic acid, and vitamin B_{12} must be present for normal erythrocyte production. However, none of these substances constitutes the signal that *regulates* production rate.

The direct control of erythrocyte production (erythropoiesis) is exerted primarily by a hormone called **erythropoietin,** which is secreted into the blood mainly by a particular group of hormone-secreting connective-tissue cells in the kidneys (the liver also secretes this hormone, but to a much lesser extent).

Erythropoietin acts on the bone marrow to stimulate the proliferation of erythrocyte progenitor cells and their differentiation into mature erythrocytes.

Erythropoietin is normally secreted in relatively small amounts, which stimulate the bone marrow to produce erythrocytes at a rate adequate to replace the usual loss. The erythropoietin secretion rate is increased markedly above basal values when there is a decreased oxygen delivery to the kidneys. Situations in which this occurs include insufficient pumping of

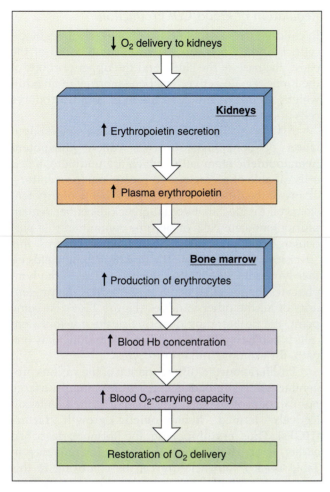

FIGURE 14–4

Reflex by which a decreased oxygen delivery to the kidneys increases erythrocyte production via increased erythropoietin secretion.

blood by the heart, lung disease, anemia (a decrease in number of erythrocytes or in hemoglobin concentration), and exposure to high altitude. As a result of the increase in erythropoietin secretion, plasma erythropoietin concentration, erythrocyte production, and the oxygen-carrying capacity of the blood all increase; therefore, oxygen delivery to the tissues returns toward normal (Figure 14–4).

Testosterone, the male sex hormone, also stimulates the release of erythropoietin. This accounts, at least in part, for the higher hemoglobin concentration in men than in women.

Anemia *Anemia* is defined as a decrease in the ability of the blood to carry oxygen due to (1) a decrease in the total number of erythrocytes, each having a normal quantity of hemoglobin, or (2) a diminished concentration of hemoglobin per erythrocyte, or (3) a combination of both. Anemia has a wide variety of causes summarized in Table 14–2.

TABLE 14–2 Major Causes of Anemia
1. Dietary deficiencies of iron (**iron-deficiency anemia**), vitamin B$_{12}$, or folic acid
2. Bone marrow failure due to toxic drugs or cancer
3. Blood loss from the body (**hemorrhage**) leading to iron deficiency
4. Inadequate secretion of erythropoietin in kidney disease
5. Excessive destruction of erythrocytes (for example, sickle-cell anemia)

Sickle-cell anemia is due to a genetic mutation that alters one amino acid in the hemoglobin chain. At the low oxygen concentrations existing in many capillaries, the abnormal hemoglobin molecules interact with each other to form fiberlike structures that distort the erythrocyte membrane and cause the cell to form sickle shapes or other bizarre forms. This results both in the blockage of capillaries, with consequent tissue damage and pain, and in the destruction of the deformed erythrocytes, with consequent anemia. Sickle-cell anemia is an example of a disease that is manifested fully only in persons homozygous for the mutated gene (Chapter 5). In heterozygotes, who are said to have sickle-cell "trait," the normal allele codes for normal hemoglobin and the mutated allele for the abnormal hemoglobin. The erythrocytes in this case contain both types of hemoglobin, but symptoms are manifest only when the oxygen concentration is unusually low, as at high altitude.

Finally, there also exist conditions in which the problem is just the opposite of anemia, namely, more erythrocytes than normal; this is termed *polycythemia.* An example, to be described in Chapter 15, is the polycythemia that occurs in high-altitude dwellers; in this case the increased number of erythrocytes is an adaptive response. As we shall see later, however, the existence of polycythemia makes the flow of blood through blood vessels more difficult.

Leukocytes

If appropriate dyes are added to a drop of blood, which is then examined under a microscope, the various classes of leukocytes (Table 14–3) are clearly visible (Figure 14–5). They are classified according to their structure and affinity for the various dyes.

The name **polymorphonuclear granulocytes** refers to the three classes of leukocytes that have multilobed nuclei and abundant membrane-surrounded granules. The granules of one group take up the red dye eosin, thus giving the cells their name **eosinophils.** Cells of a second class have an affinity for a blue dye

TABLE 14–3 Numbers and Distributions of Erythrocytes, Leukocytes, and Platelets in Normal Human Blood

Total erythrocytes = 5,000,000 per mm^3 of blood
Total leukocytes = 7000 per mm^3 of blood
  Percent of total leukocytes:
    Polymorphonuclear granulocytes
      Neurtrophils 50–70%
      Eosinophils 1–4%
      Basophils 0.1%
    Monocytes 2–8%
    Lymphocytes 20–40%
Total platelets = 250,000 per mm^3 of blood

termed a "basic" dye and are called **basophils.** The granules of the third class have little affinity for either dye and are therefore called **neutrophils.** Neutrophils are by far the most abundant kind of leukocytes.

A fourth class of leukocyte is the **monocyte,** which is somewhat larger than the granulocyte and has a single oval or horseshoe-shaped nucleus and relatively few cytoplasmic granules. The final class of leukocytes is the **lymphocyte,** which contains scanty cytoplasm and, like the monocyte, a single relatively large nucleus.

Like the erythrocytes, all classes of leukocytes are produced in the bone marrow. In addition, monocytes and many lymphocytes undergo further development and cell division in tissues outside the bone marrow. These events and the specific leukocyte functions in the body's defenses are discussed in Chapter 20.

Platelets

The circulating platelets are colorless cell fragments that contain numerous granules and are much smaller than erythrocytes. Platelets are produced when cytoplasmic portions of large bone marrow cells, termed **megakaryocytes,** become pinched off and enter the circulation. Platelet functions in blood clotting are described in Section G of this chapter.

Regulation of Blood Cell Production

In children the marrow of most bones produces blood cells. By adulthood, however, only the bones of the chest, the base of the skull, and the upper portions of the limbs remain active. The bone marrow in an adult weighs almost as much as the liver, and it produces cells at an enormous rate.

All blood cells are descended from a single population of bone marrow cells called **pluripotent hematopoietic stem cells,** which are undifferentiated cells capable of giving rise to precursors (progenitors) of any of the different blood cells. When a pluripotent stem cell divides, its two daughter cells either remain pluripotent stem cells or become committed to a particular developmental pathway; what governs this "decision" is not known. The first branching yields either lymphoid stem cells, which give rise to the lymphocytes, or so-called myeloid stem cells, the progenitors of all the other varieties (Figure 14–6). At some point, the proliferating offspring of the myeloid stem cells become committed to differentiate along only one path, for example into erythrocytes.

Proliferation and differentiation of the various progenitor cells is stimulated, at multiple points, by a large number of protein hormones and paracrine agents collectively termed **hematopoietic growth factors (HGFs).** Thus, erythropoietin, the hormone described earlier, is an HGF. Others are listed for reference in Table 14–4. (Nomenclature can be confusing in this area since the HGFs belong to a still larger general family of messengers termed "cytokines," which are described in Chapter 20.)

The physiology of the HGFs is very complex because (1) there are so many of them, (2) any given HGF is often produced by a variety of cell types throughout the body, and (3) HGFs often exert other actions in addition to stimulating blood-cell production. There are, moreover, many interactions of the HGFs on particular bone marrow cells and processes. For example, although erythropoietin is the major stimulator of erythropoiesis, at least 10 other HGFs cooperate in the

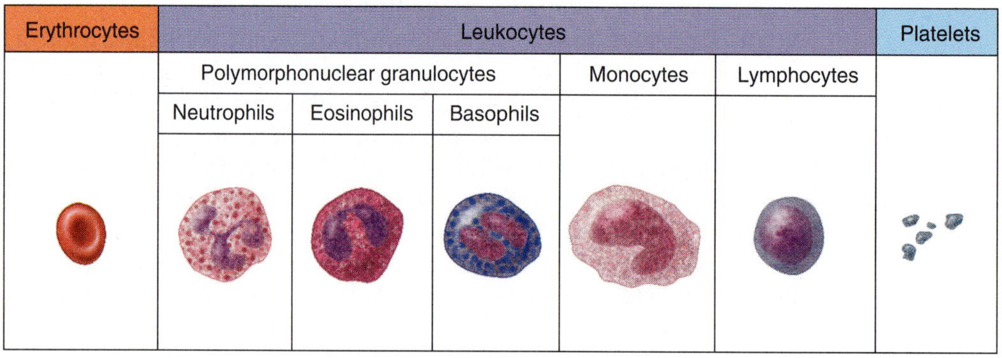

Erythrocytes	Leukocytes					Platelets
	Polymorphonuclear granulocytes			Monocytes	Lymphocytes	
	Neutrophils	Eosinophils	Basophils			

FIGURE 14–5
Classes of blood cells.

TABLE 14–4 Reference Table of Major Hematopoietic Growth Factors (HGFs)

Name	Stimulates Progenitor Cells Leading to:
Erythropoietin	Erythrocytes
Colony-stimulating factors (CSFs) (example: granulocyte CSF)	Granulocytes and monocytes
Interleukins (example: interleukin 3)	Various leukocytes
Thrombopoietin	Platelets (from megakaryocytes)
Stem cell factor	Many blood-cell types

process. Finally, in several cases the HGFs not only stimulate differentiation and proliferation of progenitor cells, they inhibit the usual programmed death (apoptosis, Chapter 7) of these cells.

The administration of specific HGFs is proving to be of considerable clinical importance. Examples are the use of erythropoietin in persons having a deficiency of this hormone due to kidney disease, and the use of granulocyte colony stimulating factor (G-CSF) to stimulate granulocyte production in individuals whose bone marrow has been damaged by anticancer drugs.

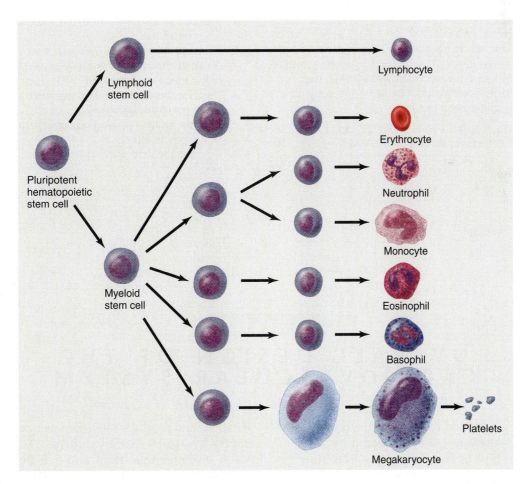

FIGURE 14–6

Production of blood cells by the bone marrow. For simplicity, no attempt has been made to differentiate the appearance of the various precursors.

Adapted form Golde and Gasson.

Blood

I. Blood is composed of cells (erythrocytes, leukocytes, and platelets) and plasma, the liquid in which the cells are suspended.

II. Plasma contains proteins (albumins, globulins, and fibrinogen), nutrients, metabolic end products, hormones, and inorganic electrolytes.

III. Erythrocytes, which make up more than 99 percent of blood cells, contain hemoglobin, an oxygen-binding protein. Oxygen binds to the iron in hemoglobin.
   a. Erythrocytes are produced in the bone marrow and destroyed in the spleen and liver.
   b. Iron, folic acid, and vitamin B_{12} are essential for erythrocyte formation.
   c. The hormone erythropoietin, which is produced by the kidneys in response to low oxygen supply, stimulates erythrocyte differentiation and production by the bone marrow.

IV. The leukocytes include three classes of polymorphonuclear granulocytes (neutrophils, eosinophils, and basophils), monocytes, and lymphocytes.

V. Platelets are cell fragments essential for blood clotting.

VI. Blood cells are descended from stem cells in the bone marrow. Their production is controlled by hematopoietic growth factors.

circulatory system	blood
blood vessel	plasma
vascular system	erythrocyte
heart	leukocyte
cardiovascular system	platelet

hematocrit	intrinsic factor
bilirubin	erythropoietin
plasma proteins	polymorphonuclear
albumin	granulocyte
globulin	eosinophil
fibrinogen	basophil
serum	neutrophil
hemoglobin	monocyte
bone marrow	lymphocyte
reticulocyte	megakaryocyte
iron	pluripotent hematopoietic
ferritin	stem cell
transferrin	hematopoietic growth
folic acid	factors (HGFs)
vitamin B_{12}	

1. Give average values for total blood volume, erythrocyte volume, plasma volume, and hematocrit.
2. Which is the most abundant class of plasma protein?
3. Which solute is found in the highest concentration in plasma?
4. Summarize the production, life span, and destruction of erythrocytes.
5. What are the routes of iron gain, loss, and distribution, and how is iron recycled when erythrocytes are destroyed?
6. Describe the control of erythropoietin secretion and the effect of this hormone.
7. State the relative proportions of erythrocytes and leukocytes in blood.
8. Diagram the derivation of the different blood cell lines.

SECTION B

OVERALL DESIGN OF THE CARDIOVASCULAR SYSTEM

The rapid flow of blood throughout the body is produced by pressures created by the pumping action of the heart. This type of flow is known as **bulk flow** because all constituents of the blood move in one direction together. The extraordinary degree of branching of blood vessels ensures that almost all cells in the body are within a few cell diameters of at least one of the smallest branches, the capillaries. Nutrients and metabolic end products move between capillary blood and the interstitial fluid by diffusion. Movements between the interstitial fluid and cell interior are accomplished by both diffusion and mediated transport.

At any given moment, approximately 5 percent of the total circulating blood is actually in the capillaries. Yet it is this 5 percent that is performing the ultimate functions of the entire cardiovascular system: the supplying of nutrients and the removal of metabolic end products. All other components of the system subserve the overall aim of getting adequate blood flow through the capillaries. This point should be kept in mind as we describe these components.

As discovered by the British physiologist William Harvey in 1628, the cardiovascular system forms a circle, so that blood pumped out of the heart through one set of vessels returns to the heart via a different set. There are actually two circuits (Figure 14–7), both originating and terminating in the heart, which is divided longitudinally into two functional halves. Each half contains two chambers: an upper chamber—the **atrium**—and a lower chamber—the **ventricle.** The atrium on each side empties into the ventricle on that side, but there is no direct communication between the two atria or the two ventricles in the adult heart.

Blood is pumped via one circuit, the **pulmonary circulation,** from the right ventricle through the lungs and then to the left atrium. It is then pumped through the **systemic circulation** from the left ventricle through all the organs and tissues of the body except the lungs and then to the right atrium. In both circuits, the vessels carrying blood *away from the heart* are called **arteries,** and those carrying blood from either the lungs or all other parts of the body (peripheral organs and tissues) back *toward the heart* are called **veins.**

In the systemic circuit, blood leaves the left ventricle via a single large artery, the **aorta** (Figure 14–8). The arteries of the systemic circulation branch off the aorta, dividing into progressively smaller vessels. The smallest arteries branch into **arterioles,** which branch into a huge number (estimated at 10 billion) of very small vessels, the **capillaries,** which unite to form larger diameter vessels, the **venules.** The arterioles, capillaries, and venules are collectively termed the **microcirculation.**

The venules in the systemic circulation then unite to form larger vessels, the veins. The veins from the various peripheral organs and tissues unite to produce two large veins, the **inferior vena cava,** which collects blood from the lower portion of the body, and the **superior vena cava,** which collects blood from the upper half of the body. It is via these two veins that blood is returned to the right atrium.

The pulmonary circulation is composed of a similar circuit. Blood leaves the right ventricle via a single large artery, the **pulmonary trunk,** which divides into the two **pulmonary arteries,** one supplying the right lung and the other the left. In the lungs, the arteries continue to branch, ultimately forming capillaries that unite into venules and then veins. The blood leaves the lungs via four **pulmonary veins,** which empty into the left atrium.

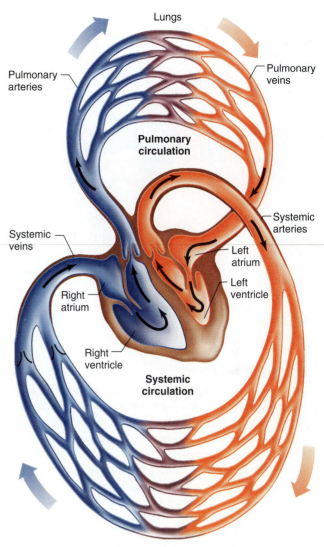

FIGURE 14–7

The systemic and pulmonary circulations. As depicted by the color change from blue to red, blood becomes fully oxygenated (red) as it flows through the lungs and then loses some oxygen (red to blue) as it flows through the other organs and tissues. For simplicity, the arteries and veins leaving and entering the heart are depicted as single vessels; in reality, this is true for the arteries but not for the veins (see Figure 14–8).

As blood flows through the lung capillaries, it picks up oxygen supplied to the adjacent lung air sacs by breathing. Therefore, the blood in the pulmonary veins, left heart, and systemic arteries has a high oxygen content. As this blood flows through the capillaries of peripheral tissues and organs, some of this oxygen leaves the blood to enter and be used by cells, resulting in the lower oxygen content of systemic venous blood.

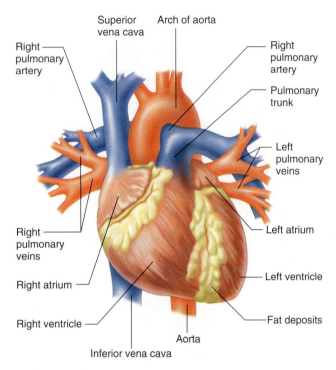

Superior vena cava

Arch of aorta

Right pulmonary artery

Right pulmonary artery

Pulmonary trunk

Left pulmonary veins

Right pulmonary veins

Left atrium

Right atrium

Left ventricle

Right ventricle

Fat deposits

Aorta

Inferior vena cava

FIGURE 14–8

Blood leaves each of the ventricles via a single artery, the pulmonary trunk from the right ventricle, and the aorta from the left ventricle. [Because the aorta and pulmonary trunk cross each other before emerging from the heart (see Figure 14–12), one gets the mistaken notion that both arise from the right ventricle.] Blood enters the right atrium via two large veins, the superior vena cava and inferior vena cava; it enters the left atrium via four pulmonary veins. ✗

Organ	Rest ml/min
Brain	650 (13%)
Heart	215 (4%)
Skeletal muscle	1030 (20%)
Skin	430 (9%)
Kidney	950 (20%)
Abdominal organs	1200 (24%)
Other	525 (10%)
Total	5000 (100%)

FIGURE 14–9

Distribution of systemic blood flow to the various organs and tissues of the body at rest.

Adapted from Chapman and Mitchell.

As shown in Figure 14–7, blood can pass from the systemic veins to the systemic arteries only by first being pumped through the lungs. Thus the blood returning from the body's peripheral organs and tissues via the systemic veins is oxygenated before it is pumped back to them.

It must be emphasized that the lungs receive all the blood pumped by the right heart, whereas each of the peripheral organs and tissues receives only a fraction of the blood pumped by the left ventricle. For reference, the typical distribution of the blood pumped by the left ventricle in an adult at rest is given in Figure 14–9.

Finally, there are several exceptions, notably the liver, kidneys, and pituitary, to the usual anatomical pattern described in this section for the systemic circulation, and these will be presented in the relevant chapters dealing with those organs.

Pressure, Flow, and Resistance

The last task in this survey of the design of the cardiovascular system is to introduce the concepts of pressure, flow, and resistance. In all parts of the system, blood flow (F) is always from a region of higher pressure to one of lower pressure. The pressure exerted by any fluid is termed a **hydrostatic pressure,** but this is usually shortened simply to "pressure" in descriptions of the cardiovascular system and denotes the force exerted by the blood. This force is generated in the blood by the contraction of the heart, and its magnitude varies throughout the system for reasons to be described in subsequent sections. The units for the rate of flow are volume per unit time, usually liters per minute (L/min). The units for the pressure difference (ΔP) driving the flow are millimeters of mercury (mmHg) because historically blood pressure was measured by determining how high a column of mercury could be driven by the blood pressure.

It must be emphasized that it is not the absolute pressure at any point in the cardiovascular system that determines flow rate but the *difference* in pressure between the relevant points (Figure 14–10).

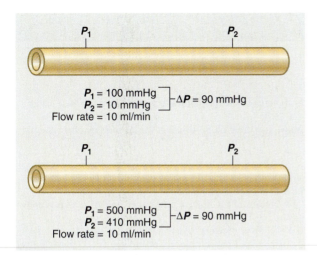

FIGURE 14–10
Flow between two points within a tube is proportional to the pressure difference between the points. The flows in these two identical tubes are the same (10 ml/min was selected arbitrarily), because the pressure differences are the same.

Knowing only the pressure difference between two points will not tell you the flow rate, however. For this, you also need to know the **resistance** (R) to flow—that is, how difficult it is for blood to flow between two points *at any given pressure difference*. Resistance is the measure of the friction that impedes flow. The basic equation relating these variables is:

$$F = \Delta P/R \qquad (14\text{–}1)$$

In words, flow rate is directly proportional to the pressure difference between two points and inversely proportional to the resistance. This equation applies not only to the cardiovascular system but to any system in which liquid or air moves by bulk flow (for example, in the urinary and respiratory systems, respectively).

Resistance cannot be measured directly, but it can be calculated from the directly measured F and ΔP. For example, in Figure 14–10 the resistances in both tubes can be calculated to be 90 mmHg ÷ 10 ml/min = 9 mmHg/ml per minute.

This example illustrates how resistance can be *calculated*, but what is it that actually *determines* the resistance? (The distinction between how a thing is calculated or measured and its determinants may seem confusing, but consider the following: By standing on a scale you *measure* your weight, but your weight is not *determined* by the scale but rather by how much you eat and exercise, and so on.) One determinant of resistance is the fluid property known as **viscosity,** which is a function of the friction between adjacent layers of a flowing fluid; the greater the friction, the greater the viscosity. The other determinants of resistance are the length and radius of the tube through which the fluid is flowing, since these characteristics determine the amount of friction between the fluid and the wall of the tube. The following equation defines the contributions of these three determinants:

$$R = (\eta L/r^4)(8/\pi) \qquad (14\text{–}2)$$

where η = fluid viscosity
L = length of the tube
r = inside radius of the tube
$8/\pi$ = a constant

In other words, resistance is directly proportional to both the fluid viscosity and the structure's length, and inversely proportional to the fourth power of the structure's radius (that is, the radius multiplied by itself four times).

Blood viscosity is not fixed but increases as hematocrit increases, and changes in hematocrit, therefore, can have significant effects on the resistance to flow in certain situations. Under most physiological conditions, however, the hematocrit and, hence, viscosity of blood is relatively constant and does not play a role in the *control* of resistance.

Similarly, since the lengths of the blood vessels remain constant in the body, length is also not a factor in the control of resistance along these vessels. In contrast, as we shall see, the radii of the blood vessels do not remain constant, and so vessel radius—the $1/r^4$ term in our equation—is the most important determinant of changes in resistance along the blood vessels. Just how important changes in radius can be is illustrated in Figure 14–11: Decreasing the radius of a tube *twofold* increases its resistance *sixteenfold.* If ΔP is held constant in this example, flow through the tube decreases sixteenfold since $F = \Delta P/R$.

Because resistance in the cardiovascular system is so often discussed in the context of blood vessels—the "tubes"—it is easy to forget that the equation relating pressure, flow, and resistance applies not only to flow through blood vessels but to the flows into and out of the various chambers of the heart. As we shall see, these flows occur through valves, and the resistance offered by a valvular opening determines the flow through the valve at any given pressure difference across it.

This completes our introductory survey of the cardiovascular system. We now turn to a description of its components and their control. In so doing, we might very easily lose sight of the forest for the trees if we do not persistently ask of each section: How does this component of the circulation contribute to adequate blood flow through the capillaries of the various organs or to an adequate exchange of materials between blood and cells? Refer to the summary in Table 14–5 as you read the description of each component to keep focused on this question.

TABLE 14–5 The Cardiovascular System

Component	Function
Heart	
Atria	Chambers through which blood flows from veins to ventricles. Atrial contraction adds to ventricular filling but is not essential for it.
Ventricles	Chambers whose contractions produce the pressures that drive blood through the pulmonary and systemic vascular systems and back to the heart.
Vascular system	
Arteries	Low-resistance tubes conducting blood to the various organs with little loss in pressure. They also act as pressure reservoirs for maintaining blood flow during ventricular relaxation.
Arterioles	Major sites of resistance to flow; responsible for the pattern of blood-flow distribution to the various organs; participate in the regulation of arterial blood pressure.
Capillaries	Sites of nutrient, metabolic end product, and fluid exchange between blood and tissues.
Venules	Sites of nutrient, metabolic end product, and fluid exchange between blood and tissues.
Veins	Low-resistance conduits for blood flow back to the heart. Their capacity for blood is adjusted to facilitate this flow.

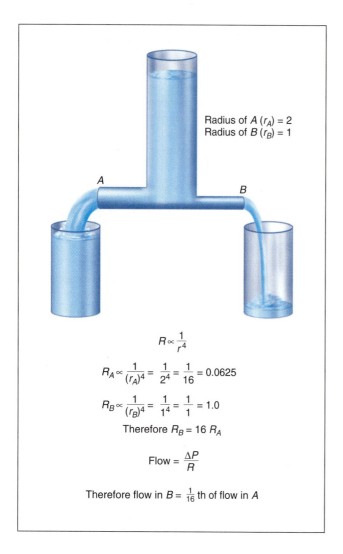

Radius of A (r_A) = 2
Radius of B (r_B) = 1

$$R \propto \frac{1}{r^4}$$

$$R_A \propto \frac{1}{(r_A)^4} = \frac{1}{2^4} = \frac{1}{16} = 0.0625$$

$$R_B \propto \frac{1}{(r_B)^4} = \frac{1}{1^4} = \frac{1}{1} = 1.0$$

Therefore $R_B = 16\ R_A$

$$\text{Flow} = \frac{\Delta P}{R}$$

Therefore flow in $B = \frac{1}{16}$ th of flow in A

FIGURE 14–11

Effect of tube radius (r) on resistance (R) and flow.

S E C T I O N **B** S U M M A R Y

I. The cardiovascular system consists of two circuits: the pulmonary circulation, from the right ventricle to the lungs and then to the left atrium, and the systemic circulation, from the left ventricle to all peripheral organs and tissues and then to the right atrium.

II. Arteries carry blood away from the heart, and veins carry blood toward the heart.

   a. In the systemic circuit, the large artery leaving the left heart is the aorta, and the large veins emptying into the right heart are the superior vena cava and inferior vena cava. The analogous vessels in the pulmonary circulation are the pulmonary trunk and the four pulmonary veins.

   b. The microcirculation consists of the vessels between arteries and veins: the arterioles, capillaries, and venules.

III. Flow between two points in the cardiovascular system is directly proportional to the pressure difference between the points and inversely proportional to the resistance: $F = \Delta P / R$.

IV. Resistance is directly proportional to the viscosity of a fluid and to the length of the tube. It is inversely proportional to the fourth power of the tube's radius, which is the major variable controlling changes in resistance.

S E C T I O N **B** K E Y T E R M S

bulk flow	venule
atrium	microcirculation
ventricle	inferior vena cava
pulmonary circulation	superior vena cava
systemic circulation	pulmonary trunk
artery	pulmonary arteries
vein	pulmonary veins
aorta	hydrostatic pressure
arteriole	resistance (R)
capillary	viscosity

S E C T I O N **B** R E V I E W Q U E S T I O N S

1. State the formula relating flow, pressure difference, and resistance.
2. What are the three determinants of resistance?

SECTION C

THE HEART

Anatomy

The heart is a muscular organ enclosed in a fibrous sac, the **pericardium,** and located in the chest (thorax). The extremely narrow space between the pericardium and the heart is filled with a watery fluid that serves as a lubricant as the heart moves within the sac.

The walls of the heart are composed primarily of cardiac muscle cells and are termed the **myocardium.** The inner surface of the walls—that is, the surface in contact with the blood within the cardiac chambers—is lined by a thin layer of cells known as **endothelial cells,** or **endothelium.** (As we shall see, endothelial cells line not only the heart chambers, but the entire cardiovascular system.)

As noted earlier, the human heart is divided into right and left halves, each consisting of an atrium and a ventricle. Located between the atrium and ventricle in each half of the heart are the **atrioventricular (AV) valves,** which permit blood to flow from atrium to ventricle but not from ventricle to atrium (Figure 14–12). The right AV valve is called the **tricuspid valve,** and the left is called the **mitral valve.**

The opening and closing of the AV valves is a passive process resulting from pressure differences across the valves. When the blood pressure in an atrium is greater than that in the ventricle separated from it by a valve, the valve is pushed open and flow proceeds from atrium to ventricle. In contrast, when a contracting ventricle achieves an internal pressure greater than that in its connected atrium, the AV valve between them is forced closed. Therefore, blood does not normally move back into the atria but is forced into the pulmonary trunk from the right ventricle and into the aorta from the left ventricle.

To prevent the AV valves from being pushed up into the atrium, the valves are fastened to muscular projections (**papillary muscles**) of the ventricular walls by fibrous strands (chordae tendinae). The papillary muscles do *not* open or close the valves. They act only to limit the valves' movements and prevent them from being everted.

The opening of the right ventricle into the pulmonary trunk and of the left ventricle into the aorta also contain valves, the **pulmonary** and **aortic valves,** respectively (Figure 14–12) (these valves are also col-

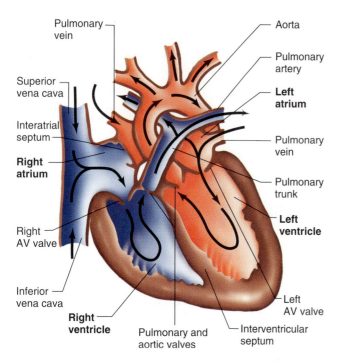

FIGURE 14–12

Diagrammatic section of the heart. The arrows indicate the direction of blood flow.

lectively referred to as the semilunar valves). These valves permit blood to flow into the arteries during ventricular contraction but prevent blood from moving in the opposite direction during ventricular relaxation (Figure 14–13). Like the AV valves, they act in a purely passive manner. Their being open or closed depends upon the pressure differences across them.

Another important point concerning the heart valves is that, when open, they offer very little resistance to flow. Accordingly, very small pressure differences across them suffice to produce large flows. In disease states, however, a valve may become narrowed so that even when open it offers a high resistance to flow. In such a state, the contracting cardiac chamber must produce an unusually high pressure to cause flow across the valve.

There are no valves at the entrances of the superior and inferior venae cavae (plural of vena cava) into the right atrium, and of the pulmonary veins into the left atrium. However, atrial contraction pumps very little blood back into the veins because atrial contraction compresses the veins at their sites of entry into the atria, greatly increasing the resistance to backflow. (Actually, a little blood is ejected back into the veins, and this accounts for the venous pulse that can often be seen in the neck veins when the atria are contracting.)

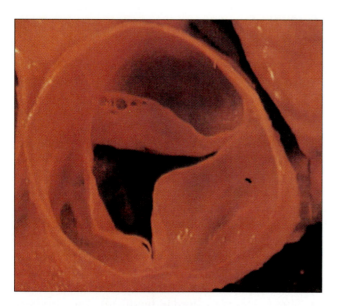

Valve partly open

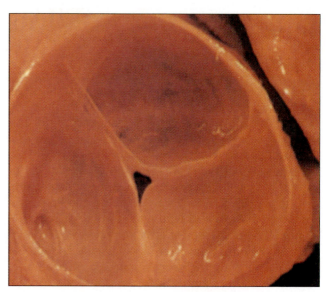

Valve almost completely closed

FIGURE 14–13

Photographs of the pulmonary valve viewed from the top—that is, from the pulmonary trunk looking down into the right ventricle. On the left the valve is in the process of opening as blood flows through it from the right ventricle into the pulmonary trunk—that is, toward the viewer. On the right, the valve is in the process of closing, the cusps being forced together by the downward pressure of the blood—that is, by the pressure of the blood in the pulmonary trunk being greater than the pressure in the right ventricle.

From R. Carola, J. P. Harley, and C. R. Noback, "Human Anatomy and Physiology," McGraw-Hill, New York, 1990 (photos by Dr. Wallace McAlpine).

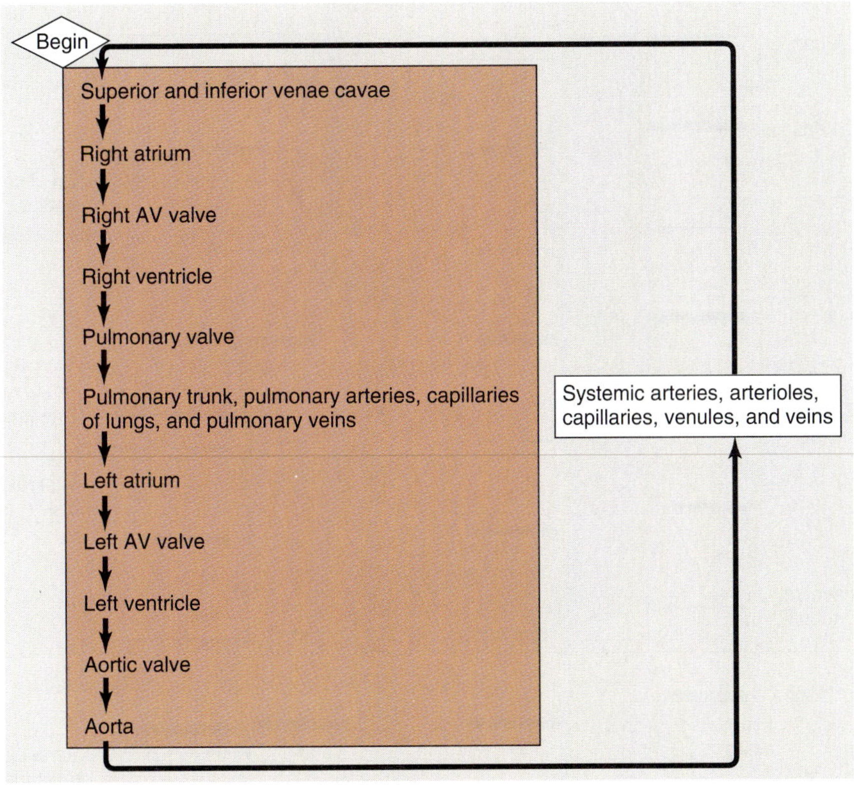

FIGURE 14–14
Path of blood flow through the entire cardiovascular system. All the structures within the colored box are located in the chest.

Figure 14–14 summarizes the path of blood flow through the entire cardiovascular system.

Cardiac Muscle

The cardiac-muscle cells of the myocardium are arranged in layers that are tightly bound together and completely encircle the blood-filled chambers. When the walls of a chamber contract, they come together like a squeezing fist and exert pressure on the blood they enclose.

Cardiac muscle combines properties of both skeletal and smooth muscle (Chapter 11). The cells are striated (Figure 14–15) as the result of an arrangement of thick myosin and thin actin filaments similar to that of skeletal muscle. Cardiac-muscle cells are considerably shorter than skeletal-muscle fibers, however, and have several branching processes. Adjacent cells are joined end to end at structures called **intercalated disks,** within which are desmosomes that hold the cells together and to which the myofibrils are attached. Adjacent to the intercalated disks are gap junctions, similar to those in many smooth muscles.

Approximately 1 percent of the cardiac-muscle cells do not function in contraction, but have specialized features that are essential for normal heart excitation. These cells constitute a network known as the conducting system of the heart and are in contact with the other cardiac-muscle cells via gap junctions. The conducting system initiates the heartbeat and helps spread the impulse rapidly throughout the heart.

One final point about the cardiac-muscle cells is that certain cells in the atria secrete the family of peptide hormones collectively called atrial natriuretic factor, described in Chapter 16.

Innervation   The heart receives a rich supply of sympathetic and parasympathetic nerve fibers, the latter contained in the vagus nerves. The sympathetic postganglionic fibers release primarily norepinephrine, and the parasympathetics release primarily acetylcholine. The receptors for norepinephrine on cardiac muscle are mainly beta-adrenergic. The hormone epinephrine, from the adrenal medulla, combines with the same receptors as norepinephrine and exerts the same actions on the heart. The receptors for acetylcholine are of the muscarinic type.

Blood Supply   The blood being pumped through the heart chambers does not exchange nutrients and metabolic end products with the myocardial cells. They, like the cells of all other organs, receive their blood supply via arteries that branch from the aorta. The arteries

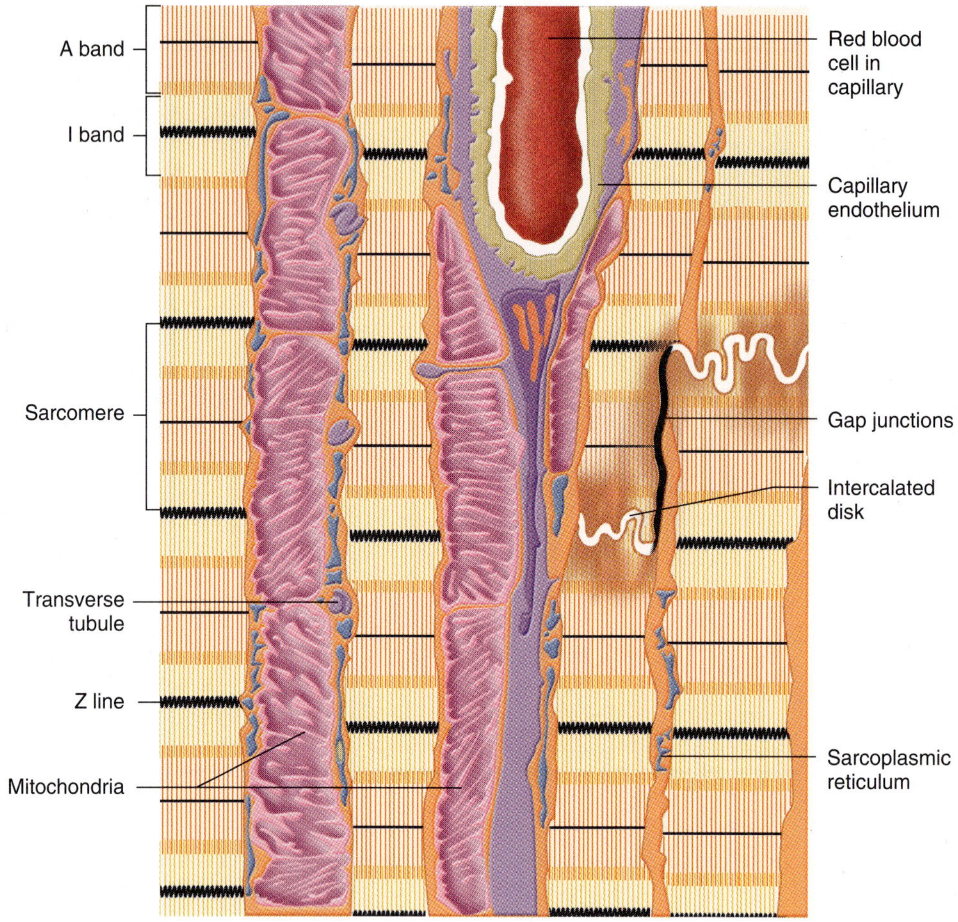

FIGURE 14–15

Diagram of an electron micrograph of cardiac muscle.

Adapted from R. M. Berne and M. N. Levy.

supplying the myocardium are the **coronary arteries,** and the blood flowing through them is termed the **coronary blood flow.** The coronary arteries exit from the very first part of the aorta and lead to a branching network of small arteries, arterioles, capillaries, venules, and veins similar to those in other organs. Most of the coronary veins drain into a single large vein, the coronary sinus, which empties into the right atrium.

Heartbeat Coordination

The heart is, in essence, a dual pump in that the atria contract first, followed almost immediately by the ventricles. Contraction of cardiac muscle, like that of skeletal muscle and many smooth muscles (Chapter 11), is triggered by depolarization of the plasma membrane. As described earlier, myocardial cells are connected to each other by gap junctions that allow action potentials to spread from one cell to another. Thus, the initial excitation of one cardiac cell eventually results in

the excitation of all cardiac cells. This initial depolarization normally arises in a small group of conducting-system cells, the **sinoatrial (SA) node,** located in the right atrium near the entrance of the superior vena cava (Figure 14–16). The action potential then spreads from the SA node throughout the atria and then into and throughout the ventricles. This pattern raises two questions: (1) What causes the SA node to "fire," and (2) precisely what is the path of spread of excitation? We'll deal initially with the second question and then return to the first question in the next section.

Sequence of Excitation

To reiterate, the SA node is the normal pacemaker for the entire heart. Its depolarization normally generates the current that leads to depolarization of all other cardiac muscle cells, and so its discharge rate determines the **heart rate,** the number of times the heart contracts per minute.

The action potential initiated in the SA node spreads throughout the myocardium, passing from cell to cell by way of gap junctions. The spread

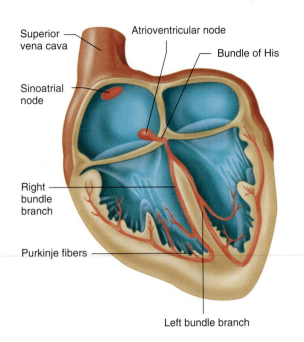

FIGURE 14–16

Conducting system of the heart. ⚡

throughout the right atrium and from the right atrium to the left atrium does not depend on fibers of the conducting system. The spread is rapid enough that the two atria are depolarized and contract at essentially the same time.

The spread of the action potential to the ventricles is more complicated and involves the rest of the conducting system (Figures 14–16 and 14–17). The link between atrial depolarization and ventricular depolarization is a portion of the conducting system called the **atrioventricular (AV) node,** which is located at the base of the right atrium. The action potential spreading through the right atrium causes depolarization of the AV node. This node manifests a particularly important characteristic: For several reasons related to the electrical properties of the AV-node cells, the propagation of action potentials through the AV node is relatively slow (requiring approximately 0.1 s). This delay allows atrial contraction to be completed before ventricular excitation occurs.

After leaving the AV node, the impulse enters the wall—the interventricular septum—between the two ventricles via the conducting-system fibers termed the **bundle of His** (or atrioventricular bundle) after its discoverer (pronounced Hiss). It should be emphasized that the AV node and the bundle of His constitute the only electrical link between the atria and the ventricles. There are no others because a layer of nonconducting connective tissue, pierced by the bundle of His, completely separates each atrium from its ventricle.

Within the interventricular septum the bundle of His divides into **right** and **left bundle branches,** which

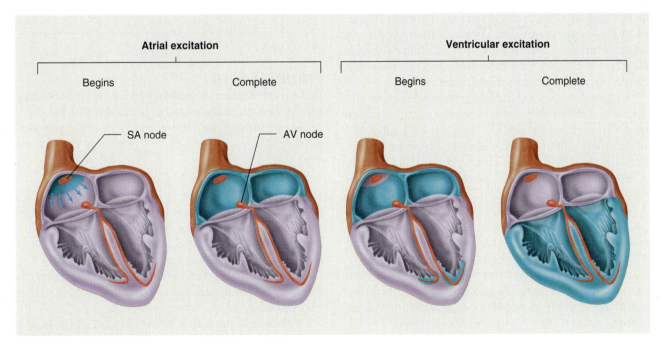

FIGURE 14–17

Sequence of cardiac excitation. The blue color denotes areas that are depolarized. Impulse spread from right atrium to left atrium is via the atrial muscle cells where the atria contact each other in their shared wall.

Adapted from Rushmer.

eventually leave the septum to enter the walls of both ventricles. These fibers in turn make contact with **Purkinje fibers,** large conducting cells that rapidly distribute the impulse throughout much of the ventricles. Finally, the Purkinje fibers make contact with nonconducting-system ventricular myocardial cells, via which the impulse spreads through the rest of the ventricles.

The rapid conduction along the Purkinje fibers and the diffuse distribution of these fibers cause depolarization of all right and left ventricular cells more or less simultaneously and ensure a single coordinated contraction. Actually, depolarization and contraction begin slightly earlier in the bottom (apex) of the ventricles and spread upward. The result is a more efficient contraction, like squeezing a tube of toothpaste from the bottom up.

Cardiac Action Potentials and Excitation of the SA Node

A typical ventricular myocardial cell action potential is illustrated in Figure 14–18a. The plasma-membrane permeability changes that underlie it are shown in Figure 14–18b. As in skeletal-muscle cells and neurons, the resting membrane is much more permeable to potassium than to sodium. Therefore, the resting membrane potential is much closer to the potassium equilibrium potential (-90 mV) than to the sodium equilibrium potential ($+60$ mV). Similarly, the depolarizing phase of the action potential is due mainly to a positive-feedback increase in sodium permeability caused by the opening of voltage-gated sodium channels; that is, the channels are opened by depolarization. At almost the same time, the permeability to potassium decreases as certain potassium channels close, and this also contributes to the membrane depolarization.

Again as in skeletal-muscle cells and neurons, the increased sodium permeability is very transient, since the sodium channels quickly close again. Unlike the case in these other excitable tissues, however, in cardiac muscle the return of sodium permeability toward its resting value is *not* accompanied by membrane repolarization. The membrane remains depolarized at a plateau of about 0 mV (Figure 14–18a). The reasons for this continued depolarization are (1) potassium permeability stays below the resting value (that is, the potassium channels mentioned above remain closed), and (2) there is a marked increase in the membrane permeability to *calcium*. The second reason is the more important of the two, and the explanation for it is as follows.

In myocardial cells, the original membrane depolarization causes voltage-gated calcium channels in the plasma membrane to open, which results in a flow of calcium ions down their electrochemical gradient into the cell. These channels are referred to as **slow**

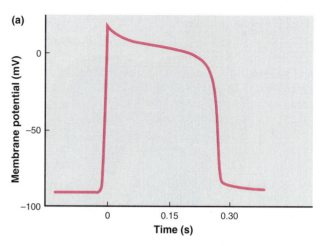

(a)

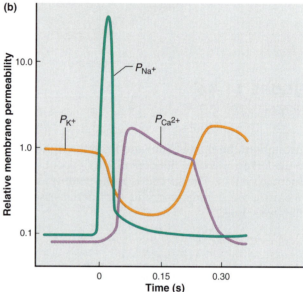

(b)

FIGURE 14–18

(a) Membrane potential recording from a ventricular muscle cell. (b) Simultaneously measured permeabilities P to potassium, sodium, and calcium during the action potential of (a).

channels because there is a delay in their opening (Figure 14–18b). The flow of positive calcium ions into the cell just balances the flow of positive potassium charge out of the cell and keeps the membrane depolarized at the plateau value.

Ultimately, repolarization does occur when the permeabilities of calcium and potassium return to their original state.

The action potentials of atrial cells, *except those of the SA node,* are similar in shape to those just described for ventricular cells, although the duration of their plateau phase is shorter.

In contrast, there are extremely important differences between action potentials of the vast majority of

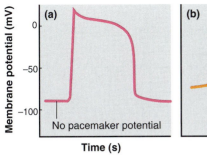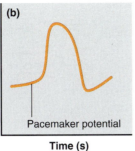

FIGURE 14–19

Comparison of action potentials in (a) a ventricular muscle cell (from Figure 14–18) and (b) a sinoatrial (SA)-nodal cell. The most important difference is the presence of the pacemaker potential in the SA node.

the atrial and ventricular myocardial cells, as just described, and those in the conducting system. Figure 14–19b illustrates the action potentials of a myocardial cell from the SA node. Note that the resting potential of the SA-node cell is not steady but instead manifests a slow depolarization. This gradual depolarization is known as a **pacemaker potential;** it brings the membrane potential to threshold, at which point an action potential occurs. Following the peak of the action potential, the membrane repolarizes, and the gradual depolarization begins again.

Thus, the pacemaker potential provides the SA node with **automaticity,** the capacity for spontaneous, rhythmical self-excitation. The slope of the pacemaker potential—that is, how quickly the membrane potential changes per unit time—determines how quickly threshold is reached and the next action potential elicited. The inherent rate of the SA node—the rate exhibited in the total absence of any neural or hormonal input to the node—is approximately 100 depolarizations per minute.

What is responsible for the pacemaker potential? There are multiple ion-permeability changes that contribute to this gradual depolarization. The most important one, however, is movement of sodium ions into the cells through a special set of voltage-gated plasma-membrane channels that are opened by the *repolarizing* phase of the *preceding* action potential. (Recall that the more common voltage-gated sodium channels of nerve, skeletal muscle, and nonconducting-system cardiac muscle are opened by *depolarization* occurring in the *on-going action potential*).

Several other portions of the conducting system are capable of generating pacemaker potentials, but the inherent rate of these other areas is slower than that of the SA node, and so they normally are "captured" by the SA node and do not manifest their own rhythm. However, they can do so under certain circumstances and are then termed *ectopic pacemakers,* an example of which is given in the next paragraph.

Recall that excitation travels from the SA node to both ventricles only through the AV node; therefore, drug- or disease-induced malfunction of the AV node may reduce or completely eliminate the transmission of action potentials from the atria to the ventricles. If this occurs, autorhythmic cells in the bundle of His, no longer driven by the SA node, begin to initiate excitation at their own inherent rate and become the pacemaker for the ventricles. Their rate is quite slow, generally 25 to 40 beats/min, and it is completely out of synchrony with the atrial contractions, which continue at the normal, higher rate of the SA node. Under such conditions, the atria are ineffective as pumps since they are often contracting against closed AV valves. Fortunately, atrial pumping, as we shall see, is relatively unimportant for cardiac function except during strenuous exercise.

The current treatment for all severe *AV conduction disorders,* as well as for many other abnormal rhythms is permanent surgical implantation of an electrical device, a *pacemaker,* that stimulates the ventricular cells at a normal rate.

The Electrocardiogram

The **electrocardiogram** (**ECG** or EKG—the *k* is from the German "kardio" for "heart") is primarily a tool for evaluating the *electrical* events within the heart. The action potentials of cardiac muscle cells can be viewed as batteries that cause charge to move throughout the body fluids. These moving charges—currents, in other words—are caused by all the action potentials occurring simultaneously in many individual myocardial cells and can be detected by recording electrodes at the surface of the skin. The top of Figure 14–20 illustrates a typical normal ECG recorded as the potential difference between the right and left wrists. The first deflection, the **P wave,** corresponds to current flows during atrial depolarization. The second deflection, the **QRS complex,** occurring approximately 0.15 s later, is the result of ventricular depolarization. It is a complex deflection because the paths taken by the wave of depolarization through the thick ventricular walls differ from instant to instant, and the currents generated in the body fluids change direction accordingly. Regardless of its form (for example, the Q and/or S portions may be absent), the deflection is still called a QRS complex. The final deflection, the **T wave,** is the result of ventricular repolarization. Atrial repolarization is usually not evident on the ECG because it occurs at the same time as the QRS complex.

A typical clinical ECG makes use of multiple combinations of recording locations on the limbs and chest so as to obtain as much information as possible

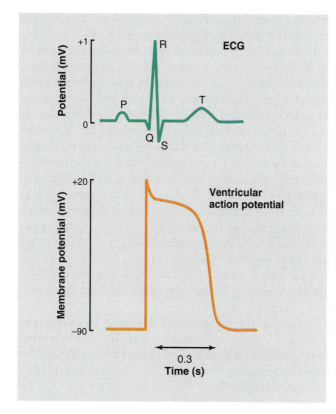

FIGURE 14–20

(Top) Typical electrocardiogram recorded from electrodes connecting the arms. P, atrial depolarization; QRS, ventricular depolarization; T, ventricular repolarization. (Bottom) Ventricular action potential recorded from a single ventricular muscle cell. Note the correspondence of the QRS complex with depolarization and the correspondence of the T wave with repolarization.

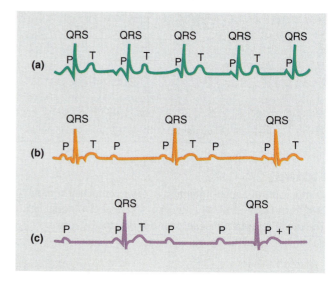

FIGURE 14–21

Electrocardiograms from a healthy person and from two persons suffering from atrioventricular block. (a) A normal ECG. (b) Partial block. Damage to the AV node permits only one-half of the atrial impulses to be transmitted to the ventricles. Note that every second P wave is not followed by a QRS and T. (c) Complete block. There is absolutely no synchrony between atrial and ventricular electrical activities, and the ventricles are being driven by a pacemaker in the bundle of His.

concerning different areas of the heart. The shapes and sizes of the P wave, QRS complex, and T wave vary with the electrode locations.

To reiterate, the ECG is not a direct record of the changes in membrane potential across individual cardiac muscle cells but is rather a measure of the currents generated in the extracellular fluid by the changes occurring simultaneously in many cardiac cells. To emphasize this point, the bottom of Figure 14–20 shows the simultaneously occurring changes in membrane potential in a single ventricular cell.

Because many myocardial defects alter normal impulse propagation, and thereby the shapes and timing of the waves, the ECG is a powerful tool for diagnosing certain types of heart disease. Figure 14–21 gives one example. It must be emphasized, however, that the ECG provides information concerning only the *electrical* activity of the heart. Thus, if something is wrong with the heart's mechanical activity, but this defect does not give rise to altered electrical activity, then the ECG will not be of diagnostic value.

Excitation-Contraction Coupling

As described in Chapter 11, the mechanism that couples excitation—an action potential in the plasma membrane of the muscle cell—and contraction is an increase in the cell's cytosolic calcium concentration. As is true for skeletal muscle, the increase in cytosolic calcium concentration in cardiac muscle is due mainly to release of calcium from the sarcoplasmic reticulum. This calcium combines with the regulator protein troponin, and cross-bridge formation between actin and myosin is initiated.

But there is a difference between skeletal and cardiac muscle in the sequence of events by which the action potential leads to increased release of calcium from the sarcoplasmic reticulum. In both muscle types, the plasma-membrane action potential spreads into the interior of muscle cells via the T tubules (the lumen of each tubule is continuous with the extracellular fluid). In skeletal muscle, as we saw in Chapter 11, the action potential in the T tubules then causes the direct opening of calcium channels in the sarcoplasmic reticulum adjacent to the T tubules. In cardiac muscle (Figure 14–22): (1) The action potential in the T tubule opens voltage-sensitive calcium channels in the T tubule membrane itself; calcium diffuses from the extracellular fluid through these channels into the cells, causing a small increase in cytosolic calcium concentration in

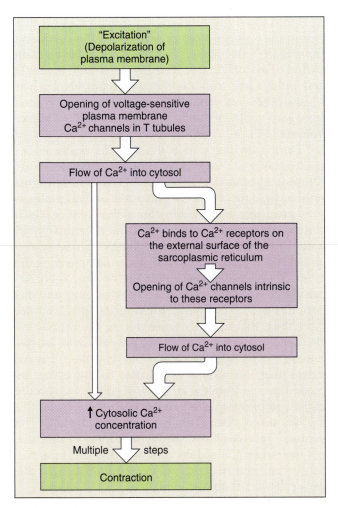

FIGURE 14–22
Excitation-contraction coupling in cardiac muscle.

the region of the T tubules and immediately adjacent sarcoplasmic reticulum. (2) This small increase in calcium concentration then causes calcium to bind to calcium receptors on the external surface of the sarcoplasmic reticulum membranes. (3) These calcium-sensitive receptors contain intrinsic calcium channels, and activation of the receptors opens the channels, allowing a large net diffusion of calcium from the calcium-rich interior of the sarcoplasmic reticulum into the cytosol (this is termed "calcium-induced calcium release"). (4) It is mainly this calcium that causes the contraction.

Thus, even though most of the calcium causing contraction comes from the sarcoplasmic reticulum, the process—unlike that in skeletal muscle—is dependent on the movement of *extracellular* calcium into the muscle, the calcium then acting as the signal for release of the sarcoplasmic-reticulum calcium.

Contraction ends when the cytosolic calcium concentration is restored to its original extremely low

value by active transport of calcium back into the sarcoplasmic reticulum. Also, an amount of calcium equal to the small amount that had entered the cell from the extracellular fluid during excitation is transported out of the cell, so that the total cellular calcium content remains constant. (The transport mechanisms involved in these movements offer an excellent review of key aspects of calcium transport described in Chapter 6. The transport into the sarcoplasmic reticulum is by primary active Ca-ATPase pumps; the transport across the plasma membrane is also by Ca-ATPase pumps plus Ca/Na exchangers.)

As we shall see, how much cytosolic calcium concentration increases during excitation is a major determinant of the strength of cardiac-muscle contraction. In this regard, cardiac muscle differs importantly from skeletal muscle, in which the increase in cytosolic calcium occurring during membrane excitation is always adequate to produce maximal "turning-on" of cross bridges by calcium binding to all troponin sites. In cardiac muscle, the amount of calcium released from the sarcoplasmic reticulum is not usually sufficient to saturate all troponin sites. Therefore, the number of active cross bridges and thus the strength of contraction can be increased still further if more calcium is released from the sarcoplasmic reticulum.

Refractory Period of the Heart

Ventricular muscle, unlike skeletal muscle, is incapable of any significant degree of summation of contractions, and this is a very good thing. Imagine that cardiac muscle were able to undergo a prolonged tetanic contraction. During this period, no ventricular filling could occur since filling can occur only when the ventricular muscle is relaxed, and the heart would therefore cease to function as a pump.

The inability of the heart to generate tetanic contractions is the result of the long absolute **refractory period** of cardiac muscle, defined as the period during and following an action potential when an excitable membrane cannot be re-excited. As described in Chapter 11, the absolute refractory periods of skeletal muscle are much shorter (1 to 2 ms) than the duration of contraction (20 to 100 ms), and a second contraction can therefore be elicited before the first is over (summation of contractions). In contrast, because of the long plateau in the cardiac-muscle action potential, the absolute refractory period of cardiac muscle lasts almost as long as the contraction (250 ms), and the muscle cannot be re-excited in time to produce summation (Figure 14–23).

In this and previous sections, we have presented various similarities and differences between cardiac and skeletal muscle. These were summarized in Table 11–6.

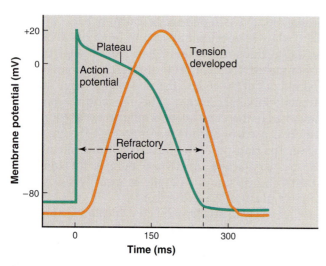

FIGURE 14–23

Relationship between membrane potential changes and contraction in a ventricular muscle cell. The refractory period lasts almost as long as the contraction.

Mechanical Events of the Cardiac Cycle

The orderly process of depolarization described in the previous sections triggers a recurring **cardiac cycle** of atrial and ventricular contractions and relaxations (Figure 14–24). For orientation, we shall first merely name the parts of this cycle and their key events. Then we shall go through the cycle again, this time describing the pressure and volume changes that cause the events.

The cycle is divided into two major phases, both named for events in the ventricles: the period of ventricular contraction and blood ejection, **systole,** followed by the period of ventricular relaxation and blood filling, **diastole.** At an average heart rate of 72 beats/min, each cardiac cycle lasts approximately 0.8 s, with 0.3 s in systole and 0.5 s in diastole.

As illustrated in Figure 14–24, both systole and diastole can be subdivided into two discrete periods. During the first part of systole, the ventricles are contracting but all valves in the heart are closed, and so no blood can be ejected. This period is termed **isovolumetric ventricular contraction** because the ventricular volume is constant. The ventricular walls are developing tension and squeezing on the blood they enclose, raising the ventricular blood pressure, but because the volume of blood in the ventricles is constant and because blood, like water, is essentially incompressible, the ventricular muscle fibers cannot shorten. Thus, isovolumetric ventricular contraction is analogous to an isometric skeletal-muscle contraction: the muscle develops tension, but does not shorten.

Once the rising pressure in the ventricles exceeds that in the aorta and pulmonary trunk, the aortic and pulmonary valves open, and the **ventricular ejection** period of systole occurs. Blood is forced into the aorta and pulmonary trunk as the contracting ventricular muscle fibers shorten. The volume of blood ejected from each ventricle during systole is termed the **stroke volume (SV).**

During the first part of diastole, the ventricles begin to relax, and the aortic and pulmonary valves close. (Physiologists and clinical cardiologists do not all agree on the dividing line between systole and diastole; as presented here, the dividing line is the point at which ventricular contraction stops and the pulmonary and aortic valves close.) At this time the AV valves are also closed. Accordingly, no blood is entering or leaving the ventricles since once again all the valves are closed. Accordingly, ventricular volume is not changing, and this period is termed **isovolumetric ventricular relaxation.** Note then, that the only times during the cardiac cycle that all valves are closed are the periods of isovolumetric ventricular contraction and relaxation. The AV valves then open, and **ventricular filling** occurs as blood flows in from the atria. *Atrial* contraction occurs at the end of diastole, after most of the ventricular filling has taken place. This is an important point: The ventricle receives blood throughout most of diastole, not just when the atrium contracts. Indeed, in a person at rest, approximately 80 percent of ventricular filling occurs *before* atrial contraction.

This completes the basic orientation. We can now analyze, using Figure 14–25, the pressure and volume changes that occur in the left atria, left ventricle, and aorta during the cardiac cycle. Events on the right side of the heart are described later. Electrical events (ECG) and heart sounds, the latter described in a subsequent section, are at the top of the figure so that their timing can be correlated with phases of the cycle.

Mid-Diastole to Late Diastole

Our analysis of events in the left atrium and ventricle, and the aorta begins at the far left of Figure 14–25 with the events of mid-diastole to late diastole. The left atrium and ventricle are both relaxed, but atrial pressure is very slightly higher than ventricular pressure. Because of this pressure difference, the AV valve is open, and blood entering the atrium from the pulmonary veins continues on into the ventricle. To reemphasize a point made earlier: All the valves of the heart offer very little resistance when they are open, and so only very small pressure differences across them are required to produce relatively large flows. Note that at this time—indeed, throughout all of diastole—the aortic valve is closed because the aortic pressure is higher than the ventricular pressure.

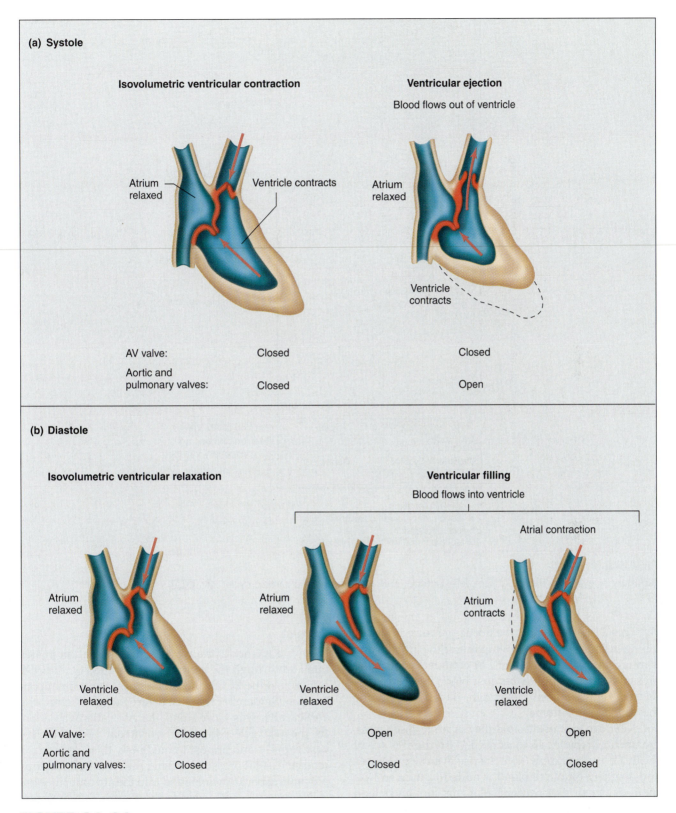

(a) Systole

Isovolumetric ventricular contraction

Atrium relaxed — Ventricle contracts

Ventricular ejection

Blood flows out of ventricle

Atrium relaxed

Ventricle contracts

AV valve:	Closed	Closed
Aortic and pulmonary valves:	Closed	Open

(b) Diastole

Isovolumetric ventricular relaxation

Atrium relaxed

Ventricle relaxed

Ventricular filling

Blood flows into ventricle

Atrial contraction

Atrium relaxed

Ventricle relaxed

Atrium contracts

Ventricle relaxed

AV valve:	Closed	Open	Open
Aortic and pulmonary valves:	Closed	Closed	Closed

FIGURE 14–24

Divisions of the cardiac cycle: (a) systole; (b) diastole. For simplicity, only one atrium and ventricle are shown. The phases of the cycle are identical in both halves of the heart. The direction in which the pressure difference *favors* flow is denoted by an arrow; note, however, that flow, although favored by a pressure difference, will not actually occur if a valve prevents it.

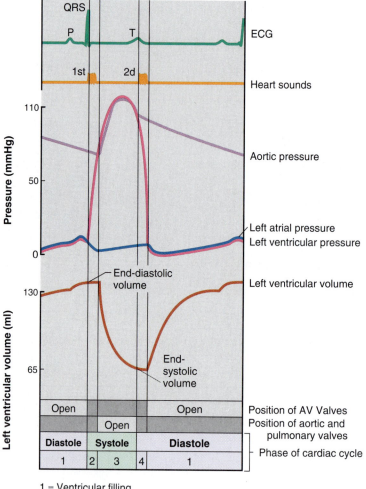

FIGURE 14–25

Summary of events in the left atrium, left ventricle, and aorta during the cardiac cycle.

Throughout diastole, the aortic pressure is slowly falling because blood is moving out of the arteries and through the vascular system. In contrast, ventricular pressure is rising slightly because blood is entering the relaxed ventricle from the atrium, thereby expanding the ventricular volume.

Near the end of diastole the SA node discharges, the atrium depolarizes (as signified by the P wave of the ECG) and contracts (note the rise in atrial pressure), and a small volume of blood is added to the ventricle (note the small rise in ventricular pressure and blood volume). The amount of blood in the ventricle at the end of diastole is called the **end-diastolic volume (EDV).**

Systole

From the AV node, the wave of depolarization passes into and through the ventricle (as signified by the QRS complex of the ECG), and this triggers ventricular contraction. Remember that just before the contraction, the aortic valve was closed and the AV valve was open. As the ventricle contracts, ventricular pressure rises very rapidly, and almost immediately this pressure exceeds the atrial pressure, closing the AV valve and thus preventing backflow of blood into the atrium. Since the aortic pressure still exceeds the ventricular pressure, the aortic valve remains closed, and the ventricle cannot empty despite its contraction.

This brief phase of isovolumetric ventricular contraction ends when the rapidly rising ventricular pressure exceeds aortic pressure. The aortic valve opens, and ventricular ejection occurs. The ventricular volume curve shows that ejection is rapid at first and then tapers off. Note that the ventricle does not empty completely. The amount of blood remaining after ejection is called the **end-systolic volume (ESV).** Thus:

Stroke volume = End-diastolic volume − End-systolic volume
     SV   EDV   ESV

As shown in Figure 14–25, normal values for an adult at rest are stroke volume = 70 ml, end-diastolic volume = 135 ml, and end-systolic volume = 65 ml.

As blood flows into the aorta, the aortic pressure rises along with the ventricular pressure. Throughout ejection, only very small pressure differences exist between the ventricle and aorta because the aortic valve opening offers little resistance to flow.

Note that peak ventricular and aortic pressures are reached before the end of ventricular ejection; that is, these pressures start to fall during the last part of systole despite continued ventricular contraction. This is because the strength of ventricular contraction and rate of blood ejection diminish during the last part of systole as shown by the ventricular volume curve. Therefore the ejection rate becomes less than the rate at which blood is leaving the aorta. Accordingly, the volume and therefore the pressure in the aorta begin to decrease.

Early Diastole

Diastole begins as ventricular contraction and ejection stop and the ventricular muscle begins to relax (recall that the T wave of the ECG corresponds to the end of the plateau phase of ventricular action potentials—that is, to the onset of ventricular repolarization). Immediately, the ventricular pressure falls significantly below aortic pressure, and the aortic valve closes. However, at this time, ventricular pressure still exceeds atrial pressure, so that the AV valve also remains closed. This early diastolic phase of isovolumetric ventricular relaxation ends as the rapidly decreasing ventricular pressure falls below atrial pressure, the AV valve opens, and rapid ventricular filling begins.

The ventricle's previous contraction compressed the elastic elements of this chamber in such a way that the ventricle actually tends to recoil outward once systole is over. This expansion, in turn, lowers ventricular pressure more rapidly than would otherwise occur and may even create a negative (subatmospheric) pressure in the ventricle, which enhances filling. Thus, some energy is stored within the myocardium during contraction, and its release during the subsequent relaxation aids filling.

The fact that ventricular filling is almost complete during early diastole is of the greatest importance. It ensures that filling is not seriously impaired during periods when the heart is beating very rapidly, and the duration of diastole and therefore total filling time are reduced. However, when rates of approximately 200 beats/min or more are reached, filling time does become inadequate, and the volume of blood pumped during each beat is decreased. The significance of this will be described in Section F.

Early ventricular filling also explains why the conduction defects that eliminate the atria as effective pumps do not seriously impair ventricular filling, at least in otherwise normal individuals at rest. This is true, for example, of *atrial fibrillation,* a state in which the cells of the atria contract in a completely uncoordinated manner and so fail to serve as effective pumps. Thus, the atrium may be conveniently viewed as merely a continuation of the large veins.

Pulmonary Circulation Pressures

The pressure changes in the right ventricle and pulmonary arteries (Figure 14–26) are qualitatively similar to those just described for the left ventricle and aorta. There are striking quantitative differences, however; typical pulmonary artery systolic and diastolic pressures are 24 and 8 mmHg, respectively, compared to systemic arterial pressures of 120 and 70 mmHg. Thus, the pulmonary circulation is a low-pressure system, for reasons to be described in a later section. This difference is clearly reflected in the ventricular architecture, the right ventricular wall being much thinner

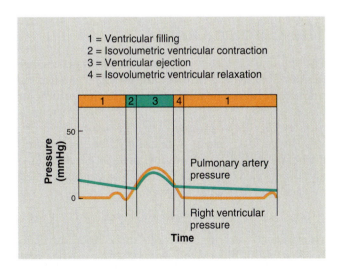

1 = Ventricular filling
2 = Isovolumetric ventricular contraction
3 = Ventricular ejection
4 = Isovolumetric ventricular relaxation

FIGURE 14–26

Pressures in the right ventricle and pulmonary artery during the cardiac cycle. This figure is done on the same scale as Figure 14–25 to facilitate comparison.

than the left. Despite its lower pressure during contraction, however, the right ventricle ejects the same amount of blood as the left over a given period of time. In other words, the stroke volumes of the two ventricles are identical.

Heart Sounds

Two sounds, termed **heart sounds,** stemming from cardiac contraction are normally heard through a stethoscope placed on the chest wall. The first sound, a soft low-pitched *lub,* is associated with closure of the AV valves at the onset of systole and isovolumetric ventricular contraction (see Figure 14–24); the second sound, a louder *dup,* is associated with closure of the pulmonary and aortic valves at the onset of diastole and isovolumetric ventricular relaxation (see Figure 14–24). These sounds, which result from vibrations caused by the closing valves, are perfectly normal, but other sounds, known as *heart murmurs,* are frequently a sign of heart disease.

Murmurs can be produced by blood flowing rapidly in the usual direction through an abnormally narrowed valve (*stenosis*), by blood flowing backward through a damaged, leaky valve (*insufficiency*), or by blood flowing between the two atria or two ventricles via a small hole in the wall separating them.

The exact timing and location of the murmur provide the physician with a powerful diagnostic clue. For example, a murmur heard throughout systole suggests a stenotic pulmonary or aortic valve, an insufficient AV valve, or a hole in the interventricular septum. In contrast, a murmur heard during diastole suggests a stenotic AV valve or an insufficient pulmonary or aortic valve.

The Cardiac Output

The volume of blood pumped by *each* ventricle per minute is called the **cardiac output (CO),** usually expressed in liters per minute. It is also the volume of blood flowing through *either* the systemic *or* the pulmonary circuit per minute.

The cardiac output is determined by multiplying the heart rate (HR)—the number of beats per minute—and the stroke volume (SV)—the blood volume ejected by each ventricle with each beat:

$$CO = HR \times SV$$

Thus, if each ventricle has a rate of 72 beats/min and ejects 70 ml of blood with each beat, the cardiac output is:

$$CO = 72 \text{ beats/min} \times 0.07 \text{ L/beat} = 5.0 \text{ L/min}$$

These values are within the normal range for a resting average-sized adult. Since, by coincidence, total blood

volume is also approximately 5 L, this means that essentially all the blood is pumped around the circuit once each minute. During periods of strenuous exercise in well-trained athletes, the cardiac output may reach 35 L/min; that is, the entire blood volume is pumped around the circuit seven times a minute. Even sedentary, untrained individuals can reach cardiac outputs of 20–25 L/min during exercise.

The following description of the factors that alter the two determinants of cardiac output—heart rate and stroke volume—applies in all respects to both the right and left heart since stroke volume and heart rate are the same for both under steady-state conditions. It must also be emphasized that heart rate and stroke volume do not always change in the same direction. For example, as we shall see, stroke volume decreases following blood loss while heart rate increases. These changes produce opposing effects on cardiac output.

Control of Heart Rate

Rhythmical beating of the heart at a rate of approximately 100 beats/min will occur in the complete absence of any nervous or hormonal influences on the SA node. This is, as we have seen, the inherent autonomous discharge rate of the SA node. The heart rate may be much lower or higher than this, however, since the SA node is normally under the constant influence of nerves and hormones.

As mentioned earlier, a large number of parasympathetic and sympathetic postganglionic fibers end on the SA node. Activity in the parasympathetic (vagus) nerves causes the heart rate to decrease, whereas activity in the sympathetic nerves increases the heart rate. In the resting state, there is considerably more parasympathetic activity to the heart than sympathetic, and so the normal resting heart rate of about 70 beats/min is well below the inherent rate of 100 beats/min.

Figure 14–27 illustrates how sympathetic and parasympathetic activity influences SA-node function. Sympathetic stimulation increases the slope of the pacemaker potential, causing the SA-node cells to reach threshold more rapidly and the heart rate to increase. Stimulation of the parasympathetics has the opposite effect—the slope of the pacemaker potential decreases, threshold is reached more slowly, and heart rate decreases. Parasympathetic stimulation also hyperpolarizes the plasma membrane of the SA-node cells so that the pacemaker potential starts from a more negative value.

How do the neurotransmitters released by the autonomic neurons change the slope of the potential? They mainly influence the special set of ion channels through which sodium ions move into the cell to cause the diastolic depolarization. Norepinephrine, the sympathetic neurotransmitter, enhances this current by

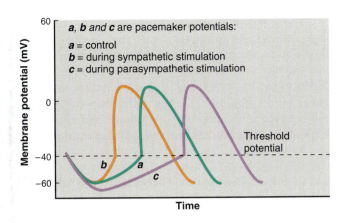

FIGURE 14–27

Effects of sympathetic and parasympathetic nerve stimulation on the slope of the pacemaker potential of an SA-nodal cell. Note that parasympathetic stimulation not only reduces the slope of the pacemaker potential but also causes the membrane potential to be more negative before the pacemaker potential begins.

Adapted from Hoffman and Cranefield.

opening more of these channels, whereas acetylcholine, the parasympathetic neurotransmitter, closes them. [This last fact is surprising since, as described earlier for synapses (Chapter 8) and motor endplates (Chapter 11), the usual effect of acetylcholine is to open, not close, channels that allow ion movement; this should reinforce the generalization that a messenger's effect on its target cells is determined by the signal transduction pathways triggered by binding of that messenger to its receptors, pathways that can differ from target to target.]

Factors other than the cardiac nerves can also alter heart rate. Epinephrine, the main hormone liberated from the adrenal medulla, speeds the heart by acting on the same beta-adrenergic receptors in the SA node as norepinephrine released from neurons. The heart rate is also sensitive to changes in body temperature, plasma electrolyte concentrations, hormones other than epinephrine, and a metabolite— adenosine—produced by myocardial cells. These factors are normally of lesser importance, however, than the cardiac nerves. Figure 14–28 summarizes the major determinants of heart rate.

As stated in the previous section on innervation, sympathetic and parasympathetic neurons innervate not only the SA node but other parts of the conducting system as well. Sympathetic stimulation also increases conduction velocity through the AV node, whereas parasympathetic stimulation decreases the rate of spread of excitation through the AV node and other portions of the conducting system.

Control of Stroke Volume

The second variable that determines cardiac output is stroke volume, the volume of blood ejected by each ventricle during each contraction. As stated earlier, the ventricles do not completely empty themselves of blood during contraction. Therefore, a more forceful contraction can produce an increase in stroke volume by causing greater emptying. Changes in the force of contraction can be produced by a variety of factors, but three are dominant under most physiological and pathophysiological conditions: (1) changes in the end-diastolic volume (that is, the volume of blood in the

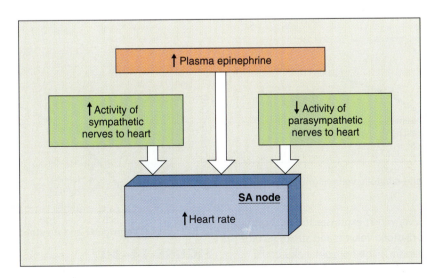

FIGURE 14–28

Major factors that influence heart rate. All effects are exerted upon the SA node. The figure shows how heart rate is increased; reversal of all the arrows in the boxes would illustrate how heart rate is decreased.

ventricles just before contraction); (2) changes in the magnitude of sympathetic nervous system input to the ventricles; and (3) **afterload** (that is, the arterial pressures against which the ventricles pump).

Relationship between Ventricular End-Diastolic Volume and Stroke Volume: The Frank-Starling Mechanism The mechanical properties of cardiac muscle are the basis for an inherent mechanism for altering stroke volume: The ventricle contracts more forcefully during systole when it has been filled to a greater degree during diastole. In other words, all other factors being equal, the stroke volume increases as the end-diastolic volume increases, as illustrated in Figure 14–29, termed a **ventricular function curve.** This relationship between stroke volume and end-diastolic volume is known as the **Frank-Starling mechanism** (also called Starling's law of the heart) in recognition of the two physiologists who identified it.

What accounts for the Frank-Starling mechanism? Basically it is simply a length-tension relationship, as described for skeletal muscle in Chapter 11, in that end-diastolic volume is a major determinant of how stretched the ventricular sarcomeres are just before contraction. Thus, the greater the end-diastolic volume, the greater the stretch, and the more forceful the contraction. However, a comparison of Figure 14–29 with Figure 11–25 reveals an important difference between the length-tension relationship in skeletal and cardiac muscle. The normal point for cardiac muscle in a resting individual is not at its optimal length for contraction, as it is for most resting skeletal muscles, but is on the rising phase of the curve; for this reason, additional stretching of the cardiac-muscle fibers by greater filling causes increased force of contraction.

The significance of the Frank-Starling mechanism is as follows: At any given heart rate, an increase in the **venous return**—the flow of blood from the veins into the heart—automatically forces an increase in cardiac output by increasing end-diastolic volume and hence stroke volume. One important function of this relationship is maintaining the equality of right and left cardiac outputs. Should the right heart, for example, suddenly begin to pump more blood than the left, the increased blood flow to the left ventricle would automatically produce an increase in left ventricular output. This ensures that blood will not accumulate in the lungs.

The Sympathetic Nerves Sympathetic nerves are distributed not only to the conducting system, as described earlier, but to the entire myocardium. The effect of the sympathetic mediator norepinephrine acting on beta-adrenergic receptors is to increase ventricular **contractility,** defined as the strength of contraction *at any given end-diastolic volume.* Plasma epinephrine acting on these receptors also increases myocardial contractility. Thus, the increased force of contraction and stroke volume resulting from sympathetic-nerve stimulation or epinephrine is *independent* of a change in end-diastolic ventricular volume.

Note that a change in contraction force due to increased end-diastolic volume (the Frank-Starling mechanism) does *not* reflect increased contractility. Increased contractility is specifically defined as an increased contraction force at any given end-diastolic volume.

The relationship between the Frank-Starling mechanism and the cardiac sympathetic nerves is illustrated in Figure 14–30. The orange ventricular function curve

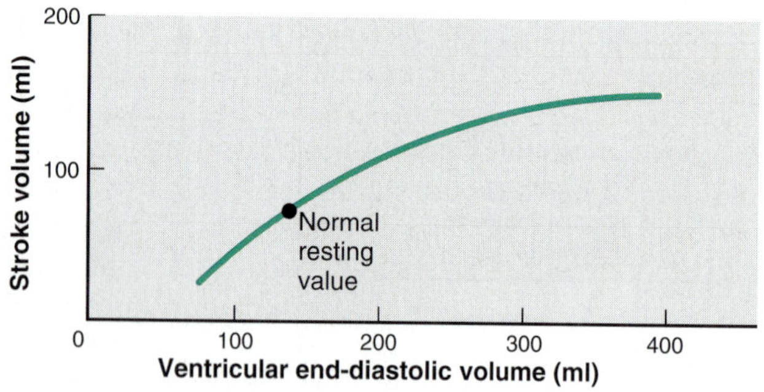

FIGURE 14–29

A ventricular function curve, which expresses the relationship between ventricular end-diastolic volume and stroke volume (the Frank-Starling mechanism). The horizontal axis could have been labeled "sarcomere length," and the vertical "contractile force." In other words, this is a length-tension curve, analogous to that for skeletal muscle ("see" Figure 11–25).

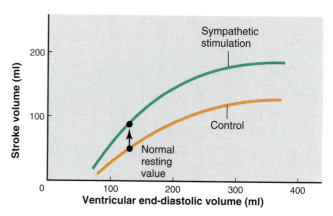

FIGURE 14–30

Effects on stroke volume of stimulating the sympathetic nerves to the heart. Stroke volume is increased at any given end-diastolic volume; that is, the sympathetic stimulation has increased ventricular contractility.

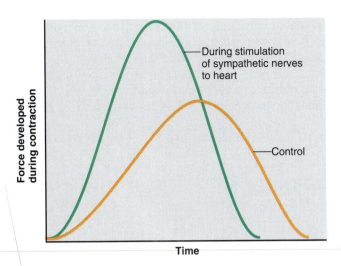

FIGURE 14–31

Effects of sympathetic stimulation on ventricular contraction and relaxation. Note that both the rate of force development and the rate of relaxation are increased, as is the maximal force developed. All these changes reflect an increased contractility.

is the same as that shown in Figure 14–29. The blue ventricular function curve was obtained for the same heart during sympathetic-nerve stimulation. The Frank-Starling mechanism still applies, but during nerve stimulation the stroke volume is greater at any given end-diastolic volume. In other words, the increased contractility leads to a more complete ejection of the end-diastolic ventricular volume.

One way of quantitating contractility is as the **ejection fraction (EF),** defined as the ratio of stroke volume (SV) to end-diastolic volume (EDV):

$$EF = SV/EDV$$

Expressed as a percentage, the ejection fraction normally averages 67 percent under resting conditions. Increased contractility causes an increased ejection fraction.

Not only does enhanced sympathetic-nerve activity to the myocardium cause the contraction to be more powerful, it also causes both the contraction and relaxation of the ventricles to occur more quickly

(Figure 14–31). These latter effects are quite important since, as described earlier, increased sympathetic activity to the heart also increases heart rate. As heart rate increases, the time available for diastolic filling decreases, but the quicker contraction and relaxation induced simultaneously by the sympathetic neurons partially compensate for this problem by permitting a larger fraction of the cardiac cycle to be available for filling.

There are multiple mechanisms by which the signal transduction pathways triggered by the binding of norepinephrine or epinephrine to their receptors causes increased contractility. These include: (1) opening more plasma-membrane calcium channels during excitation; (2) stimulating active calcium pumping into the sarcoplasmic reticulum; and (3) altering the binding of calcium by troponin. The net effect of these changes is that cytosolic calcium concentration increases to a greater value during excitation (thus facilitating contraction) and then returns to its preexcitation value more quickly following excitation (thus facilitating relaxation).

There is little parasympathetic innervation of the ventricles (in contrast to the SA node, as described in the section on control of heart rate) and so the parasympathetic system normally has only a negligible effect on ventricular contractility.

Table 14–6 summarizes the effects of the autonomic nerves on cardiac function.

Afterload   An increased arterial pressure tends to reduce stroke volume. This is because, in analogy to the situation in skeletal muscle (Chapter 11), the arterial pressure constitutes the "load" (technically termed the afterload) for contracting ventricular muscle; the greater this load, the less the contracting muscle fibers can shorten. This factor will not be dealt with further, however, since in the *normal* heart, several inherent adjustments minimize the over-all influence of arterial pressure on stroke volume. We will see, however, in the sections on high blood pressure and heart failure that long-term elevations of arterial pressure can weaken the heart and, thereby, influence stroke volume.

TABLE 14–6 Effects of Autonomic Nerves on the Heart

Area Affected	Sympathetic Nerves	Parasympathetic Nerves
SA node	Increased heart rate	Decreased heart rate
AV node	Increased conduction rate	Decreased conduction rate
Atrial muscle	Increased contractility	Decreased contractility
Ventricular muscle	Increased contractility	Decreased contractility (minor)

In summary (Figure 14–32), the two most important physiologic controllers of stroke volume are a mechanism (the Frank-Starling mechanism) dependent upon changes in end-diastolic volume, and a mechanism that is mediated by the cardiac sympathetic nerves and circulating epinephrine and that causes increased ventricular contractility. The contribution of each of these two mechanisms in specific physiological situations is described in later sections.

A summary of the major factors that determine cardiac output is presented in Figure 14–33, which combines the information of Figures 14–28 and 14–32.

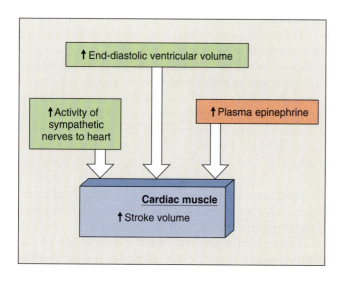

FIGURE 14–32

Major physiological controllers of stroke volume. The figure as drawn shows how stroke volume is increased. A reversal of all arrows in the boxes would illustrate how stroke volume is decreased.

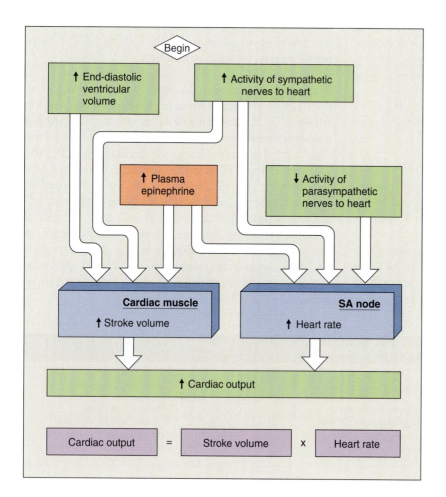

FIGURE 14–33

Major factors determining cardiac output (an amalgamation of Figures 14–28 and 14–32).

Measurement of Cardiac Function

Cardiac output in human beings can be measured by a variety of methods. Moreover, two- and three-dimensional images of the heart can be obtained throughout the entire cardiac cycle. For example, in *echocardiography,* ultrasound is beamed at the heart, and returning echoes are electronically plotted by computer to produce continuous images of the heart. This technique can detect abnormal functioning of cardiac valves or contractions of the cardiac walls and can also be used to measure ejection fraction.

Echocardiography is a "noninvasive" technique because everything used remains external to the body. Other visualization techniques are invasive. One, *cardiac angiography,* requires the temporary threading of a thin flexible tube (catheter) into the heart, via an artery or vein, under fluoroscopy. A dye is then injected through the catheter during high-speed x-ray filming. This technique is useful not only for evaluating cardiac function but also for identifying narrowed coronary arteries.

SECTION C SUMMARY

Anatomy

I. The atrioventricular (AV) valves prevent flow from the ventricles back into the atria.

II. The pulmonary and aortic valves prevent backflow from the pulmonary trunk into the right ventricle and from the aorta into the left ventricle, respectively.

III. Cardiac-muscle cells are joined by gap junctions that permit action potentials to be conducted from cell to cell.

IV. The myocardium also contains specialized muscle cells that constitute the conducting system of the heart, initiating the cardiac action potentials and speeding their spread through the heart.

Heartbeat Coordination

I. Action potentials must be initiated in cardiac-muscle cells for contraction to occur.

a. The rapid depolarization of the action potential in atrial and ventricular cells (other than those in the conducting system) is due mainly to a positive-feedback increase in sodium permeability.

b. Following the initial rapid depolarization, the cardiac-muscle cell membrane remains depolarized (the plateau phase) for almost the entire duration of the contraction because of prolonged entry of calcium into the cell through slow plasma-membrane channels.

II. The SA node generates the current that leads to depolarization of all other cardiac-muscle cells.

a. The SA node manifests a pacemaker potential, which brings its membrane potential to threshold and initiates an action potential.

b. The impulse spreads from the SA node throughout both atria and to the AV node, where a small delay occurs. The impulse then passes, in turn, into the bundle of His, right and left bundle branches, Purkinje fibers, and nonconducting-system ventricular fibers.

III. Calcium, mainly released from the sarcoplasmic reticulum (SR), functions as the excitation-contraction coupler in cardiac muscle, as in skeletal muscle, by combining with troponin.

a. The major signal for calcium release from the SR is extracellular calcium entering through voltage-gated calcium channels in the T-tubular membrane during the action potential.

b. The amount of calcium released does not usually saturate all troponin binding sites, and so the number of active cross bridges can be increased if cytosolic calcium is increased still further.

IV. Cardiac muscle cannot undergo summation of contractions because it has a very long refractory period.

Mechanical Events of the Cardiac Cycle

I. The cardiac cycle is divided into systole (ventricular contraction) and diastole (ventricular relaxation).

a. At the onset of systole, ventricular pressure rapidly exceeds atrial pressure, and the AV valves close. The aortic and pulmonary valves are not yet open, however, and so no ejection occurs during this isovolumetric ventricular contraction.

b. When ventricular pressures exceed aortic and pulmonary trunk pressures, the aortic and pulmonary valves open, and ventricular ejection of blood occurs.

c. When the ventricles relax at the beginning of the diastole, the ventricular pressures fall significantly below those in the aorta and pulmonary trunk, and the aortic and pulmonary valves close. Because the AV valves are also still closed, no change in ventricular volume occurs during this isovolumetric ventricular relaxation.

d. When ventricular pressures fall below the pressures in the right and the left atria, the AV valves open, and the ventricular filling phase of diastole begins.

e. Filling occurs very rapidly at first so that atrial contraction, which occurs at the very end of diastole, usually adds only a small amount of additional blood to the ventricles.

II. The amount of blood in the ventricles just before systole is the end-diastolic volume. The volume remaining after ejection is the end-systolic volume, and the volume ejected is the stroke volume.

III. Pressure changes in the systemic and pulmonary circulations have similar patterns, but the pulmonary pressures are much lower.

IV. The first heart sound is due to the closing of the AV valves, and the second to the closing of the aortic and pulmonary valves.

The Cardiac Output

I. The cardiac output is the volume of blood pumped by each ventricle and equals the product of heart rate and stroke volume.

   a. Heart rate is increased by stimulation of the sympathetic nerves to the heart and by epinephrine; it is decreased by stimulation of the parasympathetic nerves to the heart.

   b. Stroke volume is increased mainly by an increase in end-diastolic volume (the Frank-Starling mechanism) and by an increase in contractility due to sympathetic-nerve stimulation or to epinephrine. Afterload can also play a significant role in certain situations.

SECTION C KEY TERMS

pericardium	P wave
myocardium	QRS complex
endothelial cell	T wave
endothelium	refractory period (of cardiac
atrioventricular (AV) valve	muscle)
tricuspid valve	cardiac cycle
mitral valve	systole
papillary muscles	diastole
pulmonary valve	isovolumetric ventricular
aortic valve	contraction
intercalated disks	ventricular ejection
conducting system	stroke volume (SV)
coronary artery	isovolumetric ventricular
coronary blood flow	relaxation
sinoatrial (SA) node	ventricular filling
heart rate	end-diastolic volume (EDV)
atrioventricular (AV) node	end-systolic volume (ESV)
bundle of His	heart sounds
right and left bundle	cardiac output (CO)
branches	afterload
Purkinje fibers	ventricular function curve
slow channel	Frank-Starling mechanism
pacemaker potential	venous return
automaticity	contractility
electrocardiogram (ECG)	ejection fraction (EF)

SECTION C REVIEW QUESTIONS

1. List the structures through which blood passes from the systemic veins to the systemic arteries.
2. Contrast and compare the structure of cardiac muscle with skeletal and smooth muscle.
3. Describe the autonomic innervation of the heart, including the types of receptors involved.
4. Draw a ventricular action potential. Describe the changes in membrane permeability that underlie the potential changes.
5. Contrast action potentials in ventricular cells with SA-node action potentials. What is the pacemaker potential due to, and what is its inherent rate? By what mechanism does the SA node function as the pacemaker for the entire heart?
6. Describe the spread of excitation from the SA node through the rest of the heart.
7. Draw and label a normal ECG. Relate the P, QRS, and T waves to the atrial and ventricular action potentials.
8. Describe the sequence of events leading to excitation-contraction coupling in cardiac muscle.
9. What prevents the heart from undergoing summation of contractions?
10. Draw a diagram of the pressure changes in the left atrium, left ventricle, and aorta throughout the cardiac cycle. Show when the valves open and close, when the heart sounds occur, and the pattern of ventricular ejection.
11. Contrast the pressures in the right ventricle and pulmonary trunk with those in the left ventricle and aorta.
12. What causes heart murmurs in diastole? In systole?
13. Write the formula relating cardiac output, heart rate, and stroke volume; give normal values for a resting adult.
14. Describe the effects of the sympathetic and parasympathetic nerves on heart rate. Which is dominant at rest?
15. What are the two major factors influencing force of contraction?
16. Draw a ventricular function curve illustrating the Frank-Starling mechanism.
17. Describe the effects of the sympathetic nerves on cardiac muscle during contraction and relaxation.
18. Draw a family of curves relating end-diastolic volume and stroke volume during different levels of sympathetic stimulation.
19. Summarize the effects of the autonomic nerves on the heart.
20. Draw a flow diagram summarizing the factors determining cardiac output.

THE VASCULAR SYSTEM

The functional and structural characteristics of the blood vessels change with successive branching. Yet the entire cardiovascular system, from the heart to the smallest capillary, has one structural component in common, a smooth, single-celled layer of endothelial cells, or endothelium, which lines the inner (blood-contacting) surface of the vessels. Capillaries consist only of endothelium, whereas all other vessels have, in addition, layers of connective tissue and smooth muscle. Endothelial cells have a large number of active functions. These are summarized for reference in Table 14–7 and are described in relevant sections of this chapter or subsequent chapters.

We have previously described the pressures in the aorta and pulmonary arteries during the cardiac cycle. Figure 14–34 illustrates the pressure changes that occur along the rest of the systemic and pulmonary vascular systems. Text sections below dealing with the individual vascular segments will describe the reasons for these changes in pressure. For the moment, note only that by the time the blood has completed its journey back to the atrium in each circuit, virtually all the pressure originally generated by the ventricular contraction has been dissipated. The reason pressure at any point in the vascular system is less than that at an earlier point is that the blood vessels offer resistance to the flow from one point to the next.

Arteries

The aorta and other systemic arteries have thick walls containing large quantities of elastic tissue. Although they also have smooth muscle, arteries can be viewed most conveniently as elastic tubes. Because the arteries have large radii, they serve as low-resistance tubes conducting blood to the various organs. Their second major function, related to their elasticity, is to act as a "pressure reservoir" for maintaining blood flow through the tissues during diastole, as described below.

Arterial Blood Pressure

What are the factors determining the pressure within an elastic container, such as a balloon filled with water? The pressure inside the balloon depends on (1) the volume of water, and (2) how easily the balloon walls can be stretched. If the walls are very stretchable, large quantities of water can be added with only a small rise in pressure. Conversely, the addition of a small quantity of water causes a large pressure rise in a balloon that is difficult to stretch. The term used to denote how easily a structure can be stretched is **compliance**:

$$\text{Compliance} = \Delta \text{ volume} / \Delta \text{ pressure}$$

The *higher* the compliance of a structure, the *more easily* it can be stretched.

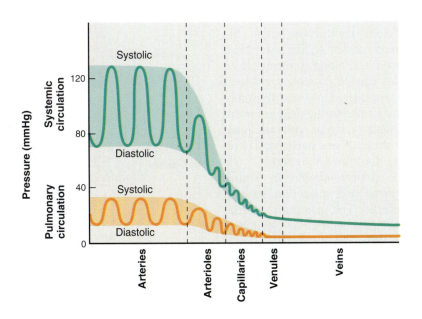

FIGURE 14–34
Pressures in the vascular system.

TABLE 14–7 Functions of Endothelial Cells

1. Serve as a physical lining of heart and blood vessels to which blood cells do not normally adhere.

2. Serve as a permeability barrier for the exchange of nutrients, metabolic end products, and fluid between plasma and interstitial fluid; regulate transport of macromolecules and other substances.

3. Secrete paracrine agents that act on adjacent vascular smooth-muscle cells; these include vasodilators—prostacyclin and nitric oxide (endothelium-derived relaxing factor, EDRF)—and vasoconstrictors—notably endothelin-1.

4. Mediate angiogenesis (new capillary growth).

5. Play a central role in vascular remodeling by detecting signals and releasing paracrine agents that act on adjacent cells in the blood vessel wall.

6. Contribute to the formation and maintenance of extracellular matrix (Chapter 1).

7. Produce growth factors in response to damage.

8. Secrete substances that regulate platelet clumping, clotting, and anticlotting.

9. Synthesize active hormones from inactive precursors (Chapter 16).

10. Extract or degrade hormones and other mediators (Chapter 15).

11. Secrete cytokines during immune responses (Chapter 20).

12. Influence vascular smooth-muscle proliferation in the disease atherosclerosis.

These principles can be applied to an analysis of arterial blood pressure. The contraction of the ventricles ejects blood into the pulmonary and systemic arteries during systole. If a precisely equal quantity of blood were to flow simultaneously out of the arteries, the total volume of blood in the arteries would remain constant and arterial pressure would not change. Such is not the case, however. As shown in Figure 14–35, a volume of blood equal to only about one-third the stroke volume leaves the arteries during systole. The rest of the stroke volume remains in the arteries during systole, distending them and raising the arterial pressure. When ventricular contraction ends, the stretched arterial walls recoil passively, like a stretched rubber band being released, and blood continues to be driven into the arterioles during diastole. As blood leaves the arteries, the arterial volume and therefore the arterial pressure slowly fall, but the next ventricular contraction occurs while there is still adequate blood in the arteries to stretch them partially. Therefore, the arterial pressure does not fall to zero.

The aortic pressure pattern shown in Figure 14–36a is typical of the pressure changes that occur in all the large systemic arteries. The maximum arterial pressure reached during peak ventricular ejection is called **systolic pressure (SP).** The minimum arterial pressure occurs just before ventricular ejection begins and is called **diastolic pressure (DP).** Arterial pressure is generally recorded as systolic/diastolic—that is, 125/75 mmHg in our example (see Figure 14–36b for average values at different ages in the population of the United States).

The difference between systolic pressure and diastolic pressure (125 − 75 = 50 mmHg in the example) is called the **pulse pressure.** It can be felt as a pulsation or throb in the arteries of the wrist or neck with each heartbeat. During diastole, nothing is felt over the

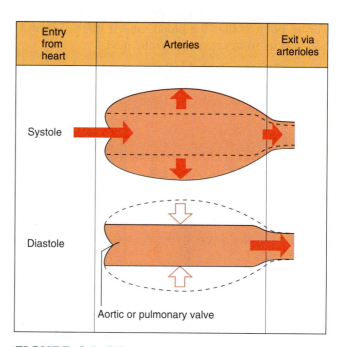

FIGURE 14–35

Movement of blood into and out of the arteries during the cardiac cycle. The lengths of the arrows denote relative quantities flowing into and out of the arteries and remaining in the arteries.

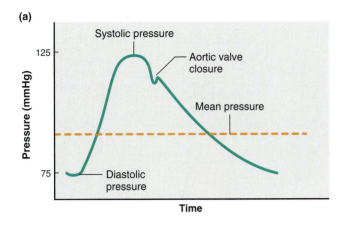

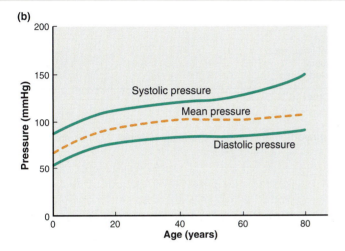

FIGURE 14–36

(a) Typical arterial pressure fluctuations during the cardiac cycle. (b) Changes in arterial pressure with age in the U.S. population.

Adapted from Guyton.

artery, but the rapid rise in pressure at the next systole pushes out the artery wall, and it is this expansion of the vessel that produces the detectable throb.

The most important factors determining the magnitude of the pulse pressure—that is, how much greater systolic pressure is than diastolic—are (1) stroke volume, (2) speed of ejection of the stroke volume, and (3) arterial compliance. Specifically, the pulse pressure produced by a ventricular ejection is greater if the volume of blood ejected is increased, if the speed at which it is ejected is increased, or if the arteries are less compliant. This last phenomenon occurs in atherosclerosis, the "hardening" of the arteries that progresses with age and accounts for the increasing pulse pressure seen so often in older people.

It is evident from Figure 14–36a that arterial pressure is continuously changing throughout the cardiac cycle. The *average* pressure (**mean arterial pressure, MAP**) in the cycle is not merely the value halfway between systolic pressure and diastolic pressure because diastole usually lasts longer than systole. The true mean arterial pressure can be obtained by complex methods, but for most purposes it is approximately equal to the diastolic pressure plus one-third of the pulse pressure (SP − DP), largely because diastole lasts about twice as long as systole:

$$MAP = DP + 1/3 \ (SP - DP)$$

Thus, in our example: MAP = 75 + 1/3 (50) = 92 mmHg.

The MAP is the most important of the pressures described because it is the pressure driving blood into the tissues averaged over the entire cardiac cycle. We can say mean "arterial" pressure without specifying to which artery we are referring because the aorta and other large arteries have such large diameters that they offer only negligible resistance to flow, and the mean pressures are therefore similar everywhere in the large arteries.

One additional important point should be made: We have stated that arterial compliance is an important determinant of *pulse* pressure, but for complex reasons, compliance does *not* influence the *mean* arterial pressure. Thus, for example, a person with a low arterial compliance (due to atherosclerosis) but an otherwise normal cardiovascular system will have a large pulse pressure but a normal mean arterial pressure. The determinants of mean arterial pressure are described in Section E.

Measurement of Systemic Arterial Pressure

Both systolic and diastolic blood pressure are readily measured in human beings with the use of a sphygmomanometer. An inflatable cuff is wrapped around the upper arm, and a stethoscope is placed in a spot on the arm just below the cuff and beneath which the major artery to the lower arm runs.

The cuff is then inflated with air to a pressure greater than systolic blood pressure (Figure 14–37). The high pressure in the cuff is transmitted through the tissue of the arm and completely compresses the artery under the cuff, thereby preventing blood flow through the artery. The air in the cuff is then slowly released, causing the pressure in the cuff and on the artery to drop. When cuff pressure has fallen to a value just below the systolic pressure, the artery opens slightly and allows blood flow for a brief time at the peak of systole. During this interval, the blood flow through the partially compressed artery occurs at a very high velocity because of the small opening and the large pressure difference across the opening. The high-velocity blood flow is turbulent and, therefore, produces vibrations that can be heard through the stethoscope. Thus, the pressure, measured on the

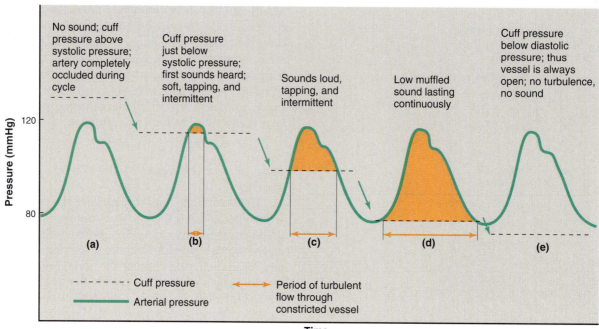

FIGURE 14–37

Sounds heard through a stethoscope while the cuff pressure of a sphygmomanometer is gradually lowered. Sounds are first heard at systolic pressure, and they disappear at diastolic pressure.

gauge attached to the cuff, at which sounds are first heard as the cuff pressure is lowered is identified as the systolic blood pressure.

As the pressure in the cuff is lowered farther, the duration of blood flow through the artery in each cycle becomes longer. When the cuff pressure reaches the diastolic blood pressure, all sound stops because flow is now continuous and nonturbulent through the open artery. Thus, diastolic pressure is identified as the cuff pressure at which sounds disappear.

It should be clear from this description that the sounds heard during measurement of blood pressure are *not* the same as the *heart* sounds described earlier, which are due to closing of cardiac valves.

Arterioles

The arterioles play two major roles: (1) The arterioles in individual organs are responsible for determining the relative blood flows to those organs at any given mean arterial pressure, and (2) the arterioles, as a whole, are a major factor in determining mean arterial pressure itself. The first function will be described in this section, and the second in Section E.

Figure 14–38 illustrates the major principles of blood-flow distribution in terms of a simple model, a fluid-filled tank with a series of compressible outflow tubes. What determines the rate of flow through each exit tube? As stated in Section B of this chapter,

$$F = \Delta P / R$$

Since the driving pressure (the height of the fluid column in the tank) is identical for each tube, differences in flow are completely determined by differences in the resistance to flow offered by each tube. The lengths of the tubes are approximately the same, and the viscosity of the fluid is constant; therefore, differences in resistance offered by the tubes are due solely to differences in their radii. Obviously, the widest tubes have the greatest flows. If we equip each outflow tube with an adjustable cuff, we can obtain various combinations of flows.

This analysis can now be applied to the cardiovascular system. The tank is analogous to the arteries, which serve as a pressure reservoir, the major arteries themselves being so large that they contribute little resistance to flow. Therefore, all the large arteries of the body can be considered a single pressure reservoir.

The arteries branch within each organ into progressively smaller arteries, which then branch into arterioles. The smallest arteries are narrow enough to offer significant resistance to flow, but the still narrower arterioles are the major sites of resistance in the vascular tree and are therefore analogous to the outflow tubes in the model. This explains the large decrease in

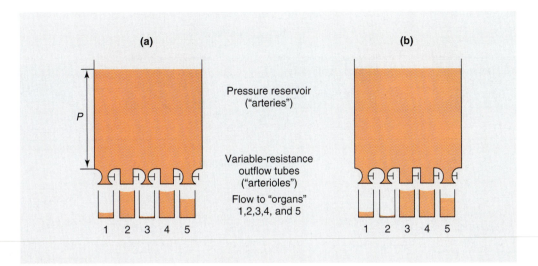

FIGURE 14–38

Physical model of the relationship between arterial pressure, arteriolar radius in different organs, and blood-flow distribution. In (a), blood flow is high through tube 2 and low through tube 3, whereas just the opposite is true for (b). This shift in blood flow was achieved by constricting tube 2 and dilating tube 3.

mean pressure—from about 90 mmHg to 35 mmHg—as blood flows through the arterioles (see Figure 14–34). Pulse pressure also diminishes to the point that flow beyond the arterioles—that is, through capillaries, venules, and veins—is much less pulsatile.

Like the model's outflow tubes (Figure 14–38), the arteriolar radii in individual organs are subject to independent adjustment. The blood flow through any organ is given by the following equation:

$$F_{organ} = (MAP - \text{venous pressure})/Resistance_{organ}$$

Since venous pressure is normally approximately zero, we may write:

$$F_{organ} = MAP/Resistance_{organ}$$

Since the MAP, the driving force for flow through each organ, is identical throughout the body, differences in flows between organs depend entirely on the relative resistances offered by the arterioles of each organ. Arterioles contain smooth muscle, which can either relax and cause the vessel radius to increase (**vasodilation**) or contract and decrease the vessel radius (**vasoconstriction**). Thus the pattern of blood-flow distribution depends upon the degree of arteriolar smooth-muscle contraction within each organ and tissue. Look back at Figure 14–9, which illustrates the distribution of blood flows at rest; these are due to differing resistances in the various locations. Such distribution can be changed markedly, as during exercise, for example, by changing the various resistances.

How can resistance be changed? Arteriolar smooth muscle possesses a large degree of spontaneous activity (that is, contraction independent of any neural,

hormonal, or paracrine input). This spontaneous contractile activity is called **intrinsic tone** (also termed basal tone). It sets a baseline level of contraction that can be increased or decreased by external signals, such as neurotransmitters. These signals act by inducing changes in the muscle cells's cytosolic calcium concentration (see Chapter 11 for a description of excitation-contraction coupling in smooth muscle). An increase in contractile force above the vessel's intrinsic tone causes vasoconstriction, whereas a decrease in contractile force causes vasodilation. The mechanisms controlling vasoconstriction and vasodilation in arterioles fall into two general categories: (1) local controls, and (2) extrinsic (or reflex) controls.

Local Controls

The term **local controls** denotes mechanisms independent of nerves or hormones by which organs and tissues alter their own arteriolar resistances, thereby self-regulating their blood flows. It does include changes caused by autocrine/paracrine agents. This self-regulation includes the phenomena of active hyperemia, flow autoregulation, reactive hyperemia, and local response to injury.

Active Hyperemia   Most organs and tissues manifest an increased blood flow (**hyperemia**) when their metabolic activity is increased (Figure 14–39a); this is termed **active hyperemia**. For example, the blood flow to exercising skeletal muscle increases in direct proportion to the increased activity of the muscle. Active hyperemia is the direct result of arteriolar dilation in the more active organ or tissue.

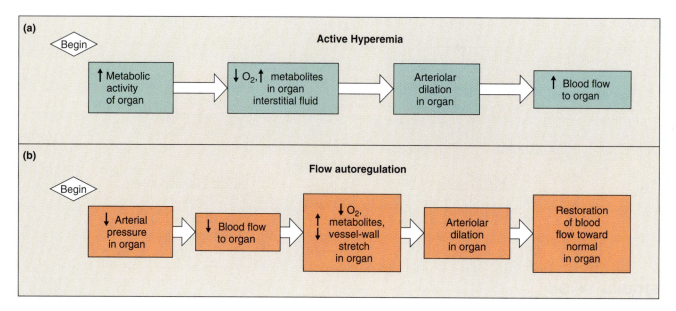

FIGURE 14-39

Local control of organ blood flow in response to (a) increases in metabolic activity, and (b) decreases in blood pressure. Decreases in metabolic activity or increases in blood pressure would produce changes opposite those shown here.

The factors acting upon arteriolar smooth muscle in active hyperemia to cause it to relax are local chemical changes in the extracellular fluid surrounding the arterioles. These result from the increased metabolic activity in the cells near the arterioles. The relative contributions of the various factors implicated vary, depending upon the organs involved and on the duration of the increased activity. Therefore, we shall name but not quantify some of these local chemical changes that occur in the extracellular fluid: decreased oxygen concentration; increased concentrations of carbon dioxide, hydrogen ion, the metabolite adenosine, potassium (as a result of enhanced potassium movement out of muscle cells during the more frequent action potentials) and eicosanoids (Chapter 7); increased osmolarity (resulting from the increased breakdown of high-molecular-weight substances); and, in some glands, increased concentration of a peptide known as **bradykinin.** The last substance is generated locally from a circulating protein called **kininogen** by the action of an enzyme, **kallikrein,** secreted by the active gland cells.

Local changes in all these chemical factors have been shown to cause arteriolar dilation under controlled experimental conditions, and they all probably contribute to the active-hyperemia response in one or more organs. It is likely, moreover, that additional important local factors remain to be discovered. It must be emphasized that all these chemical changes in the extracellular fluid act locally upon the arteriolar smooth muscle, causing it to relax. No nerves or hormones are involved.

It should not be too surprising that active hyperemia is most highly developed in skeletal muscle, cardiac muscle, and glands, tissues that show the widest range of normal metabolic activities in the body. It is highly efficient, therefore, that their supply of blood be primarily determined locally.

Flow Autoregulation During active hyperemia, increased metabolic activity of the tissue or organ is the initial event leading to local vasodilation. However, locally mediated changes in arteriolar resistance can occur when a tissue or organ suffers a change in its blood supply resulting from a change in blood pressure (Figure 14-39b). The change in resistance is in the direction of maintaining blood flow nearly constant in the face of the pressure change and is therefore termed **flow autoregulation.** For example, when arterial pressure in an organ is reduced, say, because of a partial occlusion in the artery supplying the organ, local controls cause arteriolar vasodilation, which tends to maintain flow relatively constant.

What is the mechanism of flow autoregulation? One mechanism is the same metabolic factors described for active hyperemia. When an arterial pressure reduction lowers blood flow to an organ, the supply of oxygen to the organ is diminished, and the local extracellular oxygen concentration decreases. Simultaneously, the extracellular concentrations of carbon dioxide, hydrogen ion, and metabolites all increase because they are not removed by the blood as fast as they are produced. Also, eicosanoid synthesis is increased by still unclear stimuli. Thus, the local metabolic changes occurring during decreased blood supply at constant metabolic activity are similar to those that occur during increased metabolic activity. This is because both situations reflect an initial imbalance between blood supply and level of cellular metabolic activity. Note then that the vasodilations of active hyperemia and of flow autoregulation in response to low arterial pressure do not differ in their major mechanisms, which involve local metabolic factors, but in the event—altered metabolism or altered blood pressure—that brings these mechanisms into play.

Flow autoregulation is not limited to circumstances in which arterial pressure decreases. The opposite events occur when, for various reasons, arterial pressure increases: The initial increase in flow due to the increase in pressure removes the local vasodilator chemical factors faster than they are produced and also increases the local concentration of oxygen. This causes the arterioles to constrict, thereby maintaining local flow relatively constant in the face of the increased pressure.

Although our description has emphasized the role of local *chemical* factors in flow autoregulation, it should be noted that another mechanism also participates in this phenomenon in certain tissues and organs. Some arteriolar smooth muscle responds directly to increased stretch, caused by increased arterial pressure, by contracting to a greater extent. Conversely, decreased stretch, due to decreased arterial pressure, causes this vascular smooth muscle to decrease its tone. These direct responses of arteriolar smooth muscle to stretch are termed **myogenic responses.** They are due to changes in calcium movement into the smooth-muscle cells through stretch-sensitive calcium channels in the plasma membrane.

Reactive Hyperemia When an organ or tissue has had its blood supply completely occluded, a profound transient increase in its blood flow occurs as soon as the occlusion is released. This phenomenon, known as **reactive hyperemia,** is essentially an extreme form of flow autoregulation. During the period of no blood flow, the arterioles in the affected organ or tissue dilate, owing to the local factors described above. Blood flow, therefore, is very great through these wide-open arterioles as soon as the occlusion to arterial inflow is removed.

Response to Injury Tissue injury causes a variety of substances to be released locally from cells or generated from plasma precursors. These substances make arteriolar smooth muscle relax and cause vasodilation in an injured area. This phenomenon, a part of the general process known as inflammation, will be described in detail in Chapter 20.

Extrinsic Controls

Sympathetic Nerves Most arterioles receive a rich supply of sympathetic postganglionic nerve fibers. These neurons release mainly norepinephrine, which binds to alpha-adrenergic receptors on the vascular smooth muscle to cause vasoconstriction.

In contrast, recall that the receptors for norepinephrine on heart muscle, including the conducting system, are mainly beta-adrenergic. This permits the pharmacologic use of beta-adrenergic antagonists to block the actions of norepinephrine on the heart but not the arterioles, and vice versa for alpha-adrenergic antagonists.

Control of the sympathetic nerves to arterioles can also be used to produce *vasodilation*. Since the sympathetic nerves are seldom completely quiescent but discharge at some finite rate that varies from organ to organ, they always are causing some degree of tonic constriction in addition to the vessels' intrinsic tone. Dilation can be achieved by *decreasing* the rate of sympathetic activity below this basal level.

The skin offers an excellent example of the role of the sympathetic nerves. At room temperature, skin arterioles are already under the influence of a moderate rate of sympathetic discharge. An appropriate stimulus—cold, fear, or loss of blood, for example—causes reflex enhancement of this sympathetic discharge, and the arterioles constrict further. In contrast, an increased body temperature reflexly inhibits the sympathetic nerves to the skin, the arterioles dilate, and the skin flushes.

In contrast to active hyperemia and flow autoregulation, the primary functions of sympathetic nerves to blood vessels are concerned not with the coordination of local metabolic needs and blood flow but with reflexes that serve whole body "needs." The most common reflex employing these nerves, as we shall see, is that which regulates arterial blood pressure by influencing arteriolar resistance throughout the body. Other reflexes redistribute blood flow to achieve a specific function (for example, to increase heat loss from the skin).

Parasympathetic Nerves With few exceptions, there is little or no important parasympathetic innervation of arterioles. In other words, the great majority of blood vessels receive sympathetic but not parasympathetic input.

Noncholinergic, Nonadrenergic Autonomic Neurons As described in Chapter 8, there is a population of autonomic postganglionic neurons that are labeled noncholinergic, nonadrenergic neurons because they release neither acetylcholine nor norepinephrine. Instead they release **nitric oxide,** which is a vasodilator,

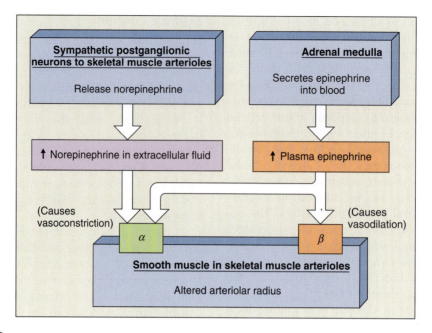

FIGURE 14–40

Effects of sympathetic nerves and plasma epinephrine on the arterioles in skeletal muscle. After its release from neuron terminals, norepinephrine diffuses to the arterioles, whereas epinephrine, a hormone, is blood-borne. Note that activation of alpha-adrenergic receptors and beta-adrenergic receptors produces opposing effects. For simplicity, norepinephrine is shown binding only to alpha-adrenergic receptors; it can also bind to beta-adrenergic receptors on the arterioles, but this occurs to a lesser extent.

and, possibly, other noncholinergic vasodilator substances. These neurons are particularly prominent in the enteric nervous system, which plays a significant role in the control of the gastrointestinal system's blood vessels (Chapter 17). These neurons also innervate arterioles in certain other locations, for example, in the penis, where they mediate erection (Chapter 19).

Hormones Epinephrine, like norepinephrine released from sympathetic nerves, can bind to alpha-adrenergic receptors on arteriolar smooth muscle and cause vasoconstriction. The story is more complex, however, because many arteriolar smooth-muscle cells possess beta-adrenergic receptors as well as alpha-adrenergic receptors, and the binding of epinephrine to these beta-adrenergic receptors causes the muscle cells to relax rather than contract (Figure 14–40).

In most vascular beds, the existence of beta-adrenergic receptors on vascular smooth muscle is of little if any importance since they are greatly outnumbered by the alpha-adrenergic receptors. The arterioles in skeletal muscle are an important exception, however. Because they have a large number of beta-adrenergic receptors, circulating epinephrine usually causes vasodilation in this vascular bed.

Another hormone important for arteriolar control is **angiotensin II,** which constricts most arterioles. This

peptide is part of the renin-angiotensin system (Chapter 16).

Yet another important hormone that, when present at high plasma concentrations, causes arteriolar constriction is **vasopressin,** which is released into the blood by the posterior pituitary gland (Chapter 10). The functions of vasopressin will be described more fully in Chapter 16.

Finally, the hormone secreted by the cardiac atria—**atrial natriuretic factor**—is a potent vasodilator. Whether this hormone, whose actions on the kidneys are described in Chapter 16, plays a widespread physiologic role in control of arterioles is unsettled.

Endothelial Cells and Vascular Smooth Muscle

It should be clear from the previous sections that a large number of substances can induce the contraction or relaxation of vascular smooth muscle. Many of these substances do so by acting directly on the arteriolar smooth muscle, but others act indirectly via the endothelial cells adjacent to the smooth muscle. Endothelial cells, in response to these latter substances as well as certain mechanical stimuli, secrete several paracrine agents that diffuse to the adjacent vascular smooth muscle and induce either relaxation or contraction, resulting in vasodilation or vasoconstriction, respectively.

One very important paracrine vasodilator released by endothelial cells is nitric oxide; note that we are dealing here with nitric oxide released from endothelial cells, not nerve endings as described in an earlier section. [Before the identity of the vasodilator paracrine agent released by the endothelium was determined to be nitric oxide, it was called **endothelium-derived relaxing factor (EDRF),** and this name is still often used because there may be substances other than nitric oxide that also fit this general definition.] Nitric oxide is released continuously in significant amounts by endothelial cells in the arterioles and contributes to arteriolar vasodilation in the basal state. In addition, its secretion is rapidly and markedly increased in response to a large number of the chemical mediators involved in both reflex and local control of arterioles. For example, nitric oxide release is stimulated by bradykinin and histamine, substances produced locally during inflammation (Chapter 20).

Another vasodilator released by endothelial cells is the eicosanoid **prostacyclin (PGI$_2$).** Unlike the case for nitric oxide, there is little basal secretion of PGI$_2$, but secretion can increase markedly in response to various inputs. The roles of PGI$_2$ in the vascular responses to blood clotting are described in Section G of this chapter.

One of the important *vasoconstrictor* paracrine agents released by endothelial cells in response to certain mechanical and chemical stimuli is **endothelin-1 (ET-1).** ET-1 is a member of the endothelin family of peptide paracrine agents secreted by a variety of cells in diverse tissues, including the brain, kidneys, and lungs. Not only does ET-1 serve as a paracrine agent but under certain circumstances it can also achieve high enough concentrations in the blood to serve as a hormone, causing widespread arteriolar vasoconstriction.

This discussion has so far focused only on arterioles. However, endothelial cells in *arteries* can also secrete various paracrine agents that influence the arteries' smooth muscle and, hence, their diameters and resistances to flow. The force exerted on the inner surface of the arterial wall, specifically on the endothelial cells, by the flowing blood is termed **shear stress;** it increases as the blood flow through the vessel increases. In response to this increased shear stress, arterial endothelium releases PGI$_2$, increased amounts of nitric oxide, and less ET-1. All these changes cause the arterial vascular smooth muscle to relax and the artery to dilate. This **flow-induced arterial vasodilation** (which should be distinguished from *arteriolar* flow autoregulation) may be important in remodeling of arteries and in optimizing the blood supply to tissues under certain conditions.

Arteriolar Control in Specific Organs

Figure 14–41 summarizes the factors that determine arteriolar radius. The importance of local and reflex controls varies from organ to organ, and Table 14–8 lists for reference the key features of arteriolar control in specific organs.

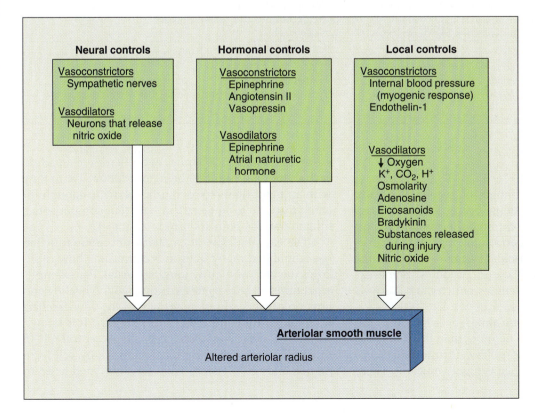

FIGURE 14–41

Major factors affecting arteriolar radius. Note that epinephrine can be a vasodilator or vasoconstrictor, depending on the tissue.

TABLE 14–8 Reference Summary of Arteriolar Control in Specific Organs

Heart
High intrinsic tone; oxygen extraction is very high at rest, and so flow must increase when oxygen consumption increases if adequate oxygen supply is to be maintained.
Controlled mainly by local metabolic factors, particularly adenosine, and flow autoregulation; direct sympathetic influences are minor and normally overridden by local factors.
Vessels are compressed during systole, and so coronary flow occurs mainly during diastole.

Skeletal Muscle
Controlled by local metabolic factors during exercise.
Sympathetic nerves cause vasoconstriction (mediated by alpha-adrenergic receptors) in reflex response to decreased arterial pressure.
Epinephrine causes vasodilation, via beta-adrenergic receptors, when present in low concentration and vasoconstriction, via alpha-adrenergic receptors, when present in high concentration.

GI Tract, Spleen, Pancreas, and Liver ("Splanchnic Organs")
Actually two capillary beds partially in series with each other; blood from the capillaries of the GI tract, spleen, and pancreas flows via the portal vein to the liver. In addition, the liver also receives a separate arterial blood supply.
Sympathetic nerves cause vasoconstriction, mediated by alpha-adrenergic receptors, in reflex response to decreased arterial pressure and during stress. In addition, venous constriction causes displacement of a large volume of blood from the liver to the veins of the thorax.
Increased blood flow occurs following ingestion of a meal and is mediated by local metabolic factors, neurons, and hormones secreted by the GI tract.

Kidneys
Flow autoregulation is a major factor.
Sympathetic nerves cause vasoconstriction, mediated by alpha-adrenergic receptors, in reflex response to decreased arterial pressure and during stress. Angiotensin II is also a major vasoconstrictor. These reflexes help conserve sodium and water.

Brain
Excellent flow autoregulation.
Distribution of blood within the brain is controlled by local metabolic factors.
Vasodilation occurs in response to increased concentration of carbon dioxide in arterial blood.
Influenced relatively little by the autonomic nervous system.

Skin
Controlled mainly by sympathetic nerves, mediated by alpha-adrenergic receptors; reflex vasoconstriction occurs in response to decreased arterial pressure and cold, whereas vasodilation occurs in response to heat.
Substances released from sweat glands and noncholinergic, nonadrenergic neurons also cause vasodilation.
Venous plexus contains large volumes of blood, which contributes to skin color.

Lungs
Very low resistance compared to systemic circulation.
Controlled mainly by gravitational forces and passive physical forces within the lung.
Constriction, mediated by local factors, occurs in response to low oxygen concentration—just opposite that which occurs in the systemic circulation.

Capillaries

As mentioned at the beginning of Section B, at any given moment, approximately 5 percent of the total circulating blood is flowing through the capillaries, and it is this 5 percent that is performing the ultimate function of the entire cardiovascular system—the exchange of nutrients and metabolic end products. Some exchange also occurs in the venules, which can be viewed as extensions of capillaries.

The capillaries permeate almost every tissue of the body. Since most cells are no more than 0.1 mm (only a few cell widths) from a capillary, diffusion distances are very small, and exchange is highly efficient. There

are an estimated 25,000 miles of capillaries in an adult, each individual capillary being only about 1 mm long with an inner diameter of 5 μm, just wide enough for an erythrocyte to squeeze through. (For comparison, a human hair is about 100 μm in diameter.)

The essential role of capillaries in tissue function has stimulated many questions concerning how capillaries develop and grow (**angiogenesis**). For example, what turns on angiogenesis during wound healing and how do cancers stimulate growth of the new capillaries required for continued cancer growth? It is known that the vascular endothelial cells play a central role in the building of a new capillary network by cell locomotion and cell division. They are stimulated to do so by a variety of **angiogenic factors** [for example, vascular endothelial growth factor (VEGF)] secreted locally by various tissue cells (fibroblasts, for example) and the endothelial cells themselves. Cancer cells also secrete angiogenic factors, and development of drugs to interfere with the secretion or action of these factors is a promising research area in anticancer therapy. For example, substances that inhibit blood-vessel growth have been found to reduce the size of almost any tumor (or eliminate the tumor completely) in mice; these agents are presently being studied in people with cancer.

Anatomy of the Capillary Network

Capillary structure varies considerably from organ to organ, but the typical capillary (Figure 14–42) is a thin-walled tube of endothelial cells one layer thick resting on a basement membrane, without any surrounding smooth muscle or elastic tissue. Capillaries in several organs (for example, the brain) have a second set of cells that adhere to the opposite side of the basement membrane and influence the ability of substances to penetrate the capillary wall.

The flat cells that constitute the endothelial tube are not attached tightly to each other but are separated by narrow water-filled spaces termed **intercellular**

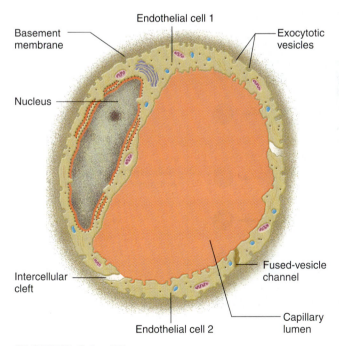

FIGURE 14–42

Capillary cross section. There are two endothelial cells in the figure, but the nucleus of only one is seen because the other is out of the plane of section. The fused-vesicle channel is part of endothelial cell 2. 𝄐

Adapted from Lentz.

clefts. The endothelial cells generally contain large numbers of endocytotic and exocytotic vesicles, and sometimes these fuse to form continuous **fused-vesicle channels** across the cell (Figure 14–42).

Blood flow through capillaries depends very much on the state of the other vessels that constitute the microcirculation (Figure 14–43). Thus, vasodilation of the arterioles supplying the capillaries causes increased capillary flow, whereas arteriolar vasoconstriction reduces capillary flow.

In addition, in some tissues and organs, blood does not enter capillaries directly from arterioles but from vessels called **metarterioles,** which connect arterioles to venules. Metarterioles, like arterioles, contain scattered smooth-muscle cells. The site at which a capillary exits from a metarteriole is surrounded by a ring of smooth muscle, the **precapillary sphincter,** which relaxes or contracts in response to local metabolic factors. When contracted, the precapillary sphincter closes the entry to the capillary completely. The more active the tissue, the more precapillary sphincters are open at any moment and the more capillaries in the network are receiving blood. Precapillary sphincters may also exist at the site of capillary exit from arterioles.

Velocity of Capillary Blood Flow

Figure 14–44 illustrates a simple mechanical model of a series of 1-cm-diameter balls being pushed down a single tube that branches into narrower tubes. Although each tributary tube has a smaller cross section than the wide tube, the *sum* of the tributary cross sections is three times greater than that of the wide tube. Let us assume that in the wide tube each ball moves 3 cm/min. If the balls are 1 cm in diameter and they move two abreast, six balls leave the wide tube per minute and enter the narrow tubes, and six balls leave the narrow tubes per minute. At what speed does each ball move in the small tubes? The answer is 1 cm/min.

This example illustrates the following important principle: When a continuous stream moves through consecutive sets of tubes, the velocity of flow *decreases* as the sum of the cross-sectional areas of the tubes *increases.* This is precisely the case in the cardiovascular system (Figure 14–45). The blood velocity is very great in the aorta, slows progressively in the arteries and arterioles, and then slows markedly as the blood passes through the huge cross-sectional area of the capillaries. The velocity of flow then progressively increases in the venules and veins because the cross-sectional area decreases. To reemphasize, flow velocity is not dependent on proximity to the heart but rather on total cross-sectional area of the vessel type.

The huge cross-sectional area of the capillaries accounts for another important feature of capillaries:

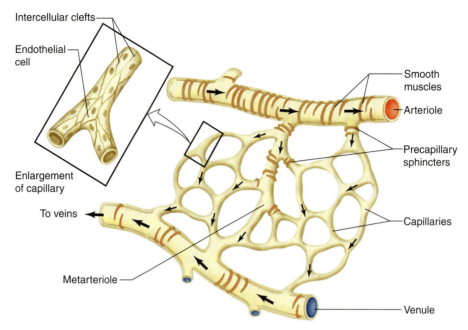

FIGURE 14–43

Diagram of microcirculation. Note the absence of smooth muscle in the capillaries.
Adapted from Chaffee and Lytle.

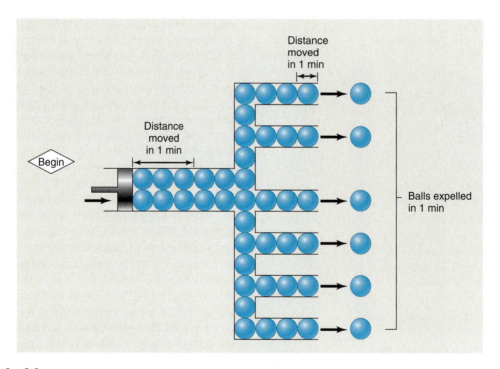

FIGURE 14–44

Relationship between total cross-sectional area and flow velocity. The total cross-sectional area of the small tubes is three times greater than that of the large tube. Accordingly, velocity of flow is one-third as great in the small tubes.

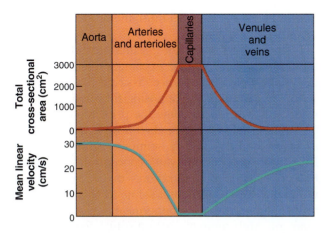

FIGURE 14–45

Relationship between total cross-sectional area and flow velocity in the systemic circulation.

Adapted from Lytle.

Because each capillary is very narrow, it offers considerable resistance to flow, but the huge total number of capillaries provides such a large cross-sectional area that the *total* resistance of *all* the capillaries is much less than that of the arterioles.

Diffusion across the Capillary Wall: Exchanges of Nutrients and Metabolic End Products

There are three basic mechanisms by which substances move across the capillary walls in most organs and tissues to enter or leave the interstitial fluid: diffusion, vesicle transport, and bulk flow. Mediated transport constitutes a fourth mechanism in the capillaries of the brain. Diffusion and vesicle transport are described in this section, and bulk flow in the next.

In all capillaries, excluding those in the brain, diffusion constitutes the only important means by which net movement of nutrients, oxygen, and metabolic end products occurs across the capillary walls. As described in the next section, there is some movement of these substances by bulk flow, but it is of negligible importance.

The factors determining diffusion rates were described in Chapter 6. Lipid-soluble substances, including oxygen and carbon dioxide, easily diffuse through the plasma membranes of the capillary endothelial cells. In contrast, ions and polar molecules are poorly soluble in lipid and must pass through small water-filled channels in the endothelial lining.

The presence of water-filled channels in the capillary walls causes the permeability of ions and small polar molecules to be quite high, although still much lower than that of lipid-soluble molecules. One location of these channels is the intercellular clefts—that is, the narrow water-filled spaces between adjacent

cells. Another set of water-filled channels is provided by the fused-vesicle channels that penetrate the endothelial cells.

The water-filled channels allow only very small amounts of protein to diffuse through them. Very small amounts of protein may also cross the endothelial cells by vesicle transport—endocytosis of plasma at the luminal border and exocytosis of the endocytotic vesicle at the interstitial side.

Variations in the size of the water-filled channels account for great differences in the "leakiness" of capillaries in different organs. At one extreme are the "tight" capillaries of the brain, which have no intracellular clefts, only tight junctions. Therefore, water-soluble substances, even those of low molecular weight, can gain access to or exit from brain interstitial space only by carrier-mediated transport through the blood-brain barrier (Chapter 8).

At the other end of the spectrum are liver capillaries, which have large intercellular clefts as well as large holes in the plasma membranes of the endothelial cells so that even protein molecules can readily pass across them. This is very important because two of the major functions of the liver are the synthesis of plasma proteins and the metabolism of substances bound to plasma proteins.

The leakiness of capillaries in most organs and tissues lies between these extremes of brain and liver capillaries.

What is the sequence of events involved in transfers of nutrients and metabolic end products between capillary blood and cells? Nutrients diffuse first from the plasma across the capillary wall into the interstitial fluid, from which they gain entry to cells. Conversely, metabolic end products from the tissues move across the cells' plasma membranes into interstitial fluid, from which they diffuse across the capillary endothelium into the plasma.

Transcapillary diffusion gradients for oxygen and nutrients occur as a result of cellular utilization of the substance. Those for metabolic end products arise as a result of cellular production of the substance. Let us take two examples: glucose and carbon dioxide in muscle.

Glucose is continuously transported from interstitial fluid into the muscle cell by carrier-mediated transport mechanisms. The removal of glucose from interstitial fluid lowers the interstitial-fluid glucose concentration below the glucose concentration in capillary plasma and creates the gradient for diffusion of glucose from the capillary into the interstitial fluid.

Simultaneously, carbon dioxide, which is continuously produced by muscle cells, diffuses into the interstitial fluid. This causes the carbon dioxide concentration in interstitial fluid to be greater than that in capillary plasma, producing a gradient for carbon dioxide diffusion from the interstitial fluid into the capillary.

Note that in both examples, metabolism—either utilization or production—of the substance is the event that ultimately establishes the transcapillary diffusion gradients.

If a tissue is to increase its metabolic rate, it must obtain more nutrients from the blood and it must eliminate more metabolic end products. One mechanism for achieving that is active hyperemia. The second important mechanism is increased diffusion gradients between plasma and tissue: Increased cellular utilization of oxygen and nutrients lowers their tissue concentrations, whereas increased production of carbon dioxide and other end products raises their tissue concentrations. In both cases the substance's transcapillary concentration difference is increased, which increases the rate of diffusion.

Bulk Flow across the Capillary Wall: Distribution of the Extracellular Fluid

At the same time that the diffusional exchange of nutrients, oxygen, and metabolic end products is occurring across the capillaries, another, completely distinct process is also taking place across the capillary—the bulk flow of protein-free plasma. The function of this process is *not* exchange of nutrients and metabolic end products but rather distribution of the extracellular fluid. As described in Chapter 1, extracellular fluid comprises the blood plasma and interstitial fluid. Normally, there is approximately three times more interstitial fluid than plasma, 10 L versus 3 L in a 70-kg person. This distribution is not fixed, however, and the interstitial fluid functions as a reservoir that can supply fluid to the plasma or receive fluid from it.

As described in the previous section, the capillary wall is highly permeable to water and to almost all plasma solutes, except plasma proteins. Therefore, in the presence of a hydrostatic pressure difference across it, the capillary wall behaves like a porous filter through which protein-free plasma moves by bulk flow from capillary plasma to interstitial fluid through the water-filled channels. (This is technically termed "ultrafiltration" but we shall refer to it simply as filtration.) *The concentrations of all the plasma solutes except protein are virtually the same in the filtering fluid as in plasma.*

The magnitude of the bulk flow is determined, in part, by the difference between the capillary blood pressure and the interstitial-fluid hydrostatic pressure. Normally, the former is much larger than the latter. Therefore, a considerable hydrostatic-pressure difference exists to filter protein-free plasma out of the cap-

illaries into the interstitial fluid, the protein remaining behind in the plasma.

Why then does all the plasma not filter out into the interstitial space? The explanation is that the hydrostatic pressure difference favoring filtration is offset by an osmotic force opposing filtration. To understand this, we must review the principle of osmosis.

In Chapter 6, we described how a net movement of water occurs across a semipermeable membrane from a solution of high water concentration to a solution of low water concentration—that is, from a region with a low concentration of solute to which the membrane is impermeable (nonpermeating solute) to a region of high nonpermeating-solute concentration. Moreover, this osmotic flow of water "drags" along with it any dissolved solutes to which the membrane is highly permeable (permeating solute). Thus, a difference in water concentration secondary to different concentrations of *nonpermeating solute* on the two sides of a membrane can result in the movement of a solution containing both water and permeating solutes in a manner similar to the bulk flow produced by a hydrostatic-pressure difference. Units of pressure are used in expressing this osmotic flow across a membrane just as for flow driven by a hydrostatic-pressure difference.

This analysis can now be applied to osmotically induced flow across capillaries. The plasma within the capillary and the interstitial fluid outside it contain large quantities of low-molecular-weight permeating solutes (also termed **crystalloids**), for example, sodium, chloride, and glucose. Since the capillary lining is highly permeable to all these crystalloids, their concentrations in the two solutions are essentially identical (as we have seen, there actually are small concentration differences for substances that are utilized or produced by the cells, but these tend to cancel each other). Accordingly, the presence of the crystalloids causes no significant difference in water concentration. In contrast, the plasma proteins (also termed **colloids**), being essentially nonpermeating, have a very low concentration in the interstitial fluid. The difference in protein concentration between plasma and interstitial fluid means that the water concentration of the plasma is very slightly lower (by about 0.5 percent) than that of interstitial fluid, inducing an osmotic flow of water from the interstitial compartment into the capillary. Since the crystalloids in the interstitial fluid move along with the water, osmotic flow of fluid, like flow driven by a hydrostatic-pressure difference, does not alter crystalloid *concentrations* in either plasma or interstitial fluid.

A key word in this last sentence is "concentrations." The *amount* of water (the *volume*) and the *amount* of

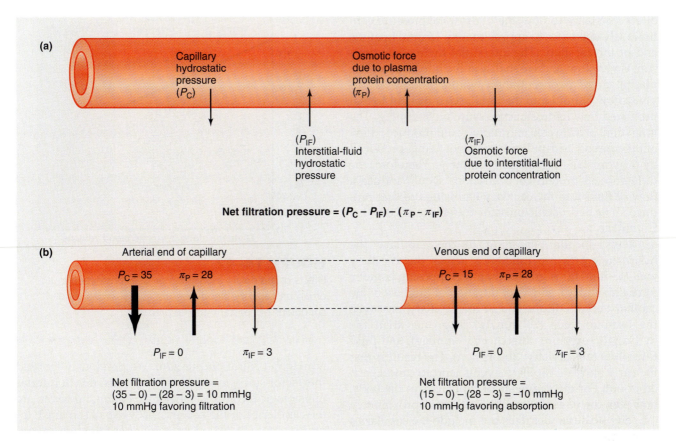

FIGURE 14–46

(a) The four factors determining fluid movement across capillaries. (b) Quantitation of forces causing filtration at the arterial end of the capillary and absorption at the venous end. Arrows in (b) denote magnitude of forces. No arrow is shown for interstitial-fluid hydrostatic pressure (P_{IF}) in (b) because it is approximately zero.

crystalloids in the two locations do change. Thus, an increased filtration of fluid from plasma to interstitial fluid increases the volume of the interstitial fluid and decreases the volume of the plasma, even though no changes in crystalloid concentrations occur.

In summary (Figure 14–46a), opposing forces act to move fluid across the capillary wall: (1) the difference between capillary blood hydrostatic pressure and interstitial-fluid hydrostatic pressure favors filtration out of the capillary; and (2) the water-concentration difference between plasma and interstitial fluid, which results from differences in protein concentration, favors the filtration of interstitial fluid *into* the capillary (filtration in this direction is termed **absorption**). Accordingly, the **net filtration pressure (NFP)** depends directly upon the algebraic sum of four variables: capillary hydrostatic pressure, P_c (favoring fluid movement out of the capillary); interstitial hydrostatic pressure, P_{IF} (favoring fluid movement into the capillary);

the osmotic force due to plasma protein concentration, π_p (favoring fluid movement into the capillary); and the osmotic force due to interstitial-fluid protein concentration, π_{IF} (favoring fluid movement out of the capillary).

$$\begin{aligned} NFP &= P_c - P_{IF} - \pi_p + \pi_{IF} \\ &= (P_c - P_{IF}) - (\pi_p - \pi_{IF}) \end{aligned}$$

The four factors that determine net filtration pressure are termed the **Starling forces** (because Starling, the same physiologist who helped elucidate the Frank-Starling mechanism of the heart, was the first to develop the ideas).

We may now consider this movement quantitatively in the systemic circulation (Figure 14–46b). Much of the arterial blood pressure has already been dissipated as the blood flows through the arterioles, so that hydrostatic pressure at the beginning of a typical capillary is about 35 mmHg. Since the capillary also offers resistance to flow, the hydrostatic pressure continuously decreases to approximately 15 mmHg at the end of the capillary. The interstitial hydrostatic pressure is very low, and we shall assume it to be zero. The plasma protein concentration would produce an osmotic flow of fluid into the capillary equivalent to that produced by a hydrostatic pressure of 28 mmHg. The interstitial protein concentration would produce a flow of fluid out of the capillary equivalent to that produced by a hydrostatic pressure of 3 mmHg. Therefore, the difference in protein concentrations would induce a flow of fluid into the capillary equivalent to that produced by a hydrostatic-pressure difference of $28 - 3 = 25$ mmHg.

Thus, in the beginning of the capillary the hydrostatic-pressure difference across the capillary wall (35 mmHg) is greater than the opposing osmotic force (25 mmHg), and a net filtration of fluid *out of* the capillary occurs. In the end of the capillary, however, the osmotic force (25 mmHg) is greater than the hydrostatic-pressure difference (15 mmHg), and fluid filters *into* the capillary (absorption). The result is that the early and late capillary events tend to cancel each other out. For the aggregate of capillaries in the body, however, there is small net filtration of approximately 4 L/day (this number does not include the capillaries in the kidneys). The fate of this fluid will be described in the section on the lymphatic system.

In our example, we have assumed a "typical" capillary hydrostatic pressure of 35 mmHg. In reality, capillary hydrostatic pressures vary in different vascular beds and, as will be described in a later section, are strongly influenced by whether the person is lying down, sitting, or standing. Moreover, a very important point is that capillary hydrostatic pressure in any vascular bed is subject to physiological regulation, mediated mainly by changes in the resistance of the arterioles in that bed. As shown in Figure 14–47, dilating the arterioles in a particular vascular bed raises capillary hydrostatic pressure in that bed because less pressure is lost overcoming resistance between the arteries and the capillaries. Because of the increased capillary hydrostatic pressure, filtration is increased, and more protein-free fluid is lost to the interstitial fluid. In contrast, marked arteriolar constriction produces decreased capillary hydrostatic pressure and hence favors net movement of interstitial fluid into the vascular compartment. Indeed, the arterioles supplying a capillary bed may be so dilated or so constricted that the capillaries manifest only filtration or only reabsorption, respectively, along their entire length.

We have presented the story of capillary filtration entirely in terms of the Starling forces, but one other factor is involved—the **capillary filtration coefficient.** This is a measure of how much fluid will filter per

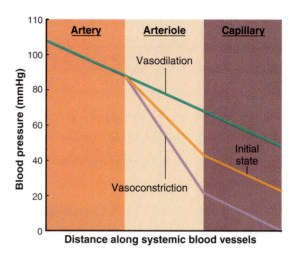

FIGURE 14–47

Effects of arteriolar vasodilation or vasoconstriction in an organ on capillary blood pressure in that organ (under conditions of constant arterial pressure).

mmHg net filtration pressure. We previously ignored this factor since, in most capillary beds, it is not under physiological control. A major exception, however, are the capillaries of the kidneys; as we shall see in Chapter 16, certain of the kidney capillaries filter huge quantities of protein-free fluid because they have a very large capillary filtration coefficient, one that can be altered physiologically.

It must be stated again that capillary filtration and absorption play no significant role in the exchange of nutrients and metabolic end products between capillary and tissues. The reason is that the total quantity of a substance, such as glucose or carbon dioxide, moving into or out of a capillary as a result of net bulk flow is extremely small in comparison with the quantities moving by net diffusion.

Finally, this analysis of capillary fluid dynamics has been in terms of the systemic circulation. Precisely the same Starling forces apply to the capillaries in the pulmonary circulation, but the values of the four variables differ. In particular, because the pulmonary circulation is a low-resistance, low-pressure circuit, the normal pulmonary capillary hydrostatic pressure—the major force favoring movement of fluid out of the pulmonary capillaries into the interstitium—is only 15 mmHg. Therefore, normally, net absorption of fluid occurs along the entire length of lung capillaries.

Veins

Blood flows from capillaries into venules and then into veins. Some exchange of materials occurs between the interstitial fluid and the venules, just as in capillaries.

Indeed, permeability to macromolecules is often greater for venules than for capillaries, particularly in damaged areas.

The veins are the last set of tubes through which blood flows on its way back to the heart. In the systemic circulation, the force driving this venous return is the pressure difference between the peripheral veins and the right atrium. The pressure in the first portion of the peripheral veins is generally quite low—only 5 to 10 mmHg—because most of the pressure imparted to the blood by the heart is dissipated by resistance as blood flows through the arterioles, capillaries, and venules. The right atrial pressure is normally approximately 0 mmHg. Therefore, the total driving pressure for flow from the peripheral veins to the right atrium is only 5 to 10 mmHg. This pressure difference is adequate because of the low resistance to flow offered by the veins, which have large diameters. Thus, a major function of the veins is to act as low-resistance conduits for blood flow from the tissue to the heart.

The veins outside the chest, the **peripheral veins,** contain valves that permit flow only toward the heart. Why are these valves necessary if the pressure gradient created by cardiac contraction pushes blood only toward the heart anyway? The answer will be given below in the section on determinants of venous pressure.

In addition to their function as low-resistance conduits, the veins perform a second important function: Their diameters are reflexly altered in response to changes in blood volume, thereby maintaining peripheral venous pressure and venous return to the heart. In a previous section, we emphasized that the rate of venous return to the heart is a major determinant of end-diastolic ventricular volume and thereby stroke volume. Thus, we now see that peripheral venous pressure is an important determinant of stroke volume.

Determinants of Venous Pressure

The factors determining pressure in any elastic tube are, as we know, the volume of fluid within it and the compliance of its walls. Accordingly, total blood volume is one important determinant of venous pressure since, as we shall see, at any given moment most blood is in the veins. Also, the walls of veins are thinner and much more compliant than those of arteries. Thus, veins can accommodate large volumes of blood with a relatively small increase in internal pressure. Approximately 60 percent of the total blood volume is present in the systemic veins at any given moment (Figure 14–48), but the venous pressure averages less than 10 mmHg. (In contrast, the systemic arteries contain less than 15 percent of the blood, at a pressure of approximately 100 mmHg.)

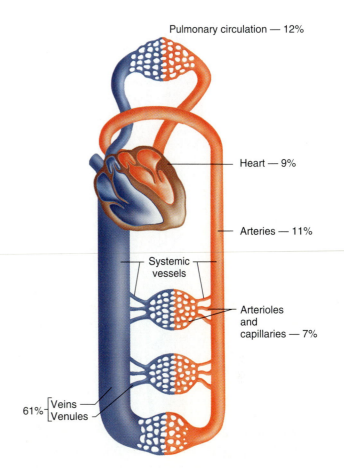

FIGURE 14–48

Distribution of the total blood volume in different parts of the cardiovascular system.

Adapted from Guyton.

The walls of the veins contain smooth muscle innervated by sympathetic neurons. Stimulation of these neurons releases norepinephrine, which causes contraction of the venous smooth muscle, decreasing the diameter and compliance of the vessels and raising the pressure within them. Increased venous pressure then drives more blood out of the veins into the right heart. Although the sympathetic nerves are the most important input, venous smooth muscle, like arteriolar smooth muscle, is also influenced by hormonal and paracrine vasodilators and vasoconstrictors.

Two other mechanisms, in addition to contraction of venous smooth muscle, can increase venous pressure and facilitate venous return. These mechanisms are the **skeletal-muscle pump** and the **respiratory pump.** During skeletal-muscle contraction, the veins running through the muscle are partially compressed, which reduces their diameter and forces more blood back to the heart. Now we can describe a major function of the peripheral-vein valves: When

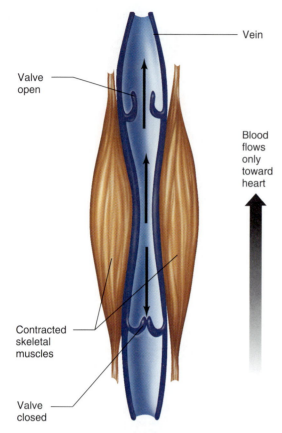

Vein

Valve
open

Blood
flows
only
toward
heart

Contracted
skeletal
muscles

Valve
closed

FIGURE 14–49

The skeletal-muscle pump. During muscle contraction, venous diameter decreases, and venous pressure rises. The resulting increase in blood flow can occur only toward the heart because the valves in the veins are forced closed by any backward flow.

the skeletal-muscle pump raises venous pressure locally, the valves permit blood flow only toward the heart and prevent flow back toward the tissues (Figure 14–49).

The respiratory pump is somewhat more difficult to visualize. As will be described in Chapter 15, during inspiration of air, the diaphragm descends, pushes on the abdominal contents, and increases abdominal pressure. This pressure increase is transmitted passively to the intraabdominal veins. Simultaneously, the pressure in the thorax decreases, thereby decreasing the pressure in the intrathoracic veins and right atrium. The net effect of the pressure changes in the abdomen and thorax is to increase the pressure difference between the peripheral veins and the heart. Accordingly, venous return is enhanced during inspiration (expiration would reverse this effect if it were not for the venous valves). The larger the inspiration, the greater the effect. Thus, breathing deeply and frequently, as in exercise, helps blood flow from the peripheral veins to the heart.

One might get the (incorrect) impression from these descriptions that venous return and cardiac output are independent entities. However, any change in venous return, due say to the skeletal-muscle pump, almost immediately causes equivalent changes in cardiac output, largely through the operation of the Frank-Starling mechanism. *Venous return and cardiac output therefore must be identical except for very brief periods of time.*

In summary (Figure 14–50), the effects of venous smooth-muscle contraction, the skeletal-muscle pump, and the respiratory pump are to facilitate venous return and thereby to enhance cardiac output by the same amount.

The Lymphatic System

The **lymphatic system** is a network of small organs (lymph nodes) and tubes (**lymphatic vessels** or simply "lymphatics") through which **lymph**—a fluid derived from interstitial fluid—flows. The lymphatic system is not technically part of the cardiovascular system, but it is described in this chapter because its vessels constitute a route for the movement of interstitial fluid to the cardiovascular system (Figure 14–51).

Present in the interstitium of virtually all organs and tissues are numerous **lymphatic capillaries** that are completely distinct from blood-vessel capillaries. Like the latter, they are tubes made of only a single layer of endothelial cells resting on a basement membrane, but they have large water-filled channels that are permeable to all interstitial-fluid constituents, including protein. The lymphatic capillaries are the first of the lymphatic vessels, for unlike the blood-vessel capillaries, no tubes flow into them.

Small amounts of interstitial fluid continuously enter the lymphatic capillaries by bulk flow (the precise mechanisms by which this occurs remain unclear). Now known as lymph, the fluid flows from the lymphatic capillaries into the next set of lymphatic vessels, which converge to form larger and larger lymphatic vessels. At various points, the lymph flows through lymph nodes, the function of which is described in Chapter 20. Ultimately, the entire network ends in two large lymphatic ducts that drain into the subclavian veins in the lower neck. Valves at these junctions permit only one-way flow from lymphatic ducts into the veins. Thus, the lymphatic vessels carry interstitial fluid to the cardiovascular system.

The movement of interstitial fluid to the cardiovascular system via the lymphatics is very important because, as noted earlier, the amount of fluid filtered out of all the blood-vessel capillaries (except those in the kidneys) exceeds that reabsorbed by approximately

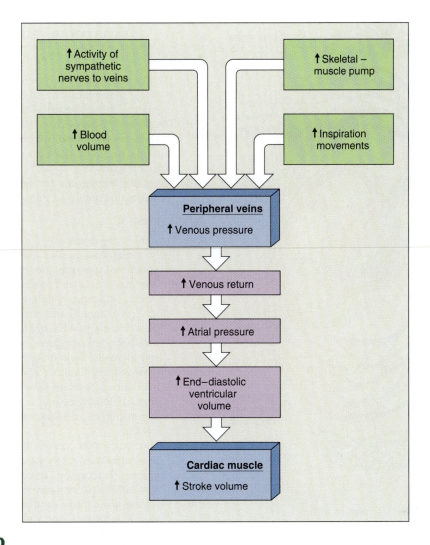

FIGURE 14–50

Major factors determining peripheral venous pressure and, hence, venous return and stroke volume. The figure shows how venous pressure and stroke volume are increased. Reversing the arrows in the boxes would indicate how these can be decreased. The effects of increased inspiration on end-diastolic ventricular volume are actually quite complex, but for the sake of simplicity, they are shown only as increasing venous pressure.

4 L each day. This 4 L is returned to the blood via the lymphatic system. In the process, the small amounts of protein that leak out of blood-vessel capillaries into the interstitial fluid are also returned to the cardiovascular system.

Failure of the lymphatic system due, for example, to occlusion by infectious organisms (as in the disease *elephantiasis*) allows the accumulation of excessive interstitial fluid. The result can be massive swelling of the involved area. The accumulation of large amounts of interstitial fluid from whatever cause (others are described in the section on heart failure) is termed *edema.*

In addition to draining excess interstitial fluid, the lymphatic system provides the pathway by which fat absorbed from the gastrointestinal tract reaches the blood (Chapter 17). The lymphatics also, unfortunately,

are often the route by which cancer cells spread from their area of origin to other parts of the body.

Mechanism of Lymph Flow

In large part, the lymphatic vessels beyond the lymphatic capillaries propel the lymph within them by their own contractions. The smooth muscle in the wall of the lymphatics exerts a pumplike action by inherent rhythmical contractions. Since the lymphatic vessels have valves similar to those in veins, these contractions produce a one-way flow toward the points at which the lymphatics enter the circulatory system. The lymphatic-vessel smooth muscle is responsive to stretch, so when there is no accumulation of interstitial fluid, and hence no entry of lymph into the lymphatics, the smooth muscle is inactive. As lymph

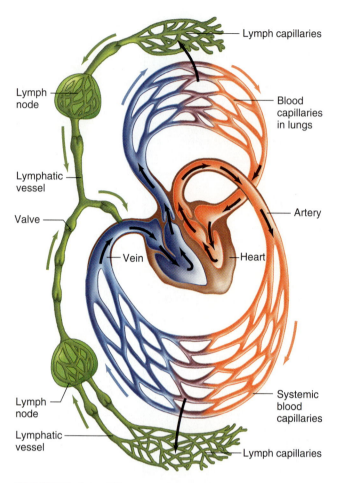

Lymph capillaries

Lymph node

Lymphatic vessel

Valve

Vein

Lymph node

Lymphatic vessel

Lymph capillaries

Blood capillaries in lungs

Artery

Heart

Systemic blood capillaries

FIGURE 14–51

The lymphatic system (green) in relation to the cardiovascular system (blue and red). The lymphatic system is a one-way system from interstitial fluid to the cardiovascular system. ⚡

formation increases, however, say as a result of increased fluid filtration out of blood-vessel capillaries, the increased fluid entering the lymphatics stretches the walls and triggers rhythmical contractions of the smooth muscle. This constitutes a negative-feedback mechanism for adjusting the rate of lymph flow to the rate of lymph formation and thereby preventing edema.

In addition, the smooth muscle of the lymphatic vessels is innervated by sympathetic neurons, and excitation of these neurons in various physiological states such as exercise may contribute to increased lymph flow. Lymph flow is also enhanced by forces external to the lymphatic vessels. These include the same external forces we described for veins—the skeletal-muscle pump and respiratory pump.

Arteries

I. The arteries function as low-resistance conduits and as pressure reservoirs for maintaining blood flow to the tissues during ventricular relaxation.

II. The difference between maximal arterial pressure (systolic pressure) and minimal arterial pressure (diastolic pressure) during a cardiac cycle is the pulse pressure.

III. Mean arterial pressure can be estimated as diastolic pressure plus one-third pulse pressure.

Arterioles

I. Arterioles, the dominant site of resistance to flow in the vascular system, play major roles in determining mean arterial pressure and in distributing flows to the various organs and tissues.

II. Arteriolar resistance is determined by local factors and by reflex neural and hormonal input.

   a. Local factors that change with the degree of metabolic activity cause the arteriolar vasodilation and increased flow of active hyperemia.

   b. Flow autoregulation, a change in resistance that maintains a constant flow in the face of changing arterial blood pressure, is due to local metabolic factors and to arteriolar myogenic responses to stretch.

   c. The sympathetic nerves, the only innervation of most arterioles, cause vasoconstriction via alpha-adrenergic receptors. In certain cases, noncholinergic, nonadrenergic neurons that release nitric oxide or other noncholinergic vasodilators also innervate blood vessels.

   d. Epinephrine causes vasoconstriction or vasodilation, depending on the proportion of alpha- and beta-adrenergic receptors in the organ.

   e. Angiotensin II and vasopressin cause vasoconstriction.

   f. Some chemical inputs act by stimulating endothelial cells to release vasodilator or vasoconstrictor paracrine agents, which then act on adjacent smooth muscle. These paracrine agents include the vasodilators nitric oxide (endothelium-derived relaxing factor) and prostacyclin, and the vasoconstrictor endothelin-1.

III. Arteriolar control in specific organs is summarized in Table 14–8.

Capillaries

I. Capillaries are the site of exchange of nutrients and waste products between blood and tissues.

II. Blood flows through the capillaries more slowly than through any other part of the vascular system because of the huge cross-sectional area of the capillaries.

III. Capillary blood flow is determined by the resistance of the arterioles supplying the capillaries and by the number of open precapillary sphincters.
IV. Diffusion is the mechanism by which nutrients and metabolic end products are exchanged between capillary plasma and interstitial fluid.
   a. Lipid-soluble substances move across the entire endothelial wall, whereas ions and polar molecules move through water-filled intercellular clefts or fused-vesicle channels.
   b. Plasma proteins move across most capillaries only very slowly, either by diffusion through water-filled channels or by vesicle transport.
   c. The diffusion gradient for a substance across capillaries arises as a result of cell utilization or production of the substance. Increased metabolism increases the diffusion gradient and increases the rate of diffusion.
V. Bulk flow of protein-free plasma or interstitial fluid across capillaries determines the distribution of extracellular fluid between these two fluid compartments.
   a. Filtration from plasma to interstitial fluid is favored by the hydrostatic-pressure difference between the capillary and the interstitial fluid. Absorption from interstitial fluid to plasma is favored by the protein concentration difference between the plasma and the interstitial fluid.
   b. Filtration and absorption do not change the concentrations of crystalloids in the plasma and interstitial fluid because these substances move together with water.
   c. There is normally a small excess of filtration over absorption.

Veins

I. Veins serve as low-resistance conduits for venous return.
II. Veins are very compliant and contain most of the blood in the vascular system.
   a. Their diameters are reflexly altered by sympathetically mediated vasoconstriction so as to maintain venous pressure and venous return.
   b. The skeletal-muscle pump and respiratory pump increase venous pressure locally and enhance venous return. Venous valves permit the pressure to produce only flow toward the heart.

The Lymphatic System

I. The lymphatic system provides a one-way route for movement of interstitial fluid to the cardiovascular system.
II. Lymph returns the excess fluid filtered from the blood-vessel capillaries, as well as the protein that leaks out of the blood-vessel capillaries.
III. Lymph flow is driven mainly by contraction of smooth muscle in the lymphatic vessels, but also by the skeletal-muscle pump and the respiratory pump.

compliance
systolic pressure (SP)
diastolic pressure (DP)
pulse pressure
mean arterial pressure (MAP)
vasodilation
vasoconstriction
intrinsic tone
local controls
hyperemia
active hyperemia
bradykinin
kininogen
kallikrein
flow autoregulation
myogenic response
reactive hyperemia
nitric oxide
angiotensin II
vasopressin
atrial natriuretic factor
endothelium-derived relaxing factor (EDRF)

prostacyclin (PGI$_2$)
endothelin-1 (ET-1)
shear stress
flow-induced arterial vasodilation
angiogenesis
angiogenic factors
intercellular cleft
fused-vesicle channel
metarteriole
precapillary sphincter
crystalloids
colloids
absorption
net filtration pressure (NFP)
Starling forces
capillary filtration coefficient
peripheral vein
skeletal-muscle pump
respiratory pump
lymphatic system
lymphatic vessel
lymph
lymphatic capillary

1. Draw the pressure changes along the systemic and pulmonary vascular systems during the cardiac cycle.
2. What are the two functions of the arteries?
3. What are normal values for systolic, diastolic, and mean arterial pressures? How is mean arterial pressure estimated?
4. What are two major factors that determine pulse pressure?
5. What denotes systolic and diastolic pressure in the measurement of arterial pressure with a sphygmomanometer?
6. What are the major sites of resistance in the systemic vascular system?
7. What are two functions of arterioles?
8. Write the formula relating flow through an organ to mean arterial pressure and the resistance to flow offered by that organ.
9. List the chemical factors thought to mediate active hyperemia.
10. Name the mechanism other than chemical factors that contributes to flow autoregulation.
11. What is the only autonomic innervation of most arterioles? What are the major adrenergic receptors influenced by these nerves? How can control of sympathetic nerves to arterioles achieve vasodilation?
12. Name four hormones that cause vasodilation or vasoconstriction of arterioles and specify their effects.

13. Describe the role of endothelial paracrine agents in mediating arteriolar vasoconstriction and vasodilation, and give three examples.
14. Draw a flow diagram summarizing the factors affecting arteriolar radius.
15. What are the relative velocities of flow through the various segments of the vascular system?
16. Contrast diffusion and bulk flow. Which is the mechanism of exchange of nutrients, oxygen, and metabolic end products across the capillary wall?
17. What is the only solute to have significant concentration differences across the capillary wall? How does this difference influence water concentration?
18. What four variables determine the net filtration pressure across the capillary wall? Give representative values for each of them in the systemic capillaries.
19. How do changes in local arteriolar resistance influence local capillary pressure?
20. What is the relationship between cardiac output and venous return in the steady state? What is the force driving venous return?
21. Contrast the compliances and blood volumes of the veins and arteries.
22. What three factors influence venous pressure?
23. Approximately how much fluid is returned to the blood by the lymphatics each day?
24. Describe the forces that cause lymph flow.

INTEGRATION OF CARDIOVASCULAR FUNCTION: REGULATION OF SYSTEMIC ARTERIAL PRESSURE

In Chapter 7 we described the fundamental ingredients of all reflex control systems: (1) an internal environmental variable being maintained relatively constant, (2) receptors sensitive to changes in this variable, (3) afferent pathways from the receptors, (4) an integrating center that receives and integrates the afferent inputs, (5) efferent pathways from the integrated center, and (6) effectors "directed" by the efferent pathways to alter their activities. The control and integration of cardiovascular function will be described in these terms.

The major cardiovascular variable being regulated is the mean arterial pressure in the systemic circulation. This should not be surprising since this pressure is the driving force for blood flow through all the organs except the lungs. Maintaining it is therefore a prerequisite for ensuring adequate blood flow to these organs.

The mean systemic arterial pressure is the arithmetic product of two factors: (1) the cardiac output and (2) the **total peripheral resistance (TPR),** which is the sum of the resistances to flow offered by all the systemic blood vessels.

Mean systemic arterial pressure (MAP)	=	Cardiac output (CO)	×	Total peripheral resistance (TPR)

These two factors, cardiac output and total peripheral resistance, set the mean systemic arterial pressure because they determine the average volume of blood in the systemic arteries over time, and it is this blood volume that causes the pressure. This relationship cannot be emphasized too strongly: *All changes in mean arterial pressure must be the result of changes in cardiac output and/or total peripheral resistance.* Keep in mind that mean arterial pressure will change only if the arithmetic product of cardiac output and total peripheral resistance changes. For example, if cardiac output doubles and total peripheral resistance goes down 50 percent, mean arterial pressure will not change because the product of cardiac output and total peripheral resistance has not changed.

That the volume of blood pumped into the arteries per unit time—the cardiac output—is one of the two direct determinants of mean arterial blood volume and hence mean arterial pressure should come as no surprise. That the total resistance of the blood vessels to flow—the TPR—is the other determinant may not be so intuitively obvious but can be illustrated by using the model introduced previously in Figure 14–38.

As shown in Figure 14–52, a pump pushes fluid into a container at the rate of 1 L/min. At steady state, fluid also leaves the container via outflow tubes at a total rate of 1 L/min. Therefore, the height of the fluid column (ΔP), which is the driving pressure for outflow, remains stable. We then disturb the steady state by loosening the cuff on outflow tube 1, thereby increasing its radius, reducing its resistance, and increasing its flow. The total outflow for the system immediately becomes greater than 1 L/min, and more fluid leaves the reservoir than enters from the pump. Therefore the volume and hence the height of the fluid column begin to decrease until a new steady state between inflow and outflow is reached. In other words, at any given pump input, a change in *total* outflow resistance must produce changes in the volume and hence the height (pressure) in the reservoir.

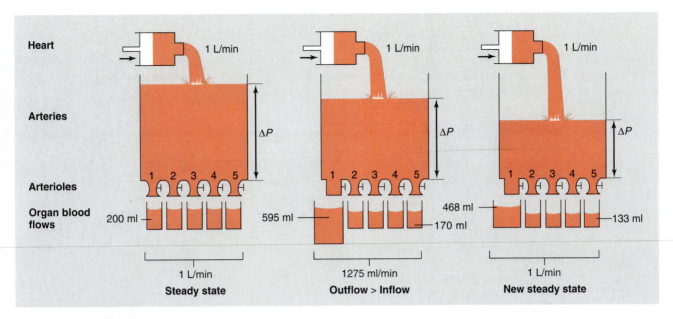

FIGURE 14–52

Dependence of arterial blood pressure upon total arteriolar resistance. Dilating one arteriolar bed affects arterial pressure and organ blood flow if no compensatory adjustments occur. The middle panel indicates a transient state before the new steady state occurs.

This analysis can be applied to the cardiovascular system by again equating the pump with the heart, the reservoir with the arteries, and the outflow tubes with various arteriolar beds. As described earlier, small arteries and capillaries offer some resistance to flow, but the major site of resistance in the systemic blood vessels is the arterioles; moreover *changes* in total resistance are normally due to changes in the resistance of arterioles. Therefore, in our discussions we equate total peripheral resistance with total arteriolar resistance.

A physiological analogy to opening outflow tube 1 is exercise: During exercise, the skeletal-muscle arterioles dilate, thereby decreasing resistance. If the cardiac output and the arteriolar diameters of all other vascular beds were to remain unchanged, the increased runoff through the skeletal-muscle arterioles would cause a decrease in systemic arterial pressure.

It must be reemphasized that it is the *total* arteriolar resistance that influences systemic arterial blood pressure. The *distribution* of resistances among organs is irrelevant in this regard. Figure 14–53 illustrates this

FIGURE 14–53

Compensation for dilation in one bed by constriction in others. When outflow tube 1 is opened, outflow tubes 2 to 4 are simultaneously tightened so that total outflow resistance, total run-off rate, and reservoir pressure all remain constant.

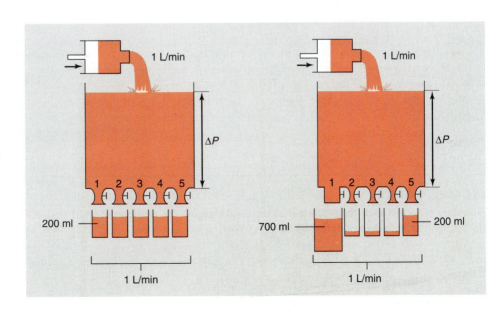

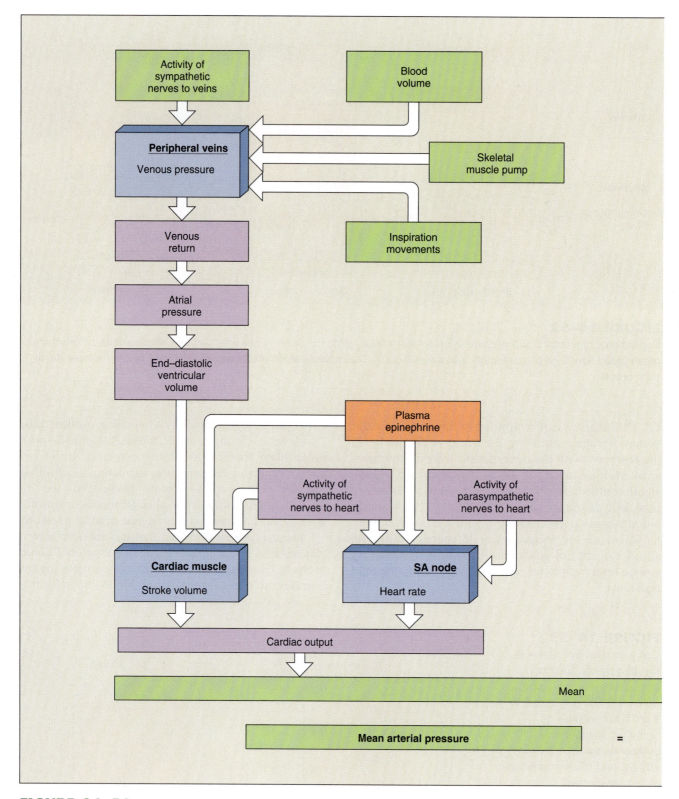

FIGURE 14–54

Summary of factors that determine systemic arterial pressure, an amalgamation of Figures 14–32, 14–41, and 14–50, with the addition of the effect of hematocrit on resistance.

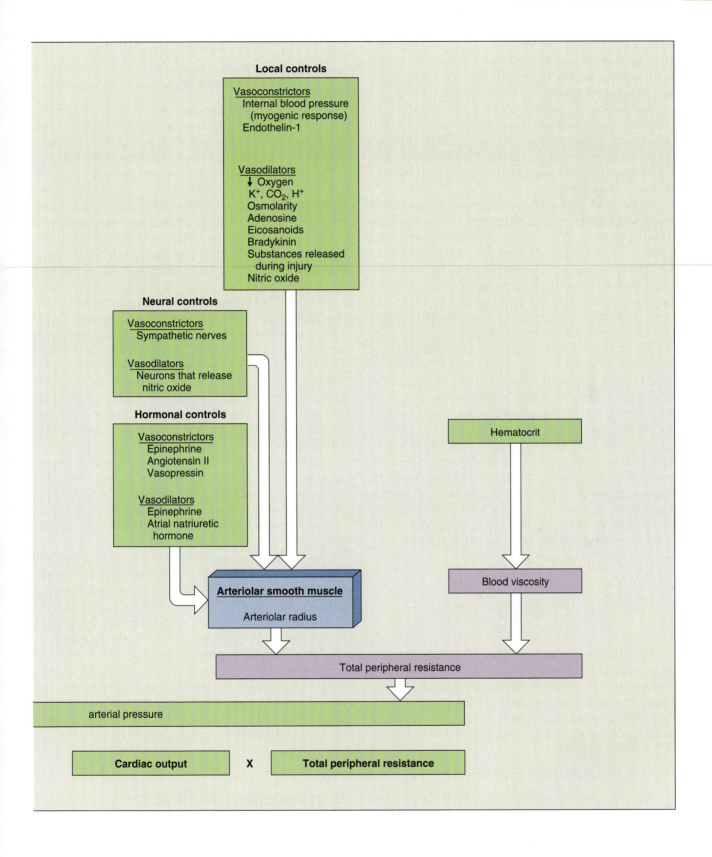

point. On the right, outflow tube 1 has been opened, as in the previous example, while tubes 2 to 4 have been simultaneously tightened. The increased resistance offered by tubes 2 to 4 compensates for the decreased resistance offered by tube 1; therefore total resistance remains unchanged, and reservoir pressure is unchanged. Total outflow remains 1 L/min, although the distribution of flows is such that flow through tube 1 is increased, that of tubes 2 to 4 is decreased, and that of tube 5 is unchanged.

Applied to the systemic circulation, this process is analogous to altering the distribution of systemic vascular resistances. When the skeletal-muscle arterioles (tube 1) dilate during exercise, the *total* resistance of the systemic circulation can still be maintained if arterioles constrict in other organs, such as the kidneys, gastrointestinal tract, and skin (tubes 2 to 4). In contrast, the brain arterioles (tube 5) remain unchanged, ensuring constant brain blood supply.

This type of resistance juggling can maintain total resistance only within limits, however. Obviously if tube 1 opens very wide, even complete closure of the other tubes cannot prevent total outflow resistance from falling. We shall see that this is actually the case during exercise.

We have thus far explained in an intuitive way why cardiac output (CO) and total peripheral resistance (TPR) are the two variables that set mean systemic arterial pressure. This intuitive approach, however, does not explain specifically why MAP is the *arithmetic product* of CO and TPR. This relationship can be derived formally from the basic equation relating flow, pressure, and resistance:

$$F = \Delta P/R$$

Rearranging terms algebraically, we have

$$\Delta P = F \times R$$

Because the systemic vascular system is a continuous series of tubes, this equation holds for the entire system—that is, from the arteries to the right atrium. Therefore, the ΔP term is mean systemic arterial pressure (MAP) minus the pressure in the right atrium, F is the cardiac output (CO), and R is the total peripheral resistance (TPR).

$$\text{MAP} - \text{Right arterial pressure} = \text{CO} \times \text{TPR}$$

Since the pressure in the right atrium is very close to 0 mmHg, we can drop this term and we are left with the equation presented earlier:

$$\text{MAP} = \text{CO} \times \text{TPR}$$

This equation is the fundamental equation of cardiovascular physiology. An analogous equation can also be applied to the pulmonary circulation:

$$\text{Mean pulmonary arterial pressure} = \text{CO} \times \text{Total pulmonary vascular resistance}$$

These equations provide a way to integrate almost all the information presented in this chapter. For example, we can now explain why mean pulmonary arterial pressure is much lower than mean systemic arterial pressure. The cardiac output through the pulmonary and systemic arteries is of course the same. Therefore, the pressures can differ only if the resistances differ. Thus, we can deduce that the pulmonary vessels offer much less resistance to flow than do the systemic vessels. In other words, the total pulmonary vascular resistance is lower than the total peripheral resistance.

Figure 14–54 presents the grand scheme of factors that determine mean systemic arterial pressure. None of this information is new, all of it having been presented in previous figures. A change in only a single variable will produce a change in mean systemic arterial pressure by altering either cardiac output or total peripheral resistance. For example, Figure 14–55 illustrates how the decrease in blood volume occurring during hemorrhage leads to a decrease in mean arterial pressure.

Conversely, any deviation in arterial pressure, such as that occurring during hemorrhage, will elicit homeostatic reflexes so that cardiac output and/or total peripheral resistance will be changed in the direction required to minimize the initial change in arterial pressure.

In the short term—seconds to hours—these homeostatic adjustments to mean arterial pressure are brought about by reflexes termed the baroreceptor reflexes. They utilize mainly changes in the activity of the autonomic nerves supplying the heart and blood vessels, as well as changes in the secretion of the hormones (epinephrine, angiotensin II, and vasopressin) that influence these structures. Over longer time spans, the baroreceptor reflexes become less important, and factors controlling blood volume play a dominant role in determining blood pressure. The next two sections analyze these phenomena.

Baroreceptor Reflexes

Arterial Baroreceptors

It is only logical that the reflexes that homeostatically regulate arterial pressure originate primarily with arterial receptors that respond to changes in pressure. High in the neck, each of the two major vessels

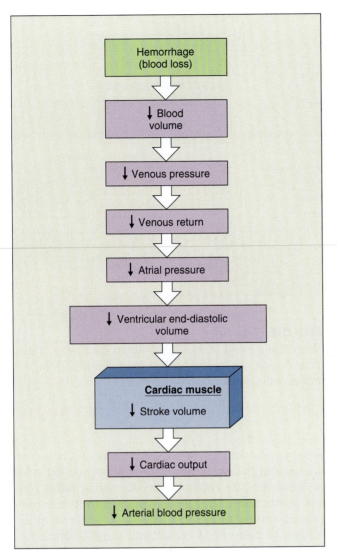

FIGURE 14–55

Sequence of events by which a decrease in blood volume leads to a decrease in mean arterial pressure.

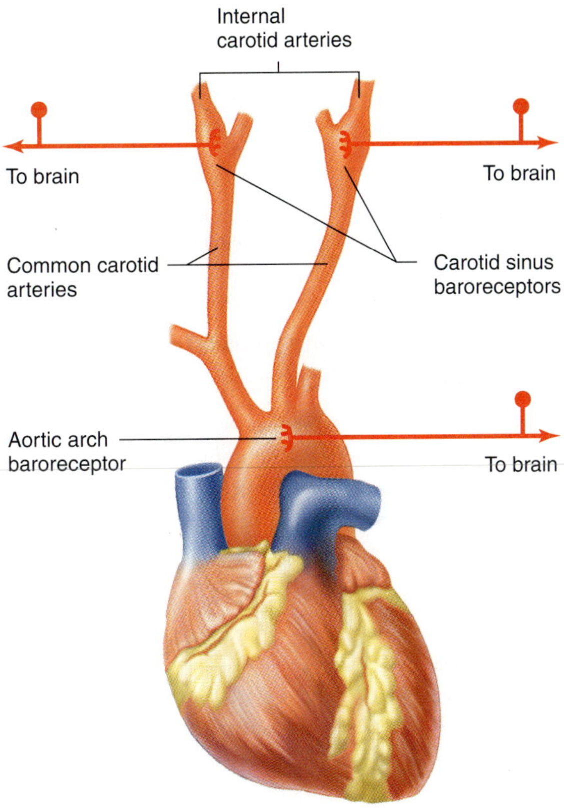

FIGURE 14–56

Locations of arterial baroreceptors.

supplying the head, the common carotid arteries, divides into two smaller arteries (Figure 14–56). At this division, the wall of the artery is thinner than usual and contains a large number of branching, vinelike nerve endings. This portion of the artery is called the **carotid sinus** (the term "sinus" denotes a recess, space, or dilated channel). Its nerve endings are highly sensitive to stretch or distortion. Since the degree of wall stretching is directly related to the pressure within the artery, the carotid sinuses serve as pressure receptors, or **baroreceptors.** An area func-

tionally similar to the carotid sinuses is found in the arch of the aorta and is termed the **aortic arch baroreceptor.** The two carotid sinuses and the aortic arch baroreceptor constitute the **arterial baroreceptors.** Afferent neurons from them travel to the brainstem and provide input to the neurons of cardiovascular control centers there.

Action potentials recorded in single afferent fibers from the carotid sinus demonstrate the pattern of baroreceptor response (Figure 14–57). In this experiment the pressure in the carotid sinus is artificially controlled so that the pressure is either steady or pulsatile—that is, varying as usual between systolic and diastolic pressure. At a particular steady pressure, for example, 100 mmHg, there is a certain rate of discharge by the neuron. This rate can be increased by raising the arterial pressure, or it can be decreased by lowering the pressure. Thus, the rate of discharge of the carotid sinus is directly proportional to the mean arterial pressure.

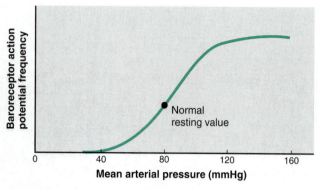

FIGURE 14–57

Effect of changing mean arterial pressure (MAP) on the firing of action potentials by afferent neurons from the carotid sinus. This experiment is done by pumping blood in a nonpulsatile manner through an isolated carotid sinus so as to be able to set the pressure inside it at any value desired.

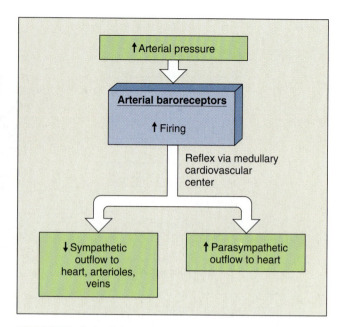

FIGURE 14–58

Neural components of the arterial baroreceptor reflex. If the initial change were a decrease in arterial pressure, all the arrows in the boxes would be reversed.

If the experiment is repeated using the same mean pressures as before but allowing pressure pulsations, it is found that at any given mean pressure, the larger the pulse pressure, the faster the rate of firing by the carotid sinus. This responsiveness to pulse pressure adds a further element of information to blood pressure regulation, since small changes in factors such as blood volume may cause changes in arterial pulse pressure with little or no change in mean arterial pressure.

The Medullary Cardiovascular Center

The primary integrating center for the baroreceptor reflexes is a diffuse network of highly interconnected neurons called the **medullary cardiovascular center,** located in the brainstem medulla oblongata. The neurons in this center receive input from the various baroreceptors. This input determines the outflow from the center along neural pathways that terminate upon the cell bodies and dendrites of the vagus (parasympathetic) neurons to the heart and the sympathetic neurons to the heart, arterioles, and veins. When the arterial baroreceptors *increase* their rate of discharge, the result is a *decrease* in sympathetic outflow to the heart, arterioles, and veins, and an *increase* in parasympathetic outflow to the heart (Figure 14–58). A decrease in baroreceptor firing rate results in just the opposite pattern.

As parts of the baroreceptor reflexes, angiotensin II generation and vasopressin secretion are also altered so as to help restore blood pressure. Thus, decreased arterial pressure elicits increased plasma concentrations of both these hormones, which raise arterial pressure by constricting arterioles. For simplicity, however, we focus in the rest of this chapter mainly on the sympathetic nervous system when discussing reflex control of arterioles. The roles of angiotensin II and vasopressin will be described further in Chapter 16 in the context of their effects on salt and water balance.

Operation of the Arterial Baroreceptor Reflex

Our description of the arterial baroreceptor reflex is now complete. If arterial pressure decreases as during a hemorrhage (Figure 14–59), this causes the discharge rate of the arterial baroreceptors to decrease. Fewer impulses travel up the afferent nerves to the medullary cardiovascular center, and this induces (1) increased heart rate because of increased sympathetic activity to the heart and decreased parasympathetic activity, (2) increased ventricular contractility because of increased sympathetic activity to the ventricular myocardium, (3) arteriolar constriction because of increased sympathetic activity to the arterioles and increased plasma concentrations of angiotensin II and vasopressin, and (4) increased venous constriction because of increased sympathetic activity to the veins. The net result is an increased cardiac output (increased heart rate and stroke volume), increased total peripheral resistance (arteriolar constriction), and return of blood pressure toward normal.

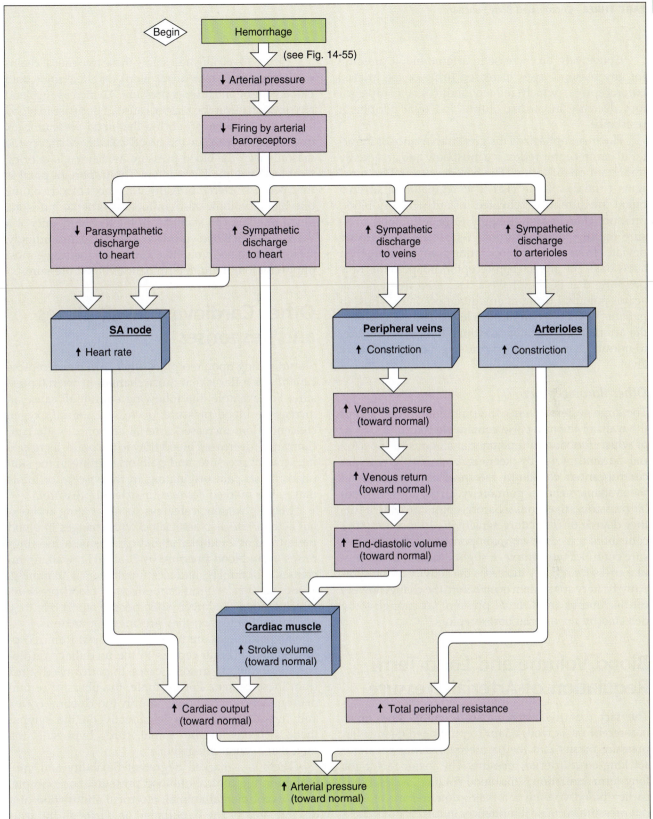

FIGURE 14–59

Arterial baroreceptor reflex compensation for hemorrhage. The compensatory mechanisms do not restore arterial pressure completely to normal. The increases designated "toward normal" are relative to prehemorrhage values; for example, the stroke volume is increased reflexly "toward normal" relative to its low point caused by the hemorrhage (that is, before the reflex occurs), but does not reach the level that existed prior to the hemorrhage. For simplicity, the fact that plasma angiotensin II and vasopressin are also reflexly increased and help constrict arterioles is not shown.

Conversely, an increase in arterial blood pressure for any reason causes increased firing of the arterial baroreceptors, which reflexly induces a compensatory decrease in cardiac output and total peripheral resistance.

Having emphasized the great importance of the arterial baroreceptor reflex, we must now add an equally important qualification. The baroreceptor reflex functions primarily as a *short-term* regulator of arterial blood pressure. It is activated instantly by any blood pressure change and attempts to restore blood pressure rapidly toward normal. Yet, if arterial pressure deviates from its normal operating point for more than a few days, the arterial baroreceptors adapt to this new pressure; that is, they have a decreased frequency of action-potential firing at any given pressure. Thus, in patients who have chronically elevated blood pressure, the arterial baroreceptors continue to oppose minute-to-minute changes in blood pressure, but at the higher level.

Other Baroreceptors

The large systemic veins, the pulmonary vessels, and the walls of the heart also contain baroreceptors, most of which function in a manner analogous to the arterial baroreceptors. By keeping brain cardiovascular control centers constantly informed about changes in the systemic venous, pulmonary, atrial, and ventricular pressures, these other baroreceptors provide a further degree of regulatory sensitivity. In essence, they contribute a feedforward component of arterial pressure control. For example, a slight decrease in ventricular pressure reflexly increases the activity of the sympathetic nervous system even before the change lowers cardiac output and arterial pressure far enough to be detected by the arterial baroreceptors.

Blood Volume and Long-Term Regulation of Arterial Pressure

The fact that the arterial baroreceptors (and other baroreceptors as well) adapt to prolonged changes in pressure means that the baroreceptor reflexes cannot set long-term arterial pressure. The major factor for long-term regulation is the blood volume. As described earlier, blood volume is a major determinant of arterial pressure because it influences in turn venous pressure, venous return, end-diastolic volume, stroke volume, and cardiac output. Thus, an increased blood volume increases arterial pressure. But the opposite causal chain also exists—an increased arterial pressure reduces blood volume (more specifically, the plasma component of the blood volume) by increasing the excretion of salt and water by the kidneys, as described in Chapter 16.

Figure 14–60 illustrates how these two causal chains constitute negative-feedback loops that determine both blood volume and arterial pressure. An increase in blood *pressure,* whatever the reason, causes a decrease in blood volume, which tends to bring the blood pressure back down. An increase in the blood *volume,* whatever the reason, raises the blood pressure, which tends to bring the blood volume back down. The important point is this: Because arterial pressure influences blood volume but blood volume also influences arterial pressure, blood pressure can stabilize, in the long run, only at a value at which blood volume is also stable. Accordingly, steady-state blood volume changes are the single most important long-term determinant of blood pressure.

Other Cardiovascular Reflexes and Responses

Stimuli acting upon receptors other than baroreceptors can initiate reflexes that cause changes in arterial pressure. For example, the following stimuli all cause an increase in blood pressure: decreased arterial oxygen concentration; increased arterial carbon dioxide concentration; decreased blood flow to the brain; increased intracranial pressure; and pain originating in the skin (in contrast, pain originating in the viscera or joints may cause marked *decreases* in arterial pressure).

Many physiological states such as eating and sexual activity are also associated with changes in blood pressure. For example, attending business meetings raises mean blood pressure by 20 mmHg, walking increases it 10 mmHg, and sleeping lowers it 10 mmHg. Mood also has a significant effect on blood pressure, which tends to be lower when people report that they are happy than when they are angry or anxious.

These changes are triggered by input from receptors or higher brain centers to the medullary cardiovascular center or, in some cases, to pathways distinct from these centers. For example, the fibers of certain neurons whose cell bodies are in the cerebral cortex and hypothalamus synapse directly on the sympathetic neurons in the spinal cord, bypassing the medullary center altogether.

There is a marked degree of flexibility and integration in the control of blood pressure. For example, in an experimental animal, electrical stimulation of a discrete area of the hypothalamus elicits all the usually observed neurally mediated cardiovascular responses to an acute emotional situation. Stimulation of other brain sites elicits cardiovascular changes appropriate to the maintenance of body temperature, feeding, or sleeping. It seems that such outputs are "preprogrammed." The complete pattern can be released by a natural stimulus that initiates the flow of information to the appropriate controlling brain center.

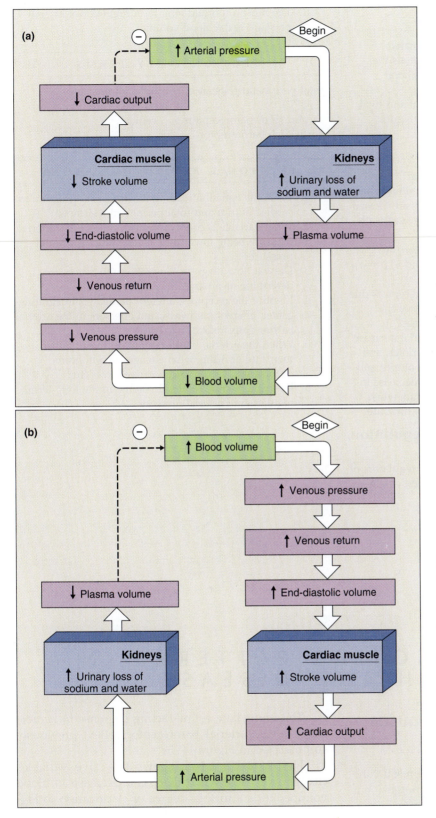

FIGURE 14–60

Causal reciprocal relationships between arterial pressure and blood volume. (a) An increase in arterial pressure due, for example, to an increased cardiac output induces a decrease in blood volume by promoting fluid excretion by the kidneys, which tends to restore arterial pressure to its original value. (b) An increase in blood volume due, for example, to altered kidney function induces an increase in arterial pressure, which tends to restore blood volume to its original value by promoting fluid excretion by the kidneys. Because of these relationships, blood volume is a major determinant of arterial pressure.

I. Mean arterial pressure, the primary regulated variable in the cardiovascular system, equals the product of cardiac output and total peripheral resistance.

II. The factors that determine cardiac output and total peripheral resistance are summarized in Figure 14–54.

Baroreceptor Reflexes

I. The primary baroreceptors are the arterial baroreceptors, including the two carotid sinuses and the aortic arch. Nonarterial baroreceptors are located in the systemic veins, pulmonary vessels, and walls of the heart.

II. The firing rates of the arterial baroreceptors are proportional to mean arterial pressure and to pulse pressure.

III. An increase in firing of the arterial baroreceptors due to an increase in pressure causes, by way of the medullary cardiovascular center, an increase in parasympathetic outflow to the heart and a decrease in sympathetic outflow to the heart, arterioles, and veins. The result is a decrease in cardiac output and total peripheral resistance and, hence, a decrease in mean arterial pressure. The opposite occurs when the initial change is a decrease in arterial pressure.

Blood Volume and Long-Term Regulation of Arterial Pressure

I. The baroreceptor reflexes are short-term regulators of arterial pressure but adapt to a maintained change in pressure.

II. The most important long-term regulator of arterial pressure is the blood volume.

total peripheral resistance (TPR)
carotid sinus baroreceptors
aortic arch baroreceptor
arterial baroreceptors
medullary cardiovascular center

1. Write the equation relating mean arterial pressure to cardiac output and total peripheral resistance.
2. What variable accounts for mean pulmonary arterial pressure being lower than mean systemic arterial pressure?
3. Draw a flow diagram illustrating the factors that determine mean arterial pressure.
4. Identify the receptors, afferent pathways, integrating center, efferent pathways, and effectors in the arterial baroreceptor reflex.
5. When the arterial baroreceptors decrease or increase their rate of firing, what changes in autonomic outflow and cardiovascular function occur?
6. Describe the role of blood volume in the long-term regulation of arterial pressure.

CARDIOVASCULAR PATTERNS IN HEALTH AND DISEASE

Hemorrhage and Other Causes of Hypotension

The term *hypotension* means a low blood pressure, regardless of cause. One general cause of hypotension is a loss of blood volume, as for example in a hemorrhage, which produces hypotension by the sequence of events shown previously in Figure 14–55. The most serious consequences of hypotension are reduced blood flow to the brain and cardiac muscle.

The immediate counteracting response to hemorrhage is the arterial baroreceptor reflex, previously summarized in Figure 14–59.

Figure 14–61, which shows how five variables change over time when there is a decrease in blood volume, adds a further degree of clarification to Figure 14–59. The values of factors changed as a *direct* result of the hemorrhage—stroke volume, cardiac output, and mean arterial pressure—are restored by the baroreceptor reflex *toward,* but not *to,* normal. In contrast, values not altered directly by the hemorrhage but

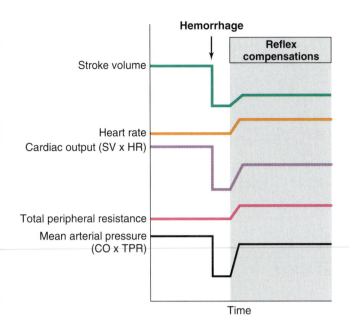

FIGURE 14–61

Five simultaneous graphs showing the time course of cardiovascular effects of hemorrhage. Note that the entire decrease in arterial pressure immediately following hemorrhage is secondary to the decrease in stroke volume and, hence, cardiac output. This figure emphasizes the relativeness of the "increase" and "decrease" arrows of Figure 14–59. All variables shown are increased relative to the state immediately following the hemorrhage, but not necessarily to the state prior to the hemorrhage.

only by the *reflex response* to hemorrhage—heart rate and total peripheral resistance—are increased above their prehemorrhage values. The increased peripheral resistance results from increases in sympathetic outflow to the arterioles in many vascular beds but not those of the heart and brain. Thus, skin blood flow may decrease markedly because of arteriolar vasoconstriction—this is why the skin becomes cold and pale. Kidney and intestinal blood flow also decrease.

A second important type of compensatory mechanism (one not shown in Figure 14–59) involves the movement of interstitial fluid into capillaries. This occurs because both the drop in blood pressure and the increase in arteriolar constriction decrease capillary hydrostatic pressure, thereby favoring absorption of interstitial fluid (Figure 14–62). Thus, the initial event—blood loss and decreased blood volume—is in large part compensated for by the movement of interstitial fluid into the vascular system. Indeed, 12 to 24 h after a moderate hemorrhage, the blood volume may be restored virtually to normal by this mechanism (Table 14–9). At this time, the entire restoration of blood volume is due to expansion of the plasma volume.

The early compensatory mechanisms for hemorrhage—the baroreceptor reflexes and interstitial fluid absorption are highly efficient, so that losses of as much as 1.5 L of blood—approximately 30 percent of total blood volume—can be sustained with only slight reductions of mean arterial pressure or cardiac output.

We must emphasize that absorption of interstitial fluid only *redistributes* the extracellular fluid. Ultimate *replacement* of the fluid lost involves the control of fluid ingestion and kidney function. Both processes are described in Chapter 16. Replacement of the lost erythrocytes requires stimulation of erythropoiesis by erythropoietin. These replacement processes require days to weeks in contrast to the rapidly occurring reflex compensations described in Figure 14–62.

Hemorrhage is a striking example of hypotension due to decrease in blood volume, but loss of large quantities of cell-free extracellular fluid rather than whole blood is also a common cause of low-blood-volume hypotension. In such cases the basic loss is of salts, particularly sodium (along with chloride or bicarbonate) and water. Such fluid loss may occur via the skin, as in severe sweating (Chapter 18) or burns. It may occur via the gastrointestinal tract, as in diarrhea or vomiting (Chapter 17), or via unusually large urinary losses (Chapter 16). Regardless of the route, the loss of fluid decreases circulating blood volume and produces symptoms and compensatory cardiovascular changes similar to those seen in hemorrhage.

Hypotension may be caused by events other than blood or fluid loss. One major cause is a depression of cardiac pumping ability (for example, during a heart attack).

Another cause is strong emotion, during which hypotension can cause fainting. Somehow, the higher brain centers involved with emotions inhibit sympathetic activity to the cardiovascular system and enhance parasympathetic activity to the heart, resulting in a markedly decreased arterial pressure and brain blood flow. This whole process is usually transient. It should be noted that the fainting that sometimes occurs in a person donating blood is usually due to hypotension brought on by emotion, not the blood loss, since losing 0.5 L of blood will not itself cause serious hypotension in normal individuals.

Massive liberation of endogenous substances that relax arteriolar smooth muscle may also cause hypotension by reducing total peripheral resistance. An important example is the hypotension that occurs during severe allergic responses (Chapter 20).

Shock

The term **shock** denotes any situation in which a decrease in blood flow to the organs and tissues damages them. Arterial pressure is usually, but not always, low

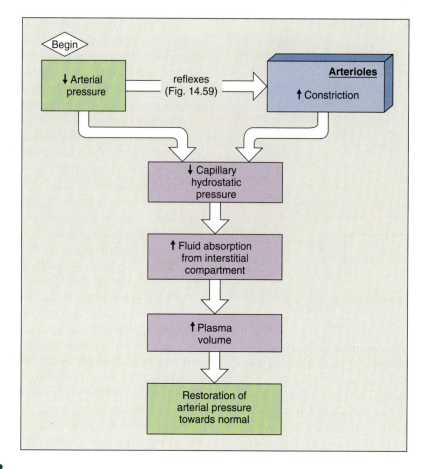

FIGURE 14–62

Mechanisms compensating for blood loss by movement of interstitial fluid into the capillaries.

in shock, and the classification of shock is quite similar to what we have already seen for hypotension: (1) *Hypovolemic shock* is caused by a decrease in blood volume secondary to hemorrhage or loss of fluid other than blood; (2) *low-resistance shock* is due to a decrease in total peripheral resistance secondary to excessive release of vasodilators, as in allergy and infection; and (3) *cardiogenic shock* is due to a marked decrease in cardiac output from a variety of factors (for example, during a heart attack).

The cardiovascular system, especially the heart, suffers damage if shock is prolonged. As the heart deteriorates, cardiac output declines markedly and shock becomes progressively worse and ultimately irreversible even though blood transfusions and other appropriate therapy may temporarily restore blood pressure.

TABLE 14–9 Fluid Shifts after Hemorrhage

| | | Percent of Normal Value | |
	Normal	Immediately after Hemorrhage	18 h after Hemorrhage
Total blood volume, ml	5000	80	98
Erythrocyte volume, ml	2300	80	80
Plasma volume, ml	2700	80	113

The Upright Posture

A decrease in the *effective* circulating blood volume occurs in the circulatory system when going from a lying, horizontal position to a standing, vertical one. Why this is so requires an understanding of the action of gravity upon the long, continuous columns of blood in the vessels between the heart and the feet.

The pressures we have given in previous sections of this chapter are for an individual in the horizontal position, in which all blood vessels are at approximately the same level as the heart. In this position, the *weight* of the blood produces negligible pressure. In contrast, when a person is vertical, the intravascular pressure everywhere becomes equal to the pressure generated by cardiac contraction *plus* an additional pressure equal to the weight of a column of blood from the heart to the point of measurement. In an average adult, for example, the weight of a column of blood extending from the heart to the feet amounts to 80 mmHg. In a foot capillary, therefore, the pressure increases from 25 (the capillary pressure resulting from cardiac contraction) to 105 mmHg, the extra 80 mmHg being due to the weight of the column of blood.

This increase in pressure due to gravity influences the effective circulating blood volume in several ways. First, the increased hydrostatic pressure that occurs in the legs (as well as the buttocks and pelvic area) when a person is quietly standing pushes outward on the highly distensible vein walls, causing marked distension. The result is pooling of blood in the veins; that is, much of the blood emerging from the capillaries simply goes into expanding the veins rather than returning to the heart. Simultaneously, the increase in capillary pressure caused by the gravitational force produces increased filtration of fluid out of the capillaries into the interstitial space. This accounts for the fact that our feet swell during prolonged standing. The combined effects of venous pooling and increased capillary filtration reduce the *effective* circulating blood volume very similarly to the effects caused by a mild hemorrhage. This explains why a person may sometimes feel faint upon standing up suddenly. This feeling is normally very transient, however, since the decrease in arterial pressure immediately causes reflex baroreceptor-mediated compensatory adjustments virtually identical to those shown in Figure 14–59 for hemorrhage.

The effects of gravity can be offset by contraction of the skeletal muscles in the legs. Even gentle contractions of the leg muscles without movement produce intermittent, complete emptying of the leg veins so that uninterrupted columns of venous blood from

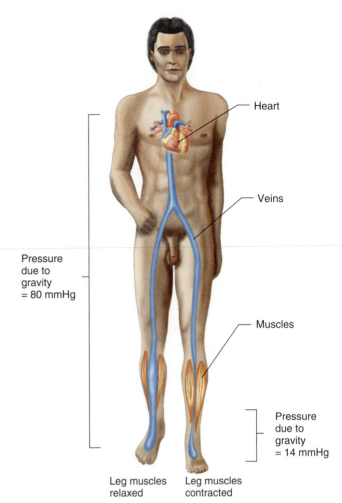

Heart

Veins

Pressure due to gravity = 80 mmHg

Muscles

Pressure due to gravity = 14 mmHg

Leg muscles relaxed   Leg muscles contracted

FIGURE 14–63

Role of contraction of the leg skeletal muscles in reducing capillary pressure and filtration in the upright position. The skeletal-muscle contraction compresses the veins, causing intermittent emptying so that the columns of blood are interrupted.

the heart to the feet no longer exist (Figure 14–63). The result is a decrease in both venous distension and pooling plus a marked reduction in capillary hydrostatic pressure and fluid filtration out of the capillaries. The importance of this phenomenon is illustrated by the fact that soldiers may faint while standing at attention for long periods of time because they have minimal contractions of the leg muscles. Here fainting may be considered adaptive in that the venous and capillary pressure changes induced by gravity are eliminated once the person is horizontal. The pooled venous blood is mobilized, and the filtered fluid is absorbed back into the capillaries. Thus, the wrong thing to do to a person who has fainted for whatever reason is to hold him or her upright.

Exercise

During exercise, cardiac output may increase from a resting value of 5 L/min to a maximal value of 35 L/min in trained athletes. The distribution of this cardiac output during strenuous exercise is illustrated in Figure 14–64. As expected, most of the increase in cardiac output goes to the exercising muscles, but there are also increases in flow to skin, required for dissipation of heat, and to the heart, required for the additional work performed by the heart in pumping the increased cardiac output. The increases in flow through these three vascular beds are the result of arteriolar vasodilation in them. In both skeletal and cardiac muscle, the vasodilation is mediated by local metabolic factors, whereas the vasodilation in skin is achieved mainly by a *decrease* in the firing of the sympathetic neurons to the skin (additional mechanisms are described in Chapter 18). At the same time that arteriolar vasodilation is occurring in these three beds, arteriolar vasoconstriction—manifested as decreased blood flow in Figure 14–64—is occurring in the kidneys and gastrointestinal organs, secondary to *increased* activity of the sympathetic neurons supplying them.

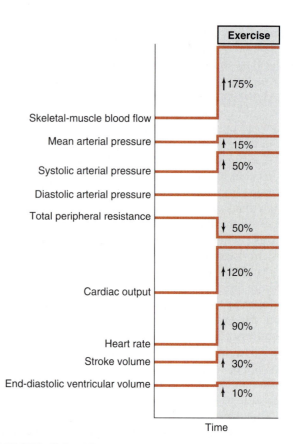

FIGURE 14–65

Summary of cardiovascular changes during mild upright exercise. The person was sitting quietly prior to the exercise.

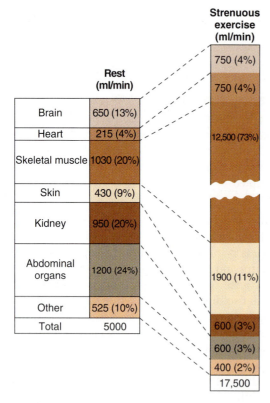

FIGURE 14–64

Distribution of the systemic cardiac output at rest and during strenuous exercise. The values at rest were previously presented in Figure 14–9.

Adapted from Chapman and Mitchell.

Vasodilation of arterioles in skeletal muscle, cardiac muscle, and skin causes a decrease in total peripheral resistance to blood flow. This decrease is partially offset by vasoconstriction of arterioles in other organs. Such resistance "juggling," however is quite incapable of compensating for the huge dilation of the muscle arterioles, and the net result is a marked decrease in total peripheral resistance.

What happens to arterial blood pressure during exercise? As always, the mean arterial pressure is simply the arithmetic product of cardiac output and total peripheral resistance. During most forms of exercise (Figure 14–65 illustrates the case for mild exercise), the cardiac output tends to increase somewhat more than the total peripheral resistance decreases, so that mean arterial pressure usually increases a small amount. Pulse pressure, in contrast, markedly increases, because of an increase in both stroke volume and the speed at which the stroke volume is ejected.

The cardiac output increase during exercise is due to a large increase in heart rate and a small increase in stroke volume. The heart rate increase is caused by a combination of decreased parasympathetic activity to the SA node and increased sympathetic activity. The increased stroke volume is due mainly to an increased ventricular contractility, manifested by an increased ejection fraction and mediated by the sympathetic nerves to the ventricular myocardium.

Note, however, in Figure 14–65 that there is a small increase (10 percent) in end-diastolic ventricular volume. Because of this increased filling, the Frank-Starling mechanism also contributes to the increased stroke volume, although not to the same degree as the increased contractility does. The increased contractility also accounts for the greater speed at which the stroke volume is ejected, as noted in the previous discussion of pulse pressure.

We have focused our attention on factors that act directly upon the heart to alter cardiac output during exercise, but it would be incorrect to leave the impression that these factors, by themselves, are sufficient to account for the elevated cardiac output. The fact is that cardiac output can be increased to high levels only if the peripheral processes favoring venous return to the heart are simultaneously activated to the same degree. Otherwise, the shortened filling time resulting from the high heart rate would lower end-diastolic volume and stroke volume by the Frank-Starling mechanism.

Factors promoting venous return during exercise are: (1) increased activity of the skeletal-muscle pump, (2) increased depth and frequency of inspiration (the respiratory pump), (3) sympathetically mediated increase in venous tone, and (4) greater ease of blood flow from arteries to veins through the dilated skeletal-muscle arterioles.

What are the control mechanisms by which the cardiovascular changes in exercise are elicited? As described previously, vasodilation of arterioles in skeletal and cardiac muscle once exercise is underway represents active hyperemia secondary to local metabolic factors within the muscle. But what drives the enhanced sympathetic outflow to most other arterioles, the heart, and the veins, and the decreased parasympathetic outflow to the heart? The control of this autonomic outflow during exercise offers an excellent example of what we earlier referred to as a preprogrammed pattern, modified by continuous afferent input. One or more discrete control centers in the brain are activated during exercise by output from the cerebral cortex, and according to this "central command," descending pathways from these centers to the appropriate autonomic preganglionic neurons elicit the firing pattern typical of exercise. Indeed, these centers begin to "direct traffic" even before the exercise begins, since a person just about to begin exercising already manifests many of these changes in cardiac and vascular function; thus, this constitutes a feedforward system.

Once exercise is underway, local chemical changes in the muscle can develop, particularly during high levels of exercise, because of imperfect matching between flow and metabolic demands. These changes activate chemoreceptors in the muscle. Afferent input from these receptors goes to the medullary cardiovascular center and facilitates the output reaching the autonomic neurons from higher brain centers (Figure 14–66). The result is a further increase in heart rate, myocardial contractility, and vascular resistance in the nonactive organs. Such a system permits a fine degree of matching between cardiac pumping and total oxygen and nutrients required by the exercising muscles. Mechanoreceptors in the exercising muscles are also stimulated and provide input to the medullary cardiovascular center.

Finally, the arterial baroreceptors also play a role in the altered autonomic outflow. Knowing that the mean and pulsatile pressures rise during exercise, you might logically assume that the arterial baroreceptors will respond to these elevated pressures and signal for increased parasympathetic and decreased sympathetic outflow, a pattern designed to counter the rise in arterial pressure. In reality, however, exactly the opposite occurs; the arterial baroreceptors play an important role in *elevating* the arterial pressure over that existing at rest. The reason is that one neural component of the central command output goes to the arterial baroreceptors and "resets" them upward as exercise begins. This resetting causes the baroreceptors to respond as though arterial pressure had decreased, and their output (decreased action-potential frequency) signals for decreased parasympathetic and increased sympathetic outflow.

Table 14–10 summarizes the changes that occur during moderate endurance exercise—that is, exercise (like jogging, swimming, or fast walking) that involves large muscle groups for an extended period of time.

In closing, a few words should be said about the other major category of exercise, which involves maintained *isometric* contractions, as in weight-lifting. Here, too, cardiac output and arterial blood pressure increase, and the arterioles in the exercising muscles undergo vasodilation due to local metabolic factors. However, there is a crucial difference: During isometric contractions, once the contracting muscles exceed 10 to 15 percent of their maximal force, the blood flow to the muscle is greatly reduced because the muscles are physically compressing the blood vessels that run through them. In other words, the arteriolar vasodilation is completely overcome by the physical compression of the blood vessels. Thus, the cardiovascular changes are ineffective in causing increased blood flow to the muscles, and isometric contractions can be

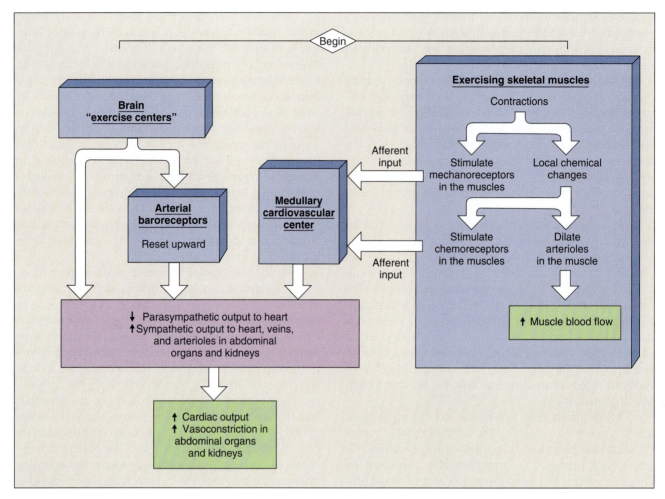

FIGURE 14–66

Control of the cardiovascular system during exercise. The primary outflow to the sympathetic and parasympathetic neurons is via pathways from "exercise centers" in the brain. Afferent input from mechanoreceptors and chemoreceptors in the exercising muscles and from reset arterial baroreceptors also influence the autonomic neurons by way of the medullary cardiovascular center.

maintained only briefly before fatigue sets in. Moreover, because of the blood-vessel compression, total peripheral resistance may go up considerably (instead of down, as in endurance exercise), contributing to a large rise in mean arterial pressure.

Maximal Oxygen Consumption and Training

As the magnitude of any endurance exercise increases, oxygen consumption also increases in exact proportion until a point is reached when it fails to rise despite a further increment in work load. This is known as **maximal oxygen consumption ($\dot{V}_{O_2}$max).** After $\dot{V}_{O_2}$max has been reached, work can be increased and sustained only very briefly by anaerobic metabolism in the exercising muscles.

Theoretically, $\dot{V}_{O_2}$max could be limited by (1) the cardiac output, (2) the respiratory system's ability to deliver oxygen to the blood, or (3) the exercising muscles' ability to use oxygen. In fact, in normal people (except for a few very highly trained athletes), cardiac output is the factor that determines $\dot{V}_{O_2}$max. With increasing workload (Figure 14–67), heart rate increases progressively and markedly until it reaches a maximum. Stroke volume increases much less and tends to level off when 75 percent of $\dot{V}_{O_2}$max has been reached (it actually starts to go back down in elderly people). The major factors responsible for limiting the rise in stroke volume and, hence, cardiac output are (1) the very rapid heart rate, which decreases diastolic filling time, and (2) inability of the peripheral factors

TABLE 14–10 Cardiovascular Changes in Moderate Endurance Exercise

Variable	Change	Explanation
Cardiac output	Increases	Heart rate and stroke volume both increase, the former to a much greater extent.
Heart rate	Increases	Sympathetic nerve activity to the SA node increases, and parasympathetic nerve activity decreases.
Stroke volume	Increases	Contractility increases due to increased sympathetic nerve activity to the ventricular myocardium; increased ventricular end-diastolic volume also contributes to increased stroke volume by the Frank-Starling mechanism.
Total peripheral resistance	Decreases	Resistance in heart and skeletal muscles decreases more than resistance in other vascular beds increases.
Mean arterial pressure	Increases	Cardiac output increases more than total peripheral resistance decreases.
Pulse pressure	Increases	Stroke volume and velocity of ejection of the stroke volume increase.
End-diastolic volume	Increases	Filling time is decreased by the high heart rate, but this is more than compensated for by the factors favoring venous return—venoconstriction, skeletal-muscle pump, and increased inspiratory movements.
Blood flow to heart and skeletal muscle	Increases	Active hyperemia occurs in both vascular beds, mediated by local metabolic factors.
Blood flow to skin	Increases	Sympathetic nerves to skin vessels are inhibited reflexly by the increase in body temperature.
Blood flow to viscera	Decreases	Sympathetic nerves to the blood vessels in the abdominal organs and the kidneys are stimulated.
Blood flow to brain	Unchanged	Autoregulation of brain arterioles maintains constant flow despite the increased mean arterial pressure

favoring venous return (skeletal-muscle pump, respiratory pump, venous vasoconstriction, arteriolar vasodilation) to increase ventricular filling further during the very short time available.

A person's $\dot{V}_{O_2}$max is not fixed at any given value but can be altered by the habitual level of physical activity. For example, prolonged bed rest may decrease $\dot{V}_{O_2}$max by 15 to 25 percent, whereas intense long-term physical training may increase it by a similar amount. To be effective, the training must be of an endurance type and must include certain minimal levels of duration, frequency, and intensity. For example, jogging 20 to 30 min three times weekly at 5 to 8 mi/h definitely produces a significant training effect in most people.

At rest, compared to values prior to training, the trained individual has an increased stroke volume and decreased heart rate with no change in cardiac output (see Figure 14–67). At $\dot{V}_{O_2}$max, he or she has an increased cardiac output, compared to pretraining values; this is due entirely to an increased maximal stroke volume since maximal heart rate is not altered by training (see Figure 14–67). The increase in stroke volume is due to a combination of (1) effects on the heart (the mechanism is unknown but may include a thicker myocardium and increased ventricular contractility), and (2) peripheral effects, including increased blood volume and increases in the number of blood vessels in skeletal muscle, which permit increased muscle blood flow and venous return.

Training also increases the concentrations of oxidative enzymes and mitochondria in the exercised muscles (Chapter 12). These changes increase the speed and efficiency of metabolic reactions in the muscles and permit large increases—200 to 300 percent—in exercise *endurance*, but they do not increase $\dot{V}_{O_2}$max because they were not limiting it in the untrained individuals.

Aging is associated with significant changes in the heart's performance during exercise. Most striking is a decrease in the maximum heart rate (and hence cardiac output) achievable.

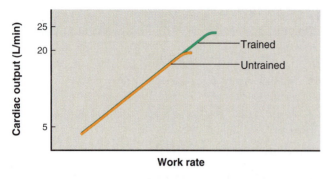

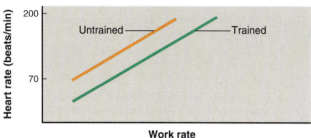

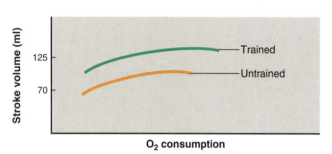

FIGURE 14–67

Changes in cardiac output, heart rate, and stroke volume with increasing work in untrained and trained persons.

Hypertension

Hypertension is defined as a chronically increased systemic arterial pressure. The dividing line between normal pressure and hypertension is set at approximately 140/90 mmHg.

Theoretically, hypertension could result from an increase in cardiac output or in total peripheral resistance, or both. In reality, however, the major abnormality in most cases of well-established hypertension is increased total peripheral resistance caused by abnormally reduced arteriolar radius.

What causes the arteriolar constriction? In only a small fraction of cases is the cause known. For example, diseases that damage a kidney or decrease its blood supply are often associated with **renal hypertension.** The cause of the hypertension is increased release of renin from the kidneys, with subsequent increased generation of the potent vasoconstrictor angiotensin II. However, for more than 95 percent of the individuals with hypertension, the cause of the

arteriolar constriction is unknown. Hypertension of unknown cause is called **primary hypertension** (formerly "essential hypertension").

Many hypotheses have been proposed to explain the increased arteriolar constriction of primary hypertension. At present, much evidence suggests that excessive sodium retention is a contributing factor in genetically predisposed ("salt-sensitive") persons. Many persons with hypertension show a drop in blood pressure after being on low-sodium diets or receiving drugs, termed diuretics, that cause increased sodium loss via the urine. Low dietary intake of calcium has also been implicated as a possible contributor to primary hypertension. Obesity and a sedentary lifestyle are definite risk factors for primary hypertension, and weight reduction and exercise are frequently effective in causing some reduction of blood pressure in persons with hypertension. Cigarette smoking, too, is a definite risk factor.

Hypertension causes a variety of problems. One of the organs most affected is the heart. Because the left ventricle in a hypertensive person must chronically pump against an increased arterial pressure (afterload), it develops an adaptive increase in muscle mass **(left ventricular hypertrophy).** In the early phases of the disease, this helps maintain the heart's function as a pump. With time, however, changes in the organization and properties of myocardial cells occur, and these result in *diminished* contractile function and heart failure (see below). The presence of hypertension also enhances the development of atherosclerosis and heart attacks (see below), kidney damage, and rupture of a cerebral blood vessel, which causes localized brain damage—a **stroke.**

The major categories of drugs used to treat hypertension are summarized in Table 14–11. These drugs all act in ways that reduce cardiac output and/or total peripheral resistance. You will note in subsequent sections of this chapter that these same drugs are also used in the treatment of heart failure and in both the prevention and treatment of heart attacks. One reason for this overlap is that these three diseases are causally interrelated; for example, as noted in this section, hypertension is a major risk factor for the development of heart failure and heart attacks. But in addition, the drugs often have multiple cardiovascular effects, which may play different roles in the treatment of the different diseases.

Heart Failure

Heart failure (also termed congestive heart failure) is a complex of signs and symptoms that occurs when the heart fails to pump an adequate cardiac output.

TABLE 14–11 Drugs Used in the Treatment of Hypertension

1. **Diuretics:** These drugs increase urinary excretion of sodium and water (Chapter 16). They tend to decrease cardiac output with little or no change in total peripheral resistance.

2. **Beta-adrenergic receptor blockers:** These drugs exert their antihypertensive effects mainly by reducing cardiac output.

3. **Calcium channel blockers:** These drugs reduce the entry of calcium into vascular smooth-muscle cells, causing them to contract less strongly. This lowers total peripheral resistance. (Surprisingly, it has been found that despite their effectiveness in lowering blood pressure, at least some of these drugs may significantly increase the risk of a heart attack. Accordingly their use as therapy for hypertension is presently under intensive review.)

4. **Angiotensin-converting enzyme (ACE) inhibitors:** As will be described in Chapter 16, the final step in the formation of angiotensin II, a vasoconstrictor, is mediated by an enzyme called angiotensin-converting enzyme. Drugs that block this enzyme therefore reduce the concentration of angiotensin II in plasma, which causes arteriolar vasodilation, lowering total peripheral resistance. The same effect can be achieved with drugs that block the receptors for angiotensin II. A reduction in plasma angiotensin II or blockage of its receptors is also protective against the development of heart-wall changes that lead to heart failure (see next section in text).

5. Drugs that antagonize one or more components of the sympathetic nervous system: The major effect of these drugs is to reduce sympathetic mediated stimulation of arteriolar smooth muscle and, thereby, reduce total peripheral resistance. Examples are drugs that inhibit the brain centers that mediate the sympathetic outflow to arterioles, and drugs that block alpha-adrenergic receptors on the arterioles.

This may happen for many reasons; two examples are pumping against a chronically elevated arterial pressure in hypertension, and structural damage due to decreased coronary blood flow. It has become standard practice to separate persons with heart failure into two categories: (1) those with diastolic dysfunction (problems with ventricular filling) and (2) those with systolic dysfunction (problems with ventricular ejection). Many persons with heart failure, however, have elements of both categories.

In *diastolic dysfunction* the wall of the ventricle has reduced compliance; that is, it is abnormally stiff and, therefore, has a reduced ability to fill adequately at normal diastolic filling pressures. The result is a reduced end-diastolic *volume* (even though the end-diastolic *pressure* in the stiff ventricle may be quite high) and, therefore, a reduced stroke volume by the Frank-Starling mechanism. Note that in pure diastolic dysfunction, ventricular *compliance* is decreased but ventricular *contractility* is normal.

There are several situations that lead ultimately to decreased ventricular compliance, but by far the most common is the existence of systemic hypertension. As noted in the previous section, the left ventricle, pumping chronically against an elevated arterial pressure, hypertrophies. The structural and biochemical changes associated with this hypertrophy make the ventricle stiff and less able to expand.

In contrast to diastolic dysfunction, *systolic dysfunction* results from myocardial damage due, for example, to a heart attack (see below) and is characterized by a decrease in cardiac *contractility*—a lower stroke volume at any given end-diastolic volume. This is manifested as a decrease in ejection fraction and, as illustrated in Figure 14–68, a downward shift of the ventricular function curve. The affected ventricle does not hypertrophy.

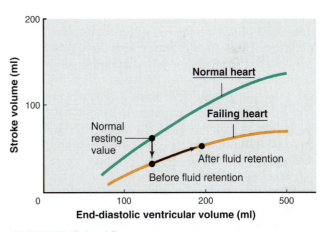

FIGURE 14–68

Relationship between end-diastolic ventricular volume and stroke volume in a normal heart and one with heart failure due to systolic dysfunction (decreased contractility). The normal curve is that shown previously in Figure 14–29. With decreased contractility, the ventricular function curve is displaced downward; that is, there is a lower stroke volume at any given end-diastolic volume. Fluid retention causes an increase in end-diastolic volume and restores stroke volume toward normal by the Frank-Starling mechanism. Note that this compensation occurs even though contractility—the basic defect—has not been altered by the fluid retention.

TABLE 14–12 Drugs Used in the Treatment of Heart Failure

1. **Diuretics:** Drugs that increase urinary excretion of sodium and water (Chapter 16). These drugs eliminate the excessive fluid accumulation contributing to edema and/or worsening myocardial function.

2. **Cardiac inotropic drugs:** Drugs (like **digitalis**) that increase ventricular contractility by increasing cytosolic calcium concentration in the myocardial cell (see Chapter 6 for mechanism). The use of these drugs is presently controversial, however, since although they clearly improve the symptoms of heart failure, they do not prolong life and, in some studies, seem to have shortened it.

3. **Vasodilator drugs:** Drugs that lower total peripheral resistance and hence the arterial blood pressure (afterload) against which the failing heart must pump. Some inhibit a component of the sympathetic nervous pathway to the arterioles, whereas others **[angiotensin-converting enzyme (ACE) inhibitors]** block the formation of angiotensin II (see Chapter 16). In addition, the ACE inhibitors prevent or reverse the maladaptive remodeling of the myocardium that is mediated by the elevated plasma concentration of angiotensin II existing in heart failure.

4. **Beta-adrenergic receptor blockers:** Drugs that block the major adrenergic receptors in the myocardium. The mechanism by which this action improves heart failure is unknown (indeed, you might have predicted that such an action, by blocking sympathetically induced increases in cardiac contractility, would be counterproductive). One hypothesis is as follows: Excess sympathetic stimulation of the heart reflexly produced by the decreased cardiac output of heart failure may cause an excessive elevation of cytosolic calcium concentration, which would lead to cell apoptosis and necrosis; beta-adrenergic receptor blockers would prevent this.

The reduced cardiac output of heart failure, regardless of whether it is due to diastolic or systolic dysfunction, triggers the arterial baroreceptor reflexes. In this situation these reflexes are elicited more than usual because, for unknown reasons, there is a decreased sensitivity of the afferent baroreceptor receptors. In other words, the baroreceptors discharge less rapidly than normal at any given mean or pulsatile arterial pressure, and the brain "interprets" this decreased discharge as a larger-than-usual fall in pressure. The results of the reflexes are (1) heart rate is increased through increased sympathetic and decreased parasympathetic discharge to the heart, and (2) total peripheral resistance is increased by increased sympathetic discharge to systemic arterioles as well as by increased plasma concentrations of the two major hormonal vasoconstrictors—angiotensin II and vasopressin.

The reflex increases in heart rate and total peripheral resistance are initially beneficial in restoring cardiac output and arterial pressure just as if the changes in these parameters had been triggered by hemorrhage.

Maintained chronically throughout the period of cardiac failure, the baroreceptor reflexes also bring about fluid retention and an expansion—often massive—of the extracellular volume. This is because, as described in Chapter 16, the neuroendocrine efferent components of the reflexes cause the kidneys to reduce their excretion of sodium and water. The retained fluid then causes expansion of the extracellular volume. Since the plasma volume is part of the extracellular fluid volume, plasma volume also increases. This in turn increases venous pressure, venous return, and

end-diastolic ventricular volume, which tends to restore stroke volume toward normal by the Frank-Starling mechanism (Figure 14–68). Thus, fluid retention is also, at least initially, an adaptive response to the decreased cardiac output.

However, problems emerge as the fluid retention progresses. For one thing, when a ventricle with systolic dysfunction (as opposed to a normal ventricle) becomes very distended with blood, its force of contraction actually decreases and the situation worsens. Second, the fluid retention, with its accompanying elevation in venous pressure, causes edema—accumulation of interstitial fluid. Why does an increased venous pressure cause edema? The capillaries, of course, drain via venules into the veins, and so when venous pressure increases, the capillary pressure also increases and causes increased filtration of fluid out of the capillaries into the interstitial fluid. Thus, most of the fluid retained by the kidneys ends up as extra interstitial fluid rather than extra plasma. Swelling of the legs and feet is usually prominent, but the engorgement occurs elsewhere as well.

Most important in this regard, failure of the *left* ventricle—whether due to diastolic or systolic dysfunction—leads to **pulmonary edema,** which is the accumulation of fluid in the interstitial spaces of the lung or in the air spaces themselves. This impairs gas exchange. The reason for such accumulation is that the left ventricle temporarily fails to pump blood to the same extent as the right ventricle, and so the volume of blood in all the pulmonary vessels increases. The resulting engorgement of pulmonary capillaries raises the capillary pressure above its normally very low value, causing increased filtration out of the capillaries.

TABLE 14–13 Major Causes of Edema

Physiological Event	Cause of Edema
Increased arterial pressure secondary to increased cardiac output*	Increased capillary pressure, leading to increased filtration
Local arteriolar dilation, as in exercise or inflammation	Increased capillary pressure, leading to increased filtration
Increased venous pressure, as in heart failure or venous obstruction	Increased capillary pressure, leading to increased filtration
Decreased plasma protein concentration, as in liver disease (decreased protein production), kidney disease (loss of protein in the urine), or protein malnutrition	Decreased force for osmotic absorption across capillary. Therefore, *net* filtration is increased.
Increased interstitial-fluid protein concentration resulting from increased capillary permeability to protein (as in inflammation)	Decreased force for osmotic absorption across capillary. Therefore, *net* filtration is increased.
Obstruction of lymphatic vessels, as in infection by filaria roundworms (elephantiasis)	Fluid filtered from the blood capillaries into the interstitial compartment is not carried away. Protein also accumulates in the interstitial fluid.

*In contrast, if arterial pressure is elevated because of increased total peripheral resistance, edema may not occur; the capillary pressure may not be elevated because the increased arteriolar resistance will prevent most of that increase in pressure from reaching the capillaries.

This situation usually worsens at night: During the day, because of the patient's upright posture, fluid accumulates in the legs; then the fluid is slowly absorbed back into the capillaries when the patient lies down at night, thus expanding the plasma volume and precipitating an attack of pulmonary edema.

Another component of the reflex response to heart failure that is at first beneficial but ultimately becomes maladaptive is the increase in total peripheral resistance, mediated by the sympathetic nerves to arterioles and by angiotensin II and vasopressin. By chronically maintaining the arterial blood pressure against which the failing heart must pump, more work must be expended by the failing heart.

One obvious treatment for heart failure is to correct, if possible, the precipitating cause (for example, hypertension). In addition, the various drugs available for treatment are summarized in Table 14–12. Finally, although cardiac transplantation is often the treatment of choice, the paucity of donor hearts, the high costs, and the challenges of postsurgical care render it a feasible option for only a very small number of patients.

In closing this section, it should be emphasized that there are causes of edema other than heart failure and the lymphatic malfunction described earlier. They are all understandable as imbalances in the Starling forces and are listed in Table 14–13.

Coronary Artery Disease and Heart Attacks

We have seen that the myocardium does not extract oxygen and nutrients from the blood within the atria and ventricles but depends upon its own blood supply via the coronary arteries. In *coronary artery disease,* changes in one or more of the coronary arteries causes insufficient blood flow (*ischemia*) to the heart. The result may be myocardial damage in the affected region and, if severe enough, death of that portion of the heart—a *myocardial infarction,* or *heart attack.* Many patients with coronary artery disease experience recurrent transient episodes of inadequate coronary blood flow, usually during exertion or emotional tension, before ultimately suffering a heart attack. The chest pain associated with these episodes is called *angina pectoris* (or, more commonly, angina).

The symptoms of myocardial infarction include prolonged chest pain, often radiating to the left arm, nausea, vomiting, sweating, weakness, and shortness of breath. Diagnosis is made by ECG changes typical of infarction and by measurement of certain proteins in plasma. These proteins are present in cardiac muscle and leak out into the blood when the muscle is damaged; the most commonly used are the enzymes creatine kinase (CK), particularly the myocardial specific isoform, CK-MB, and cardiac-specific isoforms of troponin.

Of the more than 1.5 million heart attack victims in the United States each year, approximately half are admitted to a hospital, where they can be given advanced care, and more than 80 percent of these people survive the attack and are discharged.

Sudden cardiac deaths in myocardial infarction are due mainly to *ventricular fibrillation,* an abnormality in impulse conduction triggered by the damaged myocardial cells and resulting in completely uncoordinated ventricular contractions that are ineffective in producing flow. (Note that ventricular fibrillation is fatal, whereas atrial fibrillation, as described earlier in this chapter, generally causes only minor cardiac problems.) A small fraction of individuals with ventricular fibrillation can be saved if modern emergency resuscitation procedures are applied immediately after the attack. This treatment is *cardiopulmonary resuscitation (CPR),* a repeated series of mouth-to-mouth respirations and chest compressions that circulate a small amount of blood to the brain, heart, and other vital organs when the heart has stopped. CPR is then followed by definitive treatment, including *defibrillation,* a procedure in which electric current is passed through the heart to try to halt the abnormal electrical activity causing the fibrillation.

The major cause of coronary artery disease is the presence of atherosclerosis in these vessels. *Atherosclerosis* is a disease of arteries characterized by a thickening of the portion of the arterial vessel wall closest to the lumen with (1) large numbers of abnormal smooth-muscle cells, macrophages (derived from blood monocytes), and lymphocytes, (2) deposits of cholesterol and other fatty substances both in these cells and extracellularly, and (3) dense layers of connective-tissue matrix.

The mechanisms by which atherosclerosis reduces coronary blood flow are as follows: (1) The extra muscle cells and various deposits in the wall bulge into the lumen of the vessel and increase resistance to flow; and (2) dysfunctional endothelial cells in the atherosclerotic area release excess vasoconstrictors (for example, endothelin-1) and deficient vasodilators (nitric oxide and prostacyclin). These processes are progressive, sometimes leading ultimately to complete occlusion. Total occlusion is usually caused, however, by the formation of a blood clot (*coronary thrombosis*) in the narrowed atherosclerotic artery, and this triggers the heart attack.

The processes that lead to atherosclerosis are complex and still not completely understood. It is likely that the damage is initiated by agents that injure the endothelium and underlying smooth muscle, leading to an inflammatory and proliferative response that may well be protective at first but ultimately becomes excessive. This sequence of events is described in Chapter 20 as an example of a maladaptive inflammatory response.

Cigarette smoking, high plasma concentrations of cholesterol and the amino acid homocysteine, hypertension, diabetes, obesity, a sedentary lifestyle, and stress can all increase the incidence and the severity of the atherosclerotic process. These are all termed, therefore, "risk factors" for coronary artery disease, and prevention of this disease focuses on eliminating or minimizing them through lifestyle changes (for example, changing to a diet designed to lower plasma cholesterol concentration) and/or medications. In a sense,

menopause can also be considered a risk factor for coronary artery disease since the incidence of heart attacks in women is very low until after menopause. This relationship is explained by the protective effects of estrogen, including the actions of this hormone on plasma cholesterol concentration. The control of plasma cholesterol is described in Chapter 18, and the use of sex-hormone replacement therapy in menopausal women in Chapter 19.

A few words about exercise are warranted here because of several potential confusions. It is true that a sudden burst of strenuous physical activity can sometimes trigger a heart attack. However, the risk is almost totally eliminated in individuals who perform regular physical activity, and much more important, the overall risk of heart attack at any time can be reduced as much as 35 to 55 percent by maintenance of an active lifestyle, as compared with a sedentary one. In general, the more one exercises the better is the protective effect, but anything is better than nothing in this regard. For example, even moderately paced walking 3 to 4 times a week confers significant benefit.

Regular exercise is protective against heart attacks for a variety of reasons. Among other things, it induces: (1) decreased resting heart rate and blood pressure, two major determinants of myocardial oxygen demand; (2) increased diameter of coronary arteries; (3) decreased severity of hypertension and diabetes, two major risk factors for atherosclerosis; (4) decreased total plasma cholesterol concentration with simultaneous increase in the plasma concentration of a "good" cholesterol-carrying lipoprotein (HDL, discussed in Chapter 18); and (5) decreased tendency of blood to clot and improved ability of the body to dissolve blood clots.

Another protective factor against heart attacks is supplements of *vitamin E.* This vitamin, which functions as an antioxidant, may act by preventing the oxidation of "bad" cholesterol (LDL, discussed in Chapter 18) since such oxidation makes this cholesterol more likely to promote the buildup of plaques. Supplements of *folacin* (a B vitamin also called folate or folic acid) are also protective, in this case because folacin helps reduce the blood concentration of the amino acid *homocysteine,* one of the risk factors for heart attacks. Homocysteine is cysteine with an extra CH_2, and is an intermediary in the metabolism of methionine and cysteine. In increased amounts, it exerts several pro-atherosclerotic effects, including damaging the endothelium of blood vessels. Folacin is involved in the metabolism of homocysteine in a reaction that lowers the plasma concentration of this amino acid.

Finally, there is the question of alcohol and coronary artery disease. In many studies, moderate alcohol intake has been shown to reduce the risk of dying from a heart attack. The problem is that alcohol also increases the chances of an early death from a variety of other diseases (cirrhosis of the liver, for example) and accidents. Because of these complex health effects, it is presently recommended that people should take no more than 1 drink a day.

The treatment of coronary artery disease and angina with certain drugs can be understood in terms of physiological and pathophysiological concepts described in this chapter, and we will give only a few examples. First, vasodilator drugs such as *nitroglycerin* (which is a vasodilator because it is converted in the body to nitric oxide) help in the following way: They dilate the coronary arteries and the systemic arterioles and veins. The arteriolar effect lowers total peripheral resistance, thereby lowering arterial blood pressure and the work the heart must expend in ejecting blood. The venous dilation, by lowering venous pressure, reduces venous return and thereby the stretch of the ventricle and its oxygen requirement during subsequent contraction.

To take a second example, drugs that block beta-adrenergic receptors are used to lower the arterial pressure in persons with hypertension, but more importantly, they block the effects of the sympathetic nerves on the heart; the result is reduced myocardial work, and hence oxygen demand because both heart rate and contractility are reduced.

Third, drugs that prevent or reverse clotting within hours of its occurrence are extremely important in the treatment (and prevention) of heart attacks. Use of these drugs, including aspirin, is described in Section G of this chapter.

There are several surgical treatments for coronary artery disease after the area of narrowing or occlusion is identified by cardiac angiography (described earlier in this chapter). *Coronary balloon angioplasty* is the passing of a catheter with a balloon at its tip into the occluded artery and then expanding the balloon; this enlarges the lumen by stretching the vessel and breaking up abnormal tissue deposits. A second, but related, surgical technique utilizes the permanent placing of *coronary stents* in the narrowed or occluded coronary vessel. Stents are latticelike stainless steel tubes that provide a scaffold within a vessel to open it and keep it open. A third surgical technique is *coronary bypass,* in which an area of occluded coronary artery is removed and replaced with a new vessel, often a vein taken from elsewhere in the patient's body.

We do not wish to leave the impression that atherosclerosis attacks only the coronary vessels, for such is not the case. Many arteries of the body are subject to this same occluding process, and wherever the atherosclerosis becomes severe, the resulting symptoms reflect the decrease in blood flow to the specific area. For example, occlusion of a cerebral artery due to atherosclerosis and its associated blood clotting can cause localized brain damage—a stroke. (Recall that rupture of a cerebral vessel, as in hypertension, is the other cause of stroke.) Persons with atherosclerotic cerebral vessels may also suffer reversible neurologic deficits, known as *transient ischemic attacks (TIAs),* lasting minutes to hours, without actually experiencing a stroke at the time.

Finally, it should be noted that both myocardial infarcts and strokes due to occlusion may result from the breaking off of a fragment of blood clot or fatty deposit that then lodges downstream, completely blocking a smaller vessel. The fragment is termed an *embolus,* and the process is *embolism.*

SECTION F SUMMARY

Hemorrhage and Other Causes of Hypotension

I. The physiological responses to hemorrhage are summarized in Figures 14–55, 14–59, 14–61, and 14–62.

II. Hypotension can be caused by loss of body fluids, by cardiac malfunction, by strong emotion, and by liberation of vasodilator chemicals.

III. Shock is any situation in which blood flow to the tissues is low enough to cause damage to them.

The Upright Posture

I. In the upright posture, gravity acting upon unbroken columns of blood reduces venous return by increasing vascular pressures in the veins and capillaries in the limbs.
   a. The increased venous pressure distends the veins, causing venous pooling, and the increased capillary pressure causes increased filtration out of the capillaries.
   b. These effects are minimized by contraction of the skeletal muscles in the legs.

Exercise

I. The cardiovascular changes that occur in endurance-type exercise are illustrated in Figures 14–64 and 14–65.

II. The changes are due to active hyperemia in the exercising skeletal muscles and heart, to increased sympathetic outflow to the heart, arterioles, and veins, and to decreased parasympathetic outflow to the heart.

III. The increase in cardiac output depends not only on the autonomic influences on the heart but on factors that help increase venous return.

IV. Training can increase a person's maximal oxygen consumption by increasing maximal stroke volume and hence cardiac output.

Hypertension

I. Hypertension is usually due to increased total peripheral resistance resulting from increased arteriolar vasoconstriction.

II. More than 95 percent of hypertension is termed primary in that the cause of the increased arteriolar vasoconstriction is unknown.

Heart Failure

I. Heart failure can occur as a result of diastolic dysfunction or systolic dysfunction; in both cases, cardiac output becomes inadequate.

II. This leads to fluid retention by the kidneys and formation of edema because of increased capillary pressure.

III. Pulmonary edema can occur when the left ventricle fails.

Coronary Artery Disease and Heart Attacks

I. Insufficient coronary blood flow can cause damage to the heart.

II. Sudden death from a heart attack is usually due to ventricular fibrillation.

III. The major cause of reduced coronary blood flow is atherosclerosis, an occlusive disease of arteries.

IV. Persons may suffer intermittent attacks of angina pectoris without actually suffering a heart attack at the time of the pain.

V. Atherosclerosis can also cause strokes and symptoms of inadequate blood flow in other areas.

SECTION F KEY TERMS

maximal oxygen consumption ($\dot{V}_{O_2}$max)

SECTION F REVIEW QUESTIONS

1. Draw a flow diagram illustrating the reflex compensation for hemorrhage.
2. What happens to plasma volume and interstitial-fluid volume following a hemorrhage?
3. What causes hypotension during a severe allergic response?
4. How does gravity influence effective blood volume?
5. Describe the role of the skeletal-muscle pump in decreasing capillary filtration.
6. List the directional changes that occur during exercise for all relevant cardiovascular variables. What are the specific efferent mechanisms that bring about these changes?
7. What factors enhance venous return during exercise?
8. Diagram the control of autonomic outflow during exercise.
9. What is the limiting cardiovascular factor in endurance exercise?
10. What changes in cardiac function occur at rest and during exercise as a result of endurance training?
11. What is the abnormality in most cases of established hypertension?
12. State how fluid retention can help restore stroke volume in heart failure.
13. How does heart failure lead to edema?
14. Name the major risk factors for atherosclerosis.

SECTION G

HEMOSTASIS: THE PREVENTION OF BLOOD LOSS

The stoppage of bleeding is known as **hemostasis** (don't confuse this word with homeostasis). Physiological hemostatic mechanisms are most effective in dealing with injuries in small vessels—arterioles, capillaries, venules, which are the most common source of bleeding in everyday life. In contrast, the bleeding from a medium or large artery is not usually controllable by the body. Venous bleeding leads to less rapid blood loss because veins have low blood pressure. Indeed, the drop in pressure induced by raising the bleeding part above the heart level may stop hemorrhage from a vein. In addition, if the venous bleeding is into the tissues, the accumulation of blood may in-crease interstitial pressure enough to eliminate the pressure gradient required for continued blood loss. Accumulation of blood in the tissues can occur as a result of bleeding from any vessel type and is termed a *hematoma.*

When a blood vessel is severed or otherwise injured, its immediate inherent response is to constrict (the mechanism is unclear). This short-lived response slows the flow of blood in the affected area. In addition, this constriction presses the opposed endothelial surfaces of the vessel together, and this contact induces a stickiness capable of keeping them "glued" together.

Permanent closure of the vessel by constriction and contact stickiness occurs only in the very smallest vessels of the microcirculation, however, and the staunching of bleeding ultimately is dependent upon two other processes that are interdependent and occur in rapid succession: (1) formation of a platelet plug; and (2) blood coagulation (clotting). The blood platelets—cell fragments circulating in blood that are derived from bone-marrow-cells known as megakaryocytes—are involved in both processes.

Formation of a Platelet Plug

The involvement of platelets in hemostasis requires their adhesion to a surface. Although platelets have a propensity for adhering to surfaces, they do not adhere to the normal endothelial cells lining the blood vessels. However, injury to a vessel disrupts the endothelium and exposes the underlying connective-tissue collagen molecules. Platelets adhere to collagen, largely via an intermediary called **von Willebrand factor (vWF),** a plasma protein secreted by endothelial cells and platelets. This protein binds to exposed collagen molecules, changes its conformation, and becomes able to bind platelets; thus vWF forms a bridge between the damaged vessel wall and the platelets. (vWF has several additional functions in hemostasis that are not described in this chapter.)

Binding of platelets to collagen triggers the platelets to release the contents of their secretory vesicles, which contain a variety of chemical agents. Many of these agents, including adenosine diphosphate (ADP) and serotonin, then act locally to induce multiple changes in the metabolism, shape, and surface proteins of the platelets, a process termed **platelet activation.** Some of these changes cause new platelets to adhere to the old ones, a positive-feedback phenomenon termed **platelet aggregation,** that rapidly creates a **platelet plug** inside the vessel.

Chemical agents in the platelets' secretory vesicles are not the only stimulators of platelet activation and aggregation. Adhesion of the platelets rapidly induces them to synthesize **thromboxane A_2,** a member of the eicosanoid family (Chapter 7), from arachidonic acid in the platelet plasma membrane. Thromboxane A_2 is released into the extracellular fluid and acts locally to further stimulate platelet aggregation and release of their secretory-vesicle contents (Figure 14–69).

Fibrinogen, a plasma protein whose crucial role in blood clotting is described in the next section, also plays a crucial role in the platelet aggregation produced by the factors described above. It does so by forming the bridges between aggregating platelets. The receptors (binding sites) for fibrinogen on the platelet plasma membrane become exposed and activated during platelet activation.

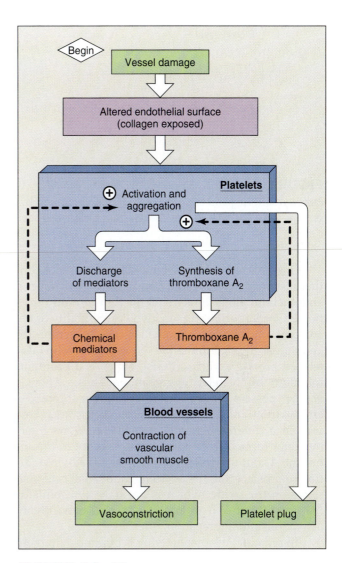

FIGURE 14–69

Sequence of events leading to formation of a platelet plug and vasoconstriction following damage to a blood-vessel wall. Note the two positive feedbacks in the pathways.

The platelet plug can completely seal small breaks in blood vessel walls. Its effectiveness is further enhanced by another property of platelets—contraction. Platelets contain a very high concentration of actin and myosin, which are stimulated to contract in aggregated platelets. This results in a compression and strengthening of the platelet plug. (When they occur in a test tube, this contraction and compression are termed "clot retraction.")

While the plug is being built up and compacted, the vascular smooth muscle in the damaged vessel is being stimulated to contract (Figure 14–69), thereby decreasing the blood flow to the area and the pressure within the damaged vessel. This vasoconstriction is the result of platelet activity, for it is mediated by thromboxane A_2 and by several chemicals contained in the platelet's secretory vesicles.

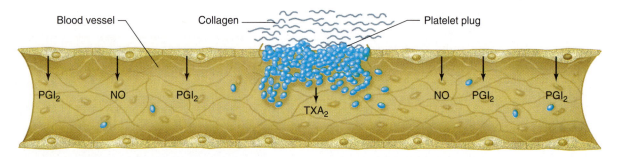

FIGURE 14–70

Prostacyclin (PGI$_2$) and nitric oxide (NO), both produced by endothelial cells, inhibit platelet aggregation and therefore prevent spread of platelet aggregation from a damaged site. TXA$_2$ = thromboxane A$_2$.

Once started, why does the platelet plug not continuously expand, spreading away from the damaged endothelium along intact endothelium in both directions? One important reason involves the ability of the adjacent undamaged endothelial cells to synthesize and release the eicosanoid known as **prostacyclin** (also termed prostaglandin I$_2$, **PGI$_2$**), which is a profound inhibitor of platelet aggregation. Thus, whereas *platelets* possess the enzymes that produce thromboxane A$_2$ from arachidonic acid, normal *endothelial cells* contain a different enzyme, one that converts intermediates formed from arachidonic acid not to thromboxane A$_2$ but to prostacyclin (Figure 14–70). In addition to prostacyclin, the adjacent endothelial cells also release **nitric oxide,** which is not only a vasodilator (Section D) but also an inhibitor of platelet adhesion, activation, and aggregation.

To reiterate, the platelet plug is built up extremely rapidly and is the primary sealer of breaks in vessel walls. In the following section, we shall see that platelets are also essential for the next, more slowly occurring, hemostatic event, blood coagulation.

Blood Coagulation: Clot Formation

Blood coagulation, or **clotting,** is the transformation of blood into a solid gel termed a **clot** or **thrombus** and consisting mainly of a protein polymer known as **fibrin.** Clotting occurs locally around the original platelet plug and is the dominant hemostatic defense. Its function is to support and reinforce the platelet plug and to solidify blood that remains in the wound channel.

The events leading to clotting are summarized, in very simplified form, in Figure 14–71. These events, like platelet aggregation, are initiated when injury to a vessel disrupts the endothelium and permits the blood to contact the underlying tissue. This contact initiates a locally occurring sequence, or "cascade," of chemical activations. At each step of the cascade, an inactive plasma protein, or "factor," is converted (activated) to a proteolytic enzyme, which then catalyzes the generation of the next enzyme in the sequence. Each of these activations results from the splitting of a small peptide fragment from the inactive plasma protein precursor, thus exposing the active site of the enzyme. It should be noted, however, that several of the plasma protein factors, following their activation, function not as enzymes but rather as cofactors for the enzymes.

To reduce the risk that you might "lose sight of the forest for the trees," Figure 14–71 gives no specifics about the cascade until the point at which the plasma protein **prothrombin** is converted to the enzyme **thrombin.** Thrombin then catalyzes a reaction in which several polypeptides are split from molecules of the large rod-shaped plasma protein **fibrinogen.** The still-large fibrinogen remnants then bind to each other to form fibrin. The fibrin, initially a loose mesh of interlacing strands, is rapidly stabilized and strengthened by enzymatically mediated formation of covalent cross-linkages. This chemical linking is catalyzed by an enzyme known as factor XIIIa, which is formed from plasma protein factor XIII in a reaction also catalyzed by thrombin.

Thus, thrombin catalyzes not only the formation of loose fibrin but also the activation of factor XIII, which stabilizes the fibrin network. But thrombin does even more than this—it exerts a profound positive-feedback effect on its own formation. It does this by activating several proteins in the cascade and also by activating platelets. Therefore, once thrombin formation has begun, reactions leading to much more thrombin generation are brought into play by this initial thrombin. We will make use of this crucial fact later when we describe the specifics of the cascade leading to thrombin.

In the process of clotting, many erythrocytes and other cells are trapped in the fibrin meshwork (Figure 14–72), but the essential component of the clot is fibrin, and clotting can occur in the absence of all

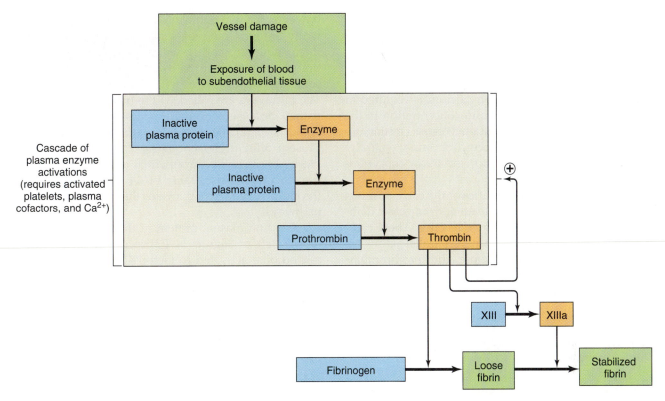

FIGURE 14–71

Simplified diagram of the clotting pathway. The pathway leading to thrombin is denoted by two enzyme activations, but the story is actually much more complex (as will be shown in Figure 14–73). Note that thrombin has three different effects—generation of fibrin, activation of factor XIII, and positive feedback on the cascade leading to itself.

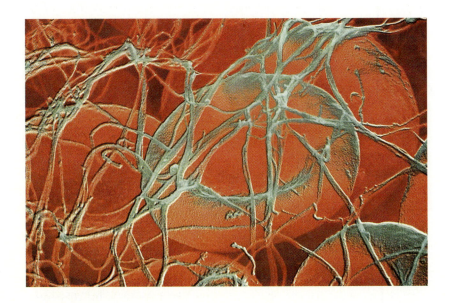

FIGURE 14–72

Electron micrograph of erythrocytes enmeshed in fibrin.

NIBSC/Science Photo Library/Custom Medical Stock.

cellular elements except platelets. Activated platelets are essential because several of the cascade reactions take place on the surface of these platelets. As noted above, platelet activation occurs early in the hemostatic response as a result of platelet adhesion to collagen,

but in addition, thrombin is an important stimulator of platelet activation. The activation causes the platelets to display specific plasma-membrane receptors that bind several of the clotting factors, and this permits the reactions to take place on the surface of the platelets. The activated platelets also display particular phospholipids, called **platelet factor (PF)**, which function as a cofactor in the steps mediated by the bound clotting factors.

One more generalization about the clotting cascade should be noted: Plasma calcium is required at various steps. However, calcium concentration in the plasma can never get low enough to cause clotting defects because death would occur from muscle paralysis or cardiac arrhythmias before such low concentrations were reached.

Now we present the specifics of the early portions of the clotting cascade, those leading from vessel damage to the prothrombin-thrombin reaction. These early portions consist of two seemingly parallel pathways that merge at the step before the prothrombin-thrombin reaction. Under physiological conditions,

however, the two pathways are not parallel but are actually brought into play *sequentially,* the link between them being thrombin. It will be clearer, however, if we first discuss the two pathways as though they were separate and then deal with their actual interaction. The pathways are called (1) the **intrinsic pathway,** so named because everything necessary for it is *in the* blood; and (2) the **extrinsic pathway,** so named because a cellular element *outside* the blood is needed. Figure 14–73 will be an essential reference for this entire discussion. Also, Table 14–14 is a reference list of the names of and synonyms for the substances in these pathways.

The first plasma protein in the intrinsic pathway (upper left of Figure 14–73) is called factor XII. It can become activated to factor XIIa when it contacts certain types of surfaces, including the collagen fibers

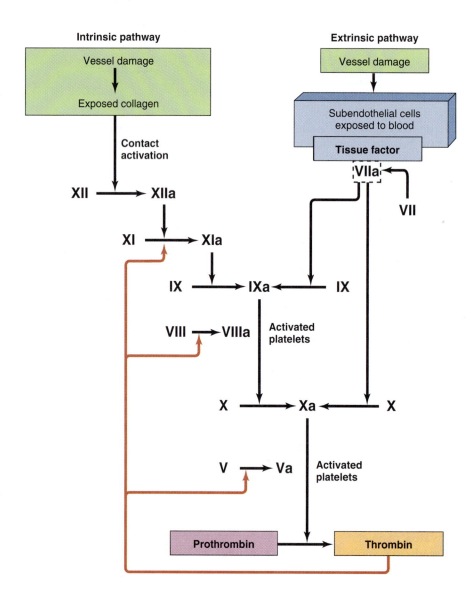

FIGURE 14–73

Two clotting pathways—called intrinsic and extrinsic—merge and can lead to the generation of thrombin. Under most physiological conditions, however, factor XII and the contact activation step that begin the intrinsic pathway probably play little, if any, roles in clotting. Rather, clotting is initiated solely by the extrinsic pathway, as described in the text. You might think that factors IX and X were accidentally transposed in the intrinsic pathway, but such is not the case; the order of activation really is XI, IX, and X. For the sake of clarity, the roles of calcium in clotting are not shown.

TABLE 14–14 Official Designations for Clotting Factors Along with Synonyms More Commonly Used

Factor I (fibrinogen)
Factor Ia (fibrin)
Factor II (prothrombin)
Factor IIa (thrombin)
Factor III (tissue factor, tissue thromboplastin)
Factor IV (Ca^{2+})
Factors V, VII, VIII, IX, X, XI, XII, and XIII are the inactive forms of these factors; the active forms add an "a" (for example, factor XIIa). There is no factor VI.
Platelet factor (PF)

underlying damaged endothelium. The contact activation of factor XII to XIIa is a complex process that requires the participation of several other plasma proteins not shown in Figure 14–73. (Contact activation also explains why blood coagulates when it is taken from the body and put in a glass tube: This has nothing whatever to do with exposure to air, but happens because the glass surface acts like collagen and induces the same activation of factor XII and aggregation of platelets as a damaged vessel surface. A silicone coating markedly delays clotting by reducing the activating effects of the glass surface.)

Factor XIIa then catalyzes the activation of factor XI to factor XIa, which activates factor IX to factor IXa. This last factor then activates factor X to factor Xa, which is the enzyme that converts prothrombin to thrombin. Note in Figure 14–73 that another plasma protein—factor VIIIa—serves as a cofactor (not an enzyme) in the factor IXa-mediated activation of factor X. The importance of factor VIII in clotting is emphasized by the fact that the disease *hemophilia,* in which excessive bleeding occurs, is usually due to a genetic absence of this factor. (In a smaller number of cases, hemophilia is due to an absence of factor IX.)

Now we turn to the extrinsic pathway for initiating the clotting cascade (upper right of Figure 14–73). This pathway begins with a protein called **tissue factor,** which is *not* a plasma protein. It is located instead on the outer plasma membrane of various tissue cells, including fibroblasts and other cells in the walls of blood vessels below the endothelium. The blood is exposed to these subendothelial cells when vessel damage disrupts the endothelial lining, and tissue factor on these cells then binds a plasma protein, factor VII, which becomes activated to factor VIIa. The complex of tissue factor and factor VIIa on the plasma membrane of the tissue cell then catalyzes the activation of factor X. In addition, it catalyzes the activation of factor IX, which can then help activate even more factor X by plugging into the intrinsic pathway.

In summary, theoretically clotting can be initiated either by the activation of factor XII or by the generation of the tissue factor–factor VIIa complex. The two paths merge at factor Xa, which then catalyzes the conversion of prothrombin to thrombin, which catalyzes the formation of fibrin. As shown in Figure 14–73, thrombin also contributes to the activation of: (1) factors XI and VIII in the intrinsic pathway; and (2) factor V, with factor Va then serving as a cofactor for factor Xa. Not shown in the figure is the fact that thrombin also activates platelets.

As stated earlier, under physiological conditions, the two pathways just described actually are brought into play sequentially. How this works can be understood by turning again to Figure 14–73; hold your hand over the first part of the *intrinsic* pathway so that you can eliminate the contact activation of factor XII and then begin the next paragraph's description at the top of the *extrinsic* pathway in the figure.

(1) The *extrinsic pathway,* with its tissue factor, is the usual way of *initiating* clotting *in the body,* and factor XII—the beginning of the full intrinsic pathway— normally plays little if any role (in contrast to its initiation of clotting in test tubes or, within the body in several unusual situations). Accordingly, thrombin is *initially* generated only by the extrinsic pathway. The amount of thrombin is too small, however, to produce adequate, sustained coagulation. (2) It *is* large enough, though, to trigger thrombin's positive-feedback effects on the *intrinsic pathway*—activation of factors XI and VIII and of platelets. (3) This is all that is needed to trigger the intrinsic pathway independently of factor XII, and this pathway then generates the large amounts of thrombin required for adequate coagulation. Thus, the extrinsic pathway, via its initial generation of small amounts of thrombin, provides the means for recruiting the more potent intrinsic pathway without the participation of factor XII. In essence, thrombin eliminates the need for factor XII. Moreover, thrombin not only recruits the intrinsic pathway but facilitates the prothrombin-thrombin step itself by activating factor V and platelets.

Finally it should be noted that the liver plays several important indirect roles in clotting (Figure 14–74), and persons with liver disease frequently have serious bleeding problems. First, the liver is the site of production for many of the plasma clotting factors. Second, the liver produces bile salts (Chapter 17), and these are important for normal intestinal absorption of the lipid-soluble substance **vitamin K.** The liver requires this vitamin to produce prothrombin and several other clotting factors.

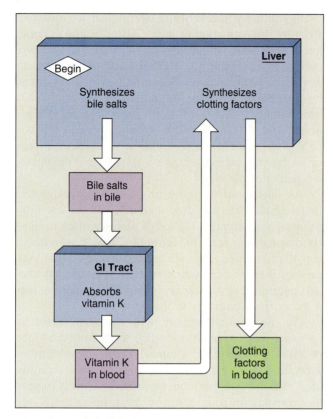

FIGURE 14–74
Roles of the liver in clotting.

Anticlotting Systems

Earlier we described how the release, by endothelial cells, of prostacyclin and nitric oxide inhibit platelet aggregation. Since this aggregation is an essential precursor for clotting, these agents reduce the magnitude and extent of clotting. In addition, however, the body has mechanisms for limiting clot formation, itself, and for dissolving a clot after it has formed.

Factors That Oppose Clot Formation

There are at least three different mechanisms that oppose clot formation, once underway, thereby helping to limit this process and prevent it from spreading excessively. Defects in any of these natural anticoagulant mechanisms are associated with abnormally high risk of clotting (*hypercoagulability*).

The first anticoagulant mechanism acts during the initiation phase of clotting and utilizes the plasma protein called **tissue factor pathway inhibitor (TFPI)**, which is secreted mainly by endothelial cells. This substance binds to tissue factor–factor VIIa complexes and inhibits the ability of these complexes to generate factor Xa. This anticoagulant mechanism is the reason that the extrinsic pathway by itself can generate only small amounts of thrombin.

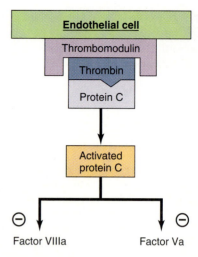

FIGURE 14–75
Thrombin indirectly inactivates factors VIIIa and Va via protein C. To activate protein C, thrombin must first bind to a thrombin receptor, thrombomodulin, on endothelial cells; this binding also eliminates thrombin's procoagulant effects.

The second anticoagulant mechanism is triggered by thrombin. As illustrated in Figure 14–75, thrombin can bind to an endothelial cell receptor known as **thrombomodulin.** This binding eliminates all of thrombin's clot-producing effects and causes the bound thrombin to bind a particular plasma protein, **protein C** (distinguish this from protein kinase C, Chapter 7). The binding to thrombin activates protein C, which, in combination with yet another plasma protein, then inactivates factors VIIIa and Va. Thus, we saw earlier that thrombin *directly activates* factors VIII and V, and now we see that it *indirectly inactivates* them via protein C. Table 14–15 summarizes the effects—both stimulatory and inhibitory—of thrombin on the clotting pathways.

A third naturally occurring anticoagulant mechanism is a plasma protein called **antithrombin III,** which inactivates thrombin and several other clotting factors. To do so, circulating antithrombin III must itself be activated, and this occurs when it binds to **heparin,** a substance that is present on the surface of endothelial cells. Antithrombin III prevents spread of a clot by rapidly inactivating clotting factors that are carried away from the immediate site of the clot by the flowing blood.

The Fibrinolytic System

TFPI, protein C, and antithrombin III all function to *limit* clot formation. The system to be described now, however, *dissolves* a clot *after* it is formed.

A fibrin clot is not designed to last forever. It is a transitory device until permanent repair of the vessel occurs. The **fibrinolytic** (or thrombolytic) **system** is the

TABLE 14–15 Actions of Thrombin

Procoagulant	Cleaves fibrinogen to fibrin Activates clotting factors XI, VIII, V, and XIII Stimulates platelet activation
Anticoagulant	Activates protein C, which inactivates clotting factors VIIIa and Va

principal effector of clot removal. The physiology of this system (Figure 14–76) is analogous to that of the clotting system: It constitutes a plasma proenzyme, **plasminogen,** which can be activated to the active enzyme **plasmin** by protein **plasminogen activators.** Once formed, plasmin digests fibrin, thereby dissolving the clot.

The fibrinolytic system is proving to be every bit as complicated as the clotting system, with multiple types of plasminogen activators and pathways for generating them, as well as several inhibitors of these plasminogen activators. In describing how this system can be set into motion, we restrict our discussion to one example—the particular plasminogen activator known as **tissue plasminogen activator (t-PA),** which is secreted by endothelial cells. During clotting, both plasminogen and t-PA bind to fibrin and become incorporated throughout the clot. The binding of t-PA to fibrin is crucial because t-PA is a very weak enzyme in the absence of fibrin. The presence of fibrin profoundly increases the ability of t-PA to catalyze the generation of plasmin from plasminogen. Thus, fibrin is an important initiator of the fibrinolytic process that leads to its own dissolution.

The secretion of t-PA is the last of the various anticlotting functions exerted by endothelial cells that we have mentioned in this chapter. They are summarized in Table 14–16.

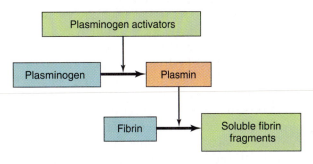

FIGURE 14–76

Basic fibrinolytic system. There are many different plasminogen activators and many different pathways for bringing them into play.

Anticlotting Drugs

Various drugs are used clinically to prevent or reverse clotting, and a brief description of their actions serves as a review of key clotting mechanisms. One of the most common uses of these drugs is in the prevention and treatment of myocardial infarction (heart attack), which, as described in Section F, is often the result of damage to endothelial cells. Such damage not only triggers clotting but interferes with the normal *anticlotting* functions of endothelial cells. For example, atherosclerosis interferes with the ability of endothelial cells to secrete nitric oxide.

TABLE 14–16 Anticlotting Roles of Endothelial Cells

Action	Result
Normally provide an intact barrier between the blood and subendothelial connective tissue	Platelet aggregation and the formation of tissue factor–factor VIIa complexes are not triggered
Synthesize and release PGI$_2$ and nitric oxide	These inhibit platelet activation and aggregation
Secrete tissue factor pathway inhibitor	Inhibits the ability of tissue factor–factor VIIa complexes to generate factor Xa
Bind thrombin (via thrombomodulin), which then activates protein C	Active protein C inactivates clotting factors VIIIa and Va
Display heparin molecules on the surfaces of their plasma membranes	Heparin binds antithrombin III, and this molecule then inactivates thrombin and several other clotting factors
Secrete tissue plasminogen activator	Tissue plasminogen activator catalyzes the formation of plasmin, which dissolves clots

Aspirin inhibits the cyclooxygenase enzyme in the eicosanoid pathways that generate prostaglandins and thromboxanes (Chapter 7). Since thromboxane A_2, produced by the platelets, is important for platelet aggregation, aspirin reduces both platelet aggregation and the ensuing coagulation. Importantly, low doses of aspirin cause a steady-state decrease in *platelet* cyclooxygenase activity but not *endothelial-cell* cyclooxygenase, and so the formation of prostacyclin—the prostaglandin that opposes platelet aggregation—is not impaired. [The reason for this difference between the responses of platelet and endothelial-cell cyclooxygenase (COX) to drugs is as follows: Platelets, once formed and released from megakaryocytes, have lost their ability to synthesize proteins, so that when their COX is blocked—the effect on any given COX molecule is irreversible—thromboxane A_2 synthesis is gone for that platelet's lifetime. In contrast, the endothelial cells produce new COX molecules to replace the ones blocked by the drug.] Aspirin is highly effective at preventing heart attacks. In addition, the administration of aspirin following a heart attack significantly reduces the incidence of sudden death and a recurrent heart attack.

A variety of new drugs that interfere with platelet function by mechanisms different from those of aspirin also have great promise in the treatment or prevention of heart attacks. In particular, certain drugs block the binding of fibrinogen to platelets and thus interfere with platelet aggregation.

Drugs that interfere with the action of vitamin K, which is required for synthesis of clotting factors by the liver, are collectively termed *oral anticoagulants.*

Heparin, the naturally occurring endothelial-cell cofactor for antithrombin III, can also be administered as a drug, which then binds to endothelial cells. In addition to its role in facilitating the action of antithrombin III, heparin also inhibits platelet function.

In contrast to aspirin, the fibrinogen blockers, the oral anticoagulants, and heparin, all of which *prevent* clotting, the fifth type of drug—plasminogen activators—dissolves a clot *after* it is formed. Use of such drugs is termed *thrombolytic therapy.* Intravenous administration of *recombinant t-PA* or a proteolytic drug called *streptokinase* within three hours after myocardial infarction significantly reduces myocardial damage and mortality. Recombinant t-PA has also been effective in reducing brain damage following a stroke caused by blood-vessel occlusion.

S E C T I O N G S U M M A R Y

I. The initial response to blood-vessel damage is vasoconstriction and the sticking together of the opposed endothelial surfaces.

II. The next events are formation of a platelet plug followed by blood coagulation (clotting).

Formation of a Platelet Plug

I. Platelets adhere to exposed collagen in a damaged vessel and release the contents of their secretory vesicles.
   a. These substances help cause platelet activation and aggregation.
   b. This process is also enhanced by von Willebrand factor, secreted by the endothelial cells, and by thromboxane A_2 produced by the platelets.
   c. Fibrinogen forms the bridges between aggregating platelets.
   d. Contractile elements in the platelets compress and strengthen the plug.
II. The platelet plug does not spread along normal endothelium because the latter secretes prostacyclin and nitric oxide, both of which inhibit platelet aggregation.

Blood Coagulation: Clot Formation

I. Blood is transformed into a solid gel when, at the site of vessel damage, plasma fibrinogen is converted into fibrin molecules, which bind to each other to form a mesh.
II. This reaction is catalyzed by the enzyme thrombin, which also activates factor XIII, a plasma protein that stabilizes the fibrin meshwork.
III. The formation of thrombin from the plasma protein prothrombin is the end result of a cascade of reactions in which an inactive plasma protein is activated and then enzymatically activates the next protein in the series.
   a. Thrombin exerts a positive-feedback stimulation of the cascade by activating platelets and several clotting factors.
   b. Activated platelets, which display platelet factor and binding sites for several activated plasma factors, are essential for the cascade.
IV. In the body, the cascade usually begins via the extrinsic clotting pathway when tissue factor forms a complex with factor VIIa. This complex activates factor X, which then catalyzes the conversion of small amounts of prothrombin to thrombin. This thrombin then recruits the intrinsic pathway by activating factor XI and factor VIII, as well as platelets, and this pathway generates large amounts of thrombin.
V. Vitamin K is required by the liver for normal production of prothrombin and other clotting factors.

Anticlotting Systems

I. Clotting is limited by three events: (1) Tissue factor pathway inhibitor inhibits the tissue factor–factor VIIa complex; (2) protein C, activated by thrombin, inactivates factors VIIIa and Va; and (3) antithrombin III inactivates thrombin and several other clotting factors.

II. Clots are dissolved by the fibrinolytic system.
  a. A plasma proenzyme, plasminogen, is activated by plasminogen activators to plasmin, which digests fibrin.
  b. Tissue plasminogen activator is secreted by endothelial cells and is activated by fibrin in a clot.

SECTION G KEY TERMS

hemostasis
von Willebrand factor (vWF)
platelet activation
platelet aggregation
platelet plug
thromboxane A_2
prostacyclin (PGI_2)
nitric oxide
blood coagulation
clotting
clot
thrombus
fibrin
prothrombin
thrombin
fibrinogen

platelet factor (PF)
intrinsic pathway
extrinsic pathway
tissue factor
vitamin K
tissue factor pathway inhibitor (TFPI)
thrombomodulin
protein C
antithrombin III
heparin
fibrinolytic system
plasminogen
plasmin
plasminogen activators
tissue plasminogen activator (t-PA)

SECTION G REVIEW QUESTIONS

1. Describe the sequence of events leading to platelet activation and aggregation, and the formation of a platelet plug. What helps keep this process localized?
2. Diagram the clotting pathway beginning with prothrombin.
3. What is the role of platelets in clotting?
4. List all the procoagulant effects of thrombin.
5. How is the clotting cascade initiated? How does the extrinsic pathway recruit the intrinsic pathway?
6. Describe the roles of the liver and vitamin K in clotting.
7. List three ways in which clotting is limited.
8. Diagram the fibrinolytic system.
9. How does fibrin help initiate the fibrinolytic system?

CHAPTER 14 CLINICAL TERMS

iron deficiency
hemochromatosis
anemia
iron-deficiency anemia
hemorrhage
sickle-cell anemia
polycythemia
ectopic pacemakers
AV conduction disorders
pacemaker
atrial fibrillation
heart murmurs
stenosis

insufficiency
echocardiography
cardiac angiography
elephantiasis
edema
hypotension
shock
hypovolemic shock
low-resistance shock
cardiogenic shock
hypertension
renal hypertension
primary hypertension

left ventricular hypertrophy
stroke
diuretics
beta-adrenergic receptor blockers
calcium-channel blockers
angiotensin-converting enzyme (ACE) inhibitors
heart failure
diastolic dysfunction
systolic dysfunction
pulmonary edema
cardiac inotropic drugs
digitalis
vasodilator drugs
coronary artery disease
ischemia
myocardial infarction
heart attack
angina pectoris
ventricular fibrillation
cardiopulmonary resuscitation (CPR)

defibrillation
atherosclerosis
coronary thrombosis
vitamin E
folacin
homocysteine
nitroglycerin
coronary balloon angioplasty
coronary stents
coronary bypass
transient ischemic attacks (TIAs)
embolus
embolism
hematoma
hemophilia
hypercoagulability
aspirin
oral anticoagulants
thrombolytic therapy
recombinant t-PA
streptokinase

CHAPTER 14 THOUGHT QUESTIONS

(Answers are given in Appendix A.)

1. A person is found to have a hematocrit of 35 percent. Can you conclude from this that there is a decreased volume of erythrocytes in the blood?
2. Which would cause a greater increase in resistance to flow, a doubling of blood viscosity or a halving of tube radius?
3. If all plasma-membrane calcium channels in contractile cardiac-muscle cells were blocked with a drug, what would happen to the muscle's action potentials and contraction?
4. A person with a heart rate of 40 has no P waves but normal QRS complexes on the ECG. What is the explanation?
5. A person has a left ventricular systolic pressure of 180 mmHg and an aortic systolic pressure of 110 mmHg. What is the explanation?
6. A person has a left atrial pressure of 20 mmHg and a left ventricular pressure of 5 mmHg during ventricular filling. What is the explanation?
7. A patient is taking a drug that blocks beta-adrenergic receptors. What changes in cardiac function will the drug cause?
8. What is the mean arterial pressure in a person whose systolic and diastolic pressures are, respectively, 160 and 100 mmHg?
9. A person is given a drug that doubles the blood flow to her kidneys but does not change the mean arterial pressure. What must the drug be doing?
10. A blood vessel removed from an experimental animal dilates when exposed to acetylcholine. After the endothelium is scraped from the lumen of the vessel, it no longer dilates in response to this mediator. Explain.

11. A person is accumulating edema throughout the body. Average capillary pressure is 25 mmHg, and lymphatic function is normal. What is the most likely cause of the edema?

12. A person's cardiac output is 7 L/min and mean arterial pressure is 140 mmHg. What is the person's total peripheral resistance?

13. The following data are obtained for an experimental animal before and after a drug. Before: Heart rate = 80 beats/min, and stroke volume = 80 ml/beat. After: Heart rate = 100 beats/min, and stroke volume = 64 ml/beat. Total peripheral resistance remains unchanged. What has the drug done to mean arterial pressure?

14. When the nerves from all the arterial baroreceptors are cut in an experimental animal, what happens to mean arterial pressure?

15. What happens to the hematocrit within several hours after a hemorrhage?

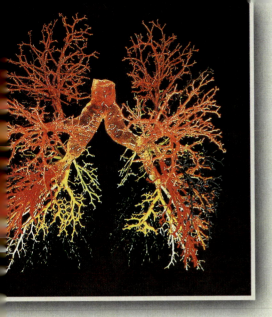

Respiration

Respiration has two quite different meanings: (1) utilization of oxygen in the metabolism of organic molecules by cells (often termed internal or cellular respiration), as described in Chapter 4, and (2) the exchanges of oxygen and carbon dioxide between an organism and the external environment. The second meaning is the subject of this chapter.

Human cells obtain most of their energy from chemical reactions involving oxygen. In addition, cells must be able to eliminate carbon dioxide, the major end product of oxidative metabolism. A unicellular organism can exchange oxygen and carbon dioxide directly with the external environment, but this is obviously impossible for most cells of a complex organism like a human being. Therefore, the evolution of large animals required the development of specialized structures to exchange oxygen and carbon dioxide for the entire animal with the external environment. In humans (and other mammals), the **respiratory system** includes the lungs, the series of tubes leading to the lungs, and the chest structures responsible for moving air into and out of the lungs during breathing.

In addition to the provision of oxygen and elimination of carbon dioxide, the respiratory system serves other functions, as listed in Table 15–1.

TABLE 15–1 Functions of the Respiratory System
1. Provides oxygen.
2. Eliminates carbon dioxide.
3. Regulates the blood's hydrogen-ion concentration (pH).
4. Forms speech sounds (phonation).
5. Defends against microbes.
6. Influences arterial concentrations of chemical messengers by removing some from pulmonary capillary blood and producing and adding others to this blood.
7. Traps and dissolves blood clots.

Organization of the Respiratory System

There are two lungs, the right and left, each divided into several lobes. **Pulmonary** is the adjective referring to "lungs." The lungs consist mainly of tiny air-containing sacs called **alveoli** (singular, **alveolus**), which number approximately 300 million in the adult. The alveoli are the sites of gas exchange with the blood. The **airways** are all the tubes through which air flows between the external environment and the alveoli.

Inspiration (inhalation) is the movement of air from the external environment through the airways into the alveoli during breathing. **Expiration** (exhalation) is movement in the opposite direction. An inspiration and an expiration constitute a **respiratory cycle.** During the entire respiratory cycle, the right ventricle of the heart pumps blood through the capillaries surrounding each alveolus. At rest, in a normal adult, approximately 4 L of fresh air enters and leaves the alveoli per minute, while 5 L of blood, the entire cardiac output, flows through the pulmonary capillaries. During heavy exercise, the air flow can increase twentyfold, and the blood flow five- to sixfold.

The Airways and Blood Vessels

During inspiration air passes through either the nose (the most common site) or mouth into the **pharynx** (throat), a passage common to both air and food (Figure 15–1). The pharynx branches into two tubes, the esophagus through which food passes to the stomach, and the **larynx,** which is part of the airways. The larynx houses the **vocal cords,** two folds of elastic tissue stretched horizontally across its lumen. The flow of air past the vocal cords causes them to vibrate, producing sounds. The nose, mouth, pharynx, and larynx are termed the upper airways.

The larynx opens into a long tube, the **trachea,** which in turn branches into two **bronchi** (singular, **bronchus**), one of which enters each lung. Within the lungs, there are more than 20 generations of branchings, each resulting in narrower, shorter, and more numerous tubes, the names of which are summarized in Figure 15–2. The walls of the trachea and bronchi contain cartilage, which gives them their cylindrical shape and supports them. The first airway branches that no longer contain cartilage are termed **bronchioles.** Alveoli first begin to appear in respiratory bronchioles, attached to their walls. The number of alveoli increases

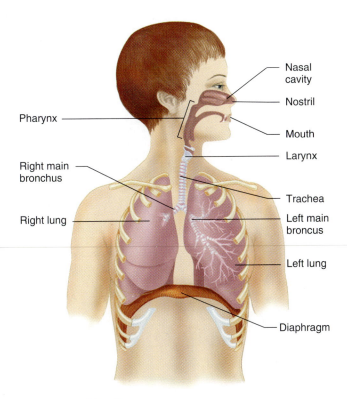

Pharynx

Right main bronchus

Right lung

Nasal cavity

Nostril

Mouth

Larynx

Trachea

Left main broncus

Left lung

Diaphragm

FIGURE 15–1

Organization of the respiratory system. The ribs have been removed in front, and the left lung is shown in a way that makes visible the airways within it. 𝄡

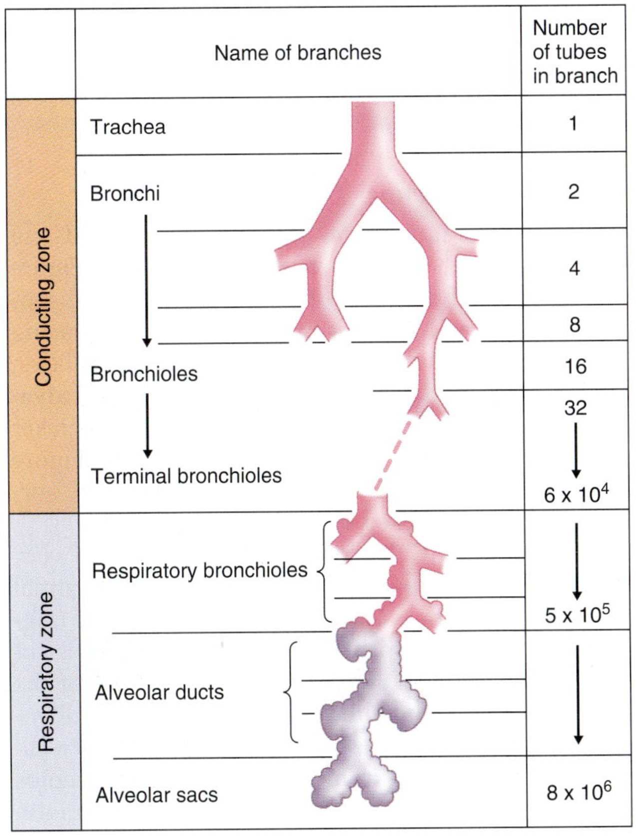

	Name of branches	Number of tubes in branch
Conducting zone	Trachea	1
	Bronchi	2
		4
		8
	Bronchioles	16
		32
	Terminal bronchioles	6×10^4
Respiratory zone	Respiratory bronchioles	5×10^5
	Alveolar ducts	
	Alveolar sacs	8×10^6

in the alveolar ducts (Figure 15–2), and the airways then end in grapelike clusters consisting entirely of alveoli (Figure 15–3). The airways, like blood vessels, are surrounded by smooth muscle, the contraction or relaxation of which can alter airway radius.

The airways beyond the larynx can be divided into two zones: (1) The **conducting zone** extends from the top of the trachea to the beginning of the respiratory bronchioles; it contains no alveoli and there is no gas exchange with the blood (Table 15–2). (2) The **respiratory zone,** which extends from the respiratory bronchioles on down, contains alveoli and is the region where gases exchange with the blood.

The epithelial surfaces of the airways, to the end of the respiratory bronchioles, contain cilia that constantly beat toward the pharynx. They also contain glands and individual epithelial cells that secrete mucus. Particulate matter, such as dust contained in the inspired air, sticks to the mucus, which is continuously and slowly moved by the cilia to the pharynx and then swallowed. This mucus escalator is important in keeping the lungs clear of particulate matter and the many bacteria that enter the body on dust particles. Ciliary activity can be inhibited by many noxious agents. For example, smoking a single cigarette can immobilize the cilia for several hours.

The airway epithelium also secretes a watery fluid upon which the mucus can ride freely. The production of this fluid is impaired in the disease *cystic fibrosis,* the most common lethal genetic disease of Caucasians, and the mucous layer becomes thick and dehydrated, obstructing the airways. The impaired secretion is due to a defect in the chloride channels involved in the secretory process (Chapter 6).

A second protective mechanism against infection is provided by cells that are present in the airways and alveoli and are termed macrophages. These cells, described in detail in Chapter 20, engulf inhaled particles and bacteria, rendering them harmless. Macrophages, like cilia, are injured by cigarette smoke and air pollutants.

The pulmonary blood vessels generally accompany the airways and also undergo numerous branchings. The smallest of these vessels branch into networks of capillaries that richly supply the alveoli (Figure 15–3). As noted in Chapter 14, the pulmonary circulation has a very low resistance compared to the systemic circulation, and for this reason the pressures within all pulmonary blood vessels are low. Therefore, the diameters of the vessels are determined largely by gravitational forces and *passive* physical forces within

FIGURE 15–2

Airway branching. 𝄡

(a)
(b)

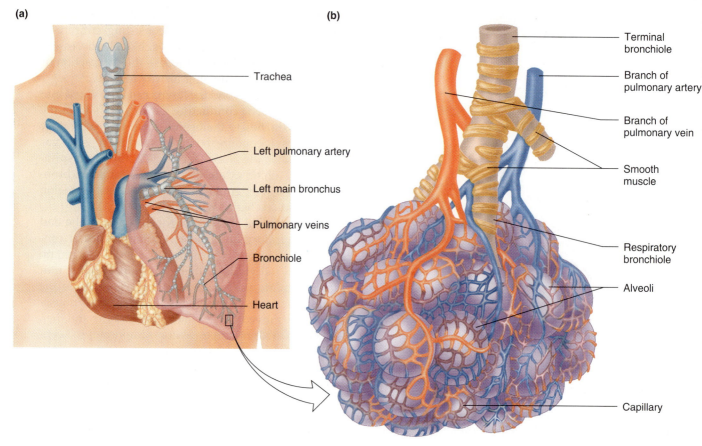

Trachea

Left pulmonary artery

Left main bronchus

Pulmonary veins

Bronchiole

Heart

Terminal bronchiole

Branch of pulmonary artery

Branch of pulmonary vein

Smooth muscle

Respiratory bronchiole

Alveoli

Capillary

FIGURE 15–3

Relationships between blood vessels and airways. (a) The lung appears transparent so that the relationships can be seen. The airways beyond the bronchiole are too small to be seen. (b) An enlargement of a small section of Figure 15–3a to show the continuation of the airways and the clusters of alveoli at their ends. Virtually the entire lung, not just the surface, consists of such clusters.

TABLE 15–2	Functions of the Conducting Zone of the Airways

1. Provides a low-resistance pathway for air flow; resistance is physiologically regulated by changes in contraction of airway smooth muscle and by physical forces acting upon the airways.

2. Defends against microbes, toxic chemicals, and other foreign matter; cilia, mucus, and phagocytes perform this function.

3. Warms and moistens the air.

4. Phonates (vocal cords).

the lungs. The net result is that there are marked regional differences in blood flow. The significance of this phenomenon will be described in the section on ventilation-perfusion inequality.

Site of Gas Exchange: The Alveoli

The alveoli are tiny hollow sacs whose open ends are continuous with the lumens of the airways (Figure 15–4a). Typically, the air in two adjacent alveoli is separated by a single alveolar wall. Most of the air-facing surface(s) of the wall are lined by a continuous layer, one cell thick, of flat epithelial cells termed **type I alveolar cells.** Interspersed between these cells are thicker specialized cells termed **type II alveolar cells** (Figure 15–4b) that produce a detergent-like substance, surfactant, to be discussed below.

The alveolar walls contain capillaries and a very small interstitial space, which consists of interstitial fluid and a loose meshwork of connective tissue (Figure 15–4b). In many places, the interstitial space is absent altogether, and the basement membranes of the alveolar-surface epithelium and the capillary-wall endothelium fuse. Thus the blood within an alveolar-wall capillary is separated from the air within the alveolus by an extremely thin barrier (0.2 μm, compared with

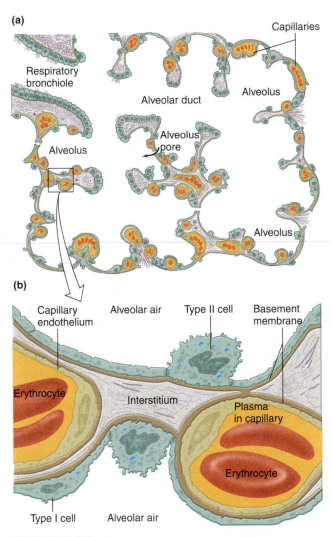

(a)

Capillaries

Respiratory
bronchiole

Alveolar duct

Alveolus

Alveolus
pore

Alveolus

Alveolus

(b)

Capillary
endothelium   Alveolar air   Type II cell   Basement
membrane

Erythrocyte

Interstitium

Plasma
in capillary

Erythrocyte

Type I cell   Alveolar air

FIGURE 15–4

(a) Cross section through an area of the respiratory zone.
There are 18 alveoli in this figure, only four of which are
labelled. Two frequently share a common wall.

From R.O. Greep and L. Weiss, "Histology," 3d ed., McGraw-Hill New York,
1973.

(b) Schematic enlargement of a portion of an alveolar wall.

Adapted from Gong and Drage.

the 7-μm diameter of an average red blood cell). The
total surface area of alveoli in contact with capillaries
is roughly the size of a tennis court. This extensive area
and the thinness of the barrier permit the rapid exchange of large quantities of oxygen and carbon dioxide by diffusion.

In some of the alveolar walls, there are pores that
permit the flow of air between alveoli. This route can
be very important when the airway leading to an alveolus is occluded by disease, since some air can still enter the alveolus by way of the pores between it and
adjacent alveoli.

Relation of the Lungs to the Thoracic (Chest) Wall

The lungs, like the heart, are situated in the **thorax,** the
compartment of the body between the neck and abdomen. "Thorax" and "chest" are synonyms. The thorax is a closed compartment that is bounded at the neck
by muscles and connective tissue and completely separated from the abdomen by a large, dome-shaped
sheet of skeletal muscle, the **diaphragm.** The wall of
the thorax is formed by the spinal column, the ribs, the
breastbone (sternum), and several groups of muscles
that run between the ribs (collectively termed the **intercostal muscles**). The thoracic wall also contains
large amounts of elastic connective tissue.

Each lung is surrounded by a completely closed
sac, the **pleural sac,** consisting of a thin sheet of cells
called **pleura.** The two pleural sacs, one on each side
of the midline, are completely separate from each
other. The relationship between a lung and its pleural
sac can be visualized by imagining what happens
when you push a fist into a balloon (Figure 15–5): The
arm represents the major bronchus leading to the lung,
the fist is the lung, and the balloon is the pleural sac.
The fist becomes coated by one surface of the balloon.
In addition, the balloon is pushed back upon itself so
that its opposite surfaces lie close together. Unlike the
hand and balloon, however, the pleural surface coating the lung (the visceral pleura) is firmly attached to

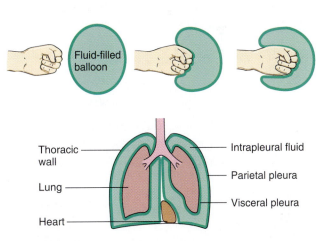

Fluid-filled
balloon

Thoracic
wall

Lung

Heart

Intrapleural fluid

Parietal pleura

Visceral pleura

FIGURE 15–5

Relationship of lungs, pleura, and thoracic wall, shown as
analogous to pushing a fist into a fluid-filled balloon. Note
that there is no communication between the right and left
intrapleural fluids. For purposes of illustration in this figure,
the volume of intrapleural fluid is greatly exaggerated. It
normally consists of an extremely thin layer of fluid between
the pleura membrane lining the inner surface of the thoracic
wall (the parietal pleura) and that lining the outer surface of
the lungs (the visceral pleura).

the lung by connective tissue. Similarly, the outer layer (the parietal pleura) is attached to and lines the interior thoracic wall and diaphragm. The two layers of pleura in each sac are so close to each other that normally they are always in virtual contact, but they are *not* attached to each other. Rather, they are separated by an extremely thin layer of **intrapleural fluid,** the total volume of which is only a few milliliters. The intrapleural fluid totally surrounds the lungs and lubricates the pleural surfaces so that they can slide over each other during breathing. More important, as we shall see in the next section, changes in the hydrostatic pressure of the intrapleural fluid—the **intrapleural pressure (P_{ip})** (also termed intrathoracic pressure)—cause the lungs and thoracic wall to move in and out together during normal breathing.

Ventilation and Lung Mechanics

An inventory of steps involved in respiration (Figure 15–6) is provided for orientation before beginning the detailed descriptions of each step, beginning with ventilation.

(1) Ventilation: Exchange of air between atmosphere and alveoli by *bulk flow*
(2) Exchange of O_2 and CO_2 between alveolar air and blood in lung capillaries by *diffusion*
(3) Transport of O_2 and CO_2 through pulmonary and systemic circulation by *bulk flow*
(4) Exchange of O_2 and CO_2 between blood in tissue capillaries and cells in tissues by *diffusion*
(5) Cellular utilization of O_2 and production of CO_2

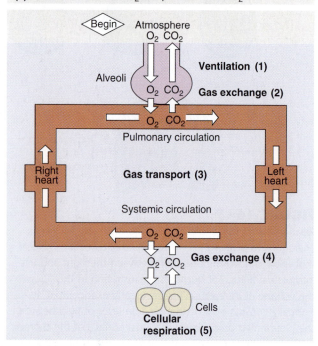

FIGURE 15–6
The steps of respiration.

Ventilation is defined as the exchange of air between the atmosphere and alveoli. Like blood, air moves by *bulk flow,* from a region of high pressure to one of low pressure. We saw in Chapter 14 that bulk flow can be described by the equation

$$F = \Delta P/R \qquad (15\text{-}1)$$

That is, flow (F) is proportional to the pressure difference (ΔP) between two points and inversely proportional to the resistance (R). For air flow into or out of the lungs, the relevant pressures are the gas pressure in the alveoli—the **alveolar pressure (P_{alv})**—and the gas pressure at the nose and mouth, normally **atmospheric pressure (P_{atm}),** the pressure of the air surrounding the body:

$$F = (P_{atm} - P_{alv})/R \qquad (15\text{-}2)$$

A very important point must be made at this point: All pressures in the respiratory system, as in the cardiovascular system, are given *relative to atmospheric pressure,* which is 760 mmHg at sea level. For example, the alveolar pressure between breaths is said to be 0 mmHg, which means that it is the same as atmospheric pressure.

During ventilation, air moves into and out of the lungs because the alveolar pressure is alternately made less than and greater than atmospheric pressure (Figure 15–7). These alveolar pressure changes are caused, as we shall see, by changes in the dimensions of the lungs.

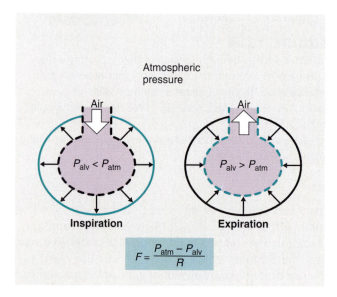

FIGURE 15–7
Relationships required for ventilation. When the alveolar pressure (P_{alv}) is less than atmospheric pressure (P_{atm}), air enters the lungs. Flow (*F*) is directly proportional to the pressure difference and inversely proportional to airway resistance (*R*). Black lines show lung's position at beginning of inspiration or expiration, and blue lines at end.

To understand how a change in lung dimensions causes a change in alveolar pressure, you need to learn one more basic concept—**Boyle's law** (Figure 15–8). At constant temperature, the relationship between the pressure exerted by a fixed number of gas molecules and the volume of their container is as follows: An increase in the volume of the container decreases the pressure of the gas, whereas a decrease in the container volume increases the pressure.

It is essential to recognize the correct causal sequences in ventilation: *During inspiration and expiration the volume of the "container"—the lungs—is made to change, and these changes then cause, by Boyle's law, the alveolar pressure changes that drive air flow into or out of the lungs.* Our descriptions of ventilation must focus, therefore, on how the changes in lung dimensions are brought about.

There are no muscles attached to the lung surface to pull the lungs open or push them shut. Rather, the lungs are passive elastic structures—like balloons—and their volume, therefore, depends upon: (1) the *difference* in pressure—termed the **transpulmonary pressure**—between the inside and the outside of the lungs; and (2) how stretchable the lungs are. The rest of this section and the next three sections focus only on transpulmonary pressure; stretchability will be discussed later in the section on lung compliance.

The pressure inside the lungs is the *air pressure* inside the alveoli (P_{alv}), and the pressure outside the lungs is the pressure of the *intrapleural fluid* surrounding the lungs (P_{ip}). Thus,

$$\text{Transpulmonary pressure} = P_{alv} - P_{ip} \qquad (15\text{-}3)$$

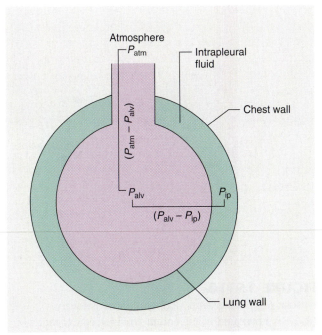

FIGURE 15–9

Two pressure differences involved in ventilation. ($P_{alv} - P_{ip}$) is a determinant of lung size; ($P_{atm} - P_{ip}$) is a determinant of air flow. (The volume of intrapleural fluid is greatly exaggerated for purpose of illustration.)

Compare this equation to Equation 15-2 (the equation that describes air flow into or out of the lungs), as it will be essential to distinguish these equations from each other (Figure 15–9).

To summarize, the muscles used in respiration are not attached to the lung surface. Rather these muscles are part of the *chest wall*. When they contract or relax, they directly change the dimensions of the *chest*, which in turn causes the transpulmonary pressure ($P_{alv} - P_{ip}$) to change. The change in transpulmonary pressure then causes a change in *lung* size, which causes changes in alveolar pressure and, thereby, in the difference in pressure between the atmosphere and the alveoli ($P_{atm} - P_{alv}$). It is this difference in pressure that causes air flow into or out of the lungs.

Let us now apply these concepts to the three phases of the respiratory cycle: the period between breaths, inspiration, and expiration.

The Stable Balance between Breaths

Figure 15–10 illustrates the situation that normally exists at the end of an unforced expiration—that is, between breaths when the respiratory muscles are relaxed and no air is flowing. The alveolar pressure (P_{alv}) is 0 mmHg; that is, it is the same as atmospheric pressure. The intrapleural pressure (P_{ip}) is approximately 4 mmHg less than atmospheric pressure—that is, −4 mmHg, using the standard convention of

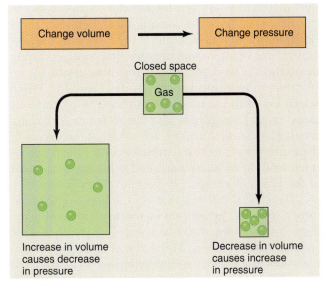

FIGURE 15–8

Boyle's law: The pressure exerted by a constant number of gas molecules in a container is inversely proportional to the volume of the container; that is, P is proportional to $1/V$ (at constant temperature).

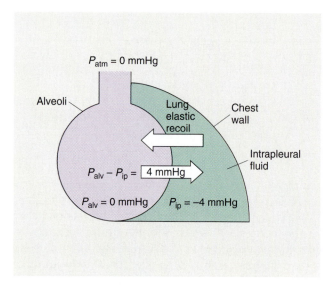

FIGURE 15–10

Alveolar (P_{alv}), intrapleural (P_{ip}), and transpulmonary ($P_{alv} - P_{ip}$) pressures at the end of an unforced expiration— that is, between breaths. The transpulmonary pressure exactly opposes the elastic recoil of the lung, and the lung volume remains stable. (The volume of intrapleural fluid is greatly exaggerated for purposes of illustration.)

giving all pressures relative to atmospheric pressure. Therefore, the transpulmonary pressure ($P_{alv} - P_{ip}$) equals [0 mmHg − (−4 mmHg)] = 4 mmHg. As emphasized in the previous section, this transpulmonary pressure is the force acting to expand the lungs; it is opposed by the elastic recoil of the partially expanded and, therefore, partially stretched lungs. **Elastic recoil** is defined as the tendency of an elastic structure to oppose stretching or distortion. In other words, inherent elastic recoil tending to collapse the lungs is exactly balanced by the transpulmonary pressure tending to expand them, and the volume of the lungs is stable at this point. As we shall see, a considerable volume of air is present in the lungs between breaths.

At the same time, there is also a pressure difference of 4 mmHg pushing *inward* on the *chest wall*— that is, tending to compress the chest—for the following reason. The pressure difference across the chest wall is the difference between atmospheric pressure and intrapleural pressure ($P_{atm} - P_{ip}$). P_{atm} is 0 mmHg, and P_{ip} is −4 mmHg; accordingly, $P_{atm} - P_{ip}$ is [0 mmHg − (−4 mmHg)] = 4 mmHg directed inward. This pressure difference across the chest wall just balances the tendency of the partially compressed elastic chest wall to move outward, and so the chest wall, like the lungs, is stable in the absence of any respiratory muscular contraction.

Clearly, the subatmospheric *intrapleural pressure* is the essential factor keeping the lungs partially expanded

and the chest wall partially compressed between breaths. The important question now is: What has caused the intrapleural pressure to be subatmospheric? As the lungs (tending to move inward from their stretched position because of their elastic recoil) and the thoracic wall (tending to move outward from its compressed position because of its elastic recoil) "try" to move ever so slightly away from each other, there occurs an infinitesimal enlargement of the fluid-filled intrapleural space between them. But fluid cannot expand the way air can, and so even this tiny enlargement of the intrapleural space—so small that the pleural surfaces still remain in contact with each other—drops the intrapleural pressure below atmospheric pressure. In this way, the elastic recoil of both the lungs and chest wall creates the subatmospheric intrapleural pressure that keeps them from moving apart more than a very tiny amount.

The importance of the transpulmonary pressure in achieving this stable balance can be seen when, during surgery or trauma, the chest wall is pierced without damaging the lung. Atmospheric air rushes through the wound into the intrapleural space (a phenomenon called *pneumothorax*), and the intrapleural pressure goes from −4 mmHg to 0 mmHg. The transpulmonary pressure acting to hold the lung open is thus eliminated, and the lung collapses. At the same time, the chest wall moves outward since its elastic recoil is also no longer opposed.

Inspiration

Figures 15–11 and 15–12 summarize the events during normal inspiration at rest. Inspiration is initiated by the neurally induced contraction of the diaphragm and the "inspiratory" intercostal muscles located between the ribs. (The adjective "inspiratory" here is a functional term, not an anatomical one; it denotes the several groups of intercostal muscles that contract during inspiration.) The diaphragm is the most important inspiratory muscle during normal quiet breathing. When activation of the nerves to it causes it to contract, its dome moves downward into the abdomen, enlarging the thorax. Simultaneously, activation of the nerves to the inspiratory intercostal muscles causes them to contract, leading to an upward and outward movement of the ribs and a further increase in thoracic size.

The crucial point is that contraction of the inspiratory muscles, by *actively* increasing the size of the thorax, upsets the stability set up by purely elastic forces between breaths. As the thorax enlarges, the thoracic wall moves ever so slightly farther away from the lung surface, and the intrapleural fluid pressure therefore becomes even more subatmospheric than it was between breaths. This decrease in intrapleural pressure *increases* the transpulmonary pressure. Therefore, the

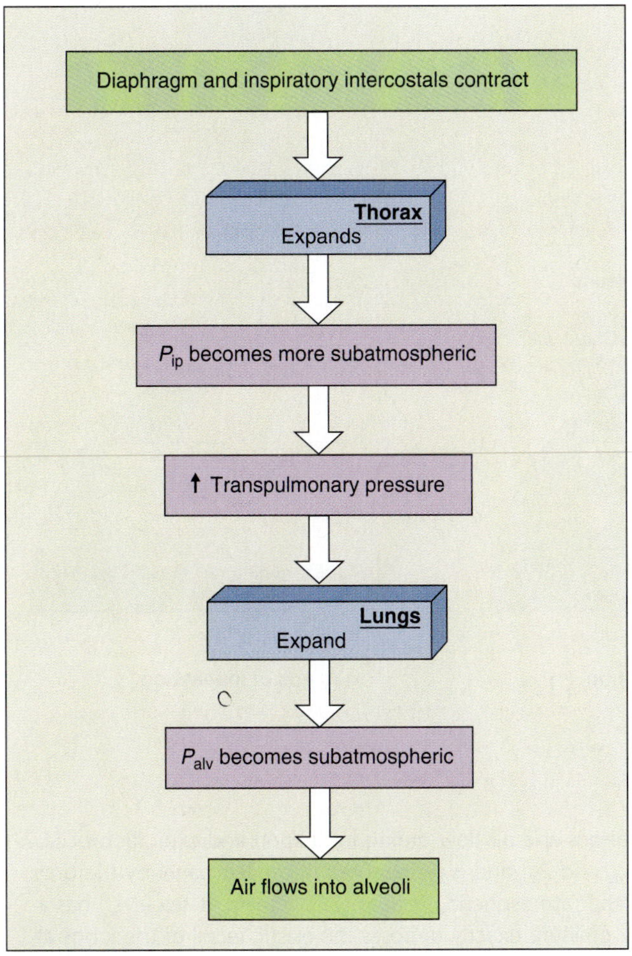

FIGURE 15–11

Sequence of events during inspiration. Figure 15–12 illustrates these events quantitatively.

force acting to expand the lungs—the transpulmonary pressure—is now greater than the elastic recoil exerted by the lungs at this moment, and so the lungs expand further. Note in Figure 15–12 that, by the end of inspiration, equilibrium *across the lungs* is once again established since the more inflated lungs exert a greater elastic recoil, which equals the increased transpulmonary pressure. In other words, lung volume is stable whenever transpulmonary pressure is balanced by the elastic recoil of the lungs (that is, after both inspiration and expiration).

Thus, when contraction of the inspiratory muscles actively increases the thoracic dimensions, the lungs are passively forced to enlarge virtually to the same degree because of the change in intrapleural pressure and hence transpulmonary pressure. The enlargement of the lungs causes an increase in the sizes of the alveoli throughout the lungs. Therefore, by Boyle's law, the pressure within the alveoli drops to less than atmospheric (see Figure 15–12). This produces the differ-

ence in pressure ($P_{alv} < P_{atm}$) that causes a bulk-flow of air from the atmosphere through the airways into the alveoli. By the end of the inspiration, the pressure in the alveoli again equals atmospheric pressure because of this additional air, and air flow ceases.

Expiration

Figures 15–12 and 15–13 summarize the sequence of events during expiration. At the end of inspiration, the nerves to the diaphragm and inspiratory intercostal muscles decrease their firing, and so these muscles relax. The chest wall is no longer being actively pulled outward and upward by the muscle contractions and so it starts to recoil inward to its original smaller dimensions existing between breaths. This immediately makes the intrapleural pressure less subatmospheric and hence *decreases* the transpulmonary pressure. Therefore, the transpulmonary pressure acting to expand the lungs is now smaller than the elastic recoil exerted by the more expanded lungs, and the lungs passively recoil to their original dimensions.

As the lungs become smaller, air in the alveoli becomes temporarily compressed so that, by Boyle's law, alveolar pressure exceeds atmospheric pressure (see Figure 15–12). Therefore, air flows from the alveoli through the airways out into the atmosphere. Thus, expiration at rest is *completely passive*, depending only upon the relaxation of the inspiratory muscles and recoil of the chest wall and stretched lungs.

Under certain conditions (during exercise, for example), expiration of larger volumes is achieved by contraction of a different set of intercostal muscles and the abdominal muscles, which *actively* decreases thoracic dimensions. The "expiratory" intercostal muscles (again a functional term, not an anatomical one) insert on the ribs in such a way that their contraction pulls the chest wall downward and inward. Contraction of the abdominal muscles increases intraabdominal pressure and forces the relaxed diaphragm up into the thorax.

Lung Compliance

To repeat, the degree of lung expansion at any instant is proportional to the transpulmonary pressure; that is, $P_{alv} - P_{ip}$. But just how much any given transpulmonary pressure expands the lungs depends upon the stretchability, or compliance, of the lungs. **Lung compliance (C_L)** is defined as the magnitude of the change in lung volume (ΔV_L) produced by a given change in the transpulmonary pressure:

$$C_L = \Delta V_L / \Delta(P_{alv} - P_{ip}) \qquad (15\text{-}4)$$

Thus, the greater the lung compliance, the easier it is to expand the lungs at any given transpulmonary pressure (Figure 15–14). A low lung compliance means

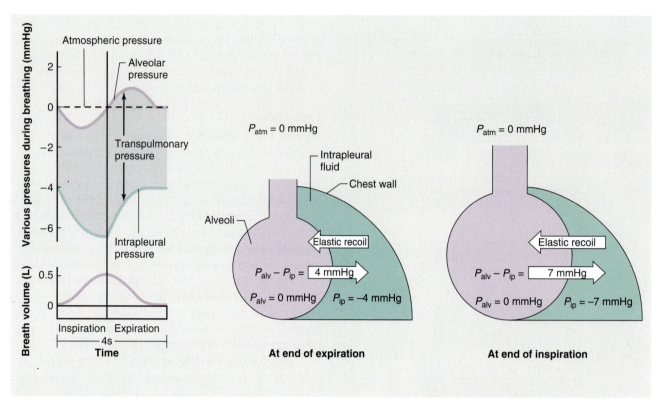

FIGURE 15–12

Summary of alveolar, intrapleural, and transpulmonary pressure changes and air flow during inspiration and expiration of 500 ml of air. The transpulmonary pressure is the difference between P_{alv} and P_{ip}, and is represented in the left panel by the gray area between the alveolar pressure and intrapleural pressure. Note that atmospheric pressure (760 mmHg at sea level) has a value of zero on the respiratory pressure scale. The transpulmonary pressure exactly opposes the elastic recoil of the lungs at the end of both inspiration and expiration. ✗

that a greater-than-normal transpulmonary pressure must be developed across the lung to produce a given amount of lung expansion. In other words, when lung compliance is abnormally low, intrapleural pressure must be made more subatmospheric than usual during inspiration to achieve lung expansion. This requires more vigorous contractions of the diaphragm and inspiratory intercostal muscles. Thus, the less compliant the lung, the more energy is required for a given amount of expansion. Persons with low lung compliance due to disease therefore tend to breathe shallowly and must breathe at a higher frequency to inspire an adequate volume of air.

Determinants of Lung Compliance There are two major determinants of lung compliance. One is the stretchability of the lung tissues, particularly their elastic connective tissues. Thus a thickening of the lung tissues decreases lung compliance. However, an equally important determinant of lung compliance is not the elasticity of the lung tissues, but the surface tension at the air-water interfaces within the alveoli.

The surface of the alveolar cells is moist, and so the alveoli can be pictured as air-filled sacs lined with water. At an air-water interface, the attractive forces between the water molecules, known as **surface tension,** make the water lining like a stretched balloon that constantly tries to shrink and resists further stretching. Thus, expansion of the lung requires energy not only to stretch the connective tissue of the lung but also to overcome the surface tension of the water layer lining the alveoli.

Indeed, the surface tension of pure water is so great that were the alveoli lined with pure water, lung expansion would require exhausting muscular effort and the lungs would tend to collapse. It is extremely important, therefore, that the type II alveolar cells secrete a detergent-like substance known as pulmonary **surfactant,** which markedly reduces the cohesive forces between water molecules on the alveolar surface. Therefore, surfactant lowers the surface tension, which increases lung compliance and makes it easier to expand the lungs (Table 15–3).

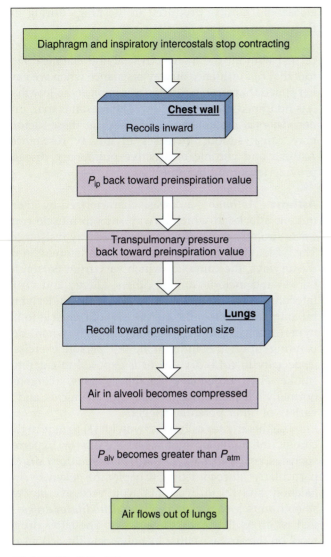

FIGURE 15–13

Sequence of events during expiration. Figure 15–12 illustrates these events quantitatively. 𝍏

$$Compliance = \frac{\Delta \text{ Lung volume}}{\Delta (P_{alv} - P_{ip})}$$

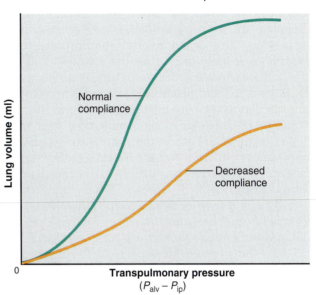

FIGURE 15–14

A graphical representation of lung compliance. Changes in lung volume and transpulmonary pressure are measured as a subject takes progressively larger breaths. When compliance is lower than normal, there is a lesser increase in lung volume for any given increase in transpulmonary pressure. 𝍏

TABLE 15–3 Some Important Facts about Pulmonary Surfactant
1. Pulmonary surfactant is a mixture of phospholipids and protein.
2. It is secreted by type II alveolar cells.
3. It lowers surface tension of the water layer at the alveolar surface, which increases lung compliance (that is, makes the lungs easier to expand).
4. A deep breath increases its secretion (by stretching the type II cells). Its concentration decreases when breaths are small.

Surfactant is a complex of both lipids and proteins, but its major component is a phospholipid that forms a monomolecular layer between the air and water at the alveolar surface. The amount of surfactant tends to decrease when breaths are small and constant. A deep breath, which people normally intersperse frequently in their breathing pattern, stretches the type II cells, which stimulates the secretion of surfactant. This is why patients who have had thoracic or abdominal surgery and are breathing shallowly because of the pain must be urged to take occasional deep breaths.

A striking example of what occurs when surfactant is deficient is the disease known as *respiratory-distress syndrome of the newborn.* This is the second leading cause of death in premature infants, in whom the surfactant-synthesizing cells may be too immature to function adequately. Because of low lung compliance, the infant is able to inspire only by the most strenuous efforts, which may ultimately cause complete exhaustion, inability to breathe, lung collapse, and death. Therapy in such cases is assisted breathing with a mechanical ventilator and the administration of natural or synthetic surfactant via the infant's trachea.

Airway Resistance

As previously stated, the volume of air that flows into or out of the alveoli per unit time is directly proportional to the pressure difference between the atmosphere and alveoli and inversely proportional to the resistance to flow offered by the airways (Equation 15-2). The factors that determine airway resistance are analogous to those determining vascular resistance in the circulatory system: tube length, tube radius, and interactions between moving molecules (gas molecules, in this case). As in the circulatory system, the most important factor by far is the tube radius: Airway resistance is inversely proportional to the fourth power of the airway radii.

Airway resistance to air flow is normally so small that very small pressure differences suffice to produce large volumes of air flow. As we have seen (see Figure 15–12), the average atmosphere-to-alveoli pressure difference during a normal breath at rest is less than 1 mmHg; yet, approximately 500 ml of air is moved by this tiny difference.

Airway radii and therefore resistance are affected by physical, neural, and chemical factors. One important physical factor is the transpulmonary pressure, which exerts a distending force on the airways, just as on the alveoli. This is a major factor keeping the smaller airways—those without cartilage to support them—from collapsing. Because, as we have seen, transpulmonary pressure increases during inspiration, airway radius becomes larger and airway resistance smaller as the lungs expand during inspiration. The opposite occurs during expiration.

A second physical factor holding the airways open is the elastic connective-tissue fibers that link the outside of the airways to the surrounding alveolar tissue. These fibers are pulled upon as the lungs expand during inspiration, and in turn they help pull the airways open even more than between breaths. This is termed **lateral traction.** Thus, putting this information and that of the previous paragraph together, both the transpulmonary pressure and lateral traction act in the same direction, reducing airway resistance during inspiration.

Such physical factors also explain why the airways become narrower and airway resistance increases during a forced expiration. Indeed, because of increased airway resistance, there is a limit as to how much one can increase the air flow rate during a forced expiration no matter how intense the effort.

In addition to these physical factors, a variety of neuroendocrine and paracrine factors can influence airway smooth muscle and thereby airway resistance. For example, the hormone epinephrine relaxes airway smooth muscle (via an effect on beta-adrenergic receptors), whereas the leukotrienes (members of the eicosanoid family) produced in the lungs during inflammation contract the muscle.

One might wonder why physiologists are concerned with all the many physical and chemical factors that *can* influence airway resistance when we earlier stated that airway resistance is *normally* so low that it is no impediment to air flow. The reason is that, under *abnormal* circumstances, changes in these factors may cause serious increases in airway resistance. Asthma and chronic obstructive pulmonary disease provide important examples.

Asthma *Asthma* is a disease characterized by intermittent attacks in which airway smooth muscle contracts strongly, markedly increasing airway resistance. The basic defect in asthma is chronic inflammation of the airways, the causes of which vary from person to person and include, among others, allergy and virus infections. The important point is that the underlying inflammation causes the airway smooth muscle to be hyperresponsive and to contract strongly when, depending upon the individual, the person exercises (especially in cold, dry air) or is exposed to cigarette smoke, environmental pollutants, viruses, allergens, normally released bronchoconstrictor chemicals, and a variety of other potential triggers.

The therapy for asthma is twofold: (1) to reduce the chronic inflammation and hence the airway hyperresponsiveness with so-called *anti-inflammatory drugs,* particularly glucocorticoids (Chapter 20) taken by inhalation; and (2) to overcome acute excessive airway smooth-muscle contraction with *bronchodilator drugs*— that is, drugs that relax the airways. The latter drugs work on the airways either by enhancing the actions of bronchodilator neuroendocrine or paracrine messengers or by blocking the actions of bronchoconstrictors. For example, one class of bronchodilator drug mimics the normal action of epinephrine on beta-adrenergic receptors.

Chronic Obstructive Pulmonary Disease The term *chronic obstructive pulmonary disease* refers to (1) emphysema, (2) chronic bronchitis, or (3) a combination of the two. These diseases, which cause severe difficulties not only in ventilation but in oxygenation of the blood, are among the major causes of disability and death in the United States. In contrast to asthma, increased smooth-muscle contraction is *not* the cause of the airway obstruction in these diseases

Emphysema is discussed later in this chapter; suffice it to say here that the cause of obstruction in this disease is destruction and collapse of the smaller airways.

Chronic bronchitis is characterized by excessive mucus production in the bronchi and chronic inflammatory changes in the small airways. The cause of obstruction is an accumulation of mucus in the airways and thickening of the inflamed airways. The same agents that cause emphysema—smoking, for example—also cause chronic bronchitis, which is why the two diseases frequently coexist.

The Heimlich Maneuver   The *Heimlich maneuver* is used to aid a person choking on a foreign body caught in and obstructing the upper airways. A sudden increase in abdominal pressure is produced as the rescuer's fists, placed against the victim's abdomen slightly above the navel and well below the tip of the sternum, are pressed into the abdomen with a quick upward thrust (Figure 15–15). The increased abdominal pressure forces the diaphragm upward into the thorax, reducing thoracic size and, by Boyle's Law, increasing alveolar pressure. The forceful expiration produced by the increased alveolar pressure often expels the object caught in the respiratory tract.

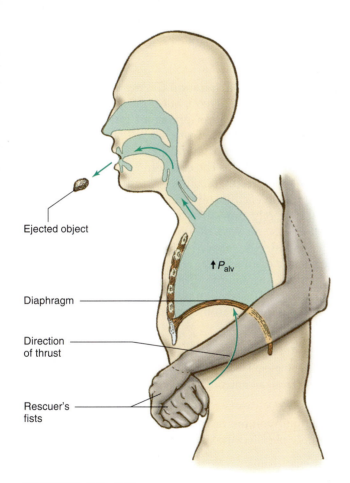

FIGURE 15–15

The Heimlich maneuver. The rescuer's fists are placed against the victim's abdomen. A quick upward thrust of the fists causes elevation of the diaphragm and a forceful expiration. As this expired air is forced through the trachea and larynx, the foreign object in the airway is often expelled.

Lung Volumes and Capacities

Normally the volume of air entering the lungs during a single inspiration is approximately equal to the volume leaving on the subsequent expiration and is called the **tidal volume.** The tidal volume during normal quiet breathing is termed the resting tidal volume and is approximately 500 ml. As illustrated in Figure 15–16, the maximal amount of air that can be increased above this value during deepest inspiration is termed the **inspiratory reserve volume** (and is about 3000 ml—that is, sixfold greater than resting tidal volume).

After expiration of a resting tidal volume, the lungs still contain a very large volume of air. As described earlier, this is the resting position of the lungs and chest wall (that is, the position that exists when there is no contraction of the respiratory muscles); it is termed the **functional residual capacity** (and averages about 2500 ml). Thus, the 500 ml of air inspired with each resting breath adds to and mixes with the much larger volume of air already in the lungs, and then 500 ml of the total is expired. Through maximal active contraction of the expiratory muscles, it is possible to expire much more of the air remaining after the resting tidal volume has been expired; this additional expired volume is termed the **expiratory reserve volume** (about 1500 ml). Even after a maximal active expiration, approximately 1000 ml of air still remains in the lungs and is termed the **residual volume.**

A useful clinical measurement is the **vital capacity,** the maximal volume of air that a person can expire after a maximal inspiration. Under these conditions, the person is expiring both the resting tidal volume and inspiratory reserve volume just inspired, plus the expiratory reserve volume (Figure 15–16). In other words, the vital capacity is the sum of these three volumes.

A variant on this method is the *forced expiratory volume in 1 s, (FEV_1),* in which the person takes a maximal inspiration and then exhales maximally *as fast as possible.* The important value is the fraction of the total "forced" vital capacity expired in 1 s. Normal individuals can expire approximately 80 percent of the vital capacity in this time.

These *pulmonary function tests* are useful diagnostic tools. For example, people with *obstructive lung diseases* (increased airway resistance) typically have a FEV_1 which is less than 80 percent of the vital capacity because it is difficult for them to expire air rapidly through the narrowed airways. In contrast to obstructive lung diseases, *restrictive lung diseases* are characterized by normal airway resistance but impaired respiratory movements because of abnormalities in the lung tissue, the pleura, the chest wall, or the neuromuscular machinery. Restrictive lung diseases are

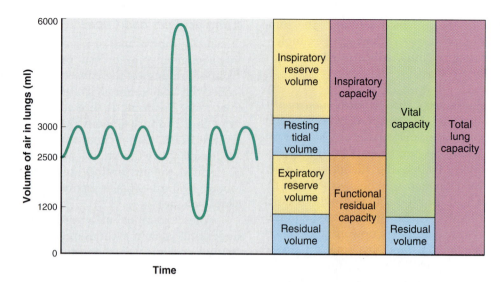

FIGURE 15–16

Lung volumes and capacities recorded on a spirometer, an apparatus for measuring inspired and expired volumes. When the subject inspires, the pen moves up; with expiration, it moves down. The capacities are the sums of two or more lung volumes. The lung volumes are the four distinct components of total lung capacity. Note that residual volume, total lung capacity, and functional residual capacity cannot be measured with a spirometer.

characterized by a reduced vital capacity but a normal ratio of FEV_1 to vital capacity.

Alveolar Ventilation

The total ventilation per minute, termed the **minute ventilation,** is equal to the tidal volume multiplied by the respiratory rate:

$$\begin{array}{l} \text{Minute ventilation} = \\ \text{(ml/min)} \end{array}$$

$$\begin{array}{cc} \text{Tidal volume} \times \text{Respiratory rate} & (15\text{-}5) \\ \text{(ml/breath)} \quad \text{(breaths/min)} \end{array}$$

For example, at rest, a normal person moves approximately 500 ml of air in and out of the lungs with each breath and takes 10 breaths each minute. The minute ventilation is therefore 500 ml/breath × 10 breaths/minute = 5000 ml of air per minute. However, because of dead space, not all this air is available for exchange with the blood.

Dead Space The conducting airways have a volume of about 150 ml. Exchanges of gases with the blood occur only in the alveoli and not in this 150 ml of the airways. Picture, then, what occurs during expiration of a tidal volume, which in this example we'll set at 450 ml instead of the 500 ml mentioned earlier. The 450 ml of air is forced out of the alveoli and through the airways. Approximately 300 ml of this alveolar air is exhaled at the nose or mouth, but approximately 150 ml still remains in the airways at the end of expiration.

During the next inspiration (Figure 15–17), 450 ml of air flows into the alveoli, but the first 150 ml entering the alveoli is not atmospheric air but the 150 ml left behind in the airways from the last breath. Thus, only 300 ml of new atmospheric air enters the alveoli during the inspiration. The end result is that 150 ml of the 450 ml of atmospheric air entering the respiratory system during each inspiration never reaches the alveoli but is merely moved in and out of the airways. Because these airways do not permit gas exchange with the blood, the space within them is termed the **anatomic dead space.**

Thus the volume of *fresh* air entering the alveoli during each inspiration equals the tidal volume *minus* the volume of air in the anatomic dead space. For the previous example:

Tidal volume = 450 ml
Anatomic dead space = 150 ml
Fresh air entering alveoli in one inspiration =
450 ml − 150 ml = 300 ml

The total volume of fresh air entering the alveoli per minute is called the **alveolar ventilation:**

$$\begin{array}{l} \text{Alveolar ventilation} = \\ \text{(ml/min)} \end{array}$$

$$\begin{array}{cc} \text{(Tidal volume} - \text{Dead space)} \times \text{Respiratory rate} & (15\text{-}6) \\ \text{(ml/breath)} \quad \text{(ml/breath)} \quad \text{(breath/min)} \end{array}$$

When evaluating the efficacy of ventilation, one should always focus on the alveolar ventilation, not

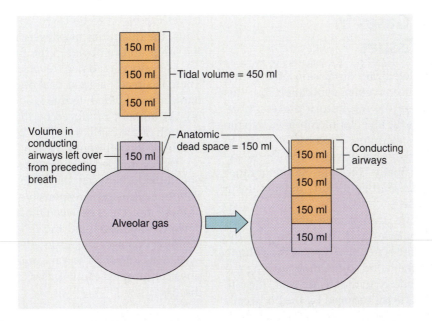

FIGURE 15–17

Effects of anatomic dead space on alveolar ventilation. Anatomic dead space is the volume of the conducting airways.

the minute ventilation. This generalization is demonstrated readily by the data in Table 15–4. In this experiment, subject A breathes rapidly and shallowly, B normally, and C slowly and deeply. Each subject has exactly the same minute ventilation; that is, each is moving the same amount of air in and out of the lungs per minute. Yet, when we subtract the anatomic dead-space ventilation from the minute ventilation, we find marked differences in alveolar ventilation. Subject A has no alveolar ventilation and would become unconscious in several minutes, whereas C has a considerably greater alveolar ventilation than B, who is breathing normally.

Another important generalization to be drawn from this example is that increased *depth* of breathing is far more effective in elevating alveolar ventilation than is an equivalent increase in breathing *rate*. Conversely, a decrease in depth can lead to a critical reduction in alveolar ventilation. This is because a fixed volume of *each* tidal volume goes to the dead space. If

the tidal volume decreases, the fraction of the tidal volume going to the dead space increases until, as in subject A, it may represent the entire tidal volume. On the other hand, any increase in tidal volume goes entirely toward increasing alveolar ventilation. These concepts have important physiological implications. Most situations that produce an increased ventilation, such as exercise, reflexly call forth a relatively greater increase in breathing depth than rate.

The anatomic dead space is not the only type of dead space. Some fresh inspired air is not used for gas exchange with the blood even though it reaches the alveoli because some alveoli, for various reasons, have little or no blood supply. This volume of air is known as **alveolar dead space.** It is quite small in normal persons but may be very large in several kinds of lung disease. As we shall see, it is minimized by local mechanisms that match air and blood flows. The sum of the anatomic and alveolar dead spaces is known as the **physiologic dead space.**

TABLE 15–4 Effect of Breathing Patterns on Alveolar Ventilation

Subject	Tidal Volume, ml/breath	×	Frequency, breaths/min	=	Minute Ventilation, ml/min	Anatomic Dead-Space Ventilation, ml/min	Alveolar Ventilation, ml/min
A	150		40		6000	150 × 40 = 6000	0
B	500		12		6000	150 × 12 = 1800	4200
C	1000		6		6000	150 × 6 = 900	5100

Exchange of Gases in Alveoli and Tissues

We have now completed our discussion of the lung mechanics that produce alveolar ventilation, but this is only the first step in the respiratory process. Oxygen must move across the alveolar membranes into the pulmonary capillaries, be transported by the blood to the tissues, leave the tissue capillaries and enter the extracellular fluid, and finally cross plasma membranes to gain entry into cells. Carbon dioxide must follow a similar path in reverse.

In the steady state, the volume of oxygen that leaves the tissue capillaries and is consumed by the body cells per unit time is exactly equal to the volume of oxygen added to the blood in the lungs during the same time period. Similarly, in the steady state, the rate at which carbon dioxide is produced by the body cells and enters the systemic blood is identical to the rate at which carbon dioxide leaves the blood in the lungs and is expired. Note that these statements apply to the steady state; *transiently,* oxygen utilization in the tissues *can* differ from oxygen uptake in the lungs and carbon dioxide production can differ from elimination in the lungs, but within a short time these imbalances automatically produce changes in diffusion gradients in the lungs and tissues that reestablish steady-state balances.

The amounts of oxygen consumed by cells and carbon dioxide produced, however, are not necessarily identical to each other. The balance depends primarily upon which nutrients are being used for energy. The ratio of CO_2 produced/O_2 consumed is known as the **respiratory quotient (RQ).** On a mixed diet, the RQ is approximately 0.8; that is, 8 molecules of CO_2 are produced for every 10 molecules of O_2 consumed. (The RQ is 1 for carbohydrate, 0.7 for fat, and 0.8 for protein.)

For purposes of illustration, Figure 15–18 presents typical exchange values during 1 min for a person at rest, assuming a cellular oxygen consumption of 250 ml/min, a carbon dioxide production of 200 ml/min, an alveolar ventilation (supply of fresh air to the alveoli) of 4000 ml/min, and a cardiac output of 5000 ml/min.

Since only 21 percent of the atmospheric air is oxygen, the total oxygen entering the alveoli per min in our illustration is 21 percent of 4000 ml, or 840 ml/min. Of this inspired oxygen, 250 ml crosses the alveoli into the pulmonary capillaries, and the rest is subsequently exhaled. Note that blood entering the lungs already contains a large quantity of oxygen, to which the new 250 ml is added. The blood then flows from the lungs to the left heart and is pumped by the left ventricle through the tissue capillaries, where 250 ml of oxygen leaves the blood to be taken up and utilized by cells. Thus, the quantities of oxygen added to the blood in the lungs and removed in the tissues are identical.

The story reads in reverse for carbon dioxide: There is already a good deal of carbon dioxide in systemic arterial blood; to this is added an additional 200 ml, the amount produced by the cells, as blood flows through tissue capillaries. This 200 ml leaves the blood as blood flows through the lungs, and is expired.

Blood pumped by the heart carries oxygen and carbon dioxide between the lungs and tissues by bulk flow, but diffusion is responsible for the net movement of these molecules between the alveoli and blood, and between the blood and the cells of the body. Understanding the mechanisms involved in these diffusional exchanges depends upon some basic chemical and physical properties of gases, which we will now discuss.

Partial Pressures of Gases

Gas molecules undergo continuous random motion. These rapidly moving molecules exert a pressure, the magnitude of which is increased by anything that increases the rate of movement. The pressure a gas exerts is proportional to (1) the temperature (because heat increases the speed at which molecules move) and (2) the concentration of the gas—that is, the number of molecules per unit volume.

As stated by **Dalton's law,** in a mixture of gases, the pressure exerted by each gas is independent of the pressure exerted by the others. This is because gas molecules are normally so far apart that they do not interfere with each other. Since each gas in a mixture behaves as though no other gases are present, the total pressure of the mixture is simply the sum of the individual pressures. These individual pressures, termed **partial pressures,** are denoted by a P in front of the symbol for the gas. For example, the partial pressure of oxygen is represented by P_{O_2}. The partial pressure of a gas is directly proportional to its concentration. Net diffusion of a gas will occur from a region where its partial pressure is high to a region where it is low.

Atmospheric air consists primarily of nitrogen (approximately 79 percent) and oxygen (approximately 21 percent), with very small quantities of water vapor, carbon dioxide, and inert gases. The sum of the partial pressures of all these gases is termed atmospheric pressure, or barometric pressure. It varies in different parts of the world as a result of differences in altitude (it also varies with local weather conditions), but at sea level it is 760 mmHg. Since the partial pressure of any gas in a mixture is the fractional concentration of that gas times the total pressure of all the gases, the P_{O_2} of atmospheric air is 0.21×760 mmHg = 160 mmHg at sea level.

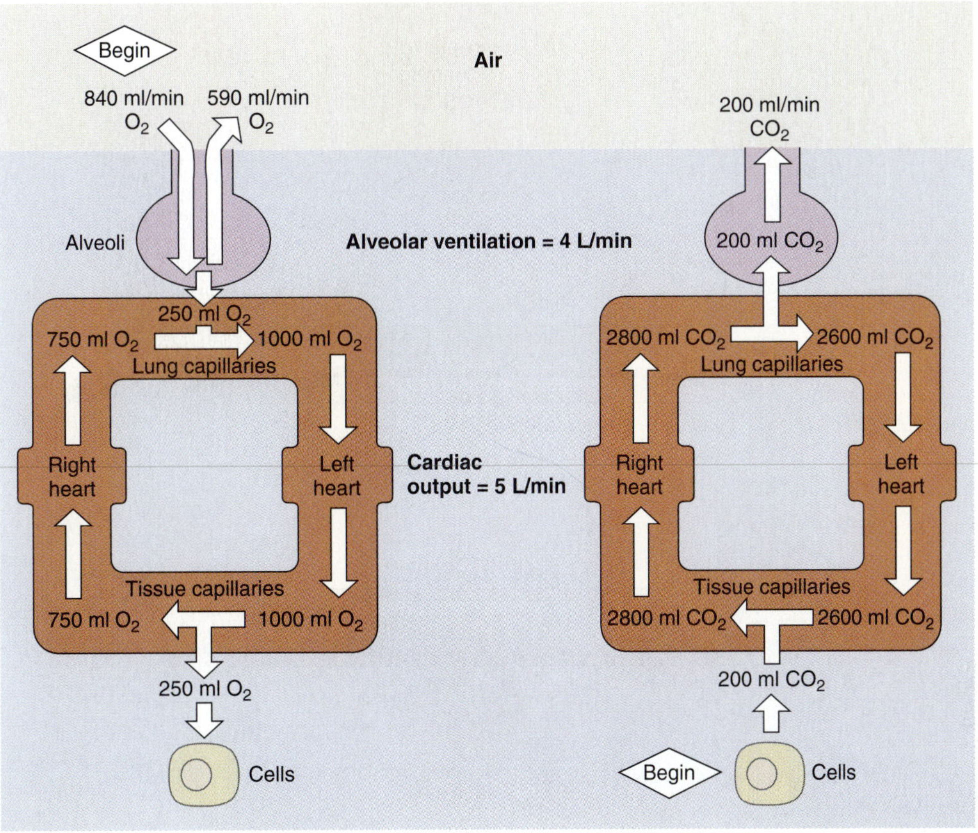

FIGURE 15–18
Summary of typical oxygen and carbon dioxide exchanges between atmosphere, lungs, blood, and tissues during 1 min in a resting individual. Note that the values given in this figure for oxygen and carbon dioxide in blood are *not* the values per liter of blood but rather the amounts transported *per minute* in the cardiac output (5 L in this example). The volume of oxygen in 1 L of arterial blood is 200 ml O_2/L of blood—that is, 1000 ml O_2/5 L of blood.

Diffusion of Gases in Liquids When a liquid is exposed to air containing a particular gas, molecules of the gas will enter the liquid and dissolve in it. **Henry's law** states that the amount of gas dissolved will be directly proportional to the partial pressure of the gas with which the liquid is in equilibrium. A corollary is that, at equilibrium, the partial pressures of the gas molecules in the liquid and gaseous phases must be identical. Suppose, for example, that a closed container contains both water and gaseous oxygen. Oxygen molecules from the gas phase constantly bombard the surface of the water, some entering the water and dissolving. Since the number of molecules striking the surface is directly proportional to the P_{O_2} of the gas phase, the number of molecules entering the water and dissolving in it is also directly proportional to the P_{O_2}. As long as the P_{O_2} in the gas phase is higher than the P_{O_2} in the liquid, there will be a net diffusion of oxygen into the liquid. Diffusion equilibrium will be reached only when the P_{O_2} in the liquid is equal to the P_{O_2} in the gas phase, and there will be no further net diffusion between the two phases.

Conversely, if a liquid containing a dissolved gas at high partial pressure is exposed to a lower partial pressure of that same gas in a gas phase, there will be a net diffusion of gas molecules out of the liquid into the gas phase until the partial pressures in the two phases become equal.

The exchanges *between* gas and liquid phases described in the last two paragraphs are precisely the phenomena occurring in the lungs between alveolar air and pulmonary capillary blood. In addition, *within* a liquid, dissolved gas molecules also diffuse from a region of higher partial pressure to a region of lower partial pressure, an effect that underlies the exchange of gases between cells, extracellular fluid, and capillary blood throughout the body.

Why must the diffusion of gases into or within liquids be presented in terms of partial pressures rather than "concentrations," the values used to deal with the diffusion of all other solutes? The reason is that the *concentration* of a gas in a liquid is proportional not only to the partial pressure of the gas but also to the solubility of the gas in the liquid; the more soluble the

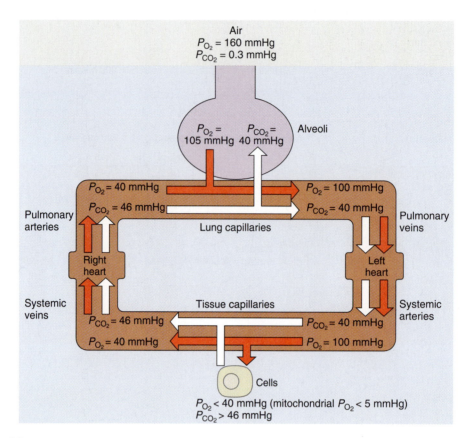

FIGURE 15–19

Partial pressures of carbon dioxide and oxygen in inspired air at sea level and various places in the body. The reason that the alveolar P_{O_2} and pulmonary vein P_{O_2} are not exactly the same is described later in the text. Note also that the P_{O_2} in the systemic arteries is shown as identical to that in the pulmonary veins; for reasons involving the anatomy of the blood flow to the lungs, the systemic arterial value is actually slightly less, but we have ignored this for the sake of clarity.

gas, the greater will be its concentration at any given partial pressure. Thus, if a liquid is exposed to two different gases having the same partial pressures, at equilibrium the *partial pressures* of the two gases will be identical in the liquid but the *concentrations* of the gases in the liquid will differ, depending upon their solubilities in that liquid.

With these basic gas properties as the foundation, we can now discuss the diffusion of oxygen and carbon dioxide across alveolar and capillary walls, and plasma membranes. The partial pressures of these gases in air and in various sites of the body are given in Figure 15–19 for a resting person at sea level. We start our discussion with the *alveolar* gas pressures because their values set those of systemic arterial blood. This fact cannot be emphasized too strongly: The alveolar P_{O_2} and P_{CO_2} determine the systemic arterial P_{O_2} and P_{CO_2}.

Alveolar Gas Pressures

Normal alveolar gas pressures are $P_{O_2} = 105$ mmHg and $P_{CO_2} = 40$ mmHg. (We do not deal with nitrogen, even though it is the most abundant gas in the alve-

oli, because nitrogen is biologically inert under normal conditions and does not undergo any net exchange in the alveoli.) Compare these values with the gas pressures in the air being breathed: $P_{O_2} = 160$ mmHg and $P_{CO_2} = 0.3$ mmHg, a value so low that we will simply assume it to be zero. The alveolar P_{O_2} is lower than atmospheric P_{O_2} because some of the oxygen in the air entering the alveoli leaves them to enter the pulmonary capillaries. Alveolar P_{CO_2} is higher than atmospheric P_{CO_2} because carbon dioxide enters the alveoli from the pulmonary capillaries.

The factors that determine the precise value of alveolar P_{O_2} are (1) the P_{O_2} of atmospheric air, (2) the rate of alveolar ventilation, and (3) the rate of total-body oxygen consumption. Although there are equations for calculating the alveolar gas pressures from these variables, we will describe the interactions in a qualitative manner (Table 15–5). *To start, we will assume that only one of the factors changes at a time.*

First, a decrease in the P_{O_2} of the inspired air (at high altitude, for example) will decrease alveolar P_{O_2}. A decrease in alveolar ventilation will do the same thing (Figure 15–20) since less fresh air is entering the

TABLE 15–5 Effects of Various Conditions on Alveolar Gas Pressures

Condition	Alveolar P_{O_2}	Alveolar P_{CO_2}
Breathing air with low P_{O_2}	Decreases	No change*
↑Alveolar ventilation and unchanged metabolism	Increases	Decreases
↓Alveolar ventilation and unchanged metabolism	Decreases	Increases
↑Metabolism and unchanged alveolar ventilation	Decreases	Increases
↓Metabolism and unchanged alveolar ventilation	Increases	Decreases
Proportional increases in metabolism and alveolar ventilation	No change	No change

*Breathing air with low P_{O_2} has no direct effect on alveolar P_{CO_2}. However, as described later in the text, people in this situation will reflexly increase their ventilation, and that will lower P_{CO_2}.

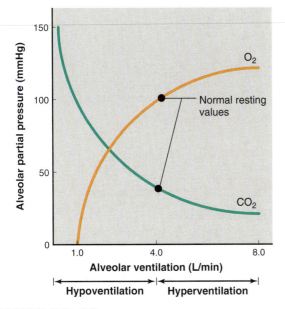

FIGURE 15–20

Effects of increasing or decreasing alveolar ventilation on alveolar partial pressures in a person having a constant metabolic rate (cellular oxygen consumption and carbon dioxide production). Note that alveolar P_{O_2} approaches zero when alveolar ventilation is about 1 L/min. At this point all the oxygen entering the alveoli crosses into the blood, leaving virtually no oxygen in the alveoli.

alveoli per unit time. Finally, an increase in the cells' consumption of oxygen will also lower alveolar P_{O_2} because a larger fraction of the oxygen in the entering fresh air will leave the alveoli to enter the blood and be used by the tissues (recall that in the steady state, the volume of oxygen entering the blood in the lungs per unit time is always equal to the volume utilized by the tissues). This discussion has been in terms of things that *lower* alveolar P_{O_2}; just reverse the direction of change of the three factors to see how to increase P_{O_2}.

The story for P_{CO_2} is analogous, again assuming that only one factor changes at a time. There is normally essentially no carbon dioxide in inspired air and so we can ignore that factor. A decreased alveolar ventilation will increase the alveolar P_{CO_2} (Figure 15–20) because there is less inspired fresh air to dilute the carbon dioxide entering the alveoli from the blood. An increased production of carbon dioxide will also increase the alveolar P_{CO_2} because more carbon dioxide will be entering the alveoli from the blood per unit time (recall that in the steady state the volume of carbon dioxide entering the alveoli per unit time is always equal to the volume produced by the tissues). Just reverse the direction of changes in this paragraph to cause a *decrease* in alveolar P_{CO_2}.

For simplicity we allowed only one factor to change at a time, but if more than one factor changes, the effects will either add to or subtract from each other. For example, if oxygen consumption and alveolar ventilation both increase at the same time, their opposing effects on alveolar P_{O_2} will tend to cancel each other out.

This last example emphasizes that, at any particular atmospheric P_{O_2}, it is the *ratio* of oxygen consumption to alveolar ventilation (that is, O_2 consumption/alveolar ventilation) that determines alveolar P_{O_2}—the higher the ratio, the *lower* the alveolar P_{O_2}. Similarly, alveolar P_{CO_2} is determined by a ratio, in this case the ratio of carbon dioxide production to alveolar ventilation (that is, CO_2 production/alveolar ventilation); the higher the ratio, the *higher* the alveolar P_{CO_2}.

Two terms that denote the adequacy of ventilation—that is, the relationship between metabolism and alveolar ventilation—can now be defined. Physiologists state these definitions in terms of carbon dioxide rather than oxygen. **Hypoventilation** exists when there is an *increase* in the ratio of carbon dioxide production to alveolar ventilation. In other words, a person is said to be hypoventilating if his alveolar ventilation cannot keep pace with his carbon dioxide production. The result is that alveolar P_{CO_2} increases above the normal value of 40 mmHg. **Hyperventilation** exists when there is a *decrease* in the ratio of carbon dioxide production to alveolar ventilation—that is, when alveolar ventilation is actually too great for the amount of carbon dioxide being produced. The result is that alveolar P_{CO_2} decreases below the normal value of 40 mmHg.

TABLE 15–6 Normal Gas Pressure

	Venous Blood	Arterial Blood	Alveoli	Atmosphere
P_{O_2}	40 mmHg	100 mmHg*	105 mmHg*	160 mmHg
P_{CO_2}	46 mmHg	40 mmHg	40 mmHg	0.3 mmHg

*The reason that the arterial P_{O_2} and alveolar P_{O_2} are not exactly the same is described later in the text.

Note that "hyperventilation" is not synonymous with "increased ventilation." Hyperventilation represents increased ventilation *relative to metabolism.* Thus, for example, the increased ventilation that occurs during moderate exercise is *not* hyperventilation since, as we shall see, in this situation the increase in production of carbon dioxide is exactly proportional to the increased ventilation.

Alveolar-Blood Gas Exchange

The blood that enters the pulmonary capillaries is, of course, systemic venous blood pumped to the lungs via the pulmonary arteries. Having come from the tissues, it has a relatively high P_{CO_2} (46 mmHg in a normal person at rest) and a relatively low P_{O_2} (40 mmHg) (Figure 15–19 and Table 15–6). The differences in the partial pressures of oxygen and carbon dioxide on the two sides of the alveolar-capillary membrane result in the net diffusion of oxygen from alveoli to blood and of carbon dioxide from blood to alveoli. As this diffusion occurs, the capillary blood P_{O_2} rises and its P_{CO_2} falls. The net diffusion of these gases ceases when the capillary partial pressures become equal to those in the alveoli.

In a normal person, the rates at which oxygen and carbon dioxide diffuse are so rapid and the blood flow through the capillaries so slow that complete equilibrium is reached well before the end of the capillaries (Figure 15–21). Only during the most strenuous exercise, when blood flows through the lung capillaries very rapidly, is there insufficient time for complete equilibration.

Thus, the blood that leaves the pulmonary capillaries to return to the heart and be pumped into the systemic arteries has essentially the same P_{O_2} and P_{CO_2} as alveolar air. (They are not exactly the same, for reasons given later.) Accordingly, the factors described in the previous section—atmospheric P_{O_2}, cellular oxygen consumption and carbon dioxide production, and alveolar ventilation—determine the alveolar gas pressures, which then determine the systemic arterial gas pressures.

Given that diffusion between alveoli and pulmonary capillaries normally achieves complete equilibration, the more capillaries participating in this process, the more total oxygen and carbon dioxide can

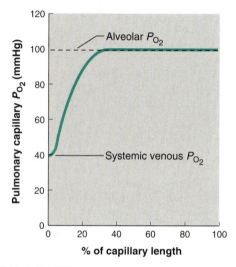

FIGURE 15–21

Equilibration of blood P_{O_2} with alveolar P_{O_2} along the length of the pulmonary capillaries.

be exchanged. Many of the pulmonary capillaries at the apex (top) of each lung are normally closed at rest. During exercise, these capillaries open and receive blood, thereby enhancing gas exchange. The mechanism by which this occurs is a simple physical one; the pulmonary circulation at rest is at such a low blood pressure that the pressure in these apical capillaries is inadequate to keep them open, but the increased cardiac output of exercise raises pulmonary vascular pressures, which opens these capillaries.

The diffusion of gases between alveoli and capillaries may be *impaired* in a number of ways, resulting in inadequate oxygen diffusion into the blood, particularly during exercise when the time for equilibration is reduced. For one thing, the surface area of the alveoli in contact with pulmonary capillaries may be decreased. In lung infections or pulmonary edema, for example, some of the alveoli may become filled with fluid. Diffusion may also be impaired if the alveolar walls become severely thickened with connective tissue, as, for example, in the disease (of unknown cause) called **diffuse interstitial fibrosis.** Pure diffusion problems of these types are restricted to oxygen and do not affect elimination of carbon dioxide, which is much more diffusible than oxygen.

Matching of Ventilation and Blood Flow in Alveoli

However, the major disease-induced cause of inadequate oxygen movement between alveoli and pulmonary capillary blood is not diffusion problems but the mismatching of the air supply and blood supply in individual alveoli.

The lungs are composed of approximately 300 million alveoli, each capable of receiving carbon dioxide from, and supplying oxygen to, the pulmonary capillary blood. To be most efficient, the right proportion of alveolar air flow (ventilation) and capillary blood flow (perfusion) should be available to *each* alveolus. Any mismatching is termed *ventilation-perfusion inequality.*

The major effect of ventilation-perfusion inequality is to lower the P_{O_2} of systemic arterial blood. Indeed, largely because of gravitational effects on ventilation and perfusion there is enough ventilation-perfusion inequality in normal people to lower the arterial P_{O_2} about 5 mmHg. This is the major explanation of the fact, given earlier, that the P_{O_2} of blood in the pulmonary veins and systemic arteries is normally 5 mmHg less than that of average alveolar air.

In disease states, regional changes in lung compliance, airway resistance, and vascular resistance can cause marked ventilation-perfusion inequalities. The extremes of this phenomenon are easy to visualize: (1) There may be ventilated alveoli with no blood supply at all, or (2) there may be blood flowing through areas of lung that have no ventilation (this is termed a *shunt*). But the inequality need not be all-or-none to be quite significant.

Carbon dioxide elimination is also impaired by ventilation-perfusion inequality but, for complex reasons, not nearly to the same degree as oxygen uptake. Nevertheless, severe ventilation-perfusion inequalities in disease states can lead to some elevation of arterial P_{CO_2}.

There are several local homeostatic responses within the lungs to minimize the mismatching of ventilation and blood flow. One of the most important operates on the blood vessels to alter blood-flow distribution. If an alveolus is receiving too little air relative to its blood supply, the P_{O_2} in the alveolus and its surrounding area will be low. This decreased P_{O_2} causes vasoconstriction of the small pulmonary blood vessels. (Note that this local effect of low oxygen on pulmonary blood vessels is precisely the opposite of that exerted on systemic arterioles.) The net adaptive effect is to supply less blood to poorly ventilated areas and more blood to well-ventilated areas.

Gas Exchange in the Tissues

As the systemic arterial blood enters capillaries throughout the body, it is separated from the interstitial fluid by only the thin capillary wall, which is highly permeable to both oxygen and carbon dioxide. The interstitial fluid in turn is separated from intracellular fluid by the plasma membranes of the cells, which are also quite permeable to oxygen and carbon dioxide. Metabolic reactions occurring within cells are constantly consuming oxygen and producing carbon dioxide. Therefore, as shown in Figure 15–19, intracellular P_{O_2} is lower and P_{CO_2} higher than in blood. The lowest P_{O_2} of all—less than 5 mmHg—is in the mitochondria, the site of oxygen utilization. As a result, there is a net diffusion of oxygen from blood into cells (and, within the cells, into the mitochondria), and a net diffusion of carbon dioxide from cells into blood. In this manner, as blood flows through systemic capillaries, its P_{O_2} decreases and its P_{CO_2} increases. This accounts for the systemic venous blood values shown in Figure 15-19 and Table 15–6.

The mechanisms that enhance diffusion of oxygen and carbon dioxide between cells and blood when a tissue increases its metabolic activity were discussed in Chapter 14.

In summary, the supply of new oxygen to the alveoli and the consumption of oxygen in the cells create P_{O_2} gradients that produce net diffusion of oxygen from alveoli to blood in the lungs and from blood to cells in the rest of the body. Conversely, the production of carbon dioxide by cells and its elimination from the alveoli via expiration create P_{CO_2} gradients that produce net diffusion of carbon dioxide from cells to blood in the rest of the body and from blood to alveoli in the lungs

Transport of Oxygen in Blood

Table 15–7 summarizes the oxygen content of systemic arterial blood (we shall henceforth refer to systemic arterial blood simply as arterial blood). Each liter normally contains the number of oxygen molecules equivalent to 200 ml of pure gaseous oxygen at atmospheric pressure. The oxygen is present in two forms: (1) dissolved in the plasma and erythrocyte water and (2) reversibly combined with hemoglobin molecules in the erythrocytes.

As predicted by Henry's law, the amount of oxygen dissolved in blood is directly proportional to the P_{O_2} of the blood. Because oxygen is relatively insoluble in water, only 3 ml can be dissolved in 1 L of blood at the normal arterial P_{O_2} of 100 mmHg. The other 197 ml of oxygen in a liter of arterial blood, more than 98 percent of the oxygen content in the liter, is transported in the erythrocytes reversibly combined with hemoglobin.

Each **hemoglobin** molecule is a protein made up of four subunits bound together. Each subunit consists of a molecular group known as **heme** and a polypeptide attached to the heme. [Hemoglobin is not the only

TABLE 15–7 Oxygen Content of Systemic Arterial Blood at Sea Level

1 liter (L) arterial blood contains
3 ml O_2 physically dissolved (1.5%)
197 ml O_2 bound to hemoglobin (98.5%)
Total   200 ml O_2
Cardiac output = 5 L/min
O_2 carried to tissues/min = 5 L/min × 200 ml O_2/L
= 1000 ml O_2/min

heme-containing protein in the body; others are the cytochromes and myoglobin (Chapters 4 and 11).] The four polypeptides of a hemoglobin molecule are collectively called **globin.** Each of the four heme groups in a hemoglobin molecule (Figure 15–22) contains one atom of **iron (Fe),** to which oxygen binds. Since each iron atom can bind one molecule of oxygen, a single hemoglobin molecule can bind four molecules of oxygen. However, for simplicity, the equation for the reaction between oxygen and hemoglobin is usually written in terms of a single polypeptide-heme chain of a hemoglobin molecule:

$$O_2 + Hb \rightleftharpoons HbO_2 \qquad (15\text{-}7)$$

Thus this chain can exist in one of two forms—**deoxyhemoglobin (Hb)** and **oxyhemoglobin (HbO_2).** In a blood sample containing many hemoglobin molecules, the fraction of all the hemoglobin in the form of oxyhemoglobin is expressed as the **percent hemoglobin saturation:**

Percent saturation =

$$\frac{O_2 \text{ bound to Hb}}{\text{Maximal capacity of Hb to bind } O_2} \times 100 \quad (15\text{-}8)$$

For example, if the amount of oxygen bound to hemoglobin is 40 percent of the maximal capacity of hemoglobin to bind oxygen, the sample is said to be 40 percent saturated. The denominator in this equation is also termed the **oxygen-carrying capacity** of the blood.

What factors determine the percent hemoglobin saturation? By far the most important is the blood P_{O_2}. Before turning to this subject, however, it must be stressed that the *total amount* of oxygen carried by hemoglobin in blood depends not only on the percent saturation of hemoglobin but also on how much hemoglobin there is in each liter of blood. For example, if a person's blood contained only half as much hemoglobin per liter as normal, then at any given percent saturation the oxygen content of the blood would be only half as much.

Effect of P_{O_2} on Hemoglobin Saturation

From inspection of Equation 15-7 and the law of mass action, one can see that raising the blood P_{O_2} should increase the combination of oxygen with hemoglobin. The experimentally determined quantitative relationship between these variables is shown in Figure 15–23,

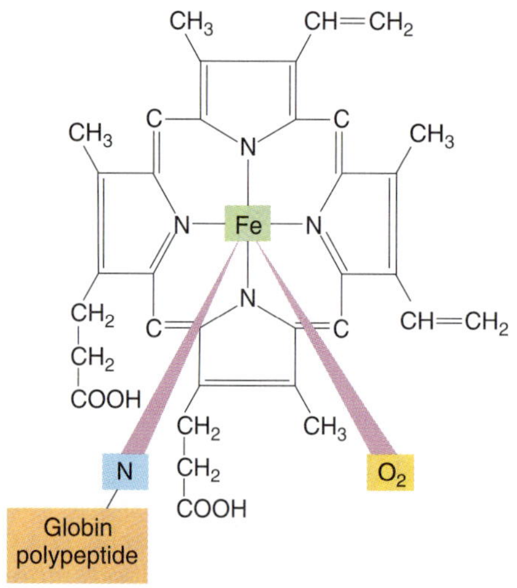

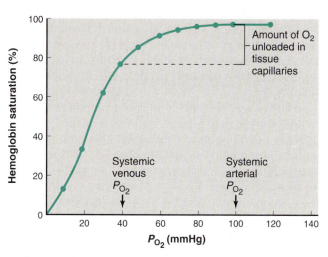

FIGURE 15–22

Heme. Oxygen binds to the iron atom (Fe). Heme attaches to a polypeptide chain by a nitrogen atom to form one subunit of hemoglobin. Four of these subunits bind to each other to make a single hemoglobin molecule.

FIGURE 15–23

Oxygen-hemoglobin dissociation curve. This curve applies to blood at 37°C and a normal arterial hydrogen-ion concentration. At any given blood hemoglobin concentration, the vertical axis could also have plotted oxygen content, in milliliters of oxygen. At 100 percent saturation, the amount of hemoglobin in normal blood carries 200 ml of oxygen.

which is called an **oxygen-hemoglobin dissociation curve.** (The term "dissociate" means "to separate," in this case oxygen from hemoglobin; it could just as well have been called an oxygen-hemoglobin association curve.) The curve is S-shaped because, as stated earlier, each hemoglobin molecule contains four subunits; each subunit can combine with one molecule of oxygen, and the reactions of the four subunits occur sequentially, with each combination facilitating the next one. (This combination of oxygen with hemoglobin is an example of cooperativity, as described in Chapter 4. The explanation in this case is as follows. The globin units of deoxyhemoglobin are tightly held by electrostatic bonds in a conformation with a relatively low affinity for oxygen. The binding of oxygen to a heme molecule breaks some of these bonds between the globin units, leading to a conformation change such that the remaining oxygen-binding sites are more exposed. Thus, the binding of one oxygen molecule to deoxyhemoglobin increases the affinity of the remaining sites on the same hemoglobin molecule, and so on.)

Note that the curve has a steep slope between 10 and 60 mmHg P_{O_2} and a relatively flat portion (or plateau) between 70 and 100 mmHg P_{O_2}. Thus, the extent to which oxygen combines with hemoglobin increases very rapidly as the P_{O_2} increases from 10 to 60 mmHg, so that at a P_{O_2} of 60 mmHg, 90 percent of the total hemoglobin is combined with oxygen. From this point on, a further increase in P_{O_2} produces only a small increase in oxygen binding.

The importance of this plateau at higher P_{O_2} values is as follows. Many situations, including high altitude and pulmonary disease, are characterized by a moderate reduction in alveolar and therefore arterial P_{O_2}. Even if the P_{O_2} fell from the normal value of 100 to 60 mmHg, the total quantity of oxygen carried by hemoglobin would decrease by only 10 percent since hemoglobin saturation is still close to 90 percent at a P_{O_2} of 60 mmHg. The plateau therefore provides an excellent safety factor in the supply of oxygen to the tissues.

The plateau also explains another fact: In a normal person at sea level, raising the alveolar (and therefore the arterial) P_{O_2} either by hyperventilating or by breathing 100 percent oxygen adds very little additional oxygen to the blood. A small additional amount dissolves, but because hemoglobin is already almost completely saturated with oxygen at the normal arterial P_{O_2} of 100 mmHg, it simply cannot pick up any more oxygen when the P_{O_2} is elevated beyond this point. But this applies only to normal people at sea level. If the person initially has a low arterial P_{O_2} because of lung disease or high altitude, then there would be a great deal of deoxyhemoglobin initially present in the arterial blood. Therefore, raising the alveolar and thereby the arterial P_{O_2} would result in significantly more oxygen transport.

We now retrace our steps and reconsider the movement of oxygen across the various membranes, this time including hemoglobin in our analysis. It is essential to recognize that the oxygen bound to hemoglobin does *not* contribute directly to the P_{O_2} of the blood. Only dissolved oxygen does so. Therefore, oxygen diffusion is governed only by the dissolved portion, a fact that permitted us to ignore hemoglobin in discussing transmembrane partial pressure gradients. However, the presence of hemoglobin plays a critical role in determining the *total amount* of oxygen that will diffuse, as illustrated by a simple example (Figure 15–24).

Two solutions separated by a semipermeable membrane contain equal quantities of oxygen, the gas pressures are equal, and no net diffusion occurs. Addition of hemoglobin to compartment B destroys this equilibrium because much of the oxygen combines with hemoglobin. Despite the fact that the total *quantity* of oxygen in compartment B is still the same, the number of *dissolved* oxygen molecules has decreased. Therefore, the P_{O_2} of compartment B is less than that of A, and so there is a net diffusion of oxygen from A to B. At the new equilibrium, the oxygen pressures are once again equal, but almost all the oxygen is in compartment B and is combined with hemoglobin.

Let us now apply this analysis to capillaries of the lungs and tissues (Figure 15–25). The plasma and erythrocytes entering the lungs have a P_{O_2} of 40 mmHg. As we can see from Figure 15–23, hemoglobin

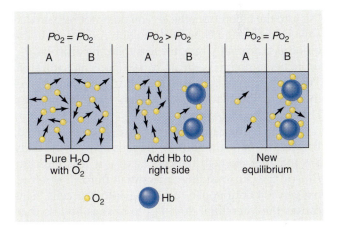

FIGURE 15–24

Effect of added hemoglobin on oxygen distribution between two compartments containing a fixed number of oxygen molecules and separated by a semipermeable membrane. At the new equilibrium, the P_{O_2} values are again equal to each other but lower than before the hemoglobin was added. However, the total oxygen, in other words, that dissolved plus that combined with hemoglobin, is now much higher on the right side of the membrane.

Adapted from Comroe.

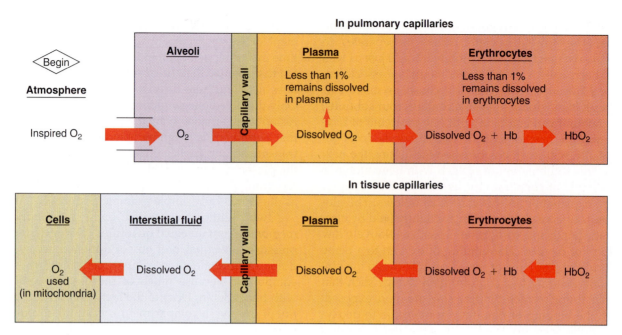

In pulmonary capillaries

Alveoli	Capillary wall	Plasma	Erythrocytes

Begin

Atmosphere

Inspired O_2 → O_2 → Dissolved O_2 → Dissolved O_2 + Hb → HbO_2

Less than 1% remains dissolved in plasma

Less than 1% remains dissolved in erythrocytes

In tissue capillaries

Cells	Interstitial fluid	Capillary wall	Plasma	Erythrocytes

O_2 used (in mitochondria) ← Dissolved O_2 ← Dissolved O_2 ← Dissolved O_2 + Hb ← HbO_2

FIGURE 15–25

Oxygen movement in the lungs and tissues. Movement of inspired air into the alveoli is by bulk-flow; all movements across membranes are by diffusion.

saturation at this P_{O_2} is 75 percent. The alveolar P_{O_2}— 105 mmHg—is higher than the blood P_{O_2} and so oxygen diffuses from the alveoli into the plasma. This increases plasma P_{O_2} and induces diffusion of oxygen into the erythrocytes, elevating erythrocyte P_{O_2} and causing increased combination of oxygen and hemoglobin. The vast preponderance of the oxygen diffusing into the blood from the alveoli does not remain dissolved but combines with hemoglobin. Therefore, the blood P_{O_2} normally remains less than the alveolar P_{O_2} until hemoglobin is virtually 100 percent saturated. Thus the diffusion gradient favoring oxygen movement into the blood is maintained despite the very large transfer of oxygen.

In the tissue capillaries, the procedure is reversed. Because the mitochondria of the cells all over the body are utilizing oxygen, the cellular P_{O_2} is less than the P_{O_2} of the surrounding interstitial fluid. Therefore, oxygen is continuously diffusing into the cells. This causes the interstitial fluid P_{O_2} to always be less than the P_{O_2} of the blood flowing through the tissue capillaries, and so net diffusion of oxygen occurs from the plasma within the capillary into the interstitial fluid. Accordingly, plasma P_{O_2} becomes lower than erythrocyte P_{O_2}, and oxygen diffuses out of the erythrocyte into the plasma. The lowering of erythrocyte P_{O_2} causes the dissociation of oxygen from hemoglobin, thereby liberating oxygen, which then diffuses out of the erythrocyte. The net result is a transfer, purely by diffusion, of large quantities of oxygen from hemo-

globin to plasma to interstitial fluid to the mitochondria of tissue cells.

To repeat, in most tissues under resting conditions, hemoglobin is still 75 percent saturated as the blood leaves the tissue capillaries. This fact underlies an important automatic mechanism by which cells can obtain more oxygen whenever they increase their activity. An exercising muscle consumes more oxygen, thereby lowering its tissue P_{O_2}. This increases the blood-to-tissue P_{O_2} gradient and hence the diffusion of oxygen from blood to cell. In turn, the resulting reduction in erythrocyte P_{O_2} causes additional dissociation of hemoglobin and oxygen. In this manner, an exercising muscle can extract almost all the oxygen from its blood supply, not just the usual 25 percent. Of course, an increased blood flow to the muscles (active hyperemia) also contributes greatly to the increased oxygen supply.

Carbon Monoxide and Oxygen Carriage *Carbon monoxide* is a colorless, odorless gas that is a product of the incomplete combustion of hydrocarbons, such as gasoline. It is one of the most common causes of sickness and death due to poisoning, both intentional and accidental. Its most striking pathophysiological characteristic is its extremely high affinity— 250 times that of oxygen—for the oxygen-binding sites in hemoglobin. For this reason, it reduces the amount of oxygen that combines with hemoglobin in pulmonary capillaries by competing for these sites. It exerts a second deleterious effect: It alters the hemoglobin molecule so as to shift the oxygen-hemoglobin dissociation curve to the left, thus decreasing the *unloading* of oxygen from hemoglobin in the tissues. As we shall see later, the situation is worsened by the fact that persons suffering from carbon monoxide poisoning do not show any reflex increase in their ventilation.

Effects of Blood P_{CO_2}, H^+ Concentration, Temperature, and DPG Concentration on Hemoglobin Saturation

At any given P_{O_2}, a variety of other factors influence the degree of hemoglobin saturation: blood P_{CO_2}, H^+ concentration, temperature, and the concentration of a substance—**2,3-diphosphoglycerate (DPG)** (also known as bisphosphoglycerate, BPG)—produced by the erythrocytes. As illustrated in Figure 15–26, an increase in any of these factors causes the dissociation curve to shift to the right, which means that, at any given P_{O_2}, hemoglobin has less affinity for oxygen. In contrast, a decrease in any of these factors causes the dissociation curve to shift to the left, which means that, at any given P_{O_2}, hemoglobin has a greater affinity for oxygen.

The effects of increased P_{CO_2}, H^+ concentration, and temperature are continuously exerted on the blood in tissue capillaries, because each of these factors is higher in tissue-capillary blood than in arterial blood: The P_{CO_2} is increased because of the carbon dioxide entering the blood from the tissues. For reasons to be described later, the H^+ concentration is elevated because of the elevated P_{CO_2} and the release of metabolically produced acids such as lactic acid. The temperature is increased because of the heat produced by tissue metabolism. Therefore, hemoglobin exposed to this elevated blood P_{CO_2}, H^+ concentration, and temperature as it passes through the tissue capillaries has its affinity for oxygen decreased, and therefore hemoglobin gives up even more oxygen than it would have if the decreased tissue-capillary P_{O_2} had been the only operating factor.

The more active a tissue is, the greater its P_{CO_2}, H^+ concentration, and temperature. At any given P_{O_2}, this causes hemoglobin to release more oxygen during passage through the tissue's capillaries and provides the more active cells with additional oxygen. Here is another local mechanism that increases oxygen delivery to tissues that have increased metabolic activity.

What is the mechanism by which these factors influence hemoglobin's affinity for oxygen? Carbon dioxide and hydrogen ions do so by combining with the globin portion of hemoglobin and altering the conformation of the hemoglobin molecule. Thus, these effects are a form of allosteric modulation (Chapter 4).

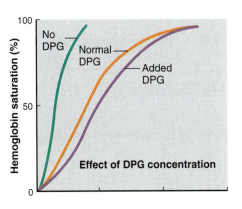
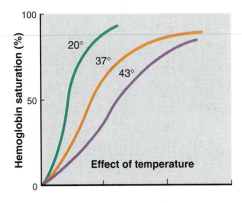
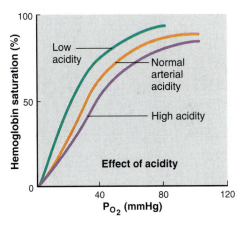

FIGURE 15–26

Effects of DPG concentration, temperature, and acidity on the relationship between P_{O_2} and hemoglobin saturation. The temperature of normal blood, of course, never diverges from 37°C as much as shown in the figure, but the principle is still the same when the changes are within the physiological range.

Adapted from Comroe.

An elevated temperature also decreases hemoglobin's affinity for oxygen by altering its molecular configuration.

Erythrocytes contain large quantities of DPG, which is present in only trace amounts in other mammalian cells. DPG, which is produced by the erythrocytes during glycolysis, binds reversibly with hemoglobin, allosterically causing it to have a lower affinity for oxygen (Figure 15–26). The net result is that whenever DPG levels are increased, there is enhanced unloading of oxygen from hemoglobin as blood flows through the tissues. Such an increase in DPG concentration is triggered by a variety of conditions associated with inadequate oxygen supply to the tissues and helps to maintain oxygen delivery. Examples include anemia and exposure to high altitude.

Transport of Carbon Dioxide in Blood

In a resting person, metabolism generates about 200 ml of carbon dioxide per minute. When arterial blood

flows through tissue capillaries, this volume of carbon dioxide diffuses from the tissues into the blood (Figure 15–27). Carbon dioxide is much more soluble in water than is oxygen, and so more dissolved carbon dioxide than dissolved oxygen is carried in blood. Even so, only a relatively small amount of blood carbon dioxide is transported in this way; only 10 percent of the carbon dioxide entering the blood remains physically dissolved in the plasma and erythrocytes.

Another 30 percent of the carbon dioxide molecules entering the blood reacts reversibly with the amino groups of hemoglobin to form **carbamino hemoglobin.** For simplicity, this reaction with hemoglobin is written as:

$$CO_2 + Hb \rightleftharpoons HbCO_2 \qquad (15\text{-}9)$$

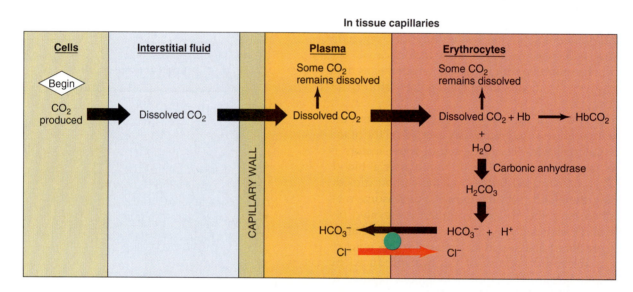

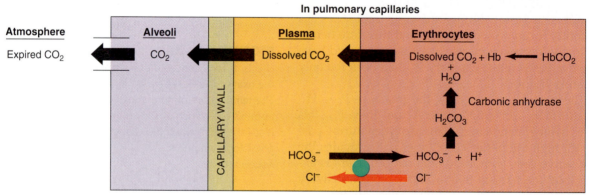

FIGURE 15–27

Summary of CO_2 movement. Expiration of CO_2 is by bulk-flow, whereas all movements of CO_2 across membranes are by diffusion. Arrows reflect relative proportions of the fates of the CO_2. About two-thirds of the CO_2 entering the blood in the tissues ultimately is converted to HCO_3^- in the erythrocytes because carbonic anhydrase is located there, but most of the HCO_3^- then moves out of the erythrocytes into the plasma in exchange for chloride ions (the "chloride shift"). See Figure 15–28 for the fate of the hydrogen ions generated in the erythrocytes.

This reaction is aided by the fact that deoxyhemoglobin, formed as blood flows through the tissue capillaries, has a greater affinity for carbon dioxide than does oxyhemoglobin.

The remaining 60 percent of the carbon dioxide molecules entering the blood in the tissues is converted to bicarbonate:

$$\underset{\substack{\text{Carbonic}\\\text{acid}}}{\underset{\text{anhydrase}}{\overset{\text{Carbonic}}{CO_2 + H_2O \rightleftharpoons H_2CO_3}}} \rightleftharpoons \underset{\text{Bicarbonate}}{HCO_3^- + H^+} \quad (15\text{-}10)$$

The first reaction in Equation 15-10 is rate-limiting and is very slow unless catalyzed by the enzyme **carbonic anhydrase.** This enzyme is present in the erythrocytes but not in the plasma; therefore, this reaction occurs mainly in the erythrocytes. In contrast, carbonic acid dissociates very rapidly into a bicarbonate ion and a hydrogen ion without any enzyme assistance. Once formed, most of the bicarbonate moves out of the erythrocytes into the plasma via a transporter that exchanges one bicarbonate for one chloride ion (this is called the "chloride shift").

The reactions shown in Equation 15-10 also explain why, as mentioned earlier, the H^+ concentration in tissue capillary blood and systemic venous blood is higher than that of the arterial blood and increases as metabolic activity increases. The fate of these hydrogen ions will be discussed in the next section.

Because carbon dioxide undergoes these various fates in blood, it is customary to add up the amounts of dissolved carbon dioxide, bicarbonate, and carbon dioxide in carbamino hemoglobin and call this sum the **total blood carbon dioxide.**

Just the opposite events occur as systemic venous blood flows through the lung capillaries (Figure 15–27). Because the blood P_{CO_2} is higher than alveolar P_{CO_2}, a net diffusion of CO_2 from blood into alveoli occurs. This loss of CO_2 from the blood lowers the blood P_{CO_2} and drives reactions 15-10 and 15-9 to the left: HCO_3^- and H^+ combine to give H_2CO_3, which then dissociates to CO_2 and H_2O. Similarly, $HbCO_2$ generates Hb and free CO_2. Normally, as fast as CO_2 is generated from HCO_3^- and H^+ and from $HbCO_2$, it diffuses into the alveoli. In this manner, all the CO_2 delivered into the blood in the tissues now is delivered into the alveoli; it is eliminated from the alveoli and from the body during expiration.

Transport of Hydrogen Ions between Tissues and Lungs

To repeat, as blood flows through the tissues, a fraction of oxyhemoglobin loses its oxygen to become deoxyhemoglobin, while simultaneously a large quantity of carbon dioxide enters the blood and undergoes the reactions that generate bicarbonate and hydrogen ions. What happens to these hydrogen ions?

Deoxyhemoglobin has a much greater affinity for H^+ than does oxyhemoglobin, and so it binds (buffers) most of the hydrogen ions (Figure 15–28). Indeed, deoxyhemoglobin is often abbreviated HbH rather than Hb to denote its binding of H^+. In effect, the reaction is $HbO_2 + H^+ \rightleftharpoons HbH + O_2$. In this manner, only a small number of the hydrogen ions generated in the blood remains free. This explains why the acidity of venous blood (pH = 7.36) is only slightly greater than that of arterial blood (pH = 7.40).

As the venous blood passes through the lungs, all these reactions are reversed. Deoxyhemoglobin becomes converted to oxyhemoglobin and, in the process, releases the hydrogen ions it had picked up in the tissues. The hydrogen ions react with bicarbonate to give carbonic acid, which dissociates to form carbon dioxide and water, and the carbon dioxide diffuses into the alveoli to be expired. Normally all the hydrogen ions that are generated in the tissue capillaries from the reaction of carbon dioxide and water recombine with bicarbonate to form carbon dioxide and water in the pulmonary capillaries. Therefore, none of these hydrogen ions appear in the *arterial* blood.

But what if the person is hypoventilating or has a lung disease that prevents normal elimination of carbon dioxide? Not only would arterial P_{CO_2} rise as a result but so would arterial H^+ concentration. Increased arterial H^+ concentration due to carbon dioxide retention is termed ***respiratory acidosis.*** Conversely, hyperventilation would lower the arterial values of both P_{CO_2} and H^+ concentration, producing ***respiratory alkalosis.***

In the course of describing the transport of oxygen, carbon dioxide, and H^+ in blood, we have presented multiple factors that influence the binding of these substances by hemoglobin. They are all summarized in Table 15–8.

One more aspect of the remarkable hemoglobin molecule should at least be mentioned—its ability to bind and transport **nitric oxide.** As described in Chapters 8, 14, and 20 respectively, nitric oxide is an important neurotransmitter and is also released by endothelial cells and macrophages. A present hypothesis is that as blood passes through the lungs, hemoglobin picks up and binds not only oxygen but nitric oxide synthesized there, carries it to the peripheral tissues, and releases it along with oxygen. Simultaneously, via a different binding site hemoglobin picks up and catabolizes nitric oxide produced in the peripheral tissues. Theoretically this cycle could play an important role in determining the peripheral concentration of nitric oxide and, thereby, the overall effect of this vasodilator agent. For example, by supplying net nitric

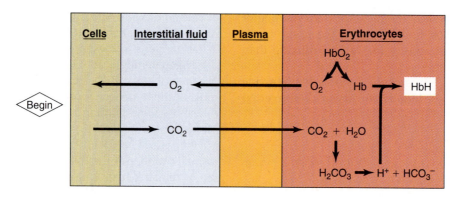

FIGURE 15–28

Binding of hydrogen ions by hemoglobin as blood flows through tissue capillaries. This reaction is facilitated because deoxyhemoglobin, formed as oxygen dissociates from hemoglobin, has a greater affinity for hydrogen ions than does oxyhemoglobin. For this reason, Hb and HbH are both abbreviations for deoxyhemoglobin.

oxide to the periphery, the process could cause additional vasodilation by systemic blood vessels; this would have effects on both local blood flow and systemic arterial blood pressure.

Control of Respiration

Neural Generation of Rhythmical Breathing

The diaphragm and intercostal muscles are skeletal muscles and therefore do not contract unless stimulated to do so by nerves. Thus, breathing depends entirely upon cyclical respiratory muscle excitation of the diaphragm and the intercostal muscles by their motor nerves. Destruction of these nerves (as in the viral disease *poliomyelitis*, for example) results in paralysis of the respiratory muscles and death, unless some form of artificial respiration can be instituted.

TABLE 15–8 Effects of Various Factors on Hemoglobin

The affinity of hemoglobin for oxygen is decreased by:

1. Increased hydrogen-ion concentration
2. Increased P_{CO_2}
3. Increased temperature
4. Increased DPG concentration

The affinity of hemoglobin for both hydrogen ions and carbon dioxide is decreased by increased P_{O_2}; that is, deoxyhemoglobin has a greater affinity for hydrogen ions and carbon dioxide than does oxyhemoglobin.

Inspiration is initiated by a burst of action potentials in the nerves to the inspiratory muscles. Then the action potentials cease, the inspiratory muscles relax, and expiration occurs as the elastic lungs recoil. In situations when expiration is facilitated by contraction of expiratory muscles, the nerves to these muscles, which were quiescent during inspiration, begin firing during expiration.

By what mechanism are nerve impulses to the respiratory muscles alternately increased and decreased? Control of this neural activity resides primarily in neurons in the medulla oblongata, the same area of brain that contains the major cardiovascular control centers. (For the rest of this chapter we shall refer to the medulla oblongata simply as the medulla.) In several nuclei of the medulla, neurons called **medullary inspiratory neurons** discharge in synchrony with inspiration and cease discharging during expiration. They provide, through either direct or interneuronal connections, the rhythmic input to the motor neurons innervating the inspiratory muscles. The alternating cycles of firing and quiescence in the medullary inspiratory neurons are generated by a cooperative interaction between synaptic input from other medullary neurons and intrinsic pacemaker potentials in the inspiratory neurons themselves.

The medullary inspiratory neurons receive a rich synaptic input from neurons in various areas of the pons, the part of the brainstem just above the medulla. This input modulates the output of the medullary inspiratory neurons and may help terminate inspiration by inhibiting them. It is likely that an area of the lower pons called the apneustic center is the major source of this output, whereas an area of the upper pons called the pneumotaxic center modulates the activity of the apneustic center.

Another cutoff signal for inspiration comes from **pulmonary stretch receptors,** which lie in the airway smooth-muscle layer and are activated by a large lung inflation. Action potentials in the afferent nerve fibers from the stretch receptors travel to the brain and inhibit the medullary inspiratory neurons. (This is known as the Hering-Breur inflation reflex.) Thus, feedback from the lungs helps to terminate inspiration. However, this pulmonary stretch-receptor reflex plays a role in setting respiratory rhythm only under conditions of very large tidal volumes, as in rigorous exercise.

One last point about the medullary inspiratory neurons should be made: They are quite sensitive to depression by drugs such as barbiturates and morphine, and death from an overdose of these drugs is often due directly to a cessation of ventilation.

Control of Ventilation by P_{O_2}, P_{CO_2}, and H^+ Concentration

Respiratory rate and tidal volume are not fixed but can be increased or decreased over a wide range. For simplicity, we shall describe the control of ventilation without discussing whether rate or depth makes the greater contribution to the change.

There are many inputs to the medullary inspiratory neurons, but the most important for the automatic control of ventilation *at rest* are from peripheral chemoreceptors and central chemoreceptors.

The **peripheral chemoreceptors,** located high in the neck at the bifurcation of the common carotid arteries and in the thorax on the arch of the aorta (Figure 15–29), are called the **carotid bodies** and **aortic bodies.** In both locations they are quite close to, but distinct from, the arterial baroreceptors described in Chapter 14 and are in intimate contact with the arterial blood. The peripheral chemoreceptors are composed of specialized receptor cells that are stimulated mainly by a decrease in the arterial P_{O_2} and an increase in the arterial H^+ concentration (Table 15–9). These cells communicate synaptically with neuron terminals from which afferent nerve fibers pass to the brainstem. There they provide excitatory synaptic input to the medullary inspiratory neurons.

The **central chemoreceptors** are located in the medulla and, like the peripheral chemoreceptors, provide excitatory synaptic input to the medullary inspiratory neurons. They are stimulated by an increase in the H^+ concentration of the brain's extracellular fluid. As we shall see, such changes result mainly from changes in blood P_{CO_2}.

Control by P_{O_2} Figure 15–30 illustrates an experiment in which healthy subjects breathe low P_{O_2} gas mixtures for several minutes. (The experiment is per-

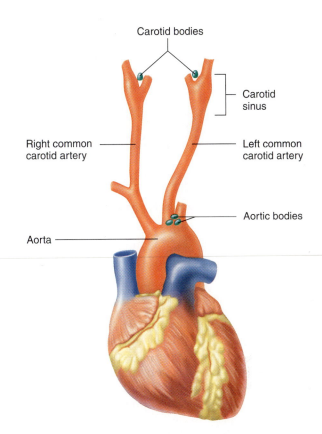

FIGURE 15–29
Location of the carotid and aortic bodies. Note that each carotid body is quite close to a carotid sinus, the major arterial baroreceptor. Both right and left common carotid bifurcations contain a carotid sinus and a carotid body.

formed in a way that keeps arterial P_{CO_2} constant so that the pure effects of changing only P_{O_2} can be studied.) Little increase in ventilation is observed until the oxygen content of the inspired air is reduced enough to lower arterial P_{O_2} to 60 mmHg. Beyond this point, any further reduction in arterial P_{O_2} causes a marked reflex increase in ventilation.

TABLE 15–9 Major Stimuli for the Central and Peripheral Chemoreceptors
Peripheral chemoreceptors—that is, carotid bodies and aortic bodies—respond to changes in the *arterial blood*. They are stimulated by: 1. Decreased P_{O_2} 2. Increased hydrogen-ion concentration
Central chemoreceptors—that is, located in the medulla oblongata—respond to changes in the *brain extracellular fluid*. They are stimulated by increased P_{CO_2}, via associated changes in hydrogen-ion concentration. (See Equation 15-10.)

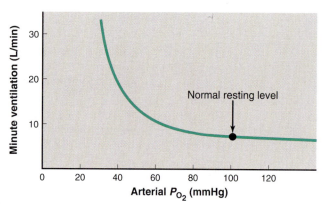

FIGURE 15–30

The effect on ventilation of breathing low-oxygen mixtures. The arterial P_{CO_2} was maintained at 40 mmHg throughout the experiment.

This reflex is mediated by the peripheral chemo-receptors (Figure 15–31). The low arterial P_{O_2} increases the rate at which the receptors discharge, resulting in an increased number of action potentials traveling up the afferent nerve fibers and stimulating the medullary inspiratory neurons. The resulting increase in ventilation provides more oxygen to the alveoli and minimizes the drop in alveolar and arterial P_{O_2} produced by the low P_{O_2} gas mixture.

It may seem surprising that we are so insensitive to smaller reductions of arterial P_{O_2}, but look again at the oxygen-hemoglobin dissociation curve (see Figure 15–23). Total oxygen transport by the blood is not really reduced very much until the arterial P_{O_2} falls below about 60 mmHg. Therefore, increased ventilation would not result in very much more oxygen being added to the blood until that point is reached.

To reiterate, the peripheral chemoreceptors respond to decreases in arterial P_{O_2}, as occurs in lung disease or exposure to high altitude. However, the peripheral chemoreceptors are *not* stimulated in situations in which there are modest reductions in the oxygen *content* of the blood but no change in arterial P_{O_2}. An example of this is anemia, where there is a decrease in the amount of hemoglobin present in the blood (Chapter 14) but no decrease in arterial P_{O_2}, because the concentration of dissolved oxygen in the blood is normal.

This same analysis holds true when oxygen content is reduced moderately by the presence of carbon monoxide, which as described earlier reduces the amount of oxygen combined with hemoglobin by competing for these sites. Since carbon monoxide does not affect the amount of oxygen that can dissolve in blood, the arterial P_{O_2} is unaltered, and no increase in peripheral chemoreceptor output occurs.

Control by P_{CO_2} Figure 15–32 illustrates an experiment in which subjects breathe air to which variable quantities of carbon dioxide have been added. The presence of carbon dioxide in the inspired air causes an elevation of alveolar P_{CO_2} and thereby an elevation of arterial P_{CO_2}. Note that even a very small increase in arterial P_{CO_2} causes a marked reflex increase in ventilation. Experiments like this have documented that small increases in arterial P_{CO_2} are resisted by the reflex mechanisms controlling ventilation to a much greater degree than are equivalent decreases in arterial P_{O_2}.

Of course we don't usually breathe bags of gas containing carbon dioxide. What is the physiological role of this reflex? If a defect in the respiratory system (emphysema, for example) causes a retention of carbon dioxide in the body, the increase in arterial P_{CO_2}

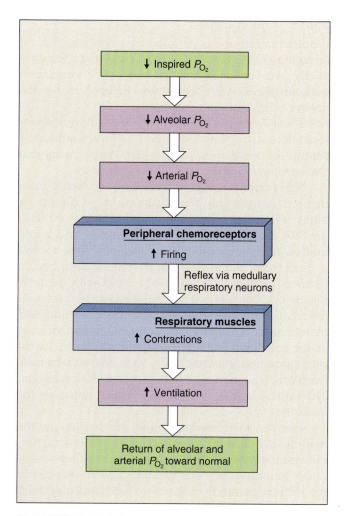

FIGURE 15–31

Sequence of events by which a low arterial P_{O_2} causes hyperventilation, which maintains alveolar (and, hence, arterial) P_{O_2} at a value higher than would exist if the ventilation had remained unchanged.

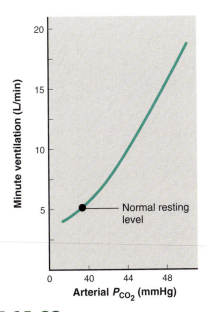

FIGURE 15–32

Effects on respiration of increasing arterial P_{CO_2} achieved by adding carbon dioxide to inspired air.

stimulates ventilation, which promotes the elimination of carbon dioxide. Conversely, if arterial P_{CO_2} decreases below normal levels for whatever reason, this removes some of the stimulus for ventilation, thereby reducing ventilation and allowing metabolically produced carbon dioxide to accumulate and return the P_{CO_2} to normal. In this manner, the arterial P_{CO_2} is stabilized near the normal value of 40 mmHg.

The ability of changes in arterial P_{CO_2} to control ventilation reflexly is largely due to associated changes in H^+ concentration (Equation 15-10). As summarized in Figure 15–33, the pathways that mediate these reflexes are initiated by both the peripheral and central chemoreceptors. The peripheral chemoreceptors are stimulated by the increased arterial H^+ concentration resulting from the increased P_{CO_2}. At the same time, because carbon dioxide diffuses rapidly across the membranes separating capillary blood and brain tissue, the increase in arterial P_{CO_2} causes a rapid increase in brain extracellular fluid P_{CO_2}. This increased P_{CO_2} increases *brain extracellular-fluid* H^+ concentration, which stimulates the central chemoreceptors. Inputs from both the peripheral and central chemoreceptors stimulate the medullary inspiratory neurons to increase ventilation. The end result is a return of arterial and brain extracellular fluid P_{CO_2} and H^+ concentration toward normal. Of the two sets of receptors involved in this reflex response to elevated P_{CO_2}, the central chemoreceptors are the more important, accounting for about 70 percent of the increased ventilation.

It should also be noted that the effects of increased P_{CO_2} and decreased P_{O_2} not only exist as independent inputs to the medulla but manifest synergistic interactions as well. Acute ventilatory response to combined low P_{O_2} and high P_{CO_2} is considerably greater than the sum of the individual responses.

Throughout this section, we have described the stimulatory effects of carbon dioxide on ventilation via reflex input to the medulla, but very high levels of carbon dioxide actually *inhibit* ventilation and may be lethal. This is because such concentrations of carbon dioxide act *directly* on the medulla to inhibit the respiratory neurons by an anesthesia-like effect. Other symptoms caused by very high blood P_{CO_2} include severe headaches, restlessness, and dulling or loss of consciousness.

Control by Changes in Arterial H^+ Concentration That Are Not Due to Altered Carbon Dioxide We have seen that retention or excessive elimination of carbon dioxide causes respiratory acidosis and respiratory alkalosis, respectively. There are, however, many normal and pathological situations in which a change in arterial H^+ concentration is due to some cause other than a primary change in P_{CO_2}. These are termed *metabolic acidosis* when H^+ concentration is increased and *metabolic alkalosis* when it is decreased. In such cases, the peripheral chemoreceptors play the major role in altering ventilation.

For example, addition of lactic acid to the blood, as in strenuous exercise, causes hyperventilation almost entirely by stimulation of the peripheral chemoreceptors (Figures 15–34 and 15–35). The central chemoreceptors are only minimally stimulated in this case because brain H^+ concentration is increased to only a small extent, at least early on, by the hydrogen ions generated from the lactic acid. This is because hydrogen ions penetrate the blood-brain barrier very slowly. In contrast, as described earlier, *carbon dioxide* penetrates the blood-brain barrier easily and changes brain H^+ concentration.

The converse of the above situation is also true: When arterial H^+ concentration is lowered by any means other than by a reduction in P_{CO_2} (for example, by loss of hydrogen ions from the stomach in vomiting), ventilation is reflexly depressed because of decreased peripheral chemoreceptor output.

The adaptive value such reflexes have in regulating arterial H^+ concentration is shown in Figure 15–35. The hyperventilation induced by a metabolic acidosis reduces arterial P_{CO_2}, which lowers arterial H^+ concentration back toward normal. Similarly, hypoventilation induced by a metabolic alkalosis results in an elevated arterial P_{CO_2} and a restoration of H^+ concentration toward normal.

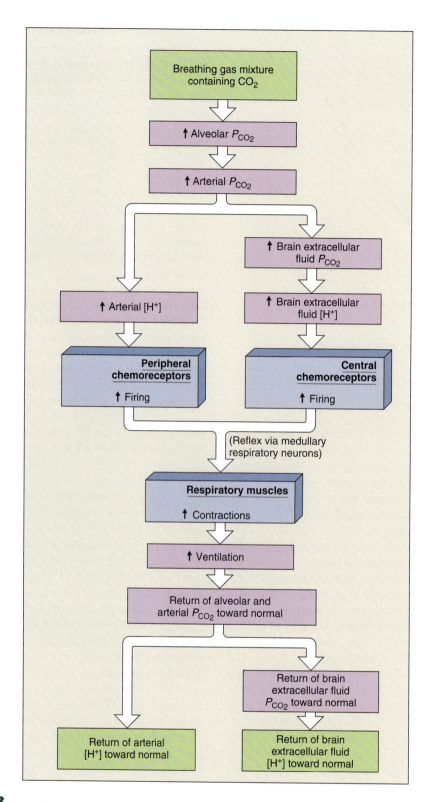

FIGURE 15–33

Pathways by which increased arterial P_{CO_2} stimulates ventilation. Note that the peripheral chemoreceptors are stimulated by an *increase* in H$^+$ concentration, whereas they are also stimulated by a *decrease* in P_{O_2} (Figure 15–31).

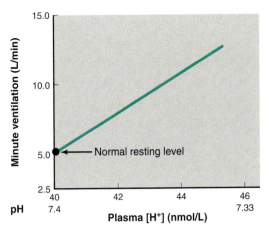

FIGURE 15–34

Changes in ventilation in response to an elevation of plasma hydrogen-ion concentration, produced by the administration of lactic acid.

Adapted from Lambertsen.

Notice that when a change in arterial H^+ concentration due to some acid unrelated to carbon dioxide influences ventilation via the peripheral chemoreceptors, P_{CO_2} is displaced from normal. This is a reflex that regulates arterial H^+ concentration at the expense of changes in arterial P_{CO_2}.

Figure 15–36 summarizes the control of ventilation by P_{O_2}, P_{CO_2}, and H^+ concentration.

Control of Ventilation during Exercise

During exercise, the alveolar ventilation may increase as much as twentyfold. On the basis of our three variables—P_{O_2}, P_{CO_2}, and H^+ concentration—it might seem easy to explain the mechanism that induces this increased ventilation. Unhappily, such is not the case, and the major stimuli to ventilation during exercise, at least that of moderate degree, remain unclear.

Increased P_{CO_2} as the Stimulus? It would seem logical that, as the exercising muscles produce more carbon dioxide, blood P_{CO_2} would increase. This is true, however, only for systemic *venous* blood but not for systemic *arterial* blood. Why doesn't arterial P_{CO_2} increase during exercise? Recall two facts from the section on alveolar gas pressures: (1) alveolar P_{CO_2} sets arterial P_{CO_2}, and (2) alveolar P_{CO_2} is determined by the *ratio* of carbon dioxide production to alveolar ventilation. During moderate exercise, the alveolar ventilation increases in exact proportion to the increased carbon dioxide production, and so alveolar and therefore arterial P_{CO_2} do not change. Indeed, for reasons described below, in very strenuous exercise the alveolar ventilation increases relatively more than carbon dioxide production. In other words, during severe exercise

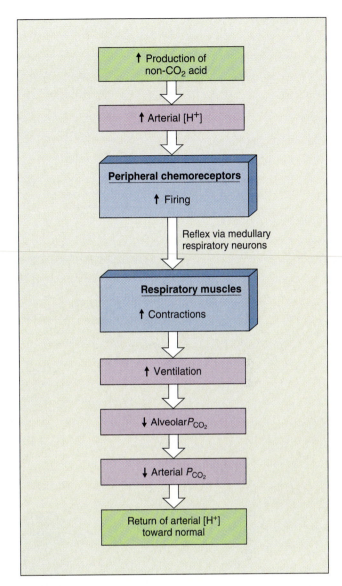

FIGURE 15–35

Reflexly induced hyperventilation minimizes the change in arterial hydrogen-ion concentration when acids are produced in excess in the body. Note that under such conditions, arterial P_{CO_2} is reflexly reduced below its normal value.

the person hyperventilates, and thus alveolar and systemic arterial P_{CO_2} actually decrease (Figure 15–37)!

Decreased P_{O_2} as the Stimulus? The story is similar for oxygen. Although systemic *venous* P_{O_2} decreases during exercise, alveolar P_{O_2} and hence systemic *arterial* P_{O_2} usually remain unchanged (Figure 15–37). This is because cellular oxygen consumption and alveolar ventilation increase in exact proportion to each other, at least during moderate exercise.

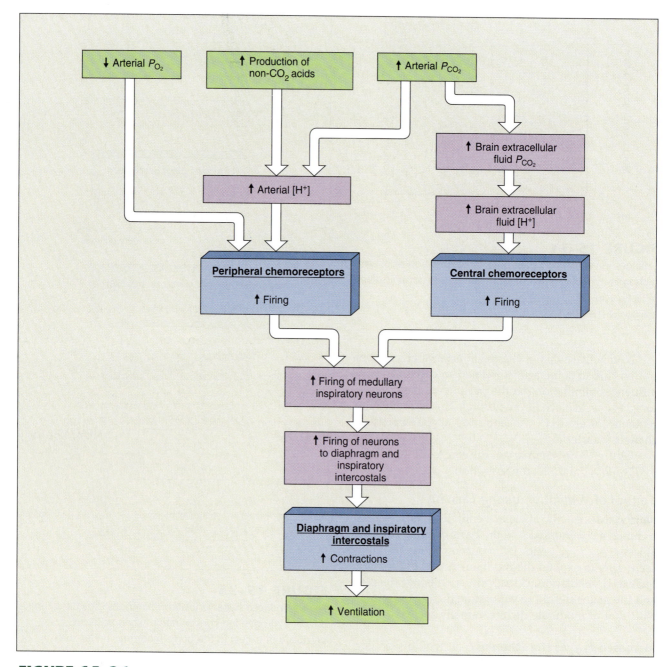

FIGURE 15–36

Summary of the major chemical inputs that stimulate ventilation. This is a combination of Figures 15–31, 15–33, and 15–35. When arterial P_{O_2} increases or when P_{CO_2} or hydrogen-ion concentration decreases, ventilation is reflexly decreased.

Note in Figure 15–37 that even though the hyperventilation of very strenuous exercise lowers arterial P_{CO_2}, it does not produce the increase in arterial P_{O_2} you would expect to occur during hyperventilation. In fact, not shown in the figure, alveolar P_{O_2} does increase during severe exercise, as expected, but for a variety of reasons, the increase in alveolar P_{O_2} is not accompanied by an increase in arterial P_{O_2}. (Indeed, in highly conditioned athletes the arterial P_{O_2} may decrease somewhat during strenuous exercise and may also contribute to stimulation of ventilation at that time.)

This is a good place to remind you of an important point made in Chapter 14: In normal individuals, ventilation is not the limiting factor in endurance exercise—cardiac output is. Ventilation can, as has just been seen, increase enough to maintain arterial P_{O_2}.

Increased H⁺ Concentration as the Stimulus? Since the arterial P_{CO_2} does not change during moderate exercise and decreases in severe exercise, there is no accumulation of excess H⁺ resulting from carbon dioxide accumulation. However, during strenuous

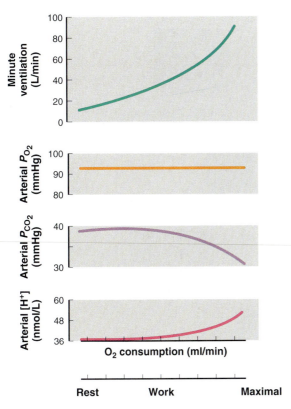

FIGURE 15–37

The effect of exercise on ventilation, arterial gas pressures, and hydrogen-ion concentration. All these variables remain constant during moderate exercise; any change occurs only during strenuous exercise when the person is actually hyperventilating.

Adapted from Comroe.

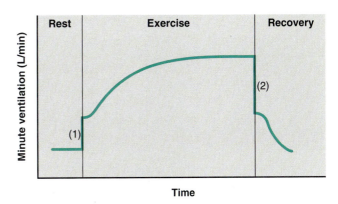

FIGURE 15–38

Ventilation changes during exercise. Note (1) the abrupt increase at the onset of exercise and (2) the equally abrupt but larger decrease at the end of exercise.

of exercise and an equally abrupt decrease at the end; these changes occur too rapidly to be explained by alteration of chemical constituents of the blood or altered body temperature.

Figure 15–39 summarizes various factors that influence ventilation during exercise. The possibility that *oscillatory* changes in arterial P_{O_2}, P_{CO_2}, or H^+ concentration occur, despite unchanged *average* levels of these variables, and play a role has been proposed but remains unproven.

Other Ventilatory Responses

Protective Reflexes   A group of responses protect the respiratory system from irritant materials. Most familiar are the cough and the sneeze reflexes, which originate in receptors located between airway epithelial cells. The receptors for the sneeze reflex are in the nose or pharynx, and those for cough are in the larynx, trachea, and bronchi. When the receptors initiating a cough are stimulated, the medullary respiratory neurons reflexly cause a deep inspiration and a violent expiration. In this manner, particles and secretions are moved from smaller to larger airways, and aspiration of materials into the lungs in also prevented.

The cough reflex is inhibited by alcohol, which may contribute to the susceptibility of alcoholics to choking and pneumonia.

Another example of a protective reflex is the immediate cessation of respiration that is frequently triggered when noxious agents are inhaled. Chronic smoking causes a loss of this reflex.

Voluntary Control of Breathing   Although we have discussed in detail the involuntary nature of most respiratory reflexes, it is quite obvious that considerable voluntary control of respiratory movements exists. Voluntary control is accomplished by descending

exercise, there *is* an increase in arterial H^+ concentration (Figure 15–37), but for quite a different reason, namely, generation and release of lactic acid into the blood. This change in H^+ concentration is responsible, in part, for stimulating the hyperventilation of severe exercise.

Other Factors   A variety of other factors play some role in stimulating ventilation during exercise. These include (1) reflex input from mechanoreceptors in joints and muscles; (2) an increase in body temperature; (3) inputs to the respiratory neurons via branches from axons descending from the brain to motor neurons supplying the exercising muscles; (4) an increase in the plasma epinephrine concentration; (5) an increase in the plasma potassium concentration (because of movement out of the exercising muscles); and (6) a conditioned (learned) response mediated by neural input to the respiratory centers. The operation of this last factor can be seen in Figure 15–38: There is an abrupt increase—within seconds—in ventilation at the onset

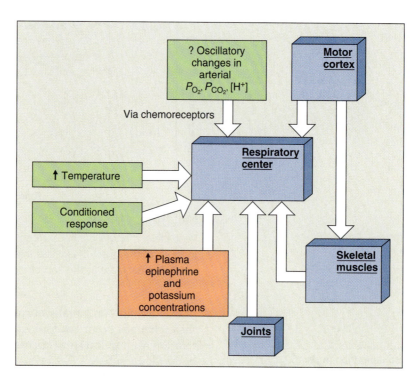

FIGURE 15–39
Summary of factors that stimulate ventilation during exercise.

pathways from the cerebral cortex to the motor neurons of the respiratory muscles. This voluntary control of respiration cannot be maintained when the involuntary stimuli, such as an elevated P_{CO_2} or H^+ concentration, become intense. An example is the inability to hold one's breath for very long.

The opposite of breath-holding—deliberate hyperventilation—lowers alveolar and arterial P_{CO_2} and increases P_{O_2}. Unfortunately, swimmers sometimes voluntarily hyperventilate immediately before underwater swimming to be able to hold their breath longer. We say "unfortunately" because the low P_{CO_2} may still permit breath-holding at a time when the exertion is lowering the arterial P_{O_2} to levels that can cause unconsciousness and lead to drowning.

Besides the obvious forms of voluntary control, respiration must also be controlled during such complex actions as speaking and singing.

Reflexes from J Receptors   In the lungs, either in the capillary walls or the interstitium, are a group of receptors called **J receptors.** They are normally dormant but are stimulated by an increase in lung interstitial pressure caused by the collection of fluid in the interstitium. Such an increase occurs during the vascular congestion caused by either occlusion of a pulmonary vessel **(pulmonary embolus)** or left ventricular heart failure (Chapter 14), as well as by strong exercise in healthy people. The main reflex effects are rapid breathing (tachypnea) and a dry cough. In addition,

neural input from J receptors gives rise to sensations of pressure in the chest and **dyspnea**—the feeling that breathing is labored or difficult.

Hypoxia

Hypoxia is defined as a deficiency of oxygen at the tissue level. There are many potential causes of hypoxia, but they can be classed in four general categories: (1) **hypoxic hypoxia** (also termed **hypoxemia**), in which the arterial P_{O_2} is reduced; (2) **anemic hypoxia**, in which the arterial P_{O_2} is normal but the total oxygen *content* of the blood is reduced because of inadequate numbers of erythrocytes, deficient or abnormal hemoglobin, or competition for the hemoglobin molecule by carbon monoxide; (3) **ischemic hypoxia** (also called hypoperfusion hypoxia), in which blood flow to the tissues is too low; and (4) **histotoxic hypoxia**, in which the quantity of oxygen reaching the tissues is normal but the cell is unable to utilize the oxygen because a toxic agent—cyanide, for example—has interfered with the cell's metabolic machinery.

The primary causes of hypoxic hypoxia in disease are listed in Table 15–10. Exposure to the reduced P_{O_2} of high altitude also causes hypoxic hypoxia but is, of course, not a "disease." The brief summaries in Table 15–10 provide a review of many of the key aspects of respiratory physiology and pathophysiology described in this chapter.

TABLE 15–10 Causes of a Decreased Arterial P_{O_2} (Hypoxic Hypoxia) in Disease

1. **Hypoventilation** may be caused (a) by a defect anywhere along the respiratory control pathway, from the medulla through the respiratory muscles, (b) by severe thoracic cage abnormalities, and (c) by major obstruction of the upper airway. The hypoxemia of hypoventilation is always accompanied by an increased arterial P_{CO_2}.

2. **Diffusion impairment** results from thickening of the alveolar membranes or a decrease in their surface area. In turn, it causes failure of equilibration of blood P_{O_2} with alveolar P_{O_2}. Often it is apparent only during exercise. Arterial P_{CO_2} is either normal, since carbon dioxide diffuses more readily than oxygen, or *reduced*, if the hypoxemia reflexly stimulates ventilation.

3. A **shunt** is (a) an anatomic abnormality of the cardiovascular system that causes mixed venous blood to bypass ventilated alveoli in passing from the right heart to the left heart, or (b) an intrapulmonary defect in which mixed venous blood perfuses unventilated alveoli (ventilation/perfusion = 0). Arterial P_{CO_2} generally does not rise since the effect of the shunt on arterial P_{CO_2} is counterbalanced by the increased ventilation reflexly stimulated by the hypoxemia.

4. **Ventilation-perfusion inequality** is by far the most common cause of hypoxemia. It occurs in chronic obstructive lung diseases and many other lung diseases. Arterial P_{CO_2} may be normal or increased, depending upon how much ventilation is reflexly stimulated.

This table also emphasizes that some of the diseases that produce hypoxia also produce carbon dioxide retention and an increased arterial P_{CO_2} **(hypercapnea)**. In such cases, treating only the oxygen deficit by administering oxygen may constitute inadequate therapy because it does nothing about the hypercapnea. (Indeed, such therapy may be dangerous. The primary respiratory drive in such patients is the hypoxia, since for several reasons the reflex ventilatory response to an increased P_{CO_2} may be lost in chronic situations; the administration of pure oxygen may cause such patients to stop breathing entirely.)

Emphysema

The pathophysiology of emphysema, a major cause of hypoxia, offers an excellent review of many basic principles of respiratory physiology. *Emphysema* is characterized by destruction of the alveolar walls, as well as atrophy and collapse of the lower airways—those from the terminal bronchioles on down. The lungs actually undergo self-destruction by proteolytic enzymes secreted by leukocytes in the lungs in response to a variety of factors. Cigarette smoking is by far the most important of these factors; it exerts its action by

stimulating the release of the proteolytic enzymes and by destroying other enzymes that normally protect the lung against the proteolytic enzymes.

As a result of alveolar-wall loss, adjacent alveoli fuse to form fewer but larger alveoli, and there is a loss of the pulmonary capillaries that were originally in the walls. The merging of alveoli, often into huge balloonlike structures, reduces the *total* surface area available for diffusion, and this impairs gas exchange. Moreover, even more important, because the destructive changes are not uniform throughout the lungs, some areas may receive large amounts of air and little blood, while others have just the opposite pattern; the result is marked ventilation-perfusion inequality.

In addition to problems in gas exchange, emphysema is associated with a marked increase in airway resistance, which greatly increases the work of breathing and, if severe enough, may cause hypoventilation. This is why emphysema is classified, as noted earlier in this chapter, as a "chronic *obstructive* pulmonary disease." The airway obstruction in emphysema is due to collapse of the lower airways. To understand this, recall that two physical factors passively holding the airways open are the transpulmonary pressure and the lateral traction of connective-tissue fibers attached to the airway exteriors. Both of these factors are diminished in emphysema because of the destruction of the lung elastic tissues, and so the airways collapse.

In summary, patients with emphysema suffer from increased airway resistance, decreased total area available for diffusion, and ventilation-perfusion inequality. The result, particularly of the ventilation-perfusion inequality, is always some degree of hypoxia. What happens to arterial P_{CO_2} in emphysema? Certainly the ventilation-perfusion inequality should cause it to increase, but the fact is that many patients with emphysema have a normal arterial P_{CO_2}. The reason is that the hypoxia and any tendency for arterial P_{CO_2} to increase will stimulate ventilation, and this increased ventilation can bring arterial P_{CO_2} back virtually to normal. In other words, any tendency for the arterial P_{CO_2} to increase due to the ventilation-perfusion inequality may be compensated by the increased ventilation. Such increased ventilation, however, does not occur in many people with emphysema, probably because the increased airway resistance prevents it, and indeed may make it very difficult to maintain even a normal ventilation; in such people, the arterial P_{CO_2} rises above normal.

Acclimatization to High Altitude

Atmospheric pressure progressively decreases as altitude increases. Thus, at the top of Mt. Everest (approximately 29,000 ft, or 9000 m), the atmospheric pressure is 253 mmHg (recall that it is 760 mmHg at sea

TABLE 15–11 Acclimatization to the Hypoxia of High Altitude

1. The peripheral chemoreceptors stimulate ventilation.

2. Erythropoietin, secreted by the kidneys, stimulates erythrocyte synthesis, resulting in increased erythrocyte and hemoglobin concentration in blood.

3. DPG increases and shifts the hemoglobin dissociation curve to the right, facilitating oxygen unloading in the tissues. However, this DPG change is not always adaptive and may be maladaptive. For example, at very high altitudes, a right shift in the curve impairs oxygen *loading* in the lungs, an effect that outweighs any benefit from facilitation of *unloading* in the tissues.

4. Increases in capillary density (due to hypoxia-induced expression of the genes that code for angiogenic factors), mitochondria, and muscle myoglobin occur, all of which increase oxygen transfer.

5. The peripheral chemoreceptors stimulate an increased loss of sodium and water in the urine. This reduces plasma volume, resulting in a concentration of the erythrocytes and hemoglobin in the blood.

level). The air is still 21 percent oxygen, which means that the P_{O_2} is 53 mmHg (0.21×253 mmHg). Obviously, the alveolar and arterial P_{O_2} must decrease as one ascends unless pure oxygen is breathed. The highest villages permanently inhabited by people are in the Andes at 19,000 ft (5700 m). These villagers work quite normally, and the only major precaution they take is that the women come down to lower altitudes during late pregnancy.

The effects of oxygen deprivation vary from one individual to another, but most people who ascend rapidly to altitudes above 10,000 ft experience some degree of **mountain sickness.** This disorder consists of breathlessness, headache, nausea, vomiting, insomnia, fatigue, and impairment of mental processes. Much more serious is the appearance, in some individuals, of life-threatening pulmonary edema, the leakage of fluid from the pulmonary capillaries into the alveolar walls and eventually the airspaces, themselves. Brain edema can also occur.

Over the course of several days, the symptoms of mountain sickness disappear, although maximal physical capacity remains reduced. Acclimatization to high altitude is achieved by the compensatory mechanisms given in Table 15–11.

Finally, it should be noted that the responses to high altitude are essentially the same as the responses to hypoxia from any other cause. Thus, a person with severe hypoxia from lung disease may show many of the same changes—increased hematocrit, for example—as a high-altitude sojourner.

Nonrespiratory Functions of the Lungs

The lungs have a variety of functions in addition to their roles in gas exchange and regulation of H^+ concentration. Most notable are the influences they have on the arterial concentrations of a large number of biologically active substances. Many substances (neurotransmitters and paracrine agents, for example) released locally into interstitial fluid may diffuse into capillaries and thus make their way into the systemic venous system. The lungs partially or completely remove some of these substances from the blood and thereby prevent them from reaching other locations in the body via the arteries. The cells that perform this function are the endothelial cells lining the pulmonary capillaries.

In contrast, the lungs may also produce and add new substances to the blood. Some of these substances play local regulatory roles within the lungs, but if produced in large enough quantity, they may diffuse into the pulmonary capillaries and be carried to the rest of the body. For example, inflammatory responses (Chapter 20) in the lung may lead, via excessive release of potent chemicals such as histamine, to profound alterations of systemic blood pressure or flow. In at least one case, the lungs contribute a hormone, angiotensin II, to the blood (Chapter 16).

Finally, the lungs also act as a "sieve" that traps and dissolves small blood clots generated in the systemic circulation, thereby preventing them from reaching the systemic arterial blood where they could occlude blood vessels in other organs.

S U M M A R Y

Organization of the Respiratory System

I. The respiratory system comprises the lungs, the airways leading to them, and the chest structures responsible for movement of air into and out of them.
   a. The conducting zone of the airways consists of the trachea, bronchi, and terminal bronchioles.
   b. The respiratory zone of the airways consists of the alveoli, which are the sites of gas exchange, and those airways to which alveoli are attached.
   c. The alveoli are lined by type I cells and some type II cells, which produce surfactant.
   d. The lungs and interior of the thorax are covered by pleura; between the two pleural layers is an extremely thin layer of intrapleural fluid.
II. The lungs are elastic structures whose volume depends upon the pressure difference across the lungs—the transpulmonary pressure—and how stretchable the lungs are.

Summary of Steps Involved in Respiration

I. The steps involved in respiration are summarized in Figure 15–6. In the steady state, the net volumes of oxygen and carbon dioxide exchanged in the lungs per unit time are equal to the net volumes exchanged in the tissues. Typical volumes per minute are 250 ml for oxygen consumption and 200 ml for carbon dioxide production.

Ventilation and Lung Mechanics

I. Bulk flow of air between the atmosphere and alveoli is proportional to the difference between the atmospheric and alveolar pressures and inversely proportional to the airway resistance: $F = (P_{atm} - P_{alv})/R$.

II. Between breaths at the end of an unforced expiration $P_{atm} = P_{alv}$, no air is flowing, and the dimensions of the lungs and thoracic cage are stable as the result of opposing elastic forces. The lungs are stretched and are attempting to recoil, whereas the chest wall is compressed and attempting to move outward. This creates a subatmospheric intrapleural pressure and hence a transpulmonary pressure that opposes the forces of elastic recoil.

III. During inspiration, the contractions of the diaphragm and inspiratory intercostal muscles increase the volume of the thoracic cage.
   a. This makes intrapleural pressure more subatmospheric, increases transpulmonary pressure, and causes the lungs to expand to a greater degree than between breaths.
   b. This expansion initially makes alveolar pressure subatmospheric, which creates the pressure difference between atmosphere and alveoli to drive air flow into the lungs.

IV. During expiration, the inspiratory muscles cease contracting, allowing the elastic recoil of the chest wall and lungs to return them to their original between-breath size.
   a. This initially compresses the alveolar air, raising alveolar pressure above atmospheric pressure and driving air out of the lungs.
   b. In forced expirations, the contraction of expiratory intercostal muscles and abdominal muscles actively decreases chest dimensions.

V. Lung compliance is determined by the elastic connective tissues of the lungs and the surface tension of the fluid lining the alveoli. The latter is greatly reduced, and compliance increased, by surfactant, produced by the type II cells of the alveoli.

VI. Airway resistance determines how much air flows into the lungs at any given pressure difference between atmosphere and alveoli. The major determinant of airway resistance is the radii of the airways.

VII. The vital capacity is the sum of resting tidal volume, inspiratory reserve volume, and expiratory reserve volume. The volume expired during the first second of a forced vital capacity (FVC) measurement is the FEV_1 and normally averages 80 percent of FVC.

VIII. Minute ventilation is the product of tidal volume and respiratory rate. Alveolar ventilation = (tidal volume − anatomic dead space) × (respiratory rate).

Exchange of Gases in Alveoli and Tissues

I. Exchange of gases in lungs and tissues is by diffusion, as a result of differences in partial pressures. Gases diffuse from a region of higher partial pressure to one of lower partial pressure.

II. Normal alveolar gas pressure for oxygen is 105 mmHg and for carbon dioxide is 40 mmHg.
   a. At any given inspired P_{O_2}, the ratio of oxygen consumption to alveolar ventilation determines alveolar P_{O_2}—the higher the ratio, the lower the alveolar P_{O_2}.
   b. The higher the ratio of carbon dioxide production to alveolar ventilation, the higher the alveolar P_{CO_2}.

III. The average value at rest for systemic venous P_{O_2} is 40 mmHg and for P_{CO_2} is 46 mmHg.

IV. As systemic venous blood flows through the pulmonary capillaries, there is net diffusion of oxygen from alveoli to blood and of carbon dioxide from blood to alveoli. By the end of each pulmonary capillary, the blood gas pressures have become equal to those in the alveoli.

V. Inadequate gas exchange between alveoli and pulmonary capillaries may occur when the alveolus-capillary surface area is decreased, when the alveolar walls thicken, or when there are ventilation-perfusion inequalities.

VI. Significant ventilation-perfusion inequalities cause the systemic arterial P_{O_2} to be reduced. An important mechanism for opposing mismatching is that a low local P_{O_2} causes local vasoconstriction, thereby diverting blood away from poorly ventilated areas.

VII. In the tissues, net diffusion of oxygen occurs from blood to cells, and net diffusion of carbon dioxide from cells to blood.

Transport of Oxygen in the Blood

I. Each liter of systemic arterial blood normally contains 200 ml of oxygen, more than 98 percent bound to hemoglobin and the rest dissolved.

II. The major determinant of the degree to which hemoglobin is saturated with oxygen is blood P_{O_2}.
   a. Hemoglobin is almost 100 percent saturated at the normal systemic arterial P_{O_2} of 100 mmHg. The fact that saturation is already more than 90 percent at a P_{O_2} of 60 mmHg permits relatively normal uptake of oxygen by the blood even when alveolar P_{O_2} is moderately reduced.
   b. Hemoglobin is 75 percent saturated at the normal systemic venous P_{O_2} of 40 mmHg. Thus, only 25 percent of the oxygen has dissociated from hemoglobin and entered the tissues.

III. The affinity of hemoglobin for oxygen is decreased by an increase in P_{CO_2}, hydrogen-ion concentration, and temperature. All these conditions exist in the tissues and facilitate the dissociation of oxygen from hemoglobin.

IV. The affinity of hemoglobin for oxygen is also decreased by binding DPG, which is synthesized by the erythrocytes. DPG increases in situations associated with inadequate oxygen supply and helps maintain oxygen release in the tissues.

Transport of Carbon Dioxide in Blood

I. When carbon dioxide molecules diffuse from the tissues into the blood, 10 percent remains dissolved in plasma and erythrocytes, 30 percent combines in the erythrocytes with deoxyhemoglobin to form carbamino compounds, and 60 percent combines in the erythrocytes with water to form carbonic acid, which then dissociates to yield bicarbonate and hydrogen ions. Most of the bicarbonate then moves out of the erythrocytes into the plasma in exchange for chloride ions.

II. As venous blood flows through lung capillaries, blood P_{CO_2} decreases because of diffusion of carbon dioxide out of the blood into the alveoli, and the above reactions are reversed.

Transport of Hydrogen Ions between Tissues and Lungs

I. Most of the hydrogen ions generated in the erythrocytes from carbonic acid during blood passage through tissue capillaries bind to deoxyhemoglobin because deoxyhemoglobin, formed as oxygen unloads from oxyhemoglobin, has a high affinity for hydrogen ions.

II. As the blood flows through the lung capillaries, hydrogen ions bound to deoxyhemoglobin are released and combine with bicarbonate to yield carbon dioxide and water.

Control of Respiration

I. Breathing depends upon cyclical inspiratory muscle excitation by the nerves to the diaphragm and intercostal muscles. This neural activity is triggered by the medullary inspiratory neurons.

II. The most important inputs to the medullary inspiratory neurons for the involuntary control of ventilation are from the peripheral chemoreceptors—the carotid and aortic bodies—and the central chemoreceptors.

III. Ventilation is reflexly stimulated, via the peripheral chemoreceptors, by a decrease in arterial P_{O_2}, but only when the decrease is large.

IV. Ventilation is reflexly stimulated, via both the peripheral and central chemoreceptors, when the arterial P_{CO_2} goes up even a slight amount. The stimulus for this reflex is not the increased P_{CO_2} itself, but the concomitant increased hydrogen-ion concentration in arterial blood and brain extracellular fluid.

V. Ventilation is also stimulated, mainly via the peripheral chemoreceptors, by an increase in arterial hydrogen-ion concentration resulting from causes other than an increase in P_{CO_2}. The result of this reflex is to restore hydrogen-ion concentration toward normal by lowering P_{CO_2}.

VI. Ventilation is reflexly inhibited by an increase in arterial P_{O_2} and by a decrease in arterial P_{CO_2} or hydrogen-ion concentration.

VII. During moderate exercise, ventilation increases in exact proportion to metabolism, but the signals causing this are not known. During very strenuous exercise, ventilation increases more than metabolism.
   a. The proportional increases in ventilation and metabolism during moderate exercise cause the arterial P_{O_2}, P_{CO_2}, and hydrogen-ion concentration to remain unchanged.
   b. Arterial hydrogen-ion concentration increases during very strenuous exercise because of increased lactic acid production. This accounts for some of the hyperventilation seen in that situation.

VIII. Ventilation is also controlled by reflexes originating in airway receptors and by conscious intent.

Hypoxia

I. The causes of hypoxic hypoxia are listed in Table 15–10.

II. During exposure to hypoxia, as at high altitude, oxygen supply to the tissues is maintained by the five responses listed in Table 15–11.

Nonrespiratory Functions of the Lungs

I. The lungs influence arterial blood concentrations of biologically active substances by removing some from systemic venous blood and adding others to systemic arterial blood.

II. The lungs also act as sieves that dissolve small clots formed in the systemic tissues.

KEY TERMS	
respiration (two definitions)	intrapleural pressure (P_{ip})
respiratory system	ventilation
pulmonary	alveolar pressure (P_{alv})
alveoli (alveolus)	atmospheric pressure (P_{atm})
airway	Boyle's law
inspiration	transpulmonary pressure
expiration	elastic recoil
respiratory cycle	lung compliance (C_L)
pharynx	surface tension
larynx	surfactant
vocal cords	lateral traction
trachea	tidal volume
bronchi (bronchus)	inspiratory reserve volume
bronchiole	functional residual capacity
conducting zone	expiratory reserve volume
respiratory zone	residual volume
type I alveolar cell	vital capacity
type II alveolar cell	minute ventilation
thorax	anatomic dead space
diaphragm	alveolar ventilation
intercostal muscle	alveolar dead space
pleural sac	physiologic dead space
pleura	respiratory quotient (RQ)
intrapleural fluid	Dalton's law

partial pressure
Henry's law
hypoventilation
hyperventilation
hemoglobin
heme
globin
iron (Fe)
deoxyhemoglobin (Hb)
oxyhemoglobin (HbO$_2$)
percent hemoglobin
  saturation
oxygen-carrying capacity
oxygen-hemoglobin
  dissociation curve

2,3-diphosphoglycerate
  (DPG)
carbamino hemoglobin
carbonic anhydrase
nitric oxide
total blood carbon dioxide
medullary inspiratory
  neuron
pulmonary stretch receptor
peripheral chemoreceptor
carotid body
aortic body
central chemoreceptor
J receptor

REVIEW QUESTIONS

1. List the functions of the respiratory system.
2. At rest, how many liters of air and blood flow through the lungs per minute?
3. Describe four functions of the conducting portion of the airways.
4. Which respiration steps occur by diffusion and which by bulk flow?
5. What are normal values for intrapleural pressure, alveolar pressure, and transpulmonary pressure at the end of an unforced expiration?
6. Between breaths at the end of an unforced expiration, in what directions are the lungs and chest wall tending to move? What prevents them from doing so?
7. State typical values for oxygen consumption, carbon dioxide production, and cardiac output at rest. How much oxygen (in milliliters per liter) is present in systemic venous and systemic arterial blood?
8. Write the equation relating air flow into or out of the lungs to atmospheric pressure, alveolar pressure, and airway resistance.
9. Describe the sequence of events that cause air to move into the lungs during inspiration and out of the lungs during expiration. Diagram the changes in intrapleural pressure and alveolar pressure.
10. What factors determine lung compliance? Which is most important?
11. How does surfactant increase lung compliance?
12. How is airway resistance influenced by airway radii?
13. List the physical factors that alter airway resistance.
14. Contrast the causes of increased airway resistance in asthma, emphysema, and chronic bronchitis.
15. What distinguishes lung capacities, as a group, from lung volumes?
16. State the formula relating minute ventilation, tidal volume, and respiratory rate. Give representative values for each at rest.
17. State the formula for calculating alveolar ventilation. What is an average value for alveolar ventilation?
18. The partial pressure of a gas is dependent upon what two factors?

19. State the alveolar partial pressures for oxygen and carbon dioxide in a normal person at rest.
20. What factors determine alveolar partial pressures?
21. What is the mechanism of gas exchange between alveoli and pulmonary capillaries? In a normal person at rest, what are the gas pressures at the end of the pulmonary capillaries, relative to those in the alveoli?
22. Why does thickening of alveolar membranes impair oxygen movement but have little effect on carbon dioxide exchange?
23. What is the major result of ventilation-perfusion inequalities throughout the lungs? Describe one homeostatic response that minimizes mismatching.
24. What generates the diffusion gradients for oxygen and carbon dioxide in the tissues?
25. In what two forms is oxygen carried in the blood? What are the normal quantities (in milliliters per liter) for each form in arterial blood?
26. Describe the structure of hemoglobin.
27. Draw an oxygen-hemoglobin dissociation curve. Put in the points that represent systemic venous and systemic arterial blood (ignore the rightward shift of the curve in systemic venous blood). What is the adaptive importance of the plateau?
28. Would breathing pure oxygen cause a large increase in oxygen transport by the blood in a normal person? In a person with a low alveolar P_{O_2}?
29. Describe the effects of increased P_{CO_2}, H$^+$ concentration, and temperature on the oxygen-hemoglobin dissociation curve. How are these effects adaptive for oxygen unloading in the tissues?
30. Describe the effects of increased DPG on the oxygen-hemoglobin dissociation curve. Under what conditions does an increase in DPG occur?
31. Draw figures showing the reactions carbon dioxide undergoes entering the blood in the tissue capillaries and leaving the blood in the alveoli. What fractions are contributed by dissolved carbon dioxide, bicarbonate, and carbamino?
32. What happens to most of the hydrogen ions formed in the erythrocytes from carbonic acid? What happens to blood H$^+$ concentration as blood flows through tissue capillaries?
33. What are the effects of P_{O_2} on carbamino formation and H$^+$ binding by hemoglobin?
34. In what area of the brain does automatic control of rhythmical respirations reside?
35. Describe the function of the pulmonary stretch receptors.
36. What changes stimulate the peripheral chemoreceptors? The central chemoreceptors?
37. Why does moderate anemia or carbon monoxide exposure not stimulate the peripheral chemoreceptors?
38. Is respiratory control more sensitive to small changes in arterial P_{O_2} or in arterial P_{CO_2}?
39. Describe the pathways by which increased arterial P_{CO_2} stimulates ventilation. What pathway is more important?

40. Describe the pathway by which a change in arterial H^+ concentration independent of altered carbon dioxide influences ventilation. What is the adaptive value of this reflex?

41. What happens to arterial P_{O_2}, P_{CO_2}, and H^+ concentration during moderate and severe exercise? List other factors that may stimulate ventilation during exercise.

42. List four general causes of hypoxic hypoxia.

43. Describe two general ways in which the lungs can alter the concentrations of substances other than oxygen, carbon dioxide, and H^+ in the arterial blood.

CLINICAL TERMS

cystic fibrosis	shunt
pneumothorax	carbon monoxide
respiratory-distress	respiratory acidosis
syndrome of the	respiratory alkalosis
newborn	poliomyelitis
asthma	metabolic acidosis
anti-inflammatory drugs	metabolic alkalosis
bronchodilator drugs	pulmonary embolus
chronic obstructive	dyspnea
pulmonary disease	hypoxia
chronic bronchitis	hypoxic hypoxia
Heimlich maneuver	hypoxemia
forced expiratory volume	anemic hypoxia
in 1 s (FEV_1)	ischemic hypoxia
pulmonary function tests	histotoxic hypoxia
obstructive lung diseases	diffusion impairment
restrictive lung diseases	hypercapnea
hypoventilation	emphysema
hyperventilation	mountain sickness
diffuse interstitial fibrosis	
ventilation-perfusion	
inequality	

THOUGHT QUESTIONS

(Answers are given in Appendix A.)

1. At the end of a normal expiration, a person's lung volume is 2 L, his alveolar pressure is 0 mmHg, and his intrapleural pressure is −4 mmHg. He then inhales 800 ml, and at the end of inspiration the alveolar pressure is 0 mmHg and the intrapleural pressure is −8 mmHg. Calculate this person's lung compliance.

2. A patient is unable to produce surfactant. In order to inhale a normal tidal volume, will her intrapleural pressure have to be more or less subatmospheric during inspiration, relative to a normal person?

3. A 70-kg adult patient is artificially ventilated by a machine during surgery at a rate of 20 breaths/min and a tidal volume of 250 ml/breath. Assuming a normal anatomic dead space of 150 ml, is this patient receiving an adequate alveolar ventilation?

4. Why must a person floating on the surface of the water and breathing through a snorkel increase his tidal volume and/or breathing frequency if alveolar ventilation is to remain normal?

5. A normal person breathing room air voluntarily increases her alveolar ventilation twofold and continues to do so until new steady-state alveolar gas pressures for oxygen and carbon dioxide are reached. Are the new values higher or lower than normal?

6. A person breathing room air has an alveolar P_{O_2} of 105 mmHg and an arterial P_{O_2} of 80 mmHg. Could hypoventilation, say, due to respiratory muscle weakness, produce these values?

7. A person's alveolar membranes have become thickened enough to moderately decrease the rate at which gases diffuse across them at any given partial pressure differences. Will this person necessarily have a low arterial P_{O_2} at rest? During exercise?

8. A person is breathing 100 percent oxygen. How much will the oxygen content (in milliliters per liter of blood) of the arterial blood increase compared to when the person is breathing room air?

9. Which of the following have higher values in systemic venous blood than in systemic arterial blood: plasma P_{CO_2}, erythrocyte P_{CO_2}, plasma bicarbonate concentration, erythrocyte bicarbonate concentration, plasma hydrogen-ion concentration, erythrocyte hydrogen-ion concentration, erythrocyte carbamino concentration, erythrocyte chloride concentration, plasma chloride concentration?

10. If the spinal cord were severed where it joins the brainstem, what would happen to respiration?

11. The peripheral chemoreceptors are denervated in an experimental animal, and the animal then breathes a gas mixture containing 10 percent oxygen. What changes occur in the animal's ventilation? What changes occur when this denervated animal is given a mixture of air containing 21 percent oxygen and 5 percent carbon dioxide to breathe?

12. Patients with severe uncontrolled diabetes mellitus produce large quantities of certain organic acids. Can you predict the ventilation pattern in these patients and whether their arterial P_{O_2} and P_{CO_2} increase or decrease?

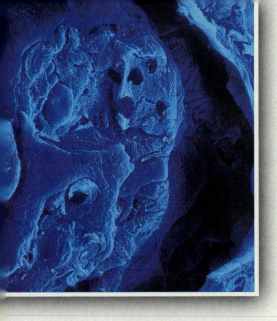

CHAPTER 16

The Kidneys and Regulation of Water and Inorganic Ions

This chapter deals with how the water and inorganic-ion composition of the internal environment is homeostatically regulated. The kidneys play the central role in these processes.

Regulation of the total-body balance of any substance can be studied in terms of the balance concept described in Chapter 7. Theoretically, a substance can appear in the body either as a result of ingestion or as a product of metabolism. On the loss side of the balance, a substance can be excreted from the body or can be metabolized. Therefore, if the quantity of any substance in the body is to be maintained at a nearly constant level over a period of time, the total amounts ingested and produced must equal the total amounts excreted and metabolized.

Reflexes that alter excretion, specifically excretion via the urine, constitute the major mechanisms that regulate the body balances of water and many of the inorganic ions that determine the properties of the extracellular fluid. The extracellular concentrations of these ions were given in Table 6–1. We will first describe how the kidneys work in general and then apply this information to how they process specific substances—sodium, water, potassium, and so on—and participate in reflexes that regulate these substances.

BASIC PRINCIPLES OF RENAL PHYSIOLOGY

Renal Functions

The adjective **renal** means "pertaining to the kidneys"; thus, for example, we refer to "renal physiology" and "renal functions."

The kidneys process the plasma portion of blood by removing substances from it and, in a few cases, by adding substances to it. In so doing, they perform a variety of functions, as summarized in Table 16–1.

First, and very importantly, the kidneys play the central role in regulating the water concentration, inorganic-ion composition, and volume of the internal environment. They do so by excreting just enough water and inorganic ions to keep the amounts of these substances in the body relatively constant. For example, if you start eating a lot of salt (sodium chloride), the kidneys will increase the amount of the salt they excrete to match the intake. Alternatively, if there is not enough salt in the body, the kidneys will excrete very little salt or virtually none at all.

Second, the kidneys excrete metabolic waste products into the urine as fast as they are produced. This keeps waste products, which can be toxic, from accumulating in the body. These metabolic wastes include **urea** from the catabolism of protein, **uric acid** from nucleic acids, **creatinine** from muscle creatine, the end products of hemoglobin breakdown (which give urine much of its color), and many others.

A third function of the kidneys is the excretion, in the urine, of some foreign chemicals, such as drugs, pesticides, and food additives, and their metabolites.

A fourth function is **gluconeogenesis.** During prolonged fasting, the kidneys synthesize glucose from amino acids and other precursors and release it into the blood. The kidneys can supply approximately 20 percent as much glucose as the liver does at such times (Chapter 18).

Finally, the kidneys act as endocrine glands, secreting at least three hormones: erythropoietin (described in Chapter 14), renin, and 1,25-dihydroxyvitamin D_3. These last two hormones are described in this chapter. (Note that renin is part of a hormonal system called the renin-angiotensin system; although renin functions as an enzyme in this system, it is customary to refer to it as a "hormone.")

TABLE 16-1 Functions of the Kidneys

1. Regulation of water and inorganic-ion balance

2. Removal of metabolic waste products from the blood and their excretion in the urine

3. Removal of foreign chemicals from the blood and their excretion in the urine

4. Gluconeogenesis

5. Secretion of hormones:
   a. Erythropoietin, which controls erythrocyte production (Chapter 14)
   b. Renin, which controls formation of angiotensin, which influences blood pressure and sodium balance (this chapter)
   c. 1,25-dihydroxyvitamin D_3, which influences calcium balance (this chapter)

Structure of the Kidneys and Urinary System

The two kidneys lie in the back of the abdominal wall but not actually in the abdominal cavity. They are retroperitoneal, meaning they are just behind the peritoneum, the lining of this cavity. The urine flows from the kidneys through the **ureters** into the **bladder,** from which it is eliminated via the **urethra** (Figure 16–1).

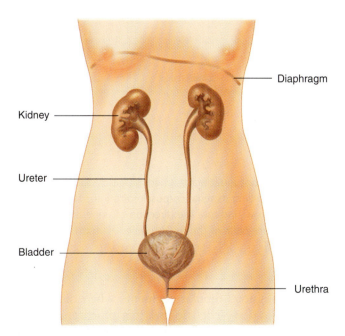

FIGURE 16–1

Urinary system in a woman. In the male, the urethra passes through the penis (Chapter 19).

Each kidney contains approximately 1 million similar subunits called **nephrons.** Each nephron (Figure 16–2) consists of (1) an initial filtering component called the **renal corpuscle,** and (2) a **tubule** that extends out from the renal corpuscle. The renal corpuscle forms a filtrate from blood that is free of cells and proteins. This filtrate then leaves the renal corpuscle and enters the tubule. As it flows through the tubule, substances are added to it or removed from it. Ultimately the fluid from all the nephrons exits the kidneys as urine.

Let us look first at the anatomy of the renal corpuscles—the filters. Each renal corpuscle contains a compact tuft of interconnected capillary loops called the **glomerulus** (plural, *glomeruli*), or **glomerular capillaries** (Figures 16–2 and 16–3). Each glomerulus is supplied with blood by an arteriole called an **afferent arteriole.** The glomerulus protrudes into a fluid-filled capsule called **Bowman's capsule.** The combination of a glomerulus and a Bowman's capsule constitutes a renal corpuscle. As blood flows through the glomerulus, a portion of the plasma filters into Bowman's capsule. The remaining blood then leaves the glomerulus by another arteriole, the **efferent arteriole.**

One way of visualizing the relationships within the renal corpuscle is to imagine a loosely clenched fist—the glomerulus—punched into a balloon—the Bowman's capsule. The part of Bowman's capsule in contact with the glomerulus becomes pushed inward but does not make contact with the opposite side of the capsule. Accordingly, a fluid-filled space—**Bowman's space**—exists within the capsule, and it is into this space that protein-free fluid filters from the glomerulus.

Blood in the glomerulus is separated from the fluid in Bowman's space by a filtration barrier consisting of three layers (Figure 16–3b): (1) the single-celled capillary endothelium, (2) a noncellular proteinaceous layer of basement membrane (also termed basal lamina) between the endothelium and the next layer, which is (3) the single-celled epithelial lining of Bowman's capsule. The epithelial cells in this region are quite different from the simple flattened cells that line the rest of Bowman's capsule (the part of the "balloon" not in contact with the "fist") and are called **podocytes.** They have an octopus-like structure in that they possess a large number of extensions, or foot processes. Fluid filters first across the endothelial cells, then through the basement membrane, and finally between the foot processes of the podocytes.

In addition to the capillary endothelial cells and the podocytes, there is a third cell type, **mesangial cells,** which are modified smooth-muscle cells that surround the glomerular capillary loops but are not part of the filtration pathway. Their function will be described later.

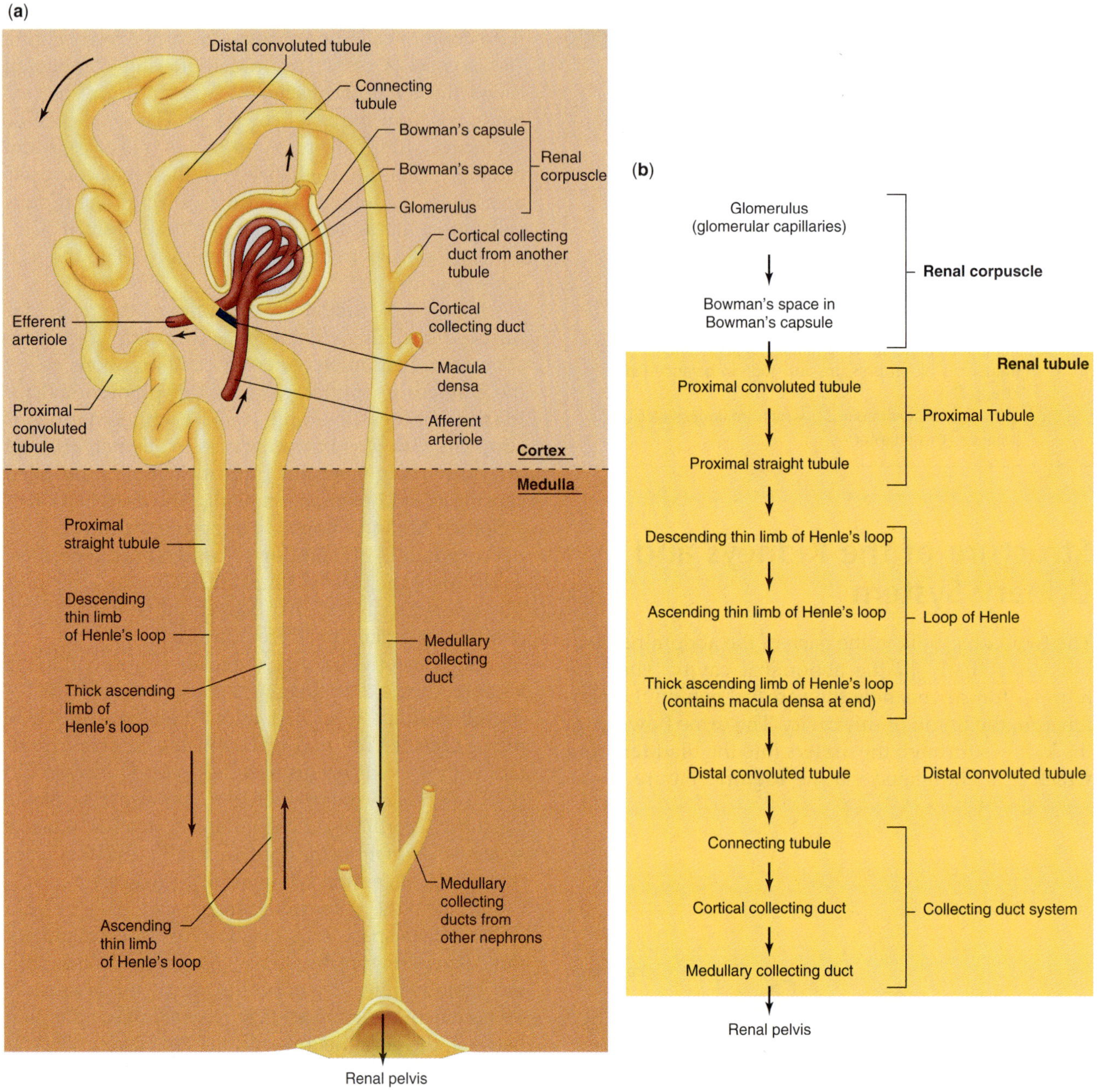

(a)

(b)

FIGURE 16–2

Basic structure of a nephron. (a) Anatomical organization. The macula densa is not a distinct segment but a plaque of cells in the ascending loop of Henle where the loop passes between the arterioles supplying its renal corpuscle of origin. The outer area of the kidney is called the cortex and the inner the medulla. The black arrows indicate the direction of urine flow. (b) Consecutive segments of the nephron. All segments in the screened area are parts of the renal tubule; the terms to the right of the brackets are commonly used for several consecutive segments. ✗

The renal tubule is continuous with a Bowman's capsule. It is a very narrow hollow cylinder made up of a single layer of epithelial cells (resting on a basement membrane). The epithelial cells differ in structure and function along the tubule's length, and 10 to 12 distinct segments are presently recognized (see Figure 16–2). It is customary, however, to group two or more contiguous tubular segments when discussing function, and we shall follow this practice. Accordingly, the segment of the tubule that drains Bowman's capsule is the **proximal tubule** (comprising the proximal convoluted tubule and the proximal straight tubule of Figure 16–2). The next portion of the tubule is the **loop of Henle**, which is a sharp hairpin-like loop

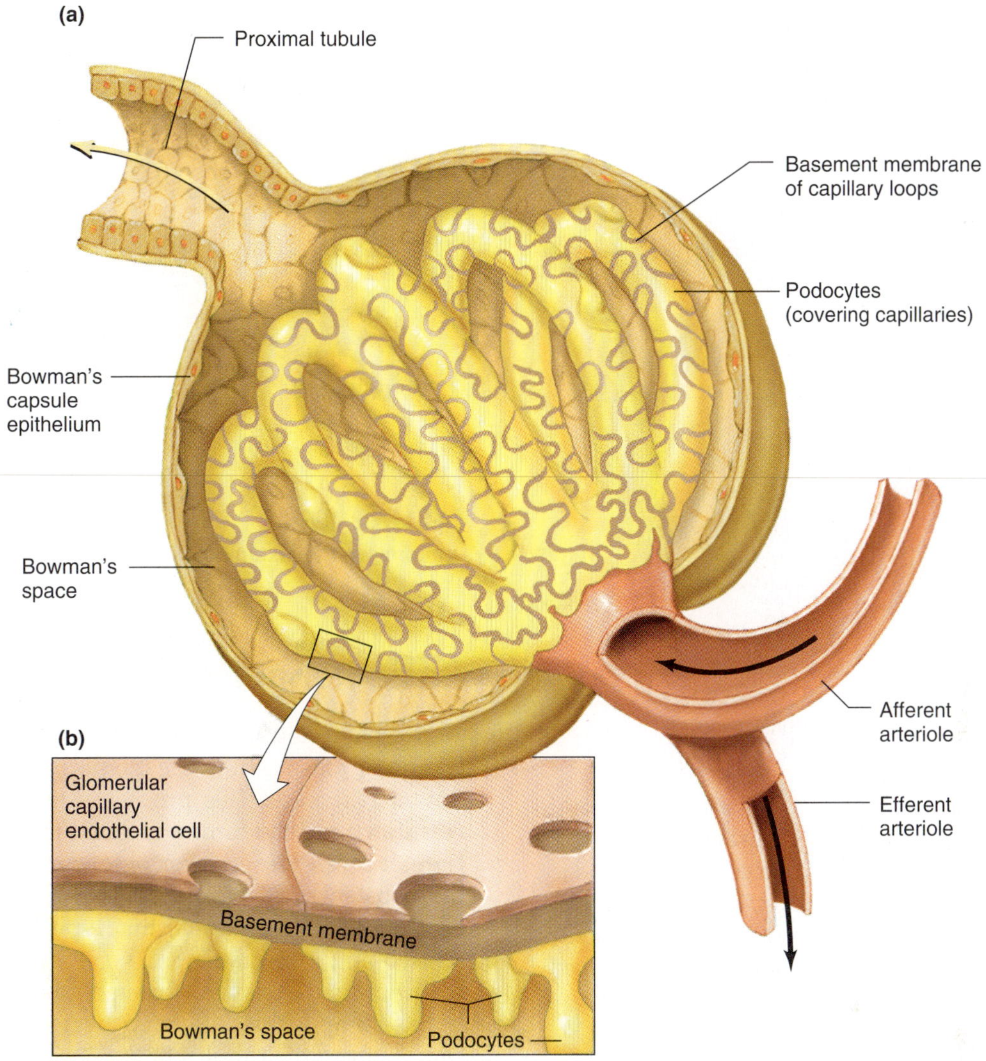

(a) Proximal tubule

Basement membrane of capillary loops

Podocytes (covering capillaries)

Bowman's capsule epithelium

Bowman's space

Afferent arteriole

Efferent arteriole

(b)

Glomerular capillary endothelial cell

Basement membrane

Bowman's space

Podocytes

FIGURE 16–3

(a) Anatomy of the renal corpuscle. Brown lines in the capillary loops indicate space between adjoining podocytes. (b) Cross section of the three corpuscular membranes—capillary endothelium, basement membrane, and epithelium (podocytes) of Bowman's capsule. For simplicity, glomerular mesangial cells are not shown in this figure.

consisting of a **descending limb** coming from the proximal tubule and an **ascending limb** leading to the next tubular segment, the **distal convoluted tubule.** Fluid flows from the distal convoluted tubule into the **collecting duct system,** the first portion of which is the connecting tubule, followed by the **cortical collecting duct** and then the **medullary collecting duct** (the reasons for the terms "cortical" and "medullary" will be apparent shortly). The connecting tubule is so similar in function to the cortical collecting duct that, in subsequent discussions, we will not describe its function separately.

From Bowman's capsule to the collecting-duct system, each nephron is completely separate from the others. This separation ends when multiple cortical collecting ducts merge. The result of additional mergings

from this point on is that the completed urine drains into the kidney's central cavity, the **renal pelvis,** via only several hundred large medullary collecting ducts. The renal pelvis is continuous with the ureter draining that kidney (Figure 16–4).

There are important regional differences in the kidney (Figures 16–2 and 16–4). The outer portion is the **renal cortex,** and the inner portion the **renal medulla.** The cortex contains all the renal corpuscles. The loops of Henle extend from the cortex for varying distances down into the medulla. The medullary collecting ducts pass through the medulla on their way to the renal pelvis.

All along its length, each tubule is surrounded by capillaries, called the **peritubular capillaries.** Note that we have now mentioned two sets of capillaries in the

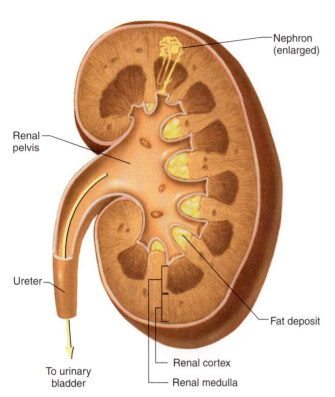

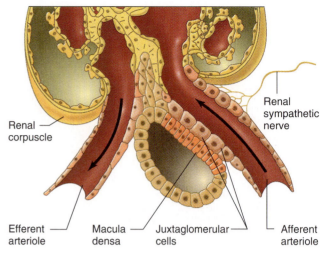

FIGURE 16–5

Anatomy of the juxtaglomerular apparatus.

FIGURE 16–4

Section of a human kidney. For clarity, the nephron illustrated to show nephron orientation is not to scale—its outline would not be clearly visible without a microscope. The outer kidney, which contains all the renal corpuscles, is the cortex, and the inner kidney is the medulla. Note that in the medulla, the loops of Henle and the collecting ducts run parallel to each other. The medullary collecting ducts drain into the renal pelvis. ꭗ

kidneys—the glomerular capillaries (glomeruli) and the peritubular capillaries. Within each nephron, the two sets of capillaries are connected to each other by an efferent arteriole, the vessel by which blood leaves the glomerulus (see Figures 16–2 and 16–3). Thus the renal circulation is very unusual in that it includes *two* sets of arterioles and *two* sets of capillaries. After supplying the tubules with blood, the peritubular capillaries then join together to form the veins by which blood leaves the kidney.

One additional anatomical detail involving both the tubule and the arterioles must be mentioned. Near its end, the ascending limb of each loop of Henle passes between the afferent and efferent arterioles of that loop's own nephron (see Figure 16–2). At this point there is a patch of cells in the wall of the ascending limb called the **macula densa,** and the wall of the afferent arteriole contains secretory cells known as **juxtaglomerular (JG) cells.** The combination of macula densa and juxtaglomerular cells is known as the **juxtaglomerular apparatus (JGA)** (Figure 16–5). The juxtaglomerular cells secrete the hormone renin.

Basic Renal Processes

As we have said, urine formation begins with the filtration of plasma from the glomerular capillaries into Bowman's space. This process is termed **glomerular filtration,** and the filtrate is called the **glomerular filtrate.** It is cell-free and except for proteins, contains all the substances in plasma in virtually the same concentrations as in plasma; this type of filtrate is also termed an ultrafiltrate.

During its passage through the tubules, the filtrate's composition is altered by movements of substances from the tubules to the peritubular capillaries and vice versa (Figure 16–6). When the direction of movement is from tubular lumen to peritubular capillary plasma, the process is called **tubular reabsorption,** or simply reabsorption. (A more accurate term for this process is absorption, but *re*absorption persists, for historical reasons, as the term more commonly used by renal physiologists.) Movement in the opposite direction—that is, from peritubular plasma to tubular lumen—is called **tubular secretion,** or simply secretion. Tubular secretion is also used to denote the movement of a solute from the cell interior to the lumen in the cases in which the kidney tubular cells themselves generate the substance.

To summarize: A substance can gain entry to the tubule and be excreted in the urine by glomerular filtration or tubular secretion. Once in the tubule, however, the substance need not be excreted but can be reabsorbed. Thus, the amount of any substance excreted in the urine is equal to the amount filtered plus the amount secreted minus the amount reabsorbed.

$$\begin{matrix} \text{Amount} \\ \text{excreted} \end{matrix} = \begin{matrix} \text{Amount} \\ \text{filtered} \end{matrix} + \begin{matrix} \text{Amount} \\ \text{secreted} \end{matrix} - \begin{matrix} \text{Amount} \\ \text{reabsorbed} \end{matrix}$$

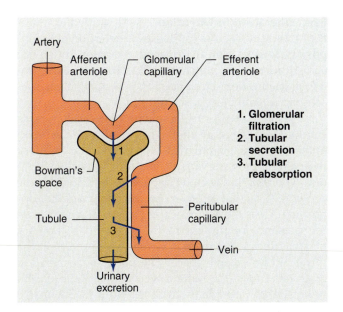

FIGURE 16–6

The three basic components of renal function. This figure is to illustrate only the *directions* of reabsorption and secretion, not specific sites or order of occurrence. Depending on the particular substance, reabsorption and secretion can occur at various sites along the tubule.

It should be stressed that not all these processes—filtration, secretion, reabsorption—apply to all substances.

To emphasize general principles, the renal handling of three hypothetical substances is illustrated in Figure 16–7. Approximately 20 percent of the plasma that enters the glomerular capillaries is filtered into Bowman's space. This filtrate, which contains X, Y, and Z in the same concentrations as in the capillary plasma, enters the proximal tubule and begins its flow through the rest of the tubule. Simultaneously, the remaining 80 percent of the plasma, with its X, Y, and Z, leaves the glomerular capillaries via the efferent arteriole and enters the peritubular capillaries.

Assume that the tubule can secrete 100 percent of the peritubular-capillary X into the tubular lumen but cannot reabsorb X. Thus, by the combination of filtration and tubular secretion, all the plasma that originally entered the renal artery loses all of its substance X, which leaves the body via the urine.

Assume that the tubule can reabsorb, but not secrete, Y and Z. The amount of Y reabsorption is small, so that much of the filtered material is not reabsorbed and escapes from the body. But for Z the reabsorptive mechanism is so powerful that all the filtered Z is transported back into the plasma. Therefore, no Z is lost from the body. Hence, for Z the processes of filtration and reabsorption have canceled each other out, and the net result is as though Z had never entered the kidney.

For each substance in plasma, a particular combination of filtration, tubular reabsorption, and tubular secretion applies. The critical point is that, for many substances, the rates at which the processes proceed are subject to physiological control. By triggering changes in the rate of filtration, reabsorption, or secretion whenever the body content of a substance goes above or below normal, homeostatic mechanisms can regulate the substance's bodily balance. For example, consider what happens when a normally hydrated person drinks a lot of water: Within 1–2 h all the excess water has been excreted in the urine, partly as a result of an increase in filtration but mainly as a result of decreased tubular reabsorption of water. In this example, the kidneys are effector organs of a reflex that maintains body water concentration within very narrow limits.

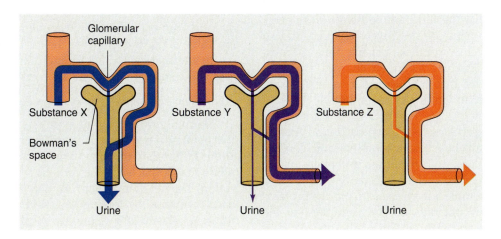

FIGURE 16–7

Renal handling of three hypothetical substances X, Y, and Z. X is filtered and secreted but not reabsorbed. Y is filtered, and a fraction is then reabsorbed. Z is filtered and completely reabsorbed.

Although renal physiologists traditionally list glomerular filtration, tubular reabsorption, and tubular secretion as the three basic renal processes, a fourth process—metabolism by the tubular cells—is also of considerable importance for some substances. In some cases, the renal tubular cells remove substances from blood or glomerular filtrate and metabolize them, resulting in their disappearance from the body. In other cases, the cells *produce* substances and add them either to the blood or tubular fluid; the most important of these, as we shall see, are ammonium, hydrogen ion, and bicarbonate.

In summary, one can study the normal renal processing of any given substance by asking a series of questions:

1. To what degree is the substance filterable at the renal corpuscle?
2. Is it reabsorbed?
3. Is it secreted?
4. What factors homeostatically regulate the quantities filtered, reabsorbed, or secreted: that is, what are the pathways by which renal excretion of the substance is altered to maintain stable body balance?

Glomerular Filtration

As stated above, the glomerular filtrate—that is, the fluid in Bowman's space—normally has no cells but contains all plasma substances except proteins in virtually the same concentrations as in plasma. This is because glomerular filtration is a bulk-flow process in which water and all low-molecular-weight substances move together. The overwhelming majority of plasma proteins—the albumins and globulins—are excluded almost entirely from the filtrate. One reason for their exclusion is that the renal corpuscles restrict the movement of such high-molecular-weight substances. A second reason is that the filtration pathways in the corpuscular membranes are negatively charged and so oppose the movement of these plasma proteins, most of which are themselves negatively charged. It should be noted, however, that low-molecular-weight plasma proteins do filter in varying degrees, but we shall ignore this fact.

The only exceptions to the generalization that all nonprotein plasma substances have the same concentrations in the glomerular filtrate as in the plasma are certain low-molecular-weight substances that would otherwise be filterable but are *bound* to plasma proteins and therefore are not filtered. For example, half the plasma calcium and virtually all of the plasma fatty acids are bound to plasma protein and so are not filtered.

Forces Involved in Filtration As described in Chapter 14, filtration across capillaries is determined by opposing forces: The hydrostatic pressure difference across the capillary wall favors filtration, while the protein concentration difference across the wall creates an osmotic force that opposes filtration.

This also applies to the glomerular capillaries, as summarized in Figure 16–8. The pressure of the blood in the glomerular capillaries—the glomerular capillary hydrostatic pressure (P_{GC})—is a force favoring filtration. The fluid in Bowman's space exerts a hydrostatic pressure (P_{BS}) that opposes this filtration. Another opposing force is the osmotic force (π_{GC}) that results from the presence of protein in the glomerular capillary plasma. Recall that there is virtually no protein in the filtrate in Bowman's space. The unequal distribution of protein causes the water concentration of the plasma to be slightly less than that of the fluid in Bowman's space, and this difference in water concentration favors fluid movement by bulk-flow from Bowman's space into the glomerular capillaries—that is, opposes glomerular filtration.

Note that in Figure 16–8 the value given for this osmotic force—29 mmHg—is larger than the value—24 mmHg—for the osmotic force given in Chapter 14 for plasma in all arteries and nonrenal capillaries. The reason is that, unlike the situation elsewhere in the body, so much water (about 20 percent of the plasma supplying the kidneys) filters out of the glomerular capillaries that the protein left behind in the plasma becomes significantly more concentrated than in arterial plasma. In other capillaries, in contrast, so little water filters that the capillary protein concentration remains essentially unchanged from its value in arterial plasma. In other words, unlike the situation in other capillaries, the plasma protein concentration and, hence, the osmotic force, increases from the beginning to the end of the glomerular capillaries. The value given in Figure 16–8 for the osmotic force is the average value along the length of the capillaries.

To summarize, the **net glomerular filtration pressure** is the sum of three relevant forces:

$$\text{Net glomerular filtration pressure} = P_{GC} - P_{BS} - \pi_{GC}$$

Normally the net filtration pressure is always positive because the glomerular capillary hydrostatic pressure is larger than the sum of the hydrostatic pressure in Bowman's space and the osmotic force opposing filtration. The net glomerular filtration pressure initiates urine formation by forcing an essentially protein-free filtrate of plasma out of the glomeruli and into Bowman's space and thence down the tubule into the renal pelvis.

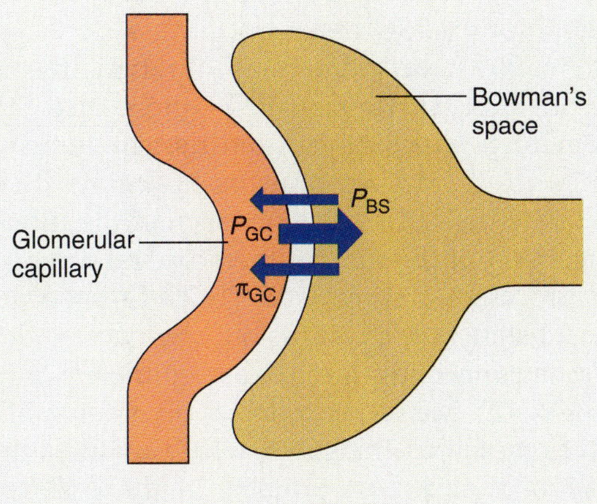

Forces	mmHg
Favoring filtration:	
Glomerular capillary blood pressure (P_{GC})	60
Opposing filtration:	
Fluid pressure in Bowman's space (P_{BS})	15
Osmotic force due to protein in plasma (π_{GC})	29
Net glomerular filtration pressure = $P_{GC} - P_{BS} - \pi_{GC}$	16

FIGURE 16–8

Forces involved in glomerular filtration. The symbol π denotes the osmotic force due to the presence of protein in glomerular capillary plasma.

Rate of Glomerular Filtration The volume of fluid filtered from the glomeruli into Bowman's space per unit time is known as the **glomerular filtration rate (GFR).** GFR is determined not only by the net filtration pressure but also by the permeability of the corpuscular membranes and the surface area available for filtration (see Chapter 14). In other words, at any given net filtration pressure, the GFR will be directly proportional to the membrane permeability and the surface area. The glomerular capillaries are so much more permeable to fluid than, say muscle or skin capillaries, that the net glomerular filtration pressure causes massive filtration. In a 70-kg person, the GFR averages 180 L/day (125 ml/min)! Contrast this number with the net filtration of fluid across all the other capillaries in the body—4 L/day, as described in Chapter 14.

When we recall that the total volume of plasma in the cardiovascular system is approximately 3 L, it follows that the entire plasma volume is filtered by the kidneys about 60 times a day. This opportunity to process such huge volumes of plasma enables the kidneys to regulate the constituents of the internal environment rapidly and to excrete large quantities of waste products.

It must be emphasized that the GFR is not a fixed value but is subject to physiological regulation. As we shall see, this is achieved mainly by neural and hormonal input to the afferent and efferent arterioles, which results in changes in net glomerular filtration pressure.

In this regard it must be emphasized that the glomerular capillaries are unique in that they are situated between two sets of arterioles—the afferent and efferent arterioles. Constriction of the afferent arterioles alone has the same effect on the hydrostatic pressure in the glomerular capillaries (P_{GC}) as does constriction of arterioles anywhere in the body on the pressure in the capillaries supplied by those arterioles: Capillary pressure *decreases* because the increased arteriolar resistance causes a greater loss of pressure between the arteries and the capillaries. In contrast, efferent arteriolar constriction alone has precisely the opposite effect on P_{GC}—it *increases* it. This occurs because the efferent arteriole lies beyond the glomerulus, so that efferent-arteriolar constriction tends to "dam back" the blood in the glomerular capillaries, raising P_{GC}. Finally, simultaneous constriction of *both* sets of arterioles tends to leave P_{GC} unchanged because of the opposing effects. The effects of arteriolar dilation are the reverse of those described for constriction.

In addition to the neuroendocrine input to the arterioles, there is also input to the mesangial cells that surround the glomerular capillaries. Contraction of these cells reduces the surface area of the capillaries, which causes a decrease in GFR at any given net filtration pressure.

It is possible to measure the total amount of any nonprotein substance (assuming also that the substance is not bound to protein) filtered into Bowman's space by multiplying the GFR by the plasma concentration of the substance. This amount is called the **filtered load** of the substance. For example, if the GFR is 180 L/day and plasma glucose concentration is 1 g/L, then the filtered load of glucose is 180 L/day × 1 g/L = 180 g/day.

Once we know the filtered load of the substance, we can compare it to the amount of the substance excreted and tell whether the substance undergoes net tubular reabsorption or net secretion. Whenever the quantity of a substance excreted in the urine is less than the filtered load, tubular reabsorption must have occurred. Conversely, if the amount excreted in the urine is greater than the filtered load, tubular secretion must have occurred.

Tubular Reabsorption

Table 16–2, which summarizes data for a few plasma components that undergo filtration and reabsorption, gives an idea of the magnitude and importance of

TABLE 16–2 Average Values for Several Components That Undergo Filtration and Reabsorption

Substance	Amount Filtered per Day	Amount Excreted per Day	Percent Reabsorbed
Water, L	180	1.8	99
Sodium, g	630	3.2	99.5
Glucose, g	180	0	100
Urea, g	54	30	44

reabsorptive mechanisms. The values in this table are typical for a normal person on an average diet. There are at least three important conclusions to be drawn from this table: (1) The filtered loads are enormous, generally larger than the amounts of the substances in the body. For example, the body contains about 40 L of water, but the volume of water filtered each day is, as we have seen, 180 L. (2) Reabsorption of waste products is relatively incomplete (for example, only 44 percent in the case of urea) so that large fractions of their filtered loads are excreted in the urine. (3) Reabsorption of most useful plasma components (for example, water, inorganic ions, and organic nutrients) is relatively complete so that the amounts excreted in the urine represent very small fractions of their filtered loads.

In this last regard, an important distinction should be made between reabsorptive processes that can be controlled physiologically and those that cannot. The reabsorption rates of most organic nutrients (for example, glucose) are always very high and are not physiologically regulated, and so the filtered loads of these substances are normally completely reabsorbed, none appearing in the urine. For these substances, like substance Z in our earlier example, it is as though the kidneys do not exist since the kidneys do not eliminate these substances from the body at all. Therefore, the kidneys do not *regulate* the plasma concentrations of these organic nutrients—that is, they do not minimize changes from normal plasma levels. Rather, the kidneys merely maintain whatever plasma concentrations already exist. (As described in Chapter 18, these concentrations are generally the result of hormonal regulation of nutrient metabolism.)

In contrast, the reabsorptive rates for water and many ions, although also very high, are regulatable. Recall, for example, the situation given earlier in which a person drinks a lot of water, which decreases tubular water reabsorption and thereby leads to increased water excretion. The critical point is that the rates at which water and many inorganic ions are reabsorbed,

and therefore the rates at which they are excreted, are subject to physiological control.

In contrast to glomerular filtration, the crucial steps in tubular reabsorption—those that achieve movement of a substance from tubular lumen to interstitial fluid—do *not* occur by bulk flow (there are inadequate pressure differences across the tubule and permeability of the tubular membranes). Instead, two other processes are involved. (1) The reabsorption of some substances is by diffusion, often across the tight junctions connecting the tubular epithelial cells (Figure 16–9). (2) The reabsorption of all other substances involves mediated transport, which requires the participation of transport proteins in the cell's plasma membranes.

Regardless of how a substance being reabsorbed goes from the lumen to the interstitial fluid, the final step in reabsorption—movement from interstitial fluid into the peritubular capillaries—is by a combination of diffusion and bulk flow, the latter driven by the capillary Starling forces. We will not mention this final step again in our discussions of reabsorption, but simply assume it occurs automatically once the substance reaches the interstitial fluid.

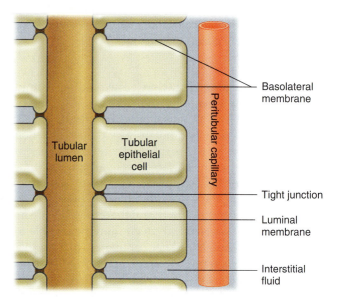

FIGURE 16–9

Diagrammatic representation of tubular epithelium. In this and subsequent figures illustrating transport in this chapter, the basement membrane of the tubule—a homogeneous proteinaceous structure that plays no significant role in transport—will not be shown. (Do not confuse the basement membrane with the basolateral membrane of the tubular cells, as illustrated in this and subsequent figures.)

Reabsorption by Diffusion Urea reabsorption by the proximal tubule provides an example of passive reabsorption by diffusion. An analysis of urea concentrations in the proximal tubule will help elucidate the mechanism. Because the corpuscular membranes are freely filterable to urea, the urea concentration in the fluid within Bowman's space is the same as that in the peritubular-capillary plasma and the interstitial fluid surrounding the tubule. Then, as the filtered fluid flows through the proximal tubule, water reabsorption occurs (by mechanisms to be described later). This removal of water increases the concentration of urea in the tubular fluid above that in the interstitial fluid and peritubular capillaries. Therefore, urea diffuses down this concentration gradient from tubular lumen to peritubular capillary. Urea reabsorption is thus dependent upon the reabsorption of water.

Reabsorption by diffusion in precisely this manner occurs for a variety of lipid-soluble organic substances, both naturally occurring and foreign (the pesticide DDT, for example).

Reabsorption by Mediated Transport Figure 16–9 highlights the fact that a substance reabsorbed by mediated transport must first cross the **luminal membrane** separating the tubular lumen from the cell interior, then diffuse through the cytosol of the cell, and finally cross the **basolateral membrane,** which begins at the tight junctions and constitutes the plasma membrane of the sides and base of the cell. This route is termed *transcellular* epithelial transport; its mechanisms were described in Chapter 6 and this material, particularly Figures 6–23 and 6–24, should be reviewed at this time.

Note in those two figures that a substance need not be actively transported across *both* the luminal and basolateral membranes in order to be actively transported across the overall epithelium—that is, to move from lumen to interstitial fluid against its electrochemical gradient. Thus, for example, sodium moves "downhill" (passively) into the cell across the luminal membrane either by diffusion (see Figure 6–23) or by facilitated diffusion (see Figure 6–24) and then moves "uphill" (actively) out of the cell across the basolateral membrane via the Na,K-ATPase pumps in this membrane. If the movement across *either* the luminal membrane or the basolateral membrane is active, then the entire process will achieve active reabsorption across the overall epithelium.

Figure 6–24 illustrates another important principle that applies to tubular reabsorption: The reabsorption of many substances is coupled to the reabsorption of sodium. Note in this figure that substance X moves uphill into the epithelial cell via a secondary active cotransporter as sodium moves downhill into the cell via this same cotransporter. The high intracellular concentration of substance X created by this active transport then drives "downhill" movement of the substance across the basolateral membrane to complete the reabsorptive process. This is precisely how glucose, many amino acids, and other organic substances undergo tubular reabsorption. The reabsorption of several inorganic ions is also coupled in a variety of ways to the reabsorption of sodium.

Many of the mediated-transport reabsorptive systems in the renal tubule have a limit, termed a **transport maximum (T_m),** to the amounts of material they can transport per unit time. This is because the binding sites on the membrane transport proteins become saturated. An important example is the secondary active-transport proteins for glucose, located in the proximal tubule and described in the previous paragraph. As noted earlier, normal persons do not excrete glucose in their urine because all filtered glucose is reabsorbed. Indeed, even by eating an extremely carbohydrate-rich meal, normal persons cannot raise their plasma glucose concentrations high enough so that the filtered load of glucose exceeds the renal glucose T_m. In contrast, in people with diabetes mellitus, a disease in which the hormonal control of the plasma glucose concentration is defective (Chapter 18), the plasma glucose concentration can become so high that the filtered load of glucose exceeds the ability of the tubules to reabsorb glucose—that is, exceeds the glucose T_m—therefore, glucose appears in the urine *(glucosuria).* In other words, the kidneys' ability to reabsorb glucose is normal in diabetes, but the tubules cannot reabsorb the markedly increased filtered load.

The pattern described for glucose is also true for a large number of other organic nutrients. For example, most amino acids and water-soluble vitamins are filtered in large amounts each day, but almost all these filtered molecules are reabsorbed by the proximal tubule. If the plasma concentration becomes high enough, however, reabsorption of the filtered load will not be as complete, and the substance will appear in larger amounts in the urine. Thus, persons ingesting very large quantities of vitamin C manifest progressive increases in their plasma concentrations of vitamin C until the filtered load exceeds the tubular reabsorptive T_m for this substance, and any additional ingested vitamin C is excreted in the urine.

Tubular Secretion

Tubular secretion moves substances from peritubular capillaries into the tubular lumen; like glomerular filtration, it constitutes a pathway into the tubule. Like reabsorption, secretion can occur by diffusion or by transcellular mediated transport. The most important substances secreted by the tubules are hydrogen ions and potassium, but a large number of normally occurring organic anions, such as choline and creatinine, are also secreted. So are many foreign chemicals, such as penicillin. Active secretion of a substance requires active transport either from the blood side (the interstitial fluid) into the cell (across the basolateral membrane) or out of the cell into the lumen (across the luminal membrane). Also as in reabsorption, tubular secretion of certain substances is coupled in one way or another to the reabsorption of sodium. We'll describe this later in the context of the secretion of potassium and hydrogen ions.

Metabolism by the Tubules

We noted earlier that, during fasting, the cells of the renal tubules synthesize glucose and add it to the blood. They can also synthesize certain substances, notably ammonium, which are then secreted into the fluid in the tubular lumen and excreted. Why the cells would make something just to have it excreted will be made clear when we discuss the role of ammonium in the regulation of plasma hydrogen-ion concentration. Also, the cells can catabolize certain organic substances (peptides, for example) taken up from either the tubular lumen or peritubular capillaries. Catabolism eliminates these substances from the body as surely as if they had been excreted into the urine.

Regulation of Membrane Channels and Transporters

We emphasized earlier that tubular reabsorption and/or secretion of many substances is under physiological control. For most of these substances, the control is achieved by regulating the activity or concentrations of the membrane proteins involved in their transport—channels and transporters. This regulation is achieved largely by hormones, neurotransmitters, and paracrine/autocrine agents.

Interestingly, the recent explosion of information concerning the structures, functions, and regulation of renal tubular-cell ion channels and transporters has made it possible to explain the underlying defects in some genetic diseases. For example, a genetic mutation can lead to an abnormality in the Na/glucose transporter that mediates active reabsorption of glucose in the proximal tubule. This abnormality leads to inadequate glucose reabsorption and loss of glucose in the urine at normal or even subnormal plasma concentrations of glucose. Contrast this condition, termed *familial renal glucosuria*, to diabetes mellitus, in which the ability to reabsorb glucose is normal, but the filtered load of glucose is greater than the maximal ability of the tubules to reabsorb this sugar.

"Division of Labor" in the Tubules

Several generalizations concerning tubular function should be kept in mind as you proceed through subsequent sections of this chapter, which deal with the renal handling of individual substances. As we have seen, in order to excrete waste products adequately, the GFR must be very large. This, though, means that the filtered volume of water and the filtered loads of all the nonwaste low-molecular-weight plasma solutes are also very large. The primary role of the proximal tubule is to reabsorb much of this filtered water and solutes. This segment has been called a "mass reabsorber" since for every substance reabsorbed by the tubule, the proximal tubule does most of the reabsorbing. Similarly, with one major exception (potassium) the proximal tubule is quantitatively the major site of solute secretion. Henle's loop also reabsorbs relatively large quantities of the major ions and, to a lesser extent, water.

Extensive reabsorption by the proximal tubule and Henle's loop ensures that the masses of solutes and the volume of water entering the tubular segments beyond Henle's loop are relatively small. These segments then do the fine-tuning for most substances, determining the final amounts excreted in the urine by adjusting their rates of reabsorption and, in a few cases, secretion. It should not be surprising, therefore, that most (but not all) homeostatic controls are exerted on these more distal segments.

The Concept of Renal Clearance

A useful way of quantitating renal functions is in terms of clearance. The renal **clearance** of any substance is the *volume of plasma* from which that substance is completely removed ("cleared") by the kidneys per unit time. Every substance has its own distinct clearance value, but the units are always in volume of plasma per time. The basic clearance formula for any substance S is

$$\text{Clearance of S} = \frac{\text{Mass of S excreted per unit time}}{\text{Plasma concentration of S}}$$

Thus, the clearance of a substance really answers the question: How much plasma had to be completely cleared of the substance to account for the mass of the substance excreted in the urine?

Since the mass of S excreted per unit time is equal to the urine concentration of S multiplied by the urine volume during that time, the formula for the clearance of S becomes

$$C_S = \frac{U_S V}{P_S}$$

where

C_S = clearance of S
U_S = urine concentration of S
V = urine volume per unit time
P_S = plasma concentration of S

Let us take the particularly important example of a polysaccharide named **inulin** (not insulin). This substance is an important research tool in renal physiology because, as will be described, its clearance is equal to the glomerular filtration rate. It is not found normally in the body, but we will administer it intravenously to a person at a rate sufficient to maintain plasma concentration constant at 4 mg/L. Urine collected over a 1-h period has a volume of 0.1 L and an inulin concentration of 300 mg/L; thus, inulin excretion equals 0.1 L/h × 300 mg/L, or 30 mg/h. How much plasma had to be completely cleared of its inulin to supply this 30 mg/h? We simply divide 30 mg/h by the plasma concentration, 4 mg/L, to obtain the volume cleared—7.5 L/h. In other words, we are calculating the inulin clearance (C_{In}) from the measured urine volume per time (V), urine inulin concentration (U_{In}), and plasma inulin concentration (P_{In}):

$$C_{In} = \frac{U_{In}V}{P_{In}}$$

$$C_{In} = \frac{300 \text{ mg/L} \times 0.1 \text{ L/h}}{4 \text{ mg/L}}$$

$$C_{In} = 7.5 \text{ L/h}$$

Now for the crucial points. From a variety of experiments, it is known that inulin is filterable at the renal corpuscle but is not reabsorbed, secreted, or metabolized by the tubule. Therefore, the mass of inulin excreted in our experiment—30 mg/h—must be equal to the mass filtered over that same time period (Figure 16–10). Accordingly, the *clearance* of inulin must equal the volume of plasma originally filtered; that is, C_{In} is equal to GFR.

It is important to realize that the clearance of *any* substance handled by the kidneys in the same way as inulin—filtered, but not reabsorbed, secreted, or metabolized—would equal the GFR. Unfortunately, there are no substances normally present in the plasma that meet these criteria. For clinical purposes, the **creatinine clearance** (C_{Cr}) is commonly used to *approximate* the GFR as follows. The waste product creatinine produced by muscle is filtered at the renal corpuscle and does not undergo reabsorption. It does undergo a small amount of secretion, however, so that some plasma is cleared of its creatinine by secretion. Accordingly, the C_{Cr} overestimates the GFR but is close enough to be highly useful.

This leads to an important generalization: When the clearance of any substance is greater than the GFR,

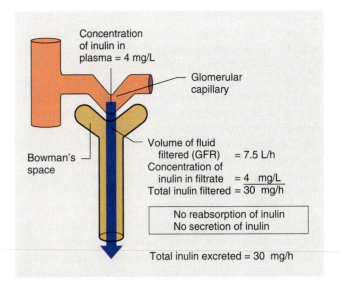

FIGURE 16–10

Example of renal handling of inulin, a substance that is filtered by the renal corpuscles but is neither reabsorbed nor secreted by the tubule. Therefore, the mass of inulin excreted per unit time is equal to the mass filtered during the same time period, and as explained in the text, the clearance of inulin is equal to the glomerular filtration rate.

as measured by the inulin clearance, that substance must undergo tubular secretion. Look back now at our hypothetical substance X (see Figure 16–7): X is filtered, and all the X that escapes filtration is secreted; no X is reabsorbed. Accordingly, all the plasma that enters the kidney per unit time is cleared of its X, and the clearance of X is therefore a measure of renal plasma flow. A substance that is handled like X is the organic anion para-amino-hippurate (PAH), which is used for this purpose (unfortunately, like inulin, it must be administered intravenously).

A similar logic leads to another important generalization: When the clearance of a filterable substance is less than the GFR, as measured by the inulin clearance, that substance must undergo reabsorption.

The remainder of this chapter describes how the kidneys function in the homeostasis of individual substances and how renal function is coordinated with that of other organs. Before turning to these individual substances, however, we complete the general story by describing the mechanisms of eliminating urine from the body—micturition.

Micturition

Urine flow through the ureters to the bladder is propelled by contractions of the ureter-wall smooth muscle. The urine is stored in the bladder and intermittently ejected during urination, or **micturition.**

The bladder is a balloon-like chamber with walls of smooth muscle collectively termed the **detrusor muscle.** The contraction of the detrusor muscle squeezes on the urine in the bladder lumen to produce urination. That part of the detrusor muscle at the base (or "neck") of the bladder where the urethra begins functions as a sphincter called the **internal urethral sphincter.** Just below the internal urethral sphincter, a ring of skeletal muscle surrounds the urethra. This is the **external urethral sphincter,** the contraction of which can prevent urination even when the detrusor muscle contracts strongly.

What factors influence these bladder structures (Figure 16–11)? (1) The detrusor muscle is innervated by parasympathetic neurons, which cause muscular contraction. Because of the arrangement of the smooth-muscle fibers, when the detrusor muscle is relaxed, the internal urethral sphincter is closed; when the detrusor muscle contracts, changes in its shape tend to pull open the internal urethral sphincter. (2) In addition, the internal sphincter receives sympathetic innervation, which causes contraction of the sphincter. (3) The external urethral sphincter, being skeletal muscle, is innervated by somatic motor neurons, which cause contraction.

As might be predicted from these inputs, while the bladder is filling, there is little parasympathetic input to the detrusor muscle but strong sympathetic input to the internal urethral sphincter and strong input by the somatic motor neurons to the external urethral sphincter. Therefore, the detrusor muscle is relaxed, and the sphincters are closed.

What happens during micturition? As the bladder fills with urine, the pressure within it increases, and this stimulates stretch receptors in the bladder wall. The afferent fibers from these receptors enter the spinal cord and *stimulate* the *parasympathetic* neurons, which then cause the detrusor muscle to contract. As noted above, this contraction facilitates the opening of the internal urethral sphincter. Simultaneously, the afferent input from the stretch receptors reflexly *inhibits* the *sympathetic* neurons to the internal urethral sphincter, which further contributes to its opening. In addition, the afferent input also reflexly *inhibits* the *somatic motor neurons* to the external urethral sphincter, causing it to relax. Both sphincters are now open, and the contraction of the detrusor muscle is able to produce urination.

We have thus far described micturition as a local spinal reflex, but descending pathways from the brain can also profoundly influence this reflex, determining the ability to prevent or initiate micturition voluntarily. Loss of these descending pathways as a result of spinal-cord damage eliminates one's ability to voluntarily control micturition. Prevention of micturition, learned during childhood, operates in the following way. As the bladder distends, the input from the bladder stretch receptors causes, via ascending pathways to the brain, a sense of bladder fullness and the urge to urinate. But in response to this, urination can be voluntarily prevented by activating descending pathways that stimulate both the sympathetic nerves to the internal urethral sphincter and the somatic motor nerves to the external urethral sphincter.

In contrast, urination can be voluntarily initiated via the descending pathways to the appropriate neurons.

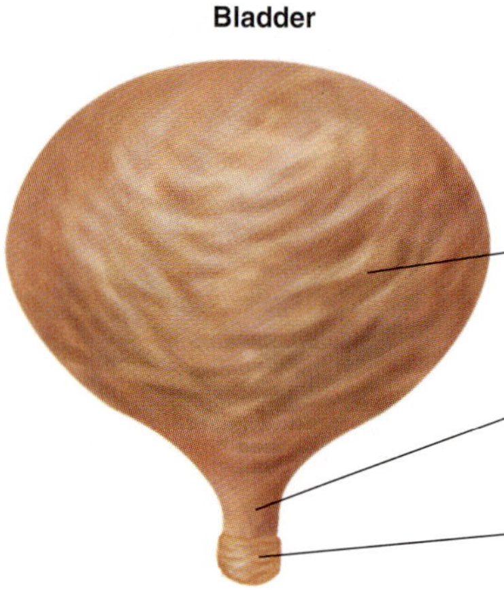

Bladder

Muscle	Innervation		
	Type	During filling	During micturition
Detrusor (smooth muscle)	Parasympathetic (causes contraction)	Inhibited	Stimulated
Internal urethral sphincter (smooth muscle)	Sympathetic (causes contraction)	Stimulated	Inhibited
External urethral sphincter (skeletal muscle)	Somatic motor (causes contraction)	Stimulated	Inhibited

FIGURE 16–11
Control of the bladder.

SECTION A SUMMARY

Functions and Structure of the Kidneys and Urinary System

I. The kidneys regulate the water and ionic composition of the body, excrete waste products, excrete foreign chemicals, produce glucose during prolonged fasting, and secrete three hormones—renin, 1,25-dihydroxyvitamin D_3, and erythropoietin. The first three functions are accomplished by continuous processing of the plasma.

II. Each nephron in the kidneys consists of a renal corpuscle and a tubule.

   a. Each renal corpuscle comprises a capillary tuft, termed a glomerulus, and a Bowman's capsule, into which the tuft protrudes.

   b. The tubule extends out from Bowman's capsule and is subdivided into many segments, which can be combined for reference purposes into the proximal tubule, loop of Henle, distal convoluted tubule, and collecting-duct system. At the level of the collecting ducts, multiple tubules join and empty into the renal pelvis, from which urine flows through the ureters to the bladder.

   c. Each glomerulus is supplied by an afferent arteriole, and an efferent arteriole leaves the glomerulus to branch into peritubular capillaries, which supply the tubule.

Basic Renal Processes

I. The three basic renal processes are glomerular filtration, tubular reabsorption, and tubular secretion. In addition, the kidneys synthesize and/or catabolize certain substances. The excretion of a substance is equal to the amount filtered plus the amount secreted minus the amount reabsorbed.

II. Urine formation begins with glomerular filtration—approximately 180 L/day—of essentially protein-free plasma into Bowman's space.

   a. Glomerular filtrate contains all plasma substances other than proteins (and substances bound to proteins) in virtually the same concentrations as in plasma.

   b. Glomerular filtration is driven by the hydrostatic pressure in the glomerular capillaries and is opposed by both the hydrostatic pressure in Bowman's space and the osmotic force due to the proteins in the glomerular capillary plasma.

III. As the filtrate moves through the tubules, certain substances are reabsorbed either by diffusion or mediated transport.

   a. Substances to which the tubular epithelium is permeable are reabsorbed by diffusion because water reabsorption creates tubule-interstitium concentration gradients for them.

   b. Active reabsorption of a substance requires the participation of transporters in the luminal or basolateral membrane.

   c. Tubular reabsorption rates are generally very high for nutrients, ions, and water, but are lower for waste products.

   d. Many of the mediated-transport systems manifest transport maximums, so that when the filtered load of a substance exceeds the transport maximum, large amounts may appear in the urine.

IV. Tubular secretion, like glomerular filtration, is a pathway for entrance of a substance into the tubule.

The Concept of Renal Clearance

I. The clearance of any substance can be calculated by dividing the mass of the substance excreted per unit time by the plasma concentration of the substance.

II. GFR can be measured by means of the inulin clearance and estimated by means of the creatinine clearance.

Micturition

I. In the basic micturition reflex, bladder distention stimulates stretch receptors that trigger spinal reflexes; these reflexes lead to contraction of the detrusor muscle, mediated by parasympathetic neurons, and relaxation of both the internal and the external urethral sphincters, mediated by inhibition of the neurons to these muscles.

II. Voluntary control is exerted via descending pathways to the parasympathetic nerves supplying the detrusor muscle, the sympathetic nerves supplying the internal urethral sphincter, and the motor nerves supplying the external urethral sphincter.

SECTION A KEY TERMS

renal	renal pelvis
urea	renal cortex
uric acid	renal medulla
creatinine	peritubular capillaries
gluconeogenesis	macula densa
ureter	juxtaglomerular (JG) cells
bladder	juxtaglomerular apparatus
urethra	(JGA)
nephron	glomerular filtration
renal corpuscle	glomerular filtrate
tubule	tubular reabsorption
glomerulus	tubular secretion
glomerular capillaries	net glomerular filtration
afferent arteriole	pressure
Bowman's capsule	glomerular filtration rate
efferent arteriole	(GFR)
Bowman's space	filtered load
podocyte	luminal membrane
mesangial cells	basolateral membrane
proximal tubule	transport maximum (T_m)
loop of Henle	clearance
descending limb	inulin
ascending limb	creatinine clearance
distal convoluted tubule	micturition
collecting duct system	detrusor muscle
cortical collecting duct	internal urethral sphincter
medullary collecting duct	external urethral sphincter

1. What are the functions of the kidneys?
2. What three hormones do the kidneys secrete?
3. Fluid flows in sequence through what structures from the glomerulus to the bladder? Blood flows through what structures from the renal artery to the renal vein?
4. What are the three basic renal processes that lead to the formation of urine?
5. How does the composition of the glomerular filtrate compare with that of plasma?
6. Describe the forces that determine the magnitude of the GFR. What is a normal value of GFR?
7. Contrast the mechanisms of reabsorption for glucose and urea. Which one shows a T_m?
8. Diagram the sequence of events leading to micturition in infants and in adults.

REGULATION OF SODIUM, WATER, AND POTASSIUM BALANCE

Total-Body Balance of Sodium and Water

Table 16–3 summarizes total-body water balance. These are average values, which are subject to considerable normal variation. There are two sources of body water gain: (1) water produced from the oxidation of organic nutrients, and (2) water ingested in liquids and so-called solid food (a rare steak is approximately 70 percent water). There are four sites from which water is lost to the external environment: skin, respiratory passageways, gastrointestinal tract, and urinary tract. Menstrual flow constitutes a fifth potential source of water loss in women.

The loss of water by evaporation from the skin and the lining of respiratory passageways is a continuous process. It is called **insensible water loss** because the person is unaware of its occurrence. Additional water can be made available for evaporation from the skin by the production of sweat. Normal gastrointestinal loss of water in feces is generally quite small, but can be severe in diarrhea. Gastrointestinal loss can also be large in vomiting.

Table 16–4 is a summary of total-body balance for sodium chloride. The excretion of sodium and chloride via the skin and gastrointestinal tract is normally quite small but may increase markedly during severe sweating, vomiting, or diarrhea. Hemorrhage can also result in loss of large quantities of both salt and water.

Under normal conditions, as can be seen from Tables 16–3 and 16–4, salt and water losses exactly equal salt and water gains, and no net change in body salt and water occurs. This matching of losses and gains is primarily the result of regulation of urinary loss, which can be varied over an extremely wide range. For example, urinary water excretion can vary from approximately 0.4 L/day to 25 L/day,

TABLE 16–3 Average Daily Water Gain and Loss in Adults

Intake	
In liquids	1200 ml
In food	1000 ml
Metabolically produced	350 ml
Total	**2550 ml**
Output	
Insensible loss (skin and lungs)	900 ml
Sweat	50 ml
In feces	100 ml
Urine	1500 ml
Total	**2550 ml**

TABLE 16–4 Daily Sodium Chloride Intake and Loss

Intake	
Food	**10.50 g**
Output	
Sweat	0.25 g
Feces	0.25 g
Urine	10.00 g
Total	**10.50 g**

depending upon whether one is lost in the desert or participating in a beer-drinking contest. Similarly, some individuals ingest 20 to 25 g of sodium chloride per day, whereas a person on a low-salt diet may ingest only 0.05 g; normal kidneys can readily alter their excretion of salt over this range to match loss with gain.

We will first present the basic renal processes for sodium and water and then describe the homeostatic reflexes that influence these processes. The renal processing of chloride is usually coupled directly or indirectly to that of sodium, and so we shall have little more to say about chloride even though it is the most abundant anion in the extracellular fluid.

Basic Renal Processes for Sodium and Water

Having low molecular weights and not being bound to protein, both sodium and water freely filter from the glomerular capillaries into Bowman's space. They both undergo considerable reabsorption—normally more than 99 percent (see Table 16–2)—but no secretion. Most renal energy utilization goes to accomplish this enormous reabsorptive task. The bulk of sodium and water reabsorption (about two-thirds) occurs in the proximal tubule, but the major hormonal controls of reabsorption are exerted on the collecting ducts.

The mechanisms of sodium and water reabsorption can be summarized by two generalizations: (1) Sodium reabsorption is an active process occurring in all tubular segments except the descending limb of the loop of Henle; and (2) water reabsorption is by diffusion and is dependent upon sodium reabsorption.

Primary Active Sodium Reabsorption

The essential feature underlying sodium reabsorption *throughout* the tubule is the primary active transport of sodium out of the cells and into the interstitial fluid, as illustrated for the cortical collecting duct in Figure 16–12. This transport is achieved by Na,K-ATPase pumps in the basolateral membrane of the cells. The active transport of sodium out of the cell keeps the intracellular concentration of sodium low compared to the luminal concentration, and so sodium moves "downhill" out of the lumen into the tubular epithelial cells. (The fact that the cell interior is negatively charged relative to the lumen also contributes to the electrochemical gradient favoring this movement from lumen to cell.)

The precise mechanism of the downhill sodium movement across the luminal membrane into the cell varies from segment to segment of the tubule depending upon which channels and/or transport proteins are present in their luminal membranes. For

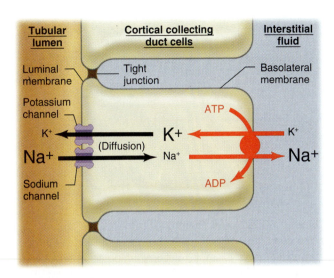

FIGURE 16–12

Mechanism of sodium reabsorption in the cortical collecting duct. Movement of the reabsorbed sodium from the interstitial fluid into peritubular capillaries is shown in Figure 16–13. The sizes of the letters for Na$^+$ and K$^+$ denote high and low concentrations of these ions. The fate of the potassium ions transported by the Na,K-ATPase pumps is discussed in the later section dealing with renal potassium handling.

example, as illustrated in Figure 16–12, the luminal entry step for sodium in the cortical collecting duct is by diffusion through sodium channels. To take another example, in the proximal tubule the luminal entry step is either by cotransport with a variety of organic molecules (glucose, for example) or by countertransport with hydrogen ions (that is, the hydrogen ions move from cell to lumen as the sodium moves into the cell). In this manner, in the proximal tubule sodium reabsorption drives the reabsorption of the cotransported substances and the secretion of hydrogen ions.

While the movement of sodium downhill from lumen into cell across the *luminal membrane* varies from one segment of the tubule to another, the *basolateral membrane* step is the same in all sodium-reabsorbing tubular segments—the primary active transport of sodium out of the cell is via Na,K-ATPase pumps in this membrane. It is this transport process that lowers intracellular sodium concentration and so makes possible the downhill luminal entry step, whatever its mechanism.

Coupling of Water Reabsorption to Sodium Reabsorption

How does active sodium reabsorption lead to passive water reabsorption? This type of coupling was described in Chapter 6 (see Figure 6-25) and is summarized again in Figure 16–13. (1) Sodium (and other

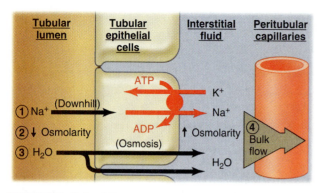

Tubular lumen	Tubular epithelial cells	Interstitial fluid	Peritubular capillaries

FIGURE 16–13

Coupling of water and sodium reabsorption. See text for explanation of numbers. The reabsorption of solutes other than sodium—for example, glucose, amino acids, and bicarbonate—also contributes to the difference in osmolarity between lumen and interstitial fluid, but the reabsorption of all these substances is ultimately dependent on direct or indirect cotransport and countertransport with sodium; therefore, they are not shown in the figure.

solutes whose reabsorption is dependent on sodium transport—for example, glucose, amino acids, and bicarbonate) is transported from the tubular lumen to the interstitial fluid across the epithelial cells, as described in the previous section. (2) This removal of solute lowers the osmolarity (that is, *raises* the water concentration) of the tubular fluid. It simultaneously raises the osmolarity (that is, *lowers* the water concentration) of the interstitial fluid adjacent to the epithelial cells. (3) The difference in water concentration between lumen and interstitial fluid causes net diffusion of water from the lumen across the tubular cells' plasma membranes and/or tight junctions into the interstitial fluid. (4) From there, water, sodium, and everything else dissolved in the interstitial fluid move together by bulk flow into peritubular capillaries as the final step in reabsorption.

Water movement across the tubular epithelium can occur, however, only if the epithelium is permeable to water. No matter how large its concentration gradient, water cannot cross an epithelium impermeable to it. Water permeability varies from tubular segment to segment and depends largely on the presence of water channels, termed **aquaporins,** in the plasma membranes (Chapter 6). The water permeability of the proximal tubule is always very high, and so water molecules are reabsorbed by this segment almost as rapidly as sodium ions. As a result, the proximal tubule always reabsorbs sodium and water in the same proportions.

We will describe the water permeability of the next tubular segments—the loop of Henle and distal convoluted tubule—later. Now for the really crucial point: The water permeability of the last portions of the tubules, the cortical and medullary collecting ducts, can be high or low because it is subject to physiological control, the only tubular segments in which water permeability is under such control.

The major determinant of this controlled permeability, and hence of water reabsorption in the collecting ducts, is a peptide hormone secreted by the posterior pituitary and known as **vasopressin,** or **antidiuretic hormone, ADH.** Vasopressin stimulates the insertion into the luminal membrane, by exocytosis, of a particular group of aquaporin water channels made by the collecting-duct cells. Accordingly, in the presence of a high plasma concentration of vasopressin, the water permeability of the collecting ducts becomes very great. Therefore, water reabsorption is maximal, and the final urine volume is small—less than 1 percent of the filtered water.

Without vasopressin, the water permeability of the collecting ducts is extremely low, and very little water is reabsorbed from these sites. Therefore, a large volume of water remains behind in the tubule to be excreted in the urine. This increased urine excretion resulting from low vasopressin is termed **water diuresis** (**diuresis** simply means a large urine flow from any cause). In a subsequent section, we will describe the reflexes that control vasopressin secretion.

The disease *diabetes insipidus,* which is distinct from the other kind of diabetes mentioned earlier in this chapter (diabetes mellitus or "sugar diabetes"), illustrates what happens when the vasopressin system malfunctions. Most people with this disease have lost the ability to produce vasopressin, usually as a result of damage to the hypothalamus. Thus, the permeability to water of the collecting ducts is low and unchanging regardless of the state of the body fluids. Therefore a constant water diuresis is present—as much as 25 L/day.

Note that in water diuresis, there is an increased urine flow, but not an increased solute excretion. In all other cases of diuresis, termed **osmotic diuresis,** the increased urine flow is the result of a primary increase in solute excretion. For example, failure of normal sodium reabsorption causes both increased sodium excretion and increased water excretion, since, as we have seen, water reabsorption is absolutely dependent on solute reabsorption. Another example of osmotic diuresis occurs in people with uncontrolled, marked diabetes *mellitus:* In this case, the glucose that escapes reabsorption because of the huge filtered load retains water in the lumen, causing it to be excreted along with the glucose. We'll talk more about the consequences of this in Chapter 18.

To summarize, any loss of solute in the urine *must* be accompanied by water loss (osmotic diuresis), but the reverse is not true; that is, water diuresis is not accompanied by equivalent solute loss.

Urine Concentration: The Countercurrent Multiplier System

Before reading this section you should review, by looking up in the glossary, several terms presented in Chapter 6—**hypoosmotic, isoosmotic,** and **hyperosmotic.**

In the section just concluded, we described how the kidneys produce a small volume of urine when the plasma concentration of vasopressin is high. Under these conditions, the urine is concentrated (hyperosmotic) relative to plasma. This section describes the mechanisms by which this hyperosmolarity is achieved.

The ability of the kidneys to produce hyperosmotic urine is a major determinant of one's ability to survive with limited amounts of water. The human kidney can produce a maximal urinary concentration of 1400 mOsmol/L, almost five times the osmolarity of plasma, which is 290 mOsmol/L (for ease of calculation, we shall round this off to 300 mOsmol/L in future discussions). The typical daily excretion of urea, sulfate, phosphate, other waste products, and ions amounts to approximately 600 mOsmol. Therefore, the minimal volume of urine water in which this mass of solute can be dissolved equals

$$\frac{600 \text{ mOsmol/day}}{1400 \text{ mOsmol/L}} = 0.444 \text{ L/day}$$

This volume of urine is known as the **obligatory water loss.** The loss of this minimal volume of urine (it would be somewhat lower if no food were available) contributes to dehydration when a person is deprived of water intake.

Urinary concentration takes place as tubular fluid flows through the *medullary* collecting ducts. The interstitial fluid surrounding these ducts is very hyperosmotic, and in the presence of vasopressin, water diffuses out of the ducts into the interstitial fluid of the medulla and then enters the blood vessels of the medulla to be carried away. The key question is: How does the medullary interstitial fluid become hyperosmotic? The answer is: through the function of Henle's loop. Recall that Henle's loop forms a hairpin-like loop between the proximal tubule and the distal convoluted tubule (see Figure 16–2). The fluid entering the loop from the proximal tubule flows down the descending limb, turns the corner, and then flows up the ascending limb. The opposing flows in the two limbs is termed a countercurrent flow, and as described next, the entire loop functions as a **countercurrent multiplier system** to create a hyperosmotic medullary interstitial fluid.

Because the proximal tubule always reabsorbs sodium and water in the same proportions, the fluid entering the descending limb of the loop from the proximal tubule has the same osmolarity as plasma—300 mOsmol/L. For the moment, let's skip the descending limb since the events in it can only be understood in the context of what the *ascending* limb is doing. Along the entire length of the ascending limb, sodium and chloride are reabsorbed into the medullary interstitial fluid. In the upper (thick) portion of the ascending limb, this reabsorption is achieved by transporters that actively cotransport sodium and chloride (as well as potassium, which we shall ignore). Such transporters are not present in the lower (thin) portion of the ascending limb, and the reabsorption there is a passive process. For simplicity, however, we shall treat the entire ascending limb as a homogeneous structure that actively reabsorbs sodium and chloride.

Very importantly, the ascending limb is relatively impermeable to water, so that little water follows the salt. The net result is that the interstitial fluid of the medulla becomes hyperosmotic compared to the fluid in the ascending limb.

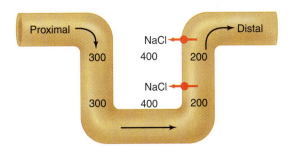

Now back to the *descending* limb. This segment, in contrast to the ascending limb, does *not* reabsorb sodium chloride and *is* highly permeable to water. Therefore, there is a net diffusion of water out of the descending limb into the more concentrated interstitial fluid until the osmolarities inside this limb and in the interstitial fluid are again equal. The interstitial hyperosmolarity is maintained during this equilibration because the ascending limb continues to pump sodium chloride to maintain the concentration difference between it and the interstitial fluid.

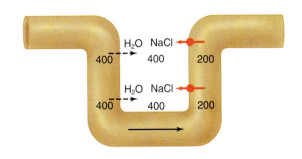

Thus, because of diffusion of water the osmolarities of the descending limb and interstitial fluid become equal, and both are higher—by 200 mOsmol/L in our example—than that of the ascending limb. This is the essence of the system: The loop countercurrent multiplier causes the interstitial fluid of the medulla to become concentrated. It is this hyperosmolarity that will draw water out of the collecting ducts and concentrate the urine. However, one more crucial feature—the "multiplication"—must be considered.

So far we have been analyzing this system as though the flow through the loop of Henle stopped while the ion pumping and water diffusion were occurring. This is not true, so let us see what happens when we allow flow in the system: As shown in the loop portion of Figure 16–14, the osmolarity difference—200 mOsmol/L—that exists at each horizontal level is "multiplied" to a much higher value—1400 mOsmol/L—at the bend in the loop. It should be

emphasized that the active sodium chloride transport mechanism in the ascending limb (coupled with low water permeability in this segment) is the essential component of the system. Without it, the countercurrent flow would have no effect whatever on loop and medullary interstitial osmolarity, which would simply remain 300 mOsmol/L throughout.

Now we have our concentrated medullary *interstitial fluid*, but we must still follow the fluid *within* the *tubules* from the loop of Henle through the distal convoluted tubule and into the collecting-duct system, using Figure 16–14 as our guide. As we have seen, the countercurrent multiplier system concentrates the descending-loop fluid, but then lowers the osmolarity in the ascending loop so that the fluid entering the distal convoluted tubule is actually more dilute (hypoosmotic)—100 mOsmol/L in Figure 16–14—than the plasma. The fluid becomes even more dilute during its passage through the distal convoluted tubule, since this tubular segment, like the ascending loop, actively transports sodium and chloride out of the tubule but is relatively impermeable to water. This hypoosmotic fluid then enters the cortical collecting duct.

As noted earlier, vasopressin increases tubular permeability to water in both the cortical and medullary collecting ducts. In contrast vasopressin does not influence water reabsorption in the parts of the tubule prior to the collecting ducts, and so, regardless of the plasma concentration of this hormone, the fluid entering the cortical collecting duct is always hypoosmotic. From there on, however, vasopressin is crucial. In the presence of high levels of vasopressin, water reabsorption occurs by diffusion from the hypoosmotic fluid in the *cortical* collecting duct until the fluid in this segment becomes isoosmotic to plasma in the peritubular capillaries of the cortex—that is, until it is once again at 300 mOsmol/L.

The isoosmotic tubular fluid then enters and flows through the *medullary* collecting ducts. In the presence of high plasma concentrations of vasopressin, water diffuses out of the ducts into the medullary interstitial fluid as a result of the high osmolarity set up there by the loop countercurrent multiplier system. This water then enters the medullary capillaries and is carried out of the kidneys by the venous blood. Water reabsorption occurs all along the lengths of the medullary collecting ducts so that, in the presence of vasopressin, the fluid at the end of these ducts has essentially the same osmolarity as the interstitial fluid surrounding the bend in the loops—that is, at the bottom of the medulla. By this means, the final urine is hyperosmotic. By retaining as much water as possible, the kidneys minimize the rate at which dehydration occurs during water deprivation.

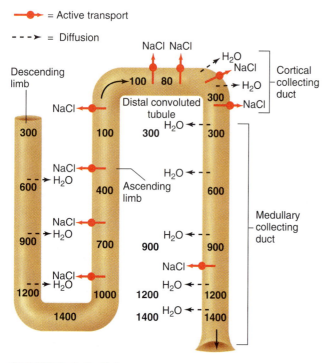

FIGURE 16–14

Generation of an interstitial fluid osmolarity gradient by the renal countercurrent multiplier system and its role in the formation of hyperosmotic urine. The transporter in the ascending limb of Henle's loop is actually a Na,K,2Cl cotransporter, but for simplicity we do not include the potassium since this ion does not contribute to the multiplier effect. Another simplification, as discussed in the text, is that the entire ascending limb is shown as actively transporting sodium chloride, whereas this is actually true only for the thick (upper) portion of the ascending limb.

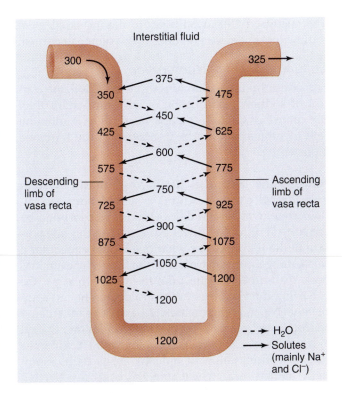

FIGURE 16–15
Function of the vasa recta. All movements of water and solutes are by diffusion. Not shown is the simultaneously occurring uptake of interstitial fluid by bulk-flow.

In contrast, as we saw earlier, when plasma vasopressin concentration is low, both the cortical and medullary collecting ducts are relatively impermeable to water. As a result, a large volume of hypoosmotic urine is excreted, thereby eliminating an excess of water in the body.

In describing how the countercurrent gradient is created, we have presented only the absolutely essential components, namely the interactions between sodium, chloride, and water. The story is actually more complex and not fully understood, but includes at least one important role for urea in determining the maximal urine concentration attainable. It also includes a unique function of the medullary circulation, to which we now turn.

The Medullary Circulation A major problem with the countercurrent system as described above is this: Why doesn't the blood flowing through medullary capillaries eliminate the countercurrent gradient set up by the loops of Henle? One would think that as plasma having the usual osmolarity of 300 mOsm/L enters the highly concentrated environment of the medulla, there would be massive net diffusion of sodium and chloride into the capillaries and water out of them, and

thus the interstitial gradient would be "washed away." However, the solution to this problem is as follows. The blood vessels in the medulla—termed **vasa recta**—form hairpin loops that run parallel to the loops of Henle and medullary collecting ducts. As shown in Figure 16–15, blood enters the top of the vessel loop at an osmolarity of 300 mOsm/L, and as the blood flows down the loop deeper and deeper into the medulla, sodium and chloride do indeed diffuse into, and water out of, the vessel. However, after the bend in the loop is reached, the blood then flows up the ascending vessel loop, where the process is almost completely reversed. Thus, the hairpin-loop structure of the vasa recta minimizes excessive loss of solute from the interstitium by *diffusion*. At the same time, both the salt and water being reabsorbed from the loops of Henle and collecting ducts are carried away in equivalent amounts by *bulk-flow*, as determined by the usual capillary Starling forces, and the steady-state countercurrent gradient set up by the loops of Henle is maintained.

Renal Sodium Regulation

Now we turn to the reflexes that act upon the basic renal processes for sodium and, in the next section, water to regulate their excretion. In normal individuals, urinary sodium excretion is reflexly increased when there is a sodium excess in the body and reflexly decreased when there is a sodium deficit. These reflexes are so precise that total-body sodium normally varies by only a few percent despite a wide range of sodium intakes and the sporadic occurrence of large losses via the skin and gastrointestinal tract.

As we have seen, sodium is freely filterable from the glomerular capillaries into Bowman's space and is actively reabsorbed, but not secreted. Therefore:

Sodium excreted = Sodium filtered − Sodium reabsorbed

The body can reflexly adjust sodium excretion by changing both processes on the right of the equation. Thus, for example, when total-body sodium decreases for any reason, sodium excretion is reflexly decreased below normal levels by lowering the GFR and simultaneously raising sodium reabsorption. Under most physiological conditions, however, the reflexly induced changes in GFR are relatively small, and sodium reabsorption is the major controlled process.

Our first problem in understanding the reflexes controlling sodium reabsorption is to determine what inputs initiate them; that is, what variables are actually being sensed by receptors? It may come as a surprise, but there are no important receptors capable of detecting either sodium concentration or the total amount of sodium in the body. Rather, the reflexes that regulate urinary sodium excretion are initiated mainly by various cardiovascular *baroreceptors,* such as the carotid sinus.

As described in Chapter 14, baroreceptors respond to pressure changes within the cardiovascular system and initiate reflexes that rapidly regulate these pressures by acting on the heart, arterioles, and veins. The new information in this chapter is that *regulation of cardiovascular pressures by baroreceptors also simultaneously achieves regulation of total-body sodium.*

Because sodium is the major extracellular solute (constituting, along with associated anions, approximately 90 percent of these solutes), changes in total-body sodium result in similar changes in extracellular volume. Since extracellular volume comprises plasma volume and interstitial volume, plasma volume is also positively related to total-body sodium. We saw in Chapter 14 that plasma volume is an important determinant of, in sequence, the blood pressures in the veins, cardiac chambers, and arteries. Thus, the chain linking total-body sodium to cardiovascular pressures is completed: Low total-body sodium leads to low plasma volume, which leads to low cardiovascular pressures, which, via baroreceptors, initiate reflexes that influence the renal arterioles and tubules so as to lower GFR and increase sodium reabsorption. These latter events decrease sodium excretion, thereby retaining sodium in the body and preventing further decreases in plasma volume and cardiovascular pressures. Increases in total-body sodium have the reverse reflex effects.

To summarize, the amount of sodium in the body determines the extracellular fluid volume, the plasma volume component of which helps determine cardiovascular pressures, which initiate the reflexes that control sodium excretion.

Control of GFR

Figure 16–16 summarizes the major mechanisms by which a lower total-body sodium, as caused by diarrhea, for example, elicits a decrease in GFR. The main direct cause of the reduced GFR—a reduced net glomerular filtration pressure—occurs both as a consequence of a lowered arterial pressure in the kidneys and, more importantly, as a result of reflexes acting on the renal arterioles. Note that these reflexes are simply the basic baroreceptor reflexes described in Chapter 14, where it was pointed out that a decrease in cardiovascular pressures causes neurally mediated reflex vasoconstriction in many areas of the body. As we shall see later, the hormones angiotensin II and vasopressin also participate in this renal vasoconstrictor response.

Conversely, an increased GFR is reflexly elicited by neuroendocrine inputs when an increased total-body sodium level causes increased plasma volume. This increased GFR contributes to the increased renal sodium loss that returns extracellular volume to normal.

Control of Sodium Reabsorption

For long-term regulation of sodium excretion, the control of sodium reabsorption is more important than the control of GFR. The major factor determining the rate of tubular sodium reabsorption is the hormone aldosterone.

Aldosterone and the Renin-Angiotensin System

The adrenal cortex produces a steroid hormone, **aldosterone,** which stimulates sodium reabsorption by the cortical collecting ducts. An action on this late portion of the tubule is just what one would expect for a fine-tuning input since most of the filtered sodium has been reabsorbed by the time the filtrate reaches the collecting-duct system. When aldosterone is completely absent, approximately 2 percent of the filtered sodium (equivalent to 35 g of sodium chloride per day) is not reabsorbed but is excreted. In contrast, when the plasma concentration of aldosterone is high, essentially all the sodium reaching the cortical collecting ducts is reabsorbed. In a normal person, the plasma concentration of aldosterone and the amount of sodium excreted lie somewhere between these extremes.

Aldosterone, like other steroids, acts by inducing the synthesis of proteins in its target cells; in the case of the cortical collecting ducts, the proteins participate in sodium transport. Look again at Figure 16–12; aldosterone induces the synthesis of all the channels and pumps shown in this figure. (By this same mechanism, aldosterone also stimulates sodium absorption from the lumens of both the large intestine and the ducts carrying fluid from the sweat glands and salivary glands. In this manner, less sodium is lost in the feces and from the surface of the skin in sweat.)

When a person is eating a lot of sodium, aldosterone secretion is low, whereas it is high when the person ingests a low-sodium diet or becomes sodium-depleted for some other reason. What controls the secretion of aldosterone under these circumstances? The answer is the hormone angiotensin II, which acts directly on the adrenal cortex to stimulate the secretion of aldosterone.

Angiotensin II is a component of the hormonal complex termed the **renin-angiotensin system,** summarized in Figure 16–17. As stated earlier, **renin** is an enzyme secreted by the juxtaglomerular cells of the juxtaglomerular apparatuses in the kidneys. Once in the bloodstream, renin splits a small polypeptide, **angiotensin I,** from a large plasma protein, **angiotensinogen,** which is produced by the liver. Angiotensin I then undergoes further cleavage to form the active agent of the renin-angiotensin

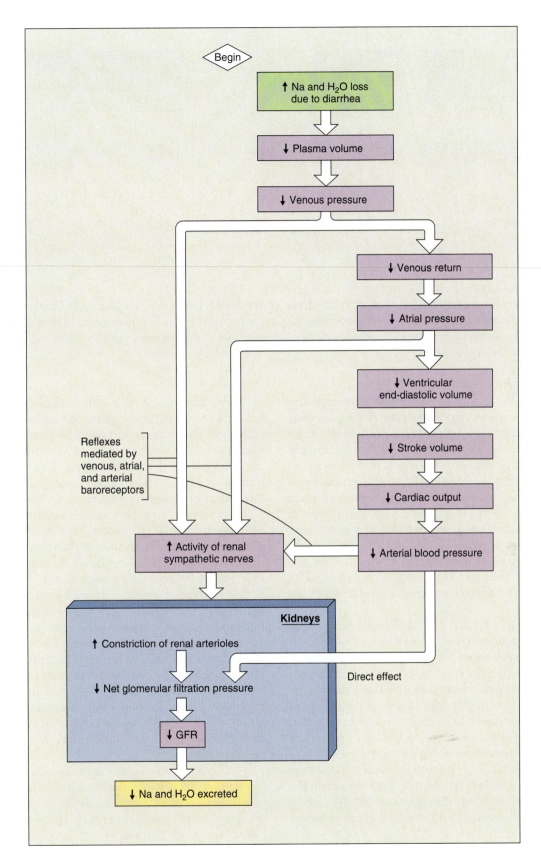

FIGURE 16–16
Direct and neurally mediated reflex pathways by which the GFR and hence sodium and water excretion are decreased when plasma volume decreases. The renal nerves also cause contraction of glomerular mesangial cells, resulting in a decreased surface area for filtration.

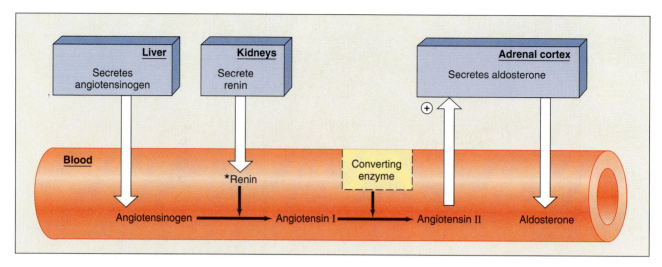

FIGURE 16–17
Summary of the renin-angiotensin system and the stimulation of aldosterone secretion by angiotensin II. Converting enzyme is located on the surface of capillary endothelial cells, particularly in the lungs. The plasma concentration of renin (as denoted by the asterisk) is the rate-limiting factor in the renin-angiotensin system; that is, it is the major determinant of the plasma concentration of angiotensin II.

system, **angiotensin II.** This conversion is mediated by an enzyme known as **angiotensin converting enzyme,** which is found in very high concentration on the luminal surface of capillary endothelial cells, particularly those in the lung. Angiotensin II exerts many effects, but the most important are its stimulation of the secretion of aldosterone and its constriction of arterioles (described in Chapter 14). Plasma angiotensin II is high during salt depletion and low when the individual is sodium replete, and it is this change in angiotensin II that brings about the changes in aldosterone secretion. Now we must ask the question: What causes the changes in plasma angiotensin II concentration with changes in salt balance?

Angiotensinogen and angiotensin converting enzyme are normally present in high and relatively unchanging concentrations, so the rate-limiting factor in angiotensin II formation is the plasma renin concentration. Normally, this concentration depends upon the rate of renin secretion by the kidneys. Thus the chain of events in salt depletion is: increased renin secretion → increased plasma renin concentration → increased plasma angiotensin concentration → increased aldosterone secretion → increased plasma aldosterone concentration. Our last task is, therefore, to explain the first link in this chain—the mechanisms by which sodium depletion causes an increase in renin secretion (Figure 16–18).

There are at least three distinct inputs to the juxtaglomerular cells: (1) the renal sympathetic nerves, (2) intrarenal baroreceptors, and (3) the macula densa. We will look at these in turn.

First, the renal sympathetic nerves directly innervate the juxtaglomerular cells, and an increase in the activity of these nerves stimulates renin secretion. This makes excellent sense since, as have seen, these nerves are reflexly activated via baroreceptors whenever a reduction in body sodium (and, hence, plasma volume) lowers cardiovascular pressures (see Figure 16–16).

The other two inputs for controlling renin release—intrarenal baroreceptors and the macula densa—are totally contained within the kidneys and require no external neuroendocrine input (although they can be influenced by such input). As noted earlier, the juxtaglomerular cells are located in the walls of the afferent arterioles; they are themselves sensitive to the pressure within these arterioles, and so function as **intrarenal baroreceptors.** When blood pressure in the kidneys *decreases,* as occurs when plasma volume is down, these cells are stretched *less* and, therefore, secrete *more* renin (Figure 16–18). Thus, the juxtaglomerular cells respond simultaneously to the combined effects of sympathetic input, triggered by baroreceptors *external* to the kidneys, and to their own pressure sensitivity.

The other completely internal input to the juxtaglomerular cells is via the macula densa, which, as noted earlier, is located near the ends of the ascending loops of Henle (see Figure 16–5). The macula densa senses the sodium and/or chloride concentration in the tubular fluid flowing past it, a *decreased* salt concentration causing *increased* renin secretion. For several reasons, including a decrease in GFR and hence tubular flow rate, macula densa sodium and

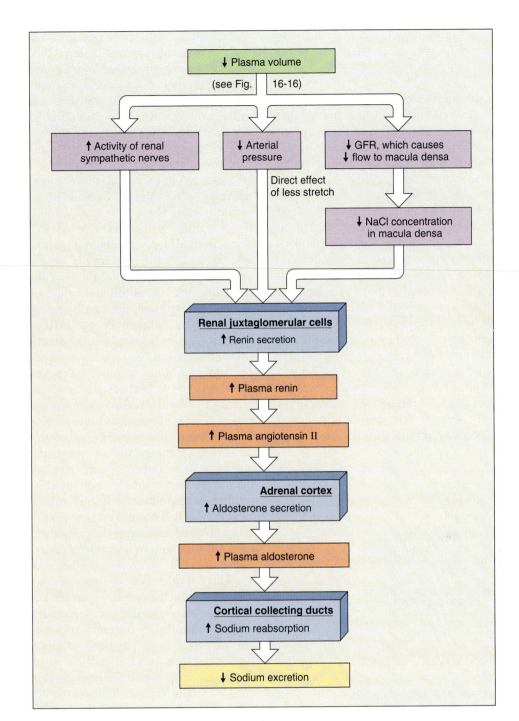

FIGURE 16–18

Pathways by which decreased plasma volume leads, via the renin-angiotensin system and aldosterone, to increased sodium reabsorption by the cortical collecting ducts and hence decreased sodium excretion.

chloride concentrations tend to decrease when a person's arterial pressure is decreased. This input therefore also signals for increased renin release at the same time that the sympathetic nerves and intrarenal baroreceptors are doing so (Figure 16–18).

Obviously, there is considerable redundancy in the control of renin secretion. As illustrated in Figure 16–18, the various mechanisms can all be participating at the same time.

By helping to regulate sodium balance and thereby plasma volume, the renin-angiotensin system contributes to the control of arterial blood pressure. However, this is not the only way in which it influences arterial pressure. Recall from Chapter 14 that angiotensin

II is a potent constrictor of arterioles all over the body and that this effect on peripheral resistance increases arterial pressure.

Other Factors Although aldosterone is the most important controller of sodium reabsorption, many other factors also play roles. For example, in addition to their *indirect* roles via control of aldosterone secretion, the renal nerves and angiotensin II also act *directly* on the tubules to stimulate sodium reabsorption.

Another input is the peptide hormone known as **atrial natriuretic factor (ANF),** which is synthesized and secreted by cells in the cardiac atria. ANF acts on the tubules (several tubular segments and mechanisms are involved) to *inhibit* sodium reabsorption. It can also act on the renal blood vessels to increase GFR, which further contributes to increased sodium excretion. As would be predicted, the secretion of ANF is increased when there is an excess of sodium in the body, but the stimulus for this increased secretion is not alterations in sodium *concentration*. Rather, using the same logic (only in reverse) that applies to the control of renin and aldosterone secretion, ANF secretion increases because of the expansion of plasma volume that accompanies an increase in body sodium. The specific stimulus is increased atrial distension (Figure 16–19).

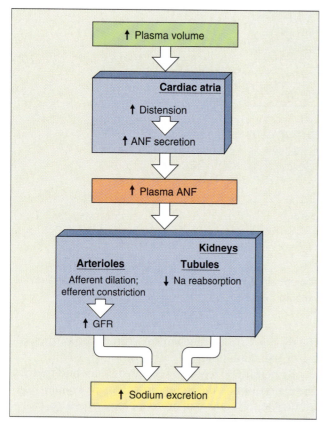

FIGURE 16–19
Atrial natriuretic factor (ANF) increases sodium excretion.

Finally, another important input controlling sodium reabsorption is the arterial blood pressure. We have previously described how the arterial blood pressure constitutes a signal for important *reflexes* (involving the renin-angiotensin system and aldosterone) that influence sodium reabsorption, but now we are emphasizing that arterial pressure also acts locally on the tubules themselves. Specifically, an *increase* in arterial pressure *inhibits* sodium reabsorption and thereby increases sodium excretion; this is termed **pressure natriuresis** ("natriuresis" means increased urinary sodium loss). Thus, an increased blood pressure reduces sodium reabsorption by two mechanisms: It reduces the activity of the renin-angiotensin-aldosterone system, and it also acts locally on the tubules. Conversely, a decreased blood pressure decreases sodium excretion both by stimulating the renin-angiotensin-aldosterone system and acting on the tubules to enhance sodium reabsorption. Now is a good time to look back at Figure 14–60, which describes the strong causal, reciprocal relationship between arterial blood pressure and blood volume, the result of which is that blood volume is perhaps the major long-term determinant of blood pressure. The direct effect of blood pressure on sodium excretion is, as shown in Figure 14–60, one of the major links in these relationships. Thus, for example, an important hypothesis is that most people who develop hypertension do so because their kidneys, for some reason, do not excrete enough sodium in response to a normal arterial pressure. Accordingly, at this normal pressure some dietary sodium is retained, which causes the pressure to rise enough to produce adequate sodium excretion to balance sodium intake, albeit at an increased body sodium content.

This completes our survey of the control of sodium excretion, which depends upon the control of two renal variables—the GFR and sodium reabsorption. The latter is controlled by the renin-angiotensin-aldosterone hormone system and by other factors, including atrial natriuretic factor and arterial blood pressure. The reflexes that control both GFR and sodium reabsorption are essentially reflexes that regulate blood pressure, since they are most frequently initiated by changes in arterial or venous pressures.

Renal Water Regulation

Water excretion is the difference between the volume of water filtered (the GFR) and the volume reabsorbed. Accordingly, the baroreceptor-initiated GFR-controlling reflexes described in the previous section tend to have the same effects on water excretion as on sodium excretion. As is true for sodium, however, the major regulated determinant of water excretion is not GFR but rather the rate of water reabsorption. As we have seen, this is determined by vasopressin, and so total-body water is regulated mainly by reflexes that alter the secretion of this hormone.

As described in Chapter 10, vasopressin is produced by a discrete group of hypothalamic neurons whose axons terminate in the posterior pituitary, from which vasopressin is released into the blood. The most important of the inputs to these neurons are from baroreceptors and osmoreceptors.

Baroreceptor Control of Vasopressin Secretion

We have seen that a decreased extracellular volume, due say to diarrhea or hemorrhage, reflexly calls forth, via the renin-angiotensin system, an increased aldosterone secretion. But the decreased extracellular volume also triggers increased vasopressin secretion. This increased vasopressin increases the water permeability of the collecting ducts, more water is reabsorbed and less is excreted, and so water is retained in the body to help stabilize the extracellular volume.

This reflex is initiated by several baroreceptors in the cardiovascular system (Figure 16–20). The baroreceptors decrease their rate of firing when cardiovascular pressures decrease, as occurs when blood volume decreases. Therefore, few impulses are transmitted from the baroreceptors via afferent neurons and ascending pathways to the hypothalamus, and the result is *increased* vasopressin secretion. Conversely, increased cardiovascular pressures cause more firing by the baroreceptors, resulting in a decrease in vasopressin secretion.

In addition to its effect on water excretion, if the plasma vasopressin concentration becomes very high, it, like angiotensin II, causes widespread arteriolar constriction. This helps restore arterial blood pressure toward normal (Chapter 14).

The baroreceptor reflex for vasopressin, as just described, has a relatively high threshold—that is, there must be a sizable reduction in cardiovascular pressures to trigger it. Therefore, this reflex, compared to the osmoreceptor reflex described next, generally plays a lesser role under most physiological circumstances, but it can become very important in pathological states such as hemorrhage.

Osmoreceptor Control of Vasopressin Secretion

We have seen how changes in extracellular volume simultaneously elicit reflex changes in the excretion of *both* sodium and water. This is adaptive since the situations causing extracellular volume alterations are very often associated with loss or gain of both sodium and water in approximately proportional amounts. In contrast, we shall see now that changes in total-body water in which no change in total-body sodium occurs are compensated for reflexly by altering water excretion *without altering sodium excretion*.

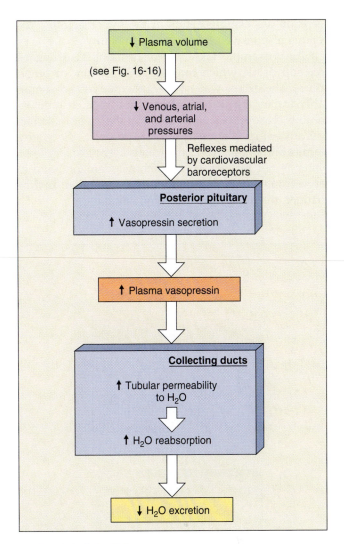

FIGURE 16–20

Baroreceptor pathway by which vasopressin secretion is increased when plasma volume is decreased. The opposite events (culminating in a decrease in vasopressin secretion) occur when plasma volume increases.

A crucial point in understanding how such reflexes are initiated is that changes in water alone, in contrast to sodium, have relatively little effect on extracellular volume. The reason is that water, unlike sodium, distributes throughout *all* the body-fluid compartments, about two-thirds entering the *intracellular* compartment rather than simply staying in the extracellular compartment as sodium does. Therefore, cardiovascular pressures and, hence, baroreceptors are little affected by pure water gains or losses. In contrast, the major change caused by water loss or gain out of proportion to sodium loss or gain is a change in the osmolarity of the body fluids. This is a key point because, under conditions due predominantly to water gain or loss, the receptors that initiate the reflexes controlling vasopressin secretion are **osmoreceptors** in the hypothalamus, receptors responsive to changes in osmolarity.

As an example, take a person drinking 2 L of sugar-free soft drink, which contains little sodium or other solute. The excess water lowers the body-fluid osmolarity (raises the water concentration), which reflexly inhibits vasopressin secretion via the hypothalamic osmoreceptors (Figure 16–21). As a result, the water permeability of the collecting ducts becomes very low, water is not reabsorbed from these segments, and a large volume of hypoosmotic urine is excreted. In this manner, the excess water is eliminated.

At the other end of the spectrum, when the osmolarity of the body fluids increases (water concentration decreases), say, because of water deprivation, vasopressin secretion is reflexly increased via the osmoreceptors, water reabsorption by the collecting ducts is increased, and a very small volume of highly concentrated urine is excreted. By retaining relatively more water than solute, the kidneys help reduce the body-fluid osmolarity back toward normal.

To summarize, regulation of body-fluid osmolarity requires separation of water excretion from sodium excretion—that is, requires the kidneys to excrete a urine that, relative to plasma, either contains more water than sodium and other solutes (water diuresis) or less water than solute (concentrated urine). This is made possible by two physiological factors: (1) osmoreceptors and (2) vasopressin-dependent dissociation of water reabsorption from sodium reabsorption in the collecting ducts.

We have now described two afferent pathways controlling the vasopressin-secreting hypothalamic cells, one from baroreceptors and one from osmoreceptors. To add to the complexity, the hypothalamic cells receive synaptic input from many other brain areas, so that vasopressin secretion, and therefore urine volume and concentration, can be altered by pain, fear, and a variety of drugs. For example, alcohol is a powerful inhibitor of vasopressin release, and this probably accounts for much of the increased urine volume produced following ingestion of alcohol, a urine volume well in excess of the volume of beverage consumed.

A Summary Example: The Response to Sweating

Figure 16–22 shows the factors that control renal sodium and water excretion in response to severe sweating. Sweat is a hypoosmotic solution containing mainly water, sodium, and chloride. Therefore, sweating causes both a decrease in extracellular volume and an increase in body-fluid osmolarity (a decrease in water concentration). The renal retention of water and sodium minimizes the deviations from normal caused by the loss of water and salt in the sweat.

Thirst and Salt Appetite

Now we turn to the other component of any balance—control of intake. Deficits of salt and water must eventually be compensated for by ingestion of these substances, because the kidneys cannot create new sodium ions or water, they can only minimize their excretion until ingestion replaces the losses.

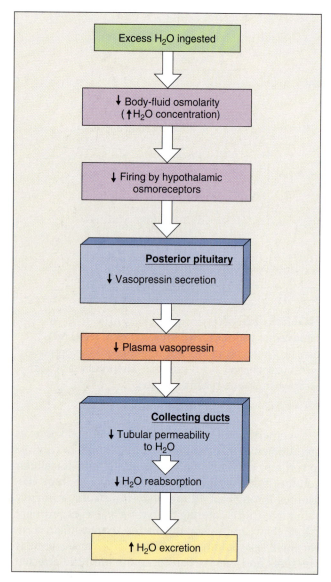

FIGURE 16–21

Osmoreceptor pathway by which vasopressin secretion is lowered and water excretion raised when excess water is ingested. The opposite events (an increase in vasopressin secretion) occur when osmolarity increases, as during water deprivation.

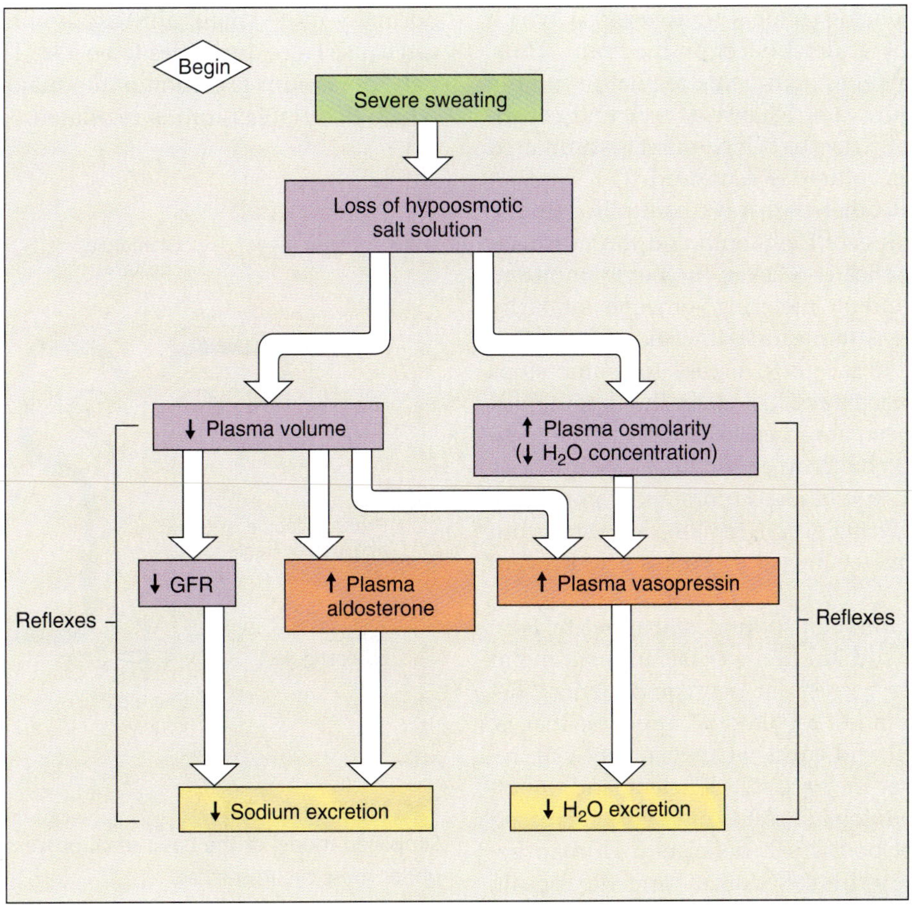

FIGURE 16–22

Pathways by which sodium and water excretion are decreased in response to severe sweating. This figure is an amalgamation of Figures 16–16, 16–18, 16–20, and the reverse of 16–21.

The subjective feeling of thirst, which leads us to obtain and ingest water, is stimulated both by a lower extracellular volume and a higher plasma osmolarity (Figure 16–23), the latter being the single most important stimulus under normal physiological conditions. Note that these are precisely the same two changes that stimulate vasopressin production, and the osmoreceptors and baroreceptors that control vasopressin secretion are identical to those for thirst. The brain centers that receive input from these receptors and mediate thirst are located in the hypothalamus, very close to those areas that produce vasopressin.

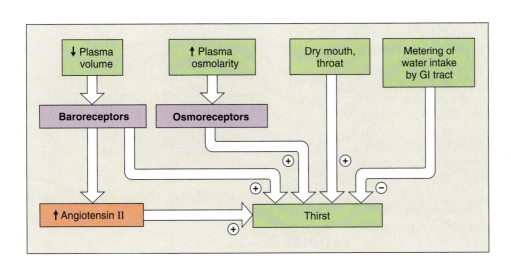

FIGURE 16–23

Inputs reflexly controlling thirst. The osmoreceptor input is the single most important stimulus under most physiological conditions. Psychosocial factors and conditioned responses are not shown.

Another influencing factor is angiotensin II, which stimulates thirst by a direct effect on the brain. Thus, the renin-angiotensin system helps regulate not only sodium balance but water balance as well and constitutes one of the pathways by which thirst is stimulated when extracellular volume is decreased.

There are still other pathways controlling thirst. For example, dryness of the mouth and throat causes profound thirst, which is relieved by merely moistening them. Some kind of "metering" of water intake by other parts of the gastrointestinal tract also occurs; that is, a thirsty individual given access to water stops drinking after replacing the lost water but before most of the water has been absorbed from the gastrointestinal tract and has a chance to eliminate the stimulatory inputs to the systemic baroreceptors and osmoreceptors. How this metering occurs remains a mystery, but one function of this feedforward process is to prevent overhydration.

The analog of thirst for sodium, **salt appetite,** is an important part of sodium homeostasis in most mammals. Salt appetite consists of two components: "hedonistic" appetite and "regulatory" appetite; that is, animals "like" salt and eat it whenever they can, regardless of whether they are salt-deficient, and, in addition, their drive to obtain salt is markedly increased in the presence of bodily salt deficiency. Human beings certainly have a strong hedonistic appetite for salt, as manifested by almost universally large intakes of salt whenever it is cheap and readily available (for example, the average American consumes 10–15 g/day despite the fact that human beings can survive quite normally on less than 0.5 g/day). However, unlike most other mammals, humans have relatively little regulatory salt appetite, at least until a bodily salt deficit becomes extremely large.

Potassium Regulation

Potassium is, as we have seen, the most abundant intracellular ion. However, although only 2 percent of total-body potassium is in the extracellular fluid, the potassium concentration in this fluid is extremely important for the function of excitable tissues, notably nerve and muscle. Recall (Chapter 8) that the resting-membrane potentials of these tissues are directly related to the relative intracellular and extracellular potassium concentrations. Accordingly, either increases or decreases in extracellular potassium concentration can cause abnormal rhythms of the heart (*arrhythmias*) and abnormalities of skeletal-muscle contraction.

A normal person remains in potassium balance by daily excreting an amount of potassium in the urine equal to the amount ingested minus the amounts eliminated in the feces and sweat. Also, like sodium, potassium losses via sweat and the gastrointestinal tract are normally quite small, although vomiting or diarrhea can cause large quantities to be lost. The control of urinary potassium excretion is the major mechanism by which body potassium is regulated.

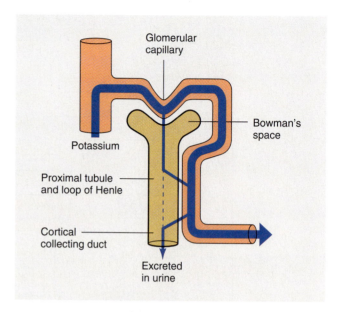

FIGURE 16–24

Simplified model of the basic renal processing of potassium under most circumstances.

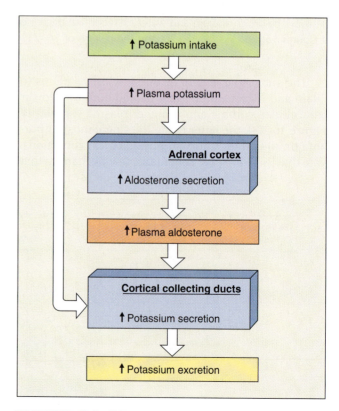

FIGURE 16–25

Pathways by which an increased potassium intake induces greater potassium excretion.

Renal Regulation of Potassium

Potassium is freely filterable in the renal corpuscle. Normally, the tubules reabsorb most of this filtered potassium so that very little of the filtered potassium appears in the urine. However, the cortical collecting ducts can *secrete* potassium, and changes in potassium *excretion* are due mainly to changes in potassium *secretion* by this tubular segment (Figure 16–24).

During potassium depletion, when the homeostatic response is to minimize potassium loss, there is no potassium secretion by the cortical collecting ducts, and only the small amount of filtered potassium that escapes tubular reabsorption is excreted. In all other situations, to the small amount of potassium not reabsorbed is added a variable amount of potassium secreted by the cortical collecting ducts, an amount necessary to maintain total-body potassium balance.

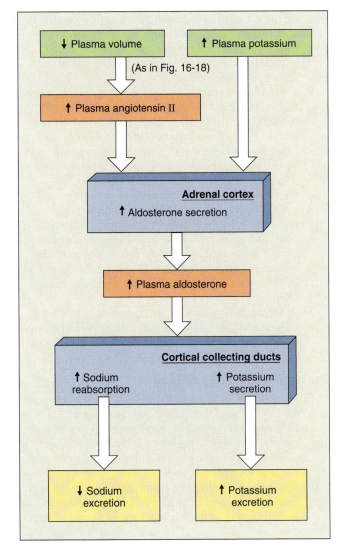

FIGURE 16–26

Summary of the control of aldosterone and its effects on sodium reabsorption and potassium secretion.

The mechanism of potassium secretion by the cortical collecting ducts was illustrated in Figure 16–12. In this tubular segment, the K^+ pumped into the cell across the basolateral membrane by Na,K-ATPases diffuses into the tubular lumen through K^+ channels in the luminal membrane. Thus, the *secretion* of potassium by the cortical collecting duct is associated with the *reabsorption* of sodium by this tubular segment. (Potassium secretion does not occur in other sodium-reabsorbing tubular segments because there are few potassium channels in the luminal membranes of their cells; rather, in these segments the potassium pumped into the cell by Na,K-ATPases simply diffuses back across the basolateral membrane through potassium channels located there.)

What factors influence potassium secretion by the cortical collecting ducts to achieve homeostasis of bodily potassium? The single most important factor is as follows: When a high-potassium diet is ingested (Figure 16–25), plasma potassium concentration increases, though very slightly, and this drives enhanced basolateral uptake via the Na,K-ATPase pumps and hence an enhanced potassium secretion. Conversely, a low-potassium diet or a negative potassium balance, for example, from diarrhea, lowers basolateral potassium uptake; this reduces potassium secretion and excretion, thereby helping to reestablish potassium balance.

A second important factor linking potassium secretion to potassium balance is the hormone aldosterone (Figure 16–25). Besides stimulating tubular sodium reabsorption by the cortical collecting ducts, aldosterone simultaneously enhances tubular potassium secretion by this tubular segment.

The reflex by which an excess or deficit of potassium controls aldosterone production (Figure 16–25) is completely different from the reflex described earlier involving the renin-angiotensin system. The aldosterone-secreting cells of the adrenal cortex are sensitive to the potassium concentration of the extracellular fluid bathing them. Thus, an increased intake of potassium leads to an increased extracellular potassium concentration, which in turn directly stimulates aldosterone production by the adrenal cortex. The resulting increased plasma aldosterone concentration increases potassium secretion and thereby eliminates the excess potassium from the body.

Conversely, a lowered extracellular potassium concentration decreases aldosterone production and thereby reduces potassium secretion. Less potassium than usual is excreted in the urine, thus helping to restore the normal extracellular concentration.

The control and major renal tubular effects of aldosterone are summarized in Figure 16–26. The fact that a single hormone regulates both sodium and potassium excretion raises the question of potential conflicts between homeostasis of the two ions. For example, if a person were sodium-deficient and therefore secreting large amounts of aldosterone, the potassium-secreting effects of this hormone would tend to cause some potassium loss even though potassium balance was normal to start with. Usually, such conflicts cause only minor imbalances because there are a variety of other counteracting controls of sodium and potassium excretion.

SECTION B SUMMARY

Total-Body Balance of Sodium and Water

I. The body gains water via ingestion and internal production, and it loses water via urine, the gastrointestinal tract, and evaporation from the skin and respiratory tract (as insensible loss and sweat).

II. The body gains sodium and chloride by ingestion and loses them via the skin (in sweat), gastrointestinal tract, and urine.

III. For both water and sodium, the major homeostatic control point for maintaining stable balance is renal excretion.

Basic Renal Processes for Sodium and Water

I. Sodium is freely filterable at the glomerulus, and its reabsorption is a primary active process dependent upon Na,K-ATPase pumps in the basolateral membranes of the tubular epithelium. Sodium is not secreted.

II. Sodium entry into the cell from the tubular lumen is always passive. Depending on the tubular segment, it is either through channels or by cotransport or countertransport with other substances.

III. Sodium reabsorption creates an osmotic difference across the tubule, which drives water reabsorption, largely through water channels (aquaporins).

IV. Water reabsorption is independent of the posterior pituitary hormone vasopressin until the collecting-duct system, where vasopressin increases water permeability. A large volume of dilute urine is produced when plasma vasopressin concentration, and hence water reabsorption by the collecting ducts, is low.

V. A small volume of concentrated urine is produced by the renal countercurrent multiplier system when plasma vasopressin concentration is high.

  a. The active transport of sodium chloride by the ascending loop of Henle causes increased osmolarity of the interstitial fluid of the medulla but a dilution of the luminal fluid.

  b. Vasopressin increases the permeability of the cortical collecting ducts to water, and so water is reabsorbed by this segment until the luminal fluid is isoosmotic to plasma in the cortical peritubular capillaries.

  c. The luminal fluid then enters and flows through the medullary collecting ducts, and the concentrated medullary interstitium causes water to move out of these ducts, made highly permeable to water by vasopressin. The result is concentration of the collecting duct fluid and the urine.

  d. The hairpin-loop structure of the vasa recta prevents the countercurrent gradient from being washed away.

Renal Sodium Regulation

I. Sodium excretion is the difference between the amount of sodium filtered and the amount reabsorbed.

II. GFR, and hence the filtered load of sodium, is controlled by baroreceptor reflexes. Decreased vascular pressures cause decreased baroreceptor firing and hence increased sympathetic outflow to the renal arterioles, resulting in vasoconstriction and decreased GFR. These changes are generally relatively small under most physiological conditions.

III. The major control of tubular sodium reabsorption is the adrenal cortical hormone aldosterone, which stimulates sodium reabsorption in the cortical collecting ducts.

IV. The renin-angiotensin system is one of the two major controllers of aldosterone secretion. When extracellular volume decreases, renin secretion is stimulated by three inputs: (1) stimulation of the renal sympathetic nerves to the juxtaglomerular cells by extrarenal baroreceptor reflexes; (2) pressure decreases sensed by the juxtaglomerular cells, themselves acting as intrarenal baroreceptors; and (3) a signal generated by low sodium or chloride concentration in the lumen of the macula densa.

V. Many other factors influence sodium reabsorption. One of these, atrial natriuretic factor, is secreted by cells in the atria in response to atrial distension; it inhibits sodium reabsorption and it also increases GFR.

VI. Arterial pressure acts locally on the renal tubules to influence sodium reabsorption, an increased pressure causing decreased reabsorption and hence increased excretion.

Renal Water Regulation

I. Water excretion is the difference between the amount of water filtered and the amount reabsorbed.

II. GFR regulation via the baroreceptor reflexes plays some role in regulating water excretion, but the major control is via vasopressin-mediated control of water reabsorption.

III. Vasopressin secretion by the posterior pituitary is controlled by cardiovascular baroreceptors and by osmoreceptors in the hypothalamus.

a. Via the baroreceptor reflexes, a low extracellular volume stimulates vasopressin secretion, and a high extracellular volume inhibits it.
b. Via the osmoreceptors, a high body-fluid osmolarity stimulates vasopressin secretion, and a low osmolarity inhibits it.

Thirst and Salt Appetite

I. Thirst is stimulated by a variety of inputs, including baroreceptors, osmoreceptors, and angiotensin II.
II. Salt appetite is not of major regulatory importance in human beings.

Potassium Regulation

I. A person remains in potassium balance by excreting an amount of potassium in the urine equal to the amount ingested minus the amounts lost in the feces and sweat.
II. Potassium is freely filterable at the renal corpuscle and undergoes both reabsorption and secretion, the latter occurring in the cortical collecting ducts and being the major controlled variable determining potassium excretion.
III. When body potassium is increased, extracellular potassium concentration increases. This increase acts directly on the cortical collecting ducts to increase potassium secretion and also stimulates aldosterone secretion, the increased plasma aldosterone then also stimulating potassium secretion.

SECTION B KEY TERMS

insensible water loss	aldosterone
aquaporins	renin-angiotensin system
vasopressin (antidiuretic hormone, ADH)	renin
	angiotensin I
water diuresis	angiotensinogen
diuresis	angiotensin II
osmotic diuresis	angiotensin converting
hypoosmotic	enzyme
isoosmotic	intrarenal baroreceptors
hyperosmotic	atrial natriuretic factor
obligatory water loss	(ANF)
countercurrent multiplier system	pressure natriuresis
	osmoreceptors
vasa recta	salt appetite

SECTION B REVIEW QUESTIONS

1. What are the sources of water gain and loss in the body? What are the sources of sodium gain and loss?
2. Describe the distribution of water and sodium between intracellular and extracellular fluids.
3. What is the relationship between body sodium and extracellular-fluid volume?
4. What is the mechanism of sodium reabsorption, and how is the reabsorption of other solutes coupled to it?
5. What is the mechanism of water reabsorption, and how is it coupled to sodium reabsorption?
6. What is the effect of vasopressin on the renal tubules, and what are the sites affected?
7. Describe the characteristics of the two limbs of the loop of Henle with regard to their transport of sodium, chloride, and water.
8. Diagram the osmolarities in the two limbs of the loop of Henle, distal convoluted tubule, cortical collecting duct, cortical interstitium, medullary collecting duct, and medullary interstitium in the presence of vasopressin. What happens to the cortical and medullary collecting-duct values in the absence of vasopressin?
9. What two processes determine how much sodium is excreted per unit time?
10. Diagram the sequence of events by which a decrease in blood pressure leads to a decreased GFR.
11. List the sequence of events leading from increased renin secretion to increased aldosterone secretion.
12. What are the three inputs controlling renin secretion?
13. Diagram the sequence of events leading from decreased cardiovascular pressures or from an increased plasma osmolarity to an increased secretion of vasopressin.
14. What are the stimuli for thirst?
15. Which of the basic renal processes apply to potassium? Which of them is the controlled process, and which tubular segment performs it?
16. Diagram the steps leading from increased plasma potassium to increased potassium excretion.
17. What are the two major controls of aldosterone secretion, and what are this hormone's major actions?

CALCIUM REGULATION

Extracellular calcium concentration normally remains relatively constant. Large deviations in either direction would cause problems. A low plasma calcium concentration increases the excitability of nerve and muscle plasma membranes, so that individuals with low plasma calcium suffer from *hypocalcemic tetany,* characterized by skeletal-muscle spasms. A high plasma calcium concentration causes cardiac arrhythmias as well as depressed neuromuscular excitability. These effects reflect, in part, the ability of extracellular calcium to bind to plasma-membrane proteins that function as ion channels, thereby altering membrane potentials. The binding alters the open or closed state of the channels. This effect of calcium on plasma membranes is totally distinct from its role as an intracellular excitation-contraction coupler, as described in Chapter 11.

Effector Sites for Calcium Homeostasis

The sections in this chapter on sodium, water, and potassium homeostasis were concerned almost entirely with the *renal* handling of these substances. In contrast, the regulation of calcium depends not only on the kidneys but also on bone and the gastrointestinal tract. The activities of the gastrointestinal tract and kidneys determine the net intake and output of calcium for the entire body and, thereby, the overall state of calcium balance. In contrast, interchanges of calcium between extracellular fluid and bone do not alter total-body balance but, rather, the *distribution* of calcium within the body. We will first describe how the effector sites handle calcium and then discuss how they are influenced by hormones in the homeostatic control of plasma calcium concentration.

Bone

Approximately 90 percent of total-body calcium is contained in bone. Therefore, deposition of calcium in bone or its removal very importantly influences plasma calcium concentration.

Bone has functions, summarized in Table 16–5, other than regulating plasma calcium concentration. It is important to recognize that its role in maintaining normal plasma calcium concentration takes precedence over the mechanical supportive role, sometimes to the detriment of the latter.

TABLE 16–5 Functions of Bone

1. Supports the body and imposed loads against gravity.

2. Provides the rigidity that permits locomotion.

3. Affords protection to the internal organs. The rib cage, vertebrae, and skull perform this function.

4. Serves as a reservoir for calcium, inorganic phosphate, and other mineral elements.

5. Produces blood cells in the bone marrow.

Bone is a special connective tissue made up of several cell types surrounded by a collagen matrix, called **osteoid,** upon which are deposited minerals, particularly the crystals of calcium and phosphate known as hydroxyapatite. In some instances, bones have central marrow cavities where blood cells are formed (Chapter 14). Typically, approximately one-third of a bone, by weight, is osteoid, and two-thirds is mineral (the bone cells contribute negligible weight).

The three types of bone cells (the blood-forming cells of the marrow are not included in this term) are osteoblasts, osteocytes, and osteoclasts (Figure 16–27). **Osteoblasts** are the bone-forming cells. They secrete collagen to form a surrounding matrix, which then becomes calcified; just how this mineralization is brought about remains controversial. Once surrounded by calcified matrix, the osteoblasts are called **osteocytes.** The osteocytes have long cytoplasmic processes that extend throughout the bone and form tight junctions with other osteocytes. **Osteoclasts** are large multinucleated cells that break down (resorb) previously formed bone by secreting hydrogen ions, which dissolve the crystals, and hydrolytic enzymes, which digest the osteoid.

The growth of bones during childhood will be discussed in Chapter 18. What is important here is that throughout life, bone is being constantly "remodeled" by the osteoblasts and osteoclasts working together. Osteoclasts resorb old bone, and then osteoblasts move into the area and lay down new matrix, which becomes calcified. This process is dependent, in part, on the stresses imposed on the bones by gravity and muscle tension, both of which stimulate osteoblastic activity. It is also influenced by many hormones, as summarized in Table 16–6, and a bewildering variety

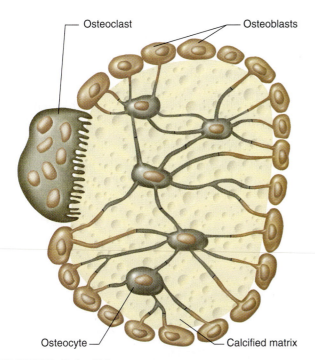

FIGURE 16–27

Cross section through a small portion of bone. The light tan area is mineralized osteoid. The osteocytes have long processes that extend through small canals and connect with each other and to osteoblasts via tight junctions.

Adapted from Goodman.

TABLE 16–6 Reference Summary of Major Hormonal Influences on Bone Mass*
Hormones that favor bone formation and increased bone mass
Insulin
Growth hormone
Insulin-like growth factor I (IGF-I)
Estrogen
Testosterone
1,25-dihydroxyvitamin D_3 (influences only mineralization, not matrix)
Calcitonin
Hormones that favor increased bone resorption and decreased bone mass
Parathyroid hormone
Cortisol
Thyroid hormones (T_4 and T_3)

*The effects of all hormones except 1,25-dihydroxyvitamin D_3 are ultimately due to a hormone-induced alteration of the balance between the activities of the osteoblasts and osteoclasts. This altered balance is not always the result of a *direct* action of the hormone on these cells, however, but may result *indirectly* from some other action of the hormone. For example, large amounts of cortisol depress intestinal absorption of calcium, which in turn causes reduced plasma calcium, increased parathyroid hormone secretion, and stimulation of osteoclasts by parathyroid hormone. Also, in addition to hormones, many paracrine agents produced by the bone cells and bone-marrow connective-tissue cells influence bone formation and resorption.

of autocrine/paracrine growth factors produced locally in the bone. Of the hormones listed, only parathyroid hormone and 1,25-dihydroxyvitamin D_3 are controlled primarily by reflexes that regulate plasma calcium concentration. Nonetheless, changes in the other listed hormones have important influences on bone mass and plasma calcium concentration.

Kidneys

About 60 percent of plasma calcium is filterable at the renal corpuscle (the rest is bound to plasma protein), and most of this filtered calcium is reabsorbed. There is no tubular secretion of calcium. Accordingly, the urinary excretion of calcium is the difference between the amount filtered and the amount reabsorbed. Like that of sodium, the control of calcium excretion is exerted mainly on reabsorption; that is, reabsorption is reflexly decreased when plasma calcium concentration goes up for whatever reason, and reflexly increased when plasma calcium goes down.

In addition, as we shall see, the renal handling of phosphate plays a role in the regulation of extracellular calcium. Phosphate, too, is handled by a combination of filtration and reabsorption, the latter being hormonally controlled.

Gastrointestinal Tract

The absorption of sodium, water, and potassium from the gastrointestinal tract normally approximates 100 percent. There is some homeostatic control of these processes, but it is relatively unimportant and so we ignored it. In contrast, a considerable amount of ingested calcium is not absorbed from the intestine and simply leaves the body along with the feces. Moreover, the active transport system that achieves calcium absorption is under important hormonal control. Accordingly, there can be large regulated increases or decreases in the amount of calcium absorbed. Indeed, hormonal control of this absorptive process is the major means for homeostatically regulating total-body calcium balance, more important than the control of renal calcium excretion.

Hormonal Controls

The two major hormones that homeostatically regulate plasma calcium concentration are parathyroid hormone and 1,25-dihydroxyvitamin D_3. A third hormone, calcitonin, plays a more limited role.

Parathyroid Hormone

All three of the effector sites described previously—bone, kidneys, and gastrointestinal tract—are subject, directly or indirectly, to control by a protein hormone called **parathyroid hormone,** produced by the parathyroid glands. These glands are in the neck, embedded in the surface of the thyroid gland, but are distinct from it. Parathyroid hormone production is controlled by the extracellular calcium concentration acting directly on the secretory cells (via a plasma-membrane calcium receptor). *Decreased* plasma calcium concentration *stimulates* parathyroid hormone secretion, and an increased plasma calcium concentration does just the opposite.

Parathyroid hormone exerts multiple actions that increase extracellular calcium concentration, thus compensating for the decreased concentration that originally stimulated secretion of this hormone (Figure 16–28).

1. It directly increases the resorption of bone by osteoclasts, which results in the movement of calcium (and phosphate) from bone into extracellular fluid.

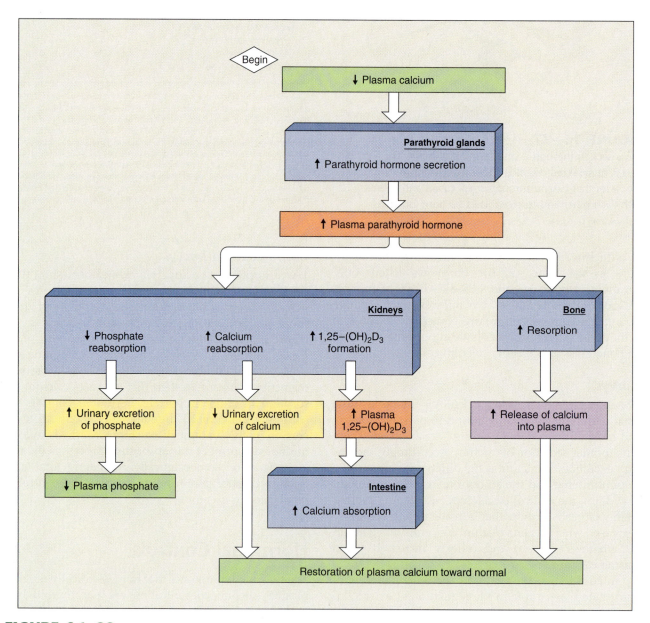

FIGURE 16–28

Reflexes by which a reduction in plasma calcium concentration is restored toward normal via the actions of parathyroid hormone. See Figure 16–29 for a more complete description of 1,25-$(OH)_2D_3$.

2. It directly stimulates the formation of 1,25-dihydroxyvitamin D₃, (discussed below), and this latter hormone then increases intestinal absorption of calcium. Thus, the effect of parathyroid hormone on the intestinal tract is an indirect one.

3. It directly increases renal tubular calcium reabsorption, thus decreasing urinary calcium excretion.

In addition, parathyroid hormone directly reduces the tubular reabsorption of phosphate, thus raising its urinary excretion. This keeps plasma phosphate from increasing at a time when parathyroid hormone is simultaneously causing increased release of both calcium and phosphate from bone.

1,25-Dihydroxyvitamin D₃

The term **vitamin D** denotes a group of closely related compounds. One of these, called **vitamin D₃** is formed by the action of ultraviolet radiation (from sunlight, usually) on a cholesterol derivative (7-dehydrocholesterol) in skin. Another form of vitamin D very similar to vitamin D₃ is ingested in food, specifically from plants. (Both forms can be found in vitamin pills and foods enriched with vitamin D.)

Because of clothing and decreased outdoor living, people are often dependent upon dietary vitamin D, and for this reason it was originally classified as a vitamin. However, regardless of source, vitamin D₃ and its similar ingested form are metabolized by addition of hydroxyl groups, first in the liver and then in certain kidney tubular cells (Figure 16–29). The end result of these changes is **1,25-dihydroxyvitamin D₃** (abbreviated **1,25-(OH)₂D₃,** also called calcitriol), the active form of vitamin D. It should be clear from this description that since 1,25-(OH)₂D₃ is made in the body, it is not, itself, a vitamin: instead, it fulfills the criteria for a hormone.

The major action of 1,25-(OH)₂D₃ is to stimulate absorption of calcium by the intestine. Thus, the major event in vitamin D deficiency is decreased intestinal calcium absorption, resulting in decreased plasma calcium.

The blood concentration of 1,25-(OH)₂D₃ is subject to physiological control. The major control point is the second hydroxylation step, the one that occurs in the kidneys. The enzyme catalyzing this step is stimulated by parathyroid hormone. Thus, as we have seen, a low plasma calcium concentration stimulates the secretion of parathyroid hormone, which in turn enhances the production of 1,25-(OH)₂D₃, and both hormones contribute to restoration of the plasma calcium toward normal.

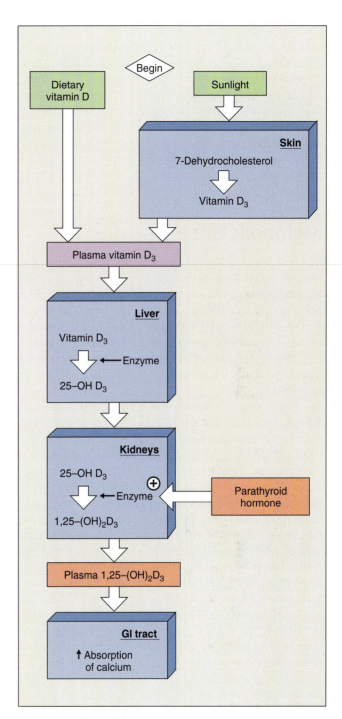

FIGURE 16–29

Metabolism of vitamin D to the active form, 1,25-(OH)₂D₃. The kidney enzyme that mediates the final step is activated by parathyroid hormone.

Calcitonin

Calcitonin is a peptide hormone secreted by cells (termed parafollicular cells) that are within the thyroid gland but are distinct from the thyroid follicles.

Calcitonin decreases plasma calcium concentration, mainly by inhibiting osteoclasts, thereby reducing bone resorption. Its secretion is stimulated by an increased plasma calcium concentration, just the opposite of the stimulus for parathyroid hormone secretion. Unlike parathyroid hormone and $1,25\text{-}(OH)_2D_3$, however, calcitonin plays little role in the normal day-to-day regulation of plasma calcium regulation, but is involved primarily in protecting the skeleton from excessive resorption during periods of "calcium stress" such as growth, pregnancy, and lactation.

Metabolic Bone Diseases

Various diseases reflect abnormalities in the metabolism of bone. *Rickets* (in children) and *osteomalacia* (in adults) are conditions in which mineralization of bone matrix is deficient, causing the bones to be soft and easily fractured. In addition, a child suffering from rickets typically is severely bowlegged due to the effect of weight-bearing on the developing leg bones. A major cause of rickets and osteomalacia is deficiency of $1,25\text{-}(OH)_2D_3$.

In contrast to these diseases, in *osteoporosis* both matrix and minerals are lost as a result of an imbalance between bone resorption and bone formation. The resulting decrease in bone mass and strength leads to an increased incidence of fractures. Osteoporosis can occur in people who are immobilized (disuse osteoporosis), in people who have an excessive plasma concentration of a hormone that favors bone resorption, and in people who have a deficient plasma concentration of a hormone that favors bone formation (Table 16–6). It is most commonly seen, however, with aging. Everyone loses bone as he or she ages, but osteoporosis is much more common in elderly women than men for several reasons: Women have a smaller bone mass to begin with, and the loss that occurs with aging occurs more rapidly, particularly after menopause removes the bone-promoting influence of estrogen.

Prevention is the focus of attention for osteoporosis. Estrogen treatment in postmenopausal women is very effective in reducing the rate of bone loss. A regular weight-bearing exercise program (brisk walking and stair-climbing, for example) is also helpful. Adequate dietary calcium (1000 mg/day before menopause and 1200–1500 mg/day after menopause) throughout life is important to build up and maintain bone mass. Several agents also provide effective therapy once osteoporosis is established. Most prominent is a group of drugs, called *bisphosphonates*, that interfere with the resorption of bone by osteoclasts. Other therapeutic agents include the hormones calcitonin, $1,25\text{-}(OH)_2D_3$, and estrogen, as well as sodium fluoride, which stimulates osteoblasts to form bone.

SECTION C SUMMARY

Effector Sites for Calcium Homeostasis

I. The effector sites for the regulation of plasma calcium concentration are bone, the gastrointestinal tract, and the kidneys.

II. Approximately 99 percent of total-body calcium is contained in bone as minerals on a collagen matrix. Bone is constantly remodeled as a result of the interaction of osteoblasts and osteoclasts, a process that determines bone mass and provides a means for raising or lowering plasma calcium concentration.

III. Calcium is actively absorbed by the gastrointestinal tract, and this process is under hormonal control.

IV. The amount of calcium excreted in the urine is the difference between the amount filtered and the amount reabsorbed, the latter process being under hormonal control.

Hormonal Controls

I. Parathyroid hormone increases plasma calcium concentration by influencing all the effector sites.
   a. It stimulates tubular reabsorption of calcium, bone resorption with release of calcium, and formation of the hormone 1,25-dihydroxyvitamin D_3, which stimulates calcium absorption by the intestine.
   b. It also inhibits the tubular reabsorption of phosphate.

II. Vitamin D_3 is formed in the skin or ingested and then undergoes hydroxylations in the liver and kidneys, in the latter stimulated by parathyroid hormone, to the active form, 1,25-dihydroxyvitamin D_3.

SECTION C KEY TERMS
---

osteoid	vitamin D
osteoblast	vitamin D_3
osteocyte	1,25-dihydroxyvitamin D_3
osteoclast	[$1,25\text{-}(OH)_2D_3$]
parathyroid hormone	calcitonin

SECTION C REVIEW QUESTIONS
---

1. List the functions of bone.
2. Describe bone remodeling.
3. Describe the handling of calcium by the kidneys and gastrointestinal tract.
4. What controls the secretion of parathyroid hormone, and what are this hormone's four major effects?
5. Describe the formation and action of $1,25\text{-}(OH)_2D_3$. How does parathyroid hormone influence the production of this hormone?

HYDROGEN-ION REGULATION

Metabolic reactions are highly sensitive to the hydrogen-ion concentration of the fluid in which they occur. This sensitivity is due to the influence on enzyme function exerted by hydrogen ions, which change the shapes of proteins. Accordingly, the hydrogen-ion concentration of the extracellular fluid is closely regulated. (At this point the reader might want to review the section on hydrogen ions, acidity, and pH in Chapter 2.)

This regulation can be viewed in the same way as the balance of any other ion—that is, as the matching of gains and losses. When loss exceeds gain, the arterial plasma hydrogen-ion concentration goes down (pH goes above 7.4), and this is termed an *alkalosis.* When gain exceeds loss, the arterial plasma hydrogen-ion concentration goes up (pH goes below 7.4), and this is termed an *acidosis.*

Sources of Hydrogen-Ion Gain or Loss

Table 16–7 summarizes the major routes for gains and losses of hydrogen ion. First, as described in Chapter 15, a huge quantity of CO_2—about 20,000 mmol—is generated daily as the result of oxidative metabolism, and these CO_2 molecules participate in the generation of hydrogen ions during passage of blood through peripheral tissues via the reactions:

$$CO_2 + H_2O \underset{}{\overset{\text{carbonic anhydrase}}{\rightleftharpoons}} H_2CO_3 \rightleftharpoons HCO_3^- + H^+$$
$$(16\text{-}1)$$

This source does not normally constitute a net gain of hydrogen ions, however, since all the hydrogen ions generated via these reactions are reincorporated into water when the reactions are reversed during passage of blood through the lungs (Chapter 15). Net retention of CO_2 does occur, however, in hypoventilation or respiratory disease and causes a net gain of hydrogen ions. Conversely, net loss of CO_2 occurs in hyperventilation, and this causes net elimination of hydrogen ions.

The body also produces acids, both organic and inorganic, from sources other than CO_2. These are collectively termed **nonvolatile acids.** They include phosphoric acid and sulfuric acid, generated mainly by the catabolism of proteins, as well as lactic acid and several other organic acids. Dissociation of all these acids yields anions and hydrogen ions. But simultaneously

TABLE 16–7 Sources of Hydrogen-Ion Gain or Loss
Gain
1. Generation of hydrogen ions from CO_2
2. Production of nonvolatile acids from the metabolism of protein and other organic molecules
3. Gain of hydrogen ions due to loss of bicarbonate in diarrhea or other nongastric GI fluids
4. Gain of hydrogen ions due to loss of bicarbonate in the urine
Loss
1. Utilization of hydrogen ions in the metabolism of various organic anions
2. Loss of hydrogen ions in vomitus
3. Loss of hydrogen ions in the urine
4. Hyperventilation

the metabolism of a variety of organic anions utilizes hydrogen ions and produces bicarbonate. Thus, metabolism of "nonvolatile" solutes both generates and utilizes hydrogen ions. In the United States, where the diet is high in protein, the generation of nonvolatile acids predominates in most people and there is an average net production of 40 to 80 mmol of hydrogen ions per day.

A third potential source of net body gain or loss of hydrogen ion is gastrointestinal secretions leaving the body. Vomitus contains a high concentration of hydrogen ions and so constitutes a source of net loss. In contrast, the other gastrointestinal secretions are alkaline; they contain very little hydrogen ion, but their concentration of bicarbonate is higher than exists in plasma. Loss of these fluids, as in diarrhea, constitutes in essence a body *gain* of hydrogen ions. This is an extremely important point: Given the mass action relationship shown in Equation 16-1, when a bicarbonate ion is lost from the body it is the same as if the body had gained a hydrogen ion. The reason is that loss of the bicarbonate causes the reactions shown in Equation 16-1 to be driven to the right, thereby generating a hydrogen ion within the body. Similarly, when the body gains a bicarbonate ion, it is the same as if the body had lost a hydrogen ion, as the reactions of Equation 16-1 are driven to the left.

Finally, the kidneys constitute the fourth source of net hydrogen-ion gain or loss; that is, the kidneys can either remove hydrogen ions from the plasma or add them.

Buffering of Hydrogen Ions in the Body

Any substance that can reversibly bind hydrogen ions is called a **buffer.** Between their generation in the body and their elimination, most hydrogen ions are buffered by extracellular and intracellular buffers. The normal extracellular-fluid pH of 7.4 corresponds to a hydrogen-ion concentration of only 0.00004 mmol/L (40 nanomols/L). Without buffering, the daily turnover of the 40 to 80 mmol of H^+ produced from nonvolatile acids generated in the body from metabolism would cause huge changes in body-fluid hydrogen-ion concentration.

The general form of buffering reactions is:

$$Buffer^- + H^+ \rightleftharpoons HBuffer \qquad (16\text{-}2)$$

HBuffer is a weak acid in that it can dissociate to $Buffer^-$ plus H^+ or it can exist as the undissociated molecule (HBuffer). When H^+ concentration increases for whatever reason, the reaction is forced to the right, and more H^+ is bound by $Buffer^-$ to form HBuffer. For example, when H^+ concentration is increased because of increased production of lactic acid, some of the hydrogen ions combine with the body's buffers, so the hydrogen-ion concentration does not increase as much as it otherwise would have. Conversely, when H^+ concentration decreases because of the loss of hydrogen ions or the addition of alkali, Equation 16-2 proceeds to the left and H^+ is released from HBuffer. In this manner, the body buffers stabilize H^+ concentration against changes in either direction.

The major extracellular buffer is the CO_2/HCO_3^- system summarized in Equation 16-1. This system also plays some role in buffering within cells, but the major intracellular buffers are phosphates and proteins. One intracellular protein buffer is hemoglobin, as described in Chapter 15.

You must recognize that buffering does not eliminate hydrogen ions from the body or add them to the body; it only keeps them "locked-up" until balance can be restored. How balance is achieved is the subject of the rest of our description of hydrogen-ion regulation.

Integration of Homeostatic Controls

The kidneys are ultimately responsible for balancing hydrogen-ion gains and losses so as to maintain a relatively constant plasma hydrogen-ion concentration.

Thus, the kidneys normally excrete the excess hydrogen ions from nonvolatile acids generated in the body from metabolism—that is, all acids other than carbonic acid. Moreover, if there is an additional net gain of hydrogen ions due to abnormally increased production of these nonvolatile acids, or to hypoventilation or respiratory malfunction, or to loss of alkaline gastrointestinal secretions, the kidneys increase their elimination of hydrogen ions from the body so as to restore balance. Alternatively, if there is a net loss of hydrogen ions from the body due to increased metabolic utilization of hydrogen ions (as in a vegetarian diet), hyperventilation, or vomiting, the kidneys replenish these hydrogen ions.

Although the kidneys are the ultimate hydrogen-ion balancers, the respiratory system also plays a very important homeostatic role. We have pointed out that hypoventilation, respiratory malfunction, and hyperventilation can *cause* a hydrogen-ion imbalance; now we emphasize that when a hydrogen-ion imbalance is due to a nonrespiratory cause, then ventilation is *reflexly altered* so as to help *compensate* for the imbalance. We described this phenomenon in Chapter 15 (see Figure 15-34): An elevated arterial hydrogen-ion concentration stimulates ventilation, and this reflex hyperventilation causes reduced arterial P_{CO_2}, and hence, by mass action, reduced hydrogen-ion concentration. Alternatively, a decreased plasma hydrogen-ion concentration inhibits ventilation, thereby raising arterial P_{CO_2} and increasing the hydrogen-ion concentration.

Thus, the respiratory system and kidneys work together. The respiratory response to altered plasma hydrogen-ion concentration is very rapid (minutes) and keeps this concentration from changing too much until the more slowly responding kidneys (hours to days) can actually eliminate the imbalance. Of course, if the respiratory system is the actual *cause* of the hydrogen-ion imbalance, then the kidneys are the sole homeostatic responder. By the same token, malfunctioning kidneys can create a hydrogen-ion imbalance by eliminating too little or too much hydrogen ion from the body, and then the respiratory response is the only one operating.

Renal Mechanisms

In the previous section we wrote of the kidneys eliminating hydrogen ions from the body or replenishing them. The kidneys perform this task by altering plasma *bicarbonate* concentration. The key to understanding how altering plasma bicarbonate concentration eliminates or replenishes hydrogen ions was stated earlier: The excretion of a bicarbonate in the urine increases the plasma hydrogen-ion concentration just as if a hydrogen ion had been added to the plasma. Similarly, the addition of a bicarbonate to the plasma lowers the plasma hydrogen-ion concentration just as if a hydrogen ion had been removed from the plasma.

Thus, when there is a lowering of plasma hydrogen-ion concentration (alkalosis) for whatever reason, the kidneys' homeostatic response is to excrete large quantities of bicarbonate. This raises plasma hydrogen-ion concentration back toward normal. In contrast, in response to a rise in plasma hydrogen-ion concentration (acidosis), the kidneys do not excrete bicarbonate in the urine, but instead kidney tubular cells produce *new* bicarbonate and add it to the plasma. This lowers the plasma hydrogen-ion concentration back toward normal.

Let us now look at the basic mechanisms by which bicarbonate excretion or addition of new bicarbonate to the plasma is achieved.

Bicarbonate Handling

Bicarbonate is completely filterable at the renal corpuscles and undergoes marked tubular reabsorption in various tubular segments (the proximal tubule, ascending loop of Henle, and cortical collecting ducts). Bicarbonate can also be secreted (in the collecting ducts). Therefore:

HCO_3^- excretion =
$\quad HCO_3^-$ filtered + HCO_3^- secreted − HCO_3^- reabsorbed

For simplicity, we will ignore the secretion of bicarbonate (because it is always quantitatively much less than tubular reabsorption) and treat bicarbonate excretion as the difference between filtration and reabsorption.

Bicarbonate reabsorption is an active process, but it is not accomplished in the conventional manner of simply having an active pump for bicarbonate ions at the luminal or basolateral membrane of the tubular cells. Instead, bicarbonate reabsorption is absolutely dependent upon the tubular secretion of hydrogen ions, which combine in the lumen with filtered bicarbonates.

Figure 16–30 illustrates the sequence of events. Start this figure inside the cell with the combination of CO_2 and H_2O to form H_2CO_3, a reaction catalyzed by the enzyme carbonic anhydrase. The H_2CO_3 immediately dissociates to yield H^+ and bicarbonate (HCO_3^-). The HCO_3^- moves down its concentration gradient across the basolateral membrane into interstitial fluid and then into the blood. Simultaneously the H^+ is secreted into the lumen; depending on the tubular segment, this secretion is achieved by some combination of primary H-ATPase pumps, primary H,K-ATPase pumps, and Na/H countertransporters.

But the secreted H^+ *is not excreted.* Instead, it combines in the lumen with a *filtered* HCO_3^- and generates

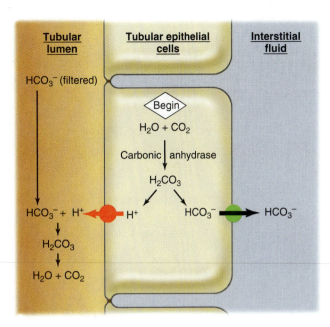

FIGURE 16–30

Reabsorption of bicarbonate. Start this figure inside the cell, with the combination of CO_2 and H_2O to form H_2CO_3. As shown in this figure, active H-ATPase pumps are involved in the movement of H^+ out of the cell across the luminal membrane; in several tubular segments, this transport step is also mediated by Na/H countertransporters and/or H,K-ATPase pumps.

CO_2 and H_2O (both of which can diffuse into the cell and be used for another cycle of hydrogen-ion generation). The overall result is that the bicarbonate filtered from the plasma at the renal corpuscle has disappeared, but its place in the plasma has been taken by the bicarbonate that was produced inside the cell, and so no net change in plasma bicarbonate concentration has occurred. It may seem inaccurate to refer to this process as bicarbonate "reabsorption," since the bicarbonate that appears in the peritubular plasma is not the same bicarbonate ion that was filtered. Yet the overall result is, in effect, the same as if the filtered bicarbonate had been more conventionally reabsorbed like a sodium or potassium ion.

Except in response to alkalosis (discussed below), the kidneys normally reabsorb all filtered bicarbonate, thereby preventing the loss of bicarbonate in the urine.

Addition of New Bicarbonate to the Plasma

It is essential to realize in Figure 16–30 that as long as there are still significant amounts of filtered bicarbonate ions in the lumen, almost all secreted hydrogen ions will combine with them. But what happens to any secreted hydrogen ions once almost all the bicarbonate has been reabsorbed and is no longer available in the lumen to combine with the hydrogen ions?

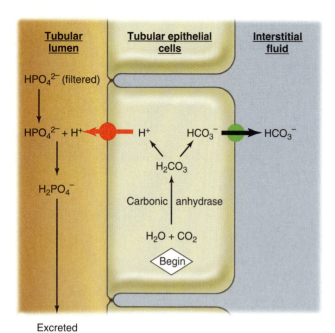

FIGURE 16–31

Renal contribution of new HCO_3^- to the plasma as achieved by tubular secretion of H^+. The process of intracellular H^+ and HCO_3^- generation, with H^+ movement into the lumen and HCO_3^- into the plasma, is identical to that shown in Figure 16–30. Once in the lumen, however, the H^+ combines with filtered phosphate (HPO_4^{2-}) rather than filtered HCO_3^- and is excreted. As described in the legend for Figure 16–30, the transport of hydrogen ions into the lumen is accomplished not only by H-ATPase pumps but, in several tubular segments, by Na/H countertransporters and/or H,K-ATPase pumps as well.

The answer, illustrated in Figure 16–31, is that the extra secreted hydrogen ions combine in the lumen with a filtered *nonbicarbonate* buffer, usually HPO_4^{2-}. (Other filtered buffers can also participate, but HPO_4^{2-} is the most important.) The hydrogen ion is then excreted in the urine as part of an $H_2PO_4^-$ ion. Now for the critical point: Note in Figure 16–31 that, under these conditions, the bicarbonate generated within the tubular cell by the carbonic anhydrase reaction and entering the plasma constitutes a *net gain* of bicarbonate by the plasma, not merely a replacement for a filtered bicarbonate. Thus, when a secreted hydrogen ion combines in the lumen with a buffer other than bicarbonate, the overall effect is not merely one of bicarbonate conservation, as in Figure 16–30, but rather of addition to the plasma of a *new* bicarbonate. This raises the bicarbonate concentration of the plasma and alkalinizes it.

To repeat, significant numbers of hydrogen ions combine with filtered nonbicarbonate buffers like HPO_4^{2-} only after the filtered bicarbonate has virtually

all been reabsorbed. The main reason is that there is such a large load of filtered bicarbonate buffers—25 times more than the load of filtered nonbicarbonate buffers—competing for the secreted hydrogen ions.

There is a second mechanism by which the tubules contribute new bicarbonate to the plasma, one that involves not hydrogen-ion secretion but rather the renal production and secretion of ammonium (NH_4^+) (Figure 16–32). Tubular cells, mainly those of the proximal tubule, take up glutamine from both the glomerular filtrate and peritubular plasma and, by a series of steps, metabolize it. In the process, both NH_4^+ and bicarbonate are formed inside the cells. The NH_4^+ is actively secreted (via Na^+/NH_4^+ countertransport) into the lumen and excreted, while the bicarbonate moves into the peritubular capillaries and constitutes new plasma bicarbonate.

A comparison of Figures 16–31 and 16–32 demonstrates that the overall result—renal contribution of new bicarbonate to the plasma—is the same regardless of whether it is achieved by: (1) H^+ secretion and excretion on nonbicarbonate buffers such as phosphate (Figure 16–31); or (2) by glutamine metabolism with NH_4^+ excretion (Figure 16–32). It is convenient, therefore, to view the latter case as representing H^+ excretion "bound" to NH_3, just as the former case constitutes H^+ excretion bound to nonbicarbonate buffers. Thus, the amount of H^+ excreted in the urine in these

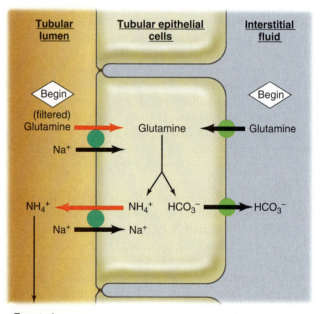

FIGURE 16–32

Renal contribution of new HCO_3^- to the plasma as achieved by renal metabolism of glutamine and excretion of ammonium (NH_4^+). Compare this figure to Figure 16–31. This process occurs mainly in the proximal tubule.

two forms is a measure of the amount of new bicarbonate added to the plasma by the kidneys. Indeed, "urinary H^+ excretion" and "renal contribution of new bicarbonate to the plasma" are really two sides of the same coin and are synonymous phrases.

The kidneys normally contribute enough new bicarbonate to the blood (excrete enough hydrogen ions) to compensate for the hydrogen ions from nonvolatile acids generated in the body.

One last point needs to be emphasized: The last paragraphs summarize the two forms in which H^+ is excreted in the urine, but to be completely accurate there is also a third form—as free H^+. However, the amount of free H^+ is always so small that it can be ignored. For example, even the most acid of urines (4.4 is the lowest pH achievable by the tubules) contains less than 0.1 mmol of free H^+, compared to several hundred mmols of H^+ bound up in nonbicarbonate buffers and NH_4^+. This emphasizes how important these two sources are in achieving the excretion of H^+.

Renal Responses to Acidosis and Alkalosis

We can now apply this material to the renal responses to the presence of an acidosis or alkalosis. These are summarized in Table 16–8.

Clearly, these homeostatic responses require that the rates of hydrogen-ion secretion, glutamine metabolism, and ammonium excretion be subject to physiological control by changes in blood hydrogen-ion concentration. The specific pathways and mechanisms that bring about these rate changes are very complex, however, and are not presented here.

Classification of Acidosis and Alkalosis

To repeat, acidosis refers to any situation in which the hydrogen-ion concentration of arterial plasma is elevated; alkalosis denotes a reduction. All such situations fit into two distinct categories (Table 16–9): (1) *respiratory acidosis* or *alkalosis*; (2) *metabolic acidosis* or *alkalosis.*

As its name implies, respiratory acidosis results from altered respiration. Respiratory acidosis occurs when the respiratory system fails to eliminate carbon dioxide as fast as it is produced. Respiratory alkalosis occurs when the respiratory system eliminates carbon dioxide faster than it is produced. As described earlier, the imbalance of arterial hydrogen-ion concentrations in such cases is completely explainable in terms of mass action. Thus, the hallmark of a respiratory acidosis is an elevation in both arterial P_{CO_2} and hydrogen-ion concentration; that of respiratory alkalosis is a reduction in both.

TABLE 16–8 Renal Responses to Acidosis and Alkalosis

Responses to Acidosis

1. Sufficient hydrogen ions are secreted to reabsorb all the filtered bicarbonate.

2. Still more hydrogen ions are secreted, and this contributes new bicarbonate to the plasma as these hydrogen ions are excreted bound to nonbicarbonate urinary buffers such as HPO_4^{2-}.

3. Tubular glutamine metabolism and ammonium excretion are enhanced, which also contributes new bicarbonate to the plasma.

Net result: More new bicarbonate ions than usual are added to the blood, and plasma bicarbonate is increased, thereby compensating for the acidosis. The urine is highly acidic (lowest attainable pH = 4.4).

Responses to Alkalosis

1. Rate of hydrogen-ion secretion is inadequate to reabsorb all the filtered bicarbonate, so that significant amounts of bicarbonate are excreted in the urine, and there is little or no excretion of hydrogen ions on nonbicarbonate urinary buffers.

2. Tubular glutamine metabolism and ammonium excretion are decreased so that little or no new bicarbonate is contributed to the plasma from this source.

Net result: Plasma bicarbonate concentration is decreased, thereby compensating for the alkalosis. The urine is alkaline (pH > 7.4).

Metabolic acidosis or alkalosis includes all situations other than those in which the primary problem is respiratory. Some common causes of metabolic acidosis are excessive production of lactic acid (during severe exercise or hypoxia) or of ketone bodies (in uncontrolled diabetes mellitus or fasting, as described in Chapter 18). Metabolic acidosis can also result from excessive loss of bicarbonate, as in diarrhea. A frequent cause of metabolic alkalosis is persistent vomiting, with its associated loss of hydrogen ions as HCl from the stomach.

What is the arterial P_{CO_2} in metabolic acidosis or alkalosis? Since, by definition, metabolic acidosis and alkalosis must be due to something other than excess retention or loss of carbon dioxide, you might have predicted that arterial P_{CO_2} would be unchanged, but such is not the case. As emphasized earlier in this chapter, the elevated hydrogen-ion concentration associated with metabolic acidosis *reflexly* stimulates ventilation and lowers arterial P_{CO_2}. By mass action, this helps restore the hydrogen-ion concentration toward normal. Conversely, a person with metabolic alkalosis will reflexly have ventilation inhibited. The result is a rise in arterial P_{CO_2} and, by mass action, an associated restoration of hydrogen-ion concentration toward normal.

TABLE 16-9 Changes in the Arterial Concentrations of Hydrogen Ion, Bicarbonate, and Carbon Dioxide in Acid-Base Disorders

Primary Disorder	H^+	HCO_3^-	CO_2	Cause of HCO_3^- Change	Cause of CO_2 Change
Respiratory acidosis	↑	↑	↑	Renal compensation	Primary abnormality
Respiratory alkalosis	↓	↓	↓		
Metabolic acidosis	↑	↓	↓	Primary abnormality	Reflex ventilatory compensation
Metabolic alkalosis	↓	↑	↑		

To reiterate, the plasma P_{CO_2} changes in metabolic acidosis and alkalosis are not the *cause* of the acidosis or alkalosis but are the *result* of compensatory reflex responses to nonrespiratory abnormalities. Thus, in metabolic, as opposed to respiratory conditions, the arterial plasma P_{CO_2} and hydrogen-ion concentration go in opposite directions, as summarized in Table 16–9.

SECTION D SUMMARY

Sources of Hydrogen-Ion Gain or Loss

I. Total-body balance of hydrogen ions is the result of both metabolic production of these ions and of net gains or losses via the respiratory system, gastrointestinal tract, and urine (Table 16–7).

II. A stable balance is achieved by regulation of urinary losses.

Buffering of Hydrogen Ions in the Body

I. Buffering is a means of minimizing changes in hydrogen-ion concentration by combining these ions reversibly with anions such as bicarbonate and intracellular proteins.

II. The major extracellular buffering system is the CO_2/HCO_3^- system, and the major intracellular buffers are proteins and phosphates.

Integration of Homeostatic Controls

I. The kidneys and the respiratory system are the homeostatic regulators of plasma hydrogen-ion concentration.

II. The kidneys are the organs that achieve body hydrogen-ion balance.

III. A decrease in arterial plasma hydrogen-ion concentration causes reflex hypoventilation, which raises arterial P_{CO_2} and, hence, raises plasma hydrogen-ion concentration toward normal. An increase in plasma hydrogen-ion concentration causes reflex hyperventilation, which lowers arterial P_{CO_2} and, hence, lowers hydrogen-ion concentration toward normal.

Renal Mechanisms

I. The kidneys maintain a stable plasma hydrogen-ion concentration by regulating plasma bicarbonate concentration. They can either excrete bicarbonate or contribute new bicarbonate to the blood.

II. Bicarbonate is reabsorbed when hydrogen ions, generated in the tubular cells by a process catalyzed by carbonic anhydrase, are secreted into the lumen and combine with filtered bicarbonate. The secreted hydrogen ions are not excreted in this situation.

III. In contrast, when the secreted hydrogen ions combine in the lumen with filtered phosphate or other nonbicarbonate buffer, they are excreted, and the kidneys have contributed new bicarbonate to the blood.

IV. The kidneys also contribute new bicarbonate to the blood when they produce and excrete ammonium.

Classification of Acidosis and Alkalosis

I. Acid-base disorders are categorized as respiratory or metabolic.
   a. Respiratory acidosis is due to retention of carbon dioxide, and respiratory alkalosis to excessive elimination of carbon dioxide.
   b. All other causes of acidosis or alkalosis are termed metabolic and reflect gain or loss, respectively, of hydrogen ions from a source other than carbon dioxide.

SECTION D KEY TERMS

nonvolatile acids   buffer

SECTION D REVIEW QUESTIONS

1. What are the sources of gain and loss of hydrogen ions in the body?
2. List the body's major buffer systems.
3. Describe the role of the respiratory system in the regulation of hydrogen-ion concentration.
4. How does the tubular secretion of hydrogen ions occur, and how does it achieve bicarbonate reabsorption?
5. How does hydrogen-ion secretion contribute to the renal addition of new bicarbonate to the blood? What determines whether a secreted hydrogen ion will achieve these events or will instead cause bicarbonate reabsorption?

6. How does the metabolism of glutamine by the tubular cells contribute new bicarbonate to the blood and ammonium to the urine?
7. What two quantities make up "hydrogen-ion excretion?" Why can this term be equated with "contribution of new bicarbonate to the plasma?"
8. How do the kidneys respond to the presence of an acidosis or alkalosis?
9. Classify the four types of acid-base disorders according to plasma hydrogen-ion concentration, bicarbonate concentration, and P_{CO_2}.

DIURETICS AND KIDNEY DISEASE

Diuretics

Drugs used clinically to increase the volume of urine excreted are known as *diuretics.* Such agents act on the tubules to inhibit the reabsorption of sodium, along with chloride and/or bicarbonate, resulting in increased excretion of these ions. Since water reabsorption is dependent upon sodium reabsorption, water reabsorption is also reduced, resulting in increased water excretion.

A large variety of clinically useful diuretics are available and are classified according to the specific mechanisms by which they inhibit sodium reabsorption. For example, one type, called loop diuretics, acts on the ascending limb of the loop of Henle to inhibit the transport protein that mediates the first step in sodium reabsorption in this segment—cotransport of sodium and chloride (and potassium) into the cell across the luminal membrane.

Except for one category of diuretics, called *potassium-sparing diuretics,* all diuretics not only increase sodium excretion but also cause increased potassium excretion, an unwanted side effect. By several mechanisms, the potassium-sparing diuretics inhibit sodium reabsorption in the cortical collecting duct, and they simultaneously inhibit potassium secretion there. This explains why they do not cause increased potassium excretion.

Diuretics are among the most commonly used medications. For one thing, they are used to treat diseases characterized by renal retention of salt and water. As emphasized earlier in this chapter, in normal persons the regulation of blood pressure simultaneously produces stability of total-body sodium mass and extracellular volume because there is a close correlation between these variables. In contrast, in several types of disease, this correlation is broken and the reflexes that maintain blood pressure can cause renal retention of sodium. Sodium excretion may fall virtually to zero despite continued sodium ingestion, leading to abnormal expansion of the extracellular fluid and formation of *edema.* Diuretics are used to prevent or reverse this renal retention of sodium and water.

The most common example of this phenomenon is *congestive heart failure* (Chapter 14). A person with a failing heart manifests (1) a decreased GFR and (2) increased aldosterone secretion, both of which, along with other important factors, contribute to the virtual absence of sodium from the urine. The net result is extracellular volume expansion and formation of edema. The sodium-retaining responses are triggered by the lower cardiac output (a result of cardiac failure) and the decrease in arterial blood pressure that results directly from this decrease in cardiac output.

Another disease in which diuretics are frequently employed is hypertension (Chapter 14). The decrease in body sodium and water resulting from the diuretic-induced excretion of these substances brings about arteriolar dilation and a lowering of the blood pressure; why decreased body sodium causes arteriolar dilation is not known.

Kidney Disease

The term "kidney disease" is no more specific than "car trouble," since many diseases affect the kidneys. Bacteria, allergies, congenital defects, kidney stones, tumors, and toxic chemicals are some possible sources of kidney damage. Obstruction of the urethra or a ureter may cause injury as the result of a buildup of pressure and may predispose the kidneys to bacterial infection.

One frequent sign of kidney disease is the appearance of protein in the urine. In normal kidneys, there is a very tiny amount of protein in the glomerular filtrate because the corpuscular membranes are not completely impermeable to proteins, particularly those with lower molecular weights. However, the cells of the proximal tubule completely remove this filtered protein from the tubular lumen, and no protein appears in the final urine. In contrast, diseased renal corpuscles may become much more permeable to protein, and diseased proximal tubules may lose their ability to remove filtered protein from the tubular lumen. The result is that protein will appear in the urine.

Although many diseases of the kidney are self-limited and produce no permanent damage, others progress if untreated. The symptoms of profound renal malfunction are relatively independent of the damaging agent and are collectively known as *uremia,* literally, "urine in the blood."

The severity of uremia depends upon how well the impaired kidneys are able to preserve the constancy of the internal environment. Assuming that the person continues to ingest a normal diet containing the usual quantities of nutrients and electrolytes, what problems arise? The key fact to keep in mind is that the kidney destruction markedly reduces the number of functioning nephrons. Accordingly, the many substances, particularly potentially toxic waste products, that gain entry to the tubule by filtration build up in the blood. In addition, the excretion of potassium is impaired because there are too few nephrons capable of normal tubular *secretion* of this ion. The person may also develop acidosis because the reduced number of nephrons fail to add enough new bicarbonate to the blood to compensate for the daily metabolic production of nonvolatile acids.

The remarkable fact is how large the safety factor is in renal function. In general, the kidneys are still able to perform their regulatory function quite well as long as 10 percent of the nephrons are functioning. This is because these remaining nephrons undergo alterations in function—filtration, reabsorption, and secretion—so as to compensate for the missing nephrons. For example, each remaining nephron increases its rate of potassium secretion so that the total amount of potassium excreted by the kidneys can be maintained at normal levels. The limits of regulation are restricted, however. To use potassium as our example again, if someone with severe renal disease were to go on a diet high in potassium, the remaining nephrons might not be able to secrete enough potassium to prevent potassium retention.

Other problems arise in uremia because of abnormal secretion of the hormones produced by the kidneys. Thus, decreased secretion of erythropoietin results in anemia (Chapter 14). Decreased ability to form 1,25-$(OH)_2D_3$ results in deficient absorption of calcium from the gastrointestinal tract, with a resulting decrease in plasma calcium and inadequate bone calcification. Both of these hormones are now available for administration to patients with uremia.

The problem with renin, the third of the renal hormones, is rarely too little secretion but rather too much secretion by the juxtaglomerular cells of the damaged kidneys. The result is increased plasma angiotensin II concentration and the development of *renal hypertension.*

Hemodialysis, Peritoneal Dialysis, and Transplantation

As described above, failing kidneys reach a point when they can no longer excrete water and ions at rates that maintain body balances of these substances, nor can they excrete waste products as fast as they are produced. Dietary alterations can minimize these problems, for example, by lowering potassium intake and thereby reducing the amount of potassium to be excreted, but such alterations cannot eliminate the problems. The techniques used to perform the kidneys' excretory functions are hemodialysis and peritoneal dialysis. The general term "dialysis" means to separate substances using a membrane.

The artificial kidney is an apparatus that utilizes a process termed *hemodialysis* to remove excess substances from the blood. During hemodialysis, blood is pumped from one of the patient's arteries through tubing that is surrounded by special dialysis fluid. The tubing then conducts the blood back into the patient by way of a vein. The tubing is generally made of cellophane that is highly permeable to most solutes but relatively impermeable to protein and completely impermeable to blood cells—characteristics quite similar to those of capillaries. The dialysis fluid is a salt solution with ionic concentrations similar to or lower than those in normal plasma, and it contains no creatinine, urea, or other substances to be completely removed from the plasma. As blood flows through the tubing, the concentrations of nonprotein plasma solutes tend to reach diffusion equilibrium with those of the solutes in the bath fluid. For example, if the plasma potassium concentration of the patient is above normal, potassium diffuses out of the blood across the cellophane tubing and into the dialysis fluid. Similarly, waste products and excesses of other substances also diffuse into the dialysis fluid and thus are eliminated from the body.

Patients with acute reversible renal failure may require hemodialysis for only days or weeks. Patients with chronic irreversible renal failure require treatment for the rest of their lives, however, unless they receive a renal transplant. Such patients undergo hemodialysis several times a week, often at home.

Another way of removing excess substances from the blood is *peritoneal dialysis,* which uses the lining of the patient's own abdominal cavity (peritoneum) as a dialysis membrane. Fluid is injected, via a needle inserted through the abdominal wall, into this cavity and allowed to remain there for hours, during which solutes diffuse into the fluid from the person's blood. The dialysis fluid is then removed by reinserting the needle and is replaced with new fluid. This procedure can be performed several times daily by a patient who is simultaneously doing normal activities.

The treatment of choice for most patients with permanent renal failure is kidney transplantation. Rejection of the transplanted kidney by the recipient's body is a potential problem with transplants, but great strides have been made in reducing the frequency of rejection (Chapter 20). Many people who might benefit from a transplant, however, do not receive one. Presently, the major source of kidneys for transplanting is recently deceased persons, and improved public understanding should lead to many more individuals giving permission in advance to have their kidneys and other organs used following their death.

SECTION E SUMMARY

Diuretics and Kidney Disease

I. Diuretics inhibit reabsorption of sodium and water, thereby enhancing the excretion of these substances. Different diuretics act on different nephron segments.

II. Many of the symptoms of uremia—general renal malfunction—are due to retention of substances because of reduced GFR and, in the case of potassium and hydrogen ion, reduced secretion. Other symptoms are due to inadequate secretion of erythropoietin and 1,25-dihydroxyvitamin D_3, and too much secretion of renin.

III. Either hemodialysis or peritoneal dialysis can be used chronically to eliminate water, ions, and waste products retained during uremia.

CHAPTER 16 CLINICAL TERMS

glucosuria	respiratory alkalosis
familial renal glycosuria	metabolic acidosis
diabetes insipidus	metabolic alkalosis
arrhythmias	diuretics
hypocalcemic tetany	potassium-sparing diuretics
rickets	edema
osteomalacia	congestive heart failure
osteoporosis	uremia
bisphosphonates	renal hypertension
alkalosis	hemodialysis
acidosis	peritoneal dialysis
respiratory acidosis	

CHAPTER 16 THOUGHT QUESTIONS

(Answers are in Appendix A.)

1. Substance T is present in the urine. Does this prove that it is filterable at the glomerulus?

2. Substance V is not normally present in the urine. Does this prove that it is neither filtered nor secreted?

3. The concentration of glucose in plasma is 100 mg/100 ml, and the GFR is 125 ml/min. How much glucose is filtered per minute?

4. A person is found to be excreting abnormally large amounts of a particular amino acid. Just from the theoretical description of T_m-limited reabsorptive mechanisms in the text, list several possible causes.

5. The concentration of urea in urine is always much higher than the concentration in plasma. Does this mean that urea is secreted?

6. If a drug that blocks the reabsorption of sodium is taken, what will happen to the reabsorption of water, urea, chloride, glucose, and amino acids and to the secretion of hydrogen ions?

7. Compare the changes in GFR and renin secretion occurring in response to a moderate hemorrhage in two individuals—one taking a drug that blocks the sympathetic nerves to the kidneys and the other not taking such a drug.

8. If a person is taking a drug that completely inhibits angiotensin converting enzyme, what will happen to aldosterone secretion when the person goes on a low-sodium diet?

9. In the steady state, is the amount of sodium chloride excreted daily in the urine by a normal person ingesting 12 g of sodium chloride per day: (a) 12 g/day or (b) less than 12 g/day? Explain.

10. A young woman who has suffered a head injury seems to have recovered but is thirsty all the time. What do you think might be the cause?

11. A patient has a tumor in the adrenal cortex that continuously secretes large amounts of aldosterone. What effects does this have on the total amount of sodium and potassium in her body?

12. A person is taking a drug that inhibits the tubular secretion of hydrogen ions. What effect does this drug have on the body's balance of sodium, water, and hydrogen ion?

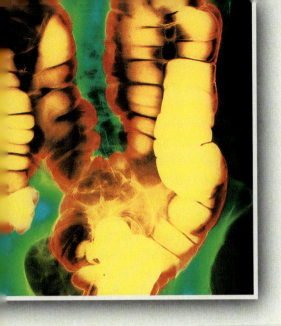

The Digestion and Absorption of Food

The **gastrointestinal (GI) system** (Figure 17–1) includes the **gastrointestinal tract** (mouth, pharynx, esophagus, stomach, small intestine, and large intestine) plus the accessory organs (salivary glands, liver, gallbladder, and pancreas) that are not part of the tract but secrete substances into it via connecting ducts. The overall function of the gastrointestinal system is to process ingested foods into molecular forms that are then transferred, along with salts and water to the body's internal environment, where they can be distributed to cells by the circulatory system.

The adult gastrointestinal tract is a tube approximately 15 ft long, running through the body from mouth to anus. The lumen of the tract, like the hole in a doughnut, is continuous with the external environment, which means that its contents are technically outside the body. This fact is relevant to understanding some of the tract's properties. For example, the large intestine is inhabited by billions of bacteria, most of which are harmless and even beneficial in this location. However, if the same bacteria enter the internal environment, as may happen, for example, in the case of a ruptured appendix, they may cause a severe infection.

Most food enters the gastrointestinal tract as large particles containing macromolecules, such as proteins and polysaccharides, which are unable to cross the intestinal epithelium. Before ingested food can be absorbed, therefore, it must be dissolved and broken down into small molecules. This dissolving and breaking-down process—**digestion**—is accomplished by the action of hydrochloric acid in the stomach, bile from the liver, and a variety of digestive enzymes that are released by the system's exocrine glands. Each of these substances is released into the lumen of the GI tract by the process of **secretion.**

The molecules produced by digestion then move from the lumen of the gastrointestinal tract across a layer of epithelial cells and enter the blood or lymph. This process is called **absorption.**

While digestion, secretion, and absorption are taking place, contractions of smooth muscles in the gastrointestinal tract wall serve two functions; they mix the luminal contents with the various secretions, and they move the contents through the tract from mouth to anus. These contractions are referred to as the **motility** of the gastrointestinal tract.

The functions of the gastrointestinal system can be described in terms of these four processes—digestion, secretion, absorption, and motility (Figure 17–2)—and the mechanisms controlling them.

The gastrointestinal system is designed to maximize absorption, and within fairly wide limits it will absorb as much of any particular substance as is ingested. With a few important exceptions (to be described later), therefore, the gastrointestinal system does *not regulate* the amount of nutrients absorbed or their concentrations in the internal environment. The regulation of the plasma concentration of the absorbed nutrients is primarily the function of the kidneys (Chapter 16) and a number of endocrine glands (Chapter 18).

Small amounts of certain metabolic end products are excreted via the gastrointestinal tract, primarily by way of the bile, but the elimination of most of the body's waste products is achieved by the lungs and kidneys. The material—**feces**—leaving the system at the end of the gastrointestinal tract consists almost entirely of bacteria and ingested material that was neither digested nor absorbed—that is, material that was never actually part of the internal environment.

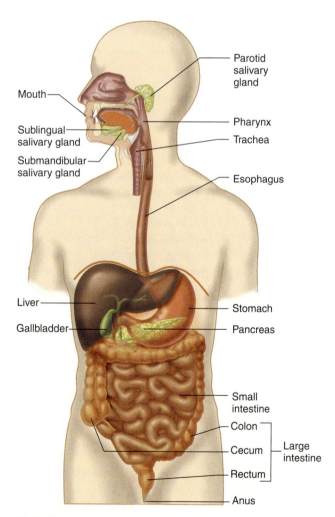

Mouth
Sublingual salivary gland
Submandibular salivary gland
Liver
Gallbladder

Parotid salivary gland
Pharynx
Trachea
Esophagus
Stomach
Pancreas
Small intestine
Colon
Cecum } Large intestine
Rectum
Anus

FIGURE 17–1

Anatomy of the gastrointestinal system. The liver overlies the gallbladder and a portion of the stomach, and the stomach overlies part of the pancreas.

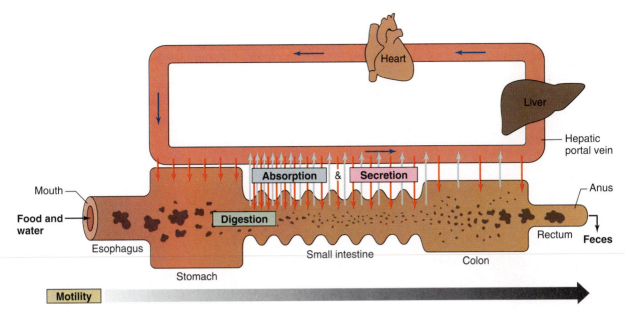

FIGURE 17–2

Four processes carried out by the gastrointestinal tract: digestion, secretion, absorption, and motility.

Overview: Functions of the Gastrointestinal Organs

Figure 17–3 presents an overview of the secretions and functions of the gastrointestinal organs. The gastrointestinal tract begins with the **mouth,** and digestion starts there with chewing, which breaks up large pieces of food into smaller particles that can be swallowed. **Saliva,** secreted by three pairs of **salivary glands** (see Figure 17–1) located in the head, drains into the mouth through a series of short ducts. Saliva, which contains mucus, moistens and lubricates the food particles before swallowing. It also contains the enzyme **amylase,** which partially digests polysaccharides. A third function of saliva is to dissolve some of the food molecules. Only in the dissolved state can these molecules react with chemoreceptors in the mouth, giving rise to the sensation of taste (Chapter 9).

The next segments of the tract, the **pharynx** and **esophagus,** contribute nothing to digestion but provide the pathway by which ingested materials reach the stomach. The muscles in the walls of these segments control swallowing.

The **stomach** is a saclike organ, located between the esophagus and the small intestine. Its functions are to store, dissolve, and partially digest the macromolecules in food and to regulate the rate at which the stomach's contents empty into the small intestine. The glands lining the stomach wall secrete a strong acid, **hydrochloric acid,** and several protein-digesting enzymes collectively known as **pepsin** (actually a precursor of pepsin known as pepsinogen is secreted and converted to pepsin in the lumen of the stomach).

The primary function of hydrochloric acid is to dissolve the particulate matter in food. The acid environment in the **gastric** (adjective for "stomach") lumen alters the ionization of polar molecules, especially proteins, disrupting the extracellular network of connective-tissue proteins that form the structural framework of the tissues in food. The proteins and polysaccharides released by hydrochloric acid's dissolving action are partially digested in the stomach by pepsin and amylase, the latter contributed by the salivary glands. A major food component that is not dissolved by acid is fat.

Hydrochloric acid also kills most of the bacteria that enter along with food. This process is not 100 percent effective, and some bacteria survive to take up residence and multiply in the gastrointestinal tract, particularly the large intestine.

The digestive actions of the stomach reduce food particles to a solution known as **chyme,** which contains molecular fragments of proteins and polysaccharides, droplets of fat, and salt, water, and various other small molecules ingested in the food. Virtually none of these molecules, except water, can cross the epithelium of the gastric wall, and thus little absorption of organic nutrients occurs in the stomach.

Digestion's final stages and most absorption occur in the next section of the tract, the **small intestine,** a tube about 1.5 inches in diameter and 9 ft in length that leads from the stomach to the large intestine. Here molecules of intact or partially digested carbohydrates, fats, and proteins are broken down by hydrolytic enzymes into monosaccharides, fatty acids, and amino acids. Some of these enzymes are on the

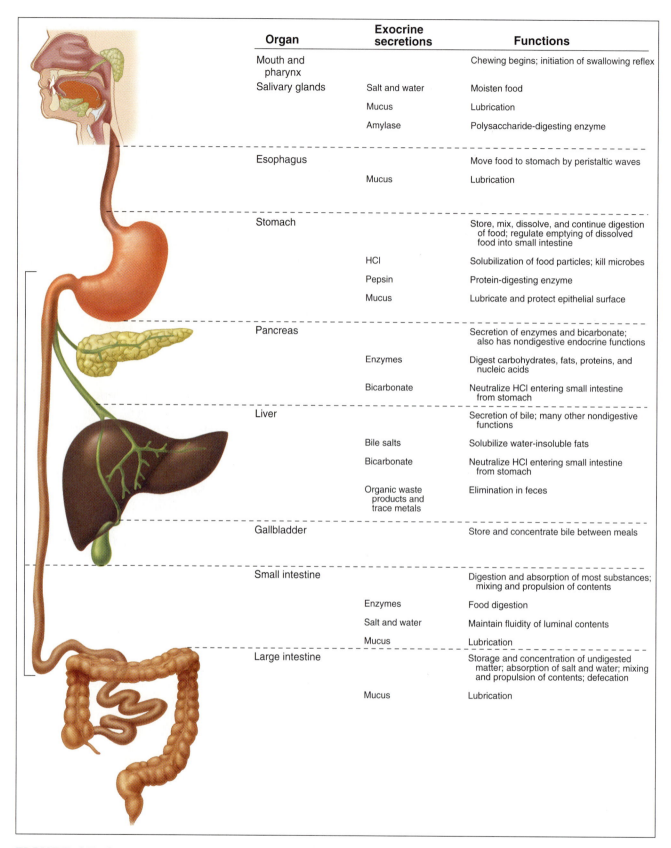

Organ	Exocrine secretions	Functions
Mouth and pharynx		Chewing begins; initiation of swallowing reflex
Salivary glands	Salt and water	Moisten food
	Mucus	Lubrication
	Amylase	Polysaccharide-digesting enzyme
Esophagus		Move food to stomach by peristaltic waves
	Mucus	Lubrication
Stomach		Store, mix, dissolve, and continue digestion of food; regulate emptying of dissolved food into small intestine
	HCl	Solubilization of food particles; kill microbes
	Pepsin	Protein-digesting enzyme
	Mucus	Lubricate and protect epithelial surface
Pancreas		Secretion of enzymes and bicarbonate; also has nondigestive endocrine functions
	Enzymes	Digest carbohydrates, fats, proteins, and nucleic acids
	Bicarbonate	Neutralize HCl entering small intestine from stomach
Liver		Secretion of bile; many other nondigestive functions
	Bile salts	Solubilize water-insoluble fats
	Bicarbonate	Neutralize HCl entering small intestine from stomach
	Organic waste products and trace metals	Elimination in feces
Gallbladder		Store and concentrate bile between meals
Small intestine		Digestion and absorption of most substances; mixing and propulsion of contents
	Enzymes	Food digestion
	Salt and water	Maintain fluidity of luminal contents
	Mucus	Lubrication
Large intestine		Storage and concentration of undigested matter; absorption of salt and water; mixing and propulsion of contents; defecation
	Mucus	Lubrication

FIGURE 17–3

Functions of the gastrointestinal organs.

luminal surface of the intestinal lining cells, while others are secreted by the pancreas and enter the intestinal lumen. The products of digestion are absorbed across the epithelial cells and enter the blood and/or lymph. Vitamins, minerals, and water, which do not require enzymatic digestion, are also absorbed in the small intestine.

The small intestine is divided into three segments: An initial short segment, the **duodenum,** is followed by the **jejunum** and then by the longest segment, the **ileum.** Normally, most of the chyme entering from the stomach is digested and absorbed in the first quarter of the small intestine, in the duodenum and jejunum.

Two major glands—the pancreas and liver—secrete substances that flow via ducts into the duodenum. The **pancreas,** an elongated gland located behind the stomach, has both endocrine (Chapter 18) and exocrine functions, but only the latter are directly involved in gastrointestinal function and are described in this chapter. The exocrine portion of the pancreas secretes (1) digestive enzymes and (2) a fluid rich in bicarbonate ions. The high acidity of the chyme coming from the stomach would inactivate the pancreatic enzymes in the small intestine if the acid were not neutralized by the bicarbonate ions in the pancreatic fluid.

The **liver,** a large gland located in the upper right portion of the abdomen, has a variety of functions, which are described in various chapters. This is a convenient place to provide, in Table 17–1, a comprehensive reference list of these **hepatic** (the term means "pertaining to the liver") functions and the chapters in which they are described. We will be concerned in this

---

TABLE 17–1 Summary of Liver Functions

A. Exocrine (digestive) functions (Chapter 17)
1. Synthesizes and secretes bile salts, which are necessary for adequate digestion and absorption of fats.
2. Secretes into the bile a bicarbonate-rich solution, which helps neutralize acid in the duodenum.

B. Endocrine functions
1. In response to growth hormone, secretes insulin-like growth factor I (IGF-I), which promotes growth by stimulating cell division in various tissues, including bone (Chapter 18).
2. Contributes to the activation of vitamin D (Chapter 16).
3. Forms triiodothyronine (T_3) from thyroxine (T_4) (Chapter 10).
4. Secretes angiotensinogen, which is acted upon by renin to form angiotensin I (Chapter 16).
5. Metabolizes hormones (Chapter 10).
6. Secretes cytokines involved in immune defenses (Chapter 20).

C. Clotting functions
1. Produces many of the plasma clotting factors, including prothrombin and fibrinogen (Chapter 14).
2. Produces bile salts, which are essential for the gastrointestinal absorption of vitamin K, which is, in turn, needed for production of the clotting factors (Chapter 14).

D. Plasma proteins
1. Synthesizes and secretes plasma albumin (Chapter 14), acute phase proteins (Chapter 20), binding proteins for various hormones (Chapter 10) and trace elements (Chapter 14), lipoproteins (Chapter 18), and other proteins mentioned elsewhere in this table.

E. Organic metabolism (Chapter 18)
1. Converts plasma glucose into glycogen and triacylglycerols during absorptive period.
2. Converts plasma amino acids to fatty acids, which can be incorporated into triacylglycerols during absorptive period.
3. Synthesizes triacylglycerols and secretes them as lipoproteins during absorptive period.
4. Produces glucose from glycogen (glycogenolysis) and other sources (gluconeogenesis) during postabsorptive period and releases the glucose into the blood.
5. Converts fatty acids into ketones during fasting.
6. Produces urea, the major end product of amino acid (protein) catabolism, and releases it into the blood.

F. Cholesterol metabolism (Chapter 18)
1. Synthesizes cholesterol and releases it into the blood.
2. Secretes plasma cholesterol into the bile.
3. Converts plasma cholesterol into bile salts.

G. Excretory and degradative functions
1. Secretes bilirubin and other bile pigments into the bile (Chapter 17).
2. Excretes, via the bile, many endogenous and foreign organic molecules as well as trace metals (Chapter 20).
3. Biotransforms many endogenous and foreign organic molecules (Chapter 20).
4. Destroys old erythrocytes (Chapter 14).

chapter only with the liver's exocrine functions that are directly related to the secretion of **bile.**

Bile contains bicarbonate ions, cholesterol, phospholipids, bile pigments, a number of organic wastes and—most important—a group of substances collectively termed **bile salts.** The bicarbonate ions, like those from the pancreas, help neutralize acid from the stomach, while the bile salts, as we shall see, solubilize dietary fat. These fats would otherwise be insoluble in water, and their solubilization increases the rates at which they are digested and absorbed.

Bile is secreted by the liver into small ducts that join to form a single duct called the common hepatic duct. Between meals, secreted bile is stored in the **gallbladder,** a small sac underneath the liver that branches from the common hepatic duct. The gallbladder concentrates the organic molecules in bile by absorbing salts and water. During a meal, the smooth muscles in the gallbladder wall contract, causing a concentrated bile solution to be injected into the duodenum via the **common bile duct** (Figure 17–4), an extension of the common hepatic duct. The gallbladder can be surgically removed without impairing bile secretion by the liver or its flow into the intestinal tract. In fact, many animals that secrete bile do not have a gallbladder.

In the small intestine, monosaccharides and amino acids are absorbed by specific transporter-mediated processes in the plasma membranes of the intestinal epithelial cells, whereas fatty acids enter these cells by diffusion. Most mineral ions are actively absorbed by transporters, and water diffuses passively down osmotic gradients.

The motility of the small intestine, brought about by the smooth muscles in its walls, (1) mixes the luminal contents with the various secretions, (2) brings the contents into contact with the epithelial surface where absorption takes place, and (3) slowly advances the luminal material toward the large intestine. Since most substances are absorbed in the small intestine, only a small volume of water, salts, and undigested material is passed on to the **large intestine.** The large intestine temporarily stores the undigested material (some of which is metabolized by bacteria) and concentrates it by absorbing salts and water. Contractions of the **rectum,** the final segment of the large intestine, and relaxation of associated sphincter muscles expel the feces—**defecation.**

The average adult consumes about 800 g of food and 1200 ml of water per day, but this is only a fraction of the material entering the lumen of the gastrointestinal tract. An additional 7000 ml of fluid from salivary glands, gastric glands, pancreas, liver, and intestinal glands is secreted into the tract each day (Figure 17–5). Of the 8 L of fluid entering the

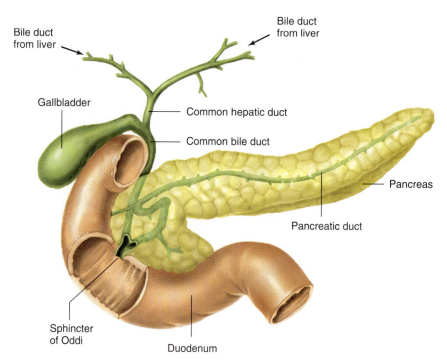

FIGURE 17–4

Bile ducts from the liver converge to form the common hepatic duct, from which branches the duct leading to the gallbladder. Beyond this branch, the common hepatic duct becomes the common bile duct. The common bile duct and the pancreatic duct converge and empty their contents into the duodenum at the sphincter of Oddi.

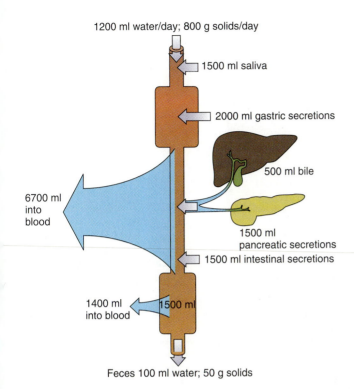

1200 ml water/day; 800 g solids/day

1500 ml saliva

2000 ml gastric secretions

500 ml bile

6700 ml into blood

1500 ml pancreatic secretions

1500 ml intestinal secretions

1400 ml into blood

1500 ml

Feces 100 ml water; 50 g solids

FIGURE 17–5

Average amounts of solids and fluid ingested, secreted, absorbed, and excreted from the gastrointestinal tract daily.

tract, 99 percent is absorbed; only about 100 ml is normally lost in the feces. This small amount of fluid loss represents only 4 percent of the total fluids lost by the body each day (most fluid loss is via the kidneys and respiratory system). Almost all the salts in the secreted fluids are also reabsorbed into the blood. Moreover, the secreted digestive enzymes are themselves digested, and the resulting amino acids are absorbed into the blood.

This completes our overview of the gastrointestinal system. Since its major task is digestion and absorption, we begin our more detailed description with these processes. Subsequent sections of the chapter will then describe, organ by organ, regulation of the secretions and motility that produce the optimal conditions for digestion and absorption. A prerequisite for this physiology, however, is a knowledge of the structure of the gastrointestinal tract wall.

Structure of the Gastrointestinal Tract Wall

From the midesophagus to the anus, the wall of the gastrointestinal tract has the general structure illustrated in Figure 17–6. Most of the tube's luminal surface is highly convoluted, a feature that greatly increases the surface area available for absorption. From the stomach on, this surface is covered by a single layer of epithelial cells linked together along the edges of their luminal surfaces by tight junctions.

Included in this epithelial layer are exocrine cells that secrete mucus into the lumen of the tract and endocrine cells that release hormones into the blood. Invaginations of the epithelium into the underlying tissue form exocrine glands that secrete acid, enzymes, water, and ions, as well as mucus.

Just below the epithelium is a layer of connective tissue, the lamina propria, through which pass small blood vessels, nerve fibers, and lymphatic ducts. (These structures are not shown in Figure 17–6 but are in Figure 17–7.) The lamina propria is separated from underlying tissues by a thin layer of smooth muscle, the muscularis mucosa. The combination of these three layers—the epithelium, lamina propria, and muscularis mucosa—is called the **mucosa** (Figure 17–6).

Beneath the mucosa is a second connective tissue layer, the submucosa, containing a network of nerve cells, termed the **submucous plexus,** and blood and lymphatic vessels whose branches penetrate into both the overlying mucosa and the underlying layers of smooth muscle called the **muscularis externa.** Contractions of these muscles provide the forces for moving and mixing the gastrointestinal contents. The muscularis externa has two layers: (1) a relatively thick inner layer of **circular muscle,** whose fibers are oriented in a circular pattern around the tube such that contraction produces a narrowing of the lumen, and (2) a thinner outer layer of **longitudinal muscle,** whose contraction shortens the tube. Between these two muscle layers is a second network of nerve cells known as the **myenteric plexus.**

Finally, surrounding the outer surface of the tube is a thin layer of connective tissue called the **serosa.** Thin sheets of connective tissue connect the serosa to the abdominal wall, supporting the gastrointestinal tract in the abdominal cavity.

Extending from the luminal surface of the small intestine are fingerlike projections known as **villi** (Figure 17–7). The surface of each villus is covered with a layer of epithelial cells whose surface membranes form small projections called **microvilli** (also known collectively as the brush border) (Figure 17–8). The combination of folded mucosa, villi, and microvilli increases the small intestine's surface area about 600-fold over that of a flat-surfaced tube having the same length and diameter. The human small intestine's total surface area is about 300 m^2, the area of a tennis court.

Epithelial surfaces in the gastrointestinal tract are continuously being replaced by new epithelial cells. In the small intestine, new cells arise by cell division from cells at the base of the villi. These cells differentiate as

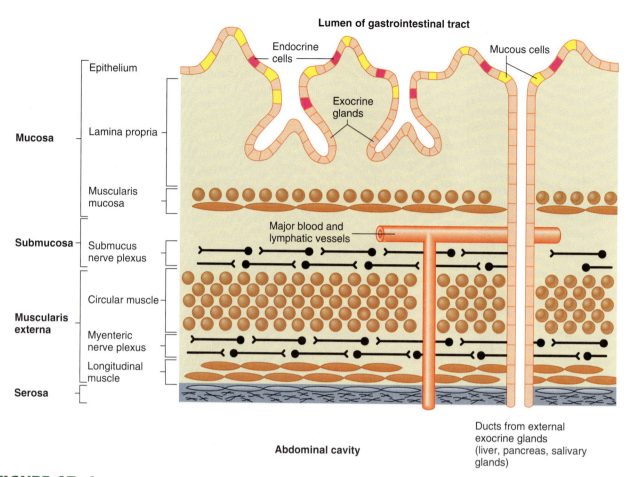

Lumen of gastrointestinal tract

Endocrine cells

Mucous cells

Epithelium

Exocrine glands

Mucosa

Lamina propria

Muscularis mucosa

Submucosa — Submucus nerve plexus

Major blood and lymphatic vessels

Circular muscle

Muscularis externa

Myenteric nerve plexus

Longitudinal muscle

Serosa

Abdominal cavity

Ducts from external exocrine glands (liver, pancreas, salivary glands)

FIGURE 17–6

Structure of the gastrointestinal wall in longitudinal section. Not shown are the smaller blood vessels and lymphatics, neural connections between the two nerve plexuses, and neural terminations on muscles, glands and epithelium.

they migrate to the top of the villus, replacing older cells that disintegrate and are discharged into the intestinal lumen. These disintegrating cells release into the lumen their intracellular enzymes, which then contribute to the digestive process. About 17 billion epithelial cells are replaced each day, and the entire epithelium of the small intestine is replaced approximately every 5 days. It is because of this rapid cell turnover that the lining of the intestinal tract is so susceptible to damage by agents, such as radiation and anticancer drugs, that inhibit cell division.

The center of each intestinal villus is occupied both by a single blind-ended lymphatic vessel termed a **lacteal** and by a capillary network (see Figure 17–7). As we will see, most of the fat absorbed in the small intestine enters the lacteals, while other absorbed nutrients enter the blood capillaries. The venous drainage from the small intestine, as well as from the large

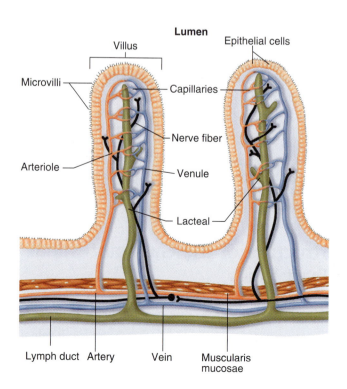

Lumen

Villus

Epithelial cells

Microvilli

Capillaries

Nerve fiber

Arteriole

Venule

Lacteal

Lymph duct   Artery   Vein   Muscularis mucosae

FIGURE 17–7

Structure of villi in small intestine.

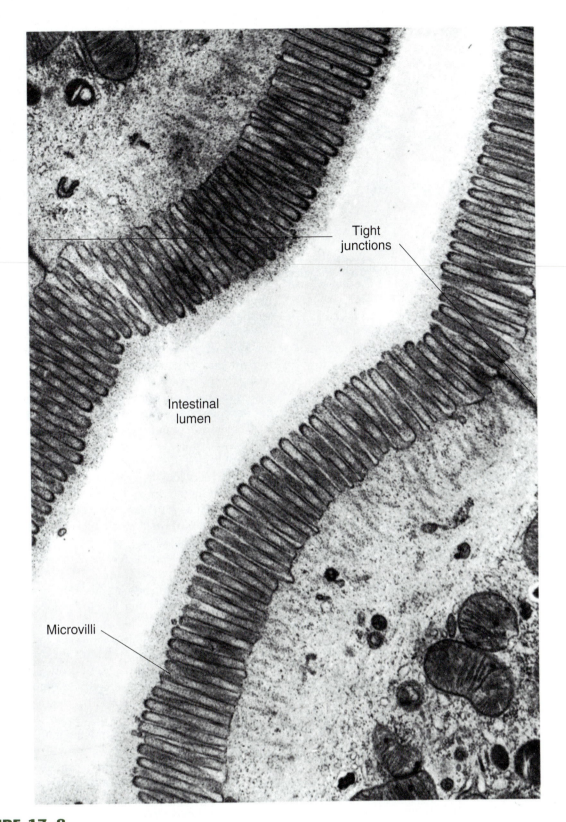

FIGURE 17–8

Microvilli on the surface of intestinal epithelial cells.

From D. W. Fawcett, J. *Histochem. Cytochem.* 13: 75–91 (1965). Courtesy of Susumo Ito.

intestine, pancreas, and portions of the stomach, does not empty directly into the vena cava but passes first, via the **hepatic portal vein,** to the liver. There it flows through a second capillary network before leaving the liver to return to the heart. Thus, material absorbed into the intestinal capillaries, in contrast to the lacteals, can be processed by the liver before entering the general circulation.

Digestion and Absorption

Carbohydrate

Carbohydrate intake per day ranges from about 250 to 800 g in a typical American diet. About two-thirds of this carbohydrate is the plant polysaccharide starch, and most of the remainder consists of the disaccharides sucrose (table sugar) and lactose (milk sugar) (Table 17–2). Only small amounts of monosaccharides are normally present in the diet. Cellulose and certain other complex polysaccharides found in vegetable matter—referred to as **fiber**—cannot be broken down by the enzymes in the small intestine and are passed on to the large intestine, where they are partially metabolized by bacteria.

Starch digestion by salivary amylase begins in the mouth and continues in the upper part of the stomach before the amylase is destroyed by gastric acid. Starch digestion is completed in the small intestine by pancreatic amylase. The products produced by both amylases are the disaccharide maltose and a mixture of short, branched chains of glucose molecules. These products, along with ingested sucrose and lactose, are broken down into monosaccharides—glucose, galactose, and fructose—by enzymes located on the luminal membranes of the small-intestine epithelial cells. These monosaccharides are then transported across the intestinal epithelium into the blood. Fructose enters the epithelial cells by facilitated diffusion, while glucose and galactose undergo secondary active transport coupled to sodium. These monosaccharides then leave the epithelial cells and enter the blood by way of facilitated diffusion transporters in the basolateral membranes of the epithelial cells. Most ingested carbohydrate is digested and absorbed within the first 20 percent of the small intestine.

Protein

Only 40 to 50 g of protein per day is required by a normal adult to supply essential amino acids and replace the amino acid nitrogen converted to urea. A typical American diet contains about 125 g of protein per day. In addition, a large amount of protein, in the form of enzymes and mucus, is secreted into the gastrointestinal tract or enters it via the disintegration of epithelial cells. Regardless of source, most of the protein in the

TABLE 17–2 Carbohydrates in Food

Class	Examples	Made of
Polysaccharides	Starch	Glucose
	Cellulose	Glucose
	Glycogen	Glucose
Disaccharides	Sucrose	Glucose-fructose
	Lactose	Glucose-galactose
	Maltose	Glucose-glucose
Monosaccharides	Glucose	
	Fructose	
	Galactose	

lumen is broken down into amino acids and absorbed by the small intestine.

Proteins are broken down to peptide fragments in the stomach by pepsin, and in the small intestine by **trypsin** and **chymotrypsin,** the major proteases secreted by the pancreas. These fragments are further digested to free amino acids by **carboxypeptidase** from the pancreas and **aminopeptidase,** located on the luminal membranes of the small-intestine epithelial cells. These last two enzymes split off amino acids from the carboxyl and amino ends of peptide chains, respectively. At least 20 different peptidases are located on the luminal membrane of the epithelial cells, with various specificities for the peptide bonds they attack.

The free amino acids then enter the epithelial cells by secondary active transport coupled to sodium. There are multiple transporters with different specificities for the 20 types of amino acids. Short chains of two or three amino acids are also absorbed by a secondary active transport that is coupled to the hydrogen-ion gradient. (This is in contrast to carbohydrate absorption, in which molecules larger than monosaccharides are not absorbed.) Within the epithelial cell, these di- and tripeptides are hydrolyzed to amino acids, which then leave the cell and enter the blood through a facilitated diffusion carrier in the basolateral membranes. As with carbohydrates, protein digestion and absorption are largely completed in the upper portion of the small intestine.

Very small amounts of intact proteins are able to cross the intestinal epithelium and gain access to the interstitial fluid. They do so by a combination of endocytosis and exocytosis. The absorptive capacity for intact proteins is much greater in infants than in adults, and antibodies (proteins involved in the immunological defense system of the body) secreted into the mother's milk can be absorbed by the infant, providing some immunity until the infant begins to produce its own antibodies.

Fat

Fat intake ranges from about 25 to 160 g/day in a typical American diet; most is in the form of triacylglycerols. Fat digestion occurs almost entirely in the small intestine. The major digestive enzyme in this process is pancreatic **lipase,** which catalyzes the splitting of bonds linking fatty acids to the first and third carbon atoms of glycerol, producing two free fatty acids and a monoglyceride as products:

Triacylglycerol $\xrightarrow{\text{Lipase}}$ Monoglyceride + 2 Fatty acids

The fats in the ingested foods are insoluble in water and aggregate into large lipid droplets in the upper portion of the stomach. Since pancreatic lipase is a water-soluble enzyme, its digestive action in the small intestine can take place only at the *surface* of a lipid droplet. Therefore, if most of the ingested fat remained in large lipid droplets, the rate of lipid digestion would be very slow. The rate of digestion is, however, substantially increased by division of the large lipid droplets into a number of much smaller droplets, each about 1 mm in diameter, thereby increasing their surface area and accessibility to lipase action. This process is known as **emulsification,** and the resulting suspension of small lipid droplets is an emulsion.

The emulsification of fat requires (1) mechanical disruption of the large fat droplets into smaller droplets, and (2) an emulsifying agent, which acts to prevent the smaller droplets from reaggregating back into large droplets. The mechanical disruption is provided by contractile activity, occurring in the lower portion of the stomach and in the small intestine, which acts to grind and mix the luminal contents. Phospholipids in food and phospholipids and bile salts secreted in the bile provide the emulsifying agents, whose action is as follows.

Phospholipids are amphipathic molecules (Chapter 2) consisting of two nonpolar fatty acid chains attached to glycerol, with a charged phosphate group located on glycerol's third carbon. Bile salts are formed from cholesterol in the liver and are also amphipathic (Figure 17–9). The nonpolar portions of the phospholipids and bile salts associate with the nonpolar interior of the lipid droplets, leaving the polar portions exposed at the water surface. There they repel other lipid droplets that are similarly coated with these emulsifying agents, thereby preventing their reaggregation into larger fat droplets (Figure 17–10).

The coating of the lipid droplets with these emulsifying agents, however, impairs the accessibility of the water-soluble lipase to its lipid substrate. To overcome this problem, the pancreas secretes a protein known as **colipase,** which is amphipathic and lodges on the lipid droplet surface. Colipase binds the lipase enzyme, holding it on the surface of the lipid droplet.

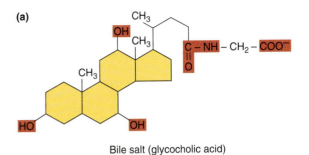

Bile salt (glycocholic acid)

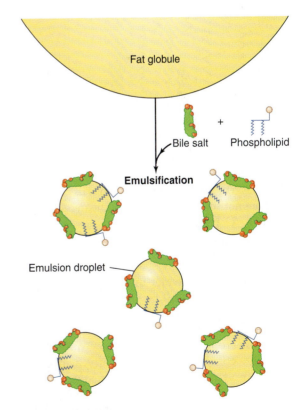

FIGURE 17–9

Structure of bile salts. (a) Chemical formula of glycocholic acid, one of several bile salts secreted by the liver (polar groups in color). (b) Three-dimensional structure of a bile salt, showing its polar and nonpolar surfaces.

FIGURE 17–10

Emulsification of fat by bile salts and phospholipids.

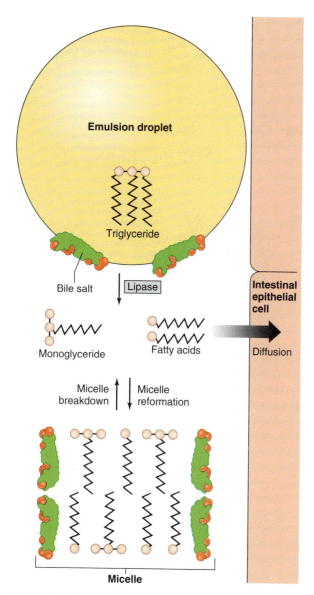

FIGURE 17–11

The products of fat digestion by lipase are held in solution in the micellar state, combined with bile salts and phospholipids. For simplicity, the phospholipids and colipase (see text) are not shown and the size of the micelle is greatly exaggerated. 🏃 ▭

Although digestion is speeded up by emulsification, absorption of the water-insoluble products of the lipase reaction would still be very slow if it were not for a second action of the bile salts, the formation of **micelles,** which are similar in structure to emulsion droplets but are much smaller—4 to 7 nm in diameter. Micelles consist of bile salts, fatty acids, monoglycerides, and phospholipids all clustered together with the polar ends of each molecule oriented toward the micelle's surface and the nonpolar portions forming the micelle's core (Figure 17–11). Also included in

the core of the micelle are small amounts of fat-soluble vitamins and cholesterol.

How do micelles increase absorption? Although fatty acids and monoglycerides have an extremely low solubility in water, a few molecules do exist in solution and are free to diffuse across the lipid portion of the luminal plasma membranes of the epithelial cells lining the small intestine. Micelles, containing the products of fat digestion, are in equilibrium with the small concentration of fat digestion products that are free in solution. Thus, micelles are continuously breaking down and reforming. When a micelle breaks down, its contents are released into the solution and become available to diffuse across the intestinal lining. As the concentrations of free lipids fall, because of their diffusion into epithelial cells, more lipids are released into the free phase as micelles break down (Figure 17–11). Thus, the micelles provide a means of keeping most of the insoluble fat digestion products in small soluble aggregates, while at the same time replenishing the small amount of products that are free in solution and are able to diffuse into the intestinal epithelium. Note that it is not the micelle that is absorbed but rather the individual lipid molecules that are released from the micelle.

Although fatty acids and monoglycerides enter epithelial cells from the intestinal lumen, it is triacylglycerol that is released on the other side of the cell into the interstitial fluid. In other words, during their passage through the epithelial cells, fatty acids and monoglycerides are resynthesized into triacylglycerols. This occurs in the agranular (smooth) endoplasmic reticulum, where the enzymes for triacylglycerol synthesis are located. This process lowers the concentration of cytosolic free fatty acids and monoglycerides and thus maintains a diffusion gradient for these molecules into the cell. Within this organelle, the resynthesized fat aggregates into small droplets coated with amphipathic proteins that perform an emulsifying function similar to that of bile salts.

The exit of these fat droplets from the cell follows the same pathway as a secreted protein. Vesicles containing the droplet pinch off the endoplasmic reticulum, are processed through the Golgi apparatus, and eventually fuse with the plasma membrane, releasing the fat droplet into the interstitial fluid. These one-micron-diameter, extracellular fat droplets are known as **chylomicrons.** Chylomicrons contain not only triacylglycerols but other lipids (including phospholipids, cholesterol, and fat-soluble vitamins) that have been absorbed by the same process that led to fatty acid and monoglyceride movement into the epithelial cells of the small intestine.

The chylomicrons released from the epithelial cells pass into lacteals—lymphatic capillaries in the

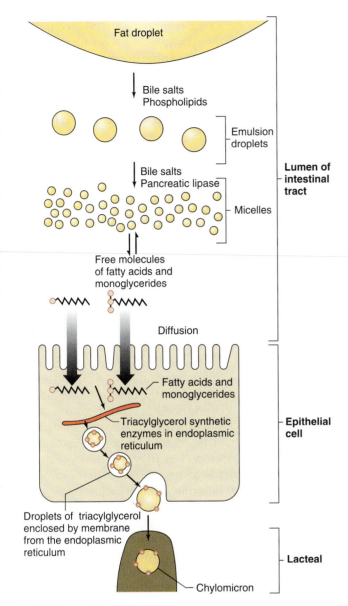

FIGURE 17–12

Summary of fat absorption across the walls of the small intestine.

Figure 17–12 summarizes the pathway taken by fat in moving from the intestinal lumen into the lymphatic system.

Vitamins

The fat-soluble vitamins—A, D, E, and K—follow the pathway for fat absorption described in the previous section. They are solubilized in micelles; thus, any interference with the secretion of bile or the action of bile salts in the intestine decreases the absorption of the fat-soluble vitamins.

With one exception, water-soluble vitamins are absorbed by diffusion or mediated transport. The exception, vitamin B_{12}, is a very large, charged molecule. In order to be absorbed, vitamin B_{12} must first bind to a protein, known as **intrinsic factor,** secreted by the acid-secreting cells in the stomach. Intrinsic factor with bound vitamin B_{12} then binds to specific sites on the epithelial cells in the lower portion of the ileum, where vitamin B_{12} is absorbed by endocytosis. As described in Chapter 14, vitamin B_{12} is required for erythrocyte formation, and deficiencies result in *pernicious anemia.* This form of anemia may occur when the stomach either has been removed (as, for example, to treat ulcers or gastric cancer) or fails to secrete intrinsic factor. Since the absorption of vitamin B_{12} occurs in the lower part of the ileum, removal of this segment because of disease can also result in pernicious anemia.

Water and Minerals

Water is the most abundant substance in chyme. Approximately 8000 ml of ingested and secreted water enters the small intestine each day, but only 1500 ml is passed on to the large intestine since 80 percent of the fluid is absorbed in the small intestine. Small amounts of water are absorbed in the stomach, but the stomach has a much smaller surface area available for diffusion and lacks the solute-absorbing mechanisms that create the osmotic gradients necessary for net water absorption. The epithelial membranes of the small intestine are very permeable to water, and net water diffusion occurs across the epithelium whenever a water-concentration difference is established by the active absorption of solutes. The mechanisms coupling solute and water absorption by epithelial cells were described in Chapter 6.

Sodium ions account for much of the actively transported solute because they constitute the most abundant solute in chyme. Sodium absorption is a primary active process, using the Na,K-ATPase pumps in a manner described in Chapter 6 and similar to that for renal tubular sodium and water reabsorption (Chapter 16). Chloride and bicarbonate ions are absorbed with the sodium ions and contribute another large fraction of the absorbed solute.

intestinal villi—rather than into the blood capillaries. The chylomicrons cannot enter the blood capillaries because the basement membrane (an extracellular glycoprotein layer) at the outer surface of the capillary provides a barrier to the diffusion of large chylomicrons. In contrast, the lacteals do not have basement membranes and have large slit pores between their endothelial cells through which the chylomicrons can pass into the lymph. The lymph from the small intestine, as from everywhere else in the body, eventually empties into systemic veins. In Chapter 18 we describe how the lipids in the circulating blood chylomicrons are made available to the cells of the body.

Other minerals present in smaller concentrations, such as potassium, magnesium, and calcium, are also absorbed, as are trace elements such as iron, zinc, and iodide. Consideration of the transport processes associated with each of these is beyond the scope of this book, and we shall briefly consider as an example the absorption of only one—iron. Calcium absorption and its regulation were described in Chapter 16.

Iron Only about 10 percent of ingested iron is absorbed into the blood each day. Iron ions are actively transported into intestinal epithelial cells, where most of them are incorporated into ferritin, the protein-iron complex that functions as an intracellular iron store (Chapter 14). The absorbed iron that does not bind to ferritin is released on the blood side where it circulates throughout the body bound to the plasma protein transferrin. Most of the iron bound to ferritin in the epithelial cells is released back into the intestinal lumen when the cells at the tips of the villi disintegrate, and it is excreted in the feces.

Iron absorption depends on the body's iron content. When body stores are ample, the increased concentration of free iron in the plasma and intestinal epithelial cells leads to an increased transcription of the gene encoding the ferritin protein and thus an increased synthesis of ferritin. This results in the increased binding of iron in the intestinal epithelial cells and a reduction in the amount of iron released into the blood. When the body stores drop, for example, when there is a loss of hemoglobin during hemorrhage, the production and hence the concentration of intestinal ferritin decreases; the amount of iron bound to ferritin decreases, thereby increasing the unbound iron released into the blood.

Once iron has entered the blood, the body has very little means of excreting it, and it accumulates in tissues. Although the control mechanisms for iron absorption just described tend to maintain the iron content of the body fairly constant, a very large ingestion of iron can overwhelm them, leading to an increased deposition of iron in tissues and producing toxic effects. This condition is termed *hemochromatosis.* Some people have genetically defective control mechanisms and therefore develop hemochromatosis even when iron ingestion is normal.

Iron absorption also depends on the type of food ingested because it binds to many negatively charged ions in food, which can retard its absorption. For example, iron in ingested liver is much more absorbable than iron in egg yolk since the latter contains phosphates that bind the iron to form an insoluble and unabsorbable complex.

The absorption of iron is typical of that of most trace metals in several respects: (1) Cellular storage proteins and plasma carrier proteins are involved, and (2) the control of *absorption,* rather than urinary excretion, is the major mechanism for the homeostatic control of the body's content of the trace metal.

Regulation of Gastrointestinal Processes

Unlike control systems that regulate variables in the internal environment, the control mechanisms of the gastrointestinal system regulate conditions in the lumen of the tract. With few exceptions, like those just discussed for iron and other trace metals, these control mechanisms are governed not by the nutritional state of the body, but rather by the volume and composition of the luminal contents.

Basic Principles

Gastrointestinal reflexes are initiated by a relatively small number of luminal stimuli: (1) distension of the wall by the volume of the luminal contents; (2) chyme osmolarity (total solute concentration); (3) chyme acidity; and (4) chyme concentrations of specific digestion products (monosaccharides, fatty acids, peptides, and amino acids). These stimuli act on receptors located in the wall of the tract (mechanoreceptors, osmoreceptors, and chemoreceptors) to trigger reflexes that influence the effectors—the muscle layers in the wall of the tract and the exocrine glands that secrete substances into its lumen.

Neural Regulation The gastrointestinal tract has its own local nervous system, known as the **enteric nervous system,** in the form of two nerve networks, the myenteric plexus and the submucous plexus (see Figure 17–6). These neurons either synapse with other neurons in the plexus or end near smooth muscles, glands, and epithelial cells. Many axons leave the myenteric plexus and synapse with neurons in the submucous plexus, and vice versa, so that neural activity in one plexus influences the activity in the other. Moreover, stimulation at one point in the plexus can lead to impulses that are conducted both up and down the tract. Thus, for example, stimuli in the upper part of the small intestine may affect smooth muscle and gland activity in the stomach as well as in the lower part of the intestinal tract.

The enteric nervous system contains adrenergic and cholinergic neurons as well as nonadrenergic, noncholinergic neurons that release neurotransmitters, such as nitric oxide, several neuropeptides, and ATP.

Many of the effectors mentioned earlier—muscle cells and exocrine glands—are supplied by neurons that are part of the enteric nervous system. This permits

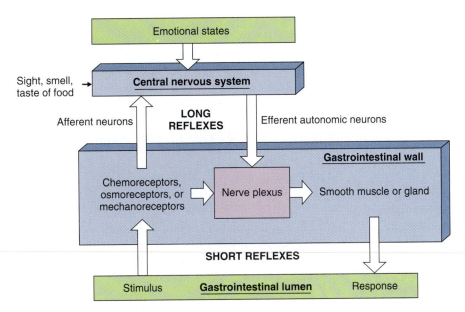

FIGURE 17–13

Long and short neural reflex pathways activated by stimuli in the gastrointestinal tract. The long reflexes utilize neurons that link the central nervous system to the gastrointestinal tract.

neural reflexes that are completely within the tract—that is, independent of the central nervous system (CNS). In addition, nerve fibers from both the sympathetic and parasympathetic branches of the autonomic nervous system enter the intestinal tract and synapse with neurons in both plexuses. Via these pathways, the CNS can influence the motility and secretory activity of the gastrointestinal tract.

Thus, two types of neural reflex arcs exist (Figure 17–13): (1) **short reflexes** from receptors through the nerve plexuses to effector cells; and (2) **long reflexes** from receptors in the tract to the CNS by way of afferent nerves and back to the nerve plexuses and effector cells by way of autonomic nerve fibers. Some controls are mediated either solely by short reflexes or solely by long reflexes, whereas other controls involve both.

Finally, it should be noted that not all neural reflexes are indicated by signals *within* the tract. The sight or smell of food and the emotional state of an individual can have significant effects on the gastrointestinal tract, effects that are mediated by the CNS via autonomic neurons.

Hormonal Regulation The hormones that control the gastrointestinal system are secreted mainly by endocrine cells scattered throughout the epithelium of the stomach and small intestine; that is, these cells are not clustered into discrete organs like the thyroid or adrenal glands. One surface of each endocrine cell is exposed to the lumen of the gastrointestinal tract. At this surface, various chemical substances in the chyme stimulate the cell to release its hormones from the opposite side of the cell into the blood. Although some of these hormones can also be detected in the lumen and may therefore act locally as paracrine agents, most of the gastrointestinal hormones reach their target cells via the circulation.

Several dozen substances are currently being investigated as possible gastrointestinal hormones, but only four—**secretin, cholecystokinin (CCK), gastrin,** and **glucose-dependent insulinotropic peptide (GIP)**—have met all the criteria for true hormones. They, as well as several candidate hormones, also exist in the CNS and in gastrointestinal plexus neurons, where they function as neurotransmitters or neuromodulators.

Table 17–3, which summarizes the major characteristics of the four established GI hormones, not only serves as a reference for future discussions but also illustrates the following generalizations: (1) Each hormone participates in a feedback control system that regulates some aspect of the GI luminal environment, and (2) each hormone affects more than one type of target cell.

These two generalizations can be illustrated by CCK. The presence of fatty acids and amino acids in the small intestine triggers CCK secretion from cells in the small intestine into the blood. Circulating CCK then stimulates secretion by the pancreas of digestive enzymes. CCK also causes the gallbladder to contract, delivering to the intestine the bile salts required for micelle formation. As fat and amino acids are absorbed,

TABLE 17–3 Properties of Gastrointestinal Hormones

	Gastrin	CCK	Secretin	GIP
Structure **Endocrine-Cell location**	Peptide Antrum of stomach	Peptide Small intestine	Peptide Small intestine	Peptide Small intestine
Stimuli for hormone release	Amino acids, peptides in stomach; parasympathetic nerves	Amino acids, fatty acids in small intestine	Acid in small intestine	Glucose, fat in small intestine
Stimuli inhibiting hormone release	Acid in stomach; somatostatin			

Target-Cell Responses

	Gastrin	CCK	Secretin	GIP
Stomach				
Acid secretion	Stimulates	Inhibits	Inhibits	
Motility	Stimulates	Inhibits	Inhibits	
Growth	Stimulates			
Pancreas				
Bicarbonate secretion		Potentiates secretin's actions	Stimulates	
Enzyme secretion		Stimulates	Potentiates CCK's actions	
Insulin secretion				Stimulates
Growth of exocrine pancreas	Stimulates	Stimulates	Stimulates	
Liver (bile ducts)				
Bicarbonate secretion		Potentiates secretin's actions	Stimulates	
Gallbladder				
Contraction		Stimulates		
Sphincter of Oddi		Relaxes		
Small intestine				
Motility	Stimulates ileum			
Growth	Stimulates			
Large intestine	Stimulates mass movement			

the stimuli (fatty acids and amino acids in the lumen) for CCK release are removed.

In many cases, a single effector cell contains receptors for more than one hormone, as well as receptors for neurotransmitters and paracrine agents, with the result that a variety of inputs can affect the cell's response. One such event is the phenomenon known as **potentiation,** which is exemplified by the interaction between secretin and CCK. Secretin strongly stimulates pancreatic bicarbonate secretion, whereas CCK is a weak stimulus of bicarbonate secretion. Both hormones together, however, stimulate pancreatic bicarbonate secretion more strongly than would be predicted by the sum of their individual stimulatory effects. This is because CCK potentiates the effect of secretin. One of the consequences of potentiation is that small changes in the plasma concentration of one gastrointestinal hormone can have large effects on the actions of other gastrointestinal hormones.

In addition to their stimulation (or in some cases inhibition) of effector-cell functions, the gastrointestinal hormones also have tropic (growth-promoting) effects on various tissues, including the gastric and intestinal mucosa and the exocrine portions of the pancreas.

Phases of Gastrointestinal Control The neural and hormonal control of the gastrointestinal system is, in large part, divisible into three phases—cephalic, gastric, and intestinal—according to stimulus location.

The **cephalic phase** is initiated when receptors in the head (*cephalic*, head) are stimulated by sight, smell, taste, and chewing. It is also initiated by various emotional states. The efferent pathways for these reflexes include both parasympathetic fibers, mostly in the vagus nerves, and sympathetic fibers. These fibers activate neurons in the gastrointestinal nerve plexuses, which in turn affect secretory and contractile activity.

Four types of stimuli in the stomach initiate the reflexes that constitute the **gastric phase** of regulation: distension, acidity, amino acids, and peptides formed during the digestion of ingested protein. The responses to these stimuli are mediated by short and long neural reflexes and by release of the hormone gastrin.

Finally, the **intestinal phase** is initiated by stimuli in the intestinal tract: distension, acidity, osmolarity, and various digestive products. The intestinal phase is mediated by both short and long neural reflexes and by the gastrointestinal hormones secretin, CCK, and GIP, all of which are secreted by endocrine cells in the small intestine.

We reemphasize that each of these phases is named for the site at which the various stimuli *initiate* the reflex and not for the sites of effector activity. Each phase is characterized by efferent output to virtually all organs in the gastrointestinal tract. Also, these phases do not occur in temporal sequence except at the very beginning of a meal. Rather, during ingestion and the much longer absorptive period, reflexes characteristic of all three phases may occur simultaneously.

Keeping in mind the neural and hormonal mechanisms available for regulating gastrointestinal activity, we can now examine the specific contractile and secretory processes that occur in each segment of the gastrointestinal system.

Mouth, Pharynx, and Esophagus

Chewing Chewing is controlled by the somatic nerves to the skeletal muscles of the mouth and jaw. In addition to the voluntary control of these muscles, rhythmical chewing motions are reflexly activated by the pressure of food against the gums, hard palate at the roof of the mouth, and tongue. Activation of these mechanoreceptors leads to reflexive inhibition of the muscles holding the jaw closed. The resulting relaxation of the jaw reduces the pressure on the various mechanoreceptors, leading to a new cycle of contraction and relaxation.

Although chewing prolongs the subjective pleasure of taste, it does not appreciably alter the rate at which the food will be digested and absorbed. On the other hand, attempting to swallow a large particle of food can lead to choking if the particle lodges over the trachea, blocking the entry of air into the lungs. A number of preventable deaths occur each year from choking, the symptoms of which are often confused with those of a heart attack so that no attempt is made to remove the obstruction from the airway. The Heimlich maneuver, described in Chapter 15, can often dislodge the obstructing particle from the airways.

Saliva The secretion of saliva is controlled by both sympathetic and parasympathetic neurons; unlike their antagonistic activity in most organs, both systems stimulate salivary secretion, with the parasympathetics producing the greater response. There is no hormonal regulation of salivary secretion. In the absence of ingested material, a low rate of salivary secretion keeps the mouth moist. In the presence of food, salivary secretion increases markedly. This reflex response is initiated by chemoreceptors (acidic fruit juices are a particularly strong stimulus) and pressure receptors in the walls of the mouth and on the tongue.

Increased secretion of saliva is accomplished by a large increase in blood flow to the salivary glands, which is mediated by both neural activity and paracrine/autocrine agents released by the active cells in the salivary gland. The volume of saliva secreted per gram of tissue is the largest secretion of any of the body's exocrine glands.

Swallowing Swallowing is a complex reflex initiated when pressure receptors in the walls of the pharynx are stimulated by food or drink forced into the rear of the mouth by the tongue. These receptors send afferent impulses to the **swallowing center** in the brainstem medulla oblongata. This center then elicits swallowing via efferent fibers to the muscles in the pharynx and esophagus as well as to the respiratory muscles.

As the ingested material moves into the pharynx, the soft palate is elevated and lodges against the back wall of the pharynx, preventing food from entering the nasal cavity (Figure 17–14b). Impulses from the swallowing center inhibit respiration, raise the larynx, and close the **glottis** (the area around the vocal cords and the space between them), keeping food from moving into the trachea. As the tongue forces the food farther back into the pharynx, the food tilts a flap of tissue, the **epiglottis,** backward to cover the closed glottis (Figure 17–14c).

The next stage of swallowing occurs in the esophagus, the foot-long tube that passes through the thoracic cavity, penetrates the diaphragm, which separates the thoracic cavity from the abdominal cavity, and joins the stomach a few centimeters below the diaphragm. Skeletal muscles surround the upper third of the esophagus, smooth muscles the lower two-thirds.

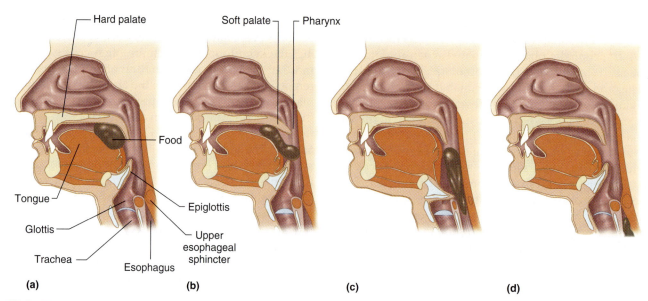

(a) (b) (c) (d)

FIGURE 17–14

Movements of food through the pharynx and upper esophagus during swallowing.

As described in Chapter 15, the pressure in the thoracic cavity is 4 to 10 mmHg less than atmospheric pressure, and this subatmospheric pressure is transmitted across the thin wall of the intrathoracic portion of the esophagus to the lumen. In contrast, the luminal pressure in the pharynx at the opening to the esophagus is equal to atmospheric pressure, and the pressure at the opposite end of the esophagus in the stomach is slightly greater than atmospheric. Therefore, pressure differences exist that would tend to force both air (from above) and gastric contents (from below) into the esophagus. This does not occur, however, because both ends of the esophagus are normally closed by the contraction of sphincter muscles. Skeletal muscles surround the esophagus just below the pharynx and form the **upper esophageal sphincter,** whereas the smooth muscles in the last portion of the esophagus form the **lower esophageal sphincter** (Figure 17–15).

The esophageal phase of swallowing begins with relaxation of the upper esophageal sphincter. Immediately after the food has passed, the sphincter closes, the glottis opens, and breathing resumes. Once in the esophagus, the food is moved toward the stomach by a progressive wave of muscle contractions that proceeds along the esophagus, compressing the lumen and forcing the food ahead of it. Such waves of contraction in the muscle layers surrounding a tube are known as **peristaltic waves.** One esophageal peristaltic wave takes about 9 s to reach the stomach. Swallowing can occur even when a person is upside down since it is not primarily gravity but the peristaltic wave that moves the food to the stomach.

The lower esophageal sphincter opens and remains relaxed throughout the period of swallowing, allowing the arriving food to enter the stomach. After the food has passed, the sphincter closes, resealing the junction between the esophagus and the stomach.

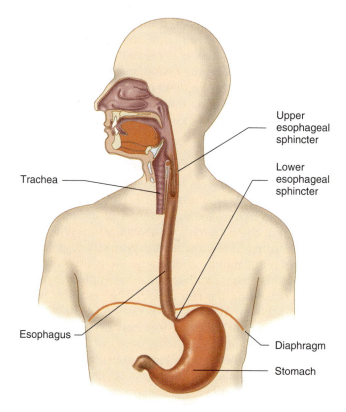

FIGURE 17–15

Location of upper and lower esophageal sphincters.

Swallowing is an example of a reflex in which multiple responses occur in a temporal sequence determined by the pattern of synaptic connections between neurons in a brain coordinating center. Since both skeletal and smooth muscles are involved, the swallowing center must direct efferent activity in both somatic nerves (to skeletal muscle) and autonomic nerves (to smooth muscle). Simultaneously, afferent fibers from receptors in the esophageal wall send information to the swallowing center that can alter the efferent activity. For example, if a large food particle does not reach the stomach during the initial peristaltic wave, the maintained distension of the esophagus by the particle activates receptors that initiate reflexes causing repeated waves of peristaltic activity (**secondary peristalsis**) that are not accompanied by the initial pharyngeal events of swallowing.

The ability of the lower esophageal sphincter to maintain a barrier between the stomach and the esophagus when swallowing is not taking place is aided by the fact that the last portion of the esophagus lies below the diaphragm, and is subject to the same abdominal pressures as is the stomach. In other words, if the pressure in the abdominal cavity is raised, for example, during cycles of respiration or by contraction of the abdominal muscles, the pressures on both the gastric contents and the terminal segment of the esophagus are raised together, preventing the formation of a pressure gradient between the stomach and esophagus that could force the stomach's contents into the esophagus.

During pregnancy the growth of the fetus not only increases the pressure on the abdominal contents but also pushes the terminal segment of the esophagus through the diaphragm into the thoracic cavity. The sphincter is therefore no longer assisted by changes in abdominal pressure. Accordingly, during the last half of pregnancy there is a tendency for increased abdominal pressure to force some of the gastric contents up into the esophagus. The hydrochloric acid from the stomach irritates the esophageal walls, producing pain known as *heartburn* (because the pain appears to be located over the heart). Heartburn often subsides in the last weeks of pregnancy as the uterus descends lower into the pelvis prior to delivery, decreasing the pressure on the stomach.

Heartburn also occurs in the absence of pregnancy. Some people have less efficient lower esophageal sphincters, resulting in repeated episodes of refluxed gastric contents into the esophagus (*gastro-esophageal reflux*), heartburn, and in extreme cases, ulceration, scarring, obstruction, or perforation of the lower esophagus. Heartburn can occur after a large meal, which can raise the pressure in the stomach enough to force acid into the esophagus. Gastro-esophageal reflux can also cause coughing and irritation of the larynx in the absence of any esophageal symptoms.

The lower esophageal sphincter not only undergoes brief periods of relaxation during a swallow but also in the absence of a swallow. During these periods of relaxation, small amounts of the acid contents from the stomach are normally refluxed into the esophagus. The acid in the esophagus triggers a secondary peristaltic wave and also stimulates increased salivary secretion, which helps to neutralize the acid and clear it from the esophagus.

Stomach

The epithelial layer lining the stomach invaginates into the mucosa, forming numerous tubular glands. Glands in the thin-walled upper portions of the stomach, the **body** and **fundus** (Figure 17–16), secrete mucus, hydrochloric acid, and the enzyme precursor **pepsinogen**. The lower portion of the stomach, the **antrum,** has a much thicker layer of smooth muscle. The glands in this region secrete little acid but contain the endocrine cells that secrete the hormone gastrin.

Mucus is secreted by the cells at the opening of the glands (Figure 17–17). Lining the walls of the glands are **parietal cells** (also known as oxyntic cells), which secrete acid and intrinsic factor, and **chief cells,** which secrete pepsinogen. Thus, each of the three major *exocrine* secretions of the stomach—mucus, acid, and pepsinogen—is secreted by a different cell type. In addition, **enterochromaffin-like (ECL) cells,** which release the paracrine agent histamine, and cells that secrete the peptide messenger somatostatin, are scattered throughout the tubular glands.

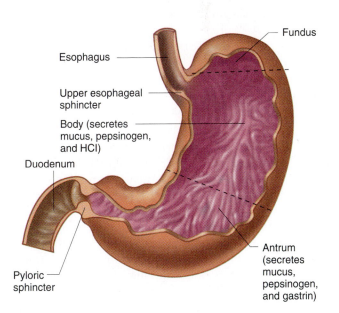

Esophagus

Upper esophageal sphincter

Body (secretes mucus, pepsinogen, and HCl)

Duodenum

Pyloric sphincter

Fundus

Antrum (secretes mucus, pepsinogen, and gastrin)

FIGURE 17–16

The three regions of the stomach: fundus, body, and antrum.

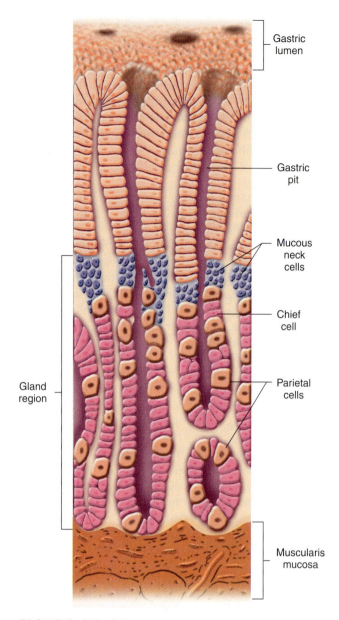

FIGURE 17–17

Gastric glands in the body of the stomach.

HCl Secretion The stomach secretes about 2 L of hydrochloric acid per day. The concentration of hydrogen ions in the stomach's lumen may reach 150 mM, 3 million times greater than the concentration in the blood.

Primary H,K-ATPases in the luminal membrane of the parietal cells pump hydrogen ions into the stomach's lumen (Figure 17–18). This primary active transporter also pumps potassium into the cell, which then leaks back into the lumen through potassium channels. Excessive vomiting can lead to potassium depletion due to this leak. As hydrogen ions are secreted into the lumen, bicarbonate ions are being secreted on the opposite side of the cell into the blood, in exchange for chloride ions. This addition of bicarbonate lowers the acidity in the venous blood from the stomach.

Increased acid secretion, stimulated by factors described in the next paragraph, is the result of the transfer of H,K-ATPase proteins from the membranes of intracellular vesicles to the plasma membrane by fusion of these vesicles with the membrane, thus increasing the number of pump proteins in the plasma membrane. This process is analogous to that described in Chapter 16 for the transfer of water channels to the plasma membrane of kidney collecting-duct cells in response to ADH.

Four chemical messengers regulate the insertion of H,K-ATPases into the plasma membrane and hence acid secretion: gastrin (a GI hormone), acetylcholine (ACh, a neurotransmitter), histamine, and somatostatin (two paracrine agents). Parietal cell membranes contain receptors for all four of these agents (Figure 17–19). Somatostatin inhibits acid secretion, while the other three stimulate secretion. Histamine is particularly important in stimulating acid secretion in that it markedly potentiates the response to the other two stimuli, gastrin and ACh. As will be discussed later when considering ulcers, this potentiating effect of histamine is the reason that drugs that block histamine receptors in the stomach suppress acid secretion. Not only do these chemical messengers act directly on the parietal cells, they also influence each other's secretion.

During a meal, the rate of acid secretion increases markedly as stimuli arising from the cephalic, gastric, and intestinal phases alter the release of the four chemical messengers described in the previous paragraph. During the cephalic phase, increased activity of the parasympathetic nerves to the stomach's enteric nervous system results in the release of ACh from the plexus neurons, gastrin from the gastrin-releasing cells, and histamine from ECL cells (Figure 17–20).

Once food has reached the stomach, the gastric phase stimuli—distension by the volume of ingested material and the presence of peptides and amino acids released by digestion of luminal proteins—produce a further increase in acid secretion. These stimuli use some of the same neural pathways used during the cephalic phase, in that nerve endings in the mucosa of the stomach respond to these luminal stimuli and send action potentials to the enteric nervous system, which in turn, can relay signals to the gastrin-releasing cells, histamine-releasing cells, and parietal cells. In addition, peptides and amino acids can act directly on the gastrin-releasing endocrine cells to promote gastrin secretion.

The concentration of acid in the gastric lumen is itself an important determinant of the rate of acid

Labels for figure:
Gastric lumen
Gastric pit
Mucous neck cells
Chief cell
Parietal cells
Gland region
Muscularis mucosa

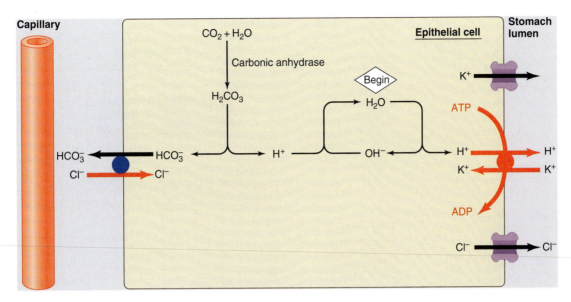

FIGURE 17–18

Secretion of hydrochloric acid by parietal cells. The hydrogen ions secreted into the lumen by primary active transport are derived from the breakdown of water molecules, leaving hydroxyl ions (OH^-) behind. These hydroxyl ions are neutralized by combination with other hydrogen ions generated by the reaction between carbon dioxide and water, a reaction catalyzed by the enzyme carbonic anhydrase, which is present in high concentrations in parietal cells. The bicarbonate ions formed by this reaction move out of the parietal cell on the blood side, in exchange for chloride ions.

secretion for the following reason. Hydrogen ions (acid) stimulate the release of somatostatin from endocrine cells in the gastric wall. Somatostatin then acts on the parietal cells to inhibit acid secretion; it also inhibits the release of gastrin and histamine. The net result is a negative-feedback control of acid secretion; as the acidity of the gastric lumen increases, it turns off the stimuli that are promoting acid secretion.

Increasing the protein content of a meal increases acid secretion. This occurs for two reasons. First, the more protein ingested, the more peptides are generated in the stomach's lumen, and these peptides, as we have seen, stimulate acid secretion. The second reason is more complicated and reflects the effects of proteins on luminal acidity. Before food enters the stomach, the H^+ *concentration* in the lumen is *high* because there are few buffers present to bind any secreted hydrogen ions; therefore, the rate of acid secretion is *low* because high acidity inhibits acid secretion. The protein in food is an excellent buffer however, and so as protein enters the stomach the H^+ concentration drops as the hydrogen ions bind to the proteins. This decrease in acidity removes the inhibition of acid secretion. The more protein in a meal, the greater the buffering of acid, and the more acid secreted.

We now come to the intestinal phase controlling acid secretion, the phase in which stimuli in the early

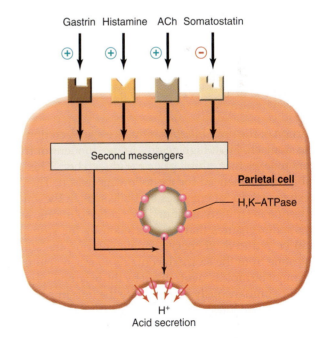

FIGURE 17–19

The four inputs to parietal cells that regulate acid secretion by controlling the transfer of the H,K-ATPase pumps in cytoplasmic vesicle membranes to the plasma membrane.

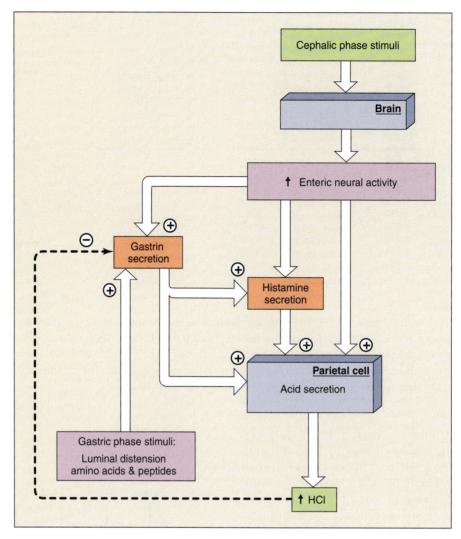

FIGURE 17–20
Cephalic and gastric phases controlling acid secretion by the stomach.

portion of the *small intestine* influence acid secretion by the stomach. First, high acidity in the duodenum triggers reflexes that inhibit gastric acid secretion. This inhibition is beneficial for the following reason. The digestive activity of enzymes and bile salts in the small intestine is strongly inhibited by acidic solutions, and this reflex ensures that acid secretion by the stomach will be reduced whenever chyme entering the small intestine from the stomach contains so much acid that it cannot be rapidly neutralized by the bicarbonate-rich fluids simultaneously secreted into the intestine by the liver and pancreas.

Acid, distension, hypertonic solutions, and solutions containing amino acids, and fatty acids in the small intestine reflexly inhibit gastric acid secretion. Thus, the extent to which acid secretion is inhibited during the intestinal phase varies, depending upon the volume and composition of the intestinal contents, but the net result is the same—balancing the secretory ac-

tivity of the stomach with the digestive and absorptive capacities of the small intestine.

The inhibition of gastric acid secretion during the intestinal phase is mediated by short and long neural reflexes and by hormones that inhibit acid secretion by influencing the four signals *directly* controlling acid secretion: ACh, gastrin, histamine, and somatostatin. The hormones released by the intestinal tract that reflexly inhibit gastric activity are collectively called **enterogastrones** and include secretin, CCK, and additional unidentified hormones.

Table 17–4 summarizes the control of acid secretion.

Pepsin Secretion   Pepsin is secreted by chief cells in the form of an inactive precursor called pepsinogen. The acidity in the stomach's lumen alters the shape of pepsinogen, exposing its active site so that this site can act on other pepsinogen molecules to break off a small chain of amino acids from their ends. This cleavage

TABLE 17–4 Control of HCL Secretion during a Meal

Stimuli	Pathways	Result
Cephalic phase Sight Smell Taste Chewing	Parasympathetic nerves to enteric nervous system	↑HCl secretion
Gastric contents (gastric phase) Distension ↑Peptides ↓H^+ concentration	Long and short neural reflexes, and direct stimulation of gastrin secretion	↑HCl secretion
Intestinal contents (intestinal phase) Distension ↑H^+ concentration ↑Osmolarity ↑Nutrient concentrations	Long and short neural reflexes; secretin, CCK, and other unspecified duodenal hormones	↓HCl secretion

converts pepsinogen to pepsin, the fully active form (Figure 17–21). Thus the activation of pepsin is an autocatalytic, positive-feedback process.

The synthesis and secretion of pepsinogen, followed by its intraluminal activation to pepsin, provides an example of a process that occurs with many other secreted proteolytic enzymes in the gastrointestinal tract. Because these enzymes are synthesized in inactive forms, collectively referred to as **zymogens,** any substrates that these enzymes might be able to act

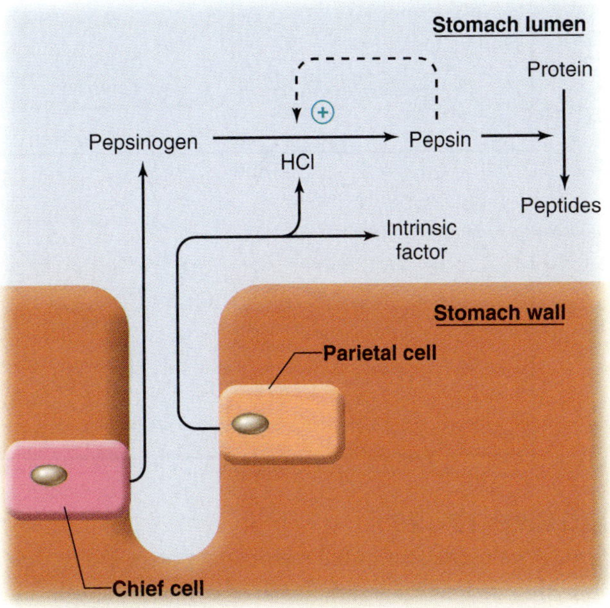

upon *inside* the cell producing them are protected from digestion, thus preventing damage to the cells.

Pepsin is active only in the presence of a high H^+ concentration. It becomes inactive, therefore, when it enters the small intestine, where the hydrogen ions are neutralized by the bicarbonate ions secreted into the small intestine.

The primary pathway for stimulating pepsinogen secretion is input to the chief cells from the enteric nervous system. During the cephalic, gastric, and intestinal phases, most of the factors that stimulate or inhibit acid secretion exert the same effect on pepsinogen secretion. Thus, pepsinogen secretion parallels acid secretion.

Pepsin is not essential for protein digestion since in its absence, as occurs in some pathological conditions, protein can be completely digested by enzymes in the small intestine.

FIGURE 17–21

Conversion of pepsinogen to pepsin in the lumen of the stomach. 🕱 ▭

Gastric Motility An empty stomach has a volume of only about 50 ml, and the diameter of its lumen is only slightly larger than that of the small intestine. When a meal is swallowed, however, the smooth muscles in the fundus and body relax before the arrival of food, allowing the stomach's volume to increase to as much as 1.5 L with little increase in pressure. This is called **receptive relaxation** and is mediated by the parasympathetic nerves to the stomach's enteric nerve plexuses, with coordination by the swallowing center in the brain. Nitric oxide and serotonin released by enteric neurons mediate this relaxation.

As in the esophagus, the stomach produces peristaltic waves in response to the arriving food. Each wave begins in the body of the stomach and produces only a ripple as it proceeds toward the antrum, a contraction

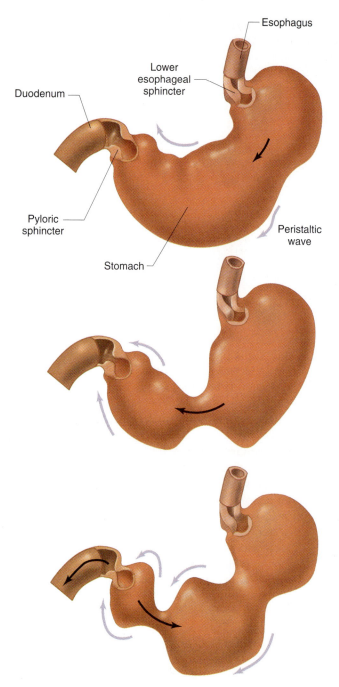

FIGURE 17–22

Peristaltic waves passing over the stomach force a small amount of luminal material into the duodenum. Black arrows indicate movement of luminal material; purple arrows indicate movement of the peristaltic wave in the stomach wall.

traction, which both mixes the luminal contents and *closes* the **pyloric sphincter,** a ring of smooth muscle and connective tissue between the antrum and the duodenum (Figure 17–22). The pyloric sphincter muscles contract upon arrival of a peristaltic wave. As a consequence of sphincter closing, only a small amount of chyme is expelled into the duodenum with each wave, and most of the antral contents are forced backward toward the body of the stomach, thereby contributing to the mixing activity in the antrum.

What is responsible for producing gastric peristaltic waves? Their rhythm (three per minute) is generated by pacemaker cells in the longitudinal smooth muscle layer. These smooth-muscle cells undergo spontaneous depolarization-repolarization cycles (slow waves) known as the **basic electrical rhythm** of the stomach. These slow waves are conducted through gap junctions along the stomach's longitudinal muscle layer and also induce similar slow waves in the overlying circular muscle layer. In the absence of neural or hormonal input, however, these depolarizations are too small to cause significant contractions. Excitatory neurotransmitters and hormones act upon the smooth muscle to further depolarize the membrane, thereby bringing it closer to threshold. Action potentials may be generated at the peak of the slow wave cycle if threshold is reached (Figure 17–23) and thus cause

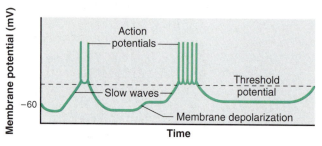

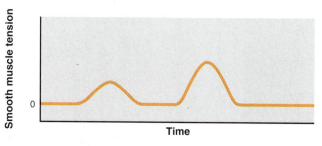

FIGURE 17–23

Slow wave oscillations in the membrane potential of gastric smooth-muscle fibers trigger bursts of action potentials when threshold potential is reached at the wave peak. Membrane depolarization brings the slow wave closer to threshold, increasing the action-potential frequency and thus the force of smooth-muscle contraction.

too weak to produce much mixing of the luminal contents with acid and pepsin. As the wave approaches the larger mass of wall muscle surrounding the antrum, however, it produces a more powerful con-

larger contractions. The number of spikes fired with each wave determines the strength of the muscle contraction.

Thus, whereas the *frequency* of contraction is determined by the intrinsic basic electrical rhythm and remains essentially constant, the *force* of contraction and therefore the amount of gastric emptying per contraction are determined reflexly by neural and hormonal input to the antral smooth muscle.

The initiation of these reflexes depends upon the contents of both the stomach and small intestine. All the factors previously discussed that regulate acid secretion (Table 17–4) can also alter gastric motility. For example, gastrin, in sufficiently high concentrations, increases the force of antral smooth-muscle contractions. Distension of the stomach also increases the force of antral contractions through long and short reflexes triggered by mechanoreceptors in the stomach wall. Therefore, the larger a meal, the faster the stomach's initial emptying rate. As the volume of the stomach decreases, the force of gastric contractions and the rate of emptying also decrease.

In contrast, distension of the *duodenum* or the presence of fat, high acidity, or hypertonic solutions in its lumen all inhibit gastric emptying (Figure 17–24) just as they inhibit acid and pepsin secretion. Fat is the most potent of these chemical stimuli.

Autonomic nerve fibers to the stomach can be activated by the CNS independently of the reflexes originating in the stomach and duodenum and can influence gastric motility. Decreased parasympathetic or increased sympathetic activity inhibits motility. Via these pathways, pain and emotions such as sadness, depression, and fear tend to decrease motility, whereas aggression and anger tend to increase it. These relationships are not always predictable, however, and different people show different gastrointestinal responses to apparently similar emotional states.

As we have seen, a hypertonic solution in the duodenum is one of the stimuli inhibiting gastric emptying. This reflex prevents the fluid in the duodenum from becoming too hypertonic since it slows the rate of entry of chyme and thereby the delivery of large molecules that can rapidly be broken down into many

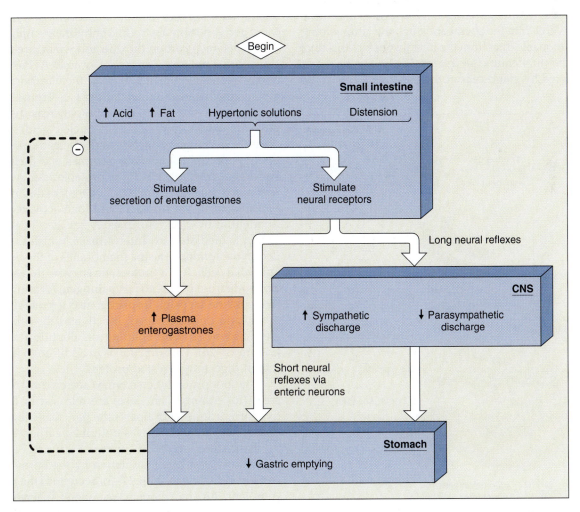

FIGURE 17–24
Intestinal-phase pathways inhibiting gastric emptying.

small molecules by enzymes in the small intestine. A patient who has had his stomach removed because of disease (for example, cancer) must eat a number of small meals. A large meal, in the absence of the controlled emptying by the stomach, would rapidly enter the intestine, producing a hypertonic solution. This hypertonic solution can cause enough water to flow (by osmosis) into the intestine from the blood to lower the blood volume and produce circulatory complications. The large distension of the intestine by the entering fluid can also trigger vomiting in these patients. All these symptoms produced by the rapid entry of large quantities of ingested material into the small intestine are known as the *dumping syndrome.*

Once the contents of the stomach have emptied over a period of several hours, the peristaltic waves cease and the empty stomach is mostly quiescent. During this time, however, there are brief intervals of peristaltic activity that will be described along with the events controlling intestinal motility.

Pancreatic Secretions

The exocrine portion of the pancreas secretes bicarbonate ions and a number of digestive enzymes into ducts that converge into the pancreatic duct, the latter joining the common bile duct from the liver just before this duct enters the duodenum (see Figure 17–4). The enzymes are secreted by gland cells at the pancreatic end of the

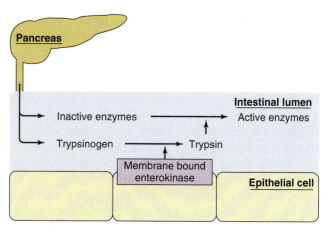

FIGURE 17–26

Activation of pancreatic enzyme precursors in the small intestine.

duct system, whereas bicarbonate ions are secreted by the epithelial cells lining the ducts (Figure 17–25).

The mechanism of bicarbonate secretion is analogous to that of hydrochloric acid secretion by the stomach, except that the directions of hydrogen-ion and bicarbonate-ion movement are reversed. Hydrogen ions, derived from a carbonic anhydrase-catalyzed reaction between carbon dioxide and water, are actively transported out of the duct cells by an H-ATPase pump and released into the blood, while the bicarbonate ions are secreted into the duct lumen (see Figure 17–18).

The enzymes secreted by the pancreas digest fat, polysaccharides, proteins, and nucleic acids to fatty acids, sugars, amino acids, and nucleotides, respectively. A partial list of these enzymes and their activities is given in Table 17–5. The proteolytic enzymes are secreted in inactive forms (zymogens), as described for pepsinogen in the stomach, and then activated in the duodenum by other enzymes. A key step in this activation is mediated by **enterokinase,** which is embedded in the luminal plasma membranes of the intestinal epithelial cells. It is a proteolytic enzyme that splits off a peptide from pancreatic **trypsinogen,** forming the active enzyme trypsin. Trypsin is also a proteolytic enzyme, and once activated, it activates the other pancreatic zymogens by splitting off peptide fragments (Figure 17–26). This function is in addition to trypsin's role in digesting ingested protein.

The nonproteolytic enzymes secreted by the pancreas (for example, amylase) are released in fully active form. Along with lipase, the pancreas secretes colipase, whose function was described earlier.

Pancreatic secretion increases during a meal, mainly as a result of stimulation by the hormones secretin and CCK (see Table 17–3). Secretin is the primary stimulant for bicarbonate secretion, whereas CCK mainly stimulates enzyme secretion. (As noted earlier, these two hormones potentiate each other's actions.)

Since the function of pancreatic bicarbonate is to

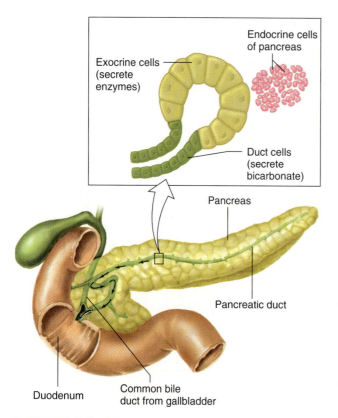

FIGURE 17–25

Structure of the pancreas. ✗

TABLE 17–5 Pancreatic Enzymes

Enzyme	Substrate	Action
Trypsin, chymotrypsin, elastase	Proteins	Breaks peptide bonds in proteins to form peptide fragments
Carboxypeptidase	Proteins	Splits off terminal amino acid from carboxyl end of protein
Lipase	Fats	Splits off two fatty acids from triacylglycerols, forming free fatty acids and monoglycerides
Amylase	Polysaccharides	Splits polysaccharides into glucose and maltose
Ribonuclease, deoxyribonuclease	Nucleic acids	Splits nucleic acids into free mononucleotides

neutralize acid entering the duodenum from the stomach, it is appropriate that the major stimulus for secretin release is increased acidity in the duodenum (Figure 17–27). In analogous fashion, since CCK stimulates the secretion of digestive enzymes, including those for fat and protein digestion, it is appropriate that the stimuli for its release are fatty acids and amino acids in the duodenum (Figure 17–28).

Luminal acid and fatty acids also act on afferent nerve endings in the intestinal wall, initiating reflexes that act on the pancreas to increase both enzyme and bicarbonate secretion. Thus, the organic nutrients in the small intestine initiate, via hormonal and neural reflexes, the secretions involved in their own digestion.

Although most of the pancreatic exocrine secretions are controlled by stimuli arising from the intestinal phase of digestion, cephalic and gastric stimuli, by way of the parasympathetic nerves to the pancreas, also play a role. Thus, the taste of food or the distension of the stomach by food, will lead to increased pancreatic secretion.

Bile Secretion

As stated earlier, bile is secreted by liver cells into a number of small ducts, the **bile canaliculi** (Figure 17–29), which converge to form the common hepatic duct (see Figure 17–4). Bile contains six major ingredients: (1) bile salts; (2) lecithin (a phospholipid); (3) bicarbonate ions and other salts; (4) cholesterol; (5) bile pigments and small amounts of other metabolic end products, and (6) trace metals. Bile salts and lecithin are synthesized in the liver and, as we have seen, help solubilize fat in the small intestine. Bicarbonate ions neutralize acid in the duodenum, and the last three ingredients represent substances extracted from the blood by the liver and excreted via the bile.

From the standpoint of gastrointestinal function, the most important components of bile are the bile salts. During the digestion of a fatty meal, most of the bile salts entering the intestinal tract via the bile are absorbed by specific sodium-coupled transporters in

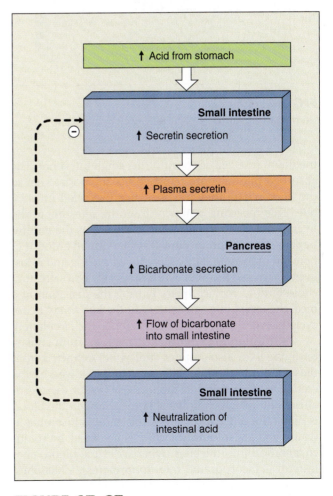

FIGURE 17–27

Hormonal regulation of pancreatic bicarbonate secretion.

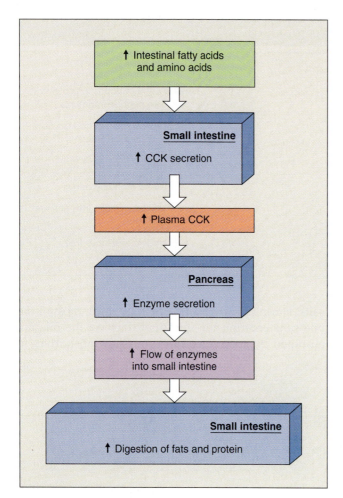

FIGURE 17–28
Hormonal regulation of pancreatic enzyme secretion.

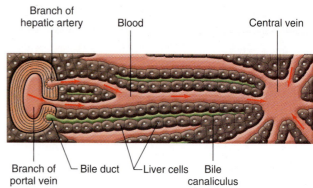

FIGURE 17–29
A small section of the liver showing location of bile canaliculi and ducts with respect to blood and liver cells.
Adapted from Kappas and Alvares.

Bile pigments are substances formed from the heme portion of hemoglobin when old or damaged erythrocytes are digested in the spleen and liver. The predominant bile pigment is **bilirubin,** which is extracted from the blood by liver cells and actively secreted into the bile. It is bilirubin that gives bile its yellow color. After entering the intestinal tract, bilirubin is modified by bacterial enzymes to form the brown

the ileum (the last segment of the small intestine). The absorbed bile salts are returned via the portal vein to the liver, where they are once again secreted into the bile. This recycling pathway from the intestine to the liver and back to the intestine is known as the **enterohepatic circulation** (Figure 17–30). A small amount (5 percent) of the bile salts escape this recycling and is lost in the feces, but the liver synthesizes new bile salts from cholesterol to replace them. During the digestion of a meal the entire bile salt content of the body may be recycled several times via the enterohepatic circulation.

In addition to synthesizing bile salts from cholesterol, the liver also secretes cholesterol extracted from the blood into the bile. Bile secretion, followed by excretion of cholesterol in the feces, is one of the mechanisms by which cholesterol homeostasis in the blood is maintained (Chapter 18). Cholesterol is insoluble in water, and its solubility in bile is achieved by its incorporation into micelles (whereas in the blood, cholesterol is incorporated into lipoproteins). Gallstones, consisting of precipitated cholesterol, will be discussed at the end of this chapter.

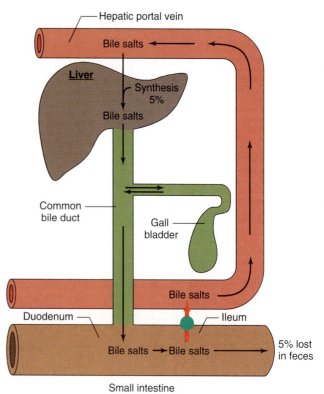

FIGURE 17–30
Enterohepatic circulation of bile salts.

pigments that give feces their characteristic color. During their passage through the intestinal tract, some of the bile pigments are absorbed into the blood and are eventually excreted in the urine, giving urine its yellow color.

Like pancreatic secretions, the components of bile are secreted by two different cell types. The bile salts, cholesterol, lecithin, and bile pigments are secreted by **hepatocytes** (liver cells), whereas most of the bicarbonate-rich salt solution is secreted by the epithelial cells lining the bile ducts. Secretion of the salt solution by the bile ducts, just like that secreted by the pancreas, is stimulated by secretin in response to the presence of acid in the duodenum.

Unlike the pancreas, whose secretions are controlled by intestinal hormones, bile salt secretion is controlled by the concentration of bile salts in the blood—the greater the plasma concentration of bile salts, the greater their secretion into the bile canaliculi. Absorption of bile salts from the intestine during the digestion of a meal leads to their increased plasma concentration and thus to an increased rate of bile salt secretion by the liver. Although bile secretion is greatest during and just after a meal, some bile is always being secreted by the liver. Surrounding the common bile duct at the point where it enters the duodenum is a ring of smooth muscle known as the **sphincter of Oddi.** When this sphincter is closed, the dilute bile secreted by the liver is shunted into the gallbladder where the organic components of bile become concentrated as NaCl and water are absorbed into the blood.

Shortly after the beginning of a fatty meal, the sphincter of Oddi relaxes and the gallbladder contracts, discharging concentrated bile into the duodenum. The signal for gallbladder contraction and sphincter relaxation is the intestinal hormone CCK—appropriately so, since as we have seen, a major stimulus for this hormone's release is the presence of fat in the duodenum. (It is from this ability to cause contraction of the gallbladder that cholecystokinin received its name: *chole,* bile; *cysto,* bladder; *kinin,* to move). Figure 17–31 summarizes the factors controlling the entry of bile into the small intestine.

Small Intestine

Secretion Approximately 1500 ml of fluid is secreted by the walls of the small intestine from the blood into the lumen each day. One of the reasons for water movement into the lumen (secretion) is that the intestinal epithelium at the base of the villi secretes a number of mineral ions, notably sodium, chloride, and bicarbonate ions into the lumen, and water follows by osmosis. These secretions, along with mucus, lubricate the surface of the intestinal tract and help protect the epithelial cells from excessive damage by the digestive

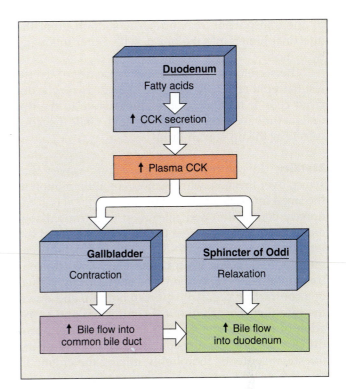

FIGURE 17–31

Regulation of bile entry into the small intestine.

enzymes in the lumen. Some damage to these cells still occurs, however, and the intestinal epithelium has one of the highest cell renewal rates of any tissue in the body.

Chloride is the primary ion determining the magnitude of fluid secretion. It exits the luminal membrane through the same chloride channel that is mutated in cystic fibrosis (Chapter 6). Various hormonal and paracrine signals—as well as certain bacterial toxins—can increase the opening frequency of these channels and thus increase fluid secretion.

As stated earlier, water movement into the lumen also occurs when the chyme entering the small intestine from the stomach is hypertonic because of a high concentration of solutes in the meal and because digestion breaks down large molecules into many more small molecules. This hypertonicity causes the osmotic movement of water from the isotonic plasma into the intestinal lumen.

Absorption Normally, virtually all of the fluid secreted by the small intestine is absorbed back into the blood. In addition, a much larger volume of fluid, which includes salivary, gastric, hepatic, and pancreatic secretions, as well as ingested water, is simultaneously absorbed from the intestinal lumen into the blood. Thus, overall there is a large net absorption of water from the small intestine. Absorption is achieved by the transport of ions, primarily sodium, from the intestinal lumen into the blood, with water following by osmosis.

The secretory and absorptive capacities of the intestinal epithelial cells become altered as newly derived cells at the base of the villi migrate to the tip. Cells at the base of the villi secrete fluid, while the older cells near the tip absorb fluid.

Motility   In contrast to the peristaltic waves that sweep over the stomach, the most common motion in the small intestine during digestion of a meal is a stationary contraction and relaxation of intestinal segments, with little apparent net movement toward the large intestine (Figure 17–32). Each contracting segment is only a few centimeters long, and the contraction lasts a few seconds. The chyme in the lumen of a contracting segment is forced both up and down the intestine. This rhythmical contraction and relaxation of the intestine, known as **segmentation,** produces a continuous division and subdivision of the intestinal contents, thoroughly mixing the chyme in the lumen and bringing it into contact with the intestinal wall.

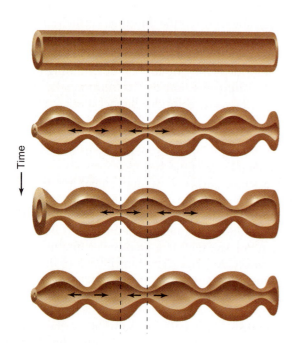

FIGURE 17–32

Segmentation movements of the small intestine in which segments of the intestine contract and relax in a rhythmical pattern but do not undergo peristalsis. This is the pattern encountered during a meal, which mixes the luminal contents.

These segmenting movements are initiated by electrical activity generated by pacemaker cells in or associated with the circular smooth-muscle layer. Like the slow waves in the stomach, this intestinal basic electrical rhythm produces oscillations in the smooth-muscle membrane potential that, if threshold is reached, trigger action potentials that increase muscle contraction. The frequency of segmentation is set by the frequency of the intestinal basic electrical rhythm, but unlike the stomach, which normally has a single rhythm (three per minute), the intestinal rhythm varies along the length of the intestine, each successive region having a slightly lower frequency than the one above. For example, segmentation in the duodenum occurs at a frequency of about 12 contractions/min, whereas in the last portion of the ileum the rate is only 9 contractions/min. Segmentation produces, therefore, a slow migration of the intestinal contents toward the large intestine because more chyme is forced downward, on the average, than upward.

The intensity of segmentation can be altered by hormones, the enteric nervous system, and autonomic nerves; parasympathetic activity increases the force of contraction, and sympathetic stimulation decreases it. Thus, cephalic phase stimuli, including emotional states, can alter intestinal motility. As is true for the stomach, these inputs produce changes in the force of smooth-muscle contraction but do not significantly change the frequencies of the basic electrical rhythms.

After most of a meal has been absorbed, the segmenting contractions cease and are replaced by a pattern of peristaltic activity known as the **migrating motility complex.** Beginning in the lower portion of the stomach, repeated waves of peristaltic activity travel about 2 ft along the small intestine and then die out. This short segment of peristaltic activity slowly migrates down the small intestine, taking about 2 h to reach the large intestine. By the time the migrating motility complex reaches the end of the ileum, new waves are beginning in the stomach, and the process is repeated.

The migrating motility complex moves any undigested material still remaining in the small intestine into the large intestine and also prevents bacteria from remaining in the small intestine long enough to grow and multiply excessively. In diseases in which there is an aberrant migrating motility complex, bacterial overgrowth in the small intestine can become a major problem. Upon the arrival of a meal in the stomach, the migrating motility complex rapidly ceases in the intestine and is replaced by segmentation.

A rise in the plasma concentration of a candidate intestinal hormone, **motilin,** is thought to initiate the migrating motility complex. The mechanisms of motilin action and the control of its release have not been determined.

The contractile activity in various regions of the small intestine can be altered by reflexes initiated at different points along the gastrointestinal tract. For example, segmentation intensity in the ileum increases during periods of gastric emptying, and this is known as the **gastroileal reflex.** Large distensions of the intestine, injury to the intestinal wall, and various bacterial infections in the intestine lead to a complete cessation of motility, the **intestino-intestinal reflex.**

As much as 500 ml of air may be swallowed during a meal. Most of this air travels no farther than the esophagus, from which it is eventually expelled by belching. Some of the air reaches the stomach, however, and is passed on to the intestines, where its percolation through the chyme as the intestinal contents are mixed produces gurgling sounds that are often quite loud.

Large Intestine

The large intestine is a tube 2.5 in. in diameter and about 4 ft long. Its first portion, the **cecum,** forms a blind-ended pouch from which extends the **appendix,** a small fingerlike projection having no known essential function (Figure 17–33). The **colon** consists of three relatively straight segments—the ascending, transverse, and descending portions. The terminal portion of the descending colon is S-shaped, forming the sigmoid colon, which empties into a relatively straight segment of the large intestine, the rectum, which ends at the anus.

Although the large intestine has a greater diameter than the small intestine, its epithelial surface area is far less, since the large intestine is about half as long as the small intestine, its surface is not convoluted, and

its mucosa lacks villi. The secretions of the large intestine are scanty, lack digestive enzymes, and consist mostly of mucus and fluid containing bicarbonate and potassium ions. The primary function of the large intestine is to store and concentrate fecal material before defecation.

Chyme enters the cecum through the **ileocecal sphincter.** This sphincter is normally closed, but after a meal, when the gastroileal reflex increases ileal contractions, it relaxes each time the terminal portion of the ileum contracts, allowing chyme to enter the large intestine. Distension of the large intestine, on the other hand, produces a reflex contraction of the sphincter, preventing fecal material from moving back into the small intestine.

About 1500 ml of chyme enters the large intestine from the small intestine each day. This material is derived largely from the secretions of the lower small intestine since most of the ingested food has been absorbed before reaching the large intestine. Fluid absorption by the large intestine normally accounts for only a small fraction of the fluid entering the gastrointestinal tract each day.

The primary absorptive process in the large intestine is the active transport of sodium from lumen to blood, with the accompanying osmotic absorption of water. If fecal material remains in the large intestine for a long time, almost all the water is absorbed, leaving behind hard fecal pellets. There is normally a net movement of potassium from blood into the large-intestine lumen, and severe depletion of total-body potassium can result when large volumes of fluid are excreted in the feces. There is also a net movement of bicarbonate ions into the lumen, and loss of this bicarbonate (a base) in patients with prolonged diarrhea can cause the blood to become acidic.

The large intestine also absorbs some of the products formed by the bacteria inhabiting this region. Undigested polysaccharides (fiber) are metabolized to short-chain fatty acids by bacteria in the large intestine and absorbed by passive diffusion. The bicarbonate secreted by the large intestine helps to neutralize the increased acidity resulting from the formation of these fatty acids. These bacteria also produce small amounts of vitamins, especially vitamin K, that can be absorbed into the blood. Although this source of vitamins generally provides only a small part of the normal daily requirement, it may make a significant contribution when dietary vitamin intake is low. An individual who depends on absorption of vitamins formed by bacteria in the large intestine may become vitamin deficient if treated with antibiotics that inhibit other species of bacteria as well as the disease-causing bacteria.

Other bacterial products include gas (**flatus**), which is a mixture of nitrogen and carbon dioxide,

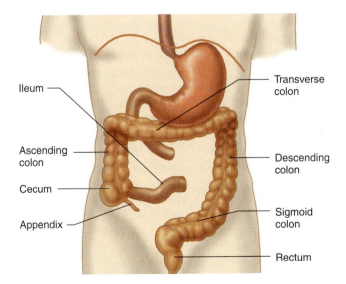

Ileum

Transverse colon

Ascending colon

Descending colon

Cecum

Sigmoid colon

Appendix

Rectum

FIGURE 17–33

The large intestine begins with the cecum.

with small amounts of the inflammable gases hydrogen, methane, and hydrogen sulfide. Bacterial fermentation of undigested polysaccharides produces these gases in the colon (except for nitrogen, which is derived from swallowed air), at the rate of about 400 to 700 ml/day. Certain foods (beans, for example) contain large amounts of carbohydrates that cannot be digested by intestinal enzymes but are readily metabolized by bacteria in the large intestine, producing large amounts of gas.

Motility and Defecation Contractions of the circular smooth muscle in the large intestine produce a segmentation motion with a rhythm considerably slower (one every 30 min) than that in the small intestine. Because of the slow propulsion of the large intestine contents, material entering the large intestine from the small intestine remains for about 18 to 24 h. This provides time for bacteria to grow and multiply. Three to four times a day, generally following a meal, a wave of intense contraction, known as a **mass movement,** spreads rapidly over the transverse segment of the large intestine toward the rectum. This usually coincides with the gastroileal reflex. Unlike a peristaltic wave, in which the smooth muscle at each point relaxes after the wave of contraction has passed, the smooth muscle of the large intestine remains contracted for some time after a mass movement.

The **anus,** the exit from the rectum, is normally closed by the **internal anal sphincter,** which is composed of smooth muscle, and the **external anal sphincter,** which is composed of skeletal muscle under voluntary control. The sudden distension of the walls of the rectum produced by the mass movement of fecal material into it initiates the neurally mediated **defecation reflex.**

The conscious urge to defecate, mediated by mechanoreceptors, accompanies distension of the rectum. The reflex response consists of a contraction of the rectum, relaxation of the internal anal sphincter, but *contraction* of the external anal sphincter (initially), and increased peristaltic activity in the sigmoid colon. Eventually, a pressure is reached in the rectum that triggers reflex *relaxation* of the external anal sphincter, allowing the feces to be expelled.

Brain centers can, however, via descending pathways to somatic nerves to the external anal sphincter, override the reflex signals that eventually would relax the sphincter, thereby keeping the external sphincter closed and allowing a person to delay defecation. In this case, the prolonged distension of the rectum initiates a reverse peristalsis, driving the rectal contents back into the sigmoid colon. The urge to defecate then subsides until the next mass movement again propels more feces into the rectum, increasing its volume and

again initiating the defecation reflex. Voluntary control of the external anal sphincter is learned during childhood. Spinal-cord damage can lead to a loss of voluntary control over defecation.

Defecation is normally assisted by a deep inspiration, followed by closure of the glottis and contraction of the abdominal and thoracic muscles, producing an increase in abdominal pressure that is transmitted to the contents of the large intestine and rectum. This maneuver (termed the Valsalva maneuver) also causes a rise in intrathoracic pressure, which leads to a transient rise in blood pressure followed by a fall in pressure as the venous return to the heart is decreased. The cardiovascular changes resulting from excessive strain during defecation may precipitate a stroke or heart attack, especially in constipated elderly individuals with cardiovascular disease.

Pathophysiology of the Gastrointestinal Tract

Since the end result of gastrointestinal function is the absorption of nutrients, salts, and water, most malfunctions of this organ system affect either the nutritional state of the body or its salt and water content. The following provide a few examples of disordered gastrointestinal function.

Ulcers

Considering the high concentration of acid and pepsin secreted by the stomach, it is natural to wonder why the stomach does not digest itself. Several of the factors that protect the walls of the stomach from being digested are: (1) The surface of the mucosa is lined with cells that secrete a slightly alkaline mucus, which forms a thin layer over the luminal surface. Both the protein content of mucus and its alkalinity neutralize hydrogen ions in the immediate area of the epithelium. Thus, mucus forms a chemical barrier between the highly acid contents of the lumen and the cell surface. (2) The tight junctions between the epithelial cells lining the stomach restrict the diffusion of hydrogen ions into the underlying tissues. (3) Damaged epithelial cells are replaced every few days by new cells arising by the division of cells within the gastric pits.

Yet these protective mechanisms can prove inadequate, and erosion (*ulcers*) of the gastric surface occur. Ulcers can occur not only in the stomach but also in the lower part of the esophagus and in the duodenum. Indeed, duodenal ulcers are about 10 times more frequent than gastric ulcers, affecting about 10 percent of the U.S. population. Damage to blood vessels in the tissues underlying the ulcer may cause bleeding into the gastrointestinal lumen. On occasion, the ulcer may penetrate the entire wall, resulting in leakage of the luminal contents into the abdominal cavity.

Ulcer formation involves breaking the mucosal barrier and exposing the underlying tissue to the corrosive action of acid and pepsin, but it is not always clear what produces the initial damage to the barrier. Although acid is essential for ulcer formation, it is not necessarily the primary factor, and many patients with ulcers have normal or even subnormal rates of acid secretion.

Many factors, including genetic susceptibility, drugs, alcohol, bile salts, and an excessive secretion of acid and pepsin, may contribute to ulcer formation. The major factor, however, is the presence of a bacterium, *Helicobacter pylori,* that is present in the stomachs of a majority of patients with ulcers or *gastritis* (inflammation of the stomach walls). Suppression of these bacteria with antibiotics usually leads to healing of the damaged mucosa.

Once an ulcer has formed, inhibition of acid secretion can remove the constant irritation and allow the ulcer to heal. Two classes of drugs are potent inhibitors of acid secretion. One class of inhibitors acts by blocking a specific class of histamine receptors found on parietal cells, which stimulate acid secretion. The second class of drugs directly inhibits the H,K-ATPase pump in parietal cells. Although both classes of drugs are effective in healing ulcers, if the *Helicobacter pylori* bacteria are not removed, the ulcers tend to recur.

Despite popular notions that ulcers are due to emotional stress and despite the existence of a potential pathway (the parasympathetic nerves) for mediating stress-induced increases in acid secretion, the role of stress in producing ulcers remains unclear. Once the ulcer has been formed, however, emotional stress can aggravate it by increasing acid secretion.

Vomiting

Vomiting is the forceful expulsion of the contents of the stomach and upper intestinal tract through the mouth. Like swallowing, vomiting is a complex reflex coordinated by a region in the brainstem medulla oblongata, in this case known as the **vomiting center.** Neural input to this center from receptors in many different regions of the body can initiate the vomiting reflex. For example, excessive distension of the stomach or small intestine, various substances acting upon chemoreceptors in the intestinal wall or in the brain, increased pressure within the skull, rotating movements of the head (motion sickness), intense pain, and tactile stimuli applied to the back of the throat can all initiate vomiting.

What is the adaptive value of this reflex? Obviously, the removal of ingested toxic substances before they can be absorbed is of benefit. Moreover, the nausea that usually accompanies vomiting may have the adaptive value of conditioning the individual to avoid the future ingestion of foods containing such toxic substances. Why other types of stimuli, such as those producing motion sickness, have become linked to the vomiting center is not clear.

Vomiting is usually preceded by increased salivation, sweating, increased heart rate, pallor, and feelings of nausea. The events leading to vomiting begin with a deep inspiration, closure of the glottis, and elevation of the soft palate. The abdominal muscles then contract, raising the abdominal pressure, which is transmitted to the stomach's contents. The lower esophageal sphincter relaxes, and the high abdominal pressure forces the contents of the stomach into the esophagus. This initial sequence of events can occur repeatedly without expulsion via the mouth and is known as *retching.* Vomiting occurs when the abdominal contractions become so strong that the increased intrathoracic pressure forces the contents of the esophagus through the upper esophageal sphincter.

Vomiting is also accompanied by strong contractions in the upper portion of the small intestine, contractions that tend to force some of the intestinal contents back into the stomach from which they can be expelled. Thus, some bile may be present in the vomitus.

Excessive vomiting can lead to large losses from the stomach of the water and salts that normally would be absorbed in the small intestine. This can result in severe dehydration, upset the body's salt balance, and produce circulatory problems due to a decrease in plasma volume. The loss of acid from vomiting results in a metabolic alkalosis (Chapter 16).

Gallstones

As described earlier, bile contains not only bile salts but also cholesterol and phospholipids, which are water-insoluble and are maintained in soluble form in the bile as micelles. When the concentration of cholesterol in the bile becomes high in relation to the concentrations of phospholipid and bile salts, cholesterol crystallizes out of solution, forming *gallstones.* This can occur if the liver secretes excessive amounts of cholesterol or if the cholesterol becomes overly concentrated in the gallbladder as a result of salt and water absorption. Although cholesterol gallstones are the most frequently encountered gallstones in the Western world, the precipitation of bile pigments can also occasionally be responsible for gallstone formation.

Why some individuals develop gallstones and others do not is still unclear. Women, for example, have about twice the incidence of gallstone formation as men, and Native Americans have a very high incidence compared with other ethnic groups in the United States.

If a gallstone is small, it may pass through the common bile duct into the intestine with no complications. A larger stone may become lodged in the opening of the gallbladder, causing painful contractile spasms of the smooth muscle. A more serious complication arises when a gallstone lodges in the common bile duct, thereby preventing bile from entering the intestine. The absence of bile in the intestine decreases the rate of fat digestion and absorption, so that approximately half of ingested fat is not digested and passes on to the large intestine and eventually appears in the feces. Furthermore, bacteria in the large intestine convert some of this fat into fatty acid derivatives that alter salt and water movements, leading to a net flow of fluid into the large intestine. The result is diarrhea and fluid loss.

Since the duct from the pancreas joins the common bile duct just before it enters the duodenum, a gallstone that becomes lodged as this point prevents both bile and pancreatic secretions from entering the intestine. This results in failure both to neutralize acid and to digest adequately most organic nutrients, not just fat. The end result is severe nutritional deficiencies.

The buildup of pressure in a blocked common bile duct inhibits further secretion of bile. As a result, bilirubin, which is normally secreted into the bile from the blood, accumulates in the blood and diffuses into tissues, where it produces the yellowish coloration of the skin and eyes known as *jaundice.*

It should be emphasized, however, that bile duct obstruction is not the only cause of jaundice. Bilirubin accumulation in the blood can occur if hepatocytes are damaged by liver disease and therefore fail to secrete bilirubin into the bile. It can also occur if the level of bilirubin in the blood exceeds the capacity of the normal liver to secrete it, as in diseases that result in an increased breakdown of red blood cells—*hemolytic jaundice.* At birth, the liver's capacity to secrete bilirubin is not fully developed. During the first few days of life this may result in jaundice, which normally clears spontaneously. Excessive accumulation of bilirubin during the neonatal period, as occurs, for example, with hemolytic disease of the newborn (Chapter 20), carries a risk of bilirubin-induced neurological damage at a time when a critical phase in the development of the nervous system is occurring.

Although surgery may be necessary to remove an inflamed gallbladder or stones from an obstructed duct, newer techniques use drugs to dissolve gallstones or noninvasive ultrasound to shatter gallstones.

Lactose Intolerance

Lactose is the major carbohydrate in milk. It cannot be absorbed directly but must first be digested into its components—glucose and galactose—which are readily absorbed by active transport. Lactose is digested by the enzyme **lactase,** which is embedded in the luminal plasma membranes of intestinal epithelial cells. Lactase is present at birth, but in approximately 25 percent of white Americans and in most Asians, its concentration begins to decline when the child is between 18 and 36 months old. This decline in lactase is genetically determined. In these individuals, as the lactase declines ingested lactose cannot be completely digested—a condition known as *lactose intolerance*—and some lactose remains in the small intestine.

Since the absorption of water requires prior absorption of solute to provide an osmotic gradient, the unabsorbed lactose in persons with lactose intolerance prevents some of the water from being absorbed. This lactose-containing fluid is passed on to the large intestine, where bacteria digest the lactose. They then metabolize the released monosaccharides, producing large quantities of gas (which distends the colon, producing pain) and short-chain fatty acids, which cause fluid movement into the lumen of the large intestine, producing diarrhea. The response to milk ingestion by adults whose lactase levels have diminished varies from mild discomfort to severely dehydrating diarrhea, according to the volume of milk and milk products ingested and the amount of lactase present in the intestine. These symptoms can be avoided if the person either drinks milk in which the lactose has been predigested or takes pills containing lactase along with the milk.

Constipation and Diarrhea

Many people have a mistaken belief that, unless they have a bowel movement every day, the absorption of "toxic" substances from fecal material in the large intestine will somehow poison them. Attempts to identify such toxic agents in the blood following prolonged periods of fecal retention have been unsuccessful, and there appears to be no physiological necessity for having bowel movements at frequent intervals. Whatever maintains a person in a comfortable state is physiologically adequate, whether this means a bowel movement after every meal, once a day, or only once a week.

On the other hand, there often are symptoms—headache, loss of appetite, nausea, and abdominal distension—that may arise when defecation has not occurred for several days or even weeks, depending on the individual. These symptoms of *constipation* are caused not by toxins but by distension of the rectum. The longer that fecal material remains in the large intestine, the more water is absorbed and the harder and drier the feces become, making defecation more difficult and sometimes painful. Thus, constipation tends to promote constipation.

Decreased motility of the large intestine is the primary factor causing constipation. This often occurs in the elderly, or it may result from damage to the colon's enteric nervous system or from emotional stress.

One of the factors increasing motility in the large intestine, and thus opposing the development of constipation, is distension. As noted earlier, dietary fiber (cellulose and other complex polysaccharides) is not digested by the enzymes in the small intestine and is passed on to the large intestine, where its bulk produces distension and thereby increases motility. Bran, most fruits, and vegetables are examples of foods that have a relatively high fiber content.

Laxatives, agents that increase the frequency or ease of defecation, act through a variety of mechanisms. Thus, fiber provides a natural laxative. Some laxatives, such as mineral oil, simply lubricate the feces, making defecation easier and less painful. Others contain magnesium and aluminum salts, which are poorly absorbed and therefore lead to water retention in the intestinal tract. Still others, such as castor oil, stimulate the motility of the colon and inhibit ion transport across the wall, thus affecting water absorption.

Excessive use of laxatives in an attempt to maintain a preconceived notion of regularity leads to a decreased responsiveness of the large intestine to normal defecation-promoting signals. In such cases, a long period without defecation may occur following cessation of laxative intake, appearing to confirm the necessity of taking laxatives to promote regularity.

Diarrhea is characterized by large, frequent, watery stools. Diarrhea can result from decreased fluid absorption, increased fluid secretion, or both. The increased motility that accompanies diarrhea probably does not *cause* most cases of diarrhea (by decreasing the time available for fluid absorption) but rather is a result of the distension produced by increased luminal fluid.

A number of bacterial, protozoan, and viral diseases of the intestinal tract cause secretory diarrhea. *Cholera,* which is endemic in many parts of the world, is caused by a bacterium that releases a toxin that stimulates the production of cyclic AMP in the secretory cells at the base of the intestinal villi. This leads to an increased frequency of opening of the chloride channels in the luminal membrane and hence increased secretion of chloride ions. There is an accompanying osmotic flow of water into the intestinal lumen, resulting in massive diarrhea that can be life-threatening due to dehydration and decreased blood volume that leads to circulatory shock. The salt and water lost by this severe form of diarrhea can be balanced by ingesting a simple solution containing salt and glucose. The active absorption of these solutes is accompanied by absorption of water, which replaces the fluid lost by diarrhea.

Traveler's diarrhea, produced by several species of bacteria, produces a secretory diarrhea by the same mechanism as the cholera bacterium, but is less severe.

In addition to decreased blood volume due to the salt and water loss, other consequences of severe diarrhea are potassium depletion and metabolic acidosis (Chapter 16) resulting from the excessive fecal loss of potassium and bicarbonate ions, respectively.

SUMMARY

I. The gastrointestinal system transfers digested organic nutrients, minerals, and water from the external environment to the internal environment. The four processes used to accomplish this function are (a) digestion, (b) secretion, (c) absorption, and (d) motility.
  a. The system is designed to maximize the absorption of most nutrients, not to regulate the amount absorbed.
  b. The system does not play a major role in the removal of waste products from the internal environment.

Overview: Functions of the Gastrointestinal Organs

I. The names and functions of the gastrointestinal organs are summarized in Figure 17–3.
II. Each day the gastrointestinal tract secretes about 6 times more fluid into the lumen than is ingested. Only 1 percent of this fluid is excreted in the feces.

Structure of the Gastrointestinal Tract Wall

I. The structure of the wall of the gastrointestinal tract is summarized in Figure 17–6.
  a. The area available for absorption in the small intestine is greatly increased by the folding of the intestinal wall and by the presence of villi and microvilli on the surface of the epithelial cells.
  b. The epithelial cells lining the intestinal tract are continuously replaced by new cells arising from cell division at the base of the villi.
  c. The venous blood from the small intestine, containing absorbed nutrients other than fat, passes to the liver via the hepatic portal vein before returning to the heart. Fat is absorbed into the lymphatic vessels (lacteals) in each villus.

Digestion and Absorption

I. Starch is digested by amylases secreted by the salivary glands and pancreas, and the resulting products, as well as ingested disaccharides, are digested to monosaccharides by enzymes in the luminal membranes of epithelial cells in the small intestine.
  a. Most monosaccharides are then absorbed by secondary active transport.

b. Some polysaccharides, such as cellulose, cannot be digested and pass to the large intestine, where they are metabolized by bacteria.

II. Proteins are broken down into small peptides and amino acids, which are absorbed by secondary active transport in the small intestine.

a. The breakdown of proteins to peptides is catalyzed by pepsin in the stomach and by the pancreatic enzymes trypsin and chymotrypsin in the small intestine.

b. Peptides are broken down into amino acids by pancreatic carboxypeptidase and intestinal aminopeptidase.

c. Small peptides consisting of two to three amino acids can be actively absorbed into epithelial cells and then broken down to amino acids, which are released into the blood.

III. The digestion and absorption of fat by the small intestine requires mechanisms that solubilize the fat and its digestion products.

a. Large fat globules leaving the stomach are emulsified in the small intestine by bile salts and phospholipids secreted by the liver.

b. Lipase from the pancreas digests fat at the surface of the emulsion droplets, forming fatty acids and monoglycerides.

c. These water-insoluble products of lipase action, when combined with bile salts, form micelles, which are in equilibrium with the free molecules.

d. Free fatty acids and monoglycerides diffuse across the luminal membranes of epithelial cells, within which they are enzymatically recombined to form triacylglycerol, which is released as chylomicrons from the blood side of the cell by exocytosis.

e. The released chylomicrons enter lacteals in the intestinal villi and pass, by way of the lymphatic system, to the venous blood returning to the heart.

IV. Fat-soluble vitamins are absorbed by the same pathway used for fat absorption. Most water-soluble vitamins are absorbed in the small intestine by diffusion or mediated transport. Vitamin B_{12} is absorbed in the ileum by endocytosis after combining with intrinsic factor secreted into the lumen by parietal cells in the stomach.

V. Water is absorbed from the small intestine by osmosis following the active absorption of solutes, primarily sodium chloride.

Regulation of Gastrointestinal Processes

I. Most gastrointestinal reflexes are initiated by luminal stimuli: (a) distension, (b) osmolarity, (c) acidity, and (d) digestion products.

a. Neural reflexes are mediated by short reflexes in the enteric nervous system and by long reflexes involving afferent and efferent neurons to and from the CNS.

b. Endocrine cells scattered throughout the epithelium of the stomach secrete gastrin, and cells in the small intestine secrete secretin, CCK, and GIP. The properties of these hormones are summarized in Table 17–3.

c. The three phases of gastrointestinal regulation—cephalic, gastric, and intestinal—are named for the location of the stimulus that initiates the response.

II. Chewing breaks up food into particles suitable for swallowing, but it is not essential for the eventual digestion and absorption of food.

III. Salivary secretion is stimulated by food in the mouth acting reflexly via chemoreceptors and pressure receptors. Both sympathetic and parasympathetic stimulation increase salivary secretion.

IV. Food moved into the pharynx by the tongue initiates swallowing, which is coordinated by the swallowing center in the brainstem medulla oblongata.

a. Food is prevented from entering the trachea by inhibition of respiration and by closure of the glottis.

b. The upper esophageal sphincter relaxes as food is moved into the esophagus, and then the sphincter closes.

c. Food is moved through the esophagus toward the stomach by peristaltic waves. The lower esophageal sphincter remains open throughout swallowing.

d. If food does not reach the stomach with the first peristaltic wave, distension of the esophagus initiates secondary peristalsis.

V. The factors controlling acid secretion by parietal cells in the stomach are summarized in Table 17–4.

VI. Pepsinogen, secreted by the gastric chief cells in response to most of the same reflexes that control acid secretion, is converted to the active proteolytic enzyme pepsin in the stomach's lumen by acid and by activated pepsin.

VII. Peristaltic waves sweeping over the stomach become stronger in the antrum, where most mixing occurs. With each wave, only a small portion of the stomach's contents are expelled into the small intestine through the pyloric sphincter.

a. Cycles of membrane depolarization, the basic electrical rhythm generated by gastric smooth muscle, determine gastric peristaltic wave frequency. Contraction strength can be altered by neural and hormonal changes in membrane potential, which is imposed on the basic electrical rhythm.

b. Distension of the stomach increases the force of contractions and the rate of emptying. Distension of the small intestine, and fat, acid, or hypertonic solutions in the intestinal lumen inhibit gastric contractions.

VIII. The exocrine portion of the pancreas secretes digestive enzymes and bicarbonate ions, all of which reach the duodenum through the pancreatic duct.

a. The bicarbonate ions neutralize acid entering the small intestine from the stomach.

b. Most of the proteolytic enzymes, including trypsin, are secreted by the pancreas in inactive forms. Trypsin is activated by enterokinase located on the membranes of the small-intestine cells and in turn activates other inactive pancreatic enzymes.

c. The hormone secretin, released from the small intestine in response to increased luminal acidity, stimulates pancreatic bicarbonate secretion. CCK is released from the small intestine in response to the products of fat and protein digestion, and stimulates pancreatic enzyme secretion.

d. Parasympathetic stimulation increases pancreatic secretion.

IX. The liver secretes bile, the major ingredients of which are bile salts, cholesterol, lecithin, bicarbonate ions, bile pigments, and trace metals.

a. Bile salts undergo continuous enterohepatic recirculation during a meal. The liver synthesizes new bile salts to replace those lost in the feces.

b. The greater the bile salt concentration in the hepatic portal blood, the greater the rate of bile secretion.

c. Bilirubin, the major bile pigment, is a breakdown product of hemoglobin and is absorbed from the blood by the liver and secreted into the bile.

d. Secretin stimulates bicarbonate secretion by the cells lining the bile ducts in the liver.

e. Bile is concentrated in the gallbladder by the absorption of NaCl and water.

f. Following a meal, the release of CCK from the small intestine causes the gallbladder to contract and the sphincter of Oddi to relax, thereby injecting concentrated bile into the intestine.

X. In the small intestine, the digestion of polysaccharides and proteins increases the osmolarity of the luminal contents, producing water flow into the lumen.

XI. Sodium, chloride, bicarbonate, and water are secreted by the small intestine. However, most of these secreted substances, as well as those entering the small intestine from other sources, are absorbed back into the blood.

XII. Intestinal motility is coordinated by the enteric nervous system and modified by long and short reflexes and hormones.

a. During and shortly after a meal, the intestinal contents are mixed by segmenting movements of the intestinal wall.

b. After most of the food has been digested and absorbed, segmentation is replaced by the migrating motility complex, which moves the undigested material into the large intestine by a migrating segment of peristaltic waves.

XIII. The primary function of the large intestine is to store and concentrate fecal matter before defecation.

a. Water is absorbed from the large intestine secondary to the active absorption of sodium, leading to the concentration of fecal matter.

b. Flatus is produced by bacterial fermentation of undigested polysaccharides.

c. Three to four times a day, mass movements in the colon move its contents into the rectum.

d. Distension of the rectum initiates defecation, which is assisted by a forced expiration against a closed glottis.

e. Defecation can be voluntarily controlled through somatic nerves to the skeletal muscles of the external anal sphincter.

Pathophysiology of the Gastrointestinal Tract

I. The factors that normally prevent breakdown of the mucosal barrier and formation of ulcers are (1) secretion of an alkaline mucus, (2) tight junctions between epithelial cells, and (3) rapid replacement of epithelial cells.

a. The bacterium *Helicobacter pylori* is a major cause of damage to the mucosal barrier leading to ulcers.

b. Drugs that block histamine receptors or inhibit the H,K-ATPase pump inhibit acid secretion and promote ulcer healing.

II. Vomiting is coordinated by the vomiting center in the brainstem medulla oblongata. Contractions of abdominal muscles force the contents of the stomach into the esophagus (retching), and if the contractions are strong enough, they force the contents of the esophagus through the upper esophageal sphincter into the mouth (vomiting).

III. Precipitation of cholesterol or, less often, bile pigments in the gallbladder forms gallstones, which can block the exit of the gallbladder or common bile duct. In the latter case, the failure of bile salts to reach the intestine causes decreased digestion and absorption of fat, and the accumulation of bile pigments in the blood and tissues causes jaundice.

IV. Lactase, which is present at birth, undergoes a genetically determined decrease during childhood in many individuals. In the absence of lactase, lactose cannot be digested, and its presence in the small intestine can result in diarrhea and increased flatus production when milk is ingested.

V. Constipation is primarily the result of decreased colonic motility. The symptoms of constipation are produced by overdistension of the rectum, not by the absorption of toxic bacterial products.

VI. Diarrhea can be caused by decreased fluid absorption, increased fluid secretion, or both.

REVIEW QUESTIONS

1. List the four processes that accomplish the functions of the gastrointestinal system.
2. List the primary functions performed by each of the organs in the gastrointestinal system.
3. Approximately how much fluid is secreted into the gastrointestinal tract each day compared with the amount of food and drink ingested? How much of this appears in the feces?
4. What structures are responsible for the large surface area of the small intestine?
5. Where does the venous blood go after leaving the small intestine?
6. Identify the enzymes involved in carbohydrate digestion and the mechanism of carbohydrate absorption in the small intestine.
7. List three ways in which proteins or their digestion products can be absorbed from the small intestine.
8. Describe the process of fat emulsification.
9. What is the role of micelles in fat absorption?
10. Describe the movement of fat digestion products from the intestinal lumen to a lacteal.
11. How does the absorption of fat-soluble vitamins differ from that of water-soluble vitamins?
12. Specify two conditions that may lead to failure to absorb vitamin B_{12}.
13. How are salts and water absorbed in the small intestine?
14. Describe the role of ferritin in the absorption of iron.
15. List the four types of stimuli that initiate most gastrointestinal reflexes.
16. Describe the location of the enteric nervous system and its role in both short and long reflexes.
17. Name the four established gastrointestinal hormones and state their major functions.
18. Describe the neural reflexes leading to increased salivary secretion.
19. Describe the sequence of events that occur during swallowing.
20. List the cephalic, gastric, and intestinal phase stimuli that stimulate or inhibit acid secretion by the stomach.
21. Describe the function of gastrin and the factors controlling its secretion.
22. By what mechanism is pepsinogen converted to pepsin in the stomach?
23. Describe the factors that control gastric emptying.
24. Describe the mechanisms controlling pancreatic secretion of bicarbonate and enzymes.
25. How are pancreatic proteolytic enzymes activated in the small intestine?
26. List the major constituents of bile.
27. Describe the recycling of bile salts by the enterohepatic circulation.
28. What determines the rate of bile secretion by the liver?
29. Describe the effects of secretin and CCK on the bile ducts and gallbladder.

30. What causes water to move from the blood to the lumen of the duodenum following gastric emptying?
31. Describe the type of intestinal motility found during and shortly after a meal and the type found several hours after a meal.
32. Describe the production of flatus by the large intestine.
33. Describe the factors that initiate and control defecation.
34. Why is the stomach's wall normally not digested by the acid and digestive enzymes in the lumen?
35. Describe the process of vomiting.
36. What are the consequences of blocking the common bile duct with a gallstone?
37. What are the consequences of the failure to digest lactose in the small intestine?
38. Contrast the factors that cause constipation with those that produce diarrhea.

CLINICAL TERMS

pernicious anemia	gallstones
hemochromatosis	jaundice
heartburn	hemolytic jaundice
gastro-esophageal reflux	lactose intolerance
dumping syndrome	constipation
ulcers	laxatives
Helicobacter pylori	diarrhea
gastritis	cholera
retching	traveler's diarrhea

THOUGHT QUESTIONS

(Answers are given in Appendix A.)

1. If the salivary glands were unable to secrete amylase, what effect would this have on starch digestion?
2. Whole milk or a fatty snack consumed before the ingestion of alcohol decreases the rate of intoxication. By what mechanism may fat be acting to produce this effect?
3. A patient brought to a hospital after a period of prolonged vomiting has an elevated heart rate, decreased blood pressure, and below-normal blood acidity. Explain these symptoms in terms of the consequences of excessive vomiting.
4. Can fat be digested and absorbed in the absence of bile salts? Explain.
5. How might damage to the lower portion of the spinal cord affect defecation?
6. One of the older but no longer used procedures in the treatment of ulcers is vagotomy, surgical cutting of the vagus (parasympathetic) nerves to the stomach. By what mechanism might this procedure help ulcers to heal and decrease the incidence of new ulcers?

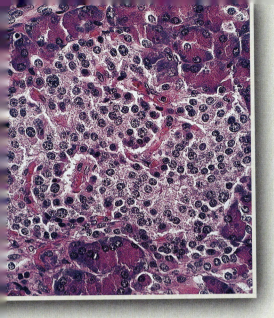

Regulation of Organic Metabolism, Growth, and Energy Balance

Chapter 4 introduced the concepts of energy and of organic metabolism at the level of the individual cell. This chapter deals with a variety of topics that are concerned in one way or another with those same concepts, but for the entire body. First we describe how the metabolic pathways for carbohydrate, fat, and protein are controlled so as to provide continuous sources of energy to the various tissues and organs at all times. The next topic is how the metabolic changes that underlie body growth occur and are controlled. Finally, we describe the determinants of total-body energy balance in terms of energy intake and output and how the regulation of body temperature is achieved.

CONTROL AND INTEGRATION OF CARBOHYDRATE, PROTEIN, AND FAT METABOLISM

Events of the Absorptive and Postabsorptive States

Mechanisms have evolved for survival during alternating periods of plenty and fasting. We speak of two functional states or periods: the **absorptive state,** during which ingested nutrients are entering the blood from the gastrointestinal tract, and the **postabsorptive state,** during which the gastrointestinal tract is empty of nutrients and energy must be supplied by the body's own stores. Since an average meal requires approximately 4 h for complete absorption, our usual three-meal-a-day pattern places us in the postabsorptive state during the late morning and afternoon and almost the entire night. We shall refer to going more than 24 h without eating as fasting.

During the absorptive period, some of the ingested nutrients supply the energy needs of the body, and the remainder are added to the body's energy stores, to be called upon during the next postabsorptive period. Total-body energy stores are adequate for the average person to easily withstand a fast of many weeks (provided that water is available).

Figures 18–1 and 18–2 summarize the major pathways to be described in this chapter. Although they may appear formidable at first glance, they should present little difficulty after we have described the component parts, and these figures should be referred to constantly during the following discussion.

Absorptive State

We shall assume, for this discussion, that an average meal contains all three of the major nutrients—carbohydrate, protein, and fat—with carbohydrate constituting most of the meal's energy content (calories). Recall from Chapter 17 that carbohydrate and protein are absorbed primarily as monosaccharides and amino acids, respectively, into the blood supplying the gastrointestinal tract. The blood leaves the gastrointestinal tract to go directly to the liver by way of the hepatic portal vein, allowing the liver to alter the nutrient composition of the blood before it returns to the heart to be pumped to the rest of the body. In contrast to carbohydrate and amino acids, fat is absorbed into the *lymph,* as triacylglycerols in chylomicrons; the lymph then drains into the systemic venous system. Thus, the liver does not get first crack at absorbed fat.

Absorbed Carbohydrate Some of the carbohydrate absorbed from the gastrointestinal tract is galactose and fructose, but since these sugars are either converted to glucose by the liver or enter essentially the same metabolic pathways as does glucose, we shall simply refer to absorbed carbohydrate as glucose.

Much of the absorbed glucose enters various body cells and is catabolized to carbon dioxide and water, providing the energy for ATP formation (Chapter 4). Indeed, and this is a key point, glucose is the body's major energy source during the absorptive state. In this regard, it should be recognized that skeletal muscle makes up the majority of body mass and is the major consumer of metabolic fuel, even at rest.

Skeletal muscle not only catabolizes glucose during the absorptive phase, but converts some of the glucose to the polysaccharide glycogen, which is then stored in the muscle.

Adipose-tissue cells (adipocytes) also catabolize glucose for energy, but the most important fate of glucose in adipocytes during the absorptive phase is its transformation to fat (triacylglycerols). Glucose is the precursor of both α-glycerol phosphate and fatty acids, and these molecules are then linked together to form triacylglycerols.

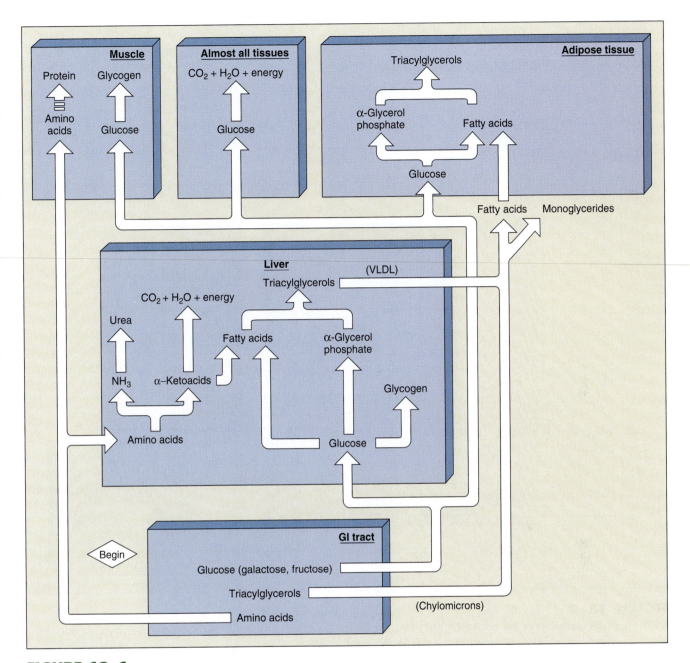

FIGURE 18–1

Major metabolic pathways of the absorptive state. The arrow from amino acids to protein in muscle is dashed to denote the fact that excess amino acids are not stored as protein (see text). All arrows between boxes denote transport of the substance via the blood. VLDL = very low density lipoproteins.

Another large fraction of the absorbed glucose enters the liver cells. This is a very important point: During the absorptive period, there is net *uptake* of glucose by the liver. It is either stored as glycogen, as in skeletal muscle, or transformed to α-glycerol phosphate and fatty acids, which are then used to synthesize triacylglycerols, as in adipose tissue. Some of the fat synthesized from glucose in the liver is stored there, but most is packaged, along with specific proteins, into molecular aggregates of lipids and proteins called lipopro-

teins. These aggregates are secreted by the liver cells and enter the blood. They are called **very low density lipoproteins (VLDL)** because they contain much more fat than protein, and fat is less dense than protein. The synthesis of VLDL by liver cells occurs by processes similar to those for synthesis of chylomicrons by intestinal mucosal cells, as described in Chapter 17.

Once in the bloodstream, VLDL complexes, being quite large, do not readily penetrate capillary walls. Instead, their triacylglycerols are hydrolyzed mainly to

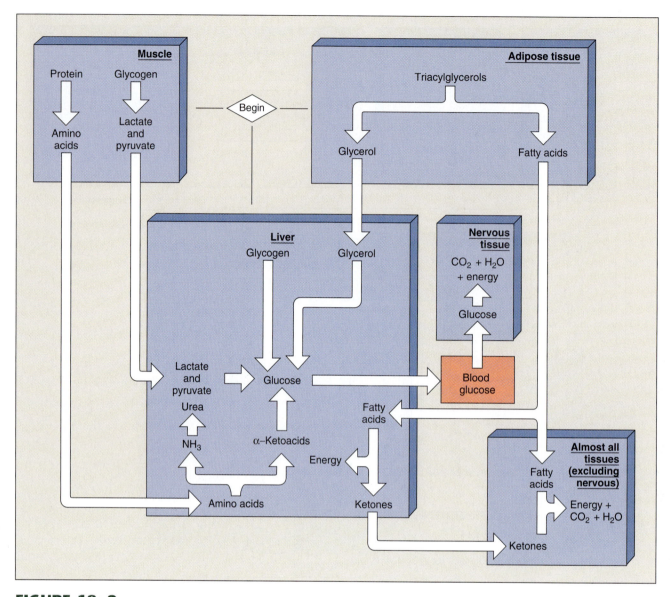

FIGURE 18–2

Major metabolic pathways of the postabsorptive state. The central focus is regulation of the blood glucose concentration. All arrows between boxes denote transport of the substance via the blood.

monoglycerides (glycerol linked to one fatty acid chain) and fatty acids by the enzyme **lipoprotein lipase,** which is located on the blood-facing surface of capillary endothelial cells, especially those in adipose tissue. In adipose-tissue capillaries, the fatty acids generated by this enzyme's action diffuse across the capillary wall and into the adipocytes. There they combine with α-glycerol phosphate, supplied by glucose metabolites, to form triacylglycerols once again. Thus, most of the fatty acids in the VLDL triacylglycerol originally synthesized from glucose by the *liver* end up being stored in triacylglycerol in *adipose tissue.* (The monoglycerides formed in the blood by the action of lipoprotein lipase in adipose-tissue capillaries circulate to the liver where they are metabolized.)

To summarize, the major fates of glucose during the absorptive phase are utilization for energy, storage as glycogen in liver and skeletal muscle, and storage as fat in adipose tissue.

Absorbed Triacylglycerols   As described in Chapter 17, almost all absorbed chylomicrons enter the lymph, which flows into the systemic circulation. The biochemical processing of these chylomicron triacylglycerols in plasma is quite similar to that just described for VLDL produced by the liver. The fatty acids of plasma chylomicrons are released, mainly within adipose-tissue capillaries, by the action of endothelial lipoprotein lipase. The released fatty acids then enter adipocytes and combine with α-glycerol phosphate, synthesized in the adipocytes from glucose metabolites, to form triacylglycerols.

The importance of glucose for triacylglycerol synthesis in adipocytes cannot be overemphasized. Adipocytes do not have the enzyme required for phosphorylation of glycerol, and so α-glycerol phosphate can be formed in these cells *only* from glucose metabolites and not from glycerol or any other fat metabolites.

In contrast to α-glycerol phosphate, there are three major sources of the fatty acids found in adipose-tissue triacylglycerol: (1) glucose that enters adipose tissue and is converted to fatty acids; (2) glucose that is converted in the liver to VLDL triacylglycerols, which are transported via the blood to the adipose tissue; and (3) ingested triacylglycerols transported to adipose tissue in chylomicrons. As we have seen, sources (2) and (3) require the action of lipoprotein lipase to release the fatty acids from the circulating triacylglycerols.

This description has emphasized the *storage* of ingested fat. For simplicity, we have not shown in Figure 18–1 that a fraction of the ingested fat is not stored but is oxidized during the absorptive state by various organs to provide energy. The relative amounts of carbohydrate and fat used for energy during the absorptive period depend largely on the content of the meal.

Absorbed Amino Acids   A minority of the absorbed amino acids enter liver cells. They are used to synthesize a variety of proteins, including liver enzymes and plasma proteins, or they are converted to carbohydrate-like intermediates known as **α-ketoacids** by removal of the amino group (deamination, Chapter 4). The amino groups are used to synthesize urea, which enters the blood and is excreted by the kidneys. The α-ketoacids can enter the Krebs tricarboxylic acid cycle and be catabolized to provide energy for the liver cells, or they can be converted to fatty acids, thereby participating in fat synthesis by the liver.

Most ingested amino acids are not taken up by the liver cells, however, but enter other cells (Figure 18–1), where they may be used to synthesize proteins. We have simplified Figure 18–1 by showing "nonliver" amino acid uptake only by muscle, because muscle contains by far the largest amount of body protein. It should be emphasized, however, that all cells require a constant supply of amino acids for protein synthesis and participate in the dynamics of protein metabolism.

Protein synthesis is represented by a dashed arrow in the muscle box in Figure 18–1 to call attention to an important fact: There is a net synthesis of protein during the absorptive period, but this basically just replaces the proteins catabolized during the postabsorptive period. In other words, excess amino acids are not *stored* as protein in the sense that glucose is stored as glycogen or that both glucose and fat are stored as fat. Rather, ingested amino acids in excess of those

TABLE 18–1 Summary of Nutrient Metabolism during the Absorptive Period

1. Energy is provided primarily by absorbed carbohydrate.

2. There is net uptake of glucose by the liver.

3. Some carbohydrate is stored as glycogen in liver and muscle, but most carbohydrate and fat in excess of that utilized for energy are stored mainly as fat in adipose tissue.

4. There is some synthesis of body proteins, but many of the amino acids in dietary protein are utilized for energy or converted to fat.

needed to maintain a stable protein turnover are merely converted to carbohydrate or fat. Therefore, eating large amounts of protein does not in itself cause increases in body protein. This discussion does not apply to growing children, who manifest a continuous increase in body protein, or to adults who are actively building body mass as, for example, by weight lifting.

Nutrient metabolism during the absorptive period is summarized in Table 18–1.

Postabsorptive State

As the absorptive period ends, net synthesis of glycogen, fat, and protein ceases, and net catabolism of all these substances begins to occur. The overall significance of these events can be understood in terms of the essential problem during the postabsorptive period: No glucose is being absorbed from the intestinal tract, yet the plasma glucose concentration must be maintained because the brain normally utilizes only glucose for energy. Too low a plasma glucose concentration can result in alterations of neural activity ranging from subtle impairment of mental function to coma and even death.

The events that maintain plasma glucose concentration fall into two categories: (1) reactions that provide sources of blood glucose, and (2) glucose sparing because of fat utilization.

Sources of Blood Glucose   The sources of blood glucose during the postabsorptive period are as follows (Figure 18–2):

1. **Glycogenolysis,** the hydrolysis of glycogen stores, occurs in the liver and skeletal muscle. In the liver, glucose is formed by this process and enters the blood. Hepatic glycogenolysis, a rapidly occurring event, is the first line of defense in maintaining plasma glucose concentration. The amount of glucose available from this source, however, can supply the body's needs for only a few hours.

Glycogenolysis also occurs in skeletal muscle, which contains approximately the same amount of glycogen as the liver. However, muscle, unlike liver, lacks the enzyme necessary to form glucose from the glucose 6-phosphate formed during glycogenolysis (Chapter 4). Instead, the glucose 6-phosphate undergoes glycolysis within the muscle to yield pyruvate and lactate. These substances enter the blood, circulate to the liver, and are converted into glucose, which can then leave the liver cells to enter the blood. Thus, muscle glycogen contributes to the blood glucose indirectly via the liver.

2. The catabolism of triacylglycerols yields glycerol and fatty acids, a process termed **lipolysis.** The major site of lipolysis is adipose tissue, and the glycerol and fatty acids then enter the blood. The glycerol reaching the liver is converted to glucose. Thus, an important source of glucose during the postabsorptive period is the glycerol released when adipose-tissue triacylglycerol is broken down.

3. A few hours into the postabsorptive period, protein becomes the major source of blood glucose. Large quantities of protein in muscle and, to a lesser extent, other tissues can be catabolized without serious cellular malfunction. There are, of course, limits to this process, and continued protein loss during a prolonged fast ultimately means functional disintegration, sickness, and death. Before this point is reached, however, protein breakdown can supply large quantities of amino acids, particularly alanine, that enter the blood and are picked up by the liver, which converts them, via the α-ketoacid pathway, to glucose.

In items 1 through 3 above, we described the synthesis by the liver of glucose from pyruvate, lactate, glycerol, and amino acids. Synthesis from any of these precursors is known as **gluconeogenesis**—that is, "new formation of glucose." During a 24-h fast, gluconeogenesis provides approximately 180 g of glucose. (The liver is not the only organ capable of gluconeogenesis; the kidneys also perform gluconeogenesis, but mainly during a prolonged fast.)

Glucose Sparing (Fat Utilization) The 180 g of glucose per day produced by gluconeogenesis in the liver (and kidneys) during fasting supplies 720 kcal. As described later in this chapter, normal total energy expenditure for an average adult equals 1500 to 3000 kcal/day. Accordingly, gluconeogenesis cannot supply all the body's energy needs. The following essential adjustment must therefore take place during the

transition from the absorptive to the postabsorptive state: Most organs and tissues markedly reduce their glucose catabolism and increase their fat utilization, the latter becoming the major energy source. This metabolic adjustment, termed **glucose sparing,** "spares" the glucose produced by the liver for use by the nervous system.

The essential step in this adjustment is lipolysis, the catabolism of adipose-tissue triacylglycerol, which liberates glycerol and fatty acids into the blood. We described lipolysis in the previous section in terms of its importance in providing *glycerol* to the liver for conversion to glucose. Now, we focus on the liberated *fatty acids*, which circulate bound to plasma albumin. [Despite this binding to protein, they are known as free fatty acids (FFA) in that they are "free" of glycerol.] The circulating fatty acids are picked up and metabolized by almost all tissues, *excluding the nervous system.* They provide energy in two ways (Chapter 4): (1) They first undergo beta-oxidation to yield hydrogen atoms (that go on to oxidative phosphorylation) and acetyl CoA; and (2) the acetyl CoA enters the Krebs cycle and is catabolized to carbon dioxide and water.

The liver is unique, however, in that most of the acetyl CoA it forms from fatty acids during the postabsorptive state does not enter the Krebs cycle but is processed into three compounds collectively called **ketones** (or ketone bodies). (Note that ketones are not the same as α-ketoacids, which, as we have seen, are metabolites of amino acids.) Ketones are released into the blood and provide an important energy source during prolonged fasting for the many tissues, *including the brain,* capable of oxidizing them via the Krebs cycle. One of the ketones is acetone, some of which is exhaled

---

TABLE 18–2 Summary of Nutrient Metabolism during the Postabsorptive Period

1. Glycogen, fat, and protein syntheses are curtailed, and net breakdown occurs.

2. Glucose is formed in the liver both from the glycogen stored there and by gluconeogenesis from blood-borne lactate, pyruvate, glycerol, and amino acids. The kidneys also perform gluconeogenesis during a prolonged fast.

3. The glucose produced in the liver (and kidneys) is released into the blood, but its utilization for energy is greatly reduced in muscle and other nonneural tissues.

4. Lipolysis releases adipose-tissue fatty acids into the blood, and the oxidation of these fatty acids by most cells and of ketones produced from them by the liver provides most of the body's energy supply.

5. The brain continues to use glucose but also starts using ketones as they build up in the blood.

and accounts for the distinctive breath odor of individuals undergoing prolonged fasting or, as we shall see, suffering from severe untreated diabetes mellitus.

The net result of fatty acid and ketone utilization during fasting is provision of energy for the body and sparing of glucose for the brain. Moreover, as just emphasized, the brain can use ketones for an energy source, and it does so increasingly as ketones build up in the blood during the first few days of a fast. The survival value of this phenomenon is very great: When the brain reduces its glucose requirement by utilizing ketones, much less protein breakdown is required to supply amino acids for gluconeogenesis. Accordingly, the protein stores will last longer, and the ability to withstand a long fast without serious tissue disruption is enhanced.

Table 18–2 summarizes the events of the postabsorptive period. The combined effects of glycogenolysis, gluconeogenesis, and the switch to fat utilization are so efficient that, after several days of complete fasting, the plasma glucose concentration is reduced by only a few percent. After 1 month, it is decreased only 25 percent.

Endocrine and Neural Control of the Absorptive and Postabsorptive States

We now turn to the endocrine and neural factors that control and integrate these metabolic pathways. We shall focus primarily on the following questions, summarized in Figure 18–3: (1) What controls net anabolism of protein, glycogen, and triacylglycerol in the absorptive phase, and net catabolism in the postabsorptive phase? (2) What induces primarily glucose utilization by cells for energy during the absorptive phase, but fat utilization during the postabsorptive phase? (3) What drives net glucose uptake by the liver during the absorptive phase, but gluconeogenesis and glucose release during the postabsorptive phase?

The most important controls of these transitions from feasting to fasting, and vice versa, are two pancreatic hormones—insulin and glucagon. Also playing a role are the hormone epinephrine, from the adrenal medulla, and the sympathetic nerves to liver and adipose tissue.

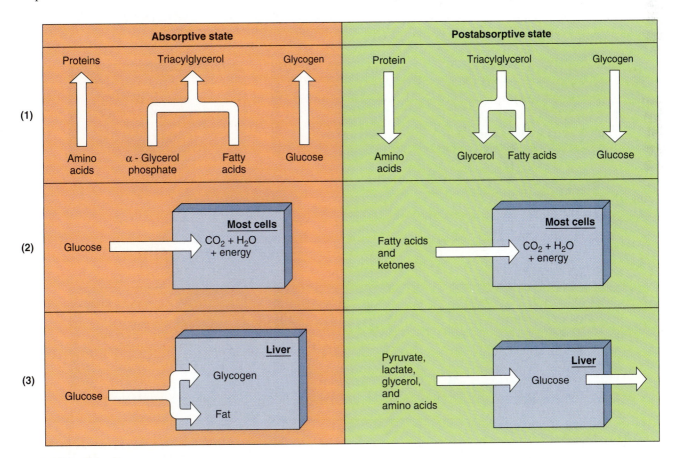

FIGURE 18–3

Summary of critical points in transition from the absorptive state to the postabsorptive state. The term "absorptive state" could be replaced with "actions of insulin," and the term "postabsorptive state" with "results of decreased insulin." The numbers at the left margin refer to discussion in the text.

Insulin and glucagon are peptides secreted by the **islets of Langerhans,** clusters of endocrine cells in the pancreas. Appropriate histological techniques reveal several distinct types of islet cells, each of which secretes a different hormone. The **beta cells** (or B cells) are the source of insulin, and the **alpha cells** (or A cells) of glucagon. (There is at least one other hormone—somatostatin—secreted by still other islet cells, but the function of this pancreatic hormone in human beings is not yet fully established.)

Insulin

Insulin is the most important controller of organic metabolism. Its secretion, and hence plasma concentration, are increased during the absorptive state and decreased during the postabsorptive state. For simplicity, insulin's many actions are often divided into two broad categories: (1) *metabolic effects* on carbohydrate,

lipid, and protein synthesis, and (2) *growth-promoting effects* on DNA synthesis, cell division, and cell differentiation. This section deals only with the metabolic effects; the growth-promoting effects are described later in this chapter.

The metabolic effects of insulin are exerted mainly on muscle cells (both cardiac and skeletal), adipose-tissue cells, and liver cells. The most important responses of these target cells are summarized in Figure 18–4. Compare the top portion of this figure to Figure 18–1 and to the left panel of Figure 18–3 and you will see that these responses to an increase in insulin are the same as the events of the absorptive-state pattern. Conversely, the effects of a reduction in plasma insulin are the same as the events of the postabsorptive pattern in Figure 18–2 and the right panel of Figure 18–3. The reasons for these correspondences is that *an increased plasma concentration of insulin is the major cause*

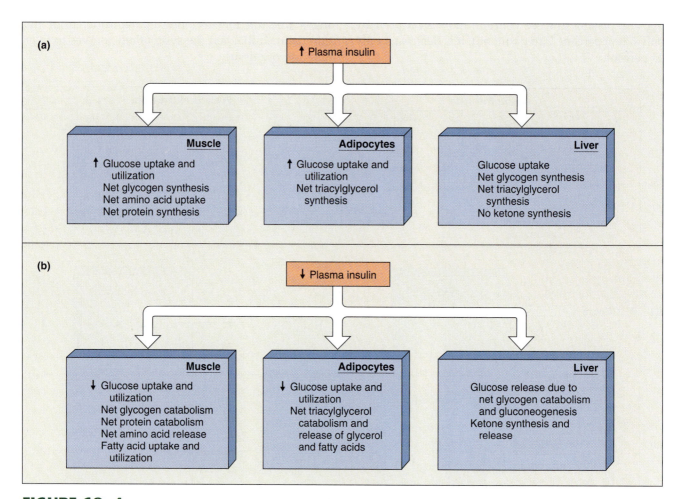

FIGURE 18–4

Summary of overall target-cell responses to (a) an increase or (b) a decrease in the plasma concentration of insulin. The responses in (a) are virtually identical to the absorptive state events of Figure 18–1 and the left panel of Figure 18–3; the responses in (b) are virtually identical to the postabsorptive state events of Figure 18–2 and the right panel of Figure 18–3. The biochemical events that underlie these responses to insulin are shown in Figure 18–6.

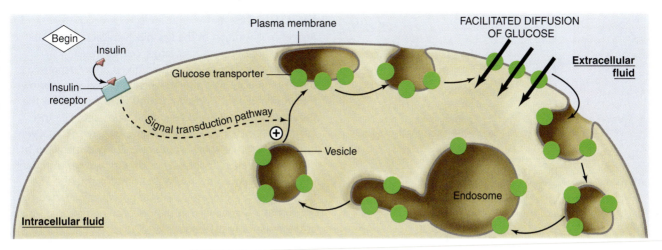

FIGURE 18–5

Stimulation by insulin of the translocation of glucose transporters from cytoplasmic vesicles to the plasma membrane in muscle cells and adipose-tissue cells. Note that these transporters are constantly recycled by endocytosis from the plasma membrane back through endosomes into vesicles. As long as insulin levels are elevated, the entire cycle continues and the number of transporters in the plasma membrane stays high. In contrast, when insulin levels decrease, the cycle is broken, the vesicles accumulate in the cytoplasm, and the number of transporters in the plasma membrane decreases.

of all the absorptive-state events, and a decreased plasma concentration of insulin is the major cause of all the postabsorptive events.

Like all peptide hormones, insulin induces its effects by binding to specific receptors in the plasma membrane of its target cells. This binding triggers a variety of signal transduction pathways that influence the target-cells' plasma-membrane transport proteins and intracellular enzymes. Thus, for example, in muscle cells and adipose-tissue cells an increased insulin concentration stimulates cytoplasmic vesicles that contain a particular type of glucose transporter (GLUT-4) in their membrane to fuse with the plasma membrane (Figure 18–5). The increased number of plasma-membrane glucose transporters resulting from this fusion then causes a greater rate of glucose movement from the extracellular fluid into the cells by facilitated diffusion. Recall from Chapter 6 that glucose enters virtually all cells of the body by facilitated diffusion; there are multiple subtypes of glucose transporters that mediate this process, however, and the subtype— GLUT-4—that is regulatable by insulin is found mainly in muscle and adipose-tissue cells.

A description of the many enzymes whose activities and/or concentrations are influenced by insulin is beyond the scope of this book, but the overall pattern is illustrated in Figure 18–6 for reference and to illustrate several principles. It is important here not to lose sight of the forest for the trees: The essential information (the "forest") to understand about insulin's actions is the target cells' ultimate responses—that is, the material summarized in Figure 18–4. Figure 18–6 merely

shows some of the specific biochemical reactions (the "trees") that underlie these responses.

A major principle illustrated by Figure 18–6 is that, in *each* of its target cells, insulin brings about its ultimate responses by multiple actions. Let us take its effects on muscle cells as an example. In these cells, insulin favors glycogen formation and storage by (1) increasing glucose transport into the cell, (2) stimulating the key enzyme (glycogen synthase) that catalyzes the rate-limiting step in glycogen synthesis, and (3) inhibiting the key enzyme (glycogen phosphorylase) that catalyzes glycogen catabolism. Thus, insulin favors glucose transformation to and storage as glycogen in muscle through three pathways. Similarly, for protein synthesis in muscle cells, insulin (1) increases the number of active plasma-membrane transporters for amino acids, thereby increasing amino acid transportation to the cells, (2) stimulates the ribosomal enzymes that mediate the synthesis of protein from these amino acids, and (3) inhibits the enzymes that mediate protein catabolism.

Control of Insulin Secretion   The major controlling factor for insulin secretion is the plasma glucose concentration. An increase in plasma glucose concentration, as occurs after a meal, acts on the pancreas to stimulate insulin secretion, whereas a decrease inhibits secretion. The feedback nature of this system is shown in Figure 18–7: Following a meal, the rise in plasma glucose concentration stimulates insulin secretion, and the insulin stimulates entry of glucose into muscle and adipose tissue, as well as net uptake, rather than net

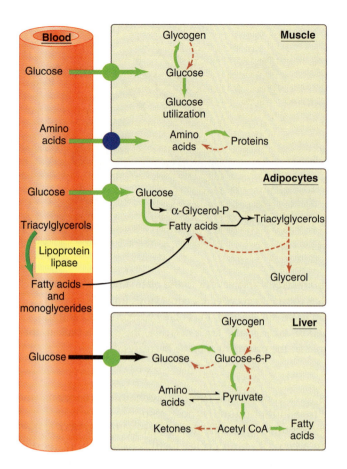

FIGURE 18–6

Reference illustration of the key biochemical events that underlie those responses of target cells to insulin summarized in Figure 18–4. Each green arrow denotes a process stimulated by insulin, whereas a dashed red arrow denotes inhibition by insulin. Except for the effects on the transport proteins for glucose and amino acids, all other effects are exerted on insulin-sensitive enzymes. The bowed arrows denote pathways whose reversibility is mediated by different enzymes (Chapter 4); such enzymes are commonly the ones influenced by insulin and other hormones. The black arrows are processes that are not *directly* affected by insulin but are enhanced in the presence of increased insulin as the result of mass-action.

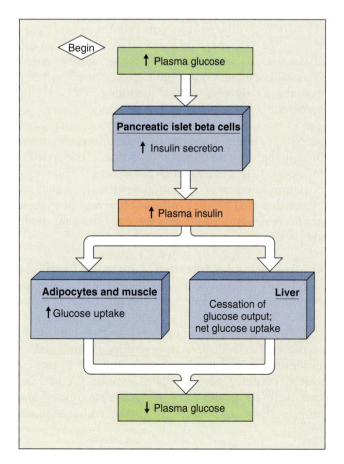

FIGURE 18–7

Nature of plasma glucose control over insulin secretion.

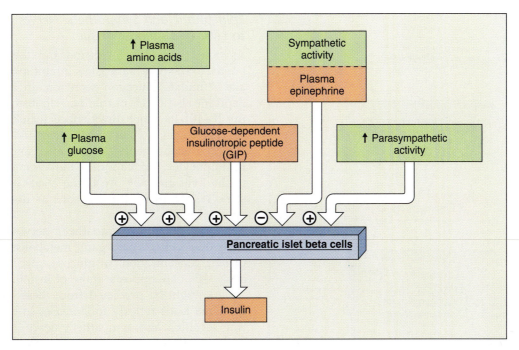

FIGURE 18–8
Major controls of insulin secretion. GIP is a gastrointestinal hormone.

output, of glucose by the liver. These effects eventually reduce the blood concentration of glucose to its premeal level, thereby removing the stimulus for insulin secretion, which returns to its previous level.

In addition to plasma glucose concentration, there are numerous other insulin-secretion controls (Figure 18–8). One is the plasma concentration of certain amino acids, an elevated amino acid concentration causing enhanced insulin secretion. This is another negative-feedback control: Amino acid concentrations increase after ingestion of a protein-containing meal, and the increased plasma insulin stimulates uptake of these amino acids by muscle (and other cells as well).

There are also important hormonal controls over insulin secretion. For example, a hormone—glucose-dependent insulinotropic peptide (GIP)—secreted by endocrine cells in the gastrointestinal tract in response to eating stimulates the release of insulin. This provides a feedforward component to glucose regulation during ingestion of a meal; thus insulin secretion rises earlier and to a greater extent than it would have if plasma glucose were the only controller.

Finally, the autonomic neurons to the islets of Langerhans also influence insulin secretion. Activation of the parasympathetic neurons, which occurs during ingestion of a meal, stimulates secretion of insulin and constitutes a second type of feedforward regulation. In contrast, activation of the sympathetic neurons to the islets or an increase in the plasma concentration of epinephrine (the hormone secreted by the adrenal medulla) inhibits insulin secretion. The significance of this relationship for the body's response to low plasma glucose (**hypoglycemia**), stress, and exercise—all situations in which sympathetic activity is increased—will be described later in this chapter.

To repeat, insulin plays the primary role in controlling the metabolic adjustments required for feasting or fasting. Other hormonal and neural factors, however, also play significant roles. They all oppose the action of insulin in one way or another and are known as **glucose-counterregulatory controls.** Of these, the most important is glucagon.

Glucagon

As noted earlier, **glucagon** is the peptide hormone produced by the alpha cells of the pancreatic islets. The major physiological effects of glucagon are all on the liver and are opposed to those of insulin (Figure 18–9): (1) increased glycogen breakdown, (2) increased gluconeogenesis, and (3) increased synthesis of ketones. Thus, the overall results of glucagon's effects are to increase the plasma concentrations of glucose and ketones, which are important for the postabsorptive period.

From a knowledge of these effects, one would logically suppose that glucagon secretion should increase during the postabsorptive period and prolonged fasting, and such is the case. The major stimulus for glucagon secretion at these times is hypoglycemia. The adaptive value of such a reflex is obvious: A decreasing

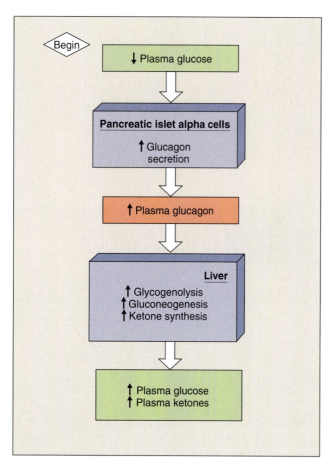

FIGURE 18–9

Nature of plasma glucose control over glucagon secretion.

plasma glucose concentration induces increased release of glucagon, which, by its effects on metabolism, serves to restore normal blood glucose concentration by glycogenolysis and gluconeogenesis while at the same time supplying (if the fast is prolonged) ketones for cell utilization. Conversely, an increased plasma glucose concentration inhibits glucagon's secretion, thereby helping to return the plasma glucose concentration toward normal. Thus, during the postabsorptive state plasma insulin concentration is low and plasma glucagon concentration is high, and this combined change accounts almost entirely for the transition from the absorptive to the postabsorptive state. Said in a different way, this shift is best explained by a rise in the glucagon: insulin ratio in the plasma.

The secretion of glucagon, like that of insulin, is controlled not only by the plasma concentration of glucose and other nutrients but also by neural and hormonal inputs to the islets. For example, the sympathetic nerves to the islets stimulate glucagon secretion—just the opposite of their effect on insulin secretion. The adaptive significance of this relationship for exercise and stress will be described subsequently.

Epinephrine and Sympathetic Nerves to Liver and Adipose Tissue

As noted earlier, epinephrine and the sympathetic nerves to the pancreatic islets inhibit insulin secretion and stimulate glucagon secretion. In addition, epinephrine also affects nutrient metabolism directly (Figure 18–10). Its major direct effects include stimulation of (1) glycogenolysis in both the liver and skeletal muscle, (2) gluconeogenesis in the liver, and (3) lipolysis in adipocytes. Activation of the sympathetic nerves to the liver and adipose tissue elicits essentially the same responses by these organs as does circulating epinephrine.

Thus, enhanced sympathetic nervous system activity exerts effects on organic metabolism—increased plasma concentrations of glucose, glycerol, and fatty acids—that are opposite those of insulin.

As might be predicted from these effects, hypoglycemia leads reflexly to increases in both epinephrine secretion and sympathetic-nerve activity to the liver and adipose tissue. This is the same stimulus that, as described above, leads to increased secretion of glucagon, although the receptors and pathways are totally different. When the plasma glucose concentration decreases, glucose receptors in the central nervous system (and, possibly, the liver) initiate the reflexes that lead to increased activity in the sympathetic pathways to the adrenal medulla, liver, and adipose tissue. The adaptive value of the response is the same as that for the glucagon response to hypoglycemia: Blood glucose returns toward normal, and fatty acids are supplied for cell utilization.

In the compensatory response to *acute* hypoglycemia, the increased activity of the sympathetic nervous system is less important than a reduced insulin concentration and an increased glucagon concentration, but nevertheless contributes. In contrast, sympathetic nervous system activity *decreases* during *prolonged* fasting or ingestion of low-calorie diets; the adaptive significance of this change—reduction of the body's rate of energy utilization—is discussed later in this chapter.

Other Hormones

In addition to the three hormones already described in this section, there are many others that have various effects on organic metabolism. The secretion of all these other hormones, however, is not primarily keyed to the transitions between the absorptive and postabsorptive states. Instead, their secretion is controlled by other factors, and these hormones are for the most part involved in homeostatic processes described elsewhere in this book. Nonetheless, the effects of two of them—cortisol and growth hormone—on nutrient metabolism are important enough to warrant description here.

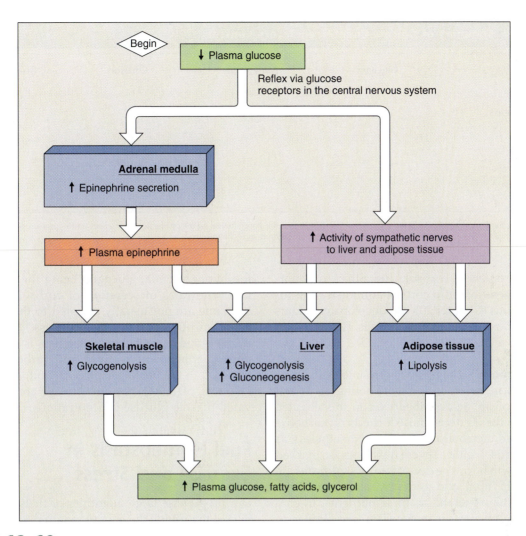

FIGURE 18-10

Participation of the sympathetic nervous system in the response to a low plasma glucose concentration (hypoglycemia). Glycogenolysis in skeletal muscle contributes to increased plasma glucose by releasing lactate and pyruvate, which are converted to glucose in the liver.

Cortisol   Cortisol, the major glucocorticoid produced by the adrenal cortex, plays an essential "permissive" role in the adjustments to fasting. We have described how fasting is associated with stimulation of both gluconeogenesis and lipolysis; however, neither of these critical metabolic transformations occurs to the usual degree in a person deficient in cortisol. In other words, the plasma cortisol level need not *rise* during fasting and usually does not, but the presence of even small amounts of cortisol in the blood somehow maintains the concentrations of the key liver and adipose-tissue enzymes required for gluconeogenesis and lipolysis. Therefore, in response to fasting, people with a cortisol deficiency develop hypoglycemia serious enough to interfere with brain function.

Moreover, cortisol can play more than a permissive role when its plasma concentration does increase, as occurs during stress (Chapter 20). In high concentration, cortisol elicits many metabolic events

ordinarily associated with fasting (Table 18–3). Clearly, here is another hormone, in addition to glucagon and epinephrine, that can exert actions opposite those of insulin. Indeed, persons with very

TABLE 18–3 Effects of Cortisol on Organic Metabolism
1. Basal concentrations are permissive for stimulation of gluconeogenesis and lipolysis in the postabsorptive state
2. Increased plasma concentrations cause: a. Increased protein catabolism   b. Increased gluconeogenesis   c. Decreased glucose uptake by muscle cells and adipose-tissue cells   d. Increased triacylglycerol breakdown
Net result: Increased plasma concentrations of amino acids, glucose, and free fatty acids

TABLE 18-4 Summary of Glucose-Counterregulatory Controls*

	Glucagon	Epinephrine	Cortisol	Growth Hormone
Glycogenolysis	√	√		
Gluconeogenesis	√	√	√	√
Lipolysis		√	√	√
Inhibition of: glucose uptake by muscle cells and adipose-tissue cells			√	√

*A √ indicates that the hormone stimulates the process; no √ indicates that the hormone has no major physiological effect on the process. Epinephrine stimulates glycogenolysis in both liver and skeletal muscle, whereas glucagon does so only in liver.

high plasma levels of cortisol, due either to abnormally high secretion or to cortisol administration for medical reasons (Chapter 20), can develop symptoms similar to those seen in individuals with insulin deficiency.

Growth Hormone The primary physiological effects of growth hormone are to stimulate both growth and protein anabolism, as described later in this chapter. Compared to these effects, those it exerts on carbohydrate and lipid metabolism are minor. Nonetheless, as is true for cortisol, either severe deficiency or marked excess of growth hormone does produce significant abnormalities in lipid and carbohydrate metabolism. Growth hormone's effects on these nutrients, in contrast to those on protein metabolism, are similar to those of cortisol and opposite those of insulin. Growth hormone (1) renders adipocytes more responsive to lipolytic stimuli, (2) increases gluconeogenesis by the liver, and (3) reduces the ability of insulin to cause glucose uptake by muscle and adipose tissue. These three effects are often termed growth hormone's "anti-insulin effects."

Summary of Hormonal Controls

To a great extent, insulin may be viewed as the "hormone of plenty." Its secretion and plasma concentration are increased during the absorptive period and decreased during postabsorption, and these changes are adequate to cause most of the metabolic changes associated with these periods. In addition, opposed in various ways to insulin's effects are the actions of four major glucose-counterregulatory controls—glucagon, epinephrine and the sympathetic nerves to the liver and adipose tissue, cortisol, and growth hormone (Table 18–4). Glucagon and the sympathetic nervous system are activated during the postabsorptive period (or in any other situation with hypoglycemia) and definitely play roles in preventing hypoglycemia, glucagon being the more important. The rates of secretion of cortisol and growth hormone are not usually coupled to the absorptive-postabsorptive pattern; nevertheless, their presence in the blood at basal concentrations is necessary for normal adjustment of lipid and carbohydrate metabolism to the postabsorptive period, and excessive amounts of either hormone cause abnormally elevated plasma glucose concentrations.

Fuel Homeostasis in Exercise and Stress

During exercise large quantities of fuels must be mobilized to provide the energy required for muscle contraction. As described in Chapter 11, these fuels include plasma glucose and fatty acids as well as the muscle's own glycogen.

The plasma glucose used during exercise is supplied by the liver, both by breakdown of its glycogen stores and by gluconeogenesis—conversion of pyruvate, lactate, glycerol, and amino acids to glucose. The glycerol is made available to the liver by a marked increase in adipose-tissue lipolysis, with a resultant release of glycerol and fatty acids into the blood, the fatty acids serving, along with glucose, as a fuel source for the exercising muscle.

What happens to plasma glucose concentration during exercise? It changes very little in short-term, mild to moderate exercise and may even increase slightly with strenuous short-term activity. However, during prolonged exercise (Figure 18–11), more than 90 min, plasma glucose concentration does decrease, but usually by less than 25 percent. Clearly, glucose output by the liver increases approximately in proportion to increased glucose utilization during exercise, at least until the later stages of prolonged exercise when it begins to lag somewhat.

The metabolic profile seen in an exercising individual—increases in hepatic glucose production, triacylglycerol breakdown, and fatty acid utilization—is similar to that seen in a fasting person, and the neuroendocrine controls are also the same. Exercise is characterized by a fall in insulin secretion and a rise in glucagon secretion (Figure 18–11), and the changes in the plasma concentrations of these two hormones are the major controls during exercise. In addition there is increased activity of the sympathetic nervous system (including increased secretion of epinephrine) and increased secretion of cortisol and growth hormone.

What triggers increased glucagon secretion and decreased insulin secretion during exercise? One signal, at least during *prolonged* exercise, is the modest decrease in plasma glucose that occurs (Figure 18–11); this is the same signal that controls the secretion of these hormones in fasting. Other inputs at all intensities of exercise are increased circulating epinephrine and enhanced activity of the sympathetic neurons supplying the pancreatic islets. Thus, the increased sympathetic nervous system activity characteristic of exercise not only contributes directly to fuel mobilization by acting on the liver and adipose tissue, but contributes indirectly by inhibiting the secretion of insulin and stimulating that of glucagon. This sympathetic output is not triggered by changes in plasma glucose concentration but is mediated by the central nervous system as part of the "programmed" neural response to exercise.

One component of the response to exercise is quite different from the response to fasting: In exercise, glucose uptake and utilization by the muscles are increased, whereas in fasting they are markedly reduced. How is it that, during exercise, the movement, via facilitated diffusion, of glucose into muscle can remain high in the presence of reduced plasma insulin and increased plasma concentrations of cortisol and growth hormone, all of which decrease glucose uptake by skeletal muscle? By an as-yet-unidentified mechanism, muscle contraction causes migration of an intracellular store of glucose transporters to the plasma membrane.

Exercise and the postabsorptive state are not the only situations characterized by the neuroendocrine profile of decreased insulin and increased glucagon, sympathetic activity, cortisol, and growth hormone. This profile also occurs in response to a variety of nonspecific stresses, both physical and emotional. The adaptive value of these neuroendocrine responses to stress is that the resulting metabolic shifts prepare the body for exercise ("fight or flight") in the face of real or threatened injury. In addition, the amino acids liberated by catabolism of body protein stores because of decreased insulin and increased cortisol not only provide energy via gluconeogenesis but also constitute a potential source of amino acids for tissue repair should injury occur. The subject of stress and the body's responses to it are further described in Chapter 20.

Diabetes Mellitus

The name "diabetes," meaning "syphon" or "running through," denotes the increased urinary volume excreted by people suffering from this disease. "Mellitus," meaning "sweet," distinguishes this urine from the large quantities of nonsweet ("insipid") urine produced by persons suffering from vasopressin deficiency. As described in Chapter 16, the latter disorder is known as diabetes insipidus, and the unmodified word "diabetes" is often used as a synonym for *diabetes mellitus,* a disease that affects nearly 15 million people in the United States.

Diabetes can be due to a deficiency of insulin or to a hyporesponsiveness to insulin, for it is not one but several diseases with different causes. Classification of these diseases rests on how much insulin the person is secreting and whether therapy requires the administration of insulin. In *insulin-dependent diabetes mellitus (IDDM),* or type 1 diabetes, the hormone is completely or almost completely absent from the islets of Langerhans and the plasma, and therapy with insulin is essential (this protein hormone cannot be given orally but must be injected because gastrointestinal enzymes would digest it). In *noninsulin-dependent diabetes mellitus (NIDDM),* or type 2 diabetes, the hormone is

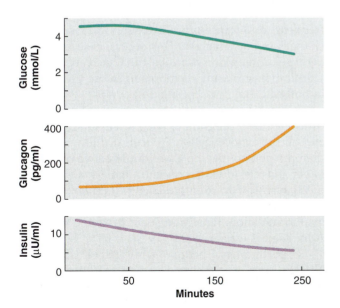

FIGURE 18–11

Plasma concentrations of glucose, glucagon, and insulin during prolonged (250 min) moderate exercise at a fixed intensity.

Adapted from Felig and Wahren.

often present in plasma at near-normal or even above-normal levels, and therapy does not require administration of insulin (although such administration may be beneficial).

IDDM is less common, affecting 15 percent of diabetic patients. It is due to the total or near-total destruction of the pancreatic beta cells by the body's own white blood cells (autoimmune disease, Chapter 20). The triggering events for this autoimmune response are not yet fully established. As noted above, treatment of IDDM always involves the administration of insulin. It is likely that in the not-too-distant future, transplantation of islet cells into the individual with IDDM will be possible.

Because of their insulin deficiency, untreated patients with IDDM always have elevated plasma glucose concentrations. This occurs both because glucose fails to enter insulin's target cells normally and because the liver continuously makes glucose—via glycogenolysis and gluconeogenesis—and releases it into the blood. Another result of the insulin deficiency is marked lipolysis with resultant elevation of plasma glycerol and fatty acids. Marked ketone formation by the liver is also present.

If extreme, these metabolic changes culminate in the acute life-threatening emergency called *diabetic ketoacidosis* (Figure 18–12). Some of the problems are due to the effects that a markedly elevated plasma glucose concentration produces on renal function. In Chapter 16, we pointed out that a normal person does not excrete glucose because all glucose filtered at the renal corpuscle is reabsorbed by the tubules. However, the elevated plasma glucose of diabetes may so increase the filtered load of glucose that the maximum tubular reabsorptive capacity is exceeded and large amounts of glucose are excreted. For the same reasons, large amounts of ketones may also appear in the urine. These urinary losses deplete the body of nutrients and lead to weight loss. Far worse, however, is the fact that these unreabsorbed solutes cause an osmotic diuresis (Chapter 16)—marked urinary excretion of sodium and water, which can lead, by the sequence of events shown in Figure 18–12, to hypotension, brain damage, and death.

The other serious abnormality in diabetic ketoacidosis is the increased plasma hydrogen-ion concentration caused by the accumulation of ketones, two of which are acids. This increased hydrogen-ion concentration causes brain dysfunction that can contribute to the development of coma and death.

Diabetic ketoacidosis is seen only in patients with untreated IDDM—that is, those with almost total inability to secrete insulin. However, 85 percent of diabetics are in the NIDDM category and never develop metabolic derangements severe enough to go into diabetic ketoacidosis. NIDDM is a disease mainly of overweight adults, typically starting in middle life. Given the earlier mention of progressive weight loss in IDDM as a symptom of diabetes, it may seem contradictory that most people with NIDDM are overweight. The paradox is resolved when one realizes that people with NIDDM, in contrast to those with IDDM, do not excrete enough glucose in the urine to cause weight loss.

There are several factors that combine to cause NIDDM. One major problem is target-cell hyporesponsiveness to insulin, termed *insulin resistance.* Obesity accounts for much of the insulin resistance in NIDDM, for obesity in any person—diabetic or not—induces some degree of insulin resistance, particularly in adipose-tissue cells. (One theory is that the excess adipose tissue overproduces a messenger that causes downregulation of insulin-responsive glucose transporters.) However, additional components of insulin resistance, not related to obesity and not yet understood, also usually occur with NIDDM.

Most people with NIDDM not only have insulin resistance but also have a defect in the ability of their beta cells to secrete insulin in response to a rise in plasma glucose concentration. In other words, although insulin resistance is the primary factor inducing hyperglycemia in NIDDM, an as-yet-unidentified defect in beta-cell function prevents these cells from responding to the hyperglycemia in normal fashion.

The major therapy for obese persons with NIDDM is weight reduction, since obesity is a major cause of insulin resistance. An exercise program is also very important, because insulin responsiveness is increased by frequent endurance-type exercise, independent of changes in body weight. This occurs, at least in part, because training causes a substantial increase in the total number of plasma-membrane glucose transporters in both skeletal muscles and adipocytes.

If plasma glucose concentration is not adequately controlled by a program of weight reduction, exercise, and dietary modification (specifically low-fat diets), then the person may be given orally active drugs that lower plasma glucose concentration by a variety of mechanisms. For example, the *sulfonylureas* lower plasma glucose by acting on the beta cells to stimulate insulin secretion. Finally, in some cases the use of insulin itself is warranted.

Unfortunately, people with either IDDM or NIDDM tend to develop a variety of chronic abnormalities, including atherosclerosis, kidney failure, small-vessel and nerve disease, susceptibility to infection, and blindness. Elevated plasma glucose contributes to most of these abnormalities either by causing the intracellular accumulation of certain glucose metabolites that exert harmful effects on cells when present in high concentrations, or by linking glucose to proteins, thereby affecting their function.

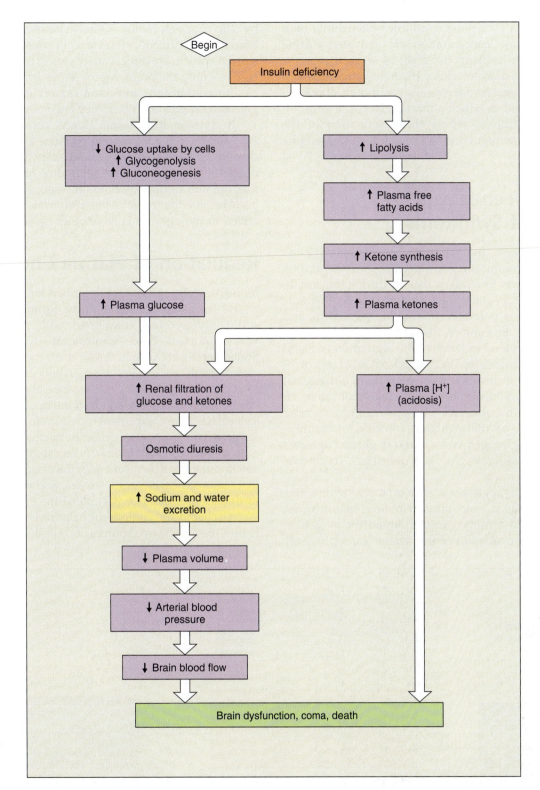

FIGURE 18–12

Diabetic ketoacidosis: Events caused by severe untreated insulin deficiency in insulin-dependent diabetes mellitus.

This discussion of diabetes has focused on insulin, but it is now clear that the hormones that elevate plasma glucose concentration may contribute to the severity of the disease. Glucagon is quite important in this regard. Most diabetics, particularly those with IDDM, have inappropriately high plasma glucagon concentrations, which contribute to the metabolic dysfunction typical of diabetes. The reason for this is that insulin normally inhibits glucagon secretion, and the low insulin of IDDM releases glucagon secretion from this inhibition.

Finally, as we have seen, all the systems that raise plasma glucose concentration are activated during stress, which explains why stress exacerbates the symptoms of diabetes. Since diabetic ketoacidosis itself constitutes a severe stress, a positive-feedback cycle is triggered in which a marked lack of insulin induces ketoacidosis, which elicits activation of the glucose-counterregulatory systems, which worsens the ketoacidosis.

Hypoglycemia as a Cause of Symptoms

As we have seen, "hypoglycemia" means a low plasma glucose concentration. Plasma glucose concentration can drop to very low values, usually during the postabsorptive state, in persons with several types of organic disorders. This is termed *fasting hypoglycemia,* and the relatively uncommon disorders responsible for it can be understood in terms of the regulation of blood glucose concentration. They include (1) an excess of insulin due to an insulin-producing tumor, a drug that stimulates insulin secretion, or the taking of too much insulin by a diabetic; and (2) a defect in one or more of the glucose-counterregulatory controls, for example, inadequate glycogenolysis and/or gluconeogenesis due to liver disease, glucagon deficiency, or cortisol deficiency.

Fasting hypoglycemia causes many symptoms. Some—increased heart rate, trembling, nervousness, sweating, and anxiety—are accounted for by activation of the sympathetic nervous system caused reflexly by the hypoglycemia. Other symptoms, such as headache, confusion, dizziness, uncoordination, and slurred speech, are direct consequences of too little glucose reaching the brain. More serious brain effects, including convulsions and coma, can occur if the plasma glucose concentration becomes low enough.

In contrast, low plasma glucose concentration has *not* been shown routinely to produce either acute or chronic symptoms of fatigue, lethargy, loss of libido, depression, or many other symptoms for which popular opinion frequently holds it responsible. Most experts believe that most of the symptoms popularly ascribed to hypoglycemia have other causes.

Regulation of Plasma Cholesterol

In the previous section, we described the flow of lipids to and from adipose tissue in the form of fatty acids and triacylglycerols complexed with proteins. One very important lipid—**cholesterol**—was not mentioned earlier because it, unlike the fatty acids and triacylglycerols, serves not as a metabolic fuel but rather as a precursor for plasma membranes, bile salts, steroid hormones, and other specialized molecules. Thus, cholesterol has many important functions in the body. Unfortunately, it can also cause problems. Specifically, high plasma concentrations of cholesterol enhance the development of **atherosclerosis,** the arterial thickening that leads to heart attacks, strokes, and other forms of cardiovascular damage (Chapter 14).

A schema for cholesterol balance is illustrated in Figure 18–13. The two sources of cholesterol are dietary

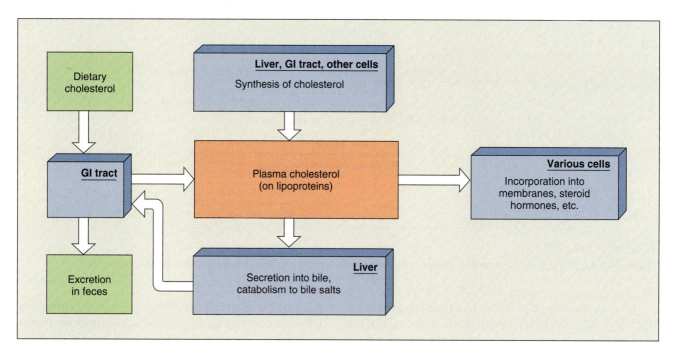

FIGURE 18–13

Cholesterol balance.

cholesterol and cholesterol synthesized within the body. Dietary cholesterol comes from animal sources, egg yolk being by far the richest in this lipid (a single egg contains about 250 mg of cholesterol). Not all ingested cholesterol is absorbed into the blood, however—much of it simply passes through the length of the gastrointestinal tract and is excreted in the feces.

What about cholesterol synthesis within the body? Almost all cells can synthesize some of the cholesterol required for their own plasma membranes, but most cannot do so in adequate amounts and depend upon receiving cholesterol from the blood. This is also true of the endocrine cells that produce steroid hormones from cholesterol. Thus, most cells *remove* cholesterol from the blood. In contrast, the liver and cells lining the gastrointestinal tract can produce large amounts of cholesterol, most of which *enters* the blood.

Now for the other side of cholesterol balance—the pathways, all involving the liver, for net cholesterol loss from the body. First of all, some plasma cholesterol is picked up by liver cells and secreted into the bile, which carries it to the intestinal tract. Here it is treated much like ingested cholesterol, some being absorbed back into the blood and the remainder being excreted in the feces. Second, much of the cholesterol picked up by the liver cells is metabolized into bile salts (Chapter 17). After their production by the liver, these bile salts, like secreted cholesterol, flow through the bile duct into the small intestine. (As described in Chapter 17, many of these bile salts are then reclaimed by absorption back into the blood across the wall of the lower small intestine.)

The liver is clearly the center of the cholesterol universe, for it can add newly synthesized cholesterol to the blood or it can remove cholesterol from the blood, secreting it into the bile or metabolizing it to bile salts. The homeostatic control mechanisms that keep plasma cholesterol relatively constant operate on all of these hepatic processes, but the single most important response involves cholesterol production. The synthesis of cholesterol by the liver is inhibited whenever dietary—and therefore plasma—cholesterol is increased. This is because cholesterol inhibits the enzyme critical for cholesterol synthesis by the liver.

Thus, as soon as the plasma cholesterol level starts rising because of increased cholesterol ingestion, hepatic synthesis is inhibited, and plasma concentration remains close to its original value. Conversely, when dietary cholesterol is reduced and plasma cholesterol begins to fall, hepatic synthesis is stimulated (released from inhibition), and this increased production opposes any further fall. The sensitivity of this negative-feedback control of cholesterol synthesis differs greatly from person to person, but it is the major reason why, for most people, it is difficult to change plasma cholesterol very much in either direction by altering only dietary cholesterol.

Thus far, the relative constancy of plasma cholesterol has been emphasized. There are, however, environmental and physiological factors that can significantly alter plasma cholesterol concentrations. Perhaps the most important of these factors are the quantity and type of dietary fatty acids. Ingesting saturated fatty acids, the dominant fatty acids of animal fat (particularly high in red meats, most cheeses, and whole milk), raises plasma cholesterol. In contrast, eating either polyunsaturated fatty acids (the dominant plant fatty acids) or monounsaturated fatty acids such as those in olive or peanut oil, lowers plasma cholesterol. The various fatty acids exert their effects on plasma cholesterol by altering cholesterol synthesis, excretion, and metabolism to bile salts.

A variety of drugs now in common use are also capable of lowering plasma cholesterol by influencing one or more of the metabolic pathways for cholesterol—for example, inhibiting the critical enzyme for hepatic cholesterol synthesis—or by interfering with intestinal absorption of bile salts.

Based on studies of the relationship between plasma cholesterol levels and cardiovascular diseases, recommendations from the National Institutes of Health call a total plasma cholesterol below 200 mg/deciliter [a deciliter (dl) is 100 ml] "desirable," 200–239 mg/dl "borderline high," and 240 mg/dl or greater "high."

The story is more complicated than this, however, since not all plasma cholesterol has the same function or significance for disease. Like most other lipids, cholesterol circulates in the plasma as part of various lipoprotein complexes. These include chylomicrons (Chapter 17), VLDL (this chapter), **low-density lipoproteins (LDL),** and **high-density lipoproteins (HDL).** LDL are the main cholesterol carriers, and they *deliver* cholesterol to cells throughout the body. LDL bind to plasma-membrane receptors specific for a protein component of the LDL, and the LDL are taken up by the cell by absorptive endocytosis. In contrast to LDL, HDL *remove* excess cholesterol from blood and tissue, including the cholesterol-loaded cells of atherosclerotic plaques (Chapter 14). They then deliver this cholesterol to the liver, which secretes it into the bile or converts it to bile salts. HDL also delivers cholesterol to steroid-producing endocrine cells. Uptake of the HDL by the liver and these endocrine cells is facilitated by the presence in their plasma membranes of large numbers of receptors specific for HDL, which bind to the receptors and then are taken into the cells.

LDL cholesterol is often designated "bad" cholesterol since high levels of it in the plasma are associated with increased deposition of cholesterol in arterial walls and higher incidences of heart attacks. (The designation "bad" should not obscure the fact that LDL are essential for supplying cells with the cholesterol they require to synthesize cell membranes and, in the case of the gonads and adrenal glands, steroid hormones.) Using the same criteria, HDL cholesterol has been designated "good" cholesterol.

The best single indicator of the likelihood of developing atherosclerotic heart disease is, therefore, not *total* plasma cholesterol but rather the *ratio* of plasma LDL-cholesterol to plasma HDL-cholesterol—the lower the ratio, the lower the risk. Cigarette smoking, a known risk factor for heart attacks, lowers plasma HDL, whereas weight reduction (in overweight persons) and regular exercise increase it. Estrogen not only lowers LDL but raises HDL, which explains, in part, why premenopausal women have so much less coronary artery disease than men. After menopause, the cholesterol values and coronary artery disease rates in women not on hormone-replacement therapy (Chapter 19) become similar to those in men.

SECTION A SUMMARY

Events of the Absorptive and Postabsorptive States

I. During absorption, energy is provided primarily by absorbed carbohydrate, and net synthesis of glycogen, triacylglycerol, and protein occurs.
   a. Some absorbed carbohydrate not used for energy is converted to glycogen, mainly in the liver and skeletal muscle, but most is converted, in liver and adipocytes, to α-glycerol phosphate and fatty acids, which then combine to form triacylglycerol. The liver releases its triacylglycerols in very low density lipoproteins, the fatty acids of which are picked up by adipocytes.
   b. The fatty acids of some absorbed triacylglycerol are used for energy, but most are rebuilt into fat in adipose tissue.
   c. Some absorbed amino acids are converted to proteins, but excess amino acids are converted to carbohydrate and fat.
   d. There is a net upake of glucose by the liver.
II. In the postabsorptive state, blood glucose level is maintained by a combination of glucose production by the liver and a switch from glucose utilization to fatty acid and ketone utilization by most tissues.
   a. Synthesis of glycogen, fat, and protein is curtailed, and net breakdown of these molecules occurs.
   b. The liver forms glucose by glycogenolysis of its own glycogen and by gluconeogenesis from lactate and pyruvate (from breakdown of muscle glycogen), glycerol (from adipose-tissue lipolysis), and amino acids (from protein catabolism).
   c. Glycolysis is decreased, and most of the body's energy supply comes from the oxidation of fatty acids released by adipose-tissue lipolysis and of ketones produced from fatty acids by the liver.
   d. The brain continues to use glucose but also starts using ketones as they build up in the blood.

Endocrine and Neural Control of the Absorptive and Postabsorptive States

I. The major hormones secreted by the pancreatic islets of Langerhans are insulin by the beta cells and glucagon by the alpha cells.

II. Insulin is the most important hormone controlling metabolism.
   a. In muscle, it stimulates glucose uptake, glycolysis, and net synthesis of glycogen and protein; in adipose tissue, it stimulates glucose uptake and net synthesis of triacylglycerol; in liver, it inhibits gluconeogenesis and glucose release and stimulates the net synthesis of glycogen and triacylglycerols.
   b. The major stimulus for insulin secretion is an increased plasma glucose concentration, but secretion is also influenced by many other factors, which are summarized in Figure 18–8.
III. Glucagon, epinephrine, cortisol, and growth hormone all exert effects on carbohydrate and lipid metabolism that are opposite, in one way or another, to those of insulin. They raise plasma concentrations of glucose, glycerol, and fatty acids.
   a. Glucagon's physiological actions are all on the liver, where it stimulates glycogenolysis, gluconeogenesis, and ketone synthesis.
   b. The major stimulus for glucagon secretion is hypoglycemia, but secretion is also stimulated by other inputs, including the sympathetic nerves to the islets.
   c. Epinephrine released from the adrenal medulla in response to hypoglycemia stimulates glycogenolysis in the liver and muscle, gluconeogenesis in liver, and lipolysis in adipocytes. The sympathetic nerves to liver and adipose tissue exert effects similar to those of epinephrine.
   d. Cortisol is permissive for gluconeogenesis and lipolysis; in higher concentrations, it stimulates gluconeogenesis and blocks glucose uptake. These last two effects are also exerted by growth hormone.

Fuel Homeostasis in Exercise and Stress

I. During exercise, the muscles use as their energy sources plasma glucose, plasma fatty acids, and their own glycogen.
   a. Glucose is provided by the liver, and fatty acids are provided by adipose-tissue lipolysis.
   b. The changes in plasma insulin, glucagon, and epinephrine are similar to those that occur during the postabsorptive period and are mediated mainly by the sympathetic nervous system.
II. Stress causes hormonal changes similar to those caused by exercise.

Diabetes Mellitus

I. Insulin-dependent diabetes is due to absolute insulin deficiency and can lead to diabetic ketoacidosis.
II. Noninsulin-dependent diabetes is usually associated with obesity and is caused by a combination of insulin resistance and a defect in beta-cell responsiveness to elevated plasma glucose concentration. Plasma insulin concentration is usually normal or elevated.

Regulation of Plasma Cholesterol

I. Plasma cholesterol is a precursor for the synthesis of plasma membranes, bile salts, and steroid hormones.

II. Cholesterol synthesis by the liver is controlled so as to homeostatically regulate plasma cholesterol concentration; it varies inversely with ingested cholesterol.

III. The liver also secretes cholesterol into the bile and converts it to bile salts.

IV. Plasma cholesterol is carried mainly by low-density lipoproteins, which deliver it to cells; high-density lipoproteins carry cholesterol from cells to the liver and steroid-producing cells. The LDL/HDL ratio correlates with the incidence of coronary heart disease.

SECTION A KEY TERMS

absorptive state
postabsorptive state
very low density lipo-
    proteins (VLDL)
lipoprotein lipase
α-ketoacids
glycogenolysis
lipolysis
gluconeogenesis
glucose sparing
ketones
islets of Langerhans

beta cells
alpha cells
insulin
hypoglycemia
glucose-counterregulatory
    controls
glucagon
cholesterol
low-density lipoproteins
    (LDL)
high-density lipoproteins
    (HDL)

SECTION A REVIEW QUESTIONS

1. Using a diagram, summarize the events of the absorptive period.
2. In what two organs does major glycogen storage occur?
3. How do the liver and adipose tissue metabolize glucose during the absorptive period?
4. How does adipose tissue metabolize absorbed triacylglycerol, and what are the three major sources of the fatty acids in adipose tissue triacylglycerol?
5. What happens to most of the absorbed amino acids when a high-protein meal is ingested?
6. Using a diagram, summarize the events of the postabsorptive period; include the four sources of

blood glucose and the pathways leading to ketone formation.
7. Distinguish between the roles of glycerol and free fatty acids during fasting.
8. List the overall responses of muscle, adipose tissue, and liver to insulin. What effects occur when plasma insulin concentration decreases?
9. List five inputs controlling insulin secretion, and state the physiological significance of each.
10. List the effects of glucagon on the liver and their consequences.
11. List two inputs controlling glucagon secretion, and state the physiological significance of each.
12. List four metabolic effects of epinephrine and the sympathetic nerves to the liver and adipose tissue, and state the net results of each.
13. List the permissive effects of cortisol and the effects that occur when plasma cortisol concentration increases.
14. List three effects of growth hormone on carbohydrate and lipid metabolism.
15. Which hormones stimulate gluconeogenesis? Glycogenolysis in liver? Glycogenolysis in skeletal muscle? Lipolysis? Blockade of glucose uptake?
16. Describe how plasma glucose, insulin, glucagon, and epinephrine levels change during exercise and stress. What causes the changes in the concentrations of the hormones?
17. Describe the metabolic disorders of severe insulin-dependent diabetes.
18. How does obesity contribute to noninsulin-dependent diabetes?
19. Hypersecretion of which hormones can induce a diabetic state?
20. Using a diagram, describe the sources of cholesterol gain and loss. Include three roles of the liver in cholesterol metabolism, and state the controls over these processes.
21. What are the effects of saturated and unsaturated fatty acids on plasma cholesterol?
22. What is the significance of the ratio of LDL cholesterol to HDL cholesterol?

CONTROL OF GROWTH

Growth is a complex process influenced by genetics, endocrine function, and a variety of environmental factors, including nutrition and the presence of infection. The process involves cell division and net protein synthesis throughout the body, but a person's height is determined specifically by bone growth, particularly of the vertebral column and legs.

Bone Growth

As described in Chapter 16, bone is a living tissue consisting of a protein (collagen) matrix upon which calcium salts, particularly calcium phosphates, are deposited. A growing long bone is divided, for descriptive purposes, into the ends, or **epiphyses,** and

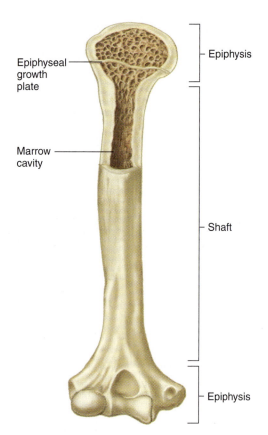

FIGURE 18–14

Anatomy of a long bone during growth. ✗

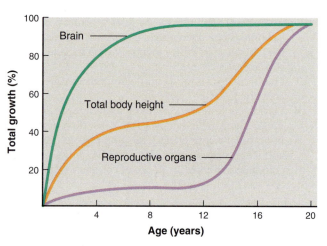

FIGURE 18–15

Relative growth in brain, total body height (a measure of long-bone and vertebral growth), and reproductive organs. Note that brain growth is nearly complete by the age 5, whereas maximal height (maximal bone lengthening) and reproductive-organ size are not reached until the late teens.

The pubertal growth spurt lasts several years in both sexes, but growth during this period is greater in boys. This, plus the fact that boys grow more before puberty because they begin puberty approximately two years later than girls, accounts for the differences in average height between men and women.

Environmental Factors Influencing Growth

Adequacy of nutrient supply and freedom from disease are the primary environmental factors influencing growth. Lack of sufficient amounts of any of the essential amino acids, essential fatty acids, vitamins, or minerals interferes with growth. Total protein and sufficient nutrients to provide energy must also be adequate.

The growth-inhibiting effects of malnutrition can be seen at any time of development but are most profound when they occur very early in life. Thus, maternal malnutrition may cause growth retardation in the fetus. Since low birth weight is strongly associated with increased infant mortality, prenatal malnutrition causes increased numbers of prenatal and early postnatal deaths. Moreover, irreversible stunting of brain development may be caused by prenatal malnutrition. During infancy and childhood, too, malnutrition can interfere with both intellectual development and total-body growth.

Following a temporary period of stunted growth due to malnutrition or illness, and given proper nutrition and recovery from illness, a child manifests a remarkable growth spurt (**catch-up growth**) that brings

the remainder, the **shaft.** The portion of each epiphysis that is in contact with the shaft is a plate of actively proliferating cartilage, the **epiphyseal growth plate** (Figure 18–14). **Osteoblasts,** the bone-forming cells (Chapter 16), at the shaft edge of the epiphyseal growth plate convert the cartilaginous tissue at this edge to bone while new cartilage is simultaneously being laid down in the interior of the plate by cells called **chondrocytes.** In this manner, the epiphyseal growth plate remains intact (indeed, actually widens) and is gradually pushed away from the center of the bony shaft as the latter lengthens.

Linear growth of the shaft can continue as long as the epiphyseal growth plates exist, but ceases when the plates are themselves ultimately converted to bone as a result of hormonal influences at puberty. This is known as **epiphyseal closure** and occurs at different times in different bones. Accordingly, a person's **bone age** can be determined by x-raying the bones and determining which ones have undergone epiphyseal closure.

As shown in Figure 18–15, children manifest two periods of rapid increase in height, one during the first two years of life and the second during puberty. Note that increase in height is not necessarily correlated with the rates of growth of specific organs.

the child up to the normal height expected for his or her age. The mechanism that accounts for this accelerated growth is unknown.

Hormonal Influences on Growth

The hormones most important to human growth are growth hormone, insulin-like growth factors I and II, thyroid hormones, insulin, testosterone, and estrogens, all of which exert widespread effects. In addition to all these hormones, there is a huge group of peptide **growth factors,** including the insulin-like growth factors, most of which act as paracrine and autocrine agents to stimulate differentiation and/or cell division of certain cell types.

The general term for a chemical that stimulates cell division is a **mitogen.** There are also peptide **growth-inhibiting factors** that modulate growth by inhibiting cell division in specific tissues. Numbering more than 60 at present, the growth factors and growth-inhibiting factors are usually produced by multiple cell types rather than by discrete endocrine glands.

The physiology of growth factors and growth-inhibiting factors is important not just for understanding control of normal growth, but also because these factors may be involved in the development of *cancer.* Thus, some *oncogenes* (genes that are involved in causing cancer, Chapter 5) code for proteins that are identical to or very similar to growth factors, growth-factor receptors, or postreceptor components of growth-factor signal transduction pathways. The problem is that these proteins have lost important regulatory constraints on their activity. For example, one oncogene codes for a version of the receptor for epidermal growth factor that is always in the activated state even in the absence of the growth factor. This activated receptor imparts a continuous growth signal to the cells containing it.

The various hormones and growth factors do not all stimulate growth at the same periods of life. For example, fetal growth is largely independent of growth hormone, the thyroid hormones, and the sex steroids, all of which importantly stimulate growth during childhood and adolescense.

Growth Hormone and Insulin-Like Growth Factors

Growth hormone, secreted by the anterior pituitary, has little or no effect on fetal growth as we have just mentioned, but is the single most important hormone for postnatal growth. Its major growth-promoting effect is stimulation (indirect, as we shall see) of cell division in its many target tissues. Thus, growth hormone promotes bone lengthening by stimulating maturation and cell division of the chondrocytes in the epiphyseal plates, thereby continuously widening the

plates and providing more cartilaginous material for bone formation.

An excess of growth hormone during childhood produces *giantism,* whereas deficiency produces *dwarfism.* When excess growth hormone is secreted in adults after epiphyseal closure, it cannot lengthen the bones further, but it does produce the disfiguring bone thickening and overgrowth of other organs known as *acromegaly.*

Importantly, growth hormone exerts its cell division-stimulating (mitogenic) effect not *directly* on cells but rather *indirectly* through the mediation of a mitogen whose synthesis and release are induced by growth hormone. This mitogen is called **insulin-like growth factor I (IGF-I)** (also known as somatomedin C). Despite its name, this messenger has its own unique effects. Under the influence of growth hormone, IGF-I is secreted by the liver, enters the blood and functions as a hormone. In addition, growth hormone stimulates the secretion of IGF-I by many other types of cells, including bone, and at these sites IGF-I functions as an autocrine or paracrine agent. The relative importance of IGF-I as a hormone versus autocrine/paracrine agent in any given organ or tissue remains controversial.

Current concepts of how growth hormone and IGF-I interact on the epiphyseal plates of bone are as follows: (1) Growth hormone stimulates the chondrocyte precursor cells (prechondrocytes) and/or young differentiating chondrocytes in the epiphyseal plates to differentiate into chondrocytes; (2) during this differentiation, the cells begin both to secrete IGF-I and to become responsive to IGF-I; (3) the IGF-I then acts as an autocrine or paracrine agent (probably along with blood-borne IGF-I) to stimulate the differentiating chondrocytes to undergo cell division.

The importance of IGF-I in mediating the major growth-promoting effect of growth hormone is illustrated by the fact that dwarfism can be due not only to decreased secretion of growth hormone but also to decreased production of IGF-I or failure of the tissues to respond to IGF-I. For example, one uncommon form of short stature (termed *growth hormone insensitivity syndrome*), is due to a genetic mutation that causes the growth hormone receptor to fail to respond to growth hormone; the result is failure to produce IGF-I in response to growth hormone.

The secretion and activity of IGF-I can be influenced by the nutritional status of the individual and by many hormones other than growth hormone. For example, malnutrition during childhood inhibits the production of IGF-I even through plasma growth hormone concentration is elevated and should be stimulating IGF-I secretion. To take another example, estrogen stimulates the secretion of IGF-I by cells of the uterus and ovaries.

TABLE 18–5 Major Effects of Growth Hormone

1. Promotes growth: Induces precursor cells in bone and other tissues to differentiate and secrete insulin-like growth factor I (IGF-I), which stimulates cell division. Also stimulates secretion of IGF-I by liver.

2. Stimulates protein synthesis, predominantly in muscle.

3. Anti-insulin effects:
   a. Renders adipocytes more responsive to lipolytic stimuli
   b. Stimulates gluconeogenesis
   c. Reduces the ability of insulin to stimulate glucose uptake

In addition to its specific growth-promoting effect on cell division via IGF-I, growth hormone directly stimulates protein synthesis in various tissues and organs, particularly muscle. It does this by increasing amino acid uptake by cells and both the synthesis and activity of ribosomes. All these events are essential for protein synthesis. This anabolic effect on protein metabolism facilitates the ability of tissues and organs to enlarge. Table 18–5 summarizes the multiple effects of growth hormone, all of which have been described in this chapter.

The control of growth hormone secretion was described in Chapter 10 (Figure 10–21). Briefly, the control system begins with two of the hypophysiotropic hormones secreted by the hypothalamus. Growth hormone secretion is stimulated by growth hormone releasing hormone (GHRH) and inhibited by somatostatin. As a result of changes in these two signals, which are virtually 180 degrees out of phase with each other (that is, one is high when the other is low), growth hormone secretion occurs in episodic bursts and manifests a striking diurnal rhythm. During most of the day, there is little or no growth hormone secreted, although bursts may be elicited by certain stimuli, including stress, hypoglycemia, and exercise. In contrast, 1 to 2 h after a person falls asleep, one or more larger, prolonged bursts of secretion may occur. (The negative-feedback controls that growth hormone and IGF-I exert on the hypothalamus and anterior pituitary are summarized in Figure 10–21.)

In addition to the hypothalamic controls, a variety of hormones—notably the sex hormones, insulin, and the thyroid hormones, as described below—influence the secretion of growth hormone. The net result of all these inputs is that the total 24-h secretion rate of growth hormone is highest during adolescence (the period of most rapid growth), next highest in children, and lowest in adults. The decreased growth hormone secretion associated with aging is responsible, in part, for the decrease in lean-body and bone mass, the expansion of adipose tissue, and the thinning of the skin that occur at that time.

The availability of large quantities of human growth hormone produced by recombinant-DNA technology has greatly facilitated the treatment of children with short stature due to deficiency of growth hormone. Controversial at present is the administration of growth hormone to short children who do not have growth hormone deficiency, to athletes in an attempt to increase muscle mass, and to normal elderly persons to reverse the growth hormone–related aging changes described in the previous paragraph.

As noted above, growth hormone plays little if any role in prenatal growth (that is, growth of the embryo and fetus). One would suppose, therefore, that this would also be true for IGF-I, but such is not the case: IGF-I is required for normal fetal total-body growth and, specifically, for normal maturation of the fetal nervous system. The stimulus for IGF-I secretion during prenatal life is unknown.

Finally, it should be noted that there is another messenger—**insulin-like growth factor II (IGF-II)**—that is closely related to IGF-I. IGF-II, the secretion of which is *independent* of growth hormone, is also a crucial mitogen during the prenatal period. It continues to be secreted throughout life, but its postnatal function is not known.

Thyroid Hormones

The thyroid hormones (TH)—thyroxine (T_4) and triiodothyronine (T_3)—are essential for normal growth because they are required for both the synthesis of growth hormone and the growth-promoting effects of that hormone. Accordingly, infants and children with *hypothyroidism* (deficient thyroid function) manifest retarded growth due to slowed bone growth.

Quite distinct from its growth-promoting effect, TH is permissive for normal development of the central nervous system during fetal life. Inadequate production of maternal and fetal TH due to severe iodine deficiency during pregnancy is one of the world's most common preventable causes of mental retardation, termed *endemic cretinism.*

This effect on brain development must be distinguished from other stimulatory effects TH exerts on the nervous system throughout life, not just during infancy. A hypothyroid person exhibits sluggishness and poor mental function, and these effects are completely reversible at any time with administration of TH. Conversely, a person with *hyperthyroidism* (excessive secretion of TH) is jittery and hyperactive.

Insulin

It should not be surprising that adequate amounts of insulin are necessary for normal growth since insulin is, in all respects, an anabolic hormone. Its inhibitory effect on protein degradation is particularly important with regard to growth.

In addition to this general anabolic effect, however, insulin exerts direct, specific growth-promoting effects on cell differentiation and cell division during fetal life (and possibly during childhood). Moreover, insulin is required for normal production of IGF-I.

Sex Hormones

As will be described in Chapter 19, sex hormone secretion (testosterone in the male and estrogen in the female) begins in earnest between the ages of 8 and 10 and progressively increases to reach a plateau over the next 5 to 10 years. A normal pubertal growth spurt, which reflects growth of the long bones and vertebrae, requires this increased production of the sex hormones. The major growth-promoting effect of the sex hormones is to stimulate the secretion of growth hormone and IGF-I.

Unlike growth hormone, however, the sex hormones not only stimulate bone growth, but ultimately *stop* it by inducing epiphyseal closure. The dual effects of the sex hormones explain the pattern seen in adolescence—rapid lengthening of the bones culminating in complete cessation of growth for life.

In addition to these dual effects on bone, testosterone, but not estrogen, exerts a direct anabolic effect on protein synthesis in many nonreproductive organs and tissues of the body. This accounts, at least in part, for the increased muscle mass of men, compared with women.

This is also why synthetic testosterone-like agents termed **anabolic steroids** [or a hormone—dehydroepi-androsterone (DHEA)—that is converted in the body to testosterone] are sometimes used by athletes—both male and female—in an attempt to increase their muscle mass and strength. However, these steroids have multiple potential toxic side effects (for example, liver damage, increased risk of prostate cancer, and infertility). Moreover, in females they can produce masculinization.

Cortisol

Cortisol, the major hormone secreted by the adrenal cortex in response to stress, can have potent *antigrowth* effects under certain conditions. When present in high concentration, it inhibits DNA synthesis and stimulates protein catabolism in many organs, and it inhibits bone growth. Moreover, it causes bone breakdown by inhibiting osteoblasts and stimulating osteoclasts (Chapter 16). It also inhibits the secretion of growth hormone. For all these reasons, in children, the elevation in plasma cortisol that accompanies infections and other stresses is, at least in part, responsible for the retarded growth that occurs with illness.

As we shall see in Chapter 20, cortisol and very similar steroids are commonly used medically in persons with arthritis or other inflammatory disorders. A side effect of such treatment is increased protein catabolism and bone breakdown. One must carefully distinguish cortisol-type steroids (glucocorticoids) from testosterone-type steroids (anabolic steroids).

This completes our survey of the major hormones that affect growth. Their actions are summarized in Table 18–6.

TABLE 18–6 Major Hormones Influencing Growth

Hormone	Principal Actions
Growth hormone	Major stimulus of postnatal growth: Induces precursor cells to differentiate and secrete insulin-like growth factor I (IGF-I), which stimulates cell division Stimulates secretion of IGF-I by liver Stimulates protein synthesis
Insulin	Stimulates fetal growth Stimulates postnatal growth by stimulating secretion of IGF-I Stimulates protein synthesis
Thyroid hormones	Permissive for growth hormone's secretion and actions Permissive for development of the central nervous system
Testosterone	Stimulates growth at puberty, in large part by stimulating the secretion of growth hormone Causes eventual epiphyseal closure Stimulates protein synthesis in male
Estrogen	Stimulates the secretion of growth hormone at puberty Causes eventual epiphyseal closure
Cortisol	Inhibits growth Stimulates protein catabolism

Compensatory Growth

We have dealt thus far only with growth during *childhood.* During adult life, a specific type of regenerative growth known as **compensatory growth,** can occur in many human organs. For example, after the surgical removal of one kidney, the cells of the other kidney begin to manifest increased cell division, and the kidney ultimately grows until its total mass approaches the initial mass of the two kidneys combined. Many growth factors and hormones participate in compensatory growth, but the precise signals that trigger the process are not known. Moreover, these signals very likely differ from organ to organ. Of particular importance is the release of angiogenic factors (Chapter 14) since availability of blood flow is a major determinant of how large an organ can become.

SECTION B SUMMARY

Bone Growth

I. A bone lengthens as osteoblasts at the shaft edge of the epiphyseal growth plates convert cartilage to bone while new cartilage is being laid down in the plates.

II. Growth ceases when the plates are completely converted to bone.

Environmental Factors Influencing Growth

I. The major environmental factors influencing growth are nutrition and disease.

II. Malnutrition during in utero life may produce irreversible stunting and mental deficiency.

Hormonal Influences on Growth

I. Growth hormone is the major stimulus of postnatal growth.

    a. It stimulates the release of IGF-I from the liver and many other cells, and IGF-I then acts locally (and perhaps also as a hormone) to stimulate cell division.

    b. Growth hormone also acts directly on cells to stimulate protein synthesis.

    c. Growth hormone secretion is highest during adolescence.

II. Because thyroid hormones are required for growth hormone synthesis and the growth-promoting effects of this hormone, they are essential for normal growth during childhood and adolescence. They are also permissive for brain development during infancy.

III. Insulin stimulates growth mainly during in utero life.

IV. Mainly by stimulating growth hormone secretion, testosterone and estrogen promote bone growth during adolescence, but these hormones also cause epiphyseal closure. Testosterone also stimulates protein synthesis.

V. Cortisol in a high concentration inhibits growth and stimulates protein catabolism.

SECTION B KEY TERMS

epiphysis	growth factor
shaft	mitogen
epiphyseal growth plate	growth-inhibiting factor
osteoblast	insulin-like growth factor I
chondrocyte	(IGF-I)
epiphyseal closure	insulin-like growth factor II
bone age	(IGF-II)
catch-up growth	compensatory growth

SECTION B REVIEW QUESTIONS

1. Describe the process by which bone is lengthened.
2. What are the effects of malnutrition on growth?
3. List the major hormones that control growth.
4. Describe the relationship between growth hormone and IGF-I and the roles of each in growth.
5. What are the effects of growth hormone on protein synthesis?
6. What is the status of growth hormone secretion at different stages of life?
7. State the effects of the thyroid hormones on growth and development.
8. Describe the effects of testosterone on growth, cessation of growth, and protein synthesis. Which of these effects are shared by estrogen?
9. What is the effect of cortisol on growth?

REGULATION OF TOTAL-BODY ENERGY BALANCE AND TEMPERATURE

Basic Concepts of Energy Expenditure

The breakdown of organic molecules liberates the energy locked in their molecular bonds. This is the energy cells use to perform the various forms of biological work—muscle contraction, active transport, and molecular synthesis. The first law of thermodynamics states that energy can be neither created nor destroyed, but can be converted from one form to another. Thus, internal energy liberated (ΔE) during breakdown of an organic molecule can either appear as heat (H) or be used to perform work (W).

$$\Delta E = H + W$$

During metabolism, about 60 percent of the energy released from organic molecules appears immediately as heat, and the rest is used for work. The energy used for work must first be incorporated into molecules of ATP, the subsequent breakdown of which serves as the immediate energy source for the work. It is essential to realize that the body is not a heat engine since it is totally incapable of converting heat to work, but the heat released in its chemical reactions is valuable for maintaining body temperature.

Biological work can be divided into two general categories: (1) **external work**—movement of external objects by contracting skeletal muscles; and (2) **internal work**—all other forms of work, including skeletal-muscle activity not used in moving external objects. As just stated, much of the energy liberated from nutrient catabolism appears immediately as heat. What may not be obvious is that all internal work, too, is ultimately transformed to heat except during periods of growth. For example, internal work is performed during cardiac contraction, but this energy appears ultimately as heat generated by the friction of blood flow through the blood vessels.

Thus, the total energy liberated when organic nutrients are catabolized by cells may be transformed into body heat, appear as external work, or be stored in the body in the form of organic molecules. The **total energy expenditure** of the body is therefore given by the equation

Total energy expenditure =
    Internal heat produced + External work + Energy stored

Metabolic Rate

The unit for energy in metabolism is the **kilocalorie (kcal),** which is the amount of heat required to raise the temperature of one liter of water one degree celsius. (In the field of nutrition, the three terms "Calorie" with a capital C, "large calorie," and "kilocalorie" are synonyms; they are all 1000 "calories," with a small c.) Total energy expenditure per unit time is called the **metabolic rate.**

Since many factors cause the metabolic rate to vary (Table 18–7), the most common method for evaluating it specifies certain standardized conditions,

---

TABLE 18–7 Some Factors Affecting the Metabolic Rate

Age (↓ with ↑ age)
Sex (women less than men at any given size)
Height, weight, and body surface area
Growth
Pregnancy, menstruation, lactation
Infection or other disease
Body temperature
Recent ingestion of food
Prolonged alteration in amount of food intake   } (An increase in any of these factors causes an increase in metabolic rate.)
Muscular activity
Emotional stress
Environmental temperature
Circulating levels of various hormones, especially
    epinephrine and thyroid hormone
Sleep (↓ during sleep)

TABLE 18–8 Major Functions of the Thyroid Hormones (TH)

1. Required for normal maturation of the nervous system in the fetus and infant

    Deficiency: Mental retardation (cretinism)

2. Required for normal bodily growth because they facilitate the secretion of and response to growth hormone

    Deficiency: Deficient growth in children

3. Required for normal alertness and reflexes at all ages

    Deficiency: Mentally and physically slow and lethargic

    Excess: Restless, irritable, anxious, wakeful

4. Major determinant of the rate at which the body produces heat during the basal metabolic state

    Deficiency: Low BMR, sensitivity to cold, decreased food appetite

    Excess: High BMR, sensitivity to heat, increased food appetite, increased catabolism of nutrients

5. Facilitates the activity of the sympathetic nervous system by stimulating the synthesis of one class of receptors (beta receptors) for epinephrine and norepinephrine

    Excess: Symptoms similar to those observed with activation of the sympathetic nervous system (for example, increased heart rate)

and measures what is known as the **basal metabolic rate (BMR).** In the basal condition, the subject is at mental and physical rest in a room at a comfortable temperature and has not eaten for at least 12 h. These conditions are arbitrarily designated "basal," even though the metabolic rate during sleep may be less than the BMR. The BMR is often termed the "metabolic cost of living," and most of it is expended by the heart, liver, kidneys, and brain. For the following discussion, it must be emphasized that the term "BMR" can be applied to a person's metabolic rate only when the specified conditions are met; thus, a person who has recently eaten or is exercising has a metabolic rate but not a *basal* metabolic rate. The next sections describe several of the important determinants of BMR and metabolic rate.

Thyroid Hormones The thyroid hormones are the single most important determinant of BMR regardless of size, age, or sex. TH increases the oxygen consumption and heat production of most body tissues, a notable exception being the brain. This ability to increase BMR is termed a **calorigenic effect.** One mechanism of this effect is that TH increases synthesis of uncoupling proteins found in the inner mitochondrial membrane of most human cells; these proteins reduce the amount of ATP produced and increase the amount of heat generated when fuels are oxidized.

Long-term excessive TH, as in persons with hyperthyroidism, induces a host of effects secondary to the calorigenic effect. For example, the increased metabolic demands markedly increase hunger and food intake; the greater intake frequently remains inadequate to

meet the metabolic needs, and net catabolism of protein and fat stores leads to loss of body weight. Also, the greater heat production activates heat-dissipating mechanisms (skin vasodilation and sweating), and the person suffers from marked intolerance to warm environments. In contrast, the hypothyroid individual complains of cold intolerance.

The calorigenic effect of TH is only one of a bewildering variety of effects exerted by these hormones. With one exception—facilitation of the activity of the sympathetic nervous system (described in Chapter 10), the major functions of the thyroid hormones have all been described in this chapter and are listed for reference in Table 18–8.

As described in Chapter 10, secretion of the thyroid hormones is stimulated by the anterior pituitary hormone, thyroid stimulating hormone (TSH), itself stimulated by the hypophysiotropic hormone, thyrotropin releasing hormone (TRH). The thyroid hormones, in turn, exert an inhibitory effect on the hypothalamo-pituitary system. What is unusual about this entire hormonal system is that there is no known stimulus that activates this system in a way that leads to the negative-feedback elimination of the stimulus (as, for example, the way changes in plasma glucose influence insulin secretion). It is as though the thyroid hormones simply set a background tone for the various parameters, like BMR, that they influence.

Epinephrine Epinephrine is another hormone that exerts a calorigenic effect. (This effect may be related to the hormone's stimulation of glycogen and triacylglycerol catabolism, since ATP splitting and energy

TABLE 18–9	Energy Expenditure during Different Types of Activity for a 70-kg (154-lb) Person

Form of Activity	Energy kcal/h
Lying still, awake	77
Sitting at rest	100
Typewriting rapidly	140
Dressing or undressing	150
Walking on level, 4.3 km/h (2.6 mi/h)	200
Bicycling on level, 9 km/h (5.5 mi/h)	304
Walking on 3 percent grade, 4.3 km/h (2.6 mi/h)	357
Sawing wood or shoveling snow	480
Jogging, 9 km/h (5.3 mi/h)	570
Rowing, 20 strokes/min	828

liberation occur in both the breakdown and subsequent resynthesis of these molecules.) Thus, when epinephrine secretion by the adrenal medulla is stimulated, the metabolic rate rises. This accounts for part of the greater heat production associated with emotional stress, although increased muscle tone also contributes.

Food-Induced Thermogenesis The ingestion of food rapidly increases the metabolic rate by 10 to 20 percent for a few hours after eating. This effect is known as **food-induced thermogenesis.** Ingested protein produces the greatest effect, carbohydrate and fat, less. Most of the increased heat production is secondary to processing of the absorbed nutrients by the liver, not to the energy expended by the gastrointestinal tract in digestion and absorption. It is to avoid the contribution of food-induced thermogenesis that BMR tests must be performed in the postabsorptive state.

To reiterate, food-induced thermogenesis is the *rapid* increase in energy expenditure in response to ingestion of a meal. As we shall see, *prolonged* alterations in food intake (either increased or decreased total calories) also have significant effects on metabolic rate but are not termed food-induced thermogenesis.

Muscle Activity The factor that can most increase metabolic rate is altered skeletal-muscle activity. Even minimal increases in muscle contraction significantly increase metabolic rate, and strenuous exercise may raise energy expenditure more than fifteenfold (Table 18–9). Thus, depending on the degree of physical activity, total energy expenditure may vary for a normal young adult from a value of approximately 1500 kcal/24 h to more than 7000 kcal/24 h (for a lumberjack). Changes in muscle activity also account in part for the changes in metabolic rate that occur during sleep

(decreased muscle contraction), during exposure to a low environmental temperature (increased muscle contraction and shivering), and with strong emotions.

Regulation of Total-Body Energy Stores

The laws of thermodynamics dictate that the total energy expenditure (metabolic rate) of the body must equal the total energy intake. We have already identified the ultimate forms of energy expenditure: internal heat production, external work, and net molecular synthesis (energy storage). The source of input is the energy contained in ingested food. Therefore:

Energy from food intake =
   Internal heat produced + External work + Energy stored

Our equation includes no term for loss of fuel from the body via excretion of nutrients because, in normal persons, only negligible losses occur via the urine, feces, and as sloughed hair and skin. In certain diseases, however, the most important being diabetes, urinary losses of organic molecules may be quite large and would have to be included in the equation.

Let us now rearrange the equation to focus on energy storage:

Energy stored =
   Energy from food intake − (Internal heat produced
                   + External work)

Thus, whenever energy intake differs from the sum of internal heat produced and external work, changes in energy storage occur; that is, the total-body energy content increases or decreases. Normally, as we have seen, energy storage, except in growing children, is mainly in the form of fat in adipose tissue.

It is worth emphasizing at this point that "body weight" and "total-body energy content" are not synonymous terms, although there is a popular tendency to equate the two. Body weight is determined not only by the amount of fat, carbohydrate, and protein in the body, but also by the amounts of water, bone, and other minerals. For example, an individual can lose body weight quickly as the result of sweating or an unusual increase in urinary output, or can gain large amounts of weight as a result of water retention, as occurs, for example, during heart failure (Chapter 14). Moreover, even focusing only on the nutrients, a constant body weight does not mean that total-body energy content is constant. The reason is that 1 g of fat contains 9 kcal, whereas 1 g of either carbohydrate or protein contains 4 kcal. Thus, for example, aging is usually associated with a gain of fat and a loss of protein; the result is that even though the person's body weight may stay constant, the total-body energy content has increased. Despite these qualifications, however, in the remainder of this chapter changes in body weight are equated with changes in total-body energy content and, more specifically, changes in body fat stores.

Body weight in adults is usually regulated around a relatively constant set point. Theoretically this constancy can be achieved by reflexly adjusting caloric intake and/or energy expenditure in response to changes in body weight. It has long been assumed that regulation of caloric intake is the only important adjustment, and this process will be described in the next section. However, there is growing evidence that energy expenditure can also be reflexly adjusted in response to changes in body weight, and we describe it first.

A typical demonstration of this process in human beings is as follows: Total daily energy expenditure was measured in nonobese subjects at their usual body weight and again after they were caused either to lose 10 percent of their body weight by underfeeding or to gain 10 percent by overfeeding. At their new body weight, the overfed subjects manifested a large (15 percent) increase in both resting and nonresting energy expenditure, and the underfed subjects a similar decrease. These changes in energy expenditure were much greater than could be accounted for simply by the altered metabolic mass of the body or having to move a larger or smaller body.

The generalization that emerges from this and other similar studies is that a dietary-induced change in total-body energy stores triggers, in negative-feedback fashion, an alteration in energy expenditure that opposes the gain or loss of energy stores. This phenomenon helps explain why some dieters lose about 5 to 10 lbs of fat fairly easily and then become stuck at a plateau. It also helps explain why some very thin people have difficulty trying to gain much weight. Another unsettled question is whether such "metabolic resistance" to changes in body weight persists indefinitely or is only a transient response to rapid changes in body weight.

Control of Food Intake

The control of food intake can be analyzed in the same way as any other biological control system. As the previous section emphasized, the variable being maintained relatively constant in this system is total-body energy content, more specifically, total fat stores. Accordingly, an essential component of such a control system is a hormone—**leptin**—synthesized by adipose-tissue cells themselves, and released from the cells in proportion to the amount of fat in the adipose tissue. This hormone acts on the hypothalamus to cause a reduction in food intake (by inhibiting the release of

neuropeptide Y, a hypothalamic neurotransmitter that stimulates eating). Leptin also stimulates the metabolic rate and, therefore, probably plays an important role in the changes in energy expenditure that occur in response to overfeeding or underfeeding, as described in the previous section. Thus, as illustrated in Figure 18–16, leptin functions in a negative-feedback system to maintain total-body energy content constant by "telling" the brain how much fat is being stored.

Leptin may turn out to exert many other effects on the hypothalamus and anterior pituitary. For example, during long-term fasting, there is a marked decrease in the secretion of the sex steroids and thyroid hormones, and an increase in the secretion of adrenal glucocorticoids. In experimental animals, these effects were almost completely eliminated by administering leptin. This suggests that leptin normally exerts a stimulatory effect on the pathways that control the secretion of these hormones (the possible role of leptin in puberty is described in Chapter 19).

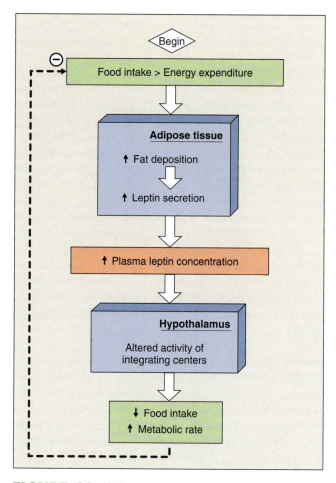

FIGURE 18–16

Role of leptin in the control of total-body energy stores.

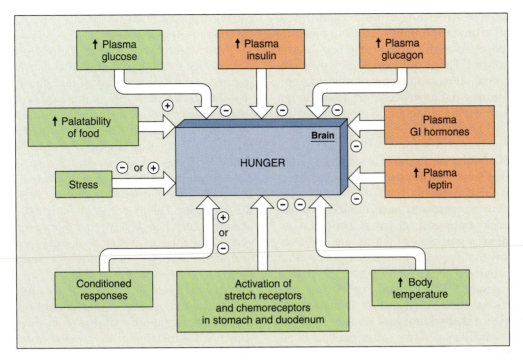

FIGURE 18–17

Short-term inputs controlling food intake. The minus signs denote hunger suppression, and the plus signs denote hunger stimulation.

It should be emphasized that leptin is crucial for *long-term* matching of caloric intake to energy expenditure. In addition, it is thought that various other signals act on the hypothalamus (and other brain areas) over short periods of time to regulate individual meal length and frequency (Figure 18–17). These **satiety signals** cause the person to cease feeling hungry and set the time period before hunger returns once again. For example, the rate of insulin-dependent glucose utilization by certain areas of the hypothalamus rises during eating, and this probably constitutes a satiety signal. Insulin, which increases during food absorption, also acts directly as a satiety signal. The increase in metabolic rate induced by eating tends to raise body temperature slightly, which acts as yet another satiety signal. Finally, there are satiety signals initiated by the presence of food within the gastrointestinal tract: These include neural signals triggered by stimulation of both stretch receptors and chemoreceptors in the stomach and duodenum, as well as by several of the hormones (cholecystokinin, for example) released from the stomach and duodenum during eating.

Food intake is also strongly influenced by the reinforcement, both positive and negative, of such things as smell, taste, and texture. In addition, the behavioral concepts of reinforcement, drive, and motivation, described in Chapter 13, must be incorporated into any comprehensive theory of food-intake control.

Another significant factor that can increase food intake is stress, as demonstrated by controlled experiments in animals.

Overweight and Obesity

The definition of *overweight* is a functional one, a state in which an increased amount of fat in the body results in a significant impairment of health from a variety of diseases, notably hypertension, atherosclerosis, heart disease, and diabetes. *Obesity* denotes a particularly large accumulation of fat—that is, being extremely overweight. The difficulty for scientists has been in establishing just how much fat constitutes "overweight"—that is, in determining at what point fat accumulation begins to constitute a health risk. This is evaluated by epidemiological studies that correlate disease rates with some measure of the amount of fat in the body. Presently, the preferred simple method for assessing the latter is not the body weight but the **body mass index (BMI),** which is calculated by dividing the weight (in kilograms) by the square of the height (in meters). For example, a 70-kg person with a height of 180 cm would have a BMI of 21.6 ($70/1.8^2$).

In 1998, the National Institutes of Health issued guidelines that categorize people with BMIs of greater than 25 as overweight (that is, as having some increased health risk because of excess fat) and those with greater than 30 as obese, with a markedly increased health risk. According to these criteria, more than half of U.S. women and men age 20 and older are now considered overweight, and nearly one-quarter are clinically obese! These guidelines, however, are quite controversial. First, the large number of epidemiological studies that have been performed do not always agree as to where along the continuum of BMIs between 25 and 30 health risks begin to occur. Second, even granting increased risk above a BMI of 25, many experts believe that the studies do not always account for confounding factors associated with being overweight or even obese, particularly a sedentary lifestyle; they believe that the increased health risk may actually be due to lack of physical activity, not body fat per se. (This question is particularly interesting and subject to study now that there are quite a few people who are overweight or even obese but are physically fit due to exercise programs.) Third, these experts feel that even if being overweight is risky, there is inadequate long-term evidence that weight loss leads to a longer life in otherwise healthy persons categorized as overweight.

To add to the complexity, there is growing evidence that not just total fat but where the fat is located has important consequences. Specifically, people with mostly abdominal fat ("apples") are at greater risk for developing serious conditions such as diabetes and cardiovascular diseases than people whose fat is mainly in the lower body ("pears")—on the buttocks and thighs. There is presently no agreement as to the explanation of this phenomenon, but it is known that there are important differences in the physiology of adipose-tissue cells in these regions. For example, adipose-tissue cells in the abdomen are much more adept at breaking down fat stores and releasing the products into the blood. Moreover, these cells are also more responsive to the hormone cortisol, which may partially explain why chronic stress seems to be associated with greater amounts of abdominal fat (stress is discussed in Chapter 20).

What is known about the underlying causes of obesity? Identical twins who have been separated soon after birth and raised in different households manifest strikingly similar body weights and incidences of obesity as adults; such studies have indicated that genetic factors play an important role in obesity. It has been postulated that natural selection favored the evolution in our ancestors of "thrifty genes," which boosted the ability to store fat from each feast in order to sustain people through the next fast. Given today's relative surfeit in many countries of the world, such an adaptation would now be a liability.

Despite the importance of genetic factors, psychological, cultural, and social factors can also be important; for example, the increasing incidence of obesity in the United States and other industrialized nations during the past few generations cannot be explained by changes in our genes.

Much recent research has focused on possible abnormalities in the leptin system as a cause of obesity. In mice that have severe hereditary obesity, the gene—*ob*—that codes for leptin is mutated so that adipose-tissue cells produce either an abnormal, inactive leptin or no leptin at all. The same is not true, however, for the vast majority of obese people: The leptin secreted by these people is normally active, and leptin concentrations in the blood are elevated, not reduced. This indicates that leptin secretion is not at fault in these people. This is not really surprising since most obesity researchers believe that there must be multiple genes that interact with one another and with environmental factors to influence a person's susceptibility to gain weight.

The methods and goals of treating obesity are presently undergoing extensive rethinking. An increase in body fat must be due to an excess of food intake over the metabolic rate, and low-calorie diets have long been the mainstay of therapy. However, it is now clear that such diets *alone* have limited effectiveness in obese people; over 90 percent regain all or most of the lost weight within 5 years. Another important reason for the ineffectiveness of such diets is that, as described earlier, the person's metabolic rate drops, sometimes falling low enough to prevent further weight loss on as little as 1000 calories a day. Related to this, many obese people continue to gain weight or remain in stable energy balance on a caloric intake equal to or less than the amount consumed by people of normal weight. These persons must either have less physical activity than normal or have lower basal metabolic rates. Finally, at least half of obese people—those who are more than 20 percent overweight—who try to diet down to desirable weights suffer medically, physically, and psychologically. This is what would be expected if the body were "trying" to maintain body weight (more specifically fat stores) at the higher set point.

Such studies, taken together, indicate that crash diets are not an effective long-term method for controlling weight. Instead one should set caloric intake at a level that can be maintained for the rest of one's life; such an intake in an overweight individual should lead to a slow steady weight loss of no more than 1 lb per week until the body weight stabilizes at a new, lower level. Most important, any program of weight loss should include increased physical activity of the endurance type. The exercise itself utilizes calories (though depressingly few), but more importantly exercise partially offsets the tendency, described earlier, for the metabolic rate to decrease during long-term caloric restriction and weight loss. Also, the combination of exercise and caloric restriction causes the person to lose more fat and less protein than with caloric

TABLE 18–10 Summary of National Research Council Dietary Recommendations

1. Reduce total fat intake to 30 percent or less of calories. Reduce saturated fatty acid intake to less than 10 percent of calories and the intake of cholesterol to less than 300 mg daily.

2. Every day eat five or more servings of a combination of vegetables and fruits, especially green and yellow vegetables and citrus fruits. Also, increase starches and other complex carbohydrates by eating six or more daily servings of a combination of breads, cereals, and legumes.

3. Maintain protein intake at moderate levels.

4. Balance food intake and physical activity to maintain appropriate body weight.

5. Alcohol consumption is not recommended. For those who drink alcoholic beverages, limit consumption to the equivalent of 1 ounce of pure alcohol in a single day.

6. Limit total daily intake of salt to 6 g or less.

7. Maintain adequate calcium intake.

8. Avoid taking dietary supplements in excess of the RDA (Recommended Dietary Allowance) in any one day.

9. Maintain an optimal intake of fluoride, particularly during the years of primary and secondary tooth formation and growth.

restriction alone. To restate the information of the previous two sentences in terms of control systems, exercise seems to lower the set point around which the body regulates total-body fat stores.

As an exercise in energy balance, let us calculate how rapidly a person can expect to lose weight on a reducing diet (assuming, for simplicity, no change in energy expenditure). Suppose an individual whose steady-state metabolic rate per 24 h is 2000 kcal goes on a 1000 kcal/day diet. How much of the person's own body fat will be required to supply this additional 1000 kcal/day? Since fat contains 9 kcal/g:

$$\frac{1000 \text{ kcal/day}}{9 \text{ kcal/g}} = 111 \text{ g/day, or } 777 \text{ g/week}$$

Approximately another 77 g of water is lost from the adipose tissue along with this fat (adipose tissue is 10 percent water), so that the grand total for 1 week's loss equals 854 g, or 1.8 lb. Thus, even on this rather severe diet, the person can reasonably expect to lose approximately this amount of weight per week, assuming no decrease in metabolic rate occurs. Actually, the amount of weight lost during the first week will probably be considerably greater since a large amount of *water* may be lost early in the diet, particularly when the diet contains little carbohydrate. This early loss is not really elimination of excess fat but often underlies the extravagant claims made for fad diets.

Eating Disorders: Anorexia Nervosa and Bulimia

The two major eating disorders are found almost exclusively in adolescent girls and young women. The typical person with *anorexia nervosa* becomes pathologically afraid of gaining weight and reduces her food intake so severely that she may die of starvation. It is not known whether the cause of anorexia nervosa is primarily psychological or biological. There are many other abnormalities associated with it—loss of menstrual periods, low blood pressure, low body temperature, altered secretion of many hormones. It is likely that these are simply the result of starvation, although it is possible that some represent signs, along with the eating disturbances, of primary hypothalamic malfunction.

Bulimia is recurrent episodes of binge eating. It is usually associated with regular employment of self-induced vomiting, laxatives, or diuretics, as well as strict dieting, fasting, or vigorous exercise to prevent weight gain. Like individuals with anorexia nervosa, those with bulimia manifest a persistent overconcern with body weight, although they are generally within 10 percent of their ideal weight. It, too, is accompanied by a variety of physiological abnormalities, but it is unknown whether they are causal or secondary.

What Should We Eat?

In the last few years, more and more dietary factors have been associated with the cause or prevention of many diseases, including not only coronary heart disease but hypertension, cancer, birth defects, osteoporosis, and a variety of other chronic diseases. These associations come mainly from animal studies, epidemiologic studies on people, and basic research concerning potential mechanisms. The problem is that the findings are often difficult to interpret and may be conflicting. To synthesize all this material in the form of simple clear recommendations to the general public is a monumental task, and all such attempts have been subjected to intense criticism. We present, in Table 18–10, one of the most commonly used sets, that issued by the National Research Council.

Regulation of Body Temperature

Animals, including people, capable of maintaining their body temperatures within very narrow limits are termed **homeothermic.** The relatively constant and high body temperature frees biochemical reactions from fluctuating with the external temperature. However, the maintenance of a relatively high body temperature (approximately 37°C, in normal persons) imposes a requirement for precise regulatory mechanisms, since further large elevations of temperature cause nerve malfunction and protein denaturation. Some people suffer convulsions at a body temperature of 41°C (106°F), and 43°C is considered to be the absolute limit for survival.

Several important generalizations about normal human body temperature should be stressed at the outset: (1) Oral temperature averages about 0.5°C less than rectal, which is generally used as an estimate of internal temperature (also known as core body temperature); thus, not all parts of the body have the same temperature. (2) Internal temperature varies several degrees in response to activity pattern and changes in external temperature. (3) There is a characteristic circadian fluctuation of about 1°C (Figure 18–18), temperature being lowest during the night and highest during the day. (4) An added variation in women is a higher temperature during the second half of the menstrual cycle.

If temperature is viewed as a measure of heat "concentration," temperature regulation can be studied by our usual balance methods. The total heat content gained or lost by the body is determined by the *net difference* between heat produced and heat loss. Maintaining a constant body temperature means that, in the

steady state, heat production must equal heat loss. The basic principles of heat production were described earlier in this chapter in the section on metabolic rate, and those of heat loss are described next. Then we will present the reflexes that play upon these processes specifically to regulate body temperature.

Mechanisms of Heat Loss or Gain

The surface of the body can lose heat to the external environment by radiation, conduction, and convection (Figure 18–19) and by the evaporation of water. Before defining each of these processes, however, it must be emphasized that radiation, conduction, and convection can under certain circumstances lead to heat *gain,* instead of heat loss.

Radiation is the process by which the surfaces of all objects constantly emit heat in the form of electromagnetic waves. The rate of emission is determined by the temperature of the radiating surface. Thus, if the body surface is warmer than the various surfaces in the environment, net heat is lost from the body, the rate being directly dependent upon the temperature difference between the surfaces.

Conduction is the loss or gain of heat by transfer of thermal energy during collisions between adjacent molecules. In essence, heat is "conducted" from molecule to molecule. The body surface loses or gains heat by conduction through direct contact with cooler or warmer substances, including the air or water.

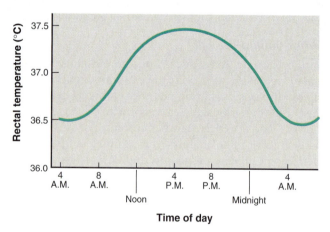

FIGURE 18–18

Circadian changes in core (measured as rectal) body temperature in normal males and in normal females in the first half of the menstrual cycle.

Adapted from Scales et al.

FIGURE 18–19

Mechanisms of heat transfer.

Convection is the process whereby conductive heat loss or gain is aided by movement of the air or water next to the body. For example, air next to the body is heated by conduction, moves away, and carries off the heat just taken from the body. The air that moved away is replaced by cooler air, which in turn follows the same pattern. Convection is always occurring because warm air is less dense and therefore rises, but it can be greatly facilitated by external forces such as wind or fans. Thus, convection aids conductive heat exchange by continuously maintaining a supply of cool air. Henceforth we shall also imply convection when we use the term "conduction."

Because of the great importance of air movement in aiding heat loss, attempts have been made to quantify the cooling effect of combinations of air speed and temperature. The most useful tool is the **wind-chill index,** which states the hypothetical temperature with *no* wind that would provide the same cooling effect as the actual temperature and wind velocity. For example, the wind-chill index would be −10°C if an object in a 5°C windy environment cooled as fast as it would if the temperature were actually −10°C and there was no wind at all.

Evaporation of water from the skin and membranes lining the respiratory tract is the other major process for loss of body heat. A very large amount of energy—600 kcal/L—is required to transform water from the liquid to the gaseous state. Thus, whenever water vaporizes from the body's surface, the heat required to drive the process is conducted from the surface, thereby cooling it.

Temperature-Regulating Reflexes

Temperature regulation offers a classic example of a biological control system (we used it as our example of such systems in Figure 7–1). The balance between heat production and heat loss is continuously being disturbed, either by changes in metabolic rate (exercise being the most powerful influence) or by changes in the external environment that alter heat loss or gain. The resulting changes in body temperature are detected by thermoreceptors, which initiate reflexes that change the output of various effectors so that heat production and/or loss are changed and body temperature is restored toward normal.

Figure 18–20 summarizes the specific components of these reflexes. There are two categories of thermoreceptors, one in the skin (**peripheral thermoreceptors,** described in Chapter 9) and the other (**central thermoreceptors)** in deep body structures, including the hypothalamus, spinal cord, and abdominal organs. Since it is the core body temperature, not the skin temperature, that is being maintained relatively constant, the central thermoreceptors provide the essential negative-feedback component of the reflexes. The peripheral thermoreceptors provide feedforward information, as described in Chapter 7, and also account for one's ability to identify a hot or cold area of the skin.

An area of the hypothalamus serves as the primary overall integrator of the reflexes, but other brain centers also exert some control over specific components of the reflexes.

Output from the hypothalamus and the other brain areas to the effectors is via: (1) sympathetic nerves to the sweat glands, skin arterioles, and the adrenal medulla; and (2) motor neurons to the skeletal muscles.

Control of Heat Production Changes in muscle activity constitute the major control of heat production for temperature regulation. The first muscle changes in response to a decrease in core body temperature are a gradual and general increase in skeletal-muscle contraction. This may lead to shivering, which consists of oscillating rhythmical muscle contractions and relaxations occurring at a rapid rate. During shivering, the efferent motor nerves to the skeletal muscles are influenced by descending pathways under the primary control of the hypothalamus. Because almost no external work is performed by shivering, virtually all the energy liberated by the metabolic machinery appears as internal heat and is known as **shivering thermogenesis.** People also use their muscles for voluntary heat-producing activities such as foot stamping and hand clapping.

Thus far, our discussion has focused primarily on the muscle response to *cold*; the opposite muscle reactions occur in response to heat. Basal muscle contraction is reflexly decreased, and voluntary movement is also diminished. These attempts to reduce heat production are relatively limited, however, both because basal muscle contraction is quite low to start with and because any increased core temperature produced by the heat acts *directly* on cells to increase metabolic rate.

Muscle contraction is not the only process controlled in temperature-regulating reflexes. In most experimental animals, chronic cold exposure induces an increase in metabolic rate (heat production) that is not due to increased muscle activity and is termed **nonshivering thermogenesis.** Its causes are an increased adrenal secretion of epinephrine and increased sympathetic activity to adipose tissue, with some contribution by thyroid hormone as well. However, nonshivering thermogenesis is quite minimal, if present at all, in adult human beings, and there is no increased secretion of thyroid hormone in response to cold. Nonshivering thermogenesis does occur in infants.

Control of Heat Loss by Radiation and Conduction For purposes of temperature control, it is convenient to view the body as a central core surrounded by a shell consisting of skin and subcutaneous tissue; we shall refer to this complex outer shell simply as skin.

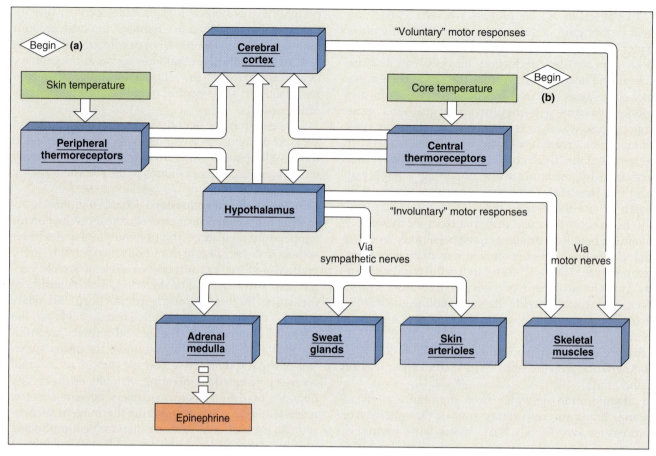

FIGURE 18-20

Summary of temperature-regulating mechanisms beginning with (a) peripheral thermoreceptors and (b) central thermoreceptors. The dashed arrow from the adrenal medulla indicates that this hormonal pathway is of minor importance in adult human beings. The solid arrows denote neural pathways. The hypothalamus influences sympathetic nerves via descending pathways.

It is the temperature of the central core that is being regulated at approximately 37°C. As we shall see, the temperature of the outer surface of the skin changes markedly.

If the skin were a perfect insulator, no heat would ever be lost from the core. The temperature of the outer skin surface would equal the environmental temperature, and net conduction would be zero. The skin is not a perfect insulator, however, and so the temperature of its outer surface generally is somewhere between that of the external environment and that of the core.

The skin's effectiveness as an insulator is subject to physiological control by a change in the blood flow to it. The more blood reaching the skin from the core, the more closely the skin's temperature approaches that of the core. In effect, the blood vessels diminish the insulating capacity of the skin by carrying heat to the surface to be lost to the external environment. These vessels are controlled largely by vasoconstrictor sympathetic nerves, the firing rate of which is reflexly increased in response to cold and decreased in response to heat. There is also a population of sympathetic neurons to the skin whose neurotransmitters (as yet unidentified) cause active *vasodilation*. Certain areas of skin participate much more than others in all these vasomotor responses, and so skin temperatures vary with location.

Finally, there are three *behavioral* mechanisms for altering heat loss by radiation and conduction: changes in surface area, changes in clothing, and choice of surroundings. Curling up into a ball, hunching the shoulders, and similar maneuvers in response to cold reduce the surface area exposed to the environment, thereby decreasing heat loss by radiation and conduction. In human beings, clothing is also an important component of temperature regulation, substituting for the insulating effects of feathers in birds and fur in other mammals. The outer surface of the clothes forms the true "exterior" of the body surface. The skin loses heat directly to the air space trapped by the clothes, which in turn pick up heat from the inner air layer and

TABLE 18–11 Summary of Effector Mechanisms in Temperature Regulation	
Desired Effect	**Mechanism**
	STIMULATED BY COLD (SEE ALSO FIGURE 7–2)
Decrease heat loss	1. Vasoconstriction of skin vessels
	2. Reduction of surface area (curling up, etc.)
	3. Behavioral response (put on warmer clothes, raise thermostat setting, etc.)
Increase heat production	1. Increased muscle tone
	2. Shivering and increased voluntary activity
	3. Increased secretion of epinephrine (minimal in adults)
	4. Increased food appetite
	STIMULATED BY HEAT
Increase heat loss	1. Vasodilation of skin vessels
	2. Sweating
	3. Behavioral response (put on cooler clothes, turn on fan, etc.)
Decrease heat production	1. Decreased muscle tone and voluntary activity
	2. Decreased secretion of epinephrine (minimal in adults)
	3. Decreased food appetite

transfer it to the external environment. The insulating ability of clothing is determined primarily by the thickness of the trapped air layer.

Clothing is important not only at low temperatures but also at very high temperatures. When the environmental temperature is greater than body temperature, conduction favors heat *gain* rather than heat loss. Heat gain also occurs by radiation during exposure to the sun. People therefore insulate themselves in such situations by wearing clothes. The clothing, however, must be loose so as to allow adequate movement of air to permit evaporation (see below). White clothing is cooler since it reflects more radiant energy, which dark colors absorb. Loose-fitting, light-colored clothes are far more cooling than going nude in a hot environment and during direct exposure to the sun.

The third behavioral mechanism for altering heat loss is to seek out warmer or colder surroundings, as for example by moving from a shady spot into the sunlight. Raising or lowering the thermostat of a house or turning on an air conditioner also fits this category.

Control of Heat Loss by Evaporation Even in the absence of sweating, there is loss of water by diffusion through the skin, which is not waterproof. A similar amount is lost from the respiratory lining during expiration. These two losses are known as **insensible water loss** and amount to approximately 600 ml/day in human beings. Evaporation of this water accounts for a significant fraction of total heat loss. In contrast to this passive water loss, sweating requires the active secretion of fluid by **sweat glands** and its extrusion into ducts that carry it to the skin surface.

Production of sweat is stimulated by sympathetic nerves to the glands. (These nerves release acetylcholine rather than the usual sympathetic neurotransmitter norepinephrine.) Sweat is a dilute solution containing sodium chloride as its major solute. Sweating rates of over 4 L/h have been reported; the evaporation of 4 L of water would eliminate almost 2400 kcal from the body!

It is essential to recognize that sweat must evaporate in order to exert its cooling effect. The most important factor determining evaporation rate is the water-vapor concentration of the air—that is, the *relative humidity*. The discomfort suffered on humid days is due to the failure of evaporation; the sweat glands continue to secrete, but the sweat simply remains on the skin or drips off.

Integration of Effector Mechanisms Table 18–11 summarizes the effector mechanisms regulating temperature, none of which is an all-or-none response but a graded, progressive increase or decrease in activity. By altering heat loss, changes in skin blood flow alone can regulate body temperature over a range of environmental temperatures (approximately 25 to 30°C or 75 to 86°F for a nude individual) known as the **thermoneutral zone.** At temperatures lower than this, even maximal vasoconstriction cannot prevent heat loss from exceeding heat production, and the body must increase its heat production to maintain temperature. At environmental temperatures above the thermoneutral zone, even maximal vasodilation cannot eliminate heat as fast as it is produced, and another heat-loss mechanism—sweating—is therefore brought strongly into play. Since at environmental temperatures above that of the body, heat is actually added to the body by radiation and conduction, evaporation is the sole mechanism for heat loss. A person's ability to tolerate such temperatures is determined by the humidity and by his/her maximal sweating rate. For example, when the air is completely dry, an individual can tolerate a temperature of 130°C (225°F) for 20 min or longer, whereas very moist air at 46°C (115°F) is bearable for only a few minutes.

Temperature Acclimatization

Changes in sweating onset, volume, and composition determine people's chronic adaptation to high temperatures. A person newly arrived in a hot environment has poor ability to do work; body temperature rises and severe weakness may occur. After several days, there is a great improvement in work tolerance, with much less increase in body temperature, and the person is said to have acclimatized to the heat (see Chapter 7 for a discussion of the concept of acclimatization). Body temperature does not rise as much because sweating begins sooner and the volume of sweat produced is greater.

There is also an important change in the composition of the sweat, namely, a marked reduction in its sodium concentration. This adaptation, which minimizes the loss of sodium from the body via sweat, is due to increased secretion of the adrenal mineralocorticoid hormone aldosterone. The sweat-gland secretory cells produce a solution with a sodium concentration similar to that of plasma, but some of the sodium is absorbed back into the blood as the secretion flows along the sweat-gland ducts toward the skin surface. Aldosterone stimulates this absorption in a manner identical to its stimulation of sodium reabsorption in the renal tubules.

Cold acclimatization has been much less studied than heat acclimatization because of the difficulty of subjecting individuals to total-body cold stress over long periods sufficient to produce acclimatization. Moreover, groups such as Eskimos that live in cold climates generally dress very warmly and so would not develop acclimatization to the cold.

Fever and Hyperthermia

Fever is an elevation of body temperature due to a "resetting of the thermostat" in the hypothalamus. A person with a fever still regulates body temperature in response to heat or cold but at a higher set point. The most common cause of fever is infection, but physical trauma and stress can also induce fever.

The onset of fever during infection is frequently gradual, but it is most striking when it occurs rapidly in the form of a chill. The brain thermostat is suddenly

raised, the person feels cold, and marked vasoconstriction and shivering occur. The person also curls up and puts on more blankets. This combination of decreased heat loss and increased heat production serves to drive body temperature up to the new set point, where it stabilizes. It will continue to be regulated at this new value until the thermostat is reset to normal and the fever "breaks." The person then feels hot, throws off the covers, and manifests profound vasodilation and sweating.

What is the basis for the thermostat resetting? Chemical messengers collectively termed **endogenous pyrogen (EP)** are released from macrophages (as well as other cell types) in the presence of infection or other fever-producing stimulus. The next steps vary depending on the precise stimulus for the release of EP. As illustrated in Figure 18–21, in some cases EP probably circulates in the blood to act upon the thermoreceptors in the hypothalamus (and perhaps other brain areas), altering the rate of firing and their input to the integrating centers. In other cases, EP may be produced by macrophage-like cells in the liver and stimulate neural receptors there that give rise to afferent neural input to the hypothalamic thermoreceptors. In both cases, the immediate cause of the resetting is a local synthesis and release of prostaglandins within the hypothalamus. *Aspirin* reduces fever by inhibiting this prostaglandin synthesis (Chapter 7).

The term "EP" was coined at a time when the identity of the chemical messenger(s) was not known. At least one peptide, **interleukin 1 (IL-1),** is now known to function as an EP, but other peptides—for example, **interleukin 6 (IL-6)**—play a role too. In addition to their effects on temperature, IL-1 and the other peptides have many other effects (described in Chapter 20) that have the common denominator of enhancing resistance to infection and promoting the healing of damaged tissue. Also as described in Chapter 20, all these peptides belong to the large family of chemical messengers called cytokines.

The story is even more complicated, however, because in response to a rising temperature the hypothalamus and other tissues release messengers that prevent excessive fever or contribute to the resetting of body temperature when the fever-causing stimulus is eliminated. Such messengers are termed **endogenous cryogens.** One known endogenous cryogen is vasopressin, functioning in this regard as a neurotransmitter.

One would expect fever, which is such a consistent concomitant of infection, to play some important protective role, and most evidence suggests that such is the case. For example, increased body temperature stimulates a large number of the body's defensive responses to infection. The likelihood that fever is a

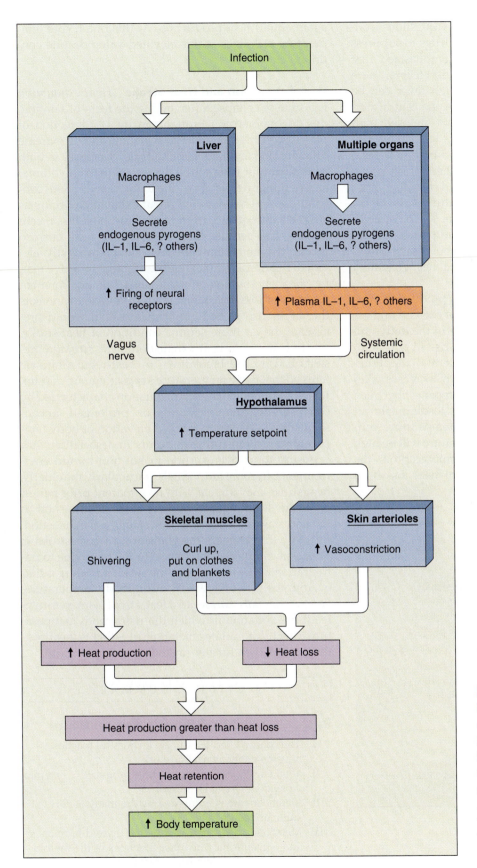

FIGURE 18–21

Pathway by which infection causes fever (IL-1 = interleukin 1, IL-6 = interleukin 6). Compare this figure to Figure 7–1: The effector responses are identical but they serve to keep body temperature *relatively constant* during exposure to cold (Figure 7–1) and *raise* body temperature during an infection (this figure). ✗

beneficial response raises important questions concerning the use of aspirin and other drugs to suppress fever during infection. It must be emphasized that these questions apply to the usual modest fevers. There is no question that an extremely high fever can be harmful, particularly in its effects on the central nervous system, and must be vigorously opposed with drugs and other forms of therapy.

To reiterate, fever is an increased body temperature caused by an elevation of the thermal set point. When body temperature is elevated for any other reason—that is, when body temperature is above the set point—it is termed **hyperthermia.** The most common cause of hyperthermia in normal people is exercise; the rise in body temperature above set point is due to retention of some of the internal heat generated by the exercising muscles.

As shown in Figure 18–22, heat production rises immediately during the initial stage of exercise and exceeds heat loss, causing heat storage in the body and a rise in the core temperature. This rise in core temperature triggers reflexes, via the central thermoreceptors, for increased heat loss; with increased skin blood flow and sweating, the discrepancy between heat production and heat loss starts to diminish but does not disappear. Therefore core temperature continues to rise. Ultimately, core temperature will be high enough to drive, via the central thermoreceptors, the heat-loss reflexes at a rate such that heat loss once

again equals heat production. At this point, core temperature stabilizes at this elevated value despite continued exercise.

Heat Exhaustion and Heat Stroke *Heat exhaustion* is a state of collapse, often taking the form of fainting, due to hypotension brought on by (1) depletion of plasma volume secondary to sweating, and (2) extreme dilation of skin blood vessels. Thus, decreases in both cardiac output (due to the decreased plasma volume) and peripheral resistance (due to the vasodilation) contribute to the hypotension. Heat exhaustion occurs as a direct consequence of the activity of heat-loss mechanisms, and because these mechanisms have been so active, the body temperature is only modestly elevated. In a sense, heat exhaustion is a safety valve that, by forcing cessation of work in a hot environment when heat-loss mechanisms are overtaxed, prevents the larger rise in body temperature that would precipitate the far more serious condition of heat stroke.

In contrast to heat exhaustion, *heat stroke* represents a complete breakdown in heat-regulating systems so that body temperature keeps going up and up. It is an extremely dangerous situation characterized by collapse, delirium, seizures, or prolonged unconsciousness—all due to marked elevation of body temperature. It almost always occurs in association with exposure to or overexertion in hot and humid environments. In some individuals, particularly the elderly, heat stroke may appear with no apparent prior period of severe sweating, but in most cases, it comes on as the end stage of prolonged untreated heat exhaustion. Exactly what triggers the transition to heat stroke is not clear—impaired circulation to the brain due to dehydration is one factor—but the striking finding is that even in the face of a rapidly rising body temperature, the person fails to sweat. Heat stroke is a positive-feedback situation in which the rising body temperature directly stimulates metabolism—that is, heat production—which further raises body temperature.

SECTION C SUMMARY

Basic Concepts of Energy Expenditure

I. The energy liberated during a chemical reaction appears either as heat or work.

II. Total energy expenditure = heat produced + external work done + energy stored.

III. Metabolic rate is influenced by the many factors summarized in Table 18–7.

IV. Metabolic rate is increased by the thyroid hormones and epinephrine. The other functions of the thyroid hormones are summarized in Table 18–8.

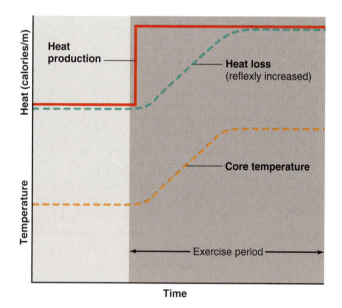

FIGURE 18–22

Thermal changes during exercise. Heat loss is reflexly increased, and when it once again equals heat production, core temperature stabilizes.

Regulation of Total-Body Energy Stores

I. Energy storage as fat can be positive or negative when the metabolic rate is less than or greater than, respectively, the energy content of ingested food.
   a. Energy storage is regulated mainly by reflex adjustment of food intake.
   b. In addition, the metabolic rate increases or decreases to some extent when food intake is chronically increased or decreased, respectively.
II. Food intake is controlled by leptin, secreted by adipose-tissue cells, and a variety of satiety factors, as summarized in Figure 18–17.
III. Being overweight or obese, the result of an imbalance between food intake and metabolic rate, increases the risk of many diseases.

Regulation of Body Temperature

I. Core body temperature shows a circadian rhythm, being highest during the day and lowest at night.
II. The body exchanges heat with the external environment by radiation, conduction, convection, and evaporation of water from the body surface.
III. The hypothalamus and other brain areas contain the integrating centers for temperature-regulating reflexes, and both peripheral and central thermoreceptors participate in these reflexes.
IV. Body temperature is regulated by altering heat production and/or heat loss so as to change total body heat content.
   a. Heat production is altered by increasing muscle tone, shivering, and voluntary activity.
   b. Heat loss by radiation, conduction, and convection depends on the difference between the skin surface and the environment.
   c. In response to cold, skin temperature is decreased by decreasing skin blood flow through reflex stimulation of the sympathetic nerves to the skin. In response to heat, skin temperature is increased by inhibiting these nerves.
   d. Behavioral responses such as putting on more clothes also influence heat loss.
   e. Evaporation of water occurs all the time as insensible loss from the skin and respiratory lining. Additional water for evaporation is supplied by sweat, stimulated by the sympathetic nerves to the sweat glands.
   f. Increased heat production is essential for temperature regulation at environmental temperatures below the thermoneutral zone, and sweating is essential at temperatures above this zone.
V. Temperature acclimatization to heat is achieved by an earlier onset of sweating, an increased volume of sweat, and a decreased sodium concentration of the sweat.
VI. Fever is due to a resetting of the temperature set point so that heat production is increased and heat loss is decreased in order to raise body temperature to the new set point and keep it there. The stimulus is endogenous pyrogen, which is interleukin 1 and other peptides as well.
VII. The hyperthermia of exercise is due to the increased heat produced by the muscles.

SECTION C KEY TERMS

external work	convection
internal work	wind-chill index
total energy expenditure	evaporation
kilocalorie (kcal)	peripheral thermoreceptor
metabolic rate	central thermoreceptor
basal metabolic rate	shivering thermogenesis
(BMR)	nonshivering thermogenesis
calorigenic effect	insensible water loss
food-induced thermo-	sweat gland
genesis	thermoneutral zone
leptin	fever
satiety signal	endogenous pyrogen (EP)
body mass index (BMI)	interleukin 1 (IL-1)
homeothermic	interleukin 6 (IL-6)
radiation	endogenous cryogens
conduction	hyperthermia

SECTION C REVIEW QUESTIONS

1. State the formula relating total energy expenditure, heat produced, external work, and energy storage.
2. What two hormones alter the basal metabolic rate?
3. State the equation for total-body energy balance. Describe the three possible states of balance with regard to energy storage.
4. What happens to the basal metabolic rate after a person has either lost or gained weight?
5. List five satiety signals.
6. List three beneficial effects of exercise in a weight-loss program.
7. Compare and contrast the four mechanisms for heat loss.
8. Describe the control of skin blood vessels during exposure to cold or heat.
9. With a diagram, summarize the reflex responses to heat or cold. What are the dominant mechanisms for temperature regulation in the thermoneutral zone and in temperatures below and above this range?
10. What changes are exhibited by a heat-acclimatized person?
11. Summarize the sequence of events leading to a fever and contrast this to the sequence leading to hyperthermia during exercise.

CHAPTER 18 CLINICAL TERMS

diabetes mellitus	sulfonylureas
insulin-dependent diabetes	fasting hypoglycemia
mellitus (IDDM)	atherosclerosis
noninsulin-dependent	cancer
diabetes mellitus	oncogene
(NIDDM)	giantism
diabetic ketoacidosis	dwarfism
insulin resistance	acromegaly

growth hormone
  insensitivity syndrome
hypothyroidism
endemic cretinism
hyperthyroidism
anabolic steroids
overweight

obesity
anorexia nervosa
bulimia
heat exhaustion
heat stroke
aspirin

CHAPTER 18 THOUGHT QUESTIONS

(*Answers are given in Appendix A.*)

1. What happens to the triacylglycerol concentrations in the plasma and in adipose tissue after administration of a drug that blocks the action of lipoprotein lipase?

2. A resting, unstressed person has increased plasma concentrations of free fatty acids, glycerol, amino acids, and ketones. What situations might be responsible and what additional plasma measurement would distinguish among them?

3. A normal volunteer is given an injection of insulin. The plasma concentrations of which hormones increase?

4. If the sympathetic preganglionic fibers to the adrenal medulla were cut in an animal, would this eliminate the sympathetically mediated component of increased gluconeogenesis and lipolysis during exercise? Explain.

5. A patient with insulin-dependent diabetes suffers a broken leg. Would you advise this person to increase or decrease his dosage of insulin?

6. A person has a defect in the ability of her small intestine to absorb bile salts. What effect will this have on her plasma cholesterol concentration?

7. A well-trained athlete is found to have a moderately elevated plasma cholesterol concentration. What additional measurements would you advise this person to have done?

8. A full-term newborn infant is abnormally small. Is this most likely due to deficient growth hormone, deficient thyroid hormones, or deficient in utero nutrition?

9. Why might the administration of androgens to stimulate growth in a small 12-year-old male turn out to be counterproductive?

10. What are the sources of heat loss for a person immersed up to the neck in a 40°C bath?

11. Lizards can regulate their body temperatures only through behavioral means. Can you predict what they do when they are infected with bacteria?

Reproduction

GENERAL TERMINOLOGY AND CONCEPTS

Before we begin detailed descriptions of male and female reproductive systems, it is worthwhile to summarize some general terminology and concepts. This brief description is for orientation, and the specifics of these processes will be dealt with in subsequent sections.

The primary reproductive organs are known as the **gonads**: the **testes** (singular, *testis*) in the male and the **ovaries** in the female. In both sexes, the gonads serve dual functions: (1) **Gametogenesis** is the production of the reproductive cells, or **gametes**; these are **spermatozoa** (singular, *spermatozoan*, usually shortened to **sperm**) by males and **ova** (singular, **ovum**) by females. (2) The gonads secrete particular steroid hormones, often termed **sex hormones**; the primary sex hormones are **testosterone** in the male and **estradiol** and **progesterone** in the female.

Several comments about sex hormone terminology and function are useful at this point. As described in Chapter 10, testosterone belongs to a group of steroid hormones that have similar masculinizing actions and are collectively called **androgens**. In the male, only the testes secrete significant amounts of testosterone. Other circulating androgens are produced by the adrenal cortex, but they are much less potent than testosterone and are unable to maintain male reproductive function if testosterone secretion is inadequate.

(A brief mention should be made of one of the adrenal androgens—dihydroepiandrosterone, DHEA, which is presently being sold as a dietary supplement and is touted as a miracle drug—one that can stop or reverse the aging process and the diseases associated with it, cure depression, strengthen the immune system, and improve athletic performance. Despite the claims, there are no long-term studies of either the benefits or risks of this hormone. DHEA is itself a weak androgen but can be converted in the body to testosterone in both men and women. Its secretion is high just before birth and again during puberty, and then falls off markedly with aging.)

Estradiol—secreted in large amounts only by the ovaries—is one of several steroid hormones that have similar actions on the female reproductive tract and are collectively termed "estrogens" or, more commonly, simply **estrogen**. In the remainder of this chapter we shall follow the common practice of using this generic term "estrogen" rather than "estradiol," pointing out where relevant the situations in which an estrogen other than estradiol is produced by an extraovarian source and becomes important.

Estrogens are not unique to females, neither are androgens to males. Plasma estrogen arises in males from secretion of small amounts by the testes and from conversion of androgens to estrogen in many nongonadal tissues (notably adipose tissue). Conversely, in females androgens are secreted, in small amounts, by the ovaries and, in larger amounts, by the adrenal cortex. (Some of these androgens are then converted to estrogen in nongonadal tissues, just as in men, and contribute to the plasma estrogen.)

As described in Chapter 7, all steroid hormones act in the same general way. They bind to intracellular receptors, and the hormone-receptor complex then binds to DNA in the nucleus, functioning as a transcription factor to alter the rate of formation of particular mRNAs. The result is a change in the rates of synthesis of the proteins coded for by the genes being transcribed. The resulting increase or decrease in the concentrations of these proteins in the target cells or their rates of secretion by the cells then account for the cells' responses to the hormone.

The systems of ducts through which the sperm or eggs are transported and the glands lining or emptying into the ducts are termed the **accessory reproductive organs**. In the female the breasts are also usually included in this category. The **secondary sexual characteristics** comprise the many external differences—hair distribution and body contours, for example—between males and females. The secondary sexual characteristics are not directly involved in reproduction.

Reproductive function is largely controlled by a chain of hormones (Figure 19–1). The first hormone in the chain is **gonadotropin releasing hormone (GnRH)**. As described in Chapter 10, GnRH is one of the hypophysiotropic hormones; that is, it is a hormone secreted by neuroendocrine cells in the hypothalamus, and it reaches the anterior pituitary via the hypothalamo-pituitary portal blood vessels. Accordingly, the brain is the primary regulator of the reproductive process.

Secretion of GnRH is triggered by action potentials in GnRH-producing hypothalamic neuroendocrine cells. These action potentials occur periodically in brief bursts, with virtually no secretion in between. This pattern of GnRH secretion is important because the cells of the anterior pituitary that secrete the gonadotropins will not respond to GnRH if the plasma concentration of this hormone remains elevated over time.

In the anterior pituitary, GnRH stimulates the release of the pituitary **gonadotropins—follicle-stimulating hormone (FSH)** and **luteinizing hormone**

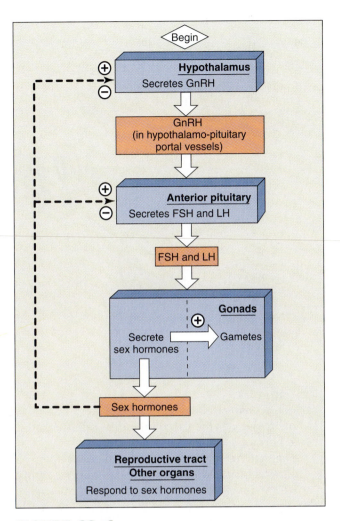

FIGURE 19-1
General pattern of reproduction control in both males and females. GnRH, like all hypothalamic hypophysiotropic hormones, reaches the anterior pituitary via the hypothalamo-pituitary portal vessels (Chapter 10). The arrow within the gonads denotes the fact that the sex hormones act locally, as paracrine agents, to influence the gametes.

(LH) (Figure 19-1). These two protein hormones, which are produced by a single pituitary cell type, were named for their effects in the female, but their molecular structures are the same in both sexes. The two hormones act upon the gonads, the result being the development of sperm or ova and sex hormone secretion. In turn, the sex hormones exert many effects on all portions of the reproductive system, including the gonads from which they come, and other parts of the body as well. In addition, the steroidal sex hormones exert feedback effects on the secretion of GnRH, FSH, and LH. A gonadal protein hormone, **inhibin,** also exerts feedback effects on the anterior pituitary.

Each link in this hormonal chain is essential, and malfunction of either the hypothalamus or the anterior pituitary can result in failure of sex hormone secretion

TABLE 19-1 Stages in the Control of Reproductive Function

1. During the initial stage, which begins during fetal life and ends in the first year of life (infancy), GnRH, the gonadotropins, and gonadal sex hormones are secreted at relatively high levels.

2. From infancy to puberty, the secretion rates of these hormones are very low, and reproductive function is quiescent.

3. Beginning at puberty, hormonal secretion rates increase markedly, showing large cyclical swings in women during the menstrual cycle, and this ushers in the period of active reproduction.

4. Finally, reproductive function diminishes later in life, largely because the gonads become less responsive to the gonadotropins, the ability to reproduce ceasing entirely in women.

and gametogenesis as surely as if the gonads themselves were diseased.

As a result of changes in the amount and pattern of hormonal secretions, reproductive function changes markedly during a person's lifetime and may be divided into the stages summarized in Table 19-1.

General Principles of Gametogenesis

At any point in gametogenesis, the developing gametes are termed **germ cells.** These cells undergo either mitosis or the type of cell division known as meiosis (described below). Because the general principles of gametogenesis are essentially the same in males and females, they are treated in this section, and features specific to the male or female are described later.

The first stage in gametogenesis is proliferation of the primordial germ cells by mitosis. Recall from our discussion of cell division in Chapter 5 that, with the exception of the gametes, the DNA of each nucleated human cell is contained in 23 pairs of chromosomes, giving a total of 46. The two corresponding chromosomes in each pair are said to be homologous to each other, with one coming from each of the person's parents. In mitosis, all the dividing cell's 46 chromosomes are replicated, and each of the two daughter cells resulting from the division receives a full set of 46 chromosomes identical to those of the original cell. Thus each daughter cell receives identical genetic information.

In this manner, mitosis of primordial germ cells, each of which contains 46 chromosomes, provides a supply of identical germ cells for the next stages. The timing of mitotic activity in germ cells differs greatly in females and males. In the female, mitosis of germ cells occurs exclusively during the individual's embryonic development. In the male, some mitosis occurs in the embryo to generate the population of germ cells present at birth, but mitosis really begins in earnest at puberty and usually continues throughout life.

The second stage of gametogenesis is **meiosis,** in which each resulting gamete receives only 23 chromosomes from a 46-chromosome germ cell, 1 chromosome from each homologous pair. Because a sperm and an about-to-be fertilized egg each has only 23 chromosomes, their union at fertilization results once again in a cell with a full complement of 46 chromosomes.

Let us see how meiosis works, using Figure 19–2 in which the letters are keyed to the text. Meiosis consists of two cell divisions in succession. The events preceding the first meiotic division are identical to those preceding a *mitotic* division, as described in Chapter 5. Recall that during the interphase preceding a mitotic division, the DNA of the chromosomes is replicated. Thus a resting interphase cell still has 46 chromosomes, but each chromosome consists of two identical strands of DNA, termed sister chromatids, which are joined together by a centromere (a).

As the first meiotic division begins, homologous chromosomes, each consisting of two identical sister chromatids, come together and line up opposite each other. Thus, (b) 23 four-chromatid groupings, called tetrads, are formed. (c) The sister chromatids of each chromosome condense into thick rodlike structures and become highly visible. Then, (d) within a tetrad, corresponding segments of homologous chromosomes come to overlap one another. Overlapping portions of the homologous chromosomes break off and exchange with each other in the process known as **crossing-over** (e). Thus, crossing-over results in the recombination of genes on homologous chromosomes.

Following crossing-over, the tetrads line up in the center of the cell (f). The orientation of each tetrad on the equator is *random,* meaning that sometimes the maternal portion points to a particular pole of the cell, and sometimes the paternal portion does so. The cell

FIGURE 19–2

Stages of meiosis in a generalized germ cell. For simplicity, the initial cell (a), which is in interphase, is given only 4 chromosomes rather than 46, the human number. Also, cytoplasm is shown only in (a), (h), and (i). Chromosomes from one parent are red, and those from the other parent are blue. The letters are keyed to descriptions in the text.

Adapted from Carlson.

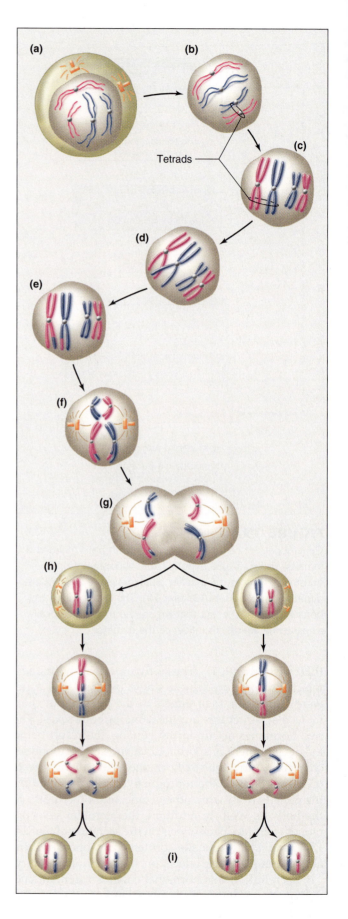

then divides (the first division of meiosis), with the maternal chromatids of any particular tetrad going to one of the two cells resulting from the division, and the paternal chromatids going to the other (g). Because of the random orientation of the tetrads at the equator, it is extremely unlikely that all 23 maternal chromatids will end up in one cell and all 23 paternal chromatids in the other. Over 8 million (2^{23}) different combinations of maternal and paternal chromosomes can result during this first meiotic division.

The second division of meiosis occurs without any further replication of DNA. The sister chromatids —both of which were originally either maternal or paternal—of each chromosome separate and move apart into the new daughter cells (h to i). The daughter cells resulting from the second meiotic division, therefore, contain 23 one-chromatid chromosomes (i).

To summarize, meiosis produces daughter cells having only 23 chromosomes, and two events during the first meiotic division contribute to the enormous genetic variability of their daughter cells: (1) crossing-over, and (2) the random distribution of maternal and paternal chromatid pairs between the two daughter cells.

SECTION **A** SUMMARY

The gonads have a dual function—gametogenesis and the secretion of sex hormones.

General Principles of Gametogenesis

I. The first stage of gametogenesis is mitosis of primordial germ cells.

II. This is followed by meiosis, a sequence of two cell divisions resulting in each gamete receiving 23 chromosomes.

III. Crossing-over and random distribution of maternal and paternal chromatids to the daughter cells during meiosis cause genetic variability in the gametes.

SECTION **A** KEY TERMS

gonad	estrogen
testis	accessory reproductive organ
ovary	secondary sexual
gametogenesis	characteristic
gamete	gonadotropin releasing
spermatozoa	hormone (GnRH)
sperm	gonadotropin
ova	follicle-stimulating hormone
ovum	(FSH)
sex hormones	luteinizing hormone (LH)
testosterone	inhibin
estradiol	germ cell
progesterone	meiosis
androgen	crossing-over

SECTION **A** REVIEW QUESTION

1. Describe the stages of gametogenesis and how meiosis results in genetic variability.

MALE REPRODUCTIVE PHYSIOLOGY

Anatomy

The male reproductive system includes the two testes, the system of ducts that store and transport sperm to the exterior, the glands that empty into these ducts, and the penis. The duct system, glands, and penis constitute the male accessory reproductive organs.

The testes are suspended outside the abdomen in the **scrotum,** which is an outpouching of the abdominal wall and is divided internally into two sacs, one for each testis. During fetal development, the testes are located in the abdomen, but during the seventh month of intrauterine development, they descend into the scrotum. This descent is essential for normal sperm production during adulthood, since sperm formation requires a temperature several degrees lower than normal internal body temperature. Cooling is achieved by air circulating around the scrotum and by a heat-exchange mechanism in the blood vessels supplying the testes. In contrast to spermatogenesis, testosterone secretion can usually occur normally at internal body temperature, and so failure of testes descent does not impair testosterone secretion.

The sites of sperm formation, or **spermatogenesis,** in the testes are the many tiny, convoluted **seminiferous tubules** (Figure 19–3), the combined length of which is 250 m (the length of almost 3 football fields). Each seminiferous tubule is bounded by a basement membrane. In the center of each tubule is a fluid-filled lumen containing spermatozoa. The tubular wall is composed of developing germ cells and another cell type, to be described later, called Sertoli cells.

The **Leydig cells,** or interstitial cells, which lie in small connective tissue spaces *between* the tubules, are

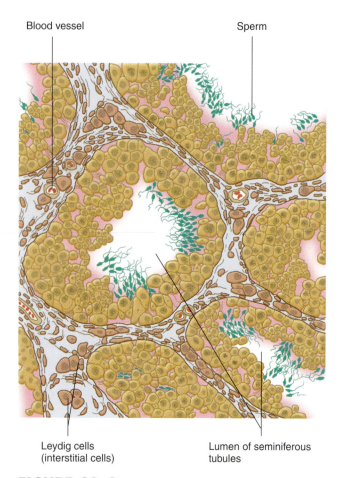

Blood vessel

Sperm

Leydig cells
(interstitial cells)

Lumen of seminiferous
tubules

FIGURE 19–3

Cross-section of an area of testis. The brown cells are in the seminiferous tubules, the sites of sperm production. This low-power magnification is only for general orientation, and the positions of the specific cells in the tubules are shown in Figure 19–8. The tubules are separated from each other by interstitial space (colored light blue) that contains Leydig cells and blood vessels.

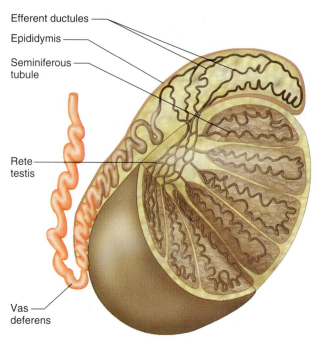

Efferent ductules

Epididymis

Seminiferous
tubule

Rete
testis

Vas
deferens

FIGURE 19–4

Section of a testis. The upper portion of the testis has been removed to show its interior.

the cells that secrete testosterone. Thus, the sperm-producing and testosterone-producing functions of the testes are carried out by different structures—the seminiferous tubules and Leydig cells, respectively.

The seminiferous tubules from different areas of a testis unite to form a network of interconnected tubes, the rete testis (Figure 19–4). Small ducts termed efferent ductules leave the rete testis, pierce the fibrous covering of the testis, and empty into a single duct within a structure called the **epididymis** (plural, *epididymides*). The epididymis is loosely attached to the outside of the testis. The duct of the epididymis is so convoluted that, when straightened out at dissection, it measures 6 m. In turn, the epididymis draining each testis leads to a **vas** (or ductus) **deferens,** a large thick-walled tube lined with smooth muscle. The vas deferens and the blood vessels and nerves supplying the testis are bound together in the **spermatic cord,** which passes to the testis through a slitlike passage, the inguinal canal, in the abdominal wall.

After entering the abdomen, the two vas deferens—one from each testis—course to the back of the urinary bladder base (Figure 19–5). The ducts from two large glands, the **seminal vesicles,** which lie behind the bladder, join the two vas deferens to form the two **ejaculatory ducts.** The ejaculatory ducts then enter the substance of the **prostate gland** and join the urethra, coming from the bladder. The prostate gland is a single donut-shaped gland below the bladder and surrounding the upper part of the urethra, into which it secretes fluid through hundreds of tiny openings in the side of the urethra. The urethra leaves the prostate gland to enter the penis. The paired **bulbourethral glands,** lying below the prostate, drain into the urethra just after it leaves the prostate.

The prostate gland and seminal vesicles secrete the bulk of the fluid in which ejaculated sperm are suspended. This fluid, plus the sperm cells, constitute **semen,** the sperm contributing only a few percent of the total volume. The glandular secretions contain a large number of different chemical substances, including nutrients, buffers for protecting the sperm against the acidic vaginal secretions, chemicals (particularly from the seminal vesicles) that increase sperm motility, and prostaglandins. The function of the prostaglandins, which are produced by the seminal vesicles, is still not

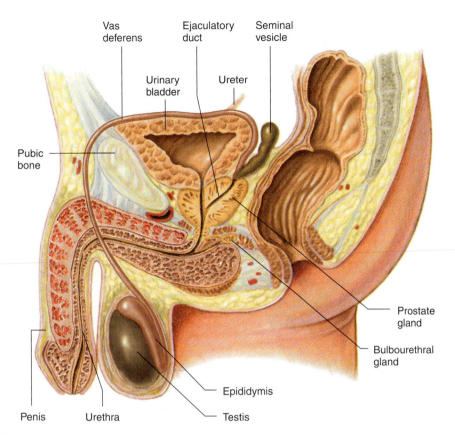

Vas
deferens

Ejaculatory
duct

Seminal
vesicle

Urinary
bladder

Ureter

Pubic
bone

Prostate
gland

Bulbourethral
gland

Epididymis

Penis   Urethra   Testis

FIGURE 19–5

Anatomic organization of the male reproductive tract. This figure shows the testis, epididymis, vas deferens, ejaculatory duct, seminal vesicle, and bulbourethral gland on only one side of the body, but they are all paired structures. The urinary bladder and a ureter are shown for orientation but are not part of the reproductive tract. Once the ejaculatory ducts join the urethra in the prostate, the urinary and reproductive tracts have merged.

clear. The bulbourethral glands contribute a small volume of lubricating mucoid secretions.

In addition to providing a route for sperm from the seminiferous tubules to the exterior, several of the duct system segments perform additional functions to be described in the section on sperm transport.

Spermatogenesis

The various stages of spermatogenesis are summarized in Figure 19–6. The undifferentiated germ cells, which are termed **spermatogonia** (singular, *spermatogonium*) begin to divide mitotically at puberty. The daughter cells of this first division then divide, and so on for a specified number of division cycles so that a clone of spermatogonia is produced from each original spermatogonium. Some differentiation occurs in addition to cell division. The cells that result from the final mitotic division and differentiation in the series are called **primary spermatocytes,** and these are the cells that will undergo the first meiotic division of spermatogenesis.

It should be emphasized that if all the cells in the clone produced by each original spermatogonium followed this pathway, the spermatogonia would disappear—that is, would all be converted to primary spermatocytes. This does not occur because, at an early point, one of the cells of each clone "drops out" of the mitosis-differentiation cycles and reverts to being a primitive spermatogonium that, at a later time, will enter into its own full sequence of divisions. In turn, one cell of the clone it produces will do likewise, and so on. Thus, the supply of undifferentiated spermatogonia does not decrease.

Each primary spermatocyte increases markedly in size and undergoes the first meiotic division (Figure 19–6) to form two **secondary spermatocytes,** each of which contains 23 two-chromatid chromosomes. Each secondary spermatocyte, in turn undergoes the second meiotic division (see Figure 19–2h to i) into **spermatids.** Thus, each primary spermatocyte, containing 46 two-chromatid chromosomes, gives rise to four spermatids, each containing 23 one-chromatid chromosomes.

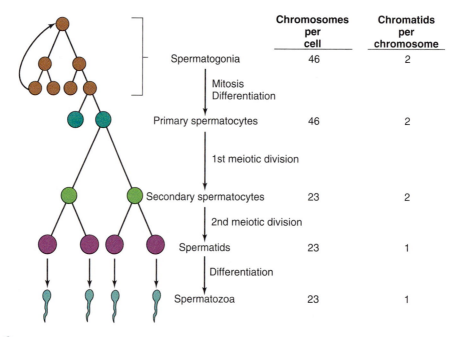

	Chromosomes per cell	Chromatids per chromosome
Spermatogonia	46	2
Mitosis Differentiation		
Primary spermatocytes	46	2
1st meiotic division		
Secondary spermatocytes	23	2
2nd meiotic division		
Spermatids	23	1
Differentiation		
Spermatozoa	23	1

FIGURE 19–6

Summary of spermatogenesis, which begins at puberty. Each spermatogonium yields, by mitosis, a clone of spermatogonia; for simplicity, the figure shows only two such cycles, with a third mitotic cycle generating two primary spermatocytes. The arrow from one of the spermatogonia back to an original spermatogonium denotes the fact that one cell of the clone does not go on to generate primary spermatocytes but reverts to being an undifferentiated spermatogonium that gives rise to a new clone. Note that each primary spermatocyte produces four spermatozoa.

The final phase of spermatogenesis is the differentiation of the spermatids into spermatozoa (sperm). This process involves extensive cell remodeling, including elongation, but no further cell divisions. The head of a sperm (Figure 19–7) consists almost entirely of the nucleus, which contains the DNA bearing the sperm's genetic information. The tip of the nucleus is covered by the **acrosome,** a protein-filled vesicle containing several enzymes that play an important role in the sperm's penetration of the egg. Most of the tail is a flagellum—a group of contractile filaments that produce whiplike movements capable of propelling the sperm at a velocity of 1 to 4 mm/min. The sperm's mitochondria form the midpiece of the sperm and provide the energy for the sperm's movement.

The entire process of spermatogenesis, from primary spermatocyte to sperm, takes approximately 64 days. The normal human male manufactures approximately 30 million sperm per day.

Thus far, we have described spermatogenesis without regard to its orientation within the seminiferous tubules or the participation of **Sertoli cells,** the second type of cell in the seminiferous tubules, with which the developing germ cells are intimately associated. As noted earlier, each seminiferous tubule is bounded by a basement membrane. Each Sertoli cell extends from the basement membrane all the way to the lumen in the center of the tubule and is joined

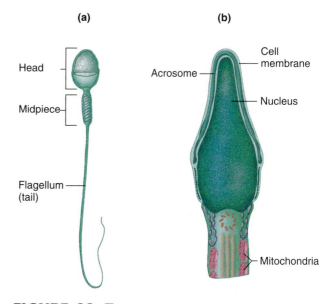

FIGURE 19–7

(a) Diagram of a human mature sperm. (b) A close-up of the head. The acrosome contains enzymes required for fertilization of the ovum.

to adjacent Sertoli cells by means of tight junctions (Figure 19–8). Thus, the Sertoli cells form an unbroken ring around the outer circumference of the seminiferous tubule, and the tight junctions divide the tubule

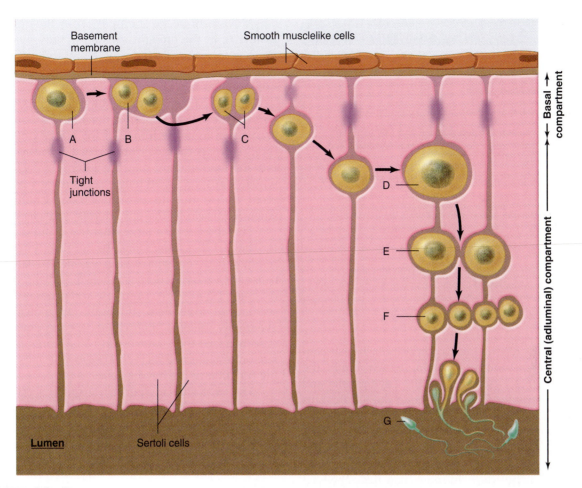

FIGURE 19–8

Relation of the Sertoli cells and germ cells. The Sertoli cells form a ring (barrier) around the entire tubule. For convenience of presentation, the various stages of spermatogenesis are shown as though the germ cells move down a line of adjacent Sertoli cells; in reality, all stages beginning with any given spermatogonium take place between the same two Sertoli cells. Spermatogonia (A and B) are found only in the basal compartment (between the tight junctions of the Sertoli cells and the basement membrane of the tubule). After several mitotic cycles (A to B), the spermatogonia (B) give rise to primary spermatocytes (C). Each of the latter crosses a tight junction, enlarges (D), and divides into two secondary spermatocytes (E), which divide into spermatids (F), which in turn differentiate into spermatozoa (G). This last step involves loss of cytoplasm by the spermatids.

Adapted from Tung.

into two compartments—a basal compartment between the basement membrane and the tight junctions, and a central compartment, beginning at the tight junctions and including the lumen.

This arrangement has several very important results: (1) The ring of interconnected Sertoli cells forms the **Sertoli-cell barrier,** which prevents the movement of many chemicals from the blood into the lumen of the seminiferous tubule and thereby ensures proper conditions for germ-cell development and differentiation in the tubules; and (2) different stages of spermatogenesis take place in different compartments and, hence, in different environments.

Mitotic cell divisions and differentiation of spermatogonia to yield primary spermatocytes take place entirely in the basal compartment. The primary spermatocytes then move through the tight junctions of the Sertoli cells (which open in front of them while at the same time new tight junctions form behind them) to gain entry into the central compartment. In this central compartment, the meiotic divisions of spermatogenesis occur, and the spermatids are remodeled into sperm while contained in recesses formed by invaginations of the Sertoli-cell plasma membranes. When sperm formation is completed, the Sertoli-cell cytoplasm around the sperm retracts, and the sperm are released into the lumen to be bathed by the luminal fluid.

Sertoli cells serve as the route by which nutrients reach developing germ cells, and they also secrete most of the fluid found in the tubule lumen. This fluid has

TABLE 19–2 Functions of Sertoli Cells

1. Provide Sertoli-cell barrier to chemicals in the plasma

2. Nourish developing sperm

3. Secrete luminal fluid, including androgen-binding protein

4. Receive stimulation by testosterone and FSH to secrete paracrine agents that stimulate sperm proliferation and differentiation

5. Secrete the protein hormone inhibin, which inhibits FSH secretion

6. Secrete paracrine agents that influence the function of Leydig cells

7. Phagocytize defective sperm

8. Secrete, during embryonic life, Müllerian inhibiting substance (MIS), which causes the primordial female duct system to regress

a highly characteristic ionic composition. It also contains **androgen-binding protein,** which binds the testosterone that is secreted by the Leydig cells and crosses the Sertoli-cell barrier to enter the tubule. This protein maintains a high concentration of testosterone in the lumen of the tubule.

Thus far we have emphasized how Sertoli cells influence the ionic and nutritional environment of the germ cells, but they do more than this: They act as intermediaries between the germ cells and the hormones—FSH from the anterior pituitary and testosterone from the Leydig cells—that stimulate spermatogenesis. In other words, these hormones do not act *directly* on the germ cells but rather on the Sertoli cells, which respond by secreting a variety of chemical messengers that function as paracrine agents to stimulate proliferation and differentiation of the germ cells.

In addition, the Sertoli cells secrete the protein hormone inhibin and paracrine agents that affect Leydig-cell function. The many functions of Sertoli cells, several of which remain to be described later in this chapter, are summarized in Table 19–2.

Transport of Sperm

From the seminiferous tubules, the sperm pass through the rete testis and efferent ductules into the epididymis and thence into the vas deferens. The vas deferens and the portion of the epididymis closest to it serve as a storage reservoir for sperm, holding them until sexual arousal leads to ejaculation. Also, in the epididymis the sperm undergo a further testosterone-dependent maturation process.

Movement of the sperm as far as the epididymis results from the pressure created mainly by the continuous formation of fluid by the Sertoli cells back in the seminiferous tubules. The sperm themselves are nonmotile at this time.

During passage through the epididymis, there occurs a hundredfold concentration of the sperm by fluid absorption from the lumen of the epididymis. Therefore, as the sperm pass from the end of the epididymis into the vas deferens, they are a densely packed mass whose transport is no longer a result of fluid movement but is due to peristaltic contractions of the smooth muscle in the epididymis and vas deferens.

The absence of a large quantity of fluid accounts for the fact that *vasectomy,* the surgical tying-off and removal of a segment of each vas deferens, does not cause the accumulation of much fluid behind the tie-off point. (The sperm, which are still produced in a vasectomized individual, do build up, however, and eventually dissolve, their chemical components being absorbed into the bloodstream.) Vasectomy has no effect on testosterone secretion because it does not alter the function of the Leydig cells.

The next step in sperm transport is ejaculation, usually preceded by erection, which permits entry of the penis into the vagina.

Erection

The penis's becoming rigid—**erection**—is a vascular phenomenon. The penis consists almost entirely of three cylindrical vascular compartments running its entire length. Normally the small arteries supplying the vascular compartments are constricted so that the compartments contain little blood and the penis is flaccid. During sexual excitation, the small arteries dilate, the three vascular compartments become engorged with blood at high pressure, and the penis becomes rigid. The vascular dilation is initiated by neural input to the small arteries of the penis. Moreover, as the vascular compartments expand, the veins emptying them are passively compressed, thus contributing to the engorgement. This entire process occurs rapidly, complete erection sometimes taking only 5 to 10 s.

What are the neural inputs to the small arteries of the penis? At rest, the dominant input is via sympathetic neurons; they release norepinephrine, which causes the arterial smooth muscle to contract. During erection this sympathetic input is inhibited, but much more important is the activation of nonadrenergic, noncholinergic autonomic neurons to the arteries (Figure 19–9). These neurons release **nitric oxide,** which relaxes the arterial smooth muscle.

Which receptors and afferent pathway initiate these reflexes? The primary stimulus comes from mechanoreceptors in the genital region, particularly in

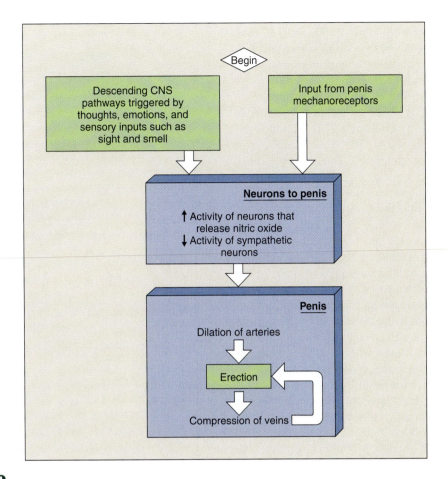

FIGURE 19–9

Reflex pathways for erection. Nitric oxide, a vasodilator, is the most important neurotransmitter to the arteries in this reflex.

the head of the penis. The afferent fibers carrying the impulses synapse in the lower spinal cord on interneurons that control the efferent outflow.

It must be stressed, however, that higher brain centers, via descending pathways, may also exert profound stimulatory or inhibitory effects upon the autonomic neurons to the small arteries of the penis. Thus, mechanical stimuli from areas other than the penis, as well as thoughts, emotions, sight, and odors, can induce erection in the complete absence of penile stimulation (or prevent erection even though stimulation is present).

Erectile dysfunction (also termed impotence), the consistent inability to achieve or sustain an erection of sufficient rigidity for sexual intercourse, is a common disorder, affecting an estimated 30 million Americans. It is very much age-dependent, increasing from an incidence of 39 percent at age 40 to 67 percent at age 70. The organic causes are multiple and include damage to or malfunction of the efferent nerves or descending pathways, endocrine disorders, various therapeutic and "recreational" drugs (alcohol, for example), and

certain diseases, particularly diabetes. Erectile dysfunction can also be due to psychological factors, which are mediated by the brain and the descending pathways.

Recently, a drug—sildenafil (better known as *Viagra*)—has been introduced that greatly improves the ability of a majority of men with erectile dysfunction to achieve and maintain an erection comparable to other men in their age group. As described above, the most important event leading to erection is the dilation of penile arteries by nitric oxide, released from autonomic neurons. Nitric oxide stimulates the enzyme guanylyl cyclase, which catalyzes the formation of cyclic GMP, as described in Chapter 7. This second messenger then continues the signal transduction pathway leading to the relaxation of the arterial smooth muscle. The sequence of events is terminated by the breakdown of cGMP by a particular phosphodiesterase enzyme. Viagra blocks the action of this enzyme and thereby permits a higher concentration of cGMP to exist during sexual excitation.

Ejaculation

The discharge of semen from the penis—**ejaculation**—is also basically a spinal reflex, the afferent pathways from penile mechanoreceptors being identical to those described for erection. When the level of stimulation is high enough, there is elicited a patterned sequence of discharge of the efferent neurons. This sequence can be divided into two phases: (1) The smooth muscles of the epididymis, vas deferens, ejaculatory ducts, prostate, and seminal vesicles contract as a result of sympathetic stimulation, emptying the sperm and glandular secretions into the urethra (**emission**); and (2) the semen (average volume 3 ml, containing 300 million sperm) is then expelled from the urethra by a series of rapid contractions of the urethral smooth muscle as well as the skeletal muscle at the base of the penis. During ejaculation, the sphincter at the base of the urinary bladder is closed so that sperm cannot enter the bladder nor can urine be expelled from it. It is worth noting that although *erection* involves *inhibition* of sympathetic nerves (to the small arteries of the penis), *ejaculation* involves *stimulation* of sympathetic nerves (to the smooth muscles of the duct system).

The rhythmical muscular contractions that occur during ejaculation are associated with intense pleasure and many systemic physiological changes, the entire event being termed an **orgasm**. Marked skeletal-muscle contractions occur throughout the body, and there is a large increase in heart rate and blood pressure. This is followed by the rapid onset of muscular and psychological relaxation.

Once ejaculation has occurred, there is a latent period during which a second erection is not possible. The latent period is quite variable but may last from minutes to hours.

As is true of erectile dysfunction, premature ejaculation or failure to ejaculate can be the result of influence by higher brain centers.

Hormonal Control of Male Reproductive Functions

Control of the Testes

Figure 19–10 summarizes the control of the testes. In a normal adult man, the GnRH-secreting neuroendocrine cells fire a brief burst of action potentials approximately every 2 h, secreting GnRH at these times. The GnRH reaching the anterior pituitary via the hypothalamo-pituitary portal vessels during each periodic pulse triggers the release from the anterior pituitary of both LH and FSH from the same cell type, although not necessarily in equal amounts. Accordingly, systemic plasma concentrations of FSH and LH

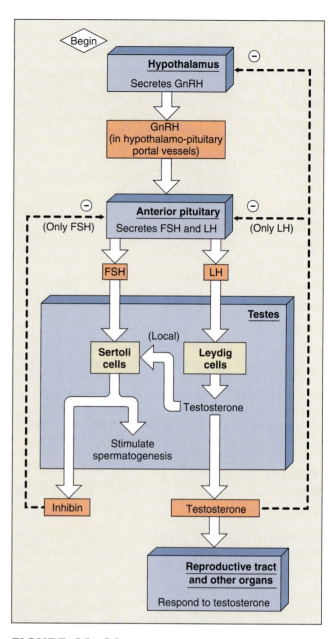

FIGURE 19–10

Summary of hormonal control of male reproductive function. Note that FSH acts only on the Sertoli cells, whereas LH acts only on the Leydig cells. The secretion of FSH is inhibited mainly by inhibin, a protein hormone secreted by the Sertoli cells, and the secretion of LH is inhibited mainly by testosterone, the hormone secreted by the Leydig cells. Testosterone, acting locally, is essential for spermatogenesis.

also show rhythmical episodic changes—rapid increases during the pulse followed by slow decreases over the next 90 min or so as the hormones are slowly removed from the plasma.

There is a clear separation of the actions of FSH and LH within the testes (Figure 19–10): FSH acts on the Sertoli cells to stimulate the secretion of paracrine agents that are essential for spermatogenesis; in contrast, LH acts on the Leydig cells to stimulate testosterone secretion. (Recall that the Leydig cells lie in the small triangular interstitial spaces between the seminiferous tubules.) In addition to its many important systemic effects as a hormone, the testosterone secreted by these cells also acts locally, as a paracrine agent, on spermatogenesis by moving from the interstitial spaces into the seminiferous tubules. There, testosterone enters Sertoli cells, and it is via these cells that it facilitates spermatogenesis. Thus, despite the absence of any *direct* effect of LH on cells in the seminiferous tubules, this hormone exerts an essential *indirect* effect because the testosterone secretion stimulated by LH is required for spermatogenesis.

The last components of the hypothalamo-pituitary control of male reproduction that remain to be discussed are the negative feedbacks exerted by testicular hormones. For this discussion it is important to realize that even though FSH and LH are produced by a single cell type, their secretion rates can be altered to different degrees by negative-feedback inputs.

Testosterone inhibits mainly LH secretion. It does so in two ways (Figure 19–10): (1) It acts on the hypothalamus to decrease the frequency of GnRH bursts, and the decreased amount of GnRH reaching the pituitary results in less secretion of the gonadotropins; and (2) it acts directly on the anterior pituitary to cause mainly less LH secretion in response to any given level of GnRH.

How does the presence of functioning testes reduce FSH secretion? The major inhibitory signal, exerted directly on the anterior pituitary, is the protein hormone **inhibin** secreted by the Sertoli cells (Figure 19–10). That the Sertoli cells, via inhibin, are the major source of feedback inhibition of FSH secretion makes sense since, as pointed out above, the facilitatory effect of FSH on spermatogenesis is exerted via the Sertoli cells. Thus, these cells are in all ways the link between FSH and spermatogenesis.

Despite all these complexities, one should not lose sight of the fact that the total amounts of GnRH, LH, FSH, testosterone, and inhibin secreted and of sperm produced are relatively constant from day to day in the adult male. This is completely different from the large cyclical swings of activity so characteristic of the female reproductive processes.

Testosterone

In addition to its essential paracrine action within the testes on spermatogenesis and its negative-feedback effects on the hypothalamus and anterior pituitary, testosterone exerts a large number of other effects, as summarized in Table 19–3.

TABLE 19–3 Effects of Testosterone in the Male
1. Required for initiation and maintenance of spermatogenesis (acts via Sertoli cells)
2. Decreases GnRH secretion via an action on the hypothalamus
3. Inhibits LH secretion via an action on the anterior pituitary
4. Induces differentiation of male accessory reproductive organs and maintains their function
5. Induces male secondary sex characteristics; opposes action of estrogen on breast growth
6. Stimulates protein anabolism, bone growth, and cessation of bone growth
7. Required for sex drive and may enhance aggressive behavior
8. Stimulates erythropoietin secretion by the kidneys

Before describing these effects, however, a word should be said concerning the metabolism of testosterone. In Chapter 10 we mentioned that hormones sometimes must undergo transformation in their target cells in order to be effective. This is true of testosterone in many (but not all) of its target cells. In some—for example, cells of the adult prostate—after its entry into the cytoplasm, testosterone undergoes an enzyme-mediated conversion to another steroid, **dihydrotestosterone,** and it is mainly this molecule that then binds to androgen receptors and elicits effects. In certain other target cells, notably in the brain, testosterone is transformed not to dihydrotestosterone but to estradiol, which is the active hormone in these cells. Note in this latter case that the "male" sex hormone must be converted to the "female" sex hormone to be active in the male. The fact that, depending on the target cells, testosterone acts as testosterone per se, dihydrotestosterone, or estradiol has important pathophysiological implications since some men lack the enzymes required to perform one or the other of the transformations of testosterone. Therefore, they will exhibit certain signs of testosterone deficiency but not others. For example, a man with absence of the enzyme for forming dihydrotestosterone will have normal differentiation of his reproductive duct structures (an effect of testosterone per se) but lack of development of his external genitalia (an effect requiring dihydrotestosterone).

Therapy of *prostate cancer* also makes use of these facts: Prostate cancer cells are stimulated by dihydrotestosterone, and so the cancer can be treated with drugs that block the enzyme catalyzing the transformation of testosterone to dihydrotestosterone.

Accessory Reproductive Organs The fetal differentiation, and later growth and function of the entire male duct system, glands, and penis all depend upon testosterone. Following *castration* (removal of the gonads) in the adult male, all the accessory reproductive organs decrease in size, the glands markedly reduce their secretion rates, and the smooth-muscle activity of the ducts is diminished. Erection and ejaculation may be deficient. These defects disappear upon the administration of testosterone.

Secondary Sex Characteristics and Growth Virtually all the male secondary sex characteristics are dependent on testosterone. For example, a male castrated before puberty does not develop a beard or either underarm or pubic hair. Other testosterone-dependent secondary sexual characteristics are deepening of the voice resulting from the growth of the larynx, thick secretion of the skin oil glands (this predisposes to acne), and the masculine pattern of fat distribution. As described in Chapter 18, testosterone also stimulates bone growth, largely indirectly through its stimulation of growth hormone secretion, but ultimately shuts off bone growth by causing closure of the bones' epiphyseal plates. Also as described in Chapter 18, testosterone is an "anabolic steroid" in that it exerts a direct stimulatory effect on protein synthesis in muscle. Testosterone is necessary for expression of the genetic determinant of the common type of baldness ("male pattern") in men. Finally, as stated in Chapter 14, testosterone stimulates the secretion of the hormone erythropoietin by the kidneys, and this is a major reason that men have a higher hematocrit than women.

Behavior Testosterone is essential in males for development of sex drive at puberty. It also plays an important role in maintaining sex drive in the adult male, although men often remain sexually active, albeit usually at a reduced level, for years after castration. (This is true only if the castration occurs during adult life.)

A controversial question is whether testosterone influences other human behaviors in addition to sexual behavior. It has proven very difficult to answer this question with respect to human beings, but there is little doubt that testosterone-dependent behavioral differences based on gender do exist in other mammals. For example, aggression is clearly greater in male animals and is testosterone-dependent.

SECTION B SUMMARY

Spermatogenesis

I. The male gonads, the testes, produce sperm in the seminiferous tubules and secrete testosterone from the Leydig cells.

II. The meiotic divisions of spermatogenesis result in each sperm containing 23 chromosomes, compared to the original 46 of the spermatogonia.

III. The developing germ cells are intimately associated with the Sertoli cells, which perform many functions, as summarized in Table 19–2.

Transport of Sperm

I. From the seminiferous tubules, the sperm pass through the epididymis, where they are concentrated.

II. The epididymis and vas deferens store the sperm, and the seminal vesicles and prostate secrete the bulk of the semen.

III. Erection of the penis occurs because of vascular engorgement accomplished by relaxation of the small arteries and passive occlusion of the veins.

IV. Ejaculation includes emission—emptying of semen into the urethra—followed by expulsion of the semen from the urethra.

Hormonal Control of Male Reproductive Functions

I. Hypothalamic GnRH stimulates the anterior pituitary to secrete FSH and LH, which then act on the testes: FSH on the Sertoli cells to stimulate spermatogenesis and inhibin secretion, and LH on the Leydig cells to stimulate testosterone secretion.

II. Testosterone, acting locally on the Sertoli cells, is essential for maintaining spermatogenesis.

III. Testosterone exerts a negative-feedback inhibition on both the hypothalamus and the anterior pituitary to reduce mainly LH secretion. Inhibin exerts a negative-feedback inhibition on FSH secretion.

IV. Testosterone maintains the accessory reproductive organs and male secondary sex characteristics and stimulates growth of muscle and bone. In many of its target cells, it must first undergo transformation to dihydrotestosterone or to estrogen.

SECTION B KEY TERMS

scrotum	seminal vesicle
spermatogenesis	ejaculatory duct
seminiferous tubule	prostate gland
Leydig cell	bulbourethral gland
epididymis	semen
vas deferens	spermatogonia
spermatic cord	primary spermatocyte

secondary spermatocyte
spermatid
acrosome
Sertoli cell
Sertoli-cell barrier
androgen-binding protein
erection

nitric oxide
ejaculation
emission
orgasm
inhibin
dihydrotestosterone

SECTION B REVIEW QUESTIONS

1. Describe the sequence of events leading from spermatogonia to sperm.
2. List the functions of the Sertoli cells.

3. Describe the path taken by sperm from the seminiferous tubules to the urethra.
4. State the roles of the prostate gland, seminal vesicles, and bulbourethral glands in the formation of semen.
5. Describe the neural control of erection and ejaculation.
6. Diagram the hormonal chain controlling the testes. Contrast the effects of FSH and LH.
7. What are the feedback controls from the testes to the hypothalamus and pituitary?
8. List the effects of testosterone on accessory reproductive organs, secondary sex characteristics, growth, protein metabolism, and behavior.

SECTION C

FEMALE REPRODUCTIVE PHYSIOLOGY

Unlike the continuous sperm production of the male, the production of the female gamete, the egg, followed by its release from the ovary—**ovulation**—is cyclical. This cyclical pattern is also true for the function and structure of virtually the entire female reproductive system. In human beings, these cycles are called **menstrual cycles.** The length of a menstrual cycle varies considerably from woman to woman and in any particular woman, but averages about 28 days. The first day of menstrual bleeding **(menstruation)** is termed day 1.

The events of the menstrual cycle are complex, and some orientation to our approach at this point may be useful. The most obvious event of a menstrual cycle in which pregnancy does not occur is, of course, menstruation, which is the result of events occurring in the uterus—the source of menstrual bleeding. However, the uterine events of the menstrual cycle are due entirely to cyclical changes in hormone secretion by the ovaries. Moreover, the ovaries, as we have stated, are the sites for production of gametes, one of which normally matures fully per menstrual cycle.

For these reasons, after summarizing the anatomy of the female reproductive tract, we begin our discussion of the menstrual cycle with the ovaries and their interactions with the anterior pituitary and hypothalamus. It is these interactions that produce the cyclical changes in the ovaries that result in (1) generation of a gamete each cycle, and (2) hormone secretions that cause cyclical changes in the entire female reproductive tract, including the uterus. The uterine changes prepare this organ to receive and nourish the gamete, and only when there is no pregnancy does menstruation occur. After describing the full menstrual cycle in the absence of pregnancy, we then describe the events of pregnancy, delivery, and lactation.

Anatomy

The female reproductive system includes the two ovaries and the female reproductive tract—two uterine tubes, a uterus, and a vagina. These structures are also termed the **female internal genitalia** (Figure 19–11a and b). In the female, unlike in the male, the urinary and reproductive duct systems are entirely separate from each other.

The ovaries are almond-sized organs in the upper pelvic cavity, one on each side of the uterus. The ends of the **uterine tubes,** (also known as oviducts or fallopian tubes) are not directly attached to the ovaries but open into the abdominal cavity close to them. The opening of each uterine tube is funnel-shaped and surrounded by long, fingerlike projections (the fimbriae) lined with ciliated epithelium. The other ends of the uterine tubes are attached to the uterus and empty directly into its cavity. The **uterus** is a hollow, thick-walled muscular organ lying between the urinary bladder and rectum. It is the source of bleeding during menstruation, and it houses the fetus during pregnancy. The lower portion of the uterus is the **cervix.** A small opening in the cervix leads to the **vagina,** the canal leading from the uterus to the outside.

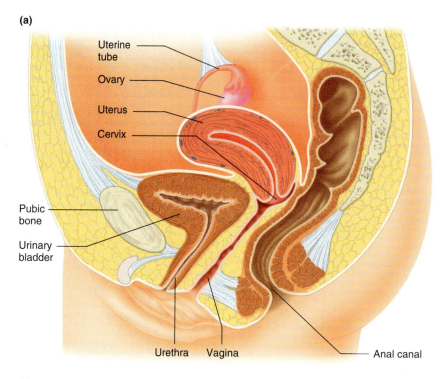

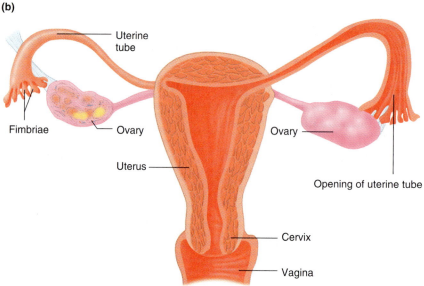

FIGURE 19–11

Female reproductive system. (a) Section through a female pelvis. (b) Diagram cut away on the right to show the continuity between the organs of the reproductive duct system—uterine tubes, uterus, and vagina.

The **female external genitalia** (Figure 19–12) include the mons pubis, labia majora, labia minora, clitoris, vestibule of the vagina, and vestibular glands. The term **vulva** is another name for all these structures. The mons pubis is the rounded fatty prominence over the junction of the pubic bones. The labia majora, the female homolog of the scrotum, are two prominent skin folds that form the outer lips of the vulva. (The terms "homolog" and "analogous" used in this chapter do not imply that one structure is dominant and its homolog or analog somehow an inferior imitation; it simply means that the two structures are derived embryologically from the same source and/or have similar functions.) The labia minora are small skin folds lying between the labia majora. They surround the urethral and vaginal openings, and the area thus enclosed is the vestibule, into which the vestibular glands empty. The vaginal opening lies behind that

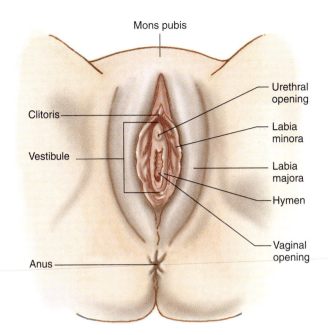

FIGURE 19–12
Female external genitalia.

of the urethra. Partially overlying the vaginal opening is a thin fold of mucous membrane, the hymen. The **clitoris,** the female homolog of the penis, is an erectile structure located at the front of the vulva.

Ovarian Function

As noted at the beginning of this chapter, the ovary, like the testis, serves a dual purpose: (1) **oogenesis,** the production of gametes—the ova; and (2) secretion of the female steroidal sex hormones, estrogen and progesterone, as well as the peptide hormone inhibin. (The ovaries secrete two other hormones—relaxin and activin—but the functions of these hormones in humans are not clear.) Before ovulation, the gametogenic and endocrine functions of the ovaries take place in a single structure, the follicle; after ovulation, the follicle, now without an egg, differentiates into a corpus luteum, which has only an endocrine function. For comparison, recall that in the testes the production of gametes and the secretion of sex steroids take place in different compartments—in the seminiferous tubules and in the Leydig cells, respectively.

This section will describe the events that occur in the ovary during a menstrual cycle, and the next section will replay these events, adding the controls.

Oogenesis

As we shall see, the female germ cells, like those of the male, have different names at different stages of development; it is helpful to use the term **egg** to refer to the germ cells at any stage.

At birth, a female's ovaries contain an estimated total of 2 to 4 million eggs, and no new ones appear after birth. Thus, in marked contrast to the male, the newborn female already has all the germ cells she will ever have. Only a few, perhaps 400, are destined to be ovulated. All the others degenerate at some point in their development so that few, if any, remain by the time a woman reaches approximately 50 years of age. One result of this developmental pattern is that the eggs that are ovulated near age 50 are 35 to 40 years older than those ovulated just after puberty. It is likely that certain defects more common among children born to older women are the result of aging changes in the egg.

During early in utero development, the primitive germ cells, or **oogonia** (singular, *oogonium*), a term analogous to spermatogonia in the male, undergo numerous mitotic divisions (Figure 19–13). Around the third month after conception, the oogonia cease dividing, and from this point on no new germ cells are generated. Still in the fetus, all the oogonia develop into **primary oocytes** (analogous to primary spermatocytes), which then begin a first meiotic division by replicating their DNA. They do not, however, complete the division in the fetus. Accordingly, all the eggs present at birth are primary oocytes containing 46 chromosomes, each with two sister chromatids. The cells are said to be in a state of **meiotic arrest.**

This state continues until puberty and the onset of renewed activity in the ovaries. Indeed, only those primary oocytes destined for ovulation will ever complete the first meiotic division, for it occurs just before the egg is ovulated. This division is analogous to the division of the primary spermatocyte, and each daughter cell receives 23 chromosomes, each with two chromatids. In this division, however, one of the two daughter cells, the **secondary oocyte,** retains virtually all the cytoplasm. The other, termed the first polar body, is very small and nonfunctional. Thus, the primary oocyte, which is already as large as the egg will be, passes on to the secondary oocyte half of its chromosomes but almost all of its nutrient-rich cytoplasm.

The second meiotic division occurs in a uterine tube *after ovulation*, but only if the secondary oocyte is fertilized—that is, penetrated by a sperm. As a result of this second meiotic division, the daughter cells each receive 23 chromosomes, each with a single chromatid. Once again, one daughter cell, now termed an ovum, retains nearly all the cytoplasm, whereas the other, termed the second polar body, is very small and nonfunctional. The net result of oogenesis is that each primary oocyte can produce only one ovum (Figure 19–13). In contrast, each primary spermatocyte produces four viable spermatozoa.

[Since the final stage of gametogenesis—formation of the ovum—occurs only after fertilization and since

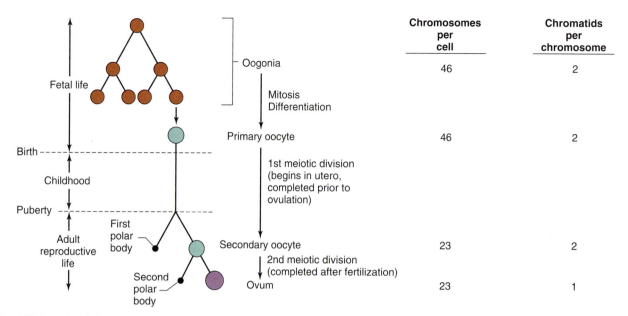

	Chromosomes per cell	Chromatids per chromosome
Oogonia	46	2
Primary oocyte	46	2
Secondary oocyte	23	2
Ovum	23	1

FIGURE 19–13

Summary of oogenesis. Compare with the male pattern of Figure 19–6. The secondary oocyte is ovulated and does not complete its meiotic division unless it is penetrated (fertilized) by a sperm. Thus, it is a semantic oddity that the egg is not termed an ovum until after fertilization occurs. Note that each primary oocyte yields only one secondary oocyte, which can yield only one ovum.

fertilization occurs *outside* the ovary (in a uterine tube, as we shall see), technically the *ovaries* do not themselves produce the fully mature gametes—the ova—but only secondary oocytes; this technicality is universally ignored by physiologists in talking about the ovarian function of gamete production.]

Follicle Growth

Throughout their life in the ovaries, the eggs exist in structures known as **follicles.** Follicles begin as **primordial follicles,** which consist of one primary oocyte surrounded by a single layer of cells called **granulosa cells.** Further development from the primordial follicle stage (Figure 19–14) is characterized by an increase in the size of the oocyte, a proliferation of the granulosa cells into multiple layers, and the separation of the oocyte from the inner granulosa cells by a thick layer of material, the **zona pellucida.** The granulosa cells secrete estrogen, small amounts of progesterone just before ovulation, and the peptide hormone inhibin.

Despite the presence of a zona pellucida, the inner layer of granulosa cells remains intimately associated with the oocyte by means of cytoplasmic processes that traverse the zona pellucida and form gap junctions with the oocyte. Through these gap junctions, nutrients and chemical messengers are passed to the oocyte. For example, the granulosa cells produce one or more factors that act on the primary oocytes to maintain them in meiotic arrest.

As the follicle grows by mitosis of granulosa cells, connective-tissue cells surrounding the granulosa cells differentiate and form layers known as the **theca,** which play an important role in estrogen secretion by the granulosa cells, as we shall see. Shortly after this, the primary oocyte reaches full size (115 μm in diameter), and a fluid-filled space, the **antrum,** begins to form in the midst of the granulosa cells as a result of fluid they secrete.

The progression of some primordial follicles to the preantral and early antral stages (Figure 19–14) occurs throughout infancy and childhood, and then during the entire menstrual cycle. Therefore, although most of the follicles in the ovaries are still primordial, there are also always present a relatively constant number of preantral and early antral follicles. At the beginning of each menstrual cycle, 10 to 25 of these preantral and early antral follicles begin to develop into larger antral follicles. About 1 week into the cycle, a further selection process occurs: Only one of the larger antral follicles, the **dominant follicle,** continues to develop, and the other follicles (in both ovaries) that had begun to enlarge undergo a degenerative process called **atresia** (an example of programmed cell death, or apoptosis). The eggs in the degenerating follicles also die.

Atresia is not limited to just antral follicles, however, for follicles can undergo atresia at all stages of development. Indeed, this process is already occurring in utero so that the 2 to 4 million follicles and eggs

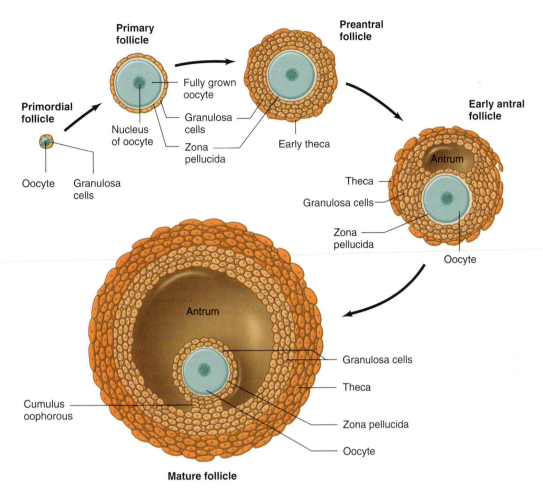

FIGURE 19–14

Development of a human oocyte and ovarian follicle. The fully mature follicle is 1.5 cm in diameter.

Adapted from Erickson et al.

present at birth represent only a small fraction of those present at an earlier time in the fetus. Atresia then continues all through prepubertal life so that only 200,000 to 400,000 follicles remain when active reproductive life begins. Of these, all but about 400 are destined for atresia during a woman's reproductive life. All told, 99.99 percent of the ovarian follicles present at birth will undergo atresia.

As the dominant follicle enlarges, mainly as a result of its expanding antrum, the granulosa cell layers surrounding the egg form a mound that projects into the antrum and is termed the cumulus oophorous (Figure 19–14). As the time of ovulation approaches, the egg (a primary oocyte) emerges from meiotic arrest and completes its first meiotic division to become a secondary oocyte. The cumulus separates from the follicle wall so that it and the oocyte float free in the antral fluid. The mature follicle (also termed a graafian follicle) becomes so large (diameter about 1.5 cm) that it balloons out on the surface of the ovary.

Ovulation occurs when the thin walls of the follicle and ovary at the site where they are joined rupture because of enzymatic digestion. The secondary oocyte, surrounded by its tightly adhering zona pellucida and granulosa cells, as well as the cumulus, is carried out of the ovary and onto the ovarian surface by the antral fluid. All this happens on approximately day 14 of the menstrual cycle.

On occasion (1 to 2 percent of all cycles), two or more follicles reach maturity, and more than one egg may be ovulated. This is the most common cause of multiple births. In such cases the siblings are fraternal, not identical, because the eggs carry different sets of genes. We will describe later how identical twins are formed.

Formation of the Corpus Luteum

After the mature follicle discharges its antral fluid and egg, its remnant in the ovary collapses around the antrum and undergoes a rapid transformation. The

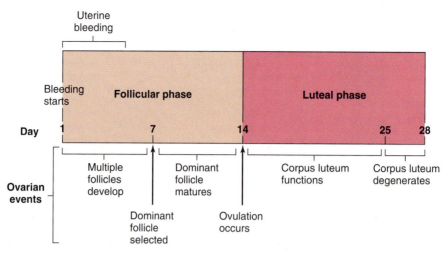

FIGURE 19–15

Summary of ovarian events during a menstrual cycle. The first day of the cycle is named for a uterine event—the onset of bleeding—even when ovarian events are used to denote the cycle phases.

granulosa cells enlarge greatly, and the entire gland-like structure is known as **corpus luteum,** which secretes estrogen, progesterone, and inhibin. If the discharged egg, now in a uterine tube, is not fertilized, the corpus luteum reaches its maximum development within approximately 10 days and then rapidly degenerates by apoptosis. As we shall see, it is the loss of corpus luteum function that leads to menstruation and the beginning of a new menstrual cycle.

In terms of ovarian function, therefore, the menstrual cycle may be divided into two phases approximately equal in length and separated by ovulation (Figure 19–15): (1) the **follicular phase,** during which a single mature follicle and secondary oocyte develop; and (2) the **luteal phase,** beginning after ovulation and lasting until the demise of the corpus luteum.

Sites of Secretion of Ovarian Hormones

This is a good place to summarize the sites of ovarian hormone secretion. Estrogen is secreted during the follicular phase mainly by the granulosa cells; following ovulation, it is secreted by the corpus luteum. Progesterone, the other major ovarian steroid hormone, is secreted in very small amounts by the granulosa and theca cells just before ovulation, but its major source is the corpus luteum. Inhibin, a peptide hormone, is secreted by both the granulosa cells and corpus luteum.

The actions of these three ovarian hormones will be described in the next sections of this chapter where we take up the control of female reproduction.

Control of Ovarian Function

The basic factors controlling ovarian function are analogous to the controls described for testicular function in that they constitute a hormonal series made up of GnRH, the anterior pituitary gonadotropins FSH and LH, and gonadal sex hormones—estrogen and progesterone. We say "basic" factors because these hormones are not the exclusive regulators of ovarian function. Several other hormones (insulin, for example) and many paracrine growth factors (for example, the insulin-like growth factors) play important but still poorly understood roles.

As in the male, the entire sequence of basic controls depends upon the secretion of GnRH from hypothalamic neuroendocrine cells in episodic pulses. In the female, however, the frequency of these pulses and hence the total amount of GnRH secreted during a 24-h period change in a patterned manner over the course of the menstrual cycle. So does the responsiveness both of the anterior pituitary to GnRH and of the ovaries to FSH and LH.

For purposes of orientation, let us look first, in Figure 19–16, at the patterns of hormone concentrations in systemic plasma during a normal menstrual cycle. (GnRH is not shown since its important concentration is not in systemic plasma but rather in the plasma within the hypothalamo-pituitary portal vessels; very few data are available on the concentration of GnRH in people, and thus, descriptions of changes in GnRH are mainly extrapolations from studies done

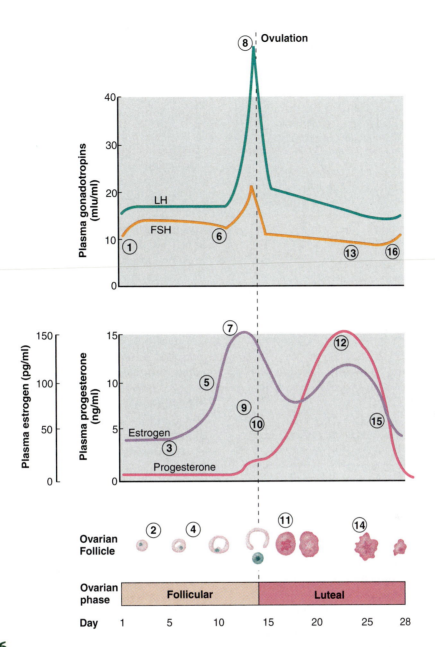

FIGURE 19–16

Summary of systemic plasma hormone concentrations and ovarian events during the menstrual cycle. The events marked by the circled numbers are described later in the text and are listed here to provide a summary. The arrows in this legend denote causality. ① FSH and LH secretion increase (because plasma estrogen concentration is low and exerting little negative feedback). → ② Multiple preantral and early antral follicles begin to enlarge and secrete estrogen. → ③ Plasma estrogen concentration begins to rise. ④ One follicle becomes dominant (cause unknown) and secretes very large amounts of estrogen. → ⑤ Plasma estrogen level increases markedly. → ⑥ FSH secretion and plasma FSH concentration decrease, but then ⑦ plasma estrogen reaches levels high enough to exert a positive feedback on gonadotropin secretion. → ⑧ An LH surge is triggered. → ⑨ The egg completes its first meiotic division and cytoplasmic maturation while the follicle secretes less estrogen accompanied by some progesterone, ⑩ ovulation occurs, and ⑪ the corpus luteum forms and begins to secrete large amounts of both estrogen and progesterone. → ⑫ Plasma estrogen and progesterone increase. → ⑬ FSH and LH secretion are inhibited and their plasma concentrations progressively fall. ⑭ The corpus luteum begins to degenerate (cause unknown) and decrease its hormone secretion. → ⑮ Plasma estrogen and progesterone concentrations fall. → ⑯ FSH and LH secretion begins to increase, and a new cycle begins.

TABLE 19–4 Summary of Major Feedback Effects of Estrogen, Progesterone, and Inhibin

1. **Estrogen**, in **low plasma concentrations**, causes the anterior pituitary to secrete less FSH and LH in response to GnRH and also may inhibit the hypothalamic neurons that secrete GnRH.

   Result: Negative-feedback inhibition of FSH and LH secretion during the early and middle follicular phase.

2. **Inhibin** acts on the pituitary to inhibit the secretion of FSH.

   Result: Negative-feedback inhibition of FSH secretion throughout the cycle.

3. **Estrogen**, in **high plasma concentrations**, causes anterior pituitary cells to secrete more LH (and FSH) in response to GnRH and also may stimulate the hypothalamic neurons that secrete GnRH.

   Result: Positive-feedback stimulation of the LH surge, which triggers ovulation.

4. High plasma concentrations of **progesterone**, in the presence of estrogen, inhibit the hypothalamic neurons that secrete GnRH.

   Result: Negative-feedback inhibition of FSH and LH secretion and prevention of LH surges during the luteal phase and pregnancy.

in experimental animals.) In Figure 19–16, the lines are plots of *average daily concentrations;* that is, the rises and falls during a single day stemming from episodic secretion have been averaged. For now, ignore both the legend and circled numbers in this figure since we are concerned here only with hormonal patterns and not the explanations of these patterns.

Note that FSH is slightly elevated in the early part of the follicular phase and then steadily decreases throughout the remainder of the cycle except for a small midcycle peak. LH is constant during most of the follicular phase but then shows a very large midcycle rise—the **LH surge**—peaking approximately 18 h *before* ovulation, followed by a rapid return toward presurge values and then a further slow decline during the luteal phase.

After remaining fairly low and stable for the first week, estrogen rises rapidly during the second week as the dominant ovarian follicle grows and secretes it. Estrogen then starts falling shortly after LH has peaked. This is followed by a second rise, due to secretion by the corpus luteum, and finally, a rapid decline during the last days of the cycle. The progesterone pattern is simplest of all: Very small amounts of progesterone are secreted by the ovaries during the follicular phase until just before ovulation, but very soon

after ovulation, the developing corpus luteum begins to secrete large amounts of progesterone, and from this point the progesterone pattern is similar to that for estrogen.

Not shown in Figure 19–16 is the plasma concentration of inhibin. Although there are quantitative differences, its pattern is similar to that of estrogen: It increases during the late follicular phase, remains high during the luteal phase, and then decreases as the corpus luteum degenerates.

The following discussion will explain how these hormonal changes are all interrelated to yield a self-cycling pattern. The numbers in Figure 19–16 are keyed to the text. The feedback effects of the ovarian hormones to be described in the text are summarized for reference in Table 19–4.

Follicle Development and Estrogen Secretion during the Early and Middle Follicular Phases

As mentioned earlier, there are always a number of preantral and early antral follicles in the adult ovary. Further development of the follicle beyond these stages requires stimulation by FSH. Prior to puberty, the plasma concentration of this gonadotropin is too low to induce such development. This all changes once sexual maturity has been reached and menstrual cycles commence: The increase in FSH secretion that occurs as one cycle ends and the next begins (number ① in Figure 19–16) provides this stimulation, and a group of preantral and early antral follicles begins to enlarge ②.

During the next week or so, there is a division of labor between the actions of FSH and LH on the follicles: FSH acts on the granulosa cells, and LH acts on the theca cells. The reason is that at this point in the cycle, granulosa cells have FSH receptors but no LH receptors, the situation for the theca cells being just the reverse. FSH stimulates the granulosa cells to multiply and produce estrogen, and it also stimulates enlargement of the antrum. Some of the estrogen produced diffuses into the blood and maintains a relatively stable plasma concentration ③, and some estrogen functions as a paracrine/autocrine agent in the follicle, where, along with FSH, it stimulates the proliferation of granulosa cells, which causes a further increase in estrogen production.

The granulosa cells, however, require help to produce estrogen because they are deficient in the enzymes required to produce the androgens that are the precursors of estrogen (Chapter 10). They are aided by the theca cells. As shown in Figure 19–17, LH acts upon the theca cells, stimulating them not only to proliferate but to synthesize androgens. The androgens diffuse into the granulosa cells and are converted to estrogen. Thus, the secretion of estrogen by the granulosa

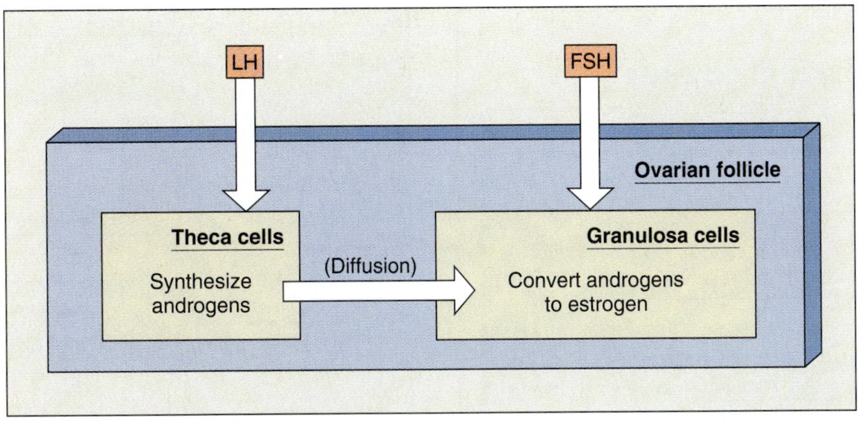

FIGURE 19–17

Control of estrogen secretion during the early and middle follicular phases. (The major androgen secreted by the theca cells is androstenedione.)

cells requires the interplay of both types of follicle cells and both pituitary gonadotropins.

At this point it is worthwhile to emphasize the similarities that the two types of follicle cells during this period of the cycle bear to cells of the testes: The granulosa cell is similar to the Sertoli cell in that it controls the microenvironment in which the germ cell develops and matures, and it is stimulated by both FSH and the major gonadal sex hormone. The thecal cell is similar to the Leydig cell in that it produces mainly androgens and is stimulated to do so by LH.

By the beginning of the second week, one follicle has become dominant (number ④ in Figure 19–16), and the other developing follicles undergo atresia by apoptosis. The reason for this apoptosis is that, as shown in Figure 19–16, the plasma concentration of FSH, a crucial "survival factor" for the follicle cells, begins to decrease, and there is no longer enough FSH to prevent such apoptosis. But why then does the dominant follicle not also undergo atresia? There are two reasons why this follicle, having gained a head start, is able to keep going: First, its granular cells have achieved a greater sensitivity to FSH because of increased numbers of FSH receptors so that less FSH is needed to stimulate them; and second, its granulosa cells now begin to be stimulated not only by FSH but by LH as well. We emphasized in the previous section that, during the first week or so of the follicular phase, LH acts only on the thecal cells; as the dominant follicle matures, however, this situation changes, and LH receptors, induced by FSH, also begin to appear in large numbers on the granulosa cells.

The dominant follicle now starts to secrete enough estrogen that the plasma concentration of this steroid begins to rise ⑤. We can now also explain why plasma FSH starts going down at this time. The reason is that estrogen, at these still relatively low concentrations, is exerting a *negative-feedback* inhibition over the secretion of gonadotropins (Table 19–4 and Figure 19–18). One site of estrogen's action is the anterior pituitary, where it reduces the amount of FSH and LH secreted in response to any given amount of GnRH. Estrogen probably also acts on the hypothalamus to decrease the amplitude of GnRH pulses and, hence, the total amount of GnRH secreted over any time period.

Therefore, as expected from this negative feedback, the plasma concentration of FSH (and LH, to a lesser extent) begins to fall as a result of the rising level of estrogen as the follicular phase continues (⑥ in Figure 19–16). One reason that FSH falls more than LH is that the granulosa cells also secrete inhibin, which, as in the male, inhibits mainly secretion of FSH (Figure 19–18).

LH Surge and Ovulation

The *inhibitory* effect of estrogen on gonadotropin secretion occurs only when plasma estrogen concentration is relatively *low*, as during the early and middle follicular phases. In contrast, high plasma concentrations of estrogen for 1 to 2 days, as occurs during the estrogen peak of the late follicular phase (⑦ in Figure 19–16), act upon the pituitary to *enhance* the sensitivity of LH releasing mechanisms to GnRH (Table 19–4 and Figure 19–19). (The high estrogen may also stimulate increased secretion of GnRH by the hypothalamus, although this remains controversial.) These effects are termed the *positive-feedback* effects of estrogen.

The net result is that, as estrogen secretion rises rapidly during the late follicular phase, its blood concentration eventually becomes high enough to cause the LH surge (⑧ in Figure 19–16). (As shown in Figure 19–16, a rise in FSH also occurs at the time of the LH surge, but it is not known whether this has a physiological role in the regulation of the cycle. The role, if any, of the

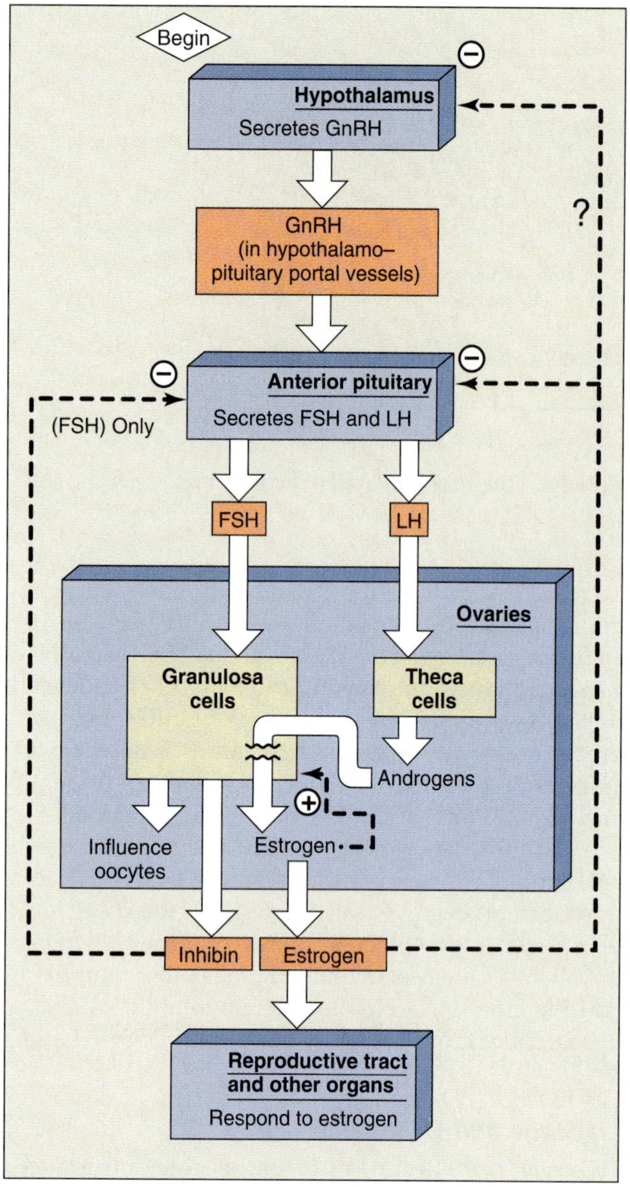

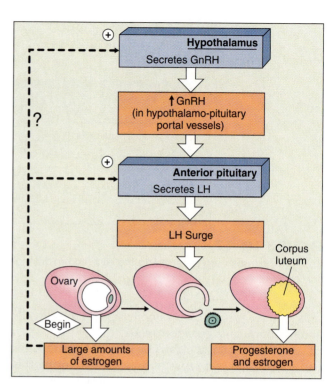

FIGURE 19–18

Summary of hormonal control of ovarian function during the early and middle follicular phases. Compare with the analogous pattern of the male (Figure 19–10). Inhibin is a protein hormone that inhibits FSH secretion. The wavy broken arrows in the granulosa cells denote the conversion of androgens to estrogen in these cells, as shown in Figure 19–17.

FIGURE 19–19

In the late follicular phase, the dominant follicle secretes large amounts of estrogen, which acts on the anterior pituitary and, possibly, the hypothalamus to cause an LH surge. The increased plasma LH then triggers both ovulation and formation of the corpus luteum. These actions of LH are mediated via the granulosa cells.

The function of the granulosa cells in mediating the effects of the LH surge is the last in the series of these cells' functions described in this chapter. They are all summarized in Table 19–6.

The Luteal Phase

The LH surge not only induces ovulation by the mature follicle but also stimulates the reactions that transform the remaining granulosa and theca cells of that follicle into a corpus luteum (⑪ in Figure 19–16). Even though its plasma concentration becomes relatively low, LH continues to stimulate the corpus luteum throughout the luteal phase, at least until the last few days of the corpus luteum's life span.

During its short life in the nonpregnant woman, the corpus luteum secretes large quantities of progesterone and estrogen ⑫, as well as inhibin. In the presence of estrogen, the high plasma concentration of progesterone causes a decrease in the secretion of the gonadotropins by the pituitary. It probably does this by acting on the hypothalamus to *suppress* the secretion of GnRH. (The progesterone also prevents any LH surges during the luteal phase despite the high

small rise in plasma progesterone at this time is also unclear.)

The midcycle surge of LH is the event that induces ovulation. The high plasma concentration of LH acts upon the granulosa cells to cause the events, presented in Table 19–5, that culminate in ovulation ⑩, as indicated by the dashed vertical line in Figure 19–16.

TABLE 19–5 Effects of the LH Surge on Ovarian Function
1. The primary oocyte completes its first meiotic division and undergoes cytoplasmic changes that prepare the ovum for implantation should fertilization occur. These LH effects on the oocyte are mediated by messengers released from the granulosa cells in response to LH.
2. Antrum size and blood flow to the follicle increase markedly.
3. The granulosa cells begin secreting progesterone and decrease their secretion of estrogen, which accounts for the midcycle drop in plasma estrogen concentration and the small rise in plasma progesterone just before ovulation.
4. Enzymes and prostaglandins, synthesized by the granulosa cells, break down the follicular-ovarian membranes. These weakened membranes rupture, allowing the oocyte and its surrounding granulosa cells to be carried out onto the surface of the ovary.
5. The remaining granulosa cells of the ruptured follicle (along with the theca cells of that follicle) are transformed into the corpus luteum, which begins to secrete progesterone and estrogen.

TABLE 19–6 Functions of Granulosa Cells
1. Nourish oocyte
2. Secrete chemical messengers that influence the oocyte (and the thecal cells)
3. Secrete antral fluid
4. Are the site of action for estrogen and FSH in the control of follicle development during early and middle follicular phases
5. Secrete estrogen from the androgens reaching them from the theca cells
6. Secrete inhibin, which inhibits FSH secretion via an action on the pituitary
7. Are the site of action for LH induction of changes in the oocyte and follicle culminating in ovulation and formation of the corpus luteum

concentrations of estrogen at this time.) The increase in plasma inhibin concentration in the luteal phase also contributes to the suppression of FSH secretion. Accordingly, during the luteal phase of the cycle, plasma concentrations of the gonadotropins are very low ⑬, which explains why no new follicles develop beyond the preantral or early antral stage during the second half of the cycle.

Recall from an earlier section that the corpus luteum degenerates within 2 weeks if pregnancy does not occur ⑭. What causes this? The most likely explanation is that the corpus luteum undergoes an age-related reduction in sensitivity to LH.

With degeneration of the corpus luteum, plasma progesterone and estrogen concentrations decrease ⑮. The secretion of FSH and LH (and probably GnRH, as well) increase (⑯ and ①) as a result of being freed from the inhibiting effects of high concentrations of progesterone, and a new group of follicles is stimulated to mature. The cycle then begins anew.

This completes our description of the control of ovarian function. It should be emphasized that, although the hypothalamus and anterior pituitary are essential links in the chain, events within the *ovary* are the real sources of *timing* for the cycle. When the ovary secretes enough estrogen, the LH surge is induced, which in turn causes ovulation. When its corpus luteum degenerates, the decrease in sex hormone secretion permits the gonadotropin levels to rise enough to

start another group of follicles developing. Thus, ovarian events, via hormonal feedbacks, "instruct" the hypothalamus and anterior pituitary.

Uterine Changes in the Menstrual Cycle

Because we have been describing ovarian function, we have so far referred to the phases of the menstrual cycle in terms of ovarian events—follicular and luteal phases, separated by ovulation. However, the phases of the menstrual cycle can also be named in terms of *uterine* events (Figure 19–20). Day 1 is, as noted earlier, the first day of menstrual bleeding, and the entire period of menstruation is known as the **menstrual phase,** which is generally about 3 to 5 days in a typical 28 day cycle. During this period, the epithelial lining of the uterus—the **endometrium**—degenerates, resulting in the menstrual flow. The menstrual flow then ceases, and the endometrium begins to thicken as it regenerates. This period of growth, the **proliferative phase,** lasts for the 10 days or so between cessation of menstruation and the occurrence of ovulation. Soon after ovulation, the endometrium begins to secrete various substances, and so the part of the menstrual cycle between ovulation and the onset of the next menstruation is called the **secretory phase.**

As shown in Figure 19–20, the *ovarian* follicular phase includes the *uterine* menstrual and proliferative phases, whereas the *ovarian* luteal phase is the same as the *uterine* secretory phase.

The uterine changes during a menstrual cycle are caused by changes in the plasma concentrations of estrogen and progesterone (Figure 19–20). During the

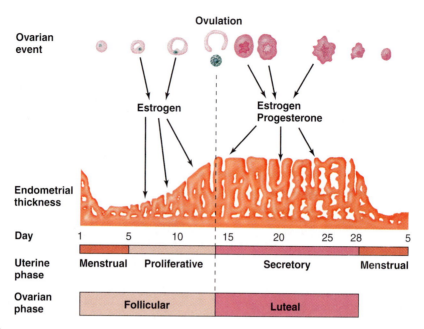

FIGURE 19–20

Relationships between ovarian and uterine changes during the menstrual cycle.

proliferative phase, a rising plasma estrogen level stimulates growth of both the endometrium and the underlying uterine smooth muscle **(myometrium).** In addition, it induces the synthesis of receptors for progesterone in endometrial cells. Then, following ovulation and formation of the corpus luteum, during the secretory phase progesterone acts upon this estrogen-primed endometrium to convert it to an actively secreting tissue: Its glands become coiled and filled with glycogen, the blood vessels become more numerous, and various enzymes accumulate in the glands and connective tissue. These changes are essential to make the endometrium a hospitable environment for an embryo.

Progesterone also inhibits myometrial contractions, in large part by opposing the stimulatory actions of estrogen and locally generated prostaglandins. This is very important to ensure that a fertilized egg, once it arrives in the uterus, will not be swept out by uterine contractions before it can implant in the wall. Uterine quiescence is maintained by progesterone throughout pregnancy and is essential to prevent premature delivery.

Estrogen and progesterone also have important effects on the secretion of mucus by the cervix. Under the influence of estrogen alone, this mucus is abundant, clear, and nonviscous. All these characteristics are most pronounced at the time of ovulation and allow sperm deposited in the vagina to move easily through the mucus on their way to the uterus and uterine tubes. In contrast, progesterone, present in significant concentrations only after ovulation, causes the

mucus to become thick and sticky—in essence a "plug" that prevents bacteria from entering the uterus from the vagina. The antibacterial blockage protects the fetus if conception has occurred.

The fall in plasma progesterone and estrogen levels that results from degeneration of the corpus luteum deprives the highly developed endometrium of its hormonal support and causes menstruation. The first event is profound constriction of the uterine blood vessels, which leads to a diminished supply of oxygen and nutrients to the endometrial cells. Disintegration starts in the entire lining, except for a thin, underlying layer that will regenerate the endometrium in the next cycle. Also, the uterine smooth muscle begins to undergo rhythmical contractions.

Both the vasoconstriction and uterine contractions are mediated by prostaglandins produced by the endometrium in response to the drop in plasma estrogen and progesterone. The major cause of menstrual cramps, *dysmenorrhea,* is overproduction of these prostaglandins, leading to excessive uterine contractions. The prostaglandins also affect smooth muscle elsewhere in the body, which accounts for the systemic symptoms (nausea, vomiting, and headache) that sometimes accompany the cramps.

After the initial period of vascular constriction, the endometrial arterioles dilate, resulting in hemorrhage through the weakened capillary walls. The menstrual flow consists of this blood mixed with endometrial debris. Typical blood loss per menstrual period is about 50 to 150 ml.

TABLE 19–7 Summary of the Menstrual Cycle	
Day(s)	**Major Events**

1–5	Estrogen and progesterone are low because the previous corpus luteum has completely regressed.
	    Therefore: (a) Endometrial lining sloughs.
	                (b) Secretion of FSH and LH is released from inhibition, and their plasma concentrations rise.
	                Therefore: Several follicles are stimulated to enlarge.

7	Dominant follicle is selected.
7–12	Plasma estrogen rises because of secretion by the dominant follicle. *Therefore:* Endometrium is stimulated to proliferate.

12–13	LH surge is induced by high plasma estrogen.
	    Therefore: (a) Oocyte is induced to complete its first meiotic division and undergo cytoplasmic maturation.
	                (b) Follicle is stimulated to secrete digestive enzymes and prostaglandin.

14	Ovulation is mediated by follicular enzymes and prostaglandin.

15–25	Corpus luteum forms and, under the influence of LH, secretes estrogen and progesterone, and so plasma concentrations of these hormones increase.
	    Therefore: (a) Secretory endometrium develops.
	                (b) Secretion of FSH and LH is inhibited, lowering their plasma concentrations.
	                Therefore: No new follicles develop.

25–28	Corpus luteum degenerates.
	    Therefore: Plasma estrogen and progesterone concentrations decrease.
	           Therefore: Endometrium begins to slough at conclusion of day 28, and a new cycle begins.

This completes our description of the menstrual cycle, the major events of which are summarized in Table 19–7. This table, in essence, combines the information in Figures 19–16 and 19–20.

Other Effects of Estrogen and Progesterone

Estrogen, in addition to its paracrine function within the ovaries, its effects on the anterior pituitary and the hypothalamus, and its uterine actions, exerts a large number of other effects, as summarized in Table 19–8 (some of these effects are discussed later in the chapter).

As noted earlier in this chapter, males also have circulating estrogen, and this hormone exerts several important effects. In particular, just as in the female, circulating estrogen provides some protection against atherosclerosis and osteoporosis, although not nearly to the same degree since plasma estrogen concentration is so much lower in the male.

Progesterone also exerts a variety of effects (Table 19–8). Since plasma progesterone is markedly elevated only after ovulation has occurred, several of these effects can be used to indicate whether ovulation has taken place. First, progesterone inhibits proliferation of the cells lining the vagina. Second, there is a small rise (approximately 0.5°C) in body temperature that usually occurs after ovulation and persists throughout

the luteal phase; this change is probably due to an action of progesterone on temperature regulatory centers in the brain.

Note that in its myometrial and vaginal effects, as well as several others in Table 19–8, progesterone exerts an "antiestrogen effect," probably by decreasing the number of estrogen receptors. In contrast, the synthesis of progesterone receptors is stimulated by estrogen in many tissues (for example, the endometrium), and so responsiveness to progesterone usually requires the presence of estrogen.

As is true for testosterone, both estrogen and progesterone act in the cell nucleus, and their biochemical mechanism of action is at the level of gene transcription.

In closing this section, brief mention should be made of the transient distressing physical and emotional symptoms that appear in many women prior to the onset of menstrual flow and disappear within a few days after the start of menstruation. The symptoms—which may include painful or swollen breasts, headache, backache, depression, anxiety, irritability, and other physical, emotional, and behavioral changes—are frequently ascribed to estrogen or progesterone excess; however, the plasma concentrations of these hormones are usually normal in women having these symptoms, and their cause is not actually known. In order of increasing severity of symptoms, the overall problem is categorized *premenstrual tension, premenstrual syndrome (PMS)*, or *premenstrual*

TABLE 19–8 Effects of Female Sex Steroids

Estrogen

1. Stimulates growth of ovary and follicles.

2. Stimulates growth of smooth muscle and proliferation of epithelial linings of reproductive tract. In addition:
   a. Uterine tubes: Increases contractions and ciliary activity.
   b. Uterus: Increases myometrial contractions and responsiveness to oxytocin. Stimulates secretion of abundant, clear cervical mucus. Prepares endometrium for progesterone's actions by inducing progesterone receptors.
   c. Vagina: Increases layering of epithelial cells.

3. Stimulates external genitalia growth.

4. Stimulates breast growth, particularly ducts and fat deposition.

5. Stimulates female body configuration development: narrow shoulders, broad hips, female fat distribution (deposition on hips, abdomen, and breasts).

6. Stimulates a more-fluid sebaceous gland secretion (this "antiacne" effect opposes the acne-producing effects of androgens).

7. Stimulates development of female pubic hair pattern (growth, as opposed to pattern, of pubic and axillary hair is androgen-stimulated).

8. Stimulates bone growth and ultimate cessation of bone growth (closure of epiphyseal plates); protects against osteoporosis; does not have an anabolic effect on skeletal muscle.

9. Vascular effects (deficiency produces "hot flashes").

10. Has feedback effects on hypothalamus and anterior pituitary (see Table 19–4).

11. Stimulates fluid retention by kidneys.

12. Stimulates prolactin secretion but inhibits prolactin's milk-inducing action on the breasts.

13. Protects against atherosclerosis by effects on plasma cholesterol (Chapter 18), blood vessels, and blood clotting (Chapter 14).

14. Exerts effects on brain neurons that may enhance learning and memory.

Progesterone

1. Converts the estrogen-primed endometrium to an actively secreting tissue suitable for implantation of an embryo.

2. Induces thick, sticky cervical mucus.

3. Decreases contractions of uterine tubes and myometrium.

4. Decreases proliferation of vaginal epithelial cells.

5. Stimulates breast growth, particularly glandular tissue.

6. Inhibits milk-inducing effects of prolactin.

7. Has feedback effects on hypothalamus and anterior pituitary (see Table 19–4).

8. Probably increases body temperature.

dysphoric disorder (PMDD), the last-named being so severe as to cause significant impairment in social functioning and work-related activities. Treatment with drugs that either block serotonin reuptake by neurons or mimic the actions of the neurotransmitter gamma-aminobutyric acid (Chapters 8 and 13) reduce the symptoms in more than 70 percent of cases. So do agents that inhibit the secretion of estrogen and progesterone by the corpus luteum. Clearly, these disorders are the result of a complex interplay between the sex steroids and brain neurotransmitters.

Androgens in Women

As noted earlier, androgens are present in the blood of normal women as a result of their production by the adrenal glands and ovaries. These androgens play several important roles in the female, including stimulation of the growth of pubic hair, axillary hair, and, possibly, skeletal muscle, and maintenance of sex drive (discussed below). In several disease states, the female adrenals may secrete abnormally large quantities of androgen, which produce *virilism:* The female fat distribution disappears, a beard appears along with the male body-hair distribution, the voice lowers in pitch, the skeletal-muscle mass enlarges, the clitoris enlarges, and the breasts diminish in size.

Female Sexual Response

The female response to sexual intercourse is characterized by marked vasocongestion and muscular contraction in many areas of the body. For example, increasing sexual excitement is associated with vascular engorgement of the breasts and erection of the nipples, resulting from contraction of smooth-muscle fibers in them. The clitoris, which is endowed with a rich supply of sensory nerve endings, increases in diameter as a result of vascular congestion. During intercourse, the vaginal epithelium becomes highly congested and secretes a mucuslike lubricant.

Orgasm in the female, as in the male, is accompanied by intense pleasure and many physical events: There is a sudden increase in skeletal-muscle activity involving almost all parts of the body; the heart rate and blood pressure increase; there is a transient rhythmical contraction of the vagina and uterus. Orgasm seems to play no essential role in ensuring fertilization, however, since conception can occur in the absence of an orgasm.

A question related to the female sexual response concerns sex drive. Incongruous as it may seem, sexual desire in women is probably more dependent upon androgens, secreted by the adrenal glands and ovaries, than estrogen, and sex drive is maintained beyond menopause, a time when estrogen levels become very low. However, the roles of sex hormones in female sexuality are much less clear than in the male.

This completes our survey of normal reproductive physiology in the nonpregnant female. In weaving one's way through this maze, it is all too easy to forget the prime function subserved by this entire system, namely, reproduction. Accordingly, we must now return to the egg we left free on the surface of the ovary, obtain a sperm for it, and carry the fertilized egg through pregnancy and delivery.

Pregnancy

For pregnancy to occur, sexual intercourse must occur no more than 5 days before ovulation or on the day of ovulation. This is because the sperm, following their ejaculation into the vagina, remain capable of fertilizing an egg for up to 5 days, and the ovulated egg remains fertile for only a few hours. (It is also possible that the rapid change in cervical mucus that occurs within a few hours after ovulation prevents entry of new sperm into the uterus.)

Egg Transport

At ovulation, the egg is extruded onto the surface of the ovary, and its first mission is to gain entry into a uterine tube. Recall that the fimbriae at the end of the uterine tubes are lined with ciliated epithelium. At ovulation, the smooth muscle of the fimbriae causes them to pass over the ovary while the cilia beat in waves toward the interior of the duct. These ciliary motions sweep the egg into the uterine tube as it emerges onto the ovarian surface.

Within the uterine tube, egg movement, driven almost entirely by uterine-tube cilia, is so slow that the egg takes about 4 days to reach the uterus. Thus, if fertilization is to occur, it must do so in the uterine tube because of the short life span of the unfertilized egg.

Sperm Transport and Capacitation

Within a minute or so after intercourse, some sperm can be detected in the uterus. The act of intercourse itself provides some impetus for transport of sperm out of the vagina through the cervix and into the uterus because of the fluid pressure of the ejaculate, and the pumping action of the penis in the vagina during ejaculation. In addition, beating of the cilia on the inner surface of the cervix probably wafts the sperm toward the uterus. Passage through the cervical mucus by the swimming sperm is dependent on the estrogen-induced changes in consistency of the mucus described earlier. Transport of the sperm the length of the uterus and into the uterine tube is mainly via the sperm's own propulsions. Another possible contributor is the posterior pituitary hormone oxytocin, which is reflexly released during intercourse and, as described later, causes contraction of the myometrium.

The mortality rate of sperm during the trip is huge. One reason for this is that the vaginal environment is acidic, a protection against yeast and bacterial infections. Another is the length and energy requirements of the trip. Of the several hundred million sperm deposited in the vagina, only a few hundred reach the uterine tube. This is one of the major reasons there must be so many sperm in the ejaculate for fertilization to occur.

Sperm are not able to fertilize the egg until they have resided in the female tract for several hours and been acted upon by secretions of the tract. This process, termed **capacitation,** causes: (1) the previously regular wavelike beats of the sperm's tail to be replaced by a more whiplike action that propels the sperm forward in strong lurches, and (2) the sperm's plasma membrane to become altered so that it will be capable of fusing with the surface membrane of the egg.

Fertilization

Fertilization is the fusion of a sperm and egg. Many sperm, after moving between the cumulus of granulosa cells still surrounding the egg, bind to the zona pellucida. This is a species-specific binding between a protein (known as ZP3) in the zona pellucida's outer surface and a complementary protein in the plasma membrane of the sperm's head. In other words, the zona pellucida proteins function as receptors for sperm surface proteins. The sperm head has many of these proteins and so becomes bound simultaneously to many sperm receptors on the zona pellucida.

This binding triggers what is termed the **acrosome reaction** in the bound sperm: The plasma membrane of the sperm head alters so that the underlying membrane-bound acrosomal enzymes become exposed to the outside—that is, to the zona pellucida. The enzymes digest a path through the zona as the

sperm, using its tail, advances through this coating. The first sperm to penetrate the entire zona and reach the egg's plasma membrane fuses with this membrane. This sperm then slowly passes into the egg's cytoplasm, penetration being achieved not by the sperm's motility but as a result of contractile elements in the egg that draw the sperm in.

Viability of the newly fertilized egg, now termed a **zygote,** depends upon preventing the entry of additional sperm. The mechanism of this "block to polyspermy" is as follows: The initial fusion of the sperm and egg plasma membranes triggers a reaction in which secretory vesicles located around the egg's periphery release their contents, by exocytosis, into the narrow space between the egg plasma membrane and the zona pellucida. Some of these molecules are enzymes that enter the zona pellucida and cause both inactivation of its sperm-binding sites and hardening of

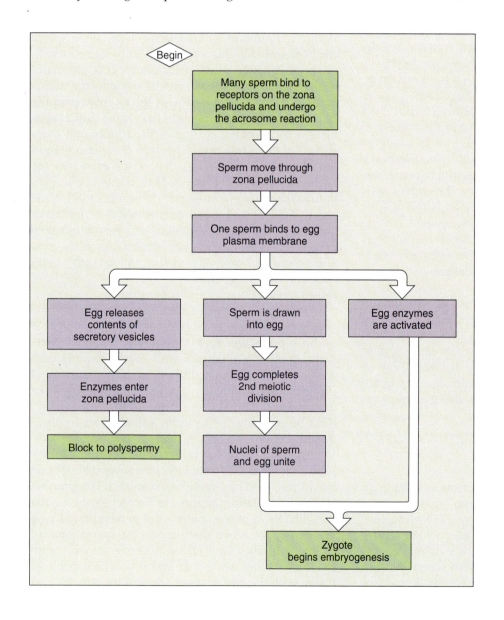

FIGURE 19–21

Events leading to fertilization, block to polyspermy, and the beginning of embryogenesis.

the entire zona. This prevents additional sperm from binding to the zona and those sperm already advancing through it from continuing.

The fertilized egg completes its second meiotic division over the next few hours, and the one daughter cell with practically no cytoplasm—the second polar body—is extruded and disintegrates. The two sets of chromosomes—23 from the egg and 23 from the sperm—are surrounded by distinct membranes and are known as pronuclei, which migrate to the center of the cell. During this period of a few hours, the DNA of the chromosomes in both pronuclei is replicated, the pronuclear membranes break down, the cell is ready to undergo a mitotic division, and fertilization is complete. Fertilization also triggers activation of the egg enzymes required for the ensuing cell divisions and embryogenesis. The major events of fertilization are summarized in Figure 19–21. If fertilization had not occurred, the egg would have slowly disintegrated and been phagocytized by cells lining the uterus.

Rarely, a fertilized egg remains in a uterine tube and embeds itself in the tube wall. Even more rarely, a fertilized egg may move backwards out of the uterine tube into the abdominal cavity, where implantation can occur. Both kinds of *ectopic pregnancies* cannot succeed, and surgery is necessary to end the pregnancy unless there is a spontaneous abortion—because of the risk of maternal hemorrhage.

Early Development, Implantation, and Placentation

The **conceptus**—a collective term for everything ultimately derived from the original zygote (fertilized egg) throughout the pregnancy—remains in the uterine tube for 3 to 4 days. The major reason is that the uterine-tube smooth muscle near where the tube enters the wall of the uterus is maintained in a contracted state by estrogen; as plasma progesterone levels rise, this smooth muscle relaxes and allows the conceptus to pass. During its stay in the uterine tube, the conceptus undergoes a number of mitotic cell divisions, a process known as **cleavage.** These divisions, however, are unusual in that no cell growth occurs before each division. Thus, the 16- to 32-cell conceptus that reaches the uterus is essentially the same size as the original fertilized egg.

Each of these cells is totipotent—that is, has the capacity to develop into an entire individual. Therefore, identical (monozygotic) twins result when, at some point during cleavage, the dividing cells become completely separated into two independently growing cell masses. In contrast, as we have seen, fraternal (dizygotic) twins result from two eggs being ovulated and fertilized.

After reaching the uterus, the conceptus floats free in the intrauterine fluid, from which it receives

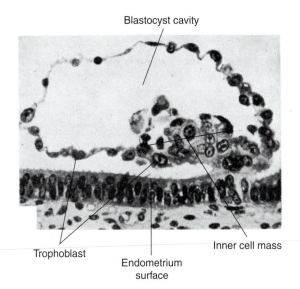

Blastocyst cavity

Trophoblast

Endometrium surface

Inner cell mass

FIGURE 19–22

Photomicrograph showing the beginning implantation of a 9-day monkey blastocyst.

From C. H. Heuser and G. L. Streeter, Carnegie Contrib. Embryol. 29:15 (1941).

nutrients, for approximately 3 days, all the while undergoing further cell divisions. Soon the conceptus reaches the stage known as a **blastocyst,** by which point the cells have lost their totipotentiality and have begun to differentiate. The blastocyst consists of an outer layer of cells, the **trophoblast,** an **inner cell mass,** and a central fluid-filled cavity (Figure 19–22). During subsequent development, the inner cell mass will give rise to the developing human—called an **embryo** during the first 2 months and a **fetus** after that—and some of the membranes associated with it. The trophoblast will surround the embryo and fetus throughout development and be involved in its nutrition as well as in the secretion of several important hormones.

The period during which the zygote develops into a blastocyst corresponds with days 14 to 21 of the typical menstrual cycle. During this period, the uterine lining is being prepared by progesterone, secreted by the corpus luteum, to receive the blastocyst. By approximately the twenty-first day of the cycle (that is, 7 days after ovulation), **implantation**—the embedding of the blastocyst in the endometrium—begins. The trophoblast cells are quite sticky, particularly in the region overlying the inner cell mass, and it is this portion of the blastocyst that adheres to the endometrium and initiates implantation.

The initial contact between blastocyst and endometrium induces rapid proliferation of the trophoblast, the cells of which penetrate between endometrial cells. Proteolytic enzymes secreted by the

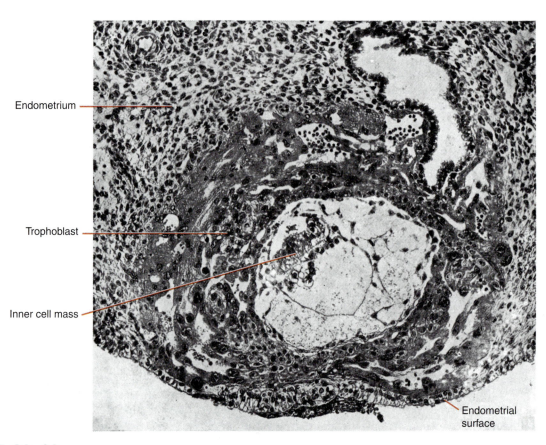

Endometrium

Trophoblast

Inner cell mass

Endometrial surface

FIGURE 19–23

Eleven-day human embryo completely embedded in the uterine lining.

From A. T. Hertig and J. Rock, Carnegie Contrib. Embryol. 29:127 (1941).

trophoblast allow the blastocyst to bury itself in the endometrial layer. The endometrium, too, is undergoing changes at the site of contact. Implantation requires an active dialogue—via a large number of paracrine agents—between the blastocyst and the cells of the endometrium. Implantation is soon completed (Figure 19–23), and the nutrient-rich endometrial cells provide the metabolic fuel and raw materials required for early growth of the embryo.

This simple nutritive system, however, is adequate to provide for the embryo only during the first few weeks, when it is very small. The structure taking over this function is the **placenta,** a combination of interlocking fetal and maternal tissues that serves as the organ of exchange between mother and fetus for the remainder of the pregnancy.

The fetal portion of the placenta is supplied by the outermost layer of trophoblast cells, the **chorion,** and the maternal portion by the endometrium underlying the chorion. Fingerlike projections of the trophoblast cells, called **chorionic villi,** extend from the chorion into the endometrium (Figure 19–24). The villi contain a rich network of capillaries linked to the embryo's circulatory system. The endometrium around the villi is

altered by enzymes and paracrine agents secreted from the cells of the invading villi so that each villus comes to be completely surrounded by a pool, or sinus, of maternal blood supplied by maternal arterioles.

The maternal blood enters these placental sinuses via the uterine artery; the blood percolates through the sinuses and then exits via the uterine veins. Simultaneously, blood flows from the fetus into the capillaries of the chorionic villi via the **umbilical arteries** and out of the capillaries back to the fetus via the **umbilical vein.** All these umbilical vessels are contained in the **umbilical cord,** a long ropelike structure that connects the fetus to the placenta.

Five weeks after implantation, the placenta has become well established, the fetal heart has begun to pump blood, and the entire mechanism for nutrition of the fetus and excretion of its waste products is in operation. A layer of epithelial cells in the villi and of endothelial cells in the fetal capillaries separate the maternal and fetal blood. Waste products move from blood in the fetal capillaries across these layers into the maternal blood, and nutrients, hormones, and growth factors move in the opposite direction. Some substances, such as oxygen and carbon dioxide, move by

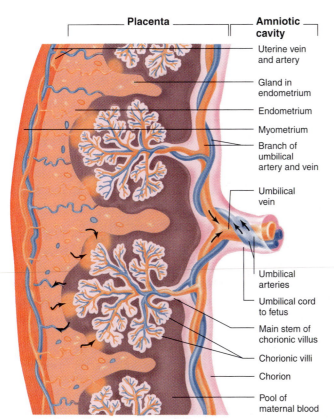

FIGURE 19–24

Interrelations of fetal and maternal tissues in the formation of the placenta. See Figure 19–25 for orientation of the placenta.

From B. M. Carlson, "Patten's Foundations of Embryology," 5th ed., McGraw-Hill, New York, 1988.

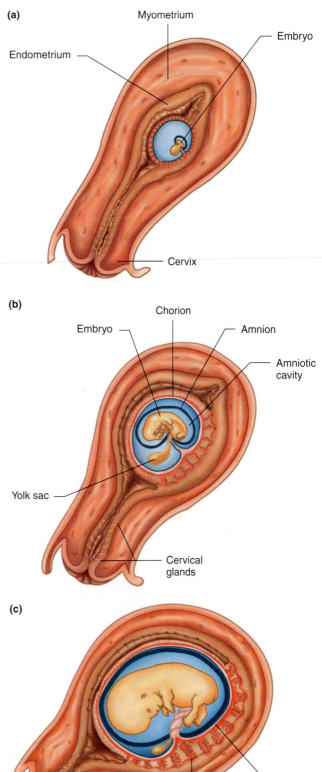

FIGURE 19–25

The uterus at (a) 3, (b) 5, and (c) 8 weeks after fertilization. Embryos and their membranes are drawn to actual size. Uterus is within actual size range.

From B. M. Carlson, "Patten's Foundations of Embryology," 5th ed., McGraw-Hill, New York, 1988.

diffusion, whereas others utilize transport proteins in the plasma membranes of the epithelial cells. Still other nutrients (for example, several amino acids) and hormones are produced by the trophoblast layers of the placenta itself and added to the fetal blood. It must be emphasized that there is an exchange of materials between the two bloodstreams but no actual mixing of the fetal and maternal blood.

Meanwhile, a space called the **amniotic cavity** has formed between the inner cell mass and the chorion (Figure 19–25). The epithelial layer lining the cavity is derived from the inner cell mass and is called the **amnion,** or **amniotic sac.** It eventually fuses with the inner surface of the chorion so that only a single combined membrane surrounds the fetus. The fluid in the amniotic cavity, the **amniotic fluid,** resembles the fetal extracellular fluid, and it buffers mechanical disturbances and temperature variations.

The fetus, floating in the amniotic cavity and attached by the umbilical cord to the placenta, develops into a viable infant during the next 8 months. Note in Figure 19–25 that eventually only the amniotic sac separates the fetus from the uterine lumen.

Amniotic fluid can be sampled (*amniocentesis*) as early as the sixteenth week of pregnancy by inserting a needle into the amniotic cavity. A large number of genetic diseases can be diagnosed by the finding of certain chemicals either in the fluid or in cells suspended in the fluid. The chromosomes of these cells can also be examined for diagnosis of certain disorders as well as to determine the sex of the fetus. Another technique for fetal diagnosis is *chorionic villus sampling.* This technique, which can be performed as early as 9–12 weeks of pregnancy, involves obtaining tissue from a chorionic villus of the placenta. This technique, however, carries a higher risk of inducing miscarriage than does amniocentesis. A third technique for fetal diagnosis is ultrasound, which provides a "picture" of the fetus without the use of x-rays. A fourth technique coming into wide use as a routine screening for various fetal abnormalities involves obtaining only *maternal* blood and measuring it for several normally occurring proteins whose concentrations change in the presence of these abnormalities. For example, particular changes in the concentrations of two hormones produced during pregnancy—chorionic gonadotropin and estriol—and alpha-fetoprotein (a major fetal plasma protein that crosses the placenta into the maternal blood) can identify 60 percent of cases with *Down's syndrome,* which includes mental retardation.

Maternal nutrition is crucial for the fetus. Malnutrition early in pregnancy can cause certain congenital abnormalities and/or the formation of an inadequate placenta. Malnutrition retards fetal growth and results in infants with higher-than-normal death rates, reduced growth after birth, and an increased incidence of learning disabilities. Specific nutrients, not just total calories, are also very important, as manifested, for example, by the increased incidence of neural defects in the offspring of mothers who are deficient in the B-vitamin folate (also called folic acid and folacin).

The developing embryo and fetus are also subject to considerable influences by a host of non-nutrient factors (noise, radiation, chemicals, viruses, and so on) to which the mother may be exposed. For example, drugs taken by the mother can reach the fetus via transport across the placenta and impair fetal growth and development. In this regard, it must be emphasized that aspirin, alcohol, and the chemicals in cigarette smoke are very potent agents, as are "street drugs" such as cocaine. Any chemical agent that can cause birth defects in the fetus is known as a *teratogen.*

It should also be recognized that, since half of the fetal genes—those from the father—differ from those of the mother, the fetus is in essence a foreign transplant in the mother. Why the fetus is not rejected as are other transplants is still not well understood and will be discussed further in Chapter 20.

Finally, how specific tissues and organs form is presently one of the most exciting and rapidly developing areas in all of biology. Before any tissues or organs form, earlier steps must occur, steps that indicate to cells "who" they are, what tissues they should form, and where. In essence, cells must be assigned "addresses" and functions in orderly sequence. Recall that initially all the cells in the embryo undergoing cleavage are identical. How is differentiation of these totipotential cells triggered? In the past few years, whole classes of genes—the *Hox genes* and *hedgehog genes,* for example—have been identified that are activated within the first few days after fertilization and guide all these processes. The proteins encoded by these genes are mainly transcription factors that are involved in proliferation and differentiation. Thus, these genes, via their encoded proteins, lay out the basic structure and organization of the body—where the head and limbs will be, and so on. Just what activates these genes in the proper sequences along the various axes of the body (top-to-bottom, right-to-left, back-to-front) remains to be determined.

Hormonal and Other Changes during Pregnancy

Throughout pregnancy, plasma concentrations of estrogen and progesterone remain high (Figure 19–26). Estrogen stimulates growth of the uterine muscle mass, which will eventually supply the contractile

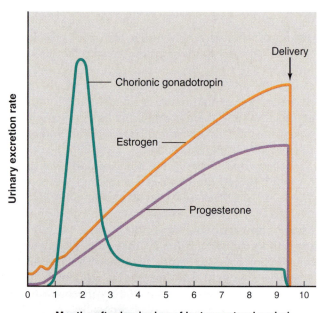

FIGURE 19–26

Urinary excretion of estrogen, progesterone, and chorionic gonadotropin during pregnancy. Urinary excretion rates are an indication of blood concentrations of these hormones.

force needed to deliver the fetus. Progesterone inhibits uterine motility so that the fetus is not expelled prematurely. During approximately the first 2 months of pregnancy, almost all the estrogen and progesterone are supplied by the extremely active corpus luteum.

Recall that if pregnancy had not occurred, this glandlike structure would have degenerated within 2 weeks after its formation. The persistence of the corpus luteum during pregnancy is due to a hormone called **chorionic gonadotropin (CG),** which starts to be secreted by the trophoblast cells around the time they start their endometrial invasion. CG gains entry to the maternal circulation, and the detection of this hormone in the mother's plasma and/or urine is used as a test of pregnancy; it is positive before the next expected menstruation. This protein hormone is very similar to LH, and it not only prevents the corpus luteum from degenerating but strongly stimulates steroid secretion by it. Thus, the signal that preserves the corpus luteum comes from the conceptus, not the mother's tissues.

The secretion of CG reaches a peak 60 to 80 days after the last menstrual period (Figure 19–26). It then falls just as rapidly, so that by the end of the third month it has reached a low, but still definitely detectable, level that remains relatively constant for the duration of the pregnancy. Associated with this falloff of CG secretion, the placenta begins to secrete large quantities of estrogen and progesterone. The very marked increases in plasma concentrations of estrogen and progesterone during the last 6 months of pregnancy are due entirely to their secretion by the trophoblast cells of the placenta, and the corpus luteum regresses after 3 months.

An important aspect of placental steroid secretion is that the placenta has the enzymes required for the synthesis of progesterone but not those for the formation of androgens, which are the precursors of estrogen. The placenta is supplied, via the fetal circulation, with these androgens, produced by the *fetal* adrenal glands and liver. The placenta converts the androgens into estrogen.

Recall that secretion of GnRH and, hence, of LH and FSH is powerfully inhibited by high concentrations of progesterone in the presence of estrogen. Since both these steroid hormones are present in high concentrations throughout pregnancy, the secretion of the pituitary gonadotropins remains extremely low.

The trophoblast cells of the placenta produce not only CG and steroids, but inhibin and many other hormones as well, all examples of hormones produced by the conceptus and influencing the mother. Some of these (for example, thyroid-stimulating hormone) are identical to hormones normally produced by other endocrine glands, whereas some are unique. One unique hormone that is secreted in very large amounts has effects similar to those of both prolactin and growth hormone. This protein hormone, **placental lactogen** (also called chorionic somatomammotropin), plays several roles in the mother—mobilizing fat for energy and stabilizing plasma glucose at relatively high levels (growth hormone–like effects) and facilitating development of the breasts (a prolactin-like effect). In the fetus, this hormone exerts growth-promoting effects, probably by stimulating the secretion of insulin-like growth factor II (IGF-II, Chapter 18).

In addition to all these messengers produced by the trophoblast, the endometrium also secretes a variety of hormones and growth factors important for maintaining the pregnancy.

Some of the numerous other physiological changes, hormonal and nonhormonal, in the mother during pregnancy are summarized in Table 19–9.

One comment about fluid balance and blood pressure during pregnancy is necessary: Approximately 5–10 percent of pregnant women retain abnormally large amounts of fluid and manifest edema, protein in the urine, and hypertension. These are the symptoms of *preeclampsia;* when convulsions also occur, the condition is termed *eclampsia.* The fetus is also affected, sometimes resulting in intrauterine growth retardation and death. The factors responsible for eclampsia are unknown, but the evidence strongly implicates abnormal vasoconstriction of the maternal blood vessels and inadequate invasion of the endometrium by trophoblast cells, resulting in poor blood perfusion of the placenta.

Pregnancy Sickness The majority of women suffer from *pregnancy sickness* (popularly called morning sickness)—nausea, vomiting, changes in the perception of food palatability, and the presence of taste aversions—during the first three months (first trimester) of pregnancy. The exact cause is unknown, but high concentrations of estrogen, progesterone, and other substances secreted at this time are thought to act on the vomiting center (Chapter 17) in the brain. It has been hypothesized that pregnancy sickness is actually an adaptive (that is, beneficial) process, one that minimizes the mother's intake of potentially toxic chemicals during the first trimester, when the embryo and fetus are particularly susceptible.

Parturition

A normal human pregnancy lasts approximately 40 weeks, counting from the first day of the last menstrual cycle, or approximately 38 weeks from the day of ovulation and conception. Safe survival of premature infants is now possible at about the twenty-fourth week of pregnancy, but treatment of these infants often requires heroic efforts at great costs.

TABLE 19–9 Maternal Responses to Pregnancy

	Response
Placenta	Secretion of estrogen, progesterone, chorionic gonadotropin, inhibin, placental lactogen, and other hormones.
Anterior pituitary	Increased secretion of prolactin and ACTH. Secretes very little FSH and LH.
Adrenal cortex	Increased secretion of aldosterone and cortisol.
Posterior pituitary	Increased secretion of vasopressin.
Parathyroids	Increased secretion of parathyroid hormone.
Kidneys	Increased secretion of renin, erythropoietin, and 1,25-dihydroxyvitamin D_3. Retention of salt and water. *Cause:* Increased aldosterone, vasopressin, and estrogen.
Breasts	Enlarge and develop mature glandular structure. *Cause:* Estrogen, progesterone, prolactin, and placental lactogen.
Blood volume	Increases. *Cause:* Total erythrocyte volume is increased by erythropoietin, and plasma volume by salt and water retention.
Calcium balance	Positive. *Cause:* Increased parathyroid hormone and 1,25-dihydroxyvitamin D_3.
Body weight	Increases by average of 12.5 kg, 60 percent of which is water.
Circulation	Cardiac output increases, total peripheral resistance decreases (vasodilation in uterus, skin, breasts, GI tract, and kidneys), and mean arterial pressure stays constant.
Respiration	Hyperventilation occurs (arterial P_{CO_2} decreases).
Organic metabolism	Metabolic rate increases. Plasma glucose, gluconeogenesis, and fatty acid mobilization all increase. *Cause:* Hyporesponsiveness to insulin due to insulin antagonism by placental lactogen and cortisol.
Appetite and thirst	Increase.
Nutritional RDAs	Increase.

During the last few weeks of pregnancy, a variety of events occur in the uterus, culminating in the delivery of the infant, followed by the placenta. All these events, including delivery, are termed **parturition.** Throughout most of pregnancy, the smooth-muscle cells of the myometrium are relatively disconnected from each other, and the uterus is sealed at its outlet by the firm, inflexible collagen fibers that constitute the cervix. These features are maintained mainly by progesterone. During the last few weeks of pregnancy, as a result of ever-increasing levels of estrogen, the smooth muscle cells synthesize *connexin,* proteins that form gap junctions between the cells, which allows the myometrium to undergo coordinated contractions. Simultaneously, the cervix becomes soft and flexible, a process termed "ripening," due to an enzymatically mediated breakup of its collagen fibers. The synthesis of the enzymes is mediated by a variety of messengers, including estrogen and placental prostaglandins, the synthesis of which is stimulated by estrogen. (The peptide hormone relaxin secreted by the ovaries may also be involved.) Estrogen has yet another important effect

on the myometrium during this period: It induces the synthesis of receptors for the posterior pituitary hormone oxytocin, which is a powerful stimulator of uterine smooth-muscle contraction, as described below.

Delivery is produced by strong rhythmical contractions of the myometrium. Actually, weak and infrequent uterine contractions begin at approximately 30 weeks and gradually increase in both strength and frequency. During the last month, the entire uterine contents shift downward so that the baby is brought into contact with the cervix. In over 90 percent of births the baby's head is downward and acts as the wedge to dilate the cervical canal when labor begins (Figure 19–27).

At the onset of labor or before, the amniotic sac ruptures, and the amniotic fluid escapes through the vagina. When labor begins in earnest, the uterine contractions become coordinated and quite strong (although usually painless at first) and occur at approximately 10- to 15-min intervals. The contractions begin in the upper portion of the uterus and sweep downward.

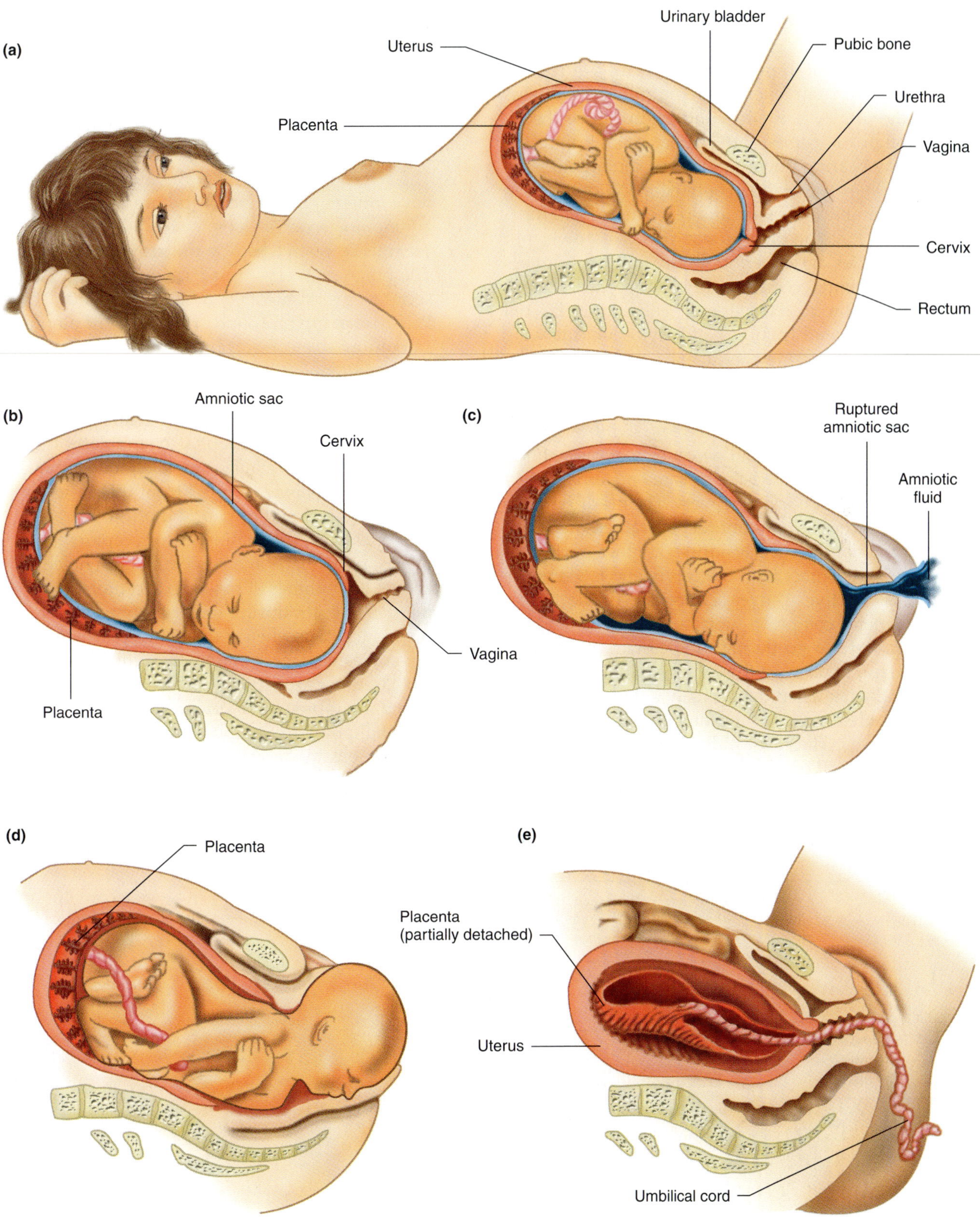

(a)

Urinary bladder

Uterus

Pubic bone

Urethra

Placenta

Vagina

Cervix

Rectum

(b)

Amniotic sac

Cervix

Vagina

Placenta

(c)

Ruptured amniotic sac

Amniotic fluid

(d)

Placenta

(e)

Placenta (partially detached)

Uterus

Umbilical cord

FIGURE 19–27

Stages of parturition. (a) Parturition has not yet begun. (b) The cervix is dilating. (c) The cervix is completely dilated, and the fetus's head is entering the cervical canal; the amniotic sac has ruptured and the amniotic fluid escapes. (d) The fetus is moving through the vagina. (e) The placenta is coming loose from the uterine wall preparatory to its expulsion.

As the contractions increase in intensity and frequency, the cervix is gradually forced open to a maximum diameter of approximately 10 cm (4 in). Until this point, the contractions have not moved the fetus out of the uterus. Now the contractions move the fetus through the cervix and vagina. At this time the mother, by bearing down to increase abdominal pressure, can help the uterine contractions to deliver the baby. The umbilical vessels and placenta are still functioning, so that the baby is not yet on its own, but within minutes of delivery both the umbilical vessels and the placental vessels completely constrict, stopping blood flow to the placenta. The entire placenta becomes separated from the underlying uterine wall, and a wave of uterine contractions delivers the placenta as the **afterbirth.**

Ordinarily, parturition proceeds automatically from beginning to end and requires no significant medical intervention. In a small percentage of cases, however, the position of the baby or some maternal defect can interfere with normal delivery. The headfirst position of the fetus is important for several reasons: (1) If the baby is not oriented headfirst, another portion of its body is in contact with the cervix and is generally a far less effective wedge. (2) Because of the head's large diameter compared with the rest of the body, if the body were to go through the cervical canal first, the canal might obstruct the passage of the head, leading to problems when the partially delivered baby attempts to breathe. (3) If the umbilical cord becomes caught between the canal wall and the baby's head or chest, mechanical compression of the umbilical vessels can result. Despite these potential problems, however, many babies who are not oriented headfirst are born normally.

What mechanisms control the events of parturition? Let us consider a set of fairly well established facts:

1. The autonomic neurons to the uterus are of little importance since anesthetizing them does not interfere with delivery.
2. The smooth-muscle cells of the myometrium have inherent rhythmicity and are capable of autonomous contractions, which are facilitated as the muscle is stretched by the growing fetus.
3. The pregnant uterus near term and during labor secretes several prostaglandins (PGE_2 and $PGF_{2\alpha}$) that are profound stimulators of uterine smooth-muscle contraction.
4. **Oxytocin,** one of the hormones released from the posterior pituitary, is an extremely potent uterine muscle stimulant. It not only acts directly on uterine smooth muscle but also stimulates it to synthesize the prostaglandins mentioned above. Oxytocin is reflexly secreted from the posterior pituitary as a result of neural input to the hypothalamus, originating from receptors in the uterus, particularly the cervix. Also, as noted above, the number of oxytocin receptors in the uterus increases during the last few weeks of pregnancy; thus, the contractile response to any given plasma concentration of oxytocin is greatly increased at parturition.
5. Throughout pregnancy, progesterone exerts an essential powerful inhibitory effect upon uterine contractions by decreasing the sensitivity of the myometrium to estrogen, oxytocin, and prostaglandins. Unlike the situation in many other species, however, the rate of progesterone

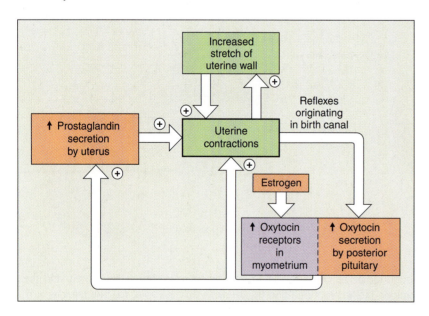

FIGURE 19–28

Factors stimulating uterine contractions during parturition. Note the positive-feedback nature of several of the inputs.

TABLE 19–10 Some Effects of Prostaglandins* on the Female Reproductive System		
Site of Production	**Action**	**Result**
Late-antral follicle	Stimulate production of digestive enzymes	Rupture of follicle
Corpus luteum	May interfere with corpus luteum's hormone secretion and function	? Death of corpus luteum
Uterus	Constrict blood vessels in endometrium	Onset of menstruation
	Cause changes in endometrial blood vessels and cells early in pregnancy	Facilitate implantation
	Increase contraction of myometrium	Help initiate both menstruation and parturition
	Cause cervical "ripening"	Facilitate cervical dilation during parturition

*The term "prostaglandins" is used loosely here, as is customary in reproductive physiology, to include all the eicosanoids.

secretion does not decrease before or during parturition in women (until after delivery of the placenta, the source of the progesterone); therefore, progesterone withdrawal does not play a role in parturition.

These facts can now be put together in a unified pattern, as shown in Figure 19–28. Once started, the uterine contractions exert a positive-feedback effect upon themselves via both local facilitation of inherent uterine contractions and reflex stimulation of oxytocin secretion. But precisely what the relative importance of all these factors is in *initiating* labor remains unclear. Because of the central role of increasing estrogen levels in parturition, one of the main candidates for the timing of delivery is a hormone from the placenta that stimulates the fetal adrenal cortex to produce androgens, which then are converted into estrogen by the placenta. [This candidate hormone, surprisingly, is identical to the hypophysiotropic hormone, corticotropin releasing hormone (CRH).]

The action of prostaglandins on parturition is the last in a series of prostaglandin effects on the female reproductive system we have described. They are summarized in Table 19–10.

Lactation

The secretion of milk by the breasts, or **mammary glands,** is termed **lactation.** The breasts contain ducts that branch all through the tissue and converge at the nipples (Figure 19–29). These ducts arise in saclike glands called **alveoli** (the same term is used to denote the lung air sacs). The breast alveoli, which are the sites of milk secretion, look like bunches of grapes with stems terminating in the ducts. The alveoli and the ducts immediately adjacent to them are surrounded by specialized contractile cells called **myoepithelial cells.**

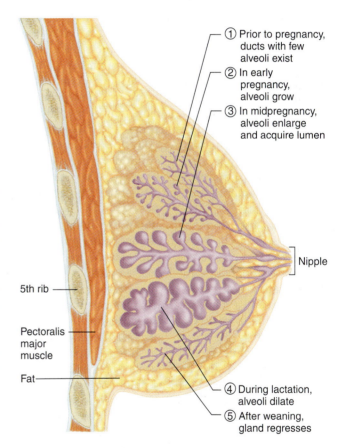

① Prior to pregnancy, ducts with few alveoli exist
② In early pregnancy, alveoli grow
③ In midpregnancy, alveoli enlarge and acquire lumen

Nipple

5th rib

Pectoralis major muscle

Fat

④ During lactation, alveoli dilate
⑤ After weaning, gland regresses

FIGURE 19–29

Anatomy of the breast. The numbers refer to the sequential changes that occur over time.

Adapted form Elias et al.

Before puberty, the breasts are small with little internal glandular structure. With the onset of puberty in females, the increased estrogen causes a marked enhancement of duct growth and branching but relatively little development of the alveoli, and much of the breast enlargement at this time is due to fat deposition. Progesterone secretion also commences at puberty during the luteal phase of each cycle, and this hormone contributes to breast growth by stimulating growth of alveoli.

During each menstrual cycle, the breasts undergo fluctuations in association with the changing blood concentrations of estrogen and progesterone, but these changes are small compared with the marked breast enlargement that occurs during pregnancy as a result of the stimulatory effects of high plasma concentrations of estrogen, progesterone, prolactin, and placental lactogen. This last hormone, as described earlier, is secreted by the placenta, whereas prolactin is secreted by the anterior pituitary. Under the influence of these hormones (and others not mentioned), both the ductal and the alveolar structures become fully developed.

As described in Chapter 10, the anterior pituitary cells that secrete prolactin are influenced by many hormones. They are inhibited by **dopamine,** which is secreted by the hypothalamus. They are probably stimulated by at least one **prolactin releasing factor (PRF),** also secreted by the hypothalamus (the chemical identity of PRF is still uncertain but may be thyrotropin releasing hormone). The dopamine and PRF secreted by the hypothalamus are hypophysiotropic hormones in that they reach the anterior pituitary by way of the hypothalamo-pituitary portal vessels. Estrogen also acts on the anterior pituitary to stimulate prolactin secretion.

Under the dominant inhibitory influence of dopamine, prolactin secretion is low before puberty. It then increases considerably at puberty in girls but not in boys, stimulated by the increased plasma estrogen concentration that occurs at this time. During pregnancy, there is a marked further increase in prolactin secretion due to stimulation by estrogen.

Prolactin is the major hormone stimulating the production of milk. However, despite the fact that prolactin is elevated and the breasts are markedly enlarged and fully developed as pregnancy progresses, there is no secretion of milk. This is because estrogen and progesterone, in large concentrations, prevent milk production by inhibiting this particular action of prolactin on the breasts. Thus, although estrogen causes an increase in the secretion of prolactin and acts with prolactin in promoting breast growth and differentiation, it, along with progesterone, is antagonistic to prolactin's ability to induce milk secretion. Delivery removes the source—the placenta—of the large amounts of estrogen and progesterone and, thereby, the inhibition of milk production.

The drop in estrogen following parturition also causes *basal* prolactin secretion to decrease from its peak late-pregnancy levels and after several months to return toward prepregnancy levels even though the mother continues to nurse. Superimposed upon this basal level, however, are large secretory bursts of prolactin during each nursing period. The episodic pulses of prolactin are signals to the breasts for maintenance of milk production, which ceases several days after the mother completely stops nursing her infant but continues uninterrupted for years if nursing is continued.

The reflexes mediating the prolactin bursts (Figure 19–30) are initiated by afferent input to the hypothalamus from nipple receptors stimulated by suckling. This input's major effect is to inhibit the hypothalamic neurons that release dopamine and, possibly, also to stimulate the neurons that secrete PRF.

One other reflex process is essential for nursing. Milk is secreted into the lumen of the alveoli, but the infant cannot suck the milk out of the alveoli. It must first be moved into the ducts, from which it can be sucked. This movement is called the **milk ejection reflex** (formerly called milk letdown) and is accomplished by contraction of the myoepithelial cells surrounding the alveoli. The contraction is under the control of oxytocin, which is reflexly released from posterior pituitary neurons in response to suckling (Figure 19–30). Higher brain centers can also exert an important influence over oxytocin release: A nursing mother may actually leak milk when she hears her baby cry or even thinks about nursing.

Yet another neuroendocrine reflex triggered by suckling (it may be mediated, in part, by prolactin) is inhibition in most women of the hypothalamo-pituitary-ovarian chain at a variety of steps, with a resultant block of ovulation. If suckling is continued at high frequency, ovulation can be delayed for years. When supplements are added to the baby's diet and the frequency of suckling is decreased, however, most women will resume ovulation even though they continue to nurse. Failure to recognize this fact may result in an unplanned pregnancy.

Initially after delivery the breasts secrete only a watery fluid called **colostrum,** which is rich in protein but poor in other nutrients. After about 24 to 48 hours the secretion of milk, itself, begins. Milk contains four major nutrient constituents: water, protein, fat, and the carbohydrate lactose (milk sugar). Although prolactin is the single most important hormone controlling milk production, insulin, growth hormone, cortisol, and still other hormones participate.

Milk also contains antibodies and other messengers of the immune system, all of which are important for the protection of the newborn, as described in Chapter 20, as well as for longer-term activation of the child's own immune system. It also contains many growth factors and hormones thought to help in tissue development and maturation, as well as a large

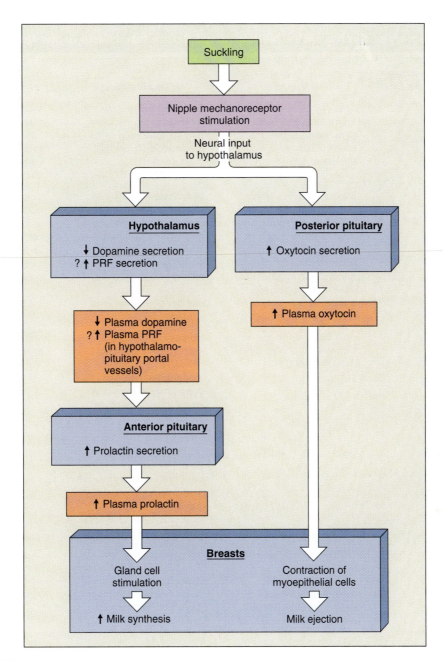

FIGURE 19–30

Major controls of the secretion of prolactin and oxytocin during nursing. The identity of PRF is not known, nor is the importance of its role in stimulating prolactin secretion during lactation.

number of neuropeptides and endogenous opioids that may subtly shape the infant's brain and behavior. Some of these substances are synthesized by the breasts themselves, not just transported from blood to milk. The reasons the milk proteins can gain entry to the newborn's blood are that (1) the low gastric acidity of the newborn does not denature them, and (2) the newborn's intestinal epithelium is much more permeable to proteins than is the adult epithelium.

Unfortunately, infectious agents, including the virus that causes AIDS (Chapter 20), can be transmitted

through breast milk, as can some drugs. For example, the concentration of alcohol in breast milk is approximately the same as in maternal plasma.

Breast feeding for the first 3 to 6 months or even longer of an infant's life is strongly advocated by the medical community. In less-developed countries, where alternative formulas are often either contaminated or nutritionally inadequate because of improper dilution or refrigeration, breast feeding significantly reduces infant sickness and mortality. In the United States, effects on infant survival are not usually

TABLE 19–11 Summary of Contraceptive Methods Used Before Intercourse in the United States*

Type	First-year Failure Rate[†]	Advantages	Disadvantages	Comments
Surgical sterilization (vasectomy in male; tubal ligation in female)	0%	Safe, highly effective, and inexpensive over long term. Frees couples from concern associated with the use of other methods.	Requires surgery. May not be reversible. Vasectomy may be associated with increased incidence of prostate cancer.	Vasectomy is performed under local anesthesia. Tubal ligation is performed vaginally or abdominally under local or general anesthesia.
Birth control pill (oral contraceptive)	2%	Most effective reversible contraceptive. Results in lighter, more regular periods. Protects against cancer of the ovaries and endometrium. Decreases risk of pelvic inflammatory disease, fibrocystic breast disease, benign ovarian cysts, and endometriosis.	Minor side effects similar to early pregnancy (nausea, breast tenderness, fluid retention) during first 3 months of use. Major complications (heart attacks and strokes) may occur in a very small number of smokers who are over 35.	Combination types contain both synthetic estrogen and progestogen. Minipill contains only progestogen and may produce irregular bleeding. Available by prescription only.
Injectable or implantable progestogens (Norplant, Depo-Provera)	1%	Long-lasting (5 yr for Norplant; 3 months for Depo-Provera).	Norplant: Prolonged menstruation; spotting. Headaches and mood changes also reported. Depo-Provera: Menstrual irregularities; ? increased risk of osteoporosis and breast cancer.	Most recently approved methods.
Intrauterine device (IUD)	5%	Once inserted, usually stays in place.	May cause bleeding and cramping. Increased risk of pelvic inflammatory disease. If pregnancy occurs, risk that it may be ectopic. Must check for placement after each period. Requires annual replacement.	Available by prescription only.
Condom (latex prophylactic, sheath)	10%	Helps protect against sexually transmitted diseases, including AIDS and herpes. May protect against cervical cancer.	Must be applied immediately before intercourse. Rare cases of allergy to rubber. May break. Blunting of sensation.	More effective when the woman uses a spermicide.
Vaginal spermicide (foams, jellies)	18% (used alone)	Available over the counter as jellies, foam, creams, and suppositories.	Messiness. Must be applied no more than 1 h before intercourse.	Best results occur when used with a condom or diaphragm.
Diaphragm	15% (with spermicide)	Can be inserted up to 6 h before intercourse.	Increased risk of urinary tract infection. Rare cases of allergy to rubber.	Use with spermicide is recommended. Available by prescription only and must be fitted. Must be left in for 6–8 h after intercourse.
Cervical cap	15% (with spermicide)	Can be inserted several days before intercourse. Is smaller than a diaphragm.	Some women find it difficult to insert. Can cause an unpleasant odor.	Must be left in place for 6–8 h after intercourse but not more than 48 h.
Vaginal sponge	15%	Easy to use because spermicide is self-contained. May be inserted as much as (but no more than) 24 h before intercourse.	May be hard to remove; may fragment. May irritate vaginal lining. Higher failure rate in women who have given birth.	Must be left in for 6 h after intercourse.

Source: Table adapted in part from University of California, Berkeley, *Wellness Letter,* 1987, and *New England Journal of Medicine,* 320:777–786, 1989.
*Coitus interruptus (withdrawal) and periodic abstinence from sexual intercourse around the time of ovulation (rhythm method) are not included because of their high failure rate (approximately 20%).
[†]For all methods, most of the failures are due to improper or inconsistent use. It is likely that oral contraceptives, the IUD, and condoms are 99 to 100 percent effective when properly used; the others are 90 to 95 percent effective.

apparent, but the practice may reduce the severity of gastrointestinal infections, has positive effects on the mother-infant interaction, is economical, and theoretically could have effects on infant development. Cow's milk has many, but not all the constituents of mother's milk, but often in very different concentrations, and it is virtually impossible to reproduce mother's milk in a commercial formula.

Contraception

About two-thirds of all American couples between the ages of 15 and 44 use some method of contraception (Table 19–11). Precise terminology is particularly important in this area. Physiologically, pregnancy is said to begin not at fertilization but after implantation is complete, approximately one week *after* fertilization. Accordingly, procedures that work prior to implantation are termed *contraceptives.* Procedures that cause death of the embryo or fetus after implantation are termed *abortifacients.*

Some forms of contraception—vasectomy, tubal ligation, vaginal diaphragms, vaginal caps, spermicides, and condoms—prevent sperm from reaching the egg. [In addition, condoms significantly reduce the risk of *sexually transmitted diseases (STDs)* such as AIDS, syphilis, gonorrhea, chlamydia, and herpes.]

Oral contraceptives are based on the fact that estrogen and progesterone can inhibit pituitary gonadotropin release, thereby preventing ovulation. One type of oral contraceptive is a combination of a synthetic estrogen and a progesterone-like substance (a progestogen or progestin). Another type is the so-called minipill, which contains only the progesterone-like substance. In actuality, the oral contraceptives, particularly the minipill, do not always prevent ovulation, but they are still effective because they have other contraceptive effects. For example, progestogens affect the composition of the cervical mucus, preventing passage of sperm through the cervix, and they also inhibit the estrogen-induced proliferation of the endometrium, making it inhospitable for implantation.

Another method of delivering a contraceptive progestogen is via tiny capsules (*Norplant*) that are implanted beneath the skin and last for 5 years. Still another method is the intramuscular injection of a different progestagen substance (*Depo-Provera*) every 3 months.

The *intrauterine device (IUD)* works beyond the point of fertilization but before implantation has begun or is complete. The presence of one of these small objects in the uterus somehow interferes with the endometrial preparation for acceptance of the blastocyst.

In addition to the methods used *before* intercourse (precoital contraception) shown in Table 19–11, there are a variety of drugs used within 72 h *after* intercourse (postcoital contraception). These most commonly interfere with ovulation, transport of the conceptus to the uterus, or implantation. One approach is a high dose of estrogen, or two large doses (12 h apart) of a combined estrogen-progestin oral contraceptive. More effective and having fewer side effects is the drug *RU 486* (mifeprestone), which has antiprogesterone activity because it binds competitively to progesterone receptors in the uterus but does not activate them. Antagonism of progesterone's effects causes the endometrium to erode and the contractions of the uterine tubes and myometrium to increase. RU 486 can also be used later in pregnancy as an abortifacient.

The rhythm method uses abstention from sexual intercourse near the time of ovulation. Unfortunately, it is difficult to time ovulation precisely, even with laboratory techniques. For example, the small rise in body temperature or change in cervical mucus and vaginal epithelium, all of which are indicators of ovulation, occur only *after* ovulation. This problem, combined with the marked variability of the time of ovulation in many women—from day 5 to day 15 of the cycle—explains why the rhythm method has a high failure rate (19 percent).

There are still no effective chemical agents for male contraception.

Infertility

Approximately 12 percent of the married couples in the United States are infertile, the number of infertile men and women being approximately equal (at least until age 30, after which infertility shifts progressively toward the woman). Careful investigation leads, in 85 percent of cases, to successful therapy with drugs, artificial insemination, or corrective surgery.

When the basic defect cannot be treated, it can sometimes be circumvented in women by the technique of *in vitro fertilization.* First, the woman is injected with drugs that stimulate multiple egg production. Immediately before ovulation, at least one egg is then removed from the ovary via a needle inserted into the ovary through the top of the vagina or the lower abdominal wall. The egg is placed in a dish for several days with sperm. After the fertilized egg has developed into a cluster of two to eight cells, it is then transferred to the woman's uterus. The success rate of this procedure, when one egg is transferred, is only about 15 percent, but recent developments may significantly increase this number.

Ovarian Function

I. The female gonads, the ovaries, produce eggs and secrete estrogen, progesterone, and inhibin.

II. The two meiotic divisions of oogenesis result in each ovum having 23 chromosomes, in contrast to the 46 of the original oogonia.

III. The follicle consists of the egg, inner layers of granulosa cells surrounding the egg, and outer layers of theca cells.

IV. At the beginning of each menstrual cycle, a group of preantral and early antral follicles begins to develop further, but soon only the dominant follicle continues its development to full maturity and ovulation.

V. Following ovulation, the remaining cells of that follicle differentiate into the corpus luteum, which lasts about 10 to 14 days if pregnancy does not occur.

VI. The menstrual cycle can be divided, according to ovarian events, into a follicular phase and a luteal phase, which last approximately 14 days each and are separated by ovulation.

Control of Ovarian Function

I. The menstrual cycle results from a finely tuned interplay of hormones secreted by the ovaries, the anterior pituitary, and the hypothalamus.

II. During the early and middle follicular phases, FSH stimulates the granulosa cells to proliferate and secrete estrogen, and LH stimulates the theca cells to proliferate and produce the androgens that the granulosa cells use to make estrogen.

  a. During this period, estrogen exerts a negative feedback on the anterior pituitary to inhibit the secretion of the gonadotropins. It probably also inhibits secretion of GnRH by the hypothalamus.

  b. Inhibin preferentially inhibits FSH secretion.

III. During the late follicular phase, plasma estrogen becomes high enough to elicit a surge of LH, which then causes, via the granulosa cells, completion of the egg's first meiotic division and cytoplasmic maturation, ovulation, and formation of the corpus luteum.

IV. During the luteal phase, under the influence of small amounts of LH, the corpus luteum secretes progesterone and estrogen. Regression of the corpus luteum results in a cessation of the secretion of these hormones.

V. Secretion of GnRH and the gonadotropins is inhibited during the luteal phase by the combination of progesterone, estrogen, and inhibin.

Uterine Changes in the Menstrual Cycle

I. The ovarian follicular phase is equivalent to the uterine menstrual and proliferative phases, the first day of menstruation being the first day of the cycle. The ovarian luteal phase is equivalent to the uterine secretory phase.

  a. Menstruation occurs when the plasma estrogen and progesterone levels fall as a result of regression of the corpus luteum.

  b. During the proliferative phase, estrogen stimulates growth of the endometrium and myometrium and causes the cervical mucus to be readily penetrable by sperm.

  c. During the secretory phase, progesterone converts the estrogen-primed endometrium to a secretory tissue and makes the cervical mucus relatively impenetrable to sperm. It also inhibits uterine contractions.

Other Effects of Estrogen and Progesterone

I. The many effects of estrogen and progesterone are summarized in Table 19–8.

Pregnancy

I. After ovulation, the egg is swept into the uterine tube, where a sperm, having undergone capacitation and the acrosome reaction, fertilizes it.

II. Following fertilization the egg undergoes its second meiotic division, and the nuclei of the egg and sperm fuse. Reactions in the ovum block penetration by other sperm and trigger cell division and embryogenesis.

III. The conceptus undergoes cleavage, eventually becoming a blastocyst, which implants in the endometrium on approximately day 7 after ovulation.

  a. The trophoblast gives rise to the fetal part of the placenta, whereas the inner cell mass develops into the embryo proper.

  b. Although they do not mix, fetal blood and maternal blood both flow through the placenta, exchanging gases, nutrients, hormones, waste products, and other substances.

  c. The fetus is surrounded by amniotic fluid in the amniotic sac.

IV. The progesterone and estrogen required to maintain the uterus during pregnancy come from the corpus luteum for the first 2 months of pregnancy, their secretion stimulated by chorionic gonadotropin produced by the trophoblast.

V. During the last 7 months of pregnancy, the corpus luteum regresses, and the placenta itself produces large amounts of progesterone and estrogen.

VI. The high levels of progesterone, in the presence of estrogen, inhibit the secretion of GnRH and thereby that of the gonadotropins, so that menstrual cycles are eliminated.

VII. Delivery occurs by rhythmical contractions of the uterus, which first dilate the cervix and then move the infant, followed by the placenta, through the vagina. The contractions are stimulated in part by oxytocin, released from the posterior pituitary in a reflex triggered by uterine mechanoreceptors, and by uterine prostaglandins.

VIII. The breasts develop markedly during pregnancy as a result of the combined influences of estrogen, progesterone, prolactin, and placental lactogen.

  a. Prolactin secretion is stimulated during pregnancy by estrogen acting on the anterior pituitary, but milk is not synthesized because high concentrations of estrogen and progesterone inhibit the milk-producing action of prolactin on the breasts.

b. As a result of the suckling reflex, large bursts of prolactin and oxytocin occur during nursing; the prolactin stimulates milk production and the oxytocin causes milk ejection.

SECTION C KEY TERMS

ovulation	fertilization
menstrual cycle	acrosome reaction
menstruation	zygote
female internal genitalia	conceptus
uterine tube	cleavage
uterus	blastocyst
cervix	trophoblast
vagina	inner cell mass
female external genitalia	embryo
vulva	fetus
clitoris	implantation
oogenesis	placenta
egg	chorion
oogonia	chorionic villi
primary oocyte	umbilical arteries
meiotic arrest	umbilical vein
secondary oocyte	umbilical cord
follicle	amniotic cavity
primordial follicle	amnion
granulosa cells	amniotic sac
zona pellucida	amniotic fluid
theca	chorionic gonadotropin (CG)
antrum	placental lactogen
dominant follicle	parturition
atresia	afterbirth
corpus luteum	oxytocin
follicular phase	mammary gland
luteal phase	lactation
LH surge	alveoli
menstrual phase	myoepithelial cell
endometrium	dopamine
proliferative phase	prolactin releasing factor (PRF)
secretory phase	milk ejection reflex
myometrium	colostrum
capacitation	

SECTION C REVIEW QUESTIONS

1. Draw the female reproductive tract.
2. Describe the various stages from oogonium to mature ovum.
3. Describe the progression from a primordial follicle to a dominant follicle.
4. Name three hormones produced by the ovaries and name the cells that produce them.
5. Diagram the changes in plasma concentrations of estrogen, progesterone, LH, and FSH during the menstrual cycle.
6. What are the analogies between the granulosa cells and the Sertoli cells and between the theca cells and the Leydig cells?
7. List the effects of FSH and LH on the follicle.
8. Describe the effects of estrogen and inhibin on gonadotropin secretion during the early, middle, and late follicular phases.
9. List the effects of the LH surge on the egg and the follicle.
10. What are the effects of the sex steroids and inhibin on gonadotropin secretion during the luteal phase?
11. Describe the hormonal control of the corpus luteum and the changes that occur in the corpus luteum in a nonpregnant cycle and if pregnancy occurs.
12. What happens to the sex steroids and the gonadotropins as the corpus luteum degenerates?
13. Compare the phases of the menstrual cycle according to uterine and ovarian events.
14. Describe the effects of estrogen and progesterone on the endometrium, cervical mucus, and myometrium.
15. Describe the uterine events associated with menstruation.
16. List the effects of estrogen on the accessory sex organs and secondary sex characteristics.
17. List the effects of progesterone on the breasts, cervical mucus, vaginal epithelium, and body temperature.
18. What are the sources and effects of androgens in women?
19. How does the egg get from the ovary to a uterine tube?
20. Where does fertilization normally occur?
21. Describe the events that occur during fertilization.
22. How many days after ovulation does implantation occur, and in what stage is the conceptus at that time?
23. Describe the structure of the placenta and the pathways for exchange between maternal and fetal blood.
24. State the sources of estrogen and progesterone during different stages of pregnancy. What is the dominant estrogen of pregnancy, and how is it produced?
25. What is the state of gonadotropin secretion during pregnancy, and what is the cause?
26. What anatomical feature permits coordinated contractions of the myometrium?
27. Describe the mechanisms and messengers that contribute to parturition.
28. List the effects of prostaglandins on the female reproductive system.
29. Describe the development of the breasts after puberty and during pregnancy, and list the major hormones responsible.
30. Describe the effects of estrogen on the secretion and actions of prolactin during pregnancy.
31. Diagram the suckling reflex for prolactin release.
32. Diagram the milk ejection reflex.

THE CHRONOLOGY OF REPRODUCTIVE FUNCTION

This section treats a variety of topics that have to do, in one way or another, with sequential changes in reproductive development or function. Genetic inheritance sets the sex of the individual, **sex determination,** which is established at the moment of fertilization. This is followed by **sex differentiation,** the multiple processes in which development of the reproductive system occurs in the fetus. Then there is the maturation of the system at puberty and the eventual decline that occurs with aging.

Sex Determination

Sex is determined by genetic inheritance of two chromosomes called the **sex chromosomes.** The larger of the sex chromosomes is called the **X chromosome** and the smaller, the **Y chromosome.** Males possess one X and one Y, whereas females have two X chromosomes. Thus, the *genetic* difference between male and female is simply the difference in one chromosome.

The reason for the approximately equal sex distribution of the population should be readily apparent: The ovum can contribute only an X chromosome, whereas half of the sperm produced during meiosis are X and half are Y. When the sperm and the egg join, 50 percent should have XX and 50 percent XY. Interestingly, however, sex ratios at birth are not exactly 1:1; rather, for unclear reasons, there tends to be a slight preponderance of male births.

An easy method exists for determining whether a person's cells contain two X chromosomes, the normal female pattern. When two X chromosomes are present, only one functions and the nonfunctional X chromosome condenses to form a nuclear mass termed the **sex chromatin,** which is readily observable with a light microscope (scrapings from the cheek mucosa are a convenient source of cells to be examined). The single X chromosome in male cells rarely condenses to form sex chromatin.

A more exacting technique for determining sex chromosome composition employs tissue culture visualization of all the chromosomes—a **karyotype.** This technique has revealed a group of genetic sex abnormalities characterized by such unusual chromosomal combinations as XXX, XXY, X, and others. The end result of such combinations is usually the failure of normal anatomical and functional sexual development.

Sex Differentiation

It is not surprising that people with abnormal genetic endowment manifest abnormal sexual development, but careful study has also revealed individuals with normal chromosomal combinations but abnormal sexual appearance and function. In these people, *sex differentiation* has been abnormal, and their appearance may even be at odds with their genetic sex—that is, the presence of XX or XY chromosomes.

It will be important to bear in mind during the following description one essential generalization: The genes directly determine only whether the individual will have testes or ovaries. All the rest of sex differentiation depends upon the presence or absence of substances produced by the genetically determined gonads, specifically the testes.

Differentiation of the Gonads

The male and female gonads derive embryologically from the same site—an area called the urogenital ridge—in the body. Until the sixth week of uterine life, there is no differentiation of this site. In the genetic male, the testes begin to develop during the seventh week. A single gene on the Y chromosome (the *SRY* **gene,** for **s**ex-determining **r**egion of the **Y** chromosome) is expressed at this time in the urogenital ridge cells and triggers this development. In the absence of a Y chromosome and, hence, the *SRY* gene, testes do not develop; instead, ovaries begin to develop in the same area at about 11 weeks.

By what mechanism does the *SRY* gene induce formation of the testes? This gene codes for a protein, SRY, which acts as a transcription factor that sets into motion a sequence of gene activations ultimately leading to formation of integrated testes from the various embryonic cells in the urogenital ridge.

There is an unusual and important fact concerning the behavior of the X and Y chromosomes during meiosis. As described earlier in this chapter, during meiosis homologous chromosomes come together, line up point for point, and then exchange fragments with each other, resulting in an exchange of genes (recombination) on these chromosomes. Such crossing-over involving the Y and X genes, however, could allow the *SRY* gene—the male-determining gene—to get unto the female's genome. To prevent this, by mechanisms still not understood, the Y and X chromosomes do not undergo recombination (except at the very tips, where the *SRY* gene is not located and the Y chromosome has the same genes as the X).

Differentiation of Internal and External Genitalia

So far as its internal duct system and external genitalia are concerned, the fetus is capable of developing into either sex. Before the functioning of the fetal gonads, the primitive reproductive tract includes a double genital duct system—**Wolffian ducts** and **Müllerian ducts**—and a common opening for the genital ducts and urinary system to the outside. Normally, most of the reproductive tract develops from only one of these duct systems: In the male, the Wolffian ducts persist and the Müllerian ducts regress, whereas in the female, the opposite happens. The external genitalia in the two sexes and the vagina do not develop from these duct systems, however, but from other structures at the body surface.

Which of the two duct systems and types of external genitalia develops depends on the presence or absence of fetal testes. These testes secrete (1) testosterone, from the Leydig cells, and (2) a protein hormone called **Müllerian-inhibiting substance (MIS)** from the Sertoli cells (Figure 19–31a). MIS, the gene for which is induced by SRY protein, acts as a paracrine agent to cause the Müllerian duct system to degenerate. Simultaneously, testosterone causes the Wolffian ducts to differentiate into the epididymis, vas deferens, ejaculatory duct, and seminal vesicle. Externally and somewhat later, under the influence of testosterone (after its conversion to dihydrotestosterone), a penis forms, and the tissue near it fuses to form the scrotum. The testes will ultimately descend into the scrotum, stimulated to do so by both MIS and testosterone.

In contrast, the female fetus, not having testes (because of the absence of the *SRY* gene), does not secrete testosterone and MIS. In the absence of MIS, the Müllerian system does not degenerate but rather develops into uterine tubes and a uterus. In the absence of testosterone, the Wolffian ducts degenerate, and a vagina and female external genitalia develop from the structures at the body surface (Figure 19–31b). Ovaries, though present in the female fetus, do not play a role in these developmental processes; in other words, female development will occur automatically unless stopped from doing so by the presence of secretions from functioning testes.

There are various conditions in which normal sex differentiation does not occur. For example, in the syndrome known as *testicular feminization* (or androgen insensitivity syndrome), the person's genetic endowment is XY and testes are present, but he has female external genitalia, a vagina, and no internal duct system at all. The problem causing this is a lack of androgen receptors due to a genetic mutation. Under the influence of SRY, the fetal testes differentiate as usual, and they secrete both MIS and testosterone. MIS causes the Müllerian ducts to regress, but the inability of the Wolffian ducts to respond to testosterone also causes them to regress, and so no duct system develops. The tissues that give rise to external genitalia (and the vagina, in the female) are also unresponsive to testosterone, and so female external genitalia and a vagina develop rather than male structures.

Sexual Differentiation of the Central Nervous System and Homosexuality

With regard to sexual behavior, differences in the brain may be formed during development. For example, genetic female monkeys given testosterone during their late fetal life manifest evidence of masculine sex behavior, such as mounting, as adults.

In this regard, a potentially important difference in brain anatomy has been reported for people: The size of a particular nucleus (neuronal cluster) in the hypothalamus is more than twice as large in men as in women. A subsequent study showed that the nucleus is also more than twice as large in heterosexual men as in homosexual men. A similar sexually dimorphic area exists in rats, in which it is known to be involved in male-type sexual behavior and is influenced during development by testosterone.

Another approach to evaluating the genetics and hormone dependency of sexual behavior and gender preference is the use of twin and family studies. The pooled data from six such studies show that 57 percent of identical twin brothers of homosexual men were also homosexual, compared to 24 percent of fraternal twins and 13 percent of nontwin brothers. The numbers for homosexual women are quite similar. On the basis of these numbers, it has been estimated that for sexual orientation, the overall heritability—that proportion of the total variability in a trait that comes from genes—is approximately 50 percent.

Puberty

Puberty is the period, usually occurring sometime between the ages of 10 and 14, during which the reproductive organs, having differentiated many years earlier in utero, mature and reproduction becomes

(a)

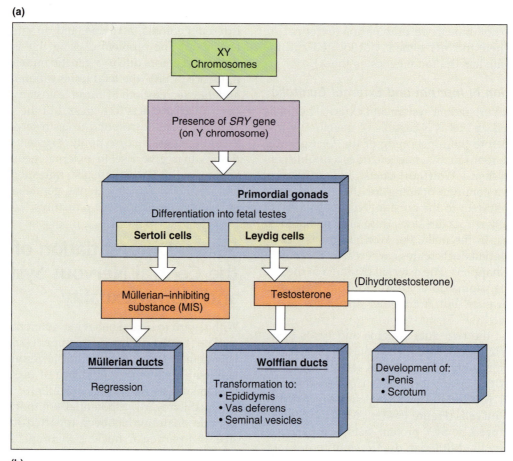

(b)

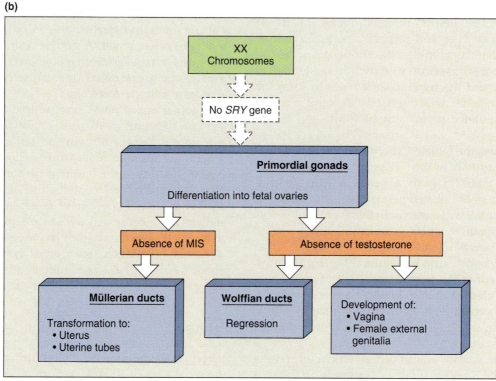

FIGURE 19–31

Sex differentiation. (a) Male. (b) Female. The *SRY* gene codes for the SRY protein.

possible. Surprisingly, the very first signs of puberty are due not to gonadal steroids but to increased secretion of adrenal androgens, under the stimulation of adrenocorticotropic hormone (ACTH). In both sexes, these androgens cause the very early development of pubic and axillary (armpit) hair, as well as the early stages of the total-body growth spurt. All other developments in puberty, however, reflect increased activity of the hypothalamic-anterior pituitary-gonadal system of hormones.

In the female, GnRH, the pituitary gonadotropins, and estrogen are all secreted at very low levels during childhood. Accordingly, there is no follicle maturation beyond the early antral stage and menstrual cycles do not occur, the female accessory sex organs remain small and nonfunctional, and there are minimal secondary sex characteristics. The onset of puberty is caused, in large part, by an alteration in brain function that raises secretion of GnRH. This hypophysiotropic hormone in turn stimulates secretion of pituitary gonadotropins, which stimulate follicle development and estrogen secretion. Estrogen, in addition to its critical role in follicle development, induces the striking changes in the accessory sex organs and secondary sex characteristics associated with puberty. **Menarche,** the first menstrual period, is a late event of puberty (average 12.3 years in the United States).

The picture for the male is analogous to that for the female. Increased GnRH secretion at puberty causes increased secretion of pituitary gonadotropins, which stimulate the seminiferous tubules and testosterone secretion. Testosterone, in addition to its critical role in spermatogenesis, induces the pubertal changes in the accessory reproductive organs, secondary sex characteristics, and sex drive.

The mechanism of the brain change that results in increased GnRH secretion at puberty remains unknown. One important event is that the brain becomes less sensitive to the negative-feedback effects of gonadal hormones at the time of puberty. Also, the adipose-tissue hormone leptin (Chapter 18) is known to stimulate the secretion of GnRH and may play a role in puberty, one that would explain the fact that the onset of puberty tends to correlate with the attainment of a certain level of energy stores, as fat, in the girl's body. The effect of leptin on GnRH could also explain why women who exercise extensively and become extremely thin lose their menstrual periods; such women would have low plasma concentrations of leptin. Whatever the mechanisms, the process is not abrupt but develops over several years, as evidenced by slowly rising plasma concentrations of the gonadotropins and testosterone or estrogen.

Some children with brain tumors or other lesions of the hypothalamus may undergo precocious puberty—that is, sexual maturation at an unusually early age—due to early secretion of increased amounts of GnRH. The youngest mother on record gave birth to a full-term, healthy infant by Cesarean section (abdominal incision) at 5 years, 8 months.

Menopause

Around the age of 50, on the average, menstrual cycles become less regular. Ultimately they cease entirely, and this cessation is known as the **menopause.** The phase of life beginning with menstrual irregularity and including the first year after cessation of menstrual flow is termed the **perimenopause,** the counterpart of puberty. It involves numerous physical and emotional changes as sexual maturity gives way to cessation of reproductive function.

Menopause and the irregular function leading to it are caused primarily by ovarian failure. The ovaries lose their ability to respond to the gonadotropins, mainly because most, if not all, ovarian follicles and eggs have disappeared by this time through atresia. That the hypothalamus and anterior pituitary are functioning relatively normally is evidenced by the fact that the gonadotropins are secreted in *greater* amounts. The main reason for this is that the decreased plasma estrogen does not exert as much negative feedback on gonadotropin secretion.

A small amount of estrogen usually persists in plasma beyond the menopause, mainly from peripheral conversion of adrenal androgens to estrogen, but the level is inadequate to maintain estrogen-dependent tissues. The breasts and genital organs gradually atrophy to a large degree. Thinning and dryness of the vaginal epithelium can cause sexual intercourse to be painful. Marked decreases in bone mass and strength, termed *osteoporosis,* may occur because of net bone resorption and can result in bone fractures (Chapter 16). Sex drive frequently stays the same and may even increase. The hot flashes so typical of menopause are caused by periodic sudden increases in body temperature, which induces a feeling of warmth, dilation of the skin arterioles, and marked sweating; how estrogen deficiency causes this is unknown. Another aspect of menopause is its relationship to cardiovascular diseases. Women have much less coronary artery disease than men until after menopause, when the incidence becomes similar in both sexes, a pattern that is due to the protective effects of estrogen: Estrogen exerts beneficial actions on plasma cholesterol (Chapter 18), and also exerts multiple direct protective actions on vessel walls. Another symptom of menopause in some women is emotional instability.

Most of the symptoms associated with menopause, as well as the increases in and death from coronary artery disease and osteoporosis, can be reduced by the administration of estrogen. Recent studies also indicate that estrogen use may reduce the risk of developing Alzheimer's disease and may also be useful in the treatment of this disease; evaluation of these possibilities must await the completion of larger studies presently in progress.

The desirability of administering estrogen to postmenopausal women is controversial, however, because of the fact that long-term estrogen administration (more than 5 years) increases the risk of developing uterine endometrial cancer and, possibly, breast cancer as well. The increased risk of endometrial cancer can be virtually eliminated by administration of a progestogen along with estrogen, but the progestogen does not influence the risk of breast cancer. The progestogen only slightly lessens estrogen's protective effect against coronary artery disease.

In conclusion, numerous studies have shown that, overall, hormone replacement therapy definitely decreases mortality in postmenopausal women, principally through estrogen's protective effects against heart disease. That is, in the average postmenopausal woman the protection against heart disease (and osteoporosis) far outweighs the negative effect of increased cancer. However, this may not be the case for individual women who have a family history of breast or endometrial cancer, or who have another known risk factor for these diseases.

Relevant to the question of hormone-replacement therapy (as well as to the hormonal treatment of breast and uterine cancer) is the development of substances (for example, *tamoxifen*) that exert some pro-estrogenic and some anti-estrogenic effects. These drugs are collectively termed *selective estrogen receptor modulators (SERMs)* because they activate estrogen receptors in certain tissues but not in others; moreover, in these latter tissues SERMs act as estrogen antagonists. Obviously, the ideal would be to have a SERM that has the pro-estrogenic effects of protecting against osteoporosis, heart attacks, and Alzheimer's disease, but opposes the development of breast and uterine cancers. What makes SERMs possible? One important contributor is that there exist two distinct forms of estrogen receptors, which are affected differentially by different SERMs.

Changes in the male reproductive system with aging are less drastic than those in women. Once testosterone and pituitary gonadotropin secretions are initiated at puberty, they continue, at least to some extent, throughout adult life. There is a steady decrease, however, in testosterone secretion, beginning at about the age of 40, which apparently reflects slow deterioration of testicular function and, as in the female, failure of the gonads to respond to the pituitary gonadotropins. Along with the decreasing testosterone levels, both sex drive and capacity diminish, and sperm become much less motile. Despite these events, many men continue to be fertile in their seventies and eighties.

With aging, some men manifest increased emotional problems, such as depression, and this is sometimes referred to as "male menopause" (or male climacteric). It is not clear, however, what role hormone changes play in this phenomenon.

SECTION D SUMMARY

Sex Determination and Sex Differentiation

I. Sex is determined by the two sex chromosomes: Males are XY, and females are XX.

II. A gene on the Y chromosome is responsible for the development of testes. In the absence of a Y chromosome, testes do not develop and ovaries do instead.

III. When a functioning male gonad is present to secrete testosterone and MIS, a male reproductive tract and external genitalia develop. In the absence of testes, the female system develops.

Puberty

I. At puberty, the hypothalamic-anterior-pituitary-gonadal chain of hormones becomes active as a result of a change in brain function that permits increased secretion of GnRH.

II. The first sign of puberty is the appearance of pubic or axillary hair.

Menopause

I. Around the age of 50, a woman's menstrual periods become less regular and ultimately disappear—menopause.

a. The cause of menopause is a decrease in the number of ovarian follicles and their hyporesponsiveness to the gonadotropins.

b. The symptoms of menopause are largely due to the marked decrease in plasma estrogen concentration.

II. Men show a steady decrease in testosterone secretion after age 40 but generally no complete cessation of reproductive function.

SECTION D KEY TERMS

sex determination	Wolffian ducts
sex differentiation	Müllerian ducts
sex chromosomes	Müllerian-inhibitin
X chromosome	substance (MIS)
Y chromosome	puberty
sex chromatin	menarche
karyotype	menopause
SRY gene	perimenopause

1. State the genetic difference between males and females and a method for identifying genetic sex.
2. Describe the sequence of events, the timing, and the control of the development of the gonads and the internal and external genitalia.
3. What is the state of gonadotropin and sex hormone secretion before puberty?
4. What is the state of estrogen and gonadotropin secretion after menopause?
5. List the hormonal and anatomical changes that occur after menopause.

CHAPTER 19 CLINICAL TERMS

vasectomy	eclampsia
erectile dysfunction	pregnancy sickness
Viagra	contraceptive
prostate cancer	abortifacient
castration	sexually transmitted disease
dysmenorrhea	(STD)
premenstrual tension	oral contraceptive
premenstrual syndrome	Norplant
(PMS)	Depo-Provera
premenstrual dysphoric	intrauterine device
disorder (PMDD)	RU 486
virilism	in vitro fertilization
ectopic pregnancy	testicular feminization
amniocentesis	osteoporosis
chorionic villus sampling	tamoxifen
Down's syndrome	selective estrogen receptor
teratogen	modulators (SERMs)
preeclampsia	

CHAPTER 19 THOUGHT QUESTIONS

(Answers are given in Appendix A.)

1. What symptom will be common to a person whose Leydig cells have been destroyed and to a person whose Sertoli cells have been destroyed? What symptom will not be common?

2. A male athlete taking large amounts of an androgenic steroid becomes sterile (unable to produce sperm capable of causing fertilization). Explain.
3. A man who is sterile is found to have no evidence of demasculinization, an increased blood concentration of FSH, and a normal plasma concentration of LH. What is the most likely basis of his sterility?
4. If you were a scientist trying to develop a male contraceptive acting on the anterior pituitary, would you try to block the secretion of FSH or that of LH? Explain the reason for your choice.
5. A 30-year-old man has very small muscles, a sparse beard, and a high-pitched voice. His plasma concentration of LH is elevated. Explain the likely cause of all these findings.
6. There are disorders of the adrenal cortex in which excessive amounts of androgens are produced. If this occurs in a woman, what will happen to her menstrual cycles?
7. Women with inadequate secretion of GnRH are often treated for their sterility with drugs that mimic the action of this hormone. Can you suggest a possible reason that such treatment is often associated with multiple births?
8. Which of the following would be a signal that ovulation is soon to occur: the cervical mucus becoming thick and sticky, an increase in body temperature, a marked rise in plasma LH?
9. The absence of what phenomenon would interfere with the ability of sperm obtained by masturbation to fertilize an egg in a test tube?
10. If a woman 7 months pregnant is found to have a marked decrease in plasma estrogen but a normal plasma progesterone for that time of pregnancy, what would you conclude?
11. What types of drugs might you work on if you were trying to develop one to stop premature labor?
12. If a genetic male failed to produce MIS during in utero life, what would the result be?
13. Could the symptoms of menopause be treated by injections of FSH and LH?

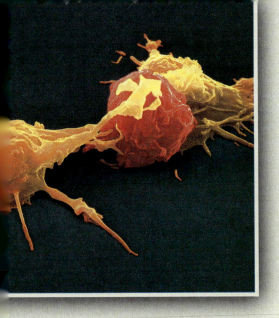

Defense Mechanisms of the Body

IMMUNOLOGY: DEFENSES AGAINST FOREIGN MATTER

Immunology is the study of the physiological defenses by which the body (the "host") destroys or neutralizes foreign matter, both living and nonliving. In distinguishing "self" from "nonself," immune defenses (1) protect against infection by *microbes*—viruses, bacteria, fungi, and parasites; (2) isolate or remove nonmicrobial foreign substances; and (3) destroy cancer cells that arise in the body, a function known as **immune surveillance.**

Immune defenses, or immunity, can be classified into two categories: nonspecific and specific, which interact with each other. **Nonspecific immune defenses** protect against foreign substances or cells without having to recognize their specific identities. The mechanisms of protection used by these defenses are not unique to the particular foreign substance or cell. **Specific immune defenses** (also called acquired immunity) depend upon specific recognition, by lymphocytes, of the substance or cell to be attacked and the launching of an attack against it that is unique for that substance or cell. As we shall see, nonspecific and specific immune responses function in tandem; in particular, components of nonspecific immunity provide instructions that enable the specific responses to select appropriate targets to attack and strategies for their elimination.

Before introducing the cells that participate in immune defenses, let us first look at the microbes we shall be most concerned with in this chapter—bacteria and viruses. These are the dominant infectious organisms in the United States and other industrialized nations. On a global basis, however, infection with parasitic eukaryotic organisms (malaria, for example) or worms are responsible for a huge amount of illness and mortality. For example, over 600 million people presently suffer from malaria.

Bacteria are unicellular organisms (prokaryotes, Chapter 3) that have an outer coating, the cell wall, in addition to a plasma membrane, but no intracellular membrane-bound organelles. Bacteria can damage tissues at the sites of bacterial replication, or they can release toxins that enter the blood and disrupt physiological functions in other parts of the body.

Viruses are essentially nucleic acids surrounded by a protein coat. Unlike bacteria, viruses lack both the enzyme machinery for metabolism and the ribosomes essential for protein synthesis. Thus, they cannot multiply by themselves but must "live" inside other cells, whose biochemical apparatus they use. The viral nucleic acid directs the host cell to synthesize the proteins required for viral replication, with the required nucleotides and energy sources also being supplied by the host cell. The effect of viral habitation and replication within a cell depends on the type of virus. Some viruses (the common cold virus, for example), after entering a cell, multiply rapidly, kill the cell, and then move on to other cells. Other viruses, such as the one that causes genital herpes, can lie dormant in infected cells before suddenly undergoing the rapid replication that causes cell damage. Finally, certain viruses cause transformation of their host cells into cancer cells.

Cells Mediating Immune Defenses

The cells that carry out immune defenses collectively make up the **immune system,** but they do not constitute a "system" in the sense of anatomically connected organs like the gastrointestinal or urinary system. Rather, they are a diverse collection of cells found both in the blood and in tissues throughout the body. In terms of cell number, the immune system ranks alongside the brain and liver.

Because of the large number of cells and the far larger number of chemical messengers that participate in immune defenses, a miniglossary defining the cells and messengers discussed in this chapter is given at the end of Section A (Table 20–11).

The most numerous of the immune system cells are the various types of white blood cells (**leukocytes**)—neutrophils, basophils, eosinophils, monocytes, and lymphocytes. (The first three types are also grouped in the general term, polymorphonuclear granulocyte.) The anatomy, production in the bone marrow, and blood concentrations of these cells were described in Chapter 14 and should be reviewed at this time. Unlike erythrocytes, the leukocytes use the blood mainly for transportation and leave the circulatory system to enter the tissues where they function. In this regard, the lymphocytes are the most complex of the leukocytes, and we shall delay until a subsequent section the description of the various classes and subclasses of lymphocytes, their origins, migrations, and major organs of residence—the lymphoid organs.

Plasma cells are not really a distinct cell line but differentiate from a particular set of lymphocytes (the B lymphocytes) during immune responses. Despite

TABLE 20–1 Cells Mediating Immune Defenses

Name	Site Produced	Functions
Leukocytes (white blood cells)		
Neutrophils	Bone Marrow	1. Phagocytosis 2. Release chemicals involved in inflammation (vasodilators, chemotaxins, etc.)
Basophils	Bone marrow	Have functions in blood similar to those of mast cells in tissues (see below)
Eosinophils	Bone marrow	1. Destroy multicellular parasites 2. Participate in immediate hypersensitivity reactions
Monocytes	Bone marrow	1. Have functions in blood similar to those of macrophages in tissues (see below) 2. Enter tissues and are transformed into macrophages
Lymphocytes	Mature in bone marrow (B cells and NK cells) and thymus (T cells); activated in peripheral lymphoid organs	Serve as "recognition cells" in specific immune responses and are essential for all aspects of these responses
B cells		1. Initiate antibody-mediated immune responses by binding specific antigens to the B cell's plasma-membrane receptors, which are immunoglobulins 2. During activation are transformed into plasma cells, which secrete antibodies 3. Present antigen to helper T cells
Cytotoxic T cells (CD8 cells)		Bind to antigens on plasma membrane of target cells (virus-infected cells, cancer cells, and tissue transplants) and directly destroy the cells
Helper T cells (CD4 cells)		Secrete cytokines that help to activate B cells, cytotoxic T cells, NK cells, and macrophages
NK cells		1. Bind directly and nonspecifically to virus-infected cells and cancer cells and kill them 2. Function as killer cells in antibody-dependent cellular cytotoxicity (ADCC)
Plasma cells	Peripheral lymphoid organs; differentiate from B cells during immune responses	Secrete antibodies
Macrophages	Almost all tissues and organs; differentiate from monocytes	1. Phagocytosis 2. Extracellular killing via secretion of toxic chemicals 3. Process and present antigens to helper T cells 4. Secrete cytokines involved in inflammation, activation and differentiation of helper T cells, and systemic responses to infection or injury (the acute phase response)
Macrophage-like cells	Almost all tissues and organs; microglia in the central nervous system	Same as macrophages
Mast cells	Almost all tissues and organs; differentiate from bone marrow cells	Release histamine and other chemicals involved in inflammation

their name, plasma cells are not usually found in the blood but rather in the tissues in which they have differentiated from their parent lymphocytes. The major function of plasma cells is to synthesize and secrete antibodies.

Macrophages are found in virtually all organs and tissues, their structures varying somewhat from location to location. They are derived from monocytes that pass out of blood vessels to enter the tissues and become transformed into macrophages. By the time of birth, this process has already supplied the tissues with a large number of macrophages that have followed this pathway, but the migration of monocytes continues throughout life. In keeping with one of their major functions, the engulfing of particles, including microbes, macrophages are strategically placed where they will encounter their targets. For example, they are found in large numbers in the various epithelia in contact with the external environment, and in several organs they line the vessels through which blood or lymph flows.

There are several cell populations that are not macrophages (and are not descended from monocytes) but exert various macrophage functions; these are collectively termed **macrophage-like cells.** They are found scattered in almost all tissues and are represented in the central nervous system by the *microglia*, a type of glial cell (Chapter 8).

Mast cells are found throughout connective tissues, particularly beneath the various epithelial surfaces of the body. They are derived from the differentiation of a unique set of bone marrow cells, which have entered the blood and then left the blood vessels to enter connective tissue, where they differentiate and undergo cell division. Thus, mature mast cells, unlike basophils, with which they share many characteristics, are not normally found in the blood. The most striking anatomical feature of mast cells is a very large number of secretory vesicles, and these cells secrete many locally acting chemical messengers.

The sites of production and functions of all these cells are briefly listed in Table 20–1 for reference and will be described in subsequent sections. Suffice it for now to emphasize two points: (1) The lymphocytes serve as "recognition cells" in specific immune defenses and are essential for all aspects of these responses; and (2) neutrophils, monocytes, macrophages, and macrophage-like cells have a variety of activities, but particularly important is their ability to secrete inflammatory mediators and to function as **phagocytes,** a term that denotes *any* cell capable of **phagocytosis,** the form of endocytosis whereby particulate matter is engulfed and usually destroyed inside the phagocyte.

Cytokines

The cells of the immune system release a multitude (more than 100, to date) of protein messengers that regulate host cell growth and function in both nonspecific and specific immune defenses. The collective term for these messengers, each of which has its own unique name, is **cytokines.** They are not produced by distinct specialized glands but rather by a variety of individual cells. The great majority of their actions occur at the site of cytokine secretion, the cytokine acting as an autocrine/paracrine agent. In some cases, however, the cytokine circulates in the blood to exert hormonal effects on distant organs and tissues involved in host defenses.

For many reasons, the physiology of the cytokines is extremely complex. (1) Most of them are secreted by more than one type of immune-system cell and by nonimmune cells as well (for example, by endothelial cells and fibroblasts). (2) There are often cascades of cytokine secretion, in which one cytokine stimulates the release of another, and so on. (3) Any given cytokine may exert actions on an extremely broad range of target cells. For example, the cytokine interleukin 2 influences the function of virtually every cell of the immune system. (4) There is great redundancy in cytokine action; that is, different cytokines can exhibit the same functions. (5) Cytokines are also involved in many nonimmunological processes, such as for example, bone formation and uterine function. Indeed, because of this last point, the precise definition of what constitutes a cytokine has become somewhat hazy. (For example, almost all the hematopoietic growth factors described in Chapter 14 are also classified as cytokines.)

We will, of necessity, limit our discussions in this chapter to just a few of the cytokines and their most important functions.

Nonspecific Immune Defenses

Nonspecific immune defenses protect against foreign cells or matter without having to recognize their specific identities. These defenses must, of course, recognize some *general* property marking the invader as foreign; the most common such identity tags are particular classes of carbohydrates or lipids that are frequent constituents of microbial cell walls. As we shall see, plasma-membrane receptors on certain immune cells, as well as a variety of circulating proteins (particularly a family called complement) are able to bind to these carbohydrates and lipids at crucial steps in nonspecific responses. This use of a system based on *carbohydrate* and *lipid* for detecting the presence of foreign cells is a key feature that distinguishes nonspecific defenses from specific ones, which, as will be described, recognize foreign cells mainly by specific *proteins* the foreign cells possess.

The nonspecific immune defenses include defenses at the body surfaces, the response to injury known as inflammation, and a family of antiviral proteins called interferons. Another nonspecific defense—the lymphocytes called natural killer (NK) cells—is best described later in the context of the specific immune defenses that commonly mobilize them.

Defenses at Body Surfaces

The body's first lines of defense against microbes are the barriers offered by surfaces exposed to the external environment. Very few microorganisms can penetrate the intact skin, and the various skin glands and the lacrymal (tear) glands all secrete antimicrobial chemicals.

The mucus secreted by the epithelial linings of the respiratory and upper gastrointestinal tracts also contains antimicrobial chemicals, but more important, mucus is sticky. Particles that adhere to it are prevented from entering the blood. They are either swept by ciliary action up into the pharynx and then swallowed, as occurs in the upper respiratory tract, or are phagocytyzed by macrophages in the various linings.

Other specialized surface defenses are the hairs at the entrance to the nose, the cough and sneeze reflexes (Chapter 15), and the acid secretion of the stomach and uterus, which kills microbes. Finally, a major defense against infection are the many relatively innocuous microbes normally found on the skin and other linings exposed to the external environment. Through a variety of mechanisms, these microbes suppress the growth of other potentially more dangerous ones.

Inflammation

Inflammation is the body's local response to infection or injury. Regardless of cause, inflammation is relatively stereotyped since a major trigger is cell or tissue injury. The functions of inflammation are to destroy or inactivate foreign invaders and to set the stage for tissue repair. The key actors are the cells that function as phagocytes; as noted earlier, the most important phagocytes are neutrophils, macrophages, and macrophage-like cells.

In this section, we describe inflammation as it occurs in the *nonspecific* defenses induced by the invasion of *microbes.* Most of the same responses can be elicited by a variety of other injuries—cold, heat, and trauma, for example. Moreover, we shall see later that inflammation is also an important component of many *specific* immune defenses in which the inflammation becomes amplified and made more effective.

The sequence of local events in a typical nonspecific inflammatory response to a bacterial infection—one caused, for example, by a cut with a bacteria-covered knife—is summarized in Table 20–2. The familiar manifestations of tissue injury and inflammation are local redness, swelling, heat, and pain.

The events of inflammation that underly these manifestations are induced and regulated by a huge number of chemical mediators, some of which are summarized for reference in Table 20–3 (not all of these will be described in this chapter). Note in this table

TABLE 20–2 Sequence of Events in a Nonspecific Local Inflammatory Response to Bacteria
1. Initial entry of bacteria into tissue
2. Vasodilation of the microcirculation in the infected area, leading to increased blood flow
3. Marked increase in protein permeability of the capillaries and venules in the infected area, with resulting diffusion of protein and filtration of fluid into the interstitial fluid
4. Chemotaxis: exit of leukocytes from the venules into the interstitial fluid of the infected area
5. Destruction of bacteria in the tissue either through phagocytosis or by mechanisms not requiring prior phagocytosis
6. Tissue repair

that some of these mediators are cytokines. Any given event of inflammation, such as vasodilation, may be induced by multiple mediators, and any given mediator may induce more than one event. Based on their origins, the mediators fall into two general categories: (1) peptides (kinins, for example) generated in the infected area by enzymatic actions on proteins that circulate in the plasma; and (2) substances secreted into the extracellular fluid from cells that either already exist in the infected area (mast cells, for example) or enter it during inflammation (neutrophils, for example).

Let us now go step by step through the process summarized in Table 20–2, assuming that the bacterial infection in our example is localized to the tissue just beneath the skin. If the invading bacteria enter the blood or lymph, then similar inflammatory responses would take place in any other tissue or organ reached by the blood-borne or lymph-borne microorganisms.

Vasodilation and Increased Permeability to Protein

A variety of chemical mediators dilate most of the microcirculation vessels in an infected and/or damaged area. The mediators also cause the local capillaries and venules to become permeable to proteins by inducing their endothelial cells to contract, opening spaces between them, through which the proteins can move.

The adaptive value of these vascular changes is twofold: (1) The increased blood flow to the inflamed area (which accounts for the redness and heat) increases the delivery of proteins and leukocytes; and (2) the increased permeability to protein ensures that the plasma proteins that participate in inflammation—many of which are normally restrained by the intact endothelium—can gain entry to the interstitial fluid.

TABLE 20-3 Some Important Local Inflammatory Mediators

Mediator	Source
Kinins	Generated from enzymatic action on plasma proteins
Complement	Generated from enzymatic action on plasma proteins
Products of blood clotting	Generated from enzymatic action on plasma proteins
Histamine	Secreted by mast cells
Eicosanoids	Secreted by many cell types
Platelet-activating factor	Secreted by many cell types
Cytokines, including chemokines	Secreted by monocytes, macrophages, neutrophils, lymphocytes, and several nonimmune cell types, including endothelial cells and fibroblasts
Lysosomal enzymes, nitric oxide, and other oxygen-derived substances	Secreted by neutrophils and macrophages

As described in Chapter 14, the vasodilation and increased permeability to protein, however, cause net filtration of plasma into the interstitial fluid and the formation of edema. This accounts for the swelling in an inflamed area, which is simply a consequence of the changes in the microcirculation and has no known adaptive value of its own.

Chemotaxis With the onset of inflammation, circulating neutrophils begin to move out of the blood across the endothelium of venules to enter the inflamed area. This multistage process is known as **chemotaxis.** It involves a variety of protein and carbohydrate **adhesion molecules** on both the endothelial cell and the neutrophil, and it is regulated by messenger molecules released by cells in the injured area, including the endothelial cells. These messengers are collectively termed **chemoattractants** (also termed **chemotaxins** or chemotactic factors).

In the first stage, the neutrophil is loosely tethered to the endothelial cells via a particular class of adhesion molecules; this event is associated with rolling of the neutrophil along the vessel surface. In essence, this initial reversible event permits the neutrophil to be exposed to chemoattractants being released in the injured area. These chemoattractants act on the neutrophil to induce the rapid appearance of another class of adhesion molecules in its plasma membrane—molecules that bind tightly to their matching molecules in the endothelial cells. In the next stage, via still other adhesion molecules, a narrow projection of the neutrophil is inserted into the space between two endothelial cells, and the entire neutrophil squeezes through the endothelial wall and into the interstitial fluid. In this way, huge numbers of neutrophils migrate into the inflamed area and move toward the microbes.

Movement of leukocytes from the blood into the damaged area is not limited to neutrophils. Monocytes follow later, and once in the tissue they undergo anatomical and functional changes that transform them to macrophages. As we shall see later, in specific immune defenses lymphocytes undergo chemotaxis, as can basophils and eosinophils under certain conditions.

An important aspect of the multistep chemotaxis process is that is provides selectivity and flexibility for the migration of the various leukocyte types. Multiple adhesion molecules that are relatively distinct for the different leukocytes are controlled by different sets of chemoattractants. Particularly important in this regard are those cytokines that function as chemoattractants for distinct subsets of leukocytes. For example, one type of cytokine simulates the chemotaxis of neutrophils, whereas another stimulates that of eosinophils. Thus, subsets of leukocytes can be stimulated to enter particular tissues at designated times during an inflammatory response, depending on the type of invader and the cytokine response it induces. The various cytokines that have chemoattractant actions are collectively referred to as **chemokines,** some of which also stimulate other steps in inflammation.

Killing by Phagocytes The initial step in phagocytosis is contact between the surfaces of the phagocyte and microbe. One of the major triggers for phagocytosis during this contact is the interaction of phagocyte receptors with certain carbohydrates or lipids in the microbial cell walls. Contact is not itself always sufficient to trigger engulfment, however, particularly with those bacteria that are surrounded by a thick, gelatinous capsule. As we shall see, chemical factors produced by the body can bind the phagocyte tightly to

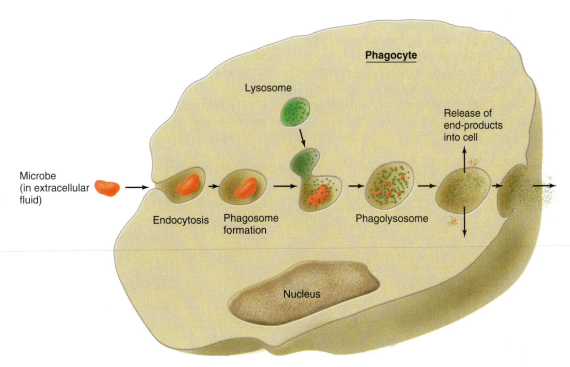

FIGURE 20–1

Phagocytosis and intracellular destruction of a microbe. After destruction has taken place in the phagolysosome, the end products are released to the outside of the cell by exocytosis or used by the cell for its own metabolism.

the microbe and markedly enhance phagocytosis. Any substance that does this is known as an **opsonin,** from the Greek word that means "to prepare for eating."

As the phagocyte engulfs the microbe (Figure 20–1), the internal, microbe-containing sac formed in this step is called a **phagosome.** A layer of plasma membrane separates the microbe from the phagocyte's cytosol. The phagosome membrane then makes contact with one of the phagocyte's lysosomes, which is filled with a variety of hydrolytic enzymes. The membranes of the phagosome and lysosome fuse, and the combined vesicles are now called the **phagolysosome.** Inside the phagolysosome, the microbe's macromolecules are broken down by the lysosomal enzymes. In addition, other enzymes in the phagolysosome membrane produce **nitric oxide** as well as **hydrogen peroxide** and other oxygen derivatives, all of which are extremely destructive to the microbe's macromolecules.

Such *intracellular* destruction is not the only way phagocytes can kill microbes. The phagocytes also release antimicrobial substances into the *extracellular* fluid, where these chemicals can destroy the microbes without prior phagocytosis. (As we shall discuss later, these chemicals can also damage normal tissue.)

Some of these substances (for example, nitric oxide) secreted into the extracellular fluid (Figure 20–2) also function as inflammatory mediators. Thus, posi-

tive feedback occurs such that when phagocytes enter the area and encounter microbes, inflammatory mediators, including chemokines, are released that bring in more phagocytes.

Complement The family of plasma proteins known as **complement** provides another means for *extracellular* killing of microbes—that is killing without prior phagocytosis. Certain of the complement proteins are always circulating in the blood in an inactive state. Upon activation (discussed below) of one of the group in response to infection or damage, there occurs a cascade in which this active protein activates a second complement protein, which activates a third, and so on. In this way, multiple active complement proteins are generated in the extracellular fluid of the infected area from inactive complement molecules that have entered from the blood. Since this system consists of at least 30 distinct proteins, it is extremely complex, and we shall identify the roles of only a few of the individual complement proteins.

Five of the active proteins generated in the complement cascade form a multiunit protein, the **membrane attack complex (MAC),** which embeds itself in the microbial plasma membrane and forms channels in the membrane, making it leaky. Water and salts enter the microbe, which disrupts the intracellular ionic environment and kills the microbe.

In addition to supplying a means for direct killing of microbes, the complement system serves other important functions in inflammation (Figure 20–3). Some of the activated complement molecules along the cascade cause, either directly or indirectly (by stimulating the release of other inflammatory mediators), vasodilation, increased microvessel permeability to protein, and chemotaxis. Also, one of the complement molecules—**C3b**—acts as an opsonin to attach the phagocyte to the microbe (Figure 20–4).

As we shall see later, antibodies, a class of proteins secreted by lymphocytes, are required to activate the very first complement protein (C1) in the full sequence known as the classical complement pathway,

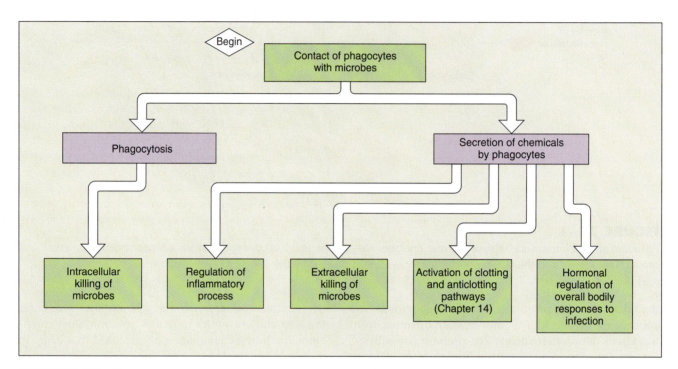

FIGURE 20–2

Role of phagocytosis in nonspecific immune defenses. Hormonal regulation of overall bodily responses to infection will be discussed later in this chapter.

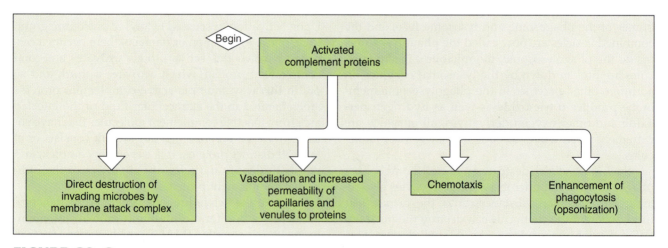

FIGURE 20–3

Functions of complement proteins. The effects on blood vessels and chemotaxis are exerted both directly by complement molecules and indirectly via other inflammatory mediators (for example, histamine) whose release the complement molecules stimulate.

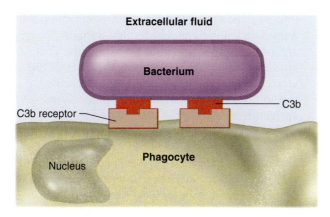

FIGURE 20–4

Function of complement C3b as an opsonin. One portion of C3b binds nonspecifically to carbohydrates on the surface of the bacterium, whereas another portion binds to specific receptor sites for C3b on the plasma membrane of the phagocyte. The structures are not drawn to scale.

but lymphocytes are not involved in *nonspecific* inflammation, our present topic. How, then, is the complement sequence initiated during nonspecific inflammation? The answer is that there is an **alternate complement pathway,** one that is not antibody-dependent and bypasses C1. The alternate pathway is initiated as the result of interactions between carbohydrates on the surface of the microbes and inactive complement molecules beyond C1. These interactions lead to formation of C3b, the opsonin described in the previous paragraph, and the activation of the subsequent complement molecules in the pathway. However, not all microbes have a surface conducive to initiating the alternate pathway.

Other Opsonins in Nonspecific Defenses In addition to complement C3b, other plasma proteins can bind nonspecifically to carbohydrates or lipids in the cell wall of microbes and facilitate opsonization. Many of these, for example **C-reactive protein,** are produced by the liver and are always found at some concentration in the plasma. Their production and plasma concentrations, however, are markedly increased during inflammation (this is discussed as part of the "acute phase response" in a subsequent section).

Tissue Repair The final stage of inflammation is tissue repair. Depending upon the tissue involved, multiplication of organ-specific cells by cell division may or may not occur during this stage. For example, liver cells multiply but skeletal muscle cells do not. In any case, fibroblasts (a type of connective-tissue cell) that reside in the area divide rapidly and begin to secrete large quantities of collagen, while blood-vessel cells proliferate in the process of angiogenesis (Chapter 14). All these events are brought about by chemical mediators, particularly a group of locally produced growth factors. Finally, remodeling occurs as the healing process winds down; the final repair may be imperfect, leaving a scar.

Interferons

Interferons are a family of cytokines that *nonspecifically* inhibit viral replication inside host cells. In response to infection by a virus, most cell types produce interferon and secrete it into the extracellular fluid. Interferon then binds to plasma-membrane receptors on the secreting cell and on other cells, whether they are infected or not (Figure 20–5). This binding triggers the

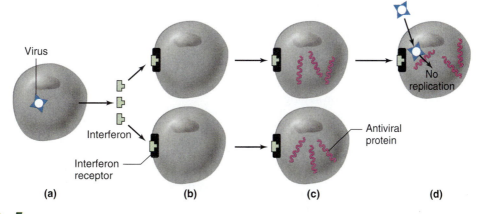

(a)　　　　(b)　　　　(c)　　　　(d)

FIGURE 20–5

Role of interferon in preventing viral replication. (a) Most cell types, when infected with viruses, secrete interferon, which enters the interstitial fluid and binds to interferon receptors on the secreting cells, themselves [autocrine function (not shown)], and (b) adjacent cells (paracrine function). In addition, some interferon enters the blood and binds to interferon receptors on far-removed cells (endocrine function). The binding of interferon to its receptors induces synthesis of proteins (c) that inhibit viral replication should viruses enter the cell (d).

synthesis of a variety of antiviral proteins by the cell. If the cell is already infected or eventually becomes infected, these proteins interfere with the ability of the viruses to replicate.

Interferon is not specific. Many kinds of viruses induce interferon synthesis, and interferon in turn can inhibit the multiplication of many kinds of viruses. We shall see in subsequent sections of this chapter that, in addition to these antiviral actions, interferons play other important roles in host defenses.

Specific Immune Defenses

Because of the complexity of specific immune defenses, it is useful to present a brief orientation before describing in more detail the various components of the response.

Overview

Lymphocytes are the essential cells in specific immune defenses. Unlike nonspecific defense mechanisms, lymphocytes must "recognize" the specific foreign matter to be attacked. Any foreign molecule that can trigger a specific immune response against itself or the cell bearing it is termed an **antigen.** (The more technically correct term "immunogen" is preferred by some immunologists.) Most antigens are either proteins or very large polysaccharides. The term "antigen" does not denote a specific structure in the way that anatomical terms like "microtubule" and "integral membrane protein" do. Rather it is a functional term; that is, any molecule, regardless of its location or function, that can induce a specific immune response against itself is by definition an antigen. It is the ability of lymphocytes to distinguish one antigen from another that confers specificity upon the immune responses in which they participate.

A common example of an antigen is ragweed pollen, the antigen that causes the specific immune response we know as hay fever. In other cases the antigen is part of the surface of a cell—microbe, virus-infected body cell, cancer cell, or transplanted cell—that appears in the body. The reason for saying "appears in" rather than "enters" is that, as we shall see, body cells that are virus-infected or cancerous produce the "foreign" molecules that act as antigens. These molecules are foreign in the sense that they are not present in normal cells.

A typical specific immune response can be divided into three stages: (1) the encounter and recognition of an antigen by lymphocytes, (2) lymphocyte activation, and (3) the attack launched by the activated lymphocytes.

1. During its development, each lymphocyte synthesizes a type of receptor that is able to bind to a specific antigen and inserts these receptors into its plasma membrane. If, at a later time, the lymphocyte ever encounters that antigen, the antigen becomes bound to the receptors. This binding is the physicochemical meaning of the word "recognize" in immunology. Accordingly, the ability of lymphocytes to distinguish one antigen from another is determined by the nature of their plasma-membrane receptors. *Each lymphocyte is specific for just one type of antigen,* and it is estimated that in a typical person the lymphocyte population expresses more than 100 million distinct antigen receptors. We do not mean to imply that every lymphocyte is different from every other one; in most cases a single type of antigen receptor may be expressed by a small number of lymphocytes, termed a clone. Thus, there are more than 100 million distinct small clones of lymphocytes in the body.

2. The binding of antigen to receptor is an absolute requirement for **lymphocyte activation** although other events may also be involved. Upon binding to an antigen, the lymphocyte undergoes a cell division, and the two resulting daughter cells then also divide (even though only one of them still has the antigen attached to it) and so on. In other words, the original binding of antigen by a single lymphocyte specific for that antigen triggers multiple cycles of cell divisions. As a result, many lymphocytes are formed that are identical to the one that started the cycles and can recognize the antigen; this is termed clonal expansion. After activation, two of the lymphocyte types—B cells and cytotoxic T cells—then function as "effector lymphocytes," which carry out the attack response. A third type of lymphocyte, called helper T cells, after activation secretes cytokines that enhance the activation and function of B cells and cytotoxic T cells.

3. The activated effector lymphocytes launch an attack against all antigens of the kind that initiated the immune response. Theoretically, it takes only one or two antigen molecules to *initiate* the specific immune response that will then result in an attack on all of the other antigens of that specific kind in the body. Activated B cells differentiate into plasma cells, which secrete antibodies into the blood, and these antibodies then recruit and guide other molecules and cells to perform the actual attack. In contrast, activated cytotoxic T cells directly attack and kill the cells bearing the antigens. Once the attack is successfully completed, the great majority of the B cells, plasma cells, helper T cells, and cytotoxic T cells

that participated in it die by apoptosis. The timely death of these cells is a homeostatic response that prevents the immune defense from becoming excessive. However, certain of the cells persist as so-called memory cells, which are ready to respond at some future time should the antigen reappear.

Lymphoid Organs and Lymphocyte Origins

Our first task is to describe the organs and tissues in which lymphocytes originate and come to reside. Then we describe the various types of lymphocytes alluded to in the overview and summarized in Table 20–1.

Lymphoid Organs   Like all leukocytes, lymphocytes circulate in the blood. At any moment, the great majority of lymphocytes are not actually in the blood, however, but in a group of organs and tissues collectively termed the **lymphoid organs.** These are subdivided into primary and secondary lymphoid organs.

The **primary lymphoid organs** are the bone marrow and thymus. These organs supply the secondary lymphoid organs with mature lymphocytes—that is, lymphocytes already programmed to perform their functions when activated by antigen. The bone marrow and thymus are not normally sites in which lymphocytes undergo activation during an immune response.

The **secondary lymphoid organs** are the lymph nodes, spleen, tonsils, and lymphocyte accumulations in the linings of the intestinal, respiratory, genital, and urinary tracts. It is in the secondary lymphoid organs that lymphocytes are activated to participate in specific immune responses.

We have stated that the bone marrow and thymus supply mature lymphocytes to the secondary lymphoid organs. Most of the lymphocytes in the secondary organs are not, however, cells that originated in the primary lymphoid organs. The explanation of this seeming paradox is that, once in the secondary organ, a mature lymphocyte coming from the bone marrow or thymus can undergo cell division to produce additional identical lymphocytes, which in turn undergo cell division and so on. In other words, all lymphocytes are *descended* from ancestors that matured in the bone marrow or thymus but may not themselves have arisen in those organs. As we mentioned earlier, all the progeny cells derived by cell division from a single lymphocyte constitute a lymphocyte clone.

(A distinction must be made between the "lymphoid organs" and the "lymphatic system," described in Chapter 14. The latter is a network of lymphatic vessels and the lymph nodes found along these vessels. Of all the lymphoid organs, only the lymph nodes also belong to the lymphatic system.)

There are no anatomical links, other than via the cardiovascular system, between the various lymphoid organs. Let us look briefly at these organs, excepting the bone marrow, which was described in Chapter 14.

The **thymus** lies in the upper part of the chest. Its size varies with age, being relatively large at birth and continuing to grow until puberty, when it gradually atrophies and is replaced by fatty tissue. Before its atrophy, the thymus consists mainly of mature lymphocytes that will eventually migrate via the blood to the secondary lymphoid organs. It also contains endocrine cells that secrete a group of hormones, collectively termed **thymopoietin,** that exert a still poorly understood regulatory effect on lymphocytes of thymic origin and descent.

Recall from Chapter 14 that the fluid flowing along the lymphatic vessels is called lymph, which is interstitial fluid that has entered the lymphatic capillaries and is being routed to the large lymphatic vessels that drain into systemic veins. During this trip, the lymph flows through **lymph nodes** scattered along the vessels. Lymph, therefore, is the route by which lymphocytes in the lymph nodes encounter the antigens that activate them. Each node is a honeycomb of lymph-filled sinuses (Figure 20–6) with large clusters of lymphocytes (the lymphatic nodules) between the sinuses. There are also many macrophages and, as is the case with other secondary lymphoid organs, macrophage-like cells.

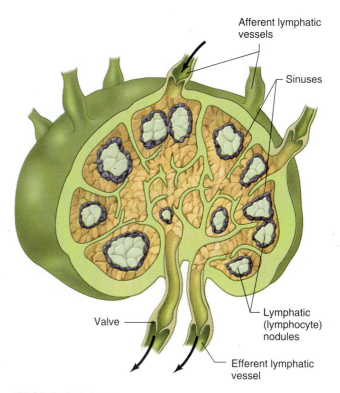

FIGURE 20–6
Anatomy of a lymph node.

The **spleen** is the largest of the secondary lymphoid organs and lies in the left part of the abdominal cavity between the stomach and the diaphragm. The spleen is to the circulating blood what the lymph nodes are to the lymph. Blood percolates through the vascular meshwork of the spleen's interior, where large collections of lymphocytes, macrophages, and macrophage-like cells are found. The macrophages of the spleen, in addition to interacting with lymphocytes, also phagocytize aging or dead erythrocytes.

The **tonsils** are a group of small, rounded organs in the pharynx. They are filled with lymphocytes, macrophages, and macrophage-like cells, and they have openings ("crypts") to the surface of the pharynx. Their lymphocytes respond to microbes that arrive by way of ingested food as well as inspired air. Similarly, the lymphocytes in the linings of the various tracts exposed to the external environment respond to infectious agents that penetrate the linings from the lumen of the tract.

Finally, we must describe the source of the lymphocytes in blood. Some are cells on their way from the bone marrow or thymus to the secondary lymphoid organs, but the vast majority are cells that are participating in lymphocyte traffic *between* the secondary lymphoid organs, blood, lymph, and all the tissues of the body. Lymphocytes from all the secondary lymphoid organs constantly enter the lymphatic vessels draining them (all lymphoid organs, not just lymph nodes, are drained by lymphatic vessels) and are carried to the blood (Chapter 14). Simultaneously, some blood lymphocytes are pushing through the endothelium of venules all over the body to enter the interstitial fluid. From there, they move into lymphatic capillaries and along the lymphatic vessels to lymph nodes. They may then leave the lymphatic vessels to take up residence in the node.

This recirculation is going on all the time, not just during an infection, although the migration of lymphocytes into an inflamed area is greatly increased by the chemotaxis process described earlier. Lymphocyte trafficking greatly increases the likelihood that any given lymphocyte will encounter the antigen it is specifically programmed to recognize. (In contrast to the lymphocytes, polymorphonuclear granulocytes and monocytes do not recirculate; once they leave the bloodstream to enter a tissue they remain there or die.)

Lymphocyte Origins The multiple populations and subpopulations of lymphocytes are summarized in Table 20–1. **B lymphocytes,** or simply **B cells,** mature in the bone marrow and then are carried by the blood to the secondary lymphoid organs (Figure 20–7). This overall process of maturation and migration continues throughout a person's life. All genera-

tions of lymphocytes that subsequently arise from these cells by cell division in the secondary lymphoid organs will be identical to the parent cells; that is, they will also be B cells.

In contrast to the B cells, other lymphocytes leave the bone marrow in an *immature* state during fetal and early neonatal life. They are carried to the thymus and mature there before moving to the secondary lymphoid organs. These cells are called **T lymphocytes** or **T cells.** Like B cells, T cells also undergo cell division in secondary lymphoid organs, the offspring being identical to the original T cells.

In addition to the B and T cells, there is another distinct population of lymphocytes, **natural killer (NK) cells.** These cells arise in the bone marrow, but their precursors and life history are still unclear. As we shall see, NK cells, unlike B and T cells, do not manifest specificity for antigens.

Functions of B Cells and T Cells

B cells, upon activation, differentiate into plasma cells, which secrete **antibodies,** proteins that travel all over the body to reach antigens identical to those that stimulated their production. The antibodies combine with these antigens and guide an attack (to be described later) that eliminates the antigens or the cells bearing them.

Antibody-mediated responses are also called humoral responses, the adjective "humoral" denoting communication by way of soluble chemical messengers (in this case, antibodies in the blood). Antibody-mediated responses have an extremely wide diversity of targets and are the major defense against bacteria, viruses, and other microbes in the extracellular fluid, and against toxic molecules (toxins).

T cells constitute a family that has two major functional subsets, termed **cytotoxic T cells** and **helper T cells.** There may also be a third subset, called suppressor T cells, which have been hypothesized to inhibit the function of both B cells and cytotoxic T cells; however, the significance of these cells and even their existence are in doubt at present, and we shall have nothing more to say about them in this chapter.

Another way to categorize T cells is not by function but rather by the presence of certain proteins, called CD4 and CD8, in their plasma membranes. Cytotoxic T cells have CD8 and so are also commonly called CD8+ cells; helper T cells have CD4 and so are also commonly called CD4+ cells.

Cytotoxic T cells are "attack" cells. They travel to the location of their target, bind to them via antigen on these targets and, following activation, directly kill them, via secreted chemicals, without the intermediation of antibodies. Responses mediated by cytotoxic T cells are directed against the body's own cells that

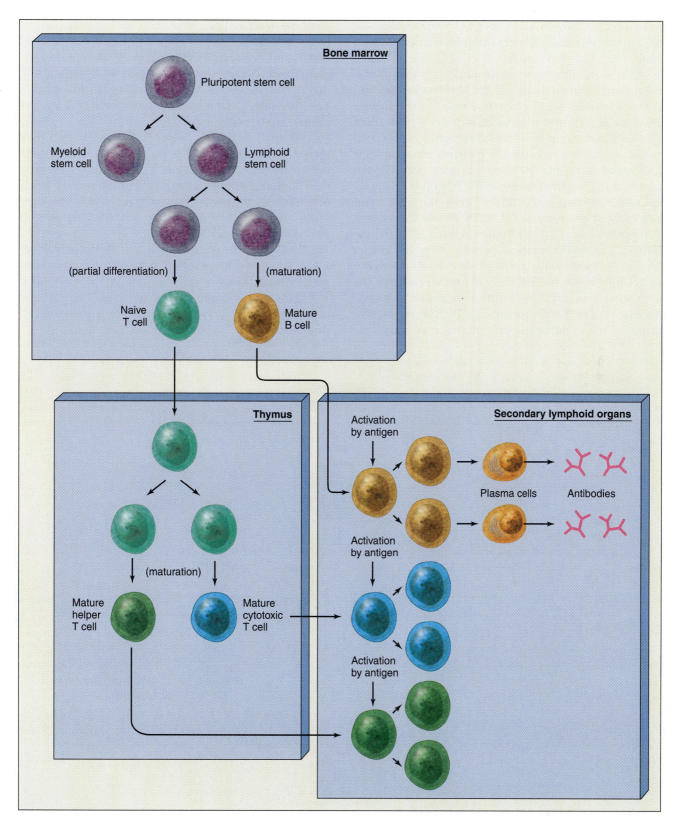

FIGURE 20–7

Derivation of B cells and T cells. NK cells are not shown because their origins in the bone marrow and their transformations, if any, after leaving it are still not clear.

have become cancerous or infected with viruses (or certain bacteria and parasites that, like viruses, take up residence inside host cells).

It is worth emphasizing the important geographical difference in antibody-mediated responses and responses mediated by cytotoxic T cells. In the former case, the B cells and plasma cells derived from them remain in whatever location the recognition and activation steps occurred, and the plasma cells send their antibodies forth, via the blood, to seek out antigens identical to those that triggered the response. In the responses mediated by cytotoxic T cells, the cells themselves must enter the blood and seek out the targets.

We have now assigned roles to the B cells and cytotoxic T cells. What role is performed by the helper T cells? As their name implies, these cells do not themselves function as "attack" cells but rather facilitate the activation and function of both B cells and cytotoxic T cells. Helper T cells go through the usual first two stages of the immune response in that they must first combine with antigen and then undergo activation. Once activated, they secrete cytokines that act on B cells and cytotoxic T cells that have also bound antigen. This is a very important point. With only a few exceptions, B cells and cytotoxic T cells cannot function adequately unless they are stimulated by cytokines from helper T cells.

We shall deal with helper T cells as though they were a homogeneous cell population, but in fact, there are several subtypes of helper T cells, distinguished by the different batteries of cytokines they secrete when activated. By means of these different cytokines, they "help" different sets of effector lymphocytes, as well as macrophages and NK cells.

Finally, it should be noted that some of the cytokines secreted by helper T cells also act as inflammatory mediators.

Figure 20–8 summarizes the basic interactions among B, cytotoxic T, and helper T cells presented in this section.

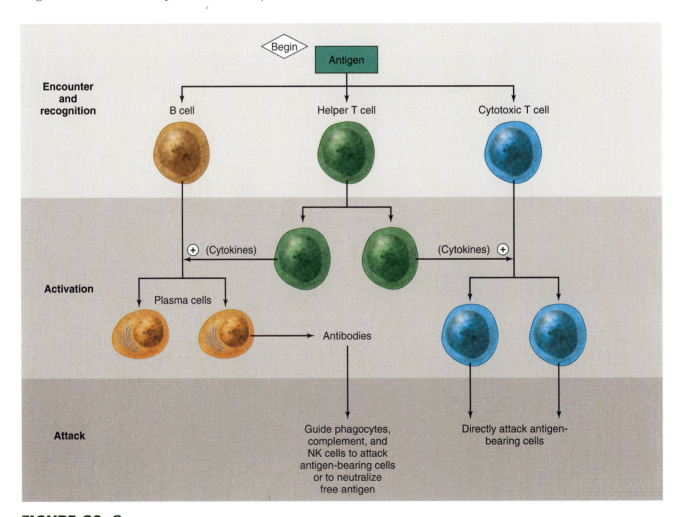

FIGURE 20–8

Summary of roles of B, cytotoxic T, and helper T cells in immune responses. Events of the attack phase are described in later sections.

Lymphocyte Receptors

To repeat, the ability of lymphocytes to distinguish one antigen from another is determined by the lymphocytes' receptors.

B-Cell Receptors   As noted earlier, after activation B cells proliferate and differentiate into plasma cells, which secrete antibodies. The plasma cells derived from a particular B cell can secrete only one particular antibody. Each B cell always displays on its plasma membrane copies of the particular antibody its plasma cell progeny are able to produce. This surface protein (glycoprotein, to be more accurate) acts as the receptor for the antigen specific to it.

B-cell receptors and antibodies constitute the family of proteins known as **immunoglobulins.** (The receptors themselves, even though they are identical to the antibodies to be secreted by the plasma cell derived from the activated B cell, are technically not antibodies since only *secreted* immunoglobulins are termed antibodies.) Each immunoglobulin molecule is composed of four interlinked polypeptide chains (Figure 20–9). The two long chains are called heavy chains, and the two short ones, light chains. There are five major classes of immunoglobulins, determined by the amino acid sequences in the heavy chains. The classes are designated by the letters A, D, E, G, and M following the symbol Ig for immunoglobulin; thus IgA, IgD, and so on.

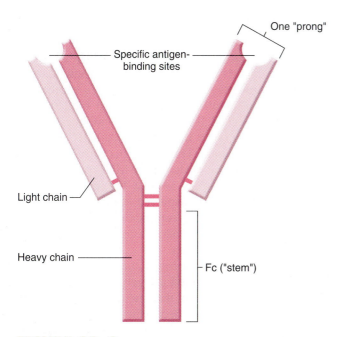

One "prong"

Specific antigen-binding sites

Light chain

Heavy chain

Fc ("stem")

FIGURE 20–9

Immunoglobulin structure. The Fc portions are the same for all immunoglobulins of a particular class. Each "prong" contains a single antigen binding site. The links between chains represent disulfide bonds.

As illustrated in Figure 20–9, immunoglobulins have a "stem," called the **Fc** portion and comprising the lower half of the two heavy chains, and two "prongs," each containing one **antigen binding site**—the amino acid sequences that bind antigen. The amino acid sequences of the Fc portion are identical for all immunoglobulins of a single class (IgA, IgD, and so on). In contrast, the amino acid sequences of the antigen binding sites vary from immunoglobulin to immunoglobulin in a given class. Thus, each of the five classes of antibody contains thousands, or millions, of unique immunoglobulins, each capable of combining with only one specific antigen (or, in some cases, several antigens whose structures are very similar). The interaction between an antigen binding site of an immunoglobulin and an antigen is analogous to the lock-and-key interactions that apply generally to the binding of ligands by proteins.

One more point should be mentioned: B-cell receptors can bind antigen whether the antigen is a free molecule dissolved in the extracellular fluid or is present on the surface of a foreign cell such as a microbe. In the latter case, the B cell becomes linked to the foreign cell via the bonds between the B-cell receptor and the surface antigen.

To summarize thus far, any given B cell or clone of identical B cells possesses unique immunoglobulin receptors—that is, receptors with unique antigen binding sites. Thus, the body has armed itself with millions of small clones of different B cells in order to ensure that receptors will exist that are specific for the vast number of different antigens the organism *might* encounter during its lifetime. The particular immunoglobulin that any given B cell displays as a receptor on its plasma membrane (and that its plasma cell progeny will secrete as antibodies) was determined during the cell's maturation in the bone marrow.

This raises a very interesting question: In the human genome there are only about 200 genes that code for immunoglobulins; how, then, can the body produce immunoglobulins having millions of different antigen binding sites, given that each immunoglobulin requires coding by a distinct gene? This diversity arises as the result of a genetic process unique to developing lymphocytes because only these cells possess the enzymes required to catalyze the process. The DNA in each of the genes that code for immunoglobulin antigen binding sites is cut into small segments, randomly rearranged along the gene, and then rejoined to form new DNA molecules. This cutting and rejoining varies from B cell to B cell, thus resulting in great diversity of the genes coding for the immunoglobulins of all the B cells taken together.

T-Cell Receptors   T cells, whether helper or cytotoxic, do not produce immunoglobulins, and so T-cell receptors for antigen are not immunoglobulins. Rather they are two-chained proteins that, like immunoglobulins, have specific regions that differ from one T-cell clone to another. As in B-cell development, multiple DNA rearrangements occur during T-cell maturation, leading to millions of distinct T-cell clones—distinct in that the cells of any given clone possess receptors of a single specificity. For T cells, this maturation occurs during their residence in the thymus.

In addition to their general structural differences, the B- and T-cell receptors differ in a much more important way: *The T-cell receptor cannot combine with antigen unless the antigen is first complexed with certain of the body's own plasma-membrane proteins.* The T-cell receptor then combines with the entire complex of antigen and body (self) protein.

The self plasma-membrane proteins that must be complexed with the antigen in order for T-cell recognition to occur constitute a group of proteins coded for by genes found on a single chromosome and known collectively as the **major histocompatibility complex (MHC).** The proteins are therefore called **MHC proteins.** Since no two persons other than identical twins have the same sets of MHC genes, no two individuals have the same MHC proteins on the plasma membranes of their cells. MHC proteins are, in essence, cellular "identity tags"—that is, genetic markers of biological individuality.

The MHC proteins are often termed "restriction elements" since, as we have seen, the ability of a T cell's receptor to recognize an antigen is restricted to situations in which the antigen is first complexed with an MHC protein. There are two classes of MHC proteins: I and II. **Class I MHC proteins** are found on the surface of virtually all cells of a person's body, excepting erythrocytes. **Class II MHC proteins** are found only on the surface of macrophages, B cells, and macrophage-like cells.

Now for another important point: The different subsets of T cells do not all have the same MHC requirements (Table 20–4): Cytotoxic T cells require antigen to be associated with class I MHC proteins, whereas helper T cells require class II MHC proteins. (One reason for this difference in requirements stems from the presence, as described earlier, of CD4 proteins on the helper T cells and CD8 proteins on the cytotoxic T cells; CD4 binds to class II MHC proteins, whereas CD8 binds to class 1 MHC proteins.)

At this point you might well ask: How do antigens, which are foreign, end up on the surface of the body's own cells complexed with MHC proteins? The answer is provided by the process known as **antigen presentation,** to which we now turn.

TABLE 20–4 MHC Restriction of the Lymphocyte Receptors

Cell Type	MHC Restriction
B	Do not interact with MHC proteins
Helper T	Class II, found only on macrophages, macrophage-like cells, and B cells
Cytotoxic T	Class I, found on all nucleated cells of the body
NK	Interaction with MHC proteins not required for activation

Antigen Presentation to T Cells

To repeat, T cells can bind antigen only when the antigen appears on the plasma membrane of a host cell complexed with the cell's MHC proteins. Cells bearing these complexes, therefore, function as **antigen-presenting cells (APCs).**

Presentation to Helper T Cells   Since helper T cells require class II MHC proteins and since these proteins are found only on macrophages, B cells, and macrophage-like cells, only these cells can function as APCs for helper T cells.

The function of the macrophage (or macrophage-like cell) as an APC for helper T cells is easier to visualize (Figure 20–10a) since the macrophage forms, in essence, a link between nonspecific and specific immune defenses. After a microbe or noncellular antigen has been phagocytized by a macrophage in a *nonspecific* response, it is partially broken down into smaller peptide fragments by the macrophage's proteolytic enzymes. The resulting fragments then bind (within endosomes) to class II MHC proteins synthesized by the macrophage (after their synthesis, MHC proteins pass into endosomes). The fragments actually fit into a deep groove in the center of the MHC proteins. The fragment-MHC complex is then transported to the cell surface where it is displayed in the plasma membrane. It is to this entire complex on the cell surface of the macrophage (or macrophage-like cell) that a specific helper T cell binds.

Note that it is not the intact antigen but rather the peptide fragments, termed antigenic determinants or **epitopes,** of the antigen that are complexed to the MHC proteins and presented to the T cell. Despite this, it is customary to refer to "antigen" presentation rather than "epitope" presentation.

How B cells process antigen and present it to helper T cells is essentially the same as the story we just described for macrophages (Figure 20–10b). It must be emphasized that the ability of B cells to

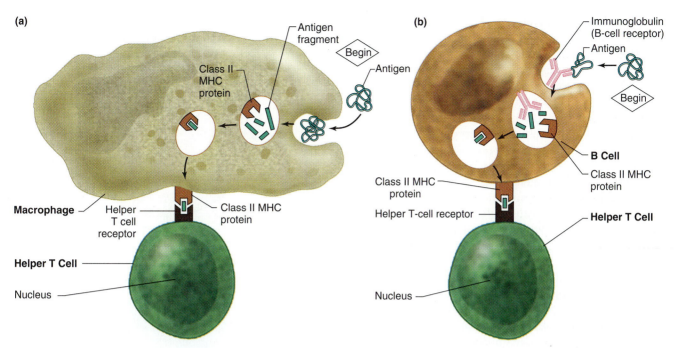

FIGURE 20–10

Sequence of events by which antigen is processed and presented to a helper T cell by (a) a macrophage or (b) a B cell. In both cases, begin the figure with the antigen in the extracellular fluid.

Adapted from Gray, Sette, and Buus.

present antigen to helper T cells is a *second* function of B cells in response to antigenic stimulation, the other being the differentiation of the B cells into antibody-secreting plasma cells.

The binding between helper T-cell receptor and antigen bound to class II MHC proteins on an APC is the essential *antigen-specific* event in helper T-cell activation. However, this binding by itself will not result in T-cell activation. In addition, *nonspecific* interactions occur between other (nonantigenic) pairs of proteins on the surfaces of the attached helper T cell and APC, and these provide a necessary **costimulus** for T-cell activation (Figure 20–11).

Finally, the antigenic binding of the APC to the T cell, along with the costimulus, causes the APC to secrete large amounts of the cytokines **interleukin 1 (IL-1)** and **tumor necrosis factor (TNF),** which act as paracrine agents on the attached helper T cell to provide yet another important stimulus for activation.

Thus, the APC participates in activation of a helper T cell in three ways: (1) antigen presentation, (2) provision of a costimulus in the form of a matching non-antigenic plasma-membrane protein, and (3) secretion of IL-1 and TNF (Figure 20–11).

The activated helper T cell itself now secretes various cytokines that have both autocrine effects on the helper T cell and paracrine effects on adjacent B cells and any nearby cytotoxic T cells, NK cells, and still other cell types; we will pick up these stories in later sections.

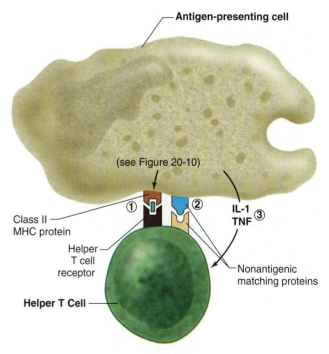

FIGURE 20–11

Three events are required for activation of helper T cells: ① presentation of the antigen bound to a class II MHC protein on an antigen-presenting cell (APC); ② the binding of matching nonantigenic proteins in the plasma membranes of the APC and the helper T cell; and ③ secretion by the APC of the cytokines interleukin 1 (IL-1) and tumor necrosis factor (TNF), which act on the helper T cell.

Presentation to Cytotoxic T Cells Because class I MHC proteins are synthesized by virtually all nucleated cells, any such cell can act as an APC for a cytotoxic T cell. This distinction helps explain the major function of cytotoxic T cells—destruction of *any* of the body's own cells that have become cancerous or infected with viruses. The key point is that the antigens that complex with class I MHC proteins arise *within* body cells. They are "endogenous" antigens, foreign proteins synthesized by a body cell.

How do such antigens arise? In the case of viruses, once a virus has taken up residence inside a host cell, the viral nucleic acid causes the host cell to manufacture viral proteins, which of course are foreign proteins. A cancerous cell, as described in Chapter 4, has had one or more of its genes altered by chemicals, radiation, or other factors; the altered genes, called **oncogenes,** code for proteins that are not normally found in the body. Such proteins act as antigens.

In both virus-infected cells and cancerous cells, some of the endogenously produced antigenic proteins are hydrolyzed by cytosolic enzymes (in proteosomes; Chapter 5) into peptide fragments, which are transported into the endoplasmic reticulum. There they are complexed with the host cell's class I MHC proteins and then shuttled by exocytosis to the plasma-membrane surface, where a cytotoxic T cell specific for the complex can bind to it (Figure 20–12).

(It is probable that, at least in some cases, full activation of cytotoxic T cells, like helper T cells, requires not only the antigen-specific binding just described but a costimulus provided by nonantigenic complementary proteins in the plasma membrane of the cytotoxic T cell and its target cell.)

NK Cells

As noted earlier, NK cells constitute a distinct class of lymphocytes. They have several functional similarities to cytotoxic T cells: (1) Their major targets are virus-infected cells and cancer cells; and (2) they attack and kill these target cells directly, after binding to them. However, unlike cytotoxic T cells, NK cells are not antigen specific; that is, each NK cell is able to attack virus-infected cells or cancer cells without any recognition of a specific antigen on the part of the NK cell. They have neither T-cell receptors nor the immunoglobulin receptors of B cells, and the exact nature of the NK-cell surface receptors that permits the cells to identify their targets is unknown (except for one case presented later). MHC proteins are not involved in the activation of NK cells.

It has been suggested that NK cells are lymphocytic "Minutemen," which can be mobilized quickly and go into action *nonspecifically* without the period of exposure and activation required by B cells and cytotoxic T cells.

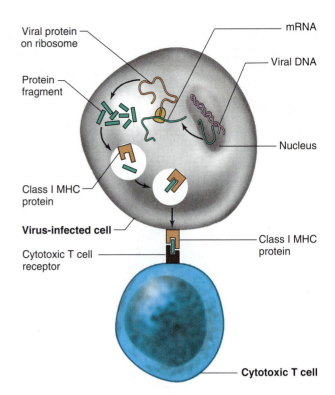

FIGURE 20–12

Processing and presentation of viral antigen to a cytotoxic T cell by an infected cell. Begin this figure with the viral DNA in the cell's nucleus. The viral DNA induces the production by the infected cell of viral protein, which is then hydrolyzed (by proteosomes) and the fragments complexed to the cell's class I MHC proteins in the endoplasmic reticulum. These complexes are then shuttled to the plasma membrane.

Adapted from Gray, Sette, and Buus.

Why then do we deal with them in the context of *specific* immune responses? Because, as will be described subsequently, their participation in an immune response is greatly enhanced either by certain antibodies or by cytokines secreted by helper T cells activated during specific immune responses.

Development of Immune Tolerance

Our basic framework for understanding specific immune responses requires consideration of one more crucial question: How does the body develop what is called **immune tolerance**—lack of immune responsiveness to *self-components?* This may seem a strange question given our definition of an antigen as a *foreign* molecule that can generate an immune response. In essence, however, we are now asking how it is that the body "knows" that its own molecules, particularly proteins, are not foreign but are "self" molecules.

Recall that the huge diversity of lymphocyte receptors is ultimately the result of multiple random DNA cutting/recombination processes. It is virtually

TABLE 20–5 Summary of Events in Antibody-Mediated Immunity Against Bacteria

1. In secondary lymphoid organs, bacterial antigen binds to specific receptors on the plasma membrane of B cells.

2. Simultaneously, antigen-presenting cells (APCs), for example, macrophages, (a) present to helper T cells processed antigen complexed to MHC class II proteins on the APCs, (b) provide a costimulus in the form of another membrane protein, and (c) secrete IL-1 and TNF, which act on the helper T cells.

3. In response, the helper T cells secrete IL-2, which stimulates the helper T cells themselves to proliferate and secrete IL-2 and other cytokines. These activate antigen-bound B cells to proliferate and differentiate into plasma cells. Some of the B cells differentiate into memory cells rather than plasma cells.

4. The plasma cells secrete antibodies specific for the antigen that initiated the response, and the antibodies circulate all over the body via the blood.

5. These antibodies combine with antigen on the surface of the bacteria anywhere in the body.

6. Presence of antibody bound to antigen facilitates phagocytosis of the bacteria by neutrophils and macrophages. It also activates the complement system, which further enhances phagocytosis and can directly kill the bacteria by the membrane attack complex. It may also induce antibody-dependent cellular cytotoxicity mediated by NK cells that bind to the antibody's Fc portion.

certain, therefore, that in each person clones of lymphocytes would have emerged with receptors that could bind to that person's own proteins. The existence and functioning of such lymphocytes would, of course, be disastrous because such binding would launch an immune attack against the cells expressing these proteins. There are at least two mechanisms, termed clonal deletion and clonal inactivation, that explain why *normally* there are no active lymphocytes that respond to self components.

First, during fetal and early postnatal life, T cells are exposed to a potpourri of self proteins in the thymus. Those T cells with receptors capable of binding self proteins are destroyed by apoptosis (Chapter 7). This process is termed **clonal deletion.** The second process, termed **clonal inactivation,** occurs not in the thymus but in the periphery and causes potentially self-reacting T cells to become nonresponsive.

What are the mechanisms of clonal deletion and inactivation during fetal and early postnatal life? Recall that full activation of a helper T cell requires not only an antigen-specific stimulus but a nonspecific costimulus (interaction between complementary nonantigenic proteins on the APC and the T cell). If this costimulus is *not* provided, the helper T cell not only fails to become activated by antigen but dies or becomes inactivated forever. This is the case during early life, although what accounts for the costimulus not being delivered is presently unclear. B cells can also undergo clonal deletion and inactivation.

This completes our framework for understanding specific immune defenses. The next two sections utilize this framework in presenting typical responses from beginning to end, fleshing out the interactions between lymphocytes and describing the attack mechanisms used by the various pathways.

Antibody-Mediated Immune Responses: Defenses against Bacteria, Extracellular Viruses, and Toxins

A classical antibody-mediated response is one that results in the destruction of bacteria. The sequence of events, which is quite similar to the response to a virus in the extracellular fluid, is summarized in Table 20–5 and Figure 20–13.

Antigen Recognition and Lymphocyte Activation

This process starts the same way as for nonspecific responses, with the bacteria penetrating one of the body's linings and entering the interstitial fluid. The bacteria then enter the lymphatic system and/or bloodstream and are carried to the lymph nodes and/or the spleen, respectively. There a B cell specific for an antigen on the bacterial surface binds to the antigen via the B-cell's plasma membrane immunoglobulin receptor.

In a few cases (notably bacteria with cell-wall polysaccharide capsules), this binding is all that is needed to trigger B-cell activation. For the great majority of antigens, however, antigen binding is not enough, and signals in the form of cytokines released into the interstitial fluid by helper T cells near the antigen-bound B cells are also needed.

For helper T cells to secrete cytokines, they must bind to a complex of antigen and class II MHC protein on an APC. Let us assume that in this case the APC is a macrophage that has phagocytized one of the bacteria, hydrolyzed its proteins into peptide fragments, complexed them with class II MHC proteins, and displayed the complexes on its surface. A helper T cell specific for the complex then binds to it, beginning the activation of the helper T cell. Moreover, the macrophage helps this activation process in two other ways: It provides a costimulus via nonantigenic plasma-membrane proteins, and it secretes IL-1 and TNF.

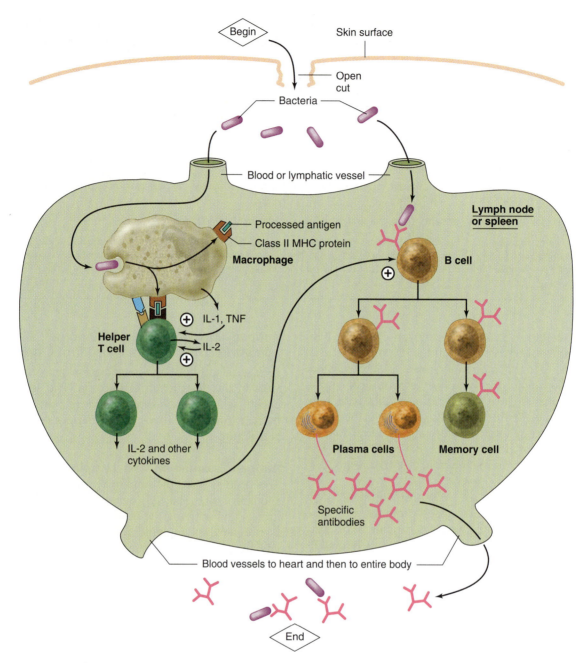

FIGURE 20–13

Summary of events by which a bacterial infection leads to antibody synthesis in peripheral lymphoid organs. The secreted antibodies travel by the blood to the site of infection where they bind to bacteria of the type that induced the response. The attack triggered by antibody's binding to bacteria is described in the text. [As illustrated in Figure 20–10b, an antigen-bound B cell, rather than a macrophage, as shown in this figure, can function as the antigen-presenting cell to the helper T cell. Also for clarity, the intracellular processing of the antigen by the macrophage (Figure 20–10a) is not shown in this figure.]

So far this has all been review. Now we begin new material. IL-1 and TNF stimulate the helper T cell to secrete another cytokine named **interleukin 2 (IL-2),** and to express the receptor for IL-2. This messenger, acting as an autocrine agent, then provides a proliferative stimulus to the activated helper T cell (Figure 20–13). The cell divides, beginning the mitotic cycles that lead to formation of a clone of activated helper T cells, and these cells then release not only IL-2 but other cytokines as well.

Certain of these cytokines provide the additional signals required to activate nearby antigen-bound B cells to proliferate and differentiate into plasma cells, which then secrete antibodies.

Thus, as shown in Figure 20–13, we are dealing with a series of protein messengers interconnecting the various cell types, the helper T cells serving as the central coordinator. The macrophage releases IL-1 and TNF, which act on the helper T cell to stimulate release of IL-2, which stimulates the helper T cell to multiply, the activated progeny then releasing still other cytokines that help activate antigen-bound B cells.

Some of the B-cell progeny, however, do not differentiate into plasma cells but instead into **memory cells,** whose characteristics permit them to respond more rapidly and vigorously should the antigen reappear at a future time (Figure 20–13).

The example we have been using employed a macrophage as the APC to helper T cells. Recall, however, that this role can also be served by B cells (see Figure 20–10). The binding of the helper T cell to the antigen-bound B cell ensures maximal stimulation of the B cell by the cytokines secreted by that helper T cell and any of its progeny that remain nearby.

Antibody Secretion After their differentiation from B cells, plasma cells produce thousands of antibody molecules per second before they die in a day or so. We mentioned earlier that there are five major classes of antibodies. The most abundant are the **IgG** antibodies, commonly called **gamma globulin,** and **IgM** antibodies; these two groups together provide the bulk of specific immunity against bacteria and viruses in the extracellular fluid. **IgE** antibodies participate in defenses against multicellular parasites and also mediate allergic responses. **IgA** antibodies are secreted by plasma cells in the linings of the gastrointestinal, respiratory, and genitourinary tracts; these antibodies generally act locally in the linings or on their surfaces. They are also secreted by the mammary glands and hence are the major antibodies in milk. The functions of **IgD** are still unclear.

In the kind of infection we've been describing, the secreted antibodies—mostly IgG and IgM—enter the blood, which carries them from the secondary lymphoid organ in which recognition and activation occurred to all tissues and organs of the body. At sites of infection, the antibodies leave the blood (recall that nonspecific inflammation had already made capillaries and venules leaky at these sites) and combine with the type of bacterial surface antigen that initiated the immune response (Figure 20–13). These antibodies then direct the attack (see below) against the bacteria to which they are now bound.

Thus, immunoglobulins play two distinct roles in immune responses: (1) During the initial recognition step, those on the surface of B cells bind to antigen brought to them; and (2) those secreted by the plasma cells (antibodies) seek out and bind to bacteria bearing the same antigens, "marking" them as the targets to be attacked.

The Attack: Effects of Antibodies The antibodies bound to antigen on the microbial surface do not directly kill the microbe but instead link up the microbe physically to the actual killing mechanisms—phagocytes (neutrophils and macrophages), complement, or NK cells. This linkage not only triggers the attack mechanism but ensures that the killing effects are restricted to the microbe. This protects adjacent normal structures from toxic effects of the chemicals employed by the killing mechanisms.

Direct enhancement of phagocytosis. Antibodies can act directly as opsonins. The mechanism is analogous to that for complement C3b (see Figure 20–4) in that the antibody links the phagocyte to the antigen. As shown in Figure 20–14, the phagocyte has membrane receptors that bind to the Fc portion of antibodies. This linkage promotes attachment of the antigen to the phagocyte and the triggering of phagocytosis.

Activation of the complement system. As described earlier in this chapter, the plasma complement system is activated in *nonspecific* inflammatory responses via the alternate complement pathway. In contrast, in *specific* immune responses, the presence of antibody of the IgG or IgM class bound to antigen activates the **classical complement pathway.** The first molecule in this pathway, C1, binds to the Fc portion of an antibody that has combined with antigen (Figure 20–15). This results in activation of the enzymatic portions of C1, thereby initiating the entire classical pathway. The end product of this cascade, the membrane attack complex (MAC), can kill the cells to which the antibody is bound by making their membranes leaky. In addition, as we saw in Figure 20–4, another activated complement molecule (C3b) functions as an opsonin to enhance phagocytosis of the microbe by neutrophils and macrophages (Figure 20–15). Thus, antibodies enhance phagocytosis both directly (Figure 20–14) and via activation of complement C3b.

It is important to note that C1 binds not to the unique antigen binding sites in the antibody's prongs but rather to complement binding sites in the Fc portion. Since the latter are the same in virtually all antibodies of the IgG and IgM classes, the complement molecule will bind to *any* antigen-bound antibodies belonging to these classes. In other words, there is only one set of complement molecules, and once activated, they do essentially the same thing regardless of the specific identity of the invader.

Antibody-dependent cellular cytotoxicity. We have seen that both a particular complement molecule (C1) and a phagocyte can bind nonspecifically to the Fc portion of antibodies bound to antigen. NK cells can also do this (just substitute a NK cell for the phagocyte in Figure 20–14). Thus, antibodies can link target cells to NK cells, which then kill the targets directly by secreting toxic chemicals. This is termed **antibody-dependent cellular cytotoxicity (ADCC)** because the killing (cytotoxicity) is carried out by cells (NK cells) but the process depends upon the presence of antibody. Note that it is the antibodies that confer specificity upon ADDC, just as they do on antibody-dependent phagocytosis and complement activation. (This mechanism for bringing NK cells into play is the one exception, mentioned earlier, to the generalization that the mechanism by which NK cells identify their targets is unclear.)

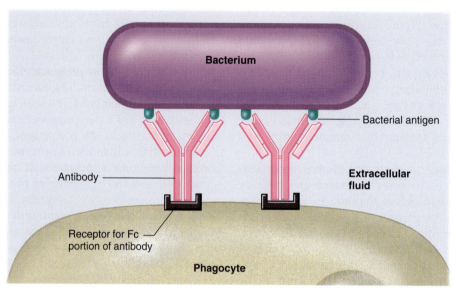

FIGURE 20–14
Direct enhancement of phagocytosis by antibody. The antibody links the phagocyte to the bacterium. Compare this mechanism of opsonization to that mediated by complement C3b (Figure 20–4).

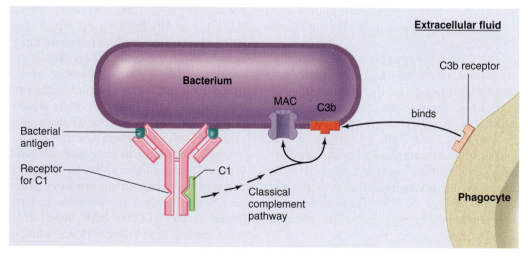

FIGURE 20–15
Activation of classical complement pathway by binding of antibody to bacterial antigen. C1 is activated by its binding to receptor for it on the Fc portion of the antibody. The membrane attack complex (MAC) is then generated, along with C3b, which acts as an opsonin by binding the bacteria to a phagocyte.

Direct neutralization of bacterial toxins and viruses.
Toxins secreted by bacteria into the extracellular fluid
can act as antigens to induce antibody production. The
antibodies then combine with the free toxins, thus preventing interaction of the toxin with susceptible cells.
Since each antibody has two binding sites for combination with antigen, clumplike chains of antibody-
antigen complexes are formed, and these clumps are
then phagocytized.

A similar binding process is the major antibody-
mediated mechanism for eliminating viruses in the extracellular fluid. Certain of the viral surface proteins
serve as antigens, and the antibodies produced against
them combine with them, preventing attachment of the
virus to plasma membranes of potential host cells. This
prevents the virus from entering a cell. As with bacterial toxins, chains of antibody-virus complexes are
formed and can be phagocytized.

Active and Passive Humoral Immunity We have
been discussing antibody formation without regard
to the course of events in time. The response of the
antibody-producing machinery to invasion by a foreign
antigen varies enormously, depending upon whether
the machinery has previously been exposed to that
antigen. Antibody production responds slowly over
several days to the first contact with an antigen, but
any subsequent infection by the same invader elicits
an immediate and marked outpouring of additional
specific antibodies (Figure 20–16). This response, which
is mediated by the memory B cells described earlier,
confers a greatly enhanced resistance toward subsequent infection with that particular microorganism.

Resistance built up as a result of the body's contact
with microorganisms and their toxins or other antigenic components is known as **active immunity.**

Until this century, the only way to develop active
immunity was to suffer an infection, but now the injection of microbial derivatives in vaccines is used. A
vaccine may consist of small quantities of living or dead
microbes, small quantities of toxins, or harmless antigenic molecules derived from the microorganism or its
toxin. The general principle is always the same: Exposure of the body to the agent results in the induction of
the memory cells required for rapid, effective response
to possible future infection by that particular organism.

A second kind of immunity, known as **passive immunity,** is simply the direct transfer of antibodies from
one person (or animal) to another, the recipient thereby
receiving preformed antibodies. Such transfers occur
between mother and fetus since IgG is selectively
moved across the placenta. Also, breast-fed children
receive IgA antibodies in the mother's milk. These
are important sources of protection for the infant
during the first months of life, when the antibody-
synthesizing capacity is relatively poor.

The same principle is used clinically when specific
antibodies (produced by genetic engineering) or
pooled gamma globulin is given to patients exposed
to or suffering from certain infections such as hepati-
tis. The protection afforded by this transfer of antibodies is relatively short-lived, usually lasting only a
few weeks or months.

Summary We can now summarize the interplay between nonspecific and specific immune defenses in re-
sisting a bacterial infection. When we encounter particular bacteria for the first time, *nonspecific* defense
mechanisms resist their entry and, if entry is gained,
attempt to eliminate them by phagocytosis and non-
phagocytic killing in the inflammatory process. Simultaneously, bacterial antigens induce the relevant
specific B-cell clones to differentiate into plasma cells
capable of antibody production. If the nonspecific defenses are rapidly successful, these slowly developing
specific immune responses may never play an important role. If the nonspecific responses are only partly
successful, the infection may persist long enough for
significant amounts of antibody to be produced. The
presence of antibody leads to both enhanced phagocytosis and direct destruction of the foreign cells, as
well as to neutralization of any toxins secreted by the
bacteria. All subsequent encounters with that type of
bacteria will be associated with the same sequence of
events, with the crucial difference that the specific responses may be brought into play much sooner and
with greater force; that is, the person may have active
immunity against that type of bacteria.

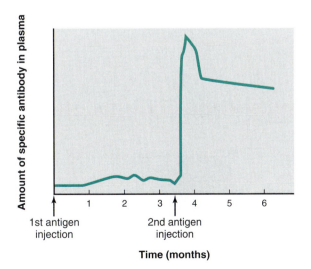

FIGURE 20–16

Rate of antibody production following initial contact with an
antigen and subsequent contact with the same antigen.

The defenses against viruses in the extracellular fluid are similar, resulting in destruction or neutralization of the virus.

Defenses against Virus-Infected Cells and Cancer Cells

The previous section described how antibody-mediated immune responses constitute the major long-term defenses against "exogenous antigens"—bacteria, viruses, and individual foreign molecules that enter the body and are encountered by the immune system in the extracellular fluid. This section now details how cytotoxic T cells, NK cells, and so-called activated macrophages (see below), all working with helper T cells, destroy the body's own cells that have become infected by viruses (or other intracellular microbes) or have been transformed into cancer cells.

What is the value of destroying virus-infected host cells? Such destruction results in release of the viruses into the extracellular fluid where they can then be directly neutralized by circulating antibody, as just described. Generally, only a few host cells are sacrificed in this way, but once viruses have had a chance to replicate and spread from cell to cell, so many virus-infected host cells may be killed by the body's own defenses that organ malfunction may occur.

Role of Cytotoxic T Cells Figure 20–17 summarizes a typical cytotoxic T-cell response triggered by viral infection of body cells. The response triggered by a cancer cell would be similar. As described earlier, a virus-infected or cancer cell produces foreign proteins, "endogenous antigens," which are processed and presented on the plasma membrane of the cell com-

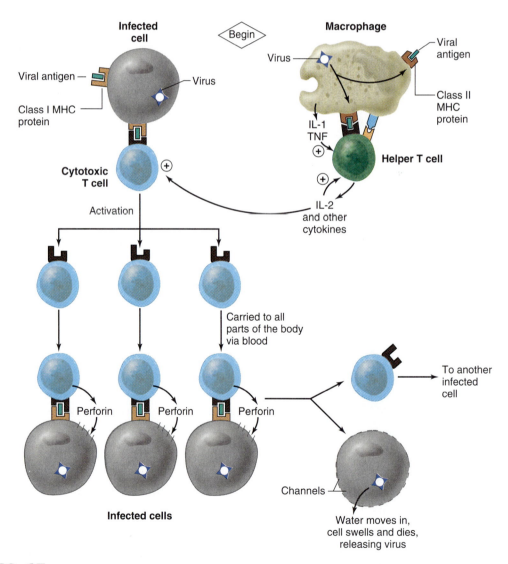

FIGURE 20–17

Summary of events in the killing of virus-infected cells by cytotoxic T cells. The released viruses can then be phagocytized. The sequence would be similar if the inducing cell were a cancer cell rather than a virus-infected cell.

plexed with MHC class I proteins. Cytotoxic T cells specific for the particular antigen can bind to the complex, but just as with B cells, binding to antigen alone does not cause activation of the cytotoxic T cell. Cytokines from adjacent activated helper T cells are also needed.

How are the helper T cells brought into play in these cases? Figure 20–17 illustrates the most likely mechanism. Macrophages phagocytize free extracellular viruses (or, in the case of cancer, antigens released from the surface of the cancerous cells) and then process and present antigen, in association with class II MHC proteins, to the helper T cells. In addition, the macrophages provide a costimulus and also secrete IL-1 and TNF. The activated helper T cell releases IL-2 and other cytokines. IL-2 then acts as an autocrine agent to stimulate proliferation of the helper T cell.

The IL-2 also acts as a paracrine agent on the cytotoxic T cell bound to the surface of the virus-infected or cancer cell, stimulating this attack cell to proliferate. Other cytokines secreted by the activated helper T cell perform the same functions. Why is proliferation important if a cytotoxic T cell has already found and bound to its target? The answer is that there is rarely just one virus-infected cell or one cancer cell. By expanding the clone of cytotoxic T cells capable of recognizing the particular antigen, proliferation increases the likelihood that the other virus-infected or cancer cells will be encountered by the specific type of cytotoxic T cell.

There are several mechanisms of target-cell killing by activated cytotoxic T cells, but one of the most important is as follows (Figure 20–17). The cytotoxic T cell releases, by exocytosis, the contents of its secretory vesicles into the extracellular space between itself and the target cell to which it is bound. These vesicles contain a protein, **perforin** (also termed pore-forming protein), which is similar in structure to the proteins of the complement system's membrane attack complex. Perforin inserts into the target cell's membrane and forms channels through the membrane. In this manner, it causes the attacked cell to become leaky and die. The fact that perforin is released into the space between the tightly attached cytotoxic T cell and the target ensures that innocent host bystander cells will not be killed since perforin is not at all specific.

Some cytotoxic T cells generated during proliferation following an initial antigenic stimulation do not complete their full activation at this time but remain as memory cells. Thus, active immunity exists for cytotoxic T cells just as for B cells.

Role of NK Cells and Activated Macrophages Although cytotoxic T cells are very important attack cells against virus-infected and cancer cells, they are not the only ones. NK cells and "activated macrophages" also destroy such cells by secreting toxic chemicals.

In the section on antibody-dependent cellular cytotoxicity (ADCC), we pointed out that NK cells can be linked to target cells by antibodies, and this certainly constitutes one potential method of bringing them into play against virus-infected or cancer cells. In most cases, however, strong antibody responses are not triggered by virus-infected or cancer cells, and the NK cell must bind *directly* to its target, without the intermediation of antibodies. As noted earlier, NK cells do not have antigen specificity; rather, they nonspecifically bind to any virus-infected and cancer cell.

The major signals for NK cells to proliferate and secrete their toxic chemicals are IL-2 and a member of the interferon family—**interferon-gamma**—secreted by the helper T cells that have been activated specifically by the targets (Figure 20–18). (Whereas essentially all body cells can produce the other interferons, as described earlier, only activated helper T cells and NK cells can produce interferon-gamma.)

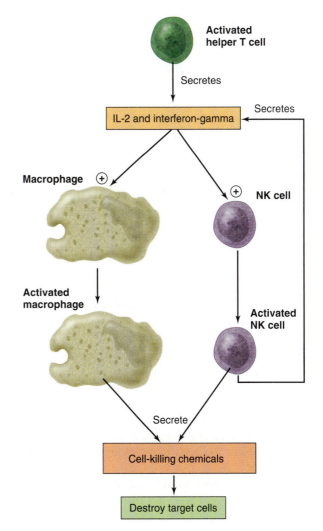

FIGURE 20–18

Role of IL-2 and interferon-gamma, secreted by activated helper T cells, in stimulating the killing ability of NK cells and macrophages.

TABLE 20–6 Summary of Host Responses to Viruses

	Main Cells Involved	Comment on Action
Nonspecific defenses		
Anatomical barriers	Body surface linings	Physical barrier; antiviral chemicals
Inflammation	Tissue macrophages	Phagocytosis of extracellular virus
Interferon	Most cell types after viruses enter them	Interferon nonspecifically prevents viral replication inside host cells
Specific defenses		
Antibody-mediated	Plasma cells derived from B cells secrete antibodies	Antibodies neutralize virus and thus prevent viral entry into cell
		Antibodies activate complement, which leads to enhanced phagocytosis of extracellular virus
		Antibodies recruit NK cells via antibody-mediated cellular cytotoxicity
Direct cell killing	Cytotoxic T cells, NK cells, and activated macrophages	Via secreted chemicals, these cells destroy host cell and thus induce release of virus into extracellular fluid where it can be phagocytized
		Activity is stimulated by IL-2 and interferon-gamma

Thus, the attack by the NK cells is nonspecific, but a specific immune response on the part of the helper T cells is required to bring the NK cells into play. Moreover, there is a positive feedback at work here since activated NK cells can themselves secrete interferon-gamma (Figure 20–18).

IL-2 and interferon-gamma act not only on NK cells but on macrophages in the vicinity to enhance their ability to kill cancer cells and cells infected with viruses and other microbes. Macrophages stimulated by IL-2 and interferon-gamma are termed **activated macrophages** (Figure 20–18). They secrete large amounts of many chemicals that are capable of killing cells by a variety of mechanisms.

Table 20–6 summarizes the multiple defenses against viruses described in this chapter.

Systemic Manifestations of Infection

There are many *systemic* responses to infection—that is, responses of organs and tissues distant from the site of infection or immune response. These systemic responses are collectively known as the **acute phase response** (Figure 20–19). It is natural to think of them as part of the disease, but the fact is that most of them actually represent the body's own adaptive responses to the infection.

The single most common and striking systemic sign of infection is fever, the mechanism of which is described in Chapter 18. Present evidence suggests that fever is often beneficial, in that an increase in body temperature enhances many of the protective responses described in this chapter.

Decreases in the plasma concentrations of iron and zinc occur in response to infection and are due to changes in the uptake and/or release of these trace elements by liver, spleen, and other tissues. The decrease in plasma iron concentration has adaptive value since bacteria require a high concentration of iron to multiply. The role of the decrease in zinc is not known. The decrease in appetite characteristic of infection may also deprive the invading organisms of nutrients—particularly minerals—they require to proliferate.

Another adaptive response to infection is the secretion by the liver of a group of proteins known collectively as **acute phase proteins.** These proteins exert many effects on the inflammatory process, many of which serve to minimize the extent of local tissue damage. In addition, they are important for tissue repair and for clearance of cell debris and the endotoxins released from the microbes. An example of an acute phase protein is C-reactive protein, one effect of which, as described earlier, is to function as a nonspecific opsonin to enhance phagocytosis.

Another response to infection, increased production and release of neutrophils and monocytes by the bone marrow, is of obvious value. There is also release of amino acids from muscle, and these amino acids provide the building blocks for the synthesis of proteins required to fight the infection and for tissue repair. Increased release of fatty acids from adipose tissue also occurs, providing a source of energy. The secretion of many hormones is increased in the acute phase response, notably that of cortisol by the adrenal cortex; the negative-feedback adaptive value of this particular hormonal response is described in Section C of this chapter.

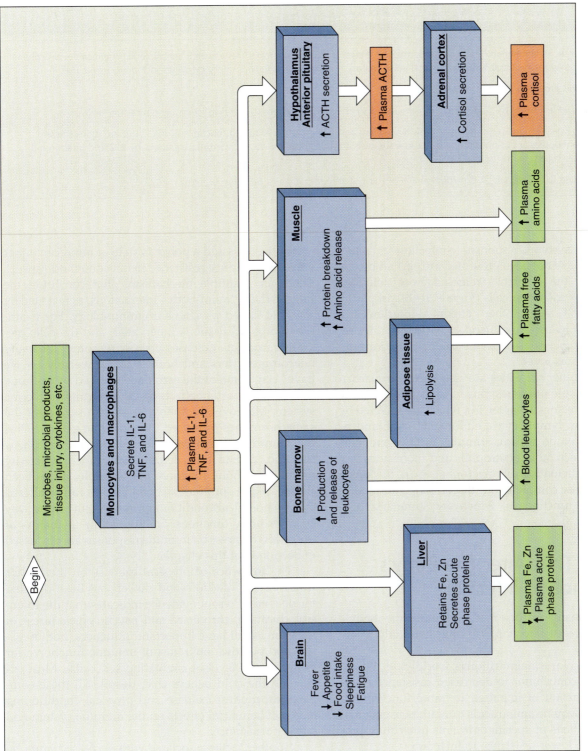

FIGURE 20–19

Systemic responses to infection or injury (the acute phase response). Other cytokines probably also participate. This figure does not include all the components of the acute phase response; for example, IL-1 and several other cytokines also stimulate secretion of insulin and glucagon. The significance of the increase in plasma cortisol is described in Section C of this chapter.

TABLE 20–7 Role of Macrophages in Immune Responses

1. In nonspecific inflammation, phagocytize particulate matter, including microbes. Also secrete antimicrobial chemicals and protein messengers (cytokines) that function as local inflammatory mediators. The inflammatory cytokines include IL-1 and TNF.

2. Process and present antigen to helper T cells.

3. The secreted IL-1 and TNF [see (1) above] stimulate helper T cells to secrete IL-2 and to express the receptor for IL-2.

4. During specific immune responses, perform same killing and inflammation-inducing functions as in (1) but are more efficient because antibodies act as opsonins and because the cells are transformed into activated macrophages by IL-2 and interferon-gamma, both secreted by helper T cells.

5. The secreted IL-1, TNF, and IL-6 mediate many of the systemic responses to infection or injury.

All these systemic responses to infection and many others that have not been presented are elicited by one or more of the cytokines released from activated macrophages and other cells (Figure 20–19). In particular IL-1, TNF, and another cytokine—**interleukin 6 (IL-6),** all of which serve *local* roles in immune responses, also serve as *hormones* to elicit far-flung responses such as fever.

Several other cytokines are also known to participate in the acute phase response. For example, colony stimulating factors (Chapter 14), which are secreted by macrophages, lymphocytes, endothelial cells, and fibroblasts, provide a major stimulus to the bone marrow to produce more neutrophils and monocytes.

The participation of macrophages in the acute phase response completes our discussion of these cells, the various functions of which are summarized in Table 20–7.

Factors That Alter the Body's Resistance to Infection

There are many factors that determine the body's capacity to resist infection. We offer here only a few examples.

Protein-calorie malnutrition is, worldwide, the single greatest contributor to decreased resistance to infection, the result of impaired immune function because of inadequate amino acids to synthesize essential proteins. Deficits of specific nutrients other than protein can also lower resistance to infection.

A preexisting disease, infectious or noninfectious, can also predispose the body to infection. People with diabetes, for example, have a propensity to numerous infections, at least partially explainable on the basis of defective leukocyte function. Moreover, any injury to a tissue lowers its resistance, perhaps by altering the chemical environment or interfering with the blood supply.

Both stress and a person's state of mind can either enhance or reduce resistance to infection (and cancer). There are multiple mechanisms that constitute the links in these "mind-body" interactions, as revealed by the field called psychoneuroimmunology. For example, lymphoid tissue is innervated, and the cells mediating immune defenses possess receptors for many neurotransmitters and hormones. Conversely, as we have seen, some of the cytokines released by immune cells have important effects on the brain and endocrine system. Moreover, lymphocytes secrete several of the same hormones produced by endocrine glands. Thus the immune system can alter neural and endocrine function, and in turn neural and endocrine activity modifies immune function. For example, it has been shown that the production of antibodies can be altered by psychological conditioning in much the same way as other body functions. Some possible mediators of the effects of stress on immune responses are discussed in Section C of this chapter.

The influence of physical exercise on the body's resistance to infection (and cancer) has been debated for decades. Present evidence indicates that the intensity, duration, chronicity, and psychological stress of the exercise all have important influences, both negative and positive, on a host of immune functions (for example, the level of circulating NK cells). Most experts in the field believe that, despite all these complexities, modest exercise and physical conditioning have net beneficial effects on the immune system and on host resistance.

Another factor associated with decreased immune function is sleep deprivation. For example, loss of half of a single night's sleep has been observed to reduce the activity of blood NK cells.

Resistance to infection will be impaired if one of the basic resistance mechanisms itself is deficient, as, for example, in people who have a genetic deficiency that impairs their ability to produce antibodies. These individuals experience frequent and sometimes life-threatening infections that can be prevented by regular replacement injections of gamma globulin. Another genetic defect is *combined immunodeficiency,* which is an absence of both B and T cells. If untreated, infants with this disorder usually die within their first year of life from overwhelming infections. Combined immunodeficiency can be cured by a bone-marrow transplantation, which supplies both B cells and cells that will migrate to the thymus and become T cells.

An environmentally induced decrease in the production of leukocytes is also an important cause of lowered resistance, as, for example, in patients given drugs specifically to inhibit rejection of tissue or organ transplants (see section on graft rejection below).

In terms of the numbers of people involved, the most important example of the lack of a basic resistance mechanism is the disease called *acquired immune deficiency syndrome (AIDS).*

Acquired Immune Deficiency Syndrome (AIDS)

AIDS is caused by the *human immunodeficiency virus (HIV),* which incapacitates the immune system. HIV belongs to the retrovirus family, whose nucleic acid core is RNA rather than DNA. Retroviruses possess an enzyme (reverse transcriptase) that, once the virus is inside a host cell, transcribes the virus's RNA into DNA, which is then integrated into the host cell's chromosomes. Replication of the virus inside the cells causes the cells' death.

Unfortunately, the cells that HIV preferentially (but not exclusively) enters are helper T cells. HIV infects these cells because the CD4 protein on the plasma membrane of helper T cells acts as a high-affinity receptor for one of the HIV's surface proteins (gp120). Thus, the helper T cell binds the virus, making it possible for the virus to then enter the cell. Very importantly, this binding of the HIV gp120 protein to CD4 is not sufficient to grant the HIV entry into the helper T cell. In addition, another surface protein on the helper T cell, one that serves normally as a receptor for certain chemokines, must serve as a coreceptor for the gp120. It has been found that persons who have a mutation in this chemokine receptor are highly resistant to infection with HIV, and much research is now focused on the possible therapeutic use of chemicals that can interact with and block this receptor.

Once in the helper T cell, the replicating virus directly kills the helper T cell and also indirectly causes its death via the body's usual immune attack, mediated in this case mainly by cytotoxic T cells, against virus-infected cells. In addition, by still poorly understood mechanisms, the HIV causes the death of many *uninfected* helper T cells by apoptosis. Without adequate numbers of helper T cells, neither B cells nor cytotoxic T cells can function normally. Thus the AIDS

patient dies from infections and cancers that ordinarily would be readily handled by the immune system.

AIDS was first described in 1981, and it has since reached epidemic proportions worldwide. The great majority of persons presently infected with HIV have no symptoms of AIDS as yet. This is an important point: One must distinguish between the presence of the symptomatic disease—AIDS—and asymptomatic infection with HIV. (The latter is diagnosed by the presence of anti-HIV antibodies in the blood.) It is thought, however, that all infected persons will eventually develop AIDS, although at highly varying rates.

The path from HIV infection to AIDS commonly takes about 10 years in untreated persons. Typically, during the first five years the rapidly replicating viruses continually kill large numbers of helper T cells in lymphoid tissues, but these are replaced by new cells. Therefore, the number of helper T cells stays normal (about 1000 cells/mm^3 of blood), and the person is asymptomatic. During the next 5 years, this balance is lost; the number of helper T cells, as measured in blood, drops to about half the normal level, but many people still remain asymptomatic. As the helper T cell count continues to fall, however, the symptoms of AIDS begin—infections with bacteria, viruses, fungi, and parasites. These are accompanied by systemic symptoms of weight loss, lethargy, and fever—all caused by high levels of the cytokines that induce the acute phase response. Certain cancers also occur with high frequency. In untreated persons, death usually ensues within two years after the onset of AIDS symptoms.

The transmission of HIV is known to occur only through (1) transfer of contaminated blood or blood products from one person to another; (2) sexual intercourse with an infected partner; or (3) transmission from an infected mother to her offspring either across the placenta during pregnancy and delivery, or via breast milk. There is no evidence that HIV is transmitted through intact skin or via food, saliva, tears, sweat, feces, urine, vomit, insects, toilet seats, or swimming pools. Even prolonged close contact with persons infected with HIV does not lead to infection in the absence of the transmission routes described above. Nor is there any risk in *donating* blood.

There are two components to the therapeutic management of HIV-infected persons: one directed against the virus itself to delay progression of the disease, and one to prevent or treat the opportunistic infections and cancers that ultimately cause death. The present recommended treatment of HIV infection itself is a simultaneous battery of at least three drugs, two which inhibit the action of the HIV enzyme (reverse transcriptase) that converts the viral RNA into the host cell's DNA, and one that inhibits the HIV enzyme (protease) that cleaves a large protein into smaller units required for the assembly of new HIV. The use of this complex and expensive regimen markedly reduces the replication of HIV in the body and should be introduced very early in the course of HIV infection, not just after the appearance of AIDS.

The ultimate hope for prevention of AIDS is development of a vaccine. For a variety of reasons related to the nature of the virus (it generates large numbers of distinct subspecies) and the fact that it infects helper T cells, which are crucial for immune responses, this is not an easy task.

Antibiotics

The most important of the drugs we employ in helping the body to resist microbes, mainly bacteria, are **antibiotics,** such as penicillin. With few exceptions (for example, the drugs used to treat HIV infections and the drug **acyclovir,** used in herpes infections), there are no antibiotics presently in common use against viruses.

Antibiotics exert a wide variety of effects, including inhibition of bacterial cell-wall synthesis, protein synthesis, and DNA replication. Fortunately, a number of the reactions involved in the synthesis of protein by bacteria, and the proteins themselves, are sufficiently different from those in human cells that certain antibiotics can inhibit them without interfering with the body's own protein synthesis. For example, the antibiotic erythromycin blocks the movement of ribosomes along bacterial messenger RNA.

Antibiotics, however, must not be used indiscriminately. For one thing, they may exert toxic effects on the *body's* cells. A second reason for judicious use is the escalating and very serious problem of drug resistance. Most large bacterial populations contain a few mutants that are resistant to the drug, and these few are capable of multiplying into large populations resistant to the effects of that particular antibiotic. Alternatively, the antibiotic can induce expression of a latent gene that confers resistance. Finally, resistance can be transferred from one resistant microbe directly to another previously nonresistant microbe by means of DNA passed between them. (One example of how drug resistance can spread rapidly by these phenomena is that many bacterial strains that were once highly susceptible to penicillin now produce an enzyme that cleaves the penicillin molecule.) A third reason for the judicious use of antibiotics is that these agents may actually contribute to a new infection by eliminating certain species of relatively harmless bacteria that ordinarily prevent growth of more dangerous ones.

Harmful Immune Responses

Until now, we have focused on the mechanisms of immune responses and their *protective* effects. In this section we shall see that immune responses can be harmful to the body.

Graft Rejection

The major obstacle to successful transplantation of tissues and organs is that the immune system recognizes the transplants, called grafts, as foreign and launches an attack against them. This is termed *graft rejection.* Although B cells and macrophages play some role, cytotoxic T cells and helper T cells are mainly responsible for graft rejection.

Except in the case of identical twins, the class I MHC proteins on the cells of a graft differ from the recipient's. So do the class II molecules present on the macrophages in the graft (recall that virtually all organs and tissues have macrophages). Accordingly, the MHC proteins of both classes are recognized as foreign by the recipient's T cells, and the cells bearing these proteins are destroyed by the recipient's cytotoxic T cells with the aid of helper T cells.

Some of the tools aimed at reducing graft rejection are radiation and drugs that kill actively dividing lymphocytes and thereby decrease the recipient's T-cell population. The single most effective drug, however, is *cyclosporin,* which does not kill lymphocytes but rather blocks the production of IL-2 and other cytokines by helper T cells. This eliminates a critical signal for proliferation of both the helper T cells themselves and the cytotoxic T cells.

Adrenal corticosteroids (Chapter 10) in large doses are also used to reduce the rejection. Their possible mechanisms of action are described in Section C of this chapter.

There are several problems with the use of drugs like cyclosporin and adrenal corticosteroids: (1) Immunosuppression with them is nonspecific, and so patients taking them are at increased risk for infections and cancer; (2) they exert other toxic side effects; and (3) they must be used continuously to inhibit rejection. An important new kind of therapy, one that may be able to avoid these problems, is under study. Recall that immune tolerance for self proteins is achieved by clonal deletion and/or inactivation, and that the mechanism for this is absence of a nonantigenic costimulus at the time the antigen is first encountered. The hope is that, at the time of graft surgery, treatment with drugs that block the complementary proteins constituting the costimulus may induce a permanent state of immune tolerance toward the graft.

TABLE 20–8 Human ABO Blood Groups

Blood Group	Percent*	Antigen on RBC	Genetic Possibilities Homozygous	Genetic Possibilities Heterozygous	Antibody in Blood
A	42	A	AA	AO	Anti-B
B	10	B	BB	BO	Anti-A
AB	3	A and B	—	AB	Neither anti-A nor anti-B
O	45	Neither A nor B	OO	—	Both anti-A and anti-B

*In the United States.

The Fetus as a Graft During pregnancy the fetal trophoblast cells of the placenta (Chapter 19) lie in direct contact with maternal immune cells. Since half of the fetal genes are paternal, all proteins coded for by these genes are foreign to the mother. Why does the mother's immune system not attack the trophoblast cells, which express such proteins, and reject the placenta? This problem is far from solved, but one critical mechanism (there are certainly others) is as follows: Trophoblast cells, unlike virtually all other nucleated cells, do not express the usual MHC class I proteins; instead they express a unique MHC class I protein that maternal immune cells do not recognize as foreign.

Transfusion Reactions

Transfusion reactions, the illness caused when erythrocytes are destroyed during blood transfusion, are a special example of tissue rejection, one that illustrates the fact that antibodies rather than cytotoxic T cells can sometimes be the major factor in rejection. Erythrocytes do not have MHC proteins, but they do have plasma-membrane proteins and carbohydrates (the latter linked to the membrane by lipids) that can function as antigens when exposed to another person's blood. There are more than 400 erythrocyte antigens, but the ABO system of carbohydrates is the most important for transfusion reactions.

Some people have the gene that results in synthesis of the A antigen, some have the gene for the B antigen, some have both genes, and some have neither gene. (Genes cannot code for the carbohydrates that function as antigens; rather they code for the particular enzymes that catalyze formation of the carbohydrates.) The erythrocytes of those with neither gene are said to have O-type erythrocytes. Accordingly, the possible blood types are A, B, AB, and O (Table 20–8).

Type A individuals always have anti-B antibodies in their plasma. Similarly, type B individuals have plasma anti-A antibodies. Type AB individuals have neither anti-A nor anti-B antibody, and type O individuals have both. These antierythrocyte antibodies are called natural antibodies. How they arise "naturally"—that is, without exposure to the appropriate antigen-bearing erythrocytes—is not presently clear.

With this information as background, we can predict what happens if a type A person were given type B blood. There are two incompatibilities: (1) The recipient's anti-B antibodies cause the transfused cells to be attacked, and (2) the anti-A antibodies in the transfused plasma cause the recipient's cells to be attacked. The latter is generally of little consequence, however, because the transfused antibodies become so diluted in the recipient's plasma that they are ineffective in inducing a response. It is the destruction of the transfused cells by the recipient's antibodies that produces the problems.

Similar analyses show that the following situations would result in an attack on the transfused erythrocytes: a type B person given either A or AB blood; a type A person given either type B or AB blood; a type O person given A, B, or AB blood. Type O people are, therefore, sometimes called universal donors, whereas type AB people are universal recipients. These terms are misleading, however, since besides antigens of the ABO system, there are a host of other erythrocyte antigens and plasma antibodies against them. Therefore, except in a dire emergency, the blood of donor and recipient must be tested for incompatibilities directly by the procedure called *cross-matching.* The recipient's serum is combined on a glass slide with the prospective donor's erythrocytes (a "major" cross-match), and the mixture is observed for rupture (hemolysis) or clumping (agglutination) of the erythrocytes; this indicates a mismatch. In addition, the recipient's erythrocytes can be combined with the prospective donor's serum (a "minor" cross-match), looking again for mismatches.

Another group of erythrocyte membrane antigens of medical importance is the Rh system of proteins. There are more than 40 such antigens, but the one most likely to cause a problem is termed Rh_o, known commonly as the **Rh factor** because it was first studied in rhesus monkeys. Human erythrocytes either have the antigen (Rh-positive) or lack it (Rh-negative). About 85 percent of the U.S. population is Rh-positive.

Antibodies in the Rh system, unlike the "natural antibodies" of the ABO system, follow the classical immunity pattern in that no one has anti-Rh antibodies unless exposed to Rh-positive cells from another person. This can occur if an Rh-negative person is subjected to multiple transfusions with Rh-positive blood, but its major occurrence involves the mother-fetus relationship. When an Rh-negative mother carries an Rh-positive fetus, some of the fetal erythrocytes may cross the placental barriers into the maternal circulation, inducing her to synthesize anti-Rh antibodies. Because this occurs mainly during separation of the placenta at delivery, a first Rh-positive pregnancy rarely offers any danger to the fetus since delivery occurs before the antibodies are made by the mother. In future pregnancies, however, these antibodies will already be present in the mother and can cross the placenta to attack and hemolyze the erythrocytes of an Rh-positive fetus. This condition, which can cause an anemia severe enough to result in death of the fetus in utero or of the newborn, is called *hemolytic disease of the newborn.* The risk increases with each Rh-positive pregnancy as the mother becomes more and more sensitized.

Fortunately, this disease can be prevented by giving an Rh-negative mother human gamma globulin against Rh-positive erythrocytes within 72 h after she has delivered an Rh-positive infant. These antibodies bind to the antigenic sites on any Rh-positive erythrocytes that might have entered the mother's blood during delivery and prevent them from inducing antibody synthesis by the mother. The administered antibodies are eventually catabolized.

You may be wondering whether ABO incompatibilities are also a cause of hemolytic disease of the newborn. For example, a woman with type O blood has antibodies to both the A and B antigens. If her fetus is type A or B, this theoretically should cause a problem. Fortunately, it usually does not, partly because the A and B antigens are not strongly expressed in fetal erythrocytes and partly because the antibodies, unlike the anti-Rh antibodies, are of the IgM type, which do not readily cross the placenta.

Allergy (Hypersensitivity)

Allergy or *hypersensitivity* refers to diseases in which immune responses to environmental antigens cause inflammation and damage to the body itself. Antigens that cause allergy are termed *allergens,* common examples of which include those in ragweed pollen and poison ivy. Most allergens themselves are relatively or

---

TABLE 20–9 Major Types of Hypersensitivity

1. **Delayed hypersensitivity**—Mediated by helper T cells and macrophages; independent of antibodies

2. **Immune-complex hypersensitivity**—Mediated by antigen-antibody complexes deposited in tissue

3. **Immediate hypersensitivity**—Mediated by IgE antibodies, mast cells, and eosinophils

---

completely harmless, and it is the immune responses to them that cause the damage. In essence, then, allergy is immunity gone wrong, for the response is inappropriate to the stimulus.

A word about terminology is useful here: As we shall see, there are three major types of hypersensitivity, as categorized by the different immunologic effector pathways involved in the inflammatory response. The term "allergy" is sometimes used popularly to denote only one of these types, that mediated by IgE antibodies. We shall follow common practice, however, of using the term "allergy" in its broader sense as synonymous with "hypersensitivity."

To develop a particular allergy, a genetically predisposed person must first be exposed to the allergen. This initial exposure causes "sensitization," and it is the subsequent exposures that elicit the damaging immune responses we recognize as the disease. The diversity of allergic responses reflects the different immunological effector pathways elicited, and the classification of allergic diseases is based on these mechanisms (Table 20–9).

In one type of allergy, the inflammatory response is independent of antibodies. It is due to marked secretion of cytokines by helper T cells activated by antigen in the area. These cytokines themselves act as inflammatory mediators and also activate macrophages to secrete their potent mediators. Because it takes several days to develop, this type of allergy is known as *delayed hypersensitivity.* The skin rash that appears after contact with poison ivy is an example.

In contrast to this are the various types of antibody-mediated allergic responses. One important type is termed *immune-complex hypersensitivity.* It occurs when so many antibodies (of either the IgG or IgM types) combine with free antigens that large numbers of antigen-antibody complexes precipitate out on the surface of endothelial cells or are trapped in capillary walls, particularly those of the renal corpuscles. These immune complexes activate complement, which then induces an inflammatory response that damages the tissues immediately surrounding the complexes. Allergy to penicillin is an example.

The more common type of antibody-mediated allergic responses, however, are those termed *immediate hypersensitivity*, because they are usually very rapid in onset. They are also called *IgE-mediated hypersensitivity* because they involve IgE antibodies.

Immediate Hypersensitivity   In immediate hypersensitivity, initial exposure to the antigen leads to some antibody synthesis and, more important, to the production of memory B cells that mediate active immunity. Upon reexposure, the antigen elicits a more powerful antibody response. So far, none of this is unusual, but the difference is that the particular antigens that elicit immediate hypersensitivity reactions stimulate, in ge-

netically susceptible persons, the production of type IgE antibodies. Production of IgE requires the participation of a particular subset of helper T cells that are activated by the allergens presented by B cells. These activated helper T cells then release cytokines that preferentially stimulate differentiation of the B cells into IgE-producing plasma cells.

Upon their release from plasma cells, IgE antibodies circulate to various parts of the body and become attached, via binding sites on their Fc portions, to connective-tissue mast cells (Figure 20–20). When subsequently the same antigen type enters the body and combines with the IgE bound to the mast cell, this triggers the mast cell to secrete many inflammatory

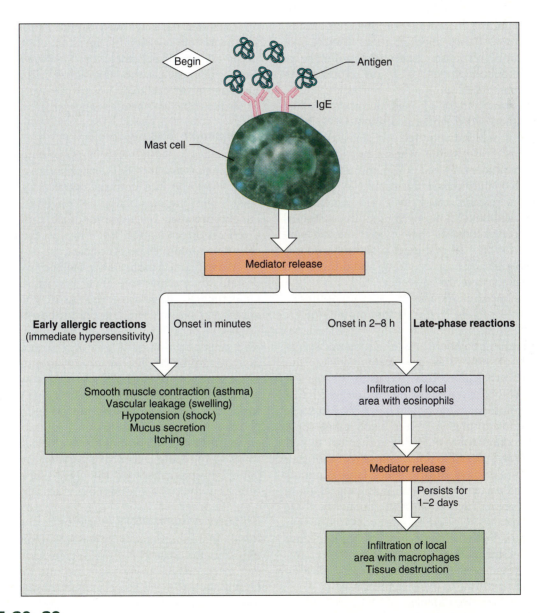

FIGURE 20–20

Sequence of events in an immediate hypersensitivity allergic response.

TABLE 20–10 *Some Possible Causes of Autoimmune Attack*

1. There may be failure of clonal deletion in the thymus or of clonal inactivation in the periphery. This is particularly true for "sequestered antigens," such as certain proteins that are unavailable to the immune system during critical early-life periods.

2. Normal body proteins may be altered by combination with drugs or environmental chemicals. This leads to an attack on the cells bearing the now-"foreign" protein.

3. In immune attacks on virus-infected bodily cells, so many cells may be destroyed that disease results.

4. Genetic mutations in the body's cells may yield new proteins that serve as antigens.

5. The body may encounter microbes whose antigens are so close in structure to certain of the body's own proteins that the antibodies or cytotoxic T cells produced against these microbial antigens also attack cells bearing the self proteins.

6. Proteins normally never encountered by lymphocytes may become exposed as a result of some other disease.

---

mediators, including **histamine,** various eicosanoids, and chemokines. All these mediators then initiate a local inflammatory response. (The entire sequence of events just described for mast cells can also occur with basophils in the circulation.)

Thus, the symptoms of IgE-mediated allergy reflect the various effects of these inflammatory mediators and the body site in which the antigen-IgE-mast cell combination occurs. For example, when a previously sensitized person inhales ragweed pollen, the antigen combines with IgE on mast cells in the respiratory passages. The mediators released cause increased secretion of mucus, increased blood flow, swelling of the epithelial lining, and contraction of the smooth muscle surrounding the airways. Thus, there follow the symptoms of congestion, running nose, sneezing, and difficulty in breathing that characterize hay fever.

Allergic symptoms are usually localized to the site of entry of the antigen. If very large amounts of the chemicals released by the mast cells (or blood basophils) enter the circulation, however, systemic symptoms may result and cause severe hypotension and bronchiolar constriction. This sequence of events, termed *anaphylaxis,* can cause death due to circulatory and respiratory failure. It can be elicited in some sensitized people by the antigen in a single bee sting.

The very rapid components of immediate hypersensitivity often proceed to a *late-phase reaction* lasting many hours or days, during which large numbers of leukocytes, particularly eosinophils, migrate into the inflamed area. The chemoattractants involved are particular cytokines released by mast cells and helper T cells activated by the allergen. The eosinophils, once in the area, secrete mediators that prolong the inflammation and sensitize the tissues, so that less allergen is needed the next time to evoke a response.

Given the inappropriateness of most immediate hypersensitivity responses, how did such a system evolve? The normal physiological function of the IgE–mast cell–eosinophil pathways is to repel invasion by multicellular parasites that cannot be phagocytized. The mediators released by the mast cells stimulate the inflammatory response against the parasites, and the eosinophils serve as the major killer cells against them by secreting several toxins. How this system also came to be inducible by harmless agents is not clear.

Autoimmune Disease

While allergy is due to an inappropriate response to an *environmental* antigen, *autoimmune disease* is due to an inappropriate immune attack triggered by the body's own proteins acting as antigens. The immune attack, mediated by autoantibodies and self-reactive T cells, is directed specifically against the body's own cells that contain these proteins.

We explained earlier how the body is normally in a state of immune tolerance toward its own cells. Unfortunately, there are situations in which this tolerance breaks down and the body does in fact launch antibody- or killer cell–mediated attacks against its own cells and tissues. A growing number of human diseases are being recognized as autoimmune in origin. Examples are *multiple sclerosis,* in which myelin is attacked; *myasthenia gravis,* in which the receptors for acetylcholine on skeletal-muscle cells are the target; rheumatoid arthritis, in which joints are damaged; and *insulin-dependent diabetes mellitus,* in which the insulin-producing cells of the pancreas are destroyed. Some possible causes for the body's failure to recognize its own cells are summarized in Table 20–10.

Excessive Inflammatory Responses

Recall that complement, other inflammatory mediators, and the toxic chemicals secreted by neutrophils and macrophages are not specific with regard to their targets. Accordingly, sometimes during an inflammatory response directed against microbes there can be so much generation or release of these substances that adjacent normal tissues may be damaged. These substances can also cause potentially lethal systemic responses. For example, macrophages release very large amounts of IL-1 and TNF, both of which are powerful inflammatory mediators (in addition to their other effects) in response to an infection with certain types of bacteria. These cytokines can cause profound vasodilation throughout the body, precipitating a type of hypotension termed *septic shock.* This is often accompanied by dangerously high fevers. In other words, it is not the bacteria themselves that cause septic shock but rather the cytokines released in response to the bacteria.

Another important example of damage produced by excessive inflammation in response to microbes is the dementia that occurs in AIDS. HIV does not itself attack neurons but it does infect microglia. Such invasion causes the microglia, which function as macrophage-like cells, to produce very high levels of inflammatory cytokines and other molecules that are toxic to neurons. (Microglia are also implicated in noninfectious brain disorders, like *Alzheimer's disease,* that are characterized by inflammation.)

Excessive long-standing inflammation can also occur in the absence of microbial infection. Thus, various major diseases, including *asthma, rheumatoid arthritis,* and *inflammatory bowel disease,* are categorized as *chronic inflammatory diseases.* The causes of these diseases, and the interplay between genetic and environmental factors, are still poorly understood. Some, like rheumatoid arthritis, are mainly autoimmune in nature, but all are associated with a marked positive-feedback increase in the production of cytokines and other inflammatory mediators.

Yet another example of excessive inflammation in a noninfectious state is the development of atherosclerotic plaques in blood vessels (Chapter 14). It is likely that, in response to endothelial cell dysfunction, the vessel wall releases inflammatory cytokines (IL-1, for example) that promote all stages of atherosclerosis—excessive clotting, chemotaxis of various leukocytes (as well as smooth-muscle cells), and so on. The endothelial-cell dysfunction is caused by initially subtle vessel-wall injury by lipoproteins and other factors, including elevated blood pressure and homocysteine (Chapter 14).

In summary, the various mediators of inflammation and immunity are a double-edged sword: In usual amounts they are essential for normal resistance, but in excessive amounts they can cause illness.

This completes the section on immunology. Table 20–11 presents a summary of immune mechanisms in the form of a miniglossary of cells and chemical mediators involved in immune responses. All the material in this table has been covered in this chapter.

TABLE 20–11 A Miniglossary of Cells and Chemical Mediators Involved in Immune Functions

Cells

Activated macrophages Macrophages whose killing ability has been enhanced by cytokines, particularly IL-2 and interferon-gamma.

Antigen-presenting cells (APC) Cells that present antigen, complexed with MHC proteins, on their surface to T cells.

B cells Lymphocytes that, upon activation, proliferate and differentiate into antibody-secreting plasma cells; provide major defense against bacteria, viruses in the extracellular fluid, and toxins; can function as antigen-presenting cells for helper T cells.

Cytotoxic T cells The class of T lymphocytes that, upon activation by specific antigen, directly attacks the cells bearing that type of antigen; are major killers of virus-infected cells and cancer cells; bind antigen associated with class I MHC proteins.

Eosinophils Leukocytes involved in destruction of parasites and in immediate hypersensitivity responses.

Helper T cells The class of T cells that, via secreted cytokines, plays a stimulatory role in the activation of B cells and cytotoxic T cells; also can activate NK cells and macrophages; bind antigen associated with class II MHC proteins.

Lymphocytes The type of leukocyte responsible for specific immune defenses; categorized mainly as B cells, T cells, and NK cells.

Macrophages Cell type that (1) functions as phagocytes, (2) processes and presents antigen to helper T cells, and (3) secretes cytokines involved in inflammation, activation of lymphocytes, and the systemic acute phase response to infection or injury.

Macrophage-like cells Several cell types that exert functions similar to those of macrophages (for example, microglia).

Mast cells Tissue cell that binds IgE and releases inflammatory mediators in response to parasites and immediate hypersensitivity reactions.

Memory cells B cells and cytotoxic T cells that differentiate during an initial immune response and respond rapidly during a subsequent exposure to the same antigen.

Monocytes A type of leukocyte; leaves the bloodstream and is transformed into a macrophage; has functions similar to those of macrophages.

Natural killer (NK) cells Class of lymphocytes that binds to cells bearing foreign antigens without specific recognition and kills them directly; major targets are virus-infected cells and cancer cells; participate in antibody-dependent cellular cytotoxicity (ADCC).

Neutrophils Leukocytes that function as phagocytes and also release chemicals involved in inflammation.

Plasma cells Cells that differentiate from activated B lymphocytes and secrete antibodies.

T cells Lymphocytes derived from precursors that differentiated in the thymus; see cytotoxic T cells and helper T cells.

TABLE 20–11 A Miniglossary of Cells and Chemical Mediators Involved in Immune Functions (Cont.)

Chemical Mediators

Acute phase proteins Group of proteins secreted by the liver during systemic response to injury or infection; stimulus for their secretion is IL-1, IL-6, and other cytokines.

Antibodies Immunoglobulins that are secreted by plasma cells; combine with the type of antigen that stimulated their production and direct an attack against the antigen or a cell bearing it.

C1 The first protein in the classical complement pathway.

Chemoattractants A general name given to any chemical mediator that stimulates chemotaxis of neutrophils or other leukocytes.

Chemokines Any cytokine that functions as a chemoattractant.

Chemotaxin A synonym for chemoattractant.

Complement A group of plasma proteins that, upon activation, kills microbes directly and facilitates the various steps of the inflammatory process, including phagocytosis; the classical complement pathway is triggered by antigen-antibody complexes, whereas the alternate pathway can operate independently of antibody.

C-reactive protein One of several proteins that function as nonspecific opsonins; production by the liver is increased during the acute phase response.

Cytokines General term for protein messengers that regulate immune responses; secreted by macrophages, monocytes, lymphocytes, neutrophils, and several nonimmune cell types; function both locally and as hormones.

Eicosanoids General term for products of arachidonic acid metabolism (prostaglandins, thromboxanes, leukotrienes); function as important inflammatory mediators.

Histamine An inflammatory mediator secreted mainly by mast cells; acts on microcirculation to cause vasodilation and increased permeability to protein.

IgA The class of antibodies secreted by the lining of the body's various "tracts."

IgD A class of antibodies whose function is unknown.

IgE The class of antibodies that mediate immediate hypersensitivity and resistance to parasites.

IgG The most abundant class of plasma antibodies.

IgM A class of antibodies that, along with IgG, provides the bulk of specific humoral immunity against bacteria and viruses.

Immunoglobulin (Ig) Proteins that function as B-cell receptors and antibodies; the five major classes are IgA, IgD, IgE, IgG, and IgM.

Interferon Group of cytokines that nonspecifically inhibits viral replication; interferon-gamma also stimulates the killing ability of NK cells and macrophages.

Interferon-gamma (See *Interferon*)

Interleukin 1 (IL-1) Cytokine secreted by macrophages (and other cells) that activates helper T cells, exerts many inflammatory effects, and mediates many of the systemic acute phase responses, including fever.

Interleukin 2 (IL-2) Cytokine secreted by activated helper T cells that causes helper T cells, cytotoxic T cells, and NK cells to proliferate, and cause activation of macrophages.

Interleukin 6 (IL-6) Cytokine secreted by macrophages (and other cells) that exerts multiple effects on immune-system cells, inflammation, and the acute phase response.

Kinins Peptides that split from kininogens in inflamed areas and facilitate the vascular changes associated with inflammation; they also activate neuronal pain receptors.

Leukotrienes A class of eicosanoids that is generated by the lipoxygenase pathway and functions as inflammatory mediators.

Membrane attack complex (MAC) Group of complement proteins that form channels in the surface of a microbe, making it leaky and killing it.

Natural antibodies Antibodies to the erythrocyte antigens (of the A or B type)

Opsonin General name given to any chemical mediator that promotes phagocytosis.

Perforin Protein secreted by cytotoxic T cells and NK cells that forms channels in the plasma membrane of the target cell, making it leaky and killing it; its structure and function are similar to that of the MAC in the complement system.

Tumor necrosis factor (TNF) Cytokine that is secreted by macrophages (and other cells) and that has many of the same actions as IL-1.

SECTION A SUMMARY

Cells Mediating Immune Defenses

I. Immune defenses may be nonspecific, in which the identity of the target is not recognized, or it may be specific, in which it is recognized.

II. The cells of the immune system are leukocytes (neutrophils, eosinophils, basophils, monocytes, and lymphocytes), plasma cells, macrophages, macrophage-like cells, and mast cells. The leukocytes use the blood for transportation but function mainly in the tissues.

III. Cells of the immune system (as well as some other cells) secrete protein messengers that regulate immune responses and are collectively termed cytokines.

Nonspecific Immune Defenses

I. External barriers to infection are the skin, the linings of the respiratory, gastrointestinal, and genitourinary tracts, the cilia of these linings, and antimicrobial chemicals in glandular secretions.

II. Inflammation, the local response to injury or infection, includes vasodilation, increased vascular permeability to protein, phagocyte chemotaxis, destruction of the invader via phagocytosis or extracellular killing, and tissue repair.

    a. The mediators controlling these processes, summarized in Table 20–3, are either released from cells in the area or generated extracellularly from plasma proteins.

    b. The main cells that function as phagocytes are the neutrophils, monocytes, macrophages, and macrophage-like cells. These cells also secrete many inflammatory mediators.

    c. One group of inflammatory mediators—the complement family of plasma proteins activated during nonspecific inflammation by the alternate complement pathway—not only stimulates many of the steps of inflammation but mediates extracellular killing via the membrane attack complex.

    d. The end result of infection or tissue damage is tissue repair.

III. Interferon stimulates the production of intracellular proteins that nonspecifically inhibit viral replication.

Specific Immune Defenses

  I. Lymphocytes mediate specific immune responses.

  II. Specific immune responses occur in three stages.

    a. A lymphocyte programmed to recognize a specific antigen encounters it and binds to it via plasma-membrane receptors specific for the antigen.

    b. The lymphocyte undergoes activation—a cycle of cell divisions and differentiation.

    c. The multiple active lymphocytes produced in this manner launch an attack all over the body against the specific antigens that stimulated their production.

  III. The lymphoid organs are categorized as primary (bone marrow and thymus) or secondary (lymph nodes, spleen, tonsils and lymphocyte collections in the linings of the body's tracts).

    a. The primary lymphoid organs are the sites of maturation of lymphocytes that will then be carried to the secondary lymphoid organs, which are the major sites of lymphocyte cell division and specific immune responses.

    b. Lymphocytes undergo a continuous recirculation among the secondary lymphoid organs, lymph, blood, and all the body's organs and tissues.

  IV. The three broad populations of lymphocytes are B, T, and NK cells.

    a. B cells mature in the bone marrow and are carried to the secondary lymphoid organs, where additional B cells arise by cell division.

    b. T cells leave the bone marrow in an immature state, are carried to the thymus, and undergo maturation there. These cells then travel to the secondary lymphoid organs and new T cells arise from them by cell division.

    c. NK cells originate in the bone marrow.

  V. B cells and T cells have different functions.

    a. B cells, upon activation, differentiate into plasma cells, which secrete antibodies. Antibody-mediated responses constitute the major defense against bacteria, viruses, and toxins in the extracellular fluid.

    b. Cytotoxic T cells directly attack and kill virus-infected cells and cancer cells, without the participation of antibodies.

    c. Helper T cells stimulate B cells and cytotoxic T cells via the cytokines they secrete. With few exceptions, this help is essential for activation of the B cells and cytotoxic T cells.

  VI. B-cell surface plasma-membrane receptors are copies of the specific antibody (immunoglobulin) that the cell is capable of producing.

    a. Any given B cell or clone of B cells produces antibodies that have a unique antigen binding site.

    b. Antibodies are composed of four interlocking polypeptide chains; the variable regions of the antibodies are the sites that bind antigen.

  VII. T-cell surface plasma-membrane receptors are not immunoglobulins, but they do have specific antigen binding sites that differ from one T-cell clone to another.

    a. The T-cell receptor binds antigen only when the antigen is complexed to one of the body's own plasma-membrane MHC proteins.

    b. Class I MHC proteins are found on all nucleated cells of the body, whereas class II MHC proteins are found only on macrophages, B cells, and macrophage-like cells. Cytotoxic T cells require antigen to be complexed to class I proteins, whereas helper T cells require class II proteins.

  VIII. Antigen presentation is required for T cell activation.

    a. Only macrophages, B cells, and macrophage-like cells function as antigen-presenting cells (APCs) for helper T cells. The antigen is internalized by the APC and hydrolyzed to peptide fragments, which are complexed with class II MHC proteins. This complex is then shuttled to the plasma membrane of the APC, which also delivers a nonspecific costimulus to the T cell and secretes interleukin 1 (IL-1) and tumor necrosis factor (TNF).

    b. A virus-infected cell or cancer cell can function as an APC for cytotoxic T cells. The viral antigen or cancer-associated antigen is synthesized by the cell itself and hydrolyzed to peptide fragments, which are complexed to class I MHC proteins. The complex is then shuttled to the plasma membrane of the cell.

  IX. NK cells have the same targets as cytotoxic T cells, but they are not antigen-specific; most of their mechanisms of target identification are not understood.

  X. Immune tolerance is the result of clonal deletion and clonal inactivation.

  XI. In antibody-mediated responses, the membrane receptors of a B cell bind antigen, and at the same time a helper T cell also binds antigen in association with a class II MHC protein on a macrophage or other APC.

    a. The helper T cell, activated by the antigen, by a nonantigenic protein costimulus, and by IL-1 and TNF secreted by the APC, secretes IL-2, which then causes the helper T cell to proliferate into a clone of cells that secrete additional cytokines.

b. These cytokines then stimulate the antigen-bound B cell to proliferate and differentiate into plasma cells, which secrete antibodies. Some of the activated B cells become memory cells, which are responsible for active immunity.

c. There are five major classes of secreted antibodies: IgG, IgM, IgA, IgD, and IgE. The first two are the major antibodies against bacterial and viral infection.

d. The secreted antibodies are carried throughout the body by the blood and combine with antigen. The antigen-antibody complex enhances the inflammatory response, in large part by activating the complement system. Complement proteins mediate many steps of inflammation, act as opsonins, and directly kill antibody-bound cells via the membrane attack complex.

e. Antibodies of the IgG class also act directly as opsonins and link target cells to NK cells, which directly kill the target cells.

f. Antibodies also neutralize toxins and extracellular viruses.

XII. Virus-infected cells and cancer cells are killed by cytotoxic T cells, NK cells, and activated macrophages.

a. A cytotoxic T cell binds, via its membrane receptor, to cells bearing a viral antigen or cancer-associated antigen in association with a class I MHC protein.

b. Activation of the cytotoxic T cell also requires cytokines secreted by helper T cells, themselves activated by antigen presented by a macrophage. The cytotoxic T cell then releases perforin, which kills the attached target cell by making it leaky.

c. NK cells and macrophages are also stimulated by helper T cell cytokines, particularly IL-2 and interferon-gamma, to attack and kill virus-infected or cancer cells.

Systemic Manifestations of Infection

I. The acute phase response is summarized in Figure 20–19.

II. The major mediators of this response are IL-1, TNF, and IL-6.

Factors That Alter the Body's Resistance to Infection

I. The body's capacity to resist infection is influenced by nutritional status, the presence of other diseases, psychological factors, and the intactness of the immune system.

II. AIDS is caused by a retrovirus that destroys helper T cells and therefore reduces the ability to resist infection and cancer.

III. Antibiotics interfere with the synthesis of macromolecules by bacteria.

Harmful Immune Responses

I. Rejection of tissue transplants is initiated by MHC proteins on the transplanted cells and is mediated mainly by cytotoxic T cells.

II. Transfusion reactions are mediated by antibodies.

a. Transfused erythrocytes will be destroyed if the recipient has natural antibodies against the antigens (type A or type B) on the cells.

b. Antibodies against Rh-positive erythrocytes can be produced following exposure of an Rh-negative person to such cells.

III. Allergy (hypersensitivity reactions), caused by allergens, are of several types.

a. In delayed hypersensitivity, the inflammation is due to the interplay of helper T cell cytokines and macrophages. Immune complex hypersensitivity is due to complement activation by antigen-antibody complexes.

b. In immediate hypersensitivity, antigen binds to IgE antibodies, which are themselves bound to mast cells. The mast cells then release inflammatory mediators such as histamine that produce the symptoms of allergy. The late phase of immediate hypersensitivity is mediated by eosinophils.

IV. Autoimmune attacks are directed against the body's own proteins, acting as antigens. Reasons for the failure of immune tolerance are summarized in Table 20–10.

V. Normal tissues can be damaged by excessive inflammatory responses to microbes.

SECTION A KEY TERMS	
immunology	C-reactive protein
immune surveillance	interferon
nonspecific immune defense	antigen
specific immune defense	lymphocyte activation
immune system	lymphoid organ
leukocyte	primary lymphoid organ
plasma cell	secondary lymphoid organ
macrophage	thymus
macrophage-like cell	thymopoietin
mast cell	lymph node
phagocyte	spleen
phagocytosis	tonsil
cytokine	B lymphocyte (B cell)
inflammation	T lymphocyte (T cell)
chemotaxis	natural killer cell (NK)
adhesion molecule	antibody
chemoattractant	antibody-mediated responses
chemotaxin	cytotoxic T cell
chemokine	helper T cell
opsonin	immunoglobulin
phagosome	Fc
phagolysosome	antigen binding site
nitric oxide	major histocompatibility
hydrogen peroxide	complex (MHC)
complement	MHC proteins (class I and
membrane attack complex	class II)
(MAC)	antigen presentation
C3b	antigen-presenting cell (APC)
alternate complement	epitope
pathway	costimulus

interleukin 1 (IL-1)
tumor necrosis factor (TNF)
oncogene
immune tolerance
clonal deletion
clonal inactivation
interleukin 2 (IL-2)
memory cell
IgG
gamma globulin
IgM
IgE
IgA
IgD

classical complement
   pathway
antibody-dependent cellular
   cytotoxicity (ADCC)
active immunity
passive immunity
perforin
interferon-gamma
activated macrophage
acute phase response
acute phase protein
interleukin 6 (IL-6)
Rh factor
histamine

SECTION A REVIEW QUESTIONS

1. What are the major cells of the immune system and their general functions?
2. Describe the major anatomical and biochemical barriers to infection.
3. Name the three cell types that function as phagocytes.
4. List the sequence of events in an inflammatory response and describe each.
5. Name the sources of the major inflammatory mediators.
6. What triggers the alternate pathway for complement activation? What roles does complement play in inflammation and cell killing?
7. Describe the antiviral role of interferon.
8. Name the lymphoid organs. Contrast the functions of the bone marrow and thymus with those of the secondary lymphoid organs.
9. Name the various populations and subpopulations of lymphocytes and state their roles in specific immune responses.
10. Contrast the major targets of antibody-mediated responses and responses mediated by cytotoxic T cells and NK cells.

11. How do the Fc and combining-site portions of antibodies differ?
12. What are the differences between B-cell receptors and T-cell receptors? Between cytotoxic T-cell receptors and helper T-cell receptors?
13. Compare and contrast antigen presentation to helper T cells and cytotoxic T cells.
14. Compare and contrast cytotoxic T cells and NK cells.
15. What two processes contribute to immune tolerance?
16. Diagram the sequence of events in an antibody-mediated response, including the role of helper T cells, interleukin 1, and interleukin 2.
17. Contrast the general functions of the different antibody classes.
18. How is complement activation triggered in the classical complement pathway, and how does complement "know" what cells to attack?
19. Name two ways in which the presence of antibodies enhances phagocytosis.
20. How do NK cells "know" what cells to attack in ADCC?
21. Diagram the sequence of events by which a virus-infected cell is attacked and destroyed by cytotoxic T cells. Include the roles of cytotoxic T cells, helper T cells, interleukin 1, and interleukin 2.
22. Contrast the extracellular and intracellular phases of immune responses to viruses, including the role of interferon.
23. List the systemic responses to infection or injury and the mediators responsible for them.
24. What factors influence the body's resistance to infection?
25. What is the major defect in AIDs, and what causes it?
26. What is the major cell type involved in graft rejection?
27. Diagram the sequences of events in immediate hypersensitivity.

SECTION B

TOXICOLOGY: THE METABOLISM OF ENVIRONMENTAL CHEMICALS

The body is exposed to a huge number of nonnutrient chemicals in the environment, many of which can be toxic. We shall refer to all these chemicals simply as "foreign" chemicals. Some are products of the natural world (lead, for example), but most are made by humans. There are now more than 10,000 chemicals being commercially synthesized, and over 1 million have been synthesized at one time or another. Virtually all foreign chemicals find their way into the body because

they are in the air, water, and food we use, or because they are purposely taken, as drugs.

As described in Section A of this chapter, foreign materials can induce inflammation and specific immune responses. These defenses do not, however, constitute the major defense mechanisms against most foreign chemicals. Rather, metabolism—molecular alteration, or **biotransformation,** and excretion—does.

The body's metabolism of foreign chemicals is summarized in Figure 20–21. First, the chemical gains entry to the body through the gastrointestinal tract, lungs, skin, or placenta in the case of a fetus. Once in the blood, the chemical may become bound reversibly to plasma proteins or to erythrocytes. Such binding lowers its free concentration and thereby its ability to alter cell function. It may accumulate in storage depots—for example, DDT in fat tissue—or it may undergo enzyme-mediated biotransformation in various organs and tissues. The metabolites resulting from biotransformation may enter the blood and follow the same pathways as the parent molecules. Finally, the chemical and its metabolites may be eliminated from the body in urine, expired air, skin secretions, or feces.

The blood concentration of any foreign chemical is determined by the interplay of all these metabolic pathways. For example, kidney function and biotransformation both tend to decrease with age, which explains why a particular dose of a drug often produces much higher blood concentrations in elderly people than in young people.

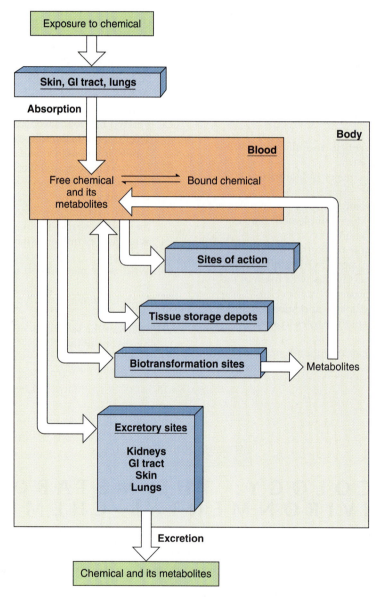

FIGURE 20–21
Metabolic pathways for foreign chemicals.

Absorption

In practice, most foreign molecules move through the lining of some portion of the gastrointestinal tract fairly readily, either by diffusion or by carrier-mediated transport. This should not be surprising since the gastrointestinal tract evolved to favor absorption of the wide variety of nutrient molecules in the environment. Foreign chemicals are the beneficiaries of these relatively nondiscriminating transport mechanisms.

The lung alveoli are highly permeable to most foreign chemicals and therefore offer an easy entrance route for those that are airborne. Lipid solubility is all-important for entry through the skin, so that this route is of little importance for charged molecules but can be used by oils, steroids, and other lipids.

The penetration of the placental membranes by foreign chemicals is important since the effects of environmental agents on the fetus during critical periods of development may be quite marked and, in many cases, irreversible. Diffusion across the placenta occurs for lipid-soluble substances, and carrier-mediated systems, which evolved for the movement of endogenous nutrients, may be usurped by foreign chemicals to gain entry into the fetus.

Storage Sites

The major storage sites for foreign chemicals are cell proteins, bone matrix, and fat. A chemical bound to cell proteins or bone or dissolved in the fat is in equilibrium with the free chemical in the blood, so that an increase in blood concentration causes more movement into storage, up to the point of saturation. Conversely, as the chemical is eliminated from the body and its blood concentration falls, movement occurs out of storage sites.

These storage sites are a source of protection, but it sometimes happens that the storage sites accumulate so much chemical that they become damaged. An example of this is lead toxicity to the kidney cells that bind lead.

Excretion

As described in Chapter 16, to appear in the urine, a chemical must either be filtered through the renal corpuscle or secreted across the tubular epithelium. Glomerular filtration is a bulk-flow process, so that all low-molecular-weight substances in plasma undergo filtration. Accordingly, there is considerable filtration of most foreign chemicals, except those bound to plasma proteins or erythrocytes. In contrast, tubular secretion is by discrete transport processes, and many foreign chemicals—penicillin, for example—utilize the mediated transport systems available for naturally occurring substances.

Once in the tubular lumen, the foreign chemical will still not be excreted if it is reabsorbed across the tubular epithelium into the blood. As the filtered fluid moves along the renal tubules, molecules that are lipid-soluble passively diffuse along with reabsorbed water through the tubular epithelium and back into the blood. The net result is that little is excreted in the urine, and the chemical is retained in the body. If these chemicals could be transformed into more polar and therefore less lipid-soluble molecules, their passive reabsorption from the tubule would be retarded and they would be excreted more readily. This type of transformation is precisely what occurs in the liver, as described in the next section.

An analogous problem exists for foreign molecules secreted in the bile. Many of these substances, having reached the lumen of the small intestine, are absorbed back into the blood, thereby escaping excretion in the feces. This cyclic enterohepatic circulation is described in Chapter 17.

Biotransformation

The metabolic alteration—biotransformation—of foreign molecules occurs mainly in the liver, but to some extent also in the kidneys, skin, placenta, and other organs. A large number of distinct enzymes and pathways are involved, but the common denominator of most of them is that they transform chemicals into more polar, less lipid-soluble substances. One consequence of this transformation is that the chemical may be rendered less toxic, but this is not always so. The second, more important, consequence is that the chemical's tubular reabsorption is diminished and urinary excretion facilitated, as described in the previous section. Similarly, for substances handled by biliary secretion, gut absorption of the metabolite is less likely so that fecal excretion is also enhanced.

The hepatic enzymes that perform these transformations are called the **microsomal enzyme system (MES)** and are located mainly in the smooth endoplasmic reticulum. One of the most important features of this enzyme system is that it is easily inducible; that is, the number of these enzymes can be greatly increased by exposure to a chemical that acts as a substrate for the system. For example, chronic overuse of alcohol results in an increased rate of alcohol catabolism because of induction of the microsomal enzymes. This accounts for much of the tolerance to alcohol—that is, the fact that increasing doses must be taken to achieve a given magnitude of effect.

The hepatic biotransformation mechanisms vividly demonstrate how an adaptive response may, under some circumstances, turn out to be maladaptive. These enzymes all too frequently "toxify" rather than "detoxify" a drug or pollutant. In fact, many foreign chemicals are quite nontoxic until the liver enzymes biotransform them. Of particular importance is the fact that many chemicals that cause cancer do so only after biotransformation. For example, a major component of cigarette smoke is transformed by the MES into a carcinogenic compound. Individuals with a very highly inducible MES have a twenty- to fortyfold higher risk of developing lung cancer from smoking.

The MES can also cause problems in another way because it evolved primarily not to defend against foreign chemicals, which were much less prevalent during our evolution, but rather to metabolize *endogenous* substances, particularly steroids and other lipid-soluble molecules. Therefore, the induction of these enzymes by a drug or pollutant increases metabolism not only of that drug or pollutant but of the endogenous substances as well. The result is a decreased concentration in the body of the normal substance, resulting in its possible deficiency.

Another important fact about the MES is that whereas certain chemicals induce it, others inhibit it. The presence of such chemicals in the environment could have deleterious effects on the system's capacity to protect against those chemicals that it transforms. Just to illustrate how complex this picture can be, note that any chemical that inhibits the microsomal enzyme system may actually confer protection against those other chemicals that must undergo transformation in order to become toxic.

SECTION B SUMMARY

I. The concentration of a foreign chemical in the body depends upon the degree of exposure to the chemical, its rate of absorption across the GI tract, lung, skin, or placenta, and its rates of storage, biotransformation, and excretion.

II. Biotransformation occurs in the liver and other tissues and is mediated by multiple enzymes, notably the microsomal enzyme system (MES). The major function of the MES is to make lipid-soluble substances more polar (less lipid-soluble), thereby decreasing renal tubular reabsorption and increasing excretion.

a. The MES can be induced or inhibited by the chemicals it processes and by other chemicals.

b. It detoxifies some chemicals but toxifies others, notably carcinogens.

SECTION B KEY TERMS

biotransformation
microsomal enzyme system (MES)

SECTION B REVIEW QUESTIONS

1. Why is the urinary excretion of lipid-soluble substances generally very low?
2. What are two functions of biotransformation mechanisms?
3. What are two ways in which activation of biotransformation mechanisms may actually cause malfunction?

RESISTANCE TO STRESS

Much of this book has been concerned with the body's response to stress in its broadest meaning of an environmental change that must be adapted to if health and life are to be maintained. Thus, any change in external temperature, water intake, and so on, sets into motion mechanisms designed to prevent a significant change in some physiological variable. In this section, however, we describe the basic stereotyped response to **stress** in the more limited sense of noxious or potentially noxious stimuli. These stimuli comprise an immense number of situations, including physical trauma, prolonged exposure to cold, prolonged heavy exercise, infection, shock, decreased oxygen supply, sleep deprivation, pain, fright, and other emotional stresses.

It is obvious that the response to cold exposure is very different from that to infection or fright, but in one respect the response to all these situations is the same: Invariably, secretion of the glucocorticoid hormone **cortisol** by the adrenal cortex is increased. Indeed, to physiologists the term "stress" has come to mean any event that elicits increased cortisol secretion.

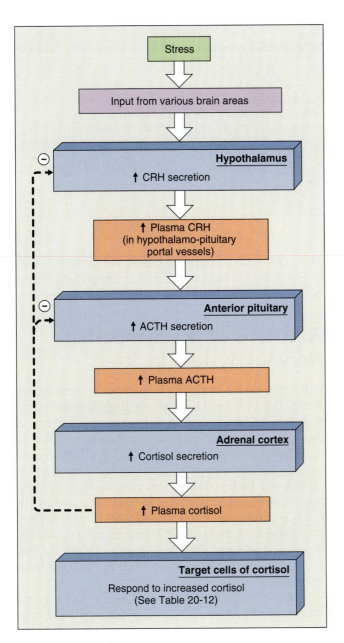

FIGURE 20–22

Pathway by which stressful stimuli elicit increased cortisol secretion. Additional stimuli, not shown in the figure, for ACTH release are vasopressin, epinephrine, and several cytokines, including IL-1, released from immune cells.

Activity of the sympathetic nervous system, including release of the hormone **epinephrine** from the adrenal medulla, is also usually increased in stress.

The increased cortisol secretion of stress is mediated mainly by the hypothalamo-anterior pituitary system described in Chapter 10. As illustrated in Figure 20–22, neural input to the hypothalamus from portions of the nervous system responding to a particular stress induces secretion of **corticotropin releasing hormone (CRH)**. This hormone is carried by the hypothalamo-pituitary portal vessels to the anterior pituitary and stimulates **adrenocorticotropic hormone (ACTH)** release. ACTH in turn circulates to the adrenal cortex and stimulates cortisol release.

The secretion of ACTH, and therefore of cortisol, is stimulated by several hormones in addition to hypothalamic CRH. These include vasopressin and epinephrine, both of which are usually increased in stress. But the most interesting recent finding, mentioned in Section A, is that several of the cytokines, including interleukin 1, also stimulate ACTH secretion (both directly and by stimulating the secretion of CRH). These cytokines provide a means for eliciting a classical stress response when the immune system is stimulated. The possible significance of this relationship for immune function is described on the following pages.

Functions of Cortisol in Stress

The major effects of increased cortisol during stress are summarized in Table 20–12. The effects on organic metabolism, as described in Chapter 18, are to mobilize fuels—to increase the plasma concentrations of amino acids, glucose, glycerol, and free fatty acids. These effects are ideally suited to meet a stressful situation. First, an animal faced with a potential threat is usually forced to forego eating, and these metabolic changes are essential for survival during fasting. Second, the amino acids liberated by catabolism of body protein not only provide a source of glucose, via gluconeogenesis, but also constitute a potential source of amino acids for tissue repair should injury occur.

A few of the medically important implications of these cortisol-induced effects on organic metabolism are as follows: (1) Any patient who is ill or is subjected to surgery catabolizes considerable quantities of body protein; (2) a diabetic who suffers an infection requires more insulin than usual; and (3) a child subjected to severe stress of any kind manifests retarded growth.

Cortisol has important effects during stress other than those on organic metabolism. It enhances vascular reactivity; that is, it increases the ability of vascular smooth muscle to contract in response to stimuli such as norepinephrine. Therefore, a patient with insufficient cortisol faced with even a moderate stress, which usually releases unknown vasodilators, may develop hypotension, due primarily to a marked decrease in total peripheral resistance caused by the vasodilators.

TABLE 20–12 Effects of Increased Plasma Cortisol Concentration during Stress

1. Effects on organic metabolism
   a. Stimulation of protein catabolism
   b. Stimulation of liver uptake of amino acids and their conversion to glucose (gluconeogenesis)
   c. Inhibition of glucose uptake and oxidation by many body cells ("insulin antagonism") but not by the brain
   d. Stimulation of triacylglycerol catabolism in adipose tissue, with release of glycerol and fatty acids into the blood

2. Enhanced vascular reactivity—that is, increased ability to maintain vasoconstriction in response to norepinephrine and other stimuli

3. Unidentified protective effects against the damaging influences of stress

4. Inhibition of inflammation and specific immune responses

As denoted by item 3 in Table 20–12, we still do not know the other reasons, in addition to the effect on vascular smooth muscle, why increased cortisol is so important for the body's optimal response to stress—that is, for its ability to resist the damaging influences of stress. What is clear is that persons exposed to severe stresses can die, usually of circulatory failure, if their plasma cortisol concentration does not increase above basal levels.

Effect 4 in the table stems originally from the known fact that administration of very large amounts of cortisol profoundly reduces the inflammatory response to injury or infection. Cortisol can also reduce the number of circulating lymphocytes and decrease both antibody production and the activity of helper T cells and cytotoxic T cells. Because of all these effects, cortisol is an invaluable tool in the treatment of allergy, arthritis, other inflammatory diseases, and graft rejection.

These anti-inflammatory and anti-immune effects have generally been classified among the various *pharmacological* effects of cortisol because it was assumed that they could be achieved only by very large doses of cortisol. It is now clear, however, that such effects also occur, albeit to a lesser degree, at the plasma concentrations achieved during stress. Thus, the increased plasma cortisol typical of infection or trauma exerts a dampening effect on the body's immune responses, protecting against possible damage from excessive inflammation.

This explains the significance of the fact, mentioned earlier, that several cytokines stimulate the secretion of ACTH and thereby cortisol. Such stimulation is part of a negative-feedback system in which the increased cortisol then partially blocks the inflammatory processes in which the cytokines participate. Moreover, cortisol normally dampens the fever caused by an infection.

It should not be assumed, however, that all the effects of stress on immune responses are due to cortisol.

Many of the other hormones released in increased quantities during stress (see below) have important effects, both inhibitory and stimulatory, on the immune system.

Functions of the Sympathetic Nervous System in Stress

Activation of the sympathetic nervous system during stress is often termed the **fight-or-flight response,** and the name is appropriate. A list of the major effects of increased sympathetic activity, including secretion of epinephrine, almost constitutes a guide to how to meet emergencies in which physical activity may be required and bodily damage may occur. Most of these actions have been discussed in other sections of the book, and they are listed in Table 20–13 with little or no comment. Actions 6 and 7, however, have not been mentioned before; they reflect stimulation by epinephrine of the brain respiratory centers and of platelet aggregation.

Also of considerable interest is the fact that the stress of birth causes a huge increase in plasma catecholamine concentrations in the newborn. They not only help in the arousal of the newborn, but perform a variety of other important functions, for example, stimulation of fluid absorption from the lung alveoli at birth.

Other Hormones Released During Stress

Other hormones that are usually released during many kinds of stress are aldosterone, vasopressin (ADH), growth hormone, glucagon, and β-endorphin (which is coreleased from the anterior pituitary with ACTH). Insulin secretion is usually decreased. The increases in vasopressin and aldosterone ensure the retention of

TABLE 20–13 Actions of the Sympathetic Nervous System, Including Epinephrine Secreted by the Adrenal Medulla, in Stress
1. Increased hepatic and muscle glycogenolysis (provides a quick source of glucose)
2. Increased breakdown of adipose tissue triacylglycerol (provides a supply of glycerol for gluconeogenesis and of fatty acids for oxidation)
3. Decreased fatigue of skeletal muscle
4. Increased cardiac output secondary to increased cardiac contractility and heart rate
5. Diverting blood from viscera to skeletal muscles by means of vasoconstriction in the former beds and vasodilation in the latter
6. Increased ventilation
7. Increased coagulability of blood

water and sodium within the body, an important adaptation in the face of potential losses by hemorrhage or sweating. Vasopressin also stimulates the secretion of ACTH, as we have seen, and may enhance learning. As described in Chapter 18, the overall effects of the changes in growth hormone, glucagon, and insulin are, like those of cortisol and epinephrine, to mobilize energy stores. The role of β-endorphin, if any, in stress, is still unclear.

This list of hormones whose secretion rates are altered by stress is by no means complete. It is likely that the secretion of almost every known hormone may be influenced by stress. For example, prolactin and thyroid hormone are often increased, whereas the pituitary gonadotropins and the sex steroids are decreased. The adaptive significance of many of these changes is unclear.

Psychological Stress and Disease

Throughout this section we have emphasized the adaptive value of the body's various responses to stress. It is now clear, however, that psychological stress, particularly if chronic, can have deleterious effects on the body, constituting important links in the mind-body interactions described on page 714. For example, it is very likely that the increased plasma cortisol associated with psychological stress can decrease the activity of the immune system enough to reduce the body's resistance to infection and, perhaps, cancer. It can also worsen the symptoms of diabetes because of its anti-insulin effects, and it can cause increased rate of death of neurons.

Similarly, it is possible that prolonged and repeated activation of the sympathetic nervous system by psychological stress may enhance the development of certain diseases, particularly atherosclerosis and

hypertension. For example, one can easily imagine that the increased blood lipid concentration and cardiac work could contribute to the former disease.

Thus, as we have seen in the case of cytokines, the body's adaptive stress responses, if excessive or inappropriate, may play a causal role in the development of diseases. Conversely, stress-related neuroendocrine responses probably play important roles in certain beneficial responses, notably the *placebo effect*—the frequent improvement or outright cure that occurs in many diseases when patients are given a pharmacologically inactive substance that they have reason to believe will produce good results.

SECTION C SUMMARY

I. Classical responses to stress, whether physical or psychological, are increased secretion of cortisol from the adrenal cortex and activation of the sympathetic nervous system, including release of epinephrine by the adrenal medulla.

II. The functions of these responses, summarized in Tables 20–12 and 20–13, can be viewed both as a preparation for fight or flight and for coping with new situations.

III. Other hormones released during stress include aldosterone, vasopressin, glucagon, growth hormone, and prolactin. Insulin secretion is usually decreased.

IV. Psychological stress can have inappropriate and deleterious effects on bodily functions and disease processes.

SECTION C KEY TERMS

stress
cortisol
epinephrine
corticotropin releasing
  hormone (CRH)

adrenocorticotropic hormone
  (ACTH)
fight-or-flight response

1. Diagram the CRH-ACTH-cortisol pathway.
2. List the functions of cortisol in stress.
3. List the major effects of activation of the sympathetic nervous system during stress.
4. List four hormones other than those listed in previous questions that increase during stress and one that decreases.

CHAPTER 20 CLINICAL TERMS

(Because of the subject matter of this chapter, it is difficult to distinguish between "physiological" key terms and "clinical" terms. This list is limited largely to disease-producing agents and disease processes.)

microbes
bacteria
viruses
oncogene
vaccine
combined immunodeficiency
acquired immune deficiency
   syndrome (AIDS)
antibiotics
acyclovir
human immunodeficiency
   virus (HIV)
graft rejection
cyclosporin
transfusion reaction
cross-matching
hemolytic disease of the
   newborn
allergy (hypersensitivity)
allergen
delayed hypersensitivity

immune-complex
   hypersensitivity
immediate hypersensitivity
IgE-mediated
   hypersensitivity
anaphylaxis
late-phase reaction
autoimmune disease
multiple sclerosis
myasthenia gravis
insulin-dependent diabetes
   mellitus
septic shock
Alzheimer's disease
asthma
rheumatoid arthritis
inflammatory bowel disease
chronic inflammatory
   diseases
placebo effect

CHAPTER 20 THOUGHT QUESTIONS

(*Answers are given in Appendix A.*)

1. If an individual failed to develop a thymus because of a genetic defect, what would happen to the immune responses mediated by antibodies and those mediated by cytotoxic T cells?
2. What abnormalities would a person with a neutrophil deficiency display? A person with a monocyte deficiency?
3. An experimental animal is given a drug that blocks phagocytosis. Will this drug prevent the animal's immune system from killing foreign cells via the complement system?
4. If the Fc portion of a patient's antibodies is abnormal, what effects could this have on antibody-mediated responses?
5. Would you predict that patients with AIDS would develop fever in response to an infection? Explain.
6. A patient with symptoms of hyperthyroidism is found to have circulating antibodies against the receptors for the thyroid hormones. Can you deduce the cause of hyperthyroidism?
7. Barbiturates and alcohol are normally metabolized by the MES. Can you deduce why the actions of an administered barbiturate last for a shorter time than normal in persons who chronically consume large quantities of alcohol?

Appendix A

Chapter 4

4-1 A drug could decrease acid secretion by (1) binding to the membrane sites that normally inhibit acid secretion, which would produce the same effect as the body's natural messengers that inhibit acid secretion; (2) binding to a membrane protein that normally stimulates acid secretion but not itself triggering acid secretion, thereby preventing the body's natural messengers from binding (competition); or (3) having an allosteric effect on the binding sites, which would increase the affinity of the sites that normally bind inhibitor messengers or decrease the affinity of those sites that normally bind stimulatory messengers.

4-2 The reason for a lack of insulin effect could be either a decrease in the number of available binding sites to which insulin can bind or a decrease in the affinity of the binding sites for insulin so that less insulin is bound. A third possibility, which does not involve insulin binding, would be a defect in the way the binding site triggers a cell response once it has bound insulin.

4-3 An increase in the concentration of compound A will lead to a decrease in the concentration of compound H by the route shown below. Sequential activations and inhibitions of proteins of this general type are frequently encountered in physiological control systems.

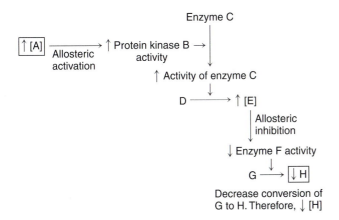

Decrease conversion of G to H. Therefore, ↓ [H]

4-4 (a) Acid secretion could be increased to 40 mmol/h by (1) increasing the concentration of compound X from 2 pM to 8 pM, thereby increasing the number of binding sites occupied; or (2) increasing the affinity of the binding sites for compound X, thereby increasing the amount bound without changing the concentration of compound X. (b) Increasing the concentration of compound X from 18 to 28 pM will not increase acid secretion because, at 18 pM, all the binding sites are occupied (the system is saturated), and there are no further binding sites available.

4-5 Phosphoprotein phosphatase removes the phosphate group from proteins that have been covalently modulated by a protein kinase. Without phosphoprotein phosphatase, the protein could not return to its unmodulated state and would remain in its activated state. The ability to decrease as well as increase protein activity is essential to the regulation of physiological processes.

4-6 The reactant molecules have a combined energy content of 55 + 93 = 148 kcal/mol, and the products have 62 + 87 = 149. Thus, the energy content of the products exceeds that of the reactants by 1 kcal/mol, and this amount of energy must be added to A and B to form the products C and D.

The reaction is reversible since the difference in energy content between the reactants and products is small. When the reaction reaches chemical equilibrium, there will be a slightly higher concentration of reactants than products.

4-7 The maximum rate at which the end product E can be formed is 5 molecules per second, the rate of the slowest—(rate-limiting)—reaction in the pathway.

4-8 Under normal conditions, the concentration of oxygen at the level of the mitochondria in cells, including muscle at rest, is sufficient to saturate the enzyme that combines oxygen with hydrogen to form water. The rate-limiting reactions in the electron transport chain depend on the available concentrations of ADP and P_i, which are combined to form ATP.

Thus, increasing the oxygen concentration above normal levels will not increase ATP production. If a muscle is contracting, it will break down ATP into ADP and P_i, which become the major rate-limiting substrates for increasing ATP production. With intense muscle activity, the level of oxygen may fall below saturating levels, limiting the rate of ATP production, and intensely active muscles must use anaerobic glycolysis to provide additional ATP. Under these circumstances, increasing the oxygen concentration in the blood will increase the rate of ATP production. As discussed in Chapter 14, it is not the concentration of oxygen in the blood that is increased during exercise but the rate of blood flow to a muscle, resulting in greater quantities of oxygen delivery to the tissue.

4-9 During starvation, in the absence of ingested glucose, the body's stores of glycogen are rapidly depleted. Glucose, which is the major fuel used by the brain, must now be synthesized from other types of molecules. Most of this newly formed glucose comes from the breakdown of proteins to amino acids and their conversion to glucose. To a lesser extent, the glycerol portion of fat is converted to glucose. The fatty acid portion of fat cannot be converted to glucose.

4-10 Fatty acids are broken down to acetyl coenzyme A during beta oxidation, and acetyl coenzyme A enters the Krebs cycle to be converted to carbon dioxide. Since the Krebs cycle can function only during aerobic conditions, the catabolism of fat is dependent on the presence of oxygen. In the absence of oxygen, acetyl coenzyme A cannot be converted to carbon dioxide, and the increased concentration of acetyl coenzyme A inhibits the further beta oxidation of fatty acids.

4-11 Ammonia is formed in most cells during the oxidative deamination of amino acids and then travels to the liver via the blood. The liver detoxifies the ammonia by converting it to the nontoxic compound urea. Since the liver is the site in which ammonia is converted to urea, diseases that damage the liver can lead to an accumulation of ammonia in the blood, which is especially toxic to nerve cells. Note that it is not the liver that produces the ammonia.

Chapter 5

5-1 Nucleotide bases in DNA pair A to T and G to C. Given the base sequence of one DNA strand as:

A-G-T-G-C-A-A-G-T-C-T

a. The complementary strand of DNA would be:

T-C-A-C-G-T-T-C-A-G-A

b. The sequence in RNA transcribed from the first strand would be:

U-C-A-C-G-U-U-C-A-G-A

Recall that uracil U replaces thymine T in RNA.

5-2 The triplet code G-T-A in DNA will be transcribed into mRNA as C-A-U, and the anticodon in tRNA corresponding to C-A-U is G-U-A.

5-3 If the gene were only composed of the triplet exon code words, the gene would be 300 nucleotides in length since a triplet of three nucleotides codes for one amino acid. However, because of the presence of intron segments in most genes, which account for 75 to 90 percent of the nucleotides in a gene, the gene would be between 1200 and 3000 nucleotides long; moreover, there are also termination codons. Thus, the exact size of a gene cannot be determined by knowing the number of amino acids in the protein coded by the gene.

5-4 Tubulin is the protein that polymerizes to form microtubules. The tubulin monomers become linked together in a spiral that forms the walls of the hollow microtubules. Without microtubules to form the spindle apparatus, the chromosomes will not separate during mitosis. There are chemical agents in the cell that cause the polymerization of tubulin at the time of mitosis and cause it to depolymerize following cell division.

5-5 A drug that inhibits DNA replication will inhibit cell division since a duplicate set of chromosomes is necessary for this process. Since one of the characteristics of cancer cells is their ability to undergo excessive uncontrolled division, a drug that inhibits DNA replication will inhibit the multiplication of cancer cells. Unfortunately, such drugs also inhibit the division of normal cells, particularly those that divide at a high rate. These include the cells that give rise to blood cells and the epithelial lining of the gastrointestinal tract. The use of such drugs must be carefully monitored to balance the damage done to normal tissues against the inhibition of tumor growth.

Chapter 6

6-1 (a) During diffusion, the net flux always occurs from high to low concentration. Thus, it will be from 2 to 1 in A and from 1 to 2 in B. (b) At equilibrium, the concentrations of solute in the two compartments will be equal: 4 mM in case A and 31 mM in case B. (c) Both will reach diffusion equilibrium at the same rate since the difference in concentration across the membrane is the same in each case, 2 mM $[(3 - 5) = -2$, and $(32 - 30) = 2]$. The two one-way fluxes will be much larger in B than in A, but the net flux has the same magnitude in both cases, although oriented in opposite directions.

6-2 The ability of one amino acid to decrease the flux of a second amino acid across a cell membrane is an example of the competition of two molecules for the same binding site, as explained in Chapter 4. The binding site for alanine on the transport protein can also bind leucine. The higher the concentration of alanine, the greater the number of binding sites that it occupies, and the fewer available for binding leucine. Thus, less leucine will be moved into the cell.

6-3 The net transport will be out of the cell in the direction from the higher-affinity site on the intracellular surface to the lower-affinity site on the extracellular surface. More molecules will be bound to the transporter on the higher-affinity side of the membrane, and thus more will move out of the cell than into it, until the concentration in the extracellular fluid becomes large enough that the number of molecules bound to transporters at the extracellular surface is equal to the number bound at the intracellular surface.

6-4 Although ATP is not used directly in secondary active transport, it is necessary for the primary active transport of sodium out of cells. Since it is the sodium concentration gradient across the plasma membrane that provides the energy for most secondary active-transport systems, a decrease in ATP production will decrease primary active sodium transport, leading to a decrease in the sodium concentration gradient and thus to a decrease in secondary active transport.

6-5 The solution with the greatest osmolarity will have the lowest water concentration. The osmolarities are:

A. $20 + 30 + (2 \times 150) + (3 \times 10) = 380$ mOsm
B. $10 + 100 + (2 \times 20) + (3 \times 50) = 300$ mOsm
C. $100 + 200 + (2 \times 10) + (3 \times 20) = 380$ mOsm
D. $30 + 10 + (2 \times 60) + (3 \times 100) = 460$ mOsm

Thus, solution D has the lowest water concentration. (Recall that NaCl forms two ions in solution and $CaCl_2$ forms three.)

Solution B is isoosmotic since it has the same osmolarity as intracellular fluid.

6-6 Initially the osmolarity of compartment 1 is $(2 \times 200) + 100 = 500$ mOsm and that of 2 is $(2 \times 100) + 300 = 500$ mOsm. The two solutions thus have the same osmolarity, and there is no difference in water concentration across the membrane. Since the membrane is permeable to urea, this substance will undergo net diffusion until it reaches the same concentration (200 mM) on the two sides of the membrane. In other words, in the steady state it will not affect the volumes of the compartments. In contrast, the higher initial NaCl concentration in compartment 1 than in compartment 2 will cause, by osmosis, the movement of water from compartment 2 to compartment 1 until the concentration of NaCl in both is 150 mM. Note that the same volume change would have occurred if there were no urea present in either compartment. It is only the concentration of nonpenetrating solutes (NaCl in this case) that determines the volume change, regardless of the concentration of any penetrating solutes that are present.

6-7 The osmolarities and nonpenetrating-solute concentrations are:

Solution	Osmolarity, mOsm	Nonpenetrating solute concentration, mOsm
A	$(2 \times 150) + 100 = 400$	$2 \times 150 = 300$
B	$(2 \times 100) + 150 = 350$	$2 \times 100 = 200$
C	$(2 \times 200) + 100 = 500$	$2 \times 200 = 400$
D	$(2 \times 100) + 50 = 250$	$2 \times 100 = 200$

Only the concentration of nonpenetrating solutes (NaCl in this case) will determine the change in cell volume. Since the intracellular concentration of nonpenetrating solute is 300 mOsm, solution A will produce no change in cell volume. Solutions B and D will cause cells to swell since they have a lower concentration of nonpenetrating solute (higher water concentration) than the intracellular fluid. Solution C will cause cells to shrink because it has a higher concentration of nonpenetrating solute than the intracellular fluid.

6-8 Solution A is isotonic because it has the same concentration of nonpenetrating solutes as intracellular fluid (300 mOsm). Solution A is also hyperosmotic since its total osmolarity is greater than 300 mOsm, as is also true for solutions B and C. Solution B is hypotonic since its concentration of nonpenetrating solutes is less than 300 mOsm. Solution C is hypertonic since its concentration of nonpenetrating solutes is greater than 300 mOsm. Solution D is hypotonic (less than 300 mOsm of nonpenetrating solutes) and also hypoosmotic (having a total osmolarity of less than 300 mOsm).

6-9 Exocytosis is triggered by an increase in cytosolic calcium concentration. Calcium ions are actively transported out of cells, in part by secondary countertransport coupled to the downhill entry of sodium ions on the same transporter. If the intracellular concentration of sodium ions were increased, the sodium concentration gradient across the membrane would be decreased, and this would decrease the secondary active transport of calcium out of the cell. This would lead to an increase in cytosolic calcium concentration, which would trigger increased exocytosis.

Chapter 7

7-1 4.4 mmol/L. If you answered 8 mmol/L, you ignored the fact that plasma potassium concentration is homeostatically regulated so that the doubling of input will lead to negative-feedback reflexes that oppose an equivalent increase in plasma concentration. If you answered 4 mmol/L, you ignored the fact that homeostatic control systems cannot *totally* prevent changes in the regulated variable when a perturbation occurs. Thus, there must be *some* rise in plasma

potassium concentration in this situation (to serve as the error signal for the compensating reflexes), and this is consistent with the answer 4.4 mmol/L. (The actual rise would have to be experimentally determined. There is no way you could have predicted that the rise would be 10 percent. All you could predict is that it would neither double nor stay absolutely unchanged.)

7-2 No. There may in fact be one, but there is another possibility—that the altered skin blood flow in the cold represents an *acclimatization* undergone by each Eskimo during his or her lifetime as a result of performing such work repeatedly.

7-3 Patient A's drug very likely acts to block phospholipase A_2, whereas patient B's drug blocks lipoxygenase (see Figure 7–6).

7-4 The chronic loss of exposure of the heart's receptors to norepinephrine causes an up-regulation of this receptor type (that is, more receptors in the heart for norepinephrine). The drug, being an agonist of norepinephrine (that is, able to bind to norepinephrine's receptors and activate them) is now more effective since there are more receptors for it to combine with.

7-5 None. Since you are told that all six responses are mediated by the cAMP system, then blockage of any of the steps listed in the question would eliminate all six of the responses. This is because the cascade for all six responses is identical from the receptor through the formation of cAMP and activation of cAMP-dependent protein kinase. Thus, the drug must be acting at a point beyond this kinase (for example, at the level of the phosphorylated protein mediating this response).

7-6 Not in most cells, since there are other physiological mechanisms by which signals impinging on the cell can increase cytosolic calcium concentration. These include (1) second-messenger–induced release of calcium from the endoplasmic reticulum and (2) voltage-sensitive calcium channels.

Chapter 8

8-1 Little change in the resting membrane potential would occur when the pump first stops because the pump's *direct* contribution to charge separation is very small. With time, however, the membrane potential would depolarize progressively toward zero because the sodium and potassium concentration gradients, which depend on the Na,K-ATPase pumps and which gives rise to the diffusion potentials that constitute most of the membrane potential, run down.

8-2 The resting potential would decrease (that is, become less negative) because the concentration gradient causing net diffusion of this positively charged ion out of the cell would be smaller. The action potential would fire more easily (that is, with smaller stimuli) because the resting potential would be closer to threshold. It would repolarize more slowly because repolarization depends on net potassium diffusion from the cell, and the concentration gradient driving this diffusion is lower. Also, the afterhyperpolarization would be smaller.

8-3 The hypothalamus was probably damaged. It plays a critical role in appetite, thirst, and sexual capacity.

8-4 The drug probably blocks cholinergic muscarinic receptors. These receptors on effector cells mediate the actions of parasympathetic nerves. Therefore, the drug would remove the slowing effect of these nerves on the heart, allowing the heart to speed up. Blocking their effect on the salivary glands would cause the dry mouth. We know that the drug is not blocking cholinergic nicotinic receptors because the skeletal muscles are not affected.

8-5 Since the membrane potential of the cells in question depolarizes (that is, becomes less negative) when chloride channels are blocked, one can predict that there was net chloride diffusion into the cells through these channels prior to the drug. Therefore, one can also predict that this passive inward movement was being exactly balanced by active transport of chloride out of the cells.

8-6 Without acetylcholinesterase, more acetylcholine would remain bound to the receptors, and all the actions normally caused by acetylcholine would be accentuated. Thus, there would be marked narrowing of the pupils, airway constriction, stomach cramping and diarrhea, sweating, salivation, slowing of the heart, and fall in blood pressure. On the other hand, in skeletal muscles, which must repolarize after excitation in order to be excited again, there would be weakness, fatigue, and finally inability to contract. In fact, lethal poisoning by high doses of cholinesterase inhibitors occurs because of paralysis of the muscles involved in respiration. Low doses of these compounds are used therapeutically.

8-7 These potassium channels, which open after a short delay following the initiation of an action potential, increase potassium diffusion out of the cell, hastening repolarization. They also account for the increased potassium permeability that causes the afterhyperpolarization. Therefore, the action potential would be broader (that is, longer in duration), returning to its resting level more slowly, and the afterhyperpolarization would be absent.

Chapter 9

9-1 (a) Use drugs to block transmission in the pathways that convey information about pain to the brain. For example, if substance P is the neurotransmitter at the central endings of the nociceptor afferent fibers, give a drug that blocks the substance P receptors. (b) Cut the dorsal root at the level of entry of the nociceptor fibers to prevent transmission of their action potentials into the central nervous system. (c) Give a drug that activates receptors in the descending pathways that block transmission of the incoming or ascending pain information. (d) Stimulate the neurons in these same descending pathways to increase their blocking activity (stimulation-produced analgesia or, possibly, acupuncture). (e) Cut the ascending pathways that transmit information from the nociceptor afferents. (f) Deal with the emotions, attitudes, memories, and so on, to decrease the sensitivity to the pain. (g) Stimulate nonpain, low-threshold afferent fibers to block transmission through the pain pathways (TENS). (h) Block transmission in the afferent nerve with a local anesthetic such as Novacaine or Lidocaine.

9-2 Information regarding temperature is carried via the anterolateral system to the brain. Fibers of this system cross to the opposite side of the body in the spinal cord at the level of entry of the afferent fibers (see Figure 9–18b). Damage to the left side of the spinal cord or any part of the left side of the brain that contains fibers of the pathways for temperature would interfere with awareness of a heat stimulus on the right. Thus, damage to the somatosensory cortex of the left cerebral hemisphere (that is, opposite the stimulus) would interfere with awareness of the stimulus. Injury to the spinal cord at the point at which fibers of the anterolateral system from the two halves of the spinal cord cross to the opposite side would interfere with the awareness of heat applied to either side of the body, as would the unlikely event that damage occurred to relevant areas of both sides of the brain.

9-3 Vision would be restricted to the rods; therefore, it would be normal at very low levels of illumination (when the cones would not be stimulated anyway), but at higher levels of illumination clear vision of fine details would be lost, and everything would appear in shades of gray. There would be no color vision. In very bright light, there would be no vision because of bleaching of the rods' rhodopsin.

9-4 (a) The individual lacks a functioning primary visual cortex. (b) The individual lacks a functioning visual association cortex.

Chapter 10

10-1 Epinephrine falls to very low levels during rest and fails to increase during stress. The sympathetic preganglionics provide the only major control of the adrenal medulla.

10-2 The increased concentration of binding protein causes more TH to be bound, thereby lowering the plasma concentration of *free* TH. This causes less negative-feedback inhibition of TSH secretion by the anterior pituitary, and the increased TSH causes the thyroid to secrete more TH until the free concentration has returned to normal. The end result is an increased *total* plasma TH—most bound to the protein—but a normal free TH. There is no hyperthyroidism because it is only the free concentration that exerts effects on TH's target cells.

10-3 Destruction of the anterior pituitary or hypothalamus. These symptoms reflect the absence of, in order, growth hormone, the gonadotropins, and ACTH (the symptom is due to the resulting decrease in cortisol secretion). The problem is either primary hyposecretion of anterior pituitary hormones or secondary hyposecretion because the hypothalamus is not secreting hypophysiotropic hormones normally.

10-4 Vasopressin and oxytocin (that is, the major posterior pituitary hormones). The anterior pituitary hormones would not be affected because the influence of the hypothalamus on these hormones is exerted not by connecting nerves but via the hypophysiotropic hormones in the portal vascular system.

10-5 The secretion of GH increases. Somatostatin, coming from the hypothalamus, normally exerts an inhibitory effect on the secretion of this hormone.

10-6 Norepinephrine and many other neurotransmitters are released by neurons that terminate on the hypothalamic neurons that secrete the hypophysiotropic hormones. Therefore, manipulation of these neurotransmitters will alter secretion of the hypophysiotropic hormones and thereby the anterior pituitary hormones.

10-7 The high dose of the cortisol-like substance inhibits the secretion of ACTH by feedback inhibition of (1) hypothalamic corticotropin releasing hormone and (2) the response of the anterior pituitary to this hypophysiotropic hormone. The lack of ACTH causes the adrenal to atrophy and decrease its secretion of cortisol.

10-8 The hypothalamus. The low basal TSH indicates either that the pituitary is defective or that it is receiving inadequate stimulation (TRH) from the hypothalamus. If the thyroid itself were defective, basal TSH would be elevated because of less negative-feedback inhibition by TH. The TSH increase in response to TRH shows that the pituitary is capable of responding to a stimulus and so is unlikely to be defective. Therefore, the problem is that the hypothalamus is secreting too little TRH.

Chapter 11

11-1 Under resting conditions, the myosin has already bound and hydrolyzed a molecule of ATP, resulting in an energized molecule of myosin ($M^* \cdot ADP \cdot P_i$). Since ATP is necessary to detach the myosin cross bridge from actin at the end of cross-bridge movement, the absence of ATP will result in rigor mortis, in which case the cross bridges become bound to actin but do not detach, leaving myosin bound to actin ($A \cdot M$).

11-2 No. The transverse tubules conduct the muscle action potential from the plasma membrane into the interior of the fiber, where it can trigger the release of calcium from the sarcoplasmic reticulum. If the transverse tubules were not attached to the plasma membrane, an action potential could not be conducted to the sarcoplasmic reticulum and there would be no release of calcium to initiate contraction.

11-3 The length-tension relationship states that the maximum tension developed by a muscle decreases at lengths below l_o. During normal shortening, as the sarcomere length becomes shorter than the optimal length, the maximum tension that can be generated decreases. With a light load, the muscle will continue to shorten until its maximal tension just equals the load. No further shortening is possible since at shorter sarcomere lengths the tension would be less than the load. The heavier the load, the less the distance shortened before reaching the isometric state.

11-4 Maximum tension is produced when the fiber is (1) stimulated by an action potential frequency that is high enough to produce a maximal tetanic tension, and (2) at its optimum length l_o, where the thick and thin filaments have overlap sufficient to provide the greatest number of cross bridges for tension production.

11-5 Moderate tension—for example, 50 percent of maximal tension—is accomplished by recruiting sufficient numbers of motor units to produce this degree of tension. If activity is maintained at this level for prolonged periods, some of the active fibers will begin to fatigue and their contribution to the total tension will decrease. The same level of total tension can be maintained, however, by recruiting new motor units as some of the original ones fatigue. At this point,

for example, one might have 50 percent of the fibers active, and 25 percent fatigued and 25 percent still unrecruited. Eventually, when all the fibers have fatigued and there are no additional motor units to recruit, the whole muscle will fatigue.

11-6 The oxidative motor units, both fast and slow, will be affected first by a decrease in blood flow since they depend on blood flow to provide both the fuel—glucose and fatty acids—and the oxygen required to metabolize the fuel. The fast-glycolytic motor units will be affected more slowly since they rely predominantly on internal stores of glycogen, which is anaerobically metabolized by glycolysis.

11-7 Two factors lead to recovery of muscle force. (1) Some new fibers can be formed by the fusion and development of undifferentiated satellite cells. This will replace some, but not all, of the fibers that were damaged. (2) Some of the restored force results from hypertrophy of the surviving fibers. Because of the loss of fibers in the accident, the remaining fibers must produce more force to move a given load. The remaining fibers undergo increased synthesis of actin and myosin, resulting in increases in fiber diameter and thus their force of contraction.

11-8 In the absence of extracellular calcium ions, skeletal muscle contracts normally in response to an action potential generated in its plasma membrane because the calcium required to trigger contraction comes entirely from the sarcoplasmic reticulum within the muscle fibers. If the motor neuron to the muscle is stimulated in a calcium-free medium, however, the muscle will not contract because the influx of calcium from the extracellular fluid into the motor nerve terminal is necessary to trigger the release of acetylcholine that in turn triggers an action potential in the muscle.

In a calcium-free solution, smooth muscles of all types would fail to respond to stimulation of the nerve supplying the muscle. However, the response to direct stimulation of the muscle's plasma membrane would depend on the type of smooth muscle. Smooth muscles that primarily depend on calcium released from the sarcoplasmic reticulum will behave like skeletal muscle and respond to direct stimulation. Smooth muscles that rely on the influx of calcium from the extracellular fluid to trigger contraction will fail to contract in response to stimulation of its plasma membrane.

11-9 The simplest model to explain the experimental observations is as follows. Upon parasympathetic nerve stimulation, a neurotransmitter is released that binds to receptors on the membranes of smooth-muscle cells and triggers contraction. The substance released, however, is not acetylcholine (ACh) for the following reason.

Action potentials in the parasympathetic nerves are essential for initiating nerve-induced contraction. When the nerves were prevented from generating action potentials by blockage of their voltage-gated sodium channels, there was no response to nerve stimulation. ACh is the neurotransmitter released from most, but not all, parasympathetic endings. When the muscarinic receptors for ACh were blocked, however, stimulation of the parasympathetic nerves still produced a contraction, providing evidence that some substance other than ACh is being released by the neurons and producing contraction.

Chapter 12

12-1   None. The gamma motor neurons are important in preventing the muscle-spindle stretch receptors from going slack, but when testing this reflex, the intrafusal fibers are not flaccid. The test is performed with a bent knee, which stretches the extensor muscles in the thigh (and the intrafusal fibers within the stretch receptors). The stretch receptors are therefore responsive.

12-2   The efferent pathway of the reflex arc (the alpha motor neurons) would not be activated, the effector cells (the extrafusal muscle fibers) would not be activated, and there would be no reflex response.

12-3   The drawing must have excitatory synapses on the motor neurons of both ipsilateral extensor and ipsilateral flexor muscles.

12-4   A toxin that interferes with the inhibitory synapses on motor neurons would leave unbalanced the normal excitatory input to these neurons. Thus, the otherwise normal motor neurons would fire excessively, which would result in increased muscle contraction. This is exactly what happens in lockjaw as a result of the toxin produced by the tetanus bacillus.

Chapter 13

13-1   Dopamine is depleted in the basal ganglia of people with Parkinson's disease, and they are therapeutically given dopamine agonists, usually L-dopa. This treatment raises dopamine levels in other parts of the brain, however, where the dopamine levels were previously normal. Schizophrenia is associated with increased brain dopamine levels, and symptoms of this disease appear when dopamine levels are high. The converse therapeutic problem can occur during the treatment of schizophrenics with dopamine-lowering drugs, which sometimes causes the symptoms of Parkinson's disease to appear.

13-2   Experiments on anesthetized animals often involve either stimulating a brain part to observe the effects of increased neuronal activity or damaging ("lesioning") an area to observe resulting deficits. Such experiments on animals, which lack the complex language mechanisms of people, cannot help with language studies. Diseases sometimes mimic these two experimental situations, and behavioral studies of the resulting language deficits in people with aphasia, coupled with study of their brains after death, have provided a wealth of information.

Chapter 14

14-1   No. Decreased erythrocyte volume is certainly one possible explanation, but there is a second: The person might have a normal erythrocyte volume but an abnormally increased plasma volume. Convince yourself of this by writing the hematocrit equation as: erythrocyte volume/(erythrocyte volume + plasma volume).

14-2   A halving of tube radius. Resistance is directly proportional to blood viscosity but inversely proportional to the *fourth power* of tube radius.

14-3   The plateau of the action potential and the contraction would be absent. You might think that contraction would persist since most calcium in excitation-contraction coupling in the heart comes from the sarcoplasmic reticulum. However, the signal for the release of this calcium is the calcium entering across the plasma membrane.

14-4   The SA node is not functioning, and the ventricles are being driven by a pacemaker in the AV node or the bundle of His.

14-5   The person has a narrowed aortic valve. Normally, the resistance across the aortic valve is so small that there is only a tiny pressure difference between the left ventricle and the aorta during ventricular ejection. In the example given here, the large pressure difference indicates that resistance across the valve must be very high.

14-6   This question is analogous to question 14-5 in that the large pressure difference across a valve while the valve is open indicates an abnormally narrowed valve—in this case, the left AV valve.

14-7   Decreased heart rate and contractility. These are effects mediated by the sympathetic nerves on beta-adrenergic receptors in the heart.

14-8   120 mmHg. MAP = DP + 1/3 (SP − DP).

14-9   The drug must have caused the arterioles in the kidneys to dilate enough to reduce their resistance by 50 percent. Blood flow to an organ is determined by mean arterial pressure and the organ's resistance to flow. Another important point can be deduced here: If mean arterial pressure has not changed even though renal resistance has dropped 50 percent, then either the resistance of some other organ or cardiac output has gone up.

14-10   The experiment suggests that acetylcholine causes vasodilation by releasing nitric oxide or some other vasodilator from endothelial cells.

14-11   A low plasma protein concentration. Capillary pressure is, if anything, lower than normal and so cannot be causing the edema. Another possibility is that capillary permeability to plasma proteins has increased, as occurs in burns.

14-12   20 mmHg/L per minute. TPR = MAP/CO.

14-13   Nothing. Cardiac output and TPR have remained unchanged, and so their product, MAP, has also remained unchanged. This question emphasizes that MAP depends on cardiac output but not on the combination of heart rate and stroke volume that produces the cardiac output.

14-14   It increases. There are a certain number of impulses traveling up the nerves from the arterial baroreceptors. When these nerves are cut, the number of impulses reaching the medullary cardiovascular center goes to zero, just as it would physiologically if the mean arterial pressure were to decrease markedly. Accordingly, the medullary cardiovascular center responds to the absent impulses by reflexly increasing arterial pressure.

14-15   It decreases. The hemorrhage causes no immediate change in hematocrit since erythrocytes and plasma are lost in the same proportion. As interstitial fluid starts entering the capillaries, however, it expands the plasma volume and decreases hematocrit. (This is too soon for any new erythrocytes to be synthesized.)

Chapter 15

15-1   200 ml/mmHg.

Lung compliance $= \Delta$ lung volume$/\Delta$ $(P_{alv} - P_{ip})$
$= 800$ ml$/[0 - (-8)]$ mmHg $-$
$\qquad\qquad\qquad [0 - (-4)]$ mmHg
$= 800$ ml$/4$ mmHg $= 200$ ml/mmHg

15-2   More subatmospheric than normal. A decreased surfactant level causes the lungs to be less compliant (that is, more difficult to expand). Therefore, a greater transpulmonary pressure $(P_{alv} - P_{ip})$ is required to expand them a given amount.

15-3   No.

Alveolar ventilation $=$ (tidal volume-dead space) $\times$
$\qquad\qquad\qquad\qquad\qquad$ breathing rate
$\qquad = (250$ ml $- 150$ ml$) \times 20$ breaths/min
$\qquad = 2000$ ml/min

whereas normal alveolar ventilation is approximately 4000 ml/min.

15-4   The volume of the snorkel constitutes an additional dead space, and so total pulmonary ventilation must be increased if alveolar ventilation is to remain constant.

15-5   The alveolar P_{O_2} will be higher than normal, and the alveolar P_{CO_2} will be lower. If you do not understand why, review the factors that determine the alveolar gas pressures.

15-6   No. Hypoventilation reduces arterial P_{O_2} but only because it reduces alveolar P_{O_2}. That is, in hypoventilation, *both* alveolar and arterial P_{O_2} are decreased to essentially the same degree. In this problem, alveolar P_{O_2} is normal, and so the person is not hypoventilating. The low arterial P_{O_2} must therefore represent a defect that causes a discrepancy between alveolar P_{O_2} and arterial P_{O_2}. Possibilities include impaired diffusion, a shunting of blood from the right side of the heart to the left through a hole in the heart wall, and mismatching of airflow and blood flow in the alveoli.

15-7   Not at rest, if the defect is not too severe. Recall that equilibration of alveolar air and pulmonary capillary blood is normally so rapid that it occurs well before the end of the capillaries. Therefore, even though diffusion may be retarded, as in this problem, there may still be enough time for equilibration to be reached. In contrast, the time for equilibration is decreased during exercise, and failure to equilibrate is much more likely to occur, resulting in a lowered arterial P_{O_2}.

15-8   Only a few percent (specifically, from approximately 200 ml O_2/L blood to approximately 215 ml O_2/L blood). The reason the increase is so small is that almost all the oxygen in blood is carried bound to hemoglobin, and hemoglobin is almost 100 percent saturated at the arterial P_{O_2} achieved by breathing room air. The high arterial P_{O_2} achieved by breathing 100 percent oxygen does cause a directly proportional increase in the amount of oxygen *dissolved* in the blood (the additional 15 ml), but this still remains a small fraction of the total oxygen in the blood. Review the numbers given in the text.

15-9   All except plasma chloride concentration. The reasons are all given in the text.

15-10   It would cease. Respiration depends on descending input from the medulla to the nerves supplying the diaphragm and the inspiratory intercostal muscles.

15-11   The 10 percent oxygen mixture will markedly lower alveolar and thus arterial P_{O_2}, but no increase in ventilation will occur because the reflex response to hypoxia is initiated solely by the peripheral chemoreceptors. The 5 percent carbon dioxide mixture will markedly increase alveolar and arterial P_{CO_2}, and a large increase in ventilation will be elicited reflexly via the central chemoreceptors. The increase will not be as large as in a normal animal because the peripheral chemoreceptors do play a role, albeit minor, in the reflex response to elevated P_{CO_2}.

15-12   These patients have profound hyperventilation, with marked increases in both the depth and rate of ventilation. The stimulus, mainly via the peripheral chemoreceptors, is the marked increase in their arterial hydrogen-ion concentration due to the acids produced. The hyperventilation causes an increase in their arterial P_{O_2} and a decrease in their arterial P_{CO_2}.

Chapter 16

16-1   No. These are possibilities, but there is another. Substance T may be secreted by the tubules.

16-2   No. It is a possibility, but there is another. Substance V may be filtered and/or secreted, but the substance V entering the lumen via these routes may be completely reabsorbed.

16-3   125 mg/min. The amount of any substance filtered per unit time is given by the product of the GFR and the filterable plasma concentration of the substance—in this case, 125 ml/min $\times$ 100 mg/100 ml $= 125$ mg/min.

16-4   The plasma concentration might be so high that the T_m for the amino acid is exceeded, and so all the filtered amino acid is not reabsorbed. A second possibility is that there is a specific defect in the tubular transport for this amino acid. A third possibility is that some other amino acid is present in the plasma in high concentration and is competing for reabsorption.

16-5   No. Urea is filtered and then partially reabsorbed. The reason its concentration in the tubule is higher than in the plasma is that relatively more water is reabsorbed than urea. Therefore, the urea in the tubule becomes concentrated. Despite the fact that urea *concentration* in the urine is greater than in the plasma, the *amount excreted* is less than the filtered load (that is, net reabsorption has occurred).

16-6   They would all be decreased. The transport of all these substances is coupled, in one way or another, to that of sodium.

16-7   GFR would not go down as much, and renin secretion would not go up as much as in a person not receiving the drug. The sympathetic nerves are a major pathway for both responses during hemorrhage.

16-8   There would be little if any increase in aldosterone secretion. The major stimulus for increased aldosterone secretion is angiotensin II, but this substance is formed from angiotensin I by the action of angiotensin-converting enzyme, and so blockade of this enzyme would block the pathway.

16-9 (b) Urinary excretion in the steady state must be less than ingested sodium chloride by an amount equal to that lost in the sweat and feces. This is normally quite small, less than 1 g/day, so that urine excretion in this case equals approximately 11 g/day.

16-10 If the hypothalamus had been damaged, there might be inadequate secretion of ADH. This would cause loss of a large volume of urine, which would tend to dehydrate the person and make her thirsty. Of course, the area of the brain involved in thirst might have suffered damage.

16-11 Because aldosterone stimulates sodium reabsorption and potassium secretion, there will be total-body retention of sodium and loss of potassium. Interestingly, the person in this situation actually retains very little sodium because urinary sodium excretion returns to normal after a few days despite the continued presence of the high aldosterone. One explanation for this is that GFR and atrial natriuretic factor both increase as a result of the initial sodium retention.

16-12 Sodium and water balance would become negative because of increased excretion of these substances in the urine. The person would also develop a decreased plasma bicarbonate concentration and metabolic acidosis because of increased bicarbonate excretion. The effects on acid-base status are explained by the fact that hydrogen-ion secretion—blocked by the drug—is needed both for bicarbonate reabsorption and for the excretion of hydrogen ion (contribution of new bicarbonate to the blood). The increased sodium excretion reflects the fact that much sodium reabsorption by the proximal tubule is achieved by Na/H countertransport. By blocking hydrogen-ion secretion, therefore, the drug also partially blocks sodium reabsorption. The increased water excretion occurs because the failure to reabsorb sodium and bicarbonate decreases water reabsorption (remember that water reabsorption is secondary to solute reabsorption), resulting in an osmotic diuresis.

Chapter 17

17-1 If the salivary glands fail to secrete amylase, the undigested starch that reaches the small intestine will still be digested by the amylase secreted by the pancreas. Thus, starch digestion is not significantly affected by the absence of salivary amylase.

17-2 Alcohol can be absorbed across the stomach wall, but absorption is much more rapid from the small intestine with its larger surface area. Ingestion of foods containing fat releases enterogastrones from the small intestine, and these hormones inhibit gastric emptying and thus prolong the time alcohol spends in the stomach before reaching the small intestine. Milk, contrary to popular belief, does not "protect" the lining of the stomach from alcohol by coating it with a fatty layer. Rather, the fat content of milk decreases the rate of absorption of alcohol by decreasing the rate of gastric emptying.

17-3 Vomiting results in the loss of fluid and acid from the body. The fluid comes from the luminal contents of the stomach and duodenum, most of which was secreted by the gastric glands, pancreas, and liver and thus is derived from the blood. The cardiovascular symptoms of this patient are the result of the decrease in blood volume that accompanies vomiting.

The secretion of acid by the stomach produces an equal number of bicarbonate ions, which are released into the blood. Normally these bicarbonate ions are neutralized by hydrogen ions released into the blood by the pancreas when this organ secretes bicarbonate ions. Because gastric acid is lost during vomiting, the pancreas is not stimulated to secrete bicarbonate by the usual high-acidity signal from the duodenum, and no corresponding hydrogen ions are formed to neutralize the bicarbonate released into the blood by the stomach. As a result, the acidity of the blood decreases.

17-4 Fat can be digested and absorbed in the absence of bile salts but in greatly decreased amounts. Without adequate emulsification of fat by bile salts and phospholipids, only the fat at the surface of large lipid droplets is available to pancreatic lipase, and the rate of fat digestion is very slow. Without the formation of micelles with the aid of bile salts, the products of fat digestion become dissolved in the large lipid droplets, where they are not readily available for diffusion into the epithelial cells. In the absence of bile salts, only about 50 percent of the ingested fat is digested and absorbed. The undigested fat is passed on to the large intestine, where the bacteria there produce compounds that increase colonic motility and promote the secretion of fluid into the lumen of the large intestine, leading to diarrhea.

17-5 Damage to the lower portion of the spinal cord produces a loss of voluntary control over defecation due to disruption of the somatic nerves to the skeletal muscle of the external anal sphincter. Damage to the somatic nerves leaves the external sphincter in a continuously relaxed state. Under these conditions, defecation occurs whenever the rectum becomes distended and the defection reflex is initiated.

17-6 Vagotomy decreases the secretion of acid by the stomach. Impulses in the parasympathetic nerves directly stimulate acid secretion by the parietal cells and also cause the release of gastrin, which in turn stimulates acid secretion. Impulses in the vagus nerves are increased during both the cephalic and gastric phases of digestion. Vagotomy, by decreasing the amount of acid secreted, decreases irritation of existing ulcers, which promotes healing and decreases the probability of acid contributing to the production of new ulcers.

Chapter 18

18-1 The concentration in plasma would increase, and the amount stored in adipose tissue would decrease. Lipoprotein lipase cleaves plasma triacylglycerols, so its blockade would decrease the rate at which these molecules were cleared from plasma and would decrease the availability of the fatty acids in them for synthesis of intracellular triacylglycerols. However, this would only reduce but not eliminate such synthesis, since the adipose-tissue cells could still synthesize their own fatty acids from glucose.

18-2 The person might be an insulin-dependent diabetic or might be a normal fasting person; plasma glucose would be increased in the first case but decreased in the second. Plasma insulin concentration would be useful because it would be decreased in both cases. The fact that the person was resting and unstressed was specified because severe stress or exercise could also produce the plasma changes mentioned in the question. Plasma glucose would be increased during stress and decreased during exercise.

18-3 Glucagon, epinephrine, and growth hormone. The insulin will produce hypoglycemia, which then induces reflex increases in the secretion of all these hormones.

18-4 It might reduce it but not eliminate it. The sympathetic effects on organic metabolism during exercise are mediated not only by circulating epinephrine but also by sympathetic nerves to the liver (glycogenolysis and gluconeogenesis), to the adipose tissue (lipolysis), and to the pancreatic islets (inhibition of insulin secretion and stimulation of glucagon secretion).

18-5 Increase. The stress of the accident will elicit increased activity of all the glucose-counterregulatory controls and will therefore necessitate more insulin to oppose these influences.

18-6 It will lower plasma cholesterol concentration. Bile salts are formed from cholesterol, and losses of these bile salts in the feces will be replaced by the synthesis of new ones from cholesterol. Chapter 17 describes how bile salts are normally absorbed from the small intestine so that very few of those secreted into the bile are normally lost from the body.

18-7 Plasma concentrations of HDL and LDL. It is the ratio of LDL cholesterol to HDL cholesterol that best correlates with the development of atherosclerosis (that is, HDL cholesterol is "good" cholesterol). The answer to this question would have been the same regardless of whether the person was an athlete or not, but the question was phrased this way to emphasize that people who exercise generally have increased HDL cholesterol.

18-8 In utero malnutrition. Neither growth hormone nor the thyroid hormones influence in utero growth.

18-9 Androgens stimulate growth but also cause the ultimate cessation of growth by closing the epiphyseal plates. Therefore, there might be a rapid growth spurt in response to the androgens but a subsequent premature cessation of growth. Estrogens exert similar effects.

18-10 Heat loss form the head, mainly via convection and sweating, is the major route for loss under these conditions. The rest of the body is *gaining* heat by conduction, and sweating is of no value in the rest of the body because the water cannot evaporate. Heat is also lost via the expired air (insensible loss), and some people actually begin to pant under such conditions. The rapid shallow breathing increases airflow and heat loss without causing hyperventilation.

18-11 They seek out warmer places, if available, so that their body temperature increases. That is, they use behavior to develop a fever. This is excellent evidence that the hyperthermia of infection is a fever (that is, a set-point change).

Chapter 19

19-1 Sterility due to lack of spermatogenesis would be the common symptom. The Sertoli cells are essential for spermatogenesis, and so is testosterone produced by the Leydig cells. The person with Leydig-cell destruction, but not the person with Sertoli-cell destruction, would also have other symptoms of testosterone deficiency.

19-2 The androgens act on the hypothalamus and anterior pituitary to inhibit the secretion of the gonadotropins. Therefore, spermatogenesis is inhibited. Importantly, even if this man were given FSH, the sterility would probably remain since the lack of LH would cause deficient testosterone secretion, and *locally* produced testosterone is needed for spermatogenesis (that is, the exogenous androgen cannot do this job).

19-3 Impaired function of the seminiferous tubules, notably of the Sertoli cells. The increased plasma FSH concentration is due to the lack of negative-feedback inhibition of FSH secretion by inhibin, itself secreted by the Sertoli cells. The Leydig cells seem to be functioning normally in this person, since the lack of demasculinization and the normal plasma LH indicate normal testosterone secretion.

19-4 FSH secretion. Since FSH acts on the Sertoli cells and LH acts on the Leydig cells, sterility would result in either case, but the loss of LH would also cause undesirable elimination of testosterone and its effects.

19-5 These findings are all due to testosterone deficiency. You would also expect to find that the testes and penis were small if the deficiency occurred before puberty.

19-6 They will be eliminated. The androgens act on the hypothalamus to inhibit the secretion of GnRH and on the pituitary to inhibit the response to GnRH. The result is inadequate secretion of gonadotropins and therefore inadequate stimulation of the ovaries. In addition to the loss of menstrual cycles, the woman will suffer some degree of masculinization of the secondary sex characteristics because of the combined effects of androgen excess and estrogen deficiency.

19-7 Such treatment may cause so much secretion of FSH that multiple dominant follicles are stimulated to develop simultaneously and have their eggs ovulated during the LH surge.

19-8 An increased plasma LH. The other two are due to increased plasma progesterone and so do not occur until *after* ovulation and formation of the corpus luteum.

19-9 The absence of sperm capacitation. When test-tube fertilization is performed, special techniques are used to induce capacitation.

19-10 The fetus is in difficulty. The placenta produces progesterone entirely on its own, whereas estriol secretion requires participation of the fetus, specifically, the fetal adrenal cortex.

19-11 Prostaglandin antagonists, oxytocin antagonists, and drugs that lower cytosolic calcium concentration. You may not have thought of the last category since calcium is not mentioned in this context in the text, but as in all muscle, calcium is the immediate cause of contraction in the myometrium.

19-12 This person would have normal male external genitals and testes, although the testes may not have descended fully, but would also have some degree of development of uterine tubes, a uterus, and a vagina. These internal female structures would tend to develop because no MIS was present to cause degeneration of the Müllerian duct system.

19-13 No. These two hormones are already elevated in menopause, and the problem is that the ovaries are unable to respond to them with estrogen secretion. Thus, the treatment must be with estrogen itself.

Chapter 20

20-1 Both would be impaired because T cells would not differentiate. The absence of cytotoxic T cells would eliminate responses mediated by these cells. The absence of helper T cells would impair antibody-mediated responses because most B cells require cytokines from helper T cells to become activated.

20-2 Neutrophil deficiency would impair nonspecific inflammatory responses to bacteria. Monocyte deficiency, by causing macrophage deficiency, would impair both nonspecific inflammation and specific immune responses.

20-3 The drug might reduce but would not eliminate the action of complement, since this system destroys cells directly (via the membrane attack complex) as well as by facilitating phagocytosis.

20-4 Antibodies would bind normally to antigen but might not be able to activate complement, act as opsonins, or recruit NK cells in ADCC. The reason for these defects is that the sites to which complement C1, phagocytes, and NK cells bind are all located in the Fc portion of antibodies.

20-5 They do develop fever, although often not to the same degree as normal. They can do so because IL-1 and other cytokines secreted by macrophages cause fever, whereas the defect in AIDS is failure of helper T cell function.

20-6 This person is suffering from an autoimmune attack against the receptors. The antibodies formed then bind to the receptors and activate them just as the thyroid hormones would have.

20-7 Alcohol over time induces a high level of MES activity, which causes an administered barbiturate to be metabolized more rapidly than normal.

Appendix B

GLOSSARY

A

A cell *see* alpha cell

absolute refractory period time during which an excitable membrane cannot generate an action potential in response to any stimulus

absorption movement of materials across an epithelial layer from body cavity or compartment toward the blood

absorptive state period during which nutrients enter bloodstream from gastrointestinal tract

accessory reproductive organ duct through which sperm or egg is transported, or a gland emptying into such a duct (in the female, the breasts are usually included)

acclimatization (ah-climb-ah-tih-ZAY-shun) environmentally induced improvement in functioning of a physiological system with no change in genetic endowment

accommodation adjustment of eye for viewing various distances by changing shape of lens

acetyl coenzyme A (acetyl CoA) (ASS-ih-teel koh-EN-zime A, koh-A) metabolic intermediate that transfers acetyl groups to Krebs cycle and various synthetic pathways

acetyl group —$COCH_3$

acetylcholine (ACh) (ass-ih-teel-KOH-leen) a neurotransmitter released by pre- and post-ganglionic parasympathetic neurons, preganglionic sympathetic neurons, somatic neurons, and some CNS neurons

acetylcholinesterase (ass-ih-teel-koh-lin-ES-ter-ase) enzyme that breaks down acetylcholine into acetic acid and choline

acid molecule capable of releasing a hydrogen ion; solution having an H^+ concentration greater than that of pure water (that is, pH less than 7); *see also* strong acid, weak acid

acidity concentration of free, unbound hydrogen ion in a solution; the higher the H^+ concentration, the greater the acidity

acidosis (ass-ih-DOH-sis) any situation in which arterial H^+ concentration is elevated above normal resting levels; *see also* metabolic acidosis, respiratory acidosis

acrosome (AK-roh-sohm) cytoplasmic vesicle containing digestive enzymes and located at head of a sperm

actin (AK-tin) globular contractile protein to which myosin cross bridges bind; located in muscle thin filaments and in micro-filaments of cytoskeleton

action potential electric signal propagated by nerve and muscle cells; an all-or-none depolarization of membrane polarity; has a threshold and refractory period and is conducted without decrement

activated macrophage macrophage whose killing ability has been enhanced by cytokines, particularly IL-2 and interferon-gamma

activation *see* lymphocyte activation

activation energy energy necessary to disrupt existing chemical bonds during a chemical reaction

active hyperemia (hy-per-EE-me-ah) increased blood flow through a tissue associated with increased metabolic activity

active immunity resistance to reinfection acquired by contact with microorganisms, their toxins, or other antigenic material; *compare* passive immunity

active site region of enzyme to which substrate binds

active transport energy-requiring system that uses transporters to move ions or molecules across a membrane against an electro-chemical difference; *see also*

primary active transport, secondary active transport

activity *see* enzyme activity

acute (ah-KUTE) lasting a relatively short time; *compare* chronic

acute phase proteins group of proteins secreted by liver during systemic response to injury or infection

acute phase response responses of tissues and organs distant from site of infection or immune response

adaptation (evolution) a biological characteristic that favors survival in a particular environment; (neural) decrease in action-potential frequency in a neuron despite constant stimulus

adenosine diphosphate (ADP) (ah-DEN-oh-seen dy-FOS-fate) two-phosphate product of ATP breakdown

adenosine monophosphate (AMP) one-phosphate derivative of ATP

adenosine triphosphate (ATP) major molecule that transfers energy from metabolism to cell functions during its breakdown to ADP and release of P_i

adenylyl cyclase (ad-DEN-ah-lil SY-klase) enzyme that catalyzes transformation of ATP to cyclic AMP

adipocyte (ad-DIP-oh-site) cell specialized for triacylglycerol synthesis and storage; fat cell

adipose tissue (AD-ah-poze) tissue composed largely of fat storing cells

adrenal cortex (ah-DREE-nal KOR-tex) endocrine gland that forms outer shell of each adrenal gland; secretes steroid hormones—mainly cortisol, aldosterone, and androgens; *compare* adrenal medulla

adrenal gland one of a pair of endocrine glands above each kidney; each gland consists of outer *adrenal cortex* and inner *adrenal medulla*

adrenal medulla (meh-DUL-ah) endocrine gland that forms inner core of each adrenal gland; secretes amine hormones, mainly epinephrine; *compare* adrenal cortex

adrenergic (ad-ren-ER-jik) pertaining to norepinephrine or epinephrine; compound that acts like norepinephrine or epinephrine

adrenocorticotropic hormone (ACTH) (ad-ren-oh-kor-tih-koh-TROH-pik) polypeptide hormone secreted by anterior pituitary; stimulates adrenal cortex to secrete cortisol; also called corticotropin

aerobic (air-OH-bik) in presence of oxygen

afferent (AF-er-ent) carrying toward

afferent arteriole vessel in kidney that carries blood from artery to renal corpuscle

afferent neuron neuron that carries information from sensory receptors at its peripheral endings to CNS; cell body lies outside CNS

afferent pathway component of reflex arc that transmits information from receptor to integrating center

affinity strength with which ligand binds to its binding site

afterbirth placenta and associated membranes expelled from uterus after delivery of infant

agonist (AG-ah-nist) chemical messenger that binds to receptor and triggers cell's response; often refers to drug that mimics action of chemical normally in the body

airway tube through which air flows between external environment and lung alveoli

albumin (al-BU-min) most abundant plasma proteins

aldosterone (al-doh-stir-OWN or al-DOS-stir-own) mineralocorticoid steroid hormone secreted by adrenal cortex; regulates electrolyte balance

alkaline having H^+ concentration lower than that of pure water (that is, having a pH greater than 7)

alkalosis (alk-ah-LOH-sis) any situation in which arterial blood H^+ concentration is reduced below normal resting levels; *see also* metabolic alkalosis, respiratory alkalosis

all-or-none pertaining to event that occurs maximally or not at all

allele (al-EEL) a gene that differs in nucleotide sequence from other copies of that same gene in the total human population

allosteric modulation (al-low-STAIR-ik) control of protein binding site properties by modulator molecules that bind to regions of the protein other than the binding site altered by them

allosteric protein protein whose binding site characteristics are subject to allosteric modulation

alpha cell glucagon-secreting cell of pancreatic islets of Langerhans

alpha-adrenergic receptor one type of plasma-membrane receptor for epinephrine and norepinephrine; also called alpha adrenoceptor; *compare* beta-adrenergic receptor

alpha-glycerol phosphate three-carbon molecule that combines with fatty acids to form tri-acylglycerol; also called glycerol 3-phosphate

alpha-ketoacid (AL-fuh KEY-toh) molecule formed from amino acid metabolism and containing carbonyl (—CO—) and carboxyl (—COOH) groups

alpha motor neuron motor neuron that innervates extrafusal skeletal-muscle fibers

alpha rhythm prominent 8- to 13-Hz oscillation on the electro-encephalograms of awake, relaxed adults with their eyes closed

alternate complement pathway sequence for complement activation that bypasses first steps in classical pathway and is not antibody dependent

alveolar dead space (al-VEE-oh-lar) volume of fresh inspired air that reaches alveoli but does not undergo gas exchange with blood

alveolar pressure (P_{alv}) air pressure in pulmonary alveoli

alveolar ventilation volume of atmospheric air entering alveoli each minute

alveolus (al-VEE-oh-lus) (lungs) thin-walled, air-filled "outpocketing" from terminal air passageways in lungs; (glands) cell cluster at end of duct in secretory gland

amine hormone (ah-MEEN) hormone derived from amino acid tyrosine; includes thyroid hormones, epinephrine, norepinephrine, and dopamine

amino acid (ah-MEEN-oh) molecule containing amino group, carboxyl group, and side chain attached to a carbon atom; molecular subunit of protein

amino group —NH_2; ionizes to —NH_3^+

aminopeptidase (ah-meen-oh-PEP-tih-dase) one of a family of enzymes located in the intestinal epithelial membrane; breaks peptide bond at amino end of polypeptide

ammonia NH_3; produced during amino acid breakdown; converted in liver to urea; ionized form is ammonium

amniotic sac (am-nee-AHT-ik) membrane surrounding fetus in utero

amphipathic (am-fuh-PATH-ik) a molecule containing polar or ionized groups at one end and nonpolar groups at the other

amplitude height, how much, magnitude of change

amylase (AM-ih-lase) enzyme that partially breaks down polysaccharides

anabolism (an-NAB-oh-lizm) cellular synthesis of organic molecules

anaerobic (an-ih-ROH-bik) in the absence of oxygen

anatomic dead space space in respiratory tract airways where gas exchange does not occur with blood

androgen (AN-dro-jen) any hormone with testosterone-like actions

anemia (ah-NEE-me-ah) reduction in total blood hemoglobin

angiogenesis (an-gee-oh-JEN-ah-sis) the development and growth of capillaries; stimulated by angiogenic factors

angiotensin I small polypeptide generated in plasma by renin's action on angiotensinogen

angiotensin II hormone formed by action of angiotensin converting enzyme on angiotensin I; stimulates aldosterone secretion from adrenal cortex, vascular smooth-muscle contraction, and thirst

angiotensin converting enzyme enzyme on capillary endothelial cells that catalyzes removal of two amino acids from angiotensin I to form angiotensin II

angiotensinogen (an-gee-oh-ten-SIN-oh-gen) plasma protein precursor of angiotensin I; produced by liver

anion (AN-eye-on) negatively charged ion; *compare* cation

antagonist (muscle) muscle whose action opposes intended movement; (drug) molecule that competes with another for a receptor and binds to the receptor but does not trigger the cell's response

anterior toward or at the front

anterior pituitary anterior portion of pituitary gland; synthesizes, stores, and releases ACTH, GH, TSH, prolactin, FSH, and LH

antibody (AN-tih-bah-dee) immunoglobulin secreted by plasma cell; combines with type of antigen that stimulated its production; directs attack against antigen or cell bearing it

antibody-dependent cellular cytotoxicity (ADCC) killing by toxic chemicals secreted by the NK cells of target cells linked to NK cells

anticodon (an-tie-KOH-don) three-nucleotide sequence in tRNA able to base-pair with complementary codon in mRNA during protein synthesis

antidiuretic hormone (ADH) (an-ty-dy-yor-ET-ik) *see* vasopressin

antigen (AN-tih-jen) any foreign molecule that stimulates a specific immune response

antigen-presenting cell (APC) cell that presents antigen, complexed with MHC proteins on its surface, to T cells

antithrombin III (an-ty-throm-bin-THREE) plasma anticlotting protein that inactivates thrombin

antrum (AN-trum) (gastric) lower portion of stomach (that is, region closest to pyloric sphincter); (ovarian) fluid-filled cavity in maturing ovarian follicle

aorta (a-OR-tah) largest artery in body; carries blood from left ventricle of heart to thorax and abdomen

aortic arch baroreceptor (a-OR-tik) *see* arterial baroreceptor

aortic body chemoreceptor chemoreceptor located near aortic arch; sensitive to arterial blood O_2 pressure and H^+ concentration

aortic valve valve between left ventricle of heart and aorta

apoptosis (a-poh-TOE-sis) self-destruction of a cell programmed by intrinsic "suicide" instructions

appendix small finger-like projection from cecum of large intestine

aquaporin (ah-qua-PORE-in) protein membrane channel through which water can diffuse

aqueous (AH-kwee-us) watery; prepared with water

arachidonic acid (ah-rak-ah-DON-ik) polyunsaturated fatty acid precursor of eicosanoids

arrhythmia (ay-RYTH-me-ah) any variation from normal heartbeat rhythm

arterial baroreceptor nerve endings sensitive to stretch or distortion produced by arterial blood pressure changes; located in carotid sinus or aortic arch; also called carotid sinus and aortic arch baroreceptors

arteriole (are-TEER-ee-ole) blood vessel between artery and capillary, surrounded by smooth muscle; primary site of vascular resistance

artery (ARE-ter-ee) thick-walled elastic vessel that carries blood away from heart to arterioles

ascending limb portion of Henle's loop of renal tubule leading to distal convoluted tubule

ascending pathway neural pathway that goes to the brain; also called sensory pathway

aspartate (ah-SPAR-tate) an excitatory neurotransmitter in CNS; ionized form of the amino acid aspartic acid

association cortex *see* cortical association areas

atmospheric pressure (P_{atm}) air pressure surrounding the body (760 mmHg at sea level)

atom smallest unit of matter that has unique chemical characteristics; has no net charge; combines with other atoms to form chemical substances

atomic nucleus dense region, consisting of protons and neutrons, at center of atom

atomic number number of protons in nucleus of atom

atomic weight value that indicates an atom's mass relative to mass of other types of atoms based on the assignment of a value of 12 to carbon atom

ATPase (aa-tea-PEE-ase) enzyme that breaks down ATP to ADP and inorganic phosphate

atrial natriuretic factor (ANF) (nat-ry-yor-ET-ik) peptide hormone secreted by cardiac atrial cells in response to atrial distension; causes increased renal sodium excretion

atrioventricular (AV) node (ay-tree-oh-ven-TRIK-you-lar) region at base of right atrium near interventricular septum, containing specialized cardiac muscle cells through which electrical activity must pass to go from atria to ventricles

atrioventricular (AV) valve valve between atrium and ventricle of heart; AV valve on right side of heart is the *tricuspid valve*, and that on the left side is the *mitral valve*

atrium (AY-tree-um) chamber of heart that receives blood from veins and passes it on to ventricle on same side of heart

atrophy (AT-roh-fee) wasting away; decrease in size

auditory (AW-dih-tor-ee) pertaining to sense of hearing

auditory cortex region of cerebral cortex that receives nerve fibers from auditory pathways

autocrine agent (AW-toh-crin) chemical messenger that is secreted into extracellular fluid and acts upon cell that secreted it; *compare* paracrine agent

automaticity (aw-toh-mah-TISS-ih-tee) capable of spontaneous, rhythmical self-excitation

autonomic nervous system (aw-toh-NAHM-ik) component of efferent division of peripheral nervous system that consists of sympathetic and parasympathetic subdivisions; innervates cardiac muscle, smooth muscle, and glands; *compare* somatic nervous system

autoreceptors receptors on a cell affected by a chemical messenger released from the same cell

autoregulation (aw-toh-reg-you-LAY-shun) ability of an individual organ to control (self-regulate) its vascular resistance independent of neural and hormonal influence; *see also* flow autoregulation

autosome chromosome that contains genes for proteins governing most cell structures and functions; *compare* sex chromosome

axon (AX-ahn) extension from neuron cell body; propagates action potentials away from cell body; also called a nerve fiber

axon terminal end of axon; forms synaptic or neuroeffector junction with postjunctional cell

axon transport process involving intracellular filaments by which materials are moved from one end of axon to other

B

B cell (immune system) lymphocyte that, upon activation, proliferates and differentiates into antibody-secreting plasma cell; (endocrine cell) *see* beta cell

B lymphocyte *see* B cell

barometric pressure *see* atmospheric pressure

baroreceptor receptor sensitive to pressure and to rate of change in pressure; *see also* arterial baroreceptor, intrarenal baroreceptor

basal (BAY-sul) resting level

basal ganglia nuclei deep in cerebral hemispheres that code and relay information associated with control of body movements; specifically, caudate nucleus, globus pallidus, and putamen

basal metabolic rate (BMR) metabolic rate when a person is at mental and physical rest but not sleeping, at comfortable temperature, and has fasted at least 12 h; also called metabolic cost of living

base (acid-base) any molecule that can combine with H^+; (nucleotide) molecular ring of carbon and nitrogen that, with a phosphate group and a sugar, constitutes a nucleotide

basement membrane thin layer of extracellular proteinaceous material upon which epithelial and endothelial cells sit

basic electrical rhythm spontaneous depolarization-repolarization

cycles of pacemaker cells in longitudinal smooth-muscle layer of stomach and intestines; coordinates repetitive muscular activity of GI tract

basilar membrane (BAS-ih-lar) membrane that separates cochlear duct and scala tympani in inner ear; supports organ of Corti

basolateral membrane (bay-so-LAH-ter-al) sides of epithelial cell other than luminal surface; also called serosal or blood side of cell

basophil (BAY-so-fill) polymorpho-nuclear granulocytic leucocyte whose granules stain with basic dyes; enters tissues and becomes mast cell

beta-adrenergic receptor (BAY-ta ad-ren-ER-jik) a type of plasma membrane receptor for epinephrine and norepinephrine; *compare* alpha-adrenergic receptor; also called beta adrenoceptor

beta cell insulin-secreting cell in pancreatic islets of Langerhans; also called B cell

beta oxidation (ox-ih-DAY-shun) series of reactions that generate hydrogen atoms (for oxidative phosphorylation) from breakdown of fatty acids to acetyl CoA

beta rhythm low, fast EEG oscillations in alert, awake adults who are paying attention to (or thinking hard about) something

bicarbonate (by-KAR-bah-nate) HCO_3^-

bile fluid secreted by liver into bile canaliculi; contains bicarbonate, bile salts, cholesterol, lecithin, bile pigments, metabolic end products, and certain trace metals

bile canaliculi (kan-al-IK-you-lee) small ducts adjacent to liver cells into which bile is secreted

bile pigment colored substance, derived from breakdown of heme group of hemoglobin, secreted in bile

bile salt one of a family of steroid molecules produced from cholesterol and secreted in bile by the liver; promotes solubilization and digestion of fat in small intestine

bilirubin (bi-eh-RUE-bin) yellow substance resulting from heme breakdown; excreted in bile as a bile pigment

binding site region of protein to which a specific ligand binds

biogenic amine (by-oh-JEN-ik ah-MEEN) one of family of neurotransmitters having basic formula $R-NH_2$; includes dopamine, norepinephrine, epinephrine, serotonin, and histamine

"biological clock" neurons that drive body rhythms

biotransformation (by-oh-trans-for-MAY-shun) alteration of foreign molecules by an organism's metabolic pathways

bladder urinary bladder; thick-walled sac composed of smooth muscle; stores urine prior to urination

blastocyst (BLAS-toh-cyst) particular early embryonic stage consisting of ball of developing cells surrounding central cavity

blood-brain barrier group of anatomical barriers and transport systems in brain capillary endothelium that controls kinds of substances entering brain extracellular space from blood and their rates of entry

blood coagulation (koh-ag-you-LAY-shun) blood clotting

blood sugar glucose

blood type blood classification according to presence of A and/or B antigens or lack of them (O)

body mass index (BMI) method for assessing degree of obesity; calculated as weight in kilograms divided by square of height in meters

bone marrow highly vascular, cellular substance in central cavity of some bones; site of erythrocyte; leukocyte, and platelet synthesis

Bowman's capsule blind sac at beginning of tubular component of kidney nephron

Boyle's law (boils) pressure of a fixed amount of gas in a container is inversely proportional to container's volume

bradykinin (braid-ee-KY-nin) protein formed by action of the enzyme kallikrein on precursor

brainstem brain subdivision consisting of medulla oblongata, pons, and midbrain and located between spinal cord and forebrain

brainstem pathways descending motor pathways whose cells of origin are in the brainstem

Broca's area (BRO-kahz) region of left frontal lobe associated with speech production

bronchiole (BRON-key-ole) small airway distal to bronchus

bronchus (BRON-kus) large-diameter air passage that enters lung; located between trachea and bronchioles

buffer weak acid or base that can exist in undissociated (Hbuffer) or dissociated (H^+ + buffer) form

buffering reversible hydrogen-ion binding by anions when H^+ concentration changes; tends to minimize changes in acidity of a solution when acid is added or removed

bulbourethral gland (bul-bo-you-WREETH-ral) one of paired glands in male that secretes fluid components of semen into the urethra

bulk flow movement of fluids or gases from region or higher pressure to one of lower pressure

C

C-reactive protein an acute phase protein that functions as a non-specific opsonin

calcitonin hormone from the thyroid gland that inhibits bone resorption

calmodulin (kal-MOD-you-lin) intracellular calcium-binding protein that mediates many of calcium's second-messenger functions

calorie (cal) unit of heat-energy measurement; amount of heat needed to raise temperature of 1 g of water 1°C; *compare* kilocalorie

calorigenic effect (kah-lor-ih-JEN-ik) increase in metabolic rate caused by epinephrine or thyroid hormones

capacitation *see* sperm capacitation

capillary smallest blood vessels

carbamino hemoglobin (kar-bah-MEEN-oh HE-moe-glow-bin) compound resulting from combination of carbon dioxide and amino groups in hemoglobin

carbohydrate substance composed of carbon, hydrogen, and oxygen

according to general formula $C_n(H_2O)_n$, where n is any whole number

carbon monoxide CO; gas that reacts with hemoglobin; decreases blood oxygen-carrying capacity and shifts oxygen-hemoglobin dissociation curve to the left

carbonic acid (kar-BAHN-ik) H_2CO_3; an acid formed from H_2O and CO_2

carbonic anhydrase (an-HY-drase) enzyme that catalyzes the reaction $CO_2 + H_2O \rightleftharpoons H_2CO_3$

carboxyl group (kar-BOX-il) —COOH; ionizes to carboxyl ion ($-COO^-$)

carboxypeptidase (kar-box-ee-PEP-tih-dase) enzyme secreted into small intestine by exocrine pancreas as precursor, procarboxypeptidase; breaks peptide bond at carboxyl end of protein

cardiac (KAR-dee-ak) pertaining to the heart

cardiac cycle one contraction-relaxation sequence of heart

cardiac muscle heart muscle

cardiac output blood volume pumped by each ventricle per minute (not total output pumped by both ventricles)

cardiovascular center neuron cluster in brainstem medulla oblongata that serves as a major integrating center for reflexes affecting heart and blood vessels

cardiovascular system heart and blood vessels

carotid (kuh-RAH-tid) pertaining to two major arteries (carotid arteries) in neck that convey blood to head

carotid body chemoreceptor chemoreceptor near main branching of carotid artery; sensitive to blood O_2 pressure and H^+ concentration

carotid sinus region of internal carotid artery just above main carotid branching; location of carotid baroreceptors

carotid sinus baroreceptor *see* arterial baroreceptor

carrier *see* transporter

catabolism (kuh-TAB-oh-lizm) cellular breakdown of organic molecules

catalyst (KAT-ah-list) substance that accelerates chemical reactions but does not itself undergo any net chemical change during the reaction

catecholamine (kat-eh-COLE-ah-meen) dopamine, epinephrine, or norepinephrine, all of which have similar chemical structures

cation (KAT-eye-on) ion having net positive charge; *compare* anion

cecum (SEE-come) dilated pouch at beginning of large intestine into which the ileum, colon, and appendix open

cell body in cells with long extensions, the part that contains the nucleus

cell differentiation *see* differentiation

cell organelle (or-guh-NEL) membrane-bound compartment, nonmembranous particle, or filament that performs specialized functions in cell

center of gravity point in a body at which body mass is in perfect balance; if the body were suspended from a string attached to this point, there would be no movement

central chemoreceptor receptor in brainstem medulla oblongata that responds to H^+ concentration changes of brain extracellular fluid

central command fatigue muscle fatigue due to failure of appropriate regions of cerebral cortex to excite motor neurons

central nervous system (CNS) brain plus spinal cord

central thermoreceptor temperature receptor in hypothalamus, spinal cord, abdominal organ, or other internal location

centriole (SEN-tree-ole) small cytoplasmic body having nine fused sets of microtubules; participates in nuclear and cell division

cephalic phase (seh-FAL-ik) (of gastrointestinal control) initiation of the neural and hormonal reflexes regulating gastrointestinal functions by stimulation of receptors in head, that is, cephalic receptors—sight, smell, taste, and chewing—as well as by emotional states

cerebellum (ser-ah-BEL-um) brain subdivision lying behind forebrain and above brainstem; deals with muscle movement control

cerebral cortex (SER-ah-brul or sah-REE-brul) cellular layer covering the cerebrum

cerebral ventricle one of four interconnected spaces in the brain; filled with cerebrospinal fluid

cerebrospinal fluid (CSF) (sah-ree-broh-SPY-nal) fluid that fills cerebral ventricles and the subarachnoid space surrounding brain and spinal cord

cerebrum (SER-ah-brum or sah-REE-brum) part of the brain that, with diencephalon, forms the forebrain

cervix (SIR-vix) lower portion of uterus; cervical opening connects uterine and vaginal lumens

channel small passage in plasma membrane formed by integral membrane proteins and through which certain small-diameter molecules and ions can diffuse; *see also* ligand-sensitive channel, voltage-gated channel, mechanosensitive channel

channel gating process of opening and closing ion channels

chaperone protein that guides the folding of large segments of a newly formed protein

charge *see* electric charge

chemical element specific type of atom

chemical equilibrium state when rates of forward and reverse components of a chemical reaction are equal, and no net change in reactant or product concentration occurs

chemical reaction breaking of some chemical bonds and formation of new ones, which changes one type of molecule to another; *see also* reversible reaction, irreversible reaction

chemical specificity *see* specificity

chemical synapse (SIN-apse) synapse at which neuro-transmitters released by one neuron diffuse across an extracellular gap to influence a second neuron's activity

chemoattractant any mediator that causes chemotaxis; also called *chemotaxin*

chemokine any cytokine that functions as a chemoattractant

chemoreceptor afferent nerve ending (or cell associated with it) sensitive to concentrations of certain chemicals

chemotaxin (kee-moh-TAX-in) *see* chemoattractant

chemotaxis (kee-moh-TAX-iss) movement of cells, particularly phagocytes, in a specific direction in response to a chemical stimulus

chief cell gastric gland cell that secretes pepsinogen, precursor of pepsin

cholecystokinin (CCK) (koh-lee-sis-toh-KY-nin) peptide hormone secreted by duodenum that regulates gastric motility and secretion, gallbladder contraction, and pancreatic enzyme secretion; possible satiety signal

cholesterol particular steroid molecule; precursor of steroid hormones and bile salts and a component of plasma membranes

cholinergic (koh-lin-ER-jik) pertaining to acetylcholine; a compound that acts like acetylcholine

chondrocyte (KON-droh-site) cell types that form new cartilage

chorionic gonadotropin (CG) (kor-ee-ON-ik go-NAD-oh-troh-pin) protein hormone secreted by trophoblastic cells of embryo; maintains secretory activity of corpus luteum during first 3 months of pregnancy

choroid plexus (KOR-oid) highly vascular epithelial structure lining portions of cerebral ventricles; responsible for much of cerebrospinal fluid formation

chromatid (KROM-ah-tid) one of two identical strands of chromatin resulting from DNA duplication during mitosis or meiosis

chromatin (KROM-ih-tin) combination of DNA and nuclear proteins; principal component of chromosomes

chromophore retinal light-sensitive component of a photopigment

chromosome highly coiled, condensed form of chromatin formed in cell nucleus during mitosis and meiosis

chronic (KRON-ik) persisting over a long time; *compare* acute

chylomicron (kye-loh-MY-kron) small droplet consisting of lipids and protein that is released from intestinal epithelial cells into the lacteals during fat absorption

chyme (kyme) solution of partially digested food in stomach and intestinal lumens

chymotrypsin enzyme secreted by exocrine pancreas; breaks certain peptide bonds in proteins and polypeptides

cilia (SIL-ee-ah) hairlike projections from specialized epithelial cells; sweep back and forth in a synchronized way to propel material along epithelial surface

circadian rhythm (sir-KAY-dee-an) occurring in an approximately 24 h cycle.

circular muscle smooth-muscle layer in stomach and intestinal walls and having muscle fibers circumferentially oriented around these organs

citric acid cycle *see* Krebs cycle

classical complement pathway antibody-dependent system for activating complement; begins with complement molecule Cl

clearance volume of plasma from which a particular substance has been completely removed in a given time

clitoris (KLIT-or-iss) small body of erectile tissue in female external genitalia; homologous to penis

clonal deletion destruction by apoptosis in the thymus of those T cells that have receptors capable of binding to self proteins

clonal inactivation process occurring in the periphery (that is, not in the thymus) that causes potentially self-reacting T cells to become nonresponsive

clone genetically identical molecules, cells, or organisms

coagulation (koh-ag-you-LAY-shun) blood clotting

cochlea (KOK-lee-ah) inner ear; fluid-filled spiral-shaped compartment that contains cochlear duct

cochlear duct (KOK-lee-er) fluid-filled membranous tube that extends length of inner ear, dividing it into compartments; contains organ of Corti

codon (KOH-don) three-base

sequence in mRNA that determines the position of a specific amino acid during protein synthesis or designates the end of the coded sequence of a protein

coenzyme (koh-EN-zime) organic cofactor; generally serves as a carrier that transfers atoms or small molecular fragments from one reaction to another; is not consumed in the reaction and can be reused

coenzyme A *see* acetyl coenzyme A

cofactor (KOH-fact-or) organic or inorganic substance that binds to a specific region of an enzyme and is necessary for the enzyme's activity

collagen (KOL-ah-jen) strong, fibrous protein that functions as extracellular structural element in connective tissue

collecting-duct system portion of renal tubules between distal convoluted tubules and renal pelvis; comprises *connecting tubule, cortical collecting duct,* and *medullary collecting duct*

colloid (KOL-oid) large molecule, mainly protein, to which capillaries are relatively impermeable

colon (KOH-lun) a portion of the large intestine, specifically that part extending from cecum to rectum

colony-stimulating factor (CSF) collective term for several hematopoietic growth factors that stimulate production of neutrophils and monocytes

commissure (KOM-ih-shur) bundle of nerve fibers linking right and left halves of the brain

common bile duct carries bile from gallbladder to small intestine

compensatory growth type of regeneration present in many organs after tissue damage

competition ability of similar molecules to combine with the same binding site or receptor

complement (KOM-plih-ment) one of a group of plasma proteins that, upon activation, kills microbes directly and facilitates the inflammatory process, including phagocytosis

compliance stretchability; *see also* lung compliance

concentration amount of material per unit volume of solution

concentration gradient gradation in concentration that occurs between two regions having different concentrations

conceptus collective term for the fertilized egg and everything derived from it

conducting system network of cardiac muscle fibers specialized to conduct electrical activity between different areas of heart

conducting zone air passages that extend from top of trachea to beginning of respiratory bronchioles and have walls too thick for gas exchange between air and blood

conduction (heat) transfer of thermal energy during collisions of adjacent molecules

cone one of two retinal receptor types for photic energy; gives rise to color vision

conformation three-dimensional shape of a molecule

connective-tissue cell cell specialized to form extracellular elements that connect, anchor, and support body structures

conscious experience things of which a person is aware; thoughts, feelings, perceptions, ideas, and reasoning during any state of consciousness

consciousness *see* conscious experience, state of consciousness

contractility (kon-trak-TIL-ity) force of heart contraction that is independent of sarcomere length

contraction operation of the force-generating process in a muscle

contraction time time between beginning of force development and peak twitch tension by the muscle

contralateral on the opposite side of the body

control system *see* homeostatic control system

convection (kon-VEK-shun) process by which a fluid or gas next to a warm body is heated by conduction, moves away, and is replaced by colder fluid or gas that in turn follows the same cycle

convergence (neuronal) many presynaptic neurons synapsing upon one postsynaptic neuron; (of eyes) turning of eyes inward (that

is, toward nose) to view near objects

cooperativity interaction between functional binding sites in a multimeric protein

core temperature temperature of inner body

cornea (KOR-nee-ah) transparent structure covering front of eye; forms part of eye's optical system and helps focus an object's image on retina

coronary pertaining to blood vessels of heart

coronary blood flow blood flow to heart muscle

corpus callosum (KOR-pus kal-LOH-sum) wide band of nerve fibers connecting the two cerebral hemispheres; a brain commissure

corpus luteum (LOO-tee-um) ovarian structure formed from the follicle after ovulation; secretes estrogen and progesterone

cortex (KOR-tex) outer layer of organ; *see also* adrenal cortex, cerebral cortex; *compare* medulla

cortical association area regions of cerebral cortex that receive input from various sensory types, memory stores, and so on, and perform further perceptual processing

corticobulbar pathway (kor-tih-koh-BUL-bar) descending pathway having its neuron cell bodies in cerebral cortex; its axons pass without synapsing to region of brainstem motor neurons

corticospinal pathway descending pathway having its neuron cell bodies in cerebral cortex; its axons pass without synapsing to region of spinal motor neurons; also called the pyramidal tract; *compare* brainstem pathways, corticobulbar pathway

corticosteroid (kor-tih-koh-STEER-oid) steroid produced by adrenal cortex or drug that resembles one

corticotropin releasing hormone (CRH) (kor-tih-koh-TROH-pin) hypophysiotropic peptide hormone that stimulates ACTH (corticotropin) secretion by anterior pituitary

cortisol (KOR-tih-sol) main glucocorticoid steroid hormone secreted by adrenal cortex; regulates various aspects of organic metabolism

cotransmitter chemical messenger released with a neurotransmitter from synapse or neuroeffector junction

cotransport form of secondary active transport in which net movement of actively transported substance and "downhill" movement of molecule supplying the energy are in the same direction

countercurrent multiplier system mechanism associated with loops of Henle that creates in renal medulla a region having high interstitial-fluid osmolarity

countertransport form of secondary active transport in which net movement of actively transported molecule is in direction opposite "downhill" movement of molecule supplying the energy

covalent bond (koh-VAY-lent) chemical bond between two atoms in which each atom shares one of its electrons with the other

covalent modulation alteration of a protein's shape, and therefore its function, by the covalent binding of various chemical groups to it

cranial nerve one of 24 peripheral nerves (12 pairs) that join brainstem or forebrain with structures outside CNS

creatine phosphate (CP) (KREE-ah-tin) molecule that transfers phosphate and energy to ADP to generate ATP

creatinine (kree-AT-ih-nin) waste product derived from muscle creatine

creatinine clearance plasma volume from which creatinine is removed by the kidneys per unit time; approximates glomerular filtration rate

critical period time during development when a system is most readily influenced by factors, sometimes irreversibly

cross bridge in muscle, myosin projection extending from thick filament and capable of exerting force on thin filament, causing the filaments to slide past each other

cross-bridge cycle sequence of events between binding of a cross bridge to actin, its release, and reattachment during muscle contraction

crossed-extensor reflex increased activation of extensor muscles contralateral to limb flexion

crossing-over process in which segments of maternal and paternal chromosomes exchange with each other during chromosomal pairing in meiosis

crystalloid low-molecular weight solute

current movement of electric charge; in biological systems, this is achieved by ion movement

cutaneous (cue-TAY-nee-us) pertaining to skin

cyclic AMP (cAMP) cyclic 3′,5′-adenosine monophosphate; cyclic nucleotide that serves as a second messenger for many "first" chemical messengers

cyclic AMP-dependent protein kinase (KY-nase) enzyme that is activated by cyclic AMP and then phosphorylates specific proteins, thereby altering their activity; also called *protein kinase A*

cyclic endoperoxide eicosanoid formed from arachidonic acid by cyclooxygenase

cyclic GMP (cGMP) cyclic 3′,5′-guanosine monophosphate; cyclic nucleotide that acts as second messenger in some cells

cyclic GMP-dependent protein kinase (KY-nase) enzyme that is activated by cyclic GMP and then phosphorylates specific proteins, thereby altering their activity; also called *protein kinase G*

cyclooxygenase (COX) (sy-klo-OX-ah-jen-ase) enzyme that acts on arachidonic acid and initiates production of cyclic endoperoxides, prostaglandins, and thromboxanes

cytochrome (SY-toh-krom) one of a series of enzymes that couples energy to ATP formation during oxidative phosphorylation

cytokine (SY-toh-kine) general term for protein extracellular messengers that regulate immune responses; secreted by macrophages, monocytes, lymphocytes, neutrophils, and several nonimmune cell types

cytokinesis (SY-toh-kin-EE-sis) stage of cell division during which cytoplasm divides to form two new cells

cytoplasm (SY-toh-plasm) region of cell interior outside the nucleus

cytosine (C) (SY-toh-seen) pyrimidine base in DNA and RNA

cytoskeleton cytoplasmic filamentous network associated with cell shape and movement

cytosol (SY-toh-sol) intracellular fluid that surrounds cell organelles and nucleus

cytotoxic T cell (SY-toh-TOX-ik) T lymphocyte that, upon activation by specific antigen, directly attacks a cell bearing that type of antigen and destroys it; major killer of virus-infected and cancer cells

D

daughter cell one of the two new cells formed when a cell divides

dead space volume of inspired air that cannot be exchanged with blood; *see also* anatomic dead space, alveolar dead space

deamination *see* oxidative deamination

declarative memory memories of facts and events

decremental decreasing in amplitude

defecation (def-ih-KAY-shun) expulsion of feces from rectum

dendrite (DEN-drite) highly branched extension of neuron cell body; receives synaptic input from other neurons

dense body cytoplasmic structure to which thin filaments of a smooth-muscle fiber are anchored

deoxyhemoglobin (Hb, HbH) (dee-ox-see-HEE-moh-gloh-bin) hemoglobin not combined with oxygen; reduced hemoglobin

deoxyribonucleic acid (DNA) (dee-ox-see-ry-boh-noo-KLAY-ik) nucleic acid that stores and transmits genetic information; consists of double strand of nucleotide subunits that contain deoxyribose

depolarize to change membrane potential value toward zero so that cell interior becomes less negative than resting level

descending limb (of Henle's loop) segment of renal tubule into which proximal tubule drains

descending pathway neural pathways that go from the brain down to the spinal cord

desmosome (DEZ-moh-some) junction that holds two cells together; consists of plasma membranes of adjacent cells linked by fibers, yet separated by a 20-nm extracellular space filled with a cementing substance

detrusor muscle (duh-TRUSS-or) the smooth muscle that forms the wall of the urinary bladder

diacylglycerol (DAG) (dy-aa-syl-GLIS-er-ol) second messenger that activates protein kinase C, which then phosphorylates a large number of other proteins

diaphragm (DY-ah-fram) dome-shaped skeletal-muscle sheet that separates the abdominal and thoracic cavities; principal muscle of respiration

diastole (dy-ASS-toh-lee) period of cardiac cycle when ventricles are relaxing

diastolic pressure (DP) (dy-ah-STAL-ik) minimum blood pressure during cardiac cycle

diencephalon (dy-en-SEF-ah-lon) core of anterior part of brain; lies beneath cerebral hemispheres and contains *thalamus* and *hypothalamus*

differentiation (dif-fer-en-she-AY-shun) process by which unspecialized cells acquire specialized structural and functional properties

diffusion (dif-FU-shun) movement of molecules from one location to another because of random thermal molecular motion; net diffusion always occurs from a region of higher concentration to a region of lower concentration

diffusion equilibrium state during which diffusion fluxes in opposite directions are equal; that is, the net flux equals zero

diffusion potential voltage difference created by net diffusion of ions

digestion process of breaking down large particles and high-molecular-weight substances into small molecules

dihydrotestosterone (dy-hy-droh-tes-TOS-ter-own) steroid formed by enzyme-mediated alteration of testosterone; active form of testosterone in certain of its target cells

1,25-dihydroxyvitamin D₃ (1,25-(OH₂D₃) (1-25-dy-hy-DROX-ee-vy-tah-min DEE-3) hormone that is formed by kidneys and is the active form of vitamin D

2,3-diphosphoglycerate (DPG) (2-3-dy-fos-foh-GLISS-er-ate) substance produced by erythrocytes during glycolysis; binds reversibly to hemoglobin, causing it to release oxygen

disaccharide (dy-SAK-er-ide) carbohydrate molecule composed of two monosaccharides

dissociation separation from

distal (DIS-tal) farther from reference point; *compare* proximal

distal convoluted tubule portion of kidney tubule between loop of Henle and collecting duct system

disulfide bond R-S-S-R

diuresis (dy-uh-REE-sis) increased urine excretion

diuretic (dy-uh-RET-ik) substance that inhibits fluid reabsorption in renal tubule; thereby increasing urine excretion

diurnal (dy-URN-al) daily; occurring in a 24-h cycle

divergence (dy-VER-gence) (neuronal) one presynaptic neuron synapsing upon many postsynaptic neurons; (of eyes) turning of eyes outward to view distant objects

dl deciliter; 0.1 L

DNA polymerase enzyme that, during DNA replication, forms new DNA strand by joining together nucleotides already base-paired with an existing DNA strand

dopamine (DOPE-ah-meen) biogenic amine (catecholamine) neurotransmitter and hormone; precursor of epinephrine and norepinephrine; *see also* prolactin inhibiting hormone

dorsal (DOR-sal) toward or at the back

dorsal root group of afferent nerve fibers that enters dorsal region of spinal cord

double bond two covalent chemical bonds formed between same two atoms; symbolized by =

down-regulation decrease in number of target-cell receptors for a given messenger in response to a chronic high concentration of that messenger; *compare* up-regulation

dual innervation (in-ner-VAY-shun) innervation of an organ or gland by both sympathetic and parasympathetic nerve fibers

duodenum (due-oh-DEE-num) first portion of small intestine (between stomach and jejunum)

E

eardrum *see* tympanic membrane

edema (ed-DEE-mah) accumulation of excess fluid in interstitial space

EEG arousal transformation of EEG pattern from alpha to beta rhythm during increased levels of attention

effector (ee-FECK-tor) cell or cell collection whose change in activity constitutes the response in a control system

efferent (EF-er-ent) carrying away from

efferent arteriole renal vessel that conveys blood from glomerulus to peritubular capillaries

efferent neuron neuron that carries information away from CNS

efferent pathway component of reflex arc that transmits information from integrating center to effector

egg female germ cell at any of its stages of development

eicosanoid (eye-KOH-sah-noid) general term for modified fatty acids that are products of arachidonic acid metabolism (cyclic endoperoxides,) prostaglandins, thromboxanes, and leukotrienes); function as paracrine/autocrine agents

ejaculation (ee-jak-you-LAY-shun) discharge of semen from penis

ejaculatory duct (ee-JAK-you-lah-tory) continuation of vas deferens after it is joined by seminal vesicle duct; joins urethra in prostate gland

ejection fraction (EF) the ratio of stroke volume to end-diastolic volume; EF = SV/EDV

electric charge particle having positivity or negativity

electric force   force that causes charged particles to move toward regions having an opposite charge and away from regions having a like charge

electric potential (*E*)   (or electric potential difference) *see* potential

electric signal   graded potential or action potential

electric synapse   (SIN-apse) synapse at which local currents resulting from electrical activity flow between two neurons through gap junctions joining them

electrocardiogram (ECG, EKG)   (ee-lek-troh-KARD-ee-oh-gram) recording at skin surface of the electric currents generated by cardiac muscle action potentials

electrochemical difference   force determining direction and magnitude of net charge movement; combination of electrical and chemical gradients

electrode   (ee-LEK-trode) probe used to stimulate electrically, or record from, the body surface or a tissue

electroencephalogram (EEG)   (eh-lek-troh-en-SEF-ah-loh-gram) recording of brain electrical activity from scalp

electrogenic pump   (elec-troh-JEN-ik) active-transport system that directly separates electric charge, thereby producing a potential difference

electrolyte   (ee-LEK-troh-lite) substance that dissociates into ions when in aqueous solution

electromagnetic radiation   radiation composed of waves with electrical and magnetic components; includes gamma rays, x-rays, and ultraviolet, visible light, infrared, and radio waves

electron   (ee-LEK-tron) subatomic particle that carries one unit of negative charge

embryo   (EM-bree-oh) organism during early stages of development; in human beings, the first 2 months of intrauterine life

emission   (ee-MISH-un) movement of male genital duct contents into urethra prior to ejaculation

emotion   *see* inner emotion, emotional behavior

emotional behavior   outward expression and display of inner emotions

emulsification   (eh-mul-suh-fah-KAY-shun) division of large lipid droplets into very small droplets

end-diastolic volume (EDV)   (dy-ah-STAH-lik) amount of blood in ventricle just prior to systole

endocrine gland   (EN-doh-krin) group of epithelial cells that secrete into the extracellular space hormones that then diffuse into bloodstream; also called a ductless gland

endocrine system   all the body's hormone-secreting glands

endocytosis   (en-doh-sy-TOH-sis) process in which plasma membrane folds into the cell, forming small pockets that pinch off to produce intracellular, membrane-bound vesicles; *see also* phagocytosis

endogenous opioid   (en-DAHJ-en-us OH-pee-oid) certain neuro-peptides—endorphin, dynorphin, and enkephalin

endogenous pyrogen (EP)   (en-DAHJ-en-us PY-roh-jen) cytokines (including interleukin 1 and interleukin 6) that act physiologically in the brain to cause fever

endometrium   (en-doh-MEE-tree-um) glandular epithelium lining uterine cavity

endoperoxide   *see* cyclic endoperoxide

endoplasmic reticulum   (en-doh-PLAS-mik reh-TIK-you-lum) cell organelle that consists of interconnected network of membrane-bound branched tubules and flattened sacs; two types are distinguished: *granular,* with ribosomes attached, and *agranular,* which is smooth-surfaced

endosome   (EN-doh-some) intracellular vesicles and tubular elements between Golgi apparatus and plasma membrane; sorts and distributes vesicles during endo- and exocytosis

endothelium   (en-doh-THEE-lee-um) thin layer of cells that lines heart cavities and blood vessels

endothelium-derived relaxing factor (EDRF)   nitric oxide and possibly other substances; secreted by vascular endothelium, it relaxes vascular smooth muscle and causes arteriolar dilation

end-plate potential (EPP)   depolarization of motor end plate of skeletal-muscle fiber in response to acetylcholine; initiates action potential in muscle plasma membrane

end-product inhibition   inhibition of a metabolic pathway by final product's action upon allosteric site on an enzyme (usually the rate-limiting enzyme) in the pathway

end-systolic volume (ESV)   (sis-TAH-lik) amount of blood remaining in ventricle after ejection

energy   ability to produce change; measured by amount of work performed during a given change

enkephalin   (en-KEF-ah-lin) peptide neurotransmitter at some synapses activated by opiate drugs; an endogenous opioid

enteric nervous system   (en-TAIR-ik) neural network residing in and innervating walls of gastro-intestinal tract

enterogastrones   (en-ter-oh-GAS-trones) collective term for hormones that are released by intestinal tract and inhibit stomach activity

enterohepatic circulation   (en-ter-oh-hih-PAT-ik) reabsorption of bile salts (and other substances) from intestines, passage to liver (via hepatic portal vein), and secretion back to intestines (via bile)

enterokinase   (en-ter-oh-KY-nase) enzyme in luminal plasma membrane of intestinal epithelial cells; converts pancreatic trypsinogen to trypsin

entrainment   (en-TRAIN-ment) adjusting biological rhythm to environmental cues

enzyme   (EN-zime) protein catalyst that accelerates specific chemical reactions but does not itself undergo net chemical change during the reaction

enzyme activity   rate at which enzyme converts reactant to product; may be measure of the properties of enzyme's active site as altered by allosteric or covalent modulation; affects rate of enzyme-mediated reaction

eosinophil   (ee-oh-SIN-oh-fil) polymorphonuclear granulocytic leukocyte whose granules take up red dye eosin; involved in parasite destruction and allergic responses

epididymis (ep-ih-DID-eh-mus) portion of male reproductive duct system located between seminiferous tubules and vas deferens

epiglottis (ep-ih-GLOT-iss) thin cartilage flap that folds down, covering trachea, during swallowing

epinephrine (E) (ep-ih-NEF-rin) amine hormone secreted by adrenal medulla and involved in regulation of organic metabolism; a biogenic amine (catecholamine) neurotransmitter; also called *adrenaline*

epiphyseal closure (ep-ih-FIZ-ee-al) conversion of epiphyseal growth plate to bone

epiphyseal growth plate actively proliferating cartilage near bone ends; region of bone growth

epiphysis (eh-PIF-ih-sis) end of long bone

epithelial cell (ep-ih-THEE-lee-al) cell at surface of body or hollow organ; specialized to secrete or absorb ions and organic molecules; with other epithelial cells, forms an *epithelium*

epithelial transport molecule movement from one extracellular compartment across epithelial cells into a second extracellular compartment

epithelium (ep-ih-THEE-lee-um) tissue that covers all body surfaces, lines all body cavities, and forms most glands

epitope (EP-ih-tope) antigenic portion of a molecule complexed to the MHC protein and presented to the T cell; also called an *antigenic determinant*

equilibrium (ee-quah-LIB-ree-um) no net change occurs in a system; requires no energy

equilibrium potential voltage gradient across a membrane that is equal in force but opposite in direction to concentration force affecting a given ion species

erection penis or clitoris becoming stiff due to vascular congestion

error signal steady-state difference between level of regulated variable in a control system and set point for that variable

erythrocyte (eh-RITH-roh-site) red blood cell

erythropoiesis (eh-rith-roh-poy-EE-sis) erythrocyte production

erythropoietin (eh-rith-roh-POY-ih-tin) peptide hormone secreted mainly by kidney cells; stimulates red blood cell production; one of the hematopoietic growth factors

esophagus (eh-SOF-uh-gus) portion of digestive tract that connects throat (pharynx) and stomach

essential amino acid amino acid that cannot be formed by the body at all (or at rate adequate to meet metabolic requirements) and must be obtained from diet

essential nutrient substance required for normal or optimal body function but synthesized by the body either not at all or in amounts inadequate to prevent disease

estradiol (es-tra-DY-ol) steroid hormone of estrogen family; major female sex hormone

estriol (ES-tree-ol) steroid hormone of estrogen family; major estrogen secreted by placenta during pregnancy

estrogen (ES-troh-jen) group of steroid hormones that have effects similar to estradiol on female reproductive tract

excitability ability to produce electric signals

excitable membrane membrane capable of producing action potentials

excitation-contraction coupling in muscle fibers, mechanism linking plasma-membrane stimulation with cross-bridge force generation

excitatory postsynaptic potential (EPSP) (post-sin-NAP-tic) depolarizing graded potential in postsynaptic neuron in response to activation of excitatory synapse

excitatory synapse (SIN-apse) synapse that, when activated, increases likelihood that postsynaptic neuron will undergo action potentials or increases frequency of existing action potentials

excretion elimination of a substance from the body

exocrine gland (EX-oh-krin) cluster of epithelial cells specialized for secretion and having ducts that lead to an epithelial surface

exocytosis (ex-oh-sy-TOH-sis) process in which intracellular vesicle fuses with plasma membrane, the vesicle opens, and its contents are liberated into the extracellular fluid

exon (EX-on) DNA gene region containing code words for a part of the amino acid sequence of a protein

expiration (ex-pur-A-shun) movement of air out of lungs

expiratory reserve volume (ex-PY-ruh-tor-ee) volume of air that can be exhaled by maximal contraction of expiratory muscles after normal resting expiration

extension straightening a joint

extensor muscle muscle whose activity straightens a joint

external anal sphincter ring of skeletal muscle around lower end of rectum

external environment environment surrounding external surface of an organism

external genitalia (jen-ih-TAH-lee-ah) (female) mons pubis, labia majora and minora, clitoris, vestibule of the vagina, and vestibular glands; (male) penis and scrotum

external work movement of external objects by skeletal-muscle contraction

external urethral sphincter ring of skeletal muscle that surrounds the urethra at base of bladder

extracellular fluid fluid outside cell; interstitial fluid and plasma

extracellular matrix (MAY-trix) a complex consisting of a mixture of proteins (and, in some cases, minerals) in which extracellular fluid is interspersed

extrafusal fiber primary muscle fiber in skeletal muscle, as opposed to modified (intrafusal) fiber in muscle spindle

extrapyramidal system *see* brainstem pathways

extrinsic (ex-TRIN-sik) coming from outside

extrinsic clotting pathway formation of fibrin clots by pathway using tissue factor on cells in interstitium; once activated, it also recruits the intrinsic clotting pathway beyond factor XII

F

facilitated diffusion (fah-SIL-ih-tay-ted) system using a transporter to move molecules from high to low concentration across a membrane; energy not required

fast fiber skeletal-muscle fiber that contains myosin having high ATPase activity

fat mobilization increased breakdown of triacylglycerols and release of glycerol and fatty acids into blood

fat-soluble vitamin *see* vitamin

fatty acid carbon chain with carboxyl group at one end through which chain can be linked to glycerol to form triacylglycerol; *see also* polyunsaturated fatty acid, saturated fatty acid, unsaturated fatty acid

Fc portion "stem" part of antibody

feces (FEE-sees) material expelled from large intestine during defecation

feedback *see* negative feedback, positive feedback

feedforward aspect of some control systems that allows system to anticipate changes in a regulated variable

female internal genitalia (jen-ih-TALE-ee-ah) ovaries, uterine tubes, uterus, and vagina

ferritin (FAIR-ih-tin) iron-binding protein that stores iron in body

fertilization union of sperm and egg

fetus (FEE-tus) human being from third month of intrauterine life until birth

fever increased body temperature due to setting of "thermostat" of temperature-regulating mechanisms at higher-than-normal level

fiber *see* nerve fiber, muscle fiber

fibrin (FY-brin) protein polymer resulting from enzymatic cleavage of fibrinogen; can turn blood into gel (clot)

fibrinogen (fy-BRIN-oh-jen) plasma protein precursor of fibrin

fibrinolytic system (fye-brin-oh-LIT-ik) cascade of plasma enzymes that breaks down clots; also called thrombolytic system

fight-or-flight response activation of sympathetic nervous system during stress

filtered load amount of any substance filtered from renal glomerular capillaries into Bowman's capsule

filtration movement of essentially protein-free plasma out across capillary walls due to a pressure gradient across the wall

first messenger extracellular chemical messenger

flatus (FLAY-tus) intestinal gas expelled through anus

flavine adenine dinucleotide (FAD) coenzyme derived from the B vitamin riboflavin; transfers hydrogen from one substrate to another

flexion (FLEK-shun) bending a joint

flow autoregulation ability of individual arterioles to alter their resistance in response to changing blood pressure so that relatively constant blood flow is maintained

fluid-mosaic model (moh-ZAY-ik) cell membrane structure consists of proteins embedded in bimolecular lipid that has the physical properties of a fluid, allowing membrane proteins to move laterally within it

flux amount of a substance crossing a surface in a unit of time; *see also* net flux

folic acid (FOH-lik) vitamin of B-complex group; essential for formation of nucleotide thiamine

follicle (FOL-ih-kel) egg and its encasing follicular, granulosa, and theca cells at all stages prior to ovulation; also called *ovarian follicle*

follicle-stimulating hormone (FSH) protein hormone secreted by anterior pituitary in males and females that acts on gonads; a gonadotropin

follicular phase (fuh-LIK-you-lar) that portion of menstrual cycle during which follicle and egg develop to maturity prior to ovulation

forebrain large, anterior brain subdivision consisting of right and left cerebral hemispheres (the cerebrum) and diencephalon

fovea centralis (FOH-vee-ah) area near center of retina where cones are most concentrated; gives rise to most acute vision

Frank-Starling mechanism the relationship between stroke volume and end-diastolic volume such that stroke volume increases as end-diastolic volume increases; also called *Starling's law of the heart*

free radical atom that has an unpaired electron in its outermost orbital; molecule containing such an atom

free-running rhythm cyclical activity driven by biological clock in absence of environmental cues

frequency number of times an event occurs per unit time

frontal lobe region of anterior cerebral cortex where motor areas, Broca's speech center, and some association cortex are located

fructose (FRUK-tose) five-carbon sugar; present in sucrose (table sugar)

functional residual capacity lung volume after relaxed expiration

functional site binding site on allosteric protein that, when activated, carries out protein's physiological function; also called active site

fused-vesicle channel endocytotic or exocytotic vesicles that have fused to form a continuous water-filled channel through capillary endothelial cell

G

G protein family of regulatory proteins that reversibly bind guanosine nucleotides; plasma-membrane G proteins interact with membrane ion channels or enzymes

gallbladder small sac under the liver; concentrates bile and stores it between meals; contraction of gallbladder ejects bile, which eventually flows into small intestine

gamete (GAM-eet) germ cell or reproductive cell; sperm in male and egg in female

gametogenesis (gah-mee-toh-JEN-ih-sis) gamete production

gamma-aminobutyric acid (GABA) major inhibitory neurotransmitter in CNS

gamma globulin immunoglobulin G (IgG), most abundant class of plasma antibodies

gamma motor neuron small motor neuron that controls intrafusal muscle fibers in muscle spindles

ganglion (GANG-glee-on) (pl. ganglia) generally reserved for cluster of neuron cell bodies outside CNS

ganglion cell retinal neuron that is postsynaptic to bipolar cells; axons of ganglion cells form optic nerve

gap junction protein channels linking cytosol of adjacent cells; allows ions and small molecules to flow between cytosols of the connected cells

gastric (GAS-trik) pertaining to the stomach

gastric phase (of gastrointestinal control) initiation of neural and hormonal gastrointestinal reflexes by stimulation of stomach wall

gastrin (GAS-trin) peptide hormone secreted by antral region of stomach; stimulates gastric acid secretion

gastroileal reflex (gas-troh-IL-ee-al) increase in contractions of ileum during gastric emptying

gastrointestinal system (gas-troh-in-TES-tin-al) gastrointestinal tract plus salivary glands, liver, gallbladder, and pancreas

gastrointestinal tract mouth, pharynx, esophagus, stomach, and small and large intestines

gating opening or closing ion channels

gene unit of hereditary information; portion of DNA containing information required to determine a protein's amino acid sequence

gene cloning process of forming identical DNA sequences using genetic engineering techniques

genetic code three-nucleotide sequence in a gene; indicates the location of a particular amino acid in the protein specified by that gene

genome complete set of an organism's genes

genotype the set of alleles present in an individual

germ cell cell that gives rise to male or female gametes (sperm and eggs)

gland *see* endocrine gland, exocrine gland

glial cell (GLEE-al) nonneuronal cell in CNS; helps regulate extracellular environment of CNS; also called *neuroglia*

globin (GLOH-bin) collective term for the four polypeptide chains of the hemoglobin molecule

globulin (GLOB-you-lin) one of a family of proteins found in blood plasma

glomerular filtration (gloh-MER-you-lar) bulk flow of an essentially protein-free plasma from renal glomerular capillaries into Bowman's capsule

glomerular filtration rate (GFR) volume of fluid filtered from renal glomerular capillaries into Bowman's capsule per unit time

glomerulus (gloh-MER-you-lus) tufts of glomerular capillaries at beginning of kidney nephron

glottis opening between vocal cords through which air passes, and surrounding area

glucagon (GLOO-kah-gahn) peptide hormone secreted by alpha cells of pancreatic islets of Langerhans; leads to rise in plasma glucose

glucocorticoid (gloo-koh-KOR-tih-koid) steroid hormone produced by adrenal cortex and having major effects on nutrient metabolism

gluconeogenesis (gloo-koh-nee-oh-JEN-ih-sis) formation of glucose by the liver or kidneys from pyruvate, lactate, glycerol, or amino acids

glucose major monosaccharide in the body; a six-carbon sugar, $C_6H_{12}O_6$; also called blood sugar

glucose-counterregulatory control neural or hormonal factors that oppose insulin's actions; glucagon, epinephrine, sympathetic nerves to liver and adipose tissue, cortisol, and growth hormone

glucose-dependent insulinotropic peptide (GIP) intestinal hormone; stimulates insulin secretion in response to glucose and fat in small intestine

glucose 6-phosphate (FOS-fate) first intermediate in glycolytic pathway

glucose sparing switch from glucose to fat utilization by most cells during postabsorptive state

glutamate (GLU-tah-mate) anion formed from the amino acid glutamic acid; a major excitatory CNS neurotransmitter

glutamine (GLOO-tah-meen) glutamate having an extra NH_3

glycerol (GLISS-er-ol) three-carbon carbohydrate; forms backbone of triacylglycerol

glycine (GLY-seen) an amino acid; a neurotransmitter at some inhibitory synapses in CNS

glycocalyx (gly-koh-KAY-lix) fuzzy coating on extracellular surface of plasma membrane; consists of short, branched carbohydrate chains

glycogen (GLY-koh-jen) highly branched polysaccharide composed of glucose subunits; major carbohydrate storage form in body

glycogenolysis (gly-koh-jen-NOL-ih-sis) glycogen breakdown to glucose

glycolysis (gly-KOL-ih-sis) metabolic pathway that breaks down glucose to two molecules of pyruvate (aerobically) or two molecules of lactate (anaerobically)

glycolytic fiber skeletal-muscle fiber that has a high concentration of glycolytic enzymes and large glycogen stores; white muscle fiber

glycoprotein protein containing covalently linked carbohydrates

Golgi apparatus (GOAL-gee) cell organelle consisting of flattened membranous sacs; usually near nucleus; processes newly synthesized proteins for secretion or distribution to other organelles

Golgi tendon organ tension-sensitive mechanoreceptor ending of afferent nerve fiber; wrapped around collagen bundles in tendon

gonad (GOH-nad) gamete-producing reproductive organ—testes in male and ovaries in female

gonadotropic hormone (goh-nad-oh-TROH-pik) hormone secreted by anterior pituitary that controls gonadal function; FSH or LH; also called *gonadotropin*

gonadotropin releasing hormone (GnRH) hypophysiotropic hormone that stimulates LH and FSH secretion by anterior pituitary in males and females

graded potential membrane potential change of variable

amplitude and duration that is conducted decrementally; has no threshold or refractory period

gradient (GRAY-dee-ent) continuous increase or decrease of a variable over distance

gram atomic mass amount of element in grams equal to the numerical value of its atomic weight

granulosa cell (gran-you-LOH-sah) cell that contributes to the layers surrounding egg and antrum in ovarian follicle; secretes estrogen, inhibin, and other messengers that influence the egg

gray matter area of brain and spinal cord that appears gray in unstained specimens and consists mainly of cell bodies and unmyelinated portions of nerve fibers

growth factor one of a group of peptides that is highly effective in stimulating cell division and/or differentiation of certain cell types

growth hormone (GH) peptide hormone secreted by anterior pituitary; stimulates insulin-like growth factor I release; enhances body growth by stimulating protein synthesis

growth hormone releasing hormone (GHRH) hypothalamic peptide hormone that stimulates growth hormone secretion by anterior pituitary

growth-inhibiting factor one of a group of peptides that modulates growth by inhibiting cell division in specific tissues

guanine (G) (GWAH-neen) purine base in DNA and RNA

guanosine triphosphate (GTP) (GWAH-noh-seen tri-FOS-fate) energy-transporting molecule similar to ATP except that it contains the base guanine rather than adenine

guanylyl cyclase (GUAN-ah-lil) enzyme that catalyzes transformation of GTP to cyclic GMP

H

H zone one of transverse bands making up striated pattern of cardiac and skeletal muscle; light region that bisects A band

habituation (hab-bit-you-A-shun) reversible decrease in response strength upon repeatedly administered stimulation

hair cell mechanoreceptor in organ of Corti and vestibular apparatus

heart rate number of heart contractions per minute

heart sound noise that results from vibrations due to closure of atrioventricular valves (first heart sound) or pulmonary and aortic valves (second heart sound)

helper T cells T cell that, via secreted cytokines, enhances the activation of B cells and cytotoxic T cells

hematocrit (heh-MAT-oh-krit) percentage of total blood volume occupied by blood cells

hematopoietic growth factor (HGF) (heh-MAT-oh-poi-ET-ik) group of protein hormones and paracrine agents that stimulate proliferation and differentiation of various types of blood cells

heme (heem) iron-containing organic molecule bound to each of the four polypeptide chains of hemoglobin or to cytochromes

hemoglobin (HEE-moh-gloh-bin) protein composed of four polypeptide chains, each attached to a heme; located in erythrocytes and transports most blood oxygen

hemoglobin saturation percent of hemoglobin molecules combined with oxygen

hemorrhage (HEM-er-age) bleeding

hemostasis (hee-moh-STAY-sis) stopping blood loss from a damaged vessel

Henle's loop *see* loop of Henle

heparin (HEP-ah-rin) anticlotting agent found on endothelial-cell surfaces; binds antithrombin III to tissues; used as an anticoagulant drug

hepatic (hih-PAT-ik) pertaining to the liver

hepatic portal vein vein that conveys blood from capillaries in the intestines and portions of the stomach and pancreas to capillaries in the liver

hertz (Hz) (hurts) cycles per second; measure used for wave frequencies

heterozygous (het-er-oh-ZY-gus) condition of having maternal and

paternal copies of a gene with slightly different nucleotide sequences (alleles); *compare* homozygous

high-density lipoprotein (HDL) lipid-protein aggregate having low proportion of lipid; promotes removal of cholesterol from cells

hippocampus (hip-oh-KAM-pus) portion of limbic system associated with learning and emotions

histamine (HISS-tah-meen) inflammatory chemical messenger secreted mainly by mast cells; monoamine neurotransmitter

homeostasis (home-ee-oh-STAY-sis) relatively stable condition of extracellular fluid that results from regulatory system actions

homeostatic control system (home-ee-oh-STAT-ik) collection of interconnected components that keeps a physical or chemical parameter of internal environment relatively constant within a predetermined range of values

homeothermic (home-ee-oh-THERM-ik) capable of maintaining body temperature within very narrow limits

homologous (hoh-MAHL-ah-gus) corresponding in origin, structure, and position

homozygous (hoh-moh-ZY-gus) condition of having maternal and paternal copies of a gene with identical nucleotide sequences (alleles); *compare* heterozygous

hormone chemical messenger synthesized by specific endocrine cells in response to certain stimuli and secreted into the blood, which carries it to target cells

hydrochloric acid (hy-droh-KLOR-ik) HCl; strong acid secreted into stomach lumen by parietal cells

hydrogen bond weak chemical bond between two molecules or parts of the same molecule, in which negative region of one polarized substance is electrostatically attracted to a positively charged region of polarized hydrogen atom in the other

hydrogen ion (EYE-on) H$^+$; single proton; H$^+$ concentration of a solution determines its acidity

hydrogen peroxide H$_2$O$_2$; chemical produced by phagosome and highly destructive to macromolecules

hydrolysis (hy-DRAHL-ih-sis) breaking of chemical bond with addition of elements of water (—H and —OH) to the products formed; also called hydrolytic reaction

hydrophilic (hy-droh-FIL-ik) attracted to, and easily dissolved in, water

hydrophobic (hy-droh-FOH-bik) not attracted to, and insoluble in, water

hydrostatic pressure (hy-droh-STAT-ik) pressure exerted by fluid

hydroxyl group (hy-DROX-il) —OH

hyper- increased

hypercalcemia increased plasma calcium

hypercapnea increased arterial P_{CO_2}

hyperemia (hy-per-EE-me-ah) increased blood flow; *see also* active hyperemia

hyperosmotic (hy-per-oz-MAH-tik) having total solute concentration greater than normal extracellular fluid

hyperpolarize to change membrane potential so cell interior becomes more negative than its resting state

hypertension chronically increased arterial blood pressure

hyperthermia increased body temperature above the set point

hypertonic (hy-per-TAH-nik) solutions containing a higher concentration of effectively membrane-impermeable solute particles than normal (isotonic) extracellular fluid

hypertrophy (hy-PER-troh-fee) enlargement of a tissue or organ due to increased cell size rather than increased cell number

hyperventilation increased ventilation adequate to reduce arterial P_{CO_2}

hypo- too little; below

hypoglycemia (hy-poh-gly-SEE-me-ah) low blood glucose (sugar) concentration

hypoosmotic (hy-poh-oz-MAH-tik) having total solute concentration less than that of normal extracellular fluid

hypophysiotropic hormone (hy-poh-fiz-ee-oh-TROH-pik) any hormone secreted by

hypothalamus that controls secretion of an anterior pituitary hormone

hypotension low blood pressure

hypothalamic releasing hormone (hy-poh-thah-LAM-ik) *see* hypophysiotropic hormone

hypothalamus (hy-poh-THAL-ah-mus) brain region below thalamus; responsible for integration of many basic neural, endocrine, and behavioral functions, especially those concerned with regulation of internal environment

hypotonic (hy-poh-TAH-nik) solutions containing a lower concentration of effectively nonpenetrating solute particles than normal (isotonic) extracellular fluid

hypoventilation decrease in ventilation that causes an increase in arterial P_{CO_2}

I

I band one of transverse bands making up repeating striations of cardiac and skeletal muscle; located between A bands of adjacent sarcomeres and bisected by Z line

IgA class of antibodies secreted by, and acting locally in, lining of gastrointestinal, respiratory, and genitourinary tracts

IgD class of antibodies whose function is unknown

IgE class of antibodies that mediate immediate hypersensitivity and resistance to parasites

IgG gamma globulin; most abundant class of antibodies

IgM class of antibodies that, along with IgG, provide major specific humoral immunity against bacteria and viruses

ileocecal sphincter (il-ee-oh-SEE-kal) ring of smooth muscle separating small and large intestines (that is, ileum and cecum)

ileum (IL-ee-um) final, longest segment of small intestine

immune defense *see* nonspecific immune defense, specific immune defense

immune surveillance (sir-VAY-lence) recognition and destruction of cancer cells that arise in body

immunity physiological mechanisms that allow body to recognize materials as foreign or abnormal and to neutralize or eliminate them; *see also* active immunity, passive immunity

immunoglobulin (Ig) (im-mun-o-GLOB-you-lin) proteins that are antibodies and antibody-like receptors on B cells (five classes are IgG, IgA, IgD, IgM, and IgE)

implantation (im-plan-TAY-shun) event during which fertilized egg becomes embedded in uterine wall

inferior vena cava (VEE-nah KAY-vah) large vein that carries blood from lower half of body to right atrium of heart

inflammation (in-flah-MAY-shun) local response to injury or infection characterized by swelling, pain, heat, and redness

inhibin (in-HIB-in) protein hormone secreted by seminiferous-tubule Sertoli cells and ovarian granulosa cells; inhibits FSH secretion

inhibitory postsynaptic potential (IPSP) hyperpolarizing graded potential that arises in postsynaptic neuron in response to activation of inhibitory synaptic endings upon it

inhibitory synapse (SIN-apse) synapse that, when activated, decreases likelihood that postsynaptic neuron will fire an action potential (or decreases frequency of existing action potentials)

initial segment first portion of axon plus the part of the cell body where axon arises

inner ear cochlea; contains organ of Corti

inner emotion emotional feelings that are entirely within a person

innervate to supply with nerves

inorganic pertaining to substances that do not contain carbon; *compare* organic

inorganic phosphate (P$_i$) (FOS-fate) $H_2PO_4^-$, HPO_4^{2-}, or PO_4^{3-}

inositol trisphosphate (IP$_3$) (in-OS-ih-tol-tris-FOS-fate) second messenger that causes release of calcium from endoplasmic reticulum into cytosol

insensible water loss water loss of which a person is unaware—that is, loss by evaporation from skin (excluding sweat) and respiratory-passage lining

inspiration air movement from atmosphere into lungs

inspiratory reserve volume maximal air volume that can be inspired above resting tidal volume

insulin (IN-suh-lin) peptide hormone secreted by beta cells of pancreatic islets of Langerhans; has metabolic and growth-promoting effects; stimulates glucose and amino acid uptake by most cells and stimulates protein, fat, and glycogen synthesis

insulin-like growth factor I (IGF-I) insulin-like growth factor that mediates mitosis-stimulating effect of growth hormone on bone and other tissues and has feedback effect on pituitary

integral membrane protein protein embedded in membrane lipid layer; may span entire membrane or be located at only one side

integrating center cells that receive one or more signals and send out appropriate response; also called an integrator

integrin (in-TEH-grin) transmembrane protein in plasma membrane; binds to specific proteins in extracellular matrix and on adjacent cells to help organize cells into tissues

intercellular cleft a narrow, water-filled space between capillary endothelial cells

intercellular fluid fluid that lies between cells; also called interstitial fluid

intercostal muscle (in-ter-KOS-tal) skeletal muscle that lies between ribs and whose contraction causes rib cage movement during breathing

interferon (in-ter-FEER-on) family of proteins that nonspecifically inhibit viral replication inside host cells; interferon-gamma also stimulates the killing ability of macrophages and NK cells

interferon-gamma *see* interferon

interleukin (in-ter-LOO-kin) a family of cytokines with many effects on immune responses and host defenses

interleukin 1 (IL-1) cytokine secreted by macrophages and other cells that activates helper T cells, exerts many inflammatory effects, and mediates many of the systemic, acute-phase responses, including fever

interleukin 2 (IL-2) cytokine secreted by activated helper T cells that causes antigen-activated helper T, cytotoxic T, and NK cells to proliferate; also causes activation of macrophages

interleukin 6 (IL-6) cytokine secreted by macrophages and other cells that exerts multiple effects on immune system cells, inflammation, and the acute-phase response

internal anal sphincter smooth-muscle ring around lower end of rectum

internal environment extracellular fluid (interstitial fluid and plasma)

internal genitalia (jen-ih-TAY-lee-ah) *see* female internal genitalia

internal urethral sphincter (you-REE-thrul) part of smooth muscle of urinary bladder wall that opens and closes the bladder outlet

internal work energy-requiring activities in body; *see also* work; *compare* external work

interneuron neuron whose cell body and axon lie entirely in CNS

interphase period of cell division cycle between end of one division and visible signs of beginning of next

interstitial fluid extracellular fluid surrounding tissue cells; excludes plasma

interstitium (in-ter-STISH-um) interstitial space; fluid-filled space between tissue cells

intestinal phase (of gastrointestinal control) initiation of neural and hormonal gastrointestinal reflexes by simulation of intestinal-tract walls

intestino-intestinal reflex cessation of contractile activity in intestines in response to various stimuli in intestine

intracellular fluid fluid in cells; cytosol plus fluid in cell organelles, including nucleus

intrafusal fiber modified skeletal-muscle fiber in muscle spindle

intrapleural fluid (in-trah-PLUR-al) thin fluid film in thoracic cavity between pleura lining the inner wall of thoracic cage and pleura covering lungs

intrapleural pressure (P_{ip}) pressure in pleural space; also called intrathoracic pressure

intrarenal baroreceptor pressure-sensitive juxtaglomerular cells of afferent arterioles, which respond to decreased renal arterial pressure by secreting more renin

intrathoracic pressure *see* intrapleural pressure

intrinsic (in-TRIN-sik) situated entirely within a part

intrinsic clotting pathway intravascular sequence of fibrin clot formation initiated by factor XII or, more usually, by the initial thrombin generated by the extrinsic clotting pathway

intrinsic factor glycoprotein secreted by stomach epithelium and necessary for absorption of vitamin B_{12} in the ileum

intron (IN-trahn) regions of noncoding nucleotides in a gene

inversely proportional relationship in which, as one factor increases by a given amount, the other decreases by a given amount

ion (EYE-on) atom or small molecule containing unequal number of electrons and protons and therefore carrying a net positive or negative electric charge

ionic bond (eye-ON-ik) strong electrical attraction between two oppositely charged ions

ionization (eye-on-ih-ZAY-shun) process of removing electrons from or adding them to an atom or small molecule to form an ion

ipsilateral (ip-sih-LAT-er-al) on the same side of the body

iris ringlike structure surrounding pupil of eye

irreversible reaction chemical reaction that releases large quantities of energy and results in almost all the reactant molecules being converted to product; *compare* reversible reaction

ischemia (iss-KEY-me-ah) reduced blood supply

islet of Langerhans (EYE-let of LAN-ger-hans) cluster of pancreatic endocrine cells; distinct islet cells secrete insulin, glucagon, somatostatin, and pancreatic polypeptide

isometric contraction (eye-soh-MET-rik) contraction of muscle under conditions in which it develops tension but does not change length

isosmotic (eye-soz-MAH-tik) having the same total solute concentration as extracellular fluid

isotonic (eye-soh-TAH-nik) containing the same number of effectively nonpenetrating solute particles as normal extracellular fluid; *see also* isotonic contraction

isotonic contraction contraction of muscle under conditions in which load on the muscle remains constant but muscle shortens

isovolumetric ventricular contraction (eye-soh-vol-you-MET-rik) early phase of systole when atrioventricular and aortic valves are closed and ventricular size remains constant

isovolumetric ventricular relaxation early phase of diastole when atrioventricular and aortic valves are closed and ventricular size remains constant

J

J receptors receptors in the lung capillary walls or interstitium that respond to increased lung interstitial pressure

JAK kinase cytoplasmic kinase bound to a receptor but not intrinsic to it

jejunum (jeh-JU-num) middle segment of small intestine

juxtaglomerular apparatus (JGA) (jux-tah-gloh-MER-you-lar) renal structure consisting of macula densa and juxtaglomerular cells; site of renin secretion and sensors for renin secretion and control of glomerular filtration rate

K

ketone (KEY-tone) product of fatty acid metabolism that accumulates in blood during starvation and in severe untreated diabetes mellitus; acetoacetic acid, acetone, or B-hydroxybutyric acid; also called ketone body

kilocalorie (kcal) (KIL-oh-kal-ah-ree) amount of heat required to change the temperature of 1 L water by 1°C; calorie used in nutrition; also called Calorie and large calorie

kinase (KY-nase) enzyme that transfers a phosphate (usually from ATP) to another molecule

kinesthesia (kin-ess-THEE-zee-ah) sense of movement derived from movement at a joint

kinin (KY-nin) peptide that splits from kininogen; facilitates vascular changes and activates pain receptors

kininogen (ky-NIN-oh-jen) plasma protein from which kinins are generated in an inflamed area

Krebs cycle mitochondrial metabolic pathway that utilizes fragments derived from carbohydrate, protein, and fat breakdown and produces carbon dioxide, hydrogen (for oxidative phosphorylation), and small amounts of ATP; also called tricarboxylic acid cycle or citric acid cycle

L

lactase (LAK-tase) small-intestine enzyme that breaks down lactose (milk sugar) into glucose and galactose

lactate ionized form of lactic acid

lactation (lak-TAY-shun) production and secretion of milk by mammary glands

lacteal (lak-TEEL) blind-ended lymph vessel in center of each intestinal villus

lactic acid (LAK-tik) three-carbon molecule formed by glycolytic pathway in absence of oxygen; dissociates to form lactate and hydrogen ions

lactose (LAK-tose) disaccharide composed of glucose and galactose; also called milk sugar

larynx (LAR-inks) part of air passageway between pharynx and trachea; contains the vocal cords

latent period (LAY-tent) period lasting several milliseconds between action-potential initiation in a muscle fiber and beginning of mechanical activity

lateral position farther from the midline

lateral inhibition method of refining sensory information in afferent neurons and ascending pathways whereby fibers inhibit each other, the most active fibers causing the greatest inhibition of adjacent fibers

lateral sac enlarged region at end of each sarcoplasmic reticulum segment; adjacent to transverse tubule

law of mass action maxim that an increase in reactant concentration causes a chemical reaction to proceed in direction of product formation; the opposite occurs with decreased reactant concentration

lecithin (LESS-ih-thin) a phospholipid

lengthening contraction contraction as an external force pulls a muscle to a longer length despite opposing forces generated by the active cross bridges

lens adjustable part of eye's optical system, which helps focus object's image on retina

leukocyte (LOO-koh-site) white blood cell

leukotrienes (LOO-koh-treens) type of eicosanoid that is generated by lipoxygenase pathway and functions as inflammatory mediator

Leydig cell (LY-dig) testosterone-secreting endocrine cell that lies between seminiferous tubules of testes; also called interstitial cell

LH surge large rise in luteinizing-hormone secretion by anterior pituitary about day 14 of menstrual cycle

ligand (LY-gand) any molecule or ion that binds to protein surface by noncovalent bonds

ligand-sensitive channel membrane channel operated by the binding of specific molecules to channel proteins

limbic system (LIM-bik) interconnected brain structures in cerebrum; involved with emotions and learning

lipase (LY-pase) enzyme that hydrolyzes triacylglycerol to monoglyceride and fatty acids; *see also* lipoprotein lipase

lipid (LIP-id) molecule composed primarily of carbon and hydrogen and characterized by insolubility in water

lipid bilayer a sheet consisting of two layers of amphipathic lipids; nonprotein part of a cell membrane

lipolysis (ly-POL-ih-sis) triacyl-glycerol breakdown

lipoprotein (lip-oh-PROH-teen) lipid aggregate partially coated by protein; involved in lipid transport in blood

lipoprotein lipase capillary endothelial enzyme that hydrolyzes triacylglycerol in lipoprotein to monoglyceride and fatty acids

lipoxygenase (ly-POX-ih-jen-ase) enzyme that acts on arachidonic acid and leads to leukotriene formation

load external force acting on muscle

local current movement of positive ions toward more negative membrane region and simultaneous movement of negative ions in opposite direction

local homeostatic response (home-ee-oh-STAT-ik) response acting in immediate vicinity of a stimulus, without nerves or hormones, and having net effect of counteracting stimulus

long-loop negative feedback inhibition of anterior pituitary and/or hypothalamus by hormone secreted by third endocrine gland in a sequence

long-term potentiation (LTP) process by which certain synapses undergo long-lasting increase in effectiveness when heavily used

loop of Henle (HEN-lee) hairpin-like segment of kidney nephron with *descending* and *ascending limbs*; situated between proximal and distal tubules

low-density lipoprotein (LDL) (lip-oh-PROH-teen) protein-lipid aggregate that is major carrier of plasma cholesterol to cells

lower esophageal sphincter smooth muscle of last portion of esophagus; can close off esophageal opening into the stomach

lumen (LOO-men) space in hollow tube or organ

luminal (LOO-min-ul) pertaining to lumen

luminal membrane portion of plasma membrane facing the lumen; also called apical or mucosal membrane

lung compliance (C_L) (come-PLY-ance) change in lung volume caused by a given change in transpulmonary pressure; the greater the lung compliance, the more stretchable the lung wall

luteal phase (LOO-tee-al) last half of menstrual cycle following ovulation; corpus luteum is active ovarian structure

luteinizing hormone (LH) (LOO-tee-en-iz-ing) peptide gonadotropic hormone secreted by anterior pituitary; rapid increase in females at midmenstrual cycle initiates ovulation; stimulates Leydig cells in males

lymph (limf) fluid in lymphatic vessels

lymph node small organ, containing lymphocytes, located along lymph vessel; a site of lymphocyte cell division and initiation of specific immune responses

lymphatic system (lim-FAT-ik) network of vessels that conveys lymph from tissues to blood and to lymph nodes along these vessels

lymphocyte (LIMF-oh-site) type of leukocyte that is responsible for specific immune defenses; B cells, T cells, and NK cells

lymphocyte activation cell division and differentiation of lymphocytes following antigen binding

lymphoid organ (LIMF-oid) bone marrow, lymph node, spleen, thymus, tonsil, or aggregate of lymphoid follicles; *see also* primary lymphoid organ, secondary lymphoid organ

lysosome (LY-soh-some) membrane-bound cell organelle containing digestive enzymes in a highly acid solution that break down bacteria, large molecules that have entered cell, and damaged components of cell

M

macrophage (MAK-roh-fahje) cell that phagocytizes foreign matter, processes it, presents antigen to lymphocytes, and secretes cytokines (monokines) involved in inflammation, activation of lymphocytes, and systemic acute phase response to infection or injury; *see also* activated macrophage, macrophage-like cell

macrophage-like cell one of several cell types that exert functions similar to those of macrophages

macula densa (MAK-you-lah DEN-sah) specialized sensor cells of renal tubule at end of loop of Henle; component of juxtaglomerular apparatus

major histocompatibility complex (MHC) group of genes that code for major histocompatibility complex proteins, which are important for specific immune function

mammary gland milk-secreting gland in breast

mass fundamental property of an object equivalent to the amount of matter in the object

mass movement contraction of large segments of colon; propels fecal matter into rectum

mast cell tissue cell that releases histamine and other chemicals involved in inflammation

maximal tubular capacity (T_m) *see* transport maximum

mean arterial pressure (MAP) average blood pressure during cardiac cycle; approximately diastolic pressure plus one-third pulse pressure

mechanoreceptor (meh-KAN-oh-re-sep-tor) sensory receptor that responds preferentially to mechanical stimuli such as bending, twisting, or compressing

mechanosensitive channel membrane ion channel that is opened or closed by deformation or stretch of the plasma membrane

median eminence (EM-ih-nence) region at base of hypothalamus containing capillary tufts into which hypophysiotropic hormones are secreted

mediate (MEE-dee-ate) bring about

mediated transport movement of molecules across membrane by binding to protein transporter; characterized by specificity, competition, and saturation; includes facilitated diffusion and active transport

medulla (meh-DUL-ah) innermost portion of an organ; *compare* cortex; *see* adrenal medulla, medulla oblongata

medulla oblongata (ob-long-GOT-ah) part of the brainstem closest to the spinal cord

medullary cardiovascular center *see* cardiovascular center

medullary inspiratory neuron *see* inspiratory neuron

megakaryocyte (meg-ah-KAR-ee-oh-site) large bone marrow cell that gives rise to platelets

meiosis (my-OH-sis) process of cell division leading to gamete (sperm or egg) formation; daughter cells receive only half the chromosomes present in original cell

melatonin (mel-ah-TOH-nin) candidate hormone secreted by pineal gland; suspected role in setting body's circadian rhythms

membrane cellular structures composed of lipids and proteins; provide selective barrier to molecule and ion movement and structural framework to which enzymes, fibers, and ligands are bound

membrane attack complex (MAC) group of complement proteins that form channels in microbe surface and destroy microbe

membrane potential voltage difference between inside and outside of cell

memory *see* procedural memory, declarative memory, working memory

memory cell B cell or T cell that differentiates during an initial infection and responds rapidly during subsequent exposure to same antigen

memory encoding processes by which an experience is transformed to a memory of that experience

menarche (MEN-ark-ee) onset, at puberty, of menstrual cycling in women

meninges (men-IN-jees) protective membranes that cover brain and spinal cord

menopause (MEN-ah-paws) cessation of menstrual cycling in middle age

menstrual cycle (MEN-stroo-al) cyclical rise and fall in female reproductive hormones and processes, beginning with menstruation

menstruation (men-stroo-AY-shun) flow of menstrual fluid from uterus; also called menstrual period

messenger RNA (mRNA) ribonucleic acid that transfers genetic information for a protein's amino acid sequence from DNA to ribosome

metabolic acidosis (met-ah-BOL-ik ass-ih-DOH-sis) acidosis due to the build up of acids other than carbonic acid (from carbon dioxide)

metabolic alkalosis (al-kah-LOH-sis) alkalosis resulting from the removal of hydrogen ions by mechanisms other than respiratory removal of carbon dioxide

metabolic end product final molecule produced by a metabolic reaction or series of reactions

metabolic pathway sequence of enzyme-mediated chemical reactions by which molecules are synthesized and broken down in cells

metabolic rate total body energy expenditure per unit time

metabolism (meh-TAB-uhl-izm) chemical reactions that occur in a living organism

metabolite (meh-TAB-oh-lite) substance produced by metabolism

metabolize change by chemical reactions

metarteriole (MET-are-teer-ee-ole) blood vessel that directly connects arteriole and venule

methyl group —CH_3

MHC protein plasma-membrane protein coded for by a major histocompatibility complex; restricts T-cell receptor's ability to combine with antigen on cell; categorized as class I and class II

micelle (MY-sell) soluble cluster of amphipathic molecules in which molecules' polar regions line surface and nonpolar regions are oriented toward center; formed from fatty acids, 2-monoglycerides, and bile salts during fat digestion in small intestine

microbe bacterium, virus, fungus, or other parasite

microcirculation blood circulation in arterioles, capillaries, and venules

microfilament rodlike cytoplasmic actin filament that forms major component of cytoskeleton

microsomal enzyme system (MES) (my-kroh-SOM-al) enzymes, found in smooth endoplasmic reticulum of liver cells, that transform molecules into more polar, less lipid-soluble substances

microtubule tubular cytoplasmic filament composed of the protein tubulin; provides internal support for cells and allows change in cell shape and organelle movement in cell

microvilli (my-kroh-VIL-eye) small fingerlike projections from epithelial-cell surface; greatly increase surface area of cell; characteristic of epithelium lining small intestine and kidney nephrons

micturition (mik-chur-RISH-un) urination

middle ear cavity air-filled space in temporal bone; contains three ear bones that conduct sound waves from tympanic membrane to cochlea

migrating motility complex pattern of peristaltic waves that pass over small segments of intestine after absorption of meal

milk ejection reflex process by which milk is moved from mammary gland alveoli into ducts, from which it can be sucked; due to oxytocin

milliliter (ml) (MIL-ih-lee-ter) volume equal to 0.001 L

millimol (mmol) (MIL-ih-mole) amount equal to 0.001 mol

millivolt (mV) (MIL-ih-volt) electric potential equal to 0.001 V

mineral inorganic substance (that is, without carbon); major minerals in body are calcium, phosphorus, potassium, sulfur, sodium, chloride, and magnesium

mineralocorticoid (min-er-al-oh-KORT-ih-koid) steroid hormone produced by adrenal cortex; has major effect on sodium and potassium balance; major mineralocorticoid is aldosterone

minute ventilation total ventilation per minute; equals tidal volume times respiratory rate

mitochondrion (my-toh-KON-dree-un) rod-shaped or oval cytoplasmic organelle that produces most of cell's ATP; site of Krebs cycle and oxidative phosphorylation enzymes

mitogen (MY-tuh-jen) chemical that stimulates cell division

mitosis (my-TOH-sis) process in cell division in which DNA is duplicated and copies of each chromosome are passed to daughter cells as the nucleus divides

mitral valve (MY-tral) valve between left atrium and left ventricle of heart

modality (moh-DAL-ih-tee) type of sensory stimulus

modulation *see* allosteric modulation, covalent modulation

modulator molecule ligand that, by acting at an allosteric regulatory site, alters properties of other binding site on a protein and thus regulates its functional activity

mol weight of a substance in grams equal to its molecular weight; 1 mol = 6×10^{23} molecules

molarity (moh-LAR-ih-tee) number of moles of solute per liter of solution

molecular weight sum of atomic weights of all atoms in molecule

molecule chemical substance formed by linking atoms together

monocyte (MAH-noh-site) type of leukocyte; leaves bloodstream and is transformed into a macrophage

monoglyceride (mah-noh-GLISS-er-ide) glycerol linked to one fatty acid side chain

monosaccharide (mah-noh-SAK-er-ide) carbohydrate consisting of one sugar molecule, which generally contains five or six carbon atoms

monosynaptic reflex (mah-noh-sih-NAP-tik) reflex in which the afferent neuron directly activates motor neurons

motilin (moh-TIL-in) candidate intestinal hormone thought to control normal GI motor activity

motivation *see* primary motivated behavior, secondary motivated behavior

motor having to do with muscles and movement

motor control hierarchy brain areas having a role in skeletal-muscle control are rank-ordered in three functional groups

motor control system CNS parts that contribute to control of skeletal-muscle movements

motor cortex strip of cerebral cortex along posterior border of frontal lobe; gives rise to many axons descending in corticospinal and multineuronal pathways; also called primary motor cortex

motor end plate specialized region of muscle-cell plasma membrane that lies directly under axon terminal of a motor neuron

motor neuron somatic efferent neuron, which innervates skeletal muscle

motor neuron pool all the motor neurons for a given muscle

motor program pattern of neural activity required to perform a certain movement

motor unit motor neuron plus the muscle fibers it innervates

mucosa (mu-KOH-sah) three layers of gastrointestinal tract wall nearest lumen—that is, *epithelium, lamina propria,* and *muscularis mucosa*

Müllerian duct (mul-AIR-ee-an) part of embryo that, in a female, develops into reproductive system ducts, but in a male, degenerates

Müllerian inhibiting substance (MIS) protein secreted by fetal testes that causes Müllerian ducts to degenerate

multimeric protein a protein composed of more than one polypeptide strand

multineuronal pathway pathway made up of chains of neurons functionally connected by synapses; also called multi-synaptic pathway

multiunit smooth muscle smooth muscle that exhibits little, if any, propagation of electrical activity from fiber to fiber and whose

contractile activity is closely coupled to its neural input

muscarinic receptor (mus-kur-IN-ik) acetylcholine receptor that responds to the mushroom poison muscarine; located on smooth muscle, cardiac muscle, some CNS neurons, and glands

muscle number of muscle fibers bound together by connective tissue

muscle fatigue decrease in muscle tension with prolonged activity

muscle fiber muscle cell

muscle-spindle stretch receptor capsule-enclosed arrangement of afferent nerve fiber endings around specialized skeletal-muscle fibers; sensitive to stretch

muscle tension force exerted by a contracting muscle on object

muscle thick filament *see* thick filament

muscle tone degree of resistance of muscle to passive stretch due to ongoing contractile activity; *see also* smooth-muscle tone

mutagen (MUTE-uh-jen) factor in the environment that increases mutation rate

mutation (mu-TAY-shun) any change in base sequence of DNA that changes genetic information

myelin (MY-uh-lin) insulating material covering axons of many neurons; consists of layers of myelin-forming cell plasma membrane wrapped around axon

myenteric plexus (my-en-TER-ik PLEX-us) nerve-cell network between circular and longitudinal muscle layers in esophagus, stomach, and intestinal walls

myo- (MY-oh) pertaining to muscle

myoblast (MY-oh-blast) embryological cell that gives rise to muscle fibers

myocardium (my-oh-KARD-ee-um) cardiac muscle, which forms heart walls

myoepithelial cell (my-oh-ep-ih-THEE-lee-al) specialized contractile cell in certain exocrine glands; contraction forces gland's secretion through ducts

myofibril (my-oh-FY-bril) bundle of thick or thin contractile filaments in cytoplasm of striated muscle; myofibrils exhibit a repeating sarcomere pattern along longitudinal axis of muscle

myogenic (my-oh-JEN-ik) originating in muscle

myoglobin (my-oh-GLOH-bin) muscle-fiber protein that binds oxygen

myometrium (my-oh-MEE-tree-um) uterine smooth muscle

myosin (MY-oh-sin) contractile protein that forms thick filaments in muscle fibers

myosin ATPase enzymatic site on globular head of myosin that catalyzes ATP breakdown to ADP and P_i, releasing the chemical energy used to produce force of muscle contraction

myosin light-chain kinase smooth-muscle protein kinase; when activated by Ca-calmodulin, phosphorylates myosin light chain

N

Na,K-ATPase pump primary active-transport protein that splits ATP and releases energy used to transport sodium out of cell and potassium in

natural killer (NK) cell type of lymphocyte that binds to virus-infected and cancer cells without specific recognition and kills them directly; participates in antibody-dependent cellular cytotoxicity

negative balance loss of substance from body exceeds gain, and total amount in body decreases, also used for physical parameters such as body temperature and energy; *compare* positive balance

negative feedback characteristic of control systems in which system's response opposes the original change in the system; *compare* positive feedback

nephron (NEF-ron) functional unit of kidney; has vascular and tubular component

nerve group of many nerve fibers traveling together in peripheral nervous system

nerve cell cell in nervous system specialized to initiate, integrate, and conduct electric signals; also called *neuron*

nerve fiber axon of a neuron

net amount remaining after deductions have been made; final amount

net flux difference between two one-way fluxes

neuroeffector junction "synapse" between a neuron and muscle or gland cell

neuroglia *see* glial cell

neurohormone chemical messenger that is released by a neuron and travels in bloodstream to its target cell

neuromodulator chemical messenger that acts on neurons, usually by a second-messenger system, to alter response to a neurotransmitter

neuromuscular junction synapse-like junction between an axon terminal of an efferent nerve fiber and a skeletal-muscle fiber

neuron (NUR-ahn) *see* nerve cell

neuropeptide family of more than 50 neurotransmitters composed of 2 or more amino acids; often also functions as chemical messenger in nonneural tissues

neurotransmitter chemical messenger used by neurons to communicate with each other or with effectors

neutrophil (NOO-troh-fil) polymorphonuclear granulocytic leukocyte whose granules show preference for neither eosin nor basic dyes; functions as phagocyte and releases chemicals involved in inflammation

neurotropic factor (neur-oh-TRO-pic) protein that stimulates growth and differentiation of some neurons

nicotinamide adenine dinucleotide (NAD$^+$) coenzyme derived from the B-vitamin niacin; transfers hydrogen from one substrate to another

nicotinic receptor (nik-oh-TIN-ik) acetylcholine receptor that responds to nicotine; primarily, receptors at motor end plate and on postganglionic autonomic neurons

nitric oxide a gas that functions as intercellular messenger, including neurotransmitters; is endothelium-derived relaxing factor; destroys intracellular microbes

nociceptor (NOH-sih-sep-tor) sensory receptor whose stimulation causes pain

node of Ranvier (RAHN-vee-ay) space between adjacent myelin-forming cells along myelinated axon where axonal plasma membrane is exposed to extracellular fluid; also called neurofibril node

nonpolar pertaining to molecule or region of molecule containing predominantly chemical bonds in which electrons are shared equally between atoms; having few polar or ionized groups

nonspecific ascending pathway chain of synaptically connected neurons in CNS that are activated by sensory units of several different types; signals general information; *compare* specific ascending pathway

nonspecific immune defense response that nonselectively protects against foreign material without having to recognize its specific identity

norepinephrine (NE) (nor-ep-ih-NEF-rin) biogenic amine (catecholamine) neurotransmitter released at most sympathetic postganglionic endings, from adrenal medulla, and in many CNS regions

NREM sleep sleep state associated with large, slow EEG waves and considerable postural-muscle tone but not dreaming; also called slow-wave sleep

nuclear envelope double membrane surrounding cell nucleus

nuclear pore opening in nuclear envelope through which molecular messengers pass between nucleus and cytoplasm

nucleic acid (noo-KLAY-ik) nucleotide polymer in which phosphate of one nucleotide is linked to the sugar of the adjacent one; stores and transmits genetic information; includes DNA and RNA

nucleolus (noo-KLEE-oh-lus) densely staining nuclear region containing portions of DNA that code for ribosomal proteins

nucleotide (NOO-klee-oh-tide) molecular subunit of nucleic acid; purine or pyrimidine base, sugar, and phosphate

nucleus (NOO-klee-us) (pl. nuclei) (cell) large membrane-bound organelle that contains cell's DNA; (neural) cluster of neuron cell bodies in CNS

O

occipital lobe (ok-SIP-ih-tul) posterior region of cerebral cortex where primary visual cortex is located

Ohm's law current (I) is directly proportional to voltage (E) and inversely proportional to resistance (R) such that I = E/R

olfaction (ol-FAK-shun) sense of smell

olfactory (ol-FAK-tor-ee) pertaining to sense of smell

olfactory epithelium mucous membrane in upper part of nasal cavity containing receptors for sense of smell

oligodendroglia (oh-lih-goh-den-droh-GLEE-ah) type of glial cell; responsible for myelin formation in CNS

oncogene (ON-koh-jeen) altered gene that can lead to cancer

oogenesis (oh-uh-JEN-ih-sis) gamete production in female

oogonium (oh-uh-GOH-nee-um) primitive germ cell that gives rise to primary oocyte

opioid (OH-pee-oid) *see* endogenous opioid

opsin (OP-sin) protein component of photopigment

opsonin (op-SOH-nin) any substance that binds a microbe to a phagocyte and promotes phagocytosis

optimal length (I$_o$) sarcomere length at which muscle fiber develops maximal isometric tension

organ collection of tissues joined in structural unit to serve common function

organ of Corti (KOR-tee) structure in inner ear capable of transducing sound-wave energy into action potentials

organ system organs that together serve an overall function

organelle *see* cell organelle

organic pertaining to carbon-containing substances; *compare* inorganic

orgasm (OR-gazm) inner emotions and systemic physiological changes that mark apex of sexual intercourse, usually accompanied in the male by ejaculation

orienting response behavior in response to a novel stimulus; that is, the person stops what he or she is doing, looks around, listens intently, and turns toward stimulus

osmol (OZ-mole) 1 mole of solute ions and molecules

osmolarity (oz-moh-LAR-ih-tee) total solute concentration of a solution; measure of water concentration in that the higher the solution osmolarity, the lower the water concentration

osmoreceptor (OZ-moh-ree-sep-tor) receptor that responds to changes in osmolarity of surrounding fluid

osmosis (oz-MOH-sis) net diffusion of water across a selective barrier from region of higher water concentration (lower solute concentration) to region of lower water concentration (higher solute concentration)

osmotic pressure (oz-MAH-tik) pressure that must be applied to a solution on one side of a membrane to prevent osmotic flow of water across the membrane from a compartment of pure water; a measure of the solution's osmolarity

osteoblast (OS-tee-oh-blast) cell type responsible for laying down protein matrix of bone; called osteocyte after calcified matrix has been set down

osteoclast (OS-tee-oh-clast) cell that breaks down previously formed bone

oval window membrane-covered opening between middle ear cavity and scala vestibuli of inner ear

ovarian follicle *see* follicle

ovary (OH-vah-ree) gonad in female

ovulation (ov-you-LAY-shun) release of egg, surrounded by its zona pellucida and granulosa cells, from ovary

ovum (pl. ova) gamete of female; egg

oxidative (OX-ih-day-tive) using oxygen

oxidative deamination (dee-am-ih-NAY-shun) reaction in which an amino group ($-NH_2$) from an amino acid is replaced by oxygen to form a keto acid

oxidative fiber muscle fiber that has numerous mitochondria and therefore a high capacity for oxidative phosphorylation; red muscle fiber

oxidative phosphorylation (fos-for-ih-LAY-shun) process by which energy derived from reaction between hydrogen and oxygen to form water is transferred to ATP during its formation

oxygen debt decrease in energy reserves during exercise that results in an increase in oxygen consumption and an increased production of ATP by oxidative phosphorylation following the exercise

oxyhemoglobin (ox-see-HEE-moh-gloh-bin) HbO_2; hemoglobin combined with oxygen

oxytocin (ox-see-TOH-sin) peptide hormone synthesized in hypothalamus and released from posterior pituitary; stimulates mammary glands to release milk and uterus to contract

P

P wave component of electro-cardiogram reflecting atrial depolarization

pacemaker neurons that set rhythm of biological clocks independent of external cues; any nerve or muscle cell that has an inherent autorhythmicity and determines activity pattern of other cells

pacemaker potential spontaneous gradual depolarization to threshold of some nerve and muscle cells' plasma membrane

paracellular pathway the space between adjacent cells of an epithelium through which some molecules diffuse as they cross the epithelium

paracrine agent (PAR-ah-krin) chemical messenger that exerts its effects on cells near its secretion site; by convention, excludes neurotransmitters; *compare* autocrine agent

paradoxical sleep *see* REM sleep

parasympathetic division (par-ah-sim-pah-THET-ik) portion of autonomic nervous system whose preganglionic fibers leave CNS from brainstem and sacral portion of spinal cord; most of its postganglionic fibers release acetylcholine; *compare* sympathetic division

parathyroid gland one of four parathyroid-hormone secreting glands on thyroid gland surface

parathyroid hormone peptide hormone secreted by parathyroid glands; regulates calcium and phosphate concentrations of extracellular fluid

parietal cell (pah-RY-ih-tal) gastric gland cell that secretes hydrochloric acid and intrinsic factor

parietal lobe region of cerebral cortex containing sensory cortex and some association cortex

partial pressure (P) that part of total gas pressure due to molecules of one gas species; measure of concentration of a gas in a gas mixture

parturition events leading to and including delivery of infant

passive immunity resistance to infection resulting from direct transfer of antibodies or sensitized T cells from one person (or animal) to another; *compare* active immunity

pepsin (PEP-sin) family of several protein-digesting enzymes formed in the stomach; breaks protein down to peptide fragments

pepsinogen (pep-SIN-ah-jen) inactive precursor of pepsin; secreted by chief cells of gastric mucosa

peptide (PEP-tide) short polypeptide chain; by convention, having less than 50 amino acids

peptide bond polar covalent chemical bond joining the amino and carboxyl groups of two amino acids; forms protein backbone

peptidergic neuron that releases peptides

percent hemoglobin saturation *see* hemoglobin saturation

perception understanding of objects and events of external world that we acquire from neural processing of sensory information

perforin protein secreted by cytotoxic T cells; forms channels in plasma membrane of target cell, which destroys it

perfusion blood flow

pericardium (per-ah-KAR-dee-um) connective-tissue sac surrounding heart

perimenopause first year after cessation of menstruation

peripheral chemoreceptor carotid or aortic body; responds to changes in arterial blood P_{O_2} and H^+ concentration

peripheral nervous system nerve fibers extending from CNS

peripheral thermoreceptor cold or warm receptor in skin or certain mucous membranes

peristaltic wave (per-ih-STAL-tik) progressive wave of smooth-muscle contraction and relaxation that proceeds along wall of a tube, compressing the tube and causing its contents to move

peritoneum (per-ih-toh-NEE-um) membrane lining abdominal and pelvic cavities and covering organs there

peritubular capillary capillary closely associated with renal tubule

permeability constant (k_p) number that defines the proportionality between a flux and a concentration gradient and depends on the properties of the membrane and the diffusing molecule

permissiveness situation whereby small quantities of one hormone are required in order for a second hormone to exert its full effects

peroxisome (per-OX-ih-some) cell organelle that destroys certain toxic products by oxidative reactions

pH expression of a solution's acidity; negative logarithm to base 10 of H^+ concentration; pH decreases as acidity increases

phagocyte (FAH-go-site) any cell capable of phagocytosis

phagocytosis (fag-uh-sy-TOH-sis) engulfment of particles by a cell

pharynx (FAIR-inks) throat; passage common to routes taken by food and air

phase shift a resetting of the internal clock due to altered environmental cues

phasic (FAYZ-ik) intermittent; *compare* tonic

phenotype (FEEN-oh-type) a particular trait expressed in a person as the result of his or her genotype

phosphate group (FOS-fate) $-PO_4^{2-}$

phosphodiesterase (fos-foh-dy-ES-ter-ase) enzyme that catalyzes cyclic AMP breakdown to AMP

phospholipase A₂ (fos-fo-LY-pase A-two) enzyme that splits arachidonic acid from plasma membrane phospholipid

phospholipase C receptor-controlled plasma-membrane enzyme that catalyzes phosphatidylinositol bisphosphate breakdown to inositol trisphosphate and diacylglycerol

phospholipid (fos-foh-LIP-id) lipid subclass similar to triacylglycerol except that a phosphate group ($-PO_4^{2-}$) and small nitrogen-containing molecule are attached to third hydroxyl group of glycerol; major component of cell membranes

phosphoprotein phosphatase (FOS-fah-tase) enzyme that removes phosphate from protein

phosphoric acid (fos-FOR-ik) acid generated during catabolism of phosphorus-containing compounds; dissociates to form inorganic phosphate and hydrogen ions

phosphorylation (fos-for-ah-LAY-shun) addition of phosphate group to an organic molecule

photopigment light-sensitive molecule altered by absorption of photic energy of certain wavelengths; consists of opsin bound to a chromophore

photoreceptor receptor sensitive to light (photic energy)

physiology (fiz-ee-OL-uh-jee) branch of biology dealing with the mechanisms by which living organisms function

pinocytosis (pin-oh-sy-TOH-sis) endocytosis when the vesicle encloses extracellular fluid or specific molecules in the extracellular fluid that have bound to proteins on the extracellular surface of the plasma membrane

pitch degree of how high or low a sound is perceived

pituitary gland (pih-TOO-ih-tar-ee) endocrine gland that lies in bony pocket below hypothalamus; constitutes anterior pituitary and posterior pituitary

pituitary gonadotropin *see* gonadotropic hormone

placenta (plah-SEN-tah) interlocking fetal and maternal tissues that serve as organ of molecular exchange between fetal and maternal circulations

placental lactogen (plah-SEN-tal LAK-toh-jen) hormone that is produced by placenta and has effects similar to those of growth hormone and prolactin

plasma (PLAS-muh) liquid portion of blood; component of extracellular fluid

plasma cell cell that differentiates from activated B lymphocytes and secretes antibodies

plasma membrane membrane that forms outer surface of cell and separates cell's contents from extracellular fluid

plasma membrane effector protein plasma-membrane protein that serves as ion channel or enzyme in signal transduction sequence

plasma protein most are albumins, globulins, or fibrinogen

plasmin (PLAZ-min) proteolytic enzyme able to decompose fibrin and thereby to dissolve blood clots

plasminogen (plaz-MIN-oh-jen) inactive precursor of plasmin

plasminogen activator any plasma protein that activates proenzyme plasminogen

plasticity (plas-TISS-ih-tee) ability of neural tissue to change its responsiveness to stimulation because of its past history of activation

platelet (PLATE-let) cell fragment present in blood; plays several roles in blood clotting

platelet aggregation (ag-reh-GAY-shun) positive-feedback process resulting in platelets sticking together

platelet factor (PF) phospholipid exposed in membranes of aggregated platelets; important in activation of several plasma factors in clot formation

pleura (PLUR-ah) thin cellular sheet attached to thoracic cage interior (*parietal pleura*) and, folding back upon itself, is attached to lung surface (*visceral pleura*); forms two enclosed *pleural sacs* in thoracic cage

pluripotent hematopoietic stem cells (plur-ih-POH-tent) single population of bone-marrow cells from which all blood cells are descended

polar pertaining to molecule or region of molecule containing polar covalent bonds or ionized groups; part of molecule to which electrons are drawn becomes slightly negative, and region from which electrons are drawn becomes slightly positive; molecule is soluble in water

polar covalent bond covalent chemical bond in which two electrons are shared unequally between two atoms; atom to which the electrons are drawn becomes slightly negative, while other atom becomes slightly positive; also called polar bond

polarized (POH-luh-rized) having two electric poles, one negative and one positive

polymer (POL-ih-mer) large molecule formed by linking together of smaller similar subunits

polymorphonuclear granulocytes (pol-ee-morf-oh-NUK-lee-er GRAN-you-loh-sites) subclass of leukocytes; consisting of eosinophils, basophils, neutrophils

polypeptide (pol-ee-PEP-tide) polymer consisting of amino acid subunits joined by peptide bonds; also called peptide or protein

polysaccharide (pol-ee-SAK-er-ide) large carbohydrate formed by linking monosaccharide subunits together

polysynaptic reflex (pol-ee-sih-NAP-tik) reflex employing one or more interneurons in its reflex arc

polyunsaturated fatty acid fatty acid that contains more than one double bond

pool the readily available quantity of a substance in the body; often equals amounts in extracellular fluid

portal vein vessel through which blood from several abdominal organs flows to the liver

portal vessel any blood vessel that links two capillary networks

positive balance gain of substance exceeds loss, and amount of that substance in body increases; *compare* negative balance

positive feedback characteristic of control systems in which an initial disturbance sets off train of events that increases the disturbance even further; *compare* negative feedback

postabsorptive state (post-ab-SORP-tive) period during which nutrients are not being absorbed by gastrointestinal tract and energy must be supplied by body's endogenous stores

posterior toward or at the back

posterior pituitary portion of pituitary from which oxytocin and vasopressin are released

postganglionic (post-gang-glee-ON-ik) autonomic-nervous-system neuron or nerve fiber whose cell body lies in a ganglion; conducts impulses away from ganglion toward periphery; *compare* preganglionic

postsynaptic neuron (post-sin-NAP-tik) neuron that conducts information away from a synapse

postsynaptic potential local potential that arises in postsynaptic neuron in response to activation of synapses upon it; *see also* excitatory postsynaptic potential, inhibitory postsynaptic potential

postural reflex reflex that maintains or restores upright, stable posture

potential (or potential difference) voltage difference between two points; *see also* graded potential, action potential

potentiation (poh-ten-she-AY-shun) presence of one agent enhances response to a second such that final response is greater than sum of the two individual responses

precapillary sphincter (SFINK-ter) smooth-muscle ring around capillary where it exits from thoroughfare channel or arteriole

preganglionic autonomic-nervous-system neuron or nerve fiber whose cell body lies in CNS and whose axon terminals lie in a ganglion; conducts action potentials from CNS to ganglion; *compare* postganglionic

presynaptic neuron (pre-sin-NAP-tik) neuron that conducts action potentials toward a synapse

presynaptic synapse relation between two neurons in which axon terminal of one neuron ends on axon terminal of second neuron; action potentials in first neuron affect neurotransmitter release from second, thereby altering effectiveness of the synapse that the second neuron makes with a third neuron

primary active transport active transport in which chemical energy is transferred directly from ATP to transporter protein

primary cortical receiving area region of cerebral cortex where specific ascending pathways end; somatosensory, visual, auditory, or taste cortex

primary lymphoid organ organs that supply secondary lymphoid organs with mature lymphocytes; bone marrow and thymus

primary motivated behavior behavior related directly to achieving homeostasis

primary motor cortex *see* motor cortex

primary oocyte (OH-uh-site) female germ cell that undergoes first meiotic division to form secondary oocyte and polar body

primary response gene (PRG) gene influenced by transcription factors generated in response to first messengers

primary RNA transcript an RNA molecule transcribed from a gene before intron removal and splicing

primary spermatocyte (sper-MAT-uh-site) male germ cell derived from spermatogonia; undergoes meiotic division to form two secondary spermatocytes

primordial follicle (FAH-lik-el) *see* ovarian follicle

procedural memory the memory of how to do things

process long extension from neuron cell body

product molecule formed in enzyme-catalyzed chemical reaction

progesterone (proh-JES-ter-own) steroid hormone secreted by corpus luteum and placenta; stimulates uterine gland secretion, inhibits uterine smooth-muscle contraction, and stimulates breast growth

program related sequence of neural activity preliminary to motor act

prohormone peptide precursor from which are cleaved one or more active peptide hormones

prolactin (pro-LAK-tin) peptide hormone secreted by anterior pituitary; stimulates milk secretion by mammary glands

prolactin-inhibiting hormone (PIH) dopamine, which serves as a hypophysiotropic hormone to inhibit prolactin secretion by anterior pituitary

proliferative phase (pro-LIF-er-ah-tive) stage of menstrual cycle between menstruation and ovulation during which endometrium repairs itself and grows

promotor specific nucleotide sequence at beginning of gene that controls the initiation of gene transcription; determines which of the paired strands of DNA is transcribed into RNA

pro-opiomelanocortin (pro-oh-pee-oh-mel-an-oh-KOR-tin) large protein precursor for ACTH, endorphin, and several other hormones

propagation (prop-ah-GAY-shun) conduction of nerve impulse

prostacyclin eicosanoid that inhibits platelet aggregation in blood clotting; also called prostaglandin I_2 (PGI_2)

prostaglandin (pros-tah-GLAN-din) one class of a group of modified unsaturated fatty acids (eicosanoids); function mainly as paracrine or autocrine agents

prostate gland (PROS-tate) large gland encircling urethra in the male; secretes seminal fluid into urethra

protein large polymer consisting of one or more sequences of amino acid subunits joined by peptide bonds

protein binding site *see* binding site

protein C plasma protein that inhibits clotting

protein kinase (KY-nase) any enzyme that phosphorylates other proteins by transferring to them a phosphate group from ATP

protein kinase C enzyme that phosphorylates certain intracellular proteins when activated by diacylglycerol

proteolytic (proh-tee-oh-LIT-ik) breaks down protein

prothrombin (proh-THROM-bin) inactive precursor of thrombin; produced by liver and normally present in plasma

proton (PROH-tahn) positively charged subatomic particle

proximal (PROX-sih-mal) nearer; closer to reference point; *compare* distal

proximal tubule first tubular component of a nephron after Bowman's capsule; comprises *convoluted* and *straight segments*

puberty attainment of sexual maturity when conception becomes possible; as commonly used, refers to 3 to 5 years of sexual development that culminates in sexual maturity

pulmonary (PUL-mah-nar-ee) pertaining to lungs

pulmonary circulation circulation through lungs; portion of cardiovascular system between pulmonary trunk, as it leaves the right ventricle, and pulmonary veins, as they enter the left atrium

pulmonary stretch receptor afferent nerve ending lying in airway smooth muscle and activated by lung inflation

pulmonary surfactant *see* surfactant

pulmonary trunk large artery that carries blood from right ventricle of heart to lungs

pulmonary valve valve between right ventricle of heart and pulmonary trunk

pulse pressure difference between systolic and diastolic arterial blood pressures

pupil opening in iris of eye through which light passes to reach retina

purine (PURE-ene) double-ring, nitrogen-containing subunit of nucleotide; adenine or guanine

Purkinje fiber (purr-KIN-gee) specialized myocardial cell that constitutes part of conducting system of heart; conveys excitation from bundle branches to ventricular muscle

pyloric sphincter (py-LOR-ik) ring of smooth muscle between stomach and small intestine

pyramidal tract *see* corticospinal pathway

pyrimidine (pi-RIM-ih-deen) single-ring, nitrogen-containing subunit of nucleotide; cytosine, thymine, or uracil

pyrogen *see* endogenous pyrogen

pyruvate (PY-roo-vayt) anion formed when pyruvic acid loses a hydrogen ion

pyruvic acid (py-ROO-vik) three-carbon intermediate in glycolysis that, in absence of oxygen, forms lactic acid or, in presence of oxygen, enters Krebs cycle

Q

QRS complex component of electrocardiogram corresponding to ventricular depolarization

R

R- in chemical formula, signifies remaining portion of molecule

rapid eye movement sleep *see* REM sleep

rate-limiting enzyme enzyme in metabolic pathway most easily saturated with substrate; determines rate of entire metabolic pathway

rate-limiting reaction slowest reaction in metabolic pathway; catalyzed by rate-limiting enzyme

reactant (ree-AK-tent) molecule that enters a chemical reaction; called the substrate in enzyme-catalyzed reactions

reaction *see* chemical reaction

reactive hyperemia (hy-per-EE-me-ah) transient increase in blood flow following release of occlusion of blood supply

receptive field (or neuron) area of body that, if stimulated, results in activity in that neuron

receptor (in sensory system) specialized peripheral ending of afferent neuron, or separate cell intimately associated with it, that detects changes in some aspect of environment; (in intercellular chemical communication) specific binding site in plasma membrane or interior of target cell with which a chemical messenger combines to exert its effects

receptor activation change in receptor conformation caused by combination of messenger with receptor

receptor potential graded potential that arises in afferent-neuron ending, or a specialized cell intimately associated with it, in response to stimulation

reciprocal innervation inhibition of motor neurons activating muscles whose contraction would oppose an intended movement

recognition binding of antigen to receptor specific for that antigen on lymphocyte surface

recombinant DNA (re-KOM-bih-nent) DNA formed by joining portions of two DNA molecules previously fragmented by a restriction enzyme

recruitment activation of additional cells in response to increased stimulus strength; increasing the number of active motor units in a muscle

rectum short segment of large intestine between sigmoid colon and anus

red muscle muscle having high oxidative capacity and large amount of myoglobin

reflex (REE-flex) biological control system linking stimulus with response and mediated by a reflex arc

reflex arc neural or hormonal components that mediate a reflex; usually includes receptor, afferent pathway, integrating center, efferent pathway, and effector

reflex response final change due to action of stimulus upon reflex arc; also called effector response

refractory period (reh-FRAK-tor-ee) time during which an excitable membrane does not respond to a stimulus that normally causes response; *see also* absolute refractory period, relative refractory period

regulatory site site on protein that interacts with modulator molecule; alters functional-site properties

relative refractory period time during which excitable membrane will produce action potential but only to a stimulus of greater strength than the usual threshold strength

releasing hormone *see* hypo-physiotropic hormone

REM sleep (rem) sleep state associated with small, rapid EEG oscillations, complete loss of tone in postural muscles, and dreaming; also called *rapid eye movement sleep, paradoxical sleep*

renal (REE-nal) pertaining to kidneys

renal corpuscle combination of glomerulus and Bowman's capsule

renal pelvis cavity at base of each kidney; receives urine from

collecting duct system and empties it into ureter

renin (REE-nin) peptide hormone secreted by kidneys; acts as an enzyme that catalyzes splitting off of angiotensin I from angiotensinogen in plasma

replicate (REP-lih-kayt) duplicate

repolarize return transmembrane potential to its resting level

residual volume air volume remaining in lungs after maximal expiration

resistance (R) hindrance to movement through a particular substance, tube, or opening

respiration (cellular) oxygen utilization in metabolism of organic molecules; (respiratory system) oxygen and carbon dioxide exchange between organism and external environment

respiratory acidosis increased arterial H^+ concentration due to carbon dioxide retention

respiratory alkalosis decreased arterial H^+ concentration when carbon dioxide elimination from the lungs exceeds its production

respiratory pump effect on venous return of changing intrathoracic and intraabdominal pressures associated with respiration

respiratory quotient (RQ) (KWOH-shunt) ratio of carbon dioxide produced to oxygen consumed during metabolism

respiratory rate number of breaths per minute

respiratory zone portion of airways from beginning of respiratory bronchioles to alveoli; contains alveoli across which gas exchange occurs

resting membrane potential voltage difference between inside and outside of cell in absence of excitatory or inhibitory stimulation; also called resting potential

restriction element *see* MHC protein

retching strong involuntary attempt to vomit but without stomach contents passing through upper esophageal sphincter

reticular formation extensive neuron network extending through brainstem core; receives and integrates information from many afferent pathways and from other CNS regions

retina   thin layer of neural tissue lining back of eyeball; contains receptors for vision

retinal   (ret-in-AL) form of vitamin A that forms chromophore component of photopigment

retrograde   opposite the usual course of events

reversible reaction   chemical reaction in which energy release is small enough for reverse reaction to occur readily; *compare* irreversible reaction

Rh factor   group of erythrocyte plasma-membrane antigens that may (Rh^+) or may not (Rh^-) be present

rhodopsin   (roh-DOP-sin) photo-pigment in rods

ribonucleic acid (RNA)   (ry-boh-noo-KLAY-ik) single-stranded nucleic acid involved in transcription of genetic information and translation of that information into protein structure; contains the sugar ribose; *see also* messenger RNA, ribosomal RNA, transfer RNA

ribosomal RNA (rRNA)   (ry-boh-SOME-al) type of RNA used in ribosome assembly; becomes part of ribosome

ribosome   (RY-boh-some) cytoplasmic particle that mediates linking together of amino acids to form proteins; attached to endoplasmic reticulum as bound ribosome, or suspended in cytoplasm as free ribosome

rigor mortis   (RIG-or MOR-tiss) stiffness of skeletal muscles after death due to failure of cross bridges to dissociate from actin because of the loss of ATP

RNA polymerase   (POL-ih-muh-rase) enzyme that forms RNA by joining together appropriate nucleotides after they have base-paired to DNA

rod   one of two receptor types for photic energy; contains the photopigment rhodopsin

S

saccade   (sah-KADE) short, jerking eyeball movement

saliva   watery solution of salts and proteins, including mucins and amylase, secreted by salivary glands

saltatory conduction   propagation of action potentials along a myelinated axon such that the action potentials jump from one node of Ranvier in the myelin sheath to the next node

sarcomere   (SAR-kuh-meer) repeating structural unit of myofibril; composed of thick and thin filaments; extends between two adjacent Z lines

sarcoplasmic reticulum   (sar-koh-PLAZ-mik reh-TIK-you-lum) endoplasmic reticulum in muscle fiber; site of storage and release of calcium ions

satiety signal   (sah-TY-ih-tee) input to food control centers that causes hunger to cease and sets time period before hunger returns

saturated fatty acid   fatty acid whose carbon atoms are all linked by single covalent bonds

saturation   occupation of all available binding sites by their ligand

scala tympani   (SCALE-ah TIM-pah-nee) fluid-filled inner-ear compartment that receives sound waves from basilar membrane and transmits them to round window

scala vestibuli   (ves-TIB-you-lee) fluid-filled inner-ear compartment that receives sound waves from oval window and transmits them to basilar membrane and cochlear duct

Schwann cell   nonneural cell that forms myelin sheath in peripheral nervous system

scrotum   (SKROH-tum) sac that contains testes and epididymides

second messenger   intracellular substance that serves as relay from plasma membrane to intracellular biochemical machinery, where it alters some aspect of cell's function

secondary active transport   active transport in which energy released during transmembrane movement of one substance from higher to lower concentration is transferred to the simultaneous movement of another substance from lower to higher concentration

secondary lymphoid organ   lymph node, spleen, tonsil, or lymphocyte accumulation in gastrointestinal, respiratory, urinary, or reproductive tract; site of stimulation of lymphocyte response

secondary peristalsis   (per-ih-STAL-sis) esophageal peristaltic waves not immediately preceded by pharyngeal phase of swallow

secondary sexual characteristics   external differences between male and female not directly involved in reproduction

secretin   (SEEK-reh-tin) peptide hormone secreted by upper small intestine; stimulates pancreas to secrete bicarbonate into small intestine

secretion   (sih-KREE-shun) elaboration and release of organic molecules, ions, and water by cells in response to specific stimuli

secretory phase   (SEEK-rih-tor-ee) stage of menstrual cycle following ovulation during which secretory type of endometrium develops

secretory vesicle   membrane-bound vesicle produced by Golgi apparatus; contains protein to be secreted by cell

segmentation   (seg-men-TAY-shun) series of stationary rhythmical contractions and relaxations of rings of intestinal smooth muscle; mixes intestinal contents

semen   (SEE-men) sperm-containing fluid of male ejaculate

semicircular canal   passage in temporal bone; contains sense organs for equilibrium and movement

seminal vesicle   one of pair of exocrine glands in males that secrete fluid into vas deferens

seminiferous tubule   (sem-ih-NIF-er-ous) tubule in testis in which sperm production occurs; lined with Sertoli cells

semipermeable membrane   (sem-eye-PER-me-ah-bul) membrane permeable to some substances but not to others

sensorimotor cortex   (sen-sor-ee-MOH-tor) all areas of cerebral cortex that play a role in skeletal-muscle control

sensory information   information that originates in stimulated sensory receptors

sensory pathway   a group of neuron chains, each chain consisting of three or more neurons connected end-to-end by synapses; carries action potentials to those parts of the brain involved in conscious recognition of sensory information

sensory receptor a cell or portion of a cell that contains structures or chemical molecules sensitive to changes in an energy form in the outside world or internal environment; in response to activation by this energy, the sensory receptor initiates action potentials in that cell or an adjacent one

sensory system part of nervous system that receives, conducts, or processes information that leads to perception of a stimulus

sensory unit afferent neuron plus receptors it innervates

serosa (sir-OH-sah) connective-tissue layer surrounding outer surface of stomach and intestines

serotonin (sair-oh-TONE-in) biogenic amine neurotransmitter; paracrine agent in blood platelets and digestive tract; also called *5-hydroxytryptamine,* or *5-HT*

Sertoli cell (sir-TOH-lee) cell intimately associated with developing germ cells in seminiferous tubule; creates blood-testis barrier, secretes fluid into seminiferous tubule, and mediates hormonal effects on tubule

serum (SEER-um) blood plasma from which fibrinogen and other clotting proteins have been removed as result of clotting

set point steady-state value maintained by homeostatic control system

sex chromatin (CHROM-ah-tin) nuclear mass not usually found in cells of males; condensed X chromosome

sex chromosome X or Y chromosome

sex determination genetic basis of individual's sex, XY determining male, and XX, female

sex differentiation development of male or female reproductive organs

sex hormone estrogen, progesterone, testosterone, or related hormones

short-loop negative feedback influence of hypothalamus by an anterior pituitary hormone

sigmoid colon (SIG-moid) S-shaped terminal portion of colon

signal sequence initial portion of newly synthesized protein (if protein is destined for secretion)

signal transduction pathway sequence of mechanisms that relay information from plasma-membrane receptor to cell's response mechanism; *see also* transduction

single-unit smooth muscle smooth muscle that responds to stimulation as single unit because gap junctions join muscle fibers, allowing electrical activity to pass from cell to cell

sinoatrial (SA) node (sy-noh-AY-tree-al) region in right atrium of heart containing specialized cardiac-muscle cells that depolarize spontaneously faster than other cells in the conducting system; determines heart rate

sister chromatids (CHROM-ah-tid) two identical DNA threads joined together during cell division

skeletal muscle striated muscle attached to bone or skin and responsible for skeletal movements and facial expression; controlled by somatic nervous system

skeletal-muscle pump pumping effect of contracting skeletal muscles on blood flow through underlying vessels

skeletomotor fiber *see* extrafusal fiber

sleep *see* REM sleep, NREM sleep

sliding-filament mechanism process of muscle contraction in which shortening occurs by thick and thin filaments sliding past each other

slow channel voltage-gated calcium channel in myocardial-cell plasma membrane; opens, after a short delay, upon depolarization

slow fiber muscle fiber whose myosin has low ATPase activity

slow-wave sleep *see* NREM sleep

smooth muscle nonstriated muscle that surrounds hollow organs and tubes; *see also* single-unit smooth muscle, multiunit smooth muscle

smooth-muscle tone smooth-muscle tension due to low-level cross-bridge activity in absence of external stimuli

sodium inactivation turning off of increased sodium permeability at action-potential peak

soft palate (PAL-et) nonbony region at back of roof of mouth

solute (SOL-yoot) substances dissolved in a liquid

solution liquid (solvent) containing dissolved substances (solutes)

solvent liquid in which substances are dissolved

somatic (soh-MAT-ik) pertaining to the body; related to body's framework or outer walls, including skin, skeletal muscle, tendons, and joints

somatic nervous system component of efferent division of peripheral nervous system; innervates skeletal muscle; *compare* autonomic nervous system

somatic receptor neural receptor in the framework or outer wall of the body that responds to mechanical stimulation of skin or hairs and underlying tissues, rotation or bending of joints, temperature changes, or painful stimuli

somatosensory cortex (suh-mat-uh-SEN-suh-ree) strip of cerebral cortex in parietal lobe in which nerve fibers transmitting somatic sensory information synapse

somatostatin (SS) (suh-mat-uh-STAT-in) hypophysiotropic hormone that inhibits growth hormone secretion by anterior pituitary; possible neurotransmitter; also found in stomach and pancreatic islets

sound wave air disturbance due to variations between regions of high air molecule density (compression) and low density (rarefaction)

spatial summation (SPAY-shul) adding together effects of simultaneous inputs to different places on a neuron to produce potential change greater than that caused by single input

specific ascending pathway chain of synaptically connected neurons in CNS, all activated by sensory units of same type

specific immune defense response that depends upon recognition of specific foreign material for reaction to it

specificity selectivity; ability of binding site to react with only one, or a limited number of, types of molecules

sperm *see* spermatozoon

sperm capacitation (kah-pas-ih-TAY-shun) process by which sperm in female reproductive tract gains ability to fertilize egg

spermatid (SPER-mah-tid) immature sperm

spermatogenesis (sper-mah-toh-JEN-ih-sis) sperm formation

spermatogonium (sper-mah-toh-GOH-nee-um) undifferentiated germ cell that gives rise to primary spermatocyte

spermatozoon (spur-ma-toh-ZOH-in) male gamete; also called sperm

sphincter (SFINK-ter) smooth-muscle ring that surrounds a tube, closing tube as muscle contracts

sphincter of Oddi (OH-dee) smooth-muscle ring surrounding common bile duct at its entrance into duodenum

sphygmomanometer (sfig-moh-mah-NOM-eh-ter) device consisting of inflatable cuff and pressure gauge for measuring arterial blood pressure

spinal nerve one of 86 peripheral nerves (43 pairs) that join spinal cord

spinal reflex reflex whose afferent and efferent components are in spinal nerves; can occur in absence of brain control

spindle fiber (muscle) *see* intrafusal fiber; (mitosis) microtubule that connects chromosome to centriole and centrioles to each other during cell division

spleen largest lymphoid organ; located between stomach and diaphragm

SRY gene gene on the Y chromosome that determines development of testes in genetic male

stable balance net loss of substance from body equals net gain, and amount of substance in body neither increases nor decreases; *compare* positive balance, negative balance

starch moderately branched plant polysaccharide composed of glucose subunits

Starling force factor that determines direction and magnitude of fluid movement across capillary wall

Starling's law of the heart *see* Frank-Starling mechanism

state of consciousness degree of mental alertness—that is, whether awake, drowsy, asleep, and so on

steady state no net change occurs; continual energy input to system is required, however, to prevent net change; *compare* equilibrium

stem cell cell that in adult body divides continuously and forms supply of cells for differentiation

stereocilia (ster-ee-oh-SIL-ee-ah) nonmotile cilia containing actin filaments

steroid (STEER-oid) lipid subclass; molecule consists of four interconnected carbon rings to which polar groups may be attached

stimulus detectable change in internal or external environment

"stop" signal three-nucleotide sequence in mRNA that signifies end of protein coding sequence

stress environmental change that must be adapted to if health and life are to be maintained; event that elicits increased cortisol secretion

stretch receptor afferent nerve ending that is depolarized by stretching; *see also* muscle-spindle stretch receptor

stretch reflex monosynaptic reflex, mediated by muscle-spindle stretch receptor, in which muscle stretch causes contraction of that muscle

striated muscle (STRY-ay-ted) muscle having transverse banding pattern due to repeating sarcomere structure; *see also* skeletal and cardiac muscle

stroke volume blood volume ejected by a ventricle during one heartbeat

strong acid acid that ionizes completely to form hydrogen ions and corresponding anions when dissolved in water; *compare* weak acid

submucous plexus (sub-MU-kus PLEX-us) nerve-cell network in submucosa of esophageal, stomach, and intestinal walls

substance P neuropeptide neurotransmitter released by afferent neurons in pain pathway as well as other sites

substrate (SUB-strate) reactant in enzyme-mediated reaction

substrate level-phosphorylation (fos-for-ih-LAY-shun) direct transfer of phosphate group from metabolic intermediate to ADP to form ATP

subsynaptic membrane (sub-sih-NAP-tik) the part of postsynaptic neuron's plasma membrane under synapse

subthreshold potential (sub-THRESH-old) depolarization less than threshold potential

subthreshold stimulus stimulus capable of depolarizing membrane but not by enough to reach threshold

sucrose (SOO-krose) disaccharide composed of glucose and fructose; also called table sugar

sulfate SO_4^{2-}

sulfhydryl group (sulf-HY-drul) —SH

sulfuric acid (sulf-YOR-ik) acid generated during catabolism of sulfur-containing compounds; dissociates to form sulfate and hydrogen ions

summation (sum-MAY-shun) increase in muscle tension or shortening in response to rapid, repetitive stimulation relative to single twitch

superior vena cava (VEE-nah KAY-vah) large vein that carries blood from upper half of body to right atrium of heart

surface tension attractive forces between water molecules at an air-water interface resulting in net force that acts to reduce surface area

surfactant (sir-FAK-tent) detergent-like phospholipid-protein mixture produced by pulmonary type II alveolar cells; reduces surface tension of fluid film lining alveoli

sympathetic division portion of autonomic nervous system whose preganglionic fibers leave CNS at thoracic and lumbar portions of spinal cord; *compare* parasympathetic division

sympathetic trunk one of paired chains of interconnected sympathetic ganglia that lie on either side of vertebral column

sympathomimetic (sym-path-oh-mih-MET-ik) produces effects similar to those of sympathetic nervous system

synapse (SIN-apse) anatomically specialized junction between two neurons where electrical activity in one neuron influences excitability of second; *see also* chemical synapse, electric synapse, excitatory synapse, inhibitory synapse

synaptic cleft narrow extracellular space separating pre- and postsynaptic neurons at chemical synapse

synaptic potential *see* postsynaptic potential

synergistic muscle (sin-er-JIS-tik) muscle that exerts force to aid intended motion

systemic circulation (sis-TEM-ik) circulation from left ventricle through all organs except lungs and back to heart

systole (SIS-toh-lee) period of ventricular contraction

systolic pressure (SP) (sis-TAHL-ik) maximum arterial blood pressure during cardiac cycle

T

T cell lymphocyte derived from precursor that differentiated in thymus; *see also* cytotoxic T cell, helper T cell

T lymphocyte *see* T cell

T tubule *see* transverse tubule

T wave component of electro-cardiogram corresponding to ventricular repolarization

target cell cell influenced by a certain hormone

taste bud sense organ that contains chemoreceptors for taste

tectorial membrane (tek-TOR-ee-al) structure in organ of Corti in contact with receptor-cell hairs

teleology (teel-ee-OL-oh-gee) explanation of events in terms of ultimate purpose served by them

template (TEM-plit) pattern

temporal lobe region of cerebral cortex where primary auditory cortex and Wernicke's speech center are located

temporal summation membrane potential produced as two or more inputs, occurring at different times, are added together; potential change is greater than that caused by single input

tendon (TEN-don) collagen-fiber bundle that connects skeletal muscle to bone and transmits muscle contraction force to the bone

tension force; *see also* muscle tension

testis (TES-tiss) (pl. testes) gonad in male

testosterone (test-TOS-ter-own) steroid hormone produced in interstitial cells of testes; major male sex hormone

tetanus (TET-ah-nus) maintained mechanical response of muscle to high-frequency stimulation; also the disease lockjaw

tetrad (TET-rad) grouping of two homologous chromosomes, each with its sister chromatid, during meiosis

thalamus (THAL-ah-mus) subdivision of diencephalon; integrating center for sensory input on its way to cerebral cortex; also contains motor nuclei

theca (THEE-kah) cell layer that surrounds ovarian-follicle granulosa cells

thermogenesis (ther-moh-JEN-ih-sis) heat generation

thermoneutral zone temperature range over which changes in skin blood flow can regulate body temperature

thermoreceptor sensory receptor for temperature and temperature changes, particularly in low (cold receptor) or high (warm receptor) range

thick filament myosin filament in muscle cell

thin filament actin filament in muscle cell

thoracic cavity (thor-ASS-ik) chest cavity

thoracic wall chest wall

thorax (THOR-aks) closed body cavity between neck and diaphragm; contains lung, heart, thymus, large vessels, and esophagus; also called the chest

threshold (THRESH-old) (or threshold potential) membrane potential to which excitable membrane must be depolarized to initiate an action potential

threshold stimulus stimulus capable of depolarizing membrane just to threshold

thrombin (THROM-bin) enzyme that catalyzes conversion of

fibrinogen to fibrin; has multiple other actions in blood clotting

thrombolytic system *see* fibrinolytic system

thrombomodulin an endothelial receptor to which thrombin can bind, thereby eliminating thrombin's clot-producing effects and causing it to bind and activate protein C

thrombosis (throm-BOH-sis) clot formation in body

thromboxane A_2 thromboxane that, among other effects, stimulates platelet aggregation in blood clotting

thrombus (THROM-bus) blood clot

thymine (T) (THIGH-meen) pyrimidine base in DNA but not RNA

thymus (THIGH-mus) lymphoid organ in upper part of chest; site of T-lymphocyte differentiation

thyroglobulin (thigh-roh-GLOB-you-lin) large protein precursor of thyroid hormones in colloid of follicles in thyroid gland; storage form of thyroid hormones

thyroid gland endocrine gland in neck; secretes thyroid hormones and calcitonin

thyroid hormones (TH) collective term for amine hormones released from thyroid gland—that is, thyroxine (T_4) and triiodo-thyronine (T_3)

thyroid-stimulating hormone (TSH) glycoprotein hormone secreted by anterior pituitary; induces secretion of thyroid hormone; also called thyrotropin

thyrotropin releasing hormone (TRH) hypophysiotropic hormone that simulates thyrotropin and prolactin secretion by anterior pituitary

thyroxine (T_4) (thigh-ROCKS-in) tetraiodothyronine; iodine-containing amine hormone secreted by thyroid gland

tidal volume air volume entering or leaving lungs with single breath during any state of respiratory activity

tight junction cell junction in which extracellular surfaces of the plasma membrane of two adjacent cells are joined together; extends around epithelial cell and restricts molecule diffusion through space between cells

tissue aggregate of single type of specialized cell; also denotes general cellular fabric of a given organ

tissue factor protein involved in initiation of clotting via the extrinsic pathway; located on plasma membrane of sub-endothelial cells

tissue plasminogen activator (t-PA) plasma protein produced by endothelial cells; after binding to fibrinogen, activates the proenzyme plasminogen

titin protein that extends from the Z line to the thick filaments and M line of skeletal-muscle sarcomere

tone maintained functional activity; *see also* muscle tone

tonic (TAH-nik) continuous activity; *compare* phasic

tonsil one of several small lymphoid organs in pharynx

total blood carbon dioxide sum total of dissolved carbon dioxide, bicarbonate, and carbamino-CO_2

total energy expenditure sum of external work done plus heat produced plus energy stored by body

total peripheral resistance (TPR) total resistance to flow in systemic blood vessels from beginning of aorta to ends of venae cavae

trace element mineral present in body in extremely small quantities

trachea (TRAY-key-ah) single airway connecting larynx with bronchi; windpipe

tract large, myelinated nerve-fiber bundle in CNS

transamination (trans-am-in-NAY-shun) reaction in which an amino acid amino group ($-NH_2$) is transferred to a ketoacid, the ketoacid thus becoming an amino acid

transcellular pathway crossing an epithelium by movement into an epithelial cell, diffusion through the cytosol of that cell, and exit across the opposite membrane

transcription formation of RNA containing, in linear sequence of its nucleotides, the genetic information of a specific gene; first stage of protein synthesis

transcription factor one of a class of proteins that act as gene switches, regulating the transcription of a particular gene by activating or repressing the initiation process

transduction process by which stimulus energy is transformed into a response

transepithelial transport *see* epithelial transport

transfer RNA (tRNA) type of RNA; different tRNAs combine with different amino acids and with codon on mRNA specific for that amino acid, thus arranging amino acids in sequence to form specific protein

transferrin (trans-FAIR-in) iron-binding protein that carries iron in plasma

translation during protein synthesis, assembly of amino acids in correct order according to genetic instructions in mRNA; occurs on ribosomes

transmural pressure pressure difference exerted on the two sides of a wall

transport maximum (T_m) upper limit to amount of material that carrier-mediated transport can move across the renal tubule

transporter integral membrane protein that mediates passage of molecule through membrane; also called carrier

transpulmonary pressure difference between alveolar and intrapleural pressures; force that holds lungs open

transverse tubule (T tubule) tubule extending from striated-muscle plasma membrane into the fiber, passing between opposed sarco-plasmic-reticulum segments; conducts muscle action potential into muscle fiber

triacylglycerol (try-ay-seel-GLISS-er-ol) subclass of lipids composed of glycerol and three fatty acids; also called fat, neutral fat, or triglyceride

tricarboxylic acid cycle *see* Krebs cycle

tricuspid valve (try-CUS-pid) valve between right atrium and right ventricle of heart

triglyceride *see* triacylglycerol

triiodothyronine (T_3) (try-eye-oh-doh-THIGH-roh-neen) iodine-containing amine hormone secreted by thyroid gland

triplet code three-base sequence in DNA and RNA that specifies particular amino acid

trophoblast (TROH-foh-blast) outer layer of blastocyst; gives rise to fetal portion of placental tissue

tropic (TROH-pik) growth promoting

tropic hormone hormone that stimulates the secretion of another hormone, and, often, growth of hormone-secreting gland

tropomyosin (troh-poh-MY-oh-sin) regulatory protein capable of reversibly converting binding sites on actin; associated with muscle thin filaments

troponin (troh-POH-nin) regulatory protein bound to actin and tropomyosin of striated-muscle thin filaments; site of calcium binding that initiates contractile activity

trypsin (TRIP-sin) enzyme secreted into small intestine by exocrine pancreas as precursor trypsinogen; breaks certain peptide bonds in proteins and polypeptides

trypsinogen (trip-SIN-oh-jen) inactive precursor of trypsin; secreted by exocrine pancreas

tubular reabsorption transfer of materials from kidney tubule lumen to peritubular capillaries

tubular secretion transfer of materials from peritubular capillaries to kidney tubule lumen

tumor necrosis factor (TNF) (neh-KROH-sis) cytokine secreted by macrophages (and other cells); has many of the same functions as IL-1

twitch mechanical response of muscle to single action potential

tympanic membrane (tim-PAN-ik) membrane stretched across end of ear canal; also called eardrum

type I alveolar cell a flat epithelial cell that with others forms a continuous layer lining the air-facing surface of the pulmonary alveoli

type II alveolar cell pulmonary cell that produces surfactant

tyrosine (TY-roh-seen) amino acid; precursor of catecholamines and thyroid hormones

tyrosine kinase protein kinase that phosphorylates tyrosine portion of proteins; may be part of plasma membrane receptor

U

ultrafiltrate (ul-tra-FIL-trate) protein-free fluid formed from plasma as it is forced through capillary walls by pressure gradient

umbilical vessel (um-BIL-ih-kul) artery or vein transporting blood between fetus and placenta

unsaturated fatty acid fatty acid containing one or more double bonds

upper esophageal sphincter (ih-sof-ih-JEE-al SFINK-ter) skeletal-muscle ring surrounding esophagus just below pharynx that, when contracted, closes entrance to esophagus

up-regulation increase in number of target-cell receptors for given messenger in response to chronic low extracellular concentration of that messenger; *see also* supersensitivity; *compare* down-regulation

uracil (U) (YOR-ah-sil) pyrimidine base; present in RNA but not DNA

urea (you-REE-ah) major nitrogenous waste product of protein breakdown and amino acid catabolism

ureter (YUR-ih-ter) tube that connects kidney to bladder

urethra (you-REE-thrah) tube that connects bladder to outside of body

uric acid (YUR-ik) waste product derived from nucleic acid catabolism

urinary bladder *see* bladder

uterine tube (YOU-ter-in) one of two tubes that carries egg from ovary to uterus; also called fallopian tube, oviduct

uterus (YOO-ter-us) hollow organ in pelvic region of females; houses fetus during pregnancy; also called womb

V

vagina (vah-JY-nah) canal leading from uterus to outside of body; also called birth canal

vagus nerve (VAY-gus) cranial nerve X; major parasympathetic nerve

van der Waals forces (walls) weak attractive forces between nonpolar regions of molecules

varicosity (vair-ih-KOS-ih-tee) swollen region of axon; contains neurotransmitter-filled vesicles; analogous to presynaptic ending

vas deferens (vas DEF-er-enz) one of paired male reproductive ducts that connect epididymis of testis to urethra; also called ductus deferens

vasa recta (VAY-zuh) blood vessels that form loops parallel to the loops of Henle in the renal medulla

vasoconstriction (vays-oh-kon-STRIK-shun) decrease in blood-vessel diameter due to vascular smooth-muscle contraction

vasodilation (vays-oh-dy-LAY-shun) increase in blood-vessel diameter due to vascular smooth-muscle relaxation

vasopressin (vas-oh-PRES-sin) peptide hormone synthesized in hypothalamus and released from posterior pituitary; increases water permeability of kidneys' collecting ducts and causes vasoconstriction; also called *antidiuretic hormone (ADH)*

vein any vessel that returns blood to heart; but *see also* portal vein

vena cava (VEE-nah KAY-vah) (pl. venae cavae) one of two large veins that returns systemic blood to heart; *see also* superior vena cava, inferior vena cava

venous return (VR) blood volume flowing *to* heart per unit time

ventilation air exchange between atmosphere and alveoli; alveolar air flow

ventral (VEN-tral) toward or at the front of body

ventral root one of two groups of efferent fibers that leave ventral side of spinal cord

ventricle (VEN-trih-kul) cavity, as in cerebral ventricle or heart ventricle; lower chamber of heart

ventricular function curve relation of the increase in stroke volume as end-diastolic volume increases, all other factors being equal

venule (VEEN-ule) small vessel that carries blood from capillary network to vein

very-low-density lipoprotein (VLDL) (lip-oh-PROH-teen) lipid-protein aggregate having high proportion of fat

vesicle (VES-ih-kul) small, membrane-bound organelle within cells

vestibular apparatus *see* vestibular system

vestibular receptor hair cell in semicircular canal, utricle, or saccule

vestibular system sense organ in temporal bone of skull; consists of three semicircular canals, a utricle, and a saccule; also called vestibular apparatus, sense organ of balance

villi (VIL-eye) finger-like projections from highly folded surface of small intestine; covered with single-layered epithelium

virus nuclei acid core surrounded by protein coat; lacks enzyme machinery for energy production and ribosomes for protein synthesis; thus cannot survive or reproduce except inside other cells whose biochemical apparatus it uses

viscera (VISS-er-ah) organs in thoracic and abdominal cavities

viscosity (viss-KOS-ih-tee) measure of friction between adjacent layers of a flowing liquid; property of fluid that makes it resist flow

visual field part of world being viewed at a given time

vital capacity maximal amount of air that can be expired, regardless of time required, following maximal inspiration

vitalism (VY-tal-ism) view that explanation of life processes requires a "life force" rather than physicochemical processes alone

vitamin organic molecule that is required in trace amounts for normal health and growth, but it is not manufactured in the body and must be supplied by diet; classified as water-soluble (vitamins C and the B complex) and fat-soluble (vitamins A, D, E, and K)

vocal cord one of two elastic-tissue bands stretched across laryngeal opening and caused to vibrate by air movement past them, producing sounds

volt (V) unit of measurement of electric potential between two points

voltage measure of potential of separated electric charges to do work; measure of electric force between two points

voltage-gated channel cell-membrane ion channel opened or closed by changes in membrane potential

vomiting center neurons in brainstem medulla oblongata that coordinate vomiting reflex

von Willebrand factor (vWF) (von-VILL-ih-brand) plasma protein secreted by endothelial cells; facilitates adherence of platelets to damaged vessel wall

vulva (VUL-vah) female external genitalia; mons pubis, labia majora and minora, clitoris, vestibule of vagina, and vestibular glands

W

waste product product from a metabolic reaction or series of reactions that serves no function

water-soluble vitamin *see* vitamin

wavelength distance between two successive wave peaks in oscillating medium

weak acid acid whose molecules do not completely ionize to form hydrogen ions when dissolved in water; *compare* strong acid

white matter portion of CNS that appears white in unstained specimens and contains primarily myelinated nerve fibers

white muscle muscle lacking appreciable amounts of myoglobin

withdrawal reflex bending of those joints that withdraw an injured part away from a painful stimulus

Wolffian duct (WOLF-ee-an) part of embryonic duct system that, in male, remains and develops into reproductive-system ducts, but in female, degenerates

work measure of energy required to produce physical displacement of matter; *see also* external work, internal work

working memory short-term memory storage process serving as initial depository of information

X

X chromosome *see* sex chromosome

Y

Y chromosome *see* sex chromosome

Z

Z line structure running across myofibril at each end of striated-muscle sarcomere; anchors one end of thin filaments and titin

zona pellucida (ZOH-nah peh-LOO-sih-dah) thick, clear layer separating egg from surrounding granulosa cells

zygote (ZY-goat) a newly fertilized egg

zymogen (ZY-moh-jen) enzyme precursor requiring some change to become active

Appendix C

	ENGLISH	METRIC
Length	1 foot = 0.305 meter	1 meter = 39.37 inches
	1 inch = 2.54 centimeters	1 centimeter (cm) = 1/100 meter
		1 millimeter (mm) = 1/1000 meter
		1 micrometer (μm) = 1/1000 millimeter
		1 nanometer (nm) = 1/1000 micrometer
°**Mass**	1 pound = 433.59 grams	1 kilogram (kg) = 1000 grams = 2.2 pounds
	1 ounce = 28.3 grams	1 gram (g) = 0.035 ounce
		1 milligram (mg) = 1/1000 gram
		1 microgram (μg) = 1/1000 milligram
		1 nanogram (ng) = 1/1000 microgram
		1 picogram (pg) = 1/1000 nanogram
Volume	1 gallon = 3.785 liters	1 liter = 1000 cubic centimeter = 0.264 gallon
	1 quart = 0.946 liter	1 liter = 1.057 quarts
	1 pint = 0.473 liter	
	1 fluid ounce = 0.030 liter	
	1 measuring cup = 0.237 liter	
		1 deciliter (dl) = 1/10 liter
		1 milliliter (ml) = 1/1000 liter
		1 microliter (μl) = 1/1000 milliliter

°A pound is actually a unit of force, not mass. The correct unit of mass in the English system is the slug. When we write 1 kg = 2.2 pounds, this means that one kilogram of *mass* will have a *weight* under standard conditions of gravity at the earth's surface of 2.2 pounds *force*.

Appendix D

ELECTROPHYSIOLOGY EQUATIONS

I. The **Nernst equation** describes the equilibrium potential for any ion species—that is, the electric potential necessary to balance a given ionic concentration gradient across a membrane so that the net passive flux of the ion is zero. The Nernst equation is

$$E = \frac{RT}{zF} \ln \frac{C_o}{C_i}$$

where E = equilibrium potential for the particular ion in question

C_i = intracellular concentration of the ion

C_o = extracellular concentration of the ion

z = valence of the ion (+1 for sodium and potassium, +2 for calcium, −1 for chloride)

R = gas constant [8314.9 J/(kg · mol · K)]

T = absolute temperature (temperature measured on the Kelvin scale: degrees centigrade +273)

F = Faraday (the quantity of electricity contained in 1 mol of electrons: 96,484.6 C/mol of charge)

ln = logarithm taken to the base e

II. A membrane potential depends on the intracellular and extracellular concentrations of potassium, sodium, and chloride (and other ions if they are in sufficient concentrations) and on the relative permeabilities of the membrane to these ions. The **Goldman equation** is used to calculate the value of the membrane potential when the potential is determined by more than one ion species. The Goldman equation is

$$V_m = \frac{RT}{F} \frac{\ln P_K \times K_o + P_{Na} \times Na_o + P_{Cl} \times Cl_i}{P_K \times K_i + P_{Na} \times Na_i + P_{Cl} \times Cl_o}$$

where V_m = membrane potential

R = gas constant [8314.9 J/(kg · mol · K)]

T = absolute temperature (temperature measured on the Kelvin scale: degrees centigrade + 273)

F = Faraday (the quantity of electricity contained in 1 mol of electrons: 96,484.6 C/mol of charge)

ln = logarithm taken to the base e

P_K, P_{Na}, and P_{Cl} = membrane permeabilities for potassium, sodium, and chloride, respectively

K_o, Na_o, and Cl_o = extracellular concentrations of potassium, sodium, and chloride, respectively

K_i, Na_i, and Cl_i = intracellular concentrations of potassium, sodium, and chloride, respectively

Appendix E

OUTLINE OF EXERCISE PHYSIOLOGY

Effects on Cardiovascular System 442–6

Atrial pumping 393
Cardiac output (increases) 400, 442–6, 464
    Distribution during exercise 429, 432, 442–3
Control mechanisms 443–5
Coronary blood flow (increases) 442–5
Gastrointestinal blood flow (decreases) 444, 445
Heart attacks (protective against) 450
Heart rate (increases) 442–5
Lymph flow (increases) 426
Maximal oxygen consumption (increases) 444–6
Mean arterial pressure (increases) 442, 443, 445
Renal blood flow (decreases) 442, 445
Skeletal-muscle blood flow (increases) 411–12, 442, 445
Skin blood flow (increases) 442, 445
Stroke volume (increases) 442, 444–6
Summary 445
Venous return (increases) 443
    Role of skeletal-muscle pump 423–4, 443
    Role of respiratory pump 423–4, 443

Effects on Organic Metabolism 606–7

Cortisol secretion (increases) 607
Diabetes mellitus (protects against) 608
Epinephrine secretion (increases) 607
Fuel homeostasis 606–7
Fuel source 78, 313, 606–7
Glucagon secretion (increases) 607
Glucose mobilization from liver (increases) 606–7
Glucose uptake by muscle (increases) 313, 607
Growth hormone secretion (increases) 607
Insulin secretion (decreases) 607
Metabolic rate (increases) 621
Plasma glucose changes 606
Plasma HDL (increases) 612
Plasma lactic acid (increases) 547
Sympathetic nervous system activity (increases) 607

Effects on Respiration 495–7

Alveolar gas pressures (no change in moderate exercise) 482
Capillary diffusion 482, 486
Control of respiration in exercise 491, 493, 495–7
Oxygen debt 313

Pulmonary capillaries (dilate) 482
Ventilation (increases) 464, 477, 493
    Breathing depth (increases) 313, 477
    Expiration 471
    Respiratory rate (increases) 313, 477
    Role of Hering-Breuer reflex 491
    Stimuli 495–7

Effects on Skeletal Muscle

Adaptation to exercise 318–9
Arterioles (dilate) 429–32
Changes with aging 319
Fatigue 313–4
Glucose uptake and utilization (increase) 313, 607
Hypertrophy 318
Local blood flow (increases) 411–12, 432, 442–4
Local metabolic rate (increases) 64
Local temperature (increases) 64
Nutrient utilization 606–7
Oxygen extraction from blood (increases) 486
Recruitment of motor units 317–8

Other Effects

Aging 156, 319
Body temperature (increases) 68, 632
Central command fatigue 314
Gastrointestinal blood flow (decreases) 442
Metabolic acidosis 547
Metabolic rate (increases) 618
Muscle fatigue 313–14
Osteoporosis (protects against) 542
Immune function 714
Soreness 315
Stress 728–30
Weight loss 624

Types of Exercise

Aerobic exercise 318–9
Endurance exercise 317, 318, 319
Long-distance running 313, 318
Moderate exercise 313
Swimming 318
Weight lifting 313, 318–19

Credits

Chapter Openers were also used in the Table of Contents

Photo Credits

Chapter 1 Opener 1: © Bruce Iverson.

Chapter 2 Opener 2: © Bruce Iverson.

Chapter 3 Opener 3: Science Photo Library / Photo Researchers, Inc.; **3.1:** From Johannes A.G. Rhodin, *"Histology, A Text & Atlas,"* Oxford University Press, New York, 1974; **3.3:** From K.R. Porter in T.W. Goodwin and O. Lindberg (eds.), *"Biological Structure and Function,"* Vol. I, Academic Press, Inc., New York, 1961; **3.6a:** From J.D. Robertson in Michael Locke (ed.), *"Cell Membranes in Development,"* Academic Press, Inc., New York, 1964; **3.10d:** M. Farquhar and G.E. Palade, *J. Cell. Biol.*, 17:375–412 (1963); **3.11:** K.R. Porter; **3.12:** D.W. Fawcett, *"The Cell, An Atlas of Fine Structure,"* W.B. Saunders Company, Philadelphia, 1966; **3.13:** W. Bloom and D.W. Fawcett, *"Textbook of Histology,"* 9th ed. W. B. Saunders Company, Philadelphia, 1968; **3.14:** K.R. Porter; **3.15:** From Roy A. Quinlan, et al., *Annals of the New York Academy of Sciences*, Vol. 455, New York, 1985.

Chapter 4 Opener 4: © Kenneth Eward / BioGrafx / Science Source / Photo Researchers, Inc.

Chapter 5 Opener 5: © Leonard Lessin / Peter Arnold, Inc.

Chapter 6 Opener 6: © Dr. Andrew Staehelin / Peter Arnold, Inc.

Chapter 7 Opener 7: © Kenneth Eward / BioGrafx / Science Source / Photo Researchers, Inc.

Chapter 8 Opener 8: © Scott Camazine / Photo Researchers, Inc.

Chapter 9 Opener 9: © Kenn Kostuk / Peter Arnold, Inc.

Chapter 10 Opener 10: Quest / Science Photo Library / Photo Researchers, Inc.

Chapter 11 Opener 11: Biology Media / Science Source / Photo Researchers, Inc.; **11.2:** From Edward K. Keith and Michael H. Ross, *"Atlas of Descriptive Histology,"* Harper & Row, New York, 1968; **11.6:** From H.E. Huxley, *J. Mol. Biol.*, 37:507–520 (1968); **11.7:** From H.E. Huxley and J. Hanson, in G.H. Bourne (ed.), *"The Structure and Function of Muscle,"* Vol. I, Academic Press, New York, 1960; **11.28a,b:** Courtesy of John A. Faulkner; **11.36 (insert):** From A.P. Somlyo, C.E. Devine, Avril V. Somlyo, and R.V. Rice, *Phil. Trans. R. Soc.* Lond. B, 265:223–229 (1973); **11.36:** From A.P. Somlyo, C.E. Devine, Avril V. Somlyo, and R.V. Rice, *Phil. Trans. R. Soc.* Lond. B, 265:223–229 (1973).

Chapter 12 Opener 12: Apostokos P. Georgopoulos, M.D.

Chapter 13 Opener 13: © Volker Steger / Peter Arnold, Inc.; **13.8:** Reprinted with permission from Malcom P. Young, *"The organization of neural systems in the primate cerebral cortex."* Proc. R. Soc. Biol. Sci., 252:13–18, 1993, Fig. 3; **13.11:** Marcus E. Raichle, MD / Washington University School of Medicine; **13.15:** Dr. Raichle / Washington University School of Medicine; **13.16:** Shaywitz, et al., 1995 NMR Research / Yale Medical School.

Chapter 14 Opener 14: Lunagrafix, Inc. / Photo Researchers, Inc.; **14.2:** © Bruce Iverson; **14.13 (right):** photos by Dr. Wallace McAlpine; **14.72:** NIBSC/Science Photo Library / Custom Medical Stock.

Chapter 15 Opener 15: Science Photo Library / Photo Researchers, Inc.

Chapter 16 Opener 16: © Manfred Kage / Peter Arnold, Inc.

Chapter 17 Opener 17: Science Photo Library / Photo Researchers, Inc.; **17.8:** From D.W. Fawcett, J. *Histochem. Cytochem.* 13:75–91 (1965). Courtesy of Susumo Ito.

Chapter 18 Opener 18: © Dr. F.C. Skvara / Peter Arnold, Inc.

Chapter 19 Opener 19: © David M. Phillips / Photo Researchers, Inc.; **19.22:** From C.H. Heuser and G.L. Streeter, *Carnegie Contrib. Embryol.* 29:15 (1941); **19.23:** From A.T. Hertig and J. Rock, *Carnegie Contrib. Embryol.* 29:127 (1941).

Chapter 20 Opener 20: © Meckes / Ottawa / Photo Researchers, Inc.

References for Figure Adaptations

Berne, R. M., and N. M. Levy: "Cardiovascular Physiology," 6th ed., Mosby, St. Louis, 1992.

Bloom, F. E., A. Lazerson, and L. Hofstadter: "Brain, Mind, and Behaviour," Freeman, New York, 1985.

Carlson, B. M.: "Patten's Foundations of Embryology," 5th ed., McGraw-Hill, New York, 1988.

Chaffee, E. E., and I. M. Lytle: "Basic Physiology and Anatomy," 4th ed., Lippincott, Philadelphia, 1980.

Chapman, C. B., and J. H. Mitchell: *Scientific American*, May 1965.

Comroe, J. H.: "Physiology of Respiration," Year Book, Chicago, 1965.

Crosby, W. H.: *Hospital Practice*, February 1987.

Davis, H., and H. R. Silverman: "Hearing and Deafness," Holt, Rinehart, and Winston, New York, 1970.

Dowling, J. E., and B. B. Boycott: *Proceedings of the Royal Society, London B*, **166:**80 (1966).

Elias, H., J. E. Pauly, and E. R. Burns: "Histology and Human Microanatomy," 4th ed., Wiley, New York, 1978.

Erickson, G. F., D. A. Magoffin, C. A. Dyer, and C. Hofeditz: *Endocrine Reviews*, Summer 1985.

Felig, P., and J. Wahren: *New England Journal of Medicine*, **293:**1078 (1975).

Gardner, E.: "Fundamentals of Neurology," 5th ed., Saunders, Philadelphia, 1968.

Golde, D. W., and J. C. Gasson: *Scientific American*, July 1988.

Goodman, H. M.: "Basic Medical Endocrinology," Raven, New York, 1988.

Gray, H. M., A. Sette, and S. Buus (drawing by G. B. Kelvin): *Scientific American*, November 1989.

Guyton, A. C.: "Functions of the Human Body," 3d ed., Saunders, Philadelphia, 1969.

Hedge, G. A., H. D. Colby, and R. L. Goodman: "Clinical Endocrine Physiology," Saunders, Philadelphia, 1987.

Hoffman, B. F., and P. E. Cranefield: "Electrophysiology of the Heart," McGraw-Hill, New York, 1960.

Hubel, D. H., and T. N. Wiesel: *Journal of Physiology*, **154:**572 (1960).

Hudspeth, A. J.: *Scientific American*, January 1983.

Kandel, E. R., and J. H. Schwartz: "Principles of Neural Science," 2d ed., Elsevier/North-Holland, New York, 1985.

Kappas, A., and A. P. Alvares: *Scientific American*, June 1975.

Lambersten, C. J.: in P. Bard (ed.), "Medical Physiological Psychology," 11th ed., Mosby, St. Louis, 1961.

Lehninger, A. L.: "Biochemistry," Worth, New York, 1970.

Lentz, T. L.: "Cell Fine Structure," Saunders, Philadelphia, 1971.

Little, R. C.: "Physiology of the Heart and Circulation," 2d ed., Year Book, Chicago, 1981.

Meyer-Franke, A., and B. Barres: *Current Biology*, **4:**847 (1994).

Moore-Ede, M. C., and F. M. Sulzman: "The Clocks That Time Us," Harvard, Cambridge (MA), 1982.

Nauta, W. J. H., and M. Fiertag: "Fundamental Neuroanatomy," Freeman, New York, 1986.

Olds, J.: *Scientific American*, October 1956.

Rasmujssen, A. T.: "Outlines of Neuroanatomy," 2d ed., Wm. C. Brown, Dubuque, IA, 1943.

Rushmer, R. F.: "Cardiovascular Dynamics," 2d ed., Saunders, Philadelphia, 1961.

Scales, W. E., A. J. Vander, M. B. Brown, and M. J. Kluger: *American Journal of Physiology*, **65:**1840 (1988).

Snyder, S. H.: "Drugs and the Brain," Freeman, New York, 1986.

Tung, K.: *Hospital Practice*, June 1988.

Young, M. P.: *Proc. R. Soc. Lond. B*, **252:**13 (1993).

von Bekesy, G.: *Scientific American*, August 1957.

Walmsley, B., F. J. Alvarez, and R. E. W. Fyffe: *Trends in Neuroscience*, **21:**81 (1998).

Wang, C. C., and M. I. Grossman: *American Journal of Physiology*, **164:**527 (1951).

Wersall, J., L. Gleisner, and P. G. Lundquist: in A. V. S. de Reuck and J. Knight (eds.), "Myostatic, Kinesthetic, and Vestibular Mechanisms," Ciba Foundation Symposium, Little Brown, Boston, 1967.

Index

Note: Page numbers in *italics* indicate illustrations. Page numbers followed by t indicate tables. Clinical terms are shown in *italics*.

A

A band, 294, *295–96, 298, 390*
Abdominal fat, 624
Abdominal pressure, 424
Abducens nerve, 213t
ABO blood group, 717, 717t
Abortifacients, 677
Absolute refractory period, 193
Absorption, 421–22
  of amino acids, 562
  of carbohydrates, 562
  epithelial transport, 136–38, *136–38*
  of fats, 563–65, *563–65*
  by gastrointestinal tract, 554, *555,* 581–83
  of iron, 566
  of minerals, 565–66
  toxicology, *726, 727*
  of vitamins, 565
Absorptive state, 594–97, *599*
  absorbed amino acids, *595,* 597
  absorbed carbohydrates, 594–96, *595*
  absorbed triacylglycerols, *595,* 596–97
  endocrine and neural control of, 599–606
  metabolic pathways of, *595*
  summary of nutrient metabolism during, 597t
Accessory nerve, 213t
Accessory reproductive organs, 636, 648
Acclimatization, 152–53
  to high altitude, 499–500, 500t
  to temperature, 630
Accommodation, 244–45, *245*
ACE inhibitors (angiotensin-converting enzyme inhibitors), 447–48t
Acetone, 598–99
Acetylcholine (ACh), 205, 205t, 208, 305–8, *306,* 389, 401
  acid secretion by stomach and, 572, *573,* 574
  in autonomic nervous system, 219, *220*
  in somatic nervous system, 216
  state of consciousness and, 354, *355*
Acetylcholine (ACh) receptors, 162, 219, 219t, 306–7, *306,* 322
Acetylcholinesterase, 205, *306,* 307–8
Acetyl coenzyme A (acetyl CoA), 73, *73,* 81–82, 598
ACh. *See* Acetylcholine (ACh)

Acid(s), 20. *See also* Hydrochloric acid
  strong, 20
  weak, 20
Acidic solutions, 21
Acidity, 20–21
Acidosis
  classification of, 547–48, 548t
  renal response to, 547, 547–48t
Acquired immune deficiency syndrome (AIDS), 715–16, 721
Acquired immunity. *See* Specific immune defenses
Acquired reflex, 147
Acromegaly, 615
Acrosome, 642, *642*
Acrosome reaction, 664
ACTH. *See* Adrenocorticotropic hormone (ACTH)
Actin, 50, *50,* 294, *295,* 298, *299,* 301, 325, 389
Action potentials, 189–96, 189t, *190–95*
  cardiac, 392–93, *392–93*
  compared to graded potentials, 196t
  direction of propagation of, 194, *194*
  excitation-contraction coupling, 302–5, *302–4,* 305t
  frequency of, 234, *235,* 236
  initiation of, 195–96, *195*
  ion-channel changes in, 191–92, *192*
  ionic basis of, 189–91
  muscle fiber, 305–8, *306*
  propagation of, 193–95, *194–95*
  refractory period, 193
  in sensory systems, 228–30, *229–30*
  smooth muscle, 328
  threshold and all-or-none response, 192–93, *193*
  T tubule, 304, *304*
  velocity of propagation of, 194
Activated macrophages, 711–12, *711,* 721t
Activation energy, 61–62, 61t
Active humoral immunity, 709, *709*
Active hyperemia, 411–12, *412,* 420, 443
Active site, 63, *63*
Active transport, 124–29, *124–28*
  primary, 125–26, *125–26,* 129t, *130*
  secondary, 125–29, *126–30,* 129t, 132, *136,* 137
  of sodium across epithelial cell, 136–37, *136*
Activin, 651

Acuity, 234
Acupuncture, 243
Acute phase proteins, 712, *713,* 722t
Acute phase response, 712–14, *713,* 714t
Acyclovir, 716
Adaptation, 152
  to stimulus, 230, *230,* 236–37, *237*
ADCC (antibody-dependent cellular cytotoxicity), 708
Adenine, 32–33, *32,* 33t, 208
Adenohypophysis, 275
Adenosine, 401, *415, 431*
Adenosine triphosphate. *See* ATP (adenosine triphosphate)
Adenylyl cyclase, 166–68, *166–69*
Adequate stimulus, 229, 233
ADH (antidiuretic hormone). *See* Vasopressin
Adhesion molecules, 692
Adipocytes, 80, 594, 596–97
Adipose tissue, 80, 265t, 596
  response to insulin, *600–602,* 601
  sympathetic nerves to, 604, *605*
ADP, 453
Adrenal cortex, 265t, 267
  hormones produced by, 268–70, *269*
Adrenal gland, 267
  steroids produced by, 152, 716
Adrenaline. *See* Epinephrine
Adrenal medulla, 220, *220,* 265t
  hormones produced by, 267, *269*
Adrenal steroids, 152, 716
Adrenergic fibers, 206
Adrenocorticotropic hormone (ACTH), 266t, 276–77, *278,* 280–81, *280–82,* 713
  memory and, 367
  at puberty, 683
  in stress response, 729–30, *729*
Adsorptive endocytosis, 134
Aerobic conditions, 72, 74
Aerobic exercise, 318–19
Affective disorders, *362*
Afferent arterioles, 507, *508–10,* 513
Afferent division, of peripheral nervous system, 216, 216t
Afferent input (information)
  alpha-gamma coactivation, 339–40, *340*
  central control of, 237–38, *237*
  length-monitoring systems, 337–39, *338–39*

  local input, 337–41, *338–41*
  tension-monitoring systems, 340, *341*
  withdrawal reflex, 340–41, *341*
Afferent neurons, 179, *179,* 180t, 216, 228–30, *228–29*
Afferent pathways, 147, *148*
Affinity
  of enzyme's active site, 65, *65*
  of protein binding sites, 55–56, *56*
  of receptors, 160, 161t
Afterbirth, 672
Afterhyperpolarization, 191, *191*
Afterload, 402–3, *403*
Aging, 108, 155–56
  blood pressure and, *409*
  caloric restriction and, 156
  exercise and, 156
  of muscles, 319
Agonists, 161, 161t, 203
Agranular endoplasmic reticulum, *39–40,* 46–47, *47*
AIDS (acquired immune deficiency syndrome), 715–16, 721
Airway(s), 464–66, *465–66*
Airway resistance, 474–75, 499
  in asthma, 474
  in chronic obstructive pulmonary disease, 474–75
  Heimlich maneuver, 475, *475*
Akinesia, 344
Albumins, 375, 376t
Alcohol, 223, 365, 365t, 668
  coronary artery disease and, 450–51
Aldosterone, 265t, 268, *268–69*
  in sodium, water, and potassium balances, 526–30, *529,* 534–35, 535
  in stress response, 730
  in temperature acclimatization, 630
Alertness, 241
Alkaline solutions, 21
Alkalosis
  classification of, 547–48, 548t
  renal response to, 547, 547–48t
Alleles, 108
Allergens, 718
Allergy (hypersensitivity), 439, 718–20, 718t, *719*
All-or-none response, action potential, 192–93, *193*
Allosteric modulation, of protein binding sites, 57–59, *58,* 65, *65,* 301
Allosteric proteins, 57–58, *58*
Alpha-adrenergic receptors, 206, 413–14, *414,* 447t
Alpha cells, 600
Alpha-fetoprotein, 668

781

ABBREVIATIONS USED IN THE TEXT

A actin, adenine
A surface area
ACE angiotensin converting enzyme
acetyl CoA acetyl coenzyme A
ACh acetylcholine
ACTH adrenocorticotropic hormone (adrenocorticotropin, corticotropin)
ADCC antibody-dependent cellular cytotoxicity
ADH antidiuretic hormone (vasopressin)
ADP adenosine diphosphate
AIDS acquired immune deficiency syndrome
alv alveoli
AMP adenosine monophosphate
ANF atrial natriuretic factor
AP action potential
APC antigen-presenting cell
atm atmosphere
ATP adenosine triphosphate
AV atrioventricular

BM basement membrane
BMI body mass index
BMR basal metabolic rate

C Celsius (centigrade), creatine, cytosine, carbon, capillary, cervical
C clearance, concentration
Ca calcium (Ca^{2+} calcium ion)
cal calorie
CAM cell adhesion molecule
cAMP cyclic 3',5'-adenosine monophosphate
CCK cholecystokinin
C$_{Cr}$ creatinine clearance
cdc kinases cell division cycle kinases
CG chorionic gonadotropin
C$_G$ glucose clearance
cGMP cyclic 3',5'-guanosine monophosphate
CGRP calcitonin gene-related peptide
C$_i$ intracellular concentration
CK creatine kinase
C$_L$ lung compliance
Cl chlorine (Cl^- chloride ion)
cm centimeter
CNS central nervous system
CO carbon monoxide, cardiac output
C$_o$ extracellular concentration
CO$_2$ carbon dioxide
CoA coenzyme A
—COOH carboxyl group (—COO$^-$ carboxyl ion)
COX cyclooxygenase

CP creatine phosphate
CPK creatine phosphokinase
CPR cardiopulmonary resuscitation
Cr creatinine
CRH corticotropin releasing hormone
CSF cerebrospinal fluid, colony-stimulating factor
CTP cytosine triphosphate
cyclic AMP cyclic 3',5'-adenosine monophosphate

d dalton
DA dopamine
DAG diacylglycerol
Δ change
Δ*E* internal energy liberated
DHEA dihydroepiandrosterone
Δ*P* pressure difference
DKA diabetic ketoacidosis
dl deciliter
DNA deoxyribonucleic acid
DP diastolic pressure
DPG 2,3-diphosphoglycerate

e$^-$ electron
E electric potential difference, voltage, internal energy
E epinephrine, enzyme
ECF extracellular fluid
ECG electrocardiogram
ECL enterochromaffin-like cell
ECT electroconvulsive therapy
EDRF endothelium-derived relaxing factor
EDV end-diastolic volume
EEG electroencephalogram
EF ejection fraction
EKG electrocardiogram
EP endogenous pyrogen
Epi epinephrine
EPP end-plate potential
EPSP excitatory postsynaptic potential
ES enzyme-substrate complex
ESV end systolic volume
ET-1 endothelin-1
η (eta) fluid viscosity

F net flux, flow
FAD flavine adenine dinucleotide
Fe iron
FEV$_1$ forced expiratory volume in 1 s
FFA free fatty acid
f$_i$ influx
f$_o$ efflux
FRC functional residual capacity
FSH follicle-stimulating hormone
ft feet
FVC forced vital capacity

G guanine
g gram
G$_0$ phase "time out" phase of cell cycle
G$_1$ phase first gap phase of cell cycle
G$_2$ phase second gap phase of cell cycle
GABA gamma-aminobutyric acid
GDP guanosine diphosphate
GFR glomerular filtration rate
GH growth hormone
GHRH growth hormone releasing hormone
G$_i$ inhibitory G protein
GI gastrointestinal
GIP glucose-dependent insulinotropic peptide
GLP-1 glucagon-like peptide-1
GMP guanosine monophosphate
GnRH gonadotropin releasing hormone
G$_s$ stimulating G protein
GTP guanosine triphosphate

H hydrogen (H^+ hydrogen ion)
H heat
h hour
Hb deoxyhemoglobin
HbH deoxyhemoglobin
HbO$_2$ oxyhemoglobin
HCl hydrochloric acid
HCO$_3^-$ bicarbonate ion
HDL high-density lipoprotein
HGF hematopoietic growth factor
HIV human immunodeficiency virus
H$_2$O$_2$ hydrogen peroxide
HPO$_4^{2-}$, H$_2$PO$_4^-$ phosphate ion, inorganic orthophosphate
HR heart rate
5-HT serotonin, 5-hydroxytryptamine
Hz hertz, or cycles per second

I current
IDDM insulin-dependent diabetes mellitus
IF interstitial fluid
Ig immunoglobulin
IGF-I insulin-like growth factor I
IGF-II insulin-like growth factor II
IL-1 interleukin 1
IL-2 interleukin 2
IL-6 interleukin 6
In insulin
in inch
IP$_3$ inositol trisphosphate
IPSP inhibitory postsynaptic potential
IUD intrauterine device